Recent Advances in Hydraulic Fracturing

John L. Gidley, PhD
Editor-in-Chief
Senior Technical Advisor (retired)
Exxon Co. U.S.A.

Stephen A. Holditch, PhD
Associate Editor
Professor, Petroleum Engineering
Texas A&M U.

Dale E. Nierode, PhD
Associate Editor
Senior Research Associate
Exxon Production Research Co.

Ralph W. Veatch Jr., PhD
Associate Editor
Section Supervisor
Amoco Production Co.

SPE Monograph Series, Volume 12

Henry L. Doherty Memorial Fund of AIME
Society of Petroleum Engineers
Richardson, TX USA

Disclaimer

This book was prepared by members of the Society of Petroleum Engineers and their well-qualified colleagues from material published in the recognized technical literature and from their own individual experience and expertise. While the material presented is believed to be based on sound technical knowledge, neither the Society of Petroleum Engineers nor any of the authors or editors herein provide a warranty either expressed or implied in its application. Correspondingly, the discussion of materials, methods, or techniques that may be covered by letters patents implies no freedom to use such materials, methods, or techniques without permission through appropriate licensing. Nothing described within this book should be construed to lessen the need to apply sound engineering judgment nor to carefully apply accepted engineering practices in the design, implementation, or application of the techniques described herein.

Printed in the United States of America.

ISBN 978-1-55563-020-1

11 12 13 14 15 16 / 13 12 11 10 9 8

Society of Petroleum Engineers
222 Palisades Creek Drive
Richardson, TX 75080-2040 USA

http://store.spe.org/
service@spe.org
1.972.952.9393

Preface

During the almost two decades that have elapsed since publication of the SPE monograph *Hydraulic Fracturing* by G.C. Howard and C.R. Fast, the science and technology of hydraulic fracturing have undergone an almost explosive growth. More than 2,000 publications have appeared in the technical literature, providing an improved simulation of the process, better materials and methods for its application, and a growing sophistication in its evaluation in the field.

While this outpouring of knowledge has generally been welcomed by the industry, it has been overwhelming to the individual and has posed somewhat of a problem to the Society. How should this body of knowledge be digested to its essence and condensed to a form that can be assimilated by an engineer faced with designing fracturing treatments?

More than 5 years ago, when the Monograph Committee decided that a new monograph on this topic might meet the need, it became obvious to those of us charged with the project that a conventional monograph written by two or three authors would be inadequate for the job at hand. The problem, simply stated, was that no two or three people within the industry were, in our opinion, sufficiently knowledgeable in all areas of the subject to cover the technology in the depth required.

As we saw it, the Society's need was to create a monograph on hydraulic fracturing that would comprehensively cover current technology. The solution appeared to be a book written by those at the forefront of each area of expertise of the evolving technology, people with a first-hand knowledge of the technical problems and their proposed solutions. Our goal was to create a comprehensive reference tool useful to the individual faced with designing, analyzing, and improving hydraulic fracturing treatments.

With some apprehension, we submitted our plan for a book of 17 chapters written by 23 authors to the SPE Board of Directors. While the plan was ultimately approved (as is obvious from the product in your hands) as a special Monograph Committee project, we suspect that it was not without some misgivings, for this approach had not previously been attempted in the Monograph Series.

The 23 authors selected represent a diversity of talent, background, and experience. All have published extensively in their fields of specialty. Perhaps of even more importance, each has demonstrated an ability in clear technical exposition. Having said that, we recognize that many other equally capable people were passed over in the selection process.

Looking at the final product, we believe that the book is unique in both the depth and breadth of its coverage in each technical area. An overview at the beginning of each chapter describes its scope and summarizes the area covered. To those already experienced in hydraulic fracturing, each chapter may be viewed as standing alone, although cross-referencing between chapters permits identification of related areas.

While *Recent Advances in Hydraulic Fracturing* is not a textbook in the normal sense, the authors frequently use illustrative problems to demonstrate application of the technology. Each author has attempted to make the material as instructive as possible.

Finally, even though every book at the time of its publication is already partially out of date, we believe that the authors of this work, by their unfailing diligence to keep abreast of new technology, have truly captured the significant recent advances in hydraulic fracturing.

John L. Gidley
Editor in Chief
Feb. 1989

Acknowledgments

Many people have contributed to the completion of this monograph in addition to the authors and editors prominently listed on its pages. Any effort to acknowledge all of them is fraught with the possibility of unintentional omission. Despite that, we wish to begin the list by acknowledging the employers of the listed authors. Their principal contributions were time, materials, and talent. Inevitably, that portion of the writing chore not accomplished after hours or on weekends took place on company time. In addition, many companies provided the authors with secretarial and graphical services for the final product.

Among the organizations we wish to recognize for their contributions to this effort are (in alphabetical order) Amoco Production Co., Brown U., Conoco Inc., Dowell Schlumberger Inc., Exxon Co. U.S.A., Exxon Production Research Co., Fast Engineering, John L. Gidley & Assocs. Inc., Guydon Software Services, Halliburton Services, S.A. Holditch & Assocs. Inc., Koninklijke/Shell E&P Laboratorium, NSI Technologies Inc., Princeton U., Stim-Lab Inc., Terra Tek Inc., Texas A&M U., and The Western Co.

We also wish to recognize those individuals whose names do not appear here who contributed their typing and word-processing skills in developing the manuscript and their creative talents for providing the illustrations to eludicate the technology discussed.

A debt of gratitude is owed to the many fine members of the SPE staff who dutifully edited the manuscript, handled the myriad chores of coordinating the effort to produce this monograph, and are responsible for both its appearance on the printed page and its adherence to recognized grammatical standards. Particularly to be recognized among this group are Christy Magargee and Georgeann Bilich.

We are also obliged to the SPE Monograph Committee, who concurred with the multiauthor approach used in the creation of this monograph and whose actions in resolving problems clearly supported the effort.

Finally, we would be negligent indeed if we did not acknowledge the contributions of our spouses. Their load was materially increased by our absence from the family during this book's preparation. We hope that the final product provides a measure of pride in collective accomplishment that justifies the sacrifices often made on its behalf.

John L. Gidley
June 1989

SPE Monograph Series

The Monograph Series of the Society of Petroleum Engineers was established in 1965 by action of the SPE Board of Directors. The Series is intended to provide authoritative, up-to-date treatment of the fundamental principles and state of the art in selected fields of technology. The Series is directed by the Society's Monograph Committee. A committee member designated as Monograph Editor provides technical evaluation with the aid of the Review Committee. Below is a listing of those who have been most closely involved with the preparation of this monograph.

Monograph Committee (1989)

Contents

Chapter 1
An Overview of Hydraulic Fracturing

Ralph W. Veatch Jr., SPE, Amoco Production Co.
Zissis A. Moschovidis, SPE, Amoco Production Co.
C. Robert Fast, SPE, Fast Engineering Co.

1.1 Introduction

The Hydraulic Fracturing Process. Hydraulic fracturing plays a major role in enhancing petroleum reserves and daily production. Fig. 1.1 portrays a conceptual version of the "typical" fracturing process. It consists of blending special chemicals to make the appropriate fracturing fluid and then pumping the blended fluid into the pay zone at high enough rates and pressures to wedge and extend a fracture hydraulically. First, a neat fluid, called a "pad," is pumped to initiate the fracture and to establish propagation. This is followed by a slurry of fluid mixed with a propping agent (often called a "proppant"). This slurry continues to extend the fracture and concurrently carries the proppant deeply into the fracture. After the materials are pumped, the fluid chemically breaks back to a lower viscosity and flows back out of the well, leaving a highly conductive propped fracture for oil and/or gas to flow easily from the extremities of the formation into the well. Note that the fracture has two wings extending in opposite directions from the well and is oriented more or less in the vertical plane. Other fracture configurations (e.g., "horizontal" fractures) are known to exist; some have been observed at relatively shallow depths (e.g., $<2,000$ ft [610 m]). But they constitute a relatively low percentage of the situations experienced to date. Hence, most of the discussion in this chapter will be in the context of "vertical" fractures, like that depicted in Fig. 1.1.

Fracturing has made a significant contribution in enhancing oil and gas producing rates and recoverable reserves. The fracturing process, introduced to the industry in 1947, is a standard operating practice. By 1981, more than 800,000 treatments had been performed.[1] As of 1988, this has grown to exceed 1 million. About 35 to 40% of all currently drilled wells are hydraulically fractured, and about 25 to 30% of the total U.S. oil reserves have been made economically producible by the process. Fracturing is responsible for increasing North America's oil reserves by 8 billion bbl [1.3×10^9 m^3].

History and Development of Hydraulic Fracturing. The first fracturing treatment specifically designed to stimulate well production was conducted in the Hugoton gas field, July 1947, on Kelpper Well 1 located in Grant County, KS. The well was completed with four gas-productive limestone pay zones from 2,340 to 2,580 ft [715 to 790 m]. The bottomhole pressure was approximately 420 psi [2.9 MPa]. This well, originally completed with a downhole acid treatment, was chosen for hydraulic fracturing because it had a low deliverability and would offer a direct comparison between acidizing and fracturing.

The mechanical pumping equipment used consisted of a centrifugal pump for mixing the gasoline-based napalm-gelled fracturing fluid and a duplex, positive-displacement piston pump for pumping the fluid into the well. Because of the fire hazard, all units, including the mixing tanks, were placed 150 ft [45 m] apart, which complicated this first operation.

This particular "hydrafrac," as it is commonly called, operation actually involved four separate treatments (one on each of the three perforated zones and one on the bottom openhole section) conducted through tubing equipped with a cup-type straddle packer. The treatment of each zone consisted of 1,000 gal [3.8 m^3] of napalm-thickened gasoline followed by 2,000 gal [7.6 m^3] of gasoline containing 1% of a cationic emulsion breaker that acted as a viscosity reducer.

By the mid-1960's, the primary method of stimulation in this field was hydraulic fracturing. The use of large volumes of low-cost, water-based fluid pumped at very high rates had proved to be an effective, economical procedure for fracturing Hugoton wells.

Since its inception, hydraulic fracturing has developed from a simple, low-volume, low-rate fracture stimulation method to a very highly engineered, complex procedure that is used for many purposes. It can be used to improve well productivity by overcoming drilling and completion damage near the bore; it can also be used to make deeply penetrating, high-conductivity fractures in low-permeability reservoirs. The fracturing of disposal and injection wells to increase injectivity is common. Fracturing has been used in secondary and tertiary recovery processes, such as water-, fire-, and steamflood operations, to improve injectivity and sweep efficiency. Hydraulic fracturing is currently the most widely used process for stimulating oil and gas wells.

In retrospect, one might say that hydraulic fracturing has been so successful that in the past we did not have to design treatments with a high degree of precision for them to work, and work extremely well. As we moved toward applications where larger quantities of more sophisticated materials were required, the economics demanded more rigorous designs. This required reliable determination of both the potential of a well to respond to fracture stimulation and the effectiveness of a treatment design to create the desired fracture.

Over the years, the technology associated with fracturing has improved significantly. A host of fracturing fluids has been developed for reservoirs ranging from shallow, low-temperature formations to those in deep, hot areas. Many different types of proppants have been developed, ranging from silica sand, the "standard," to high-strength materials, like sintered bauxite, for use in deep formations where fracture closure stresses exceed the ranges of sand capabilities. New design models and analytical and diagnostic methods have emerged, and the service industry has continually developed new equipment to meet the emerging challenges.

Fracturing treatments typically have varied in size from the small (e.g., 500 gal [1.9 m^3]) mini-hydraulic fracturing treatments for short fracture lengths to the deeply penetrating massive hydraulic fracturing (MHF) treatments which now exceed 1 million gal [3.8×10^3 m^3] fracturing fluid and 3 million lbm [1.4×10^6 kg] of propping agent. MHF treatments have played a significant role in developing otherwise uneconomical tight (i.e., low-permeability) gas formations. The design difficulties and high cost of MHF have made obvious the need to enhance our fracture design and treatment capabilities.

Fracture design still involves much judgment as well as engineering. After 40 years of fracturing experience and research, our abilities to determine in-situ fracture shapes, dimensions (lengths,

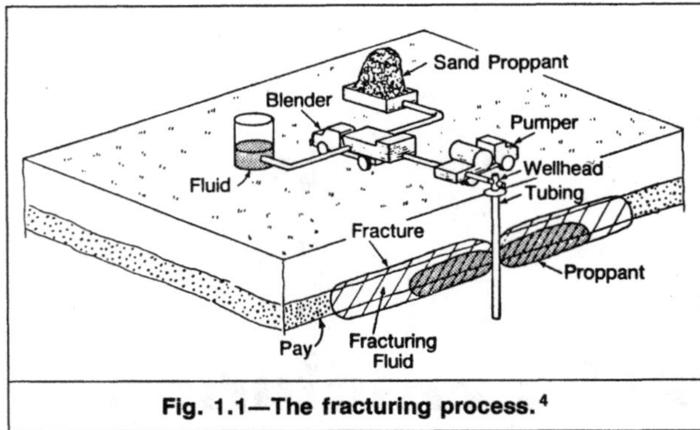

Fig. 1.1—The fracturing process.[4]

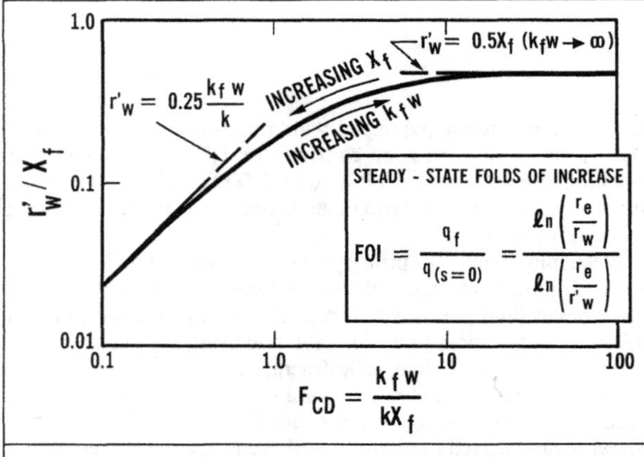

Fig. 1.2—Prats'[6] producing-rate folds-of-increase curve.

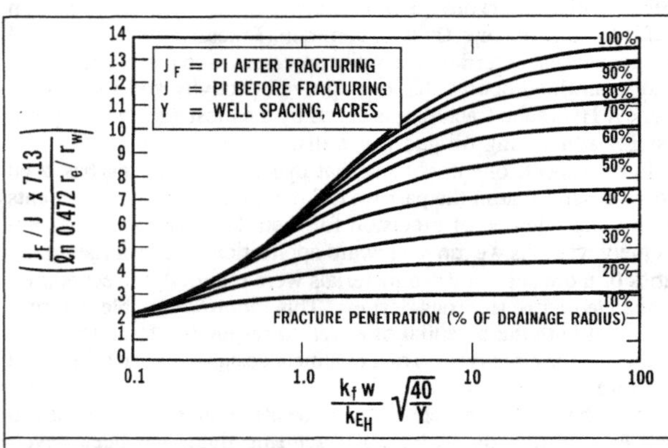

Fig. 1.3—McGuire and Sikora's[7] producing-rate folds-of-increase curves.

widths, and heights), symmetry about the wellbore, azimuths, and fracture conductivities are still not totally developed. In addition, our abilities to measure in-situ rock properties and stress fields that significantly affect fracture propagation are not perfected. Consequently, our abilities to optimize treatment designs and economics precisely are often limited. However, technology in fracturing is advancing significantly.

Scope. The discussion in this chapter covers much of the currently developing technology and the future needs for technology advances. It is by design a brief synopsis of the many aspects of fracturing. It is hoped that this chapter will provide an overview of previous survey publications[2–5] and the remaining chapters in this monograph.

The sections in this chapter are presented in the same sequence as the chapters in this monograph, except that the discussions pertinent to Chaps. 2 (Pretreatment Formation Evaluation) and 15 (Post-

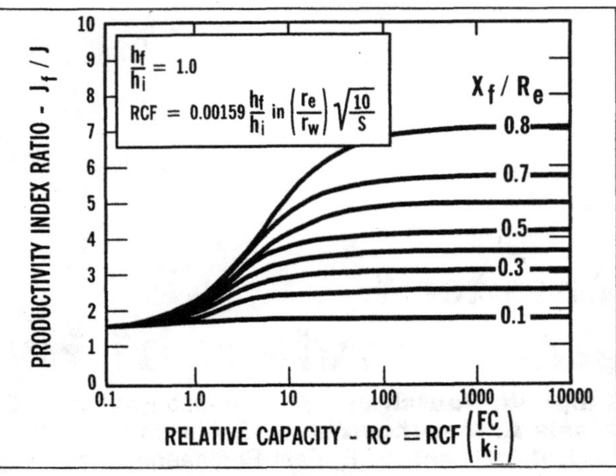

Fig. 1.4—Productivity-index-ratio increase from fracturing—steady-state flow, vertical propped fractures, $h_f/h = 1.0$.[8]

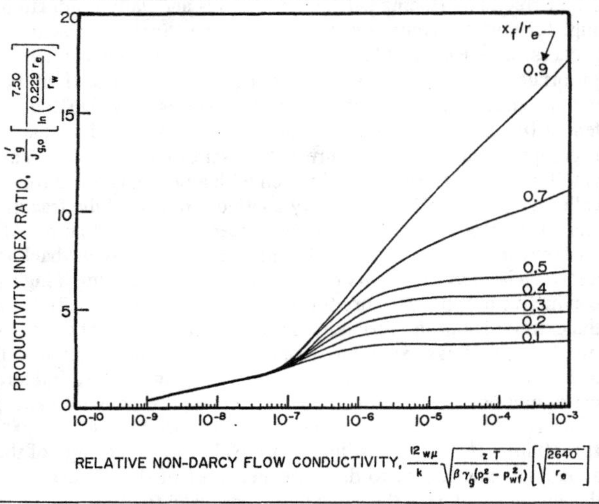

Fig. 1.5—Tannich and Nierode's[9] producing-rate folds-of-increase curves.

fracture Formation Evaluation) have been combined into one section (Section 1.2).

1.2 Formation Evaluation—The Fracturing Aspects

A wide variety of methods (both graphical and computerized) are available to estimate the effects of fracture length and fracture conductivity on well productivity for a particular formation. If the reservoir has a relatively high permeability where steady-state or pseudosteady-state flow develops relatively quickly, methods provided by Prats,[6] McGuire and Sikora,[7] Tinsley *et al.*,[8] or Tannich and Nierode[9] as shown in Figs. 1.2 through 1.5 may be used to predict or estimate the productivity improvement from a fracturing treatment. These methods do not apply if there is severe near-wellbore skin damage or transient flow.

If the reservoir has such low permeability that transient flow is the dominant regime throughout a major portion of a well's life, it is necessary to use a transient-flow reservoir computer simulator or type curves like those provided by Agarwal *et al.*[10] (Fig. 1.6) or Holditch *et al.*[11] (Fig. 1.7). In Fig. 1.6, dimensionless producing rates, q_D, are related to dimensionless times, t_{Dxf}, and dimensionless flow capacities, F_{CD}. In Fig. 1.7, dimensionless cumulative production values, Q_D, are related to dimensionless times and fracture conductivities. From these graphs, one can estimate either the producing rate or the cumulative production performance from a given reservoir for different fracture half-lengths, x_f, and fracture conductivities, k_fw. They provide considerable in-

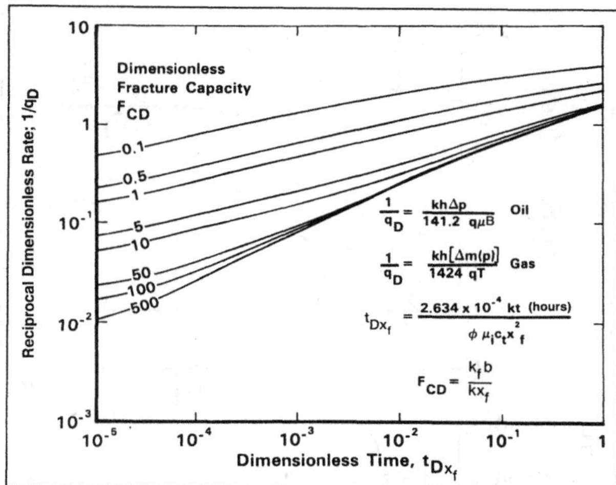

Fig. 1.6—Producing-rate type curves with propped vertical fractures—transient flow, constant wellbore pressure.[10]

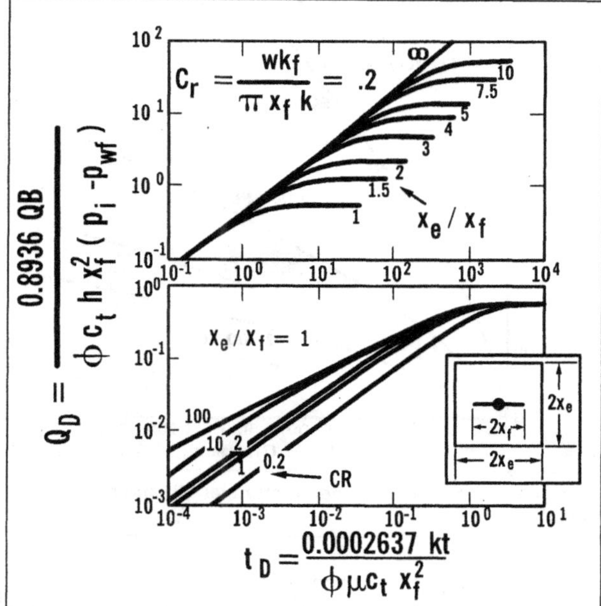

Fig. 1.7—Dimensionless cumulative-production type curves.[11]

sight into the effects of length and conductivity as they relate to formation permeability during unsteady-state flow. To maximize producing rate, one would like to achieve a fracture where F_{CD} values approach the 100 to 500 range.

In higher-permeability reservoirs, fracture stimulation will increase early-life producing rates (which increases cash flow) but will usually not increase ultimate cumulative recovery. In low-permeability formations, however, fracturing also can significantly increase ultimate recovery. The overall benefits derived from deeply penetrating fractures in low-permeability formations were investigated in a 1980 Natl. Petroleum Council study[12] on tight-gas reservoirs. The results as summarized by Baker[13] and Veatch[14] indicate that advanced technology will increase recoverable gas from tight formations by 40 to 75%. In this study, "advanced technology" implies more deeply penetrating and/or higher-conductivity fractures as required by formation permeability levels and efficiently patterned wells consistent with the azimuthal trend of long fractures for effective reservoir drainage.

In reservoir-response studies, one should be aware of how well a given reservoir model represents in-situ formation conditions. Some complex reservoirs may require equally complex reservoir simulators or methods for analyzing and/or predicting performance. Improved techniques are emerging to cope with the more complex problems. Numerous authors[15-22] have made significant contributions to improve analysis and modeling of well flow performance in hydraulically fractured formations.

1.3 Rock Mechanics and Fracture Geometry

Rock mechanics plays an important role in governing the geometry of propagating fractures. Some of the theoretically identified factors that affect fracture propagation are (1) variations of in-situ stresses existing in different layers of rock, (2) relative bed thickness of formations in the vicinity of the fracture, (3) bonding between formations, (4) variations in mechanical rock properties (including elastic modulus, Poisson's ratio, toughness, or ductility), (5) fluid pressure gradients in the fracture, and (6) variations in pore pressure from one zone to the next.

Local stress fields and variations in stresses between adjacent formations are often thought to dominate fracture orientation and vertical fracture growth. Regional stresses can impact the azimuthal trend of hydraulically created fractures. Fractures will usually propagate perpendicularly to the direction of the minimum principal stress.

Fig. 1.8 depicts the effects of differences in the magnitudes of horizontal and vertical stresses on the plane of orientation of a fracture. Here the stress magnitudes are proportional to the arrow sizes. At shallow depths, horizontal fractures have been reported.[2] These might result from a condition like that depicted in Fig. 1.8c. Experience leads us to believe that at depths below 1,000 to 2,000 ft

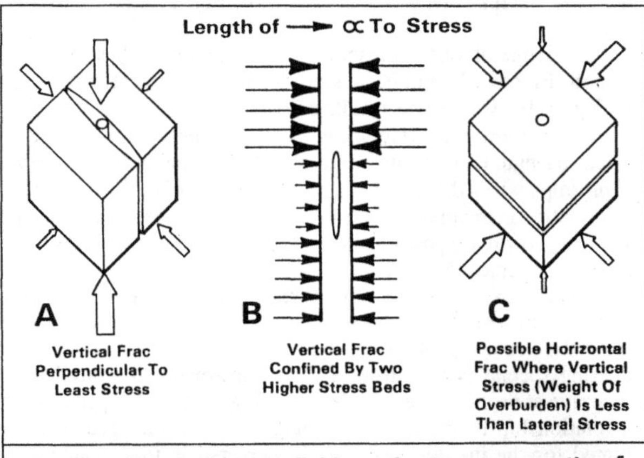

Fig. 1.8—Effect of stress fields on fracture propagation.[5]

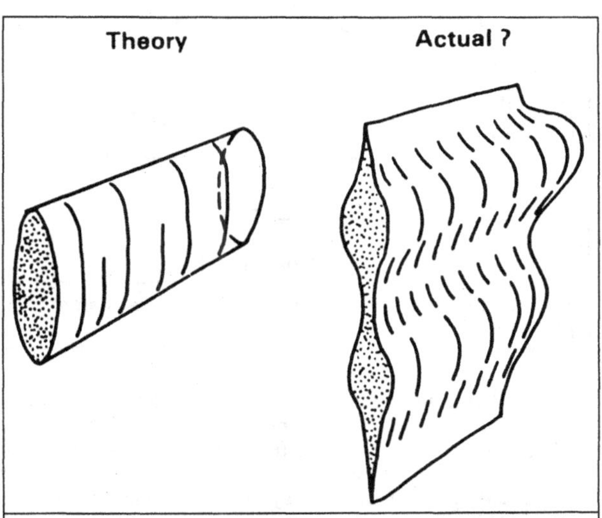

Fig. 1.9—Theoretical fracture-propagation models vs. possible actual in-situ behavior.[5]

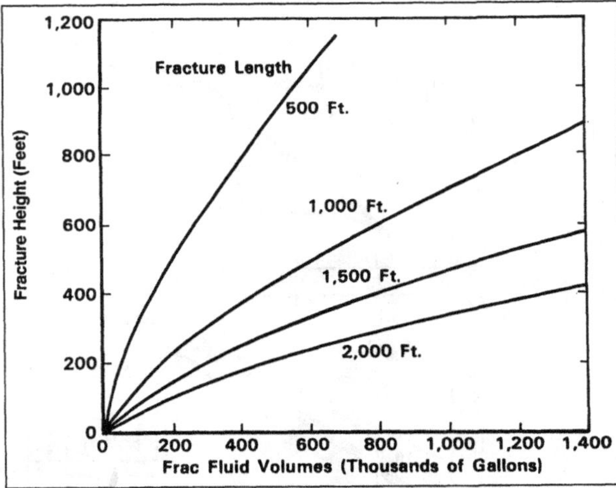

Fig. 1.10—Simulation model fracture length and height calculations.[5]

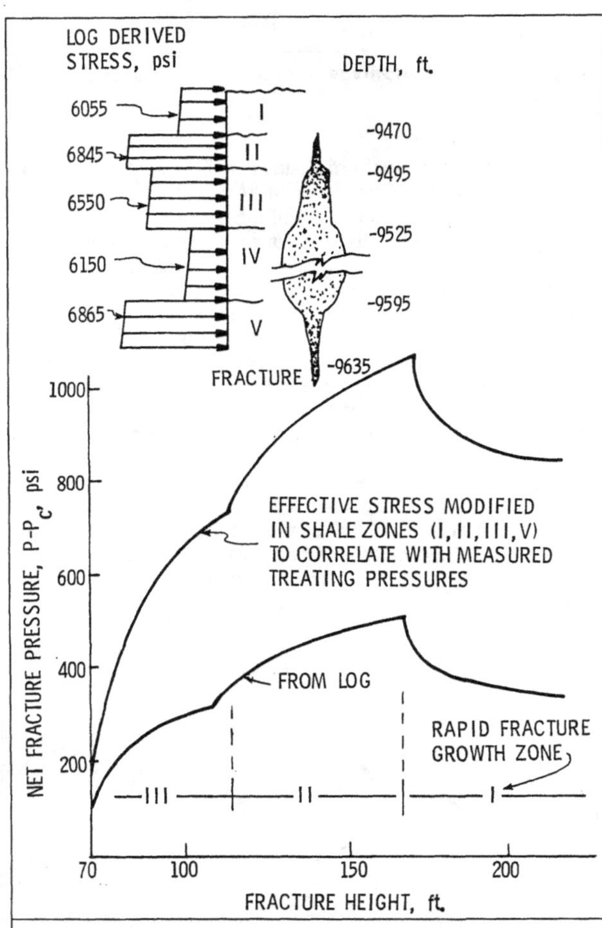

Fig. 1.11—In-situ stress profiles; fracture height vs. pressure.[5]

[305 to 610 m], fractures are usually oriented vertically, as shown in Fig. 1.8a. Vertical fracture growth can be inhibited or stopped by higher lateral stresses in the formations above and below the fracture initiation zone, as shown in Fig. 1.8b.

Stress changes between rock layers can influence fracture configuration significantly. Our commonly used equations for theoretical fracture propagation presume a rather simple fracture configuration, such as that shown on the left side of Fig. 1.9. Ex-

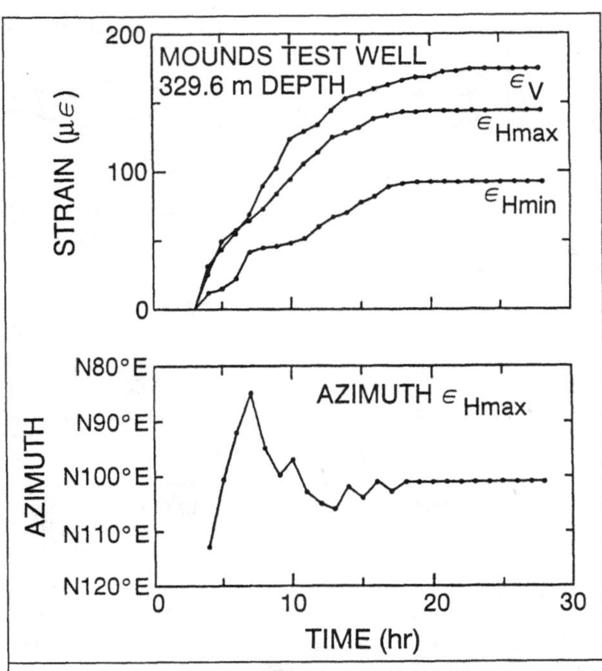

Fig. 1.12—Plot of temperature-corrected principal strains and azimuth of the maximum horizontal strain as a function of time.[33]

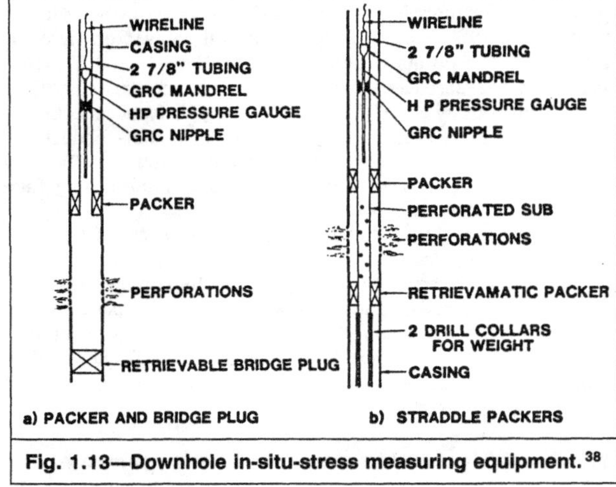

Fig. 1.13—Downhole in-situ-stress measuring equipment.[38]

perience leads us to believe that more complicated fracture configurations, such as that on the right side of Fig. 1.9, are more often the case.

Knowledge of vertical fracture height is extremely important in design. Fracture height has a significant effect on fracture length. In Fig. 1.10, we see the results of fracture-length calculations for a number of different fracture heights. For these data, we see that fracture length is essentially inversely proportional to fracture height. This emphasizes the importance of having reliable fracture-height information in treatment designs. Common methods for investigating vertical growth are posttreatment temperature-decay profiles and/or radioactive-tracer profiles.

In many cases, fracture heights may grow instead of remaining constant throughout treatment. When this occurs, methods to estimate the growth profiles must be developed. This requires conducting special in-situ-stress and fracture-mechanics studies[23–26] to arrive at profiles like that shown in Fig. 1.11. This graph shows a relationship between fracture height vs. fracture pressure estimated for the in-situ-stress profile depicted in the upper portion of Fig. 1.11. Data like these can be used to improve fracturing treat-

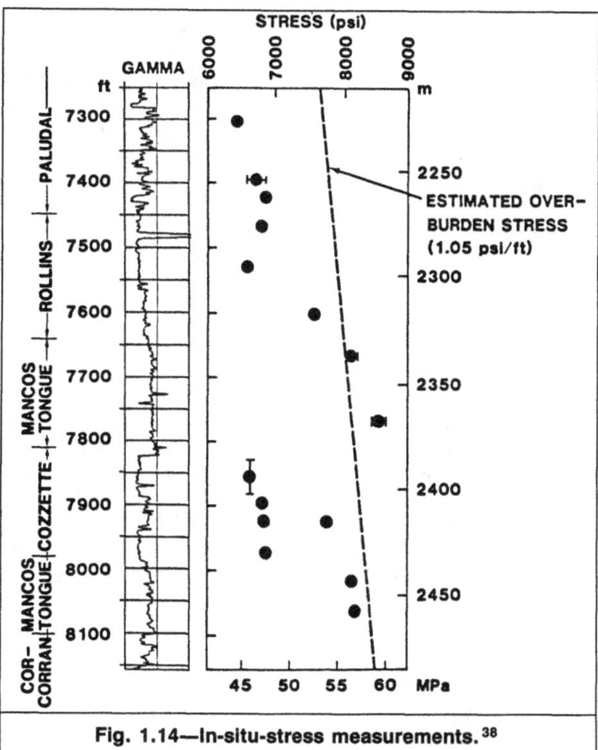

Fig. 1.14—In-situ-stress measurements.[38]

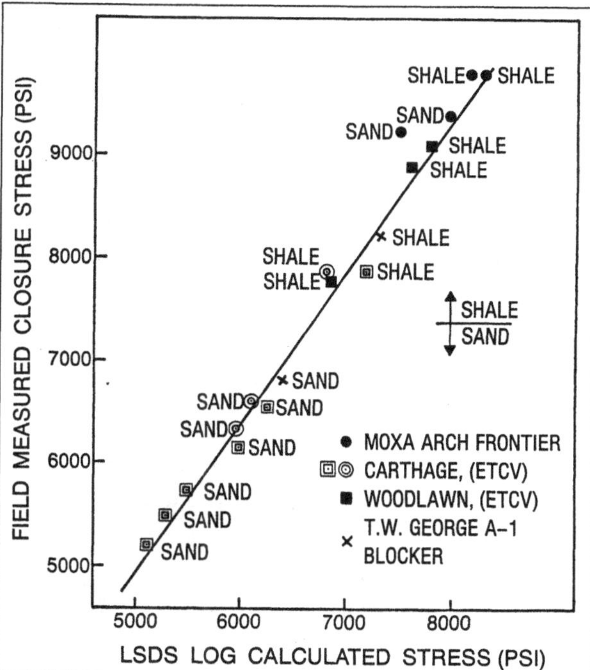

Fig. 1.15—Comparison of in-situ stresses calculated from long-spaced sonic logs vs. data measured by pump-in tests.[3]

ment effectiveness in formations known to have undesirable vertical fracture growth tendencies.

There has been significant progress in the measurement and profiling of in-situ stresses to extend the technology introduced by previous investigators.[27-31] Methods have been published for determining both the magnitude and direction of in-situ stresses using anelastic strain recovery data from oriented cores.[32-37] Fig. 1.12 shows an example of the type of test results recorded by these methods. Teufel[33] concluded that the technique generally is reliable for estimating the direction, but not as reliable as pump-in tests for determining the magnitude of principal stresses.

Pump-in methods [e.g., pump-in/shut-in pressure decline, instantaneous shut-in pressure (ISIP), pump-in/flowback, step-rate tests, etc.] have become the most prevalent procedures for measuring in-situ stresses. Techniques like those proposed by Warpinski *et al.*[38] have refined the method to yield relatively reliable results. Fig. 1.13 shows an example of the wellbore downhole closure tools they used for testing. Pumping small volumes (e.g., 1 to 2 bbl [0.16 to 0.32 m^3]) and shutting in the well downhole minimizes wellbore volume effects and improves the potential for definitive measurements. Another method tested by Veatch and Moschovidis[3] as an alternative to downhole shut-in equipment uses a constant-rate flowback control device at the surface to improve in-situ-stress measurements from the pump-in/flowback method. Daneshy *et al.*[39] presented a technique for conducting in-situ-stress measurements during drilling operations. This method uses a packer in the openhole section very near the bottom of the hole. In addition to stress data, oriented cores from the formation immediately below the fractured openhole section also provide information about the azimuthal tendencies of a fracture.

It has become apparent that in-situ stresses can vary significantly between adjacent formations. Data by Warpinski *et al.*,[38] depicted in Fig. 1.14, show in-situ-stress differences >2,000 psi [13.8 MPa] occurring over relatively small vertical intervals (e.g., <100 ft [30 m]). Large stress differences have also been reported (see Fig. 1.15) by Veatch and Moschovidis[3] in the east Texas Cotton Valley (ETCV), Wyoming Moxa Arch, and Colorado Wattenberg tight-formation gas plays. In view of the major effect that in-situ-stress profiles have on fracture-propagation geometry, it is very important to have methods to determine them reliably. There have been some successful efforts with acoustical wavetrain (i.e., shear and compressional velocity) measurements to profile in-situ stresses. Laboratory work[40-42] and field data[43] verified that these methods have some potential for use in in-situ-stress profiling. In-situ stresses are estimated from observations of acoustical velocity changes resulting from stress changes on cores. Some results obtained with long-spaced digital sonic (LSDS) logs (i.e., acoustical wavetrain data) corroborated with pump-in stress tests showed a good correlation between these two methods. Fig. 1.15 shows a comparison of stresses calculated from acoustical wavetrain data obtained with an LSDS log vs. those measured by pump-in stress tests. These data include results from both sands and shales in the Moxa Arch formations and the Blocker, Carthage, and Woodlawn fields in the ETCV play. The excellent correlations observed here may result partially from some inherent geological similarity between these particular tight-gas formations. Warpinski *et al.*[38] did not observe as close an agreement in studies of the Colorado Mesaverde group formations shown in Fig. 1.16. Here, fracturing gradients measured from pump-in tests were compared with those calculated from LSDS logs. However, acoustical wavetrain measurements still appear to be useful for inferring in-situ-stress profiles for many applications. For reliable use, they will probably have to be developed for each particular field or geological horizon in a given area.

A relatively comprehensive investigation of the effect of the in-situ-stress profile on vertical fracture growth in the Mission Canyon Ratcliffe formations in North Dakota was conducted by Begnaud and Claiborne,[44] who presented several methods of determining horizontal stress differences between the pay zones and their bounding formations. These included in-situ measurements (using acid treatments, minifracture, and pump-in/flowbacks), differential-strain-curve analyses, conventional core analyses, and LSDS logs. Their study demonstrated good agreement between field techniques and theoretical calculations used in analysis of vertical fracture growth. Their findings indicated that vertical fractures were not confined within the pay zone during either Mission Canyon or Ratcliffe fracture treatments.

1.4 Fracture-Propagation Models

Two-Dimensional Models. In two-dimensional (2D) fracture models, during propagation fracture height is assumed to be constant and the dimensions that change are width and length (or radius).

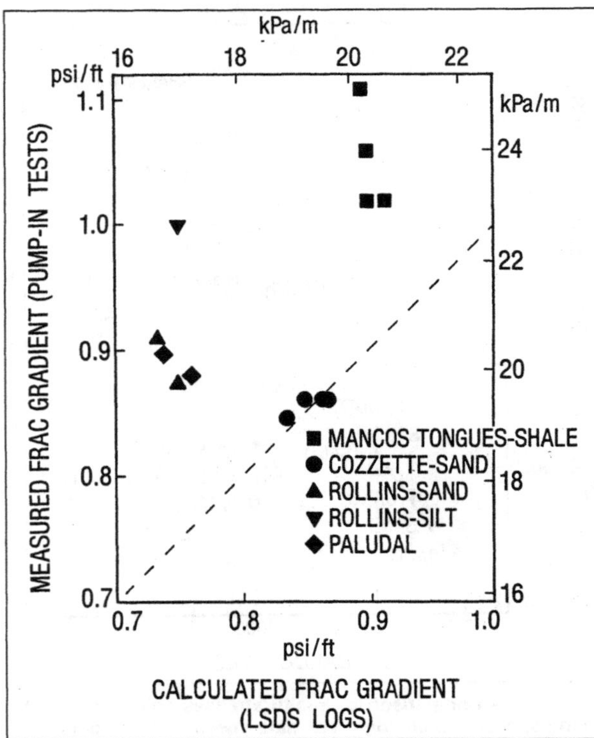

Fig. 1.16—Comparison of measured and calculated fracture gradients.[38]

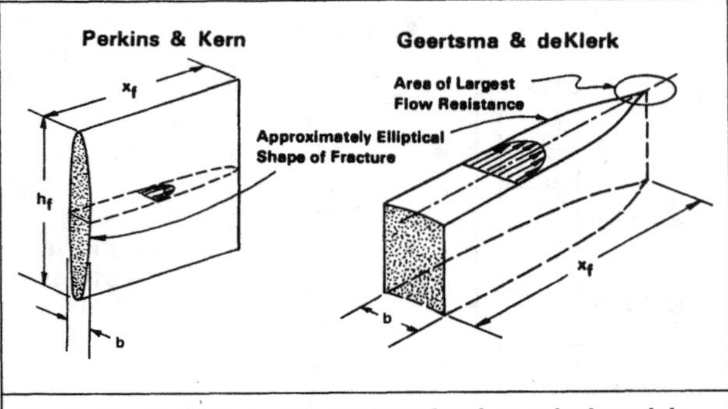

Fig. 1.17—Fracture configurations for theoretical models—PK vs. KGD.[5]

Two basic types of approaches are commonly used in 2D fracture-propagation simulators: one presented by Perkins and Kern[45] using premises published by Sneddon[46] and one by Geertsma and de Klerk,[47] who published an approach based on earlier work by Khristianovitch and Zheltov[48] and by Barenblatt.[49] As depicted in Fig. 1.17, the two approaches differ basically in that the Perkins-Kern model is developed from the premise that the cross section of the fracture in the vertical plane, perpendicular to the long axis of the fracture, generally maintains an elliptical configuration. The Geertsma-de Klerk approach presumes an approximately elliptical configuration in the horizontal plane and a rectangular shape in the vertical plane. Development of the Perkins-Kern model begins with fracture width expressed in terms of fracture height:

$$w \sim \frac{h_f p}{E} \qquad (1.1)$$

The Geertsma-de Klerk development is based on width expressed in relation to fracture length:

$$w \sim \frac{x_f p}{E} \qquad (1.2)$$

Most of the computerized fracturing simulator codes developed to date incorporate the fluid-leakoff expressions proposed by Carter[50] for both the Perkins-Kern and Geertsma-de Klerk fracture-propagation models.

Using approaches similar to Nordgren's,[51] both Smith[52] and Nolte[26] observed that the Newtonian flow equation that relates fracture pressure to injection rate and fluid viscosity yields

$$p \sim \frac{(E^3 \mu q_i x_f)^{1/4}}{h_f} \qquad (1.3)$$

for the Perkins-Kern approach, and

$$p \sim \frac{(E^3 \mu q_i)^{1/4}}{h_f^{1/4} x_f^{1/2}} \qquad (1.4)$$

for the Geertsma-de Klerk model. Thus, for a given set of conditions, the Perkins-Kern model predicts that wellbore fracturing will increase proportionally with fracture length raised to approximately the one-fourth power, and the Geertsma-de Klerk method indicates that pressures decrease proportionally to fracture length raised to approximately the one-half power.

Widths calculated from the Perkins-Kern model are generally smaller than those computed by the Geertsma-de Klerk model; hence, the Perkins-Kern results predict a significantly longer fracture length for a given amount of injected fluid at a given rate, all other parameters being the same. Geertsma and Haafkens[53] presented a study comparing the two theories; the results are summarized in Table 1.1. The study included results computed by the two approaches and by two other modified approaches. One method[54,55] is similar in concept to the Geertsma-de Klerk method. The other was developed by Nordgren[51] from a Perkins-Kern-type approach.

The data in Table 1.1 raise a number of questions about differences in results computed by the various methods. Many are addressed by Geertsma and Haafkens[53] and are not discussed in detail here. It is obvious that differences do exist as a result of the basic premises used to develop the models. These premises should be considered so that the appropriate model fits the in-situ fracturing conditions best.

TABLE 1.1—COMPARISON OF FRACTURE-DESIGN CALCULATIONS* FOR DIFFERENT FRACTURING MODELS[53]

	Geertsma and de Klerk	Daneshy	Perkins and Kern	Nordgren
Pad volume, bbl	750	320	1,350	1,650
Proppant-laden fluid volume, bbl	1,250	1,680	650	350
Average sand concentration, lbm/gal	3	2.5	2.5	3.5
Total amount of sand, lbm	157,500	176,000	68,000	51,000
Viscosity after pad, cp	36	36	36	36
Created fracture length, ft	698	670	804	845
Effective fracture length, ft	486	453	240	185
Created fracture width, in.	0.22	0.43	0.17	0.16
Effective fracture width, in.	0.20	0.31	0.16	0.16
Effective fracture height, ft	98	97	94	85
Average fracture conductivity, darcy-ft	7.1	9.8	6.5	6.5

*Results calculated on the basis of different theories for predicting fracture dimensions.

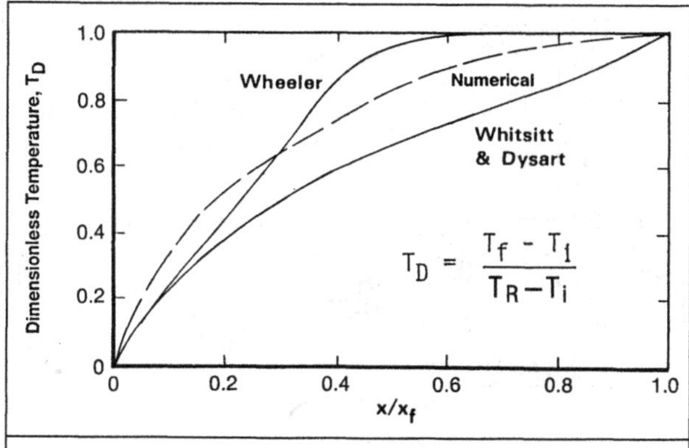

Fig. 1.18—Fluid-temperature profile predictions in fractures during treatment.[56]

Computerized Fracturing Simulators. These simulators, commonly used throughout the industry today, vary in complexity from rather simple models that handle only constant fracture height and constant fluid properties to very sophisticated methods where these parameters change throughout the treatment. The sophisticated models can incorporate vertical growth during treatment; variations in fluid rheological properties with temperature, shear rate, and time; and variations in fluid loss with pressure and temperature. The higher degrees of sophistication require more comprehensive sets of input data on formation properties and fluid behavior.

Most models use power-law theory for determining pipe friction losses, fracture width resulting from friction losses in the fracture, and hydraulic horsepower requirements. Most of today's fracturing fluids are non-Newtonian. For these, it is common practice to substitute an apparent viscosity, μ_a, value for the Newtonian viscosity term, μ, in the classic flow expressions. Here, apparent viscosity is computed by

$$\mu_a = \frac{47{,}880K}{\dot{\gamma}^{(1-n)}}, \quad \text{(1.5)}$$

where μ_a, $\dot{\gamma}$ (shear rate), and K (consistency index) are expressed in cp, seconds^{-1}, and lbf·sec/ft^2, respectively, and n (flow-behavior index) is dimensionless. The expressions used for pipe, perforation, and fracture-fluid rheology are discussed in Sec. 1.8.

Fluid-Temperature Profiles. A variety of methods have been presented to incorporate computations to predict fluid-temperature profiles in the fracture during treatment. Some of the more commonly used approaches are covered by Settari[56] in a discussion of methods by Sinclair,[57] Harrington *et al.*,[58] Wheeler,[59] and Whitsitt and Dysart.[60] The results are shown in Fig. 1.18.

Sinclair characterizes leakoff in terms of fluid efficiency (defined as the volume of the fracture divided by the total injected volume at the end of injection). At low fluid efficiency, the fluid temperature in the fracture, T_f, remains close to the injection temperature, T_i, and with increasing efficiency, the temperature along the fracture tends exponentially to the reservoir temperature, T_R. The numerical solution gives faster heat-up at the entrance than both analytical models and falls between the two toward the fracture tip. These differences result mainly from coupled leakoff calculations as opposed to assumed leakoff rate distributions (constant for Wheeler and linearly increasing for Whitsitt and Dysart).

Three-Dimensional Models. The majority of the 3D models used today assume that the fracture is planar and remains planar during propagation. The general problem of a curved hydraulic fracture in a layered formation (fully 3D model) is so computationally demanding that it is not practically feasible at this time. However, some models with curved fractures in the sense of 2D elasticity have been presented by Ingraffea *et al.*[61] and Narendran and Cleary.[62]

The planar three-dimensional (3D) hydraulic fracturing models presented in the literature include (1) lumped-parameter models, (2) pseudo-3D (P3D) models, and (3) general 3D fracture models.

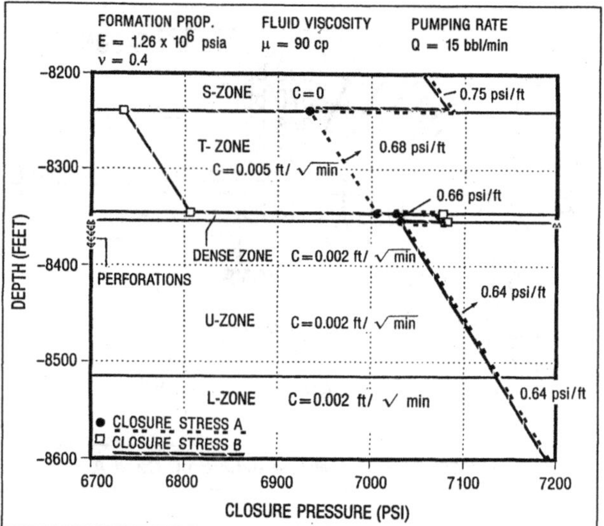

Fig. 1.19—Estimated in-situ-stress profiles and other data for 3D simulation of Field Case Studies A and B.[3]

These models are so complex that a detailed description lies beyond the scope of this chapter, but Mendelsohn[63,64] gives an excellent treatment of their main features.

The basic elements of 3D fracturing models are (1) a crack-opening model, (2) a fluid-flow model, (3) a crack-propagation criterion, and (4) (when numerical solutions are performed) a fracture-propagation algorithm. The fracture-propagation algorithm combines the fluid-flow and fracture-opening interaction into a highly nonlinear coupled problem that satisfies the fracture-propagation criterion while furnishing a numerical solution. All models treat the fracture process in a "quasistatic" sense: inertial terms are neglected both in the fracture-opening and fluid-momentum equations. The formation is assumed to be linearly elastic, and the fracturing criterion is formulated with Griffith's[65,66] approach in terms of the formation fracture toughness (i.e., critical stress-intensity factor, K_c). For most models, flow inside the fracture is approximated with equations for laminar flow of a Newtonian or power-law fluid between parallel plates. Leakoff is usually considered as one-dimensional (1D) and perpendicular to the surface of the fracture. Leakoff velocity, v_L, is assumed to be given by Carter's[50] formula. Recently, temperature and pore-pressure effects on closure stress have also been estimated with simple 1D models, such as those suggested by Keck *et al.*[67]

Crack Opening. The general crack-opening problem can be formulated and solved with finite-element or boundary-integral-equation techniques. The finite-element method can be applied to determine fracture width for any fracture shape (planar or curved) and for both homogeneous and inhomogeneous (layered) formations. It has been successfully applied in a planar 3D fracture model without fluid flow by Morita *et al.*[68] and in a model for curved 2D hydraulic fractures with branches (in the 2D elasticity sense) by Ingraffea *et al.*[61] Finite-element methods, however, generally are computationally demanding compared with boundary-integral-equation methods.

The boundary-integral-equation method is based on the influence (Green's) function approach and reduces the problem to singular integral equations on the fracture plane.[69,70] These equations can be solved numerically by discretization on the domain of the fracture by finite-element,[71] collocation,[70,72] or finite-difference methods. The boundary-integral-equation method can be practically applied for homogeneous formations for which the Green's function[73] is well known. It has been used for the majority of the fracturing models, both 2D (simple[74] and complex[62]) and 3D.[70–72] Integral equations derived from dislocation[75,76] or elastic potential theories[69,77,78] are commonly used. A different boundary-integral equation formulation has also been used by Mastrojannis *et al.*[70,79]; however, this model does not have fluid flow and uses hydrostatic pressure to calculate crack opening.

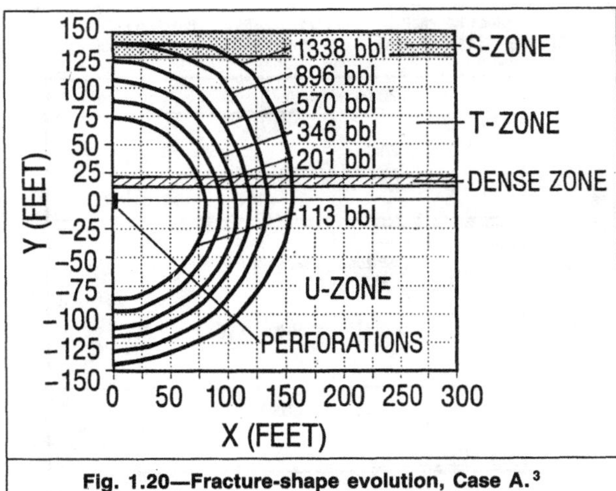

Fig. 1.20—Fracture-shape evolution, Case A.[3]

Efforts to model crack-opening behavior in layered inhomogeneous formations have also been undertaken.[80-83]

Fluid Flow. The common approach to fluid flow as related to hydraulic fracturing is to integrate the continuity and momentum equations across the width of the fracture and to derive 2D equations in the fracture plane. This assumes laminar flow of a Newtonian or power-law fluid between parallel plates. Integration procedures by Nemat-Nasser and Ohtsubo[84] were used by Clifton and Abou-Sayed[85] to develop 2D flow equations for a 3D model. Numerical solutions can be obtained either with finite-element or finite-difference methods. The finite-element method was applied both for 3D simulators[71,72,85,86] and simpler ones in which 1D flow (simple-geometry, lumped, and P3D models) is assumed.[87] The finite-difference method is more commonly applied for 1D fluid-flow models in 2D fracture-propagation simulators.

Fracture-Propagation Algorithm. Virtually all crack-propagation algorithms are iterative in nature, using an implicit or explicit finite-difference approximation of time derivatives. Timestep and crack advancement are usually related by a crack-propagation criterion expressed from linear fracture mechanics for a stable crack. A simple yet effective way to enforce a critical stress-intensity factor at the crack front is to estimate a critical crack opening, w_c, at a given distance behind the crack front. The actual crack opening for a stable crack should not exceed this estimate. Values for w_c can be estimated in terms of K_c from 2D crack displacements[88] by

$$w_c=4(1-\nu)(K_c/G)\sqrt{r/2\pi}\,, \quad\quad (1.6)$$

where

r = specific distance behind the crack front,
G = shear modulus, and
ν = Poisson's ratio of the medium.

Application of 3D Model to Field Cases. The application of 3D simulators is important primarily for complex reservoir conditions—i.e., where there are multiple zones with varying elastic properties and leakoff characteristics and where closure-stress profiles dictate complicated fracture geometries. For such in-situ conditions, the fracture shape is unknown *a priori* and, depending on in-situ parameters, can be drastically different from the shape the P3D simulators[89-99] can predict. For these complex simulations, a 3D fracturing model is required. Abou-Sayed *et al.*[100,101] presented a case study using such a 3D model that quantifies the influence of various in-situ conditions on fracture geometry.

3D-Simulation Examples. The two examples are from an actual field case study. Fig. 1.19 shows the two closure-stress profiles and the other field parameters used in the study. Case A represents the base case; Case B has a closure stress in Zone T that is 200 psi [1.4 MPa] lower relative to Case A (attributed to a pressure-drawdown scenario after production) and a closure stress 50 psi

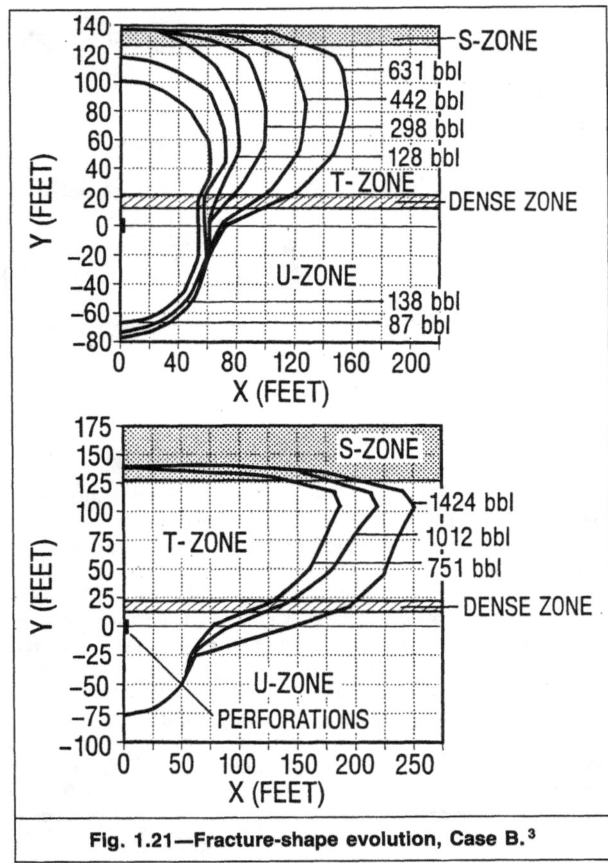

Fig. 1.21—Fracture-shape evolution, Case B.[3]

[345 kPa] higher in a dense streak ("dense zone") at the upper portion of Zone U.

Completion experience in the field established that the target zone (Zone T) should not be directly perforated because of severe solids-production problems. Zone U, located directly below Zone T, is perforated instead. Treatments initiated in Zone U have the dual purpose of stimulating Zone U and communicating with Zone T.

In Figs. 1.20 through 1.22, Point 0 on the y axis corresponds to the midpoint of the perforations in Fig. 1.19—a depth of 8,366 ft [2550 m]. Fig. 1.20 shows the computed fracture-shape evolution for the Case A stress profile for injected volumes ranging from 113 to 1,338 bbl [18 to 213 m^3]. Note that the fracture essentially remains approximately "penny-shaped," although some confinement can be observed at the interface of Zones S and T.

Fig. 1.21 shows the simulated fracture-shape evolution for the Case B stress profile where injected volumes ranged from 87 to 1,424 bbl [14 to 226 m^3]. The resulting shape is drastically different from that of Case A. The fracture grows mainly in Zone T, where closure stress is low. This type of behavior can be quantified only by numerical simulation and represents a delicate balance between in-situ values of closure stress, closure gradients, leakoff, perforation location, and fluid rheology.

Fig. 1.22 compares the fracture-width profiles along the wellbore for Cases A and B. These are the pictures one would see by looking at the fracture from the wellbore. In Case A, the maximum fracture width occurs close to the perforations. In Case B, the fracture grows unsymmetrically with respect to the perforations, and a point of reduced width, referred to as a width "pinch point," develops there. Width pinching near the perforations may cause an undesirable screenout during the early stages of the treatment. Fig. 1.23 shows the fracture-width history for both cases as a function of injected volume. The maximum fracture width (wherever it might occur) and the fracture width at the perforations (i.e., at y=0.0 ft [0.0 m] in Figs. 1.20 through 1.22) are plotted vs. the total injected volume. In Case A, we see no significant difference between these two values. The width at the perforations approxi-

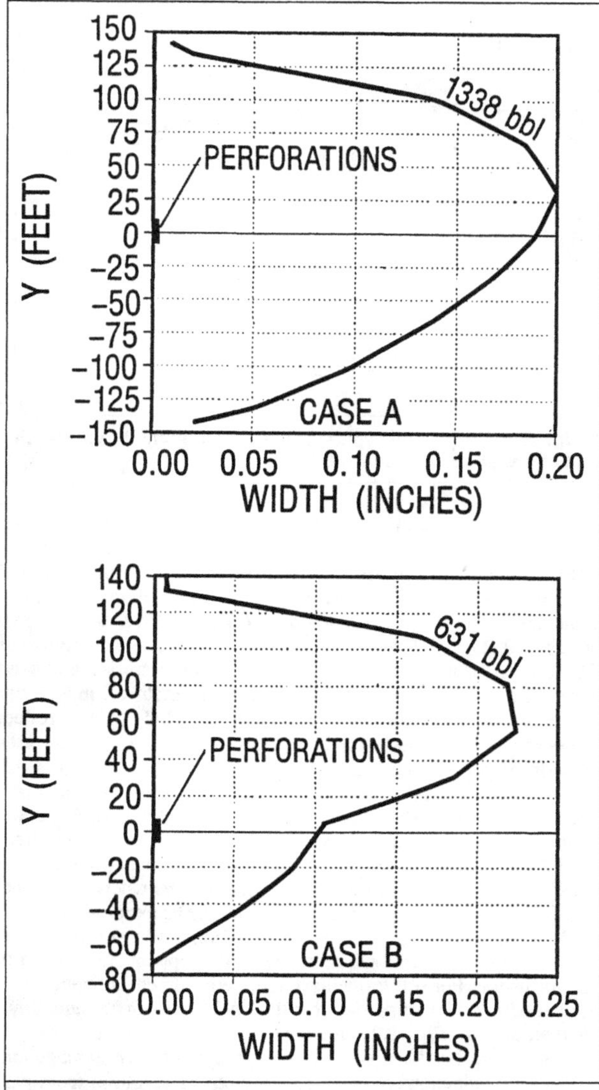

Fig. 1.22—Fracture-width profile at the wellbore, Cases A and B.[3]

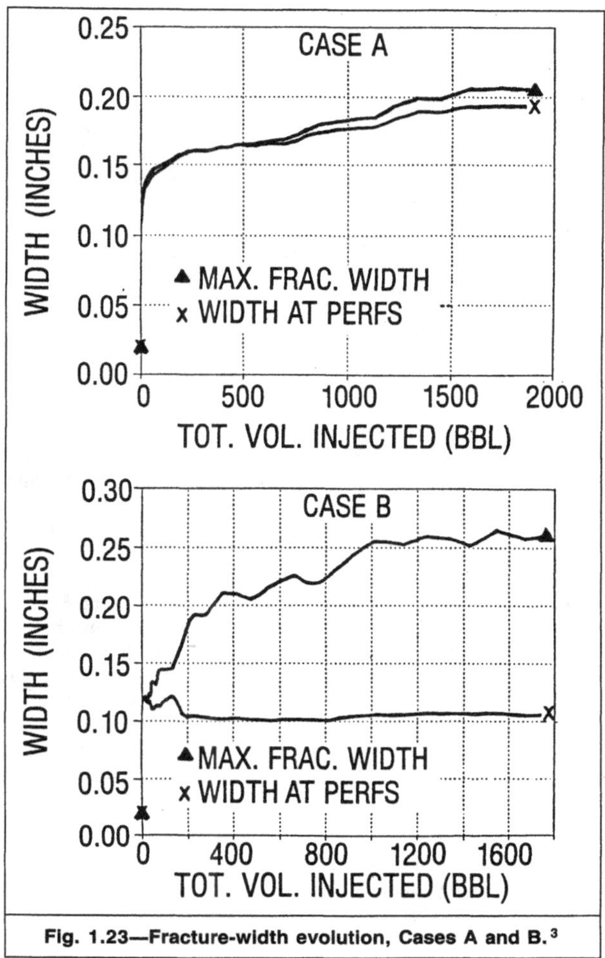

Fig. 1.23—Fracture-width evolution, Cases A and B.[3]

mately equals the maximum width, and they both increase with treatment volume. As expected, in Case B, the maximum width occurs in Zone T and increases with the volume injected. However, the width at the perforations initially increases (while the fracture is still penny-shaped) and subsequently decreases after about 200 bbl [32 m^3] injection, to remain constant at approximately 0.10 in. [0.25 cm] for the remainder of the treatment. For Case B's in-situ conditions, an increased pad volume does not diminish the danger of screenout. A higher-viscosity fluid and small-diameter proppant may be required to pump the fracturing treatment successfully without an undesirable screenout. Note that the width at the perforations can actually decrease during pumping of the treatment, especially when unconfined, unsymmetric fracture growth occurs. Such behavior can be quantified only through a 3D fracture simulation.

The width-history plot may be used to estimate the pad volume and the total treatment volume so that proppant may be introduced at the time the fracture has attained sufficient width. The maximum proppant size may also be estimated. For example, Case B allows at most a 20/40-mesh proppant with a maximum proppant diameter of 0.0331 in. [0.084 cm] to be pumped.

In summary, planar 3D simulators are very valuable for many aspects of hydraulic fracturing analysis and design. They can be used (1) to determine the fracture shape for given in-situ and pumping conditions; (2) to estimate proppant size, pad volume, and treatment volume from the fracture-width and fracture-dimension histories; (3) to study the effect of the location of the perforations and the associated problems of width pinching; and (4) to diagnose in-situ closure-stress features by comparing the actual minifracture pressure with simulated pressure.

1.5 Propping Agents and Fracture Conductivity

Propping Agents. The function of a propping agent (proppant) is to hold the fracture open after fluid injection is stopped and the fracturing fluid has been removed. The reservoir fluids can then flow from the extremities of the fracture to the wellbore through the highly conductive proppant pack.

Silica sand is currently the most commonly used proppant material in the U.S. The ready availability and low cost of high-quality sand that can provide good fracture conductivity for a wide range of conditions make it very attractive for use in fracture stimulation. The American Petroleum Inst. (API) has established sand-quality specifications for use in fracturing treatments.[102] These basically cover size distribution, sphericity and roundness, solubility in acid, silt and clay content, and crush resistance. The size designations for fracturing given in Table 1.2 relate to the corresponding size openings of sieves used in processing the sands. The remainder of the specifications are described in the API publication[102] for the interested reader.

Another commonly used material is sintered bauxite (aluminum oxide). It is significantly stronger than sand and is used in deep formations where high fracture closure stresses severely crush sand.

Fig. 1.24, a comparison of fracture conductivity vs. depth (which can be related to closure stress) for sand and sintered bauxite, illustrates the improvement achievable with sintered bauxite in the deeper formations. This improvement is especially pronounced at depths below the 8,000- to 10,000-ft [2440- to 3050-m] range where

TABLE 1.2—API FRACTURING SAND SIZE DESIGNATION [102]

Mesh Range Designation	Range (μm)
Primary Sizes	
12/20	850 to 1,700
20/40	425 to 850
40/70	212 to 425
Alternate Sizes	
6/12	1,700 to 3,350
8/16	1,180 to 2,360
16/30	600 to 1,180
30/50	300 to 600
70/140	106 to 212

sand crushes under the high stresses. Sintered bauxite has a higher density than sand (a specific gravity of 3.5 to 3.7, compared with 2.65 for sand); thus, it is not easily transported by a fracturing fluid.

A variety of materials—alumina, cordierite, mullite, silicon carbide, and some ceramic oxides[103-106]—have been introduced for use as proppants. Certain ceramics have been developed commercially to meet an intermediate-density requirement between sand and sintered bauxite. Zirconium oxide is available on the European market as an alternative to sintered bauxite, but it has not entered the U.S. market significantly. Resin-coated proppants[107] (i.e., sand coated with a polymeric material) have been introduced to the market. One purpose of the coating is to relieve high stresses caused by grain-to-grain contact and thus to improve the load-carrying capacity of the proppant pack. Another is to consolidate the proppant pack with proppant particles adhering to one another.

Fracture Conductivity. Studies[108-114] and comprehensive data sets recently published by the stimulation service companies and proppant manufacturers[115-121] provide extensive resources for fracture-conductivity laboratory test results.

To estimate fracture-conductivity values that might be expected from a proppant, (1) type of proppant, (2) proppant size distribution, (3) proppant concentration in the fracture, (4) the stress load on the proppant pack (usually related to depth and reservoir pore pressure), (5) formation embedment characteristics, (6) potential plugging from fracturing-fluid residue, and (7) long-term degradation under the in-situ environment should be considered.

Work is currently under way to quantify conductivity reduction as a result of the long-term in-situ effects. Long-term deterioration under in-situ environmental conditions can significantly reduce fracture conductivity after initial placement. Reductions are possible as a result of (1) corrosion stress cracking and/or dissolution of the proppant by the fluid/stress/temperature environment, (2) the long-term effects of stress and/or temperature, and (3) fines migration and redistribution. At this writing, definitive data are still in the making, but a conservative estimate would be a 5- to 10-fold reduction below short-term, ambient-temperature test values.

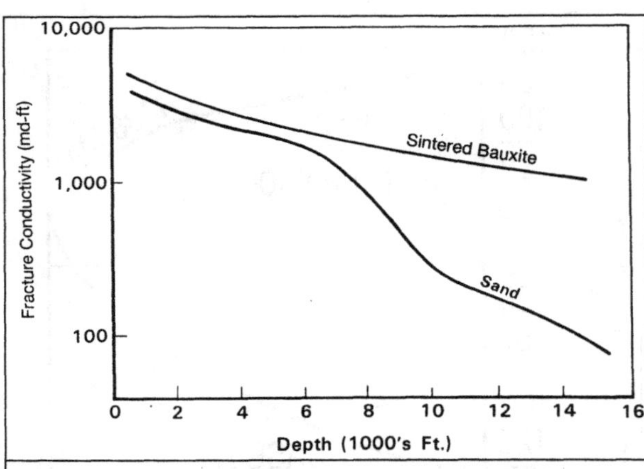

Fig. 1.24—Expected typical fracture conductivity vs. depth (closure-stress gradient = 0.7 psi/ft depth).[4]

Fracture-conductivity laboratory flow tests performed in radial or linear flow cells with the proppant subjected to a fracture-simulated load are used to develop design data such as those shown in Fig. 1.25.[117] This figure shows fracture conductivity of 20/40-mesh sand for a wide variety of closure stresses and sand concentrations in the fracture. The peaks in fracture conductivity at concentrations around 100 lbm/1,000 ft^2 [0.49 kg/m^2] reflect the existence of a partial monolayer. As sand concentration increases to a fully packed monolayer in the 200- to 400-lbm/1,000-ft^2 [0.98- to 1.96-kg/m^2] range, conductivity declines. The increase for concentrations above 500 lbm/1,000 ft^2 [2.4 kg/m^2] results from the multiple proppant layers in wider fractures. In most fracture designs, it is advisable to try to achieve concentrations of at least 1 lbm/ft^2 [4.9 kg/m^2]. In vertical fractures where proppants can fall to the lower part of the fracture, it may be extremely difficult to design a treatment to achieve a partial monolayer. Another problem usually associated with monolayers is proppant embedment in the formation. For soft formations where embedment is severe, fracture conductivities of partial or full monolayers can be much lower than shown. Fig. 1.25 does not reflect this.

Some work has been conducted on proppant-pack damage and plugging. Reductions in fracture conductivity resulting from par-

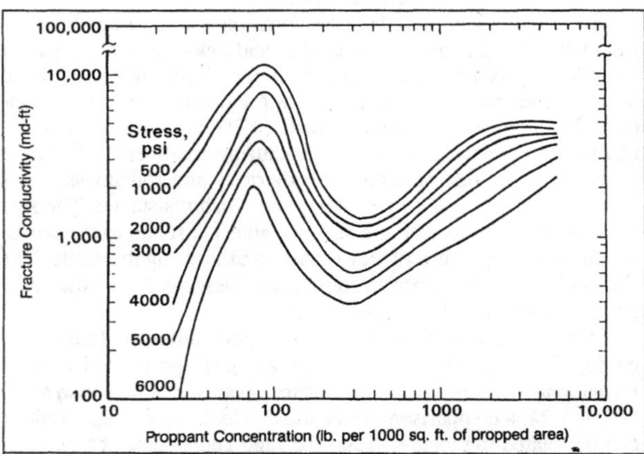

Fig. 1.25—Fracture conductivity, closure stress, and proppant concentration—20/40 fracturing sand.[117]

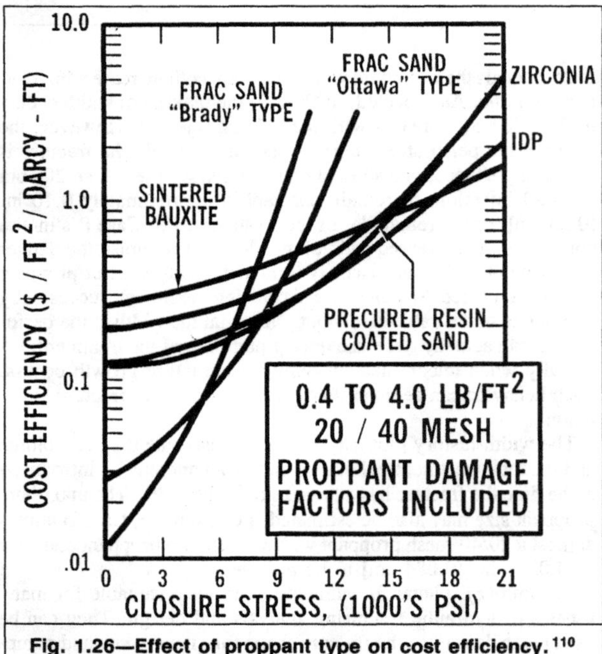

Fig. 1.26—Effect of proppant type on cost efficiency.[110]

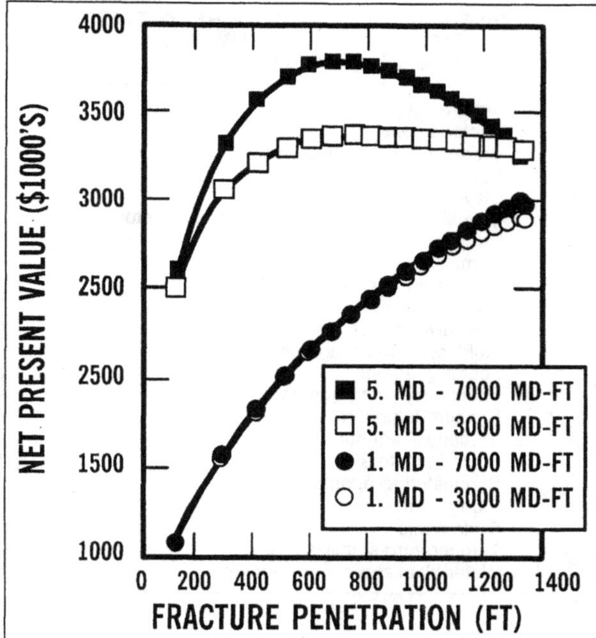

Fig. 1.27—Net present value vs. fracture length—1- and 5-md permeabilities and 3,000- and 7,000-md-ft fracture conductivities.[126]

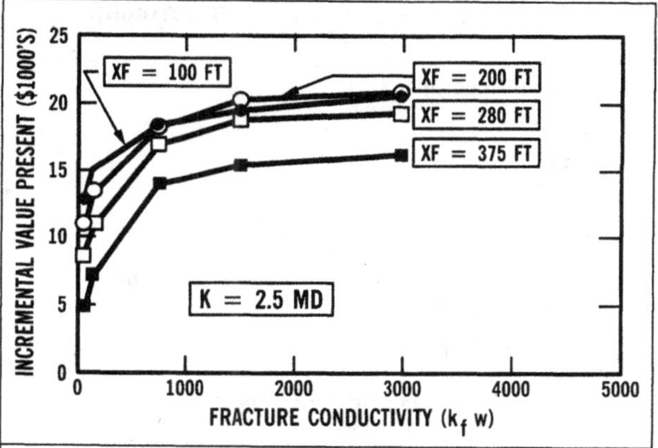

Fig. 1.28—Incremental present worth vs. fracture conductivity for fracture lengths of 100 to 375 ft, 2.5-md formation.[126]

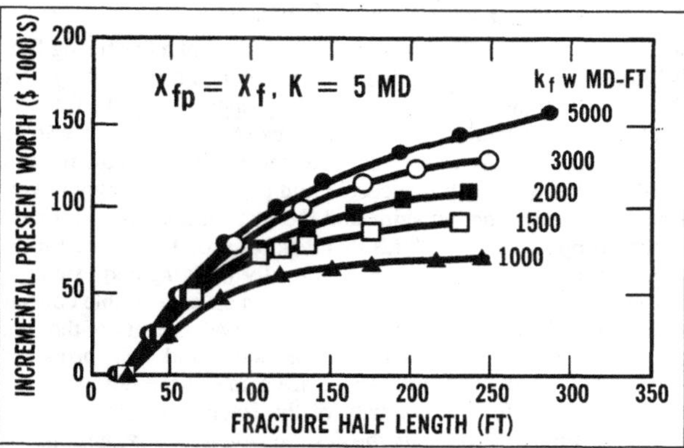

Fig. 1.29—Incremental present worth vs. fracture half-length for fracture conductivities from 500 to 5,000 md-ft, 5.0-md formation permeability.[126]

tial plugging by fracturing-fluid residue are discussed by Cooke[109] and Almond.[112] Kim and Losacano[122] conducted proppant-pack-damage tests on 20/40-mesh sand for different fracture fluids over a wide range of closure stresses and at different temperatures. The results showed that fracture conductivities could be reduced by 40 to 60% just from plugging by the gel residue. Cheung[123] reported that various concentrations of HCl/HF-acid solutions can dissolve a significant amount of proppant, which would reduce conductivity. Unpublished work sponsored by Norten-Alcoa Co. suggests that highly siliceous proppants may degrade severely in brines at high temperatures. Almond and Bland[124] reported various ways in which break temperature and break mechanism (i.e., oxidizers and enzymes) play an important role in 20/40-mesh-sand proppant-pack flow impairment from guar, derivatized guar, and cellulose-based fluids. Phillips and Wilson[125] showed that using a solvent in the pad fluids with a surfactant in the rest of the fluid reduces water blocking in the fracture and significantly enhances fracturing-fluid recovery and production.

Phillips and Anderson[110] demonstrated a method to modify the traditional conductivity-vs.-closure-stress data to include the cost of various types of proppants, as shown by the curves in Fig. 1.26. These curves represent the cost per unit fracture area per unit of conductivity over a wide range of closure stresses. They account for a proppant-pack deterioration factor of about 20 to 25% below short-term ambient test values. Phillips and Anderson acknowledged the importance of having realistic estimates of in-situ conductivity values. Graphs like these can be constructed for a given formation and field with proppant price schedules, fracture-conductivity data (adjusted for long-term in-situ environmental effects), and hydraulic fracturing propagation simulators.

Work presented by Britt[126] and Veatch[127] demonstrated a few of the many economic approaches to optimize fracture-conductivity designs. The example in Fig. 1.27 compares net present values for two different fracture conductivities (i.e., 3,000 and 7,000 md-ft [915 and 2135 md·m]) in formations of 1- and 5-md permeability. Here, the cost of proppant (sand) for the 3,000-md-ft [915-md·m] case is $8.00/100 lbm [$0.18/kg]. For the 7,000-md-ft [2135-md·m] case, proppant (sintered bauxite) cost is $70.00/100 lbm [$1.54/kg]. The results indicate that a 3,000-md-ft [915-md·m] -conductivity fracture with a sand proppant design is adequate for the 1-md formation case. For the 5-md formation, use of the more expensive ($70/100-lbm [$1.54/kg]) sintered bauxite to achieve a 7,000-md-ft [2135-md·m] -conductivity fracture will yield a higher dollar return.

One should be very careful to avoid general rules of thumb and use results such as those presented to infer that sand is the optimum proppant for all 1-md formations or that sintered bauxite is optimum for all 5-md formations. In some formations, especially the deeper ones, it may not be possible to achieve the fracture conductivities used in the above examples. Also, as discussed in Chap. 17, many factors can affect the economic results, and one should investigate their range for each formation and their effects on economics.

A comprehensive fracture-conductivity/reservoir-performance study[126] addresses optimization of fracture conductivity for an oil reservoir under both primary and secondary depletion. It showed the economic benefits of high-conductivity short fractures for moderately permeable (i.e., 1- to 10-md) formations. The results depicted in Figs. 1.28 and 1.29 show the impact of different fracture lengths and conductivities on incremental present worth. By developing curves like these, one can determine the appropriate fracture-conductivity/length relationship required to maximize economic returns for a given reservoir.

Studies by Elbel[128] and Montgomery and Steanson[129] address methods for using reservoir-performance type curves and computerized simulators to determine the appropriate fracture-conductivity-design requirements for various reservoir permeability levels. Elbel's results support previous findings by Bennett *et al.*[130] that a varying conductivity in the fracture from the wellbore to the tip can significantly affect production rates. Studies like these can be very important in the determination of proppant placement and scheduling programs for a treatment design to ensure that the appropriate distribution of conductivity in the fracture is achieved.

TABLE 1.3—COMMONLY USED FRACTURING-FLUID SYSTEMS[4]

Water-based polymer solutions
  Natural guar gum (guar)*
  HPG*
  HEC
  Carboxymethyl HEC*
Polymer water-in-oil emulsions
  ⅔ hydrocarbon** + ⅓ water-based polymer solution†
Gelled hydrocarbons
  Petroleum distillate, diesel, kerosene, crude oil
Gelled alcohol (methanol)
Gelled CO_2
Gelled acid (HCl)
Aqueous foams
  Water phase—guar, HPG solutions
  Gas phase—nitrogen, CO_2

*Can be crosslinked to increase viscosity.
**Petroleum distillate, diesel, kerosene, crude oil.
†Usually guar or HPG.

TABLE 1.4—TYPICAL FUNCTIONS OR TYPES OF ADDITIVES AVAILABLE FOR FRACTURING-FLUID SYSTEMS[4]

Antifoaming agents
Bacteria-control agents
Breakers for reducing viscosity
Buffers
Clay-stabilizing agents
Crosslinking or chelating agents (activators)
Defoamers
Demulsifying agents
Dispersing agents
Emulsifying agents
Flow-diverting or flow-blocking agents
Fluid-loss-control agents
Foaming agents
Friction-reducing agents
Gypsum inhibitors
pH-control agents
Scale inhibitors
Sequestering agents
Sludge inhibitors
Surfactants
Temperature-stabilizing agents
Water-blockage-control agents

1.6 Fracturing Fluids and Additives

Fluid's Function in Fracturing. The purpose of a fracturing fluid is basically two-fold: to wedge open and extend a fracture hydraulically and to transport and distribute the proppant along the fracture.

The fluid(s) selected for a treatment can have a significant influence on the resulting effectively propped fracture length, fracture conductivity, and treatment cost. Fluid properties strongly govern fracture-propagation behavior and the distribution and placement of the propping agents. Fluids that leak off rapidly into the formation have a low efficiency in hydraulically wedging and extending a fracture. Fluid leakoff may also result in an undesirable concentration of residue in the fracture. The effective viscosity of the fluid controls the internal fracturing pressure and the proppant-transporting characteristics. These are some of the desirable features of a fluid for the majority of the fracturing treatments.

1. Low fluid loss to obtain the desired penetration with minimum fluid volumes.
2. Sufficient effective viscosity to create the necessary fracture width and to transport and distribute the proppant in the fracture as required.
3. No excessive friction in the fracture.
4. Good temperature stability for the formation being treated.
5. Good shear stability.
6. Minimal damaging effects on formation permeability.
7. Minimal plugging effects on fracture conductivity.
8. Low friction-loss behavior in the pipe.
9. Good posttreatment breaking characteristics.
10. Good posttreatment cleanup and flowback behavior.
11. Low cost.

Fluid Systems. Many different types of fluids are currently available for fracturing treatments. It is outside the scope of this chapter to give comprehensive details on all the fluids and their design information; what is given here is extremely general. For any given specific application, one must consult published engineering data or laboratory test data relevant to the conditions to which the fluid is subjected during treatment.

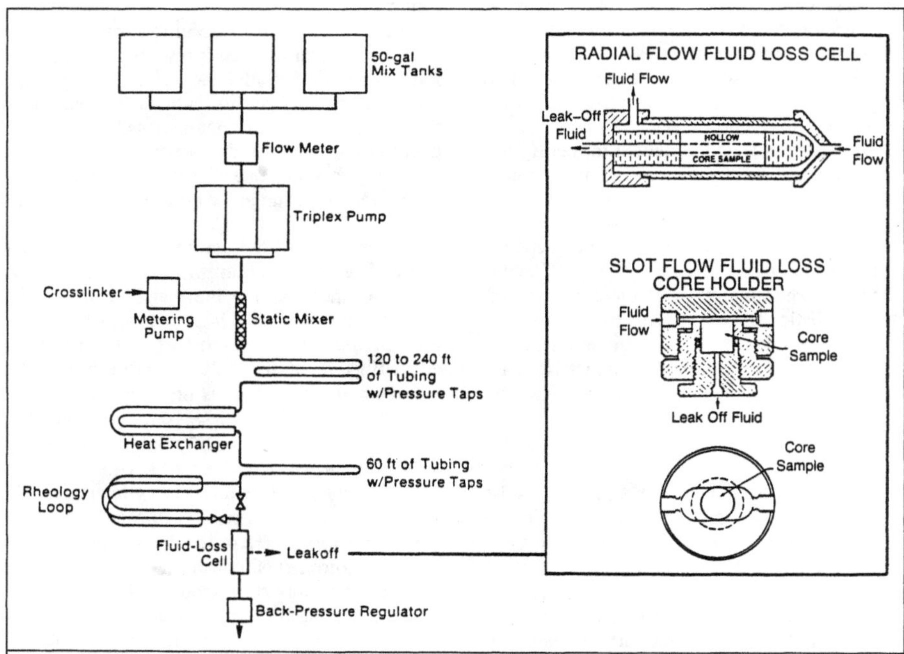

Fig. 1.30—Schematic of flow test and dynamic fluid-loss apparatus[142]; slot-flow fluid-loss core holder and radial-flow fluid-loss cell.[143]

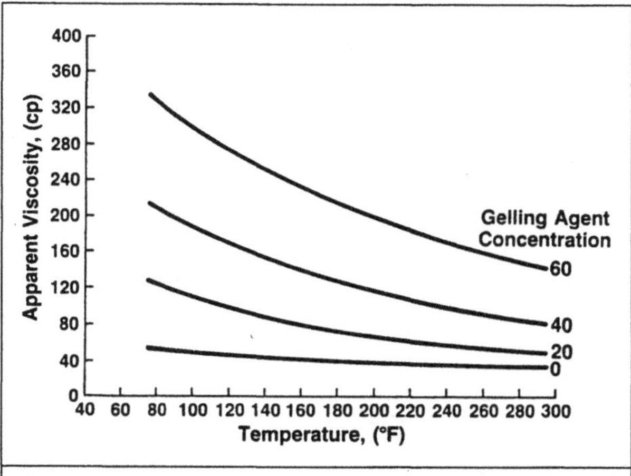

Fig. 1.31—Apparent viscosity at 170 seconds $^{-1}$ for 0 to 60 lbm HPG/1,000 gal water foamed to 70 quality.[136]

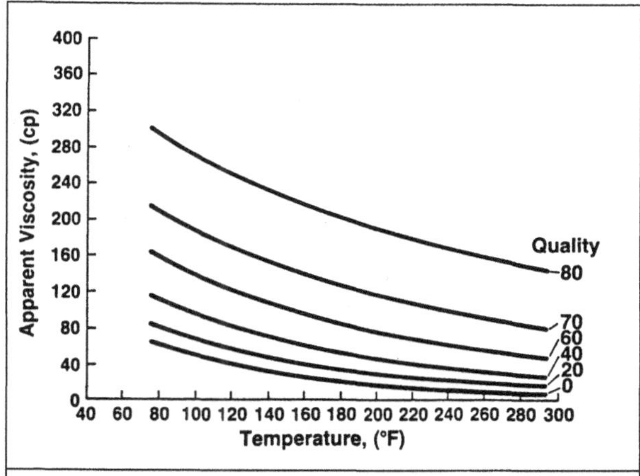

Fig. 1.32—Apparent viscosity at 170 seconds $^{-1}$ for 40 lbm HPG/1,000 gal water foamed from 0 to 80 quality.[136]

A list of the various types of fluids, or systems, that are commonly used today in the U.S. is given in Table 1.3. In addition to the basic fluid, many additives that perform different functions are available. A list of these and the functions they perform is given in Table 1.4. Some of the major considerations in designing a system include the following.

1. Formation temperature, fluid-temperature profile, and duration in the fracture.
2. Proposed treatment volume and pumping rates.
3. Type of formation (sandstone or limestone).
4. Potential fluid-loss-control requirements.
5. Formation sensitivity to fluids.
6. Pressure.
7. Depth.
8. Type of proppant to be pumped.
9. Fluid breaking requirements.

The water-based polymers, which constitute the majority of the applications, cover a wide range of formation types, depths, pressures, and temperatures and are relatively low in cost. Some [guar, hydroxypropyl guar (HPG), and carboxymethyl hydroxyethyl cellulose (HEC)], can be crosslinked to give them added viscosity and to expand their range of temperature application. Temperature stability has been enhanced by the addition of oxygen scavengers (e.g., thiosulfate salts and/or methanol). Significant fluid-loss enhancement is achieved by the addition of 5% dispersed hydrocarbon or with various concentrations and types of solid particulates.

The polymer emulsions in general provide somewhat better fluid-loss behavior, less potential for formation or fracture-conductivity damage, and possibly better proppant transportability. But they have upper temperature limits of about 250°F [120°C]. Also, they can be difficult to break at low temperatures and can be somewhat costly.

The gelled hydrocarbons are used primarily in water-sensitive formations where aqueous fluids may damage formation permeability. Gelled alcohols are also considered for formations that are subject to water blockage in the formation pores. Gelled CO_2 is claimed to have minimal formation-damage potential with good well-flowback characteristics. Gelled acids have proved to be very effective for stimulating carbonate reservoirs. One problem common to the gelled hydrocarbons, alcohols, CO_2, or acid systems is that they are considerably more costly than the water-based polymers.

In some instances, stages of different types of fluids might be used in the same treatment. For example, in limestone or dolomite formations, one might alternately inject small stages of gelled acid and gelled water. In low-pressure formations, it is not uncommon to spearhead a gelled water system with nitrogen, foam, or gelled CO_2 to enhance cleanup.

Foamed Fracturing Fluids. The use of foamed fracturing fluids has received considerable interest in recent years. Aqueous foams usually exhibit very good postfracture cleanup performance when used in stimulating abnormally low-pressured reservoirs or reservoirs that exhibit postfracture cleanup problems with denser fracturing-fluid systems. Foam systems are becoming more common in current fracturing practices. In deep formations, where pumping pressures are unusually high, gas volume requirements can become excessive. Therefore, applications in deep formations can be costly.

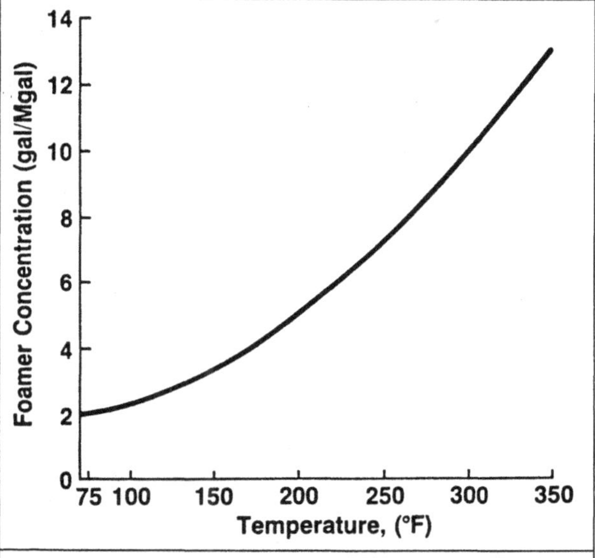

Fig. 1.33—Typical foamer concentration required to maintain a dynamically stable foam.[136]

Several laboratories have constructed equipment especially for testing foam rheology and fluid loss. Results of recent tests from these various sources have significantly extended the data base established by previous investigators.[131-133] Particular works[134-141] have shown foams to exhibit extremely good rheological and fluid-loss performance under a fairly wide range of conditions. Most of the laboratory systems used to test foams are similar to that described by Wendorff and Earl.[139] Basically, they are high-pressure systems with foam generators, foam-viewing chambers, heated rheology loops, inline fluid-loss cells, and fracture-simulation chambers, quite similar to the equipment depicted in Fig. 1.30.[142,143] Harris[134] recently conducted a comprehensive study to investigate how foam texture relates to rheology. Some of the conclusions resulting from this work are as follows: (1) foams are shear-history-dependent fluids; (2) the viscosity of foam is determined primarily by its quality and liquid-phase properties and to a lesser extent by its texture; (3) higher surfactant concentrations produce finer-texture foams; (4) viscosity measurements at low pressure may not ade-

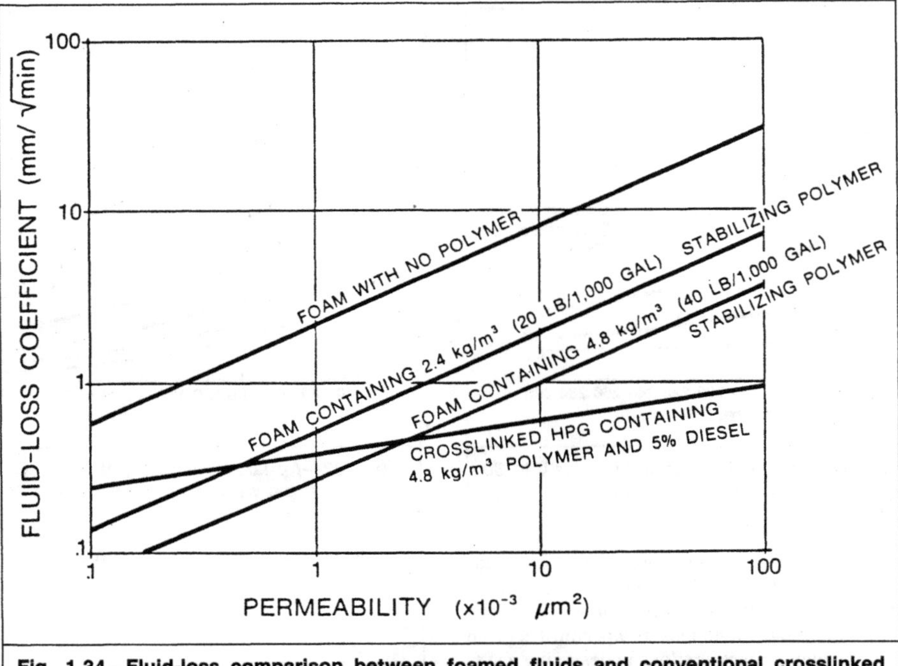

Fig. 1.34—Fluid-loss comparison between foamed fluids and conventional crosslinked gels. [138]

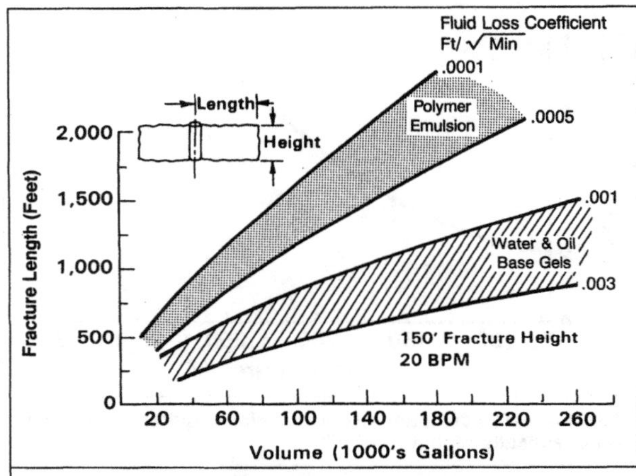

Fig. 1.35—Fluid loss vs. fracture length for low- (polymer emulsion) and high- (water- and oil-based gels) fluid-loss behavior. [4]

quately simulate field usage at high pressure; (5) the chemical type of the liquid phase influences texture; and (6) the larger bubbles of hydrocarbon and methanol foams result in sensitivity to degradation at high shear rates.

Examples of the effects of foam quality and gel concentration on apparent viscosity during tests by Harris and Reidenbach[136] are shown in Figs. 1.31 and 1.32. Here we see very good viscosities even at high temperatures for high-quality foams with 40- to 60-lbm/1,000-gal [4793- to 7190-g/m³] gel concentrations. Typical foamer concentrations required to maintain a stable foam with these good viscosities at different temperatures are shown in Fig. 1.33. Note that the requirements are not too severe even at high temperatures. Data from dynamic fluid-loss tests by Watkins *et al.*,[138] such as those in Fig. 1.34, show that in low-permeability formations, leakoff coefficients for some foams can be lower than those of crosslinked aqueous fracturing fluids. Craighead *et al.*[141] conducted proppant-settling studies on a foam generated with delayed crosslinked gels and found that the settling rate in foamed crosslinked gel was almost two orders of magnitude higher than in foamed linear (uncrosslinked) gel. They also found that foamed crosslinked gels were affected less by changes in foam quality.

Foam has developed a definite place in fracturing applications. It is particularly advantageous in low-pressure formations where limited reservoir energy is available to clean up a well after fracturing. One disadvantage that remains with foams is the limitation on achieving high proppant concentrations in the final gas/liquid mixture. However, recently developed technology using proppant concentrators has alleviated this problem to a degree.

1.7 Fracturing-Fluid Loss

Static Fluid Loss. Fluid-loss behavior can, in some cases, affect fracture penetration as much as vertical fracture-height growth, as illustrated in Fig. 1.35. This shows the comparative behavior of fracture length vs. treatment volume computed with a 2D hydraulic fracturing simulator for two systems with different fluid-loss behaviors. Here, the fluid-loss coefficients used for the polymer-emulsion system calculations were on the order of 0.0001 to 0.0005 ft/min$^{1/2}$ [30.4×10^{-6} to 152×10^{-6} m/min$^{1/2}$]. For the water- and oil-based systems, the values ranged from 0.001 to 0.003 ft/min$^{1/2}$ [0.3×10^{3} to 0.9×10^{-3} m/min$^{1/2}$]. All other parameters used in the calculations were identical. Significantly longer fractures were computed for the more efficient polymer-emulsion system. This demonstrates both the impact that fluid loss has on a fracturing treatment and the need for reasonably accurate values.

Fluid-loss behavior depends on a number of factors: (1) type and quantity of gelling agent, (2) type and quantity of fluid-loss additive (FLA), (3) pressure differential between the fracture and the formation, (4) permeability and porosity of the formation, (5) formation-fluid viscosity, temperature, and compressibility behavior, (6) fracturing-fluid and fluid-filtrate viscosity and temperature behavior, and (7) formation (or fluid) temperature. Total fluid loss generally has two parts: spurt loss and fluid-loss coefficient. Spurt loss pertains to the "instantaneous" loss when the fluid is first exposed to the fracture face; fluid-loss coefficient represents the long-term behavior over the duration of the exposure. Spurt-loss data, such as those shown in Fig. 1.36, indicate that it is highly dependent on gel concentration and fluid temperature and somewhat dependent on formation permeability. Spurt loss can be reduced significantly by the addition of solid particulate FLA material. Numerous types of these products are available. The more common components include fine-mesh silica (e.g., silica flour) and/or fine-mesh resin particles.

Fluid-loss coefficient is usually represented by three components: fracturing-fluid viscosity and relative permeability effects, C_I;

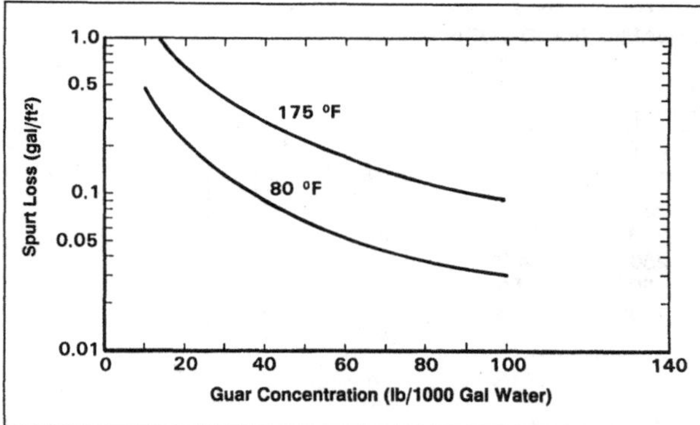

Fig. 1.36—Spurt loss vs. uncrosslinked guar concentration and temperature (formation permeability range from 5 to 10 md).[4]

reservoir-fluid viscosity/compressibility effects, C_{II}; and fracturing-fluid wall-building effects, C_{III}. Values for C_I and C_{II} can be computed from rock and fluid properties; values for C_{III} are normally determined from laboratory tests. Several different methods for computing a combined fluid-loss coefficient, C, from C_I, C_{II}, and C_{III} are discussed by Howard and Fast[2] and Settari[144]; the reader should choose the appropriate method. Fig. 1.37 shows typical C_{III} data. Note the dependence on gel concentration, FLA, and temperature. The FLA materials may have significantly different effects in the conventional-permeability formations than in the tight rocks. Solid particulate FLA may be very effective for a wide range (1 to 150 md) of high- or conventional-permeability formations but much less effective in the tight, low-permeability rocks. A 5% hydrocarbon phase, such as that discussed by Penny,[145] dispersed in an HPG system can reduce fluid loss significantly in tight formations but has little or no effect in conventional- or high-permeability rocks.

Dynamic Fluid Loss. Most of the fluid-loss data currently available for design (Figs. 1.36 and 1.37) were obtained from API-type[146] static fluid-loss tests at a pressure differential of 1,000 psi [6.9 MPa]; however, values may vary significantly with pressure differential. If pressure differentials during fracturing are expected to be significantly different from 1,000 psi [6.9 MPa], consistent laboratory data should be investigated. Fluid-loss behavior

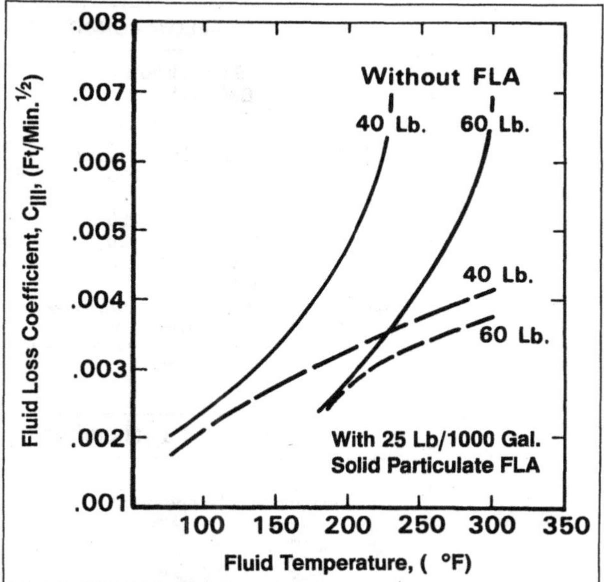

Fig. 1.37—Fluid-loss additives and temperature effects on wall-building fluid-loss coefficient, C_{III}—titanate crosslinked 40 and 60 lbm HPG.[4]

under dynamic conditions, as would actually occur in a fracturing treatment, may be quite different than static tests would imply.[131,147–149]

Several researchers[142,143,150–153] indicated that dynamic fluid-loss tests can yield different results than static tests and that shear rate and shear history can affect the tests significantly. Many of these data were obtained using systems with the fluid-loss cells, rheology loops, and heating capabilities shown in Fig. 1.30. Fig. 1.38 shows the different fluid-loss behaviors observed by Gulbis[142] for different shear rates, shear histories, and temperatures. These tests were run on the same fluid, an HPG fluid crosslinked with a titanium compound. The test conditions shown in the legend of Fig. 1.38 are given in Tables 1.5 and 1.6. The results demonstrate the significant effect that dynamics can have on fluid-loss behavior for fluids flowing in a fracture. The studies by Roodhart[152] and

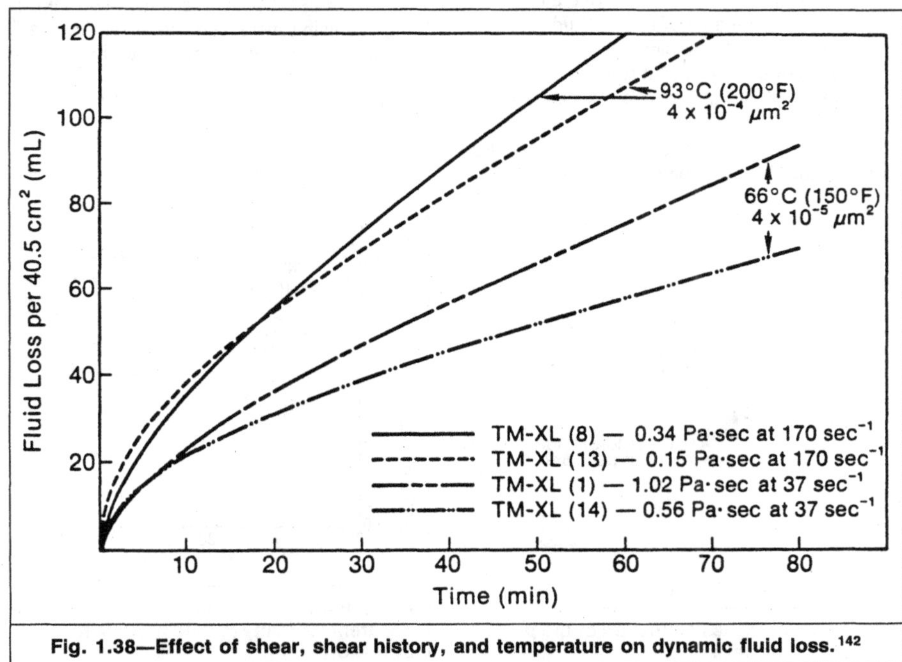

Fig. 1.38—Effect of shear, shear history, and temperature on dynamic fluid loss.[142]

TABLE 1.5—FLOW CONDITIONS FOR DYNAMIC FLUID-LOSS TESTS[142]

	Shear History Before Heating		Shear History After Heating				
Designation	Seconds*	Shear Rate (seconds $^{-1}$)	Seconds	Shear Rate (seconds $^{-1}$)	Velocity in Cell (cm/sec)	Shear Rate in Cell (seconds $^{-1}$)	Cell Type**
B	120	250	60	250	24.9	160	1
B′	17	3,380	60	250	24.9	160	1
	120	250	60	250	24.9		
C	240	120	120	120	2.5	24	2
C′	34	1,690	120	120	2.5	24	2
	240	120	120	120	2.5	24	2

*Shear times are listed in the order in which they occur.
**Cell 1 = fluid flows in 1.27-cm hole through core.
Cell 2 = fluid flows in 0.64-cm-wide annular space around core.

TABLE 1.6—FLUID RHEOLOGY FOR FLUID-LOSS TEST[142]

Test	Fluid Type	Temperature (°F)	Shear Type	n'	K'	μ at 37 seconds $^{-1}$ (cp)	μ at 170 seconds $^{-1}$ (cp)
1	TM-XL**	150	C	0.50	0.13	1,020	480
8	TM-XL	200	B	0.41	0.15	850	340
13	TM-XL	200	B′	0.59	0.026	280	150
14	TM-XL	150	C′	0.47	0.081	560	250

*See Table 1.5.
**TM-XL = crosslinked transition metal.

Harris and Penny[143] also showed that fluid-loss (i.e., both spurt-loss and fluid-loss coefficient) behavior is affected by fluid-flow dynamics.

Other interesting observations have been reported. Roodhart's[152] tests demonstrated a significant effect of pressure differential on wall-building fluid-loss coefficient, C_{III}. This is shown in Fig. 1.39 for both a crosslinked HPG fluid with 5% diesel and an HEC-based fluid with silica flour. Here we see a significant increase in C_w at higher pressure differentials.

Harris and Penny[143] observed an effect of increased viscosity in the flowing fluid caused by dehydration from fluid loss. This phenomenon is shown in Fig. 1.40 by the continually increasing viscosity for a test in a radial flow cell (Fig. 1.30) where fluid loss is occurring. It suggests that the gel thickens because of fluid loss. The other curve shows that viscosity decreases from shear and temperature degradation when fluid leakoff is prevented by replacing the core with an impermeable blank. Observations like these emphasize the need for a better understanding of in-situ fluid-loss behavior and its effects on rheology and proppant transport. This is particularly important because of the significant role that fluid loss plays as one of the more dominant parameters controlling the fracturing process.

Gulbis'[142] work indicated that at shear rates below 80 seconds $^{-1}$, dynamic and static fluid-loss behaviors were similar. Penny *et al.*[151] had corresponding results at shear rates below 40 seconds $^{-1}$. At high shear rates, however, they suggested that fluid loss follows a $t^{3/4}$ trend rather than the commonly observed $t^{1/2}$ for static tests.

Work by these investigators supported earlier studies that showed that a hydrocarbon phase (e.g., 5% diesel) could significantly reduce fluid loss, especially if mixed with silica flour (or other fine-mesh particulates) and a surfactant. This was especially effective in fractured cores, as shown by Penny *et al.*[151] in Fig. 1.41. Gulbis also reported that the effects of the hydrocarbon-phase/silica-flour additives on reducing fluid loss were less pronounced in dynamic than static tests.

Fluid Loss From Field Data. Several investigators have supplemented the literature with methods to infer fluid loss from field data since Nolte[154] introduced the pressure-decline method in 1979. Nierode[155] proposed a different approach for determining fluid loss using measurements of increasing ISIP data during the treatment. This work is based on the relationship

$$g_f(t_2)=g_f(t_1)[1+A(C\sqrt{t_2-t_1})^B], \quad (1.7)$$

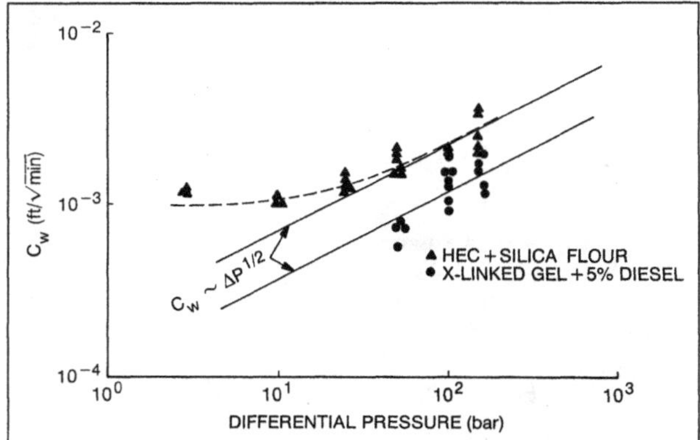

Fig. 1.39—Leakoff coefficient, C_w, as a function of the pressure differential over the filter cake.[152]

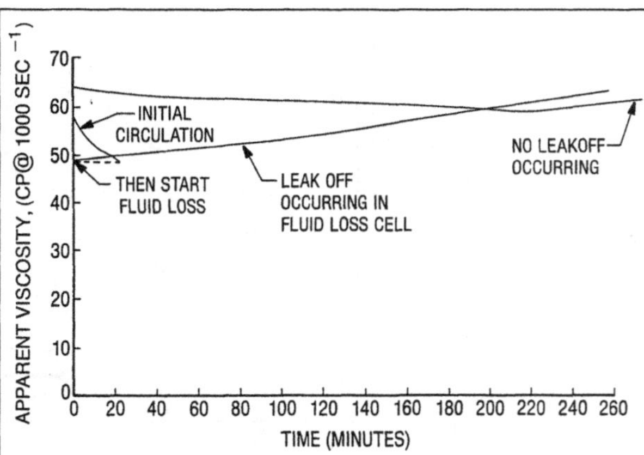

Fig. 1.40—Effect of fluid loss on fluid viscosity in a fracture—apparent viscosity of fluid within a recirculation loop vs. time of leakoff.[143]

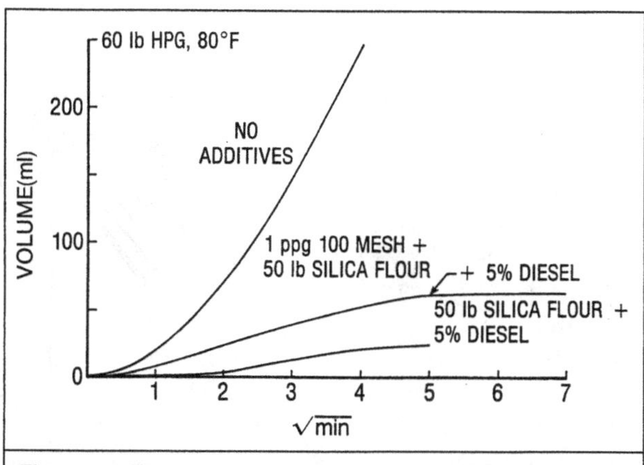

Fig. 1.41—Fractured-core leakoff volume at 25-psi pressure differential with various fluid-loss additives.[151]

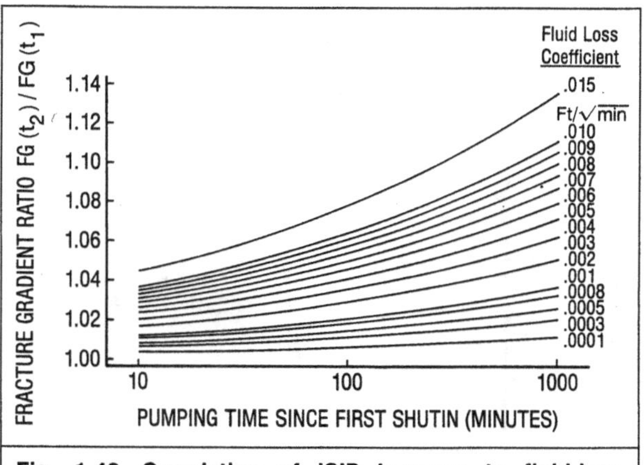

Fig. 1.42—Correlation of ISIP increase to fluid-loss coefficient.[155]

where

g_f = ISIP fracture gradient (ISIP/depth) at time t,
t_1 = time of first shut-in pressure measurement,
t_2 = time of later shut-in pressure measurement,
A,B = empirical constants, and
C = fluid-loss coefficient.

Nierode proposed the values $A=0.19043$ and $B=0.46767$ for a Kristianovitch-Geertsma-de Klerk[53] (KGD, sometimes called Kristianovitch-Zheltov[48]) -shaped fracture, and $A=0.20233$ and $B=0.47850$ for a Perkins-Kern-Nordgren[45,51] (PKN) -shaped fracture. These values served as the basis for the curves shown in Fig. 1.42 for using ISIP increase and pumping time since the first shut-in to estimate fluid-loss coefficient.

Cooper *et al.*[156] presented the results of a comprehensive field study comparing the methods of Nolte[154] and Nierode[155] with theoretical expressions of Smith[157] and Williams *et al.*[150] The theoretical expressions use the three types of linear flow leakoff mechanisms.

1. Fluid-viscosity- and permeability-controlled coefficient,

$$C_{\mathrm{I}}=0.0469\left(\frac{\Delta p\phi k}{\mu}\right)^{0.5} \quad \text{(1.8)}$$

2. Reservoir-fluid-compressibility-controlled coefficient,

$$C_{\mathrm{II}}=0.0374\Delta p\left(\frac{\phi k c_f}{\mu_{fo}}\right)^{0.5} \quad \text{(1.9)}$$

3. Wall-building-controlled coefficient,

$$C_{\mathrm{III}}=0.0164\frac{m}{A_{fl}}. \quad \text{(1.10)}$$

To compute a total fluid-loss coefficient, C_t, Smith[157] combined the terms in the form

$$C_t=\left(\frac{1}{C_{\mathrm{I}}}+\frac{1}{C_{\mathrm{II}}}+\frac{1}{C_{\mathrm{III}}}\right)^{-1}, \quad \text{(1.11)}$$

and Williams *et al.*[150] proposed the form

$$C_t=\frac{2C_{\mathrm{I}}C_{\mathrm{II}}C_{\mathrm{III}}}{C_{\mathrm{I}}C_{\mathrm{III}}+[C_{\mathrm{III}}^2C_{\mathrm{I}}^2+4C_{\mathrm{II}}^2(C_{\mathrm{I}}^2+C_{\mathrm{III}}^2)]^{0.5}}. \quad \text{(1.12)}$$

Cooper *et al.*'s[156] results are given in Table 1.7. In general, it appeared that the theoretical values (Eqs. 11 and 12) were lower than those computed by either Nolte's[154] or Nierode's[155] methods. There were several cases where close agreement was obtained for the 0.1-md formations; however, some of the other data exhibited a wide divergence. If one is faced with this dilemma (and it cannot be statistically remedied), it may be necessary to conduct sensitivity studies with a wide range of leakoff values to investigate the impact that the different values will have on a fracture-treatment design.

TABLE 1.7—COMPARISON OF FLUID-LOSS-COEFFICIENT VALUES OBTAINED FROM THEORETICAL, AMOCO-NOLTE PRESSURE DECLINE, AND NIERODE ISIP ANALYSIS[156]

Formation	Location	Fluid Type*	Permeability (md)	Theoretical C_t (ft/min ½) Eq. 1.12	Theoretical C_t (ft/min ½) Eq. 1.11	Amoco-Nolte C_t (ft/min ½)	Nierode ISIP Analysis C_t (ft/min ½) PKN	Nierode ISIP Analysis C_t (ft/min ½) KGD (KZ)
Almond	Sweetwater, WY	1	0.01	0.00231	0.00194S**	0.00321	0.00498	0.00513
Cotton Valley (Lower)	Cherokee, TX	2	0.1	0.00063	0.00050	0.00040	0.00040	0.00040
J sand	Adams, CO	3	0.1	0.00056	0.00051S	0.00161	†	†
Mission Canyon	McKenzie, ND	4	0.1	0.00030	0.00027	0.00042	0.00029	0.00029
Cotton Valley (Upper)	Panola, TX	5,8	0.1	0.00092	0.00079S	0.00315	0.00395F‡	0.00414F
Seger sand	Washita, OK	4,8	0.1	0.00091	0.00086	0.00037	0.00044	0.00044
Springer sand	Caddo, OK	2	0.3	0.00195	0.00286	0.00359	0.00182F	0.00186F
Berea	Kanawha, WV	6	18.0	0.00119	0.00098	0.00128	0.00206	0.00209
Kuparuk	North Slope, AK	7	40.0	0.00114	0.00136	0.00179	†	†

*1 = 40 lbm gel/1,000 gal crosslinked HPG.
2 = 50 lbm gel/1,000 gal delayed crosslinked HPG.
3 = high-temperature 40 lbm gel/1,000 gal crosslinked HPG.
4 = 50 lbm gel/1,000 gal crosslinked HPG.
5 = high-temperature 50 lbm gel/1,000 gal crosslinked HPG.
6 = 25 lbm gel/1,000 gal uncrosslinked HPG.
7 = 10 gal gelling agent/1,000 gal hydrocarbon, crosslinked.
8 = plus 5% hydrocarbon.

**S = Screenout problems occurred from theoretical C_t, whereas designs based on pressure-decline C_t were successful.
† = Final ISIP lower than initial ISIP.
‡F = Taken from ISIP's during the main proppant fracture treatment following minifracture.

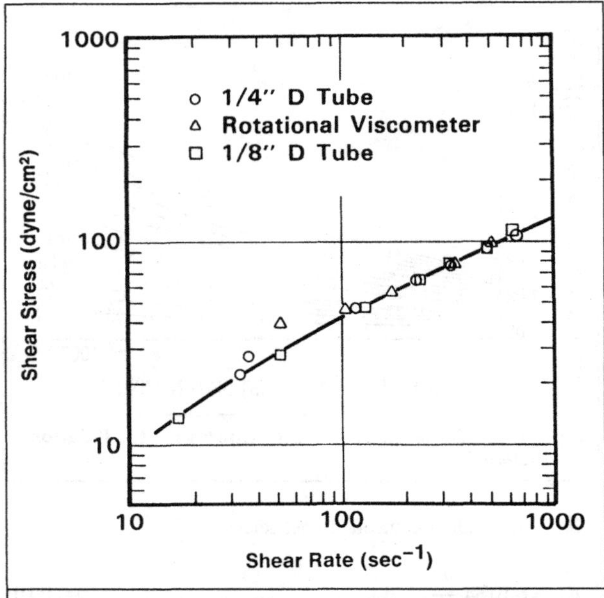

Fig. 1.43—Flow curve for uncrosslinked 40-lbm HPG, 149°F.[158]

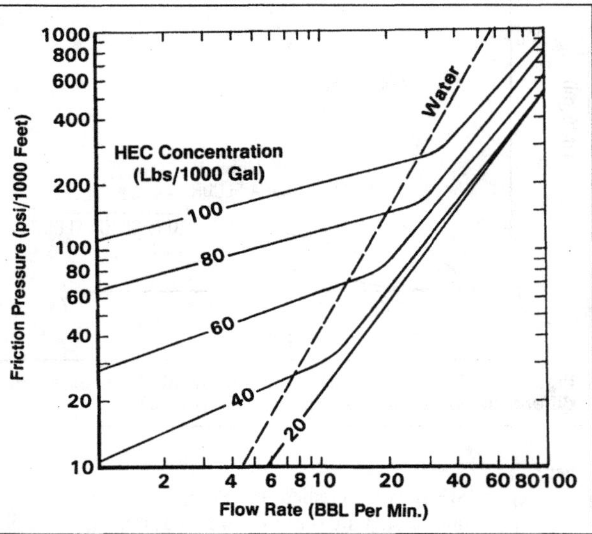

Fig. 1.44—Pipe friction vs. flow rate and gel concentration—HEC/water-based fluids (4½-in.-O.D., 11.6-lbf/ft casing).[4]

1.8 Rheology of Fracturing Fluids

The wide variety of different types of fracturing fluids is accompanied by wide differences in rheological behavior. Some of the fluids exhibit a Newtonian flow behavior. Others are non-Newtonian but essentially behave as power-law fluids. Still others, such as the cross-linked polymers commonly used for MHF treatments, can be very complex, non-power-law mixtures. For those fluids with Newtonian or power-law behavior, the rheological design data currently available are relatively reliable.

Power-Law Fluids. Viscous behavior is often determined according to API specifications[146] using a Couette (i.e., coaxial cylinder or rotational bob and sleeve) type of viscometer. For the power-law fluids, wall shear stress and shear rate are related by

$$\tau_w = K\dot{\gamma}_w^n . \qquad (1.13)$$

Rheograms are run to develop flow curves (log-log plots of shear stress vs. shear rate), such as the one shown in Fig. 1.43. From these curves, values for flow-behavior index, n, and consistency index, K, are determined from the slope and intercept (at unity shear rate), respectively. Apparent viscosity, μ_a, is then computed by Eq. 5.

Pipe Flow and Perforation Friction. For pipe flow, a form commonly used to model rheological behavior is expressed by

$$\frac{d_p \Delta p}{4L_p} = K'(8v/d_p)^{n'}, \qquad (1.14)$$

where $8v/d_p$ is the apparent shear rate in pipe and K' is related to K for a power-law fluid by

$$K' = K\left(\frac{3n+1}{4n}\right)^n . \qquad (1.15)$$

Fig. 1.44 illustrates typical pipe-flow data that are generally available for a wide range of pipe configurations. This set shows both the laminar (slope ~ ½ cycle/cycle) and turbulent (slope ~ 2

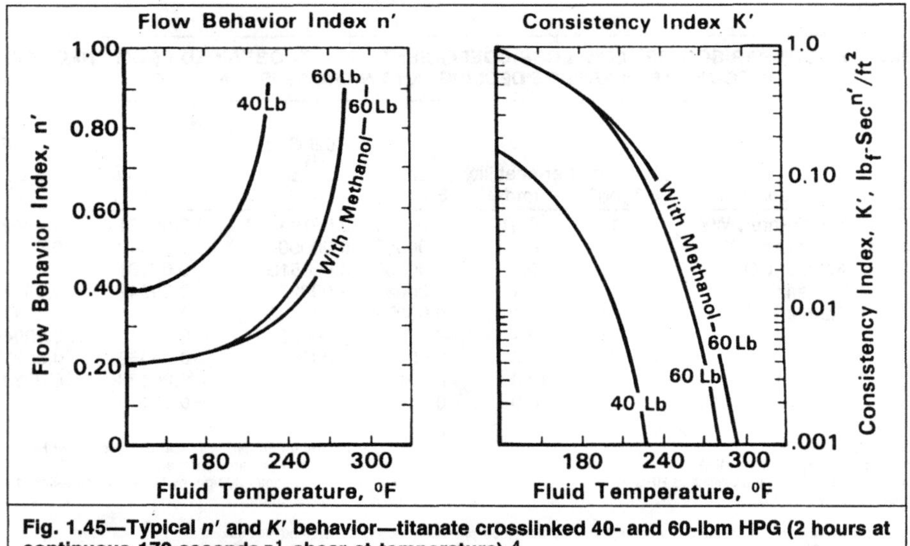

Fig. 1.45—Typical n' and K' behavior—titanate crosslinked 40- and 60-lbm HPG (2 hours at continuous 170-seconds^{-1} shear at temperature).[4]

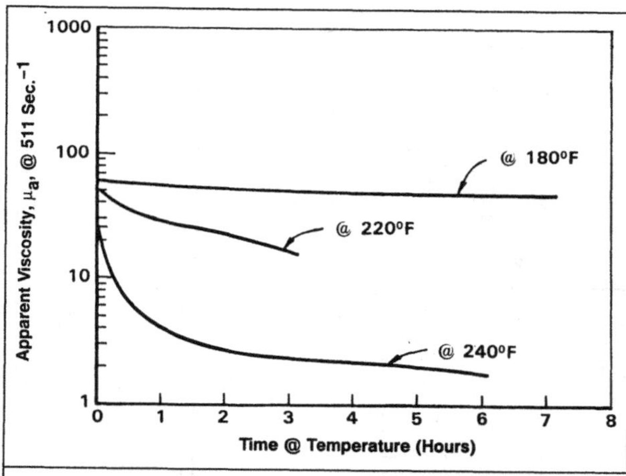

Fig. 1.46—Temperature and time effects on viscosity—40-lbm titanate crosslinked HPG.[4]

cycle/cycle) regimes for various HEC gel concentrations in 4½-in. [11.5-cm] casing. Experience has indicated excellent agreement between design data and field practice for uncrosslinked fluids. With crosslinked fluids, however, variations in the crosslinking mechanism at different points in the pipe during pumping can create fluctuations in friction pressure. Such fluctuations have been observed from downhole treating pressure measurements on MHF jobs.

Perforation friction is normally computed by

$$p_{fp}=\frac{0.2369q_i^2\rho}{\alpha^2 n_p^2 d_p^4}. \quad (1.16)$$

Values for α usually range from 0.8 to 0.9.

Flow in a Fracture. For flow in a fracture, rheological behavior is customarily modeled by

$$\frac{w\Delta p}{2x}=K_f'(6v/w)n_f', \quad (1.17)$$

where $6v/w$=apparent shear rate in the fracture and

$$K_f'=K\left(\frac{2n+1}{3n}\right)^n. \quad (1.18)$$

Even though some of the fluids do not conform strictly to a power-law behavior, they follow a near-power-law behavior sufficiently within the applicable shear-rate range that for practical purposes their behavior can be predicted adequately with a simple set of n and K values. This is the case for the example shown in Fig. 1.43, which indicates a slight deviation from power-law behavior in the lower shear-rate range, but seems to be nearly power-law through the major portion of the shear-rate range of interest.

Temperature Effects on Viscosity. All fracturing fluids exhibit a temperature dependency. The nature and degree depend, of course, on the type of fluid system. For many Newtonian liquids, viscosity can be related to temperature by the Arrhenius relationship:

$$\mu=Fe^{(E_a/RT)}. \quad (1.19)$$

This may hold for some power-law, and even some non-power-law, fluids such that apparent viscosity behavior follows the above relation over certain temperature ranges. Hence, it is not uncommon to plot $\log(\mu_a)$ vs. $1/T$ to investigate temperature effects.

Many of the power-law-type fluids will degrade with time, and this degradation is accelerated at elevated temperatures. Most will also show signs of gel degradation at elevated shear. This again depends on the type of system. However, behavior can usually be identified fairly well from a sufficient set of tests over the range of interest.

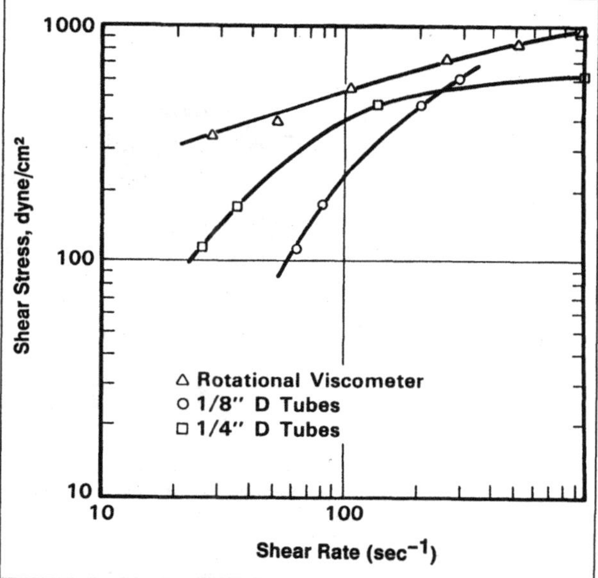

Fig. 1.47—Flow curves from pipe and rotational viscometer—40-lbm crosslinked HPG at 176°F.[158]

Crosslinked (Viscoelastic) Fluids. The crosslinked fluids, commonly used in fracturing today because of their purported better proppant-carrying and temperature-stability performances, have some very complex rheological properties. These fluids for the most part are guar, HPG, or carboxymethyl HEC solutions that are crosslinked with some type of metallic compound (e.g., a borate or titanate compound). The crosslinkers are generally claimed to be unique and proprietary to each fracturing service company. The gel behavior is affected by a number of things, including temperature, temperature history, shear rate and history, time degradation, and chemical contamination. At this time, techniques for characterizing gel rheological behavior are not well-established. Flow curves or rheograms are not always linear or repeatable. They can demonstrate significant dependence on temperature and shear rate. Here, n and K values are usually estimated with a tangent line to the flow curve at a given desired temperature and shear rate to generate data sets like that shown in Fig. 1.45. This set is quite probably unique to the shear rate (170 seconds^{-1}) and time conditions of the test. A variation in either shear rate or time could possibly result in a significantly different data set. This greatly magnifies the testing requirements to characterize fluid behavior for practical application to fracture design.

Common practice within the industry for crosslinked fluids is to estimate apparent viscosity, μ_a, with equations developed for power-law fluids (Eq. 1.5). An example of computed viscosity behavior is given in Fig. 1.46. This shows the effect of temperature on computed apparent viscosity for a crosslinked HPG fracturing fluid. It can be seen that the fluid degrades with time for a given shear rate and that for elevated temperatures, the effects of degradation are drastically increased.

No standard procedures have been developed (with which the industry is entirely comfortable) to predict behavior of the crosslinked fluids or foams accurately. One of the current problems is repeatability; scaling difficulties can also occur. To illustrate this problem, Fig. 1.47, from work by Rogers *et al.*,[158] shows comparative Couette and pipe viscometer data on "identical" crosslinked fluids. Attempts were made to test the fluids under identical conditions, yet the data did not scale. Reasons for this type of behavior are not yet totally resolved. Several studies[159-166] have addressed some of the needs in this area.

Rheological characterization of crosslinked fracturing fluids remains a difficult and elusive challenge, but some additional insights have been developed to extend the work of previous investigators. Studies by Guillot and Dunand[167] and Prud'homme[168] have demonstrated the use of laser anemometry to observe velocity profiles for investigating wall-slip phenomena. Using a circular cross-

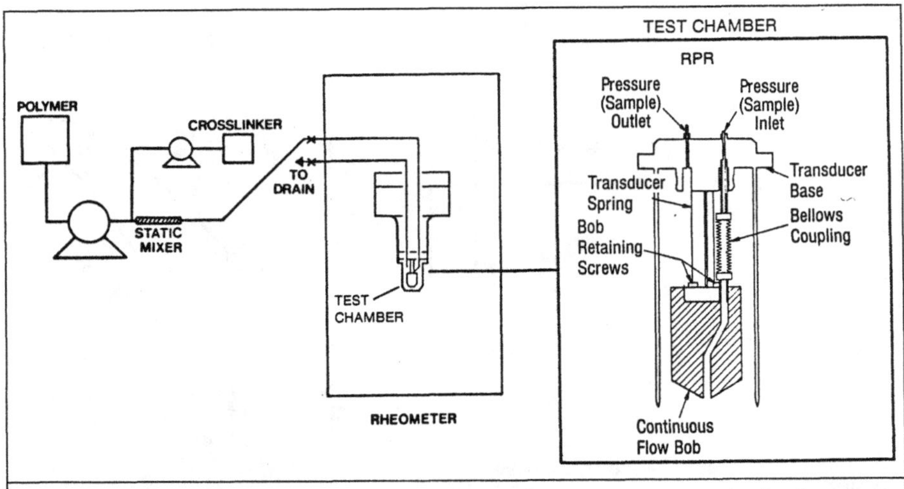

Fig. 1.48—Schematic of dynamic fluid preparation and flow through rheology test chamber.[169]

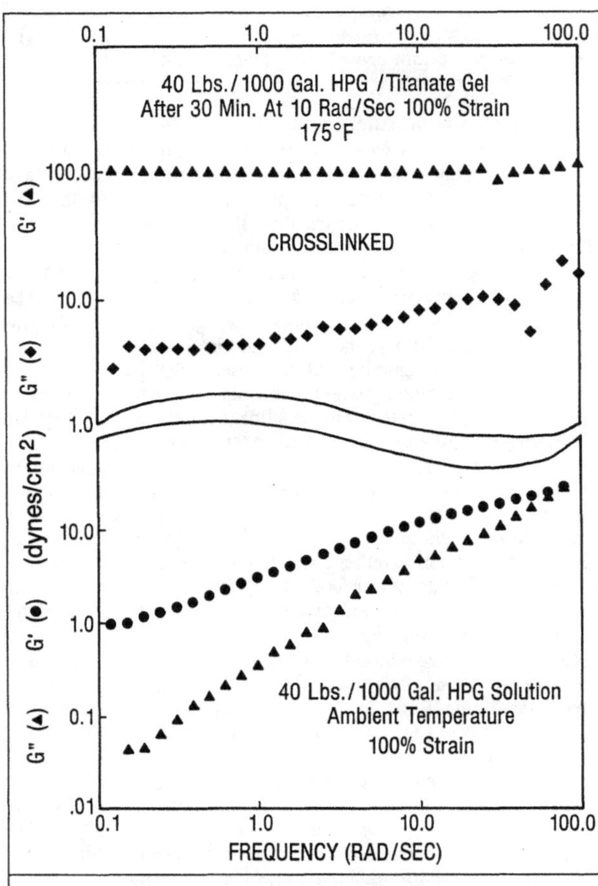

Fig. 1.49—Elastic, G′, and viscous, G″, moduli vs. oscillation frequency of an HPG solution and a crosslinked HPG gel.[169]

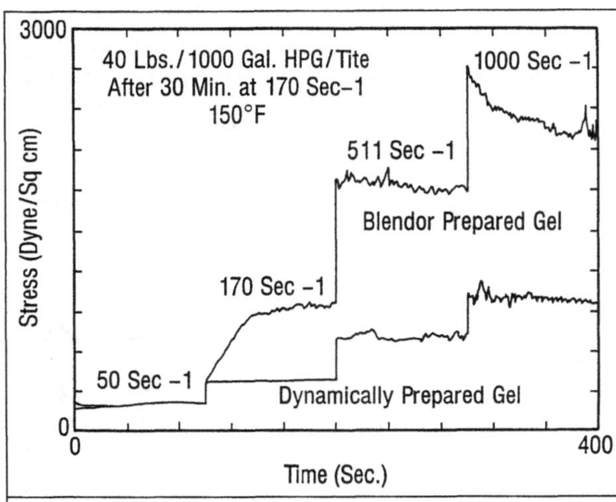

Fig. 1.50—Effect of fluid preparation on shear-stress response at various shear rates of a crosslinked HPG gel.[169]

sectional flow apparatus, Guillot and Dunand reported that, at low shear rates, aqueous HPG solutions exhibited velocity profiles much different from what known power-law-parameter calculations would indicate. Prud'homme's work in a coaxial cylinder apparatus exhibited behavior anomalous to conventionally known flow models. Further work is necessary to resolve or explain the occurrence of these anomalies.

Laboratory studies with oscillatory viscometers by Prud'homme[168] and Knoll[169] provided insight into methods for investigating gel structure, wall slip, and the significant role that mixing procedures play in testing of crosslinked fluids. Fig. 1.48 shows a schematic of Knoll's apparatus, including a pressure rheometer that is capable of both steady and oscillatory shear. Prud'homme and Knoll indicated that the physical nature of fluids can be demonstrated experimentally by oscillatory shear measurements that evaluate the elastic and viscous behavior of a fluid or gel. The elastic or storage modulus, G', as developed from classic network theory of macromolecules, indicates crosslink density. The viscous or loss modulus, G'', describes polymer behavior for these materials. By determining G' and G'' behavior as a function of strain (deformation) and frequency (rate), the structure of a material can be analyzed. Thus, it was possible to investigate the viscoelastic nature of a fluid. The rheometer apparatus was also capable of dynamic mixing and crosslinking of polymer. Fig. 1.49 shows the measured differences in G' and G'' for an uncrosslinked HPG solution and an HPG gel crosslinked with a titanium-based crosslinker. Test procedures, such as Knoll's and Prud'homme's, have shed considerable light on investigation of the conditions under which gels will form and their degree of crosslinking.

Knoll also demonstrated the variations expected between tests on blender-prepared gels and those that are dynamically prepared (i.e., crosslinked while flowing). The example in Fig. 1.50 shows the different stresses for various shear rates that resulted from use of different preparation procedures. This supports other reports and emphasizes the complex nature of characterizing fracturing-fluid rheology.

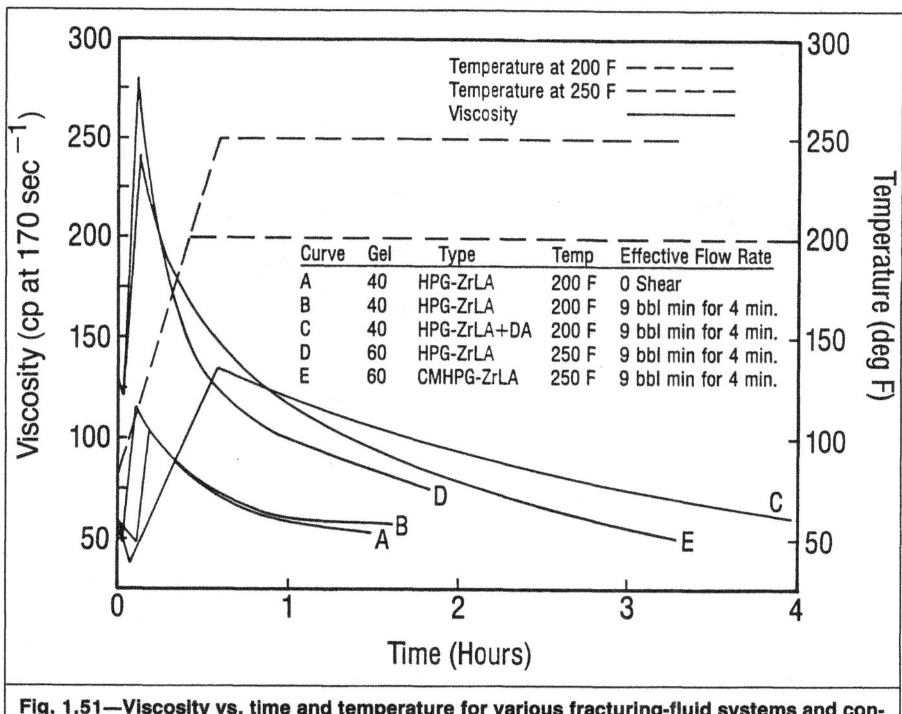

Fig. 1.51—Viscosity vs. time and temperature for various fracturing-fluid systems and conditions.[172]

Recent studies[168,170–172] in pipe-flow or capillary equipment yielded additional data on the effects of shear, temperature, and time for different fluids and crosslinking systems. These studies showed that a high-shear environment could destroy a gel if it was sheared severely after crosslinking. Observations by Gardner and Eikerts[172] indicate that high levels of shear before crosslinking have little effect on overall performance and that temperature will activate the crosslinking mechanism. Fig. 1.51 shows the composite of a series of their tests on HPG and carboxymethyl HPG systems crosslinked with a zirconium compound. Curve C shows the improvement in viscosity performance of a delayed system over that of comparable nondelayed systems (Curves A and B). Other investigations have supported similar phenomena. As a result of such findings, the industry is moving toward the use of delayed crosslink systems formulated to activate after the fluid has been pumped down the tubulars and through the perforations. This is one of the significant developments in fracturing-fluid technology.

In-Situ Fracture Flow Tests. Warpinski[173] conducted experiments on fluid flow through actual in-situ fractures created at the U.S. DOE's Nevada Test Site. The fracture was instrumented from a tunnel at a depth of 1,400 ft [425 m]. Theoretical friction factors, f_{th}, were computed by

$$f_{th}=\frac{64\mu_a}{d_H v\rho}. \qquad (1.20)$$

These values were compared with measured friction factors, f_m, computed by Eq. 1.19 from measured pressure losses:

$$f_m=\frac{\pi^3 h_f^2 w^3 \Delta p}{64\rho Q^2 \Delta x}. \qquad (1.21)$$

The results given in Table 1.8 show that pressure losses along the fracture were much larger than what would be predicted by viscous theory, which is currently used in most of the simulation models throughout the industry. The reasons for this underprediction are not identified to the degree that one can do more than make empirical corrections but are thought to be tortuosity, secondary flow, multiple fracture strands, sharp turns (corners), etc., resulting from the irregularity of the fracture faces.

TABLE 1.8—COMPARISON OF MEASURED vs. THEORETICAL FRICTION FACTORS IN A HYDRAULIC FRACTURE[173]

Test	Fluid	Flow Rate (gal/min)	Height (ft)	Apparent Viscosity (cp)	f_m/f_{th}
4	Water	10	10	1.0	1.39
5	Water	20	12	1.0	2.45
6	Water	30	12	1.0	2.75
7	Water	40	15	1.0	3.11
9	50 lbm gel	20	20	20	2.15
10	50 lbm gel	40	20	22	2.16
11	50 lbm gel	20	20	45	1.48

1.9 Proppant Transport

Models that include proppant-transport predictions generally use expressions developed from Stokes' law for laminar flow of Newtonian fluids and Newton's law for turbulent flow.

Clark and Quadir[174] present a rather comprehensive review of the various approaches[175–181] proposed to compute particle-settling velocities. Except for Harrington *et al.*'s[180] method, the expressions in general pertain to Newtonian or power-law fluids.

For Newtonian fluids, settling velocities are a function of the gravitational acceleration, drag force, fluid and particle densities, particle diameter, fluid viscosity, and surface roughness. If the assumption of single uniform spherical particles that do not exhibit electrostatic interactions is made, Govier and Aziz[181] indicate the settling velocities, v_s, for the laminar (Stokes' law), transition, and turbulent (Newton's law) regions to be

$$v_s=\frac{g(\rho_p-\rho)^2}{18\mu}, \qquad (1.22)$$

$$v_s=0.20\left[g\frac{(\rho_p-\rho)}{\rho}\right]^{0.72}\frac{d^{1.18}}{(\mu/\rho)^{0.45}}, \qquad (1.23)$$

and

$$v_s=1.74\left[\frac{q(\rho_p-\rho)d}{\rho}\right]^{0.5}, \qquad (1.24)$$

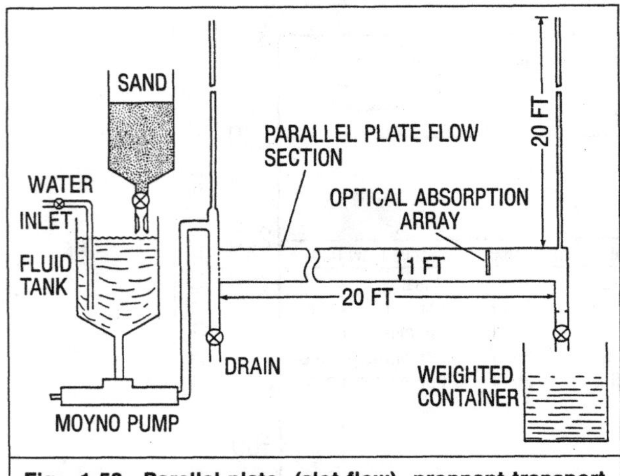

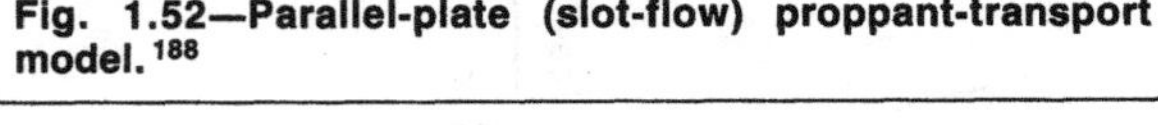
Fig. 1.52—Parallel-plate (slot-flow) proppant-transport model.[188]

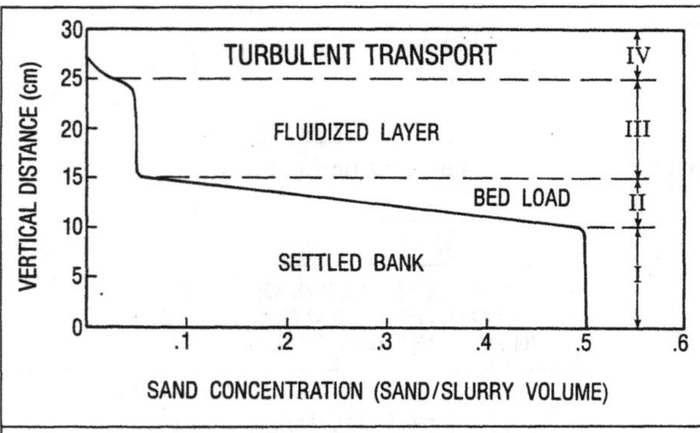

Fig. 1.53—Typical sand concentration regions observed in proppant-transport experiments.[187]

respectively. Eqs. 1.22 through 1.24 are for single particles. In slurries, settling behaves somewhat differently because of particle interference and/or clumping. Recent works by Zigrang and Sylvester,[177] which expresses the work of Zanker[182] in explicit form, and by Clark and Guler[183] address particle transport in slurries.

For particle settling in non-Newtonian fluids, the Newtonian viscosity, μ, is commonly replaced by a computed value of apparent viscosity, μ_a. Govier and Aziz[181] suggest that for some uncrosslinked fluids, this may be an adequate approximation. Harrington *et al.*[180] suggest that it may not be applicable for crosslinked fluids. There is a significant need for better proppant-transport prediction methods for the crosslinked fluids commonly used today.

Several recent studies[184–186] supplement the technology of other previous investigators in the area of proppant transport and settling for both power-law and viscoelastic fracturing fluids. Biot and Medlin[187] and Medlin *et al.*[188] conducted a comprehensive theoretical and experimental investigation on proppant transport in thin

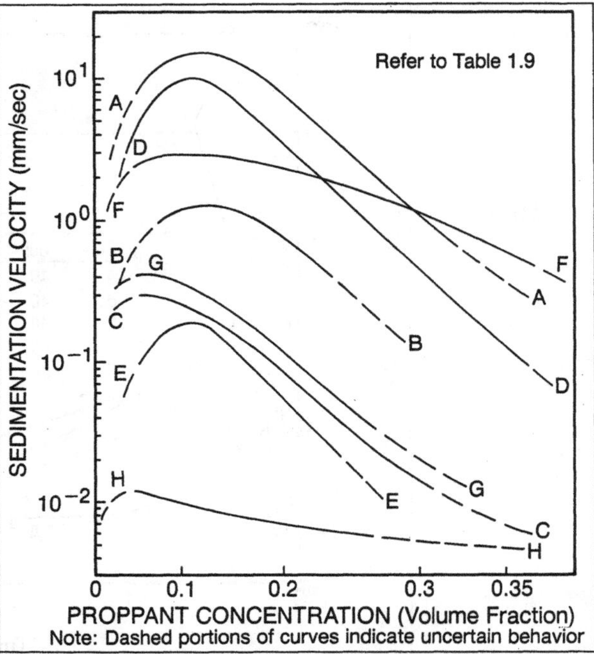

Fig. 1.54—Settling velocity at 20/25-mesh Ottawa sand.[191]

(uncrosslinked) fluids. With the apparatus shown in Fig. 1.52, they observed four regions of transport phenomena, as depicted in Fig. 1.53. Here, Region I is a settled bank where the concentration is a function of the proppant-packing characteristics; Region II, called the bed load, is a fluidized layer of relatively small height; Region III is a zone of viscous-drag transport where the proppant concentration is more or less constant; and Region IV is a zone of turbulent transport through which the concentration declines to zero. Their theoretical approaches closely modeled experimental findings, and their conclusions indicate that nearly all transport for thin fluids is by viscous drag.

Work by Roodhart[189] and Acharya[190] addressed proppant transport and settling in flowing viscoelastic fracturing fluids; Kirkby and Rockefeller[191] investigated settling in nonflowing slurries of both viscous and viscoelastic fluids. Both Roodhart and Acharya used vertical parallel-plate-type equipment somewhat similar to Medlin *et al.*'s[188] apparatus, and both developed theoretical expressions for settling velocities under different flow conditions. Some of the conclusions of Acharya's work are summarized below.

1. Correlations were developed for proppant-settling rate in inelastic (power-law) and viscoelastic fracturing fluids for low- and intermediate-Reynolds-number, N_{Re}, flow regimes.
2. In the intermediate-Reynolds-number region ($2 < N_{Re} < 500$), a power-law model would not be adequate for viscoelastic fluids.
3. No published normal stress data exist for viscoelastic characterization of commonly used fracturing fluids (both crosslinked and uncrosslinked). Such normal stress data will have to be measured for fracture design.

Roodhart's[189] work substantiated the need for separate expressions for different shear-rate regimes. His work incorporates terms

TABLE 1.9—FLUIDS AND RHEOLOGIES FOR FIG. 1.54[191]

Fluid	Polymer Loading (wt%)	Rheology at Low Shear Rates
A	0.48 HPG	Nearly Newtonian
B	0.72 HPG	Nearly Newtonian
C	0.18 Borate-crosslinked HPG	Viscoelastic; yield stress
D	0.30 Xanthan gum	Yield stress
E	0.48 Xanthan gum	High yield stress
F	0.48 Polyacrylamide	Pseudoplastic
G	Proprietary Fluid A-1	Viscoelastic; pseudoplastic
H	Proprietary Fluid A-2	Viscoelastic; very high yield stress

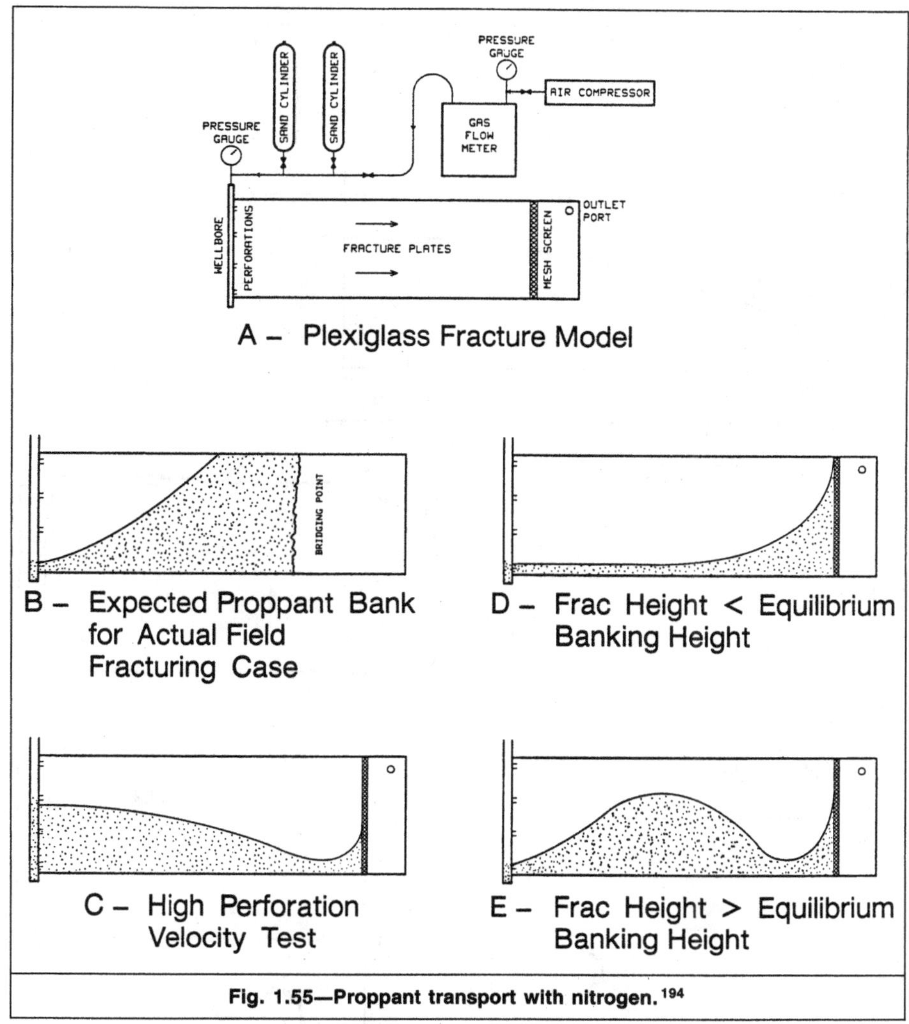

Fig. 1.55—Proppant transport with nitrogen.[194]

involving values of viscosities extrapolated to zero and infinite shear and apparent yield stresses to describe proppant settling.

Studies in a cylindrical apparatus by Kirkby and Rockefeller[191] on proppant settling under stagnant conditions (such as might occur during shut-in after a fracturing treatment is pumped) showed a strong dependence on proppant concentration. The results shown in Fig. 1.54 include both crosslinked and uncrosslinked fluids as listed in Table 1.9. The effects that proppant clustering has on increasing settling velocity with concentrations up to 0.1 to 0.2 vol/vol are shown. At higher concentrations, settling may be hindered, which reduces velocity. Also, one can observe the much lower settling velocities for the viscoelastic fluids. Work by Clark *et al.*[192] with equipment very similar to Kirkby and Rockefeller's yielded results that indicated improved proppant suspension with xanthan gum/HPG mixtures over that achieved with HPG alone. Dunand and Soucémarianadin[193] investigated both single-particle and suspension settling in quiescent fracturing fluids. Their observations indicate more rapid settling for suspensions in HPG solutions than would be computed for Newtonian fluids.

Gottschling *et al.*[194] conducted proppant-transport experiments under simulated fracture conditions using only nitrogen gas as the transport medium. Some of the results shown in Fig. 1.55 depict the proppant bank for various injection rates and equilibrium-banking conditions. Field fracturing treatments were conducted in the Devonian shale formations, where nitrogen would alleviate clay swelling or migration or oil/water emulsion problems that might occur with aqueous fluids. Although production information to date is not sufficient to evaluate whether nitrogen-gas/sand treatments are better than other types, it was found that nitrogen gas at high enough rates would effectively create a fracture and efficiently transport 20/40-mesh sand into it.

1.10 Fracture Design

Many factors influence the effectiveness and cost of a fracturing treatment. In essence, we have very little control over where and how fractures will ultimately propagate in subsurface strata. Our current efforts are limited to selecting (1) the appropriate types of materials (e.g., fluids, additives, and proppants), (2) the appropriate volumes of materials, (3) the injection rates for pumping these materials, and (4) the schedule for injecting the materials. Some success has been achieved in vertical growth control by controlling viscosity and/or by using diverting-type additives in the fracturing fluid.

With today's technology, the complete design process may use a data set like that listed below to assess reservoir producing potential and to specify appropriate design information pertinent to the fracturing treatment.

1. Well drainage area and drainage configuration.
2. Vertical distribution of formation net pay.
3. Formation permeability, porosity, and hydrocarbon saturation and the vertical distribution profile of these parameters.
4. Formation fluid properties, including viscosity and FVF's.
5. Static reservoir pressure.
6. Formation temperature.
7. Thermal conductivities of formations penetrated by the fracture, as well as in the vicinity of the fracture.
8. Fracture height or vertical growth extent that will occur during treatment.
9. Fracture extension and/or closure stress profiles.
10. Critical net fracturing pressure.
11. Formation effective modulus, Poisson's ratio, and density profiles.

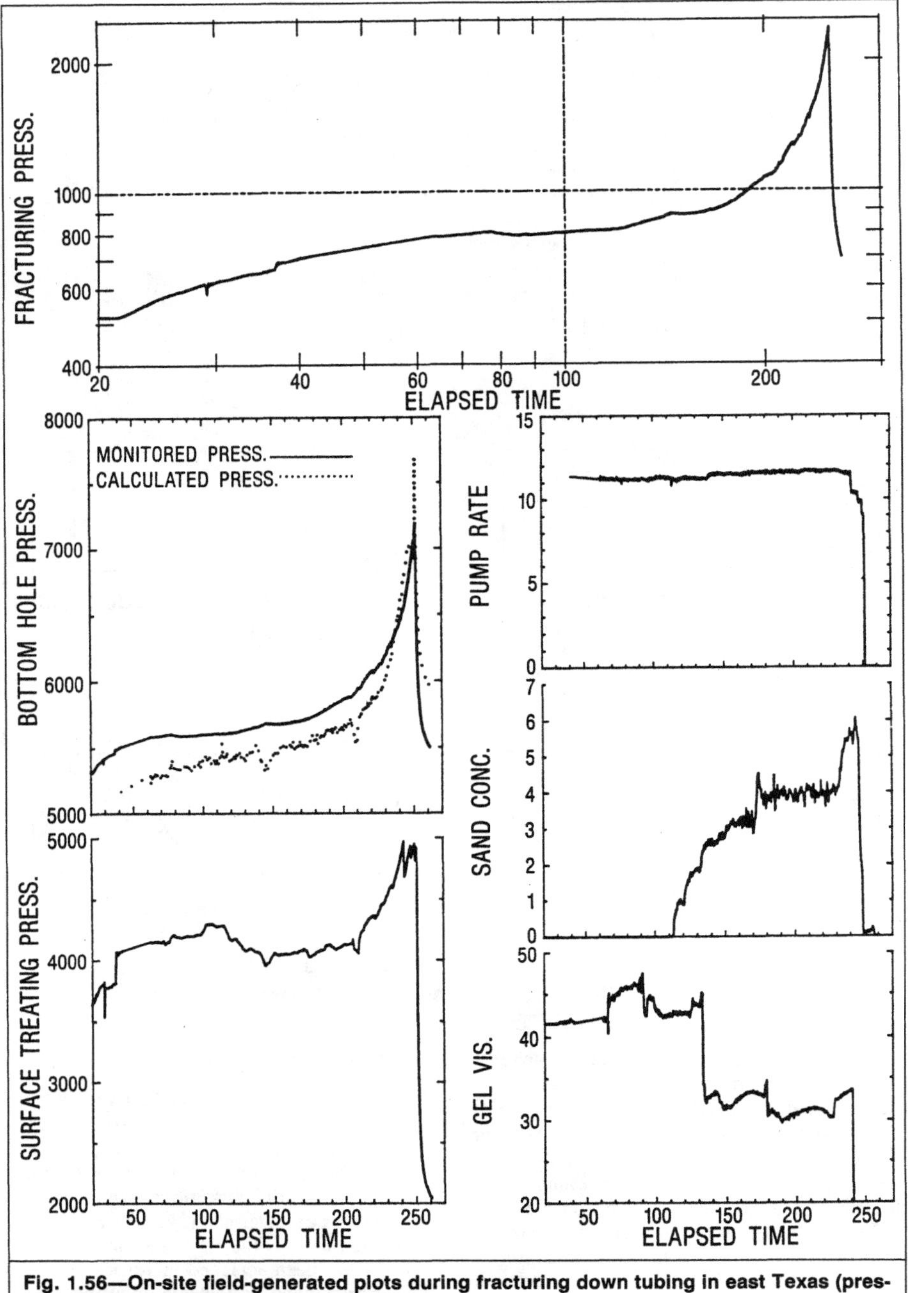

Fig. 1.56—On-site field-generated plots during fracturing down tubing in east Texas (pressure monitored in static annulus).[212]

12. Fracturing-fluid apparent viscosity or rheological n and K values. It may also be necessary to specify these values as functions of shear rate, time, and temperature.
13. Fracturing-fluid pipe- and perforation-friction data.
14. Fracturing-fluid spurt loss and, if necessary, its functional dependence on temperature.
15. Fracturing-fluid combined leakoff coefficient and, if necessary, its behavior as a function of pressure differential and temperature.
16. Vertical extent of net leakoff height.
17. Fluid thermal properties.
18. Proppant size distribution.
19. Proppant density.
20. Proppant fracture conductivity as a function of fracture closure stress, proppant type, proppant size distribution, proppant concentration in the fracture, and embedment into the formation.
21. Formation embedment pressure.
22. Perforation configuration (intervals, shots per foot, and size of holes).
23. Tubular goods and wellhead configuration, sizes, and pressure ratings.

Items 1 through 4 pertain primarily to reservoir performance, Items 5 and 6 to both reservoir performance and fracturing, and Items 7 through 23 primarily to the fracture-treatment design. Although the list appears to be quite comprehensive, it still does not reflect a complete picture of the many factors that can possibly affect fracture design.

The sensitivity of the predicted/actual results to the quality of the design depends on both the relative cost of the treatments and the nature of experience in an individual formation. In some areas, it may be typical for operators to progress through an extensive process of trying a number of alternative fluids, treatment sizes, and injection procedures to arrive at a set of standard treatments that provide acceptable results. In those areas where fracturing treat-

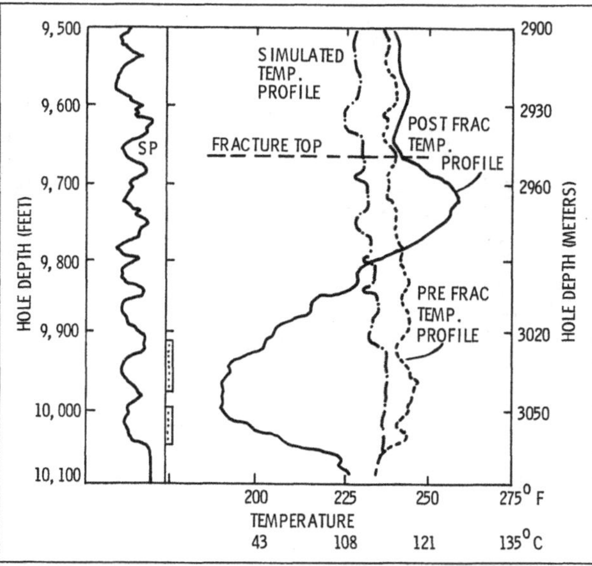

Fig. 1.57—Temperature profiles: prefracture, postfracture, and prefracture model simulated.[219]

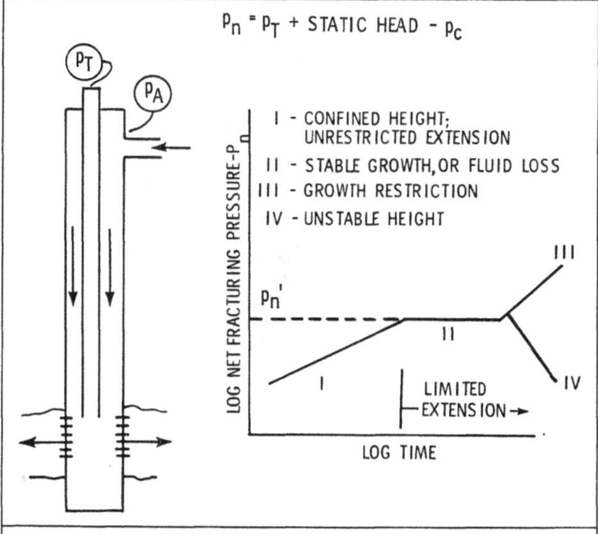

Fig. 1.58—Wellbore net fracturing pressure behavior during treatment—vertically confined fractures.[24]

ments constitute a relatively small portion of the total drilling and completion costs (e.g., high-permeability formations where short fractures are adequate), this approach is often used to obtain relatively quick and effective results. However, in low-permeability formations where deeply penetrating fractures are required, the resolution of the necessary fracturing parameters is very important. In areas where MHF treatments account for roughly one-half of the total well costs, the importance of fracturing is equal to, or greater than, that of development drilling for increasing recoverable reserves. Here it is essential to take the necessary steps to determine the required data with a high degree of resolution.

Existing methods for accurately quantifying some of the essential fracturing parameters—such as fracture length, width, conductivity, height, azimuth, shape, or symmetry about the wellbore—are still very much in the experimental stage. This makes it extremely difficult to assess how accurately we can predict fracturing behavior and effectiveness for a given set of design conditions. The input-data problem is not limited to in-situ formation or rock fracturing parameters. Current laboratory procedures and data for predicting fracturing-fluid and proppant behavior during a treatment are sometimes inadequate. The industry is, however, making significant progress in many of these areas. Recent works[25,52,195-211] describe programs and approaches to provide better information for fracture-treatment design.

1.11 Field Implementation and Equipment

One significant recent advancement in fracturing technology has been the development of on-site data-gathering and monitoring equipment and treating equipment designed for computer control.

Cooper *et al.*,[212] Hannah *et al.*,[213] and Harrington[214] describe some of the on-site computerized plotting and analysis capabilities and monitoring systems. These capabilities include an on-site, field-durable, transportable computer system; software for real-time analysis and graphical display of fracturing, pumping, and post-shut-in decline-pressure data; and an on-site rheology test system interfaced with the computer for determining rheological flow data pertinent to the treatment. Fig. 1.56 shows one example of the type of real-time, on-site data displays currently available from the fracturing and treatment-monitoring service companies.

Enhancements and advancements in computer hardware, software, microprocessors, servocontrol of blending equipment, proppant densitometers, and on-site rheological test equipment have significantly improved the design and execution of fracturing treatments. The computer age has truly come for fracturing!

1.12 Fracture Diagnostics

Comprehensive Programs. Much of the significant advancement in fracturing-treatment diagnostic and design technology relates to the interpretation of downhole fracturing pressures during pumping and the analysis of shut-in decline pressures after pumping is stopped. It includes methods applied to both minifracture calibration treatments and stimulation treatments.

Comprehensive field-data-collection programs to investigate methods for improving fracture design capabilities were reported by Veatch and Crowell.[25] Programs like this have effectively improved our ability to obtain better estimates of fluid-loss coefficient, elastic modulus, net fluid-loss interval, fracture-height growth behavior, critical net fracturing pressure, fracture closure stress, and fracture extension pressure, which are often essential for fracture design. These programs have provided significant insight into methods for controlling undesirable vertical fracture growth and for designing fluid and proppant schedules to improve fracture conductivity.

A very comprehensive set of diagnostic tests was conducted at the U.S. DOE's Multiwell Experiment Site near Rifle, CO. The experiment is still in progress; findings to date can be found in Refs. 215 through 218. As Northrop *et al.*[216] stated, one of the purposes of the work is to investigate the effectiveness of stimulation technology with diagnostic instrumentation and production performance testing. Features of the Multiwell Experiment Site include (1) three closely spaced (115 to 215 ft [35 to 66 m]) wells for reservoir characterization, interference testing, well-to-well geophysical profiling, and placement of diagnostic instrumentation adjacent to the fracture treatment; (2) complete core taken through the formations of interest; (3) a comprehensive core-analysis program; (4) an extensive logging program with conventional and experimental logs; (5) determination of in-situ stresses in sands and bounding shales; (6) use of various seismic surveys and sedimentological analyses to determine lens morphology and extent; (7) use of seismic, electrical-potential, and tilt-diagnostic techniques for hydraulic fracture characterization; and (8) a series of stimulation experiments to address key questions. Many of the techniques developed from this experiment are being incorporated into practice throughout the industry.

Fracture-Height Measurement. Postfracturing temperature-decay profile surveys are currently one of the more widely applicable techniques for determining fracture height at the wellbore. However, interpretation is sometimes masked by temperature anomalies, usually in the form of a "warm nose," as depicted near the 9,700-ft [2960-m] -depth mark in Fig. 1.57. The use of post-water-circulation temperature-decay surveys taken before perforating can significantly improve the interpretation of these data for fracture height. As shown in Figure 1.57, the prefracture surveys serve as a baseline for temperature behavior where no fluid has entered the formation. The procedures and results from many of the early tests were reported by Dobkins[219]; subsequent tests[25] supported his conclusions.

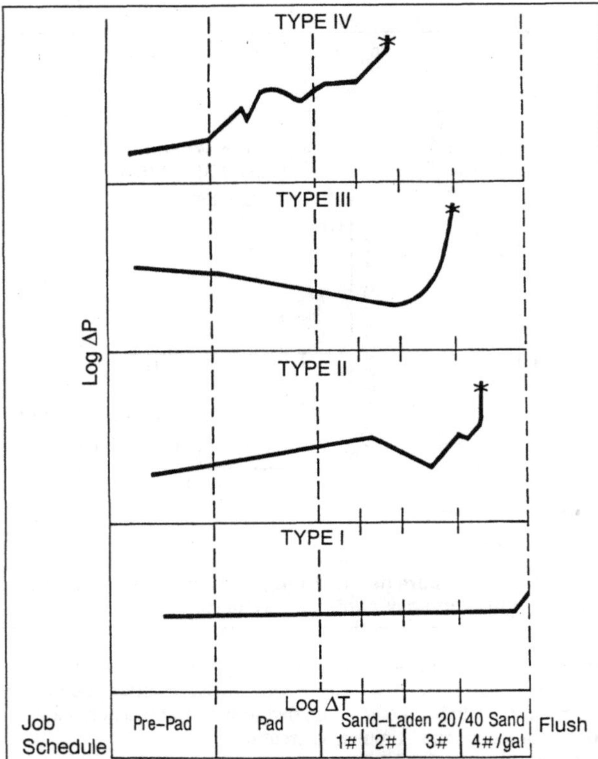

Fig. 1.59—Downhole fracturing pressure behavior for Types I through IV fractures in the order of screenout potential.[220]

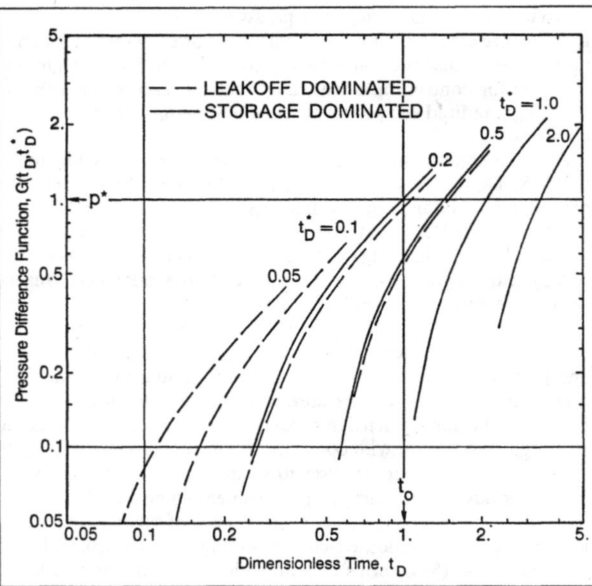

Fig. 1.60—Generalized postminifracture pressure-decline type curves.[222]

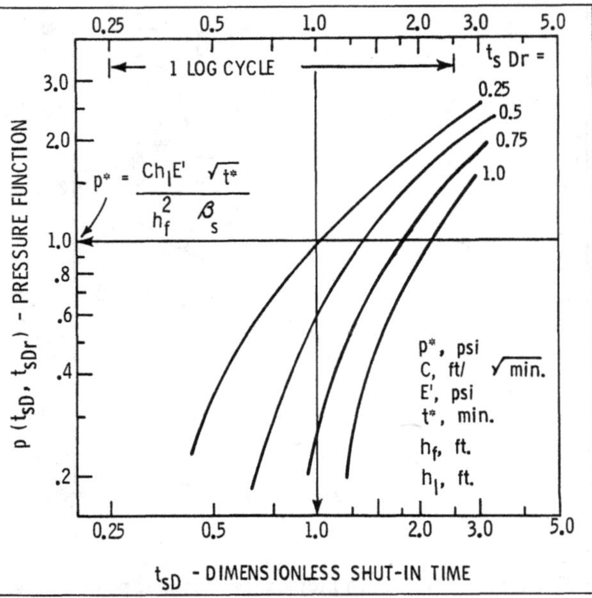

Fig. 1.61—Postfracturing treatment shut-in pressure-master-decline type curves.[154]

One possible cause of the warm nose may be warmed fluid in the fracture flowing back past the wellbore from one portion of the fracture to another after pumping has ceased at the end of the fracturing job. It is postulated that continued fracture extension at one or more points in the fracture after shut-in can cause a redistribution of fluid in the fracture, carrying heat back across the wellbore after shut-in to cause the warm nose to develop.

Radioactive surveys run with postfracturing temperature-decay profile surveys after fracturing have also been found to enhance fracture-height interpretation. They can be especially helpful in confirmation of fracture height when the warm nose appears on the surveys. They are also very effective for investigating the bottom of the fracture. Sand fill often precludes using postfracturing temperature-decay profiles for this purpose.

Downhole Fracturing Pressure. The procedures and findings from early tests were discussed by Nolte and Smith[24]; many subsequent tests[25] confirmed their observations. Lateral fracture extension rates, critical net fracturing pressures, and vertical growth behavior can be inferred from downhole fracturing pressure. This is depicted conceptually in Fig. 1.58 by a logarithmic plot of net fracturing pressure, p_n, vs. time. Mode 1 represents confined height and free lateral fracture extension; Mode 2 indicates a reduced fracture penetration rate when pressure reaches the p_n' value. One of several possible explanations is that at this point (i.e., $p = p_n'$), side fissures develop and take fluid from the main fracture. Another explanation of Mode 2 behavior is the occurrence of stabilized low-rate vertical fracture growth. Mode 3 can be interpreted as a storage mode having restricted vertical or lateral extension and implying "ballooning" width. Mode 4 indicates rapid vertical growth.

In the area of treating pressures, Conway *et al.*[220] suggested that five basic types of fracture behavior could be identified from downhole fracturing pressures during pumping. A large number of treatment-pressure charts were evaluated, grouped by similar behavior, and correlated with various design models or propagation modes. Fig. 1.59 shows four types: (I) KGD, (II) PKN, (III) penny-shaped, and (IV) Medlin and Fitch. These are described on the basis of plots of the net downhole fracturing pressure (i.e., downhole fracturing pressure minus closure stress) vs. pumping time on a logarithmic scale. Type I exhibits a constant net downhole fracturing pressure or declines with a slope of 0.05. Type II depicts Modes 1, 3, and 4 of Fig. 1.58 plus a screenout mode at the end. Type III declines steadily and then increases rather rapidly with a 2 : 1 slope as the treatment goes into a screenout mode. Type IV behavior, investigated by Medlin and Fitch,[221] is characterized by large pressure increases early in the treatment and usually approaches a screenout mode by the time viscous slurry reaches the formation, resulting in very little proppant entering the fracture. Consequently, well performance is relatively poor. Type V is depicted in Fig. 1.58.

In Fig. 1.59, the curves for Types I through IV are arranged in order of screenout tendencies, with Type I being the lowest and Type IV the highest. Conway *et al.*[220] suggest the importance of identifying characteristic behavior patterns early in the life of a fracturing-treatment program to improve downhole-fracturing-

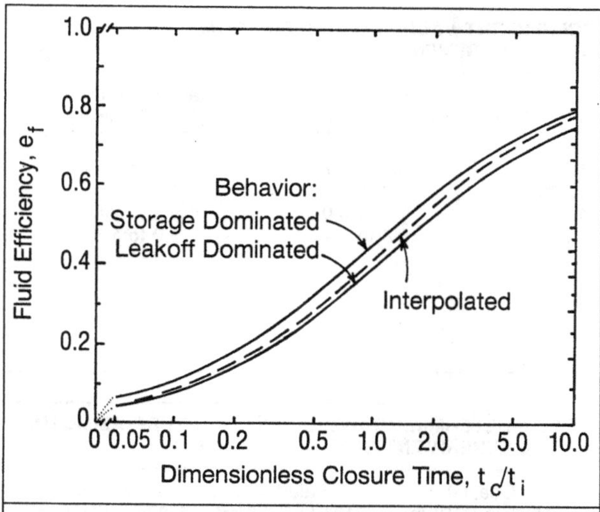

Fig. 1.62—Relationship of fracturing-fluid efficiency and fracture closure time.[208,222]

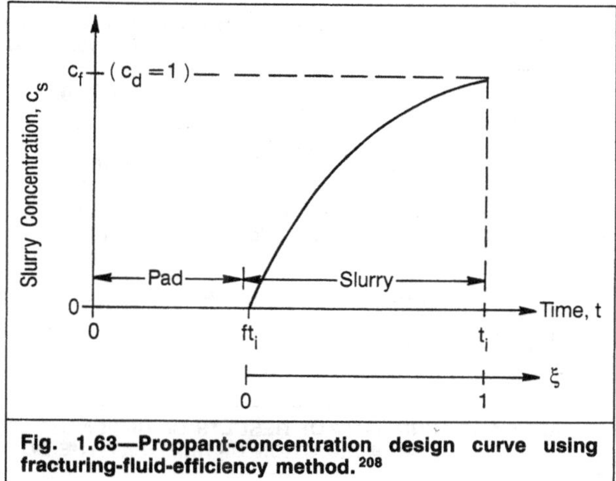

Fig. 1.63—Proppant-concentration design curve using fracturing-fluid-efficiency method.[208]

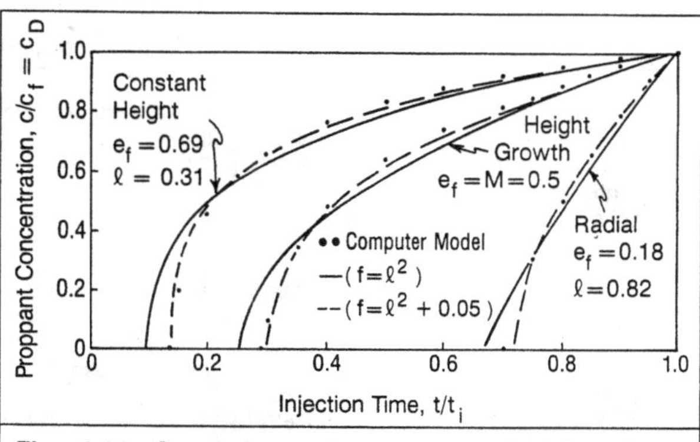

Fig. 1.64—Comparison of proppant schedules—fluid-efficiency method vs. simulation model design.[208]

pressure design and execution of future treatments. Methods have been investigated for estimating downhole fracturing pressure from surface pressure data.

Postfracturing Pressure Decline. A number of design parameters can be inferred from pressure decline after the ISIP, including p_c, h_ℓ, C, E, h_f, and x_f. This analysis can be particularly useful with a minifracture calibration test (discussed later) to obtain C, E, h_ℓ, p_c, and p_f values on a given well before MHF for use in designing its treatment. The postfracturing pressure-decline data are extremely easy and relatively inexpensive to obtain. These data do not require high-resolution pressure-measuring equipment (accuracy of ±10 psi [±70 kPa] will yield satisfactory results). It involves merely leaving a pressure-recording chart connected to the well for a shut-in period of usually two to three times the length of the pumping period.

Nolte[222] extended his original type-curve pressure-decline analysis for general application. Development and application of the theory and type-curve techniques for postfracturing pressure-decline data obtained during the initial portion of the program were presented. Procedures for analysis, using the pressure-decline type curves shown in Figs. 1.60 and 1.61, are documented in Chap. 14. This analysis covers a wide range of conditions from high-leakoff formations[52] to the very-low-leakoff tight-gas formations. The analysis was also developed for use with either the PKN, KGD, or penny fracturing models. Fig. 1.60 can be used for the general case over a wide range of dimensionless times, t_D, and dimensionless time reference values t_D^*, where type curves are given for the dimensionless pressure difference function, $G(t_D,t_D^*)$.

Martins and Harper[223] developed a type-curve approach for a fracture in a long perforated interval where it is assumed that the fracture evolves as a family of confocal ellipses and the created fracture length is on the same order of magnitude as the perforated interval. Lee[224] also developed type curves specifically for the KGD and radial geometry models that conform closely with those presented by Nolte.

Using concepts similar to those presented previously by Harrington *et al.*[206] and Harrington and Hannah,[207] Nolte[208] developed a method for using pressure-decline data to design proppant and fluid schedules for fracturing treatments with fluid-volume efficiencies.

Fig. 1.62 is used to estimate the fluid efficiency, e_f, for the slurry from dimensionless closure time (i.e., the ratio of closure time to pumping time, t_c/t_i). Curves similar to those shown in Fig. 1.63 can be constructed to optimize proppant concentration.

Fig. 1.64 shows very close agreement between optimizing proppant schedules with this technique and using those derived from computer simulator models for three types of geometry (constant height, growing height, and radial growth). This approach allows the design of proppant schedules from field minifracture pressure-decline data with very little *a priori* knowledge of the fracture geometry. Crawford[210] also discusses the impact of fluid efficiency on proppant scheduling for treatment design.

Minifracture Calibration Test. This test is aimed at measuring pertinent data directly on a well before the fracturing treatment is designed. This procedure can yield excellent values of p_c, p_f, and C, as well as some indication of the E/h_f ratio. Analysis of these data can imply an expected $h_f(p_n)$ relation during treatment. The program is usually conducted with routine perforation breakdown operations before flow testing.

First, a closure-stress test to obtain p_c and/or p_f is conducted during perforation breakdown with conventional, nondamaging breakdown fluids (KCl water, acids, etc.). The procedure can include conventional step-rate tests, repeated pump-in/flowback operations with small fluid volumes (100 to 5,000 gal [0.4 to 19 m^3]), and shut-in pressure-decline tests. A minifracture calibration test is then performed with moderate volumes (5,000 to 40,000 gal [19 to 151 m^3]) of the same fracturing fluid that will be used throughout the major portion of the MHF treatment. This fluid must be proppant-free to allow the fracture to close unrestricted. Downhole fracturing pressure is recorded during pumping, and postfracturing pressure-decline data are taken after shut-in. During this shut-in period, postfracturing temperature-decay profile surveys are taken to measure h_f.

These tests have been found to yield a good calibration of the formation. In some cases, it may also be possible to infer p_n' values and $h_f(p_n)$ behavior from the data. This approach is especially effective for investigating static in-situ C values for the fracturing fluid planned for the treatment. This, of course, requires estimates of h_f and h_ℓ. Table 1.10[25] shows a good comparison of minifracture-calibration-calculated C values vs. laboratory data.

TABLE 1.10—COMPARISON OF FLUID-LOSS-COEFFICIENT VALUES OBTAINED FROM FIELD TESTS vs. LABORATORY MEASUREMENTS[25]

Formation	Area	Permeability (μd)	Fluid Type*	Field Data (10^{-3} ft/min$^{1/2}$)	Laboratory Data (10^{-3} ft/min$^{1/2}$)
Cotton Valley	Texas	1 to 100	1,2,4	0.3 to 0.7	0.7 to 1.0
Muddy J	Colorado	1 to 100	1,3	0.5 to 0.7	0.3 to 0.7
Frontier	Wyoming	1 to 300	1,2	1.0 to 1.2	1.0 to 1.5
Mesa Verde	Wyoming	1 to 100	1,2	1.0 to 6.0	0.5 to 2.0
Dakota	New Mexico	10 to 1,000	1,2	0.8 to 1.2	1.0 to 1.5

*Fluid Types:
1 = 40 or 50 lbm gel/1,000 gal crosslinked HPG.
2 = 1 + 5% hydrocarbon.
3 = polymer emulsion.
4 = 50 lbm/gel/1,000 gal crosslinked cellulose derivative.

TABLE 1.11—SUMMARY OF RESULTS OF DIFFERENT METHODS FOR DETERMINING HYDRAULIC FRACTURE AZIMUTH AS A FUNCTION OF DEPTH AT THE DOE MULTIWELL EXPERIMENT SITE, RIFLE, CO[231]

	Upper Fluvial (1330 to 1585 m)	Lower Fluvial (1585 to 1835 m)	Coastal (1835 to 2010 m)	Paludal (2101 to 2270 m)	Marine (2270 to 2450 m)
Predictive Methods					
Paleo-Stress Directions—Geologic Observations					
Normal faults	N70W ± 10°	+	N70W ± 10°	+	N70W ± 10°
Surface fractures	N80W ± 15°	+	N80W ± 15°	+	N80W ± 15°
Oriented-core fractures	N74W ± 11°	+	N74W ± 11°	+	N74W ± 11°
Calcite strain analysis	N78W ± 7°	+	N78W ± 7°	+	N78W ± 7°
In-Situ Stress Directions—Oriented-Core Analysis					
Strain recovery	N86W ± 15°	N82W ± 12°	N71W ± 10°	N62W ± 13°	N68W ± 9°
Differential strain analysis	S85°W	NA	N89°W	NA	N75°W
In-Situ Stress Directions—Wellbore Breakouts					
Caliper logs	S86W ± 10°	N84W ± 9°	N75W ± 10°	N51W ± 21°	N64W ± 14°
Televiewer	N83W ± 11°	N78W ± 12°	N66 ± 13°	N59W ± 22°	NA
In-Situ Stress Directions—Computer Modeling of Topographic Loading					
No tectonic stress	S74°W	S75°W	S80°W	S85°W	N65°W
Tectonic stress of (N70°W)	N87°W	N83°W	N79°W	N73°W	N69°W
Observation Methods					
Geophysical Mapping Borehole Seismic	NA	NA	NA	N67W ± 8°	NA
Openhole Hydraulic Fracture Impression Packer	NA	NA	NA	NA	N50° to 70°W

1.13 Fracture Azimuth and Geometry

Knowledge of fracture azimuth and symmetry is especially important in tight formations where well locations should be selected to minimize interference of the long fractures. A full complement of tests was run in the 8,000-ft [2440-m] -deep, low-permeability Wattenberg, Colorado, area. The results indicated that a generally uniform azimuthal trend prevailed in several parts of the area. Details of the equipment, procedures, and findings may be found in Refs. 225 through 228.

Subsequent tests in deeper horizons in Wyoming (11,000 ft [3350 m]) and Texas (10,000 ft [3050 m]) have been more questionable than those from Wattenberg. The use of tiltmeters may have some depth limitations. Other methods, however, including borehole-directional-geophone measurements, earth-tidal-strain data,[229] rock-mechanics data from oriented cores,[230] borehole-ellipticity data, and magnetometer measurements,[231] may provide possible means for measuring or inferring azimuth.

A number of relatively significant investigations have enhanced the work of early investigators for mapping the azimuthal trends of hydraulically induced fractures. These included (1) the U.S. DOE Multiwell Experiment[231,232] in the Piceance basin near Rifle, CO; (2) the experiment funded primarily by the Gas Research Inst. and conducted jointly with Dowell Schlumberger and Amoco[233,234] at Amoco's Mounds Test Site near Tulsa, OK; (3) investigations[235] from multiple wells in several fields in east Texas and Alaska; and (4) studies[236] in the Kuparuk River formation on the Alaskan North Slope. The large number of tests has allowed comparison of a wide variety of azimuth-mapping methods.

The results of the methods used in the Multiwell Experiment azimuth study are summarized in Table 1.11. Maps of the borehole seismic events during fracturing in the Paludal zone are shown in Figs. 1.65 and 1.66, which depict the azimuth trends and vertical growth tendencies, respectively. The relatively close agreement between the borehole-seismic-mapping data and the oriented-core strain-recovery data in Table 1.11 is encouraging for the potential of these two methods.

The tests at Mounds consisted of seven fracture-mapping methods in a 1,000-ft [305-m] -deep sandstone formation. The results of these tests are summarized in Table 1.12. The "true" fracture azimuth[234] was N95E, as suggested by borehole television camera observations, surface tiltmeters, and strain-relaxation measurements. The difference in data from the differential strain-curve analysis and differential wave-velocity analysis was attributed to paleostress regimes combined with current stress regimes. Caliper logs and remote seismic sensing did not yield definitive results.

Lacy's[235] work included active seismic measurements from tiltmeters and a triaxial-borehole-seismic tool plus predictive methods

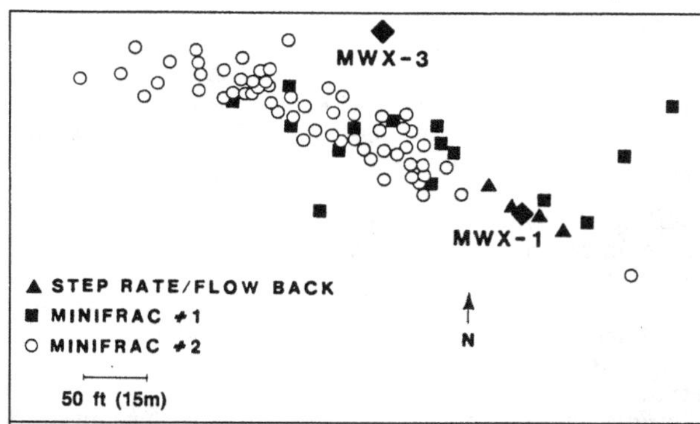

Fig. 1.65—Multiwell Experiment fracture azimuth from borehole-seismic-event locations projected on a horizontal plane.[232]

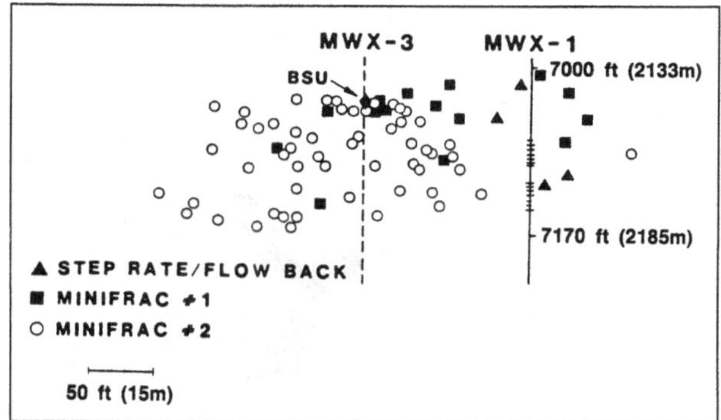

Fig. 1.66—Multiwell Experiment fracture side view from borehole-seismic-event locations projected on a vertical plane parallel to the fracture plane.[232]

TABLE 1.12—SUMMARY OF AZIMUTH RESULTS, MOUNDS TEST SITE, OK[234]

Procedure	True Azimuth
Borehole logs	
Downhole television	N95°E
Caliper logs	—
Tiltmeters	N95°E
Core Analysis	
Anelastic strain recovery	N86°E
Laboratory core analysis	N30°W
Differential strain curve	
Differential wave velocity	
Seismic Monitoring	
Borehole seismic	N70°E
Remote seismic	—

of stress-relief, thermal-expansion, and sonic-velocity measurements on oriented sandstone cores. Test depths ranged from 8,500 to 12,000 ft [2590 to 3660 m]. The results indicated good comparisons of azimuth trends among tiltmeter, triaxial-borehole-seismic, and stress-relaxation data. Fig. 1.67 shows an example of tiltmeter data, and Fig. 1.68 the triaxial-borehole-seismic results on the same well. Note in Fig. 1.67 how much the interpretation was improved by increasing the number of tiltmeters in an array from 8 to 18.

Griffin[236] investigated azimuth measurements from wellbore ellipticity, on-site core strain relaxation, differential strain-curve analysis, differential wave-velocity analysis, triaxial-borehole-seismic tools, impression packers, and borehole-televiewer studies. The results indicated that all these methods yielded azimuth information, but in these tests, the triaxial-borehole-seismic method was preferred from both a definitive and economic standpoint.

Other investigators reported the results of azimuth studies using mainly one type of instrumentation, e.g., tiltmeters[237–240] and borehole-seismic techniques.[241,242] All have obtained definitive signal responses from their instrumentation that yielded azimuth interpretations. From all the work to date, it appears that techniques are now available that can provide azimuth information. In view of the uncertainty involved with any single given method, however, one should use a sufficient number of different methods to corroborate results.

Several investigators have presented results of special design applications. Kim *et al.*[243] concluded that it may be possible to use fracturing pressure, pressure-decline data, and postfracturing temperature surveys to speculate on inferences of fracture orientation relative to the azimuth of a deviated wellbore in certain areas. Other investigations have discussed special designs for geothermal reservoirs,[200] fracture acidizing,[201] soft, unstable formations,[202] and multiple-zone stimulation.[204,205]

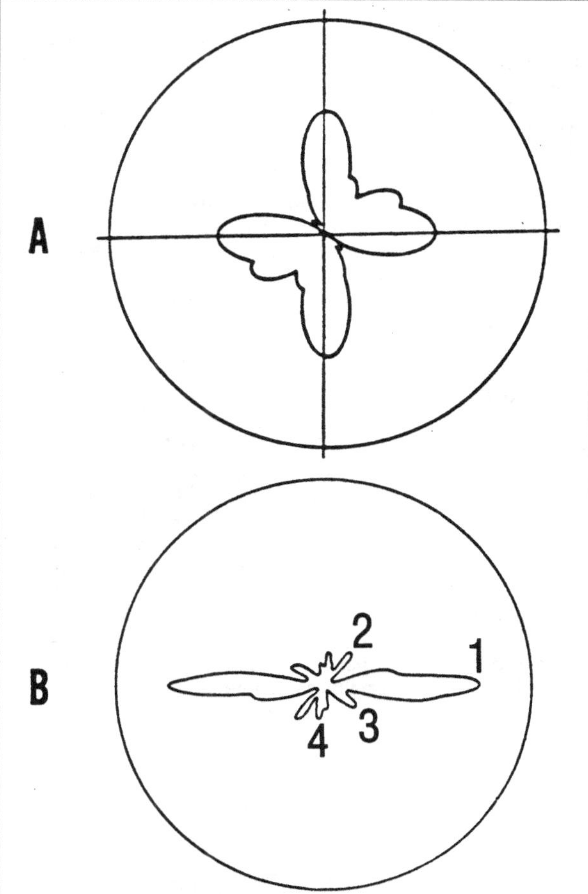

Fig. 1.67—Comparison of fracture-azimuth interpretation from (a) an earlier field test result using eight tiltmeters with (b) the latest field test result using 18 tiltmeters and other improvements.[235]

1.14 Fracturing Economics

The design of fracture treatments generally has three basic requirements: (1) to determine what oil and/or gas producing rates and

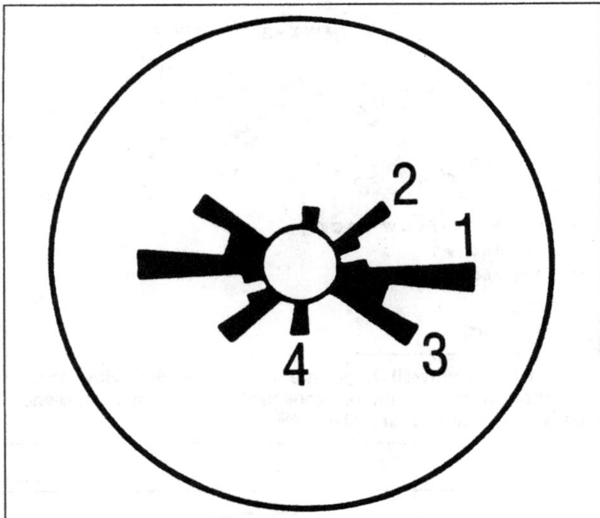

Fig. 1.68—Fracture azimuth from triaxial-borehole-seismic test in the same well and zone as the tiltmeter test shown in Fig. 1.67b. [The primary hydraulic (1) and natural (2 through 4) fracture directions agree within a few degrees.][235]

recoveries might be expected from various fracture lengths and fracture conductivities for a given reservoir, (2) to determine the fracture-treatment design requirements to achieve the desired fracture lengths and conductivities, and (3) to combine Requirements 1 and 2 to maximize economic returns. This concept is illustrated in Fig. 1.69.[5,126]

Ideally, a reservoir performance simulator will provide predictions of the production rates and recoveries for various fracture lengths and conductivities. From these data, a monetary revenue estimate can be developed for various fracture lengths. As can be seen in the upper portion of Fig. 1.69, the estimate of revenue as a function of fracture length is usually not a linear relationship. The rate of revenue increase diminishes with increasing fracture length and eventually reaches a relatively flat slope.

A hydraulic fracturing simulator is usually required to compute the treatment volumes, types of materials, and pumping schedules necessary to achieve various fracture lengths and conductivities. With these data, a relationship between fracture length (and conductivity) and treatment cost can be generated. An example of this is depicted in the lower portion of Fig. 1.69. As can be seen, the treatment costs will usually accelerate with increasing fracture length.

The final step to investigate the total net revenue—i.e., discounted revenue minus cost—is shown on the right side of Fig. 1.69. The net-revenue curve will generally exhibit some optimal point at which the cost to achieve longer fractures exceeds the revenue generated by production from the additional length. Thus, a range of treatment designs that maximizes economics (i.e., optimum treatments) can be selected.

The specific procedures for determining the optimum fracturing-treatment design for a given formation may not always conform precisely to these conceptual steps. But they will always involve some type of balance between treatment costs and revenues generated from the production response associated with a treatment.

A major factor in optimization involves achieving the appropriate balance between the fracture characteristics and the formation properties that govern reservoir performance. High-permeability reservoirs require high fracture conductivities but do not need deeply penetrating fractures; low-permeability formations require deeply penetrating fractures but can tolerate lower fracture conductivities. Some typical length requirements are illustrated in Fig. 1.70.[244] It is presumed here that adequate fracture conductivity exists for all cases. Fig. 1.70 shows that fracture half-length (i.e., wellbore to tip) requirements typically are less than 1,000 ft [305 m] for conventional-permeability ($k>1.0$ md) reservoirs, but the low-permeability (e.g., $k=0.0001$ md) formations can require half-lengths as long as 3,500 to 4,500 ft [1070 to 1370 m].

Optimal economic design is particularly important for the MHF treatments, which can make up a large portion of the total well cost. An example of the relative treatment cost as a percent of total well cost in three major U.S. tight-gas basins is given in Fig. 1.71. As can be seen, for the 500,000-gal [1890-m^3] or higher treatments, fracturing costs can approach one-half the total well cost (including fracturing).

It has generally been recognized that the fracture-length requirements depend greatly on reservoir permeability and fracture conductivity. Fracture conductivity economics presented by Phillips and Anderson[110] and Britt[126] were discussed previously (see Sec. 1.5). However, length and conductivity may not be the only parameters affecting fracture design optimization. This is sometimes not obvious in parametric fracturing studies, where the primary focus is on formation-permeability, fracture-penetration, and conductivity requirements. In some cases, other factors (e.g., net pay and fracture height) can become important considerations in fracturing economics. Their incremental effects can be very significant. For example, consider the effect of net pay on fracture-penetration requirements to optimize the net present worth of a treatment (i.e., the present worth of the hydrocarbon production for the fractured formation minus the present worth of the hydrocarbon production for the unfractured formation minus treatment costs). The results of an example case presented by Veatch[127] (Fig. 1.72) depict the percent increase in net present worth (i.e., net present worth for

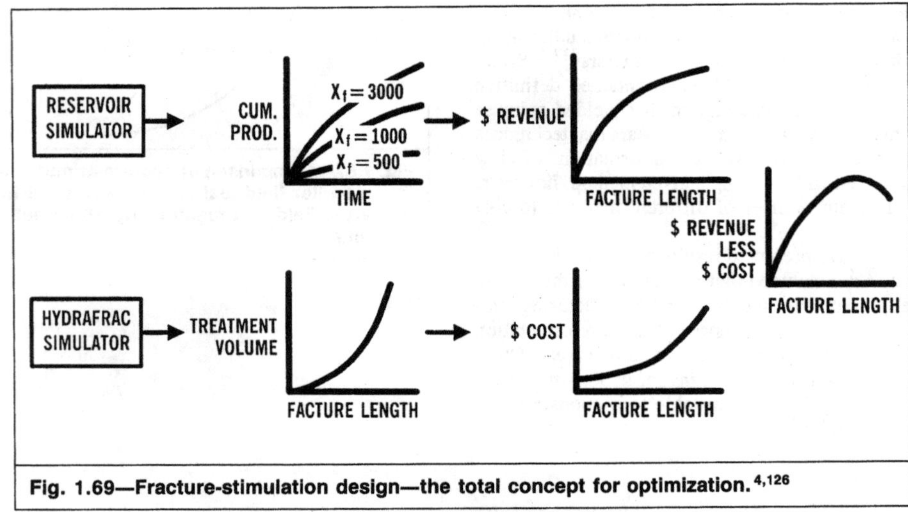

Fig. 1.69—Fracture-stimulation design—the total concept for optimization.[4,126]

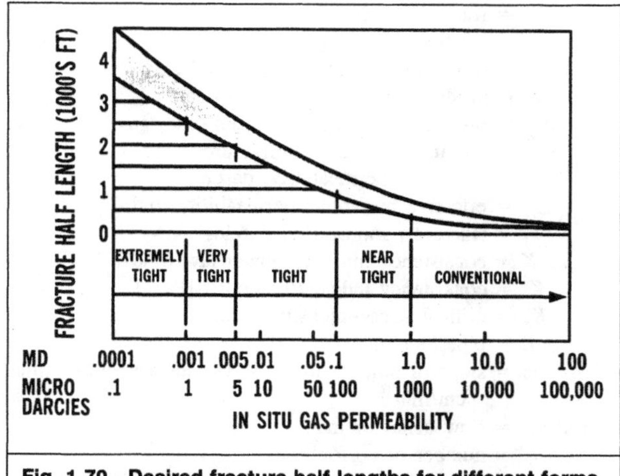

Fig. 1.70—Desired fracture half-lengths for different formation permeabilities.[244]

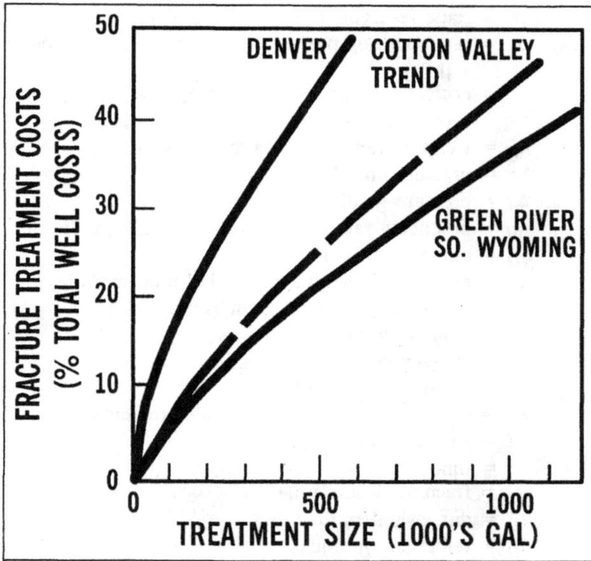

Fig. 1.71—Relative MHF costs/total drilling and completion costs vs. MHF treatment volumes.[5]

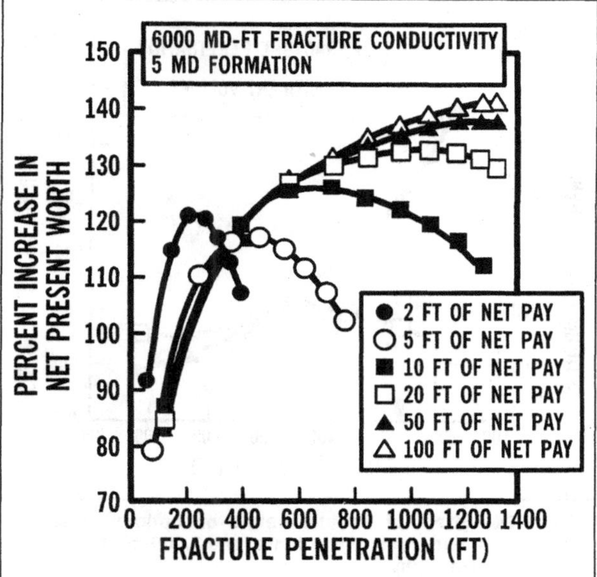

Fig. 1.72—Effect of net pay on percent increase in net present worth vs. fracture penetration for a 6,000-md-ft fracture in a 5-md formation.[127]

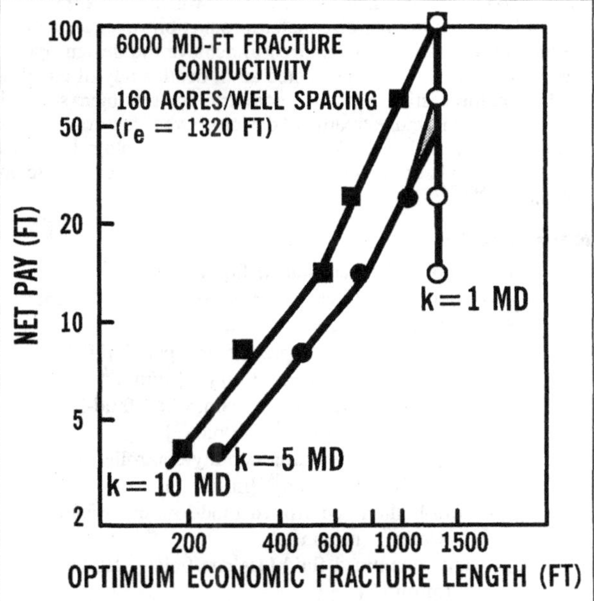

Fig. 1.73—Effect of net pay and permeability on optimum fracture penetration for a 6,000-md-ft fracture in 1-, 5-, and 10-md formations.[127]

the fractured case expressed as a percent of the unfractured case present worth) vs. fracture penetration for net pays ranging from 2 to 100 ft [0.6 to 30 m] in a 5-md formation. Here, fracture conductivity is 6,000 md-ft [1830 md·m] and the wells are on 160-acre/well [65-ha/well] spacing. Fig. 1.72 shows that the optimum fracture penetration (i.e., the penetration at which the maximum net-present-worth increase occurs) gets longer as net pay increases. The results for this case and two other formation permeability levels (1 and 10 md) are summarized in Fig. 1.73, which shows the optimum fracture penetration plotted vs. net pay. Here we see optimum fracture lengths ranging from 200 to 1,320 ft [60 to 400 m] for the 5- and 10-md formations and an almost constant optimum length for the 1-md formation. This shows that optimum lengths can vary widely for a given permeability and fracture conductivity, depending on the net-pay magnitude.

Addressing fracture height from an economic standpoint reinforces the need for reliable height data in treatment design. In addition to the obvious increase in costs, fracture height can have a significant impact on optimum economic penetration, which in turn could affect well-spacing requirements. As an example, cases were run for a 1-md formation with 10 ft [3 m] of net pay, a 2,000-md-ft [610-md·m] fracture, and 160-acre/well [65-ha/well] spacing. Fracture heights from 180 to 720 ft [55 to 219 m] were investigated. The resulting optimum fracture-length and treatment-volume requirements are shown in Fig. 1.74. The optimum values were those that yielded the maximum net present worth for each given height. As can be seen, the optimum treatment-volume requirements did not change dramatically over the wide range of fracture heights, but the optimum lengths did. At a height of 180 ft [55 m], the optimum fracture penetration approaches the drainage boundary (i.e., 1,320 ft [400 m]); at heights of about 600 to 700 ft [180 to 215 m], the optimum lengths were 300 to 400 ft [90 to 120 m]. This suggests that one may need to investigate the economics for closer well spacing for such situations.

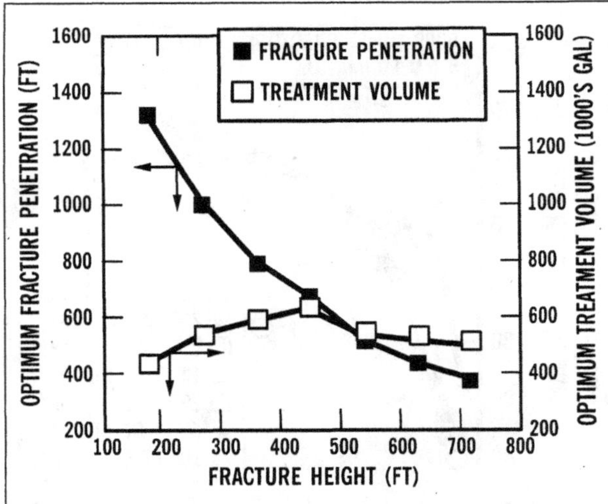

Fig. 1.74—Effect of fracture height on optimum fracture penetration and treatment volume for a 2,000-md-ft fracture in a 10-ft, 1-md pay.[127]

Warembourg *et al.*[199] presented an economic study of three examples and addressed several other important factors that should be considered for optimizing treatment designs. These included (1) the duration of the production forecast from which net present worth is calculated, (2) the net discounted production revenue, and (3) the amount of investment required to achieve the design option. Meng and Brown[245] also presented an in-depth study of coupling production forecasting, fracture geometry, and treatment scheduling to optimize fracture-treatment designs. Other factors—such as hydrocarbon price, interest (discount) factors, technology level, and risk—have also been shown to play a critical role in economic optimization.[246,247]

Nomenclature

A = empirical fit constants, Eq. 1.7
A_{fl} = area of fluid-loss paper or core, Eq. 1.10, cm^2
B = empirical fit constants, Eq. 7
c_f = reservoir fluid compressibility, psi^{-1} [kPa^{-1}]
C = fluid-loss coefficient, ft/min$^{1/2}$ [m/min$^{1/2}$]
C_I = fracturing-fluid-viscosity-controlled fluid-loss coefficient, ft/min$^{1/2}$ [m/min$^{1/2}$]
C_{II} = reservoir-fluid-compressibility-controlled fluid-loss coefficient, ft/min$^{1/2}$ [m/min$^{1/2}$]
C_{III} = wall-building-controlled fluid-loss coefficient, ft/min$^{1/2}$ [m/min$^{1/2}$]
C_t = total combined fluid-loss coefficient, ft/min$^{1/2}$ [m/min$^{1/2}$]
d = particle diameter, cm
d_H = hydraulic diameter of fracture
d_p = pipe diameter
e = base of natural logarithms
e_f = fluid efficiency for slurry
E = Young's modulus of elasticity
E' = effective in-situ Young's modulus of elasticity
E_a = fluid activation energy per mole
f_m = measured friction factor
f_{th} = theoretical friction factor
F = thermoviscous constant characteristic of fluid
g = gravitational acceleration, 980.7 cm/s^2
$g_f(t)$ = fracture gradient at time t, psi/ft [kPa/m]
G = shear modulus
G' = fluid elastic modulus
G'' = fluid viscous modulus
$G(t_D,t_D^*)$ = dimensionless pressure difference function
h = formation net pay, ft [m]
h_f = fracture height, ft [m]
h_ℓ = net fluid-loss height, ft [m]
J = productivity index, unfractured case
J_f = productivity index, fractured case
J'_g = productivity index, fractured case, gas
$J_{g,o}$ = productivity index, unfractured case, gas
k = formation permeability, darcies
k_{eH} = effective horizontal permeability, md
k_f = fracture permeability, darcies
K = consistency index of power-law fluid
K' = consistency index, (lbf-sec)/ft [Pa·s]
K_c = critical stress-intensity factor
L = pipe length, ft
m = slope of plot of fluid loss vs. square root of time, cm/min$^{1/2}$
n,n',n_f' = flow behavior index
n_p = number of perforations
N_{Re} = Reynolds number
p = pressure, psi [kPa]
p_c = fracture closure pressure (wellbore), psi [kPa]
p_f = fracture extension pressure, psi [kPa]
p_i = static reservoir pressure, psi [kPa]
p_n = net fracturing pressure (wellbore)$=(p-p_c)$, psi [kPa]
p_n' = critical net fracturing pressure; pressure capacity (wellbore), psi [kPa]
p_{wf} = wellbore flowing pressure, psi [kPa]
Δp = pressure drop, psi [kPa]
$\Delta p/\Delta x$ = pressure gradient
q = producing rate (unfractured), bbl/min [m^3/min]
q_D = dimensionless flow rate
q_f = producing rate (fractured), bbl/min [m^3/min]
q_i = injection rate, bbl/min [m^3/min]
Q = volumetric flow rate
Q_D = cumulative flow rate
r = distance from crack front
r'_w = equivalent wellbore radius after fracturing
R = gas constant
t = time
t_c = fracture closure time
t_D = dimensionless time
t_D^* = reference t_D values
t_{Dx_f} = dimensionless time based on x_f
t_i = total treatment pumping time
t_1 = time of first shut-in pressure measurement, minutes
t_2 = time of later shut-in pressure measurement, minutes
T = temperature, °R [K]
T_D = $(T_f-T_i)/(T-T_i)$=dimensionless temperature in a fracture
T_f = fluid temperature in a fracture, °R [K]
T_i = injection temperature, °R [K]
T_R = reservoir temperature, °R [K]
v = fluid velocity, ft/sec [m/s]
v_L = leakoff velocity, ft/sec [m/s]
w = fracture width, ft [m]
w_c = critical crack-opening width, ft [m]
x = distance from the wellbore to some point in a fracture
x_f = fracture length, ft [m]
α = constant, Eq. 1.14
$\dot{\gamma}$ = shear rate, seconds^{-1}
$\dot{\gamma}_w$ = wall shear rate
μ = viscosity, cp [Pa·s]
μ_a = apparent viscosity, cp [Pa·s]
ν = Poisson's ratio
ρ = density, g/cm^3

ρ_p = proppant density, g/cm^3
τ_w = wall shear stress
ϕ = porosity

References

General Overview*

1. Waters, A.B. "Hydraulic Fracturing—What Is It?" *JPT* (Aug. 1981) 1416.
2. Howard, G.C. and Fast, C.R.: *Hydraulic Fracturing*, Monograph Series, SPE, Richardson, TX (1970) **2.**
3. Veatch, R.W. Jr. and Moschovidis, Z.A.: "An Overview of Recent Advances in Hydraulic Fracturing Technology," paper SPE 14085 presented at the 1986 SPE Intl. Meeting on Petroleum Engineering, Beijing, March 17–20.
4. Veatch, R.W. Jr.: "Overview of Current Hydraulic Fracturing Design and Treatment Technology—Part 2," *JPT* (May 1983) 853–64.
5. Veatch, R.W. Jr.: "Overview of Current Hydraulic Fracturing Design and Treatment Technology—Part 1," *JPT* (April 1983) 677–87.

Formation Evaluation—The Fracturing Aspects

6. Prats, M.: "Effect of Vertical Fractures on Reservoir Behavior—Incompressible Fluid Case," *SPEJ* (June 1961) 105–18; *Trans.*, AIME, **222.**
7. McGuire, W.J. and Sikora, V.J.: "The Effect of Vertical Fractures on Well Productivity," *Trans.*, AIME (1960) **219,** 401–03.
8. Tinsley, J.M. *et al.*: "Vertical Fracture Height—Its Effect on Steady-State Production Increase," *JPT* (May 1969) 633–38; *Trans.*, AIME, **246.**
9. Tannich, J.D. and Nierode, D.E.: "The Effect of Vertical Fractures on Gas Well Productivity," paper SPE 15902 available at SPE, Richardson, TX (June 1986).
10. Agarwal, R.G., Carter, R.D., and Pollock, C.B. "Evaluation and Performance Prediction of Low-Permeability Gas Wells Stimulated by Massive Hydraulic Fracturing," *JPT* (March 1979) 362–72; *Trans.*, AIME, **267.**
11. Holditch, S.A. *et al.*: "The Optimization of Well Spacing and Fracture Length in Low Permeability Gas Reservoirs," paper SPE 7496 presented at the 1978 SPE Annual Technical Conference and Exhibition, Houston, Oct. 1-4.
12. "NPC—Unconventional Gas Sources—Volume V—Tight Gas Reservoirs—Part I—December, 1980," Tight Gas Reservoir Task Group of the Unconventional Gas Committee of the Natl. Petroleum Council (1980).
13. Baker, C.O.: "Effect of Price and Technology on Tight Gas Resources of the United States," paper 819584 presented at the 1981 ASME Intersociety Energy Conversion Conference, Atlanta, Aug. 9–14.
14. Veatch, R.W. Jr.: "A Brief Survey of the Technology Challenge to Improve Recovery from Tight Gas Reservoirs," paper 819582 presented at the 1981 ASME Intersociety Energy Conversion Conference, Atlanta, Aug. 9–14.
15. Cinco-Ley, H. and Samaniego-V., F.: "Transient Pressure Analysis for Fractured Wells," *JPT* (Sept. 1981) 1749–66.
16. Lee, W.J. and Holditch, S.A.: "Fracture Evaluation With Pressure Transient Testing in Low-Permeability Gas Reservoirs," *JPT* (Sept. 1981) 1776–92.
17. Cinco-Ley, H.: "Evaluation of Hydraulic Fracturing by Transient Pressure Analysis Methods," paper SPE 10043 presented at the 1982 SPE Intl. Petroleum Exhibition and Technology Symposium, Beijing, March 19–22.
18. Bennett, C.O. *et al.*: "Performance of Finite-Conductivity, Vertically Fractured Wells in Single-Layer Reservoirs," *SPEFE* (Aug. 1986) 399–412; *Trans.*, AIME, **281.**
19. Bennett, C.O., Reynolds, A.C., and Raghavan, R.: "Analysis of Finite-Conductivity Fractures Intercepting Multilayer Commingled Reservoirs," *SPEFE* (June 1986) 259–74; *Trans.*, AIME, **281.**
20. Guppy, K.H., Cinco-Ley, H., and Ramey, H.J. Jr.: "Pressure Buildup Analysis of Fractured Wells Producing at High Flow Rates," *JPT* (Nov. 1982) 2656–66.
21. Tison, J.K. *et al.*: "A Method for Selecting Potential Infill Locations in the East Texas Cotton Valley Tight Gas Play," paper SPE 11022 presented at the 1982 SPE Annual Technical Conference and Exhibition, New Orleans, Sept. 26–29.
22. Verbeek, C.M.J.: "Analysis of Production Tests of Hydraulically Fractured Wells in a Tight Solution Gas Drive Reservoir," paper SPE 11084 presented at the 1982 SPE Annual Technical Conference and Exhibition, New Orleans, Sept. 26–29, 1982.

Rock Mechanics and Fracture Geometry

23. Rosepiler, M.J.: "Determination of Principal Stresses and Confinement of Hydraulic Fractures in Cotton Valley," paper SPE 8405 presented at the 1979 SPE Annual Technical Conference and Exhibition, Las Vegas, Sept. 23-26.
24. Nolte, K.G. and Smith, M.B.: "Interpretation of Fracturing Pressures," *JPT* (Sept. 1981) 1767–75.
25. Veatch, R.W. Jr. and Crowell, R.F.: "Joint Research/Operations Programs Accelerate Massive Hydraulic Fracturing Technology," *JPT* (Dec. 1982) 2763–75.
26. Nolte, K.G.: "Principles for Fracture Design Based on Pressure Analysis," *SPEPE* (Feb. 1988) 22–30.
27. Aron, J., Murray, J., and Seeman, B.: "Formation Compressional and Shear Interval Transit-Time Logging by Means of Long Spacings and Digital Techniques," paper SPE 7446 presented at the 1978 SPE Annual Technical Conference and Exhibition, Houston, Oct. 1–4.
28. Fertl, W.H.: "Evaluation of Fractured Reservoir Rocks Using Geophysical Well Logs," paper SPE 8938 presented at the 1980 SPE/DOE Unconventional Gas Recovery Symposium, Pittsburgh, PA, May 18–21.
29. Voegele, M.D. and Jones, A.H.: "A Wireline Hydraulic Fracturing Tool for the Determination of In-Situ Stress Contrasts," paper SPE 8937 presented at the 1980 SPE/DOE Unconventional Gas Recovery Symposium, Pittsburgh, PA, May 18–21.
30. Koerperich, E.A.: "Shear Wave Velocities Determined From Long- and Short-Spaced Borehole Acoustical Devices," *SPEJ* (Oct. 1980) 317–26.
31. Teufel, L.W.: "Determination of In-Situ Stress From Anelastic Strain Recovery Measurements of Oriented Cores," paper SPE 11649 presented at the 1983 SPE/DOE Low-Permeability Gas Reservoirs Symposium, Denver, March 14–16.
32. Teufel, L.W.: "Prediction of Hydraulic Fracture Azimuth from Anelastic Strain Recovery Measurements of Oriented Core," *Proc.*, 23rd U.S. National Rock Mechanics Symposium (1982) 238–46.
33. Teufel, L.W.: "In-Situ Stress State in the Mounds Test Well as Determined by the Anelastic Strain Recovery Method," paper SPE 13896 presented at the 1985 SPE/DOE Low-Permeability Gas Reservoirs Symposium, Denver, May 19–22.
34. Blanton, T.L.: "The Relation Between Recovery Deformation and In-Situ Stress Magnitudes," paper SPE 11624 presented at the 1983 SPE/DOE Low-Permeability Gas Reservoirs Symposium, Denver, March 14–16.
35. Blanton, T.L. and Teufel, L.W.: "A Field Test of the Strain Recovery Method of Stress Determination in Devonian Shales," paper SPE 12304 presented at the 1983 SPE Eastern Regional Meeting, Pittsburgh, PA, Nov. 9–11.
36. Blanton, T.L. and Teufel, L.W.: "In-Situ Stress Determination From Wellbore Elongation Measurements," paper SPE 13877 presented at the 1985 SPE/DOE Low-Permeability Gas Reservoirs Symposium, Denver, May 19–22.
37. Teufel, L.W. and Warpinski, N.R.: "Determination of In-Situ Stress from Anelastic Strain Recovery Measurements of Oriented Core: Comparison to Hydraulic Fracture Stress Measurements in the Rollins Sandstone," *Proc.*, 25th U.S. Symposium on Rock Mechanics, Evanston (June 1984) 176–85.
38. Warpinski, N.R., Branagan, P., and Wilmer, R.: "In-Situ Stress Measurements at U.S. DOE's Multiwell Experiment Site, Mesaverde Group, Rifle, Colorado," *JPT* (March 1985) 527–36.
39. Daneshy, A.A. *et al.*: "In-Situ Stress Measurements During Drilling," *JPT* (Aug. 1986) 891–98; *Trans.*, AIME, **281.**
40. Lin, W.: "Ultrasonic Velocities and Dynamic Elastic Moduli of Mesaverde Rocks," Report UCID-20273, Lawrence Livermore Natl. Laboratory (Nov. 1984).
41. Mao, N.H. and Sweeney, J.J.: "Estimation of In-Situ Stresses From Ultrasonic Measurements," *SPEFE* (Oct. 1986) 532–38.
42. Newberry, B.M., Nelson, R.F., and Ahmed, U.: "Prediction of Vertical Hydraulic Fracture Migration Using Compression and Shear Wave Slowness," paper SPE 13895 presented at the 1985 SPE/DOE Low-Permeability Gas Reservoirs Symposium, Denver, May 19–22.
43. Johnson, P.A. and Albright, J.N.: "In-Situ Physical Properties Using Crosswell Acoustic Data," paper SPE 13881 presented at the 1985 SPE/DOE Low-Permeability Gas Reservoirs Symposium, Denver, May 19–22.
44. Begnaud, W.J. and Claiborne, E.B. Jr.: "Vertical Fracture Growth Considerations in the Mission Canyon/Ratcliffe Formations of the North Alexander Area," paper SPE 14375 presented at the 1985 SPE Annual Technical Conference and Exhibition, Las Vegas, Sept. 22–25.

Fracture-Propagation Models

45. Perkins, T.K. Jr. and Kern, L.R.: "Widths of Hydraulic Fractures," *JPT* (Sept. 1961) 937–49; *Trans.*, AIME, **222.**
46. Sneddon, I.N.: "The Distribution of Stress in the Neighborhood of a Crack in an Elastic Solid," *Proc.*, Royal Soc. of London (1946) **187,** 229.
47. Geertsma, J. and de Klerk, F.: "A Rapid Method of Predicting Width and Extent of Hydraulically Induced Fractures," *JPT* (Dec. 1969) 1571–81; *Trans.*, AIME, **246.**

*Other recent publications are *Reservoir Stimulation*, M.J. Economides and K.G. Nolte (eds.), Schlumberger Educational Services, Houston (1987) 1-01—12-17; and *Petroleum Engineering Handbook*, H.B. Bradley (ed.), SPE, Richardson, TX (1987) Chap. 55.

48. Khristianovitch, S.A. and Zheltov, Y.P.: "Formation of Vertical Fractures by Means of Highly Viscous Fluids," *Proc.*, Fourth World Pet. Cong., Rome (1955) **II,** 579.
49. Barenblatt, G.I.: "Mathematical Theory of Equilibrium Cracks," *Advances in Applied Mechanics* (1962) **7,** 55.
50. Carter, R.D.: Appendix I to paper by C.C. Howard and C.R. Fast, "Optimum Fluid Characteristics for Fracture Extension," presented at the 1957 ASME Spring Meeting, Mid-Continent District, Div. of Production, Tulsa, OK, April.
51. Nordgren, R.P.: "Propagation of a Vertical Hydraulic Fracture," *SPEJ* (Aug. 1972) 306–14; *Trans.*, AIME, **253.**
52. Smith, M.B.: "Stimulation Design for Short, Precise Hydraulic Fractures," *SPEJ* (June 1985) 371–79.
53. Geertsma, J. and Haafkens, R.: "A Comparison of Theories for Predicting Width and Extent of Vertical Hydraulically Induced Fractures," *Trans.*, ASME (1979) **101,** 8–19.
54. Daneshy, A.A.: "On the Design of Vertical Hydraulic Fractures," *JPT* (Jan. 1973) 83–93; *Trans.*, AIME, **255.**
55. Daneshy, A.A. *et al.*: "Effect of Treatment Parameters on the Geometry of a Hydraulic Fracture," paper SPE 3507 presented at the 1971 SPE Annual Meeting, New Orleans, Oct. 3–6.
56. Settari, A.: "Simulation of Hydraulic Fracturing Processes," *SPEJ* (Dec. 1980) 487–500.
57. Sinclair, A.R.: "Heat Transfer Effects in Deep Well Fracturing," *JPT* (Dec. 1971) 1484–92; *Trans.*, AIME, **251.**
58. Harrington, L.J., Hannah, R.R., and Beirute, R.: "Post-Fracturing Temperature Recovery and Its Implication for Stimulation Design," paper SPE 7560 presented at the 1978 SPE Annual Technical Conference and Exhibition, Houston, Oct. 1–4.
59. Wheeler, J.A.: "Analytical Calculations of Heat Transfer From Fractures," paper SPE 2494 presented at the 1969 SPE Improved Oil Recovery Symposium, Tulsa, OK, April 13–15.
60. Whitsitt, N.F. and Dysart, G.R.: "The Effect of Temperature on Stimulation Design," *JPT* (April 1970) 493–502; *Trans.*, AIME, **249.**
61. Ingraffea, A.R., Shaffer, R.J., and Heuze, F.E.: "FEFFLAP: A Finite Element Program for Analysis of Fluid-Driven Fracture Propagation in Jointed Rock," Unconventional Gas Program Reports UCID-20368 and UCID-20369 (March 1985).
62. Narendran, V.M. and Cleary, M.P.: "Analysis of Growth and Interaction of Multiple Hydraulic Fractures," paper SPE 12272 presented at the 1983 SPE Reservoir Simulation Symposium, San Francisco, Nov. 15–18.
63. Mendelsohn, D.A.: "A Review of Hydraulic Fracture Modeling—Part I: General Concepts, 2D Models, Motivation for 3D Modeling," *J. Energy Res. Tech.* (Sept. 1984) **106,** 369.
64. Mendelsohn, D.A.: "A Review of Hydraulic Fracture Modeling—Part II: 3D Modeling and Vertical Growth in Layered Rock," *J. Energy Res. Tech.* (Dec. 1984) **106,** 543.
65. Griffith, A.A.: "The Phenomena of Rupture and Flow in Solids," *Phil. Trans.*, Royal Soc. of London (1920) Ser. A, **221,** 163–98.
66. Griffith, A.A.: "Fracture Dynamics," *Fracturing of Metals (A Seminar on the Fracturing of Metals),* American Soc. for Metals (1948) 147–66.
67. Keck, R.G., Cleary, M.P., and Crockett, A.: "A Lumped Numerical Model for the Design of Hydraulic Fractures," paper SPE 12884 presented at the 1984 SPE/DOE/GRI Unconventional Gas Recovery Symposium, Pittsburgh, PA, May 13–15.
68. Morita, N., Whitfill, D.L., and Wahl, H.A.: "Stress Intensity Factor and Fracture Cross-Sectional Shape Predictions From 3D Model for Hydraulically Induced Fractures," *JPT* (Oct. 1988) 1329–42.
69. Bui, H.D.: "An Integral Equations Method for Solving the Problem of a Plane Crack of Arbitrary Shape," *J. Mech. Phys. Solids* (1977) **25,** 29–39.
70. Mastrojannis, E.N., Keer, L.M., and Mura, T.: "Stress Intensity Factor for a Plane Crack Under Normal Pressure," *Intl. J. Fracture* (June 1985) **15,** 3.
71. Clifton, R.J. and Abou-Sayed, A.S.: "A Variational Approach to the Prediction of the Three-Dimensional Geometry of Hydraulic Fractures," paper SPE 9879 presented at the 1981 SPE/DOE Low-Permeability Gas Reservoirs Symposium, Denver, May 27–29.
72. Lam, K.Y. and Cleary, M.P.: "Development of a Fully Three-Dimensional Simulator for Analysis and Design of Hydraulic Fracturing," MIT-UFRAC Project Report, Resource Extraction Laboratory (June 1985).
73. Mura, T.: *Micromechanics of Defects in Solids,* Martinus Nijhoff Publishers (1982).
74. Cleary, M.P. *et al.*: "Theoretical and Laboratory Simulation of Underground Fracturing Operations," MIT-UFRAC First Annual Report (Aug. 1981).
75. Mura, T.: "The Continuum Theory of Dislocations," *Advances in Materials Research,* H. Herman (ed.), Interscience Publishers (1968) **3.**
76. Hirth, J.P. and Lothe, J.: *Theory of Dislocations,* McGraw-Hill Book Co. Inc., New York City (1968).
77. Kossecka, E.: "Defects as Surface Distributions of Double Forces," *Arch. Mech.* (1971) **23,** 481–94.
78. Bui, H.D.: "Application des potentiels élastique à l'étude des fissures planes de forme arbitraire en milieu tridimensionnel," *Centre Res. Acad. Sci. Ser. A280* (1968) **1975,** 1157–60.
79. Mastrojannis, E.N., Keer, L., and Mura, T.: "Growth of Planar Cracks Induced by Hydraulic Fracture," *Intl. J. Numerical Methods Eng.* (1980) **15,** 41–54.
80. Annigeri, B.S. and Cleary, M.P.: "Surface Integral Finite Element Hybrid (SIFEH) Method for Fracture Mechanics," *Intl. J. Numerical Methods Eng.* (1984) **20,** 869–85.
81. Lee, J.C. and Keer, L.M.: "Study of a Three-Dimensional Crack Terminating at an Interface," *J. Applied Mech.*
82. Clifton, R.J.: "Recent Advances in the Three-Dimensional Simulation of Hydraulic Fracturing," *Proc.*, 19th Midwestern Mechanics Conference, Columbus, OH (Sept. 1985).
83. Barree, R.D.: "A Practical Numerical Simulator for Three-Dimensional Fracture Propagation in Heterogeneous Media," paper SPE 12273 presented at the 1983 SPE Reservoir Simulation Symposium, San Francisco, Nov. 15–18.
84. Nemat-Nasser, S. and Ohtsubo, H.: "Fluid Flow and Heat Transfer Through Hydraulically Induced Fractures in Hot, Dry Rock Masses," *J. Pressure Vessel Tech.* (Aug. 1978) **100,** 277–84.
85. Clifton, R.J. and Abou-Sayed, A.S.: "On the Computation of the Three-Dimensional Geometry of Hydraulic Fractures," paper SPE 7943 presented at the 1979 SPE/DOE Low-Permeability Gas Reservoirs Symposium, Denver, May 20–22.
86. Cleary, M.P., Kavvadas, M., and Lam, K.Y.: "Development of a Fully Three-Dimensional Simulator for Analysis and Design of Hydraulic Fracturing," paper SPE 11631 presented at the 1983 SPE/DOE Low-Permeability Gas Reservoirs Symposium, Denver, March 14–16.
87. Advani, S.H. *et al.*: "Fluid Flow and Structural Response Modeling Associated With the Mechanics of Hydraulic Fracturing," *SPEFE* (June 1986) 309–18.
88. Broek, D.: *Elementary Fracture Mechanics,* Sijhoff and Noordhoff Publishers (1978).
89. Settari, A. and Cleary, M.P.: "Development and Testing of Pseudo-Three-Dimensional Model of Hydraulic Fracture Geometry," *SPEPE* (Nov. 1986) 449–66; *Trans.*, AIME, **281.**
90. Palmer, I.D. and Craig, H.R.: "Modeling of Axisymmetric Vertical Growth in Elongated Hydraulic Fractures and Application to First MWX Stimulations," paper SPE 12879 presented at the 1984 SPE/DOE/GRI Unconventional Gas Recovery Symposium, Pittsburgh, PA, May 13–15.
91. Palmer, I.D. and Luiskutty, C.T.: "A Comparison of Highly Elongated Fracture Models of Variable Heights," paper SPE 13864 presented at the 1985 SPE/DOE Low-Permeability Gas Reservoirs Symposium, Denver, May 19–22.
92. Palmer, I.D. and Carroll, H.B. Jr.: "Three-Dimensional Hydraulic Fracture Propagation in the Presence of Stress Variations," *SPEJ* (Dec. 1983) 870–78.
93. Palmer, I.D. and Carroll, H.B. Jr.: "Numerical Solution for Height of Elongated Hydraulic Fractures," paper SPE 11627 presented at the 1983 SPE/DOE Low-Permeability Gas Reservoirs Symposium, Denver, March 14–16.
94. Meyer, B.R.: "Frac Model in 3D—1: New Simulator Makes Fracture Design Routine," *Oil & Gas J.* (June 17, 1985) 87.
95. Meyer, B.R.: "Frac Model in 3D—2: Proppant Placement Analyzed," *Oil & Gas J.* (July 1, 1985) 65.
96. Meyer, B.R.: "Frac Model in 3D—3: Hydraulic Fracturing Simulator's Capabilities Examined," *Oil & Gas J.* (July 22, 1985) 65.
97. Advani, S.H., Khattab, H., and Lee, J.K.: "Hydraulic Fracture Geometry Modeling, Prediction, and Comparisons," paper SPE 13863 presented at the 1985 SPE/DOE Low-Permeability Gas Reservoirs Symposium, Denver, May 19–22.
98. Thiercelin, M.J., Ben-Naceur, K., and Lemanczyk, Z.R.: "Simulation of Three-Dimensional Propagation of a Vertical Hydraulic Fracture," paper SPE 13861 presented at the 1985 SPE/DOE Low-Permeability Gas Reservoirs Symposium, Denver, May 19–22.
99. Settari, A.: "Quantitative Analysis of Factors Influencing Vertical and Lateral Fracture Growth," *SPEPE* (Aug. 1988) 310–22.
100. Abou-Sayed, A.S., Clifton, R.J., and Sinha, K.P.: "Evaluation of the Influence of In-Situ Reservoir Conditions on the Geometry of

Hydraulic Fractures Using a Three-Dimensional Simulator: Part I—Technical Approach,'' paper SPE 12877 presented at the 1984 SPE/DOE/GRI Unconventional Gas Recovery Symposium, Pittsburgh, PA, May 13–15.

101. Abou-Sayed, A.S. *et al.*: ''Evaluation of the Influence of In-Situ Reservoir Conditions on the Geometry of Hydraulic Fractures Using a Three-Dimensional Simulator: Part 2—Case Studies,'' paper SPE 12878 presented at the 1984 SPE/DOE/GRI Unconventional Gas Recovery Symposium, Pittsburgh, PA, May 13–15.

Propping Agents, Fracture Conductivity, and Economics

102. *RP 56, Recommended Practices for Testing Sand Used in Hydraulic Fracturing Operations,* API, Dallas (1983).
103. Cutler, R.A. *et al.*: ''New Proppants for Deep Gas Well Stimulation,'' paper SPE 9869 presented at the 1981 SPE/DOE Low-Permeability Gas Reservoirs Symposium, Denver, May 27–29.
104. Neal, E.A., Parmley, J.L., and Colpoys, P.J.: ''Oxide Ceramic Proppants for Treatment of Deep Well Fractures,'' paper SPE 6816 presented at the 1977 SPE Annual Technical Conference and Exhibition, Denver, Oct. 9–12.
105. Callanan, M.J., Cipolla, C.L., and Lewis, P.E.: ''The Application of a New Second-Generation High-Strength Proppant in Tight Gas Reservoirs,'' paper SPE 11633 presented at the 1983 SPE/DOE Low-Permeability Gas Reservoirs Symposium, Denver, March 13–16.
106. Cutler, R.A. *et al.*: ''Fracture Conductivity Comparison of Ceramic Proppants,'' *SPEJ* (April 1985) 157–70.
107. Sinclair, A.R. and Graham, J.W.: ''A New Proppant for Hydraulic Fracturing,'' paper presented at the 1978 ASME Energy Technology Conference, Houston, Nov. 5–9.
108. Cooke, C.E. Jr.: ''Effect of Fracturing Fluids on Fracture Conductivity,'' *JPT* (Oct. 1975) 1273–82; *Trans.*, AIME, **259.**
109. Cooke, C.E. Jr.: ''Fracturing With a High-Strength Proppant,'' *JPT* (Oct. 1977) 1222–26.
110. Phillips, A.M. and Anderson, R.W.: ''Use of Proppant Selection Models to Optimize Fracturing Treatment Designs in Low Permeability Reservoirs,'' paper SPE 13855 presented at the 1985 SPE/DOE Low-Permeability Gas Reservoirs Symposium, Denver, May 19–22.
111. Larsen, D.G. and Smith, L.J.: ''New Conductivity Found in Angular Blends of Fracturing Sand,'' paper SPE 13814 presented at the 1985 SPE Production Operations Symposium, Oklahoma City, OK, March 10–12.
112. Almond, S.W.: ''Factors Affecting Gelling Agent Residue Under Low Temperature Conditions,'' paper SPE 10658 presented at the 1982 SPE Formation Damage Control Symposium, Lafayette, LA, March 24–25.
113. Becq, D.F., Roque, C., and Sarda, J.P.: ''High-Strength Proppant Behavior Under Extreme Conditions,'' paper SPE 12487 presented at the 1984 SPE Formation Damage Control Symposium, Bakersfield, CA, Feb. 13–14.
114. Norman, M.E., Cipolla, C.L., and Webb, M.L.: ''Application of Manufactured Proppants in Moderately Permeable Oil Reservoirs,'' paper SPE 12357 presented at the 1983 SPE Production Technology Symposium, Lubbock, TX, Nov. 14–15.
115. ''Proppants, Permeability and Conductivity,'' data book, BJ-Titan Inc., Tomball, TX (May 20, 1983).
116. ''Proppant Selection Guide,'' Dowell Schlumberger, Tulsa, OK (Sept. 1985).
117. ''The Fracbook™ Design/Data Manual for Hydraulic Fracturing,'' Halliburton Services, Duncan, OK (1971).
118. Holditch, S.A.: *Criteria for Propping Agent Selection,* second edition, Norton Alcoa Proppants, Dallas (1984).
119. ''The Technical Literature File,'' Standard Oil Proppants Div., Standard Oil Co., Dallas, TX.
120. *Proppants,* second edition, The Western Co. of North America, R&D, Fort Worth, TX (1984).
121. ''Propped Fracture Flow Capacity,'' technical newsletter, The Western Co. of North America, R&D, Fort Worth, TX (1985).
122. Kim, C.M. and Losacano, J.A.: ''Fracture Conductivity Damage Due to Cross-Linked Gel Residue and Closure Stress on Propped 20/40 Mesh Sand,'' paper SPE 14436 presented at the 1985 SPE Annual Technical Conference and Exhibition, Las Vegas, Sept. 22–25.
123. Cheung, S.K.: ''Effects of Acids on Gravels and Proppants,'' *SPEPE* (May 1988) 201–04.
124. Almond, S.W. and Bland, W.E.: ''Effect of Break Mechanisms on Gelling Agent Residue and Flow Impairment in 20/40 Mesh Sand,'' paper SPE 12485 presented at the 1984 SPE Formation Damage Control Symposium, Bakersfield, CA, Feb. 13–14.
125. Phillips, A.M. and Wilson, W.J.: ''Improved Drainage of Sand Pack Enhances Fracturing Fluid Recovery and Increases Production,'' paper SPE 12924 presented at the 1984 SPE Rocky Mountain Regional Meeting, Casper, WY, May 21–23.
126. Britt, L.K.: ''Optimized Oilwell Fracturing of Moderate-Permeability Reservoirs,'' paper SPE 14371 presented at the 1985 SPE Annual Technical Conference and Exhibition, Las Vegas, Sept. 22–25.
127. Veatch, R.W.: ''Economics of Fracturing: Some Methods, Examples, and Case Studies,'' paper SPE 15509 presented at the 1986 SPE Annual Technical Conference and Exhibition, New Orleans, Oct. 5–8.
128. Elbel, J.L.: ''Considerations for Optimum Fracture Geometry Design,'' *SPEPE* (Aug. 1988) 323–27.
129. Montgomery, C.T. and Steanson, R.E.: ''Proppant Selection—The Key to Successful Fracture Stimulation,'' *JPT* (Dec. 1985) 2163–72.
130. Bennett, C.O. *et al.*: ''Influence of Fracture Heterogeneity and Wing Length on the Response of Vertically Fractured Wells,'' *SPEJ* (April 1983) 219–30.

Foamed Fracturing Fluids

131. King, G.E.: ''Factors Affecting Dynamic Fluid Leakoff with Foam Fracturing Fluids,'' paper SPE 6817 presented at the 1977 SPE Annual Technical Conference and Exhibition, Denver, Oct. 9–12.
132. Harris, P.C.: ''Dynamic Fluid-Loss Characteristics of Nitrogen Foam Fracturing Fluids,'' *JPT* (Oct. 1985) 1847–52.
133. Ainley, B.R. and Charles, J.G.: ''Fracturing Shale and Tight Sands With Stabilized Foam as the Pad Fluid and Stimulating Coal Seams With Stabilized Foam as the Sole Fluid,'' paper SPE 10825 presented at the 1982 SPE/DOE Unconventional Gas Recovery Symposium, Pittsburgh, PA, May 16–18.
134. Harris, P.C.: ''Effects of Texture on Rheology of Foam Fracturing Fluids,'' paper SPE 14257 presented at the 1985 SPE Annual Technical Conference and Exhibition, Las Vegas, Sept. 22–25.
135. Harris, P.C.: ''Dynamic Fluid-Loss Characteristics of CO_2-Foam Fracturing Fluids,'' *SPEPE* (May 1987) 89–94; *Trans.*, AIME, **283.**
136. Harris, P.C. and Reidenbach, V.G.: ''High-Temperature Rheological Study of Foam Fracturing Fluids,'' *JPT* (May 1987) 613–19; *Trans.*, AIME, **283.**
137. Reidenbach, V.G. *et al.*: ''Rheological Study of Foam Fracturing Fluids Using Nitrogen and Carbon Dioxide,'' *SPEPE* (Jan. 1986) 31–41; *Trans.*, AIME, **281.**
138. Watkins, E.K., Wendorff, C.L., and Ainley, B.R.: ''A New Crosslinked Foamed Fracturing Fluid,'' paper SPE 12027 presented at the 1983 SPE Annual Technical Conference and Exhibition, San Francisco, Oct. 5–8.
139. Wendorff, C.L. and Earl, R.B.: ''Foam Fracturing Laboratory,'' paper SPE 12025 presented at the 1983 SPE Annual Technical Conference and Exhibition, San Francisco, Oct. 5–8.
140. Craighead, M.S., Watson, R.W., and Hossaini, M.: ''Foamed Anhydrous Methanol Stimulation Technique,'' paper SPE 12315 presented at the 1983 SPE Eastern Regional Meeting, Pittsburgh, PA, Nov. 9–11.
141. Craighead, M.S., Hossaini, M., and Freeman, E.R.: ''Foam Fracturing Utilizing Delayed Crosslinked Gels,'' paper SPE 14437 presented at the 1985 SPE Annual Technical Conference and Exhibition, Las Vegas, Sept. 22–25.

Fracturing-Fluid Loss

142. Gulbis, J.: ''Dynamic Fluid Loss of Fracturing Fluids,'' paper SPE 12154 presented at the 1983 SPE Annual Technical Conference and Exhibition, San Francisco, Oct. 5–8.
143. Harris, P.C. and Penny, G.S.: ''Influence of Temperature and Shear History on Fracturing Fluid Efficiency,'' paper SPE 14258 presented at the 1985 SPE Annual Technical Conference and Exhibition, Las Vegas, Sept. 22–25.
144. Settari, A.: ''A New General Model of Fluid Loss in Hydraulic Fracturing,'' *SPEJ* (Aug. 1985) 491–501.
145. Penny, G.S.: ''Nondamaging Fluid Loss Additives for Use in Hydraulic Fracturing of Gas Wells,'' paper SPE 10659 presented at the 1982 SPE Formation Damage Control Symposium, Lafayette, LA, March 24–25.
146. *RP 39, Recommended Practice for Standard Procedure for the Evaluation of Hydraulic Fracturing Fluids,* API, Dallas (1983).
147. McDaniel, R.R. *et al.*: ''An Improved Method for Measuring Fluid Loss at Simulated Fracture Conditions,'' *SPEJ* (Aug. 1985) 482–90.
148. Hall, C.D. Jr. and Dollarhide, F.E.: ''Performance of Fracturing Fluid Loss Agents Under Dynamic Conditions,'' *JPT* (July 1968) 763–69; *Trans.*, AIME, **243.**
149. Williams, B.B.: ''Fluid Loss from Hydraulically Induced Fractures,'' *JPT* (July 1970) 882–88; *Trans.*, AIME, **249.**
150. Williams, B.B., Gidley, J.L., and Schechter, R.S.: *Acidizing Fundamentals,* Monograph Series, SPE, Richardson, TX (1979) **6.**
151. Penny, G.S., Conway, M.W., and Lee, W.S.: ''Control and Modeling of Fluid Leakoff During Hydraulic Fracturing,'' *JPT* (June 1985) 1071–81.
152. Roodhart, L.P.: ''Fracturing Fluid: Fluid-Loss Measurements Under Dynamic Conditions,'' *SPEJ* (Oct. 1985) 629–36.

153. Zigrye, J.L., Whitfill, D.L., and Sievert, J.A.: "Fluid-Loss Control Differences of Crosslinked and Linear Fracturing Fluids," *JPT* (Feb. 1985) 315–20.
154. Nolte, K.G.: "Determination of Fracturing Parameters from Fracturing Pressure Decline," paper SPE 8341 presented at the 1979 SPE Annual Technical Conference and Exhibition, Las Vegas, Sept. 23–26.
155. Nierode, D.E.: "Comparison of Hydraulic Fracture Design Methods to Observed Field Results," *JPT* (Oct. 1985) 1831–39.
156. Cooper, G.D., Nelson, S.G., and Schopper, M.D.: "Comparison of Methods for Determining In-Situ Leakoff Rate Based on Analysis With an On-Site Computer," paper SPE 13223 presented at the 1984 SPE Annual Technical Conference and Exhibition, Houston, Sept. 16–19.
157. Smith, J.E.: "Design of Hydraulic Fracture Treatments," paper SPE 1286 presented at the 1964 SPE Annual Meeting, Denver, Oct. 3–6.

Rheology of Fracturing Fluid

158. Rogers, R.E., Veatch, R.W. Jr., and Nolte, K.G.: "Pipe Viscometer Study of Fracturing Fluid Rheology," *SPEJ* (Oct. 1984) 575–81.
159. Baumgartner, S.A. *et al.*: "High-Efficiency Fracturing Fluids for High-Temperature, Low-Permeability Reservoirs," paper SPE 11615 presented at the 1983 SPE/DOE Low-Permeability Gas Reservoirs Symposium, Denver, March 14–16.
160. Cloud, J.E. and Clark, P.E.: "Stimulation Fluid Rheology III. Alternatives to the Power Law Fluid Model for Crosslinked Gels," paper SPE 9332 presented at the 1980 SPE Annual Technical Conference and Exhibition, Dallas, Sept. 21–24.
161. Buechley, T.C. and Lord, D.L.: "Hydraulic Fracturing Fluid Mechanics—State of the Art," *AIChE J.* (1973) **69,** No. 135, 199–200.
162. Conway, M.W. and Harris, E.: "A Laboratory and Field Evaluation of a Technique for Hydraulic Fracturing Stimulation of Deep Wells," paper SPE 10964 presented at the 1982 SPE Annual Technical Conference and Exhibition, New Orleans, Sept. 26–29.
163. Conway, M.W. *et al.*: "Chemical Model for the Rheological Behavior of Crosslinked Fluid Systems," *JPT* (Feb. 1983) 315–20.
164. Craigie, L.J.: "A New Method for Determining the Rheology of Crosslinked Fracturing Fluids Using Shear History Simulation," paper SPE 11635 presented at the 1983 SPE/DOE Low-Permeability Gas Reservoirs Symposium, Denver, March 14–16.
165. Gardner, D.C. and Eikerts, J.V.: "The Effects of Shear and Proppant on the Viscosity of Cross-Linked Fracturing Fluids," paper SPE 11066 presented at the 1982 SPE Annual Technical Conference and Exhibition, New Orleans, Sept. 26–29.
166. Lescarboura, J.A., Sifferman, T.R. and Wahl, H.A.: "Evaluation of Fracturing Fluid Stability by Using a Heated, Pressurized Flow Loop," *SPEJ* (June 1984) 249–55.
167. Guillot, D. and Dunand, A.: "Rheological Characteristics of Fracturing Fluids by Using Laser Anemometry," *SPEJ* (Feb. 1985) 39–45.
168. Prud'homme, R.K.: "Rheological Characterization of Fracturing Fluids," PRAC Project 45 Final Reports 82-45 and 84-45, API, Dallas (April 1984, Aug. 1985).
169. Knoll, S.K.: "Wall Slip Evaluation in Steady Shear Viscosity Measurements of Hydraulic Fracturing Fluids," paper SPE 13904 presented at the 1985 SPE/DOE Low-Permeability Gas Reservoirs Symposium, Denver, May 19–22.
170. Royce, T.N., Rickards, A.R., and Beck, L.M.: "Rheological Characteristics of Adjustable Crosslinked Fracturing Fluids," paper SPE 13178 presented at the 1984 SPE Annual Technical Conference and Exhibition, Houston, Sept. 16–19.
171. Shah, S.N. and Watters, L.T.: "Time and Shear Effects on Rheological Properties of Crosslinked Fluids—An Evaluation Method," *SPEPE* (Jan. 1986) 55–61.
172. Gardner, D.C. and Eikerts, J.V.: "Rheological Characterization of Crosslinked and Delayed Crosslinked Fracturing Fluids by Using a Closed-Loop Pipe Viscometer," paper SPE 12028 presented at the 1983 SPE Annual Technical Conference and Exhibition, San Francisco, Oct. 5–8.
173. Warpinski, N.R.: "Measurement of Width and Pressure in a Propagating Hydraulic Fracture," *SPEJ* (Feb. 1985) 46–54.

Proppant Transport

174. Clark, P.E. and Quadir, J.A.: "Proppant Transport in Hydraulic Fractures: A Critical Review of Particle Settling Velocity Equations," paper SPE 9866 presented at the 1981 SPE/DOE Low-Permeability Gas Reservoirs Symposium, Denver, May 27–29.
175. Novotny, E.J.: "Proppant Transport," paper SPE 6813 presented at the 1977 SPE Annual Technical Conference and Exhibition, Denver, Oct. 9–12.
176. Swanson, V.F.: "The Development of a Formula for Direct Determination of Free Settling Velocity of Any Size Particle," *Trans.*, SME (June 1967) 160–66.
177. Zigrang, D.J. and Sylvester, N.D.: "An Explicit Equation for Particle Settling Velocities in Solid-Liquid Systems," *AIChE J.* (Nov. 1981) **27,** 1043–44.
178. Barnea, E. and Mednick, R.L.: "Correlations for Minimum Fluidization Velocity," *Trans.*, Inst. of Chemical Engineers (1975) **3,** 278–81.
179. Daneshy, A.A.: "Numerical Solution of Sand Transport in Hydraulic Fracturing," *JPT* (Jan. 1978) 132–40.
180. Harrington, L.J., Hannah, R.R., and Williams, D.: "Dynamic Experiments and Proppant Settling in Crosslinked Fracturing Fluids," paper SPE 8342 presented at the 1979 SPE Annual Technical Conference and Exhibition, Las Vegas, Sept. 23–26.
181. Govier, G.W. and Aziz, K.: *The Flow of Complex Mixtures in Pipes,* Van Nostrand Reinhold Co., New York City (1972).
182. Zanker, A.: "Nomographs Determine Settling Velocities for Solid-Liquid Systems," *Chem. Eng.* (May 19, 1980) 147.
183. Clark, P.E. and Guler, N.: "Proppant Transport in Vertical Fractures: Settling Velocity Correlations," paper SPE 11636 presented at the 1983 SPE/DOE Low-Permeability Gas Reservoirs Symposium, Denver, March 14–16.
184. Babcock, R.E., Prokop, C.L., and Kehle, R.O.: "Distribution of Propping Agents in Vertical Fractures," *Drill. & Prod. Prac.*, API (1967).
185. Kern, L.R., Perkins, T.K., and Wyant, R.E.: "The Mechanics of Sand Movement in Fracturing," *Trans.*, AIME (1959) **216,** 403–05.
186. Shah, S.N.: "Proppant Settling Correlations for Non-Newtonian Fluids Under Static and Dynamic Conditions," *SPEJ* (April 1982) 164–70.
187. Biot, M.A. and Medlin, W.L.: "Theory of Sand Transport in Thin Fluids," paper SPE 14468 presented at the 1985 SPE Annual Technical Conference and Exhibition, Las Vegas, Sept. 22–25.
188. Medlin, W.L., Sexton, J.H., and Zumwalt, G.L.: "Sand Transport Experiments in Thin Fluids," paper SPE 14469 presented at the 1985 SPE Annual Technical Conference and Exhibition, Las Vegas, Sept. 22–25.
189. Roodhart, L.P.: "Proppant Settling in Non-Newtonian Fracturing Fluids," paper SPE 13905 presented at the 1985 SPE/DOE Low-Permeability Gas Reservoirs Symposium, Denver, May 19–22.
190. Acharya, A.: "Particle Transport in Viscous and Viscoelastic Fracturing Fluids," *SPEPE* (March 1986) 104–10.
191. Kirkby, L.L. and Rockefeller, H.A.: "Proppant Settling Velocities in Nonflowing Slurries," paper SPE 13906 presented at the 1985 SPE/DOE Low-Permeability Gas Reservoirs Symposium, Denver, May 19–22.
192. Clark, P.E. *et al.*: "Proppant Transport by Xanthan and Xanthan-Hydroxypropyl Guar Solutions: Alternatives to Crosslinked Fluids," paper SPE 13907 presented at the 1985 SPE/DOE Low-Permeability Gas Reservoirs Symposium, Denver, May 19–22.
193. Dunand, A. and Soucémarianadin, A.: "Concentration Effects on the Settling Velocities of Proppant Slurries," paper SPE 14259 presented at the 1985 SPE Annual Technical Conference and Exhibition, Las Vegas, Sept. 22–25.
194. Gottschling, J.C., Royce, T.N., and Shuck, L.Z.: "Nitrogen Gas and Sand: A New Technique for Stimulation of Devonian Shale," *JPT* (May 1985) 901–07.

Fracture Design

195. White, J.L. and Daniel, E.F.: "Key Factors in MHF Design," *JPT* (Aug. 1981) 1501–12.
196. Abou-Sayed, A.S., Ahmed, U., and Jones, A.: "Systematic Approach to Massive Hydraulic Fracturing Treatment Design," paper SPE 9877 presented at the 1981 SPE/DOE Low-Permeability Gas Reservoirs Symposium, Denver, May 27–29.
197. Schlottman, B.W., Miller, W.K. II, and Lueders, R.K.: "Massive Hydraulic Fracture Design for the East Texas Cotton Valley Sands," paper SPE 10133 presented at the 1981 SPE Annual Technical Conference and Exhibition, San Antonio, Oct. 4–7.
198. Ahmed, U. *et al.*: "State-of-the-Art Hydraulic Fracture Stimulation Treatment for a Western Tight Sand Reservoir," paper SPE 11184 presented at the 1982 SPE Annual Technical Conference and Exhibition, New Orleans, Sept. 26–29.
199. Warembourg, P.A. *et al.*: "Fracture Stimulation Design and Evaluation," paper SPE 14379 presented at the 1985 SPE Annual Technical Conference and Exhibition, Las Vegas, Sept. 22–25.
200. Rowley, J.C. *et al.*: "Fracturing Operations in a Dry Geothermal Reservoir," paper SPE 12100 presented at the 1983 SPE Annual Technical Conference and Exhibition, San Francisco, Oct. 5–8.
201. Bailey, D.E. and Wickham, J.F.: "Sand Fracturing vs. Fracture Acidizing," paper SPE 12898 presented at the 1984 SPE Rocky Mountain Regional Meeting, Casper, WY, May 21–23.

202. Smith, M.B., Miller, W.K. II, and Haga, J.: "Tip Screenout Fracturing: A Technique for Soft, Unstable Formations," *SPEPE* (Feb. 1987) 95–103; *Trans.*, AIME, **283.**
203. Smith, M.B., Rosenberg, R.J., and Bowen, J.F.: "Fracture Width: Design vs. Measurement," paper SPE 10965 presented at the 1983 SPE Eastern Regional Meeting, Pittsburgh, PA, Nov. 9–11.
204. Ahmed, U., Newberry, B.M., and Cannon, D.E.: "Hydraulic Fracturing Treatment Design of Wells With Multiple Zones," paper SPE 13857 presented at the 1985 SPE/DOE Low-Permeability Gas Reservoirs Symposium, Denver, May 19–22.
205. Al-Khatib, A.M., King, A.R., and Wilson, M.S.: "Hydraulic Fracturing Design and Evaluation: A Case History, Lost Hills Field, CA," paper SPE 12482 presented at the 1984 SPE California Regional Meeting, Long Beach, April 11–13.
206. Harrington, L.J., Whitsitt, N.F., and Hannah, R.R.: "Prediction of the Location and Movement of Fluid Interfaces in a Fracture," *Proc.*, Southwestern Petroleum Short Course, Lubbock, TX (April 26–27, 1973).
207. Harrington, L.J. and Hannah, R.R.: "Fracturing Design Using Perfect Support Fluids for Selected Fracture Proppant Concentrations in Vertical Fractures," paper SPE 5642 presented at the 1975 SPE Annual Technical Conference and Exhibition, Dallas, Sept. 28–Oct. 1.
208. Nolte, K.G.: "Determination of Proppant and Fluid Schedules From Fracturing-Pressure Decline," *SPEPE* (July 1986) 255–65; *Trans.*, AIME, **281.**
209. Economides, M.J. *et al.*: *Reservoir Stimulation*, Schlumberger Education Services, Houston (1987).
210. Crawford, H.R.: "Proppant Scheduling and Calculation of Fluid Loss During Fracturing," paper SPE 12064 presented at the 1983 SPE Annual Technical Conference and Exhibition, San Francisco, Oct. 5–8.
211. McLeod, H.O. Jr.: "A Simplified Approach to Design of Fracturing Treatments Using High-Viscosity Crosslinked Fluids," paper SPE 11614 presented at the 1983 SPE/DOE Low-Permeability Gas Reservoirs Symposium, Denver, March 13–16.

Field Implementation and Equipment

212. Cooper, G.D., Nelson, S.G., and Schopper, M.D.: "Improving Fracturing Design Through the Use of an On-Site Computer System," paper SPE 12063 presented at the 1983 SPE Annual Technical Conference and Exhibition, San Francisco, Oct. 5–8.
213. Hannah, R.R., Harrington, L.J., and Lance, L.C.: "Real-Time Calculation of Accurate Bottomhole Fracturing Pressure From Surface Measurements With Measured Pressures as a Base," paper SPE 12062 presented at the 1983 SPE Annual Technical Conference and Exhibition, San Francisco, Oct. 5–8.
214. Harrington, L.J.: "Computers and Microprocessors—Valuable Tools in Well Completions," *The Oil Man*, 1985 SPE Offshore Europe Exhibition, Aberdeen, Sept. 10–13.

Fracture Diagnostics

215. Crawley, A.B., Northrop, D.A., and Sattler, A.R.: "The Department of Energy's Western Gas Sands Project Multiwell Experiment Update," paper SPE 11183 presented at the 1982 SPE Annual Technical Conference and Exhibition, New Orleans, Sept. 26–29.
216. Northrop, D.A. *et al.*: "Current Status of the Multiwell Experiment," paper SPE 12868 presented at the 1984 SPE/DOE/GRI Unconventional Gas Recovery Symposium, Pittsburgh, PA, May 13–15.
217. Warpinski, N.R. *et al.*: "Fracturing and Testing Case Study of Paludal, Tight, Lenticular Gas Sands," *SPEFE* (Dec. 1987) 535–45.
218. Sattler, A.R., Raible, C.J., and Gall, B.L.: "Integration of Laboratory and Field Data for Insight on the Multiwell Experiment Paludal Stimulation," paper SPE 13891 presented at the 1985 SPE/DOE Low-Permeability Gas Reservoirs Symposium, Denver, May 19–22.
219. Dobkins, T.A.: "Improved Methods To Determine Hydraulic Fracture Height," *JPT* (April 1981) 719–26.
220. Conway, M.W. *et al.*: "Prediction of Formation Response From Fracture Pressure Behavior," paper SPE 14263 presented at the 1985 SPE Annual Technical Conference and Exhibition, Las Vegas, Sept. 22–25.
221. Medlin, W.L. and Fitch, J.L.: "Abnormal Treating Pressures in Massive Hydraulic Fracturing Treatments," *JPT* (May 1988) 633–42.
222. Nolte, K.G.: "A General Analysis of Fracturing Pressure Decline With Application to Three Models," *SPEFE* (Dec. 1986) 571–83; *Trans.*, AIME, **281.**
223. Martins, J.P. and Harper, T.R.: "Mini-Frac Pressure Decline Analysis for Fractures Evolving From Long Perforated Intervals and Unaffected by Confining Strata," paper SPE 13869 presented at the 1985 SPE/DOE Low-Permeability Gas Reservoirs Symposium, Denver, May 19–22.
224. Lee, W.S.: "Pressure Decline Analyses With the Christianovich and Zheltov and Penny-Shaped Geometry Model of Fracturing," paper SPE 13872 presented at the 1985 SPE/DOE Low-Permeability Gas Reservoirs Symposium, Denver, May 19–22.

Fracture Azimuth and Geometry

225. Smith, M.B. *et al.*: "The Azimuth of Deep, Penetrating Fractures in the Wattenberg Field," *JPT* (Feb. 1978) 185–93.
226. Smith, M.B., Logan, J.M., and Wood, M.D.: "Fracture Azimuth—A Shallow Experiment," *Trans.*, ASME (June 1980) **102,** 99–105.
227. Wood, M.D., Pollard, D.D., and Raleigh, C.B.: "Determination of In-Situ Geometry of Hydraulically Generated Fractures Using Tiltmeters," paper SPE 6091 presented at the 1976 SPE Annual Technical Conference and Exhibition, New Orleans, Oct. 3–6.
228. Schuster, C.L.: "Detection Within the Wellbore of Seismic Signals Created by Hydraulic Fracturing," paper SPE 7448 presented at the 1978 SPE Annual Technical Conference and Exhibition, Houston, Oct. 1–4.
229. Hanson, J.M. and Owen, L.B.: "Fracture Orientation Analysis by the Solid Earth Tidal Strain Method," paper SPE 11070 presented at the 1982 SPE Annual Technical Conference and Exhibition, New Orleans, Sept. 26–29.
230. Clark, J.A.: "The Prediction of Hydraulic Fracture Azimuth Through Geological, Core, and Analytical Studies," paper SPE 11611 presented at the 1983 SPE/DOE Low-Permeability Gas Reservoirs Symposium, Denver, March 14–16.
231. Teufel, L.W. *et al.*: "Determination of Hydraulic Fracture Azimuth by Geophysical, Geological, and Oriented-Core Methods at the Multiwell Experiment Site, Rifle, Colorado," paper SPE 13226 presented at the 1984 SPE Annual Technical Conference and Exhibition, Houston, Sept. 16–19.
232. Hart, C.M. *et al.*: "Fracture Diagnostics Results for the First Multiwell Experiment's Paludal Zone Stimulation," *SPEFE* (Sept. 1987) 320–26; *Trans.*, AIME, **283.**
233. Fitz-Patrick, R.P., Karr, G., and O'Shea, P.: "A Comprehensive Fracture Diagnostics Experiment: Part 1—An Overview," *SPEPE* (Nov. 1986) 411–22; *Trans.*, AIME, **281.**
234. Smith, M.B. *et al.*: "A Comprehensive Fracture Diagnostics Experiment: Part 2—Comparison of Fracture Azimuth Measuring Procedures," *SPEPE* (Nov. 1986) 423–31; *Trans.*, AIME, **281.**
235. Lacy, L.L.: "Comparison of Hydraulic-Fracture Orientation Techniques," *SPEFE* (March 1987) 66–76; *Trans.*, AIME, **283.**
236. Griffin, K.W.: "Induced Fracture Orientation Determination in the Kuparuk Reservoir," paper SPE 14261 presented at the 1985 SPE Annual Technical Conference and Exhibition, Las Vegas, Sept. 22–25.
237. Davis, P.M.: "Surface Deformation Associated with Dipping Hydrofracture," *J. Geophy. Res.* (1983) **88,** 5826.
238. Evans, K.: "On the Development of Shallow Hydraulic Fractures as Viewed Through the Surface Deformation Field: Part 1—Principles," *JPT* (Feb. 1983) 406–10.
239. Evans, K. and Holzhausen, G.: "On the Development of Shallow Hydraulic Fractures as Viewed Through the Surface Deformation Field: Part 2—Case Histories," *JPT* (Feb. 1983) 411–20.
240. Holzhausen, G. *et al.*: "Hydraulic-Fracture Growth During Steam Stimulation in a Single-Well Test," paper SPE 13619 presented at the 1985 SPE California Regional Meeting, Bakersfield, March 27–29.
241. Dobecki, T.L.: "Hydraulic Fracture Orientation by Use of Passive Borehole Seismics," paper SPE 12110 presented at the 1983 SPE Annual Technical Conference and Exhibition, San Francisco, Oct. 5–8.
242. Batchelor, A.S., Baria, R., and Hearn, K.: "Monitoring Effects of Hydraulic Stimulation by Microseismic Event Location: A Case Study," paper SPE 12109 presented at the 1983 SPE Annual Technical Conference and Exhibition, San Francisco, Oct. 5–8.
243. Kim, C.M., Champion, J.H., and Cooper, G.D.: "Evaluation of Fracturing Results in Deviated Wellbores Through On-Site Pressure Measurement and Post-Fracture Temperature Survey," paper SPE 14373 presented at the 1985 SPE Annual Technical Conference and Exhibition, Las Vegas, Sept. 22–25.

Fracturing Economics

244. Elkins, L.E.: "Western Tight Sands Major Research Requirements," *Proc.*, Gas Research Inst./American Gas Assn./U.S. DOE Intl. Gas Research Conference, Chicago (June 9–12, 1980).
245. Meng, H.Z. and Brown, K.E.: "Coupling of Production Forecasting, Fracture Geometry Requirements, and Treatment Scheduling in the Optimum Hydraulic Fracture Design," paper SPE 16435 presented at the 1987 SPE/DOE Low-Permeability Reservoirs Symposium, Denver, May 18–19.

246. Rosenberg, J.I. *et al.*: "A Sensitivity Analysis of the Natl. Petroleum Council Study of Tight Gas," paper SPE 11645 presented at the 1983 SPE/DOE Low-Permeability Gas Reservoirs Symposium, Denver, March 14–16.
247. Brashear, J.P., Rosenberg, J.I., and Mercer, J.: "Tight Gas Resource and Technology Appraisal: Sensitivity Analyses of the National Petroleum Council Estimates," paper SPE 12862 presented at the 1984 SPE/DOE/GRI Unconventional Gas Recovery Symposium, Pittsburgh, PA, May 13–15.
(See also Refs. 2, 5, 110, 118, 126, 127, and 129.)

SI Metric Conversion Factors

acres	× 4.046 873	E+03	=	m^2
bar	× 1.0*	E+02	=	kPa
bbl	× 1.589 873	E−01	=	m^3
cp	× 1.0*	E+00	=	mPa·s
degrees	× 1.745 329	E−02	=	rad
dynes/cm^2	× 1.0*	E−01	=	Pa
ft	× 3.048*	E−01	=	m
ft^2	× 9.290 304*	E−02	=	m^2
°F	(°F−32)/1.8		=	°C
gal	× 3.785 412	E−03	=	m^3
in.	× 2.54*	E+00	=	cm
lbf	× 4.448 222	E+00	=	N
lbm	× 4.535 924	E−01	=	kg
lbm/ft^2	× 4.882 428	E+00	=	kg/m^2
lbm/gal	× 1.198 264	E+02	=	kg/m^3
psi	× 6.894 757	E+00	=	kPa

*Conversion factor is exact.

Chapter 2
Pretreatment Formation Evaluation

Stephen A. Holditch, SPE, Texas A&M U.

2.1 Overview

This chapter describes how geologic, petrophysical, and well testing expertise should be combined to provide a complete reservoir description. We begin by discussing several aspects of geology that must be considered by the engineer designing a fracture treatment. This is followed by discussions concerning well logging and core analysis. Finally, the subject of well testing is introduced and discussed. The well testing section covers the design and analysis of well tests so that both reservoir flow and rock mechanical properties can be calculated. This chapter demonstrates that the key to obtaining an accurate formation evaluation is to correlate all phases of the evaluation into a logical, consistent explanation of reservoir behavior.

2.2 Geologic Considerations

Drainage Area. To optimize the size of a hydraulic fracture treatment, one must optimize the ratio of fracture length, L_f, to drainage radius, r_e.[1] In blanket reservoirs, the entire L_f/r_e ratio can be optimized. By projecting flow rate vs. time as a function of fracture length and drainage radius, it is possible to determine both the optimum fracture length and the optimum drainage radius. In lenticular reservoirs, however, drainage radius is a fixed parameter and is usually not a function of fracture treatment size. Therefore, in lenticular reservoirs, it is necessary to rely on geologic expertise to determine the most probable value for drainage radius in a particular situation. After determining a probable value for drainage radius, the engineer can optimize the propped fracture half-length by optimizing the L_f/r_e ratio.

The engineer should begin the fracture design process by discussing the expected reservoir characteristics with a geologist. The discussions should center around the depositional characteristics and the most probable shape of the reservoir.

A number of papers in the literature detail geologic studies in low-permeability areas.[2-5] Fig. 2.1 presents a diagrammatic cross section showing a general distribution of water and gas in conventional, tight lenticular, and tight blanket sandstone reservoir intervals. The strata labeled "L" in Fig. 2.1 are intervals with lenticular sandstones, while strata labeled "B" represent a blanket-like reservoir. Notice in Fig. 2.1 that the blanket sand interval in the shallower portion of the basin has conventional traps with gas on top of water. Deeper into the basin, many times the low-permeability reservoirs do not contain gas/water contacts (GWC's).

The lenticular intervals show sand lenses embedded in massive shales. In the shallower, more conventional portion of the strata, most of the lenses are water-bearing or contain GWC's. As one progresses downdip into the lower-permeability, deeper portion of the basin, most of the sand lenses are gas-bearing. A few lenses in the predominantly gas-bearing region, however, may contain mostly water and would not be productive.

Understanding the complexity of the geologic deposition patterns is important before a fracture treatment is designed. Not only is it important to understand whether a formation is blanket or lenticular, gas-bearing, or water-bearing, but it is also important to determine the probable size of the reservoir before the stimulation treatment is designed.

In summary, if an engineer is designing a fracture treatment in a blanket reservoir, the engineer must determine the optimum values of both the fracture half-length and the drainage radius. In lenticular reservoirs, however, it is often necessary first to determine the probable size and shape of the reservoir and then to determine the optimum fracture length on the basis of the most probable reservoir size.

Lithology. Another geologic characteristic important to the engineer designing a hydraulic fracture treatment is the lithologic characteristics of the reservoir to be treated. Of primary importance is knowledge of whether the formation is mainly a sandstone or carbonate reservoir. If the formation is a sandstone reservoir, a water-based or oil-based fracture fluid will probably be selected for the hydraulic fracture treatment. In shallow carbonate reservoirs, however, the use of acid-based fluids is sometimes feasible. In addition to the selection of fracturing fluids, the basic lithology of a reservoir is an important factor when openhole geophysical logs are analyzed. A knowledge of the type of minerals that compose the formation of interest is basic to the understanding and interpretation of well logs.

Other considerations concerning the lithologic characteristics of a reservoir also need to be determined by the design engineer. For example, the cementing material in a reservoir can be extremely important. In some shallow, low-permeability sandstones, the sand grains are cemented with different types of clay material. As long as the formation is producing only oil or gas, the compressive strength of the formation can be satisfactory. If the reservoir begins to produce formation water or if water is used as a stimulation fluid, however, the reservoir may lose strength and could collapse into the wellbore. Similar problems have occurred in deep, soft sandstones, such as the Wilcox, when the formation contains mostly carbonate cement. In such reservoirs, formation collapse can occur if the sandstones are stimulated with acid and large pressure gradients are applied during the cleanup period following the stimulation treatment. In situations where carbonate cement is holding together a fairly soft rock, acid should not be used to break down the perforations or to stimulate the reservoir.

Clay Content. Most geologic studies of low-permeability reservoirs confirm that the major cause of low permeability is the filling of the pores of a typical formation with precipitants over geologic time. It is important to determine the types of materials that are in the pore space and exactly how that material is distributed.

It is well known that many low-permeability reservoirs contain large amounts of clay material in the pore space.[6] The clay material can be either detrital or authigenic. Detrital clays are introduced into a sandstone by physical processes at the time of deposition or by biogenic processes shortly after deposition. Clays that develop by direct precipitation from solution or by regeneration of detrital clays are referred to as authigenic clays. Most clays in ancient sandstones are authigenic because detrital clays in most sandstones are altered after burial to form regenerated authigenic clays.

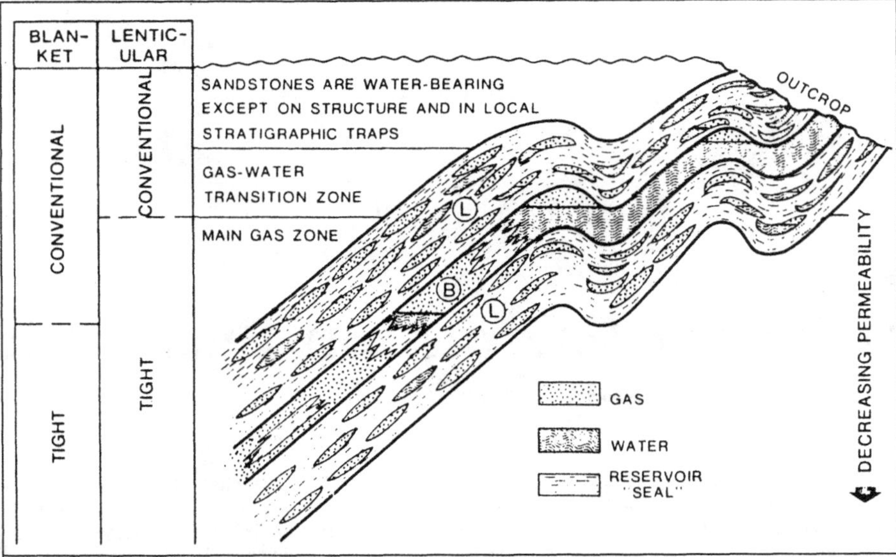

Fig. 2.1—Diagrammatic cross section showing general distribution of water and gas in conventional and tight lenticular (L) and blanket (B) sandstone reservoir intervals.[2]

To predict lateral and vertical distribution of clays, knowledge of the origin of the clay and the factors that controlled its occurrence is necessary. Therefore, geologic studies that include core descriptions, use of scanning electron microscopes (SEM's), and X-ray diffraction analysis can be quite important to the understanding of a particular formation.

The most common types of clay are kaolinite, chlorite, illite, and mixed-layer clays. Fig. 2.2 schematically illustrates how these pore-filling clays affect the permeability in a typical sandstone reservoir.[6] Notice that the type of clay will affect the permeability of the reservoir. It is also well known that the location of the clay material is quite important. If the clay is a pore-filling clay, it will reduce permeability to a higher degree than will pore-lining clay. Therefore, when one is investigating the type and amount of clay material in a formation, it is also important to observe under the SEM where the clay material is located. Clay material in or around pore throats can be much more detrimental to the permeability than clay that is simply lining the pores.

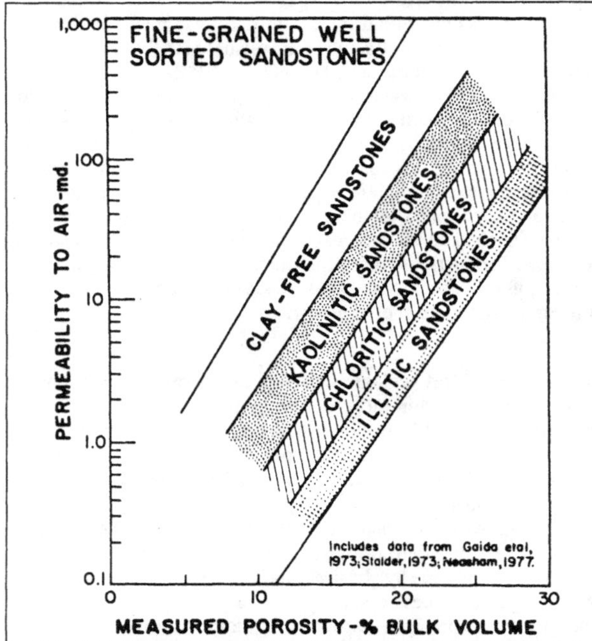

Fig. 2.2—Porosity/permeability relationships of clay-free and clay-bearing sandstones.[6]

One of the better techniques for determining clay content from openhole geophysical logs is natural gamma ray reading. Correlations exist whereby one can determine the gamma ray reading for a clean shale, a clean sandstone, or limestone and then use the actual gamma ray to interpolate and to calculate the amount of clay contained in the sandstone or limestone formation. One of the minerals found in abundance in a typical shale is potassium. Normally, large amounts of potassium are not found in sandstones; however, it is not uncommon for potassium feldspar to be mixed with sand grains in a tight reservoir. When this occurs, the gamma ray reading in a productive sandstone could actually be higher than the gamma ray reading in a clean shale.

In summary, when performing a lithologic description of the formation material, the geologist should describe the main formation components as well as the types of clays and other minerals that fill the pores of the formation. The types of minerals and their location in the rock matrix can be extremely important to the interpretation of well logs and reservoir behavior. Understanding the lithologic makeup of a reservoir is also important when one is determining the base fracturing fluid and the additives necessary to minimize formation damage from a stimulation treatment.

Fault Patterns. A geologic study in a low-permeability area would not be complete without an investigation into the regional and local stress patterns in an area. Knowledge of in-situ stresses is important in the design of fracturing treatments. One way to begin a study of stresses in an area is to examine the regional and localized fault systems. Hubbert and Willis[7] explained that localized and regional stress patterns in an area are controlling factors in determining the orientation of hydraulic fractures and that the state of stress underground is not hydrostatic but depends on tectonic conditions. In tectonically relaxed areas characterized by normal faulting, the least stress will be approximately horizontal, whereas in areas of tectonic compression characterized by folding and thrust faulting, the least stress will usually be vertical. They further concluded that hydraulically induced fractures will be formed approximately perpendicular to the least principal stress. Therefore, in tectonically relaxed areas, hydraulic fractures should be vertical, while in tectonically compressed areas, they may be horizontal.

It is clear from Hubbert and Willis, as well as from many others, that one can study the fault system in a particular area and learn a great deal about the state of stress underground. When such a study is performed, it is important to determine both the strike and the nature of the fault system. If the faults are normal and one assumes that the state of stress underground that caused these faults is still dominating the area, then one would expect a hydraulic fracture created near the fault to parallel the fault plane. If the faults near the wellbore are thrust faults, however, then a hydraulic fracture created near the fault plane would either be horizontal or, if vertical, run perpendicular to the fault plane.

Without question, reservoir inhomogeneity is one of the largest problems facing all aspects of the exploration and production industry. Such geologic discontinuities as faults, joints, and bedding planes, in conjunction with changes in material properties, permeability, and porosity in a typical reservoir, can significantly affect the hydraulic fracture treatment design and results. Mineback experiments conducted at the U.S. DOE's Nevada Test Site have revealed some of these effects.[8] A complete geologic description would include outcrop studies (if possible) to determine the probability and distribution of possible reservoir discontinuities expected in a particular area.

Summary. An accurate geologic description is quite important to an engineer trying to design and to optimize a stimulation treatment. In blanket reservoirs, the optimization process is normally straightforward. Using an appropriate reservoir model, one can optimize both the fracture half-length and the drainage area for any given set of economic criteria.

In lenticular reservoirs or blanket sand reservoirs containing numerous faults, the optimization process is not so straightforward. If the reservoir is lenticular, one must attempt to describe the size and shape of the reservoir. If such a description is possible, one can then design the optimum size fracture treatment. In highly faulted areas, one must determine the strike of the fault system and the nature of the faults in the area to estimate the strike of any hydraulic fracture that will be created.

Knowledge of the fault system and the shape of the various lenses can also affect both the analysis of prefracture pressure transient tests and the projections of postfracture well performance. For example, the predominant flow pattern in many lenticular reservoirs is linear rather than radial. Therefore, a well drilled in a channel sand must be simulated by a model that can describe a long, rectangular reservoir. Also, if a well is drilled near a sealing fault, it will behave like "half" a well. If the proper geologic conditions are not used in the reservoir flow calculations, one can easily overpredict the amount of oil and gas that will be produced after a stimulation treatment.

2.3 Logging Considerations

A very important part of obtaining an accurate formation evaluation is an accurate analysis of the geophysical logs run in a particular well. A conventional log analysis normally provides values for porosity, water saturation, and net hydrocarbon pay. These data, coupled with the PVT properties of the formation fluid, can be used to calculate the amount of oil and gas in place per acre. Then, if the areal size of the reservoir can be estimated, one can determine the amount of oil and gas in place in the entire reservoir. Small errors in porosity or water saturation can cause large errors in the estimates of the deliverability from a particular reservoir; therefore, accurate log analysis is extremely important.

Quality control of logging operations is necessary to ensure that the data being analyzed are accurate. Special attention should be paid to the wellbore size and any sudden irregularities in the borehole walls. Washouts and enlarged sections of the borehole can cause error in the measurements of porosity and resistivity that can lead to poor decisions concerning the completion method.

During a log analysis, the analyst should pay special attention to the mudcake, which can usually be detected by examining the caliper log. If a formation contains enough permeability to accept mud filtrate and to build a filter cake, that zone will be permeable enough to produce oil and gas, usually at commercial flow rates. In many areas, mud cake can be used to pick net pay.

In reservoirs that contain very low porosity and high values of matrix density, some of the logging tools currently used are operating at the low end of the accuracy range. Most evaluation problems are not caused by logging tool inaccuracies, however, but by the analyst's not using modern log analysis concepts to correct the log readings for shale content, fluid content, and borehole irregularities.

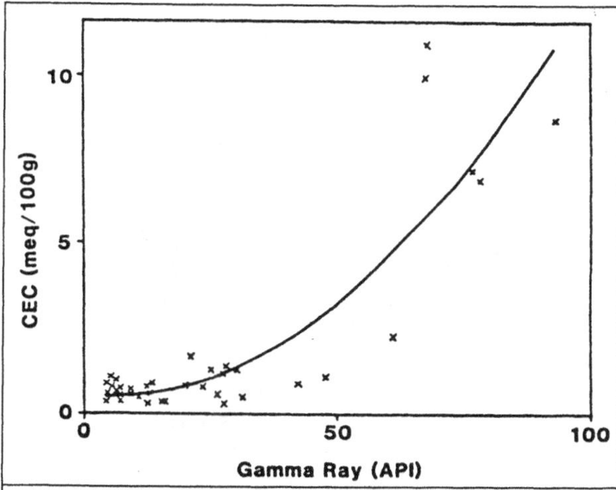

Fig. 2.3—Functional relationship between CEC and gamma ray activity.[13]

Shaly Sand Analysis. To describe most tight-gas reservoirs properly, one must perform a shaly sand analysis. To obtain all the information necessary to analyze the formation, one should run gamma ray, spontaneous potential, dual-induction, neutron, density, and acoustic logs as a minimum suite of openhole logs. Also desirable are either whole or sidewall cores to measure the clay content, pore geometry, cation exchange capacity (CEC), and lithology directly.

Several methods are available to perform a shaly sand analysis, including the modified Archie, Waxman-Smits,[9] Alger *et al.*,[10] and dual-water model[11] methods.

Probably the two best shaly sand analysis techniques are the Waxman-Smits and the dual-water model. When Archie first developed the empirical equations for analyzing logs by use of electrical resistivity measurements, the formation material was assumed to be an insulator to the flow of current in the reservoir.[12] In the Waxman-Smits and dual-water methods, however, the conductivity of the formation materials, specifically the clay material in the pores, is taken into account. The following illustrates the difference between the Archie technique and the Waxman-Smits analysis technique.

Archie's equation,

$$S_w^n=\frac{AR_w}{\phi^m R_t}, \qquad (2.1)$$

is based on the assumption that 100% of the current from a resistivity logging tool is transmitted through the fluids in the pore space.

Waxman-Smits' equation,

$$S_w^n=\frac{AR_w}{\phi^m R_t(1+R_wBQ_V/S_w)}, \qquad (2.2)$$

accounts for the current conducted by the fluids as well as any current that may be transmitted through the clays that fill the pore space. The parameter Q_V is a measure of the quantity of cation-exchangeable clays present in the pore space.

The best method for calculating Q_V is to measure the CEC, Q_{Vt}, from cores. In certain instances, it may be possible to determine values for Q_{Vt} from log data. Rosepiler[13] developed a correlation between Q_{Vt} and gamma ray activity in the Cotton Valley formation. Rosepiler's correlation, presented in Fig. 2.3, should not be used except in the Cotton Valley formation and when no direct measurements of Q_{Vt} are available. Values of Q_{Vt} can be used in Eq. 2.3 to calculate values of Q_V:

$$Q_V=Q_{Vt}(1-\phi)(\rho/\phi). \qquad (2.3)$$

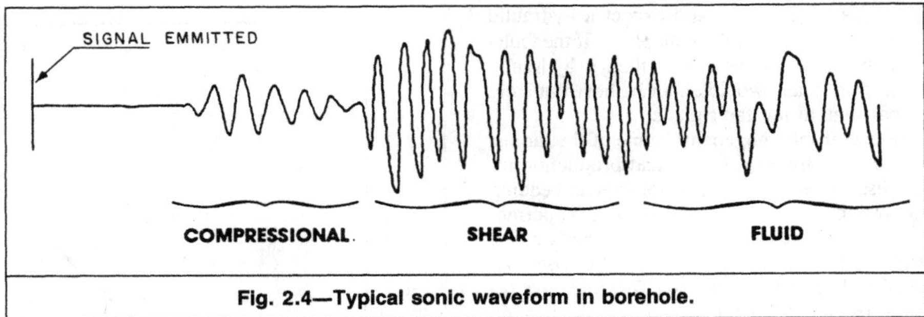

Fig. 2.4—Typical sonic waveform in borehole.

The analysis of geophysical well logs is obviously the key to performing an accurate formation evaluation. Improved well logging tools and better well logging techniques will allow the analyst to evaluate low-permeability formations better. One improvement recently presented to the industry was an improved correlation of sonic transit time vs. porosity.[14] The Wylie equation,

$$v=(1-\phi)v_{ma}+\phi v_f, \quad (2.4)$$

has always been used in the petroleum industry to evaluate porosity from compressional velocities measured with acoustic logs. The Raymer-Hunt-Gardner[14] equation,

$$v=(1-\phi)^m v_{ma}+\phi v_f, \quad (2.5)$$

where v_{ma}=17,850 ft/sec [5440 m/s] or 56 μsec/ft [184 μs/m] for sandstone, 20,500 ft/sec [6250 m/s] or 49 μsec/ft [161 μs/m] for limestone, and 22,750 ft/sec [6935 m/s] or 44 μsec/ft [144 μs/m] for dolomite and v_f=5,300 ft/sec [1615 m/s] or 189 μsec/ft [620 μs/m] for water, was empirically determined with data from a wide range of porosities and lithologies, however, and appears to estimate the true velocity response of the compressional wave in a reservoir better. It was found that the Wiley time-average equation accurately described the porosity from acoustic log data only when the porosity ranged between 25 and 30%. If the porosity was much lower than 25%, the Wiley time-average equation usually underestimated porosity. The new Raymer-Hunt-Gardner equation appears to approximate the true porosity better in the low-porosity range usually associated with low-permeability reservoirs. Therefore, in tight reservoirs, the new velocity correlation should prove to be much more accurate.

In summary, new logging tools and better log analysis techniques are available for performing shaly sand and complex lithology analysis. If one applies these techniques and if well logs can be obtained in boreholes that are not excessively washed out, accurate log evaluations can normally be obtained. The ultimate objective of any formation evaluation project is to use cores, well tests, and geologic descriptions to correlate the openhole logs better. If accurate empirical correlations can be derived, then well logs will provide the maximum amount of information for the minimum cost.

Mechanical Properties. A typical reservoir consists of multiple layers of different rock material. In a typical sand dispersal system, the various layers of formation material will usually consist of sandstones, silty sandstones, siltstones, mudstones, and shales. Depending on the diagenesis of the rock material, many of these sandstones contain carbonate cements or some carbonate layers in certain depositional environments. If the formation were deposited in a deltaic environment, it is also possible to have interbedded coal seams in the formation of interest.

In conventional reservoirs, formation evaluation normally deals with only the pay zones. Because little or no hydraulic fracturing is needed to stimulate high-permeability reservoirs, the formation evaluation problem is concerned mainly with the flow characteristics of the producing interval. In tight-gas reservoirs, however, a complete formation evaluation must include all layers of formation surrounding the potentially productive intervals. The mechanical properties of the bounding layers of formation can be as important to the design engineer as the flow properties of the producing zone. The mechanical properties are used to predict the shape and to calculate the dimensions of the hydraulic fracture that will be created. Accurate predictions of the shape and extent of a hydraulic fracture in a low-permeability reservoir often determine the success or failure of the venture.

Therefore, we need to know not only the porosity, water saturation, and net pay for the productive interval, but also such mechanical properties as Young's modulus, shear modulus, Poisson's ratio, and bulk compressibility for all layers of rock material near the productive interval. We can calculate these mechanical properties if accurate values for bulk density and acoustic travel times are recorded. The end use of these data will be to determine a profile of the in-situ stresses and the moduli of the reservoir to be fracture treated.

The following equations can be used to calculate the mechanical properties of a formation. Detailed discussions concerning these rock properties are deferred to Chap. 3. The purpose of the following information is to illustrate how the rock properties can be calculated from log data.

Poisson's ratio:

$$\nu=\frac{0.5R_v^2-1}{R_v^2-1} \quad (2.6)$$

and

$$R_v=\frac{v_c}{v_s}=\frac{\Delta t_s}{\Delta t_c}. \quad (2.7)$$

Shear modulus:

$$G=1.34\times10^{10}\rho_b/\Delta t_s^2. \quad (2.8)$$

Young's modulus:

$$E=2G(1+\nu). \quad (2.9)$$

Bulk modulus:

$$K=1.34\times10^{10}\rho_b\left(\frac{1}{\Delta t_c^2}-\frac{4}{3\Delta t_s^2}\right). \quad (2.10)$$

Compressibility:

$$c_b=1/K. \quad (2.11)$$

The values for shear-wave velocity, v_s, and compressional-wave velocity, v_c, can best be determined by recording a full waveform sonic signal from a downhole acoustic transmitter. Fig. 2.4 illustrates the type of signal that can be recorded from a long-spaced sonic tool. Notice that after the signal is emitted, the sound wave that arrives first is a compressional wave followed by the shear wave and, finally, by the fluid wave. By directly measuring the compressional-wave and shear-wave velocities with full waveform analysis and then combining that information with an accurate measurement of bulk density, one can use Eqs. 2.6 through 2.11 to determine the necessary mechanical properties of the formation. The key to accurate mechanical properties determination is accurate measurement of the shear travel time in the formation.

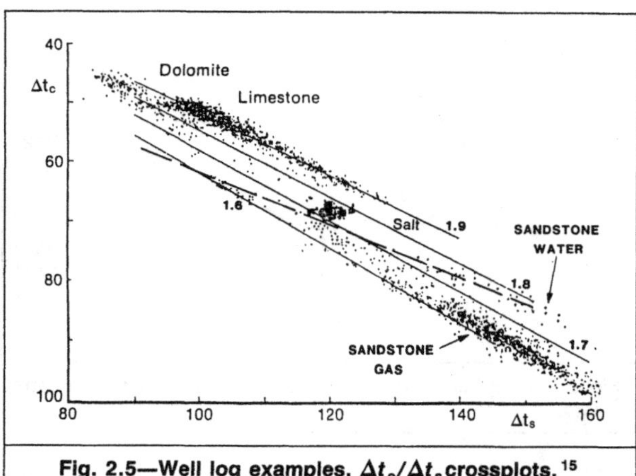

Fig. 2.5—Well log examples, $\Delta t_c/\Delta t_s$ crossplots.[15]

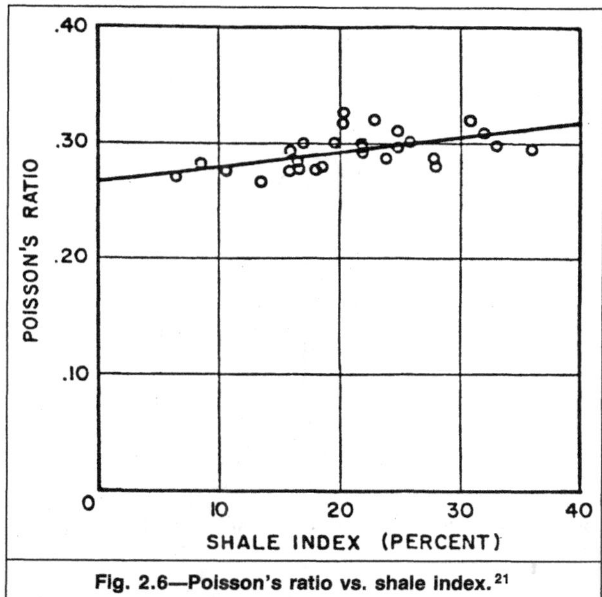

Fig. 2.6—Poisson's ratio vs. shale index.[21]

A number of papers in the petroleum industry discuss the calculation of mechanical properties from well log data.[15-21] A study of the proposed calculation techniques shows that the only truly accurate method to determine the formation mechanical properties is to measure the shear wave velocity directly. A large number of wells, however, have been drilled and logged with both sonic and density logs where full waveform sonic analysis cannot be performed. In these instances, estimation of the rock mechanical properties is possible by use of the measured compressional-wave velocity, bulk density, and estimated lithologic profile of the reservoir.

Pickett[15] was the first to suggest that the ratio of shear-wave travel time to compressional travel time was a function of lithology. Fig. 2.5 illustrates the relationship between the compressional travel time and shear-wave travel time for a number of different lithologies and fluid saturations. The velocity ratios from this graph are summarized in Table 2.1.

It is obvious from this relationship that if one can determine the amount of dolomite, limestone, sandstone, and shale in a reservoir and the most probable fluid content, an estimation of the shear-wave travel time is possible from the compressional-wave travel time. Once a velocity ratio is estimated, then values for Poisson's ratio and modulus can be computed.

Fig. 2.6 presents a correlation that illustrates how Poisson's ratio varies with shale content in a shaly sand. The shale index is defined in Eq. 2.12. The shale index was first defined by Alger *et al.*[10] and the correlation presented in Fig 2.6 was developed by Anderson *et al.*[21]

$$I_{sh}=\frac{\phi_S-\phi_D}{\phi_S}. \quad \text{(2.12)}$$

It must be re-emphasized that calculating mechanical properties on the basis of estimated values of lithology is acceptable only when the direct measurement of the shear-wave velocity has not been performed. The logging industry, however, is rapidly developing new tools and techniques that should allow the measurement of both compressional- and shear-wave velocities through casing. Therefore, in areas where profiling of mechanical properties has not been previously performed, it may be possible to enter existing boreholes and to run cased-hole logs to determine these values.[22,23]

Stress Profile. One of the most important uses of the mechanical properties data is to determine a stress profile in a formation containing multiple layers. The knowledge of in-situ stresses and an accurate stress profile are necessities if one wishes to design a fracture treatment that has the maximum opportunity of being contained in the productive interval. Even when containment is not probable, an accurate estimate of the in-situ stresses can be used to design and to pump the fracture treatment to obtain maximum effectiveness. Hubbert and Willis[7] first presented the formulation to calculate horizontal stresses in a reservoir. Detailed discussions concerning in-situ stresses are presented in Chap. 3. Here, we present only Hubbert's equation (as modified) to illustrate the use of rock mechanical properties to determine in-situ stresses in the various rock layers:

$$\sigma_x=\left(\frac{\nu}{1-\nu}\right)(\sigma_z-p)+p+\sigma_E. \quad \text{(2.13)}$$

Eq. 2.13 illustrates that the total horizontal stress, σ_x, can be calculated if one knows the value of Poisson's ratio, ν, total overburden stress, σ_z, reservoir pressure, p, and any externally generated stresses, σ_E, that may be acting on the formation. The first two terms of Eq. 2.13, which are made up of Poisson's ratio, overburden stress, and reservoir pressure, are derived from elastic stress/strain theory and can be calculated if one can accurately estimate the three required parameters. The third term, however, which is an externally generated stress, must be determined empirically. This term, σ_E, is used to depict a stress that would be caused by an outside generated force. One would expect such forces in a tectonically compressed area near a large thrust fault. Large horizontal stresses can also be found near mountain ranges or in areas where deep-seated salt or shale intrusions occur. Other factors, such as the burial history of a basin or thermal effects, can create stresses that cannot be calculated but must be measured.

As Hubbert and Willis first explained, in tectonically relaxed areas, the externally generated stresses are usually quite minimal and the elastic components of the rock can normally be used to estimate the fracture gradients. In such areas, most fractures are vertical. In tectonically compressed areas or areas with complex tectonic forces, it is often very difficult to calculate accurate horizontal stress gradients. These stress gradients must be measured by use of injection tests so that the value of any externally generated stress, σ_E, can be empirically determined with field data.

A much more thorough discussion of in-situ stresses, their causes, and uses of the data in fracture treatment design are covered in Chap. 3.

TABLE 2.1—VELOCITY RATIOS FROM FIG. 2.5

Lithology	$\Delta t_s/\Delta t_c$
Sandstone/water	1.78
Sandstone/gas	1.60
Dolomite	1.80
Limestone	1.90

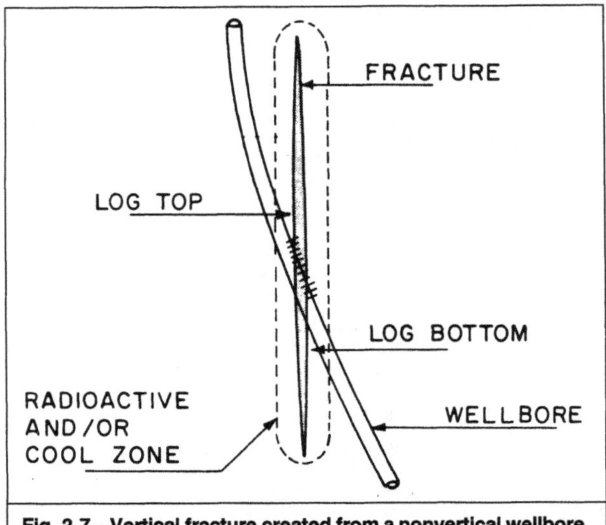

Fig. 2.7—Vertical fracture created from a nonvertical wellbore.

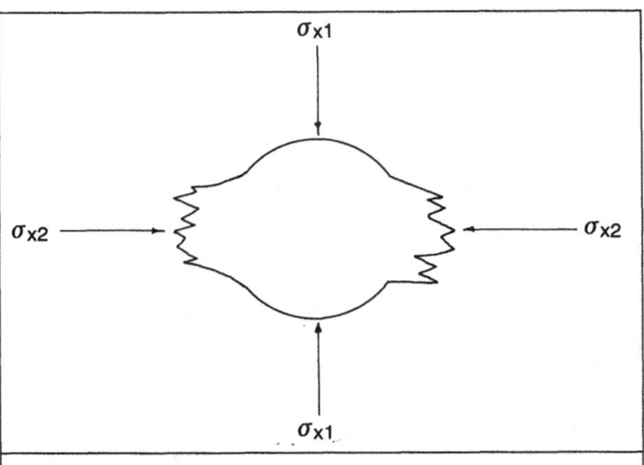

Fig. 2.8—Elliptical borehole caused by horizontal stress contrasts.

Temperature Log Base Profiles. Temperature logs are sometimes used to determine the injection profile before a fracture treatment.[24,25] Temperature logs, in combination with gamma ray logs, can usually be used to determine where fluid enters or exits from the casing. These logs may also provide useful information concerning flow in channels behind the casing. Many engineers also try to use gamma ray/temperature logs to determine the created fracture height after a stimulation treatment; however, the estimation of fracture height from temperature log data may be quite misleading.

Fig. 2.7 is a schematic of a vertical fracture created from a nonvertical wellbore. In reality, this situation should occur virtually 100% of the time. It is highly improbable that the fracture will perfectly parallel the wellbore for its entire height. The dashed zone around the fracture is the portion of the formation that will be affected either by the radioactive materials pumped into the fracture or by cooling caused by fluid pumped down the fracture. It can be easily demonstrated that once the fracture is several inches from the wellbore, a production logging tool in the wellbore will not be capable of sensing any increase in gamma ray activity or any decrease in temperature as a result of fluid movement in the fracture. In Fig. 2.7, one would estimate the created fracture height to be from "log top" to "log bottom" when, in reality, the actual fracture height can be much greater than the value estimated from the gamma ray/temperature log.

From the above discussion, it is obvious that prefracture injection surveys, along with gamma ray/temperature surveys, can be most useful when one is attempting to determine which sets of perforations have actually accepted stimulation fluid. A prefracture survey can also identify such problems as split casing and channels behind the pipe that prevent the fracture treatment from entering the zone to be stimulated.[25] In most cases, gamma ray/temperature logs should be used only to determine a minimum value of fracture height near the wellbore. One must always remember that the radius of investigation of all cased-hole logging tools is quite small and that these tools can illustrate only what is occurring very near the wellbore.

Hole Ellipticity. Hole ellipticity has been used in certain situations as an aid in determining the direction of the least compressive horizontal stress. Fig. 2.8 illustrates the concept by depicting a plan view of a borehole. The horizontal stresses are labeled as σ_{x1} and σ_{x2}. If hole ellipticity were caused only by elastic deformation, the deformation would be virtually immeasurable in a typical wellbore. Because the horizontal stress in the formation is larger than the stress exerted by the mud in the borehole, the rock may tend to spall into the wellbore along planes of weakness. Therefore, when one uses a four-arm caliper tool to determine hole ellipticity, it is usually believed that the direction of the major axis of the ellipse can be used to determine the orientation of the least principal horizontal stress. This subject is discussed more thoroughly in Chap. 16.

Fracture Height. Of all the parameters needed to design a fracture treatment, perhaps the most important and the most difficult to measure is created fracture height. As discussed in several places in this monograph, fracture height can be calculated if one can obtain complete descriptions of all layers in the reservoir and use a reliable three-dimensional (3D) fracture design model. Of course, the industry should strive to improve its ability to design fracture treatments routinely by use of 3D methods; for practical purposes, however, one must currently rely on existing two-dimensional (2D) design methods. To design a fracture treatment correctly with a 2D model, one must correctly estimate the created fracture height.

For most situations, one should consider only (1) thick, clean shales, (2) thick, dense formations, and (3) coal seams as potential barriers to fracture growth.

Clean shales will usually have higher values of Poisson's ratio and, thus, higher horizontal stresses than the siltstones and sandstones located near the shale. If the clean shale is thick enough, it will normally be a barrier to fracture propagation.

Dense, low-porosity sandstones and carbonates are usually hard and have large values for Young's modulus. As the value of modulus increases, the value of created fracture width decreases. Therefore, even if the in-situ stresses are comparable, the fracture width in hard formations will be substantially smaller than the fracture width in soft formations. Because the flow of fluids down a fracture is proportional to fracture width cubed, it is obvious that high-modulus materials can be effective barriers to fluid flow in the fracture, which in effect will cause the dense zone to be an effective barrier.

A coal seam can prevent fracture growth either by shear failure at the formation interface or by excessive leakoff of fracture fluids into the coal seam.

The best method of estimating created fracture height from a log is to start at the perforated interval and search up and down the log until a shale or dense streak is found that appears thick enough to be a barrier to fracture growth.

Fig. 2.9 is a log of a Wilcox sand in south Texas. The density log and R_{wa} curve indicate that the net gas pay is located primarily between 9,175 and 9,225 ft [2797 and 2812 m]. However, the formation is silty from 9,110 to 9,175 ft [2777 to 2797 m] and from 9,225 to 9,300 ft [2812 to 2835 m]. A clean Wilcox shale is normally recognized as an interval with low density porosity and a resistivity of less than 2 Ω·m. Therefore, even though this sand was perforated only from 9,175 to 9,225 ft [2797 to 2812 m], the fracture would be expected to grow up and down through the siltstone until it reaches a thick, clean shale. For design purposes, one should use about 200 to 220 ft [61 to 67 m] as the value of created fracture height for the Wilcox interval in Fig. 2.9.

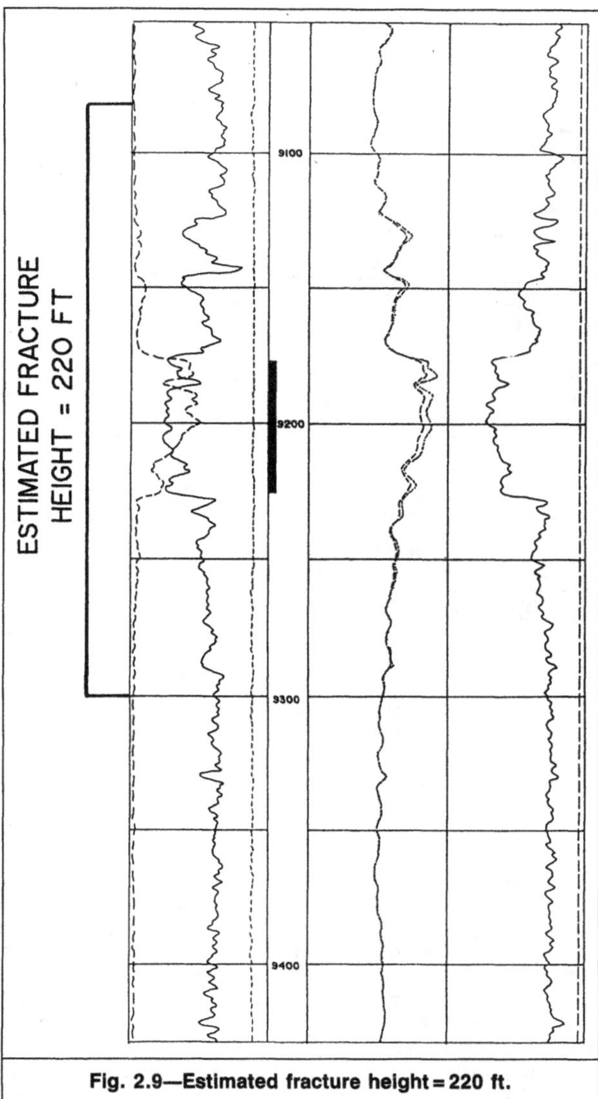

Fig. 2.9—Estimated fracture height = 220 ft.

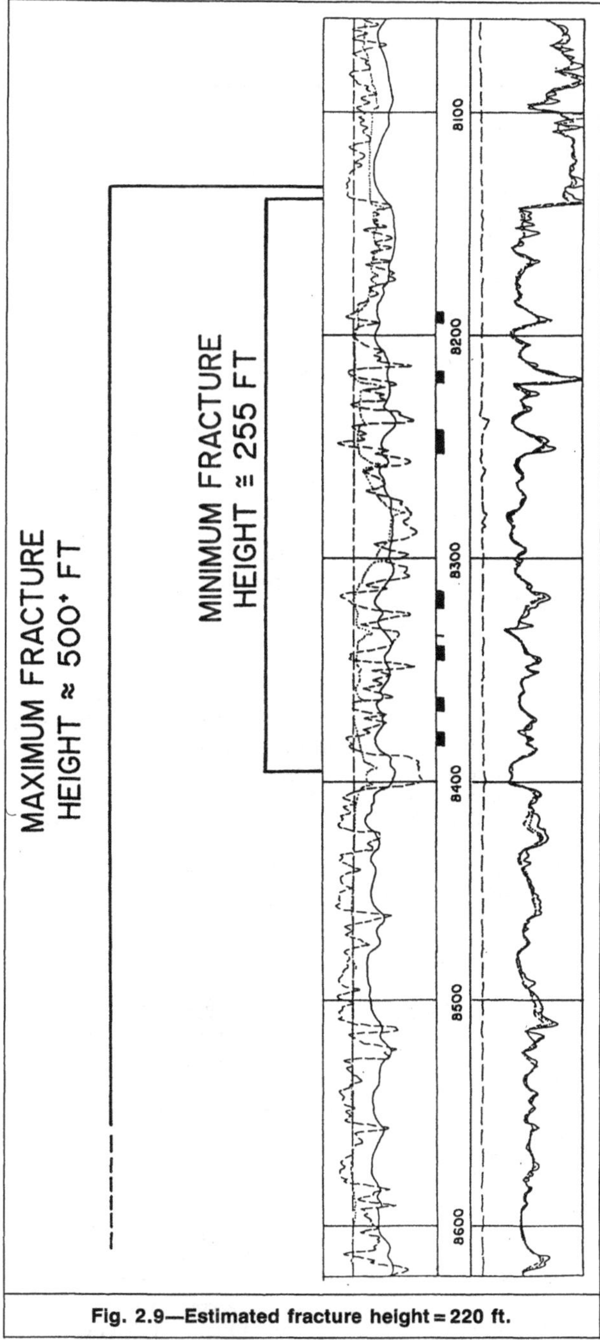

Fig. 2.9—Estimated fracture height = 220 ft.

Fig. 2.10 presents a log for a Travis Peak section in east Texas. The perforations are marked from 8,190 to 8,384 ft [2496 to 2555 m]. The top of the Travis Peak is at 8,142 ft [2482 m]. The interval from 8,060 to 8,142 ft [2457 to 2482 m] is the Sligo carbonate. One can see that the porosity of the Sligo is close to zero. Because of the high modulus of this zone, the Sligo is usually a barrier to upward propagation of a hydraulically created fracture in the Travis Peak formation.

The Travis Peak formation in Fig. 2.10 is composed of numerous layers of sandstone, siltstone, mudstone, and shale. Only one shale, located from 8,388 to 8,402 ft [2557 to 2561 m], is clean enough to be considered as a potential barrier; however, because it is only about 10 ft [3 m] thick, it probably will not prevent downward propagation of the hydraulic fracture.

To design a fracture treatment for a zone like to the one illustrated in Fig. 2.10, one should use a minimum value for fracture height of 255 ft [78 m] and a possible value of >500 ft [>152 m].

It is recognized that the size of the fracture treatment, the viscosity of the fracture fluid, and the injection rate will influence the value of created fracture height. But assuming that an adequate treatment volume will be pumped to create a long fracture and that the injection rate and fluid viscosity will be sufficient to transport proppant adequately several hundred or thousand feet into the fracture, one would expect the technique of selecting created fracture height from logs described above to be an acceptable method.

More sophisticated fracture-height estimation techniques are available to some organizations, usually on a research basis, and we expect that these more sophisticated techniques will become more widely applied as technology develops. To design a fracture treatment now, however, with current technology, one must estimate fracture height from logs.

Perhaps the most common error introduced in the fracture design process is that most engineers underestimate the value of created fracture height. When this occurs, the designed fracture treatment usually will be too small and the cash flow and profit from a well will be substantially decreased from the optimum values.

2.4 Core Analysis

In the design of a hydraulic fracture treatment, it is important that all the various layers in a reservoir be described adequately. These layers usually consist of sandstones, siltstones, limestones, and shales, and the properties of each of these rock types can be important to the hydraulic fracture treatment design. Therefore, when a well in a low-permeability area is cored, one should be sure to

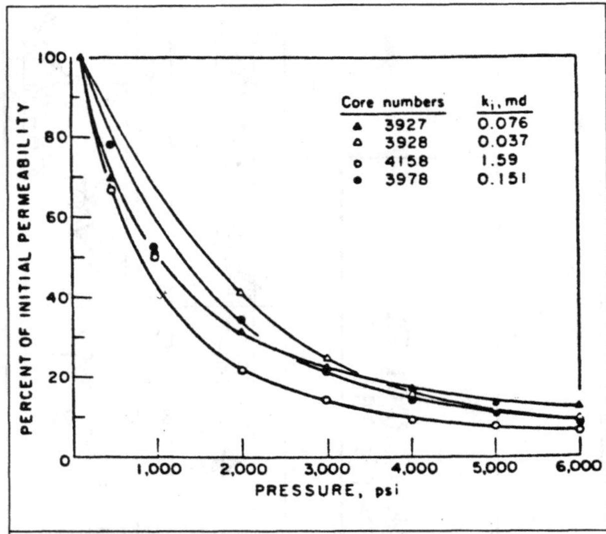

Fig. 2.11—Effect of overburden pressure on gas permeability of Gasbuggy cores.[26]

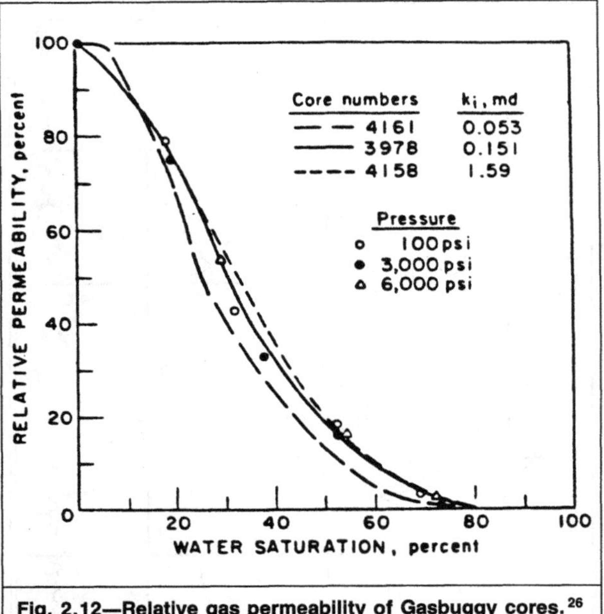

Fig. 2.12—Relative gas permeability of Gasbuggy cores.[26]

core some of the shales, siltstones, and mudstones that can be found both above and below the potentially productive intervals. In many cases, coring these potential barriers to fracture growth is more important than coring the productive interval.

The main purposes of obtaining core material normally are to evaluate the amount of oil and gas in place in the reservoir, to determine effective values of permeability, and to obtain correlations between the log readings and the core material. Whole cores are most useful when one attempts to obtain a complete data set. Sidewall cores can be useful if whole cores cannot be obtained from a particular well. The sidewall cores can be used to measure values of CEC and to determine mineral content, clay content, and clay location in the pores.

Conventional Core Analysis. A conventional core analysis is usually performed to determine values of porosity, permeability, and water saturation at essentially atmospheric conditions. Typically, a core is leached of its liquid hydrocarbons and dried in an oven. Then measurements are made at room temperature under moderate pressure (±100 psi [±690 kPa]). Such core analysis techniques have proved to be quite useful in conventional reservoirs; however, in tight-gas reservoirs, conventional core analysis is not very useful.

Thomas and Ward[26] demonstrated that in low-permeability reservoirs, the permeability can decrease by one to two orders of magnitude when the core is tested under restored-state conditions. Figs. 2.11 and 2.12 illustrate some of the core measurements published by Thomas and Ward. Notice in Fig. 2.11 that the permeability of the Gasbuggy cores decreased by an order of magnitude when the overburden stress was increased from 100 to 5,000 psi [0.69 to 34.5 MPa]. These cores were dry and did not contain any water saturation. When water-saturation effects are included, the permeability can be decreased further. Fig. 2.12 illustrates that for initial water saturations ranging between 40 and 60%, the permeability to gas can be decreased further by a factor of 5 to 10. Therefore, the values of permeability measured from conventional core analysis in low-permeability reservoirs can easily be too optimistic by a factor of 50 to 100.

Special Core Analyses. Since the publication of Thomas and Ward's[26] paper, other papers have confirmed that formation permeability decreases substantially in low-permeability reservoirs when the confining stress increases.[27,28] The industry has recognized that special core analysis procedures are needed to describe low-permeability reservoirs properly and to obtain the data necessary for evaluating the formation and designing fracture treatments. Techniques are being used in the industry to measure virtually any reservoir property under restored-state (in-situ) conditions.

Special core analysis can be used to determine values for permeability; porosity; capillary pressure; relative permeabilities to oil, gas and water; saturation exponent; and cementation factor under simulated reservoir conditions. Other parameters, such as compressional travel time, shear travel time, and formation density, can also be measured in the laboratory. Each of these values can be extremely important to the overall formation evaluation.

Core-measured values for porosity, cementation factor, and saturation exponent, along with measurements for CEC, can be very useful during attempts to determine the proper empirical water-saturation correlation for analyzing openhole geophysical logs. The measured values of permeability, relative permeability, and capillary pressure will be important when one analyzes pressure-transient data or wishes to predict well performance as a function of various forms of stimulation. Shear travel time, compressional travel time, and bulk density can be used to compute mechanical properties and to correlate these properties to values measured from openhole geophysical logs and in-situ stress tests.

In addition to the special core analysis techniques described, the cores should also be analyzed with the SEM and X-ray diffraction analysis equipment. The results from such studies can be used to determine the quantity and type of clays that are present in the pore space of the rock and their exact location. Also, by quantifying the mineral content in each core sample, one can determine the proper correlations for calculating lithology from well logs.

Along with the increased awareness of the need for special core analysis, a great deal of research and progress has been made on laboratory procedures for measuring core properties under restored-state conditions.[29,30] Gas-permeability measurements performed on tight sandstones are influenced by the quantity and morphology of interstitial, authigenic clay. Clay occurring in discrete-particle and pore-lining forms exerts little or no effect on the permeability; however, clay present in the pore-bridging form constricts or blocks pore throats and reduces rock permeability. Drying tight-sand samples under a vacuum for extended periods causes pore-bridging fibrous clay to disintegrate. The clay collapses into fine particles and opens previously blocked pore throats, thereby increasing permeability. When these same core samples are dried in humidity-controlled conditions, the clays remain unaltered and the resulting permeabilities are much lower and more realistically represent the true formation conditions. Numerous recent papers have led to a much improved understanding of the causes of low permeability and the importance of measuring formation properties under restored-state conditions.[27-33]

In addition to measuring formation properties under restored-state conditions, it is also useful at times to determine the possible interaction of fracturing fluids and fracturing proppants with the formation.[34] Such tests are difficult to run under restored-state conditions but can provide useful information concerning possible adverse reactions between the fluids used during a stimulation treatment and materials in the core. In some cases, the formation can be unusually soft or unusually hard, and the effects of proppant embedment or proppant crushing should be studied in the laboratory. Therefore, in addition to the special core analyses performed to evaluate the fluid flow properties and mechanical properties of the formation, special tests should also be run to confirm the choice of fracturing fluid and fracturing proppant in special instances.

Oriented Coring. To determine the direction of certain reservoir characteristics, such as natural fractures or stress patterns, oriented coring techniques are often useful. To obtain an oriented core, special coring equipment is used. The core barrel can be oriented with respect to magnetic north by use of standard borehole surveying tools. A scribe line is etched onto the face of the core, and periodic surveys can be run to confirm that the core orientation is properly maintained.

Knowing core orientation after the core is retrieved at the surface can be quite useful when planning the location of development wells in blanket reservoirs. Fig. 2.13 illustrates that when fracture orientation is known, the wells can be drilled to obtain adequate drainage in a reservoir; however, if the well spacing and locations with respect to the fracture azimuth are not well planned, the resulting drainage pattern could be incomplete and will not sufficiently drain a particular reservoir.[35,36]

The use of oriented cores to determine the stress state under the earth and the expected azimuth of both natural fractures and hydraulically induced fractures has been well documented in the industry.[36-38] In particular, the use of anelastic strain recovery measurements from oriented core to determine the direction of in-situ stresses has proved quite useful to the industry. Fig. 2.14 illustrates one such measurement of a core from Well MWX-2.[38] By measuring the strain in three directions, one can estimate the magnitude of the in-situ stresses in three directions.

2.5 Well Testing Considerations

Once the geology, log, and core data have been thoroughly analyzed and the decision has been made that oil and gas are present in commercial quantities in the reservoir and the well should be completed, a series of prefracture well tests should be designed and analyzed to evaluate the formation further. The main purpose of these well tests is to determine an in-situ estimate for reservoir permeability, a value for skin, the initial reservoir pressure, and other properties, such as the in-situ stresses and the effective fluid-loss coefficient.

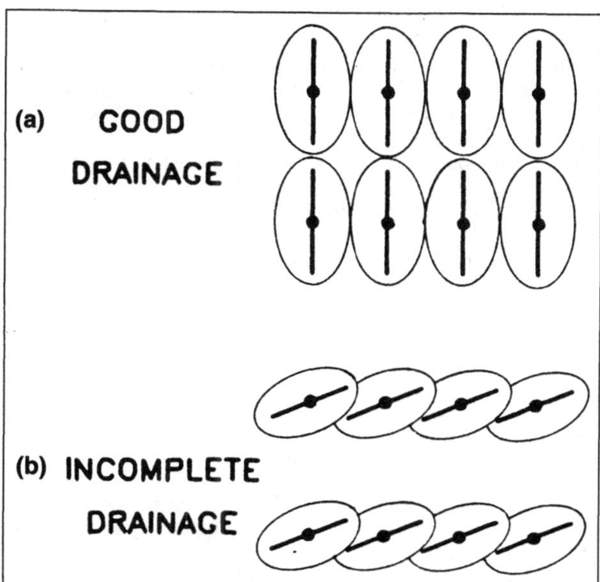

Fig. 2.13—Optimum selection of infill well locations depends on the orientation of propped fractures in low-permeability reservoirs—(a) optimum recovery and (b) inefficient recovery.[35,36]

It is possible that some prefracture well tests can be performed before casing is set. If openhole drillstem tests are run in low-permeability areas, however, one must be extremely cautious in using such data as the basis for any major decision.[39]

In most situations, all prefracture testing will be performed through casing. Therefore, the method of perforating and the perforating efficiency can become quite important. This subject will be discussed more thoroughly in Chap. 12. For this discussion of cased-hole well testing, it will be assumed that the perforations are adequate and do not affect the results of the well test analyses.

Reservoir Performance. One of the main purposes of performing a prefracture well test is to determine the reservoir flow potential from a particular formation. Even though techniques are available to analyze early-time production data,[40] the best method for determining the reservoir flow parameters is to run a pressure-buildup test. If one is attempting to define reservoir inhomogeneities (such as sand pinchouts or nearby faults) with the prefracture well tests, then the flow period before the pressure-buildup period needs to

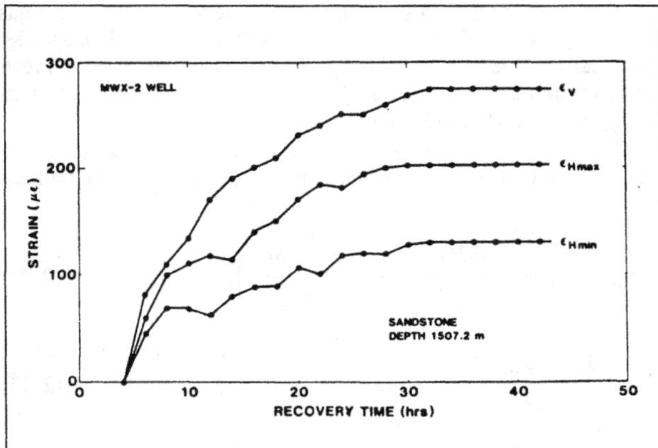

Fig. 2.14—Strain recovery-time curves of vertical and principal horizontal strains.[38]

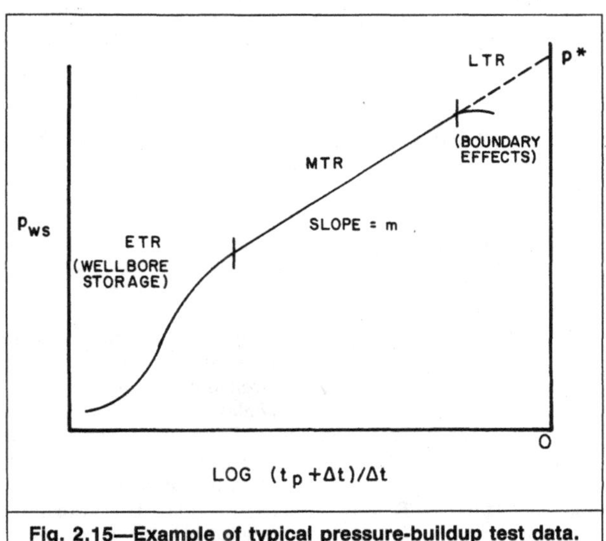

Fig. 2.15—Example of typical pressure-buildup test data.

be long enough to allow the discontinuity to be felt at the wellbore before the well is shut in for the pressure-buildup test.

The radial-flow equation for constant-rate oil production in an infinite-acting reservoir provides the basis for analyzing prefracture well test data. A complete derivation of the radial-flow equation can be found in Ref. 41.

A pressure-buildup test is simply an extension of a pressure drawdown test with a flow rate of zero. When the principle of superposition is applied, the following equations can be derived[41]:

$$p_i - p_{ws} = \frac{162.6 QB\mu}{kh} \log\left(\frac{t_p + \Delta t}{\Delta t}\right) \quad \text{(2.14)}$$

and

$$s = 1.151\left[\frac{p_{1\,\mathrm{hr}} - p_{ws}}{m} \log\left(\frac{k}{\phi \mu c_t r_w^2}\right) + 3.23\right]. \quad \text{(2.15)}$$

Eqs. 2.14 and 2.15 can be used to determine the formation permeability-thickness product and a skin factor from a graph of the pressure-buildup data.

Fig. 2.15 illustrates a typical graph of shut-in wellbore pressure vs. the log of a parameter commonly called the Horner time group. The slope, m, is equal to 162.6 $QB\mu/kh$. There is an early-time region (ETR), dominated by wellbore storage, a late-time region (LTR), influenced by boundaries, and a middle-time region (MTR), where the slope of the pressure-buildup data, m, is inversely proportional to the permeability-thickness product of the reservoir. When the reservoir is infinite acting, the MTR can be extrapolated to a value called p^* that represents the average reservoir pressure at that time. If the formation is not infinite and information concerning the shape of the reservoir is available, it is possible to use existing correlations to convert values of p^* to the correct values for reservoir pressure.[41]

Wellbore Storage. As illustrated in Fig. 2.15, early-time data measured during well tests will be distorted because of wellbore storage effects. When a well first begins to produce or when a well is initially shut in for a bottomhole-pressure-buildup test, the sandface flow rate is not immediately altered by changes at the surface. Even when the rate is set equal to zero at the surface by shutting in a valve on the Christmas tree, the well continues to produce at the perforations. As fluid flows into the wellbore and as the pressure in the wellbore increases, the fluid flow rate at the sandface decreases with time.

Agarwal *et al.*[42] solved the wellbore storage problem and presented the results graphically as dimensionless pressure vs. dimensionless time for fixed values of skin factor, s, and a dimensionless wellbore storage constant, C_D:

$$C_D = \frac{0.894 C}{\phi c_t h r_w}, \quad \text{(2.16)}$$

$$t_D > (60 + 3.5s) C_D, \quad \text{(2.17)}$$

and

$$t > \frac{(200{,}000 + 12{,}000s) C}{(kh/\mu)}. \quad \text{(2.18)}$$

As can be seen in Eq. 2.16, C_D is composed of certain wellbore and reservoir properties. To determine the duration of wellbore storage, Agarwal determined that wellbore storage effects are diminished once the value of dimensionless time exceeds the value defined in Eq. 2.17. Converting Eq. 2.17 from dimensionless units to real units results in Eq. 2.18. This equation can be used to determine the duration of the ETR, which is also equivalent to the beginning of the MTR. It is extremely important when pressure-drawdown and -buildup tests are designed in a low-permeability reservoir that the well test be run long enough that the effects of wellbore storage are diminished and the test truly reaches the MTR during both the flow and buildup periods.

Radius of Investigation. Of practical importance to any prefracture well test is designing a test that will investigate a significant portion of the reservoir. The radius of investigation from a well test, where the well is flowing under radial-flow conditions, can be approximated as

$$r_i = \left(\frac{kt}{948 \phi \mu c_t}\right)^{1/2}. \quad \text{(2.19)}$$

Eq. 2.19 can be used to determine the duration of the well test required to investigate a certain distance into the reservoir. Obviously, one would prefer to investigate a large percentage of the reservoir during a well test; however, in very-low-permeability reservoirs, the value for radius of investigation may be quite small, even for relatively long-duration well tests. Eq. 2.19 can also be used to determine the distance to a reservoir heterogeneity if that heterogeneity can be recognized from transient well test data.

Pseudopressure and Pseudotime. Eqs. 2.14 and 2.15 were derived for flow of a fluid with small and constant compressibility and constant viscosity. For most oil reservoirs, such assumptions are valid; however, when gas wells are analyzed, the liquid fluid-flow equations may not be as accurate as necessary. It has been demonstrated that the liquid fluid-flow equations can be altered with the parameters called real-gas pseudopressure and real-gas pseudotime.[43,44] The parameter real-gas pseudopressure, described in Eq. 2.20, takes into account that the viscosity and gas compressibility factor are functions of the pressure in the reservoir:

$$\psi p = 2\int_0^p \frac{p\,dp}{\mu z}. \quad \text{(2.20)}$$

The parameter group called real-gas pseudotime, described in Eq. 2.21, is used to account for the changes in viscosity and total compressibility in the reservoir system as a function of time:

$$t_a = \int_0^t \frac{dt}{\mu c_t}. \quad \text{(2.21)}$$

To compute the functions of real-gas pseudopressure and real-gas pseudotime, one must first obtain the PVT properties of the fluid. These can normally be obtained from estimates of gas gravity and bottomhole temperature for a sweet-gas reservoir. If the reservoir is an acid-gas reservoir or contains a large amount of liquid hydrocarbon, such as a gas-condensate reservoir, a complete gas analysis may be required to develop correctly the PVT properties necessary to compute the values of real-gas pseudopressure and real-gas pseudotime. The most common way to calculate these functions is by use of numerical integration techniques.[41]

The definitions of real-gas pseudopressure and real-gas pseudotime are used to develop the equations needed to analyze a buildup test:

$$p_{pws} = p_p^* - \frac{1{,}637 Q_{gs} T_r}{kh} \log\left(\frac{t_{pa} + \Delta t_a}{\Delta t_a}\right), \quad \text{(2.22)}$$

$$k = \frac{1{,}637 Q_{gs} T_r}{mh}, \quad \text{(2.23)}$$

and

$$s' = 1.151\left[\frac{(p_{pws} - p_{pwf})}{m} - \log\left(\frac{k \Delta t_a}{\phi r_w^2}\right) + 3.23 + \log\left(\frac{t_{pa} + 1}{t_{pa}}\right)\right]. \quad \text{(2.24)}$$

TABLE 2.2—DATA FOR WELL A

Net gas pay, ft	19
Wellbore radius, ft	0.255
Initial reservoir pressure, psia	6,180
Reservoir temperature, °F	279
Average porosity, %	17.6
Average water saturation, %	60
Separator gas gravity	0.707
Gas-condensate ratio, scf/STB	29,700
Condensate gravity, °API	52.3
Drainage area, acres	160
Flowing wellhead temperature, °F	80
Tubing ID, in.	2.441
Length of tubing, ft	9,700

Pressure (psi)	Pseudopresure (psi³/cp)	B_g (scf/res ft³)	μ_g (cp)	z factor	Compressibility (psi⁻¹)
250	0.4803×10^{7}	12.23	0.1330×10^{-1}	0.9786	0.408×10^{-2}
500	0.1918×10^{8}	24.97	0.1363×10^{-1}	0.9587	0.208×10^{-2}
750	0.4281×10^{8}	38.17	0.1417×10^{-1}	0.9405	0.141×10^{-2}
1,000	0.7536×10^{8}	51.78	0.1463×10^{-1}	0.9244	0.106×10^{-2}
1,250	0.1166×10^{9}	65.70	0.1507×10^{-1}	0.9107	0.854×10^{-3}
1,500	0.1661×10^{9}	79.79	0.1560×10^{-1}	0.8999	0.709×10^{-3}
1,750	0.2232×10^{9}	93.92	0.1615×10^{-1}	0.8919	0.600×10^{-3}
2,000	0.2872×10^{9}	107.92	0.1671×10^{-1}	0.8871	0.515×10^{-3}
2,250	0.3574×10^{9}	121.65	0.1745×10^{-1}	0.8854	0.446×10^{-3}
2,500	0.4326×10^{9}	134.98	0.1819×10^{-1}	0.8866	0.388×10^{-3}
2,750	0.5121×10^{9}	147.82	0.1895×10^{-1}	0.8905	0.340×10^{-3}
3,000	0.5952×10^{9}	160.10	0.1975×10^{-1}	0.8969	0.300×10^{-3}
3,250	0.6811×10^{9}	171.79	0.2055×10^{-1}	0.9056	0.265×10^{-3}
3,500	0.7695×10^{9}	182.97	0.2135×10^{-1}	0.9161	0.236×10^{-3}
3,750	0.8598×10^{9}	193.35	0.2216×10^{-1}	0.9284	0.211×10^{-3}
4,000	0.9517×10^{9}	203.25	0.2296×10^{-1}	0.9420	0.189×10^{-3}
4,250	0.1045×10^{10}	212.60	0.2374×10^{-1}	0.9569	0.171×10^{-3}
4,500	0.1139×10^{10}	221.42	0.2452×10^{-1}	0.9728	0.155×10^{-3}
4,750	0.1233×10^{10}	229.74	0.2531×10^{-1}	0.9897	0.141×10^{-3}
5,000	0.1328×10^{10}	237.60	0.2609×10^{-1}	1.0073	0.129×10^{-3}
5,250	0.1423×10^{10}	245.04	0.2688×10^{-1}	1.0256	0.118×10^{-3}
5,500	0.1519×10^{10}	252.07	0.2758×10^{-1}	1.0444	0.109×10^{-3}
5,750	0.1614×10^{10}	258.74	0.2826×10^{-1}	1.0638	0.100×10^{-3}
6,000	0.1710×10^{10}	265.06	0.2894×10^{-1}	1.0835	0.930×10^{-4}
6,250	0.1806×10^{10}	271.07	0.2961×10^{-1}	1.1037	0.864×10^{-4}

The value of p_{pws} used in the skin-factor equation must be a value in the MTR or its extrapolation at time Δt_a.

Producing Pseudotime. To conform to theory, one should keep the flow rate before a buildup exactly constant. Because it is very difficult to produce a well at constant rate for any extended period, Horner suggested that an effective, producing pseudotime could be used to compute the Horner time group. This producing pseudotime is defined as

$$t_p=\frac{Q}{q}. \quad (2.25)$$

The producing pseudotime, t_p, is equal to cumulative production, Q, divided by the last stabilized flow rate, q, from the well before the well is shut in for the pressure-buildup test.

Effective Pseudotime. Another concept that needs to be discussed is the use of effective pseudotime.[45] The following equation presents the relationship for computing effective pseudotime:

$$\Delta t_{ae}=\frac{\Delta t_a}{1+\dfrac{\Delta t_a}{t_{ap}}}. \quad (2.26)$$

The concepts of pseudopressure, pseudotime, producing pseudotime, and effective pseudotimes allow one to apply the pressure-transient analysis techniques for gas wells properly. The importance of using real-gas pseudotime and effective pseudotimes has been illustrated in the literature.[46,47]

Adjusted Pressures and Adjusted Times. Although the parameter groups defined as real-gas pseudopressures and real-gas pseudotimes can be effectively used to apply the liquid-flow equations in gas-producing situations, the units of real-gas pseudopressure and real-gas pseudotime can be somewhat confusing to a practicing engineer. In an effort to develop a parameter group that correctly adjusts for changes in the gas fluid properties with pressure and still results in parameter values that appear normal to the analyst, the terms "adjusted pressure" and "adjusted time" can be used. These parameter groups are similar to the normalized pressure and time functions previously described.[48]

To calculate adjusted pressure and adjusted time, one can use the following equations:

$$p_a=2\bar{\mu}\bar{z}\int_0^p \frac{p\,dp}{\mu z} \quad (2.27)$$

and

$$t_a=\bar{\mu}\bar{c}_t\int_0^t \frac{dt}{\mu c_t}. \quad (2.28)$$

Notice that adjusted pressure is simply real-gas pseudopressure multiplied by the constant $\bar{\mu}\bar{z}$. Adjusted time is real-gas pseudotime multiplied by the constant $\bar{\mu}\bar{c}_t$. The values of $\bar{\mu}\bar{z}$ and $\bar{\mu}\bar{c}_t$ should be evaluated at $\bar{p}=(p_i+p_{wf})/2$.

With these definitions, the following equations can be used to analyze a pressure-buildup test:

$$p_{ai}-p_{aws}=\frac{162.6Q\bar{B}\bar{\mu}}{kh}\log\left(\frac{t_{ap}-\Delta t_a}{\Delta t_a}\right), \quad (2.29)$$

TABLE 2.3—BUILDUP TEST DATA FOR WELL A

Δt (hours)	p_{ws} (psia)
0	1,347
0.0008	1,348
0.0022	1,350
0.0039	1,355
0.0067	1,364
0.0092	1,376
0.0117	1,387
0.0142	1,399
0.0169	1,412
0.0264	1,456
0.0403	1,523
0.0556	1,593
0.0681	1,650
0.0792	1,698
0.0903	1,746
0.1125	1,841
0.1403	1,953
0.1681	2,062
0.1958	2,168
0.2264	2,279
0.2542	2,376
0.3792	2,775
0.5458	3,199
0.7125	3,527
0.8792	3,771
1.046	3,947
1.813	4,327
2.646	4,491
3.479	4,585
4.313	4,649
5.229	4,740
6.313	4,757
7.146	4,792
7.979	4,824
8.813	4,852
9.646	4,879
10.31	4,898

$$k=\frac{162.6Q\bar{B}\bar{\mu}}{mh}, \quad \text{(2.30)}$$

and

$$s=1.151\left[\frac{p_{a1hr}-p_{awf}}{m}-\log\left(\frac{k}{\phi\bar{\mu}\bar{c}_t r_w{}^2}\right)+3.23+\log\left(\frac{t_{ap}+1}{t_{ap}}\right)\right]. \quad \text{(2.31)}$$

Example Problem. To illustrate how the prefracture well test data can be analyzed in several ways, the data for Well A in Table 2.2 will be used to work an example problem.

Well A was produced at a dry-gas rate of 2,500 Mcf/D [71×10^3 m^3/d], and after 12,604 Mcf [357×10^3 m^3] was produced, the flowing tubing pressure was 1,200 psia [8.3 MPa]. Although the skin factor for the well is not known exactly, it is known that the perforations were balled out with KCl and that such treatments typically result in a skin factor of -1 in this reservoir.

With these data and the following equations, an estimate of the formation permeability can be obtained.

$$k=\left[\ln\left(\frac{r_d}{r_w}\right)-0.75+s'\right]\frac{1{,}422qT}{h(p_{pi}-p_{pwf})} \quad \text{(2.32)}$$

and

$$r_d=\left(\frac{kt}{948\phi\mu c_t}\right)^{1/2}. \quad \text{(2.33)}$$

This technique is called the one-point (ONEPT) method of analysis.[40] To use the ONEPT technique, one first assumes a value for k and solves for r_d in Eq. 2.33. Then, with the value of r_d, one solves Eq. 2.32 for k. This new estimate of k can be used in Eq. 2.33, and the calculation process is repeated until it converges on a value of permeability.

For the data given in Table 2.2, a permeability value of 0.575 md can be calculated with the ONEPT technique.

Following the early-time, single-rate flow test, Well A continued to produce. After about 45 days, the well had produced 108,952 Mcf [3.1×10^6 m^3] of gas and was flowing at a rate of 2,170 Mcf/D [61×10^3 m^3/d]. Well A was shut in at this point, and the pressure-buildup data are presented in Table 2.3.

To analyze the buildup test properly, one should prepare both a semilog Horner graph and a full logarithmic type-curve graph of the pressure-buildup data. The first step should be to perform a qualitative type-curve analysis of the data. The best type curve for this analysis is usually that published by Gringarten *et al.*[49]

The type-curve graph of log Δp vs. log Δt can be analyzed to determine the early-time slope. If the early-time slope equals 1, the early pressure data are dominated by wellbore storage. If the early slope of the type-curve graph equals one-half or less, then the buildup indicates that natural fractures may be present in the reservoir. The type-curve graph of the field data should be used to overlay the Gringarten *et al.* type curve. By matching the shape of the field data curve to the shape of an analytic solution, one can determine when the MTR begins and can observe the onset of boundary effects, if they are occurring.

Fig. 2.16 presents a type-curve graph of the buildup data for Well A using the functions called delta adjusted pressure, Δp_a, and effective adjusted time, Δt_{ae}. Effective adjusted time is computed using Eq. 2.26 and the values of adjusted time.

Notice that the early time data do approximate a unit slope and that the start of the MTR occurs at an effective adjusted time of about 2.2 hours. Also, no evidence of boundary effects can be seen from the late-time data.

After the qualitative type-curve match has been completed, the Horner analysis should be performed with the data in the MTR according to the type-curve match. The Horner analysis can provide quantitative estimates of formation permeability and skin factor, and initial reservoir pressure can be evaluated.

Fig. 2.17 presents the Horner graph of the buildup data for Well A. Tables 2.4 and 2.5 present the plotting functions used to plot Figs. 2.16 and 2.17.

By drawing a straight line through the MTR, the slope was determined to be 643 psia/cycle[4.4 MPa/cycle] and p_{a1hr} was 2,830 psia [19.5 MPa]. Table 2.6 presents detailed calculations of this prefracture well test. Notice that from the Horner graph, the formation permeability was calculated to be 0.60 md with a skin of -0.42. This permeability value obtained from the Horner analysis is essentially identical to the permeability estimate obtained from the ONEPT analysis.

After evaluating the Horner graph, the analyst should re-evaluate the type-curve match and compute values for permeability and skin factor from the match point of the type-curve graph. If the values for permeability obtained from both the Horner and the type-curve methods are not in agreement, the analyst must assume that either an incorrect type-curve match was obtained or the selected MTR on the Horner graph was not correct. The analyst should be satisfied with the resulting calculations only when essentially identical answers can be obtained from both analysis techniques.

Because the Horner graph did give accurate results for Well A, one can take the estimated value of permeability and calculate a pressure match point for the type curve. In Table 2.6, one can see that for a value of $p_D=0.1$, the value of Δp_a is 55.5. This match point can be used to fix the type curve in the vertical direction and the data moved horizontally to find an excellent match. When such a match occurs, it confirms the validity of the Horner analysis.

In many cases, the analysis of prefracture well tests can be quite straightforward and can result in adequate estimates of the in-situ formation parameters. As one might expect, however, several problems must be addressed in the field to test and to analyze low-permeability formations properly.

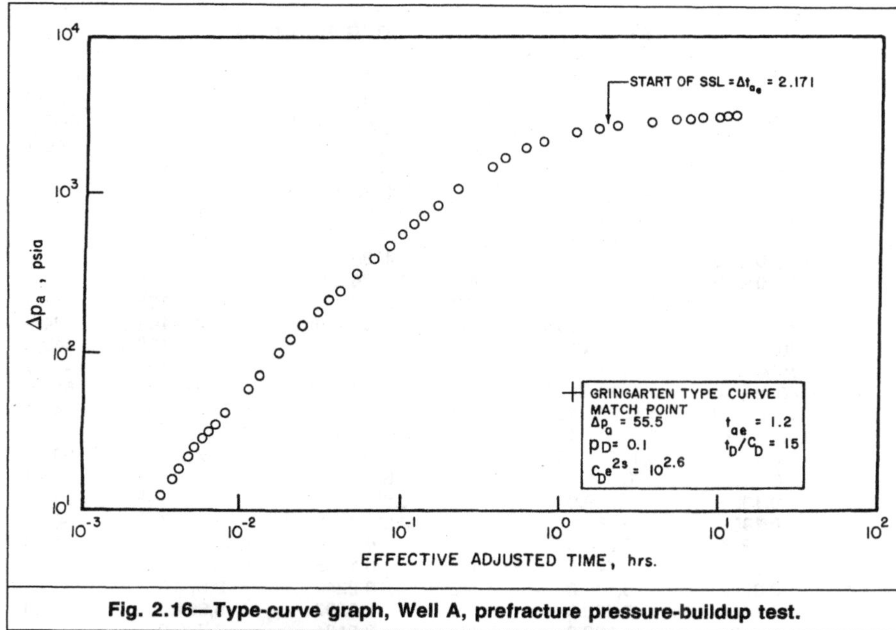

Fig. 2.16—Type-curve graph, Well A, prefracture pressure-buildup test.

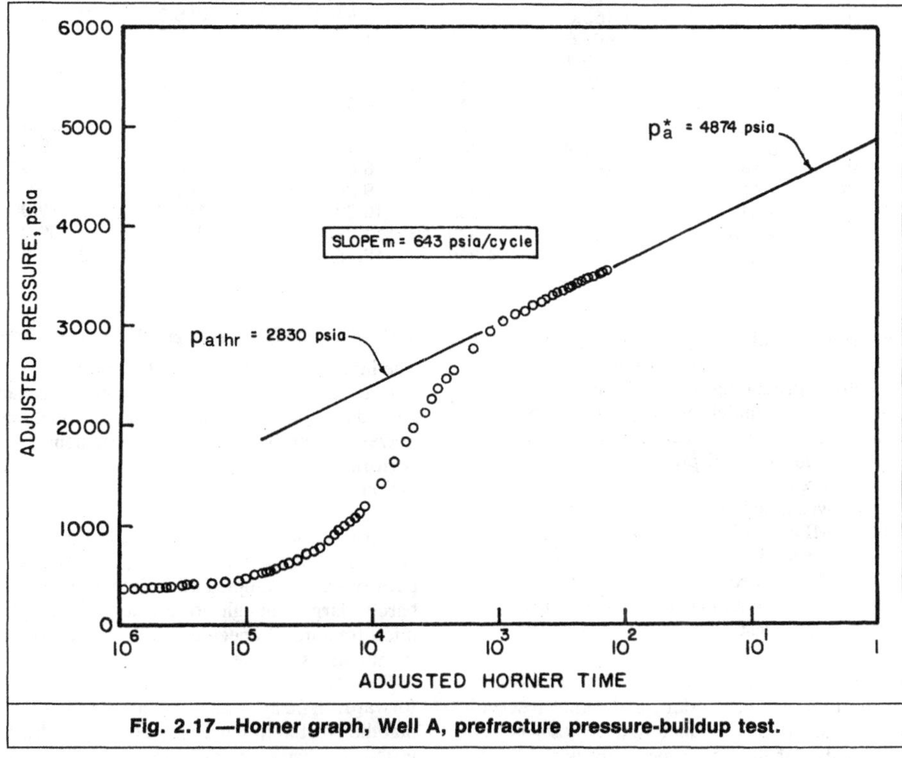

Fig. 2.17—Horner graph, Well A, prefracture pressure-buildup test.

First, in many situations, the well completion and perforations do not effectively connect the formation with the wellbore. In some cases, the well will simply not flow at sustained, measurable rates. If the well does not flow at a high enough rate to unload liquids, the pressure-drawdown phase of the test may be quite misleading. Another problem is that if the well will produce at a sustained measurable rate, but not all the perforations are open, then the resulting pressure-buildup test may not truly represent the entire formation flow capacity. It is common to pick net gas pay from logs and to use that value to determine permeability from the *kh* product. If some of the perforations are not completely open and vertical permeability is low, however, one will probably use the wrong value for net pay when analyzing prefracture well tests.

For this reason, it is usually a sound practice to break down the perforations with a nondamaging fluid before a prefracture well test is performed. In many cases, 2% KCl water with surfactants can be used during the breakdown of the perforations. Breakdown tests should not be designed to stimulate the reservoir; they should be designed simply to establish sufficient communication between the formation and wellbore for a prefracture well test to be performed and for the subsequent fracture treatment to be pumped successfully.

When an injection test is performed on a well to improve communication between the wellbore and the formation, it is quite likely that moderate stimulation will occur around the wellbore. Skin factors of −1 to −3 would be commonly expected after a ballout treatment to open up perforations. As discussed in the literature,[50] even a small negative skin can distort the pressure buildup in a low-permeability well to the point where it may be difficult to determine the correct MTR.

TABLE 2.4—HORNER PLOTTING FUNCTIONS

	Time (hours)	Horner Time	Pressure (psia)	Adjusted Time (hours)	Adjusted Horner Time	Adjusted Pressure (psia)
1	0.8000×10^{-3}	0.1506×10^{7}	1,348	0.3220×10^{-3}	0.5987×10^{7}	350.3
2	0.2200×10^{-2}	0.5477×10^{6}	1,350	0.8897×10^{-3}	0.2167×10^{7}	351.4
3	0.3900×10^{-2}	0.3090×10^{6}	1,355	0.1580×10^{-2}	0.1220×10^{7}	353.9
4	0.6700×10^{-2}	0.1799×10^{6}	1,364	0.2723×10^{-2}	0.7079×10^{6}	358.4
5	0.9200×10^{-2}	0.1310×10^{6}	1,376	0.3750×10^{-2}	0.5140×10^{6}	364.4
6	0.1170×10^{-1}	0.1030×10^{6}	1,387	0.4785×10^{-2}	0.4029×10^{6}	369.9
7	0.1420×10^{-1}	0.8486×10^{5}	1,399	0.5826×10^{-2}	0.3309×10^{6}	375.9
8	0.1690×10^{-1}	0.7130×10^{5}	1,412	0.6959×10^{-2}	0.2770×10^{6}	382.4
9	0.2640×10^{-1}	0.4564×10^{5}	1,456	0.1101×10^{-1}	0.1751×10^{6}	404.5
10	0.4030×10^{-1}	0.2990×10^{5}	1,523	0.1713×10^{-1}	0.1125×10^{6}	440.2
11	0.5560×10^{-1}	0.2167×10^{5}	1,593	0.2413×10^{-1}	0.7990×10^{5}	481.9
12	0.6810×10^{-1}	0.1770×10^{5}	1,650	0.3004×10^{-1}	0.6418×10^{5}	515.9
13	0.7920×10^{-1}	0.1522×10^{5}	1,698	0.3543×10^{-1}	0.5442×10^{5}	544.4
14	0.9030×10^{-1}	0.1335×10^{5}	1,746	0.4095×10^{-1}	0.4708×10^{5}	573.0
15	0.1125	0.1071×10^{5}	1,841	0.5238×10^{-1}	0.3680×10^{5}	633.1
16	0.1403	8,590.0	1,953	0.6740×10^{-1}	0.2860×10^{5}	709.4
17	0.1681	7,169.0	2,062	0.8317×10^{-1}	0.2318×10^{5}	783.6
18	0.1958	6,155.0	2,168	0.9960×10^{-1}	0.1936×10^{5}	861.0
19	0.2264	5,323.0	2,279	0.1185	0.1626×10^{5}	945.0
20	0.2542	4,741.0	2,376	0.1365	0.1413×10^{5}	1,018.0
21	0.3792	3,179.0	2,775	0.2249	8,573.0	1,350.0
22	0.5458	2,209.0	3,199	0.3597	5,361.0	1,733.0
23	0.7125	1,692.0	3,527	0.5101	3,780.0	2,046.0
24	0.8792	1,372.0	3,771	0.6725	2,867.0	2,287.0
25	1.046	1,153.0	3,947	0.8440	2,285.0	2,464.0
26	1.813	665.6	4,327	1.691	1,141.0	2,856.0
27	2.646	456.4	4,491	2.664	724.6	3,027.0
28	3.479	347.4	4,585	3.663	527.2	3,126.0
29	4.313	280.4	4,649	4.680	412.9	3,194.0
30	5.229	231.4	4,704	5.810	332.8	3,253.0
31	6.313	191.9	4,757	7.162	270.2	3,309.0
32	7.146	169.6	4,792	8.210	235.8	3,346.0
33	7.979	152.0	4,824	9.265	209.1	3,380.0
34	8.813	137.7	4,852	10.33	187.7	3,410.0
35	9.646	125.9	4,879	11.39	170.2	3,439.0
36	10.31	117.9	4,898	12.25	158.4	3,459.0

To maximize the chances of obtaining an accurate prefracture pressure-transient analysis, it is very useful to measure the initial reservoir pressure before producing from or injecting into a reservoir. If the correct value for initial reservoir pressure is known and the last few data points on the pressure-transient test cannot be extrapolated to that value of initial pressure, then the correct MTR may not have been reached.

In the previous data, it was noted that the initial pressure in Well A was 6,180 psia [42.6 MPa], which is equivalent to a value of adjusted pressure of 4,874 psia [33.6 MPa]. From Fig. 2.17, it can be seen that the MTR is extrapolated back to the original pressure; therefore, no boundaries were felt and no significant depletion occurred during the 45-day well test.

Summary. To develop low-permeability reservoirs successfully, the values for formation permeability, skin, and initial reservoir pressure are needed. Analysis of postfracture well tests is very difficult if the correct value of in-situ permeability is not known from a prefracture well test. It is also very difficult to optimize fracture length and to design the optimum proppant for a fracture treatment if the value of in-situ permeability is not known from a prefracture well test. In some situations, a low-productivity well may not actually need a stimulation treatment. The results from a prefracture well test of a low-productivity well may indicate that a particular zone contains either a very high skin factor or a low reservoir pressure. It may be decided that some other form of stimulation, such as acidization, is really needed in a particular reservoir.

To determine accurate values of permeability, skin, and reservoir pressure, a prefracture transient well test should be considered a necessity. It is very difficult to design or to analyze a hydraulic fracture treatment intelligently unless adequate prefracture well testing has been performed.

2.6 Mechanical Properties Testing

The importance of mechanical properties has been discussed in this chapter, and techniques for calculating mechanical properties from core data and log data have been mentioned. It is important, however, that log and core measurements be confirmed with tests performed in situ to correlate all methods into a consistent evaluation.

In-Situ Stress Tests. Perhaps the most important variable to be determined is the value of in-situ stress. An in-situ stress test is performed by pumping fluid into a zone at an injection rate just barely large enough to create a hydraulic fracture. Once a microfracture has been created, the pumps can be shut down and in-situ stress can be measured.[51,52]

The measurement of in-situ stress in the field is not straightforward. In many instances, the effects of perforations and the fact that the created fracture may not stay within the zone that is being tested can distort the measurements. By paying attention to detail and by performing small-volume injection/flowback tests, however, one can usually measure accurate values of in-situ stresses. The ideal formation evaluation would be one where the in-situ stresses calculated from logs and from cores and measured with injection/flowback tests all result in a consistent stress profile.

Leakoff Tests. Another test that can provide valuable information is a minifracture or an injection/leakoff test. These tests are normally run with larger volumes of fluid than used in an in-situ stress test. The fluids used in leakoff tests should be similar to those used during the main hydraulic fracture treatment. If fluids are injected into the wellbore at fracturing rates, the pumps then shut down, and the pressure decline measured with time, it is theoretically possible to compute a value for fluid-loss coefficient by assuming a value for fracture height (or vice versa). Therefore, any prefracture test designed to compute leakoff coefficient should be coupled with a test to determine an estimated fracture height, or at least, likely combinations of these two variables should be determined.

As discussed earlier, a minimum value of fracture height can usually be determined from gamma ray/temperature logs after an injection test. Because cased-hole logs cannot detect the fracture when it deviates from the borehole, however, production log results after an injection test cannot be used indiscriminately to design the large fracture treatment. If prefracture injection tests are designed and analyzed properly, however, they can be useful in the design. Bundy[24] illustrated how prefracture injection tests can be used in formations containing multiple pay intervals to help design large limited-entry stimulation techniques.

Fracture Azimuth. The expected azimuth of a hydraulic fracture treatment can also be determined from prefracture injection tests.[36,37,52] One method that appears promising is to use a triaxial borehole seismic tool in conjunction with a small prefracture injection test to map the orientation of the created fracture. Recent field studies have shown acceptable reproducibility in expected fracture azimuth when various tools—such as oriented core, the triaxial borehole geophone systems, and tiltmeters—have been compared to predict a common direction for fracture azimuth.

2.7 Summary of Prefracture Formation Evaluation Concepts

A large amount of information has been presented in this chapter. It is virtually impossible to explain each subject completely in an in-depth manner that can be applied by a practicing engineer. I have tried to present the pertinent comments, equations, and references to allow an interested party to research the various topics and to develop the necessary calculation procedures.

The most important aspect of a successful prefracture formation evaluation is that a consistent picture of the formation be developed by use of all the available analysis techniques. For example, if one cannot correlate the permeabilities from core analysis to the permeability obtained from pressure-transient analysis, the prefracture formation evaluation may not be accurate. Also, it is extremely important that one be able to determine a consistent stress profile from the logs, cores, and well tests.

Examples of how geologic information, well logs, cores, and well test information have been used to develop an accurate and acceptable prefracture formation evaluation are presented in the literature.[53-58] These papers and others illustrate that the various analysis techniques discussed in this chapter can be applied successfully to evaluate and to design stimulation treatments in low-permeability reservoirs.

TABLE 2.5—TYPE-CURVE PLOTTING FUNCTIONS

	Effective Time (hours)	Δp (psia)	Effective Adjusted Time (hours)	Δp_a (psia)
1	0.8000×10^{-3}	1	0.3220×10^{-3}	0.5013
2	0.2200×10^{-2}	3	0.8897×10^{-3}	1.504
3	0.3900×10^{-2}	8	0.1580×10^{-2}	4.010
4	0.6700×10^{-2}	17	0.2723×10^{-2}	8.521
5	0.9200×10^{-2}	29	0.3750×10^{-2}	14.54
6	0.1170×10^{-1}	40	0.4785×10^{-2}	20.05
7	0.1420×10^{-1}	52	0.5826×10^{-2}	26.06
8	0.1690×10^{-1}	65	0.6959×10^{-2}	32.58
9	0.2640×10^{-1}	109	0.1101×10^{-1}	54.64
10	0.4030×10^{-1}	176	0.1713×10^{-1}	90.39
11	0.5560×10^{-1}	246	0.2413×10^{-1}	132.1
12	0.6810×10^{-1}	303	0.3004×10^{-1}	166.0
13	0.7919×10^{-1}	351	0.3543×10^{-1}	194.6
14	0.9029×10^{-1}	399	0.4095×10^{-1}	223.2
15	0.1125	494	0.5238×10^{-1}	283.2
16	0.1403	606	0.6740×10^{-1}	359.5
17	0.1681	715	0.8317×10^{-1}	433.7
18	0.1958	821	0.9959×10^{-1}	511.1
19	0.2264	932	0.1185	595.2
20	0.2541	1,029	0.1365	668.6
21	0.3791	1,428	0.2249	1,000.0
22	0.5456	1,852	0.3596	1,383.0
23	0.7121	2,180	0.5100	1,696.0
24	0.8786	2,424	0.6723	1,937.0
25	1.045	2,600	0.8436	2,114.0
26	1.810	2,980	1.690	2,506.0
27	2.640	3,144	2.660	2,677.0
28	3.469	3,238	3.656	2,777.0
29	4.298	3,302	4.669	2,844.0
30	5.206	3,357	5.793	2,903.0
31	6.280	3,410	7.136	2,959.0
32	7.104	3,445	8.176	2,996.0
33	7.927	3,477	9.221	3,030.0
34	8.749	3,505	10.27	3,060.0
35	9.569	3,532	11.33	3,089.0
36	10.22	3,551	12.17	3,110.0

Nomenclature

A = area
B = formation volume factor
c_b = compressibility
c_t = total compressibility
C = wellbore storage constant
C_D = dimensionless wellbore storage constant
E = Young's modulus
G = shear modulus
h = thickness
I_{sh} = shale index
k = permeability
K = bulk modulus
L_f = fracture length
m = slope
p = reservoir pressure
p^* = average reservoir pressure at MTR
p_a = adjusted pressure
p_{ai} = adjusted average static pressure at start of buildup test
p_{awf} = adjusted flowing bottomhole pressure
p_{aws} = adjusted shut-in bottomhole pressure
p_{a1hr} = adjusted pressure at 1 hour
p_D = dimensionless pressure
p_i = average static pressure at start of buildup test
p_p = pseudopressure
p_p^* = average reservoir pseudopressure at MTR
p_{pi} = initial pseudopressure
p_{pwf} = flowing bottomhole pseudopressure
p_{pws} = shut-in bottomhole pseudopressure
p_{wf} = flowing bottomhole pressure
p_{ws} = shut-in bottomhole pressure
p_{1hr} = pressure at 1 hour
Δp = pressure change
Δp_a = adjusted pressure change
q = flow rate
Q = cumulative production
Q_V = quantity of cation-exchangeable clays in pore space
Q_{Vt} = total CEC
r_e = external drainage radius
r_d = drainage of radius
r_i = radius of investigation
r_w = wellbore radius
R_t = true resistivity
R_v = velocity ratio
R_w = water resistivity
R_{wa} = apparent water resistivity
s = skin factor
S_w = water saturation
t = time
t_{ap} = pseudoproducing time

TABLE 2.6—GAS WELL TEST ANALYSIS, PREFRACTURE PRESSURE BUILDUP

Basic Properties

q, Mcf/D	2,196	r_w, ft	0.255
T_r, °R	739	S_w	0.60
h, ft	19	t_p, hours	1,205
ϕ_t	0.176	t_{ap}, hours	1,975
$\bar{p}$, psi	3,763.5	$\bar{z}$	0.9291
$\bar{\mu}$, cp	0.0222	$\bar{c}_t$, psi^{-1}	8.99×10^{-5}
$\bar{B}_g$, RB/Mcf	0.9185		

Semilog Analysis

Horner slope, psi/cycle 643

$$k=\frac{162.6q\bar{B}_g\mu}{mh}=\frac{(162.6)(2,196)(0.9185)(0.0222)}{(643)(19)}=0.60\text{ md}$$

p_{a1hr}, psia 2,830

p_{awf}, psi 372.1

$$s=1.151\left[\frac{p_{a1hr}-p_{awf}}{m}-\log\left(\frac{k}{\phi\bar{\mu}\bar{c}r_w^{\,2}}\right)+3.23+\log\left(\frac{t_{ap}+1}{t_{ap}}\right)\right]$$

$$=1.151\left\{\frac{2,830-372.1}{643}-\log\left[\frac{0.60}{(0.176)(0.0222)(9\times10^{-5})(0.255)^2}\right]\right.$$

$$\left.+3.23+\log\left(\frac{1,975+1}{1,975}\right)\right\}=-0.42$$

Radius of Investigation

At end of production

$$r_i=\left(\frac{kt_{ap}}{948\phi\overline{\mu c}}\right)^{1/2}=\left[\frac{(0.6)(1,975)}{(948)(0.176)(0.0222)(9\times10^{-5})}\right]^{1/2}=1,866\text{ ft}$$

At end of test

$$r_i=\left(\frac{k\Delta t_a}{948\phi\overline{\mu c}}\right)^{1/2}=\left[\frac{(0.6)(12.34)}{(948)(0.176)(0.0222)(9\times10^{-5})}\right]^{1/2}=149\text{ ft}$$

A type-curve analysis should also be performed to determine the MTR and to confirm the values calculated from the semilog analysis.

t_D = dimensionless time
t_p = producing time
Δt = shut-in time
Δt_a = pseudo-shut-in time
Δt_{ae} = effective pseudo-shut-in time
Δt_c = compressional-wave travel time
Δt_s = shear-wave travel time
T = temperature
T_r = reservoir temperature
v = velocity
v_c = compressional-wave velocity
v_f = velocity of sound in fluid
v_{ma} = velocity of sound in matrix
v_s = shear-wave velocity
z = gas compressibility factor
μ = viscosity
ν = Poisson's ratio
ρ = density
ρ_b = bulk density
σ_E = externally generated stresses
σ_x = total horizontal stress
σ_{x1},σ_{x2} = total horizontal stress in Directions 1 and 2
σ_z = total overburden stress
ϕ = porosity
ϕ_D = density porosity
ϕ_S = sonic porosity

Subscripts

a = adjusted
D = dimensionless
g = gas
t = total

Superscripts

m = cementation factor
n = saturation exponent
$^{-}$ = average

References

1. Holditch, S.A., Jennings, J.W., and Neuse, S.H.: "The Optimization of Well Spacing and Fracture Length in Low Permeability Gas Reservoirs," paper SPE 7496 presented at the 1978 SPE Annual Technical Conference and Exhibition, Houston, Oct. 1–4.
2. Spencer, C.W.: "Geologic Aspects of Tight Gas Reservoirs in the Rocky Mountain Region," *JPT* (July 1985) 1308–14.
3. Peterson, R.E. and Kohout, J.: "An Approximation of Continuity of Lenticular Mesaverde Sandstone Lenses Utilizing Close-Well Correlations, Piceance Basin, Northwest Colorado," paper SPE 11610 presented at the 1983 SPE/DOE Symposium on Low Permeability Gas Reservoirs, Denver, March 14–16.
4. Finley, R.J. and O'Shea, P.A.: "Geologic and Engineering Analysis of Blanket-Geometry Tight Gas Sandstones," paper SPE 11607 presented at the 1983 SPE/DOE Symposium on Low Permeability Gas Reservoirs, Denver, March 14–16.

5. Hodges, L.T. and Knutson, C.F.: "Tight Gas Sandstone Channel Continuity and Directivity, Upper Cretaceous and Paleocene, Greater Green River Basin, Wyoming," paper SPE 9844 presented at the 1981 SPE/DOE Symposium on Low Permeability Gas Reservoirs, Denver, May 27–29.
6. Wilson, M.D.: "Origins of Clays Controlling Permeability in Tight Gas Sands," *JPT* (Dec. 1982) 2871–76.
7. Hubbert, M.K. and Willis, D.G.: "Mechanics of Hydraulic Fracturing," *Trans.*, AIME (1957) **210,** 153–66.
8. Warpinski, N.R. and Teufel, L.W.: "Influence of Geologic Discontinuities on Hydraulic Fracture Propagation," *JPT* (Feb. 1987) 209–20.
9. Waxman, M.H. and Smits, L.J.M.: "Electrical Conductivities in Oil-Bearing Shaly Sands," *SPEJ* (June 1968) 107–21; *Trans.*, AIME, **243**.
10. Alger, R.P. *et al.*: "Formation Density Log Applications in Liquid-Filled Holes," *JPT* (March 1963) 321–32; *Trans.*, AIME, **228**.
11. Clavier, C., Coates, G. and Dumanoir, J.: "Theoretical and Experimental Bases for the Dual-Water Model for the Interpretation of Shaly Sands," *SPEJ* (April 1984) 153–68.
12. Archie, G.E.: "The Electrical Resistivity Log as an Aid in Determining Some Reservoir Characteristics," *Trans.*, AIME (1942) **146,** 54–62.
13. Rosepiler, M.J.: "Calculation and Significance of Water Saturations in Low Porosity Shaly Gas Sands," paper SPE 10910 presented at the 1982 SPE Cotton Valley Symposium, Tyler, May 20.
14. Raymer, L.L., Hunt, E.R., and Gardner, J.S.: "An Improved Sonic Transit Time-to-Porosity Transform," paper P presented at the 1980 SPWLA Annual Logging Symposium, July 8–11.
15. Pickett, G.R.: "Acoustic Character Logs and Their Applications in Formation Evaluation," *JPT* (June 1963) 659–67; *Trans.*, AIME, **228**.
16. Sethi, D.K.: "Well Log Applications in Rock Mechanics," paper SPE 9833 presented at the 1981 SPE/DOE Symposium on Low Permeability Gas Reservoirs, Denver, May 27–29.
17. Leslie, H.D. and Mons, F.: "Sonic Waveform Analysis: Applications," paper GG presented at the 1982 SPWLA Annual Logging Symposium, July 6–9.
18. Coates, G.R. and Denoo, S.A.: "Log Derived Mechanical Properties and Rock Stress," paper U presented at the 1980 SPWLA Annual Logging Symposium, July 8–11.
19. Kowalski, J.: "Formation Strength Parameters from Well Logs," paper N presented at the 1975 SPWLA Annual Logging Symposium, June 4–7.
20. Tixier, M.P., Loveless, G.W., and Anderson, R.A.: "Estimation of Formation Strength from the Mechanical Properties Log," *JPT* (March 1975) 253–58.
21. Anderson, R.A., Ingram, D.S., and Zanier, A.M.: "Determining Fracture Pressure Gradients from Well Logs," *JPT* (Nov. 1973) 1259–68.
22. Wu, P.T.: "Comparison of Digital and Analog Techniques for Determining Sonic Velocity from Borehole Acoustic Waveforms," paper SPE 13286 presented at the 1984 SPE Annual Technical Conference and Exhibition, Houston, Sept. 16–19.
23. Morris, C.R., Little, T.M., and Letton, W. III: "A New Sonic Array Tool for Full-Waveform Logging," paper SPE 13285 presented at the 1984 SPE Annual Technical Conference and Exhibition, Houston, Sept. 16–19.
24. Bundy, T.E.: "Prefracture Injection Surveys: A Necessity for Successful Fracture Treatments," *JPT* (May 1982) 995–1001.
25. Dobkins, T.A.: "Improved Methods to Determine Hydraulic Fracture Height," *JPT* (April 1981) 719–26.
26. Thomas, R.D. and Ward, D.C.: "Effect of Overburden Pressure and Water Saturation on Gas Permeability of Tight Sandstone Cores," *JPT* (Feb. 1972) 120–24.
27. Jones, F.O. Jr. and Owens, W.W.: "A Laboratory Study of Low-Permeability Gas Sands," *JPT* (Sept. 1980) 1631–40.
28. Walls, J.D., Nur, A.M., and Bourbie, T.: "Effects of Pressure and Partial Water Saturation on Gas Permeability in Tight Sands: Experimental Results," *JPT* (April 1982) 930–36.
29. Sattler, A.R.: "The Multiwell Experiment Core Program, II," paper SPE 12854 presented at the 1984 SPE Unconventional Gas Recovery Symposium, Pittsburgh, May 13–15.
30. Soeder, D.J.: "Laboratory Drying Procedures and the Permeability of Tight Sandstone Core," *SPEFE* (Feb. 1986) 16–22.
31. Randolph, P.L., Soeder, D.J., and Chowdiah, P.: "Porosity and Permeability of Tight Sands," paper SPE 12836 presented at the 1984 Unconventional Gas Recovery Symposium, Pittsburgh, May 13–15.
32. Randolph, P.L.: "Porosity and Permeability of Mesaverde Sandstone Core from the U.S. DOE Multiwell Experiment, Garfield County, Colorado," paper SPE 11765 presented at the 1983 SPE/DOE Symposium on Low Permeability Gas Reservoirs, Denver, March 14–16.
33. Soeder, D.J. and Randolph, P.L.: "Porosity, Permeability, and Pore Structure of the Tight Mesaverde Sandstone, Piceance Basin, CO," *SPEFE* (June 1987) 129–36.
34. Ahmed, U., Abou-Sayed, A.S., and Jones, A.H.: "Experimental Evaluation of Fracturing Fluid Interaction with Tight Reservoir Rocks and Propped Fractures," paper SPE 7922 presented at the 1979 SPE/DOE Symposium on Low Permeability Gas Reservoirs, Denver, May 20–22.
35. Smith, M.B.: "Effect of Fracture Azimuth on Production with Applications to the Wattenburg Gas Field," paper SPE 8298 presented at the 1979 SPE Annual Technical Conference and Exhibition, Las Vegas, Sept. 23–26.
36. Lacy, L.L.: "Comparison of Hydraulic-Fracture Orientation Techniques," *SPEFE* (March 1987) 66–76.
37. Teufel, L.W. *et al.*: "Determination of Hydraulic Fracture Azimuth by Geophysical, Geological, and Oriented-Core Methods at the Multiwell Experiment Site, Rifle, CO," paper SPE 13226 presented at the 1984 SPE Annual Technical Conference and Exhibition, Houston, Sept. 16–19.
38. Teufel, L.W.: "Determination of In-Situ Stress From Anelastic Strain Recovery Measurements of Oriented Core," paper SPE 11649 presented at the 1983 SPE/DOE Symposium on Low Permeability Gas Reservoirs, Denver, March 14–16.
39. Holditch, S.A. *et al.*: "Effect of Mud Filtrate Invasion on Apparent Productivity in Drillstem Tests in Low-Permeability Gas Formations," *JPT* (Feb. 1983) 299–305.
40. Lee, W.J. *et al.*: "Estimating Formation Permeability from Single-Point Flow Data," paper SPE 12847 presented at the 1984 SPE Unconventional Gas Recovery Symposium, Pittsburgh, May 13–15.
41. Lee, W.J.: *Well Testing,* Textbook Series, SPE, Richardson, TX (1982) **1**.
42. Agarwal, R.G., Al-Hussainy, R., and Ramey, H.J. Jr.: "An Investigation of Wellbore Storage and Skin Effect in Unsteady State Liquid Flow: I. Analytical Treatment," *SPEJ* (Sept. 1970) 279–90.
43. Al-Hussainy, R., Ramey, H.J. Jr., and Crawford, P.B.: "The Flow of Real Gases Through Porous Media," *JPT* (May 1966) 624–36; *Trans.*, AIME, **237**.
44. Agarwal, R.G.: "Real Gas Pseudo-Time—A New Function for Pressure Buildup Analysis of MHF Gas Wells," paper SPE 8279 presented at the 1979 SPE Annual Technical Conference and Exhibition, Las Vegas, Sept. 23–26.
45. Agarwal, R.G.: "A New Method to Account for Producing Time Effects when Drawdown Type Curves are Used to Analyze Pressure Buildup and Other Test Data," paper SPE 9289 presented at the 1980 SPE Annual Technical Conference and Exhibition, Dallas, Sept. 21–24.
46. Lee, W.J. and Holditch, S.A.: "Application of Pseudotime to Buildup Test Analysis of Low Permeability Gas Wells with Long-Duration Wellbore Storage Distortion," *JPT* (Dec. 1982) 2877–87.
47. Lee, W.J. *et al.*: "Analyzing Gas Well Buildup Tests with Changing Wellbore Storage in Tight Gas," paper SPE 12845 presented at the 1984 SPE Unconventional Gas Recovery Symposium, Pittsburgh, May 13–15.
48. Meunier, D.F., Kabir, C.S., and Wittmann, M.J.: "Gas Well Test Analysis: Use of Normalized Pressure and Time Functions," *SPEFE* (Dec. 1987) 629–36.
49. Gringarten, A.C. *et al.*: "A Comparison Between Different Skin and Wellbore Storage Type-Curves for Early-Time Transient Analysis," paper SPE 8205 presented at the 1979 SPE Annual Technical Conference and Exhibition, Las Vegas, Sept. 23–26.
50. Bostic, J. N. and Graham, J. A.: "Prefracture Pressure Transient Testing: East Texas Cotton Valley Tight Gas Play," paper SPE 7941 presented at the 1979 SPE/DOE Symposium on Low-Permeability Gas Reservoirs, Denver, May 20–22.
51. Warpinski, N.R., Branagan, P., and Wilmer, R.: "In-Situ Stress Measurements at U.S. DOE's Multiwell Experiment Site, Mesaverde Group, Rifle, Colorado," *JPT* (March 1985) 527–36.
52. Daneshy, A.A. *et al.*: "In-Situ Stress Measurements During Drilling," *JPT* (Aug. 1986) 891–98.
53. Hunt, E.R. *et al.*: "Application of New Well Logs and Geology to Fracturing and Producibility in Tight Gas Sands, Cotton Valley Group," paper SPE 9832 presented at the 1981 SPE/DOE Symposium on Low Permeability Gas Reservoirs, Denver, May 27–29.
54. Robinson, B.M., Holditch, S.A., and Lee, W.J.: "A Case Study of the Wilcox (Lobo) Trend in Webb and Zapata Counties, Texas," *JPT* (Dec. 1986) 1355–64.

55. Schlottman, B.W., Miller, W.K., and Lueders, R.K.: "Massive Hydraulic Fracture Design for the East Texas Cotton Valley Sands," paper SPE 10133 presented at the 1981 SPE Annual Technical Conference and Exhibition, San Antonio, Oct. 5–7.
56. Kozik, H.G. and Holditch, S.A.: "A Case History for Massive Hydraulic Fracturing the Cotton Valley Lime Matrix, Fallon and Personville Fields," *JPT* (Feb. 1981) 229–44.
57. Wyman, R.E., Holditch, S.A., and Randolph, P.L.: "Analyses of an Elmworth Hydraulic Fracture in Alberta," *JPT* (Sept. 1980) 1621–30.
58. McLennan, J.D., Roegiers, J.C., and Marx, W.P.: "The Mancos Formation: An Evaluation of the Interaction of Geological Conditions, Treatment Characteristics, and Production," paper SPE 11606 presented at the 1983 SPE/DOE Symposium on Low Permeability Gas Reservoirs, Denver, March 14–16.

SI Metric Conversion Factors

acres	× 4.046 873	E−01	=	ha
°API	141.5/(131.5+°API)		=	g/cm^3
bbl	× 1.589 873	E−01	=	m^3
cp	× 1.0*	E−03	=	Pa·s
ft	× 3.048*	E−01	=	m
°F	(°F−32)/1.8		=	°C
in.	× 2.54*	E+00	=	cm
psi	× 6.894 757	E+00	=	kPa
psi^{-1}	× 1.450 377	E−01	=	kPa^{-1}
°R	× 5/9		=	K
scf/bbl	× 1.801 175	E−01	=	std m^3/m^3

*Conversion factor is exact.

Chapter 3
Rock Mechanics and Fracture Geometry

N.R. Warpinski, SPE, Sandia Natl. Laboratories
Michael Berry Smith, SPE, NSI Technologies Inc.

3.1 Overview

This chapter discusses basic rock mechanics considerations required to model the hydraulic fracture process. It begins with a discussion of in-situ stresses, stress measurement techniques, and factors influencing the stress state. This is followed by a review of linear elasticity, poroelasticity, and fracture mechanics; important parameters are defined and their contributions to the fracturing process noted. Finally, calculations of fracture height are discussed for layered formations.

3.2 Introduction

Rock mechanics is the theoretical and applied science of the mechanical behavior of rock, that branch of mechanics concerned with the response of rock to the force fields of its physical environment.[1,2] In hydraulic fracturing, rock mechanics is important in the determination of mechanical properties and the in-situ stress state of reservoir rock, the calculation of deformation and failure behavior of the rock mass caused by the treatment, and the determination of the fracture's final geometry.

Mechanical properties usually of concern for treatment design and analysis are (1) elastic properties, such as Young's modulus (or shear modulus) and Poisson's ratio; (2) strength properties, such as fracture toughness and tensile and compressive strength; (3) ductility; (4) friction; and (5) poroelastic parameters describing the compressibility of the rock matrix compared with the compressibility of the bulk rock under specific fluid flow (or migration) conditions. The most important factor for overall fracture design, however, is the in-situ stress field. Stress not only controls or influences most aspects of fracture behavior, but also influences the values of both reservoir properties and mechanical properties of the rock. For example, increased confining stress will generally result in increased strength,[3] decreased permeability[4] and porosity, and mixed results for Young's modulus and Poisson's ratio. This behavior will be discussed later.

While fracturing models are discussed fully in Chaps. 4 and 5, a brief examination of a fundamental model can be used to illustrate the relative importance of the various parameters. The Khristianovitch-Zheltov[5]/Geertsma-de Klerk[6] model is generally considered applicable for fractures with a height/length ratio greater than one,[7] and for this model, width is related to modulus (Chap. 4) by

$$w \sim (1/E)^{1/4}, \quad (3.1)$$

where E is Young's modulus for the formation. Thus, fracture width is relatively insensitive to modulus. Propagation pressure, on the other hand, is related by

$$P-\sigma_c \sim E^{3/4}, \quad (3.2)$$

so a high-modulus, "stiff" rock causes higher pressures, which can alter fracture geometry.

For fractures with a height/length ratio less than one, sections of the fracture can be viewed as plane strain, as seen in Fig. 3.1, and the Khristianovitch-Zheltov/Geertsma-de Klerk model can be used to calculate height growth. From Chap. 4, the following equations for a unit length of fracture,

$$w \sim (\mu i 4a^2/E)^{1/4} \quad (3.3a)$$

and

$$P-\sigma_c \sim Ew/2a, \quad (3.3b)$$

can be combined (assuming a constant pressure) to give

$$\frac{da}{dt}=\frac{a(P-\sigma_c)^3}{\mu E^2}. \quad (3.4)$$

Thus, the rate of height growth is most strongly influenced by the driving or net fracturing pressure. Because this term depends on the confining stress, σ_c, variations of σ_c in the upper or lower barrier layers will have the most significant impact on fracture height growth. Relations of this type are developed in Chap. 5.

Other rock properties, such as strength, ductility, and friction, generally have only second-order effects on the hydraulic fracturing process, but cases arise where these factors become important. In addition, poroelasticity may be significant in higher-permeability formations or drawn-down reservoirs.

3.3 In-Situ Stress

The in-situ stress, as it affects hydraulic fracturing, is the local stress state in a given rock mass at depth. The three principal stress components of the local stress state—typically compressive, anisotropic, and nonhomogeneous—are influenced strongly by the weight of the overburden, pore pressure, temperature, rock properties, diagenesis, tectonics, and viscoelastic relaxation. In addition, tampering with the in-situ conditions by drilling, fracturing, or production can alter some of these parameters, changing the local stress field.

In-situ stresses are clearly the single most important factor controlling hydraulic fracturing. The stresses control the fracture azimuth[8] and orientation (vertical or horizontal), vertical height growth, surface treating pressures, proppant crushing and embedment, fracture cross-sectional width profiles, and many other facets of fracture behavior. For effective fracture design, it is advantageous for the stresses to be known and for their variations with respect to completion and production techniques to be understood.

Stress Definition. Before in-situ stresses and their role in hydraulic fracturing are discussed, it is helpful to define some commonly used terms.

Closure Pressure and Closure Stress. Nolte[9] defines closure pressure as the fluid pressure required to initiate the opening of an existing fracture. This pressure is equal to and counteracts the stress in the rock perpendicular to the fracture plane. This stress is the minimum principal in-situ stress and is often called the closure stress.

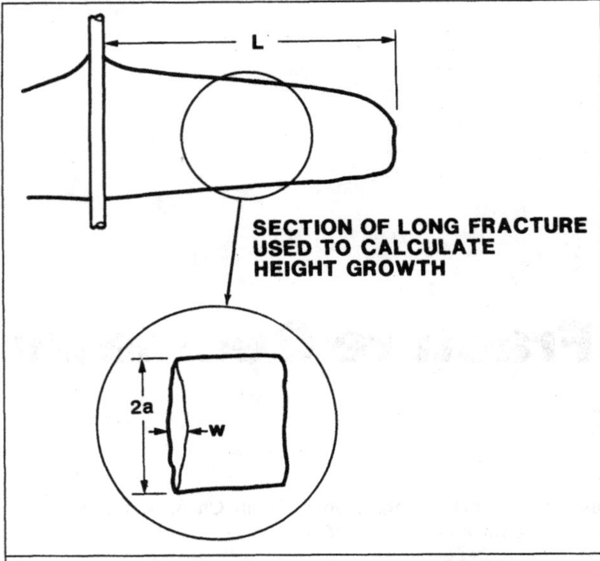

Fig. 3.1—Plane-strain geometry of a long hydraulic fracture.

Fracture Extension Pressure. The fracture extension pressure is the pressure required to extend an existing fracture. It is generally greater than the closure pressure and depends on the size of the fracture and specifics of the treatment.

Instantaneous Shut-In Pressure. The instantaneous shut-in pressure (ISIP) is the pressure in the hydraulic fracture immediately after shut-in. This pressure may vary from several psi to several hundred psi above the closure pressure, depending on the treatment and the rock. The large drop in pressure may be a result of several factors, including the loss of pressure drop across perforations or other flow-entrance restrictions and partial loss of the viscous pressure drop in the fracture in the near-wellbore region. The ISIP is generally greater than the closure stress, but in very small treatments in low-permeability rocks, it will approach closure stress.

Effective Stress. The concept of an effective stress is based on soil mechanics research[10] that showed that a uniform pore pressure, p, affected both the mechanical properties and the behavior of soils. It was shown that when the compressive stress, σ, was corrected for this fluid pressure by

$$\sigma_e=\sigma-sp, \quad \ldots (3.5)$$

where $s\sim 1$ for soils, the resultant soil behavior as a function of σ' (effective stress) was independent of p. The same behavior occurs for rocks,[11-16] but the value of s is thought to be slightly less than one. Nevertheless, the value of one is generally used. The effective stress concept must be used in any situation where there is a significant pore pressure, either naturally occurring or artificially induced. The effective stress acts to control modulus, permeability, and other properties that are stress-sensitive.

Virgin Stresses. The "virgin" stresses are the in-situ stresses that exist in a reservoir before drilling, completion, and production activities. One of the three principal stresses, the overburden stress, is generally vertical and, to a good approximation, is equal to the weight of the overlying rocks. This is usually calculated as

$$\sigma_v=\int_0^z \rho(z)g\,dz, \quad \ldots (3.6)$$

where in the general rock mechanics literature, positive stresses indicate compression. The overburden stress can be estimated by integrating the bulk density log from the surface to the depth of interest. Overburden stresses are typically in the range of 1.0 to 1.1 psi/ft [23 to 25 kPa/m] of depth in sedimentary basins.

The values of the other two principal stresses, which are horizontal, are variable and have been the subject of much research and debate. Hubbert and Willis[8] performed simple laboratory experiments and theoretical analyses showing that the horizontal effective stresses could have any value between about one-third and three times the effective overburden stress. Using a simple sandbox experiment, they showed that normal faulting would occur for values less than one-third and thrust faulting would occur for values greater than three.

The present in-situ stress state in a rock at depth is a complex interaction of rock and reservoir properties, tectonics, and burial history. Prats[17] showed that the differential horizontal effective stress induced by changes in depth, temperature, strain, or pressure could be written as

$$d\sigma_{eH_i}=\frac{\nu}{1-\nu}d(\sigma_z-p)+\frac{E\alpha}{1-\nu}dT+\frac{Ed\epsilon_i}{1-\nu^2}+\frac{\nu Ed\epsilon_j}{1-\nu^2}, \quad \ldots (3.7)$$

where the first term on the right side accounts for the effective overburden stress, the second term accounts for thermal stresses, and the last two terms account for tectonic strains. If variations in the overburden stress, σ_z, pore pressure, p, temperature, T, tectonic strains, ϵ_i and ϵ_j, and the material properties (Young's modulus, E, Poisson's ratio, ν, and coefficient of thermal expansion, α) are known as functions of depth or time, Eq. 3.7 can be integrated to develop a stress history. The viscoelastic response of the rock can also be added by incorporation of a relaxation function into the integration. Even the effect of diagenesis is incorporated in Eq. 3.7 through the form of E, ν, and α. For example, consolidation and cementation with depth of burial can be modeled as an increasing E and decreasing ν with depth.

In general, none of the variations in these parameters are known, so the calculation is currently more of academic than practical interest. Nevertheless, the form of Eq. 3.7, the incorporation of a relaxation function, and the addition of diagenesis effects through the material property terms provide an analysis that illustrates most of the important factors influencing the state of stress at depth. Prats[17] performed an analysis by assuming a form for the material properties and relaxation function and produced some example calculations to show the stress variations expected with depth of burial.

Measurement of In-Situ Stress. At present, the only reliable method of measuring the in-situ stress state at depth is the hydraulic fracturing technique. Two variations of the technique are currently in use: the standard hydraulic fracture measurement and the step-rate/flowback procedure.

Hydraulic Fracture Stress-Test Procedure. The hydraulic fracturing technique as a stress measurement tool is based on the analysis of Hubbert and Willis.[8] In openhole tests, the technique has been widely used and discussed[16,18-25]; when the test is adequately conducted, it can yield an accurate, reproducible estimate of the minimum principal in-situ stress and a somewhat less reliable estimate of the maximum horizontal in-situ stress. The procedure is to isolate the interval of interest with packers, to pump a small volume of low viscosity fluid into the formation to break it down, and to shut in and measure the ISIP. Under these conditions, it is generally accepted that the minimum in-situ stress is essentially

$$\sigma_{min}=p_{is}, \quad \ldots (3.8)$$

and for a vertical fracture,

$$\sigma_{min}=\sigma_{Hmin}=p_{is}. \quad \ldots (3.9)$$

In principle, the maximum horizontal stress can be determined, though this determination is more complicated and usually requires an analysis of the poroelastic response of the rock.[16,22]

For most oil and gas applications, however, it is impossible or impractical to conduct these tests in an openhole environment. Performing a stress measurement in a cased and perforated hole causes additional complications because of the effects of the casing, cement annulus, explosive perforation damage, and random perforation orientation. Some tests[26] and recent results,[27-29] however, show

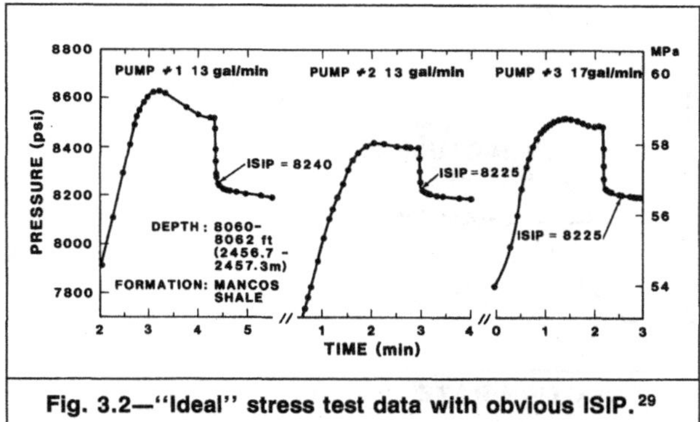

Fig. 3.2—"Ideal" stress test data with obvious ISIP.[29]

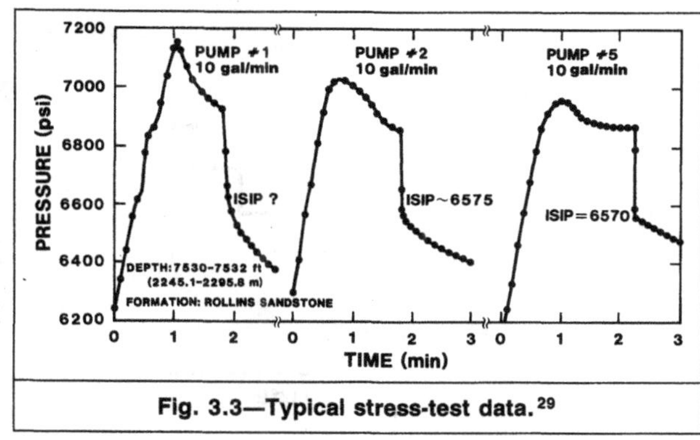

Fig. 3.3—Typical stress-test data.[29]

that accurate measurements of σ_{min} can be made through perforations. No determination of σ_{Hmax} is possible under these conditions.

To perform an accurate, reproducible stress test, several factors need to be carefully considered, including the zone to be tested, the perforations, the pressure measurement system, type of fluid, flow rate, volume injected, and the interpretation procedure.

Generally, the best zones for stress testing are uniform, thick formations. Small-scale layering can cause difficult interpretation problems because individual layers may have widely different stress magnitudes. It is not known whether the stress-test ISIP is a measure of the smallest of these stresses, the largest, or some average. In addition, multiple layers tend to smear the shut-in pressure behavior, resulting in less-defined ISIP values. Zone thickness should be at least 6 to 8 ft [1.8 to 2.4 m], preferably greater. A good cement job in the zone is essential.

The perforations may be the most important factor for obtaining stress data. The purpose of the perforations is to provide an open, undamaged flow path into the rock with minimum disturbance of the rock. Long penetration depths are not necessary and can be detrimental if the perforations also severely crush and stress the rock. Optimum perforations are generally medium-sized charges (10 to 25 g), although larger perforations may be required in heavy casing for tests at large depths. A typical perforation schedule might be a 2-ft [0.61-m] section perforated with 4 shots/ft [13 shots/m] at 90° phasing with 15-g charges. This gives eight chances of getting one good perforation (all that is needed). Field tests have not shown a large effect of perforation direction (phasing) on test quality, but the best case is clearly that where the perforation is aligned with the fracture direction. Small perforation phasing angles (excluding 0° phasing, of course) are desirable.

Test zones are isolated with straddle packers or a bridge plug and a packer, but before the top packer is set, fluid should be circulated down the tubing to clear out any gas. Large gas volumes will increase the system compliance, resulting in long pressurization times and possible masked shut-in data. The preferred pressure gauge location is downhole on a wireline with surface readout capabilities for real-time decisions. Data should be obtained at least once per second for optimum analysis capability. For diagnostic purposes, a pressure gauge should be used to monitor the casing, and another pressure bomb should be set below the lower packer or bridge plug. These will provide data on possible communication between adjacent perforated zones or leaks past the packers. Although it may complicate the test procedures somewhat, a bottomhole closure mechanism is recommended. Bottomhole closure improves the data by reducing the volume of fluid in the tubing that is in communication with the fracture at shut-in and by separating the fracture and pressure transducer from any water-hammer pulses.

The test fluid is typically KCl water, but more viscous fluids may be required in higher-permeability formations (these tests are usually not reliable in formations with permeability much greater than 1 md). Low flow rates, in the range of 1 to 20 gal/min [4×10^{-3} to 76×10^{-3} m^3/min], are normally used. Optimum rates depend on depth, fluid, size of zone, perforation, and previous tests; no set rule can be offered *a priori*. Pumping should continue until the formation breaks down and the pressure stabilizes (not always possible) before shut-in is attempted. Volumes should be on the order of 1 to 20 gal [4×10^{-3} to 76×10^{-3} m^3] in the fracture; compliance of the fluid in the tubing may cause a much larger pumped volume to be essentially stored in the tubing. Tests should be repeated at least three times. Often six or seven tests will be needed before reproducible, understandable results are obtained.

The pressure response is often abnormal, and some diagnostic procedures may aid the interpretation. If the pressures during fracturing are high and the pressure drops at shut-in are large, the problem may be a flow restriction entering the fracture. This may be caused by either the perforations or the cement job. A good diagnostic test for this is a rate test. The fracture is only mildly sensitive to flow rate (¼ to ½ power; see Chap. 5), but a restriction will be highly rate sensitive (1 to 2 power).

In some tests, the ISIP may appear to change with every test, but usually monotonically. This seems to be typical in complex lithologies and may be a result of an ever-enlarging fracture contacting more layers. Often a volume-and-bleed test can help diagnose this situation. After several stress tests are conducted in which the ISIP has changed, the pressure should be lowered to a value much less than the ISIP so that the fracture can fully close and bleed back as much of the injected fluid as possible. This often takes about 30 minutes. Now, a very small stress test will usually give the same ISIP as the first test. This value can be verified by bleeding again and repumping. Alternatively, larger-volume tests should give different ISIP values and the effect of volume on ISIP can be investigated. The ISIP obtained from the smallest-volume tests is most representative of the stress in the perforated interval, but this complex behavior also warns that the stress value cannot be extrapolated over a large zone.

Interpretation of the pressure records is often the most difficult part of the stress-test procedure. Fig. 3.2 shows an ideal stress-test record where the ISIP is clear from visual inspection. This test was conducted in a shale at a depth of 8,060 to 8,062 ft [2456.7 to 2457.3 m]. Eight 6-g perforations were used and four tests conducted with KCl water (only the first three are shown). The ISIP here is 8,230 psi [56.7 MPa], with an uncertainty of only ±20 psi [±138 kPa].[29] While these data are the desirable product of a stress test, such clear results are obtained only about 25% of the time.

A more typical record[29] is shown in Fig. 3.3. This particular example is for a sandstone at a depth of 7,530 ft [2295 m] and water was injected through eight 14-g perforations at 0.25 bbl/min [39.7×10^{-3} m^3/min]. The value of the ISIP is 6,570 psi [45.3 MPa], which is equivalent to the in-situ stress. The ISIP becomes sharper and clearer with successive pumping; this appears to be typical of stress tests conducted through perforations.

Determination of the exact point of the ISIP for a pressure record like Fig. 3.3 can be attempted many ways. These include searching for changes in the first and second derivatives of the pressure record, plots of pressure vs. log of time or square root of time, log-log plots of pressure vs. time, and graphical tangent methods. None of these is clearly superior to the others, and in ambiguous situations all should be tried.

Step-Rate/Flowback Test Procedure. The second stress measurement technique, the step-rate/flowback test used by Nolte[9,30] and Smith,[31] is more applicable for determining stress over a larger permeable interval. An upper bound for the minimum stress can be found by a step-rate test. Fluid is injected into a previously in-

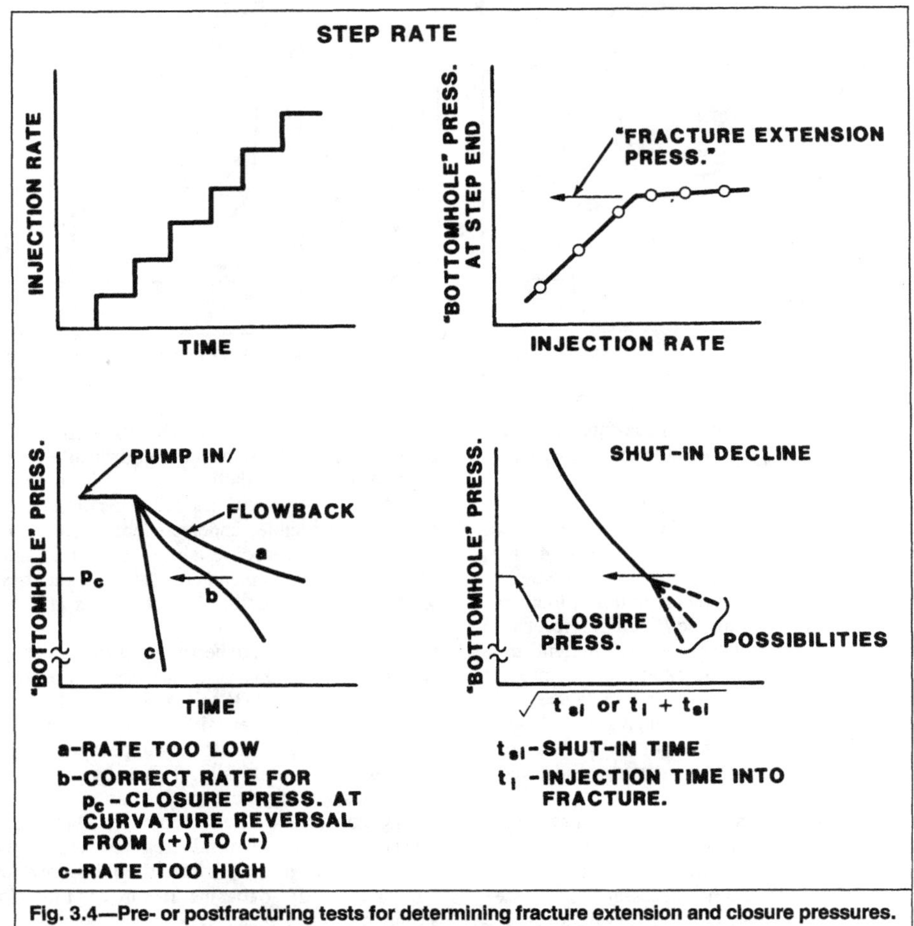

Fig. 3.4—Pre- or postfracturing tests for determining fracture extension and closure pressures.

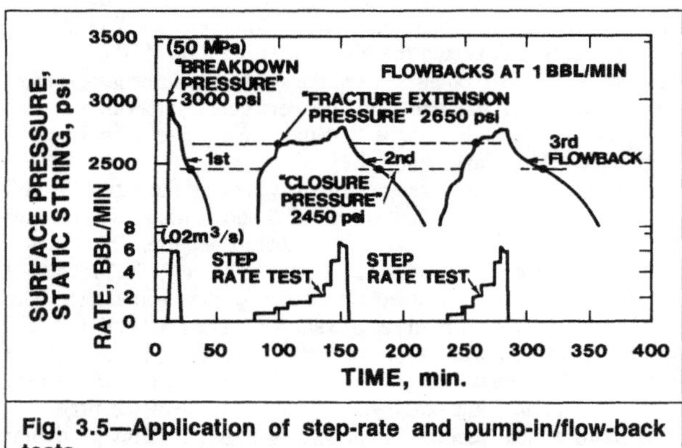

Fig. 3.5—Application of step-rate and pump-in/flow-back tests.

itiated fracture at various flow rates, and a "stabilized" pressure for each rate is recorded. The pressure is plotted vs. flow rate as shown in Fig. 3.4. The break point of the step-rate test is the extension pressure, which is larger than the closure pressure because of fluid friction in the fracture and a finite resistance to extension. The step-rate test can also be performed without stabilized pressure at each rate. Each injection rate can be maintained for a fixed period of time (5 to 10 minutes). Fig. 3.5 shows actual step-rate data; the extension pressure here is about 200 psi [1.4 MPa] above the closure stress.

While the step-rate test measures an extension pressure, and thus an upper bound for closure stress, the flowback portion of the test is the preferred method of determining the closure pressure, using the interpretation shown in Fig. 3.4. For this procedure, fluid of sufficient quantity and rate is injected to create a fracture (e.g., step-rate test) followed by an immediate flowback at a constant rate

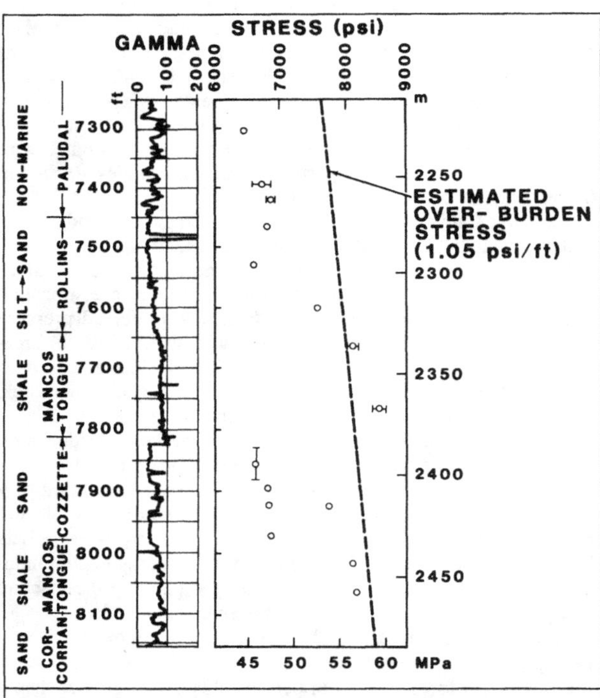

Fig. 3.6—Stress data in marine sandstones and shales, multiwell experiment, Piceance basin, CO.[29]

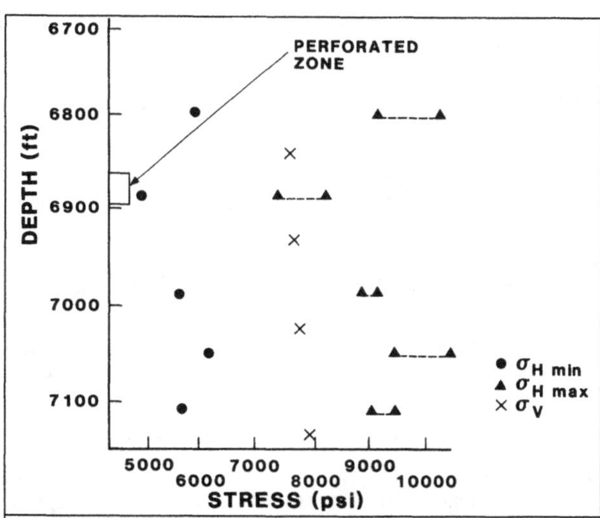

Fig. 3.7—Principal stresses in upper interval of Esso Kelly C-16-I 93-P1, Deep basin, Alta.

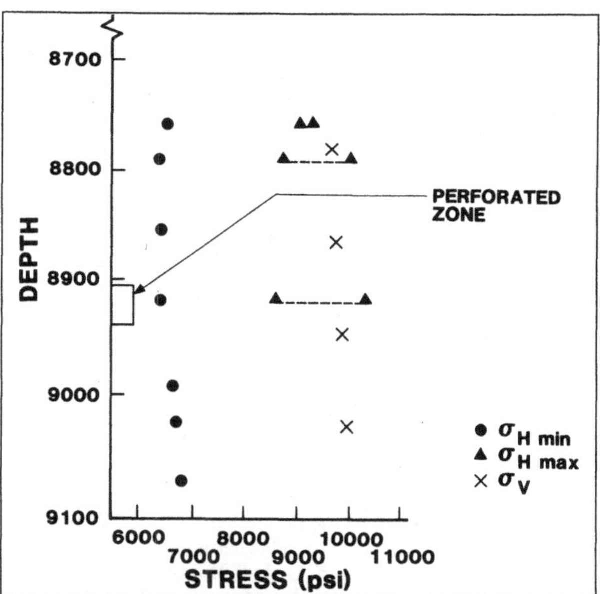

Fig. 3.8—Principal stresses in lower interval in Esso Kelly C-16-I 93-P1, Deep basin, Alta.

controlled by an adjustable valve or choke and recorded by an accurate low-rate flowmeter. A large valve is preferred over an adjustable choke to avoid plugging by debris in the fluid. If the flowback rate is within the correct range, the resultant pressure decline will show a characteristic reversal of curvature (must be from positive to negative) at the closure pressure. The accelerated pressure decline at the curvature reversal is caused by the flow restriction introduced when the fracture closes. The correct range of flowback rates must be determined by trial and error for any specific field; however, the range is on the order of 1/10 to 1/4 of the average injection rate. The effect of flow rates outside the correct range is shown in Fig. 3.4. Ungelled water can be used as the fluid for formations requiring massive hydraulic fracturing, while higher-permeability formations may require fracturing fluids to reduce the rate of fluid loss and to extend the fracture closure time during flowback. When applied in the field, the procedure should be repeated several times to ensure a consistent indication of closure pressure. Fig. 3.5 shows actual flowback data obtained during a series of tests.

Determination of the point of curvature reversal is often difficult to perform visually, so a preferred analysis technique is to fit the data with a least-squares polynomial routine. The reversal point is easily found when the first or second derivatives are calculated. A problem that may develop in permeable formations is an increase in closure stress with repeated testing because of poroelastic behavior.

The flowback test is the preferred stress measurement technique for determining the closure stress to be used in the various fracture pressure analyses (Chap. 14). It is a stress measurement over the entire interval, as is appropriate for pressure analyses.

Calculated Stresses. Finally, there have been many attempts to correlate the minimum principal in-situ stress with rock properties, particularly Poisson's ratio. This is usually performed by determining Poisson's ratio as a function of depth with a sonic log and calculating the minimum stress from[32]

$$\sigma_{H\min}=\frac{\nu}{1-\nu}(\sigma_z-sp)+sp. \qquad (3.10)$$

When Eq. 3.10 is compared with the more complete Eq. 3.7 and the possible effects of relaxation and diagenesis are considered, it is clear that Eq. 3.10 will seldom provide an accurate estimate of the in-situ stress. Nevertheless, Eq. 3.10 indicates that the horizontal in-situ stresses in a relaxed, normally pressured basin will typically be 0.6 to 0.7 psi/ft [13.6 to 15.8 kPa/m], and this is often observed; some success has been reported in using this approach.[33]

Stresses in Sedimentary Rocks. Most measurements of stresses in sand/shale sequences show that the shales generally have higher stresses than sandstones. Stress data[29] shown in Fig. 3.6 indicate that in some marine rock environments, the stresses in the shales may be 1,000 to 2,000 psi [6.9 to 13.8 MPa] greater than the stresses in the sands. Other results show both behavior similar to that in Fig. 3.7 and occasionally a nearly uniform stress throughout a section,[27] as shown in Fig. 3.8. In both Figs. 3.7 and 3.8, the perforated interval is a sand and the surrounding zones are thick shales. It should not immediately be assumed that the presence of shale bounding layers is indicative of large stress contrasts.

Effect of Reservoir Pressure Changes. Reservoir pressure changes are known to affect the in-situ stresses in the reservoir rock. This impacts hydraulic fracturing in two important ways: reservoir drawdown and leakoff of fracturing fluid.

Reservoir Drawdown. Salz[34] showed the relationship between fracture propagation pressure and pore pressure, and Cleary[35] and Geertsma[36] described the effect of a drawdown on the stresses in the reservoir rock. When the reservoir is drawn down and the pore pressure decreases, the drained portion tries to contract or to undergo volumetric shrinkage.[35] It is laterally constrained by the impermeable barrier layers above and below, however, and thus the attempted decrease in strain is converted to a decrease in stress. The result is a reduction of the lateral confining stress by an amount $\chi(p_i-p)$, where p_i is the initial reservoir pressure, p is the current reservoir pressure, and χ is a poroelastic coefficient, which depends in a most general sense on the porous-media parameters, the geometry, and the moduli of the various rock layers. Geertsma[36] simplified χ by assuming a constant vertical stress and zero lateral strain to estimate the change in closure stress as

$$\Delta\sigma_c=2\bar{\chi}(p_i-p), \qquad (3.11)$$

where $\bar{\chi}$ is given by

$$\bar{\chi}=\frac{(K_s-K)(1-2\nu)}{2K_s(1-\nu)}, \qquad (3.12)$$

with K being the bulk modulus of the reservoir rock and K_s the bulk modulus of the pure solid component of the rock. Typical $\bar{\chi}$

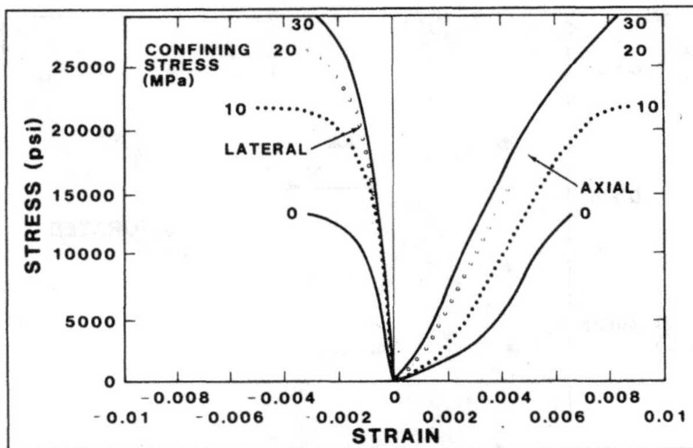

Fig. 3.9—Axial stress difference vs. axial and lateral strain for triaxial compression of sandstone for four confining stresses.

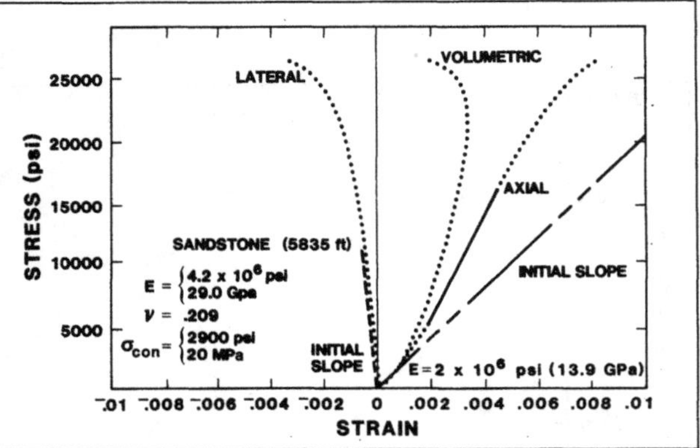

Fig. 3.10—Complete stress/strain data for triaxial compression of sandstone at 2,900-psi [20-MPa] confining stress.

values for sandstone are on the order of 0.3. The development of Eqs. 3.11 and 3.12 will be given in the discussion of poroelasticity.

Leakoff of Fracturing Fluid. Leakoff of fracturing fluid into the reservoir pore space will produce the opposite effect of a drawdown. Smith[31] and Cleary[37] described the increase in closure stress observed during a treatment and accounted for it through different approaches. Smith[31] considered a fracture in an oil well where the leakoff of a fracturing fluid will perturb the reservoir pressure over large distances. He decoupled the problem by assuming a form for the pore-pressure distribution in terms of an elliptical pressure distribution and then solved the resulting poroelastic equations. This resulted in a stress increase given by

$$\Delta\sigma_c = \frac{A\Delta P}{2+\zeta}, \quad \text{(3.13)}$$

where
A = poroelastic constant,
ΔP = treatment extension pressure minus the initial reservoir pressure, and
ζ = pressure distribution parameter calculated as

$$\zeta = \frac{ac'\sqrt{\pi}+\sqrt{a^2c'^2\pi+4}}{2}. \quad \text{(3.14)}$$

Here, a is the fracture half-height and c' is a porous-medium coefficient given by

$$c' = \sqrt{\frac{\phi\mu c}{(10.536\times10^{-4}\, kt)}}, \quad \text{(3.15)}$$

where ϕ=porosity and μ=viscosity.

Cleary[37] considered the effect of a treatment in a gas reservoir. While he suggests that a reservoir simulator should be used for accurate calculation of the stress change,[38] or back-stress as he calls it, he develops a relation for the back-stress as a function of the treatment pressure for the case where the pore-pressure distribution in the reservoir reaches a steady state:

$$\Delta\sigma = \overline{\chi}(P-p_i), \quad \text{(3.16)}$$

where $\overline{\chi}$ is given in Eq. 3.12 and P is the treatment pressure. Cleary found that this relation held for any shape of fracture and any pressure distribution in a homogeneous isotropic reservoir. The steady-state distribution is unlikely to develop for most treatments, however, and higher values of χ should be expected for shorter times.

3.4 Basic Rock Mechanics and Rock Properties

Elastic Properties. The primary rock properties of interest for hydraulic fracturing calculations are the elastic properties, particularly the stiffness of the rock. This usually defaults to the modulus of elasticity because most calculations are based on linear elasticity.

Linear Elasticity. The assumption that rock behaves as a linear elastic material provides a major simplification of the theory of elasticity with the result that many fracturing problems are tractable and have analytic solutions. As seen later, these solutions have been essential for the development of hydraulic fracturing theory. It should be remembered, however, that many rocks show considerable nonlinear behavior over the stress-loading range of interest, and the effect of the nonlinearity should be considered in certain instances.

The basic assumption of the theory of linear elasticity is that the components of stress are linear functions of the components of strain[2] and are expressed by the relations

$$\sigma_x = (\lambda+2G)\epsilon_x + \lambda\epsilon_y + \lambda\epsilon_z, \quad \text{(3.17a)}$$

$$\sigma_y = \lambda\epsilon_x + (\lambda+2G)\epsilon_y + \lambda\epsilon_z, \quad \text{(3.17b)}$$

and

$$\sigma_z = \lambda\epsilon_x + \lambda\epsilon_y + (\lambda+2G)\epsilon_z, \quad \text{(3.17c)}$$

where λ and G are known as Lamé's parameters. The equations are expressed in this form because one parameter, $\lambda+2G$, relates stress and strain in the same direction while another one, λ, relates stress and strain in orthogonal directions. In this context, G is the familiar shear modulus, but λ is seldom used in most engineering applications. Other parameters, such as Young's modulus (modulus of elasticity) and Poisson's ratio, are usually used.

Young's Modulus. Young's modulus is the ratio of stress to strain for uniaxial stress.[2] If σ_x is the direction of the uniaxial stress such that $\sigma_y=\sigma_z$=constant, it can easily be derived from Eqs. 3.17 that

$$E = \frac{\sigma_x}{\epsilon_x} = \frac{G(3\lambda+2G)}{\lambda+G}. \quad \text{(3.18)}$$

Young's modulus can be determined in the laboratory by use of triaxial compression tests with right cylindrical specimens, preferably with a length/diameter ratio of two or greater. The test is performed by applying a confining hydraulic pressure to the specimen, then loading it axially and measuring the displacement or strain. For determining modulus for fracturing calculations, the confining pressure is normally set equal to the mean effective stress acting on the reservoir rock. The value of E can be determined from the resultant stress/strain curve as $E=\Delta\sigma/\Delta\epsilon$. Examples of stress/strain curves for a sandstone under several confining stresses are shown in Fig. 3.9 (Simultaneously, lateral strains are also measured and will be used later to compute Poisson's ratio.) The value of E is obviously sensitive to the confining stress on the sample, and care must be taken to estimate it correctly. In addition, such factors as moisture content of the rock, strain rate of the test, and sample preparation procedures can result in significant deviations of the rock behavior.[39] Finally, the true in-situ modulus may be considerably different from this laboratory measurement because of the presence of joints. Generally, the laboratory measurement may be viewed as an upper bound for modulus.

As seen in Fig. 3.9, real rocks are not linear and there is no unique modulus. If the modulus of the rock at a particular stress level is desired, the tangent modulus, which is the tangent to the stress/strain curve at a given axial stress difference, can be computed. For rock behavior over a range of stresses or strains, a secant modulus is often used. In this case, a straight line is drawn through the stress/strain curve, intersecting it at the top and bottom of the stress range of interest. Alternatively, a best-fit straight line can be drawn through the data over the stress or strain range of interest. This last procedure is normally sufficient for rocks subjected to confining stresses greater than 1,000 or 2,000 psi [7 or 14 MPa].

An example of how Young's modulus might be calculated is given in Fig. 3.10. This data set is for the same sandstone shown in Fig. 3.9 at a confining pressure of 2,900 psi [20 MPa]. Young's modulus was first calculated by performing a least-squares fit of the axial stress/strain data between 20 and 60% of the failure stress. This resulted in a value of 4.2×10^6 psi [29 GPa]. Because the curve is so linear over this range, a tangent or secant modulus should have nearly the same value. For example, at 60% of failure stress (15,850 psi [109 MPa]), the strain is 0.0045; at 20% of failure stress (5,290 psi [36.5 MPa]), the strain is 0.0019. The secant modulus is then

$$E=\frac{\Delta\sigma}{\Delta\epsilon}=\frac{15{,}850-5{,}290}{0.0044-0.0019}=\frac{10{,}560\ \text{psi}}{0.0025}$$

$$=4.2\times10^6\ \text{psi [29 GPa]},$$

which agrees with the least-squares fit.

For hydraulic fracture applications, the pressures generated during the fracturing process are usually elevated only slightly above the in-situ stress (or confining pressure), and an initial tangent modulus might be more representative of the modulus needed for fracture design. In Fig. 3.10, an initial tangent line is drawn and the tangent modulus is calculated as

$$E_o=\frac{\Delta\sigma}{\Delta\epsilon}=\frac{20{,}300-0}{0.01-0}=2\times10^6\ \text{psi [14 GPa]}.$$

This rock is clearly not linear over the entire stress range. Such behavior is typical of many rocks and shows that care must be taken in using laboratory Young's modulus data. A factor-of-two difference in the Young's modulus used for design purposes can make a considerable difference in the treatment volume. Finally, Fig. 3.11a shows the tangent modulus calculated along the entire stress range. An ideal material would have a flat line so that the modulus is independent of stress.

One possible problem with the use of initial tangent Young's modulus data is anomalous behavior resulting from microcracks created during the coring process. Subjecting the sample to confining pressure may not close these microcracks entirely, and the initial Young's modulus may be abnormally low because of the large compliance of these open cracks. If this is the case, Poisson's ratio (discussed in the next section) may help diagnose this behavior; a rapidly changing Poisson's ratio at the start of the test will often be characteristic of closing cracks (Fig. 3.11b).

For a variety of elastic rocks, the initial tangent Young's modulus can be calculated as a function of confining pressure by[39]

$$E_0=K_0 p_a(p/p_a)^n, \quad\ldots\ldots\ldots\ldots\ldots\ldots\ldots\ldots\ldots (3.19)$$

where

K_0 = empirical "modulus number,"
p = confining pressure, and
p_a = atmospheric pressure.

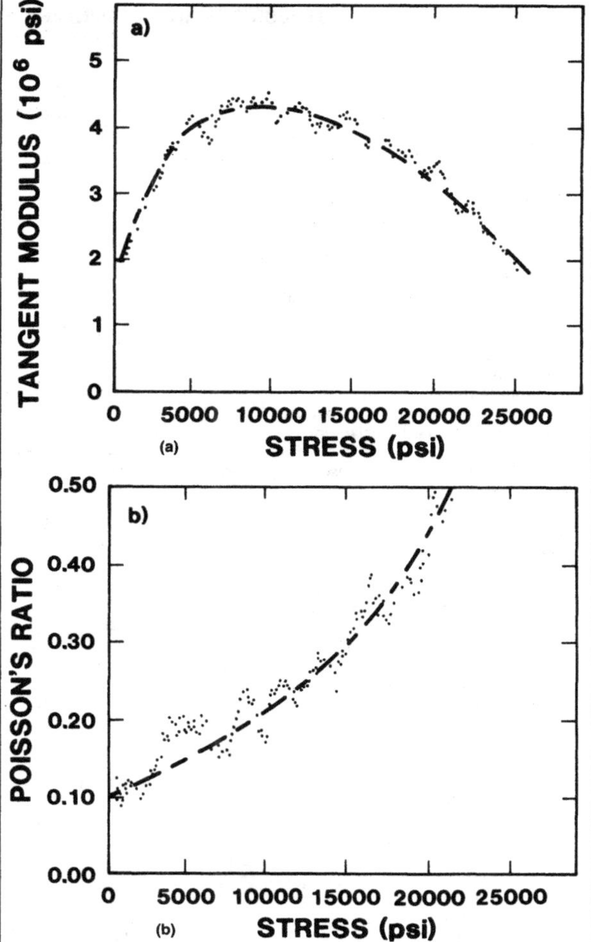

Fig. 3.11—Tangent modulus and incremental Poisson's ratio for triaxial compression of sandstone at 2,900-psi [20-MPa] confining stress.

Values for K_0 and n for several elastic rocks are included in Table 3.1.

For example, suppose a value of Young's modulus was needed for Berea sandstone at 2,000-psi [13.8-MPa] confining pressure. Table 3.1[39] gives a modulus number, K_0, of 43,600 and an exponent, n, of 0.25. According to Eq. 3.19,

$$E_0=43{,}600\times14.7\left(\frac{2{,}000}{14.7}\right)^{0.25}=2.2\times10^6\ \text{psi [15 GPa]}.$$

Young's modulus is most important for computation of fracturing pressures and the width profile of the hydraulic fracture. In addition, differences in the Young's modulus between the reservoir rock and barrier rocks may affect the height growth of the fracture.

Poisson's Ratio. Poisson's ratio is the ratio of lateral expansion to longitudinal contraction for a rock under a uniaxial stress condition. This is easily determined from Eqs. 3.17 as

$$\nu=\frac{\lambda}{2(\lambda+G)}. \quad\ldots\ldots\ldots\ldots\ldots\ldots\ldots\ldots\ldots (3.20)$$

In the laboratory, Poisson's ratio is also determined in a triaxial stress test by measuring the circumferential strain, the volumetric strain, or the lateral strain, as well as the axial strain, and then computing the lateral/axial strain ratio. Examples of volumetric and lateral strain data are shown in Fig. 3.10 for one of the samples used in Fig. 3.9. Poisson's ratio is also a function of the stress or strain (nonlinear), and a value at a particular stress or over a range of stresses needs to be computed.

TABLE 3.1—TRIAXIAL STRESS/STRAIN PARAMETERS FOR SEDIMENTARY (CLASTIC AND CHEMICAL) ROCK TYPES[39]

Description	Density (g/cm^3)	Specific Gravity	Porosity (%)	Modulus Number (in. × 10^3)	Exponent	Failure Ratio	Cohesion (MN/m^2)	Angle of Friction	Range of Confining Pressure (MN/m^2)
Stockton shale breccia, waxy to earthy	2.48	—	—	19.4	0.26	—	1.45	36 to 50	0.1 to 12.4
Berea sandstone, medium-grained, well-cemented	—	2.66	18.2	43.6	0.25	0.73	27.2	27.8	0 to 200.0
Week's Island sandstone, massive, hard, friable, fine-grained, well-cemented	—	—	—	5.96	0.39	0.66	55.2	27.5	0 to 203.0
Oil Creek sandstone, massive, very hard, very fine-grained, well-cemented	—	—	—	161.8	0.07	0.42	22.1	44.5	0 to 203.0
Bartlesville sandstone, massive, fine-grained, well-cemented	—	—	—	58.6	0.27	0.64	8.0	37.2	0 to 203.0
Pottsville sandstone, unweathered, almost pure silica	2.28	2.64	14.0	106.0	0.07	0.25	14.9	45.2	0 to 68.9
Boise sandstone, well-cemented	1.90	—	27.0	73.6	0.12	—	—	—	1.4 to 34.4
Mase sandstone, uniform, medium-grained	2.69	—	0.9	135.4	0.08	0.62	41.9	41.5	0.1 to 150.0
Mutenberg sandstone	—	—	—	127.0	0.07	—	—	—	0 to 243.0
Barnes sandstone, massive, fine-grained, well-cemented—parallel to bedding	—	—	—	48.0	0.07	0.60	8.6	34.0	0 to 203.0
perpendicular to bedding	—	—	—	62.4	0.10	0.52	8.2	34.0	0 to 203.0
Repetto siltstone, hard, fissile, dry	—	2.58	5.6	76.8	0.0	0.55	34.7	32.1	51.0 to 203.0
Repetto siltstone, hard, fissile, saturated	—	2.58	5.6	25.8	0.16	0.72	34.7	32.1	0 to 200.0
Stockton Northview shale, dense, silty, fine-grained	2.47	—	—	2.07	0.23	—	0.69	21.0	0.8 to 4.1
Stockton shale, soft, waxy	2.38	—	—	4.06	0.08	—	0.34	22.0	0.8 to 4.1
Muddy shale, hard, fine-grained, dry	—	2.67	4.7	3.57	0.43	0.36	35.4	55.5	0 to 203.0
Muddy shale, hard, fine-grained, saturated	—	2.67	4.7	85.6	0.01	0.36	38.4	14.4	0 to 203.0
5,900 ft Sands formation shale, hard, fissile	—	—	—	94.6	0.12	0.78	24.7	23.8	0 to 203.0
Edmonton clay shale, $w=20\%$	—	—	—	0.18	0.80	0.55	0.1	28.0	0.1 to 0.8
Edmonton bentonite shale, $w=30\%$	—	—	—	0.22	0.68	0.78	0.3	7.5	0.1 to 3.1
Green River shale, hard, calcareous— parallel to bedding	—	—	—	0.06	1.22	0.84	0.0	46.6	0 to 203.0
perpendicular to bedding	—	—	—	54.9	0.11	0.81	24.9	22.6	0 to 203.0
Green River Shale-1, fine-grained, brittle, calcitic and dolomitic, interbedded with kerogen— parallel to bedding	—	—	—	131.2	0.10	0.66	73.1	29.0	6.9 to 172.0
15° to bedding	—	—	—	92.5	0.15	0.64	62.0	30.0	6.9 to 172.0
20° to bedding	—	—	—	115.6	0.08	0.46	53.8	30.0	6.9 to 172.0
30° to bedding	—	—	—	85.0	0.14	0.50	44.1	30.5	6.9 to 172.0
45° to bedding	—	—	—	97.9	0.12	0.57	55.8	30.5	6.9 to 172.0
60° to bedding	—	—	—	117.0	0.11	0.49	59.3	30.0	6.9 to 172.0
75° to bedding	—	—	—	61.8	0.21	0.45	59.3	30.5	6.9 to 172.0
perpendicular to bedding	—	—	—	85.6	0.13	0.38	62.7	30.5	6.9 to 172.0
Green River Shale-2, fine-grained, plastic, calcitic and dolomitic, interbedded with kerogen— parallel to bedding	—	—	—	50.3	0.13	0.64	44.5	21.0	6.9 to 172.0
10° to bedding	—	—	—	67.3	0.04	0.74	41.4	20.5	6.9 to 172.0
20° to bedding	—	—	—	56.5	0.07	0.75	34.5	19.9	6.9 to 172.0
30° to bedding	—	—	—	29.9	0.11	0.63	29.0	18.0	6.9 to 172.0
40° to bedding	—	—	—	27.9	0.15	0.65	31.7	19.0	6.9 to 172.0
60° to bedding	—	—	—	13.9	0.25	0.69	36.5	20.6	6.9 to 172.0
perpendicular to bedding	—	—	—	78.9	0.03	0.80	38.6	20.7	6.9 to 172.0
Devonian limestone, heterogeneous, coarse-grained	—	—	—	123.1	0.12	0.96	20.6	33.6	0 to 203.0
Fusselman limestone, heterogeneous, coarse-grained	—	—	—	63.3	0.20	0.60	11.1	32.7	0 to 203.0
Wolf Camp limestone, heterogeneous, fine-grained	—	—	—	77.8	0.28	0.63	23.6	34.8	0 to 203.0
Marianna limestone, massive, friable, dry	—	2.70	13.0	107.5	0.0	0.34	26.8	26.4	0 to 203.0
Marianna limestone, massive, friable, saturated	—	2.70	13.0	59.9	0.09	0.80	12.6	37.6	0.3 to 4.4
Wells Station limestone, heterogeneous, fine-grained	—	—	—	455.0	0.0	—	—	—	20.6 to 98.0
Solenhofen limestone, homogeneous	—	—	—	594.0	0.0	—	—	—	0.1 to 1,013.0
Solenhofen limestone, homogeneous at 25°C	—	—	—	315.0	0.06	—	—	—	0 to 500.0
Limestone	—	2.64	—	544.0	0.02	—	—	—	0.1 to 100.0
Indiana limestone, oolitic	2.20	2.70	19.4	44.5	0.18	0.54	6.72	42.0	0 to 9.6
							29.6	7.0	9.6 to 68.9
Crown Point limestone	—	—	—	52.1	0.25	0.58	86.0	21.3	20.0 to 180.0
AEC Nevada site limestone, dense, fine-grained	270	2.72	0.5	742.0	0.0	0.33	14.5	44.0	0 to 27.6
Blair dolomite, homogeneous, fine-grained	—	—	—	168.6	0.16	0.50	35.9	39.0	0 to 203.0
Clear Fork dolomite, coarse to fine-grained	—	—	—	196.5	0.22	0.39	73.1	35.0	0 to 203.0
Fusselman dolomite, heterogeneous, fine-grained, calcitic	—	—	—	86.9	0.26	0.60	48.4	39.5	0 to 203.0
Glorieta dolomite, heterogeneous, medium-grained, calcitic	—	—	—	60.5	0.29	0.74	25.8	35.0	0 to 203.0
Luning dolomite, fine-grained, calcitic	—	—	—	101.3	0.21	0.88	23.7	34.0	0 to 203.0
Hasmark dolomite, homogeneous, dry, coarse-grained—parallel to foliation	—	2.91	3.5	176.2	0.11	0.86	23.1	32.1	101.0 to 203.0
perpendicular to foliation	—	2.91	3.5	153.6	0.13	0.81	45.6	30.6	0 to 203.0
Hasmark dolomite, homogeneous, coarse-grained, saturated	—	2.91	3.5	88.0	0.17	0.61	22.8	35.5	0.8 to 5.9
Stockton dolomite and dolomite breccia, calcareous, medium- to fine-grained	2.56	—	—	231.0	0.02	—	3.45	61.0	0.05 to 12.4
Stockton dolomite with shale seams, laminated	2.56	—	—	56.6	0.27	—	0.69	51.0	0.4 to 12.4
Stockton dolomite with stylolites, clay-filled	2.56	—	—	68.9	0.32	—	0.76	56.0	0.8 to 4.1
Chalk, 95% $CaCO_3$	1.62	2.72	40.0	0.13	0.67	0.54	0.0	31.5	10 to 90.0
Blaine anhydrite, fine-grained	—	—	—	93.2	0.10	0.78	43.4	29.4	0 to 203.0

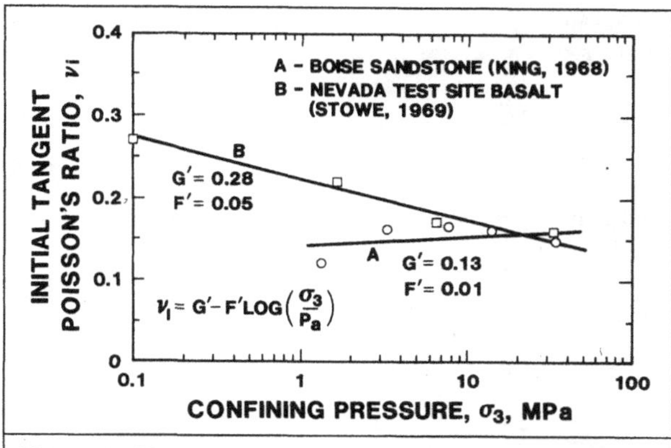

Fig. 3.12—Variations of initial tangent Poisson's ratio with confining pressure.[39]

Poisson's ratio can be calculated from the stress/strain data in Fig. 3.10, but both the lateral and axial strains are required. Again, least-squares fits of both curves were performed between 20 and 60% of the failure stress, resulting in a value of 0.209. Because the lateral stress/strain curve is also fairly linear in this range, a secant or tangent Poisson's ratio should be similar. For example, at 60% of failure stress, the lateral strain is −0.0007, while at 20% of failure stress, the lateral strain is −0.00025. From the axial strain data given in the previous section, a secant Poisson's ratio is calculated as

$$\nu=\frac{-\Delta\epsilon_y}{\Delta\epsilon_x}=\frac{-[-0.00075-(-0.00025)]}{0.0045-0.0018}=0.2.$$

The initial tangent Poisson's ratio can be found by comparing values at some point on the two initial slope lines. At 10,000 psi [69 MPa], the value of apparent axial strain is 0.0043 and the apparent lateral strain is about −0.0005, while both are zero at zero stress. Therefore, the initial modulus is given by

$$\nu=\frac{\Delta\epsilon_y}{\Delta\epsilon_x}=-\frac{(-0.0005-0)}{0.0045-0}=0.12.$$

This again shows that the rock is nonlinear and that considerably different values of Poisson's ratio can be generated, depending on the stress range chosen. Fig. 3.12 shows tangent Poisson's ratio calculated along the entire stress range.

The effect of confining pressure on Poisson's ratio is illustrated in Fig. 3.12.[39] Boise sandstone shows only a small dependence of Poisson's ratio on confining pressure. However, other sandstones often show a much greater effect.

Poisson's ratio, though less important than Young's modulus, is needed for calculation of the fracture-width distribution. Additionally, it is important for all calculations of the in-situ stress distribution in the reservoir, whether virgin stresses or an altered stress state.

Shear and Bulk Moduli. Two other moduli useful for rock mechanics modeling are the shear modulus, G, and the bulk modulus, K. The shear modulus arises naturally from linear elasticity, as seen in Eqs. 3.17, but it is not easily measured in the laboratory and is generally computed from E and ν by

$$G=\frac{E}{2(1+\nu)}. \quad (3.21)$$

In many hydraulic fracture width models, G and E are often interchanged through Eq. 3.21.

The bulk modulus, K, is the ratio of hydrostatic pressure to the volumetric strain it produces. K is related to λ and G through

$$K=\lambda+2/3G \quad (3.22)$$

TABLE 3.2—SURFACE ENERGIES OF ROCKS

	At Atmospheric Pressure	
	γ (10^4 ergs/cm^2)	Source
Carthage limestone	3.8	Perkins and Bartlett[43]
Lueders limestone	1.9	Perkins and Bartlett[43]
Indiana limestone	4.2	Perkins and Bartlett[43]
Arizona sandstone	12.7	Perkins and Bartlett[43]
Lueders limestone	1.7	Perkins and Krech[44]
Carthage limestone	3.5	Perkins and Krech[44]
Tennessee sandstone	8.8	Perkins and Krech[44]
Lueders limestone	1.1	Friedman *et al.*[42]
Indiana limestone	1.8	Friedman *et al.*[42]
Coconino sandstone	2.5	Friedman *et al.*[42]
Tennessee sandstone	3.8	Friedman *et al.*[42]
Chilhowee quartzite	5.0	Friedman *et al.*[42]
Chelmsford granite	5.0	Forootan-Rad and Moavenzadeh[45]
Danby marble	5.0	Forootan-Rad and Moavenzadeh[45]
	At 2,900-psi Confining Stress	
Lueders limestone	21.9	Perkins and Krech[44]
Carthage limestone	15.8	Perkins and Krech[44]
Tennessee sandstone	20.1	Perkins and Krech[44]

and is often calculated from E as

$$K=\frac{E}{3(1-2\nu)}, \quad (3.23)$$

but it can be measured in the laboratory by measuring the volume change during a hydrostatic compression. The reciprocal of K is called the compressibility. Most poroelastic calculations require K because the compressibility of the rock is a major factor in the response of the rock.

Poroelasticity. The previous discussion covered the response of rocks to external forces; in most physical situations, however, rocks are also subjected to internal (or body) forces, such as pore pressure and temperature. Pore pressure effects on elastic deformations of porous materials have been discussed by several authors,[11-15] with a recent work by Rice and Cleary[16] giving a thorough development, including the most general case of coupling the elasticity and diffusion equations. For a complete analysis, this coupling is required because, as mentioned earlier, permeability is sensitive to changes in stress. However, this coupling will not be discussed here.

For an isotropic, porous material, Eqs. 3.17 can be generalized as

$$\epsilon_x=\frac{1}{E}\sigma_x-\frac{\nu}{E}(\sigma_y+\sigma_z)-\frac{1-2\nu}{E}(1-K/K_s)p-\alpha T, \quad (3.24)$$

where, as before, compressive stress and strain are treated as positive and two additional equations are derived by a cyclic perturbation of the x, y, and z coordinates. In addition to the elastic constants discussed above, K_s is the bulk modulus of the minerals making up the matrix of the rock, α is the coefficient of linear thermal expansion, and T is temperature. Eqs. 3.17 can also be rewritten in terms of effective stress, $\sigma'=\sigma-sp$, where s is given by $(1-K/K_s)$.

Comparing Eq. 3.24 with the usual elastic constants shows that one additional laboratory measurement, K_s, is required. K_s is usually determined by subjecting samples to a hydrostatic confining pressure, allowing the pressurizing fluid to penetrate the rock, and measuring the resulting volume change of the sample.[2]

Physically, the poro/thermal-elasticity relations imply that for an increase in pore pressure, a volumetric expansion will occur because of the reduction in effective stress caused by the increased pressure, with the magnitude of the expansion reduced by a volume reduction of the rock grains caused by the increased pressure. It is also interesting to note the similarity between pore-pressure and temperature effects.

Under in-situ conditions, volume changes cannot occur because the reservoir is bounded by impermeable rocks where pore pressure is not changing. Therefore, the in-situ stresses must change for Eq. 3.24 to be satisfied. As an example, consider the simple case of pore pressure changing over an area that is large compared with the reservoir thickness.

For this case, the lateral strains are zero and Eq. 3.24 can be used directly to give

$$\Delta\sigma_x=\Delta\sigma_y=\frac{1-2\nu}{1-\nu}(1-K/K_s)\Delta p, \quad \text{(3.25)}$$

where x and y are assumed to be the lateral coordinates of the formation. More complex problems obviously require more sophisticated calculations. Perkins and Gonzalez[40] gave excellent examples of how changes in reservoir pressure and temperature alter in-situ stresses.

Fracture Mechanics. The theory of hydraulic fracturing depends on an understanding of crack behavior in a rock mass at depth. Because rock is predominantly a brittle material, most efforts to understand the behavior of crack equilibrium and growth in rocks have relied on elastic, brittle fracture theories.

Griffith's Theory and Surface Energy. Griffith[41] advanced the first plausible theory of crack behavior while studying the reasons for the low tensile strength observed in brittle materials, such as glass. He suggested that the low tensile strength observed in glass was a result of the presence of a population of cracks, and he attempted to analyze their behavior under tensile-loading conditions.

Griffith assumed that the microcracks were elliptic with a small minor axis and used an energy analysis to equate the work performed during extension of the crack to an energy ascribed to the newly created crack surface—the surface energy. For an elliptic crack in plane strain under a simple tensile-loading condition, the work to extend a crack of half-height a by an amount da is given by

$$dW=\frac{-\pi\sigma^2(1-\nu^2)}{E}ada, \quad \text{(3.26)}$$

and this was equated to the newly released surface energy (for two new faces),

$$dW=2\gamma da, \quad \text{(3.27)}$$

where γ is the surface energy. Now a critical value of stress for crack growth could be solved by

$$\sigma_c=\sqrt{\frac{2E\gamma}{\pi(1-\nu^2)a}}. \quad \text{(3.28)}$$

Friedman *et al.*[42] gave a good review of surface energies for rocks. Some of their data are excerpted in Table 3.2.

Barenblatt's Theory and Cohesion. Barenblatt[46] felt that Griffith's theory was inadequate because an overall elliptic fracture shape leads to unrealistic infinite stresses at the crack tip for uniformly loaded cracks in equilibrium. He proposed a model leading to the same crack-extension criterion while eliminating the singularity at the crack tip. Barenblatt recognized that there would be very large, attractive molecular forces at the crack tip that he called cohesive forces. He postulated that these forces act only in small zones near the crack tip and would tend to pull the crack faces together. Neglecting for a moment any external loading, these cohesive forces would result in a stress singularity at the tip, but one that is compressive in nature. In his theory, he deduces that this compressive stress singularity is exactly equal to the tensile stress singularity at the edge of the cohesion zone caused by external loads so that the two effects cancel and no singularity occurs. A simple rederivation of Barenblatt's analysis is given in Ref. 47.

Barenblatt suggested that there is a limit to the cohesive forces that could be characterized as a material property that he called the modulus of cohesion, K'. He ascribed all the work of the cohesive forces to surface energy as

$$K'=\sqrt{\frac{\pi\gamma E}{(1-\nu^2)}}. \quad \text{(3.29)}$$

Under simple loading conditions of uniform pressure in the crack,

$$\frac{K'}{\pi}=\sigma\sqrt{\frac{a}{2}}, \quad \text{(3.30)}$$

so that cracking will occur when

$$\sigma=\sigma_c=\sqrt{\frac{2}{a}\frac{K'}{\pi}}=\sqrt{\frac{2\gamma E}{\pi(1-\nu^2)\alpha}}. \quad \text{(3.31)}$$

This is exactly equivalent to the Griffith criteria.

Analysis of the displacements near the crack tip showed that when the compressive (cohesive) and tensile (load) stress singularities cancelled, smooth closure of the crack tip occurred. This is an important aspect of Barenblatt's solution and explains why the singularity is not present under equilibrium conditions. This condition is usually written as

$$\left.\frac{\partial w}{\partial x}\right|_{x=L}=0. \quad \text{(3.32)}$$

Barenblatt's model with the smooth closure of the crack tip will be invoked for several of the hydraulic fracture analyses discussed in later sections.

Linear Elastic Fracture Mechanics and Fracture Toughness. Linear elastic fracture mechanics (LEFM) is related to Griffith's theory, but was modified by Orowan[48] and restated by Irwin[49] to include dissipative energy processes, such as plastic flow and microcracking. These factors are incorporated through the stress intensity factors, K_I, K_{II}, and K_{III}, which quantify the intensity of the stress singularity at a crack tip. LEFM states that a fracture will advance when its stress intensity reaches a critical value, K_{Ic}, assuming that the crack tip is in a state of plane strain. K_{Ic} is known as the plane-strain fracture toughness and has been shown to be a measurable material property for metals, glasses, ceramics, polymers, and many other engineering materials. Schmidt[50] and Schmidt and Huddle[51] have shown that LEFM is well suited to the study of crack behavior in rocks as well.

Irwin classified three different singular stress fields according to the displacement. Mode I is opening, Mode II is in-plane sliding, and Mode III is antiplane sliding of the crack surfaces. For most problems in hydraulic fracturing, only the opening mode is used and this section will be restricted to the effect of K_I. It can be shown[52] that the stress-intensity factor, K_I, near the crack tip is related to the applied stresses through

$$\begin{Bmatrix}\sigma_x\\ \\ \tau_{xy}\\ \\ \sigma_y\end{Bmatrix}=\frac{K_I}{\sqrt{2\pi r}}\cos\left(\frac{\theta}{2}\right)\begin{Bmatrix}1-\sin\frac{\theta}{2}\sin\frac{3\theta}{2}\\ \sin\frac{\theta}{2}\cos\frac{3\theta}{2}\\ 1+\sin\frac{\theta}{2}\sin\frac{3\theta}{2}\end{Bmatrix}, \quad \text{(3.33)}$$

where θ is the angle measured from the crack axis. The singularity is always $\sqrt{r}$ regardless of the applied stresses, and K_I is usually thought of as the strength of that singularity.

For a crack extending from $-a$ to $+a$ on the x axis, Rice[52] shows that the Mode I stress-intensity factor can be calculated by

$$K_I=\frac{1}{\sqrt{\pi a}}\int_{-a}^{a}p(t)\sqrt{\frac{a+t}{a-t}}\,dt, \quad \text{(3.34)}$$

TABLE 3.3—FRACTURE TOUGHNESS OF ROCK

	K_{Ic} (psi-$\sqrt{\text{in.}}$)	Source
Cozzette sandstone	1,430	Unpublished
Mesaverde fluvial sandstones	1,230	Unpublished
Mancos shale	1,300	Unpublished
Indiana limestone	845	Schmidt[50]
Westerly granite	2,365	Schmidt and Lutz[55]
Devonian shale	750 to 1,200	Jones *et al.*[53]
Green River oil shale	730 to 1,000	Costin[56]
Benson sand	1,440 to 1,580	Brechtel *et al.*[54]
Benson shale	530	Brechtel *et al.*[54]

an expression that is very useful for hydraulic fracture problems. In the vicinity of a uniform stress field, σ, in the y direction (equivalent to constant pressure in the crack), the equation easily reduces to

$$K_I=\sqrt{\pi a}\,\sigma, \quad (3.35)$$

and for failure to occur, this becomes

$$\sigma_c=\frac{K_{Ic}}{\sqrt{\pi a}}. \quad (3.36)$$

Irwin showed that this is identical to the Griffith theory because

$$K_{Ic}=\sqrt{\frac{2E\gamma_{\text{eff}}}{1-\nu^2}}, \quad (3.37)$$

except that γ_{eff} includes the original surface energy, γ, plus any other plastic or microcrack work. Thus, in the simplest case of purely brittle failure, all three theories are identical. The most general and extensively used one now is LEFM, however, for which failure occurs when

$$K_I=K_{Ic}. \quad (3.38)$$

Table 3.3 gives some representative values of fracture toughness for several rock types.

Irwin[49] actually formulated his solution in terms of the strain-energy release rate, $\mathcal{G}$, which is related to the fracture toughness through

$$K_I^2=\mathcal{G}\frac{E}{1-\nu^2}. \quad (3.39)$$

$\mathcal{G}$ is also used in some hydraulic fracture modeling applications because it is entirely equivalent to K_I and it is often interchanged.

Fracture toughness can be measured in the laboratory with any one of several techniques described in Refs. 51, 52, and 56 through 58.

General Fracture Mechanics Comments. The failure criteria given in Eqs. 3.28, 3.31, and 3.35 are not suitable for hydraulic fracture applications because of the unrealistic loading condition of constant pressure throughout the entire crack. Only cracks with very short lengths or very low pressures would be stable under these conditions. All three models can be modified to account for more reasonable loads, however, resulting in more stable crack behavior. Such a modification of the Barenblatt theory forms the basis for the Khristianovitch-Zheltov[5]/Geertsma-de Klerk[6] models discussed in Chap. 5. A similar modification to Eq. 3.35 could be made by incorporating a more reasonable pressure distribution into Eq. 3.34.

Friction. While most models of hydraulic fractures assume that the rock mass at depth is a homogeneous, isotropic solid, the true state often is significantly different from the ideal one because of such features as faults, joints, and bedding planes. Many of these features can be characterized as having a nearly complete lack of tensile strength. Under this condition, compressive stresses can be transmitted across the discontinuity, tensile stresses cannot be transmitted, and various amounts of shear stress can be transmitted, depending on the shear strength and frictional properties of the interface. Thus, it is important to understand the role that friction plays in the overall rock mass behavior.

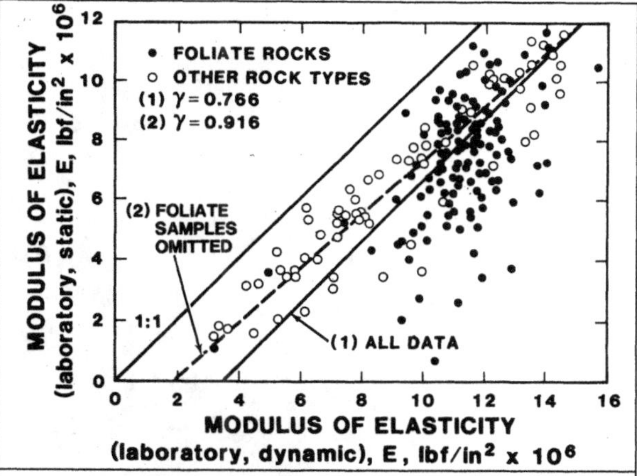

Fig. 3.13—Comparison of laboratory static and dynamic models.[39]

For most problems in rock mechanics, the shear strength of a surface can be represented to sufficient accuracy by

$$\tau=\tau_0+\mu_f\sigma_n, \quad (3.40)$$

where

τ_0 = inherent shear strength of surface,
μ_f = coefficient of friction, and
σ_n = effective normal stress on interface.

In many cases dealing with sliding interfaces, particularly where σ_n is large, τ_0 is small and can be neglected. Now the coefficient of friction is the only material property required to describe rock sliding behavior.

The coefficient of friction for most rocks has values ranging from 0.4 to 1.0, but this is caused mainly by surface roughness. The coefficient of friction increases with increasing surface roughness,[59] particularly at low effective normal stresses, because the roughness/friction relationship can be attributed to the interlocking of asperities along the surface.

If the two sliding surfaces are separated by a layer of unconsolidated or low-shear-strength material (gouge), then the frictional strength decreases considerably. Engelder *et al.*[60] found about a 20% reduction in the coefficient of friction for a quartz gouge and Shimamoto[59] showed that clay can reduce μ_f by an order of magnitude.

Friction may play an important role in naturally fractured reservoirs. It may also be significant for height growth in bedded media.

Ductility. Ductility is the ability of a material to undergo irreversible deformation while maintaining resistance to external loads. Ductility, or plastic behavior, is important for the study of cracks in metals,[52] but is generally ignored in brittle materials. Under very high compressive loadings, ductility is common in such rocks as salt, carbonates, coal, and shale, but under the tensile loads found near crack tips, ductility is generally confined to a very small region near the tip. Because no complete analysis of the role of ductility in a plane-strain Mode I (opening) fracture is available, the effects of ductility on crack growth are speculation based on the plastic analyses of Mode III cracks. Van Eekelen[61] suggests that some plastic energy dissipation will occur and may result in an apparent

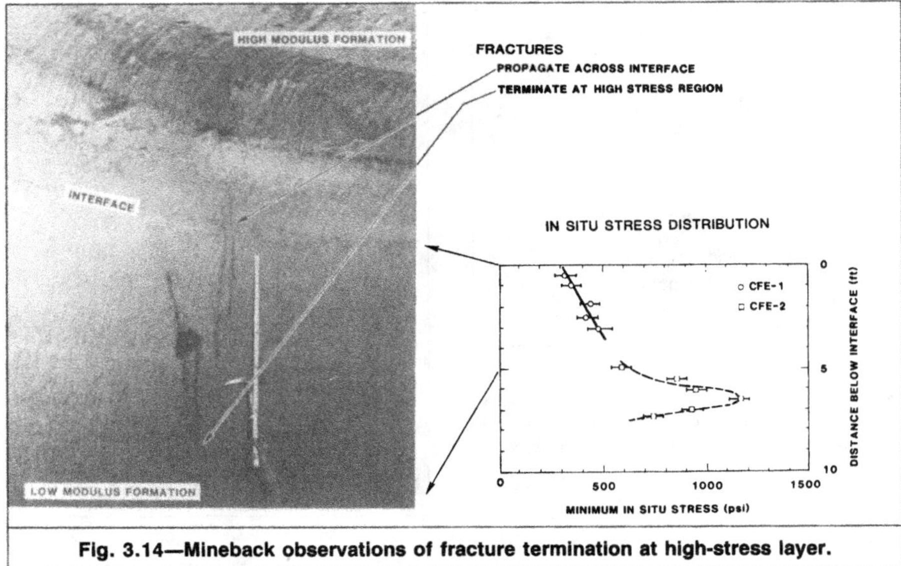

Fig. 3.14—Mineback observations of fracture termination at high-stress layer.

increase in the fracture toughness of a running crack (the resistance to fracture propagation increases with crack length until some maximum value is attained). He also thinks it likely, however, that the plastic or microcrack zone is small enough that it is already included in the measured value of K_{Ic} and therefore need not be accounted for further.

Medlin and Massé[62] take a somewhat different approach to the effect of plasticity. They suggest that Griffith's original surface energy should be reformulated as

$$\mathcal{E}=2\gamma+\alpha_p a, \quad\quad (3.41)$$

where

$\mathcal{E}$ = separation energy,
α_p = plasticity coefficient, and
a = crack half-height.

However, they were unable to determine reliable values for α_p for field size fractures. Their results suggest that highly plastic materials (possibly some shales) will blunt fracture growth.

Sonic Log Determination of Rock Properties. An attractive method of determining important rock mechanical constants is to calculate them from dynamic properties measured by logging techniques. If the compressional- and shear-wave velocities through the rock can be measured accurately with a long-spaced sonic log and if a density log can be obtained, then the elastic constants can be calculated from expressions given by Clark.[63]

For Poisson's ratio, Clark gives

$$\nu=\tfrac{1}{2}\frac{v_c^2-2v_s^2}{v_c^2-v_s^2}, \quad\quad (3.42)$$

where ν depends only on the velocities. Shear modulus is given by

$$G=\rho v_s^2, \quad\quad (3.43)$$

where ρ is the density. Young's modulus is equivalent to

$$E=\rho v^2\left(\frac{3v_c^2-4v_s^2}{v_c^2-v_s^2}\right), \quad\quad (3.44)$$

and the bulk modulus is

$$K=\rho(v_c^2-\tfrac{4}{3}v_s^2). \quad\quad (3.45)$$

Unfortunately, laboratory tests have shown that there is often a wide discrepancy between these dynamic moduli and static moduli measured in load frames, with the dynamic modulus generally higher, as seen in Fig. 3.13. This technique does have the advantage of measuring an in-situ property, however, rather than a core property that has been disturbed from its true in-situ state.

This technique is further complicated by the difficulty of measuring the shear-wave velocity. This is not a simple procedure, and small errors can lead to significant variations in the calculated elastic properties.

3.5 Effect of In-Situ Stresses and Rock Properties on Fracture Geometry

The purpose of hydraulic fracture design models is to calculate the volume of fluid and sand required to create a fracture of desired size and conductivity. To effect an acceptable fracture design, reasonable estimates or calculations of fracture height, width, and length must be attempted. In addition, the azimuth of the fracture is important for optimum well layout.[64] All these fracture-geometry parameters are influenced or controlled by the stress state and rock properties. This section reviews the current understanding of the relative importance of stress and rock properties on fracture geometry.

Fracture Azimuth. The effect of the in-situ stress field on fracture azimuth is well understood. Hubbert and Willis[8] showed that whenever the stress field is anisotropic, there is a preferred fracture azimuth perpendicular to the minimum, compressive, principal in-situ stress. Put simply, the fracture prefers to take the path of least resistance and therefore opens up against the smallest stress. Only under near-isotropy conditions in the three stresses will the rock fabric possibly be the dominant factor controlling fracture growth. Some laboratory results[65] suggest that a 200-psi [1.4-MPa] stress difference is sufficient to force fracture propagation in a preferred direction. A probable exception to this is fracture behavior in a jointed reservoir; under some reservoir conditions and joint orientations, the joint systems may strongly influence growth direction and overall geometry.[66]

Even when the stress difference is greater than 200 psi [1.4 MPa], problems may arise if the treatment pressures become greater than the intermediate stress. A possibility exists for secondary fractures to grow orthogonal to the principal fracture plane. This behavior may be enhanced by the presence of faults, fractures, or weak bedding planes.

Fracture Height. Conventional two-dimensional design models (see Chap. 5) require a value for fracture height so that width and length can be calculated with volume and flow considerations. More complex pseudo-three-dimensional models calculate fracture height, but some rationale is required for performing the height calculation. Several factors have been identified that contribute to the containment of hydraulic fractures.

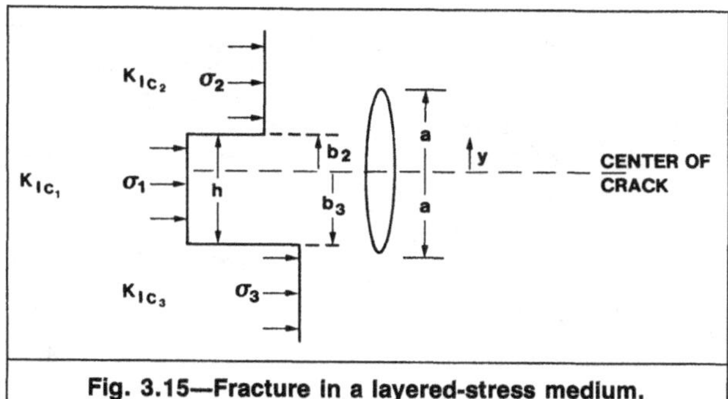

Fig. 3.15—Fracture in a layered-stress medium.

Evidence from production logs and other evaluation techniques has suggested that hydraulic fractures often terminate before propagating far into the bounding, impermeable (usually shale) layers, particularly in smaller treatments. These results have led to the conclusion that some property of the shale or the interface, or some property difference between the reservoir rocks and the abutting materials, hindered fracture growth. Much research has been focused on defining and quantifying this behavior.

In-Situ Stress Differences. Perkins and Kern[67] and Harrison *et al.*[68] suggested early on that stress differences between the pay zone and the bounding materials would have an important effect on fracture containment or restriction. Their intuition has been supported by theoretical,[38,69-75] laboratory,[76-81] and field data,[82,83] which show that the in-situ stress difference is the most important factor controlling fracture height.

An experiment clearly showing the dominant effect of the in-situ stress contrasts as opposed to rock properties was conducted in a tunnel where the fractures could be mined back for observation.[83] Hydraulic fractures with dyed water were initiated in horizontal holes in the vicinity of material property interfaces and stress contrasts. As seen in Fig. 3.14, fractures propagated upward into hard, high-strength, high-modulus materials, but they would not propagate downward through a thin, high-stress layer (stresses were measured with stress tests described in Sec. 3.2). The importance of stress contrasts was clearly seen in 20 separate fracture tests, while no similar major influence of material properties (modulus and strength) could be discerned.

Laboratory tests have demonstrated the same behavior. Tests of fracture behavior near material interfaces have not shown any obvious effect of property differences,[76] while stress contrasts have been shown to stop or to impede fracture growth.[78,80]

This is a favorable situation for hydraulic fracture containment because the bounding layers are often soft, clay-rich materials like shale with large stresses. Such materials should have high stresses because the material will be in near hydrostatic equilibrium[68]; the horizontal stresses should be close to the overburden stress.

The importance of the in-situ stress contrasts can be illustrated with a simple force balance. If Fig. 3.15 is considered as a symmetric case with $\sigma_2=\sigma_3$ and $b_2=b_3=h/2$ and the elasticity and strength of the rock are ignored, fracture height can be estimated by assuming that the fracture is in equilibrium so that the internal force caused by pressure is equal to the external force of the stresses. This results in

$$2Pa=\sigma_1 h+\sigma_2(2a-h), \quad \text{(3.46)}$$

which can be rearranged to give

$$2a=h\Delta\sigma/(\Delta\sigma-P_{net}), \quad \text{(3.47)}$$

where $\Delta\sigma=\sigma_2-\sigma_1$ and $P_{net}=P-\sigma_1$. Thus, a net pressure equal to half the stress difference would double the fracture height. While this simplistic approach is useful for illustration or mental calculations, a more complete analysis must be used for fracture design.

A simple yet important calculation of fracture height in a layered-stress medium can be made if material property variations are neglected and the vertical pressure distribution in the hydraulic fracture is assumed to be constant. This analysis was first proposed by Simonson *et al.*[69] for a symmetric geometry, but is easily generalized to more complex situations. Basically, the analysis is a calculation of the equilibrium height of a hydraulic fracture for a given internal pressure in a layered-stress environment. The stress-intensity factor is calculated at the top and bottom tips of the fracture and set equal to the fracture toughness of the materials, resulting in a unique height and position, or centering, of the crack with respect to the stress field.

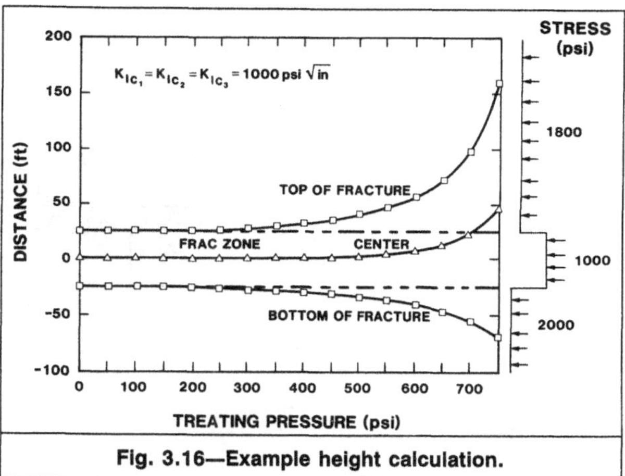

Fig. 3.16—Example height calculation.

For the geometry shown in Fig. 3.15, the stress-intensity factor at the top of the crack can be determined by

$$K_{Itop}=\frac{1}{\sqrt{\pi a}}\int_{-a}^{a}P(y)\sqrt{\frac{a+y}{a-y}}\,dy, \quad \text{(3.48)}$$

as given by Rice.[52] Here, a is the crack half-height and $P(y)$ is the net pressure distribution opening the crack. The net pressure distribution is

$$P(y)=P-\sigma_3 \text{ for } -a\leq y\leq -b_3, \quad \text{(3.49a)}$$

$$P(y)=P-\sigma_1 \text{ for } -b_3\leq y\leq b_2, \quad \text{(3.49b)}$$

and

$$P(y)=P-\sigma_2 \text{ for } b_2\leq y\leq a, \quad \text{(3.49c)}$$

with an additional geometry constraint of

$$b_3=h-b_2. \quad \text{(3.50)}$$

The integration of Eq. 3.48 and a similar one for the bottom tip yields two equations that can be solved for fracture height. After the two equations are added and subtracted, the final form is given as

$$\frac{\sqrt{\pi}\left(K_{Ictop}+K_{Icbottom}\right)}{2\sqrt{a}}=(\sigma_2-\sigma_1)\sin^{-1}\left(\frac{b_2}{a}\right)+(\sigma_3-\sigma_1)\times\sin^{-1}\left(\frac{b_3}{a}\right)-(\sigma_2+\sigma_3-2P)\frac{\pi}{2} \quad \text{(3.51)}$$

and

$$\frac{\sqrt{\pi}\left(K_{Icbottom}-K_{Ictop}\right)}{2}=(\sigma_2-\sigma_1)\sqrt{a^2-b_2^2}-(\sigma_3-\sigma_1)\sqrt{a^2-b_3^2}. \quad \text{(3.52)}$$

One note on the solution of these two equations is required. In general, one would like to calculate the height of the fracture given a pressure, P, but this would require iterative solution of the two equations. On the other hand, examination of Eqs. 3.51 and 3.52

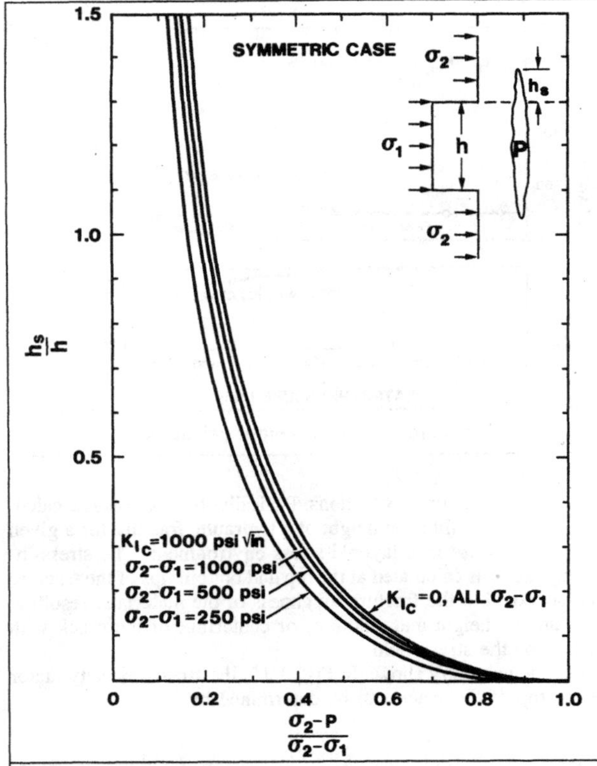

Fig. 3.17—Generic curve for height growth through barriers, symmetric case.

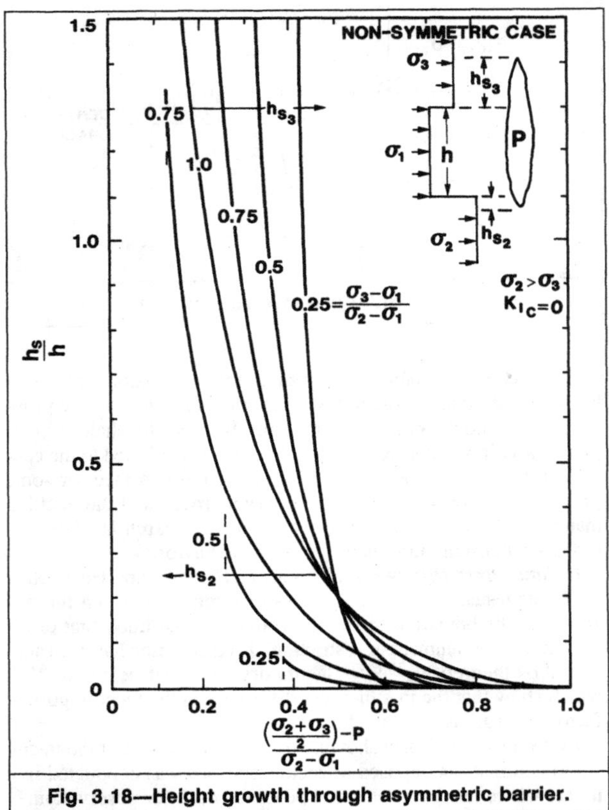

Fig. 3.18—Height growth through asymmetric barrier.

shows that the inverse calculation is much less involved. If a crack half-length, a, is assumed, the geometry factor, b_2, can be directly calculated from Eq. 3.52. Now the pressure can be calculated directly from Eq. 3.51. A few trial-and-error calculations with various a's—easily performed on a computer—result in the correct pressure and the solution to the problem.

Symmetric Case. For the symmetric case, $b_2=b_3$, $\sigma_2=\sigma_3$, and $K_{Ic\,top}=K_{Ic\,bottom}$, and we obtain the single equation given by Simonson *et al.*[69]:

$$\frac{\sqrt{\pi}\,K_{Ic}}{2\sqrt{a}}(\sigma_2-\sigma_1)\sin^{-1}\left(\frac{b_2}{a}\right)-(P-\sigma_2)\frac{\pi}{2}. \quad (3.53)$$

Eq. 3.53 is easily solved, because $b_2=h/2$.

Fig. 3.16 shows an example calculation for the nonsymmetric case, a plot of fracture height as a function of the treatment pressure above the closure stress. With the large confining stresses in the barrier zone, the fracture is restricted to a narrow height for treatment pressures less than 500 to 600 psi [3.4 to 4.1 MPa]. For greater pressures, the top of the fracture begins to grow excessively and becomes unbounded at 800 psi [5.5 MPa].

For the symmetric case, Eq. 3.53 can be rearranged as

$$\frac{\sigma_2-P}{\sigma_2-\sigma_1}=-\frac{2}{\pi}\sin^{-1}\left(\frac{h/2}{h+h_s}\right)+\frac{K_{Ic}}{\sqrt{\pi/(h/2+h_s)}\,(\sigma_2-\sigma_1)}, \quad (3.54)$$

where h_s is the distance the fracture has propagated into the bounding materials (see Fig. 3.17). If the fracture toughness of the material (or strength) is small compared with the stress contrasts, then the depth of penetration, h_s, normalized by the thickness of the zone, is a function of only the dimensionless pressure-contrast ratio, and a generic curve can be developed for any stress contrast or zone thickness. This is shown in the top curve of Fig. 3.17. Also shown is a family of curves for the case where K_{Ic} is not neglected. A different family of curves is needed for different K_{Ic} values, but the value used in Fig. 3.17 (1,000 psi-$\sqrt{\text{in.}}$ [1.1 MPa·$\sqrt{\text{m}}$]) is representative of many sedimentary rocks. The following example shows how Fig. 3.17 can be applied.

Example Problem. Suppose a 50-ft [15-m] -thick sandstone is to be fractured. The pay zone has a closure stress of 3,000 psi [20.7 MPa], while the shales above and below are both estimated or measured to be 3,500 psi [24.1 MPa]. However, 30 ft [9 m] above the pay is an aquifer that must be avoided. What kind of pressure in the crack can be supported without breaking through the upper shale into the aquifer?

Solution. In this case, $h_s/h=30/50=0.6$. If K_{Ic} is neglected, 0.6 intersects the $K_{Ic}=0$ curve in Fig. 3.17 at 0.3 so that

$$\frac{\sigma_2-p}{\sigma_2-\sigma_1}=0.3.$$

Because $\sigma_2-\sigma_1=500$ psi [3.4 MPa],

$$\sigma_2-p=0.3\times500 \text{ psi}=150 \text{ psi [1 MPa]}$$

and

$$p=\sigma_2-150=3{,}500-150=3{,}350 \text{ psi [23.1 MPa]}.$$

A bottomhole pressure (BHP) of 3,350 psi [23.1 MPa] can be supported without fracturing into the aquifer. If $K_{Ic}=1{,}000$ psi-$\sqrt{\text{in.}}$ [1.1 MPa·$\sqrt{\text{m}}$], the result is only slightly different because

$$\frac{\sigma_2-P}{\sigma_2-\sigma_1}=0.28.$$

The value of P here is 3,360 psi [23.2 MPa]. Generally K_{Ic} can be neglected.

The reverse case can also be calculated from Fig. 3.17. If a maximum pressure is calculated for a given fracture design, then

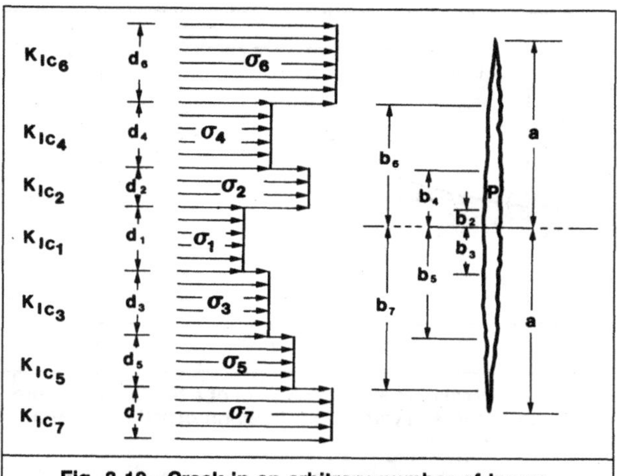

Fig. 3.19—Crack in an arbitrary number of layers.

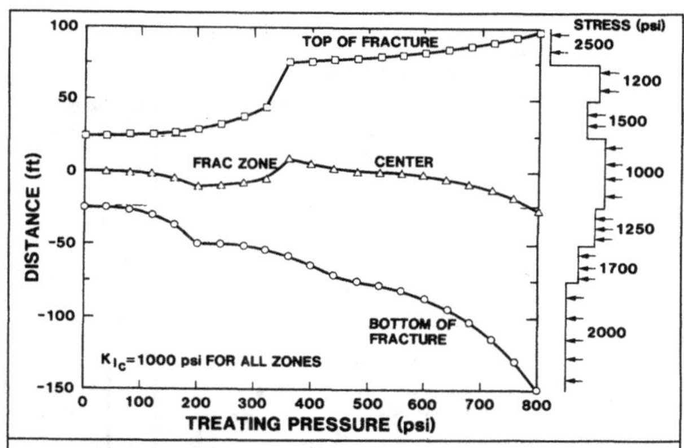

Fig. 3.20—Example calculation of height growth for a seven-layer model.

an estimate of fracture penetration can be made and the design refined.

Nonsymmetric Case. In general, the stresses in the layers above and below the pay zone will not be symmetric, and then Eqs. 3.51 and 3.52 must be solved. If K_{Ic} can be neglected again, then a family of curves can be developed that show the penetration for different ratios of asymmetry, as shown in Fig. 3.18. In this figure, h_{s2} is the penetration into the higher-stress layer and h_{s3} is the penetration into the lower-stress layer. While the diagram shows the lower-stress region on top, the opposite case is equivalent. Another example will show how Fig. 3.18 can be used.

Example Problem. Consider the same pay zone as the previous example with 50-ft [15-m] thickness and 3,000-psi [20.7-MPa] stress. The upper layer is again 3,500 psi [24.1 MPa], but the lower layer is 4,000 psi [27.6 MPa]. As before, an aquifer is 30 ft [9 m] above the pay. What pressure can be supported without breaking into the aquifer?

Solution. With $h_{s3}=30$ ft [9 m] and $h=50$ ft [15 m], $h_{s3}/h=0.6$. The ratio is

$$\frac{\sigma_3-\sigma_1}{\sigma_2-\sigma_1}=\frac{3{,}500-3{,}000}{4{,}000-3{,}000}=\frac{500}{1{,}000}=0.5.$$

A value of $h_s/h=0.6$ intercepts the 0.5 curve for h_{s3} at 0.39 so that

$$\frac{\dfrac{\sigma_2+\sigma_3}{2}-P}{\sigma_2-\sigma_1}=0.39.$$

With $\sigma_2-\sigma_1=1{,}000$ psi [6.9 MPa] and $(\sigma_2+\sigma_3)/2=3{,}750$ psi [25.9 MPa],

$$P=3{,}750-0.39\times1{,}000=3{,}750-390=3{,}360 \text{ psi [23.2 MPa]}.$$

A BHP of 3,360 psi [23.2 MPa] can be supported in this case. The penetration downward can also be estimated by the intercept of the 0.5 h_{s2} curve with 0.39. This occurs at an h_s/h value of about 0.2, so the downward penetration is $0.2\times50=10$ ft [3 m].

Occasionally, there may be circumstances where a more complex calculation requiring more layers is needed. This can be done for an arbitrary number of layers in the same way. Referring to the geometry in Fig. 3.19 (showing only seven of the layers), the same procedure as previously developed yields

$$\frac{\sqrt{\pi}\,(K_{Icm}+K_{Icn})}{2\sqrt{a}}=S_0\frac{\pi}{2}+\sum_{i=2,2}^{m} S_i\sin^{-1}\left(\frac{b_i}{a}\right)+\sum_{j=3,2}^{n} S_j\sin^{-1}\left(\frac{b_j}{a}\right) \quad\ldots\ldots(3.55)$$

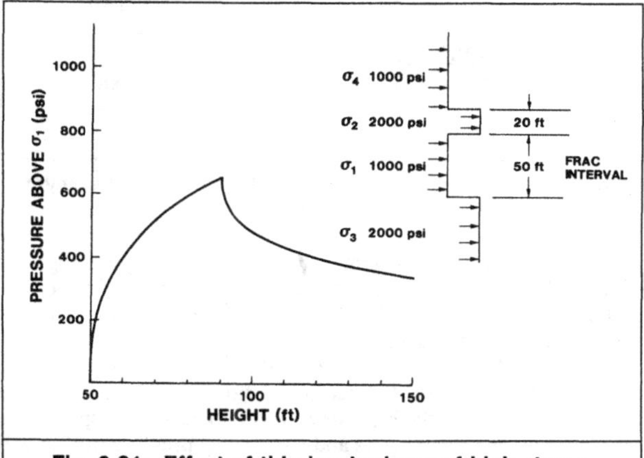

Fig. 3.21—Effect of thin barrier layer of high stress.

and

$$\frac{\sqrt{\pi a}\,(K_{Icn}+K_{Icm})}{2}=\sum_{i=2,2}^{m} S_i\sqrt{a^2-b_i^2}-\sum_{j=3,2}^{n} S_j\sqrt{a^2-b_j^2}. \quad\ldots\ldots(3.56)$$

In these equations, m is the layer in which the top crack tip resides and n is the layer in which the bottom crack tip resides; additionally, $S_2=\sigma_2-\sigma_1$, $S_3=\sigma_3-\sigma_1$, $S_i=\sigma_i-\sigma_{i-2}$ for $i>2$, $S_j=\sigma_j-\sigma_{j-2}$ for $j>3$, $\sigma_0=2P-\sigma_m-\sigma_n$, and $i=2,2$ (or $j=3,2$) means that i (or j) is incremented by 2. For example, if $m=6$, then the summation is for $i=2$, 4, and 6. These equations are solved exactly the same way as the simpler three-layer model. Fig. 3.20 shows an example calculation for the multilayer geometry.

If desired, the calculation can be performed with stress gradients rather than constant stresses in individual layers, but the problem becomes much more complex and the effect is usually small. Additionally, the available stress data will seldom warrant such detail. Palmer and Craig[74] described the model for a nonsymmetric three-layer case.

Performing such calculations for multiple layers does allow consideration of an additional factor. The three-layer models discussed above assume infinite barrier zones; however, for zones of finite thickness, some portion of the containment effect will be lost. This is illustrated by the example calculations in Fig. 3.21.

For this case, even though a barrier zone exists with higher closure stress, the finite thickness of the upper barrier limits the containment to a pressure level of only about one-half the stress difference. In

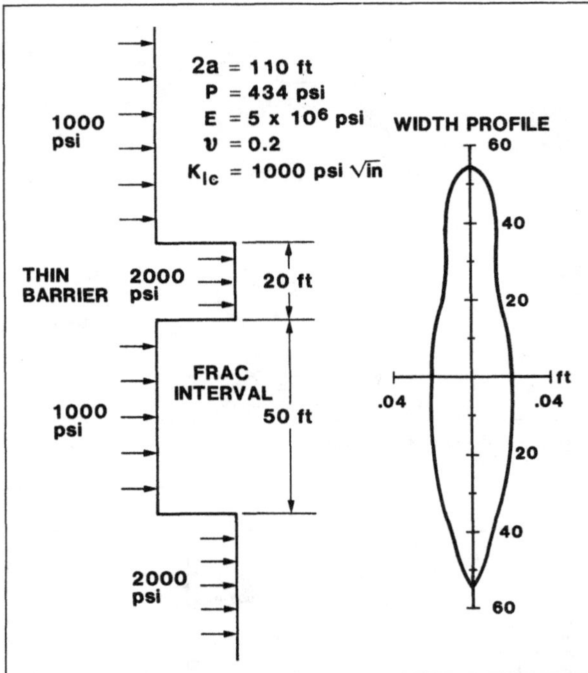

Fig. 3.22—Width distribution caused by a thin barrier layer.

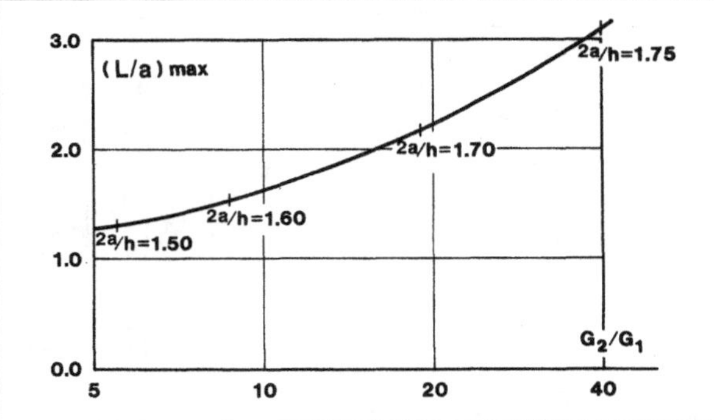

Fig. 3.23—Maximum elongation (L/a) of a hydraulic fracture in a three-layered system as a function of stiffness ratio.

addition to unbounded height growth for further pumping, the narrow interval can lead to a "pinch" point as seen in the cross-sectional width distribution in Fig. 3.22. This can lead to a premature screenout as discussed in Ref. 9.

These calculations for fracture height are conservative because they do not account for the additional flow resistance caused by the reduced width in the high-stress layers. Pressure drops will be large through the shale layers and the vertical pressure profile will not be uniform. Treatment of these conditions requires an analysis of the flow in the fracture.

One approach to flow-dependent height calculations is to use time-constant models that relate vertical growth to lateral propagation of the fracture. Cleary[37,71,72] developed one such model that showed the effect of the high-stress layers as

$$\frac{dL}{da} \propto \left(\frac{\tau_c^H}{\tau_c^L}\right)^{n'} = \left(\frac{\gamma_2}{\Gamma_2}\frac{P-\sigma_1}{P-\sigma_{2,3}}\right)^{n'+2}, \quad \ldots\ldots\ldots\ldots\ldots (3.57)$$

where

τ_c^H = time constant for height growth,
τ_c^L = time constant for length growth,
γ_2, Γ_2 = parameters of the Cleary model (see Chap. 5), and
n' = flow-consistency index of a power-law fluid.

The time-constant ratio is essentially equivalent to dL/da so that the penetration into the bounding layers (growth of a) can be estimated from the length propagation of the treatment.

Eq. 3.57 has one major drawback. While pressure, P, must always be greater than σ_1 (the minimum in-situ stress), fracture growth will extend into barrier zones for pressures less than the closure stress in these zones. For this case, $p-\sigma_{2,3}$ is negative and the analysis breaks down. The simplistic form of such time-constant models is inappropriate for calculating certain conditions—e.g., determining when a fracture may penetrate through a relatively thin barrier. For such cases, a characteristic time could be calculated with more sophisticated computer models, or vertical flow may be considered negligible and the equilibrium models used.

Young's Modulus. Young's modulus, while less important than stress contrasts, can restrict fracture growth in two ways. First, a fracture approaching an interface between two dissimilar materials will have its growth retarded slightly by the interface. The reason is that the stress-intensity factor at the tip of the crack will decrease somewhat as the interface is approached and crossed (alternatively, the tensile stresses that fracture the rock at the tip of the crack will decrease). However, this effect is small and will have little importance on field-size hydraulic fractures.

Second, Young's modulus can restrict fracture growth if the modulus of the boundary layer is greater than the modulus of the pay zone. The fracture width will be smaller in the high-modulus material and flow resistance will be higher, making fracturing more difficult. This is not usually the case in the field; moduli of sandstones are generally greater than those of shales and penetration is encouraged. As shown earlier, this trend is negated by the high stresses usually associated with shale layers.

Cleary[37,71,72] also developed a time-constant model for height growth as a function of modulus differences. He estimated that

$$\frac{dL}{da} \propto \left(\frac{\tau_c^H}{\tau_c^L}\right)^{n'} = \left(\frac{\gamma_2^F}{\gamma_2}\frac{E_{2,3}'}{E_1'}\right)^{n'+1}, \quad \ldots\ldots\ldots\ldots\ldots\ldots (3.58)$$

where γ_2^F and γ_2 are described in Ref. 71, E' is the plane-strain modulus, $E/(1-\nu^2)$, and τ_c^H and τ_c^L are as given before. Thus, higher moduli in Layers 2 and 3 result in a larger dL/da and reduced height growth.

Van Eekelen[61] formulated an approximate model to estimate vertical crack propagation in a layered material with modulus contrasts. He assumed symmetric layers with respect to modulus, $E_2'=E_3'$, and no leakoff of the fracturing fluid. He assumed a constant pressure in any vertical cross section, allowed only small penetration into the bounding layers, and approximated the cross-sectional shape of the fracture by fractional ellipses. He developed an effective modulus for the bounding layers by

$$E_2''=\left(E_1'E_2'\frac{1+\Delta h_{2h}}{1+\dfrac{E_1'\Delta h}{2E_2'h}}\right)^{1/2}, \quad \ldots\ldots\ldots\ldots\ldots\ldots\ldots (3.59)$$

where Δh is the fractional growth into both the upper and lower layers. This effective modulus would describe the fracture width if the material were homogeneous.

By use of a Khristianovitch-Zheltov/Geertsma-de Klerk[6] model for vertical growth, the ratio of vertical spreading to lateral spreading was found to be

$$\frac{da}{dL}=\frac{19}{3}\frac{La}{h^2}\left[\frac{E_1'\left(1+\dfrac{E_1\Delta h}{2E_2'h}\right)}{E_2'\left(1+\dfrac{\Delta h}{2h}\right)}\right]. \quad \ldots\ldots\ldots\ldots (3.60)$$

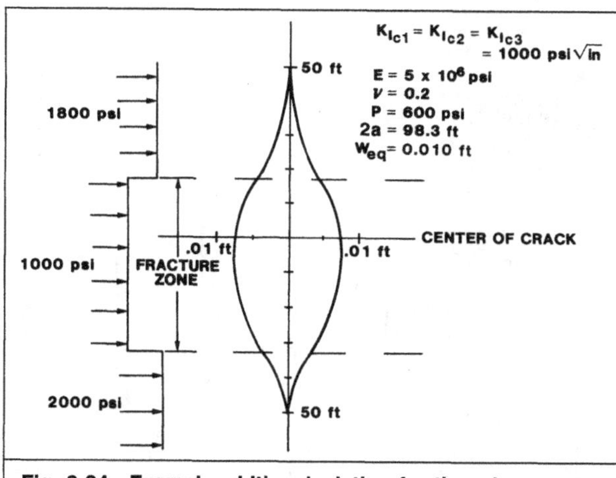

Fig. 3.24—Example width calculation for three-layer case.

If desired, integration of Eq. 3.60 yields

$$2L=h\left\{1+\frac{12}{19}\frac{E_2}{E_1}\left[\log\left(\frac{2a}{h}\right)+¼\left(3+\frac{E_1}{E_2}\right)\left(\frac{a}{h}-1\right)\right]\right\}^{½}. \quad \text{(3.61)}$$

Fig. 3.23 shows an example calculation of the modulus contrast required to keep a fracture relatively contained. In this example, shear modulus is used (see Eq. 3.21) instead of Young's modulus.

Fracture Toughness or Strength. As Fig. 3.17 or the associated example shows, the effect of toughness or strength is small, except in cases where the stress contrasts are minimal. Fracture toughness or strength contributions are often neglected.

Interface Slippage. In-situ stress contrasts can restrict fracture growth by clamping the fracture tip and reducing fracture width in the high-stress regions; moduli contrasts can slightly retard out-of-zone propagation by reducing the fracture width and thus restricting fluid flow. In both cases, the fracture must penetrate into the barrier region before any significant effect is felt. Interface slippage, when it occurs, can result in immediate termination of fracture growth and is obviously the most favorable containment barrier when it occurs.

Anderson,[77] Teufel and Clark,[78] and Teufel[79] studied the interface containment problem and found that it is controlled by the frictional shear stress acting on the interface plane. When the frictional forces are small, the tensional stress ahead of the crack cannot be easily transmitted across the interface and slippage is likely to occur. This effectively terminates fracture growth. When the frictional forces are large, the interface is transparent because stresses are readily transmitted across it. Because the frictional shear stress, as described by the linear friction law,

$$\tau=\mu_f\sigma_n, \quad \text{(3.62)}$$

depends on the effective normal stress, σ_n, acting on the interface, shear slippage is likely to occur only where this stress is small or the coefficient of friction, μ_f, is very small.

Under normal situations, this is likely only at very shallow depths where the overburden stress is small. Because the interfaces are generally horizontal, the normal stress is usually equal to the overburden. Unusual situations may occur, however, where the shear stress will be small at great depths. The two most obvious are overpressured reservoirs where the effective stress is low because the pore pressure is high, and interfaces with clay or gouge in them so that the coefficient of friction is negligible.

Fracture Width. While fracture width is generally calculated by the design model, some consideration should be given to width variations caused by stress and modulus contrasts. In one experiment,[82] a television camera showed that the width of the fracture in the pay zone was much greater than the width in the abutting shales because the higher stress in the shale clamps down on the fracture, resulting in narrower cracks.

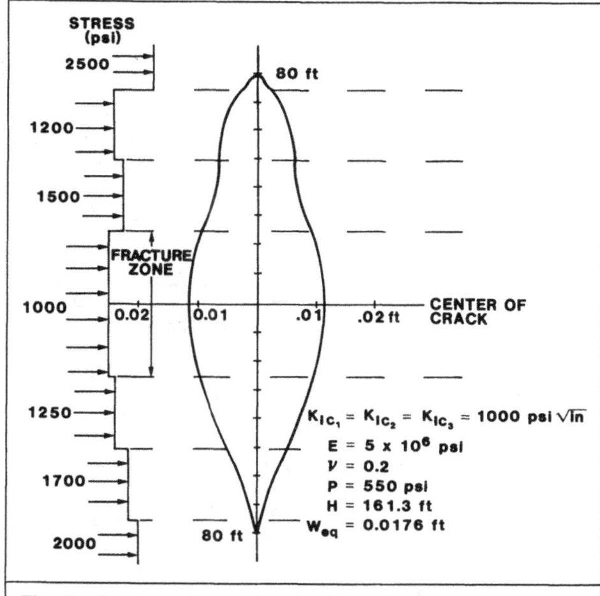

Fig. 3.25—Example width calculation for seven-layer case.

A somewhat involved calculation can be made to estimate width variations in a crack in a layered-stress medium. The analysis, which applies strictly to a crack with no slip at top and bottom, is based on England and Green's[84] formulation of crack width as an integral of even and odd functions. They showed that the width at any point y in a fracture could be determined as

$$w=\frac{-16(1-\nu^2)}{E}\int_{|y|}^{a}\frac{F(t)+yG(t)}{\sqrt{t^2-y^2}}\,dt, \quad \text{(3.63)}$$

where $F(t)$ and $G(t)$ are given by

$$F(t)=-\frac{t}{2\pi}\int_0^t\frac{f(u)du}{\sqrt{t^2-u^2}} \quad \text{(3.64)}$$

and

$$G(t)=-\frac{1}{2\pi t}\int_0^t\frac{ug(u)du}{\sqrt{t^2-u^2}}. \quad \text{(3.65)}$$

The net pressure opening the crack is

$$p(y)=-f(y)-g(y), \quad \text{(3.66)}$$

where f is an even function and g is an odd function.

The integration of Eq. 3.63 is tedious even for the three-layer case and requires considerable bookkeeping. An example calculation for the stress data used in Fig. 3.15 is shown in Fig. 3.24. The fracture is quite wide in the low-stress fracture zone, but it rapidly narrows in the higher-stress barrier zones. The maximum width of 0.0144 ft [4.4 mm] is considerably less than an elliptic width based on the height (0.0226 ft [6.9 mm]).

Once a width distribution is calculated, an equivalent width given by

$$w_{eq}=\frac{2}{\pi a}\int_{-a}^{a}w\,dy \quad \text{(3.67)}$$

is easily obtained and provides a method for updating fracture design

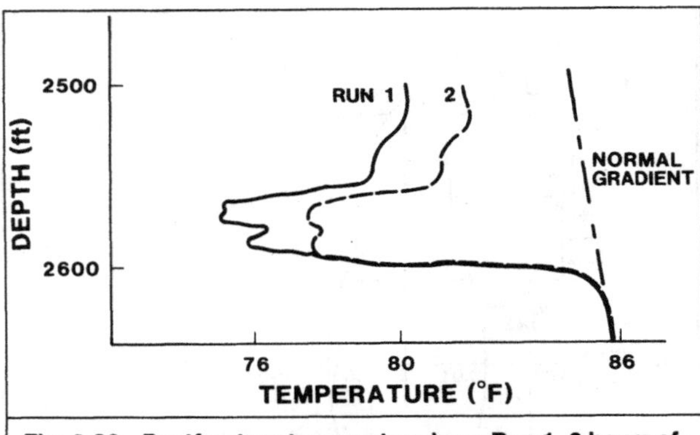

Fig. 3.26—Postfracture temperature logs: Run 1, 8 hours after injection; Run 2, 18 hours after injection.

models to reflect a more realistic width. For a Perkins and Kern[67] model (see Chap. 5), w_{eq} can be used directly. For a Khristianovitch-Zheltov[5]/Geertsma-de Klerk[6] model, the best approach is probably to develop a width ratio as

$$\frac{w_{eq}}{w_{max}}=\frac{w_{eq}E}{4(1-\nu^2)(P-\sigma_1)a}. \qquad (3.68)$$

This ratio of the reduced width to the homogeneous stress width can be used as a multiplicative factor to adjust the parallel plate width in Khristianovitch-Zheltov/Geertsma-de Klerk models.

Calculations for situations with more than three stress layers can be performed, but the details of the solution are complex. An example of such a calculation for the stress data given in Fig. 3.19 is shown in Fig. 3.25. The high-stress regions result in narrower crack widths and probably hinder fluid flow and proppant transport. In this case, the equivalent width is 0.0176 ft [5.4 mm] compared with a maximum width of 0.0226 ft [6.9 mm] and an elliptic width of 0.034 ft [10.4 mm] on the basis of the total height.

3.6 Fracture-Height Measurement

Previous sections have discussed factors that control fracture geometry, how these variables may be measured, and procedures for using such data to calculate fracture height. While these theoretical analyses capture the physics of the problem, they are necessarily idealized, and independent measurements of fracture geometry are needed whenever possible.

A good example of this is seen in a mineback experiment conducted in a tunnel at the U.S. DOE's Nevada Test Site discussed in the previous section.[83] Before a series of small hydraulic fractures, several stress tests were run to develop the stress profile in Fig. 3.14 (solid curve). On the basis of this profile, the small fractures were expected to be essentially radial or "penny-shaped" cracks because no apparent barriers existed to fracture growth. However, exposing the fracture revealed that downward growth of the fracture was limited, as seen in Fig. 3.14. Following this observation, additional stress tests were conducted, exposing a small, high-stress interval as seen in the second stress profile in Fig. 3.14 and explaining the observations. While such a small zone would probably not affect fracture geometry for a typical well-stimulation hydraulic fracture, the test emphasizes that situations exist where it is impossible (at least impractical) to collect sufficient data to predict fracture geometry fully.

For such cases, or to determine when such situations might exist, fracture height should be measured to verify and/or to modify the calculation procedures. This section discusses some of the currently available tools for making such measurements. While these techniques can often yield accurate and valuable information, they are subject to restrictions, and the result from any individual measurement should always be treated with some reservation.

Temperature Logs. The most common procedure for measuring fracture height is the use of postfracture temperature logs.[85] The basic procedure is to run a base (prefracture) temperature log to determine the temperature gradient in the formations, then run one or more logs following the stimulation as seen in Fig. 3.26. Above the treated zone, heat transfer will occur by radial heat conduction, while over the fracture faces, heat transfer will be by linear flow. Thus, the temperature will recover at different rates after pumping, which will cause the development of a temperature anomaly, identifying the fractured zone.

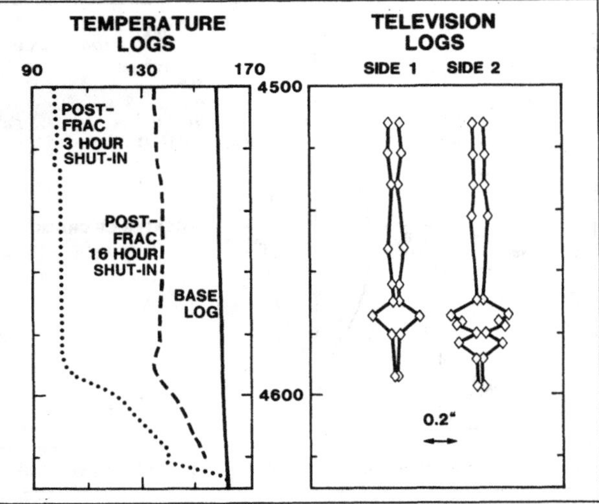

Fig. 3.27—Comparison of postfracture temperature and televiewer logs.

Unfortunately, ideal logs, such as Fig. 3.26, are rare and may be in error when they do occur. Smith *et al.*[82] presented a case of an oil field where postfracture temperature logs run in perforated completions had consistently been interpreted as indicating confined fracture height. Later, a comprehensive logging program, including downhole television, showed that extensive height growth was occurring, as seen in Fig. 3.27. The misinterpretation in this case was probably related to the fractures deviating from the wellbore, and this illustrates the primary restriction on postfracture temperature logs.

Temperature logs are shallow investigative tools and can detect fractures only very near the wellbore. Ref. 86 discusses three-dimensional finite-element analyses, which show that a fracture 1 ft [0.3 m] from the wellbore was totally transparent to postfracture logs. For typical wellbore dimensions, a difference in inclination of only 1° [0.02 rad] between the wellbore and fracture could result in only 40 to 50 ft [12 to 15 m] of the fracture being "visible" on temperature logs.

Another interpretation problem is that postfracture temperature logs are affected by fracture width and fracture height. Theoretical analyses presented earlier showed that low-stress or low-modulus zones can have significantly greater fracture width and will accept the majority of the fluid flow. Because more fluid is entering this region, there will be more cooling, and the largest temperature anomaly will be adjacent to the widest portion of the fracture. This is both a strength and a weakness: a strength because the log is indicating the variable of primary interest, e.g., where the bulk of the stimulation is going, and a weakness because the large anomaly can mask height growth, leading to false conclusions regarding fracture geometry.

A final factor to be considered in analyzing temperature logs is the effect of wellbore condition. For example, pumping down tubing will create a temperature anomaly immediately below the tubing because of the difference in radial heat flow rates for a tubing/casing configuration compared with fluid flow just inside the casing. Smith and Steffensen[87] discussed examples of this and other factors, such as washouts, in a study of temperature profiles in water injection wells.

While the problem of the fracture and wellbore deviating from one another is irreversible, this simply implies that the results of temperature logging as a single tool should be treated with reservation. In many instances, the logs can yield valuable information, and several analysis techniques have been developed to improve their interpretation.

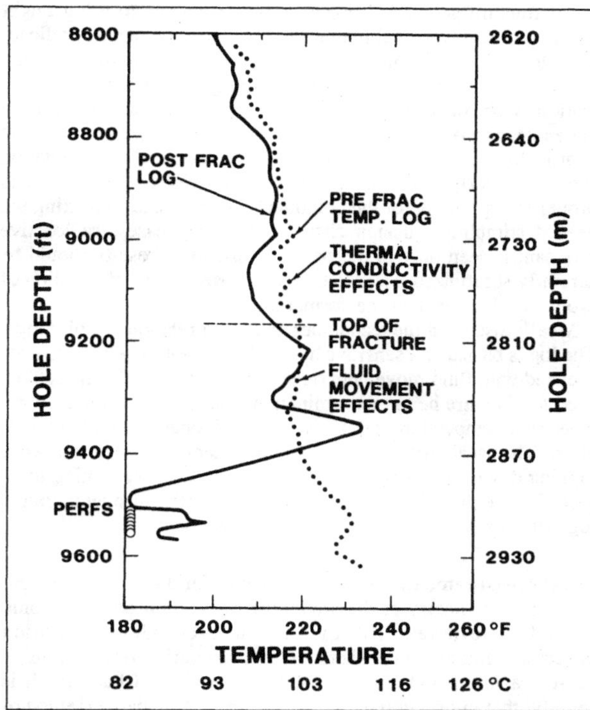

Fig. 3.28—Example of post-cold-water-circulation test log with postfracture log.

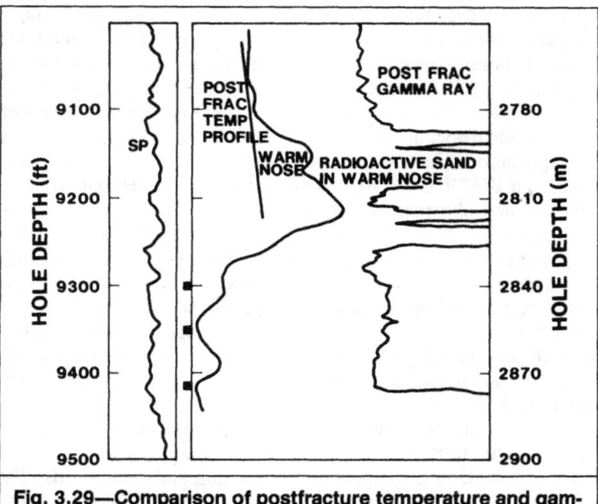

Fig. 3.29—Comparison of postfracture temperature and gamma ray logs.

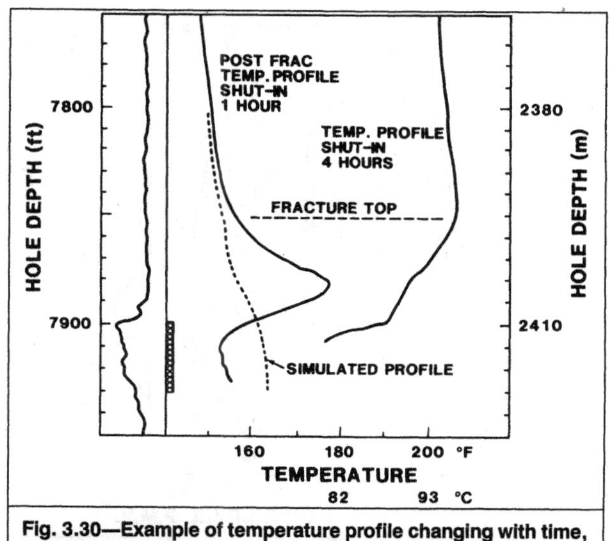

Fig. 3.30—Example of temperature profile changing with time, later logs showing a "clearer" interpretation.

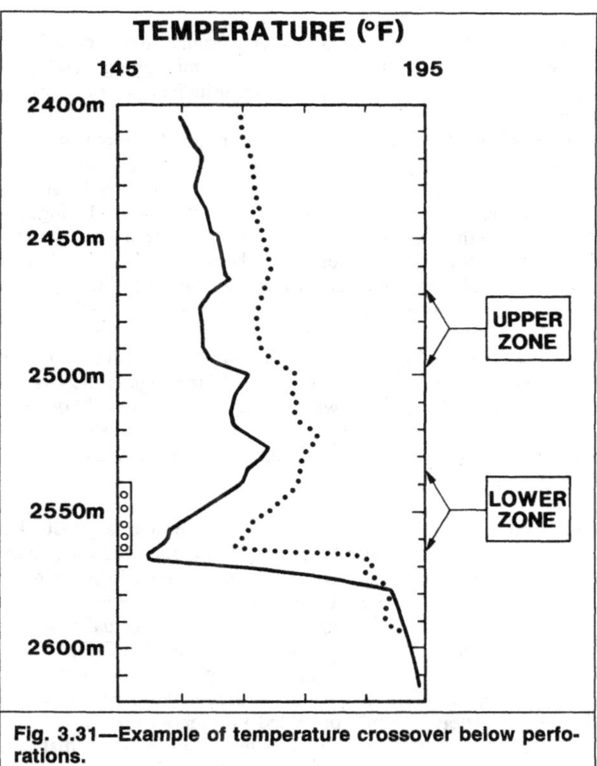

Fig. 3.31—Example of temperature crossover below perforations.

Dobkins[88] gave several examples of "nonideal" temperature logs and suggested the use of a cold-water circulating test to assist in analysis. For this test, water is circulated down tubing and up the annulus to cool the wellbore without creating a fracture. Post-circulation logs then indicate perturbations caused by thermal conductivity changes and wellbore effects, such as washouts. This prefracture measurement is then used to generate an expected post-fracture response, and any deviation from this expected response indicates fluid movement outside the pipe and thus the presence of fracture height growth. An example of this is seen in Fig. 3.28. For the particular case of a "warm nose" above perforations, further evidence of the correctness of this interpretation was given by comparing postfracture temperature and radioactivity logs, as shown in Fig. 3.29.

Postfracture temperature-log interpretation is also improved by the running of more than one log. Fig. 3.30 shows an example where the nature of the temperature profile changed with later logs, giving a much easier interpretation. This was also discussed by Wages,[89] who further suggested using the time rate of temperature recovery to identify fractured intervals, though no quantitative example was given of how this should be done.

Additional logs can also yield more realistic estimates of downward fracture growth. Typically, at the conclusion of a stimulation, the wellbore fluid below the treated interval is very near static reservoir temperature. Thus, a temperature log will show a sharp break immediately below the perforations. This is often interpreted as the fracture bottom when in fact the sharp increase in temperature only indicates stagnant fluid in the rathole. Because the wellbore fluid is stagnant and hot, any downward fracture growth will result in fluid outside the wellbore being cooler than inside. Thus, while temperatures over (and above) the treated interval increase with time, one may observe postfracture cooling below the

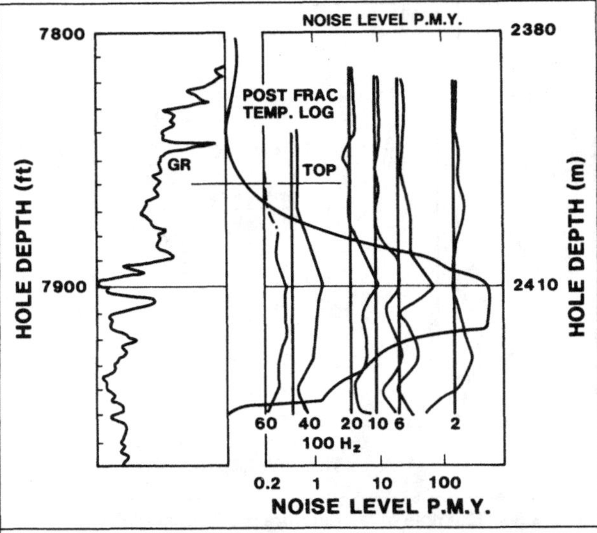

Fig. 3.32—Comparison of postfracture temperature and noise logs.

treated interval. This can result in a temperature crossover, as seen in Fig. 3.31. When wellbore conditions permit logging below perforations, such a crossover can give a clear indication of downward fracture height growth.

While specific logging practices will vary for different situations, some guidelines are generally applicable. A base log should be run to measure the prefracture temperature profile, followed by at least two postfracture logs (with best results usually obtained by logging down, so that the temperature sensor is always entering undisturbed fluid). The first postfracture log should be run shortly (1 to 3 hours) after the stimulation, followed by a second log several hours later. No backflow from the well should be allowed. If backflow does occur, however, it is usually possible to obtain a good log by allowing a couple of hours for temperatures to stabilize after the backflow. One exception to the timing of the logs is the case of underpressured formations, where fluid level in the wellbore will fall after the hydraulic fractures. For this case, the wellbore fluid level must be allowed to stabilize before postfracture temperature logs are run.

Radioactivity Logs and Noise Logs. Two other normal production logs often used for fracture height determination are postfracture gamma ray and noise logs. Of these, radioactivity logs are more common and are conducted by inducing artificial radioactivity in the fracture by including tagged sand with the normal proppant (tagged material typically added at a rate of about 1 mCi/1,000 lbm [81.6 kBq/kg] of proppant), followed by postfracture gamma ray logs. For the most definitive results with regard to fracture height, the tagged material should be added throughout the stimulation.

One advantage of gamma ray over temperature logs is that they need not be run immediately after a stimulation, allowing wellbore fill below perforations to be removed before logging. The other restrictions on temperature logs, however, apply equally to radioactivity logs; i.e., they are shallow investigative tools (shallower even than temperature logs) and the response is proportional to fracture width. Thus, while the two logs are often used in combination, the potential exists for them to confirm one another and still not yield reliable results.

One disadvantage of radioactivity logs is their inability to distinguish between a fracture and a small channel behind the casing. The temperature response caused by a small amount of flow in a channel or annular space behind the casing may not alter the radial-flow heat conduction around unfractured portions of the wellbore[85] and thus will not affect the temperature logs. In other cases, proper care may allow interpretation of the effect of such flow in the temperature logs[87]; however, any radioactive material deposited in such a channel is indistinguishable from tagged material in a fracture.

In other cases, the combination can yield very valuable results about fracture geometry. For example, a temperature log might show significant vertical height growth, but postfracture radioactivity logs show the gamma ray response restricted to the main pay interval. This could be interpreted as height growth through formations with higher stress or modulus such that fracture width in these regions would not allow proppant transport. Thus, the proppant is probably concentrated in the target formation. Such a situation can lead to early screenouts as discussed by Nolte,[9] however, and proper interpretation of the combined logs can lead to making the needed stimulation design changes. In many cases, radioactive proppant is seen only in the perforations. Such results should be carefully scrutinized because they may provide no information of proppant placement in the formation.

Mea[90] discussed the techniques and interpretation of noise logs. The log is basically a sensitive hydrophone used to detect noise associated with fluid movement outside of the casing. While seldom used for fracture height determination, the tool can often be combined with temperature logs to acquire additional data with minimal effort. The usual procedure is to record the temperature profile while logging down, then stop every 10 ft [3 m] or so on coming up to record noise data. Fig. 3.32 compares a postfracture temperature log with a noise log recorded in this manner.

Direct Measurement. Three methods exist for direct measurement of fracture geometry at the wellbore: impression packers, sonic borehole televiewers (BHTV's),[91] and downhole television cameras.[82] Impression packers are used primarily for determining fracture azimuth and are discussed more fully in Chap. 16. It is doubtful that impression packers could provide a viable method of measuring fracture height. In general, direct methods yield the most complete data; however, special planning and effort are required, so such measurements cannot be made routinely.

The BHTV is a sonar device introduced by Zemanek *et al.*[91] as a fracture detection tool. A crystal inside the tool emits a high-frequency sonic pulse and then records the reflection of the pulse from the borehole wall. Lack of any reflected signal may indicate the presence of a fracture with a width significantly larger than the wave length of the sonic signal. In principle, this should be an excellent fracture detection tool; however, results have often not lived up to expectations. In one example from the Mondak field in Montana,[92] a BHTV failed to register natural fractures that were recovered and observed in core samples.

A comparison of BHTV's with downhole television logs often indicates that the televiewer responds best where significant spalling exists along the edge of the fracture. Because the amount of spalling could be affected by mechanical properties of the formations, the BHTV could be misinterpreted with respect to fracture-height growth. In one published comparison,[82] however, the BHTV correctly measured the height of a hydraulic fracture and showed the fracture deviating from the wellbore.

Certainly the best method of direct measurement is downhole closed-circuit television. This tool requires very special conditions, however, including an openhole completion and clean fluids. The process is, of course, very simple—just inject water (or water gelled with hydroxyethylcellulose) and record the fracture geometry on videotape. When conditions warrant the effort, such as areas where many stimulations are planned or where unusual problems have been encountered, television logs can supply a great deal of data about fracture geometry in a particular formation. Such information can be used to verify or to modify the calculation procedures used to design stimulations in that formation.

New Techniques. Undoubtedly, new techniques will be developed as the need for more detailed fracture descriptions increases. One process currently being tested is the use of a borehole geophone package to "triangulate" on microseismic events triggered by the hydraulic fracturing process.

While this procedure has been used primarily to measure fracture azimuth,[93] in principle it can be extended to determine vertical fracture extent. One example[94] is illustrated in Fig. 3.33, where the location of the seismic events indicated confined fracture height, which was confirmed by postfracture temperature logs and bottomhole-treating-pressure behavior (see Chap. 14). This particular example used geophones located in offset wells, but presumably the fracture azimuth procedure of locating the geophones in the injection well can also be used. In this case, however, data can be recorded only during shut-in periods. This will limit the number of events that can be monitored and may weaken the analysis.

One potential problem is related to the phenomena creating the microseismic events. It is often speculated that the events are related not to the tensile cracking at the tip of the fracture, but to fluid leakoff that increases pore pressure and triggers small shear failures around the periphery of the fracture. In that case, fracture-height growth into impermeable layers, such as a shale, might not emit seismic signals (because there would be no leakoff) and thus lead to misinterpretation.

One overriding advantage of this technique is the ability to monitor fracture geometry away from the wellbore; therefore the procedure may have great potential. Another advantage is that the technique would be relatively easy to implement because of the availability of vertical seismic profiling tools.

Summary of Height Measurement Techniques. This section described several techniques available for routine measurement of fracture height. Of these, temperature logs are and will probably remain the most prevalent because of their ease of use and minimal impact on operations. While such logs are subject to some strict restrictions, they can often yield valuable data if the fracture remains adjacent to the wellbore. On the other hand, temperature logs can easily be misleading and should be treated with some reservations. For example, a temperature log showing fracture height limited to the perforated interval in the absence of any significant changes in lithology should be treated with some skepticism.

More definitive logs are available, including sonic BHTV's and downhole television cameras. The use of these tools often requires special planning, however, and is thus limited to particular situations where detailed data are most valuable.

In general, some type of fracture geometry measurement is needed because it may not be possible to measure sufficient data to predict fracture geometry. All available techniques have limitations, however, and their restrictions should be remembered during interpretation. Whenever possible, postfracture logs should be used in conjunction with and in addition to theoretical analysis and other types of monitoring, such as the pressure diagnostics discussed in Chap. 14, and not used as stand-alone measurements.

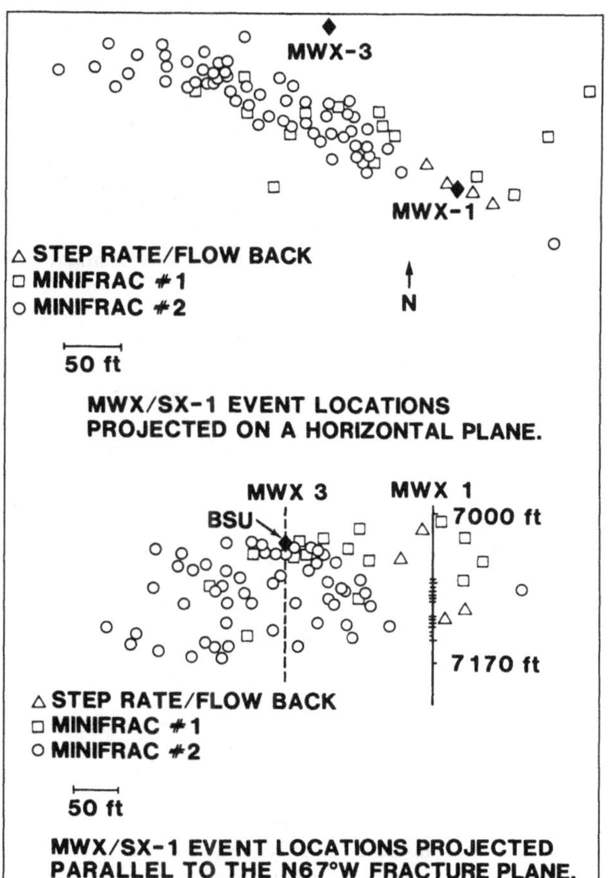

Fig. 3.33—Example of locating microseismic events to determine fracture height.

Nomenclature

a = fracture half-height, ft [m]
A = poroelastic constant
$b_2,b_3 \ldots b_k$ = geometry coefficients for layered-stress fracture analysis, ft [m]
c = fluid compressibility, psi^{-1} [MPa^{-1}]
c' = porous-medium coefficient, ft^{-1} [m^{-1}]
E = Young's modulus, psi [MPa]
E_0 = initial tangent modulus, psi [MPa]
E_1,E_2 = Young's modulus for layers, psi [MPa]
E_1',E_2',E_3' = plane-strain moduli for layers$=E_i/(1-\nu^2)$, psi [MPa]
E_2'' = effective Young's modulus in layered medium, psi [MPa]
$\mathcal{E}$ = separation energy per unit depth, ft-lbf/ft^3 [J/m^3]
$f(y)$ = even function of net fracture pressure, psi [MPa]
F = Poisson's ratio parameter
$F(y)$ = even function for width determination, psi [MPa]
$g(y)$ = odd function of net fracture pressure, psi [MPa]
G = shear modulus, psi [MPa]
G_1,G_2 = shear moduli for layered medium, psi [MPa]
$G(y)$ = odd function for width determination, psi [MPa]
$\mathcal{G}$ = strain-energy release rate, psi-in. [MPa·m]
h = fracture-zone thickness, ft [m]
Δh = fractional growth out of zone, ft [m]
h_s = fracture growth out of zone, symmetric case, ft [m]
h_{s2} = fracture growth into highest-stress layer (three-layer case), ft [m]
h_{s3} = fracture growth into lowest-stress layer (three-layer case), ft [m]
i = injection rate, bbl/min [m^3/s]
k = permeability, md
k' = fluid consistency for power-law fluid, lbf-sec$^{n'}$/ft^2 [N·s$^{n'}$/m^2]
K = bulk modulus, psi [MPa]
K' = modulus of cohesion, psi-$\sqrt{\text{in.}}$ [MPa·$\sqrt{\text{m}}$]
K_s = bulk modulus of solid rock matrix, psi [MPa]
K_0 = modulus number, psi [MPa]
$K_{\text{I}},K_{\text{II}},K_{\text{III}}$ = stress-intensity factors for three opening modes, psi-$\sqrt{\text{in.}}$ [MPa·$\sqrt{\text{m}}$]
$K_{\text{I}c}$ = fracture toughness or critical stress-intensity factor, psi-$\sqrt{\text{in.}}$ [MPa·$\sqrt{\text{m}}$]
$K_{\text{I}c1},K_{\text{I}c2},K_{\text{I}c3}$ = fracture toughness for layered medium, psi-$\sqrt{\text{in.}}$ [MPa·$\sqrt{\text{m}}$]
$K_{\text{Itop}},K_{\text{Ibottom}}$ = fracture toughness for top and bottom layers, psi-$\sqrt{\text{in.}}$ [MPa·$\sqrt{\text{m}}$]
$K_{\text{I1}},K_{\text{I2}},K_{\text{I3}}$ = Mode I stress-intensity factors of various layers, psi-$\sqrt{\text{in.}}$ [MPa·$\sqrt{\text{m}}$]
L = fracture length, ft [m]
n = modulus exponent

n' = consistency index for power-law fluid
p = pore pressure, psi [MPa]
Δp = change in pore pressure, psi [MPa]
p_a = atmospheric pressure, psi [MPa]
p_i = initial reservoir pressure, psi [MPa]
p_{is} = instantaneous shut-in pressure, psi [MPa]
P = pressure in hydraulic fracture, psi [MPa]
ΔP = net fracturing pressure, psi [MPa]
P_c = closure pressure, psi [MPa]
P_{net} = net fracturing pressure above closure stress, psi [MPa]
r = distance from tip of crack, ft [m]
s = poroelastic parameter defining effect of pore pressure
S_i, S_j = differential stresses in multilayered medium, psi [MPa]
$S_0, S_2, S_3 \ldots S_n$ = differential stresses in three-layered medium, psi [MPa]
t = time, seconds
t_i = injection time, seconds
t_{si} = shut-in time, seconds
T = temperature, °F [°R]
u = dummy variable for integration
v_c = compressional-wave velocity, ft/sec [m/s]
v_s = shear-wave velocity, ft/sec [m/s]
w = width, ft [m]
w_{eq} = equivalent width, ft [m]
w_{max} = maximum width, ft [m]
W = work resulting from creating crack surface per unit depth, ft-lbf/ft [J/m]
x = arbitrary direction, ft [m]
y = distance along fracture height, ft [m]
z = vertical direction, ft [m]
α = coefficient of linear thermal expansion, $°F^{-1}$ [$°C^{-1}$]
α_p = plasticity coefficient per unit depth, ft-lbf/ft^4 [J/m^4]
γ = surface energy per unit depth, ft-lbf/ft^3 [J/m^3]
γ_{eff} = effective surface energy per unit depth, ft-lbf/ft^3 [J/m^3]
γ_2 = slope of pressure drop in fracture (KGD), psi/ft [MPa/m]
γ_2^F = crack coefficient
Γ_2 = slope of pressure drop in fracture (PKN), psi/ft [MPa/m]
ϵ_i, ϵ_j = strains in i, j directions, $i, j = 1,2$
$\epsilon_x, \epsilon_y, \epsilon_z$ = strains in x, y, z directions
$\Delta\epsilon_x, \Delta\epsilon_y$ = change in strains in x and y directions
$\Delta\epsilon$ = change in strain
ζ = pressure distribution parameter
θ = arbitrary angle, degrees
λ = Lamé coefficient, psi [MPa]
μ = viscosity, cp
μ_f = coefficient of friction
ν = Poisson's ratio
ν_i = initial Poisson's ratio
ρ = density, g/cm^3
σ = stress, psi [MPa]
$\Delta\sigma$ = differential of change in stress, psi [MPa]
σ_e = effective stress, psi [MPa]
σ_c = closure stress, psi [MPa]
σ_{con} = confining stress, psi [MPa]
$\Delta\sigma_c$ = change in closure stress, psi [MPa]
σ_H = horizontal in-situ stress (both equal), psi [MPa]
σ_{eHi} = effective horizontal in-situ stress, i direction, psi [MPa]
σ_{Hmin} = minimum horizontal in-situ stress, psi [MPa]
σ_{Hmax} = maximum horizontal in-situ stress, psi [MPa]
σ_{min} = minimum in-situ stress, psi [MPa]
σ_n = normal stress, psi [MPa]
σ_v = overburden stress, psi [MPa]
$\sigma_x, \sigma_y, \sigma_z$ = stress in x, y, and z directions, psi [MPa]
$\Delta\sigma_x, \Delta\sigma_y$ = differential or change in stress in x and y directions, psi [MPa]
$\sigma_1, \sigma_2, \sigma_3$ = stresses in layers, psi [MPa]
τ = shear stress, psi [MPa]
τ_c^H = time constant for height growth, seconds
τ_c^L = time constant for length growth, seconds
τ_{xy} = shear stress in x,y plane, psi [MPa]
τ_0 = inherent shear strength, psi [MPa]
ϕ = porosity, fraction
χ = poroelastic coefficient
$\bar{\chi}$ = averaged poroelastic coefficient

Subscripts

i,j = dummy variables for x or y direction
m,n = dummy variables indicating layer

References

1. Judd, W.R.: "Rock Stress, Rock Mechanics, and Research," *State of Stress in the Earth's Crust,* Elsevier Science Publishing Co., New York City (1964) 5–51.
2. Jaeger, J.C. and Cook, N.G.W.: *Fundamentals of Rock Mechanics,* Halsted Press, New York City (1976).
3. Senseny, P.E.: "Laboratory Measurements of Mechanical Properties of Sandstones and Shales," paper SPE 11762 presented at the 1983 SPE/DOE Symposium on Low Permeability Gas Reservoirs," Denver, March 14–16.
4. Thomas, R.D. and Ward, D.C.: "Effect of Overburden Pressure and Water Saturation on Gas Permeability of Tight Sandstone Cores," *JPT* (Feb. 1972) 120–24.
5. Khristianovitch, S.A. and Zheltov, Y.P.: "Formation of Vertical Fractures by Means of Highly Viscous Fluids," *Proc.,* World Pet. Cong., Rome (1955) **2,** 579–86.
6. Geertsma, J. and de Klerk, F.: "A Rapid Method of Predicting Width and Extent of Hydraulically Induced Fractures," *JPT* (Dec. 1969) 1571–81.
7. Geertsma, J. and Haafkens, R.: "A Comparison of the Theories for Predicting Width and Extent of Vertical Hydraulically Induced Fractures," *Trans.,* ASME (March 1979) **101,** 8–19.
8. Hubbert, M.K. and Willis, D.G.: "Mechanics of Hydraulic Fracturing," *Trans.,* AIME (1957) **210,** 153.
9. Nolte, K.G.: "Fracture Design Considerations Based on Pressure Analysis," *SPEPE* (Feb. 1988) 22–30.
10. Terzaghi, K.: *Theoretical Soil Mechanics,* John Wiley and Sons Inc., New York City (1943) 51.
11. Biot, M.A.: "General Theory of Three Dimensional Consolidation," *J. Applied Physics* (1941) **12,** 155–65.
12. Biot, M.A.: "Theory of Elasticity and Consolidation for a Porous Anisotropic Solid," *J. Applied Physics* (1955) **26,** 182–85.
13. Biot, M.A.: "General Solutions of the Equations of Elasticity and Consolidation for a Porous Material," *J. Applied Mech.* (1956) **78,** 91–96.
14. Nur, A. and Byerlee, J.D.: "An Exact Effective Stress Law for Elastic Deformation of Rock with Fluids," *J. Geophysical Res.* (Sept. 1971) **76,** No. 26, 6414–19.
15. Cornet, F.H. and Fairhurst, C.: "Influence of Pore Pressure on the Deformation Behavior of Saturated Rocks," *Proc.,* Third Congress of the Intl. Soc. for Rock Mechanics, Natl. Academy of Science, Washington, DC (Sept. 1974) **1,** Part B, 638–44.
16. Rice, J.R. and Cleary, M.P.: "Some Basic Stress Diffusion Solutions for Fluid-Saturated Elastic Porous Media with Compressible Constituents," *Rev. of Geophysics and Space Physics* (May 1976) **14,** No. 2, 227–41.
17. Prats, M.: "Effect of Burial History on the Subsurface Horizontal Stresses of Formations Having Different Material Properties," *SPEJ* (Dec. 1981) 658–62.
18. Scheidegger, A.E.: "Stresses in Earth's Crust as Determined from Hydraulic Fracturing Data," *Geologie und Bauwesen* (1962) **27,** 45.
19. Kehle, R.O.: "Determination of Tectonic Stress Through Analysis of Hydraulic Well Fracturing," *J. Geophysical Res.* (1964) **69,** 259.
20. Fairhurst, C.: "Measurement of In Situ Rock Stresses, with Particular Reference to Hydraulic Fracturing," *Rock Mech. Eng. Geology* (1964) **2,** 129.
21. Haimson, B.C. and Fairhurst, C.: "Initiation and Extension of Hydraulic Fractures in Rocks," *SPEJ* (Sept. 1967) 310–18.

22. Haimson, B. and Fairhurst, C.: "Hydraulic Fracturing in Porous Permeable Materials," *JPT* (July 1969) 811–17.
23. Haimson, B.C.: "The State of Stress in the Earth's Crust," *Rev. of Geophysics and Space Physics* (July 1975) **13,** No. 3, 350–52 and 381–83.
24. Haimson, B.C.: "The Hydrofracturing Stress Measuring Method and Recent Field Results," *Intl. J. Rock Mech. Min. Sci.* (1978) **25,** 167–78.
25. Haimson, B.C.: "Confirmation of Hydrofracturing Results Through Comparisons with Other Stress Measurements," *Proc.*, 22nd U.S. Rock Mechanics Symposium, Massachusetts Inst. of Technology, Boston (June 1981) 379–85.
26. Warpinski, N.R.: "Investigation of the Accuracy and Reliability of In Situ Stress Measurements Using Hydraulic Fracturing in Perforated, Cased Holes," *Proc.*, 24th U.S. Symposium on Rock Mechanics, College Station, TX (June 1983) 773–86.
27. Kry, R. and Gronseth, M.: "In Situ Stresses and Hydraulic Fracturing in the Deep Basin," paper 82-3321 presented at the 1982 Petroleum Soc. of CIM Annual Meeting, Calgary, Alta., June 6–9.
28. Greenfield, H. *et al.*: "Resource Evaluation and Production Research on Tight Sands in the Pinedale Unit, Sublette County, Wyoming," GRI Report 81/0049, Gas Research Inst., Chicago (Dec. 1981) 59–73.
29. Warpinski, N.R.: "In-Situ Stress Measurements at U.S. DOE's Multiwell Experiment Site, Mesaverde Group, Rifle, Colorado," *JPT* (March 1985) 527–36.
30. Nolte, K.G.: "Determination of Fracture Parameters from Fracturing Pressure Decline," paper SPE 8341 presented at the 1979 SPE Annual Technical Conference and Exhibition, Las Vegas, Sept. 23–26.
31. Smith, M.B.: "Stimulation Design for Short, Precise Hydraulic Fractures," *SPEJ* (June 1985) 371–79.
32. Voight, B.: "Stress History and Rock Stress," *Proc.*, Third Cong. of the Intl. Soc. for Rock Mechanics, Natl. Academy of Science, Washington, DC (Sept. 1974) **2,** Part A, 580–82.
33. Rosepiler, M.J.: "Determination of Principal Stresses and Confinement of Hydraulic Fractures in Cotton Valley," paper SPE 8405 presented at the 1979 SPE Annual Technical Conference and Exhibition, Las Vegas, Sept. 23–26.
34. Salz, L.B.: "Relationship Between Fracture Propagation Pressure and Pore Pressure," paper SPE 6870 presented at the 1977 SPE Annual Technical Conference and Exhibition, Denver, Oct. 7–12.
35. Cleary, M.P.: "Rate and Structure Sensitivity in Hydraulic Fracturing of Fluid-Saturated Porous Formations," *Proc.*, 20th U.S. Symposium on Rock Mechanics, Austin, TX (June 4–6, 1978) 127–42.
36. Geertsma, J.: "Some Rock-Mechanical Aspects of Oil and Gas Well Completions," paper EUR-38 presented at the 1978 SPE European Offshore Petroleum Conference and Exhibition, London, Oct. 24–27.
37. Cleary, M.P.: "Analysis of Mechanisms and Procedures for Producing Favorable Shapes of Hydraulic Fractures," paper SPE 9260 presented at the 1980 SPE Annual Technical Conference and Exhibition, Dallas, Sept. 21–24.
38. Cleary, M.P.: "Comprehensive Design Formulae for Hydraulic Fracturing," paper SPE 9259 presented at the 1980 SPE Annual Technical Conference and Exhibition, Dallas, Sept. 21–24.
39. Lama, R.D. and Vutukuri, V.S.: *Handbook on Mechanical Properties of Rocks,* Trans Tech Publications, Clausthal, Germany (1978) **2,** 86–89, 145, 234.
40. Perkins, T.K. and Gonzalez, J.A.: "Changes in Earth Stresses Around a Wellbore Caused by Radially Symmetrical Pressure and Temperature Gradients," *SPEJ* (April 1984) 129–40.
41. Griffith, A.A.: "The Phenomenon of Rupture and Flow in Solids," *Phil. Trans. Roy. Soc. London* (1921) **A221,** 163–98.
42. Friedman, M., Handin, J., and Alani, G.: "Fracture-Surface Energy of Rocks," *Intl. J. Rock Mech. Min. Sci.* (1972) **9,** 757–66.
43. Perkins, T.K. and Bartlett, L.E.: "Surface Energies of Rocks Measured During Cleavage," *SPEJ* (Dec. 1963) 307–17.
44. Perkins, T.K. and Krech, W.W.: "Effect of Cleavage Rate and Stress Level on Apparent Surface Energies of Rocks," *SPEJ* (Dec. 1966) 308–12.
45. Forootan-Rad, D. and Moavenzadeh, F.: "Crack Initiation and Propagation in Rock," Report R68-29, Massachusetts Inst. of Technology, Dept. of Civil Engineering (1968) 126.
46. Barenblatt, G.I.: "Mathematical Theory of Equilibrium Cracks," *Advances in Applied Mech.* (1962) **7,** 55.
47. Goodier, J.N.: "Mathematical Theory of Equilibrium Cracks," *Fracture,* Liebowitz (ed.), Academic Press, New York City (1968) **2.**
48. Orowan, E.: *Fatigue and Fracture of Metals,* John Wiley & Sons Inc., New York City (1952) 139.
49. Irwin, G.R.: "Analysis of Stresses and Strains Near the End of a Crack Traversing a Plate," *J. App. Mech.* (1957) **24,** 361.
50. Schmidt, R.A.: "Fracture Toughness Testing of Limestone," *Experimental Mech.* (May 1976) **16,** No. 5, 161–67.
51. Schmidt, R.A. and Huddle, C.W.: "Effect of Confining Pressure on Fracture Toughness of Indiana Limestone," *Intl. J. Rock Mech. Min. Sci.* (1977) **14,** 289–93.
52. Rice, J.R.: "Mathematical Analysis in the Mechanics of Fracture," *Fracture,* H. Liebowitz (ed.), Academic Press, New York City (1968) **2.**
53. Jones, A.H., Abou-Sayed, A.S., and Rogers, L.A.: "Rock Mechanics Aspects of MHF Design in Eastern Devonian Shale Gas Reservoirs," Report TR 77-83, Terra Tek, Dallas (1977).
54. Brechtel, C.E., Abou-Sayed, A.S., and Jones, A.H.: "Fracture Containment Analysis Conducted on the Benson Pay Zone in Columbia Well 20538-T," *Proc.*, Second Eastern Gas Shales Symposium, Morgantown, WV (Oct. 1978) **1,** 264.
55. Schmidt, R.A. and Lutz, T.J.: "K_{Ic} and J_{Ic} of Westerly Granite—Effect of Thickness and In-Plane Dimensions," *Fracture Mechanics Applied to Brittle Materials,* STP 678, ASTM, Philadelphia (1979) 166–82.
56. Costin, L.S.: Static and Dynamic Fracture Behavior of Oil Shale," *Fracture Mechanics Methods for Ceramics, Rocks, and Concrete,* STP 745, ASTM, Philadelphia (1981) 169–84.
57. Clifton, R.J. *et al.*: "Determination of the Critical Stress-Intensity Factor K_{Ic} in a Circular Ring," *Exp. Mech.* (1976) **16,** 223–38.
58. Barker, L.M.: "Theory for Determining K_{Ic} from Small, Non-LEFM Specimens, Supported by Experiments on Aluminum," Report TR 78-6R, Terra Tek, Dallas (1978).
59. Shimamoto, T.: "Effects of Fault-Gauge on the Frictional Properties of Rocks: An Experimental Study," PhD dissertation, Texas A&M U., College Station (1977).
60. Engelder, J.T., Logan, J.M., and Handin, J.: "The Sliding Characteristics of Sandstone on Quartz Fault-Gauge," *Pure and Applied Geophysics* (1975) **113,** 68–86.
61. Van Eekelen, H.A.: "Hydraulic Fracture Geometry: Fracture Containment in Layered Formations," *SPEJ* (June 1982) 341–49.
62. Medlin, W.L. and Massé, L.: "Plasticity Effects in Hydraulic Fracturing," *JPT* (Sept. 1986) 995–1006.
63. Clark, S.P.: *Handbook of Physical Constants,* Geological Soc. of America Inc., New York City (1966) 100.
64. Smith, M.B.: "Effect of Fracture Azimuth on Production With Application to the Wattenberg Gas Field," paper SPE 8298 presented at the 1979 SPE Annual Technical Conference and Exhibition, Las Vegas, Sept. 23–26.
65. Komar, C.A. and Frohne, K.H.: "Factors Controlling Fracture Orientation in Sandstone," paper SPE 4567 presented at the 1973 SPE Annual Technical Conference and Exhibition, Las Vegas, Sept. 30–Oct. 3.
66. Warpinski, N.R. and Teufel, L.W.: "Influence of Geologic Discontinuities on Hydraulic Fracture Propagation," *JPT* (Feb. 1987) 209–20.
67. Perkins, T.K. and Kern, L.R.: "Widths of Hydraulic Fractures," *JPT* (Sept. 1961) 937–49; *Trans.*, AIME, **222.**
68. Harrison, E., Kieschnick, W.F., and McGuire, W.J.: "The Mechanics of Fracture Induction and Extension," *Trans.*, AIME (1954) **201,** 252–63.
69. Simonson, E.R., Abou-Sayed, A.S., and Clifton, R.J.: "Containment of Massive Hydraulic Fractures," *SPEJ* (Feb. 1978) 27–32.
70. Nolte, K.G. and Smith, M.B.: "Interpretation of Fracturing Pressures," *JPT* (Sept. 1981) 1767–75.
71. Cleary, M.P.: "Primary Factors Governing Hydraulic Fractures in Heterogeneous Stratified Porous Formations," paper 78-Pet-47 presented at the 1978 ASME ETC Conference, Houston, Nov. 5–9.
72. Cleary, M.P. and Keck, R.G.: "Microcomputer Models for the Design of Hydraulic Fractures," paper SPE 11628 presented at the 1983 SPE/DOE Symposium on Low Permeability Reservoirs, Denver, March 14–16.
73. Abou-Sayed, A.S. *et al.*: "Evaluation of the Influence of In-Situ Reservoir Conditions on the Geometry of Hydraulic Fractures Using a Three-Dimensional Simulator—Part 2: Case Studies," paper SPE 12878 presented at the 1984 SPE/DOE/GRI Unconventional Gas Recovery Symposium, Pittsburgh, May 13–15.
74. Palmer, I.D. and Craig, H.R.: "Modelling of Asymmetric Vertical Growth in Elongated Hydraulic Fractures and Application to First MWX Stimulation," paper SPE 12879 presented at the 1984 SPE/DOE/GRI Unconventional Gas Recovery Symposium, Pittsburgh, May 13–15.
75. Cleary, M.P.: "Comprehensive Design Formulae for Hydraulic Fracturing," paper SPE 9259 presented at the 1980 SPE Annual Technical Conference and Exhibition, Dallas, Sept. 21–24.
76. Daneshy, A.A.: "Hydraulic Fracture Propagation in Layered Formations," *SPEJ* (Feb. 1978) 33–41.
77. Anderson, G.D.: "Effects of Friction on Hydraulic Fracture Growth Near Unbonded Interfaces in Rocks," *SPEJ* (Feb. 1981) 21–29.
78. Teufel, L.W. and Clark, J.A.: "Hydraulic Fracture Propagation in Layered Rock: Experimental Studies of Fracture Containment," *SPEJ* (Feb. 1984) 19–32.
79. Teufel, L.W.: "An Experimental Study of Hydraulic Fracture Propagation in Layered Rock," PhD dissertation, Texas A&M U., College Station (Aug. 1979).
80. Warpinski, N.R. *et al.*: "Laboratory Investigation on the Effect of In-Situ Stresses on Hydraulic Fracture Containment," *SPEJ* (June 1982) 333–40.

81. Biot, M.A., Medlin, W.L., and Massé, L.: "Fracture Penetration Through an Interface," *SPEJ* (Dec. 1983) 857–69.
82. Smith, M.B., Rosenberg, R.J., and Bowen, J.F.: "Fracture Width—Design vs. Measurement," paper SPE 10965 presented at the 1982 SPE Annual Technical Conference and Exhibition, New Orleans, Sept. 26–29.
83. Warpinski, N.R., Schmidt, R.A., and Northrop, D.A.: "In-Situ Stresses: The Predominant Influence on Hydraulic Fracture Containment," *JPT* (March 1982) 653–64.
84. England, A.H. and Green, A.E.: "Some Two-Dimensional Punch and Crack Problems in Classical Elasticity," *Proc.*, Cambridge Phil Soc., London (1963) **59,** 489.
85. Agnew, B.G.: "Evaluation of Fracture Treatments with Temperature Surveys," *JPT* (July 1966) 892–98.
86. Smith, R.C. *et al.*: "Fracture Height from Temperature Logs," paper SPE 7559 presented at the 1978 SPE Annual Technical Conference and Exhibition, Houston, Oct. 1–4.
87. Smith, R.C. and Steffensen, R.J.: "Interpretation of Temperature Profiles in Water-Injection Wells," *JPT* (June 1975) 777–84.
88. Dobkins, T.A.: "Improved Methods to Determine Hydraulic Fracture Height," *JPT* (April 1981) 719–26.
89. Wages, P.E.: "Interpretation of Postfracture Temperature Surveys," paper SPE 11189 presented at the 1982 SPE Annual Technical Conference and Exhibition, New Orleans, Sept. 26–29.
90. McKinley, R.M. Bower, F.M., and Rumble, R.C.: "The Structure and Interpretation of Noise from Flow Behind Cemented Casing," *JPT* (March 1973) 329–38.
91. Zemanek, J. *et al.*: "The Borehole Televiewer—A New Logging Concept for Fracture Location and Other Types of Borehole Inspection," *JPT* (June 1969) 762–74.
92. Hirsch, J.M. *et al.*: "Recent Experience with Wireline Fracture Detection Logs," paper SPE 10333 presented at the 1981 SPE Annual Technical Conference and Exhibition, San Antonio, Oct. 5–7.
93. Schuster, C.L.: "Detection Within the Wellbore of Seismic Signals Created by Hydraulic Fracturing," paper SPE 7448 presented at the 1978 SPE Annual Technical Conference and Exhibition, Houston, Oct. 1–3.
94. Hart, C.M. *et al.*: "Fracture Diagnostics Results for the First Multiwell Experiment's Paludal Zone Stimulation," *SPEFE* (Sept. 1987) 320–26.

SI Metric Conversion Factors

bbl	× 1.589 873	E−01	=	m^3
ft	× 3.048*	E−01	=	m
°F	(°F−32)/1.8		=	°C
gal	× 3.785 412	E−03	=	m^3
in.	× 2.54*	E+00	=	cm
psi	× 6.894 757	E+00	=	kPa

*Conversion factor is exact.

Chapter 4
Two-Dimensional Fracture-Propagation Models

J. Geertsma,* SPE

4.1 Introduction

A mathematical fracture propagation model is indispensable to relate injection rate, q, time of treatment, t, and fluid leakoff, q_ℓ, with fracture dimensions—i.e., width, w, and length, L. Together, fracture dimensions and leakoff as a function of time form the basis for proppant and fracturing fluid scheduling.

So-called two-dimensional (2D) models were born out of necessity in the early 1960's to obtain closed-form solutions for a complex solid/fluid-mechanics interaction problem. Such models require that the fracture boundary in the plane of propagation be specified in advance. Models that assume a rectangular extension mode are widely used. The consequences of using a radially expanding fracture mode have also been examined.

4.2 Hydraulically Induced Fracture-Mechanics Principles

Fracture-width theories are based on the assumption that the fracture surface deforms in a linear elastic manner. This seems justified because of the usually large in-situ prestress, on which only small additional stress systems are superimposed, excluding the tip area. For plane-strain conditions, England and Green derived an equation for the width of a line crack between $x=-L$ and $x=+L$ (or $z=-\frac{1}{2}h_f$ and $z=+\frac{1}{2}h_f$) opened by an equal and opposite normal pressure distribution, p, on each side of the crack as exerted by a fluid. Assuming a symmetrically distributed in-situ normal stress, σ_h, opposing p, this equation is

$$w(x)=\frac{4(1-\nu)L}{\pi G}\int_{x/L}^{1}\frac{f_{L2}\mathrm{d}f_{L2}}{\sqrt{(f_{L2}^2-\ell^2)}}\int_{0}^{f_{L2}}\frac{\Delta p(f_{L1})\mathrm{d}f_{L1}}{\sqrt{(f_{L2}^2-f_{L1}^2)}}, \qquad (4.1)$$

where G and ν represent the rock's elastic properties, i.e., shear modulus and Poisson's ratio, respectively. Eq. 1 can also be written in terms of any other combination of linear elastic constants. For instance, instead of G, Young's modulus, E, can be used because

$$E=2(1+\nu)G. \qquad (4.2)$$

The pressure distribution $\Delta p(f_{L1})=p(f_{L1})-\sigma_H(f_{L1})$ over the fracture is the difference between hydraulic fluid pressure, p, in the fracture and in-situ normal stress component, σ_H, in the rock before fracturing and perpendicular to the fracture walls. In other words, σ_H represents the in-situ minimum horizontal stress. Furthermore, x/L, f_{L2}, and f_{L1} are all fractions of fracture half-length. This unpleasant-looking double integral (Eq. 4.1) provides much information on fracture width and fracture equilibrium conditions.

The most simple case is a uniformly distributed load, Δp=constant, over the full fracture length ($2L$). Then

$$w(x)=\frac{2(1-\nu)L\Delta p}{G}\sqrt{(1-x^2)}. \qquad (4.3)$$

One can substitute z for x and h_f for $2L$. In the first case, one considers fracture length as measured from the well for plane-strain conditions. In the second case, one considers total fracture height and assumes plane-strain conditions in the plane perpendicular to propagation. Coupling of both conditions is not allowed with Eq. 4.1. Basic principles of linear fracture mechanics, such as Barenblatt's[1] equilibrium condition, require wedge-like closing at the fracture tip; thus,

$$\left(\frac{\partial w}{\partial x}\right)_{x/L=1}=0, \qquad (4.4)$$

as depicted in Fig. 4.1. This leads to

$$\int_0^L\frac{\Delta p(x)\mathrm{d}x}{\sqrt{(L^2-x^2)}}=\frac{K}{\sqrt{2L}} \qquad (4.5)$$

if cohesion of the rock material is taken into account. K, then, is the cohesion modulus, a material property or constant.

For Δp=constant, Eq. 4.5 provides

$$\frac{\pi}{2}\Delta p_{cr}=\frac{K}{\sqrt{2L}}. \qquad (4.6)$$

This equals Griffith's static fracture equilibrium condition (see also Chap. 3)

$$\Delta p_{cr}=\left[\frac{2EE_s}{\pi L(1-\nu^2)}\right]^{1/2} \qquad (4.7)$$

because Barenblatt's cohesion modulus, K, is related to Griffith's surface energy, E_s:

$$K^2=\frac{\pi EE_s}{1-\nu^2}. \qquad (4.8a)$$

Because Eq. 4.2 relates E and G,

$$K^2=\frac{2\pi GE_s}{1-\nu}. \qquad (4.8b)$$

*Retired from Koninklijke/Shell E&P Laboratorium.

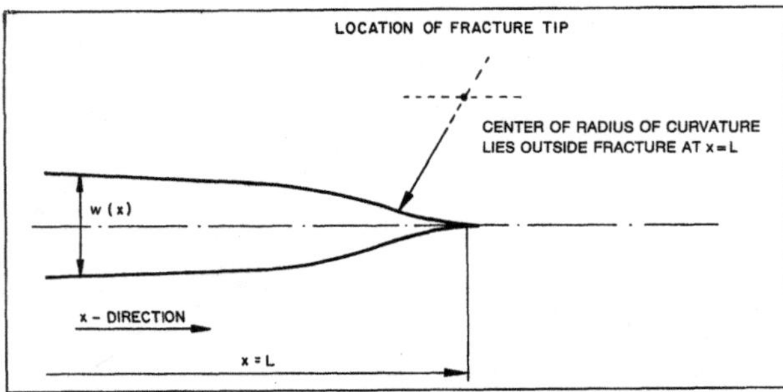

Fig. 4.1—Schematic representation of the wedge-like closing at the fracture tip.

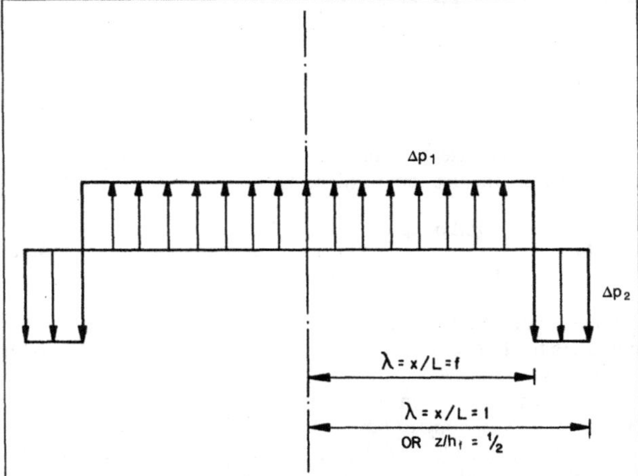

Fig. 4.2—Static fracture equilibrium resulting from penetration in the layer of higher in-situ stress, starting at λ = x/L = f.

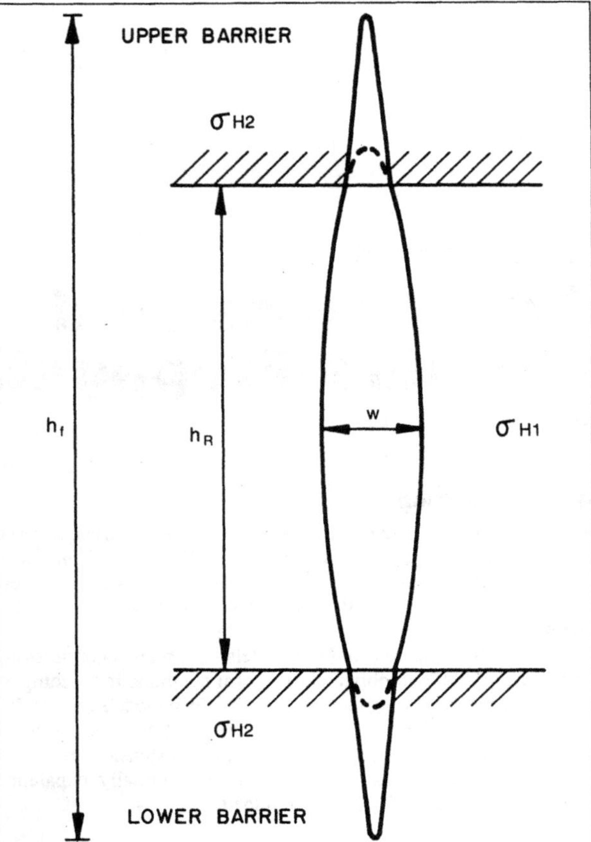

Fig. 4.3—Penetration of fracture in adjacent higher-stressed rock layers.

K is also related to the critical stress intensity factor as defined by Irwin:

$$K_c=\frac{\sqrt{2}}{\pi}K, \qquad (4.9a)$$

and thus

$$\Delta p_{cr}=K_c/\sqrt{L}. \qquad (4.9b)$$

Typical values for K_c of reservoir rocks are between 500 and 1,000 psi/$\sqrt{\text{in.}}$ [0.5 and 1.0 MPa/$\sqrt{\text{m}}$] (see table in Chap. 3). By definition, γ, K_c, and K represent material properties and thus have to be treated as constants not influenced by scale. Therefore, Eqs. 4.6 and 4.7 imply that for a uniformly distributed Δp from crack tip to its origin, the crack becomes unstable once a certain length is surpassed. So-called Griffith cracks are short. For $K_c=1{,}000$ psi/$\sqrt{\text{in.}}$, the uniform Δp required to extend a plane-strain fracture with a wing length, L, of 100 ft [30 m] is only 16 psi [110 kPa]. Hydraulically induced fractures of a size of interest for well stimulation derive their equilibrium condition from other loading conditions. In practice, two simplified cases are of fundamental interest.

Static equilibrium is obtained for low or even negligible K_c values, provided a step function for the loading condition prevails as follows:

Δp_1 for $0<\lambda<A_{f\ell}$,

and

Δp_2 for $A_{f\ell}<\lambda<1$.

Fig. 4.2 introduces the dimensionless fracture-length coordinate $\lambda=x/L$.

For this loading condition, Eq. 4.5 ensures tip stress equilibrium if

$$\Delta p_1\int_0^f\frac{d\lambda}{\sqrt{(1-\lambda^2)}}+\Delta p_2\int_f^1\frac{d\lambda}{\sqrt{(1-\lambda^2)}}$$

$$=\frac{K}{\sqrt{2L}}=\frac{\pi K_c}{2\sqrt{L}}$$

or

$$\Delta p_1=\Delta p_2+\frac{\dfrac{K_c}{\sqrt{L}}-\Delta p_2}{\dfrac{2}{\pi}\text{arc sin} f}. \qquad (4.10)$$

The condition of Eq. 4.10 extends Griffith's crack-stability requirement for fracture penetration into higher-stressed layers. Of course, it simplifies matters because it neglects gravity effects and assumes the same stress level in top and bottom layers, but it keeps the mathematics simple. If $L\rightarrow \frac{1}{2}h_f$ (half of fracture height), $\Delta p_1=p-\sigma_{H1}$, and $\Delta p_2=p-\sigma_{H2}$, such that $\sigma_{H2}>p>\sigma_{H1}$, an equilibrium crack height is obtained for a given fluid pressure level, p, for the condition

$$\sigma_{H2}-\sigma_{H1}=\frac{\pi}{2}\,\frac{\left[\dfrac{K_c\sqrt{2}}{\sqrt{h_f}}+\sigma_{H2}-p\right]}{\text{arc sin} f}$$

with $f=h_R/h_f$ (see Fig. 4.3). This provides

$$h_R=h_f\sin\left\{\frac{\pi}{2}\left[\frac{K_c\sqrt{2}+\pi(\sigma_{H2}-p)\sqrt{h_f}}{(\sigma_{H2}-\sigma_{H1})\sqrt{h_f}}\right]\right\}. \qquad (4.11a)$$

For sufficiently large h_f, this expression simplifies in practical terms to approximately

$$\frac{h_R}{h_f} \sim \sin\left(\frac{\pi}{2}\frac{\sigma_{H2}-p}{\sigma_{H2}-\sigma_{H1}}\right) \quad \text{(4.11b)}$$

or

$$\frac{h_R}{h_f} \sim \sin\left[\frac{\pi}{2}\left(1-\frac{p-\sigma_{H1}}{\sigma_{H2}-\sigma_{H1}}\right)\right]. \quad \text{(4.11c)}$$

It follows from this relationship that h_R/h_f, the inverse of which is the percentage of fracture-height penetration into adjacent layers, is governed primarily by the in-situ stress contrast $(\sigma_{H2}-\sigma_{H1})$, hence the interest in the determination of actual stress-contrast levels.

This result can be applied to predict fracture height in a Perkins-Kern[2] (PK) model, provided that the basement/overburden stress level in relation to that prevailing in the reservoir is known.

This theory implies that for a given reservoir height, a higher stress level in adjacent layers than in the reservoir rock stops fracture growth in the vertical direction. Because basement and overburden are frequently stiffer than the reservoir rock and stiff layers are stressed more than weak layers for the same tectonic strain imposed, this theory seems to be realistic for explaining fracture containment.

Barenblatt's equilibrium condition (Eq. 4.5) also explains *mobile equilibrium* in the propagation direction. Above the critical pressure, a constant overpressure, $\Delta p=p-\sigma_H$, in the fracture to the tip cannot ensure equilibrium conditions for slow propagation—i.e., in the absence of kinetic energy contributions. Mobile equilibrium is ensured, however, if the fracturing fluid pressure lags behind. For this purpose, the Zheltov-Khristianovitch[3] condition applies in impermeable rock:

$$\Delta p_1=p-\sigma_H, \text{ for } 0<\lambda<\lambda_0 \text{ as before,}$$

and

$$\Delta p_2=-\sigma_H, \text{ for } \lambda_0<\lambda<1.$$

In permeable rock, the loading between λ_0 and 1 has to be written as

$$\Delta p_2=p_t-\sigma_H,$$

where p_t takes a value between zero and reservoir pore pressure, depending on how fast the pressure sink at the tip is filled with respect to the speed of fracture propagation.[4] In practical calculations, the latter aspect is usually left out. This seems justified in view of additional assumptions that must be made to obtain workable design formulas. Therefore, taking $p_t=0$, using Eq. 4.5, and neglecting the effect of $K_c/\sqrt{L}$ for simplicity, one obtains the condition

$$\lambda_0=\sin\left(\frac{\pi}{2}\frac{\sigma_H}{p}\right) \quad \text{(4.12)}$$

for the "wetted" fracture length relative to the total fracture length. This ensures mobile equilibrium even for λ_0 close to unity, where p is only slightly higher than σ_H.

Thus, with the help of a simple step-function schematic, both concepts of static and mobile (vs. dynamic or high-speed) equilibrium can be explained.

Fracture mechanics specifies the fracture shape as a function of fluid pressure in that fracture. On the other hand, fluid mechanics specifies the fluid pressure distribution in that fracture if the shape is known. For one-dimensional transient fluid flow in a rectangular fracture, the continuity equation

$$\frac{\partial q}{\partial x}+q_\ell+\frac{\partial A}{\partial t}=0$$

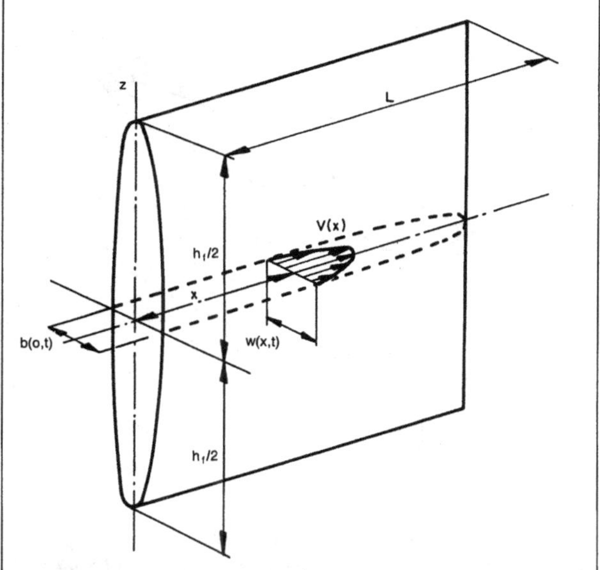

Fig. 4.4—Schematic representation of linearly propagating fracture with laminar fluid flow according to Perkins and Kern.[2]

applies, where

$q(x,t)$ = flow rate (volume per unit time) through a cross section,
x = constant of the fracture,
q_ℓ = volume rate of fluid loss to the formation per unit length of fracture, and
$A(x,t)$ = cross-sectional area of the fracture.

The flow rate, q, is related to the fluid pressure gradient. q_ℓ is determined by the flow field of the losses into the surrounding porous medium as governed by Darcy's law. These relationships have to be solved for pressures, fracture widths, and fracture lengths while the fracture-mechanics conditions are satisfied simultaneously. Even for the most simplifying assumptions, the combined set of equations is too complicated to be solved analytically. Hence, the following methods have been introduced to overcome this problem.

4.3 2D Models Without Leakoff

Two more or less complementary models to describe hydraulically induced fracture propagation in rocks have emerged for design purposes. Both models include a rectangular and a radial (circular) propagation mode. But because the majority of hydraulically induced fractures propagate in a vertical plane fed by injection over considerable intervals, the rectangular fracture is analyzed in most detail. The contrasting models are known as the PK[2] and the Geertsma-de Klerk[5] (GdK) models. (See Ref. 5 for England and Green's equation.) The literature on these concepts is extensive. In addition to the original papers,[2,5] there are numerous others. An important forerunner to the GdK study is the work by Zheltov and Khristianovitch,[3] who introduced the concept of mobile equilibrium—i.e., slow-moving fracture propagation as a result of hydraulic action. Le Tirant and Dupuy[6] were the first to present a calculation procedure for both rectangular and radial fracture propagation that makes use of the theory of mobile equilibrium, but they did not present much detail about mathematical handling of the scheme. Nordgren[7] improved the PK model considerably, resulting in the PKN model. More recently Abé *et al.*[8] published a solution scheme for the radially expanding propagation mode similar to the GdK model. Finally, Biot *et al.*[9] introduced a classic mathematical procedure to refine the solution of the GdK model concept.

This section starts with a prediction of fracture dimensions for an assumed rectangular contour according to PKN and GdK models.

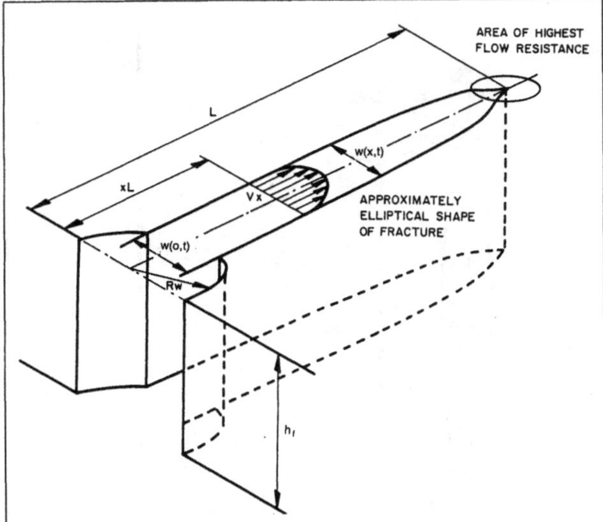

Fig. 4.5—Schematic representation of linearly propagating fracture with laminar fluid flow according to Geertsma and de Klerk.[5]

Then the prediction of the dimensions of a radially expanding fracture is discussed. Fluid loss is introduced in Secs. 4.4 and 4.5. The treatment of non-Newtonian fluid flow is the subject of Sec. 4.6, and examples of application are given in Sec. 4.7.

The PKN Model. In the PK model for vertical linear fracture propagation (Fig. 4.4), the assumptions are as follows.

1. The fracture has a fixed height, h_f, independent of fracture length.
2. The fracturing fluid pressure, p, is constant in vertical cross sections perpendicular to the direction of propagation.
3. Reservoir rock stiffness, its resistance to deformation under the action of p, prevails in the vertical plane. In other words, each vertical cross section deforms individually and is not hindered by its neighbors.
4. Accordingly, in these cross sections, Eq. 4.3 relates height, h_f, fluid pressure, p, and local fracture width. The cross sections obtain an elliptic shape with maximum width in the center:

$$w(x,t)=\frac{(1-\nu)h_f(p-\sigma_H)}{G}. \qquad (4.13)$$

5. The fluid pressure gradient in the propagating or x direction is determined by the flow resistance in a narrow, elliptical flow channel. For Newtonian flow behavior (with viscosity, μ, and flow rate, q),

$$\frac{\partial(p-\sigma_H)}{\partial x}=-\frac{64}{\pi}\frac{q\mu}{w^3h_f}. \qquad (4.14)$$

6. The fluid pressure in the fracture falls off toward the tip or leading edge such that at $x=L$, $p=\sigma_H$ for unspecified reasons.

The original theory neglected the influence of the growth rate of fracture width on the flow rate; i.e., it was assumed that in the absence of fluid losses,

$$\frac{\partial q}{\partial x}=0.$$

Although this assumption is reasonable if fluid loss dominates the material balance, it causes a significant numerical error in the case of little or no leakoff. Nordgren[7] corrected this growth-rate effect and rewrote the continuity equation in the form

$$\frac{\partial q}{\partial x}=-\frac{\pi h_f}{4}\frac{\partial w}{\partial t}. \qquad (4.15)$$

Elimination of $(p-\sigma_H)=\Delta p$ and q from Eqs. 4.13 through 4.15 provides a nonlinear partial-differential equation in terms of $w(x,t)$:

$$\frac{G}{64(1-\nu)h_f\mu}\frac{\partial^2 w^2}{\partial x^2}-\frac{\partial w}{\partial t}=0, \qquad (4.16)$$

subject to the initial condition

$w(x,0)=0$ for $t=0$,

and boundary conditions

$w(x,t)=0$ for $x>L(t)$,

$q(0,t)=q_0$ for a one-sided fracture, or

$q(0,t)=½q_0$ for a two-sided fracture.

The shape of the fracture then takes the form

$$w(x,t)=w(x,0)(1-x/L)^{1/4},$$

and the fracture volume amounts to

$$V=\frac{\pi}{5}Lh_f w(0,t)=q_0t.$$

The GdK Model. The fracture propagation model for vertical rectangular fracture propagation (Fig. 4.5) forms, to a degree, the counterpart of the PK theory.

The assumptions now include the following.

1. Again a fixed fracture height, h_f, is assumed.
2. Rock stiffness is taken into account in the horizontal plane only. As a result, fracture width does not depend on fracture height, except through the boundary condition at the wellbore that specifies a constant total injection rate, q. Of course, the flow rate per unit fracture height, q/h_f, influences fracture width, but width is constant in the vertical direction because the theory is based on the plane-strain condition, which was applied to derive a mechanically satisfying model in individual horizontal planes. However, applied in practice over full productive intervals, the model yields relatively large fracture widths that seem to be closer to reality in many field cases than the narrower fractures predicted by the PK theory. The reasons for this tendency are not yet fully explained.
3. The fluid pressure gradient in the propagating direction is determined by the flow resistance in a narrow rectangular slit of variable width in the vertical direction:

$$p(0,t)-p(x,t)=p_f-p=\frac{12\mu q_0}{h_f}\int_0^x\frac{dx}{w^3(x,t)}. \qquad (4.17)$$

The equilibrium condition dictated by applied mechanics is

$$\int_0^L\frac{p(x,t)dx}{\sqrt{L^2-x^2}}=\frac{\pi}{2}\sigma_H+\frac{K}{\sqrt{2L}}. \qquad (4.18)$$

Eq. 4.17 applies to a one-sided fracture taking the full injection rate q_0. As in the original PK theory, the assumption $\partial q/\partial x=0$ is made. The x correction term has been introduced by Biot *et al.*[9]

Following the proposition made by Zheltov and Khristianovitch,[3] the fluid pressure distribution that satisfies Eq. 4.18 can be approximated by

$p_f=p_0$ for $0<\lambda<L_0/L$,

TABLE 4.1—EQUATIONS FOR FRACTURE LENGTH, MAXIMUM FRACTURE WIDTH, AND INJECTION PRESSURE FOR CONSTANT INJECTION RATE

$L(t)$	$w(0,t)$	$p(0,t)-\sigma_H$
PKN Model		
$C_1\left[\dfrac{Gq_0^3}{(1-\nu)\mu h_f^4}\right]^{1/5} t^{4/5}$	$C_2\left[\dfrac{(1-\nu)q_0^2\mu}{Gh_f}\right]^{1/5} t^{1/5}$	$\dfrac{C_3}{H_f}\left[\dfrac{Gq_0^3\mu L}{(1-\nu)^3}\right]^{1/4}$
GdK Model		
$C_4\left[\dfrac{Gq_0^3}{(1-\nu)\mu h_f^3}\right]^{1/6} t^{2/3}$	$C_5\left[\dfrac{(1-\nu)q_0^3\mu}{Gh_f^3}\right]^{1/6} t^{1/3}$	$\dfrac{C_6}{2H_f}\left[\dfrac{Gq_0\mu h_f^3}{(1-\nu)^3L^2}\right]^{1/4}$

Observe that p_0 increases with fracture length and thus treatment time for PKN models and decreases with fracture length for GdK-type models.

and

$$p_f=0 \text{ for } L_0/L<\lambda<1,$$

with $L_0/L=\lambda_0$ close to unity. This provides for the condition of "wetted" fracture length

$$\lambda_0=L_0/L=\sin\frac{\pi}{2}\left(\frac{\sigma_H}{p_f}-\frac{K_c}{p_f\sqrt{L}}\right) \quad\text{.................(4.19)}$$

to start the calculation. This approximation is frequently good enough to prevent further refinements, such as that examined by Daneshy.[10] Furthermore, the contribution of K_c to L_0/L is neglected in the GdK formulation, as in Eq. 4.12. With the exception of the immediate surroundings of the fracture tip, the shape of the fracture in the horizontal plane is elliptic with maximum fracture width at the wellbore:

$$w(0,t)=\frac{2(1-\nu)L(p_f-\sigma_H)}{G}. \quad\text{......................(4.20)}$$

This relationship is similar to Eq. 4.13 in the PK theory, taking $L=\frac{1}{2}h_f$. What remains is the determination of the p_f level, dictated by the sharp drop in fluid pressure near the fracture tip that closes in a wedge-like shape. A good approximation to determine the fluid flow resistance in the fracture turned out to be

$$\int_0^{\lambda_0}\frac{w^3(0,t)}{w^3(x,t)}d\lambda\sim\frac{7}{4}(1-\lambda_0^2)^{-1/2}. \quad\text{...................(4.21)}$$

TABLE 4.2—VALUES FOR C_1 THROUGH C_6 IN TABLE 4.1

	One Wing	Two Wings
PK		
C_1	0.60	0.395
C_2	2.64	2.00
C_3	3.00	2.52
PKN		
C_1	0.68	0.45
C_2	2.50	1.89
C_3	2.75	2.31
GdK		
C_4	0.68	0.48
C_5	1.87	1.32
C_6	2.27	1.19

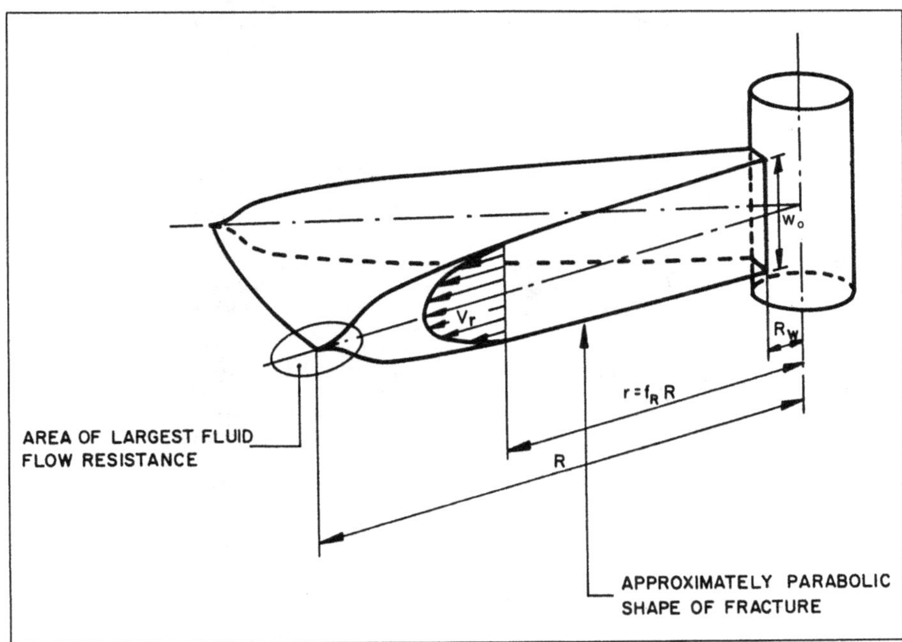

Fig. 4.6—Schematic representation of radially propagating fracture with laminar fluid flow (GdK model).

TABLE 4.3—VALUES OF $[e^{\alpha^2} \operatorname{erfc} \alpha + (2\alpha/\sqrt{\pi}) - 1]$ (from Ref. 12)

α	$e^{\alpha^2} \operatorname{erfc} \alpha$	$[e^{\alpha^2} \operatorname{erfc} \alpha + (2\alpha/\sqrt{\pi}) - 1]$
0.00	1.00000	0.00000
0.02	0.97783	0.00039
0.04	0.95642	0.00155
0.06	0.93574	0.00344
0.08	0.91576	0.00603
0.10	0.89646	0.00929
0.12	0.87779	0.01320
0.14	0.83974	0.01771
0.16	0.84228	0.02282
0.18	0.82538	0.02849
0.20	0.80902	0.03470
0.22	0.79318	0.04142
0.24	0.77784	0.04865
0.26	0.76297	0.05635
0.28	0.74857	0.06451
0.30	0.73460	0.07311
0.32	0.72106	0.08214
0.34	0.70792	0.09157
0.36	0.69517	0.10139
0.38	0.68280	0.11158
0.40	0.67079	0.12214
0.42	0.65912	0.13304
0.44	0.64779	0.14428
0.46	0.63679	0.15584
0.48	0.62609	0.16771
0.50	0.61569	0.17988
0.52	0.60588	0.19234
0.54	0.59574	0.20507
0.56	0.58618	0.21807
0.58	0.57687	0.23133
0.60	0.56780	0.24483
0.62	0.55898	0.25858
0.64	0.55039	0.27256
0.66	0.54203	0.28676
0.68	0.53387	0.30117
0.70	0.52593	0.31580
0.72	0.51819	0.33062
0.74	0.51064	0.34564
0.76	0.50328	0.36085
0.78	0.49610	0.37624
0.80	0.48910	0.39180
0.82	0.48227	0.40754
0.84	0.47560	0.42344
0.86	0.46909	0.43950
0.88	0.46274	0.45571
0.90	0.45653	0.47207
0.92	0.45047	0.48858
0.94	0.44455	0.50523
0.96	0.43876	0.52201
0.98	0.43311	0.53892
1.00	0.42758	0.55596
1.05	0.41430	0.59910
1.10	0.40173	0.64295
1.15	0.38983	0.68746
1.20	0.37854	0.73259
1.25	0.36782	0.77830
1.30	0.35764	0.82454
1.35	0.34796	0.87127
1.40	0.33874	0.91847
1.45	0.32996	0.96611

TABLE 4.3—VALUES OF $[e^{\alpha^2} \operatorname{erfc} \alpha + (2\alpha/\sqrt{\pi}) - 1]$ (from Ref. 12) (continued)

α	$e^{\alpha^2} \operatorname{erfc} \alpha$	$[e^{\alpha^2} \operatorname{erfc} \alpha + (2\alpha/\sqrt{\pi}) - 1]$
1.50	0.32159	1.01415
1.55	0.31359	1.06258
1.60	0.30595	1.11136
1.65	0.29865	1.16048
1.70	0.29166	1.20991
1.75	0.28497	1.25964
1.80	0.27856	1.30964
1.85	0.26241	1.35991
1.90	0.26651	1.41043
1.95	0.26084	1.46118
2.00	0.25540	1.51215
2.05	0.23016	1.56334
2.10	0.24512	1.61472
2.15	0.24027	1.66628
2.20	0.23559	1.71803
2.25	0.23109	1.76994
2.30	0.22674	1.82201
2.35	0.22255	1.87424
2.40	0.21850	1.92661
2.45	0.21459	1.97912
2.50	0.21081	2.03175
2.60	0.20361	2.13740
2.70	0.19687	2.24350
2.80	0.19055	2.35001
2.90	0.18460	2.45690
3.00	0.17900	2.56414
3.10	0.17372	2.67169
3.20	0.16873	2.77954
3.30	0.16401	2.88766
3.40	0.15954	2.99602
3.50	0.15529	3.10462
3.60	0.15127	3.21343
3.70	0.14743	3.32244
3.80	0.14379	3.43163
3.90	0.14031	3.54099
4.00	0.13700	3.65052
4.10	0.13383	3.76019
4.20	0.13081	3.87000
4.30	0.12791	3.97994
4.40	0.12514	4.09001
4.50	0.12248	4.20019
4.60	0.11994	4.31048
4.70	0.11749	4.42087
4.80	0.11514	4.53136
4.90	0.11288	4.64194
5.00	0.11070	4.75260
5.20	0.10659	4.97417
5.40	0.10277	5.19602
5.60	0.09921	5.41814
5.80	0.09589	5.64049
6.00	0.09278	5.86305
6.20	0.08986	6.08581
6.40	0.08712	6.30874
6.60	0.08453	6.53184
6.80	0.08210	6.75508
7.00	0.07980	6.97845
7.20	0.07762	7.20195
7.40	0.07556	7.42557
7.60	0.07361	7.64929
7.80	0.07175	7.87311

TABLE 4.3—VALUES OF $[e^{\alpha^2}\text{erfc}\,\alpha+(2\alpha/\sqrt{\pi})-1]$ (from Ref.12) (continued)

α	$e^{\alpha^2}\text{erfc}\,\alpha$	$[e^{\alpha^2}\text{erfc}\,\alpha+(2\alpha/\sqrt{\pi})-1]$
8.00	0.06999	8.09702
8.20	0.06830	8.32101
8.40	0.06670	8.54508
8.60	0.06517	8.76923
8.80	0.06371	8.99344
9.00	0.06231	9.21772
9.20	0.06097	9.44206
9.40	0.05969	9.66645
9.60	0.05846	9.89090
9.80	0.05727	10.11539
10.00	0.05614	10.33993

After substituting this expression into Eq. 4.17 and realizing further that the fracture volume of a one-sided fracture amounts approximately to

$$V=h_fLw(0,t)\int_0^1(1-\lambda^2)^{1/2}d\lambda=\frac{\pi}{4}h_fLw(0,t)=q_0t, \quad\text{.....(4.22)}$$

one finally obtains

$$L(t)=0.68\left[\frac{Gq_0^3}{\mu(1+\nu)h_f^3}\right]^{1/6}t^{2/3} \quad\text{.....(4.23)}$$

and

$$w(0,t)=1.87\left[\frac{(1+\nu)\mu q_0^3}{Gh_f^3}\right]^{1/6}t^{1/3}. \quad\text{.....(4.24)}$$

Summary of Results for Both Models. Tables 4.1 and 4.2 summarize the results for fracture length, maximum fracture width, and injection pressure for the PK, PKN, and GdK models. For details, consult the original papers.[2,5,7]

Radially Propagating Fractures. The modeling of radial fracture propagation by the PK and GdK models differs only because of the assumed hydraulic-fluid pressure distribution (see Fig. 4.6). In both concepts, the fluid pressure travels logarithmically from pressure p_0 at the entrance $(r=r_w)$ as a result of viscous flow resistance:

$$p=p_0-\frac{6\mu q}{\pi\bar{w}^3}\ln\frac{r}{r_w}. \quad\text{.....(4.25)}$$

In the PK concept, the pressure at the fracture tip is assumed to equal the rock stress perpendicular to the fracture plane, σ_H. In the GdK concept, p reduces without fluid influx to zero at $r=R_0$, a value close to R, the actual fracture radius. Stress equilibrium is pursued at $r=R$ with Barenblatt's theory:

$$\int_{\rho_w}^1\frac{\rho p(\rho)d\rho}{\sqrt{(1-\rho^2)}}=\sigma_H \quad\text{.....(4.26)}$$

(plus a cohesion term), where $\rho=r/R$.

To have an idea about the value of $R_0/R=\rho_0$, the following approximation can be used:

$$(1-\rho_0)\sim\frac{G}{10(1-\nu)R}\left(\frac{\mu q}{\sigma_H{}^4}\right)^{1/3}. \quad\text{.....(4.27)}$$

The total fluid pressure drop in the fracture is much larger in the GdK approach than in the PK modeling. The lever effect is more pronounced in the GdK model. As a result, wider fractures are predicted for the GdK radial model than with the PK model, although a similar dependence on variables is found. One obtains

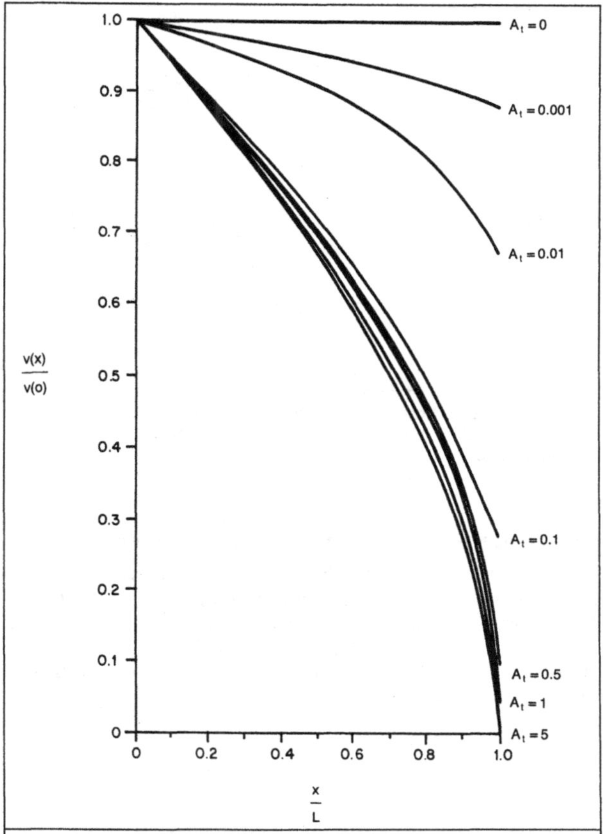

Fig. 4.7—Fracturing fluid velocity distribution between parallel walls as a function of $A_t=h_fLK_t^2/q_0w$.

$$w(0)=C_7\left[\frac{(1-\nu)\mu q_0R}{G}\right]^{1/4}, \quad\text{.....(4.28)}$$

where the PK approach obtains a value of $C_7=1.4$ and the GdK approach, a value of 2.15.

In the PK model, the average fracture width is taken to be

$$\bar{w}=\frac{2}{3}w(0), \quad\text{.....(4.29a)}$$

but in the GdK model,

$$\bar{w}=\frac{8}{15}w(0),$$

because

$$w^2=w_0^2(1-R^2). \quad\text{.....(4.29b)}$$

Because $V=qt$, for GdK,

$$R=\left[\frac{15qt}{8\pi w(0)}\right]^{1/2}, \quad\text{.....(4.30a)}$$

and for PK,

$$R=\left[\frac{3qt}{2\pi w(0)}\right]^{1/2}. \quad\text{.....(4.30b)}$$

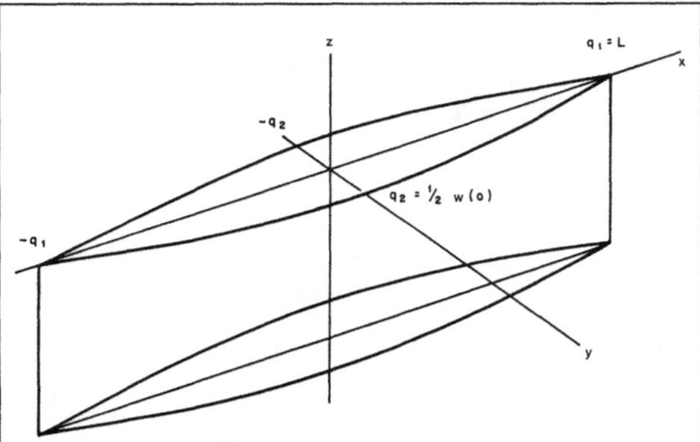

Fig. 4.8—Definition of generalized coordinates q_1 and q_2 in Biot *et al.*'s[9] variational formulation of hydraulic fracture propagation.

4.4 Effect of Leakoff on Fracture Dimensions

The Carter Equation. The basic elements to describe fluid-loss effects on fracture propagation are from Carter.[11] In his theory, fracture width and height are assumed to be constant; only fracture length, $L(t)$, is a variable. Injection rate is also assumed to be constant. The relevant fluid-loss velocity function, normal to the fracture faces, is assumed to take the form

$$v_\ell = \frac{K_\ell}{\sqrt{t-\tau}}, \quad (4.31)$$

where K_ℓ is the overall fluid-loss coefficient as measured in laboratory filtration tests and τ represents the time at which filtration starts. For details of the leakoff mechanism, see Chap. 8. For the geometry as specified above, the material balance is

$$\frac{dV}{dt} = wh_f\frac{dL}{dt} = q - q_\ell, \quad (4.32)$$

where

$$q_\ell = \int_0^{A_{f\ell}} v_\ell dA_{f\ell} = \int_0^{A_{f\ell}} v_\ell \frac{dA_{f\ell}}{d\tau} d\tau = K_\ell \int_0^t \frac{dA_{f\ell}}{d\tau} \frac{d\tau}{\sqrt{t-\tau}} \quad (4.33)$$

and $A_{f\ell} = 2h_fL$. This formulation provides

$$\frac{q}{h_f} = 2K_\ell \int_0^t \frac{dL}{d\tau} \frac{d\tau}{\sqrt{t-\tau}} + w\frac{dL}{dt}, \quad (4.34)$$

which leads to

$$V = \frac{qt}{\alpha^2}\left(e^{\alpha^2} \text{ erfc } \alpha + \frac{2\alpha}{\sqrt{\pi}} - 1\right) = wh_fL, \quad (4.35)$$

where

$$\alpha = \frac{2K_\ell}{w}\sqrt{\pi t}.$$

A similar mathematical approach has been used by Marx and Langenheim[12] to predict heat-conduction losses to cap and base rock during hot-fluid injection into a reservoir. They presented tabulated values for the expression in parentheses in Eq. 4.35 for α values ranging from 0 to 10 (see Table 4.3). For $\alpha > 4$,

$$e^{\alpha^2} \text{ erfc } \alpha \sim \frac{1}{\alpha\sqrt{\pi}},$$

and thus for large α values,

$$wh_fL \rightarrow \frac{2q_0t}{\alpha\sqrt{\pi}}, \quad (4.36)$$

which leads to

$$L \rightarrow \frac{q_0\sqrt{t}}{\pi h_f K_\ell}. \quad (4.37)$$

This seems a proper place to call attention to the low fracturing fluid efficiency for large α values. If fluid efficiency is defined as the ratio between fracture volume, wh_fL, and injected volume, q_0t, for large α values, then

$$\eta_f = \frac{wh_fL}{q_0t} \rightarrow \frac{2}{\alpha\sqrt{\pi}}. \quad (4.38)$$

Eq. 4.38 stresses the importance of keeping α as low as possible. For $\alpha = 10$, this efficiency is only 11%. The major part of the injected fluids is squeezed into the formation, causing only potential problems.

Furthermore, the velocity distribution in the fracture depends strongly on the fluid efficiency, thereby influencing the fluid pressure distribution. Fig. 4.7 shows the distribution of fracture fluid velocity between parallel walls as a function of the dimensionless parameter $A_\ell = h_fLK_\ell^2/q_0w$. For $A_\ell > 1$, theory shows that the flow rate falls off rapidly according to

$$\frac{q(x)}{q_0} = \frac{v(x)}{v(0)} = 1 - \frac{2}{\pi}\text{arc sin } \lambda. \quad (4.39)$$

For these conditions, the factor A_ℓ relates to the fluid efficiency according to

$$A_\ell = \frac{wh_fL}{q_0t}\frac{K_\ell^2t}{w^2} = \eta_f\frac{\alpha^2}{4\pi} = \frac{1}{\pi^2\eta_f}. \quad (4.40)$$

Spurt loss, V_{sp}, can be included in this context by

$$\frac{dV}{dt} = q - q_\ell - V_{sp}\frac{dA_{f\ell}}{dt},$$

and thus

$$\frac{q}{h_f} = (w + 2V_{sp})\frac{dL}{dt} + 2K_\ell\int_0^t \frac{dL}{d\tau}\frac{d\tau}{\sqrt{t-\tau}}. \quad (4.41)$$

Comparison of Eqs. 4.33 and 4.40 shows that spurt losses can be accounted for by replacing w with an effective fracture width $(w+2V_{sp})$; thus,

$$\alpha=\frac{2K_\ell}{(w+2V_{sp})}\sqrt{\pi t}\,.$$

4.5 Fracture Dimensions With Leakoff—Incorporation Into the PK Theory by Nordgren

Perkins and Kern did not consider fluid losses into the formation, a shortcoming remedied by Nordgren. The local-continuity equation now becomes (compare with Eq. 4.15)

$$\frac{\partial q}{\partial x}+\frac{\pi h_f}{4}\frac{\partial w}{\partial t}+q_\ell=0. \quad \text{(4.42)}$$

The local q_ℓ in this context is specified as the loss per unit fracture length:

$$q_\ell=\frac{2h_fK_\ell}{\sqrt{t-\tau(x)}}. \quad \text{(4.43)}$$

The extended form of Eq. 4.16 then becomes

$$\frac{G}{64(1-\nu)h_f\mu}\frac{\partial^2 w^4}{\partial x^2}-\frac{\partial w}{\partial t}-\frac{8K_\ell}{\pi\sqrt{t-\tau(x)}}=0, \quad \text{(4.44)}$$

subject to initial condition

$w(x,0)=0$

and boundary conditions

$w(x,t)=0$ for $x>L$

and

$$-\frac{\partial w^4(0,t)}{\partial x}=\frac{256\mu(1-\nu)\,q}{\pi G} \text{ for a one-sided fracture.}$$

This set of equations can be solved numerically (see Ref. 7). Because the PK model seems most applicable for long-duration treatments, the approximation

$$\frac{\partial q}{\partial x}+q_\ell=0$$

seems appropriate under these circumstances. This leads to

$$L\int_0^1\frac{d\lambda}{\sqrt{t-\tau(x)}}=\frac{q_0}{2h_fK_\ell},$$

where τ depends on $\lambda=x/L$ such that $\tau[L(t')]=t'$ for $0<t'<t$. The solution is then similar to Carter's:

$$L\to\frac{1}{\pi}\frac{q_0\sqrt{t}}{h_fK_\ell}$$

and

$$\frac{q(\lambda)}{q_0}=1-\frac{2}{\pi}\text{arc sin }\lambda.$$

This is a good approximation for

$h_fLK_\ell^2>q_0w(0)$.

For this condition,

$$\Delta p^4(0)=\frac{4^4}{\pi}\frac{\mu G^3q_0L}{h_f^4(1-\nu)^3}\int_0^1\left(1-\frac{2}{\pi}\text{arc sin }\lambda\right)d\lambda.$$

This provides

$$w(0)=4\left(\frac{2}{\pi^3}\right)^{1/4}\left[\frac{\mu(1-\nu)q_0{}^2}{Gh_fK_\ell}\right]^{1/4}t^{1/8} \quad \text{(4.45)}$$

and

$$\Delta p(0)=4\left(\frac{2}{\pi^3}\right)^{1/4}\left[\frac{\mu Gq_0^2}{h_f^5(1-\nu)^3K_\ell}\right]^{1/4}t^{1/8}. \quad \text{(4.46)}$$

Incorporation Into the GdK Model. Contrary to the PK model, the original GdK model is assumed to simulate field conditions best for small α values—i.e., for low loss coefficients and for small treatment times. The material balance is considered in the overall form as Carter did; thus, as in Eq. 4.32, inclusive of spurt loss,

$$\frac{dV}{dt}=q_0-q_\ell-2V_{sp}h_f\frac{dL}{dt}.$$

The fracture volume differs from Carter's because of the elliptic shape in the horizontal cross sections. Thus, as mentioned earlier,

$$\frac{dV}{dt}=\frac{\pi}{4}h_f\left[w(0)\frac{dL}{dt}+L\frac{dw(0)}{dt}\right]=\frac{\pi}{4}h_f\left[w(0)+L\frac{dw(0)}{dL}\right]\frac{dL}{dt}$$

and

$$q_\ell=2K_\ell h_f\int_0^t\frac{dL}{d\tau}\frac{d\tau}{\sqrt{t-\tau}}.$$

For the GdK model, it was found earlier that

$$L\frac{dw(0)}{dL}=½w(0).$$

Note that from $L=C_1t^{2/3}$ and $w(0)=C_2t^{1/3}$, it follows that

$$w(0)=\frac{C_2}{\sqrt{C_1}}L^{½},$$

$$\frac{dw(0)}{dL}=½\frac{C_2}{\sqrt{C_1}}L^{-½},$$

and thus

$$L\frac{dw(0)}{dL}=½\frac{C_2}{\sqrt{C_1}}L^{½}=½w(0).$$

The material balance now reads

$$\frac{q_0}{h_f}=\left[\frac{3\pi}{8}w(0,t)+2V_{sp}\right]\frac{dL}{dt}+2K_\ell\int_0^t\frac{dL}{d\tau}\frac{d\tau}{\sqrt{t-\tau}}. \quad \text{(4.47)}$$

If w were independent of fracture length, L, in Eq. 4.47, the solution would accord with Carter's. This solution can still be used because in the product $w(dL)/dt$, the variations in w as functions of time are much smaller than those in dL/dt. However, the "constant" w value has to be chosen such that the true solution for no leakoff is met. Therefore, $w(0,t)$ in Eq. 4.47 has to be replaced by $2/3w(0,t_p)$, when t_p represents the time when the pumps stop. The factor 2/3 makes the approximate solution numerically equal to the "exact" one for $K_\ell = V_{sp} = 0$.

For constant $w(0,t_p)$, Eq. 4.47 is similar to Eq. 4.33 and thus leads to

$$\frac{dL}{dt}=\frac{q_0}{h_f}\left[\frac{4}{\pi w(0,t_p)+8V_{sp}}\right]e^{\alpha_L^2}\ \text{erfc}\ \alpha_L, \quad \ldots\ldots\ldots\ldots (4.48)$$

where

$$\alpha_L=\frac{8K_\ell\sqrt{\pi t}}{\pi w(0,t_p)+8V_{sp}}.$$

Integration of Eq. 4.48 shows that

$$L=\frac{q_0}{16\pi h_f K_\ell^2}[\pi w(0,t_p)+8V_{sp}]\left(\frac{2\alpha_L}{\sqrt{\pi}}-1+e^{\alpha_L^2}\ \text{erfc}\ \alpha_L\right) \quad \ldots\ldots\ldots\ldots (4.49)$$

for a one-sided fracture. We can use

$$w(0,t_p)=2.27\left[\frac{(1-\nu)\mu q_0 L^2}{Gh_f}\right]^{1/4}$$

as a second relationship between L and w and to solve for both unknowns, either graphically or numerically by computer.

Biot's Variational Formulation To Obtain a Solution of the GdK Model Incorporating Large Fluid Losses. The approximations made by the GdK model to arrive at closed-form results can be analyzed with the Lagrangian variational method of Biot *et al.*[9] A prerequisite is that a suitable assumption has to be made about the fracture shape. The method is historically a forerunner of the finite-element method (FEM). It bridges the gap between purely numerical and purely analytic computational procedures in applied mechanics. The numerical FEM is based on variational principles, where the boundary is subdivided into many elements. A displacement function is introduced in each element. Displacement continuity is introduced at so-called nodal points. In an analytic variational method like Biot *et al.*'s, a shape function for the entire deformed boundary, instead of a displacement function per element, is introduced to start the calculations. If the result is not very sensitive to the proper choice of shape function, the calculation procedure is considerably simplified compared with the FEM, a bonus in the precomputer age and still an advantage today. The variational approach is based on the set of equations

$$\frac{\partial E_p}{\partial q_i}+\frac{\partial F_d}{\partial \dot{q}_i}=Q_i, \quad \ldots\ldots\ldots\ldots (4.50)$$

where

- E_p = elastic potential energy of the system,
- F_d = dissipation function that may incorporate fluid-loss effects,
- Q_i = driving or braking forces not derivable from a potential,
- q_i = generalized coordinates, and
- $\dot{q}_i$ = time derivative of q_i, $\partial q_i/\partial t$.

For a 2D problem, the generalized coordinates are chosen as fracture length, L, and fracture width at the origin, $w(0)$. To relate $w(0)$ with $w(x)$ for the GdK model, an elliptic shape function is introduced. With $q_1 = L$ and $q_2 = ½w(0)$ (see Fig. 4.8),

$$E_p=\frac{\pi}{16}\left(\frac{G}{1-\nu}\right)w^2(0),$$

$$F_d=6\mu\int_0^L\frac{q^2(x)}{w^3(x)}dx,$$

$$Q_1=\frac{\pi}{4}p(0)w(0)-E_s,$$

and

$$Q_2=\frac{\pi}{2}p(0)L.$$

The variational formulation is an attractive method to obtain an analytic solution for the GdK model for large fluid losses. Under these circumstances, the increase in fracture length becomes almost independent of fracture width:

$$L\rightarrow\frac{q_0}{\pi h_f K_\ell}t^{½}.$$

Similarly, the flow rate becomes almost insensitive of fracture-width variations, because

$$q(x)\rightarrow q_0\left(1-\frac{2}{\pi}\text{arc sin}\ \lambda\right),$$

as in Nordgren's solution for the PK model under these conditions.

The above implies that the dissipation function, F_d, for these conditions takes the form

$$F_d=\frac{6\mu}{h_f^2}\int_0^L\frac{q^2(x)}{w^3(x)}dx=\frac{6\mu L q_0^2}{w^3(0)h_f^2}\int_0^1\frac{\left(1-\frac{2}{\pi}\text{arc sin}\ \lambda\right)^2}{f^3(\lambda)}d\lambda$$

if $\lambda = x/L$. For $f(\lambda)$, try an elliptic shape function $f(\lambda)=(1-\lambda^2)^{½}$ such that one obtains a shape parameter

$$F_1=\int_0^1\frac{\left(1-\frac{2}{\pi}\text{arc sin}\ \lambda\right)^2}{(1-\lambda^2)^{3/2}}d\lambda.$$

Numerical evaluation shows that $F \cong 0.75$. Next, express q_0/h_f in terms of L, $w(0)$, and their time derivatives. From Eq. 4.37,

$$\frac{q_0^2}{h_f^2}=4\pi^2K_\ell^2\dot{L}^2t=2\pi^2K_\ell^2L\dot{L}.$$

Because these approximations are valid for $\alpha > 4$, the dissipation function becomes

$$F_d=12\pi^2\frac{\mu K_\ell^2 F_1}{w(0)^3}L^2\dot{L}.$$

From the variational formulation, Eq. 4.50, and after differentiation,

$$48\pi\mu K_\ell^2 F_1 L^2 = p_0 w^4(0)$$

and

$$Gw(0)=2(1-\nu)p_0L.$$

Elimination of p_0 yields

$$\frac{Gw^5(0)}{(1-\nu)}=96\pi\mu K_\ell^2F_1L^3.$$

Resubstituting Eq. 4.37 into the above relationship provides the desired expression for fracture width:

$$w(0)=1.58\left[\frac{(1-\nu)\mu q_0^3F_1}{Gh_f^3K_\ell}\right]^{1/5}t^{3/10}. \quad (4.51)$$

The relationship between L and $w(0)$ becomes

$$w(0)=3.13\left[\frac{(1-\nu)\mu K_\ell^2F_1}{G}\right]^{1/5}L^{3/5}. \quad (4.52)$$

Note that for this solution,

$$p_0=0.79\pi\left[\frac{G^4h_f^2K_\ell^4\mu F_1}{q_0^2(1-\nu)^4}\right]^{1/5}t^{-1/5}.$$

Therefore, for large α values (long-duration or massive hydraulic fracturing, involving large fluid losses), $w(0)$ is no longer proportional to $L^{1/2}$, but to $L^{3/5}$, and the entrance pressure still declines in time.

Radially Expanding Fractures with Fluid Loss. For a radially expanding fracture, Eq. 4.47 becomes

$$q_0=\left[\frac{3\pi}{5}w(0,t)+2\pi V_{sp}\right]\frac{dR^2}{dt}+2\pi K_\ell\int_0^t\frac{dR^2}{d\tau}\frac{d\tau}{\sqrt{t-\tau}}. \quad (4.53)$$

To accommodate the "true" solution for no fluid loss, one can replace $w(0,t)$ in Eq. 4.53 by $8/9w(0,t_p)$ and use the solution

$$R\frac{dR}{dt}=\frac{q_0}{4\pi}\left[\frac{15}{4w(0,t_p)+15V_{sp}}\right]e^{\alpha_r^2}\operatorname{erfc}\alpha_r,$$

where

$$\alpha_r=\frac{15K_\ell\sqrt{\pi t}}{4w(0,t_p)+15V_{sp}}.$$

Integration leads to

$$R^2=\frac{q_0}{30\pi^2K_\ell^2}[4w(0,t_p)+15V_{sp}]\left(\frac{2\alpha_r}{\sqrt{\pi}}-1+e^{\alpha_r^2}\operatorname{erfc}\alpha_r\right). \quad (4.54)$$

Because a radially fractured propagation mode will usually occur only during the early stages of fracture growth—i.e., for small t values such as those used in minifracture testing—one can use Eq. 4.28 as a second equation relating R and (0):

$$w(0)\sim2.15\left[\frac{\mu q_0R(1-\nu)}{G}\right]^{1/4}.$$

4.6 Effect of Non-Newtonian Fluid Flow on Fracture Dimensions

All the computational models described so far assume Newtonian fluid flow of the fracturing fluid during fracture propagation. However, most fracturing fluids used today exhibit non-Newtonian behavior in some way (see Chap. 9). Most popular is a flow law originally proposed by Ostwald and de Waele,[13] which is usually referred to as the "power law." This formulation replaces Newton's one-parameter flow law with a two-parameter one that still allows for rather simple mathematical manipulations. The major technical drawback is that this formulation predicts an infinitely high "apparent" viscosity once the shear rate goes to zero. Thus, the danger exists that the flow resistance is overestimated at low flow rates, unless a cutoff or truncation procedure, like the so-called Sprigg's model, is used.[14]

For the power law, the relationship between pressure gradient and flow rate in flow between parallel walls amounts to

$$-\frac{dp}{dx}=2K_a\left(\frac{6q}{h_f}\right)^{n'}\frac{1}{w(2n'+1)}, \quad (4.55)$$

where

$$K_a = K'\left(\frac{2n'+1}{3n'}\right)^{n'},$$

K' = consistency index, and
n' = flow behavior index.

For Newtonian fluids, $n'=1$ and $K'=\mu$; for gels, $n'<1$. Eq. 4.55 is suitable for applications of Biot's variational formulation because the dissipation function, F_d, is actually

$$F_d=-½\int_0^L q(x)\frac{\partial p}{\partial x}dx.$$

Thus, according to Eq. 4.55, $\partial p/\partial x$ can be translated again in terms of q. In particular, an analytic solution for the GdK model for large fluid losses can thus be obtained for power-law fluids:

$$F_d=\frac{6^{n'}K_a}{h_f^{n'}}\int_0^L\frac{q(x)^{(n'+1)}}{w(x)^{(2n'+1)}}dx.$$

4.7 Example Calculations

Effect of Cohesion. To become more familiar with the effects of rock cohesion, take, for example, a plausible critical stress-intensity factor, K_c, of 455 psi/$\sqrt{\text{in.}}$ [5,000 kPa/$\sqrt{\text{cm}}$] and a static shear modulus, G, of this rock of 1.45×10^6 psi [10^7 kPa].

For a Poisson's ratio of $\nu=0.20$, Eq. 4.9a provides a cohesion modulus of

$$K=1{,}013\ \text{psi}\sqrt{\text{in.}}\ [11\ 135\ \text{kPa}\sqrt{\text{cm}}].$$

Eq. 4.8a predicts a surface energy of

$$E_s=90.2\times10^{-3}\ \text{psi/in.}\ [1.58\ \text{kPa/cm}].$$

Such rock properties imply that for a fracture length of only 39.4 in. [100 cm], the critical extension pressure is only, according to Eq. 4.9b, $\Delta p_c=72.5$ psi [500 kPa]. This critical overpressure with respect to the least principal rock stress reduces to 7.25 psi [50 kPa] for a fracture length of 328 ft [100 m].

Static Equilibrium. Assume a total lateral and minimum rock stress of 2,750 psi [19 MPa] in the target rock, a slightly higher stress level of 3,045 psi [21 MPa] in the adjacent top and bottom layers, and a fracturing pressure in between of 2,900 psi [20 MPa]. With the help of Eq. 4.11b, one obtains h_R/h_f or $\sin(\pi/2\times0.50)=0.71$ for the fracture penetration into the adjacent layers. More accurately, if for the same stress and pressure data, h_f=1,181 in. [3×10³ cm], one obtains, from Eq. 4.11a and neglecting cohesion, $h_R/h_f = \sin(\pi/2\times0.56)=0.77$.

Thus, the fracture will penetrate into the adjacent formations such that the actual height, h_f, is 30 to 40% larger than formation thickness. If this is not realized in the fracture-extension calculations, one would predict a longer fracture than actually will be obtained.

Mobile Equilibrium. With the same values for formation stress and fracturing pressure as above, Eq. 4.12 predicts for the ratio between "wetted" fracture length and total fracture length:

$$\lambda_0=\sin(\pi/2\times0.95)=0.997.$$

This demonstrates that λ_0 is usually close to unity.

Prediction of Fracture Dimensions.
Summary of assumed rock data:

G = 1.45×10⁶ psi [10⁷ kPa].
ν = 0.20.
K_c = 455 psi$\sqrt{\text{in.}}$ [5000 kPa$\sqrt{\text{cm}}$].
h_R = 100 ft [30 m].
σ_H = 2,750 psi [1.9×10⁴ kPa].
σ_{Ha} = 3,045 psi [2.1×10⁴ kPa].

Fracture treatment data:

q_0 = 10 bbl/min [1.6 m³/min].
t = 200 minutes.
μ = 100 cp [1.67×10⁻⁶ kPa·min].
K_ℓ = 0.0015 ft/min^½ [0.00046 m/min^½].
V_{sp} = 0.01 gal/ft² [0.41 L/m²].

Data for the non-Newtonian case:

n' = 0.63.
K' = 0.0044 lbf-sec^n/ft² [0.211 N·sec^n/m²].

The No-Fluid-Loss Case. Assume a two-winged fracture. The results of Table 4.1 can be used to determine the length of each wing and the wellbore fracture width.

For the PKN model, we use for the fracture height the static equilibrium result $h_R/h_f=0.77$; thus, h_f=30/0.77=39 m [128 ft]. The length of each wing after 200 minutes becomes[7] (C_1=0.45) (the individual parametric values in the equations below are entered in SI units)

$$L=0.45\left[\frac{10^7\times(1.6)^3}{0.8\times1.67\times10^{-6}\times(39)^4}\right]^{1/5}(200)^{4/5}$$

$$=830 \text{ m [2,723 ft]},$$

and the wellbore fracture width becomes

$$w(0)=1.89\left[\frac{0.8\times(1.6)^2\times1.67\times10^{-6}}{10^7\times39}\right]^{1/5}(200)^{1/5}$$

$$=8.43\times10^{-3} \text{ m [0.33 in.]}.$$

For the GdK model, the fracture height is taken as 98 ft [30 m], because it is a plane-strain configuration with C_4=0.48 (two-wing concept, Table 4.1):

$$L=0.48\left[\frac{10^7\times(1.6)^3}{0.8\times1.67\times10^{-6}\times30^3}\right]^{1/6}(200)^{2/3}$$

$$=530 \text{ m [1,738 ft]},$$

and the wellbore fracture width (C_5=1.32) is

$$w(0)=1.32\left[\frac{0.8\times(1.6)^3\times1.67\times10^{-6}}{10^7\times30^3}\right]^{1/6}(200)^{1/3}$$

$$=12.48\times10^{-3} \text{ m [0.49 in.]}.$$

Incorporation of Fluid Loss. If $\alpha \gg 1$ is assumed, then Eq. 4.37 applies (two-wing case); for the PKN model,

$$L\rightarrow\frac{\tfrac{1}{2}q_0t^{1/2}}{\pi h_fK_\ell}=\frac{1}{2}\,\frac{1.6\times(200)^{1/2}}{\pi\times0.00046\times39}=200 \text{ m [656 ft]},$$

and for the GdK model,

$$L\rightarrow\frac{39}{30}\times200=260 \text{ m [853 ft]}.$$

For the PKN case, it is assumed that the fluid loss also takes place in the adjacent layers, which is usually not the case. In addition, the effect of spurt loss will be neglected in this simple example. For the incorporation of both effects in the calculations, see Chap. 11.

With the above-mentioned simplifications, the limiting case of large α values provides these estimates.

With the PKN model for the estimate of wellbore fracture width,

$$w(0)=4\left(\frac{2}{\pi^3}\right)^{1/4}\left[\frac{0.8\times1.67\times10^{-6}\times(0.8)^2}{10^7\times39\times0.00046}\right]^{1/4}(200)^{1/8}$$

$$=5.8\times10^{-3} \text{ m [0.23 in.]},$$

and with the GdK model and Eq. 4.51 for the same dimension,

$$w(0)=1.58\left[\frac{0.8\times1.67\times10^{-6}\times(0.8)^3\times0.75}{10^7\times30\times0.00046}\right]^{1/5}(200)^{3/10}$$

$$=10.2\times10^{-3} \text{ m [0.40 in.]}.$$

A check on the value of α [$\alpha=(2K_\ell/w)\sqrt{\pi t}$] provides, for the PKN model,

$$\alpha=\frac{2\times0.00046}{5.8\times10^{-3}}(\pi\times200)^{1/2}=3.96,$$

and for the GdK model,

$$\alpha=\frac{2\times0.00046}{10.2\times10^{-3}}(\pi\times200)^{1/2}=2.26.$$

Therefore, a more detailed calculation is required with a computer program. But as a first approximation, the α value gives useful indications about the expected dimensions.

Returning to the rough values for α and $w(0)$ in the GdK model, Eq. 4.49 provides (still neglecting spurt losses)

$$L=\frac{0.8\times\pi\times10.2\times10^{-3}}{16\times\pi\times30\times(0.00046)^2}\left[\frac{5.74}{\sqrt{\pi}}-1+0.18\right]$$

$$=194 \text{ m [636 ft]}.$$

Radially Expanding Fractures. Basic data:

q_0 = 1.6 m³/min [10 bbl/min].
t = 50 minutes.

μ = 1.67×10^{-6} kPa·min [100 cp].
K_ℓ = 0.00046 m/min$^{1/2}$ [0.0015 ft/min$^{1/2}$].
G = 10^7 kPa [1.45×10^6 psi].
ν = 0.25.

Starting with the no-fluid-loss case, for the GdK model, Eq. 4.28 provides

$$w(0)=2.15\left(\frac{0.8\times1.67\times10^{-6}\times1.6\times R}{10^7}\right)^{1/4}\text{ m},$$

and Eq. 4.30b provides

$$R=\left(\frac{1.6\times50}{\pi\bar{w}}\right)^{1/2}\text{ m},$$

where, according to Eq. 4.29b, $\bar{w}$=0.53 $w(0)$.
Combining these three equations yields

$w(0)$=4.63×10^{-3} m [0.18 in.],

$\bar{w}$=2.45×10^{-3} m [0.096 in.], and

R=101.9 m [334 ft].

Because radially expanding fractures are usually expected for testing purposes only (at least in oil/gas well stimulation), the large-α case does *not* provide useful information if fluid losses are taken into account. GdK modeling suggests the use of Eq. 4.54 in combination with Eq. 4.28 in a dimensionless form. This involves the introduction of the dimensionless numbers

$$K_u=\frac{K_\ell\sqrt{t}}{w(0)}$$

and

$$K_R=\frac{R^2K_\ell}{q_0\sqrt{t}},$$

together with

$$K_s=\frac{K_\ell\sqrt{t}}{V_{sp}}$$

if spurt losses are taken into account (see Ref. 5). For example, without spurt loss, Eq. 4.54 changes into

$$K_R=\frac{1}{\pi^2}\left\{1-\frac{2}{15K_u}\left[1-\exp\left(\frac{15\sqrt{\pi}K_u}{4}\right)^2\right.\right.$$

$$\left.\left.\times\operatorname{erfc}\left(\frac{15\sqrt{\pi}K_u}{4}\right)\right]\right\},$$

and Eq. 4.28 can be written in the same form, using

$$K_{\mu R}=256\left(\frac{q_0}{K_{f\ell}{}^3\sqrt{t}}\right)^3\left(\frac{\mu}{Gt}\right)^2$$

and

$$K_u^8K_R=\frac{1}{K_{\mu R}}.$$

A plot of K_u vs. K_R with $K_{\mu R}$ as a parameter provides the information needed. A simple computer program can do the same.
In the example, $K_{\mu R}$=35.84, which leads to K_R=0.088 and K_μ=0.85. Accordingly,

$$w(0)=\frac{K_\ell\sqrt{t}}{K_u}=\frac{0.46\sqrt{50}}{0.85}\times10^{-3}$$

$$=3.83\times10^{-3}\text{ m [0.15 in.]}$$

and

$$R=\left(\frac{0.088\times1.6\times\sqrt{50}}{0.46\times10^{-3}}\right)^{1/2}=46.5\text{ m [152 ft]}.$$

Nomenclature

$A_{f\ell}$ = $2h_fL$=fracture surface area exposed to fluid loss, m [ft]
A_ℓ = dimensionless fluid-loss parameter defined as $h_fLK_\ell^2/q_0w$ for Carter's model
E = Young's modulus of rock formation, kPa [psi]
E_p = elastic potential energy of system, kPa·m^2 [lbf]
E_s = surface energy as defined by Barenblatt, kPa/cm [psi/in.]
f_{L1} = fraction of fracture length
f_{L2} = fraction of fracture length
$f(\lambda)$ = shape function of fracture, function of λ
F_d = dissipation function defined by Biot

$$F_1=\int_0^1\frac{\left(1-\frac{2}{\pi}\sin^{-1}\lambda\right)^2}{(1-\lambda^2)^{3/2}}d\lambda,\text{ or in more general terms}$$

$$\int_0^1\frac{[q(x)/q(0)]^2}{f^3(\lambda)}d\lambda$$

G = shear modulus of rock formation, kPa [psi]
h_f = fracture height, m [ft]
h_R = reservoir interval height, m [ft]
K = cohesion modulus, kPa/$\sqrt{\text{cm}}$ [psi/$\sqrt{\text{in.}}$]
K' = power-law consistency index, Pa·s [lbm-sec^n/ft^2]

$$K_a=K'\left(\frac{2n'+1}{3n'}\right)^{n'},\text{ Pa·s [lbm-sec}^n\text{/ft}^2]$$

K_c = critical stress-intensity factor, kPa/$\sqrt{\text{cm}}$ [psi/$\sqrt{\text{in.}}$]
K_ℓ = overall fluid-loss coefficient, m/$\sqrt{\text{min}}$ [ft/$\sqrt{\text{min}}$]
L = fracture length, m [ft]
$\dot{L}$ = $\partial L/\partial t$, m/s [ft/min]
n' = power-law flow behavior index
p = fluid pressure, kPa [psi]
p_t = fracture pressure in tip region, $0<p_t<p_r$, kPa [psi]
Δp = $p-\sigma_H$, kPa [psi]
q = flow rate, m^3/min [bbl/min]
q_i = generalized coordinates (i=1,2...)
$\dot{q}_i$ = time derivative of q_i
q_ℓ = fluid-loss rate, m^3/min [bbl/min]
q_0 = flow rate at fracture entrance, $x/L=\lambda=0$, m^3/min [bbl/min]
Q_i = driving or braking forces, not derivable from a potential
r = radial coordinate, m [ft]
r_w = wellbore radius, m [ft]
R = fracture radius, m [ft]
R_0 = wetted fracture radius, m [ft]
t = time, minutes
t_p = time at which pumping stops, minutes
v_ℓ = fluid-loss velocity function
$v(x)$ = fluid velocity at x, m/s [ft/sec]

$v(0)$ = fluid velocity at fracture entrance, m/s [ft/sec]
V = fracture volume, m^3 [ft^3]
V_{sp} = spurt loss, m^3 [ft^3]
w = fracture width, m [in.]
$w(0)$ = fracture width at injection point, m [in.]
$\bar{w}$ = average width of radially propagating fracture, m [ft]
$\dot{w}$ = $\partial w/\partial t$, m/min [in./min]
x = fracture length coordinate, m [ft]
y = fracture height coordinate, m [ft]
α = dimensionless fluid-loss parameter defined as

$$\frac{2K_\ell}{w}\sqrt{\pi t} \text{ for Carter's model}$$

α_L = dimensionless fluid-loss parameter including spurt loss defined as

$$\frac{8K_\ell\sqrt{\pi t}}{\pi w(0,t_p)+8V_{sp}}, \text{ linearly expanding fracture}$$

α_R = dimensionless fluid-loss parameter including spurt loss defined as

$$\frac{15K_\ell\sqrt{\pi t}}{4w(0,t_p)+15V_{sp}}, \text{ radially expanding fracture}$$

η_f = fracturing-fluid efficiency, ratio between fracture volume and total amount of fluid injected
λ = x/L=dimensionless fracture coordinate
λ_0 = dimensionless "wetted" fracture length=L_0/L
μ = Newtonian fracturing-fluid viscosity, kPa·min [cp]
ν = Poisson's ratio of rock formation
ρ = r/R=dimensionless radial fracture coordinate
ρ_0 = dimensionless radial "wetted" fracture length, R_0/R
σ_H = in-situ normal rock stress, perpendicular to fracture face
τ = time at which fluid loss starts

References

1. Barenblatt, G.I.: "Mathematical Theory of Equilibrium Cracks," *Advances in Applied Mechanics* (1962) **7**, 55.
2. Perkins, T.K. and Kern, L.R.: "Widths of Hydraulic Fractures," *JPT* (Sept. 1961) 937–49; *Trans.*, AIME, **222**.
3. Zheltov, Y.P. and Khristianovitch, S.A.: "On the Mechanism of Hydraulic Fracturing of an Oil-Bearing Stratum," *Izvest. Akad. Nauk SSR, OTN* (1955) **5**, 3–41 (in Russian).
4. Warpinski, N.R.: "Measurement of Width and Pressure in a Propagating Hydraulic Fracture," *SPEJ* (Feb. 1985) 46–54.
5. Geertsma, J. and de Klerk, F.A.: "Rapid Method of Predicting Width and Extent of Hydraulically Induced Fractures," *JPT* (Dec. 1969) 1571–81; *Trans.*, AIME, **246**.
6. Le Tirant, P. and Dupuy, M.: "Fracture Dimensions Obtained During Hydraulic Fracturing Treatments of Oil Reservoirs," *Rev. Inst. Français du Pétrole* (1967) 44–98 (in French).
7. Nordgren, R.P.: "Propagation of a Vertical Hydraulic Fracture," *SPEJ* (Aug. 1972) 306–14; *Trans.*, AIME, **253**.
8. Abé, H., Mura, T., and Keer, L.M.: "Growth Rate of a Penny-Shaped Crack in Hydraulic Fracturing of Rocks," *J. Geophys. Res.* (1976) **81**, 5335.
9. Biot, M.A., Massé, L., and Medlin, W.L.: "A Two-Dimensional Theory of Fracture Propagation," *SPEPE* (Jan. 1986) 17–30.
10. Daneshy, A.A.: "On the Design of Vertical Hydraulic Fractures," *JPT* (Jan. 1973) 83–93; *Trans.*, AIME, **255**.
11. Carter, R.D.: Appendix to "Optimum Fluid Characteristics for Fracture Extension," by G.C. Howard and C.R. Fast, *Drill. and Prod. Prac.*, API (1957) 267.
12. Marx, J.W. and Langenheim, R.H.: "Reservoir Heating by Hot Fluid Injection," *Trans.*, AIME (1960) **219**, 312–15.
13. Bird, R.B. *et al.*: *Transport Phenomena*, John Wiley & Sons Publishing Co., New York City (1960) 11.
14. Bird, R.B.: "Polymer Fluid Dynamics," *Selected Topics in Transport Phenomena*, Chemical Engineering Progress Symposium Series 58 (1965) **61**, Chap. 6.

Chapter 5
Three-Dimensional Fracture-Propagation Models

R.J. Clifton, Brown U. and Terra Tek Inc.

5.1 Overview

This chapter discusses three-dimensional (3D) mathematical models developed to describe the fracturing process. It begins with a presentation of 3D models in which fracture height is allowed to vary with fluid injection and the vertical components of fluid flow are included. This is followed by a discussion of pseudo-3D models that have some of the attributes of fully 3D models but require less computational effort. Insight into the range of validity of the various types of fracture propagation models is provided by a series of example problems.

For both the fully 3D models and the pseudo-3D models, the overall approach is to subdivide the fracture into discrete elements and to solve the governing equations for these elements on a digital computer. These governing equations consist of (1) elasticity equations that relate the pressure on the crack faces to the crack opening, (2) fluid-flow equations that relate the flow of the fluid in the fracture to the pressure gradients in the fluid, and (3) a fracture criterion that relates the intensity of the stress state ahead of the crack front to the critical intensity necessary for tensile fracture of the rock. For each type of model, the governing equations are presented first, followed by a discussion of the method of solution. Finally, numerical examples are presented and comparisons made with predictions of other fracture-propagation models.

Because of the complexity of the fully 3D modeling of hydraulic fracturing, more background in solid mechanics and mathematics is used in developing the models here than in other chapters. Readers lacking this background should still find the chapter helpful, however, because the principal features of the model are presented in a manner that allows understanding of the main concepts without the need for working through lengthy derivations. Original papers that provide the necessary background and detailed derivations are listed in the References.

5.2 3D Models

Elasticity. As in the two-dimensional (2D) models discussed in Chap. 4, the formation is assumed to behave as a linear elastic solid in response to changes in crack-face pressures introduced by hydraulic fracturing. Because hydraulic fracturing is done in low-permeability reservoirs, changes in elastic response resulting from fluid migration within the formation are restricted to a sufficiently thin layer adjacent to the crack faces that these changes are assumed to have a negligible influence on the effective stiffness of the formation. Thus, poroelasticity effects are neglected and, strictly speaking, the elastic moduli are regarded to be the moduli for the ''undrained'' response of the rock. Isotropic elasticity is assumed.

For hydraulic fracturing of gas and oil reservoirs, the depths of the formations to be fractured are generally sufficiently large, compared with fracture dimensions, for the effects of the ground-level free surface to be negligible. Thus, the formation is assumed to be infinite in extent. The fracture is assumed to develop as a plane, vertical fracture oriented perpendicular to the direction of the minimum in-situ compressive stress. The elasticity problem solved is that of the change in stress and displacement fields caused by increasing the normal compressive stress $\sigma_{zz}(x,y,0)$ on Crack Surface A (Fig. 5.1) from its initial values $\sigma^0_{zz}(x,y,0)$ to the current pressure $p(x,y,t)$ in the fracturing fluid. Shear tractions on the crack surface are assumed to be zero. This fully 3D elasticity problem for an infinite medium is reduced to a 2D problem for a finite region by use of a surface-integral formulation. In this formulation, the change in normal stress on the crack plane is related to the crack opening, $w(x,y)$, by an integral of the form[1]

$$\Delta p(x,y) \equiv p(x,y) - \sigma^0_{zz}(x,y,0) = E_e \iint_A \vec{\nabla}' w \cdot \vec{\nabla}'(1/R) dA', \quad \ldots (5.1)$$

where the effective elastic modulus, E_e, is

$$E_e = \frac{G}{4\pi(1-\nu)}, \quad \ldots (5.2)$$

the gradient operator, $\vec{\nabla}'$, is

$$\vec{\nabla}' \equiv \frac{\partial}{\partial x'}\vec{i} + \frac{\partial}{\partial y'}\vec{j}, \quad \ldots (5.3)$$

and the distance, R, between point (x',y') at which the integrand is evaluated and point (x,y) at which the pressure is evaluated is

$$R = [(x-x')^2 + (y-y')^2]^{1/2}. \quad \ldots (5.4)$$

The integrand of Eq. 5.1 is obtained from the fundamental elasticity solution for the stress, normal to the plane $z=0$, resulting from a dislocation line segment on the plane $z=0$ with Burgers vector in the z direction. The dislocation segment is at point (x',y'), called the source point, and the stress is evaluated at point (x,y), the field point.

Eq. 5.1 is the form used in developing a numerical method for evaluating the crack openings for a given pressure distribution as long as the elastic moduli are uniform. Such uniformity of the moduli will be assumed in presenting the principal features of the numerical method; adjustments to account for differences in elastic moduli from layer to layer will be described once the principal features are presented.

A numerical method for finding an approximate solution $w(x,y)$ is obtained by covering the crack surface with a quadrilateral mesh, as shown in Fig. 5.2. The crack opening, $w(x,y)$, is represented in terms of local trial functions, $\phi_i(x,y)$, by

$$w(x,y) = \sum_i w_i \phi_i(x,y), \quad \ldots (5.5)$$

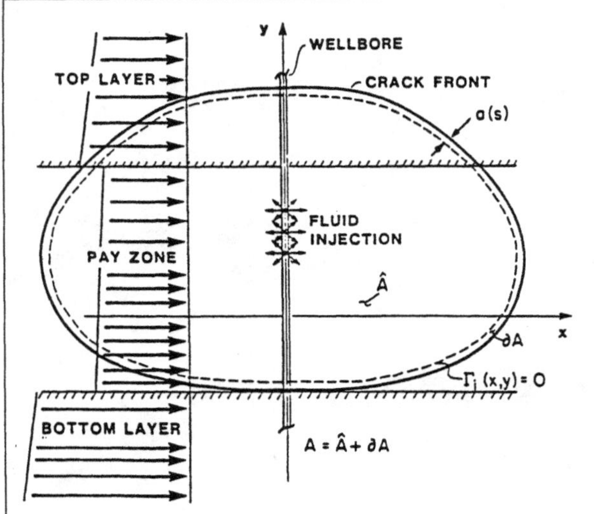

Fig. 5.1—Schematic of the hydraulic fracturing configuration.

where w_i is the displacement of the ith node. The function $\phi_i(x,y)$ has the value of unity at the ith node, varies linearly over adjacent triangles, vanishes along the opposite side of these triangles, and vanishes outside the adjacent triangles. For a node i along the inner boundary of the near-crack-tip zone of width a in Fig. 5.1, the function $\phi_i(x,y)$ varies as the square root of distance from the crack tip for points (x,y) in the near-crack-tip region, in agreement with the variation of crack openings near crack tips in elastic solids. The differential pressure is represented similarly:

$$\Delta p(x,y)=\sum_i \Delta p_i\tilde{\phi}_i(x,y), \quad (5.6)$$

where the local trial function, $\tilde{\phi}_i(x,y)$, is the same as $\phi_i(x,y)$ except for points (x,y) in the crack-tip region. In this region, the function $\tilde{\phi}_i(x,y)$ is taken to be uniform (at the value unity) with the understanding that the boundary conditions on the pressure are not known but are to be determined as part of the solution. Multiplying Eq. 5.1 by $\phi_i(x,y)$ and integrating over Fracture Surface A, we obtain a system of linear algebraic equations of the form

$$K\vec{w}=T\Delta\vec{p}, \quad (5.7)$$

where

$\vec{w}$ = column vector of nodal displacements,
K = stiffness matrix, and
$T\Delta\vec{p}$ = column vector of nodal forces, with $\Delta\vec{p}$=column vector of nodal pressures and T=sparse, symmetric matrix of element areas.

The matrix K is symmetric, positive definite, and well-conditioned as long as the aspect ratios for the triangular elements remain within reasonable limits.

Evaluation of the elements of the stiffness matrix K has been described previously on the basis of a variational approach[2]; only the essential ideas are given here. When points (x,y) and (x',y') are in different triangles, the contribution ΔK_{jk} to K_{jk} obtained from (x,y) sweeping over the jth triangle and (x',y') sweeping over the kth triangle is

$$\Delta K_{jk}=\frac{E_e}{2}\iint_{A_j}\phi_j(x,y)\left[(\vec{\nabla}'\phi_k)\cdot\oint_{S_k}\frac{\vec{n}_k ds}{R}\right]dxdy, \quad (5.8)$$

where A_j is the area of the jth triangle and $(\vec{\nabla}'\phi_k)$ is the gradient of the trial function for the kth triangle, which has unit exterior normal $\vec{n}_k$ and perimeter S_k. The integral in Eq. 5.8 is evaluated

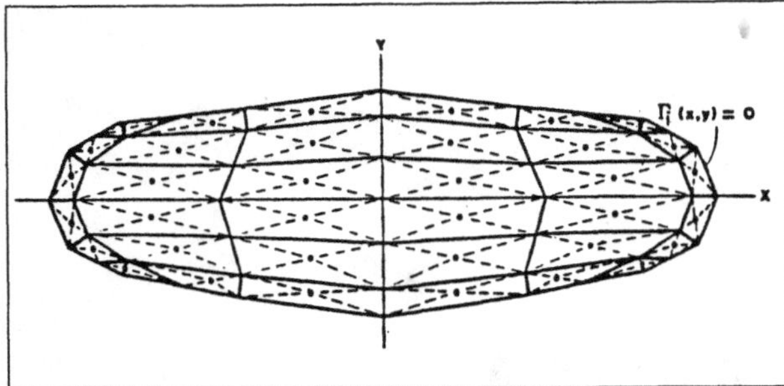

Fig. 5.2—Quadrilateral elements with triangular subelements.

by exact integration of the line integral, which depends on the coordinates (x',y'), followed by a three-point quadrature to evaluate the area integral. When points (x,y) and (x',y') are in the same triangle, it is advantageous to evaluate the integrals in Eq. 5.7 by using exact values of these integrals for isosceles triangles and making first-order corrections to account for deviations from isosceles triangles. Coefficients required for these integrals can be tabulated as functions of the height-to-base ratio of the triangle to eliminate re-evaluation of the integrals. When (x,y), (x',y'), or both are in the near-crack-tip Region ∂A, then the integral over this region is evaluated by substituting expressions for $\phi_j(x,y)$ or $\vec{\nabla}'\phi_k(x',y')$ on the basis of trial functions for the crack-tip region of the form

$$\phi_j(x,y)=\phi_j(s)[\zeta/a(s)]^{1/2}, \quad (5.9)$$

where ζ is the distance from crack tip in the direction of inward normal, $\zeta=a(s)$ at the inner boundary of the crack-tip region, and $\phi_j(s)=1$ at the jth node and decreases linearly to zero at the adjacent nodes along the crack front.

The matrix T on the right side of Eq. 5.7 is obtained by integration of the respective trial functions for $w(x,y)$ and $\Delta p(x,y)$. The contribution ΔT_{jk} from the jth triangle is

$$\Delta T_{jk}=-\frac{A_j}{6}a_k, \quad (5.10)$$

where

a_k = 1 for $k=j$,
a_k = ½ for k corresponding to other two vertices of jth triangle, and
a_k = 0 otherwise.

When (x,y) is in the near-crack-tip zone, the contribution to T_{jk} is obtained by taking $w(x,y)$ to have the form of Eq. 5.9 and assuming $p(x,y)$ to be uniform over the width $a(s)$.

Differences in elastic moduli from layer to layer are accounted for by the fundamental solution for an infinitesimal loop of unit crack opening in an infinite region consisting of two bonded, homogeneous half-spaces. This fundamental solution was obtained in closed form by Lee and Keer.[3] For use in the numerical method described here, it is convenient to express this solution as a factor f times the fundamental solution for a homogeneous infinite medium. Then the contribution to the stiffness matrix from points (x,y) in triangle j and points (x',y') in triangle k can be approximated by multiplying the contribution ΔK_{jk} for a homogeneous medium by the f applicable to the normal stress on the crack plane at (x_j,y_j) resulting from an infinitesimal disk of crack opening at (x_k',y_k'). The factor f has the form (see Eqs. 5.14)

$$f=f(x-x',y-y'',y'-y''; G_2/G_1,\nu_1,\nu_2), \quad (5.11)$$

where y'' is the y coordinate of the interface and G_2/G_1 is the ratio of the elastic shear moduli of the two layers. For definiteness, G_1 is the modulus of the layer containing the source point (x',y'); G_2 is the modulus of the other layer. The factor f is finite and is a smooth function of its arguments, except for a jump across the interface. When the distance R between (x,y) and (x',y') becomes small compared with the distance from the interface, f approaches unity.

Explicit expressions for f in Eq. 5.11 can be given in terms of the distance R (see Eq. 5.4), the distance $\hat{R}$,

$$\hat{R}=[(x-x')^2+(\bar{y}+\bar{y}')^2]^{1/2}, \quad (5.12)$$

where $\bar{y}=y-y''$, $\bar{y}'=y'-y''$, and the following material parameters:

$$\Gamma=G_2/G_1,\ \kappa_i=3-4\nu_i,\ S=(1-\Gamma)/(1+\Gamma),$$

$$\alpha=(1-\Gamma)/(1+\Gamma\kappa_1),\quad \beta=(\kappa_2-\Gamma\kappa_1)/(\Gamma+\kappa_2). \quad (5.13)$$

Two cases must be distinguished. Expressions for f for these two cases are as follows.[3]

Case 1 (source point and field point in same half space):

$$f=1+\frac{R^3}{\hat{R}^3}\left[S(1+\kappa_1)-\frac{3}{2}\alpha(\kappa_1^2-2\kappa_1+3)\right]$$

$$+\frac{3}{2}\frac{R^3}{\hat{R}^5}[12\alpha\bar{y}\bar{y}'-\alpha(3-\kappa_1)(\kappa_1-1)(\bar{y}+\bar{y}')^2]$$

$$+\frac{3R^3}{2\hat{R}(\hat{R}+|\bar{y}+\bar{y}'|)^2}(\beta-2S+2\alpha\kappa_1+\alpha\kappa_1^2-2S\kappa_1). \quad (5.14a)$$

Case 2 (source point and field point in different half spaces):

$$f=1+\frac{1}{2}[\alpha(\kappa_1-3)+\beta(\kappa_2-3)+(\kappa_1-\kappa_2)-2S(1+\kappa_1)]$$

$$+\frac{3R^2}{2(R+|\bar{y}-\bar{y}'|)^2}[-\kappa_1(1+\alpha)+\kappa_2(1-\beta)+2S(1+\kappa_1)]$$

$$+\frac{3(2R+|\bar{y}-\bar{y}'|)}{(R+|\bar{y}-\bar{y}'|)^2}|\alpha\bar{y}'-\beta\bar{y}|. \quad (5.14b)$$

Because of the smoothness of f, the approximation involved in introducing the factor f is very good as long as points (x,y) and (x',y') are not so near the interface that the triangles to which they belong cross the interface. Even when triangles cross the interface, this approximation (Eq. 5.11) is expected to be adequate for obtaining stiffness coefficients that give satisfactory approximations for the crack opening.

Fundamental solutions have not been obtained for cases of multiple interfaces separating layers of different elastic moduli. Consequently, in these cases, the stiffness coefficients cannot be modified in the straightforward way described for the case of a single interface. Nevertheless, modification procedures can be introduced that involve weighted averages of fundamental solutions for each interface regarded as the only interface. The procedures become rigorously valid when one interface becomes of dominant importance because of a combination of the strength of the modulus contrast across the interface and the proximity of the source and field points to the interface. In many cases, where one interface is not dominant, either the source and field points are so widely separated that an accurate modification of the stiffness coefficient is not required or the modulus contrast is so weak that the modification is small. Although such arguments can be introduced to support the use of weighted averages of single-interface solutions, the validity of this approach needs to be examined by comparisons with full numerical solutions for cases of multiple interfaces.

2D Fluid Flow. The fluid flow is generally idealized as that of the laminar flow of an incompressible, power-law fluid. The fluid is assumed to flow between essentially parallel porous walls. Leakoff through the walls occurs at a rate determined by the difference between the pressure in the fracturing fluid and the remote pore fluid pressure, as well as by the time elapsed since the local fracture surface was first exposed to the fracturing fluid. Integration of the governing equations through the width of the opening gives the 2D flow equations[4] consisting of the continuity equation

$$\frac{\partial q_x}{\partial x}+\frac{\partial q_y}{\partial y}=-q_L-\frac{\partial w}{\partial t}+q_I \quad (5.15)$$

and the pressure-gradient equations (cf. Table 9.1)

$$\frac{\partial p}{\partial x}+\eta'\left(\frac{|q|}{w^2}\right)^{n'-1}\frac{q_x}{w^3}=0 \quad (5.16a)$$

and

$$\frac{\partial p}{\partial y}+\eta'\left(\frac{|q|}{w^2}\right)^{n'-1}\frac{q_y}{w^3}=\rho F_y, \quad (5.16b)$$

where

q_x (q_y) = volume flow rate in x direction (y direction) per unit length in y direction (x direction),
$|q|$ = $(q_x^2+q_y^2)^{1/2}$ = resultant flow rate,
q_L = volume leakoff rate per unit fracture area, and
q_I = volume injection rate per unit fracture area (zero except for regions near the wellbore and adjacent to perforations).

The body force caused by the weight of the fluid, per unit volume, is represented by ρF_y. The viscosity parameter, η', is related to the usual power-law coefficients, K' and n', by

$$\eta'=2^{(n'+1)}K'(2+1/n')^{n'}. \quad (5.17)$$

The principal assumptions made in deriving Eqs. 5.15 and 5.16 are that inertia effects are negligible and that velocity gradients in the x-y plane are negligible compared with the gradients through the fracture width. These assumptions are expected to be quite realistic except near the perforations, where the details of the flow are not modeled. At sufficiently small values of the shearing rate ($|q|/w^2$), the power-law fluid is replaced by a Newtonian fluid (i.e., $n'=1$) to avoid singular coefficients in Eqs. 5.16.

The fluid-loss term, q_L, is obtained from a time-dependent leakoff relation of the form

$$q_L(x,y,t)=\frac{2C_L(p-p_f)}{\sqrt{t-\tau(x,y)}}, \quad (5.18)$$

where $\tau(x,y)$ is the time at which the position (x,y) was first exposed to the fracturing fluid. The coefficient C_L is the leakoff coefficient, K_ℓ, of Eq. 4.31, normalized with respect to the stress state at depth by dividing by the difference between the minimum in-situ compressive stress and the in-situ pore fluid pressure. The dependence of the fluid-loss rate on the pressure difference $p-p_f$ is consistent with the solution for one-dimensional (1D) flow into a semi-infinite porous medium with far-field pore pressure, p_f, and a constant pressure, p, maintained at the injection plane.[5] Because the change in pressure, $p-\sigma_{zz}^0$, is generally much less than the pressure difference, $p-p_f$, the leakoff obtained from Eq. 5.18 is approximately the same as that obtained from Eq. 4.31. However, use of Eq. 5.18 allows the leakoff to be distributed over the fracture area in a way that accounts, for example, for higher leakoff rates where the difference between the pressure at the injection face and the far-field pore pressure is higher. The particular normalization used—i.e., division by $[\sigma_{zz}^0(0,0)-p_f]$—does not imply any particular dependence of the leakoff coefficient on the state of stress and/or pore-fluid pressure. Instead, it simply states that the leakoff predicted by Eq. 5.18 will agree with that measured in experiments used for determining K_ℓ (see Eq. 4.31) when these experiments are done under in-situ conditions of stress and pore pressure and the pressure at the injection face of the sample is $\sigma_{zz}^0(0,0)$.

For computational purposes, it is advantageous to reformulate Eqs. 5.15 and 5.16 in terms of a single equation for the pressure distribution. Such an equation can be obtained by substituting q_x and q_y from Eq. 5.16 into Eq. 5.15. It is more convenient, however, to introduce a variational approach in which the equations for the unknown nodal pressures are obtained as the conditions for a function of these pressures to be at its minimum value. A functional $J(p)$ for such a variational approach can be written in the form[6]

$$J(p)=\iint_A F\left(p,\frac{\partial p}{\partial x},\frac{\partial p}{\partial y}\right)\mathrm{d}x\mathrm{d}y+\int_{\Gamma_i}\Big[aC_L(p-p_f)^2$$

$$+p\left(2vC_s+vw_a+\frac{2a}{3}\frac{\mathrm{d}w_a}{dt}\right)\Big]\mathrm{d}s, \qquad (5.19)$$

where

$$F\left(p,\frac{\partial p}{\partial x},\frac{\partial p}{\partial y}\right)=C_L(p-p_f)^2+p\left(\frac{\partial w}{\partial t}-q_I\right)+\left(\frac{n'}{n'+1}\right)w^2$$

$$\times\left(\frac{w}{\eta'}\right)^{1/n'}\left[\left(\frac{\partial p}{\partial x}\right)^2+\left(\frac{\partial p}{\partial y}-\rho F_y\right)^2\right]^{(n'+1)/2n'} \qquad (5.20)$$

The functional $J(p)$ has the units of power. The integrands represent power fluxes. The first three terms in F represent the power fluxes caused by injecting the fluid into the reservoir through the fracture walls, opening the fracture, and injecting fluid into the fracture in the vicinity of the perforations. The final term in F is the power dissipation caused by the flow of the viscous fluid in the fracture. The first two terms in the line integral account for the power flux of the pressure acting on the fluid being lost in the near-crack-tip zone of width a. The second of these terms corresponds to the spurt loss, which is assumed to occur instantaneously when the rock is exposed to the fracturing fluid. The spurt-loss coefficient, C_s, is a specified value for the spurt loss per unit area; v is the velocity of the crack front in the direction of its outward normal; and w_a is the opening at distance a from the crack front.

The form of Eq. 5.19 can be verified by confirming that any pressure distribution, $p(x,y)$, that minimizes $J(p)$ must satisfy Eqs. 5.15 and 5.16. If $p(x,y)$ is not prescribed on the inner boundary, $\Gamma_i(x,y)$, of the crack-tip region, then $J(p)$ is a minimum only if

$$q_n-vw_a-2C_La(p-p_f)-2vC_s-\frac{2a}{3}\frac{\mathrm{d}w_a}{dt}=0 \qquad (5.21)$$

on $\Gamma_i(x,y)=0$,

where q_n is the flow rate in the direction normal to the boundary. This "natural boundary condition" is simply the condition that the net rate of flow into the near-crack-tip region is balanced by the leakoff rate and volume expansion of this region. To include further details of the flow in the near-crack-tip region would require consideration of the velocity component perpendicular to the crack plane. Initial results for flow fields that rigorously satisfy the boundary conditions for an advancing crack, without leakoff, indicate that pressures and pressure gradients are well-behaved in the vicinity of the crack tip.[7] The absence of any singularities in the flow or pressure fields near the crack tip suggests that satisfactory solutions can be obtained without including the details of the flow in the immediate neighborhood of the crack tip.

A numerical method for obtaining the pressure $p(x,y)$ at the nodes of the mesh (Fig. 5.2) can be obtained by imposing the requirement that the functional $J(p)$ be minimized for $p(x,y)$ that vary linearly over each triangular element. Integration of the integrals in Eq. 5.19 over the area, A, and the boundary curve, $\Gamma_i(x,y)=0$, respectively, followed by differentiation with respect to the nodal pressures, gives a system of equations of the form

$$(D+C_LT)(\vec{p}-\vec{p}_f)=-\frac{T}{2}\frac{\partial\vec{w}}{\partial t}+\hat{\vec{b}}, \qquad (5.22)$$

where

$\vec{p}$ = vector of nodal pressures,

$D=D\left(\frac{\partial p}{\partial x},\frac{\partial p}{\partial y}-\rho F_y,w\right)$ = matrix that accounts for viscous resistance to fluid flow,

T = matrix of areas obtained previously in the elasticity problem, and

$\hat{\vec{b}}$ = vector comprising effects of fluid injection at wellbore, spurt loss at crack tip, and vw_a associated with moving inner boundary of crack-tip zone.

For a Newtonian fluid (i.e., $n'=1$), the coefficient D is independent of the pressure gradients $\partial p/\partial x$ and $\partial p/\partial y-\rho F_y$. In this case, Eq. 5.22 is linear in p for given values of $\vec{w}$ and $\partial\vec{w}/\partial t$. For a non-Newtonian fluid, the matrix D depends on the nodal pressures, so that the system of equations (Eq. 5.22) is nonlinear in the nodal pressures, even when $\vec{w}$ and $\partial\vec{w}/\partial t$ are given. In any case, p and w are coupled strongly and nonlinearly through the elasticity conditions (Eq. 5.7) and the fluid-flow conditions (Eq. 5.22). The simultaneous solution of these equations at successive timesteps is outlined later.

Crack Advance. Advance of the crack is controlled by the fracture criterion of linear elastic fracture mechanics. That is, crack advance occurs in such a way that the stress-intensity factor, K_I, is kept nearly equal to the critical stress-intensity factor, K_{Ic}, during crack extension at each node. Because the crack opening, $w_a(s)$, at the inner boundary of the near-crack-tip region is proportional to the stress-intensity factor at the boundary, the condition for crack advance can be expressed in terms of $w_a(s)$ as

$$w_a(s)<w_c \qquad (5.23a)$$

for no crack advance and

$$w_a(s)>w_c \qquad (5.23b)$$

for crack advance, where[8]

$$w_c=\frac{2(1-\nu)K_{Ic}}{G}\left[\frac{2a(s)}{\pi}\right]^{1/2} \qquad (5.24)$$

K_{Ic} is a measure of the intensity of the elastic stress field near the crack tip that is required for crack extension. For a brittle elastic solid, this value can be obtained from laboratory testing of precracked specimens as long as the specimens are sufficiently large for the inelastic, microcracked region near the crack tip to be small relative to all characteristic lengths of the specimen. Fracture-toughness experiments on rocks indicate that K_{Ic} is on the order of 10^3 psi-in.$^{1/2}$ [$\approx 10^3$ MPa·m$^{1/2}$] for many competent rocks.[9,10] Such values are commonly used in the simulation of hydraulic fracturing by means of the 2D analyses of Chap. 4 and the 3D and pseudo-3D analyses of this chapter. Be aware, however, that direct confirmation of the applicability of these values to the characterization of the critical conditions for crack extension in massive hydraulic fracturing is lacking. Indeed, there is evidence that the characteristic values for the massive fracturing of reservoirs may be significantly different from the values obtained from laboratory experiments. Hydrostatic pressure has been shown to increase the value of K_{Ic}.[11] Propagation of cracks over long distances may result in increased process zone sizes, resulting in increased values for K_{Ic}.[12] Such increases in K_{Ic} may explain the observation that the pressures measured at the wellbore are often larger than those predicted by fracture analyses. On the other hand, the presence of natural fractures and the increased probability of the crack being able to follow a path of reduced resistance caused by the presence of cracked and weakened regions tend to reduce the effective value of K_{Ic}. Fortunately, for many applications, the predicted fracture geometry is relatively insensitive to the value of K_{Ic}. However, for cases with relatively small pressure gradients in the fracturing fluid (e.g., caused by large crack openings resulting from small elastic moduli or the use of low-viscosity fluids), the predicted geometry can depend significantly on the value of K_{Ic}. Research on the values of K_{Ic} to use under reservoir conditions would help clarify matters.

Ideally, the values for the velocity, v, at each node along the crack front are those for which the crack opening at a distance a from the crack front remains at the critical value, w_c. However, such values of v can only be obtained iteratively because of the need to know the crack openings at time t_{n+1}. These iterations would be extremely time-consuming because each change in the crack geometry would require a re-evaluation of the stiffness matrix, K. A much more efficient approach is to use estimated values for v at each step and to solve Eqs. 5.7 and 5.22 for the pressure and crack opening at t_{n+1} for the estimated change in crack-front location during the timestep. Estimated values for v can be obtained by using the crack-tip opening at time t_n to indicate whether the crack velocity should be increased or decreased during the next timestep to bring the crack-tip opening at time t_{n+1} nearer to the critical value, w_c.

During the timestep, the crack is advanced a distance for which the projected volume in an annular element near the crack front will become equal to the volume corresponding to a crack-opening profile with $w_a(s)=w_c$. The normal velocity of a boundary segment is computed from a volume-conservation equation, analogous to Eq. 5.21, for the annular element having a width d equal to the sum of the width of the near-crack-tip zone and the annulus. This equation, when solved for the normal velocity, v, is

$$v=\frac{[\bar{q}_n-2C_L d(\bar{p}-\bar{p}_f)]\Delta s-\mathrm{d}V/\mathrm{d}t}{(\bar{w}_d+2C_{SL})\Delta s}, \quad \ldots\ldots (5.25)$$

where $\bar{w}_d$ is the average crack opening at the inside of the annular element and $\mathrm{d}V/\mathrm{d}t$ is the required rate of change of volume of the annular element if the volume after crack advance is to be equal to the desired volume corresponding to a crack opening $w_a(s)=w_c$. Overbars indicate quantities averaged over the length Δs of the annular element. Eq. 5.25 can be viewed as an equation for computing a control parameter v such that the volume correction, $\mathrm{d}V$, is kept near zero. Once the crack velocity is computed, the advance of each node is computed by multiplying v by the time increment.

Solution of Coupled Equations. The solution to the coupled system of equations (Eqs. 5.7 and 5.22) is to be obtained at time $t_{n+1}=t_n+\Delta t$ from the solution at time t_n, the known injection rate, and the crack advance corresponding to the crack velocity obtained from Eq. 5.25. Integration of Eq. 5.22 over the timestep Δt can be approximated by replacing the derivative $\partial\vec{w}/\partial t$ by the difference approximation,

$$\frac{\partial\vec{w}(x,y,t)}{\partial t}\simeq\frac{\vec{w}(x,y,t_{n+1})-\vec{w}(x,y,t_n)}{\Delta t}, \quad \ldots\ldots (5.26)$$

and evaluating the remaining terms at an intermediate time, $t_n+\theta\Delta t$, where θ is a parameter satisfying $0<\theta<1$. The crack openings in Eq. 5.26 are evaluated at nodes with the same (x,y) coordinates at the two times. For definiteness, these nodes are viewed as the nodal positions at time $t_n+\theta\Delta t$. Crack openings at these positions at time t_n are obtained by interpolation from the nodal crack openings at time t_n. Integration of Eq. 5.22 over the timestep Δt gives a relation of the form

$$(D+C_LT)\hat{\vec{p}}_{n+\theta}+\frac{1}{2\theta\Delta t}T\vec{w}_{n+\theta}=\vec{b}, \quad \ldots\ldots (5.27)$$

where

$$\vec{b}=\hat{\vec{b}}+C_LT\hat{\vec{p}}_f+\frac{1}{2\theta\Delta t}T\vec{w}_n$$

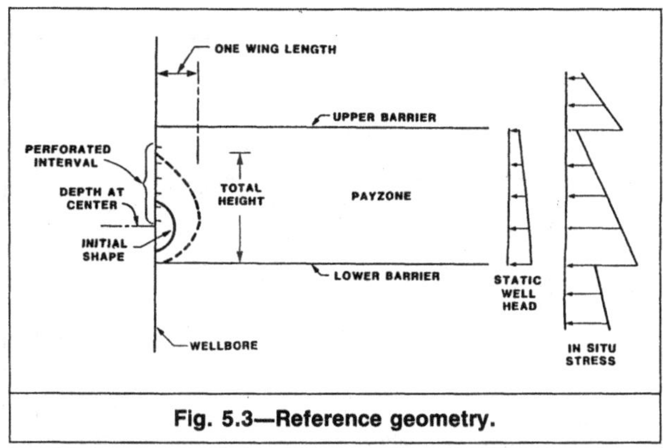

Fig. 5.3—Reference geometry.

and

$$\hat{\vec{p}}_{n+\theta}=\vec{p}_{n+\theta}-\vec{\sigma}^0_{zz}(0),\ \hat{\vec{p}}_f=\vec{p}_f-\vec{\sigma}^0_{zz}(0),$$

where $\vec{\sigma}^0_{zz}(0)$ is a constant vector with nodal values equal to the in-situ stress at the origin $x=0$, $y=0$, and the subscript $n+\theta$ denotes evaluation at time $t_n+\theta\Delta t$.

Eq. 5.7, applied at time $t_{n+\theta}$, can be used to eliminate the unknown crack openings, $\vec{w}_{n+\theta}$, in Eq. 5.27. For pressure $\hat{\vec{p}}_{n+\theta}$, this gives[7]

$$\left[\frac{1}{2\theta\Delta t}TK^{-1}T+D+C_LT\right]\hat{\vec{p}}_{n+\theta}=\vec{b}+\frac{1}{2\theta\Delta t}TK^{-1}T\hat{\vec{\sigma}}_z, \quad \ldots\ldots (5.28)$$

where K^{-1} is the inverse of the matrix K and $\hat{\vec{\sigma}}_z$ is the vector of nodal in-situ stresses σ^0_{zz} reduced by the stress $\sigma^0_{zz}(0,0,0)$. The matrix [ ] is symmetric and, as long as the mesh is satisfactory, can be made suitably well-conditioned by controlling the timestep Δt. Eq. 5.28 can be solved iteratively beginning with $\hat{\vec{p}}_{n+\theta}\simeq\hat{\vec{p}}_n$.

Numerical Examples. The availability of 3D models makes it possible to examine the limits of validity of the 2D approximate models presented in Chap. 4 and to simulate fracture growth for cases outside these limits. As an indication of some of the features of the solutions obtained with 3D models, numerical examples are presented here for the reference geometry of Fig. 5.3. For each of the examples, the crack-front positions are shown at successive stages of crack growth. The evolution of four principal quantities is also shown: (1) the crack length, L (twice the one wing length shown in Fig. 5.3); (2) the total crack height, h; (3) the crack opening at the wellbore at the center of the perforated interval; and (4) the pressure difference $\Delta p=p-\sigma^0_{zz}$ at the same position. In all examples, the center of the perforated interval is at the center of the initial fracture.

First, consider Case A of Table 5.1. This case is intended to provide a direct comparison with predictions of the 2D approximate models of Chap. 4. For this reason, strong stress barriers are introduced to maintain a constant crack height, and the fluid is assumed to be Newtonian. Furthermore, to simplify the initial comparison, leakoff is assumed to be negligible. The effects of differences between the hydrostatic pressure gradient of the fracturing fluid and the in-situ stress gradient are eliminated by setting both quantities to zero. The perforation interval is taken to be the full height of the pay zone to simulate horizontal flow at the wellbore.

The crack fronts at various times are shown in Fig. 5.4A. Except near the crack tip along the x axis, the height of the fracture is nearly equal to the height of the pay zone. The evolution of various characteristic parameters of the fracture is shown in Fig. 5.4B, in which predictions of the fully 3D model (curves without symbols) are compared with those of the constant-height models (curves with symbols) of Chap. 4. Logarithmic scales with the same number of cycles per inch are used for both horizontal and vertical axes so that a line with slope m corresponds to a quantity that is varying as t^m. In all figures, the time histories at early times (e.g., $t<0.25$ minutes

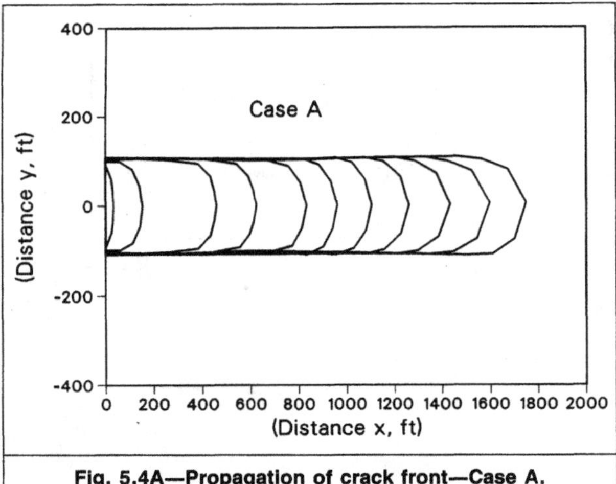

Fig. 5.4A—Propagation of crack front—Case A.

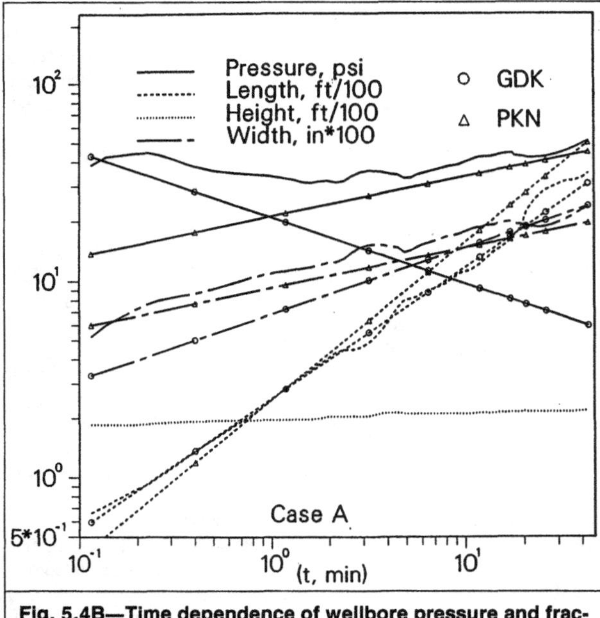

Fig. 5.4B—Time dependence of wellbore pressure and fracture geometry—Case A.

in Fig. 5.4B) are transients associated with beginning the injection into an initial static fracture that is filled with a quiet, pressurized fluid. These transients are associated with the initial conditions of the 3D simulation and are unimportant to the understanding of the principal features of hydraulic fracturing.

The waviness of the solutions obtained for the 3D simulations is a consequence of the approximate nature of the predictions of the crack advance during a given timestep, as discussed in connection with Eq. 5.25. This waviness tends to be greater for cases like Case A, where there is a strong jump in the minimum in-situ compressive stress across the interfaces between the pay zone and the bounding layers. For these cases, small errors in the predicted advance into the bounding layers result in significant changes in the pressures and crack openings obtained for the predicted position. Also, for these cases, the cracks become highly elongated, as shown in Fig. 5.4A, and few nodes are available for describing the crack front in the pay zone. Consequently, the front becomes pointed and errors in the predicted advance of a single node at the center of the pay zone can have a significant effect on the length, pressure, and opening plotted in Fig. 5.4B. The waviness can be reduced by reducing the timestep and the mesh size; however, when the trends of the solution are clearly evident (as in Fig. 5.4B), the additional computing effort required is probably not justified. Relatively little waviness occurs in 3D simulations of cases involving small stress contrast and appreciable vertical migration, resulting in fractures of modest aspect ratio, L/h.

As shown in Fig. 5.4B, once the crack length becomes greater than the height, the Perkins-Kern-Nordgren[13,14] (PKN) model predictions for the pressure and crack opening at the wellbore agree quite well with the predictions of the fully 3D model. In particular, the effective elastic stiffness at the wellbore (i.e., pressure divided by crack opening) is almost exactly the same for the two models. This agreement confirms the applicability at the wellbore of the model for elastic stiffness introduced by Perkins and Kern[13] and Nordgren[14] (namely, the stiffness corresponding to the opening of a plane-strain crack of height h subjected to uniform pressure). For $L/h >> 1$, however, the crack length predicted by the PKN model becomes significantly greater than that predicted by the 3D model. A qualitative understanding of this difference is the following. The PKN model overestimates crack openings in the region near the crack front because the openings at a position, say x_1, are computed from the pressure difference acting on the crack plane at x_1, whereas the pressure differences at nearby positions $x > x_1$ are considerably less (often negative within the fracture and certainly negative for $x > L/2$). Because the openings are overestimated, the pressure gradient is strongly underestimated (e.g., Eq. 4.14). Integration of such underestimated pressure gradients gives wellbore pressures that are less than those obtained from the 3D model, as shown in Fig. 5.4B. Consequently, the PKN model results in openings at the wellbore and over some interval $x << L/2$ that are less than those predicted by the 3D model (again, see Fig. 5.4B). These smaller openings in regions nearer the wellbore result in less fluid being stored in these regions than predicted by the 3D model. This underestimation of fracture volume near the wellbore is greater than the additional volume that results from the overestimation of crack openings near the crack front. Consequently, additional crack length is required in the PKN model to accommodate the fluid injected.

At early times when $L/h << 1$, the principal elastic stiffness is associated with the x direction, as assumed by Geertsma and de Klerk (GdK).[15] However, the crack does not propagate with a straight, vertical front, as assumed in the GdK model. Instead, the crack expands from its initially elliptical shape to approximately a circular shape with radius $h/2$ before the crack begins to propagate in the x direction with essentially an unchanged crack-front shape. The initial expansion of the crack from an ellipse to a circle can be understood by noting that, for an elliptical crack subjected to uniform pressure, the stress-intensity factor is greatest at the minor axis.[16] Consequently, the crack tends to advance along the minor axis until the crack becomes approximately circular.

During early stages when $L/h << 1$, the GdK model exhibits an effective elastic stiffness at the wellbore (again, the ratio of the pressure to the crack opening) that is comparable to that obtained from the fully 3D model. In the GdK model, however, the pressure decreases monotonically with increasing time and the effective elastic stiffness at the wellbore becomes much less than that of the 3D model. This behavior results because, for $L/h >> 1$, the elastic stiffness is associated primarily with the fracture height, not its length, as assumed in the GdK model. In spite of this relatively poor modeling of the elastic stiffness at late times, the crack length predicted by the GdK model agrees remarkably well with that predicted by the 3D model. Although this good agreement may be fortuitous, it should be noted that close agreement holds over two decades of crack advance. Furthermore, other simulations have shown reasonably good agreement between the lengths predicted by the GdK and the 3D models. Thus, it appears that the GdK model may give quite a good indication of the fracture length in well-contained fractures.

A second example, shown as Case B in Table 5.1, is designed to represent the case of axisymmetric crack growth analyzed by Perkins and Kern,[13] Geertsma and de Klerk,[15] and Abe *et al.*[17] The input parameters for this case are the same as for Case A, except the stress barriers are removed and the perforation interval is reduced to a small region at the origin to simulate a point source. As expected, the crack fronts advance in essentially an axisymmetric pattern (see Fig. 5.5A). Good agreement (Fig. 5.5B) is obtained between the values of crack length, wellbore pressure, and crack opening obtained from the 3D simulation and the corresponding values obtained from the elementary models of Chap. 4 (i.e., Tables

TABLE 5.1—INPUT DATA FOR NUMERICAL EXAMPLES

	Case						
	A	B	C	D	E	F	G
Formation Properties							
Young's modulus, psi	1,000,000						6,440,000
Poisson's ratio	0.2						0.17
Stress gradient, psi/ft	0.0			0.8			0.75
Stress contrast, psi	200		20			20	200
Fracture toughness, psi-in.$^{1/2}$	1,000						
Fluid Properties							
K', lbf-sec$^{n'}$/ft^2	0.00002					0.025	0.00438
n'	1					0.5	0.658
Weight density, psi/ft	0.0			0.43			0.069
Fluid loss, ft/min$^{1/2}$	0.0				0.0008		0.002
Spurt loss, gal/ft^2	0.0						
Perforations							
Perforated interval, ft	200	1					50
Number of perforations	200	1					100
Injection Scheme							
Pumping rate, bbl/min	40						25
Total volume, bbl	1,727	1,183	1,428	98	1,648	943	1,886
Pay-Zone Details							
Pay-zone thickness, ft	200	NA		NA			250
Initial Fracture Geometry							
Total height, ft	180	90					200
One wing length, ft	30	45					100

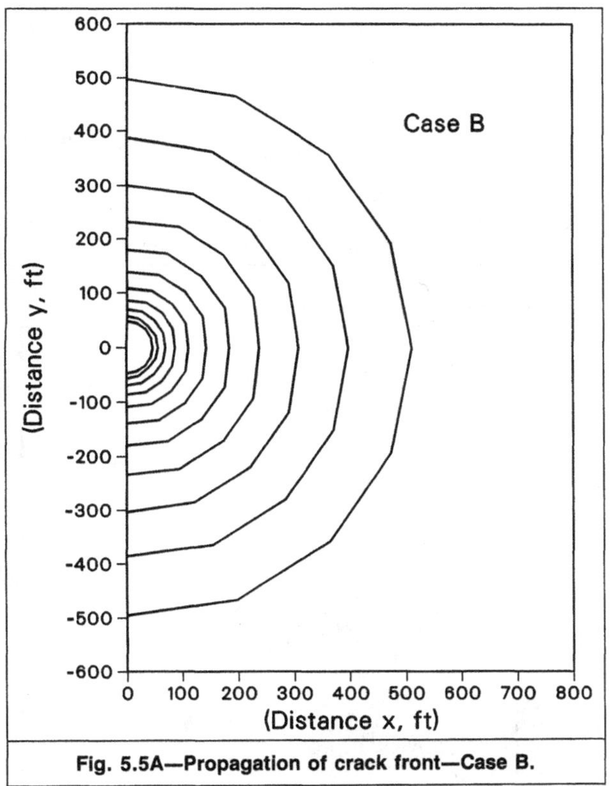

Fig. 5.5A—Propagation of crack front—Case B.

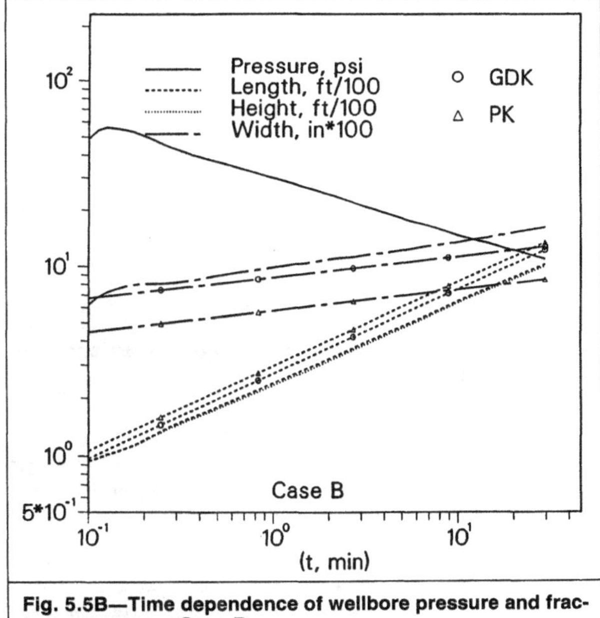

Fig. 5.5B—Time dependence of wellbore pressure and fracture geometry—Case B.

4.1 and 4.2). Better agreement in both radius and crack opening is obtained for the GdK model.

Case C in Table 5.1 is an example in which the fracture is confined largely to the pay zone but the vertical growth of the fracture has a significant effect on the fracture geometry. This case is the same as Case A, except the stress barrier is reduced from 200 to 20 psi [1379 to 138 kPa]. Fracture positions at various times are shown in Fig. 5.6A. The corresponding time histories of wellbore pressure and characteristic crack dimensions are shown in Fig. 5.6B. Comparison of Figs. 5.4A and 5.6A shows that vertical migration of the fracture strongly reduces the fracture-length growth rate. Vertical migration of the fracture has a much weaker effect on the crack opening. Also, when significant vertical migration occurs, the pressure in the later stages does not increase as strongly as for the well-contained case, Case A.

In Case C, significant vertical migration occurs, but the h/L ratio remains sufficiently small that so-called pseudo-3D models offer an attractive alternative to fully 3D models. In the pseudo-3D models, crack-height variations are accounted for approximately. Such models are described in Sec. 5.3. Predictions of such a model for Case C are discussed in detail after the formulation of pseudo-3D models is described. An important advantage of the pseudo-3D models is the greatly reduced computing time required. When these models are applicable, the computational time can often be reduced to a few percent of that required for the fully 3D models.

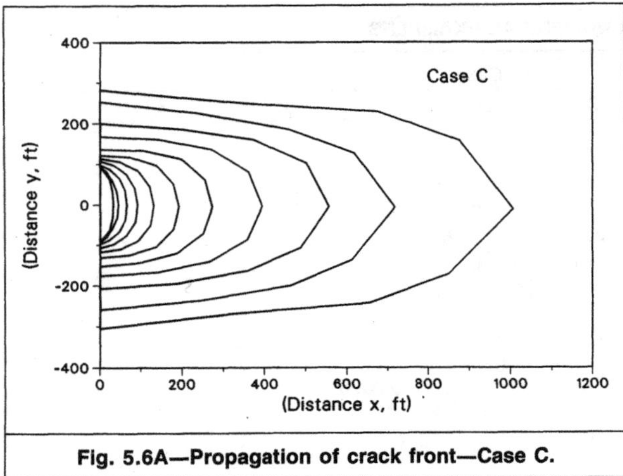

Fig. 5.6A—Propagation of crack front—Case C.

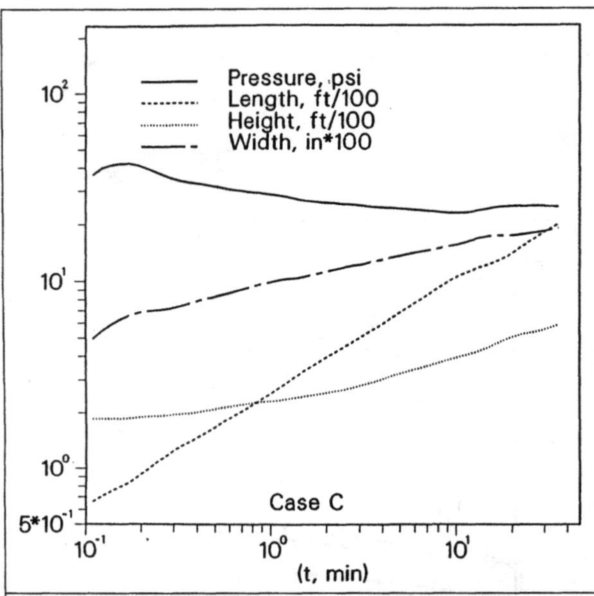

Fig. 5.6B—Time dependence of wellbore pressure and fracture geometry—Case C.

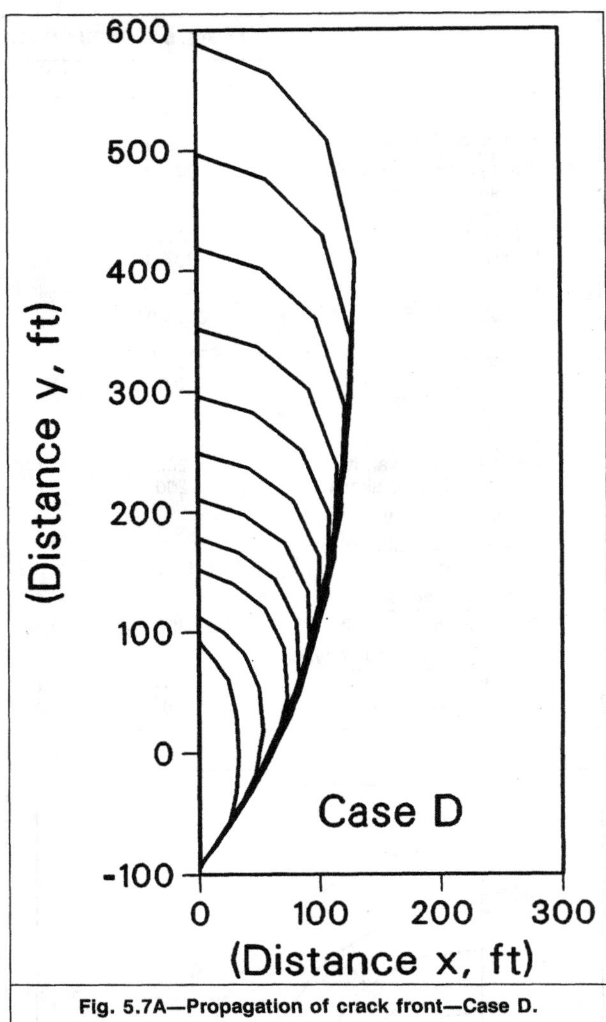

Fig. 5.7A—Propagation of crack front—Case D.

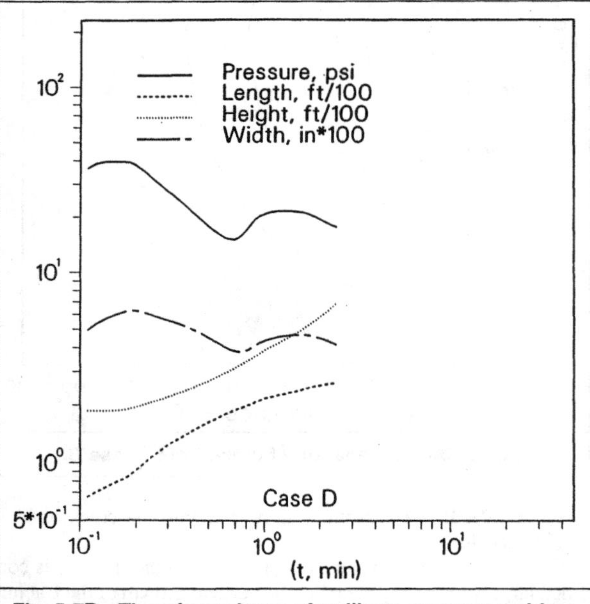

Fig. 5.7B—Time dependence of wellbore pressure and fracture geometry—Case D.

An example of upward migration of the fracture owing to a strong stress gradient is given by Case D in Table 5.1. The stress gradient (0.80 psi/ft [18 kPa/m]) for the minimum in-situ stress in the formation is much greater than the hydrostatic pressure gradient (0.43 psi/ft [9.7 kPa/m]) of the fracturing fluid. With no other barrier present, the crack migrates strongly upward, as shown in Fig. 5.7A. Time histories of the pressure and the height, length, and width of the fracture for this case of runaway vertical migration of the fracture are shown in Fig. 5.7B. At early times, the pressure decreases as the crack migrates vertically into regions of lower in-situ stress. However, the crack eventually becomes so elongated vertically that frictional pressure gradients in the vertical direction become sufficiently large that the pressure at the wellbore must increase to maintain the injection rate. Qualitatively, the pressure/time histories for the strongly contained case (Case A) and the uncontained case (Case C) are similar in that the pressure decreases while the crack advances along a large fraction of its front but increases as the crack becomes highly elongated, and extension occurs only along a small, remote part of the crack front. Correspondingly, at late times, the vertical crack advance for the uncontained fracture becomes analogous to the horizontal crack advance for the well-contained fracture.

Cases A, B, and C provide some indication of the relationship of fully 3D models to simpler, more idealized models. Examination of the limitations of idealized models is certainly one application

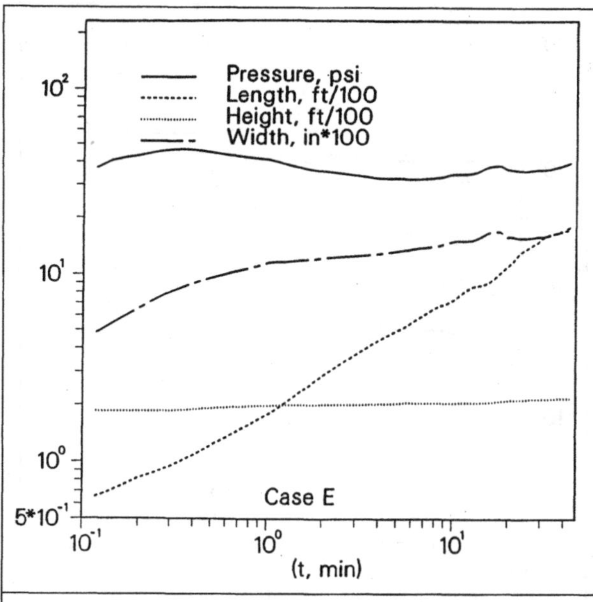

Fig. 5.8—Time dependence of wellbore pressure and fracture geometry—Case E.

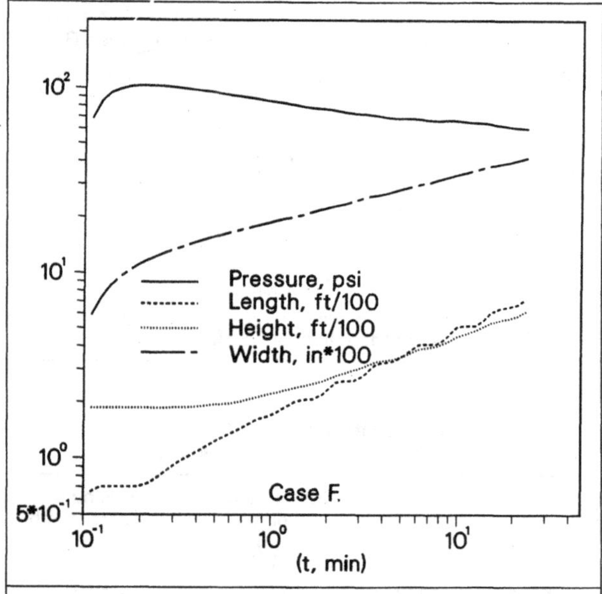

Fig. 5.9—Time dependence of wellbore pressure and fracture geometry—Case F.

of 3D models, but the primary application of 3D models is to cases for which the elementary models do not apply. These 3D models have been extended to include multiple fluids, proppant transport, and the effects of temperature variations on the fracturing fluids.[18] A full presentation of these features of the model is beyond the scope of this chapter. However, three additional examples are included in Table 5.1 to illustrate some of the effects of non-Newtonian fluids and fluid leakoff.

Case E is the same as Case A, except significant fluid loss is assumed to occur. When leakoff occurs, the principal effect is a reduced rate of crack advance in the horizontal direction (see Fig. 5.8). At late times, the pressure and crack opening at the wellbore also increase more slowly than for the case of no leakoff (Case A).

Case F is the same as Case C, except a more viscous, non-Newtonian fluid is used. Because of the higher effective viscosity, the pressures and crack openings are much larger (see Fig. 5.9). The larger pressures cause increased vertical migration of the fracture. With larger crack openings and greater height, the crack length is greatly reduced for the same volume of injected fluid.

Case G is the small-job case considered in detail in Chap. 11. Additional details regarding the pumping schedule and the proppant injection schedule are given in Table 11.2. The actual fluid/sand schedule specified in this table was used for the 3D simulation; however, because inclusion of proppant transport in fully 3D models[18] is beyond the scope of this chapter, the computed proppant distribution is not presented. The evolution of the fracture is shown in Fig. 5.10A. The fracture height grows slowly, even though a 200-psi [1379-kPa] stress barrier was put at the assumed height of 250 ft [76 m], which is a fixed height in the 2D models. A comparison of the final fracture dimensions with the values obtained from the 2D models used in Chap. 11 is given in Table 5.2. The labeling of the 2D models is the same as in Table 11.3. The KZ computer model refers to the model introduced by Khristianovitch and Zheltov.[19] Again, the PKN models tend to predict a longer fracture than that obtained from the 3D model. Of the 2D models listed, the best agreement with the 3D model is obtained for the GdK simplified model. The final crack opening at the center of the perforations (see Fig. 5.10B) agrees reasonably well with the predictions of both the PKN and GdK simplified models.

The effect of having a much larger fracture toughness is considered in Case H, which is the same as Case G in Table 5.1 except that the fracture toughness, K_{Ic}, is increased by a factor of 10 to 10^4 psi-in.$^{1/2}$. The propagation of the crack for this case is shown in Fig. 5.10C. Comparison of Figs. 5.10A and 5.10C shows that increasing the fracture toughness reduces the fracture length significantly. Note also that the crack fronts are smoother for the larger values of K_{Ic}. This occurs because larger values of K_{Ic} result in larger crack openings and smaller pressure gradients near the crack front. Both of these conditions are favorable for obtaining better estimates for the crack advance during each timestep. Consequently, fluctuations in the advance at each node are smaller and the crack fronts are smoother.

5.3 Pseudo-3D Models

2D Elasticity. In pseudo-3D models, formation elasticity is approximated by assuming that the crack length is sufficiently large relative to the height that the effective elastic stiffness (i.e., the relationship between the pressure and the crack opening), at all cross sections x=constant, is independent of the crack length and the horizontal distance from the cross section to the crack front. Under these conditions, the crack opening $w(x,y)$ at each x is obtained from the plane-strain-elasticity solution for a crack of height $h(x)$ subjected to a pressure distribution $\Delta p(x,y)$. The plane-strain-elasticity problem involves deformation in the y-z plane only; the x coordinate plays the role of a parameter used to account for variations in crack height and pressure along the fracture length. For the case of homogeneous, isotropic, linear elasticity, the crack opening $w(x,y)$ is related to the pressure difference by[16]

$$w(x,y)=\int_{-h/2}^{h/2}\Delta p(x,y_o+y')\left\{\frac{(1-\nu)}{\pi G}\ln[R_y(y,y')]\right\}dy', \quad \ldots(5.29)$$

where

$$R_y(y,y')=\frac{(h/2+y')^{1/2}(h/2+y_o-y)^{1/2}+(h/2-y')^{1/2}(h/2+y-y_o)^{1/2}}{\left|(h/2+y')^{1/2}(h/2+y_o-y)^{1/2}-(h/2-y')^{1/2}(h/2+y-y_o)^{1/2}\right|}.$$

The elevation y is measured from the midheight of the pay zone; the elevation y' is measured from the midheight of the fracture. The term in braces is the fundamental solution for the crack opening resulting from a pair of opposing unit line forces acting on the crack faces at y'. The fracture height $h(x)=H+\Delta h^a(x)+\Delta h^b(x)$, where H is the pay-zone height, and Δh^a, Δh^b are the distances that the fracture extends into the layers above and below the pay zone, respectively. $y_o=(\Delta h^a-\Delta h^b)/2$ is the location of the midheight of the fracture relative to the midheight of the pay zone. The pressure difference is generally taken to have the form

$$\Delta p(x,y)=p(x)-\sigma(y), \quad \ldots(5.30)$$

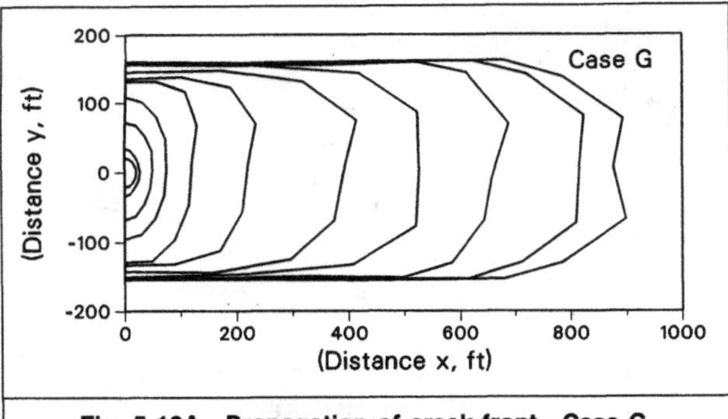

Fig. 5.10A—Propagation of crack front—Case G.

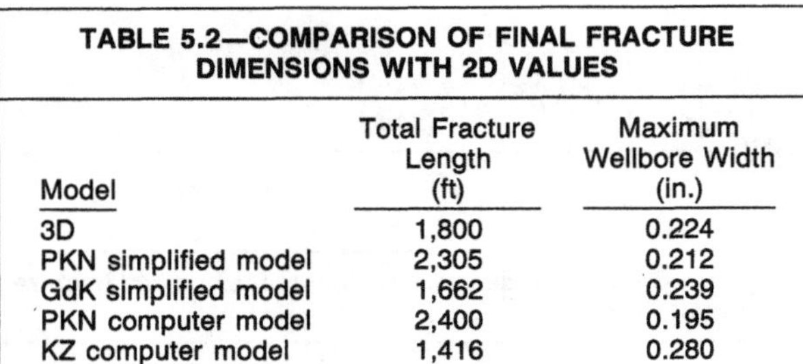

TABLE 5.2—COMPARISON OF FINAL FRACTURE DIMENSIONS WITH 2D VALUES

Model	Total Fracture Length (ft)	Maximum Wellbore Width (in.)
3D	1,800	0.224
PKN simplified model	2,305	0.212
GdK simplified model	1,662	0.239
PKN computer model	2,400	0.195
KZ computer model	1,416	0.280

where $p(x)$ is the pressure in the fracturing fluid, assumed to be constant over the fracture height, and $\sigma(y)$ is the in-situ stress normal to the crack plane. Pressure gradients in the y direction, required for the vertical flow of the fluid to fill the vertical extensions of the crack, are often neglected as being small. This assumption appears to be appropriate for the intended applications in which the vertical extension of the crack is slow relative to the horizontal extension. Approximate means for including pressure gradients in the y direction have been introduced in some pseudo-3D models.[20] Within each layer of the formation, the in-situ stress $\sigma(y)$ in Eq. 5.30 is usually taken either to be constant or to vary linearly over the depth of the layer. Larger in-situ stresses in the layers bounding the pay zone provide the mechanism for vertical fracture containment.

Explicit integration of Eq. 5.29 is possible for simple forms of the pressure difference $p(x,y)$. The general form of the resulting expressions is[20]

$$w(x,y)=f[y;p(x),h(x)]. \quad (5.31)$$

For the commonly assumed case of uniform in-situ stress within each layer, Eq. 5.31 can be written as[21]

$$w(x,y)=w_{\mathrm{I}}-w_{\mathrm{II}}-w_{\mathrm{III}}, \quad (5.32)$$

where

$$w_{\mathrm{I}}=\frac{4}{E'}p\,[h^2/4-(y-y_o)^2]^{1/2},$$

$$w_{\mathrm{II}}=\frac{4(\sigma^a-\sigma^p)}{E'\pi}\left\{-\left(\frac{h}{2}-\Delta h^a-y+y_o\right)\times\cosh^{-1}\left[\frac{\frac{h^2}{4}-\left(\frac{h}{2}-\Delta h^a\right)(y-y_o)}{\frac{h}{2}\left|y-y_o-\frac{h}{2}+\Delta h^a\right|}\right]+\left[\frac{\pi}{2}-\sin^{-1}\left(\frac{h/2-\Delta h^a}{h/2}\right)\right][h^2/4-(y-y_o)^2]^{1/2}\right\},$$

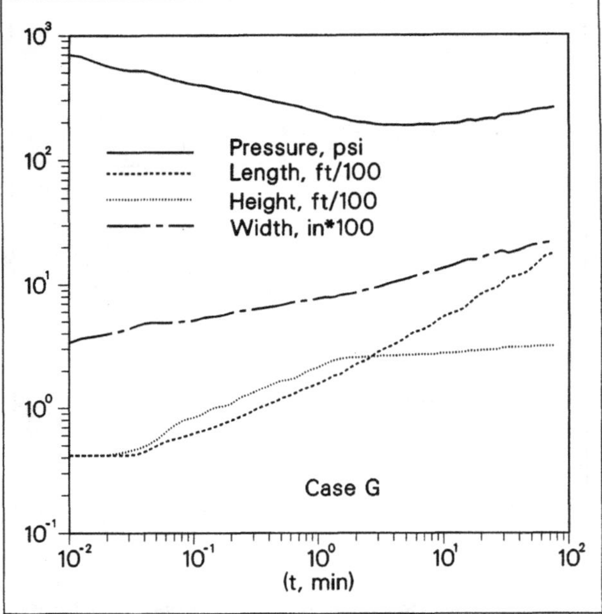

Fig. 5.10B—Time dependence of wellbore pressure and fracture geometry—Case G.

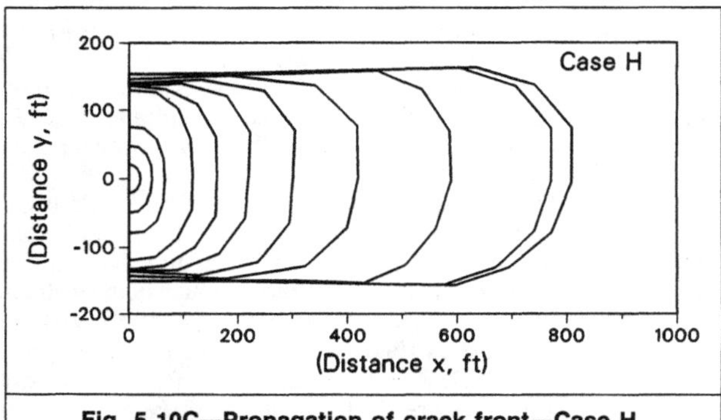

Fig. 5.10C—Propagation of crack front—Case H.

and

$$w_{\mathrm{III}}=\frac{4(\sigma^b-\sigma^p)}{E'\pi}\left\{-\left(\frac{h}{2}-\Delta h^b-y+y_o\right)\times\cosh^{-1}\left[\frac{\frac{h^2}{4}+\left(\frac{h}{2}-\Delta h^b\right)(y-y_o)}{\frac{h}{2}\left|y-y_o+\frac{h}{2}-\Delta h^b\right|}\right]+\left[\frac{\pi}{2}-\sin^{-1}\left(\frac{\frac{h}{2}-\Delta h^b}{h/2}\right)\right]\left[\frac{h^2}{4}-(y-y_o)^2\right]^{1/2}\right\},$$

in which $E'=E/(1-\nu^2)$ and σ^a, σ^p, σ^b are the in-situ stresses in the upper layer, pay zone, and lower layer, respectively. For more complicated forms of in-situ stress variation with depth, it may be necessary to integrate Eq. 5.29 numerically. In any event, the formation elasticity in the pseudo-3D formulations is modeled by a relationship of the form given in Eq. 5.31.

The vertical extension of the fracture is controlled by the requirement that the stress-intensity factor, K_I, for the stress field near the crack tip be equal to the critical value, K_{Ic}. At each cross section x, this requirement is imposed at the top and bottom of the fracture. The stress-intensity factor at the upper and lower crack tips can be expressed as[16]

$$K_I^{a,b}=\frac{1}{\sqrt{\pi h/2}}\int_{-h/2}^{h/2}\Delta p(x,y_o+y')\left(\frac{h\pm 2y'}{h\mp 2y'}\right)^{1/2}dy'. \quad \text{......(5.33)}$$

Substitution of Eq. 5.30 into Eq. 5.33, with K_I^a and K_I^b equal to the critical values for crack advance, gives a system of two nonlinear equations for the distances $\Delta h^a(x)$ and $\Delta h^b(x)$. These equations can be simplified by noting that the contributions of the pressure differences near one tip of the fracture have relatively little effect on the stress-intensity factor at the other tip. When the in-situ stress is uniform within each layer, this simplification allows Eq. 5.33 to be written as[22]

$$K_I^{a,b}=(p-\sigma^p)\left(\frac{\pi h}{2}\right)^{1/2}\left[1-\frac{2}{\pi}\left(\frac{\sigma^{a,b}-\sigma^p}{p-\sigma^p}\right)\cos^{-1}\left(\frac{H}{h}\right)\right]. \quad \text{......(5.34)}$$

Eq. 5.34 is an exact integration of Eq. 5.33 for the symmetrical case in which the in-situ stresses and the distances of crack penetration are the same for the two layers bounding the pay zone. For asymmetrical cases, Eq. 5.34 remains a useful approximation as long as the difference in the products $(p-\sigma^a)\Delta h^a$ and $(p-\sigma^b)\Delta h^b$ is small compared with their sum. When this is not the case, it is necessary to integrate Eq. 5.33 for the prescribed in-situ stress distribution $\sigma(y)$. In any case, a system of two equations is obtained for relating $\Delta h^a(x)$ and $\Delta h^b(x)$ to the fluid pressure $p(x)$. These equations can be solved for $\Delta h^a(x)$ and $\Delta h^b(x)$ by a search technique. Alternatively, $\Delta h^a(x)$ and $\Delta h^b(x)$ can be obtained by the integration of ordinary differential equations obtained from the differentiation of Eq. 5.33 with respect to x, for $K_I^{a,b}$ set equal to constant critical values. These equations are integrated from the crack front $x=L/2$ back to the wellbore $x=0$; initial conditions $\Delta h^a=0$ and $\Delta h^b=0$ are used at $x=L/2$. Negative values are viewed as indicating that the crack has not propagated into the corresponding bounding layer; such values are replaced by zero because it is assumed that for pseudo-3D models the crack height is always at least as large as the pay-zone height h.

1D Fluid Flow. In pseudo-3D models, the fluid flow is idealized as being a 1D flow along the fracture length. The equation governing the flow rate $Q(x,t)$ is obtained by integrating the flow rate $q(x,y,t)$ per unit height over the fracture height. From Eq. 16a, the resulting expression for the total flow rate at x is

$$Q(x,t)=\int_{-h/2}^{h/2}\left(\frac{w^{2n'+1}}{\eta'}\left|\frac{\partial p}{\partial x}\right|\right)^{1/n'}dy', \quad \text{......(5.35)}$$

where again the fluid is assumed to be a power-law fluid characterized by the parameters n' and η'. This equation can be solved for the pressure gradient to obtain

$$\frac{\partial p}{\partial x}=\frac{\eta'[Q(x,t)]^{n'}}{\int_{-h/2}^{h/2}w^{[(2n'+1)/n']}dy'}, \quad \text{......(5.36)}$$

which has the form

$$\frac{\partial p(x,t)}{\partial x}=g[p(x,t),Q(x,t)] \quad \text{......(5.37)}$$

once equations of the form of Eqs. 5.31 and 5.33 are used to express the opening $w(x,y,t)$ and the height $h(x,t)$ in terms of the pressure $p(x,t)$. [Or $\Delta h^a(x,t)$ and $\Delta h^b(x,t)$ are obtained by integrating their governing ordinary differential equations obtained from Eq. 5.33.]

The remaining physical condition to impose is the conservation of the mass of the fracturing fluid, assumed to be incompressible. This condition gives

$$\frac{-\partial Q(x,t)}{\partial x}=Q_L(x,t)+\frac{\partial A_c(x,t)}{\partial t}, \quad \text{......(5.38)}$$

where $Q_L(x,t)$ is the rate of leakoff at x per unit length in the x direction. Leakoff is normally assumed to occur only over the pay-zone height. Then, the classic leakoff relation has the form

$$Q_L(x,t)=\frac{2K_\ell H}{\sqrt{t-\tau(x)}}, \quad \text{......(5.39)}$$

where K_ℓ is the leakoff coefficient and $\tau(x)$ is the time at which the fracture reaches position x. The last term in Eq. 5.38 is the "storage" term corresponding to the rate of change of the cross-sectional area $A_c(x,t)$, where

$$A_c(x,t)=\int_{-h/2}^{h/2}w(x,y_o+y',t)dy'. \quad \text{......(5.40)}$$

Differentiation of Eq. 5.40 with respect to time and the use of Eq. 5.29 and the expression for w_I following Eq. 5.32 allow Eq. 5.38 to be written as

$$-\frac{\partial Q(x,t)}{\partial x}=Q_L(x,t)+\frac{\pi h^2}{2E'}\frac{\partial p(x,t)}{\partial t}. \quad \text{......(5.41)}$$

Eqs. 5.36 and 5.41 constitute a system of two coupled, nonlinear partial-differential equations for the pressure $p(x,t)$ and the flow rate $Q(x,t)$.

Solution of the Coupled Equations. Solutions of Eqs. 5.36 and 5.41 are sought to satisfy the boundary conditions

$$Q(0,t)=Q_I(t)/2 \quad \text{......(5.42)}$$

and

$$\Delta p[L(t)/2,t]=p_L, \quad \text{......(5.43)}$$

where $Q_I(t)$ is the injection rate and p_L is the stress differential required to open the crack a nominal width at the crack front. Eq. 5.42 follows immediately from the assumed symmetry of the crack with respect to $x=0$. Eq. 5.43 is more difficult to motivate because the assumptions of the pseudo-3D model do not apply well near the crack front at $x=L/2$. If the crack opening is assumed to be zero at the "crack front" $x=L/2$, then the fluid-flow equation, Eq. 5.36, cannot be used as x approaches $L/2$ because of the singularity at $w=0$. This difficulty can be overcome by regarding the crack as extending a small distance beyond $x=L/2$. Then, the pressure at $x=L/2$ can be set at a value that is approximately the right level for the stress-intensity factor along the supposed crack front to be comparable to the critical value, K_{Ic}. A convenient choice is the pressure, say p_L, at which the stress-intensity factors obtained from Eq. 5.33 (or equivalent), for $\Delta h^a(L/2)$ and $\Delta h^b(L/2)$ equal to zero, are equal to K_{Ic}.

Various methods have been used to solve the coupled equations of the pseudo-3D model. One satisfactory method is to advance

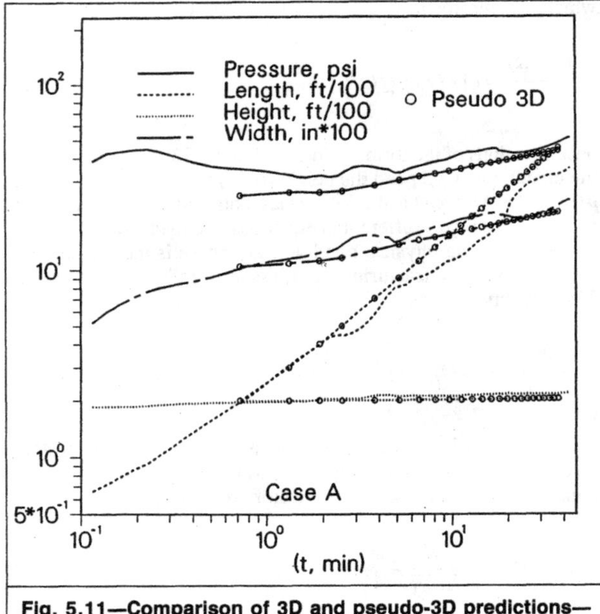

Fig. 5.11—Comparison of 3D and pseudo-3D predictions—Case A.

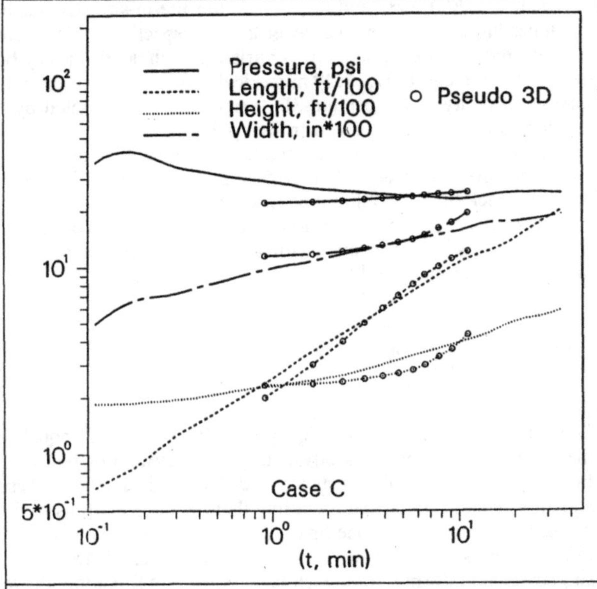

Fig. 5.12—Comparison of 3D and pseudo-3D predictions—Case C.

the crack front a distance $\Delta L/2$ during the timestep from t to $t+\Delta t$. The timestep Δt is regarded as an additional unknown to be determined by the boundary condition, Eq. 5.42. For an assumed timestep Δt, the flow at the crack front $x=L/2$ at time t can be obtained from the requirement that the net rate of flow across $x=L(t)/2$ must be equal to the leakoff rate in the crack-front region beyond $x=L/2$. This requirement gives (cf. Eq. 5.21)

$$Q[L(t)/2,t]-vA_c-2\pi K_\ell\sqrt{v}(H/2)^{3/2}-2C_sHv=0, \quad \ldots\ldots(5.44)$$

where v is the crack-front velocity and C_s is the spurt-loss coefficient introduced in Sec. 5.2. The third term in Eq. 5.44 is obtained for a crack front that is parabolic in the x-y plane and extends a distance $(H/2)$ in front of $x=L/2$. Eqs. 5.43 and 5.44 provide initial values for $p(x,t)$ and $Q(x,t)$ for the evaluation of these functions by the integration of Eqs. 5.37 and 5.41 from the crack front $x=L/2$ back to the wellbore $x=0$. The term $Q_L(x,t)$ on the right side of Eq. 5.41 can be evaluated for an assumed timestep Δt. The term involving $\partial p/\partial t$ can be replaced by the difference relation

$$\frac{\partial p}{\partial t}(x,t)=[p(x,t+\Delta t)-p(x,t)]/\Delta t, \quad \ldots\ldots(5.45)$$

because $p(x,t)$ is known from the solution at the previous timestep. Eq. 5.41 becomes an ordinary differential equation with x as the independent variable. Eqs. 5.37 and 5.41 can be integrated by a Runge-Kutta method and the value of $Q(0,t)$ obtained can be compared with the required value $Q_I(t)/2$. If the two values do not agree, then a new value is assumed for Δt. Selection of new values Δt that give improved agreement is relatively easy because the computed value for $Q(0,t)$ decreases monotonically with increasing Δt.

Numerical Examples. Comparisons of the predictions of a pseudo-3D model with the predictions of the fully 3D model are shown in Figs. 5.11 and 5.12 for Cases A and C of Table 5.1, respectively. For the well-contained case, Case A, the pseudo-3D model predictions are similar to those of the PKN model. Good overall agreement is obtained in the trends of pressure and crack opening. However, as for the PKN model, the predicted crack length is greater than that obtained for the fully 3D model.

For Case C, the pseudo-3D model provides a reasonably accurate model of the vertical migration of the fracture during the early stages of the treatment. At later times, however, when the pressure difference at the wellbore exceeds the 20-psi [138-kPa] stress barrier, the height predicted by the pseudo-3D model becomes unbounded. For this reason, the pseudo-3D calculations had to be terminated at the time shown in Fig. 5.12. When the stress barrier is large enough for the pseudo-3D model to predict stable height growth, a pseudo-3D model provides a convenient, useful computational tool. Attempts to extend the range of application of pseudo-3D models to cases with smaller stress barriers (by using approximations for a vertical pressure gradient in the fracture) have, so far, not provided reliable correlations with fully 3D models.

5.4 Concluding Remarks

Fully 3D modeling of hydraulic fracturing provides a framework for more realistic predictions of fracture geometry, proppant distribution, and treatment performance. The elastic response of the formation is modeled as a fully 3D problem in elasticity, thereby eliminating the need to introduce an idealization of 2D plane-strain deformation in either a horizontal plane (GdK model) or a vertical plane (PKN and pseudo-3D models). From the examples presented here, it is evident that the former idealization underestimates the elastic stiffness for long fractures (i.e., $L/h>>1$) and predicts, unrealistically, that injection pressures decrease with increasing crack length. The latter idealization overestimates the crack opening at distances from the crack front that are less than a few multiples of the crack height. The larger crack openings near the crack front reduce the pressure gradients, resulting in smaller pressures at the wellbore. These smaller pressures result in predictions of smaller fluid capacity for a fracture of a given length or, correspondingly, greater length for a fracture that contains a given volume of injected fluid. Inclusion of the vertical components of flow in 3D calculations makes it possible to model the flow in fractures with significant vertical migration and to account for the effects of vertical velocity components on proppant transport.[18] Enforcement of a fracture criterion for crack advance at all positions along the crack front allows the fracture shape to evolve naturally in response to the loading conditions. Such evolution contrasts with that of 2D models in which a fracture criterion is imposed only for horizontal crack extension in the GdK model and not at all in the PKN model.

The attractive features of fully 3D modeling are obtained at the increased costs of obtaining additional information on formation properties and additional computing time. The primary additional information required is the variation of the minimum in-situ stress with depth, including the pay zone and its bounding layers. Information on changes in elastic moduli from layer to layer is also helpful when these changes are substantial. The need to know variations in the in-situ stress is implicit in the application of 2D models (because of the importance of assessing the applicability of a constant-height model) and explicit in the application of pseudo-3D models. Thus, the determination of in-situ stress variations is an important aspect of improved fracture modeling, regardless of which mathematical models are used. The additional computing time required for 3D modeling, as described here, is largely due to the time required for the evaluation of the elements of the stiffness matrix, K. Because the calculation of these elements can be done in parallel, the time required can be expected to decrease dramatically as computers with many parallel processors become more widely available. Treatments of the type illustrated by the numerical examples already can be calculated in a few minutes on a supercomputer and in an hour or less on a common mainframe or an advanced minicomputer.

Overall, it appears that fully 3D modeling can be expected to play an increasingly important role in the simulation of hydraulic fracturing treatments. Currently, its role is largely that of enabling research groups at major production companies and service companies to understand hydraulic fracturing better and to design treatments for difficult cases in which 2D modeling is expected to be inadequate. With the continuation of the trend of increased computing power at lower cost, however, 3D modeling can be expected to become a common tool of fracture designers.

Nomenclature

a = width of near-crack-tip zone, ft [m]
a_k = weight factors, Eq. 5.10
A = total fracture area, ft^2 [m^2]
$\hat{A}$ = area of discretized region of fracture, ft^2 [m^2]
A_c = area of vertical cross section of fracture, ft^2 [m^2]
A_j = area of jth triangle, ft^2 [m^2]
$\overleftrightarrow{b}, \vec{b}$ = vectors in Eqs. 5.22 and 5.27, ft^3/sec [m^3/s]
C_L = normalized leakoff coefficient, ft/(sec$^{1/2}$-psi) [m/(s$^{1/2}$·kPa)]
C_s = spurt-loss coefficient, ft/sec [m/s]
d = width of annular region in Fig. 5.2, ft [m]
D = dissipation matrix, ft^3/(sec-psi) [m^3/(s·kPa)]
E = Young's modulus, psi [kPa]
E' = elastic modulus in plane strain, $E/(1-\nu^2)$, psi [kPa]
E_e = effective elastic modulus, Eq. 5.2, psi [kPa]
f,g = functions
F = integrand for functional J
G = elastic shear modulus, psi [kPa]
G_1,G_2 = elastic shear moduli of Layers 1 and 2, psi [kPa]
h = fracture height, ft [m]
$\Delta h^a,\Delta h^b$ = vertical migration of fracture above and below pay zone, ft [m]
H = pay-zone thickness, ft [m]
J = functional in variational formulation of fluid flow
K = elastic stiffness matrix, psi-ft [kPa·m]
K' = power-law viscosity coefficient, psi-sec$^{n'}$ [kPa·s$^{n'}$]
K^{-1} = inverse of K, (psi-ft)$^{-1}$ [kPa·m]$^{-1}$
K_I,K_I^a,K_I^b = stress intensity factor, psi-ft$^{1/2}$ [kPa·m$^{1/2}$]
K_{Ic} = critical value of stress-intensity factor required for crack advance, psi-ft$^{1/2}$ [kPa·m$^{1/2}$]
K_ℓ = leakoff coefficient, ft/sec$^{1/2}$ [m/s$^{1/2}$]
L = total fracture length, ft [m]
m = slope of log-log plots
n = timestep index
n' = power-law viscosity exponent
$\vec{n}_k$ = unit exterior normal to kth triangle
p = fluid pressure in fracture, psi [kPa]
$\vec{p}$ = vector of fluid pressure at nodes, psi [kPa]
p_f = pore-fluid pressure in reservoir, psi [kPa]
p_i = fluid pressure at ith node, psi [kPa]
p_L = pressure differential required to open fracture at crack front in pseudo-3D analysis, psi [kPa]
q = flow rate, $(q_x^2+q_y^2)^{1/2}$, ft^2/sec [m^2/s]
q_I = injection rate per unit fracture area for line source along perforations, ft/sec [m/s]
q_L = leakoff rate through both fracture faces per unit fracture area, ft/sec [m/s]
q_n = flow rate in direction normal to crack front per unit length of front, ft^2/sec [m^2/s]
q_x (q_y) = volume flow rate in x or y direction per unit length in y or x direction, ft^2/sec [m^2/s]
Q = horizontal flow rate in pseudo-3D model, ft^2/sec [m^2/s]
Q_I = total injection rate, ft^3/sec [m^3/s]
Q_L = leakoff rate in pseudo-3D model, ft^2/sec [m^2/s]
R = distance from source point (x',y') to field point (x,y), ft [m]
$\hat{R}$ = distance in calculation of the effects of differences in elastic moduli from layer to layer, ft [m]
R_y = height quotient used for calculating crack opening in pseudo-3D model, Eq. 5.29
Δs = distance along crack front, ft [m]
S_k = perimeter of kth triangle, ft [m]
Δt = timestep, seconds
T = matrix of weighted areas of triangular elements, ft^2 [m^2]
v = normal velocity of the crack front, ft/sec [m/s]
V = volume of annular region of fracture per unit length of crack front, ft^2 [m^2]
w = crack opening, ft [m]
$\vec{w}$ = vector of crack openings at nodes, ft [m]
w_a = crack opening at distance a from crack front, ft [m]
w_c = critical crack opening required for crack advance, ft [m]
w_i = crack opening at ith node, ft [m]
$w_{\rm I},w_{\rm II},w_{\rm III}$ = crack opening resulting from uniform pressure over heights h, Δh^a, Δh^b, respectively, ft [m]
y_o = fracture midheight relative to pay-zone midheight, ft [m]
α,β,Γ,S = dimensionless parameters defined in Eq. 5.13
Γ_i = boundary of discretized region of fracture surface
ζ = distance from crack tip in direction of inward normal, ft [m]
η' = viscosity parameter, psi-sec$^{n'}$ [kPa·s$^{n'}$]
θ = dimensionless parameter characterizing fractional timestep at which governing equations are solved
κ_1,κ_2 = dimensionless parameters defined in Eq. 5.13
ν = Poisson's ratio
ν_1,ν_2 = Poisson's ratio for Layers 1 and 2
ρF_y = weight density of fluid, psi/ft [kPa/m]
$\sigma^a,\sigma^b,\sigma^p$ = in-situ compressive stresses in upper layer, lower layer, and pay zone, psi [kPa]
σ_{zz}^0,σ = compressive stress perpendicular to crack plane before fracturing treatment, psi [kPa]
τ = time when position in fracture plane is first exposed to fracturing fluid, seconds
$\phi_i,\tilde{\phi}_i$ = local trial functions in Eqs. 5.5 and 5.6
∂A = area of near-crack-tip zone, ft^2 [m^2]
$\vec{\nabla}'$ = gradient operator

Acknowledgments

The author is grateful to J.J. Wang and H. Morales of Terra Tek Inc. for providing the numerical examples for the 3D and pseudo-3D models, respectively. Results for the 3D model were obtained with TerraFrac®; results for the pseudo-3D model were obtained with FRACTEK©.

References

1. Bui, H.D.: "An Integral Equations Method for Solving the Problem of a Plane Crack of Arbitrary Shape," *J. Mech. Phys. Solids* (1977) **25,** 29–39.
2. Clifton, R.J. and Abou-Sayed, A.S.: "A Variational Approach to the Prediction of the Three-Dimensional Geometry of Hydraulic Fractures," paper SPE 9879 presented at the 1981 SPE/DOE Low Permeability Symposium, May 27–29.
3. Lee, J.C. and Keer, L.M.: "Study of a Three-Dimensional Crack Terminating at an Interface," *J. Appl. Mechs.* (1986) **53,** 311–16.
4. Clifton, R.J. and Abou-Sayed, A.S.: "On the Computation of the Three-Dimensional Geometry of Hydraulic Fractures," paper SPE 7943 presented at the 1979 SPE Low Permeability Gas Reservoirs Symposium, Denver.
5. Kurashige, M.: "Transient Response of a Fluid-Saturated, Poro-Elastic Layer Subjected to a Sudden Pressure Rise," *J. Applied Mech.* (1982) **53,** 492–96.
6. Clifton, R.J.: "Recent Advances in the Three Dimensional Simulation of Hydraulic Fracturing," *Developments in Mechanics, Proc.*, 19th Midwestern Mechanics Conference, Ohio State U., Columbus (1985) **13,** 311–19.
7. Bui, H.D. and Parnes, R.: "A Re-Examination of the Pressure at the Tip of Fluid-Filled Crack," *Intl. J. Eng. Sci.* (1982) **20,** 1215–20.
8. Rice, J.R.: "Mathematical Analysis in the Mechanics of Fracture," *Fracture, An Advanced Treatise (Vol. II, Mathematical Fundamentals)*, H. Liebowitz (ed.), Academic Press, New York City (1968) Chap. III, 191–311.
9. Schmidt, R.A.: "Fracture Toughness Testing of Limestone," *Experimental Mechanics* (1976) **16,** 161–67.
10. Clifton, R.J., Simonson, E.R., Jones, A.H., and Green, S.J.: "Determination of the Critical Stress Intensity Factor K_{Ic} from Internally Pressurized Thick-Walled Vessels," *Experimental Mechanics* (1976) **16,** 233–38.
11. Abou-Sayed, A.S.: "An Experimental Technique for Measuring the Fracture Toughness of Rocks under Downhole Stress Conditions," *Proc.*, Sixth Intl. Conference on Experimental Stress Analysis, Munich (1978) VDI-Verlag, 819–24.
12. Botsis, J., Chudnovsky, A., and Moet, A.: "Fatigue Crack Layer Propagation in Polystyrene," *Intl. J. Fracture* (1987) **33,** 263–83.
13. Perkins, T.K. Jr. and Kern, L.R.: "Widths of Hydraulic Fractures," *JPT* (Sept. 1961) 937–49; *Trans.*, AIME, **222.**
14. Nordgren, R.P.: "Propagation of a Vertical Hydraulic Fracture," *SPEJ* (Aug. 1972) 306–14; *Trans.*, AIME, **253.**
15. Geertsma, J. and de Klerk, F.: "A Rapid Method of Predicting Width and Extent of Hydraulically Induced Fractures," *JPT* (Dec. 1969) 1572–81; *Trans.*, AIME, **246.**
16. Tada, H., Paris, P., and Irwin, G.: *The Stress Analysis of Crack Handbook,* Del Research Corp., Hellertown, PA (1973).
17. Abe, H., Mura, T., and Keer, L.M.: "Growth Rate of a Penny-Shaped Crack in Hydraulic Fracturing of Rocks," *J. Geophys. Res.* (1976) **81,** 5335–40.
18. Clifton, R.J. and Wang, J.J.: "Multiple Fluids, Proppant Transport, and Thermal Effects in Three-Dimensional Simulation of Hydraulic Fracturing," paper SPE 18198 presented at the 1988 SPE Annual Technical Conference and Exhibition, Houston, Oct. 2–5.
19. Khristianovitch, S.A. and Zheltov, J.P.: "Formation of Vertical Fractures by Means of a Highly Viscous Fluid," Fourth World Pet. Cong., Rome (1955) **II,** 579–86.
20. Palmer, I.D. and Carroll, H.B. Jr.: "Three-Dimensional Hydraulic Fracture Propagation in the Presence of Stress Variations," *SPEJ* (Dec. 1983) 870–78.
21. Morales, R.H.: "Microcomputer Analysis of Hydraulic Fracture Behavior With a Pseudo-Three-Dimensional-Simulator," *SPEPE* (Feb. 1989) 69–74.
22. Simonson, E.R., Abou-Sayed, A.S., and Clifton, R.J.: "Containment of Massive Hydraulic Fractures," *SPEJ* (Feb. 1978) 27–32.

SI Metric Conversion Factors

bbl	×	1.589 873	E−01	=	m^3
ft	×	3.048*	E−01	=	m
ft^2	×	9.290 304*	E−02	=	m^2
gal	×	3.785 412	E−03	=	m^3
in.	×	2.54*	E+00	=	cm
(lbf-sec)/ft^2	×	4.788 026	E+01	=	Pa·s
psi	×	6.894 757	E+00	=	kPa

*Conversion factor is exact.

Chapter 6
Propping Agents and Fracture Conductivity

Robert W. Anderson, SPE, The Western Co.
Claude E. Cooke Jr., SPE, Consultant
Charles L. Wendorff, Consultant

6.1 Overview

This chapter describes the role of proppants in the fracturing process, and gives a brief history of their development by the industry. The relationship between fracture permeability and well productivity is described, and laboratory techniques currently used for measuring fracture conductivity, along with a short history of the development of these techniques, are outlined. Representative data from the measurement of fracture flow capacity by techniques generally accepted within the industry are presented for the commonly used proppants. This information should be useful in the design of fracturing treatments. Factors that may decrease fracture conductivity that are not normally evaluated in routine laboratory measurements are discussed in terms of how they might influence the effectiveness of a fracturing treatment.

6.2 Introduction

At this point it is worthwhile to recall that the goal of hydraulic fracturing is to increase well productivity by altering the flow pattern in the formation near the wellbore from one that is radial with flowlines converging to the wellbore to one that is linear with flow to a conductive fracture that intersects the wellbore. For the effort to be successful, the fracture must be much more conductive than the formation. To obtain a high-permeability fracture, a granular propping agent must be added to the fracturing fluid.

The purpose of the propping agent (proppant) is to keep the walls of the fracture apart so that a conductive path to the wellbore is retained after pumping has stopped and fluid pressure has dropped below that required to hold the fracture open. Ideally, the proppant will provide flow conductivity large enough to make negligible any pressure losses in the fracture during fluid production. In practice, this ideal might not be achieved because the selection of a proppant involves many compromises imposed by economic and practical considerations.

The propped fracture must have a conductivity at least high enough to eliminate most of the radial flow path that exists in an unfractured well and to allow linear flow from the reservoir into the fracture. This requires relatively unimpeded linear flow within the fracture to the wellbore. To accomplish this, the proppant must enable the propped fracture to have a permeability several orders of magnitude larger than that of the reservoir rock.

6.3 Effect of Fracture Conductivity on Well Productivity

The McGuire-Sikora Model. The effect of fracture conductivity on well productivity is usually expressed in terms of a mathematical or graphical model. One of the earliest of these, and one still useful for analyzing high-permeability formations, is that proposed by McGuire and Sikora.[1] This model was developed from an electric-analog study and published in graphical form. The McGuire-Sikora model assumes laminar flow of an incompressible fluid and pseudo-steady-state flow conditions. The results from the model, shown in Fig. 6.1, are in terms of the predicted productivity index (PI) ratio that results from fractures of different conductivity ratio (defined on the x axis of the graph), where

J = PI of well after fracturing,
J_0 = PI of well when undamaged and unstimulated (skin=0),
r_e = drainage radius of the well,
r_w = wellbore radius,
w = propped width of fracture,
k_f = permeability of proppant,
$\bar{k}$ = average formation permeability based on gross thickness,
A_d = well spacing for square drainage area, and
L = length of one wing of fracture.

The curves presented in Fig. 6.1 are representative of a 3-in. [7.6-cm] wellbore radius and a well spacing of 40 acres [16 ha]. The scaling factors on the axes of the graph permit converting the conductivity ratio and the PI ratio to other values of well spacing and wellbore radius. Note that the PI ratio shown is the PI of the well after the fracturing treatment, J, divided by the PI of the well before the treatment, J_0. This quantity is also called the stimulation ratio.

Proppant Selection—McGuire-Sikora. The McGuire-Sikora model is suited for determining the required fracture conductivity in an oil well producing from high- or medium-permeability formation. Fig. 6.1 can be used to determine the conductivity ratio needed to achieve a given PI ratio. The selection of a proppant consists of determining the optimum economic value of wk_f, fracture conductivity, for a given well. All other variables in the curves are independent of proppant properties, except that proppant density can affect bed geometry when fluids that do not totally suspend the proppant during pumping are used.

An examination of Fig. 6.1 shows that the conductivity ratio must be about 1,000 before a stimulation ratio of 2 (a minimum design value) is achieved. For fractures in the range of 0.1 in. [0.25 cm] wide, the permeability of the proppant must be 10,000 times the permeability of the reservoir rock to meet this goal. As an example, for a formation of 10-md permeability, proppant permeability must be about 100 darcies. This value can be achieved with 20/40-mesh particles that have sustained a small degree of crushing, as shown by data in Sec. 6.4.

If the reservoir rock permeability is only 0.1 md and proppant permeability remains at 100 darcies, the conductivity ratio becomes 100,000. This is about the upper limit of fracture conductivity that

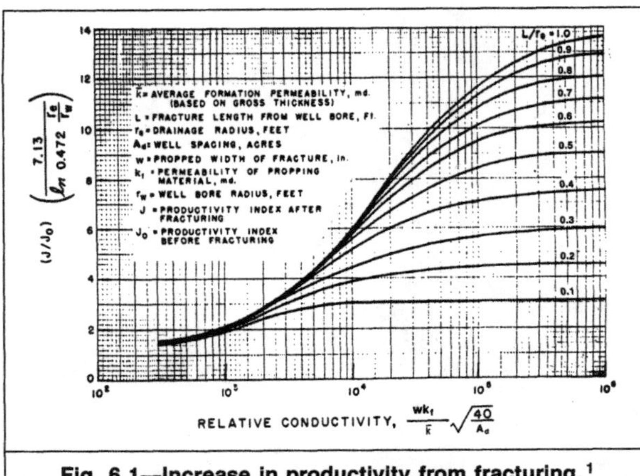

Fig. 6.1—Increase in productivity from fracturing.[1]

is beneficial in increasing the stimulation ratio of a fractured well (see Fig. 6.1). With the higher values of conductivity ratio, improved productivity depends primarily on the ratio of fracture length to the drainage radius of the well, L/r_e. A larger fracture treatment, by providing greater propped fracture length, L, would increase the stimulation ratio, but higher permeability in the fracture would have little benefit. In fact, a stimulation ratio of 2 in rock of this permeability (i.e., 0.1 md) can be attained with a proppant permeability as low as 1 darcy. The latter value is exceeded by nearly any commercial proppant, even at high stress (assuming that it is not damaged by some other mechanism, such as those discussed in Sec. 6.5). The initial PI of an oil well in a 0.1-md rock is so low, however, that even a high stimulation ratio will produce only a small incremental production rate.

Using the McGuire-Sikora chart for proppant selection can be generalized as follows. Given the predicted reservoir permeability, calculate the range of needed fracture conductivity to achieve a desired stimulation ratio. Then estimate the stress on the proppant (with the methods in Chap. 4 or 11), and use data such as those presented in the following sections of this chapter to determine which proppants will provide the needed conductivity ratio under existing stress. A comprehensive design procedure is discussed in Chap. 11.

Other Widely Used Models. Several other models to predict the effect of fracture conductivity on PI have been published since that of McGuire and Sikora. Especially notable are those of Tinsley *et al.*[2], Prats,[3] and Mao.[4] Each assumes either steady-state or pseudosteady-state flow conditions; i.e., pressures are stabilized at the wellbore and at the drainage boundary of the well. Darcy's law is also assumed to apply in the fracture.

Tinsley attempted to provide a means of correcting for fractures only partially filled with proppant. Prats provided an analytic model for the special case of incompressible fluids, and Mao used a numerical simulator to analyze the case of finite-conductivity fractures. After correcting a scaling error in the Tinsley model, Soliman[5] compared the PI ratios predicted by each of the three models and found them in good agreement.

Non-Darcy Flow and Unsteady-State Flow. The models that assume that Darcy's law applies in the fracture are generally adequate for flow of oil in the fracture, although for a low-viscosity oil at rates of thousands of barrels per day, this assumption should be re-examined. Non-Darcy flow usually exists in the fracture of a gas well, however, and models that do not consider non-Darcy flow will overestimate the stimulation to be achieved from a given fracturing treatment.[6,7] Predictions of fracture stimulation that include the effects of turbulence have been published.[8,9] A set of correlation curves that can be used much like the McGuire-Sikora curves for proppant selection is discussed in Chaps. 11 and 15.

If the effects of non-Darcy flow can be neglected, the proppant conductivity needed to attain a given increase in production rate can be predicted with the type-curve results of Agarwal *et al.*,[9] which include the effects of unsteady-state flow conditions. Fracture conductivity is expressed in a dimensionless form that depends on the width of the fracture and permeability of the proppant, as in the McGuire-Sikora model.

To include the effects of both non-Darcy flow in the fracture and unsteady-state flow, both of which are usually important in low permeability gas reservoirs, a numerical simulation of flow in the reservoir is necessary (see Chap. 11). Proppant is selected through calculations for each of several different proppants with various amounts of proppant in the fracture. Use of different proppants in the same fracture can also be investigated with a numerical simulator.[10]

6.4 Commercial Proppants

Early History. One of the first proppants used in the early days of hydraulic fracturing during the late 1940's was sand dredged from the Arkansas River. Initially, the sand was not cleaned and screened as today's standards require[11-15] but as the need became evident, steps were taken to process the sand more thoroughly.

TABLE 6.1—TYPICAL PHYSICAL PROPERTIES OF BRADY-TYPE FRACTURING SAND[17]*

API Property	Recommended Limits	API Mesh Size 6/12**	8/16	12/20	16/30	20/40
Particle diameter range, μm	Standard	3,350 to 1,700	2,360 to 1,180	1,700 to 850	1,180 to 600	850 to 425
Sieve analysis, wt% retained						
Top sieve	0.1 maximum	0.0	0.0	0.0	0.0	0.1
Between primary sieves	90.0 minimum	95.7	93.1	91.0	98.5	91.6
Second and sixth sieves		4.2	6.6	8.5	1.0	8.0
Pan	1.0 maximum	0.1	0.3	0.5	0.5	0.4
Total		100.0	100.0	100.0	100.0	100.0
Krumbein shape factor						
Roundness	0.6 minimum	0.6	0.6	0.6	0.6	0.6
Sphericity	0.6 minimum	0.6	0.6	0.6	0.6	0.6
12/3 HCl/HF solubility, 30 minutes at 150°F, wt%	3.0 maximum	0.4	1.0	1.0	0.8	0.8
Silt and fine particle, FTU†	250 maximum	20	95	120	45	115
Crush resistance, % fines	Variable with size	17.9	13.4	15.5	8.3	11.4
generated at closure stress, psi		2,000	2,000	3,000	3,000	4,000
Particle density, lbm/gal	22.1 maximum	22.1	22.1	22.1	22.1	22.1
Bulk density, lbm/ft³	105.0 maximum	95.5	98.0	99.9	101.1	100.5
Clustering, wt%	1.0 maximum	<1.0	<1.0	<1.0	0.0	0.0

*All tests performed according to Ref. 11 or 12. Sources include Hickory sandstone, aeolian dune sand, and Bidahochi formation. Values shown are averages of multiple production samples over a 4-year period.
**Not commercially available at this time.
†FTU = formazine turbidity units.

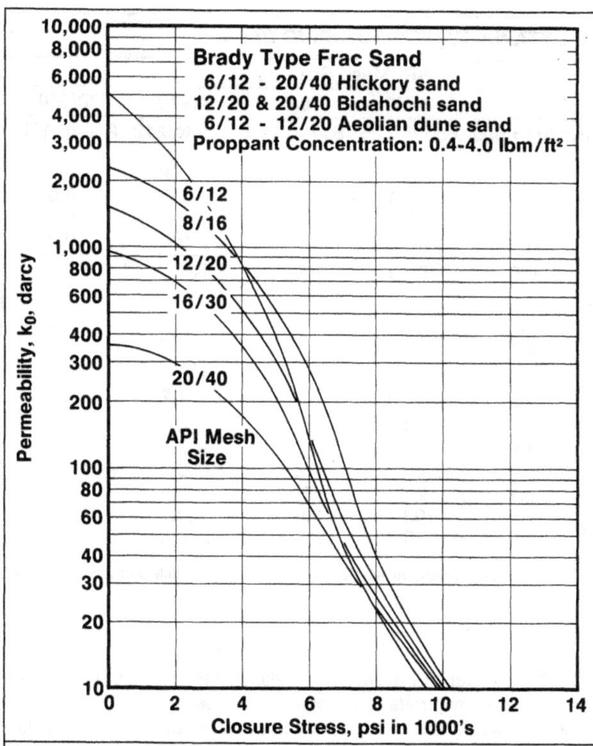

Fig. 6.2—Effect of Brady-type sand mesh size on pack permeability.[17]

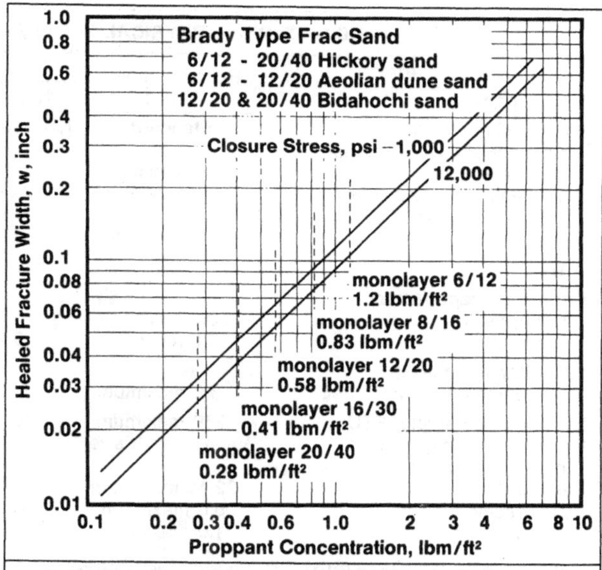

Fig. 6.3—Effect of closure stress on pack width for Brady-type fracturing sand.[20]

During the mid 1950's, sand from the Saint Peter sandstone formation near Ottawa, IL, entered the market.

As the need for a more economical and readily available fracturing sand grew, mines were opened near Brady, TX, in 1958, and production from the Hickory sandstone formation began to be marketed. This sand, as well as most other high-quality sand used today, is mined from consolidated sandstone formations. The mining process includes crushing, screening, and washing to separate the sandstone matrix into its individual sand grains. A wide range of particle sizes is found in the deposits. Typically, only 20 to 30% of such deposits is found to be in a size range useful for hydraulic fracturing applications.

The explosive growth of the hydraulic fracturing industry from the mid-1970's to the early 1980's created shortages of fracturing sand. Supplies from the Saint Peter sandstone of Illinois were supplemented by high-quality material from the Jordan, Ironton, and Galesville sandstones of Minnesota and Wisconsin.[16] Similarly, sand from the Bidahochi formation in Arizona and aeolian dune sand of Colorado augmented proppant production from the Hickory sandstone in Texas. Finally, new sand-processing plants were constructed in Minnesota and Wisconsin specifically to produce fracturing sand and to replace plants designed to supply sand for other applications.

Commercial Fracturing Sand. ***Brady-Type Sand.*** This rounded quartz sand, also known as brown or Texas sand, is mined from the Hickory sandstone in central Texas near the town of Brady. The Hickory sandstone was deposited during the Upper Cambrian Age some 500 million years ago. The color of this sand results from small amounts of iron oxide contamination in the crystal structure. Color variation has no bearing on the strength of this sand or on any other sand discussed here.

As mined, the sand is polycrystalline; i.e., each whole grain is composed of more than one quartz crystal bonded together, leaving cleavage planes in the whole grain. In terms of fines generated, the API crush resistance test[11] typically yields from <50 to as much as 85% of the API permissible fines. The deposit yields acceptable fracturing sand in the 20/40-mesh size range and larger. Production in sizes smaller than 20/40 mesh is not sized to meet API recommendations. Typical physical properties, fracture permeability, and pack-width data for this sand are presented in Table 6.1 and Figs. 6.2 and 6.3.[17-20]

The Bidahochi formation sand is mined from shallow, lightly consolidated lenses in eastern Arizona. It was deposited during the Pliocene or Tertiary Age some 6 million years ago. This distinctly colored sand contains grains of chert, which is stronger than quartz, along with rose and smoky quartz. Fracturing sand from this formation is available in limited quantities in 12/20, 20/40, and 40/70 mesh only.

The aeolian dune sand is mined in central Colorado from shallow, lightly consolidated lenses. This sand was deposited during the Holocene Age less than 1 million years ago. The large sizes, 6/12 through 12/20 mesh, are as high in quality as those from the Hickory formation, but the small sizes, 16/30 through 70/140 mesh, contain so much feldspar that they produce excessive fines in the API crush resistance test.

Ottawa-Type Sand. This well-rounded, very pure quartz sand exceeds API recommendations. In terms of fines generated, the API crush resistance test typically yields less than half of the maximum acceptable fines on this sand. The sand also is monocrystalline. Crushed particles are primarily large chipped grains rather than individual quartz crystals. Color variation is widespread in this sand but has no impact on its performance characteristics as a proppant. For the most part, the sand is well processed and of high quality for fracturing applications. Typical physical properties, permeability, and pack-width data of this sand are presented in Table 6.2 and in Figs. 6.4 and 6.5.[17-20]

The Saint Peter sandstone, commonly known as Ottawa sand, was deposited in the Ottawa district of Illinois during the Middle Ordovician Age some 460 million years ago. This sand is available in 20/40 mesh and smaller sizes only. Color variation runs from white through gray-white to pale yellow.

The Jordan sandstone was deposited in south central Minnesota and western Wisconsin during the Upper Cambrian Age some 500 million years ago. Jordan fracturing sand is available only in 12/20-mesh and smaller sizes. The color varies from white through gray-white to pale yellow to brown.

The Galesville and Ironton sandstones were deposited in south central Minnesota and western Wisconsin during the Upper Cambrian Age some 500 million years ago. Ironton fracturing sand is available in 12/20 mesh and smaller; the Galesville sand is available in 20/40-mesh and smaller sizes only. Its color varies from white to light tan.

TABLE 6.2—TYPICAL PHYSICAL PROPERTIES OF OTTAWA-TYPE FRACTURING SAND[17]*

API Property	Recommended Limits	API Mesh Size					
		12/20**	16/30	20/40	30/50	40/70	70/140
Particle diameter range, μm	Standard	1,700 to 850	1,180 to 600	850 to 425	600 to 300	425 to 212	212 to 106
Sieve analysis, wt% retained							
Top sieve	0.1 maximum	0.0	0.0	0.0	0.0	0.1	0.1
Between primary sieves	90.0 minimum	93.2	97.9	91.5	93.1	91.8	90.0
Second and sixth sieves		6.6	2.1	8.0	6.5	7.6	9.1
Pan	1.0 maximum	0.2	0.0	0.5	0.4	0.6	0.8
Total		100.0	100.0	100.0	100.0	100.0	100.0
Krumbein shape factor							
Roundness	0.6 minimum	0.7	0.7	0.7	0.7	0.7	0.6
Sphericity	0.6 minimum	0.7	0.7	0.8	0.8	0.7	0.7
12/3 HCl/HF solubility, 30 minutes at 150°F, wt%	3.0 maximum	1.5	1.0	1.0	0.9	1.2	2.5
Silt and fine particle, FTU	250 maximum	68	110	80	60	40	130
Crush resistance, % fines	Variable with size	5.4	1.6	4.0	3.3	3.4	2.5
generated at closure stress, psi		3,000	3,000	4,000	4,000	5,000	5,000
Particle density, lbm/gal	22.11 maximum	22.1	22.1	22.1	22.1	22.1	22.1
Bulk density, lbm/ft³	105.0 maximum	95.5	98.6	102.7	103.0	102.7	103.0
Clustering, wt%	1.0 maximum	0.0	0.0	0.0	0.0	0.0	0.0

*All tests performed according to Ref. 11 or 12. Sources include Saint Peter, Jordan, Galesville, and Ironton sandstones. Values shown are averages of multiple production samples over a 4-year period.
**Available in limited quantities on special order only.

Efforts To Improve on Fracturing Sand. Because of the well-recognized limitations of fracturing sand, especially at high stress levels, efforts have been made to find a different proppant with improved performance characteristics. Many of the deficiencies of sand relate to its brittle failure from point loading under high stress levels. Likewise, much effort has concentrated on materials with more strength and deformability than sand. Such materials as iron shot, aluminum pellets, quenched-glass beads, walnut hulls, plastic beads, and a vast array of high-strength and deformable particles were manufactured in the 1960's and evaluated as potential proppants. With the single exception of glass beads, none survived until the early 1970's because each of these proppants failed to achieve the desired results in actual field applications.

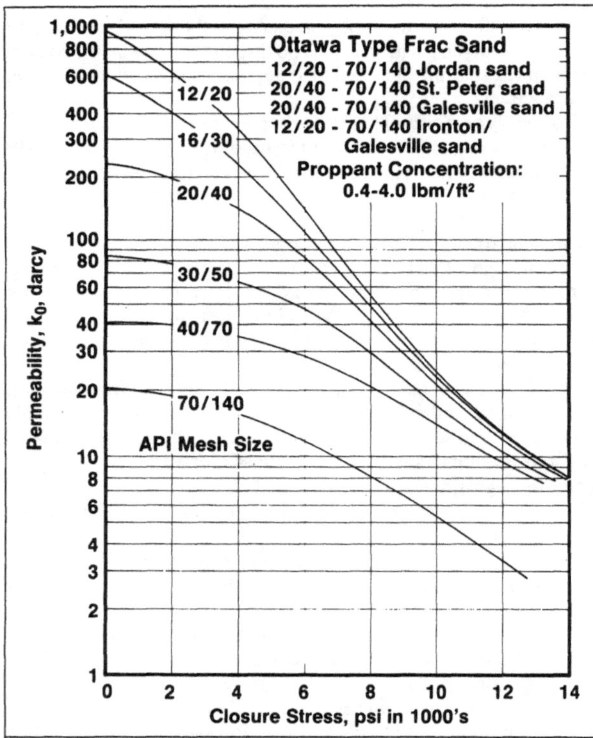

Fig. 6.4—Effect of Ottawa-type sand mesh size on pack permeability.[17]

With the drilling of deeper wells, the shortcomings of glass beads and quartzitic materials as proppants became apparent.[6] Such materials are weakened by hot formation brines and tend to fail catastrophically under high closure stress. These factors accelerated the search for improved materials, and in the mid-1970's, a high-strength ceramic proppant, sintered bauxite,[21] was introduced. The inertness and strength of sintered bauxite are caused by its major constituent, corundum, a form of aluminum oxide. Although expensive, sintered bauxite retains permeability under very high stress and severe reservoir conditions better than any other proppant available today.

The expense of sintered bauxite motivated efforts to find less costly but useful substitutes.[22-27] Under development at the same time as sintered bauxite, curable resin-coated sand[28] was the first such product to find application.

Research and development[29-35] on other ceramic proppants during the early 1980's produced a less expensive proppant containing mullite, another form of aluminum oxide, in addition to corundum.[23] It has helped to bridge the cost-performance gap be-

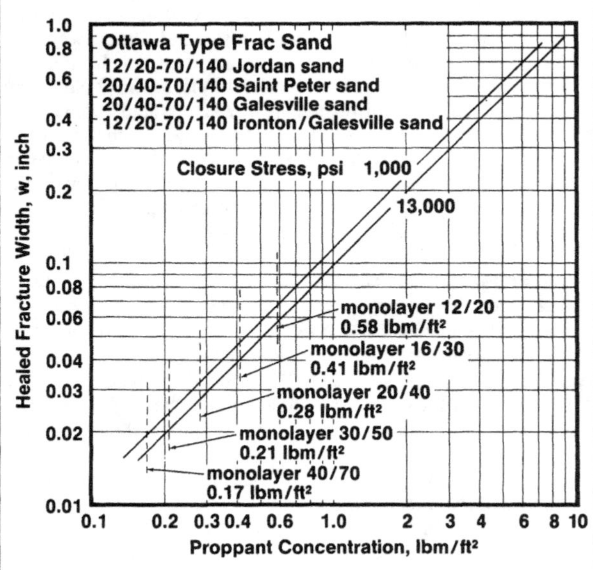

Fig. 6.5—Effect of closure stress on pack width for Ottawa-type fracturing sand.[20]

TABLE 6.3—TYPICAL PHYSICAL PROPERTIES OF SINTERED BAUXITE—HIGH-STRENGTH, SINTERED CERAMIC PROPPANT[17]*

API Property	Recommended Limits	API Mesh Size 12/20	16/20	20/40	40/70
Particle diameter range, μm	Standard	1,700 to 850	1,180 to 850	850 to 425	452 to 212
Sieve analysis, wt% retained					
Top sieve	0.1 maximum	0.0	0.0	0.0	0.0
Between primary sieves	90.0 minimum	96.3	95.3	94.0	95.4
Second and sixth sieves		3.7	4.7	6.0	4.6
Pan	1.0 maximum	0.0	0.0	0.0	0.0
Total		100.0	100.0	100.0	100.0
Krumbein shape factor					
Roundness	0.7 minimum	0.8	0.8	0.8	0.8
Sphericity	0.7 minimum	0.9	0.9	0.9	0.9
12/3 HCl/HF solubility, 30 minutes at 150°F, wt%	7.5 maximum	2.0	2.0	2.0	2.0
Silt and fine particle, FTU	250 maximum	80	100	100	120
Crush resistance, % fines generated	Variable with size and stress				
at 7,500 psi		5.4	6.4	2.6	1.7
at 10,000 psi		10.6	12.2	4.3	3.0
at 12,500 psi		16.8	18.0	6.8	5.2
at 15,000 psi		22.5	23.2	10.7	7.3
Particle density, lbm/gal	28.4 maximum	30.88	30.88	30.88	30.88
Bulk density, lbm/ft^3	140.0 maximum	140.0	140.0	140.0	140.0
Clustering, wt%	1.0 maximum	<1.0	0.0	0.0	0.0

*All tests performed according to Ref. 11 or 12. Values shown are averages of multiple production samples over a 4-year period.

tween sand and bauxite. Because of its lower cost and high performance, this material has enjoyed widespread use since its introduction.

Improved Commercial Proppants. ***Sintered Bauxite.*** As previously described, sintered bauxite is an inert, high-strength ceramic proppant. Patented by Cooke *et al.*,[21] this high-density proppant is produced by the same manufacturing techniques as refractory ceramics and metal-working abrasive grits. The raw material is primarily high-alumina bauxite ore from South America. The ore is first ground to a particle size less than 15 μm, shaped into small ceramic pellets using water and a binder, and, after drying and screening, fired in a kiln to bind the edges of the individual particles that make up each pellet. After the sintering process, the color of the product varies from black to brown or gray. Typical physical properties, pack permeability, and width data for this proppant are presented in Table 6.3 and in Figs. 6.6 and 6.7.[17-20]

Sintered bauxite draws its strength from the unique manufacturing process and from the materials present in the bauxite ore. Corundum, the major component of sintered bauxite, is one of the hardest materials known to man. It measures 9 on Moh's hardness scale. For comparison, quartz is 7 and diamond is 10. When crushed, bauxite does not shatter as completely as the sands; it simply splits into large pieces that are still capable of providing flow capacity.[17,36,37] This crush resistance is caused partially by sintered bauxite's elastic properties, which allow slight deformation before failure under high stresses.

Fig. 6.6—Effect of sintered bauxite mesh size on pack permeability.[17]

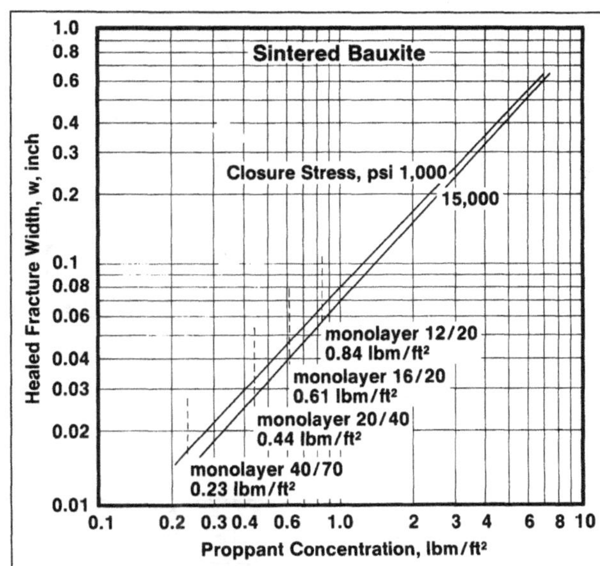

Fig. 6.7—Effect of closure stress on pack width for sintered bauxite.[20]

TABLE 6.4—TYPICAL PHYSICAL PROPERTIES OF HIGH-STRENGTH, INTERMEDIATE-DENSITY, SINTERED CERAMIC PROPPANT[17]*

API Property	Recommended Limits	API Mesh Size 12/20	16/20	20/40	40/70**
Particle diameter range, μm	Standard	1,700 to 850	1,180 to 850	850 to 425	452 to 212
Sieve analysis, wt% retained					
Top sieve	0.1 maximum	0.0	0.0	0.0	0.0
Between primary sieves	90.0 minimum	98.0	92.4	93.7	95.2
Second and sixth sieves		2.0	7.6	6.3	4.8
Pan	1.0 maximum	0.0	0.0	0.0	0.0
Total		100.0	100.0	100.0	100.0
Krumbein shape factor					
Roundness	0.7 minimum	0.8	0.8	0.8	0.7
Sphericity	0.7 minimum	0.8	0.8	0.9	0.9
12/3 HCl/HF solubility, 30 minutes at 150°F, wt%	7.5 maximum	4.5	4.8	6.2	5.6
Silt and fine particle, FTU	250 maximum	100	100	100	120
Crush resistance, % fines generated	Variable with size and stress				
at 7,500 psi		6.4	10.3	3.2	1.4
at 10,000 psi		13.6	19.4	6.0	2.7
at 12,500 psi		19.3	27.4	9.8	4.6
at 15,000 psi		26.9	33.9	14.3	7.4
Particle density, lbm/gal	28.4 maximum	26.29	25.95	25.62	26.12
Bulk density, lbm/ft³	114.0 maximum	113.0	107.0	106.0	113.0
Clustering, wt%	1.0 maximum	<1.0	<1.0	<1.0	<1.0

*All tests performed according to Ref. 11 or 12. Values shown are averages of multiple production samples over a 3-year period.
**Currently available in limited quantities on special order only.

The first sintered bauxite proppants were angular in shape, which could cause increased abrasion and failure of pumping equipment, treating lines, wellhead equipment, and chokes. Process improvements have produced a material with roundness and sphericity values better than the best fracturing sand and thus less abrasive than its predecessor. This proppant has become the standard against which all other proppants are measured.[38-48]

Intermediate-Density Proppant. Even though this material is often called "intermediate-strength proppant," a more appropriate term is "intermediate-density proppant (IDP)." The strength of this type of proppant is much closer to that of sintered bauxite than to sand. While neither as strong nor as inert as sintered bauxite, this material has an advantage over sintered bauxite in that it has a lower density (approaching that of sand) than bauxite.

The search for a more economical replacement of sintered bauxite revealed that high-alumina, domestic bauxitic ores could be used to produce a high-performance, sintered proppant with properties approaching those of sintered bauxite.[23] In addition to corundum, this proppant contains mullite, a less-dense mixed form of aluminum oxide. The result is a dark brown to tan proppant of lower bulk density and lower specific gravity than bauxite. This new material is produced by manufacturing techniques similar to those used for sintered bauxite. Typical physical properties, pack permeability, and width data for this proppant are presented in Table 6.4 and in Figs. 6.8 and 6.9.[17-20]

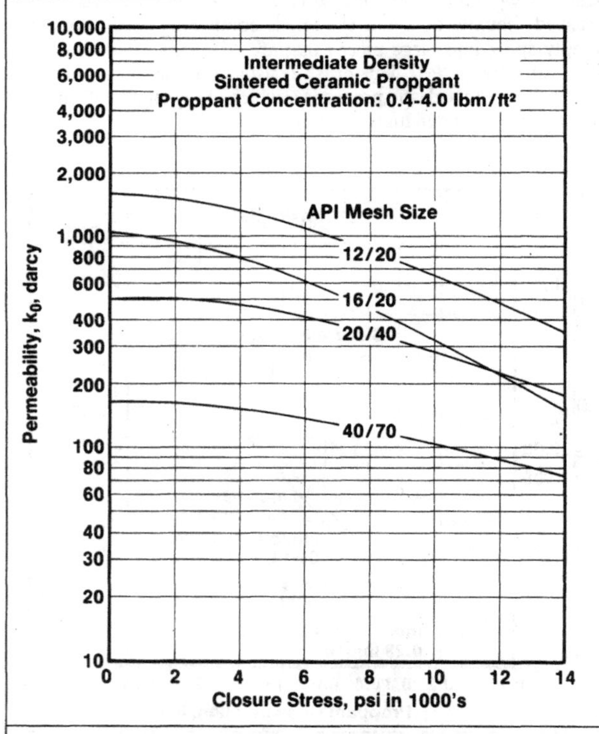

Fig. 6.8—Effect of intermediate-density proppant mesh size on pack permeability.[17]

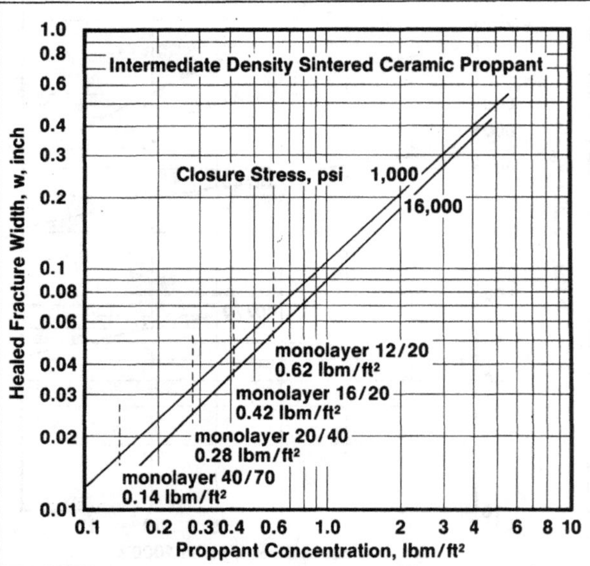

Fig. 6.9—Effect of closure stress on pack width for intermediate-density proppant.[20]

TABLE 6.5—TYPICAL PHYSICAL PROPERTIES OF CURABLE RESIN-COATED SAND—LOW-DENSITY, INTERMEDIATE-STRENGTH PROPPANT[17]*

API Property	Recommended Limits	API Mesh Size 12/20**	16/30	20/40
Particle diameter range, μm	Standard	1,700 to 850	1,180 to 600	850 to 425
Sieve analysis, wt% retained				
Top sieve	0.1 maximum	0.0	0.0	0.0
Between primary sieves	90.0 minimum	95.5	98.0	94.4
Second and sixth sieves		4.3	2.0	5.6
Pan	1.0 maximum	0.2	0.0	0.0
Total		100.0	100.0	100.0
Krumbein shape factor				
Roundness	0.7 minimum	0.8	0.8	0.8
Sphericity	0.7 minimum	0.9	0.9	0.8
12/3 HCl/HF solubility, wt%				
30 minutes at 150°F, wt%	7.5 maximum	0.5	0.6	0.5
Compressive strength, after 100 hours at 195°F, psi	Variable with size	1,400	2,000	2,800
Tensile strength after 3 minutes at 450°F, psi	Variable with size	180.0	220.0	270.0
Resin content, wt%	3.6 to 4.4	3.7	4.0	3.8
Coating continuity, count %	98.0 minimum	99.5	99.0	98.5
Uncoated particles, wt%	0.5 maximum	0.2	0.3	0.2
Particle density, lbm/gal	21.7 maximum	21.3	21.2	21.3
Bulk density, lbm/ft³	100.0 maximum	96.0	95.5	96.0
Clustering, wt%	0.5 maximum	<1.0	<1.0	<1.0

*All tests performed according to Ref. 11 or 12. Values shown are averages of multiple production samples over a 2-year period.
**Currently available on special order only.

While the durability and strength of intermediate-density proppant are some what less than those of sintered bauxite, performance is virtually equivalent in all but the deepest and hottest wells. At high stress levels, the proppant breaks into large particles capable of providing good flow capacity. The proppant particles have good resistance to corrosion by hot formation brines; their roundness and sphericity are better than those of the best fracturing sands, while their bulk density is only slightly higher.

Despite higher cost, intermediate-density proppants may replace sand at intermediate well depths because of their improved performance.[30,49,50] Within the next few years a variety of sources may be developed to make this material widely available at lower costs.[35]

Resin-Coated Proppants. The most commonly available resin-coated proppants are resin-coated sands. These low-density, intermediate-strength proppants are available in two forms: curable and precured resin-coated Ottawa-type fracturing sands.[28,51-54] Both are manufactured by a process similar to that used to produce coated sand for the foundry industry. Curable resin-coated sand was originally patented by Graham *et al.*[28] for use in gravel-packing operations. Precured resin-coated sand became available in 1982, about 7 years after the first curable product was used in fracturing operations.[55-59]

The emergence of a high-quality, curable resin-coated sand, along with the availability of a precured type, has led to a wide variety

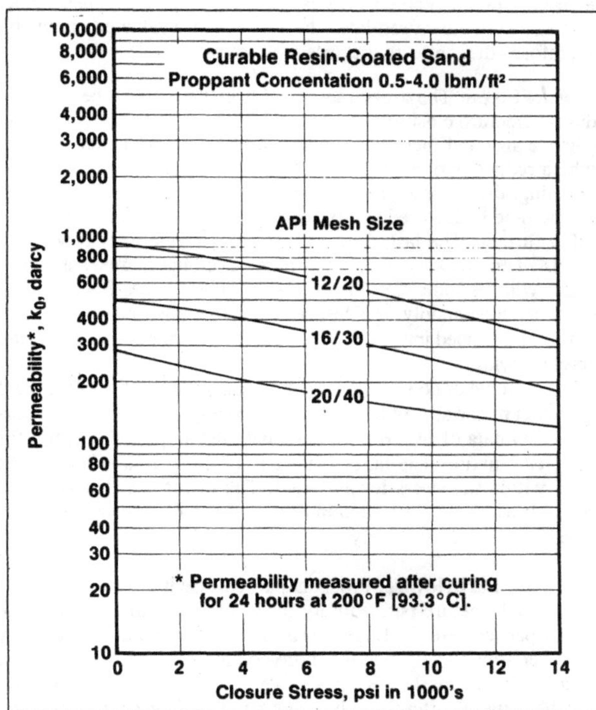

Fig. 6.10—Effect of mesh size on pack permeability for curable resin-coated Ottawa-type fracturing sand.[17]

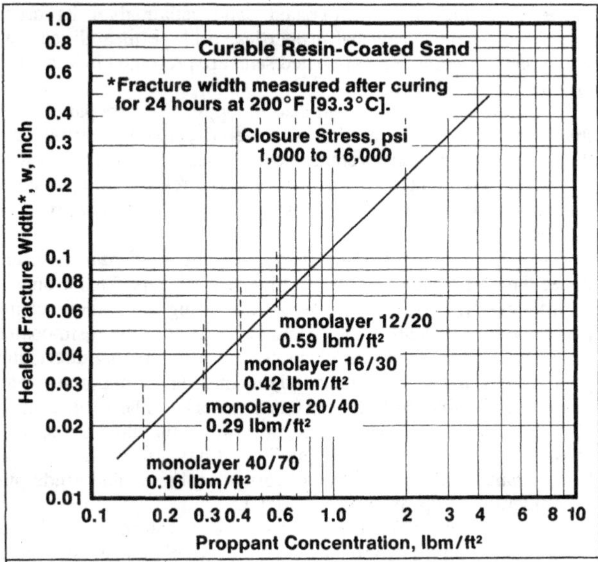

Fig. 6.11—Effect of concentration on pack width for curable resin-coated Ottawa-type sand.[20]

TABLE 6.6—TYPICAL PHYSICAL PROPERTIES OF PRECURED RESIN-COATED FRACTURING SAND—LOW-DENSITY, INTERMEDIATE-STRENGTH PROPPANT[17]*

API Property	Recommended Limits	API Mesh Size 12/20**	16/30	20/40
Particle diameter range, μm	Standard	1,700 to 850	1,180 to 600	850 to 425
Sieve analysis, wt% retained				
Top sieve	0.1 maximum	0.0	0.0	0.0
Between primary sieves	90.0 minimum	96.4	98.0	94.4
Second and sixth sieves		3.5	2.0	5.6
Pan	1.0 maximum	0.0	0.0	0.0
Total		100.0	100.0	100.0
Krumbein shape factor				
Roundness	0.7 minimum	0.8	0.8	0.8
Sphericity	0.7 minimum	0.9	0.9	0.8
12/3 HCl/HF solubility, 30 minutes at 150°F, wt%	7.5 maximum	0.3	0.3	0.4
Silt and fine particle, FTU	250 maximum	40	40	50
Crush resistance, % fines generated	Variable with size and stress			
at 7,500 psi		—	3.6	0.8
at 10,000 psi		11.2	7.0	3.0
at 12,500 psi		—	24.3	7.2
at 15,000 psi		—	39.6	11.2
Resin content, wt%	3.6 to 4.4	3.7	3.9	4.2
Coating continuity, count %	98.0 minimum	99.5	99.0	99.7
Uncoated particles, wt%	0.5 maximum	0.2	0.3	0.2
Particle density, lbm/gal	21.7 maximum	21.2	21.3	21.3
Bulk density, lbm/ft^3	100.0 maximum	97.4	98.0	98.6
Clustering, wt%	1.0 maximum	<1.0	<1.0	<1.0

*All tests performed according to Ref. 11 or 12. Values shown are averages of multiple production samples over a 3-year period.
**Currently available on special order only.

of fracturing applications. Although this proppant is not as strong nor as tough as the ceramic proppants, it is a significant improvement over uncoated sand. The plastic coating distributes point loads over a wider area on the sand grain and retards brittle failure. As such, the product is useful at higher stress levels (e.g., in deeper wells) than conventional fracturing sand.

The major application of the *curable resin-coated sand* is as a tail-in material to retain the sand in producing zones that will not retain ordinary fracturing sand. The curable coating bonds the sand grains together after they are in place in the fracture. This in-situ consolidation often prevents proppant flowback, subsequent productivity loss, and damage to well equipment. Because of the consolidated nature of the proppant pack formed with resin-coated sand, compressive or tensile strength is often used as the critical physical property to describe resin-coated sand rather than its crush resistance. Typical physical properties, pack permeabilities, and width data for curable resin-coated sand are presented in Table 6.5 and in Figs. 6.10 and 6.11.[17-20]

A curable resin coating can also be applied to proppants other than sand, and such materials as sintered bauxite, intermediate-density proppant, and zirconia have all been coated and used in fracturing treatments. The use of a curable resin coating in these applications is largely the same as with sand—to prevent proppant flowback.

The major application of *precured resin-coated sand* is to enhance the performance of sand at high stress levels. This proppant is produced by heat curing the coating during the manufacturing process rather than allowing curing to occur after the resin-coated sand has been pumped into place.[51] The resin coating distributes point loading and permits the underlying material to support a greater load than possible with the uncoated material. The resin coating also encapsulates the sand grains, thus preventing the migration of crushed fines during fluid production. It has also been shown to be resistant to destruction by hot formation brines and crude oils at temperatures up to 300°F [150°C].

At low stress levels, the performance of this material is not materially different from that of sand. At higher stress levels, however, performance of the resin-coated sand is improved considerably over the original uncoated sand. Table 6.6 shows typical physical properties of this material. Packed fracture permeability and width data are shown in Figs. 6.12 and 6.13.[17-20]

6.5 Factors Affecting Fracture Conductivity

This section discusses five factors that significantly affect the fracture flow capacity developed with proppants used in hydraulic fracturing. These factors can be readily evaluated in the laboratory, and their effect on fracture conductivity is relatively well established. Other factors, to be discussed later, have not been evaluated routinely; therefore, their effects are less well known.

Closure Stress. The stress transmitted from the earth to the proppant during fracture closure causes crushing of the proppant, reducing particle size and increasing surface area of the proppant, both of which reduce permeability of the propped fracture. In addition to crushing, the stress applied to the proppant pack serves to compact the particle bed, to reduce its porosity, and to reduce its permeability further. The last effect occurs even at relatively low stress levels when breakage is not important. Cycling of stress, as would occur with periodic shut-ins of a well, also reduces fracture conductivity irreversibly.[60] Closure stress may also cause proppant particles to embed into the walls of a soft formation, thus decreasing fracture width.

An example of how closure stress affects permeability of different proppant materials can be seen by comparing the slopes of the permeability data of Fig. 6.4 for sand with the slopes of Fig. 6.6 for sintered bauxite. The latter material is clearly less affected than sand within the stress levels tested. The effects of closure stress on pack width are illustrated in Figs. 6.5 and 6.7 for sand and sintered bauxite, respectively.

Proppant Particle Size. Proppant particle size has a material effect on packed fracture permeability, as can be seen in Fig. 6.2. The larger particles—e.g., 12/20 mesh—provide a greater conductivity at lower stress levels than the more commonly used smaller sizes, such as 20/40 mesh. As stress levels increase and particles are crushed, these differences in conductivity decrease because particle size distribution, porosity, and surface areas become similar despite initial particle-size differences. At this point, other factors

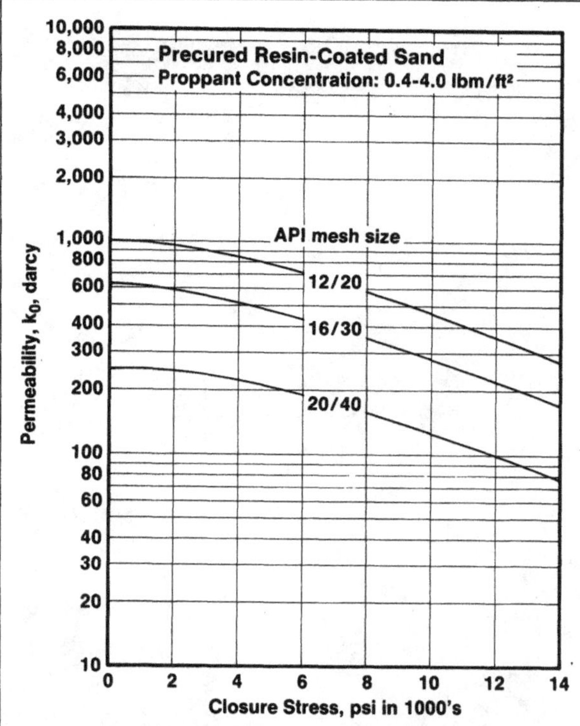

Fig. 6.12—Effect of mesh size on pack permeability for precured resin-coated Ottawa-type fracturing sand.[17]

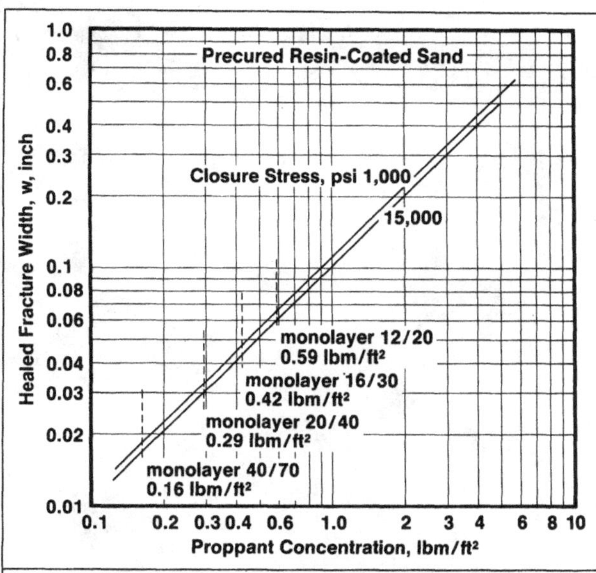

Fig. 6.13—Effect of closure stress on pack width for precured resin-coated Ottawa-type fracturing sand.[20]

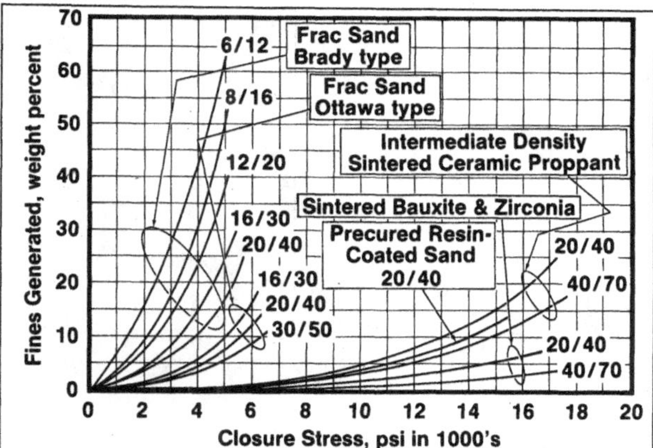

Fig. 6.14—Effect of proppant type and size on crush resistance as measured by a modification of the procedure outlined in Refs. 11 and 12. The greater the fines generated, the lower the flow capacity of the proppant-packed fracture.[20]

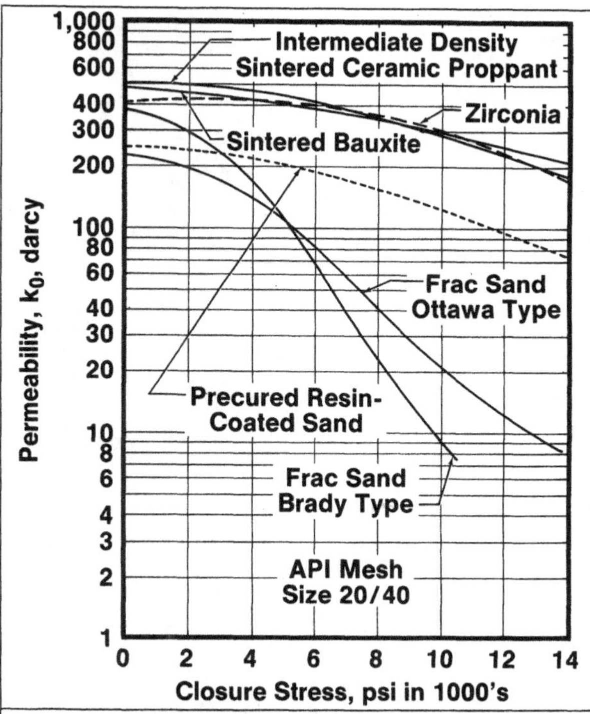

Fig. 6.15—Effect of proppant type on pack permeability. Relative performance of the various proppants is demonstrated for the 20/40-mesh size.[20]

often play a more dominating role in proppant size selection than conductivity considerations.

Consideration of proppant size is important in the design of fracturing treatments because a minimum fracture width is needed to allow the proppant to enter the fracture. The generally accepted values for this so-called admittance criterion require fracture widths in the range of two to three times the largest grain diameter. An admittance criterion based on twice the largest grain diameter requires fracture widths of 0.187, 0.066, and 0.033 in. [0.475, 0.168, and 0.084 cm] for 8/16-, 20/40-, and 40/70-mesh proppants, respectively. The largest of these values may be difficult to achieve in very deep wells with formations having high bottomhole fracturing pressures and usually requires the use of smaller proppant for successful completion of the fracturing treatment.

Additionally, it should be thoroughly understood that proppant transport must be considered during the selection of the size of the propping agent. Even though a 12/20-mesh proppant may be much more conductive than a 20/40-mesh proppant, the smaller proppant is much easier to transport deeply into a fracture than the larger proppant. In many cases, proppant-transport considerations dominate the choice of the proppant size. One should never use large-mesh proppants (even 20/40 mesh) unless they can be successfully placed in the fracture. See Chap. 10 for a detailed discussion of proppant-transport considerations.

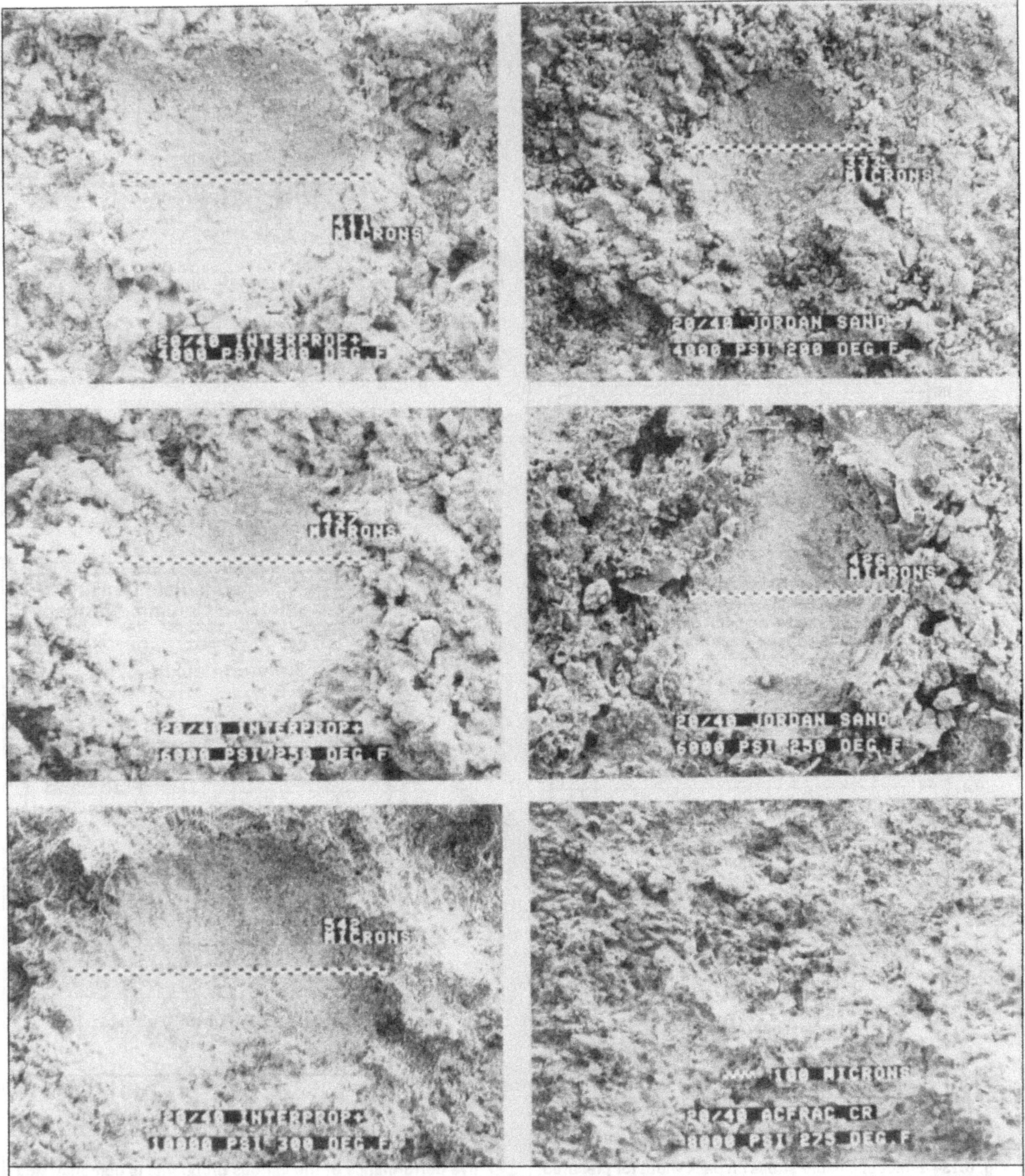

Fig. 6.16—Comparison of embedment of 20/40-mesh proppants into Ohio sandstone following the flow of 2% KCl for 100 hours at the indicated closure stress and temperature (courtesy Stim-Lab Inc.).

Proppant Concentration. The term "proppant concentration" refers to the amount of proppant per unit area of fracture wall (measured on one side only). In customary units, it is expressed in pounds of proppant per square foot of one wall of the fracture. (In SI units, it is expressed in kilograms per square meter.) If proppant settles to the bottom of a vertical fracture as it enters, the concentration will be determined by the width of the fracture at the time of entry (i.e., during pumping). If the proppant is suspended in the fracturing fluid until the fracture closes, concentration will be determined by both the width during pumping and the concentration of proppant in the fluid.

Fracture conductivity increases with increasing concentration of proppant in the fracture. This relationship is not direct for concentrations of less than about ½ lbm/ft^2 [2.44 kg/m^2] because of wall effects.

During the early days of fracturing, there was an interest in the industry in placing a "partial monolayer" of proppant in the fracture to achieve theoretically maximum conductivity.[61] This process has not proved to be widely successful. Among the reasons given for its failure are the lack of ability to obtain uniform and complete coverage of the fracture with a monolayer, partial embedment of the proppant into the fracture walls, non-Darcy flow in the very thin fracture created in this manner, and insufficient proppant strength to support the load. This method now receives little attention.[6,62-65]

Proppant Strength. The strength of proppants is of major concern in the design of propped fractures.[36,37,66] Historically, this strength has been expressed in terms of the load required to crush a single grain of proppant divided by the diameter squared of its contact area at the point of crushing.

Results from another, perhaps more appropriate, representation of the strength of proppants are shown in Fig. 6.14.[20] This relatively new testing technique, the API crush resistance test,[11] was designed to determine the relative strength of proppants in packs and has been tested and adopted by API for testing sands to be used in hydraulic fracturing. The API test uses an apparatus for imposing a sustained load on a proppant pack. The degree of size reduction sustained by the proppant is taken as an inverse measure of proppant strength.

The API crush resistance test is a more complex measure of strength than that described above for single particles. The values obtained are influenced by grain shape, particle-size distribution, packing arrangement, and other attributes of the particle pack. Although these factors are thought to make the test more representative of proppant performance under field conditions than the single-particle test, sensitivity of the measurement to several pack attributes makes the test more difficult to reproduce, and small variations in results (e.g., 2 or 3%) are considered insignificant in critical comparisons.

Fig. 6.15 shows the relationship of closure stress to flow capacity of various proppants, which is determined primarily by proppant strength. A comparison of this figure with Fig. 6.14 reveals the sensitivity of this test to small variations in particle size and shape. For example, Brady sand has a predominance of larger 20/30-mesh particles, while Ottawa sand has a predominance of smaller 30/40-mesh particles and is more rounded. More fines are generated from the larger and more angular Brady sand because the point loading per particle is greater. In addition, the larger Brady sand supplies higher proppant-pack permeability at closure stresses up to 5,000 psi [34.5 MPa] than the slightly smaller Ottawa sand.

Above 5,000-psi [34.5-MPa] closure stress, some of the largest grains have been broken into smaller particles. Thus, at higher stresses, Ottawa sand, which had not broken as much as Brady sand, was found to have the higher proppant-pack permeability.

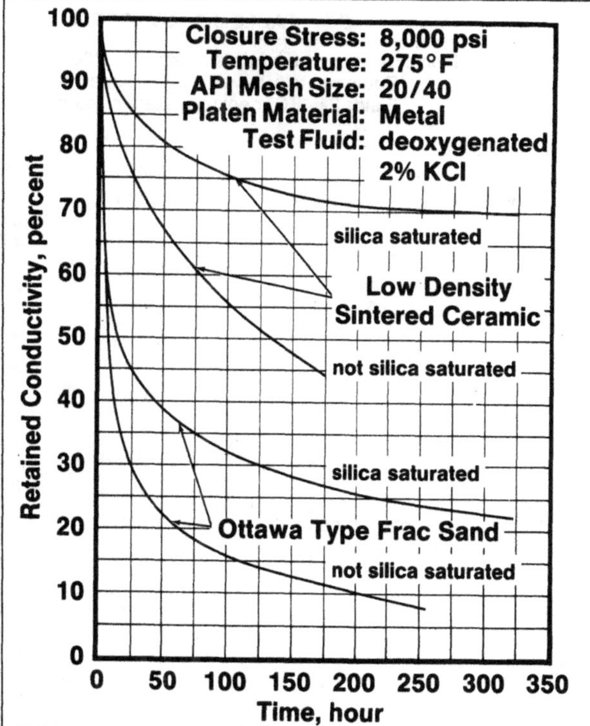

Fig. 6.17—McDaniel[73] shows an excellent example of the effects of stress corrosion for exposure of two commercial proppants to fluids in which they have a slight solubility. The initial silica content of the test fluid was about 40 mg/L for the fluid that was not presaturated and about 100 mg/L for the presaturated fluid.

While the conductivity measurements on which these results are based are very sensitive to proppant-pack attributes and difficult to reproduce, the comparison cited is from measurements made in the same laboratory and therefore are as comparable as current measurement techniques permit.[20]

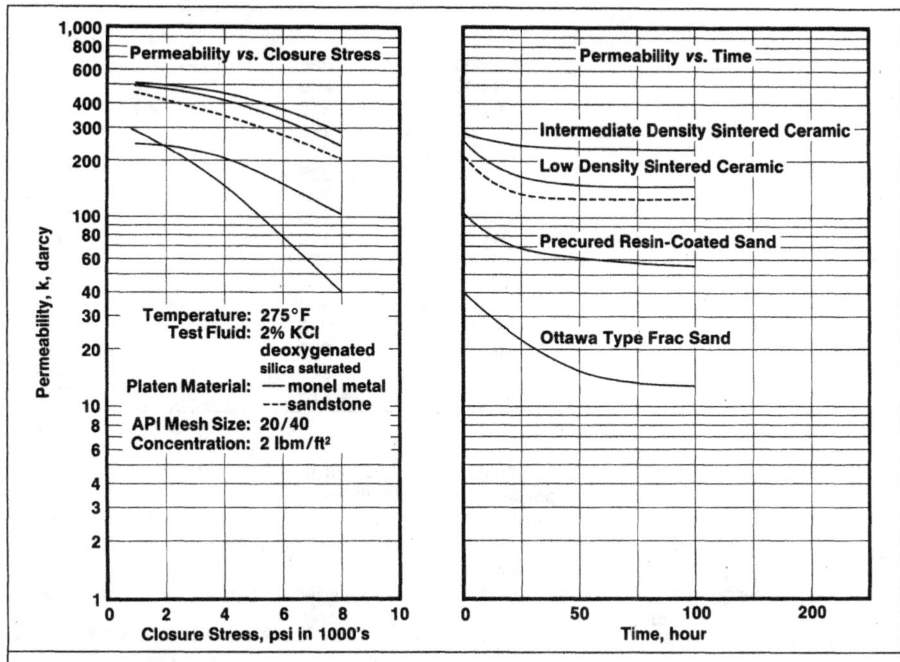

Fig. 6.18—This example of Penny's data shows the general shape of the long-term data for all proppants tested.[68]

TABLE 6.7—SYNOPSIS OF LONG-TERM PROPPANT TESTING[68,78]

Test Conditions: Time at maximum stress: 50 to 100 hours
Equipment: API-type 10-in.^{2} linear flow cell
Fluid: 2% KCl deoxygenated, silica saturated

Proppant	Mesh Size	Closure Stress (psi)	Temperature (°F)	Retention of API Short-Term Permeability Values (%) 2 lbm/ft^2 Monel Metal	2 lbm/ft^2 Ohio Sandstone	1 lbm/ft^2 Ohio Sandstone
Brady-type fracturing sand	12/20	2,000	125	—	84	85
		3,000	150	—	63	—
		4,000	175	—	56	59
		5,000	225	—	63	—
		6,000	250	—	—	68
	16/30	2,000	125	—	79	66
		3,000	150	—	72	—
		4,000	175	—	54	68
		5,000	225	—	55	53
	20/40	2,000	125	—	96	96
		3,000	150	—	75	—
		4,000	175	67	—	78
		5,000	225	34	—	55
		5,000	225	44	44	—
Ottawa-type fracturing sand	12/20	2,000	125	—	90	90
		3,000	150	—	92	69
		4,000	175	74	58	62
		5,000	225	—	61	46
	16/30	2,000	125	—	66	67
		4,000	175	—	72	65
		5,000	225	—	65	68
		6,000	250	—	67	71
	20/40	2,000	125	—	98	83
		3,000	150	—	91	85
		4,000	175	76	74	76
		4,000	200	—	74	—
		5,000	225	39	—	62
		5,000	225	50	52	—
		6,000	250	—	32	—
		8,000	225	—	56	—
		8,000	225	—	54 (Berea)	—
		8,000	275	32	48	—
Curable resin-coated sand	12/20	4,000	175	—	96	—
		8,000	275	—	74	33
	16/30	4,000	175	—	95	—
		8,000	275	—	73	—
	20/40	4,000	175	—	99	—
		6,000	250	—	91	—
		8,000	275	—	98	77
		10,000	300	—	30	—
Precured resin-coated sand	12/20	2,000	125	—	—	81
		4,000	175	—	81	94
		5,000	235	—	—	74
		6,000	250	—	78	70
		8,000	275	—	66	73
	16/30	4,000	175	—	96	92
		6,000	250	—	52	52
		8,000	225	—	58	—
		8,000	225	—	58 (Berea)	—
		8,000	275	—	69	79
	20/40	4,000	175	—	84	88
		5,000	225	—	79	—
		6,000	250	—	79	87
		7,000	250	—	46	—
		8,000	225	—	63	—
		8,000	225	—	48 (Berea)	—
		8,000	225	—	50	—
		8,000	275	52	59	52

Proppant Grain Shape. Roundness and sphericity are proppant particle properties that affect performance. Their importance depends somewhat on the stress level at which the proppant is to be used.

Because the surface stresses are more uniform, a well-rounded, spherical particle is capable of carrying higher loads without crushing than a less-rounded particle. Therefore, at high stress levels, a high degree of roundness and sphericity contribute to higher proppant-pack conductivity.

At lower stress levels, however, this may not be the case. An angular proppant particle does not pack as well as a well-rounded particle and thus has more porosity and correspondingly greater permeability.[67] An example of this phenomenon was described previously. Brady sand, which is somewhat more angular than Ottawa sand, has slightly better flow capacity below about 5,000 psi [34.5 MPa] than Ottawa sand, although the more rounded Ottawa sand is superior in proppant-pack permeability at higher stress levels.

6.6 Other Factors Affecting Fracture Conductivity

This section discusses six additional factors, some of which have been investigated in detail only recently. While most of them influence fracture conductivity adversely, the laboratory investigation

TABLE 6.7—SYNOPSIS OF LONG-TERM PROPPANT TESTING[68,78] (continued)

Test Conditions: Time at maximum stress: 50 to 100 hours
Equipment: API-type 10-in.^{2} linear flow cell
Fluid: 2% KCl deoxygenated, silica saturated

Proppant	Mesh Size	Closure Stress (psi)	Temperature (°F)	Retention of API Short-Term Permeability Values (%) 2 lbm/ft² Monel Metal	2 lbm/ft² Ohio Sandstone	1 lbm/ft² Ohio Sandstone
Sintered bauxite	20/40	4,000	175	98	—	—
		6,000	250	95	—	—
		8,000	275	90	—	—
		10,000	300	83	—	—
Intermediate-density sintered ceramic	12/20	4,000	200	—	82	87
		5,000	225	52	52	—
		6,000	250	—	90	69
		8,000	275	—	76	54
	16/20	4,000	175	88	—	93
		5,000	225	71	70	—
		6,000	250	—	93	91
		8,000	275	—	78	61
		10,000	300	—	71	—
	20/40	4,000	175	87	—	93
		5,000	225	85	82	—
		6,000	250	—	79	92
		8,000	275	80	78	64
		10,000	300	67	50	—
Low-density sintered ceramic	16/20	4,000	175	—	91	95
		6,000	250	—	79	83
		8,000	275	—	64	73
		10,000	300	—	51	—
	20/40	4,000	175	—	98	94
		6,000	250	—	88	85
		8,000	275	61	74	75
		10,000	300	—	58	—

of these factors is more difficult and time-consuming and therefore has been accomplished for only a limited number of conditions. The full effect of these factors on future treatment design is yet to be determined.

Embedment. If proppant particles penetrate the walls of the fracture, the effective width of the fracture, and thereby the conductivity, is decreased. Fig. 6.16 illustrates embedment in a series of scanning electron micrographs of the surface of a sandstone after 100 hours of exposure to simulated bottomhole stress and temperature. The indentations made by the proppants appear as fine-grained circular patterns of crushed sandstone (Young's modulus of 5×10^6 psi [34 GPa]).[68] Not only is the width of the fracture decreased by embedment, but fine particles are generated by failure of the formation rock. These fine particles may also contribute to the loss of fracture conductivity.

An attempt to assess the severity of embedment has been made by ball-point penetrometer tests of formation rock.[66,69] These tests are not as important as was earlier thought because in most modern fracture designs the proppant pack is many particles (e.g., 10 to 40) thick in the fracture. The intrusion of the proppant into the fracture wall represents only a small fraction of the proppant-to-proppant interaction.[70]

Environmental Effects. The stress transmitted from the earth to the proppant, described earlier, causes breakage of particles, rearrangement and compacting of the particle bed, and decreased permeability of the proppant pack.[71]

The strength of inorganic proppant particles is affected by their extended exposure to corrosive fluids (salt water). The most striking example of this behavior is provided by the glass-bead proppants used formerly.[6] Glass beads fail at a much lower stress level in the presence of water than in air. A stress-corrosion phenomenon operates in the presence of fluids in which the particle has some small solubility.[68,72] The effect is reduced somewhat when silica-type proppants are exposed to formation waters that are naturally saturated with silica from the surrounding rock.[68,73] An excellent example of this phenomenon is illustrated in Fig. 6.17.[73]

The long-term stability of proppants is continually being investigated in the presence of water or brine at reservoir temperatures.[68,73-81] It appears as though the majority of change in proppant-pack conductivity occurs within the first 100 hours even at 300°F [150°C]. Many factors contribute to the decline in performance of even the best proppants. Fig. 6.18 illustrates the effect of time on proppant flow capacity.[68]

A synopsis of the data available is presented in Table 6.7. These results are a tabulation of retention/regain factors that can be applied to short-term API-type permeability data to arrive at useful long-term values. Because all the long-term data available to date follow the general trend of the major reduction occurring during the first 50 to 100 hours of exposure, the indication is that the values thus obtained will be applicable over much longer periods of time.

As Table 6.7 shows, the retention factors range from 98 to about 30% simply as a result of time and temperature. Using sandstone in place of the metal platen material has had little, if any, effect, at least for a sandstone of this type with a Young's modulus of about 5×10^6 psi [34 GPa]. These factors may be used by selecting appropriate conditions from Table 6.7 and multiplying the retention factor by the appropriate short-term permeability value. For example, Ottawa-type fracturing sand has a retention factor of 32% at 8,000 psi and 275°F [55.2 MPa and 135°C]. The appropriate short-term permeability from Fig. 6.4 is 41 darcies; hence, the expected permeability is $41\times0.32=13$ darcies. The actual long-term experimental data[68] show a permeability of 13 darcies at about 100 hours and 275°F [135°C].

Fracturing-Fluid Residues. The pore space of proppants packed in a fracture is sometimes decreased by the deposition of a residue from water-based fracturing fluids.[68,72,82-84] Such residue may cause a drastic decrease in fracture conductivity under certain conditions. The problem is most pronounced when the volume of residue from the polymer is higher, when polymer concentration is higher, when the concentration of proppant in the closed fracture is lower, and when stress on the fracture is higher (causing lower porosity).

The most common residue is a product of the degradation of water-soluble polymers used to build viscosity in fracturing fluids. Fig. 6.19 shows a cross section of a fracture containing intermediate-density proppant after application of 8,000-psi [55.2-MPa] closure stress.[68] The proppant used in the test was 20/40 mesh, and the

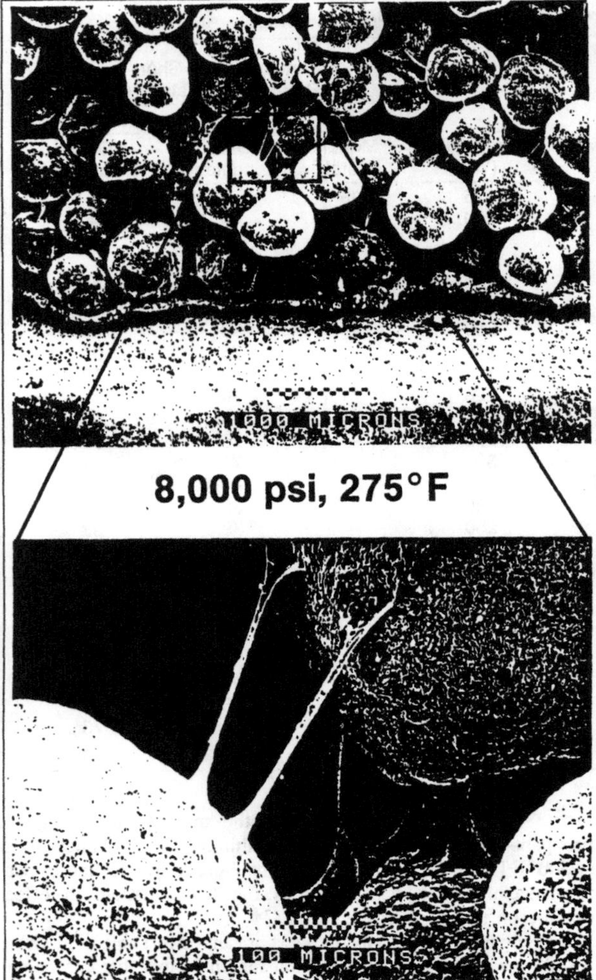

Fig. 6.19—Scanning electron photomicrograph of a core/proppant interface showing proppant pack with gel residue, filter cake, and sandstone surface. The lower frame is about an 8× magnification of the indicated area of the upper frame (courtesy Stim-Lab Inc.).

formation permeability was approximately 0.2 md. Proppant was placed in the test unit by pumping a slurry in a fracturing gel into the space between sandstone slabs. Fluid loss occurred through the sandstone.

Service companies have devoted much effort to reducing polymer residues in fracturing fluids. Recent research has focused on developing more efficient thickeners with more soluble degradation products. Some of the detrimental effects of résidue deposition can be alleviated by minimizing polymer concentrations, using higher proppant concentrations in fluids that suspend the proppant, using foam or emulsion fluids, and avoiding conditions of extreme proppant crushing. Table 6.8 is a tabulation of available data comparing retained/regained permeabilities after breaking and cleanup of various generic fracturing-fluid types.[78] The data are presented as retention factors that can be applied to API-type short-term permeability data to obtain a usable value of proppant-pack flow capacity. The retained permeability includes the effects of time, temperature, and fluid residues.

Close inspection of Table 6.8 reveals a direct comparison of the effects of increasing gellant loading for a titanate crosslinked hydroxypropyl guar gum (HPG) -type fluid. Increasing gellant from 40 to 50 lbm/1,000 gal [4793 to 5991 g/m^3] decreases retained permeability by an additional 15%. Further reduction is encountered by a gellant increase from 50 to 60 lbm/1,000 gal [5991 to 7190 g/m^3] of about 15%.

Another comparison of fluid effects, i.e., guar gum vs. HPG, shows little difference even though the guar has about three times as much insoluble residue as the HPG. Virtually no difference is seen between titanate crosslinked fluids and those linked with zirconates.

Fracture closure, fluid leakoff, and viscosity breaking processes have a dramatic effect on cleanup and regained permeability of the proppant pack. Breaking times of 2, 10, and 24 hours are compared in Table 6.8 for a generic crosslinked fluid. Slow and fast breaks are compared for gelled oil. The trend is the same: more rapid breaks tend to be more effective in terms of regained permeability.

In a comparison of the damaging effects of different types of generic fracturing fluids, one type stands out as being the least damaging: foam fracturing fluids. These fluids, composed mainly of a gas and minor amounts of gelled water, permit a proppant pack to regain 70 to 90% of its potential flow capacity.

Another type of residue problem that has been less fully investigated is that resulting from the use of particulate fluid-loss-control additives.[68] Table 6.9 is a synopsis of the available data,[78] comparing various generic fluid-loss control additives and mechanisms. The comparison is made in terms of retained/recovered permeability as a percentage of that obtained in API short-term testing of proppants. The fluid-loss agents tested generally have a great effect on regained permeability of the proppant pack. Water-soluble particulates (starch) and volatile oils (toluene) are probably the cleanest of the group. Much more work is needed in this area to evaluate the extent of the problem.

Fines Movement. The fine particles created by grain failure at higher stress levels lead to lower proppant-pack permeability. The particle-size distributions resulting from such crushed particles have been investigated by several authors[36,85] and their effects on fracture conductivity reported. Fine particles have been shown to migrate through the propped fracture and to plug the pore throats, thereby reducing fracture conductivity.[85] The long-term decrease in permeability of sand proppant reported[6] may be caused at least partially by movement of pre-existing fines with continued flow through the sand.

Non-Darcy Flow. For non-Darcy flow, the pressure drop in the fracture can be expressed by

$$\Delta p/\Delta L_f = \mu v/k_f + (\beta)(\rho)v^2, \qquad (6.1)$$

where

Δp = pressure,
ΔL_f = length of proppant pack in direction of flow,
μ = viscosity,
v = velocity,
k_f = permeability,
β = turbulence factor, and
ρ = fluid density.

The second term of the equation, with coefficient β, expresses the increased pressure gradient as a result of deviations from Darcy's law. Values of β have been measured for a variety of sand sizes at different values of stress.[6,86,87]

Non-Darcy effects can substantially reduce the effective fracture conductivity in high-flow-rate gas wells.[88] This reduction in conductivity will decrease the well's PI and can complicate the analysis of pressure-transient tests. To analyze wells properly where non-Darcy flow affects the pressure distribution in and around the fracture, a reservoir simulator that includes non-Darcy flow must be used by the analyst.

6.7 Laboratory Measurements of Fracture Conductivity

Historical Development. Within a few years after the introduction and acceptance of fracturing as a useful procedure for stimulating oil and gas well production, operators, service companies, and suppliers became involved in evaluating the properties of proppants.

TABLE 6.8—SYNOPSIS OF THE EFFECTS OF GENERIC FRACTURING FLUIDS ON LONG-TERM PROPPANT (20/40) PERFORMANCE[68,78]

Gellant[a]	Crosslinker	Proppant Type	Proppant Concentration (lbm/ft²)	Closure Stress (psi)	Temperature (°F)	Retained Permeability (%)	Remarks
Crosslinked, gelled 2% KCl water							
40 HPG[b]	Ti	IDP	1	8,000	275	51	Effect of gellant concentration
50 HPG	Ti	IDP	1	8,000	275	34	
60 HPG	Ti	IDP	1	8,000	275	19	
40 HPG	Ti	Ottawa	2	5,000	225	31	2% insoluble residue
40 guar gum[c]	Ti	Ottawa	2	5,000	225	27	6% insoluble residue
60 HPG	Ti	IDP	2	10,000	300	25	Effect of crosslinker type
60 HPG	Zr	IDP	2	10,000	300	23	
40 HPG	Ti	IDP	1	4,000	175	43	Fastest break
40 HPG	Ti	IDP	1	4,000	175	19	Slow break
40 HPG	Ti	IDP	1	4,000	175	2	Very slow break
40 guar gum	$[BO_4]$[g]	Ottawa	2	3,000	150	76	Fastest break
40 CMHPG[d]	Al	Ottawa	2	3,000	150	42	Fast break
40 HPG	Ti	Ottawa	2	3,000	150	52	Fast break
40 HPG	Ti	Ottawa	1	3,000	150	17	Fast break
40 HPG	Ti	Ottawa	1	3,000	150	55	Fastest break
40 HPG	$[BO_4]$	Ottawa	1	4,000	175	48	Fastest break
40 HPG	$[BO_4]$	Ottawa	1	5,000	225	42	Fastest break
40 HPG	Ti	Ottawa	2	4,000	175	36	No breaker mentioned
40 HPG	Ti	IDP	2	6,000	250	41	No breaker mentioned
40 HPG	Ti	IDP	2	4,000	175	52	Fastest break
50 HPG	Ti	IDP	2	8,000	275	39	No breaker mentioned
40 HPG	Ti	IDP	1	6,000	250	46	Fastest break
40 HPG	Ti	IDP	1	8,000	275	59	Fastest break
40 HPG	Ti	IDP	1	6,000	250	31	Slow break
40 HPG	Ti	IDP	1	8,000	275	41	Slow break
Polyemulsion, 67% oil in 33% gelled 2% KCl water							
40 HPG	NA	IDP	2	4,000	175	25	Cationic emulsifier
40 HPG	NA	IDP	2	4,000	175	54	Anionic emulsifier
40 HEC[e]	NA	IDP	2	4,000	175	29	Cationic emulsifier
Gelled oil, Diesel Fuel No. 2							
Phosphate ester	Al	IDP	2	4,000	175	49	Slow break
Phosphate ester	Al	IDP	2	4,000	175	74	Fast break
Foam fracture, 70% gas in 30% gelled 2% KCl water							
30 HPG	NA	IDP	2	4,000	175	71	Cationic foamer, N_2
20 XG[f]	NA	IDP	2	4,000	175	85	Cationic foamer, N_2
30 HPG	NA	IDP	2	4,000	175	79	Anionic foamer, CO_2

[a] Loading in lbm/1,000 gal plus gellant symbol or name.
[b] Hydroxypropyl guar gum.
[c] Low-residue guar gum.
[d] Carboxymethylhydroxypropyl guar gum.
[e] Hydroxyethyl cellulose.
[f] Xanthan gum.
[g] Borate ion.
NA = Not applicable.

Several concerns motivated these efforts. Oil and gas operators were interested in relating proppant properties to production results. Service companies felt the need for these properties in designing fracturing treatments. Proppant suppliers wanted data and specifications so that they could effectively market their materials. In addition, while sand was the least expensive and most readily available proppant, all concerned with its use recognized its limitations. Thus, many laboratories became involved in efforts to research and develop new proppants.

Early efforts to determine the permeabilities of proppant packs employed modified API procedures developed for determining the permeability of rock samples. It was obvious, however, that these procedures did not simulate downhole conditions and thus the permeability data had limited usefulness in predicting production results, designing treatments, setting specifications, or developing new proppants.

The downhole condition thought to have the major influence on proppant permeability was overburden stress. The change in proppant permeability resulting from continual exposure to closure stress at downhole temperature for long periods of time was recognized as important in assessing the longevity of the fracturing treatment. Hence, equipment was needed to permit imposing a load on the proppant pack while measuring its permeability as a function of time.

Modified Hassler Sleeve Permeameter. Early testing equipment designed to simulate downhole closure stresses was based on a modified rubber sleeve (Hassler) permeameter as shown in Fig. 6.20. Permeabilities, k_f, were calculated with the use of Darcy's equation:

$$k_f = q\mu L/A_p\Delta p, \quad \ldots \quad (6.2)$$

where

q = flow rate,
L_f = length of proppant pack in direction of flow, and
A_p = cross-sectional area of proppant pack open to flow.

A weighed amount of proppant was placed in a cylindrical rubber sleeve containing screens near each end. The proppant was settled by vibration, and metal end caps were inserted in the rubber sleeve. The length of the proppant pack, L_f (i.e., the dimension between the end caps), and the average proppant pack diameter, d, were measured. The sleeve was placed in a pressure chamber, closure stress was applied to the proppant pack by pumping hydraulic fluid into the chamber outside of the rubber sleeve, and stress was transmitted through the rubber sleeve to the proppant particles.

The permeability of the proppant was measured at ambient conditions by flowing air, gas, or liquids through the proppant pack

TABLE 6.9—SYNOPSIS OF THE EFFECTS OF VARIOUS GENERIC FLUID-LOSS-CONTROL ADDITIVES AND MECHANISMS ON LONG-TERM PROPPANT (20/40) PERFORMANCE[67,78]

Gellant	Crosslinker	Closure Stress (psi)	Temperature (°F)	Retained Permeability (%)	Remarks
Crosslinked, gelled 2% KCl water					
50 HPG	Zr	8,000	275	15	With 30 lbm (vegetable flour + silica)/1,000 gal
50 HPG	Zr	8,000	275	34	With 5% toluene + anionic dispersant
50 HPG	Zr	8,000	275	33	With 30 lbm starch/1,000 gal
60 HPG	Zr	10,000	300	7	With 5% diesel + cationic dispersant
60 HPG	Zr	10,000	300	23	No additives
60 HPG	Ti	10,000	300	25	No additives
60 HPG	Ti	10,000	300	21	With 30 lbm silica flour/1,000 gal
60 HPG	Ti	10,000	300	12	With 30 lbm silica flour + 5% diesel/1,000 gal

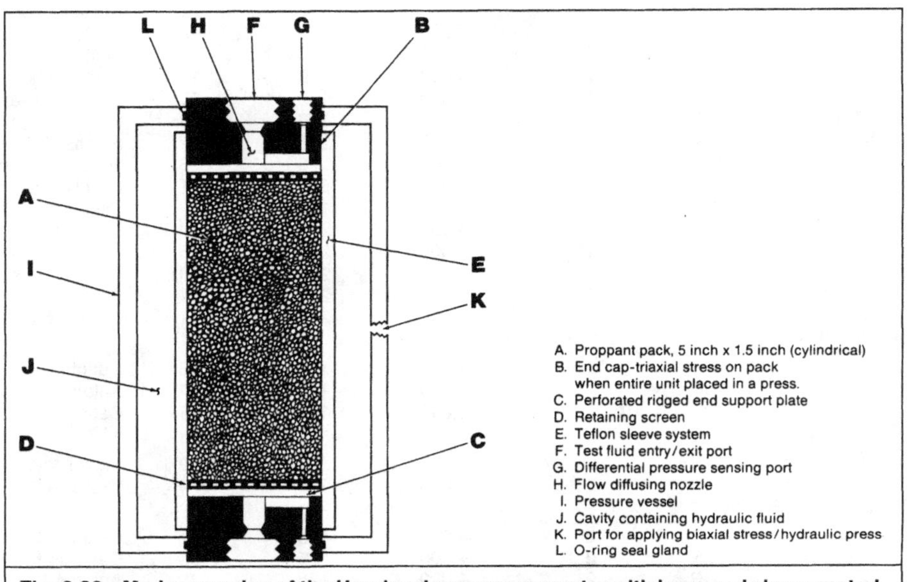

Fig. 6.20—Modern version of the Hassler sleeve permeameter with improved sleeve material and design. Provisions for triaxial loading of the pack are included.

contained in the sleeve. Fluid viscosity, μ, of the test fluid at test temperature was recorded. The differential pressure, Δp, between the pack inlet and outlet was recorded at various flow rates, q. Flow rates were usually determined by measuring effluent volume vs. time. Liquid volumes were measured in graduated volumetric glassware, while gas volumes were measured with a gas volume meter. Usually, test liquids were Newtonian fluids—i.e., water, brine, or refined oil—so that fluid viscosities were obtained from the literature.

At the start of a test, a relatively low closure stress (500 to 1,000 psi [3.5 to 6.9 MPa]) was applied. This stress was held for a few minutes or until a semisteady-state condition existed. This could usually be determined by a reduction in noise level. When a given stress level was initially applied, cracking and breaking of many particles occurred. But as the stress was held at that level, the number of particles being broken declined until eventually breakage became inaudible. At that point, fluid flow was started and continued until consistent reproducible values of pressure drop and flow rate were obtained. Closure stresses were raised incrementally, and the permeability of the proppant at various stresses was obtained.

The effect of temperature on proppants under stress could also be determined by placing the apparatus in a thermostatted bath. Long-term effects could be determined with this apparatus because laboratories could afford to tie up this relatively inexpensive equipment for long periods of time.

Several problems were recognized with this configuration, however. A major concern involved measurement of the cross-sectional area. When stress was applied, the cross-sectional area did not remain uniform throughout the pack length. Another concern was the inability to determine the effect of proppant/rock interaction—i.e., proppant embedment. Last, a cylindrical configuration where the cross-sectional area was tens to hundreds of particle diameters across did not simulate a crack whose width was frequently less than 20 particle diameters.

Early Fracture Conductivity Procedures. Several laboratories developed equipment and procedures to simulate fracturing conditions more closely. These efforts were aimed at determining how proppant embedment influenced conductivity. Notable among these were procedures developed by Pan American (Amoco) Production Research, Gulf R&D Co., and Socony Mobil Field Research.

In the Amoco and Gulf procedures, samples of proppant were placed between pieces of rock cored from actual producing zones, while Mobil tested proppants between metal platens of various hardness to simulate the fracture face. In all these procedures, bed thickness ranged from partial monolayers—a layer of single grains with grains separated so that they did not touch each other—to proppant packs having multiple layers of proppant.

Differences in flow path in the test apparatus were also significant. In the procedure developed by Amoco, test fluid followed a radial flow path through the proppant-supported crack, while in the Gulf and Mobil apparatus, it followed a linear path. As in the case of the earlier Hassler sleeve apparatus, these procedures, which are discussed in more detail in the following sections, had both advantages and limitations.

Amoco Radial Conductivity Test Unit. The radial flow cell developed by Amoco was adapted and used by several laboratories[18,69,88,89] to determine proppant conductivity. In this procedure, a cored sample 2 to 6 in. [5 to 15 cm] in diameter was machined into slices ½ to 1 in. [1.3 to 2.5 cm] thick, and appropriate holes for pressure taps and fluid flow were drilled into the core. Each core slice was then mounted into a steel cup and held in place with a low-melting alloy, such as Wood's metal. This equipment is illustrated in Fig. 6.21. A loose-fitting screen was placed around the periphery of the bottom core. A known amount of proppant was placed on the bottom core slice and distributed uniformly on its surface. A second core slice mounted in a steel cup was carefully positioned directly on top of the proppant layer. The assembled unit was placed in the frame of a press. Closure stress was then applied by the press. Flow rates and differential pressures were recorded. The dimension between platens of the pressure frame was recorded and used to determine fracture width.

The advantages of this method included a relatively easy procedure with low equipment costs, short sample preparation time, and the ability to stress proppants between rock samples in a fracture-like configuration. This apparatus is still used in some laboratories.[18,90]

In spite of these obvious advantages, the radial flow pattern had serious drawbacks that severely limited the test results. Under steady-state flow conditions, the differential pressure required to move fluid at a constant rate through porous media is inversely proportional to the cross-sectional area of the media. In radial flow, the cross-sectional area of the proppant pack is increased by a factor of 2π with each unit increase of radius. Similarly, differential pressure is decreased by a corresponding factor of 2π for each unit increase in test-cell radius. Because of the large difference in cross section of the flow path from the entrance to the exit, any change in proppant conductivity near the entrance has a much greater effect than a similar change near the exit. If flow rates are kept low to ensure a laminar flow regime (required to determine proppant permeability), high-permeability proppants produce very low differential pressures that are difficult to measure accurately. If higher flow rates are employed to produce more readily measurable differential pressures, non-Darcy flow occurs at the inlet near the center of the core. Finally, the results obtained are very sensitive to the uniformity of proppant loading, particularly at the center of the apparatus.

Gulf Linear Conductivity Test Unit. Gulf recognized the disadvantages of a radial flow tester and developed a unit having a linear flow path (see Fig. 6.22). A 4- to 5½-in. [10- to 14-cm] section of a 3-in. [7.6-cm] -diameter core was suspended in a rectangular box-shaped mold. The mold was designed so that there would be 2 to 3 in. [5 to 7.6 cm] of set plastic between each end of the core and the end of the casting. Epoxy resin with a high aluminum-powder content was poured into the mold and allowed to set. The high aluminum content both strengthened the casting and helped reduce heat buildup. After setting and cooldown, the block was cut in half longitudinally with a diamond saw. One half, which was to be the bottom portion of the test unit, was machined to have both inlet and outlet ports. Uniform flow across the propped surface was achieved through the use of both quieting basins and an open flow path to and from the propped core. Fig. 6.22 illustrates how the unit was assembled. A gasket of closed-cell polyurethane foam surrounded the flow path and propped area. At the very low fluid pressures used in this apparatus, test fluids were contained by the gasket.

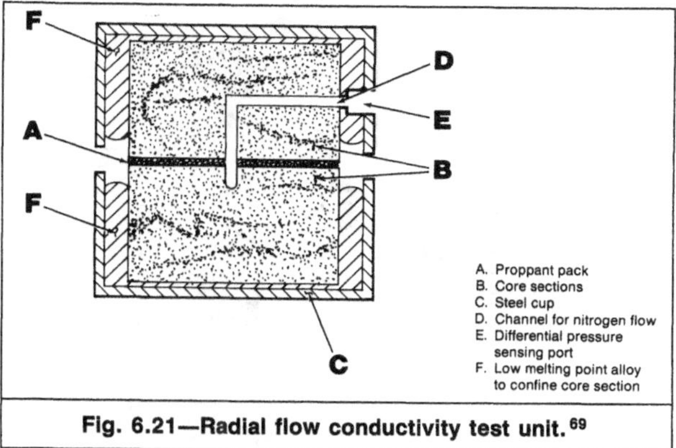

Fig. 6.21—Radial flow conductivity test unit.[69]

While this unit had a linear flow path between rock faces supported by proppant, it had several drawbacks. Sample preparation was expensive and time-consuming. The large test area, about 50% larger than that of the API unit, required a very large press to attain typical closure stresses. The major disadvantage, however, was in not knowing how much of the stress supplied by the press was supported by the proppant and how much was supported at the gasket. At moderate to high stresses, deformation of the plastic caused it to creep around the core so that stress concentrations occurred at the outer edge, thus relieving stress on the propped area.

Mobil Linear Conductivity Test Unit. Mobil recognized many of the problems associated with the Amoco and Gulf conductivity cells and designed a linear unit in which proppant was stressed between metal platens rather than rock. Metals varying in hardness,

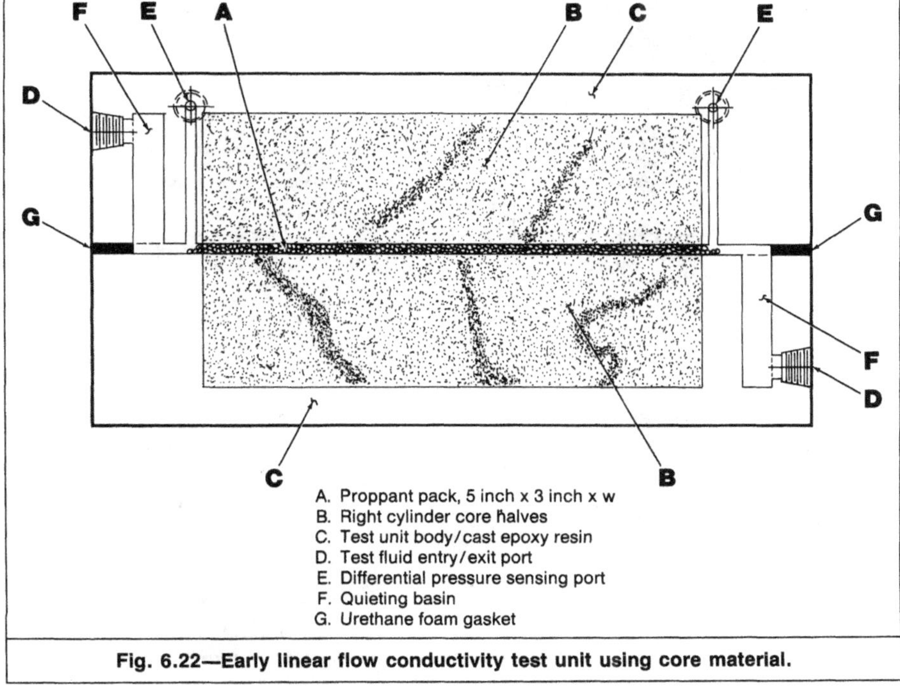

Fig. 6.22—Early linear flow conductivity test unit using core material.

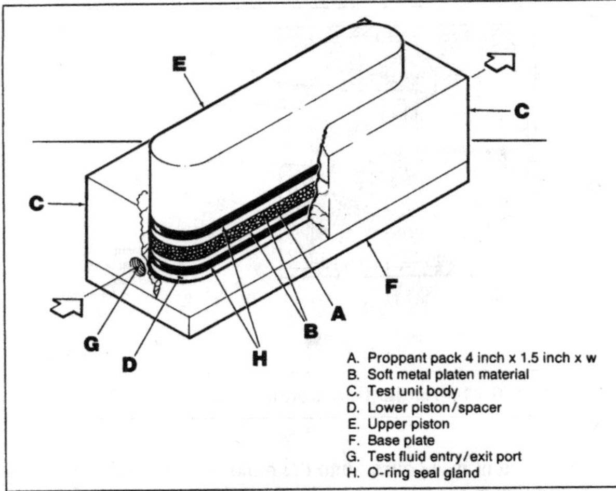

Fig. 6.23—Early linear flow conductivity test unit using soft metal platen material.

such as lead, aluminum, brass, and stainless steel, were chosen as typical of formation rocks on the basis of embedment tests. The test unit was relatively small, having a test area of approximately 6 in.^{2} [39 cm^2]. This was an advantage because a moderate-size press could be used to achieve closure.

The apparatus, shown in Fig. 6.23, was relatively inexpensive to manufacture and easy to use. It had the advantages of a linear configuration and fracture widths that could be easily measured.

While there was some interest within the industry for this design, it did not gain widespread acceptance. Perhaps at the time, most laboratories were trying to design equipment where proppant was stressed between rock surfaces. Also, it was generally believed that conductivity tests should include the production of fines such as occurs at a proppant/rock interface under stress. Thus, the use of metal platens instead of rock platens had only a very limited acceptance. Although the Mobil design overcame many of the problems associated with the other methods of conductivity testing, the basic design was not generally accepted until some years later, when work using a similar conductivity cell was described by Cooke.[6]

Split Core in Hassler Sleeve. Several laboratories, including Arco, B.J. Hughes, Dowell, and The Western Co., used a cell employing a split core in a Hassler sleeve. More recently, this design was improved by Terra Tek. In the earlier work, a unit similar to that shown in Fig. 6.24 was used. For this apparatus, a core plug was cut longitudinally into two equal parts and placed in the sleeve. A spacer of the desired thickness was used to separate the core pieces and to control the opening of the simulated fracture. The spacer had the same width as the core halves and was shaped like a tuning fork. To prepare for a test, a tight-fitting screen was placed at one end of the Hassler sleeve. The cores separated by the spacer were placed in the sleeve. A weighed amount of proppant was carefully poured into the open crack between core halves. As proppant was poured in, the spacer was carefully pulled out, leaving a proppant-filled crack with approximately the same fracture width as the spacer thickness. Some workers carefully tamped the proppant to improve test reproducibility; others believed that untamped tests are more representative of actual field conditions.

People using this apparatus experienced a number of problems and often disagreed on procedures. The first example of the disagreement on procedures has been mentioned above—i.e., the tamping of the tests. Problems in measuring fracture width were encountered, and core breakage was a problem. While fracture width could be measured initially as well as finally, it could not be

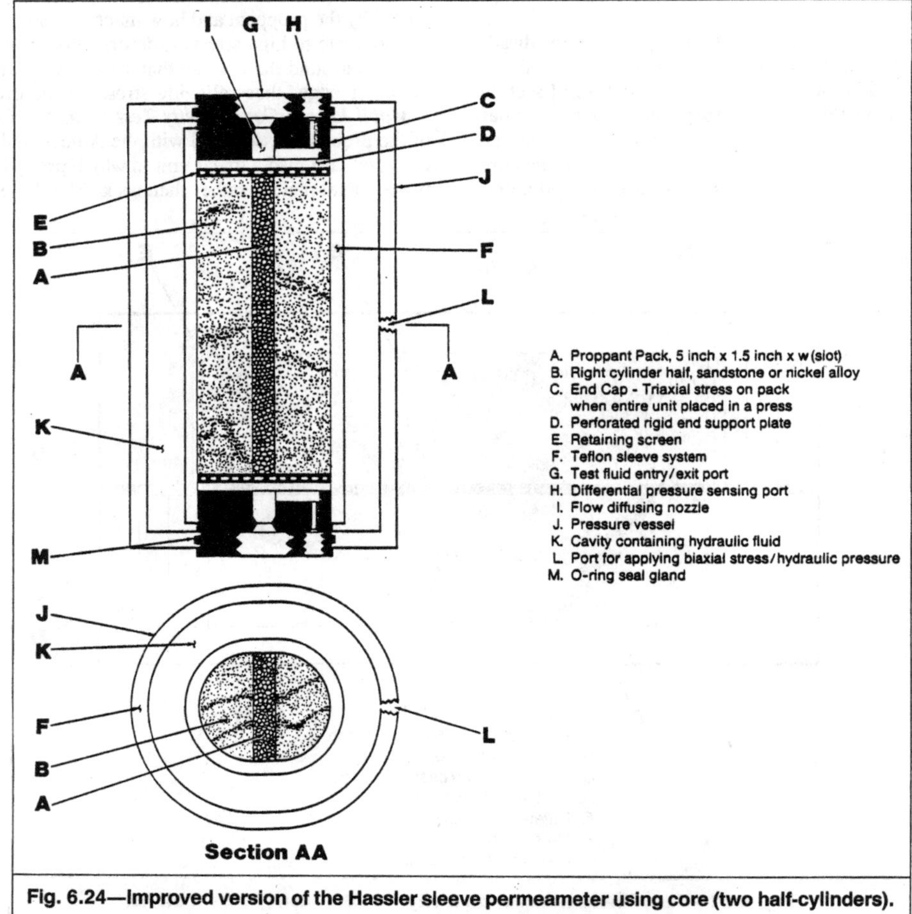

Fig. 6.24—Improved version of the Hassler sleeve permeameter using core (two half-cylinders).

measured easily during the test. Thus, fracture permeabilities were obtained by estimating what the fracture width was during the test.

Tests were usually made by applying stress biaxially to the core halves through the Hassler sleeve. The third stress caused by the confining action of the end caps on the proppant pack was unknown but usually thought to be much lower than the biaxial stress. At moderate to high stress differentials (>5,000 psi [>34.5 MPa]), the core halves experienced sufficient differences between stresses to cause rock failure. This resulted in nonuniform unknown closure stresses being placed on the propped fracture.

An improvement to eliminate breakage of the core halves consisted of replacing the rock core halves with a steel right cylinder cut longitudinally. Thin slices of formation rock ⅛ to ¼ in. [0.3 to 0.6 cm] thick were glued to the flat surface of each steel half-cylinder. Proppant was then placed between the rock surfaces with a spacer as described earlier. When biaxial stresses were applied, core breakage did not occur because the tensile strength of the steel was sufficiently high to tolerate the stress differential. While this improvement greatly reduced core breakage, the problem of measuring fracture width remained unsolved; thus, proppant permeabilities could only be estimated. Several laboratories currently use this apparatus with major improvements in the confining sleeve for high-temperature use. Some of the units permit the application of triaxial stresses.[77,78]

Terra Tek Conductivity Test Unit. In this unit, several changes in the Hassler sleeve design were made to improve its usefulness for testing proppants. A schematic of the apparatus is shown in Fig. 6.24. With this equipment, known triaxial stresses can be applied to the proppant. The entire unit is placed in a press. Hydraulic pressure applied to Port L supplies biaxial stresses to the proppant. The press is used to apply stress to the end caps, thus exerting a differing triaxial stress to the proppant pack. Improved flow-distribution, pressure-sealing, and pressure-sensing devices made this equipment easier to use and more reliable than some of the earlier units.

When comparative tests are made on proppants, a right cylinder made of an inert, very hard material, such as Hastelloy™, is used. When the interaction of proppant with rock or the effect of differing triaxial stresses is desired, a right cylinder is cut from a rock sample. The cylinder is then cut in half longitudinally, and rock cylinder halves are used in place of the metal cylinder halves. Proppant is loaded between the cylinder halves in a manner similar to that described earlier. The same careful loading procedure is used in all tests to provide reproducible results.

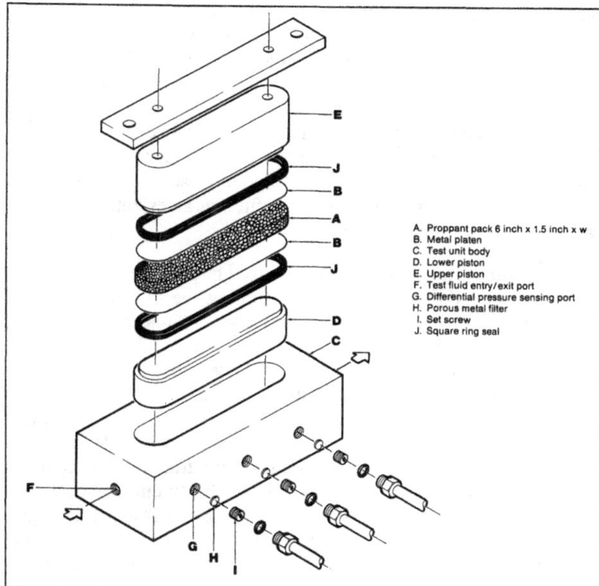

Fig. 6.25—API fracture conductivity test unit, exploded view.

Fracture Conductivity. The difficulty of measuring fracture width to permit calculating proppant permeability has led to the widespread use of the term "fracture conductivity." Fracture conductivity, the product of permeability and fracture width, has utility in both laboratory tests and field treatments. It represents an overall resistance the proppant pack presents to fluid flow. This term is also useful for evaluating well productivity after a fracturing treatment when fracture geometry is assumed to be rectangular.

In three of the four early test procedures, fracture width was found to be difficult to measure, but it is even more difficult to determine in well treatments. Fracture height (at least at the wellbore) can be estimated from well logs and from drilling records. It was assumed that induced fractures had a rectangular cross section where the fracture width was usually only a fraction of an inch in dimension, while the vertical dimension ranged from a few to hundreds of feet. In a fracture, the cross-sectional area, A, has a width w and a height h so that Eq. 6.3 can be modified as follows:

$$k_f = q\mu L_f / wh\Delta p. \quad \ldots\ldots (6.3)$$

Further, because fracture width often cannot be determined independently, it is useful to combine permeability and width and to solve the equation for fracture conductivity:

$$k_f w = q\mu L_f / h\Delta p. \quad \ldots\ldots (6.4)$$

To use this equation (which applies to linear systems in laminar flow) in laboratory conductivity tests, a weighed amount of proppant is placed in a fracture with a known surface area. Closure stress

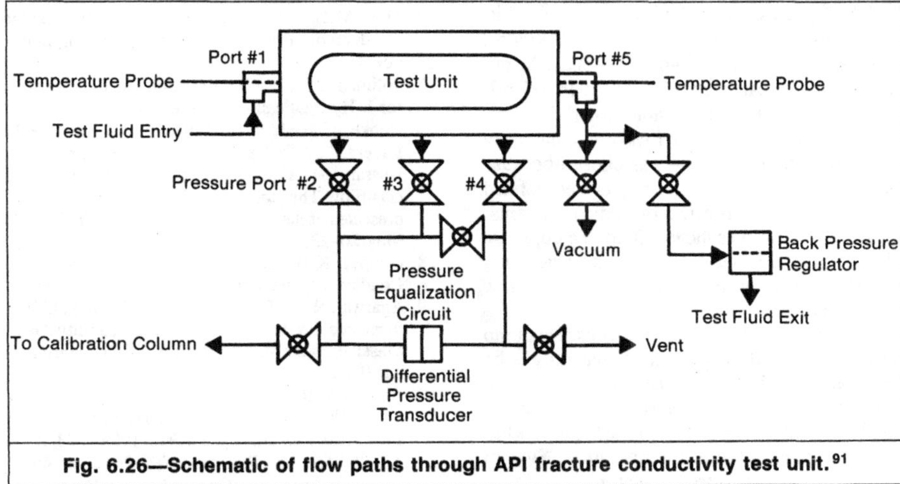

Fig. 6.26—Schematic of flow paths through API fracture conductivity test unit.[91]

is applied and fracture conductivity is calculated with Eq. 6.4. The reasoning is that if one has a fracture filled with the same weight of proppant at the same ratio of proppant mass per unit area, the fracture conductivity should be the same. If this assumption is correct, then the term "fracture conductivity" so obtained is a valuable descriptor. However, the advent of newer testing procedures where fracture width can be evaluated independently and the development of more realistic knowledge of fracture geometry make the proppant permeability a more useful characterization of proppant-pack fluid conductivity.

API Fracture Conductivity Test Unit. The API fracture conductivity test procedure[91] is rapidly becoming the industry standard for determining proppant permeability. This test, under development by an API committee for more than 8 years, had as its goal a scientifically sound, reproducible procedure acceptable to both the users and suppliers of fracture proppants. A test that would simulate the critical downhole conditions while keeping testing costs within practical limits was needed. The downhole conditions incorporated into the test procedure are closure stress, temperature, and time to reach a semisteady-state flow condition. Permeability and fracture conductivity values are obtained by measuring the flow of a single-phase liquid under laminar flow conditions through a uniform bed of proppant stressed at selected closure stresses. The major downhole factor that has not been incorporated into the test is rock hardness under *in-situ* conditions. This cell has been modified by some testing laboratories for use with rock-covered platens.[68]

In the API conductivity test unit, which was adapted from an apparatus described by Cooke[6] and illustrated in Fig. 6.25, a weighed amount of proppant is placed in a uniform layer between stainless-steel platens. Fig. 6.26 is a schematic of the apparatus. The unit incorporates two movable pistons with replaceable metal platens in a body. Overburden stress applied to the two pistons is exerted on the pack of proppant. Stress is applied to the test unit with a press in a frame having parallel platens. Fracture widths at stress can be determined by measuring the distance between platens and subtracting the dimension of the test unit components. Liquid of known viscosity is passed lengthwise through the proppant pack and the flow rate and differential pressure are measured. Proppant permeability can then be calculated using Eq. 6.2 because both the cross-sectional area of the test unit and the length between the pressure ports are known. Tests can be made at elevated temperatures if heating elements are incorporated into the platens of the press. Changes in the test-unit platen material may be required if high-chloride brines, typifying formation fluids or fracturing fluids, are used in the test.[68,76,79-83,91-93]

The API test cell has a relatively large test area of 10 in.^{2} [65 cm^2]. Its width is 1½ in. [3.8 cm], the length is 7 in. [17.8 cm], and the distance between pressure ports is 5 in. [12.7 cm]. These dimensions were picked for ease of obtaining reliable data economically. Larger-size test units would possibly reduce operator and equipment error but would greatly increase the cost of the required equipment (particularly the press). Presses having a capacity of 150,000 lbf [667×10^3 N] are moderately priced and will supply up to 15,000-psi [103.4-MPa] stress to the 10-in.^{2} [65-cm^2] unit. Closure stress gradients usually run less than 0.75-psi/ft [17-kPa/m] depth; thus, 15,000-psi [103.4-MPa] stress would normally be encountered in wells with depths greater than 20,000 ft [6100 m].

In addition to the test unit and press, a suitable test fluid is required. Although both gas and liquid systems were evaluated by the API committee, a liquid system—water—was chosen because of its ubiquitous presence in hydrocarbon reservoirs, its well-defined physical and chemical properties, and its effect on some proppants, an effect not revealed when gas is used as the test fluid. To duplicate the reactive chemical content of formation waters and fracturing fluids as closely as possible, the water used in the test must be low in oxygen content and presaturated with silica.[68,81]

Testing with liquid requires a low flow-rate, high-pressure pump and a differential-pressure gauge. It is imperative that no gas be present in the system during a test. Degassed distilled water is used as the test fluid. To ensure that residual gases remain in solution, system pressure during the test should be at least 10 atm [1013 kPa] above the vapor pressure of water at the test temperature. System pressure is controlled with a backpressure regulator.

API Test Procedure. A sample of proppant is weighed, carefully poured into the test unit, and leveled so that the proppant pack has a uniform thickness. The top piston is placed on the proppant pack and the unit is placed in the press. A small stress is applied to the test unit and the entire flow system is evacuated and then filled with water. The initial fracture width is measured at both ends of the test unit (it must be the same at both ends for a valid test). The closure stress desired is applied for the time period prescribed in the procedure.

When tests are run at elevated temperatures, the heating system is allowed to stabilize at test temperature before flow tests are started. Flow tests are run against a backpressure sufficiently high to prevent outgassing during the test. Flow tests are started only after the majority of particle crushing has occurred and the temperature is stable at each stress level. Fracture width and differential pressure are measured at three different flow rates. The permeability of the proppant at the desired stress and temperature is calculated from the experimental data using Eq. 6.3. The test is described in more detail in Ref. 91.

Nomenclature

A_d = well spacing for square drainage area, acres [ha]
A_p = cross-sectional area of proppant pack open to flow, m^2 [ft^2]
h = height, ft [m]
J = PI of well after fracturing
J_0 = PI of well when undamaged and unstimulated (skin=0)
$\bar{k}$ = average formation permeability based on gross thickness, md
k_f = permeability of proppant, md
L_f = length of proppant pack in direction of flow, m [ft]
L = length of one wing of fracture, ft [m]
p = pressure, atm [kPa]
q = flow rate, cm^3/s [ft^3/sec]
r_e = drainage radius of well, ft [m]
r_w = wellbore radius, ft [m]
v = velocity, cm/s [ft/sec]
w = propped width of fracture, in. [cm]
β = turbulence factor, atm·sec^2/g [kPa-s^2/g]
μ = viscosity, mPa·s [cp]
ρ = fluid density, g/cm^3 [lbm/gal]

References

1. McGuire, W.J. and Sikora, V.J.: "The Effect of Vertical Fractures on Well Productivity," *Trans.*, AIME (1960) **219**, 401–04.
2. Tinsley, J.M. *et al.*: "Vertical Fracture Height—Its Effect on Steady-State Production Increase," *JPT* (May 1969) 633–38; *Trans.*, AIME, **246**.
3. Prats, M.: "Effect of Vertical Fracture on Reservoir Behavior—Incompressible Fluid Case," *SPEJ* (June 1961) 105–17; *Trans.*, AIME, **222**.
4. Mao, M.L.: "Performance of Vertically Fractured Wells with Finite Conductivity Fractures," PhD dissertation, Stanford U., Stanford, CA (1977).
5. Soliman, M.Y.: "Modifications to Production Increase Calculations for a Hydraulically Fractured Well," *JPT* (Jan. 1983) 170–72.
6. Cooke, C.E. Jr.: "Conductivity of Fracture Proppants in Multiple Layers," *JPT* (Sept. 1973) 1101–07; *Trans.*, AIME, **255**.
7. Hossain, M., Cady, G.V., and Honarpour, M.M.: "Simulation of Real Gas Flow Through Finite-Conductivity Fractures," paper SPE 12919 presented at the 1984 Rocky Mountain Regional Meeting, Casper, WY, May 20–23.
8. Guppy, K.H. *et al.*: "Non-Darcy Flow in Wells With Finite-Conductivity Vertical Fractures," *SPEJ* (Oct. 1982) 681–98.
9. Agarwal, R.G., Carter, R.D., and Pollock, C.B.: "Evaluation and Performance Prediction of Low-Permeability Gas Wells Stimulated by Massive Hydraulic Fracturing," *JPT* (March 1979) 362–72; *Trans.*, AIME, **267**.
10. Anderson, R.W. and Phillips, A.M.: "Practical Application of Economic Well-Performance Criteria to the Optimization of Fracturing Treatment Design," *JPT* (Feb. 1988) 223–28.
11. *API RP-56, API Recommended Practices for Testing Sand Used in Hydraulic Fracturing Operations,* API Production Dept., Dallas (1983).

12. *API RP-60, API Recommended Practices for Testing High Strength Proppants Used in Hydraulic Fracturing Operations,* API, Dallas (Feb. 1, 1989).
13. Fickling, J. and Mack, D.: "The Effects of Sand Quality on Fracture Stimulation Treatments," *Northeast Oil Reporter* (July 1985) 33–35.
14. Wendorff, C.L., Burke, R.F., and Garbis, S.J.: "The Effect Sand Handling Equipment Has on Prop Sand Quality," paper SPE 8402 presented at the 1979 SPE Annual Technical Conference and Exhibition, Las Vegas, Sept. 23–26.
15. Wendorff, C.L.: "Frac Sand Quality Control—A Must for Good Frac Treatments," paper presented at the 1978 ASME Annual Meeting, Houston, Nov.
16. Hoaberg, R.K. and Koerner-Moore, J.: "Silica-Sand Proppants Used in Well Stimulation," paper SME-80-345 presented at the 1980 SME Fall Meeting, Minneapolis, MN, Oct. 22–24.
17. *Proppants*, second edition, The Western Co., Fort Worth, TX (1984).
18. *Proppants (F-3235),* Halliburton Services, Duncan, OK (1986).
19. *Proppant Selection Guide*, Dowell Schlumberger, Tulsa, OK (1985).
20. Hannah, R.R. and Anderson, R.W.: *Propped Fracture Flow Capacity—Treatment Design Data*, The Western Co., Fort Worth, TX (Jan. 1985).
21. Cooke, C.E. Jr., Hedden, W.A., and Chard, W.C.: "Hydraulic Fracturing Method using Sintered Bauxite Propping Agent," U.S. Patent No. 4,068,718 (1978).
22. Beck, W.R. and Castle, R.B.: "Proppant for Well Fractures and Method of Making Same," U.S. Patent No. 4,493,875 (Jan. 15, 1985).
23. Fitzgibbon, J.J.: "Sintered Spherical Pellets Containing Clay as a Major Component Useful for Gas and Oil Well Proppants," U.S. Patent No. 4,427,068 (1984).
24. Lunghofer, E.P.: "Hydraulic Fracturing Propping Agent," European Patent Application No. 87,852 (1983).
25. Urffer, D.: "Zircon and Silica Based Propping Agent for Deep Geologic Fractures," French Patent No. 2,493,910 (1982).
26. Urffer, D.: "Propping Agent Based on Zirconia and Silica for Deep Geological Fractures," U.S. Patent No. 4,607,697 (Aug. 26, 1986).
27. Watson, D.R., Carithers, V.G., and McDaniel, L.T.: "Alumino Silicate Ceramic Proppant for Gas and Oil Well Fracturing and Method of Forming Same," U.S. Patent No. 4,555,493 (Nov. 26, 1985).
28. Graham, J.W. *et al.*: "Method for Treating Subterranean Formations," U.S. Patent No. 3,929,191 (Dec. 30, 1975).
29. Callahan, M.J., Cipolla, C.L., and Lewis, P.E.: "The Application of a New Second-Generation High-Strength Proppant in Tight Gas Reservoirs," paper SPE 11633 presented at the 1983 SPE/DOE Low-Permeability Gas Reservoirs Symposium, Denver, March 13–16.
30. Norman, M.E., Cipolla, C.L., and Webb, M.L.: "Application of Manufactured Proppants in Moderately Permeable Oil Reservoirs," paper SPE 12357 presented at the 1983 SPE Production Technology Symposium, Lubbock, TX, Nov. 14–15.
31. Cutler, R.A. *et al.*: "Fracture Conductivity Comparison of Ceramic Proppants," *SPEJ* (April 1985) 157–70.
32. Cutler, R.A. *et al.*: "Lightweight Proppants for Deep Gas Well Stimulation," final report, Contract No. DOE/BC/100038-29, U.S. DOE (Jan. 1984).
33. Cutler, R.A. *et al.*: "New Proppants for Deep Gas Well Stimulation," paper SPE 9869 presented at the 1981 SPE/DOE Low Permeability Symposium, Denver, May 27–29.
34. Sarda, J.P.: "Use of High-Strength Ceramic Beads for Propping Deep Hydraulic Fractures," *JPT* (Jan. 1981) 55–56.
35. Sparlin, D.D. and Hagen, R.: "The New Intermediate Strength Fracture Proppants," *Drilling* (June 1983) **44,** 78, 79, 81, 83, 84, 86.
36. Carroll, H.B. Jr. and Baker, B.H.: "Particle Size Distributions Generated by Crushed Proppants and Their Effects on Fracture Conductivity," paper SPE 7923 presented at the 1979 SPE Low Permeability Symposium, Denver, May 20–22.
37. Swanson, S.R. and Cutler, R.A.: "Fracture Analysis of Ceramic Proppants," *J. Energy Res. Tech.* (June 1983) **105,** 128–33.
38. Tucker, R.L.: "Practical Pressure Analysis in Evaluation of Proppant Selection for the Low-Permeability, Highly Geopressured Reservoirs of the McAllen Ranch (Vicksburg) Field," paper SPE 7925 presented at the 1979 SPE Low Permeability Gas Reservoirs Symposium, Denver, May 20–22.
39. Atteberry, R.D., Tucker, R.L., and Ritz, J.W.: "Application of Sintered Bauxite Proppants to Stimulation of Low Permeability South Texas Gas Reservoirs," paper 7924 presented at the 1979 SPE Low Permeability Gas Reservoirs Symposium, Denver, May 20–22.
40. Bleakley, W.B.: "Mobil AG Scores with Massive Frac," *Pet. Eng.* (Jan. 1984) 72, 74, 78, 80, 83.
41. Brinkman, F.W.: "Status Report on Fracturing of Deep and Low Permeability Formations in West Germany," paper SPE 9852 presented at the 1981 SPE/DOE Low Permeability Gas Reservoirs Symposium, Denver, May 27–29.
42. Cooke, C.E. Jr. and Gidley, J.L.: "High-Strength Proppant Extends Deep Well Fracturing Capabilities," *Proc.*, 10th World Pet. Cong., Bucharest (1980) **3,** 89–96.
43. Cooke, C.E. Jr., Gidley, J.L., and Mutti, D.H.: "Use of High-Strength Proppant for Fracturing Deep Wells," paper SPE 6440 presented at 1977 SPE Deep Drilling and Production Symposium, Amarillo, TX April 17–19.
44. Hickey, J.W., Brown, W.E., and Crittendon, S.J.: "The Comparative Effectiveness of Propping Agents in the Red Fork Formation of the Anadarko Basin," paper SPE 10132 presented at the 1981 SPE Annual Technical Conference and Exhibition, San Antonio, Oct. 5–7.
45. Hickey, J. W.: "The Comparative Effectiveness of Propping Agents in the Red Fork Formation of the Anadarko Basin," paper SPE 11577 presented at the 1983 SPE Production Operations Symposium, Oklahoma City, OK, Feb. 27–March 1.
46. "Sintered Bauxite Unlocks Gas Well," *Drilling Cont.* (Sept. 1980) 94.
47. Lindley, B.W. and McGhee, B.F.: "An Investigation of a High-Strength Proppant Tail-In at McAllen Ranch Field," paper SPE 11935 presented at the 1983 SPE Annual Technical Conference and Exhibition, San Francisco, Oct. 5–8.
48. Seccombe, J.C. and Anderson, C.E.: "Selection of a Fracture Proppant in a Tight Gas Field, Bauxite vs. Sand, Wamsutter Area, Wyoming," paper SPE 10827 presented at the 1982 SPE/DOE Unconventional Gas Recovery Symposium, Pittsburgh, PA, May 16–18.
49. Kohlhaas, C.A.: "Field Study—Steinle Ranch, An Intermediate Depth Oil Field, Shows Significant Benefit From Bauxite Proppants," paper SPE 10875 presented at the 1982 SPE Rocky Mountain Regional Meeting, Billings, MT, May 19–21.
50. Pearce, K.W.: "Application of High Strength Proppants in Relatively Shallow and Hard Formations," *Proc.*, 30th Annual Southwestern Petroleum Short Course, Lubbock, TX (1983) 153–62.
51. Johnson, C.K. and Armbruster, D.R.: "Particles Covered with a Cured Infusable Thermoset Film and Process for Their Production," U.S. Patent No. 4,439,489 (March 27, 1984).
52. Underdown, D.R. and Glaze, H.: "Proppant Charge and Method," U.S. Patent No. 4,443,347 (April 17, 1984).
53. Graham, J.W. and Sinclair, A.R.: "Well Completion Method," U.S. Patent No. 4,533,596 (Nov. 19, 1985).
54. Graham, J.W. and Sinclair, A.R.: "High Strength Particulates," U.S. Patent No. 4,585,064 (April 29, 1986).
55. Sinclair, A.R. and Graham, J.W.: "A New Proppant to Sustain Fracture Conductivity and Maximize Stimulated Production," paper presented at the 1977 Oklahoma U. Massive Hydraulic Fracturing Symposium, Norman, March.
56. Sinclair, A.R. and Graham, J.W.: "A New Proppant for Hydraulic Fracturing," paper 78-PET-34 presented at the 1978 ASME Energy Technology Conference, Houston, Nov. 5–9.
57. Sinclair, A.R., Graham, J.W., and Sinclair, C.P.: "Improved Well Stimulation With Resin-Coated Proppants," paper SPE 11579 presented at the 1983 SPE Production Operations Symposium, Oklahoma City, Feb. 27–March 1.
58. Underdown, D.R. and Das, K.: "New Proppant for Deep Hydraulic Fracturing," *JPT* (Jan. 1985) 98-104.
59. Underdown, D.R. and Das, K.: "Stability of Gravel-Packing Materials for Thermal Wells," *JPT* (Nov. 1985) 2006–12.
60. Kim, C.M. and Willingham, J.R.: "Flow Response of Propped Fractures to Repeated Production Cycles," paper SPE 16912 presented at the 1987 SPE Annual Technical Conference and Exhibition, Dallas, Sept. 27–30.
61. Darin, S.R. and Huitt, J.L.: "Effect of Partial Monolayer of Propping Agent on Fracture Flow Capacity," *Trans.*, AIME (1960) **219**, 31–37.
62. Coulter, G.R. and Wells, R.D.: "The Advantages of High Proppant Concentration in Fracture Stimulations," *JPT* (June 1972) 643–50.
63. Ely, J.W. and Holditch, S.A.: "Deep Well Stimulation Utilizing High Concentrations of Proppant," paper SPE 4118 presented at the 1972 SPE Annual Fall Meeting, San Antonio, Oct. 8–11.
64. McDaniel, R.R. and Willingham, J.R.: "The Effect of Various Proppants and Proppant Mixtures on Fracture Permeability," paper SPE 7573 presented at the 1978 SPE Annual Technical Conference and Exhibition, Houston, Oct. 1–3.
65. Watkins, H.: "Higher Sand Concentration in East Texas," paper SPE 10913 presented at the 1982 SPE Cotton Valley Symposium, Tyler, TX (May 20).
66. McGlothlin, B.B. and Huitt, J.L.: "Relation of Formation Rock Strength to Propping Agent Strength in Hydraulic Fracturing," *JPT* (March 1966) 377–84; *Trans.*, AIME, **237**.
67. Larsen, D.G. and Smith, L.J.: "New Conductivity Found in Angular Blends of Fracturing Sand," paper SPE 13814 presented at the 1985 SPE Production Operations Symposium, Oklahoma City, March 10–12.
68. Penny, G.S.: "Evaluation of the Effects of Environmental Conditions and Fracturing Fluids on the Long-Term Conductivity of Proppants,"

paper SPE 16900 presented at the 1987 Annual Technical Conference and Exhibition, Dallas, Sept. 27–30.
69. Howard, G.C. and Fast, C.R.: *Hydraulic Fracturing*, Monograph Series, SPE, Richardson, TX (1970) **2**, 61–63, 67.
70. Volk, L.J. *et al.*: "Embedment of High Strength Proppant into Low-Permeability Reservoir Rock," paper SPE 9867 presented at the 1981 SPE/DOE Low Permeability Gas Reservoirs Symposium, Denver, May 27–29.
71. Cooke, C.E. Jr.: "Fracturing With a High-Strength Proppant," *JPT* (Oct. 1977) 1222–26.
72. Kim, C.M. and Losacano, J.A.: "Fracture Conductivity Damage Due to Crosslinked Gel Residue and Closure Stress on Propped 20/40 Mesh Sand," paper SPE 14436 presented at the 1985 SPE Annual Technical Conference and Exhibition, Las Vegas, Sept. 22–25.
73. McDaniel, B.W.: "Conductivity Testing of Proppants at High Temperature and Stress," paper SPE 15067 presented at the 1986 SPE California Regional Meeting, Oakland, April 2–4.
74. Becq, D.F., Roque, C., and Sarda, J.P.: "High-Strength Proppant Behavior Under Extreme Conditions," paper SPE 12487 presented at the 1984 SPE Formation Damage Symposium, Bakersfield, Feb. 13–14.
75. Brown, W.E. and Much, M.G.: "An Evaluation of Four Commonly Used Proppants," Norton-Alcoa Proppants, Dallas (Feb. 1986).
76. Cobb, S.L. and Farrell, J.J.: "Evaluation of Long-Term Proppant Stability," paper SPE 14133 presented at the 1986 SPE Intl. Meeting on Petroleum Engineering, Beijing, March 17–20.
77. "Investigation of the Effects of Fracturing Fluids Upon the Conductivity of Proppants," final report, Stim-Lab Inc., Duncan, OK (Jan. 29, 1987).
78. "Investigation of the Effects of Fracturing Fluids Upon the Conductivity of Proppants," final report, Stim-Lab Inc., Duncan, OK (Jan. 18, 1988).
79. McDaniel, B.W.: "Realistic Fracture Conductivities of Proppants as a Function of Reservoir Temperature," paper SPE 16453 presented at the 1987 SPE/DOE Low Permeability Reservoirs Symposium, Denver, May 18–19.
80. Much, M.G. and Penny, G.S.: "Long Term Performance of Proppants Under Simulated Reservoir Conditions," paper SPE 16415 presented at the 1987 SPE/DOE Low Permeability Reservoirs Symposium, Denver, May 18–19.
81. Katagiri, K., Ott, W.K., and Nutley, B.G.: "Hydraulic Fracturing Aids Geothermal Field Development," *World Oil* (Dec. 1980) 75, 76, 78, 80, 84, 88.
82. Cooke, C.E. Jr.: "Effect of Fracturing Fluids on Fracture Conductivity, *JPT* (Oct. 1975) 1273–82; *Trans.*, AIME, **259**.
83. Almond, S.W. and Bland, W.E.: "Effect of Break Mechanism on Gelling Agent Residue and Flow Impairment in 20/40 Mesh Sand," paper SPE 12485 presented at the 1985 SPE Formation Damage Symposium, Bakersfield, Feb. 13–14.
84. Roodhart, L., Kuiper, T.O., and Davies, D.R.: "Proppant Pack and Formation Impairment During Gas Well Hydraulic Fracturing," paper SPE 15629 presented at the 1986 SPE Annual Technical Conference and Exhibition, New Orleans, Oct. 5–8.
85. Gruesbeck, C. and Collins, R.E.: "Entrainment and Deposition of Fine Particles in Porous Media," *SPEJ* (Dec. 1982) 847–56.
86. Holditch, S.A. and Morse, R.S.: "The Effects of Non-Darcy Flow on the Behavior of Hydraulically Fractured Wells," *JPT* (Oct. 1976) 1169–79.
87. Maloney, D.R., Gall, B.L., and Raible, C.J.: "Non-Darcy Gas Flow Through Propped Fractures: Effects of Partial Saturation, Gel Damage and Stress," paper SPE 16899 presented at the 1987 SPE Annual Technical Conference and Exhibition, Dallas, Sept. 27–30.
88. Pursell, D.A.: "Laboratory Investigation of Inertial Flow in High Strength Fracture Proppants," MS thesis, Texas A&M U., College Station (Dec. 1987).
89. Jennings, A.R. Jr. and Lord, D.L.: "Fracture Flow Capacity—A Key to Sustained Production After Hydraulic Fracturing," paper SPE 6127 presented at the 1976 SPE Annual Technical Conference and Exhibition, New Orleans, Oct. 3–6.
90. Parker, M.A. and McDaniel, B.W.: "Fracturing Treatment Design Improved by Conductivity Measurements Under In-Situ Conditions," paper SPE 16901 presented at the 1987 SPE Annual Technical Conference and Exhibition, Dallas, Sept. 27–30.
91. *API RP-61, API Recommended Practices for Evaluating Short Term Proppant Pack Conductivity*, eighth draft, API, Production Dept., Dallas (Feb. 1989).
92. Much, M.G.: "Proppant Evaluation Test Improved," *Oil & Gas J.* (April 6, 1987) 33–36.
93. Wendorff, C.L.: "Fracture Porosimeter—A New Tool for Determining Fracture Conductivity Under Downhole Stress," paper SPE 10963 presented at the 1982 SPE Annual Technical Conference and Exhibition, New Orleans, Sept. 26–29.

SI Metric Conversion Factors

acres	×	4.046 873	E−01	=	ha
atm	×	1.013 250*	E+02	=	kPa
°F		(°F−32)/1.8		=	°C
in.	×	2.54*	E+00	=	cm
in.^{2}	×	6.451 6*	E+00	=	cm^2
lbm/ft^2	×	4.882 428	E+00	=	kg/m^2
lbm/ft^3	×	1.601 846	E+01	=	kg/m^3
lbm/gal	×	1.198 264	E+02	=	kg/m^3
psi	×	6.894 757	E+00	=	kPa

*Conversion factor is exact.

Chapter 7
Fracturing Fluids and Additives

John W. Ely, SPE, S.A. Holditch & Assocs.

7.1 Overview

This chapter discusses the various fracturing fluids and the additives necessary to achieve certain properties. Included is a discussion of the properties desired in an ideal fracturing fluid. This is followed by a discussion of the numerous fluids available, including water-based fluids, oil-based fluids, alcohol-based fluids, emulsion fluids, and foam-based fluids.[1-7] Use of additives in the fracturing fluid system is also detailed.

The information in this chapter should provide the design engineer with the basic knowledge needed to choose a base fluid and the important additives.

7.2 Properties of a Fracturing Fluid

Fracturing fluids are pumped into underground formations to stimulate oil and gas production. To achieve successful stimulation, the fracturing fluid must have certain physical and chemical properties.

1. It should be compatible with the formation material.
2. It should be compatible with the formation fluids.
3. It should be capable of suspending proppants and transporting them deep into the fracture.
4. It should be capable, through its inherent viscosity, to develop the necessary fracture width to accept proppants or to allow deep acid penetration.
5. It should be an efficient fluid (i.e., have low fluid loss).
6. It should be easy to remove from the formation.
7. It should have low friction pressure.
8. Preparation of the fluid should be simple and easy to perform in the field.
9. It should be stable so that it will retain its viscosity throughout the treatment.
10. The fracturing fluid should be cost-effective.

The first characteristic listed may be the most critical. If the chemical nature of the fracturing fluid causes swelling of naturally occurring clays in the formation, thereby plugging pore channels, the treatment will be a failure. If the fracturing fluid causes migration of fines and/or clays, the success of the treatment will be nullified.[9-19] If the fracturing fluid creates emulsions and/or sludging of the crude oil, then plugging rather than stimulation will occur.[20-26] If the fracturing fluid dissolves the cementing material that holds the grains of the sandstone together, spalling of the formation can occur and failure will result. The fracturing fluid should not cause scaling or paraffin problems.[27-33] Compatibility is therefore a critical and necessary characteristic of a fracturing fluid.

Another important characteristic of a fracturing fluid is its ability to transport the proppants down the tubular goods through perforations and deep into the fracture.[35-41] Depending on the nature of the fracturing fluid, it may perfectly suspend the proppant or, as is the case of linear fluids, allow for some settling and banking of the proppant in the fracture. Large viscosities are needed to transport proppants and to develop the fracture width needed to create and to prop long fractures. It is well known that insufficient fracture width from insufficient viscosity will not allow proppants to be transported very far into the fracture.

The ideal fracturing fluid should be moderately efficient. A high percentage of the fluid should remain in the fracture and not be lost to the formation. Fluid efficiency is normally attained by combining high fluid viscosity with fluid-loss additives.[42-54] These fluid-loss additives may consist of plastering agents, bridging agents, microemulsions, or emulsified gas. A low-efficiency fracturing fluid would not create the desired fracture volume, carry the proppant, or achieve the desired formation penetration if most of the fracturing fluid leaks off during treatment.

Another important characteristic of a fracturing fluid is its ability to revert from high viscosity to low viscosity upon residing in the formation.[55-57] Viscosity reduction is necessary so that the treating fluid can be removed from the formation easily. High fluid viscosities in the fracture or in the formation near the fracture can reduce hydrocarbon production. Fracture-fluid viscosity is normally reduced by thermal degradation in high-temperature wells or by controlled degradation through the use of such breaking agents as enzymes, oxidizers, or weak acids. Controlled degradation is essential for the fluid to maintain its viscosity during the treatment but to degrade and lose its viscosity after the treatment.

Many fracturing-fluid systems used during the 1950's and 1960's had high viscosity and the ability to degrade; however, they were extremely difficult to pump down small tubular goods. Modern fracturing-fluid systems have been developed that allow for high viscosity but have reduced friction properties.[58-74] In fact, most of these fluids will pump at pressures lower than low-viscosity base fluids, such as water or oil, through turbulence suppression by long-chain polymer systems. If a fluid cannot be pumped easily, it normally is not acceptable as a fracturing fluid. Exceptions to this case are high-viscosity crude oils used in shallow casing jobs. These high-viscosity, Newtonian fluids, however, are not acceptable for pumping down small tubular goods.

Fracture-fluid stability at high temperature is a critical aspect of any fluid.[75-79] A fluid that rapidly loses its viscosity because of thermal thinning or degradation is not applicable for treatment of high-temperature wells. A fracturing fluid should be able to maintain the designed viscosity with minimal viscosity loss vs. time at bottomhole temperature (BHT).

Finally, fracturing fluids should be cost-effective[80] and easy to mix in the field. One of the most important and realistic selection criteria for a fracturing fluid is cost-effectiveness in treating the formation under study. Quite obviously, a fluid that has all these attributes but will not yield cost-effective stimulation will not be an ideal fluid.

A fairly comprehensive list of service company products and processes is given in Appendix A, and fracturing-fluid-selection flow charts are presented in Appendix B. These flow charts should assist engineers in selecting fracturing fluids and processes that meet their particular requirements.

7.3 Water-Based Fluids

Water-based fracturing fluids are used in the majority of hydraulic fracturing treatments today. This was not the case in the early days

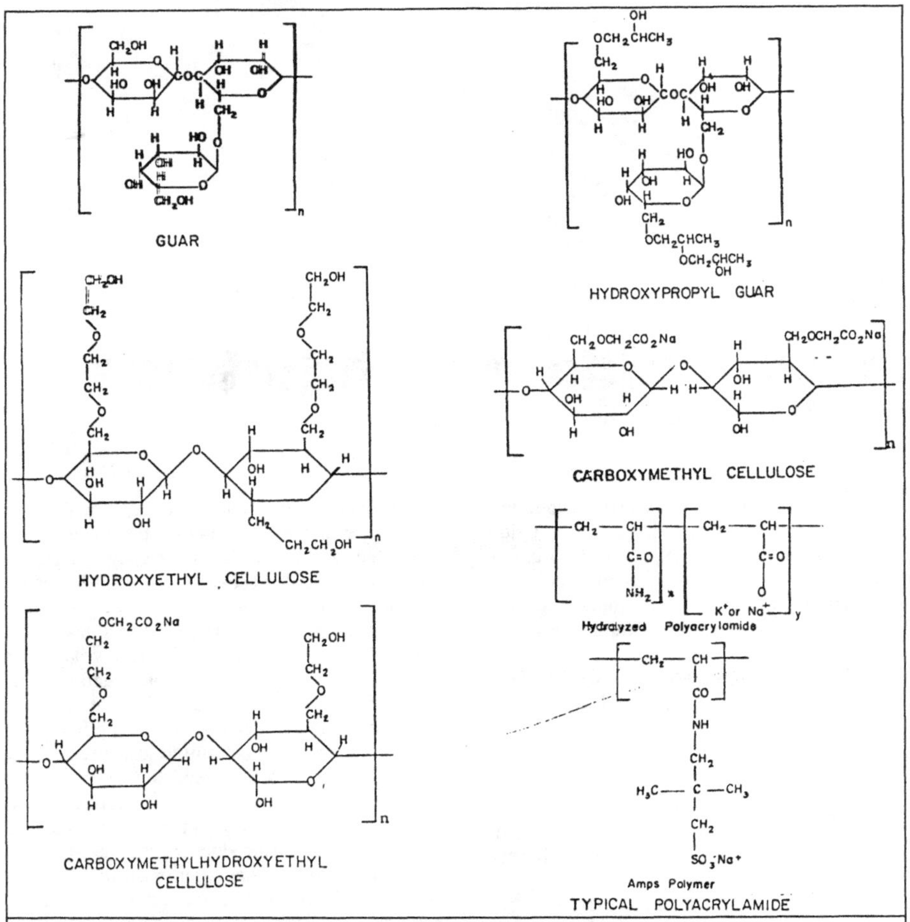

GUAR

HYDROXYPROPYL GUAR

HYDROXYETHYL CELLULOSE

CARBOXYMETHYL CELLULOSE

Hydrolyzed Polyacrylamide

CARBOXYMETHYLHYDROXYETHYL CELLULOSE

Amps Polymer

TYPICAL POLYACRYLAMIDE

Fig. 7.1—Chemical structures of guar, HPG, HEC, CMHEC, and polyacrylamides.

of fracturing, when oil-based fluids were the fluids selected for virtually all the treatments. Water-based fracturing fluids have many advantages over oil-based fluids.

1. Water-based fluids are economical. The base fluid, water, is much cheaper than oil, condensate, methanol, and acid.
2. Water-based fluids yield increased hydrostatic head compared with oil, gases, and methanol.
3. These fluids are incombustible; hence they are not a fire hazard.
4. Water-based fluids are readily available.
5. This type of fluid is easily viscosified and controlled.

In the early days of fracturing, there was a great deal of reluctance to pump water into an oil-bearing formation. In retrospect, many of the early water-based fracturing treatments were conducted on formations that were not very water-sensitive. Also, the numerous waterfloods being performed in the 1950's and 1960's added confidence that a water-based fluid could be used as a stimulation fluid.

The availability, cost-effectiveness, hydrostatic head, and lack of fire danger provided incentives for the service companies to develop such additives as potassium chloride, clay stabilizers, surfactants, and nonemulsifiers that make water-based fluids more versatile.

Although many improvements have been made in oil-based fluids, most recent technical development has occurred in water-based fluid technology, partly because of the recognition that water-based fluids are chosen for fracturing the vast majority of reservoirs.

Linear Fracturing Fluids. The need to thicken water to help transport proppant, to decrease fluid loss, and to increase fracture width was apparent to early investigators. The first water viscosifier was starch. Starch had been used to thicken and to decrease the fluid loss in drilling muds. This particular fluid was short-lived because of shear sensitivity, lack of temperature stability, and bacterial degradation. In the early 1960's, guar gum was found to be a ready replacement.[81-83] Guar polymer comes from a bean that when added to water thickens and viscosifies the fluid. The structure of unmodified guar gum is shown in Fig. 7.1. Guar is a naturally occurring polymer that undergoes hydration upon contact with water. The polymer uncoils, with water molecules attaching themselves to the polymer chain. This creates a viscous fluid by interaction of the polymer coils, one to another, in the water-based system.

The first guar polymers developed in the early 1960's are still in use today. Some of the suppliers have improved the product by removing more of the hulls and inert material to give a lower-residue material, but the product remains basically the same as that used in the early days of fracturing. The guar molecule, as a point of reference, is widely used in viscosifying such things as ice cream and other food items.

Other linear gels used today as fracturing fluids are hydroxypropyl guar (HPG), hydroxyethylcellulose (HEC), carboxymethyl HPG (CMHPG), xanthan gum, and in some rare cases, polyacrylamides.[84-86] The chemical structures of various polymers are presented in Fig. 7.1.

Developed in the early 1970's, HPG has become the most widely used viscosifier for water-based fracturing treatments.[87] HPG is obtained by the reaction of propylene oxide with the guar molecule, creating a more-temperature-stable, somewhat-higher-viscosity polymer. It was developed primarily to reduce the residue obtained from guar gum and to achieve greater temperature stability.

The residue after degradation of a particular viscosifying polymer is defined as the material that remains as insoluble product upon complete degradation of the polymer.[88-93] Guar products typically vary in nondegradable residue from 8 to 12%. HPG will vary from 1 to 4% residue. This residue was construed as extremely important to some within the industry who felt that the residue could act as

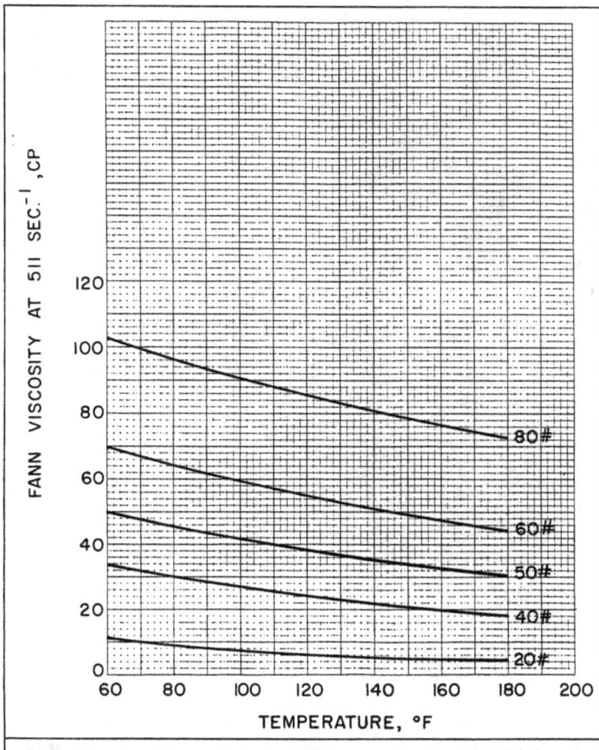

Fig. 7.2—High-viscosity guar gum, Fann viscosity vs. temperature for various concentrations of polymer (2% KCl).

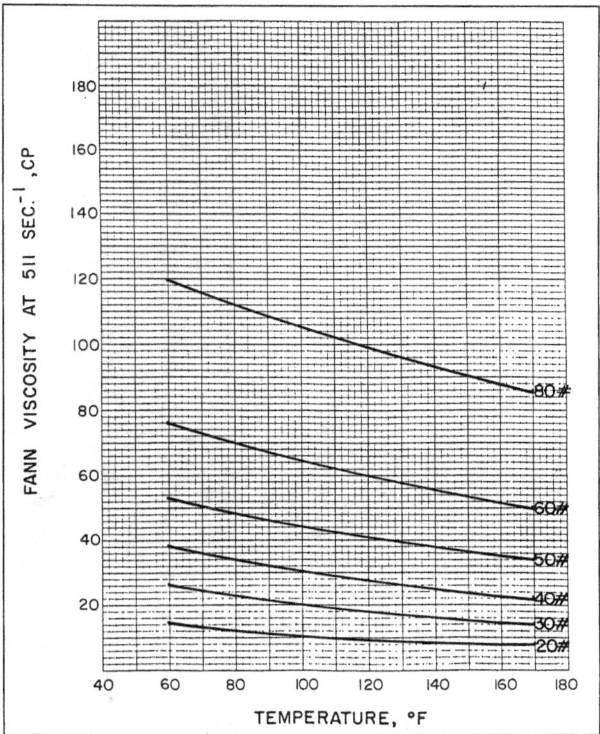

Fig. 7.3—High-viscosity HPG, Fann viscosity vs. temperature for various concentrations of polymer (2% KCl).

a plugging agent in the fracture system or fracture proppant pack or as a plugging material in the formation pore spaces.

Perhaps of even more importance was the enhanced temperature stability achieved with HPG. The propylene oxide groups basically tend to stabilize the polymer against thermal degradation. Figs. 7.2 and 7.3 illustrate comparative viscosity vs. temperature at various concentrations of guar and HPG.

The reduction in residue between HPG and guar gum can be a disadvantage in fluid-loss control in high-permeability intervals. The residue consisting of colloidal solids is a good fluid-loss-control agent. Therefore, where temperature stability is not required and the formation is highly permeable, the best fluid may in fact be guar gum. This choice is particularly beneficial when one must add large quantities of nondegradable fluid-loss additive to HPG to fracture-treat a low-temperature well with moderate to high permeability successfully.

A further derivative of guar gum, CMHPG is formed by the reaction of HPG with sodium monochloroacetate. This double-derivatized product is used only in crosslinked gel applications. The CMHPG has little application in linear gel systems because it is even higher in cost (see Table 7.1) than HPG. Although CMHPG allegedly has enhanced properties in crosslinked systems, its current usage is relatively limited, particularly in the case of linear gels. This product has somewhat lower residue than HPG.

Other viscosifiers used in linear gel systems include HEC, carboxymethylcellulose (CMC), and carboxymethylhydroxyethylcellulose (CMHEC). These cellulose polymers are usually considered synthetics; guar is usually considered a natural polymer. Cellulose derivatives are formed by reacting natural cellulose from cotton or wood products to form the derivative. HEC is formed by treating cellulose with sodium hydroxide and reacting it with ethylene oxide. Hydroxyethyl groups are introduced to yield a hydroxyethyl ether. CMC is produced by reacting alkali cellulose with sodium monochloroacetate under very controlled conditions. Again, CMHEC is created by a double derivitization, combining the reactions mentioned earlier for the creation of HEC followed by the reaction of sodium monochloroacetate with HEC. These products yield high-viscosity polymers (Figs. 7.4 through 7.6) that have no residue whatsoever upon degradation. They form clear fluid solutions (see Figs. 7.7 through 7.9) that show the turbidity of HPG vs. the nearly opaque nature of guar and the very clean, clear solutions of cellulose derivatives. These relatively-high-viscosity synthetic polymers are also high in cost (Table 7.1); therefore, the use of HEC is somewhat limited. HEC has relatively low cost-effectiveness compared with guar and HPG. Also, HEC suffers from extreme difficulty in crosslinking; few metal or metal-chelant techniques are currently available to crosslink HEC.

TABLE 7.1—COMPARATIVE COSTS OF POLYMERS*

	Service Company				
	A	B	C	D	E
Guar	1.13	0.93	0.94	1.04	1.0
HPG	1.41	1.31	1.18	1.35	1.29
CMHPG	—	—	1.41	—	1.40
CMC	—	1.38	—	—	1.62
HEC	1.5	1.65	1.38	1.85	1.62
CMHEC	1.59	1.44	—	—	1.62
Xanthan	2.62	2.35	—	2.43	2.65

*Relative costs shown are based on the average price for guar gum. Because Service Company E's published price for guar was the exact average, it would have an index of 1.0 for guar. All other polymer prices are then based on relative costs. Company A's published price for xanthan is 2.62 times higher than guar. The comparison of relative costs, instead of price per pound, should withstand the test of time.

HEC was used widely in the late 1960's and early 1970's,[77,79] but the primary use for HEC today has been in gravel-pack applications, where a nonresidual high-viscosity fluid is required. The use of a retarded (glyoxylated) secondary gel of HEC allowed a large number of high-temperature fracturing treatments to be conducted with relatively high concentrations of HEC.

CMC has little or no use today in hydraulic-fracturing applications because of its sensitivity to salt concentrations (see Fig. 7.5). It was used briefly in the mid-1970's as a crosslinked fluid because of the ease of crosslinking the carboxymethylene group with heavy metals. Its high cost (Table 7.1) and sensitivity to salts have basically eliminated it from the product line of most fracturing service

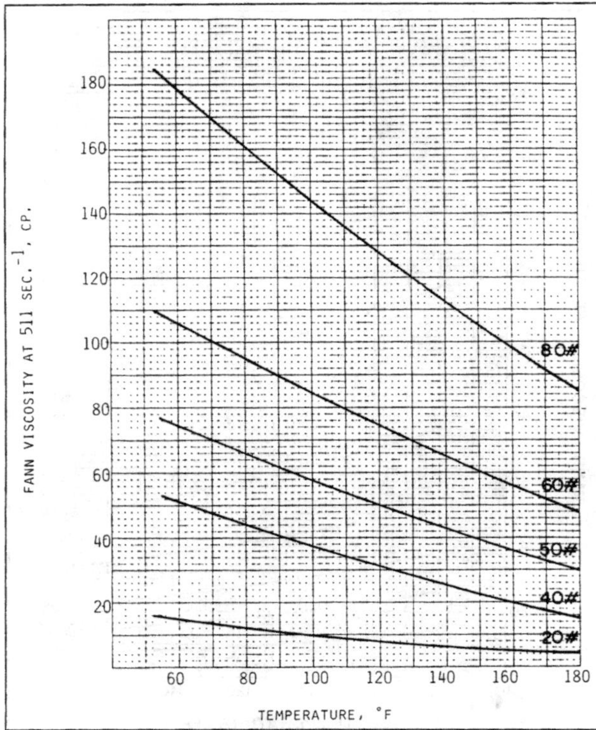

Fig. 7.4—High-viscosity HEC, Fann viscosity vs. temperature for various concentrations of polymer (2% KCl).

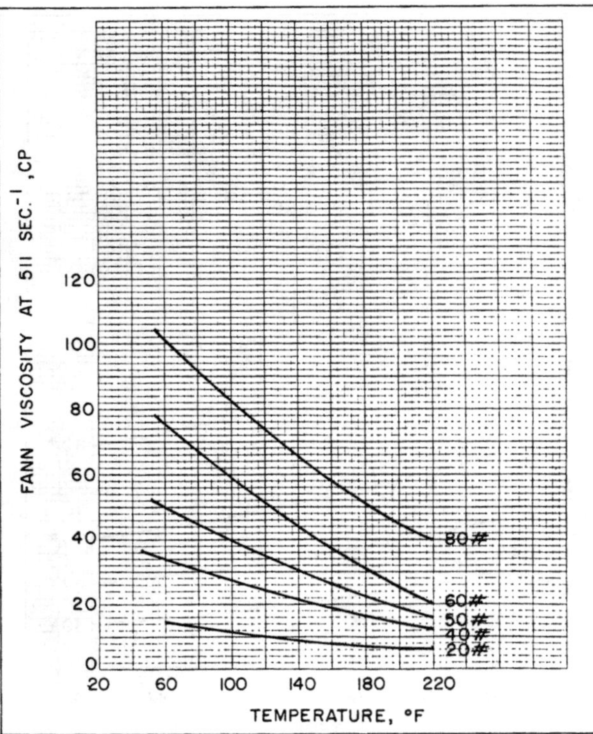

Fig. 7.6—High-viscosity CMHEC, Fann viscosity vs. temperature for various concentrations of polymer (2% KCl).

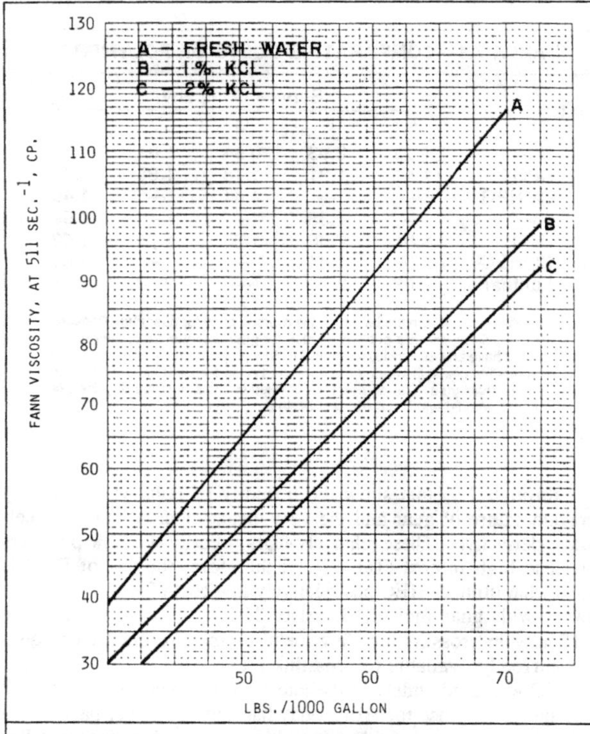

Fig. 7.5—High-viscosity CMHEC, Fann viscosity vs. polymer concentrations for various base fluids (80°F).

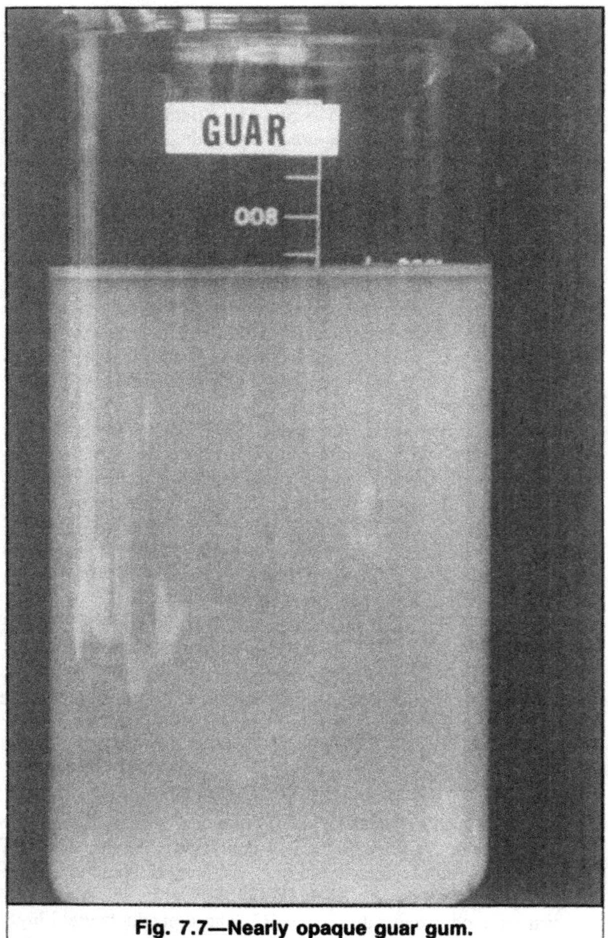

Fig. 7.7—Nearly opaque guar gum.

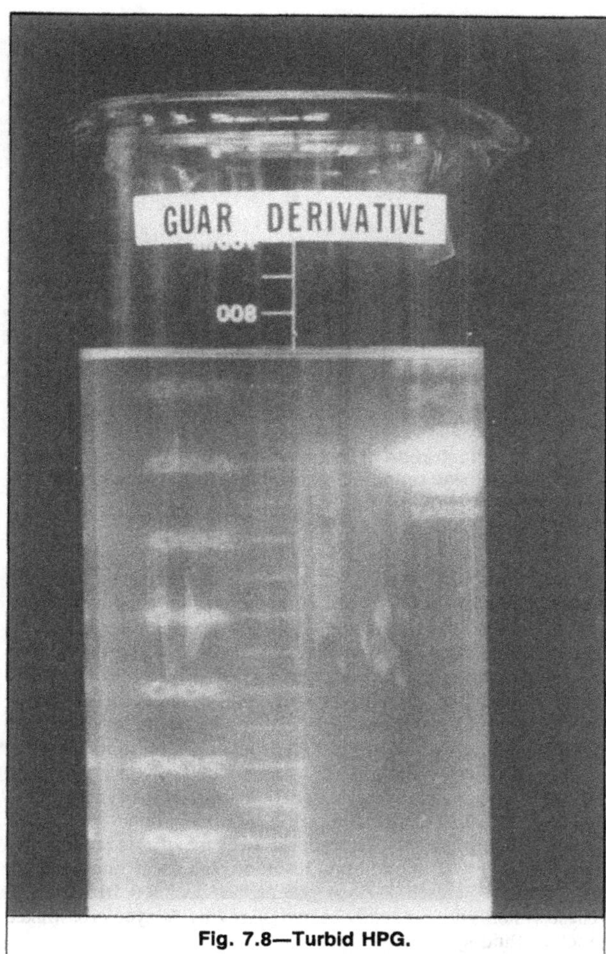

Fig. 7.8—Turbid HPG.

Fig. 7.9—Clear cellulose derivative.

companies. An improvement on CMC is CMHEC. The CMHEC product, while retaining the ease in crosslink ability, does not have CMC's sensitivity to salt. This product is used relatively sparingly in linear gel systems and moderately in crosslinked acid systems in low-temperature applications.[94]

Another viscosifier finding some use as a thickener for fracturing fluids is xanthan gum, which is used both as a linear gel system and as a crosslinked fluid. It is used as a thickener more often in drilling fluids than in fracturing fluids. Xanthan is a relatively-high-cost product (Table 7.1). Its major use in stimulation during the past few years has been as a thickener for hydrochloric acid. Its use is limited to acid concentrations up to 15% and temperatures of 200°F [93°C] or less.

The last linear viscosifiers to be discussed are polyacrylamides. There are many examples of polymers and copolymers of acrylamides. Fig. 7.1 presents an example of an acrylamide structure. The primary use for acrylamides is not as a linear gel viscosifier but as a friction reducer; these products yield excellent friction reduction at very low concentrations. The temperature stability of acrylamides is usually adequate, but the cost is sometimes prohibitive. Acrylamides also are somewhat difficult to degrade controllably under low-temperature conditions.[95] Typically, these products are readily crosslinkable but, because of their very high cost compared with guar or cellulose derivatives, have found very limited use in the oil field other than as friction reducers for water and acid. Copolymers of acrylamide are finding fairly wide usage as stable linear and crosslinked gels in fracture-acidizing applications.[96]

Acidizing and fracture acidizing require extensive theoretical and practical explanation. Completing a literature search on hydraulic fracturing uncovered several references[97-116] that, combined with *Acidizing Fundamentals,*[117] should be used for furthering knowledge of acidizing and fracture acidizing.

Linear gels are relatively simple fluids to use and to control. Excellent reproducible data are available on the viscosity of these fluids. The problem with linear fluids is their poor proppant suspension capability. Also, the linear gel has less temperature stability than a similar crosslinked fluid.[96,118-132]

If one is designing a damage-removal treatment or banking-type proppant pack for high fracture conductivity near the wellbore, then a linear gel may be the ideal fluid. One needs to look at cost-effectiveness, amount of proppant placed, and type of proppant pack needed. If one is trying to achieve deep penetration of proppant or of acid away from the wellbore, then quite obviously, one needs to choose the higher-viscosity crosslinked fracturing fluids.

Crosslinked Fracturing Fluids. Crosslinked fracturing fluids, first used in the late 1960's, were considered a major advancement in hydraulic-fracturing technology. With linear gels, the only means to obtain increased viscosity is to increase the polymer concentration. Often, 80 to 100 lbm polymer/1,000 gal [9586 to 11 983 g/m^3] water is needed to yield the viscosities necessary to fracture-treat a well successfully with linear gel systems. Adding proppant and dispersing fluid-loss additives into such concentrated solutions of linear fluids is difficult.

The development of crosslinked fluids eliminated many of the problems that occurred when linear gels were used to fracture-treat deep, hot reservoirs. The crosslinking reaction is one where the molecular weight of the base polymer is substantially increased by tying together the various molecules of the polymer into a structure through metal or metal-chelate crosslinkers.

The earliest crosslinkers were borates and antimony metal crosslinkers. The metals are dispersed between the polymer strands, and an attraction occurs between the metals and the hydroxyl groups or, in the case of carboxymethyl derivatives, the carboxy groups.

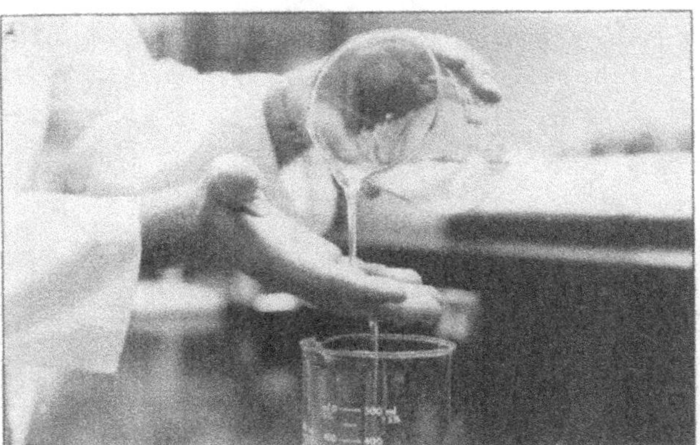

Fig. 7.10—Water viscosified with guar gum. Viscosity will vary depending on the concentration of the gel used. A loading of about 0.5% guar polymer is shown. This yields about 34 cp at 511 seconds^{-1} shear rate measured with a Model 35 Fann viscometer.

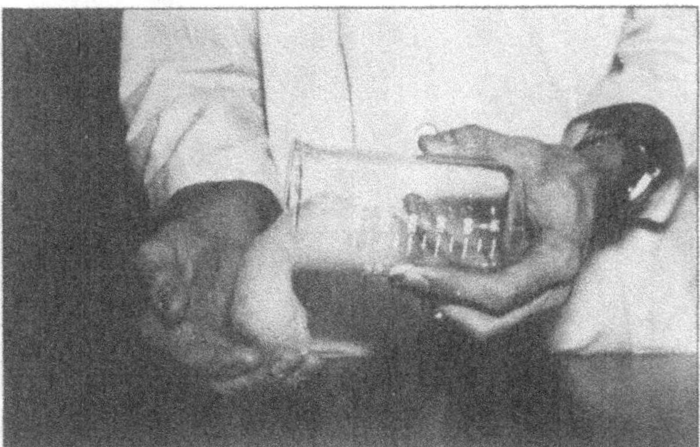

Fig. 7.11—Illustration of the rigid structure of a crosslinked gel. The gel was prepared by adding a metal crosslinker to the linear guar gel of Fig. 7.10.

This interaction takes a gel system from a true liquid (Fig. 7.10) to a pseudoplastic fluid (Fig. 7.11).

The first crosslinked fluid was a guar gum system. A typical crosslinked gel in the late 1960's consisted of 60 to 80 lbm/1,000 gal [9586 g/m^3] of guar crosslinked with antimony or borate. The antimony system was a relatively-low-pH fracturing fluid. The borate frac-turing fluid was a high-pH system, typically in the pH 10 range, while the antimony was approximately pH 3 to 5. These early crosslinked fracture-fluid systems suffered from some distinct disadvantages relating to pumpability because of the rapid crosslink reaction and the high base-gel viscosity. Both the antimony crosslinked and the borate crosslinked systems suffered quite a few field problems with incomplete gel degradation.

Incomplete gel degradation after fracturing treatments would result, at the very least, in producing back very viscous gel that could possibly carry proppant back out of the fracture. The worst scenario could result in a temporarily or, perhaps, permanently plugged fracture proppant pack. The antimony system experienced fewer gel degradation problems than the borate system because the operating pH range was such that enzyme breakers would function. Field personnel typically did not account for the effects of formation fluids in raising the pH, which thereby necessitated higher loadings of enzyme breakers.

In the early days, borate systems had degradation problems in low-temperature wells (i.e., below 125°F [52°C]). In the high pH range of the borate crosslinked gel, the standard enzyme breaker systems simply do not function. Many treatments were overflushed with acid in an attempt to break the borate crosslink. Catalyzed, oxidizer breaker systems can controllably degrade borate crosslinked fluids. Antimony crosslink is used with both guar and HPG through higher loadings of the enzyme breakers, taking into account any increase in the pH resulting from formation-water contamination.

Many other crosslinker systems have been developed, such as aluminum, chromium, copper, and manganese. Additionally, in the late 1960's and early 1970's, crosslinked CMC and some crosslinked HEC were used, although the difficulty in crosslinking HEC limited its use. With the development of the HPG and CMHEC polymers, a new generation of crosslinkers was also developed. The first and most widely used of the new crosslinkers were the titanium chelates[102] and aluminum compounds in the case of the CMHEC polymer. The use of titanium chelates was initiated in the early 1970's with HPG. Through the early 1980's, titanium chelates were the most widely used crosslinkers, although zirconium chelates found increasing use as a result of enhanced temperature stability and cost-effectiveness.

Crosslinking the polymer molecule tends to increase the temperature stability of the base polymer. It is theorized that this temperature stability is derived from less thermal agitation of the molecule because of its rigid nature and some shielding from hydrolysis, oxidation, or other depolymerization reactions that can occur.

Crosslinking of the polymer, although increasing the apparent viscosity of the fluid by several orders of magnitude, does not necessarily cause friction pressures to increase to any degree in the pumping operation. It was originally theorized that the ease in pumping of crosslinked fluids was caused by the formation of a water ring similar to that physically prepared for the Superfrac process.[133,134] This hypothesis was disproved by observing crosslinked fluids pumped down test wells and back up to the surface.

Even though crosslinked HPG systems can be pumped into deep, hot reservoirs, severe shear degradation occurs if the fluid is crosslinked at the surface and pumped at high rates down the tubular goods and through the perforations. Because of this tendency to lose viscosity permanently as a result of high shear rates, the use of "standard" crosslinked gel systems has declined dramatically. These systems have recently been replaced by delayed crosslinked fracture-fluid systems.

Delayed Crosslink Systems. A noteworthy development of the 1980's has been the use of fracturing fluid with a controlled crosslink time, or a delayed crosslink reaction.[135-137] Crosslink time is defined as the time for the base fluid to take on a rigid structure. This is evaluated in the field by many techniques. Some visually observe the fluid in a blender. The crosslink time occurs when the vortex in the blender disappears under a set shear rate. Others simply collect a sample and observe the fluid until it becomes rigid enough to become unpourable (i.e., lipping out of a jar and coming back into the jar itself) (Fig. 7.11). Obviously, crosslink time is the time required to observe a very large increase in viscosity as the fluid becomes rigid.

A significant amount of research has been performed to understand the importance of using the delayed crosslink fluid systems. Recent research results indicate that a delayed crosslink system allows better dispersion of the crosslinker, yields more viscosity, and improves fracturing-fluid temperature stability. Although dispersion is felt to be a major factor in stabilizing crosslinked gels, further research[127] has proved that the final stability of the crosslinked gel is directly related to the shear history of the gel at the time of crosslinking. The gel stability is a direct function of crosslinking at low shear rates. The popular explanation is that at low shear rates the polymer strands are uniformly laid out and crosslinking actually occurs in a very uniform, structured manner, yielding ultimately much higher viscosities and better stability to temperature, hydrolytic, and oxidative degradation.

Another advantage of delayed crosslink systems is lower pumping friction because of lower viscosity in the tubular goods. Although the crosslinked gels are pumpable down tubular goods, a certain amount of energy is required to shear the crosslink back toward the base gel, and this viscosity is exhibited as higher pumping friction; therefore, the use of delayed crosslink fluids yields a higher ultimate viscosity downhole and gives much more efficient use of

TABLE 7.2—4-HOUR BREAKER CONCENTRATION FOR BATCH-MIXED CROSSLINKED GEL (lbm or gal/1,000 gal)
Base fluid: 1% KCl water
Additives: Indicated breaker and lbm HPG/1,000 gal HPG gelling agent/1,000 gal base fluid

Temperature	20 lbm			30 lbm			40 lbm			50 lbm			60 lbm		
(°F)	E*	O**	HTO†	E	O	HTO	E	O	HTO	E	O	HTO	E	O	HTO
80	0.50	—	—	0.75	—	—	1.00	—	—	4.0	—	—	<4.0	—	—
100	0.30	—	—	0.40	—	—	0.60	—	—	<2.0	—	—	<2.5	—	—
120	0.25	—	—	0.35	—	—	0.50	—	—	0.50	—	—	2.0	—	—
140	—	0.30	—	0.30	—	—	0.15	0.50	—	—	0.50	—	1.0	1.0	—
160	—	0.15	—	—	<0.20	—	—	0.30	—	—	0.50	—	—	0.60	—
180	—	0.05	—	—	0.10	—	—	0.10	—	—	0.20	—	—	<0.20	—
200	—	<0.05	—	—	0.05	—	—	0.05	—	—	0.10	—	—	0.15	—
220			—	—	—	<0.40	—	—	<0.5	—	—	<0.5	—	—	<0.75
240							—	—	0.25	—	—	<0.25	—	—	<0.2
260							—	—	0.20	—	—	0.15	—	—	<0.10

*E = enzyme breaker.
**O = oxidizer breaker.
†HTO = high temperature oxidizer breaker.

available horsepower on location. As a result, delayed crosslink systems are used more than conventional crosslink systems. The main advantages of using a crosslinked fluid vs. a linear fluid are summarized below.

1. One can achieve much higher viscosity in the fracture with a comparable gel loading.
2. The system is more efficient from the standpoint of fluid-loss control.
3. A crosslinked fluid has better proppant-transport capabilities.
4. A crosslinked fluid has better temperature stability.
5. A crosslinked fluid is more cost-effective per pound of polymer.

If one requires high viscosity at high temperatures and deep penetration into the fracture, the ideal fluid is definitely a zirconium or titanium delayed crosslink system. If one is working with small zones, low pressure, and low temperature, however, then one can consider the use of linear fluids. The most cost-effective approach must be used to obtain the needed fracture width and to stay within the productive zone.

7.4 Oil-Based Fracturing Fluids

The most common oil-based fracturing gel available today is a reaction product of aluminum phosphate ester and a base, typically sodium aluminate.[138,139] Reaction of the ester and the base creates an association reaction, which in turn creates a sol that yields viscosity in diesels or moderate- to high-gravity crude systems. The aluminum phosphate ester gels have been improved to gel more crude oils and to enhance temperature stability. The earliest viscosified oils were napalm-type fluids of aluminum octoate. Later fluids were reaction products of caustic and tall oil fatty acids; in fact, some of these fluids are still in use.[140,141] These fatty-acid soaps, although useful as fracturing fluids, frequently cause pumpability problems. Aluminum phosphate esters can be used to create fluids with enhanced stability at high temperature and good proppant-carrying capacity for use on wells with BHT's in excess of 260°F [127°C].

Using gelled hydrocarbons is advantageous in certain situations to avoid formation damage to water-sensitive oil-producing formations that may be caused by the use of water-based fluids. If the produced crude has high enough gravity, typically above 35° [0.85 g/cm^3], then produced crude oil can be used to fracture the formation. The primary disadvantage of using gelled oil systems is the fire hazard.

In most cases, the pumping friction of an oil-based fluid is higher than a delayed, crosslinked water-based fluid system. Pumping pressures are also higher because of a lack of hydrostatic head of the hydrocarbon compared with water. Additionally, when one fractures a high-temperature well (above 260°F [127°C]), the temperature stability of a delayed, crosslinked water-based system is more predictable, and such a system is less costly than the typical oil-based fluid system.

It should also be mentioned that preparation of oil-based fracturing fluids requires a great deal of technical capability and quality control.[142] The preparation of water-based fracturing fluids is relatively straightforward by comparison. In particular, the preparation and quality control of gelling crude oil require much more care than those of water-based fluids.

7.5 Alcohol-Based Fluids

Methanol and isopropanol have been used for many years either as a component of water- and acid-based fracturing fluids or, in some cases, as the sole fracturing fluid. Alcohol, which reduces the surface tension of water, has frequently been used for the removal of water blocks. In fracturing fluids, alcohol has found

TABLE 7.3—4-HOUR BREAKER CONCENTRATION FOR BATCH-MIXED CROSSLINKED GEL (lbm or gal/1,000 gal)
Base fluid: 20% methanol in 1% KCl water
Additives: Indicated breaker and lbm HPG/1,000 gal base fluid
HPG gelling agent/1,000 gal base fluid

Temperature	20 lbm			30 lbm			40 lbm			50 lbm			60 lbm		
(°F)	E*	O**	HTO†	E	O	HTO	E	O	HTO	E	O	HTO	E	O	HTO
80	<1.50	—	—	<1.50	—	—	2.00	—	—	<2.25	—	—	<4.00	—	—
100	1.15	—	—	<1.00	—	—	1.85	—	—	<2.00	—	—	<3.00	—	—
120	<1.25	<3.00	—	0.75	5.00	—	<0.75	<7.00	—	1.75	5.00	—	<3.00	<6.00	—
140	<1.25	<2.00	—	<0.75	<3.00	—	1.00	5.00	—	<1.75	<5.00	—	<3.00	<5.00	—
160	—	2.00	—	—	<2.00	—	—	3.00	—	—	3.50	—	—	3.00	—
180	—	0.85	<3.00	—	1.00	—	—	2.50	—	—	2.50	—	—	1.50	—
200	—	—	<1.00	—	—	<4.00	—	—	<7.00	—	—	<7.00	—	—	<7.00
220	—	—	0.75	—	—	<1.00	—	—	<1.50	—	—	<4.00	—	—	<2.00
240	—	—		—	—	0.75	—	—	<1.00	—	—	<1.25	—	—	<1.50
260				—	—	0.25	—	—	<0.30	—	—	<0.50	—	—	<0.60

*E = enzyme breaker.
**O = oxidizer breaker
†HTO = high-temperature oxidizer breaker.

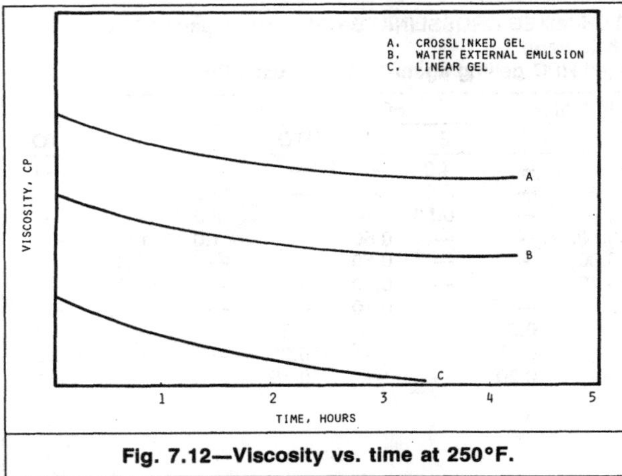

Fig. 7.12—Viscosity vs. time at 250°F.

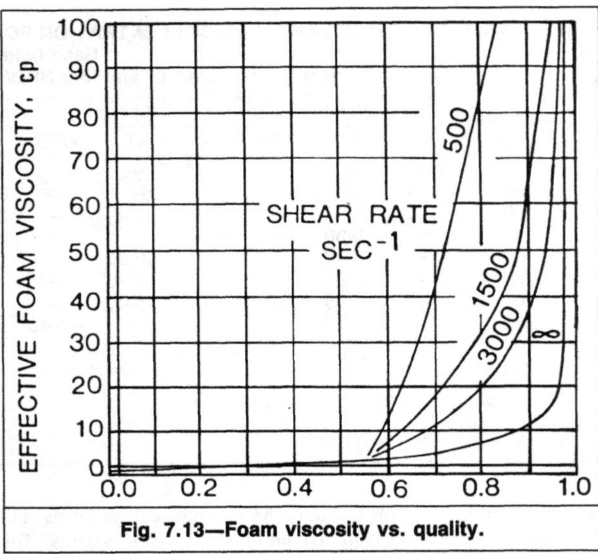

Fig. 7.13—Foam viscosity vs. quality.

wide use as a temperature stabilizer because it acts as an oxygen scavenger. Polymers are available that will viscosify pure methanol or isopropanol. These polymers include hydroxypropylcellulose and HPG, which is substituted frequently. Guar gum will viscosify up to 25% methanol or isopropanol, but above this, the guar will precipitate. HPG will gel up to approximately 60% methanol.

In a water-sensitive oil reservoir, hydrocarbon-based fluids are generally preferred over alcohol-based fluids. Use of alcohol-based fluids creates several drawbacks, especially the inherent danger to personnel who breathe alcohol fumes and the ever-present danger of combustion. As fracturing fluids, methanol-based fluids, particularly at higher concentrations, present difficulty in the controlled degradation of the base fluid. Very high concentrations of any type of breaker are required for complete degradation. Tables 7.2 and 7.3 illustrate the necessary breaker concentrations of a typical crosslinked gel and one containing 20% methanol.

Numerous papers have described the benefits of methanol in water-based linear and crosslinked fracturing fluids and in acid-based fracturing fluids.[143-147] The primary benefits relate to low surface tension, miscibility with water, removal of water blocks, and compatibility with formations that are water-sensitive.

7.6 Emulsion Fracturing Fluids

Emulsion fracturing fluids have been used for many years. In fact, some of the first oil-based fluids were oil-external emulsions. These products had many drawbacks, and their use was greatly limited because of extremely high friction pressure resulting from their high inherent viscosity and lack of friction reduction. Water-external emulsion fracturing fluids were introduced in the mid-1970's.[147] These fracturing fluids, although yielding somewhat higher friction pressure than comparable water-based gels, were indeed a breakthrough in our industry and continue to be used widely as a very cost-effective, functional fracturing system. These water-external emulsions require only one-third the quantity of gelling agent, surfactant, and other additives, with two-thirds oil making up the bulk of fluid.

The cost-effectiveness of an oil emulsion implies that the load oil can be produced back and sold. A water-external emulsion was a very popular fluid when crude oil and condensate sold for \$3 to \$5/bbl [\$19 to \$31/m^3]. The use of oil-in-water emulsions has decreased recently with the increased cost of crude oil. An oil-in-water emulsion has good fluid-loss control, exhibits excellent proppant-carrying capacity, and tends to clean up very well.

The emulsion is broken in the formation when the surfactant that created the emulsion is absorbed into the formation. Water-external emulsions are relatively simple fracturing fluids that are easy to mix and pump into most reservoirs.

The two basic types of oil/water emulsions are oil external and water external. An oil-external emulsion is a two-phase system where oil is the continuous phase and water is emulsified in the oil. A water-external emulsion is one where water is continuous and oil is the discontinuous phase. In an oil-external emulsion, the fluid has viscous properties very similar to the base oil. As a result, oil-external emulsions yield high friction pressures related to high oil viscosity.

On the other hand, because of the low viscosity of water compared with oil, water-external emulsions have lower friction pressures. Also, there is some tendency to achieve friction reduction with the polymers in the water phase of water-external emulsions. Attempts have been made to place friction reducers in the oil phase of an oil-external emulsion, but these efforts have met with little or no success. The pumping pressures of the water-external emulsions, however, are somewhat higher than for typical, conventional crosslinked fracturing fluids but much lower than the oil-external emulsions. The water-external-emulsion friction properties seem to lie between the friction properties of linear gels and crosslinked fluids.

Fig. 7.12 presents a comparison of the viscosities of linear gels, emulsion fluids, and crosslinked fluids at various temperatures. The water-external emulsions are relatively cost-effective fluids for use when good viscosity is needed but the temperature stability or the ultrahigh viscosity of crosslinked fracturing fluids is not. One needs to take into account the possibility that the load oil might not be recovered in very-low-pressure or depleted reservoirs and that pumping any oil-based system has potential hazards.

7.7 Foam-Based Fluids

The introduction of foam fracturing fluids was a significant achievement in fluid technology,[148-166] but foam was not widely used as a fracturing fluid until the mid-1970's. Foam fracturing fluids are simply a gas-in-liquid emulsion. The gas bubbles provide high viscosity and excellent proppant-transport capabilities.

Stable foam has viscous properties similar to a gelled, water-based fluid. Fig. 7.13 illustrates that the volume of gas necessary to create a stable foam is approximately 60 to 90% of the total volume at a given pressure and temperature. As Fig. 7.13 shows, foam stability and viscosity increase as foam quality increases from 60 to 90%. The foam reverts to a mist above 90%. It is extremely important to stay within the stable foam-quality range during treatment. Typical fracturing treatments are designed to achieve a 70-75-80-quality foam. This means that 70, 75, or 80% of the fracturing fluid is gas.

Gas bubbles are created by turbulence when liquid and gas are mixed. The bubbles emulsified in the liquid create a foam that will break out slowly with time. The half-life of a foam is the time necessary for one-half of the liquid used to generate the foam to break out of the foam under atmospheric conditions.

The gas-in-water emulsion can be stabilized by adding a surfactant to coat the gas bubbles. The addition of polymers to the liquid also affects foam stability.

A foam mixed to a quality of 70 to 80% with a good-quality foamer but without such stabilizers as guar, HPG, or xanthan

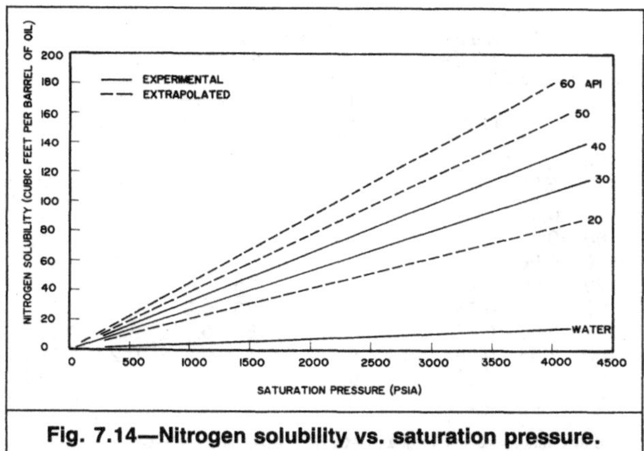

Fig. 7.14—Nitrogen solubility vs. saturation pressure.

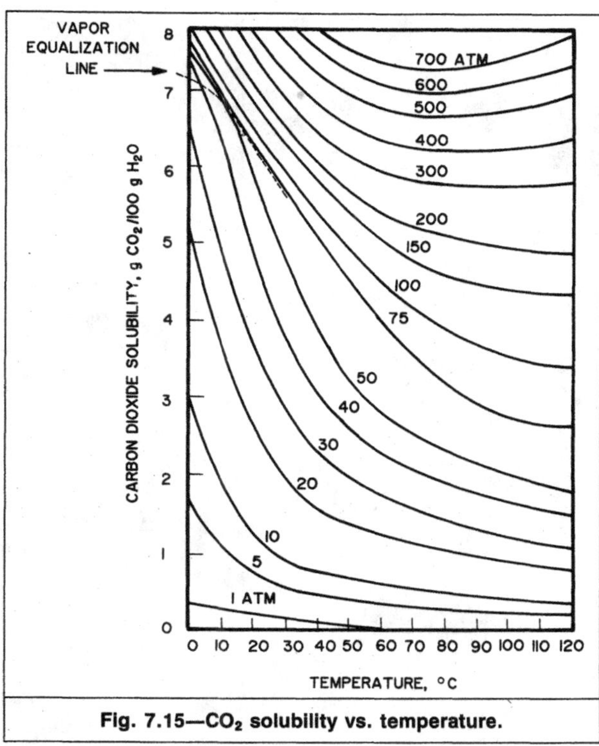

Fig. 7.15—CO_2 solubility vs. temperature.

generally yields a half-life of 3 to 4 minutes. Many early fracturing treatments, as well as some current treatments, were conducted with this type of foam. Addition of polymer stabilizers increases half-life to 20 to 30 minutes. Half-life measurements are used only as qualitative indicators of foam stability in the laboratory. In the fracture and under high pressure, foam half-life is much longer than measured at atmospheric conditions.

When CO_2 is used as the energizing medium for foam, liquid CO_2 is pumped instead of dry N_2 gas. What is formed at the point of intersection is not a true gas/liquid foam. The resulting emulsion turns into a foam only if the liquid CO_2 turns into a gas at reservoir conditions. When N_2 is used to foam a fluid, however, a true foam exists whenever the quality is between 60 and 90%.

Using foam as a fracturing fluid has several advantages. The two most obvious are minimizing the amount of fluid placed on the formation and improving recovery of fracturing fluid by the inherent energy in the gas. In preparing a foam, one typically uses 65 to 80% less water than in conventional treatments. The inherent energizing capabilities of the fluid caused by entrained gas assist in rapid cleanup or simply promote cleanup in low-pressure formations.

However, using foam as a fracturing fluid has several disadvantages. Much more care must be taken in running a foam fracturing treatment from a mechanical point of view. Small variations in the water or gas mixing rates can cause the loss of foam stability. N_2 foam is not very dense; therefore, pumping pressures will be large compared with gelled water. Another disadvantage of foam is that it is very difficult to get high sand concentrations in foam fracturing.[164] The highest achievable sand concentration downhole is typically about 8 lbm/gal [957 kg/m^3] because all the sand must be added to about one-quarter of the fluid. For example, to run a 75-quality foam, one has to run 20 lbm/gal [2397 kg/m^3] in the liquid phase to achieve a 5-lbm/gal [599-kg/m^3] slurry concentration.

Virtually any liquid can be foamed. One can foam methanol, methanol/water mixtures, hydrocarbons, and water. Foam fracturing is a technique that has many inherent advantages and disadvantages; therefore, it is the ideal fluid in certain formations, while it should not even be considered in others. Recent improvements in equipment to handle high sand concentrations and in metering capabilities of gas and liquid have made foam fracturing more controllable. The greatest application for foam fracturing is probably in shallow, low-pressure wells that require an energized fluid or in those wells that are so water-sensitive that foam must be used.

7.8 Energized Fracturing Fluids

The use of high-pressure hydrocarbon gases to assist in acid-treatment flowback is documented as far back as the 1950's. The use of CO_2 and N_2 was initiated in the late 1960's.[167-174] The advantages of energizing fracturing fluids are quite obvious, particularly for a formation with low bottomhole pressure. The energy imparted by the gases enables more rapid removal of the stimulation fluid and may be cost-effective compared with long-term swabbing or pumping operations. Previous research has shown that entrained gas is also beneficial for fluid-loss control. The incorporation of inert gases into a fracturing fluid will yield proportionally better fluid efficiency than the same fluid without the entrained gases.

The type of gas used for energizing a fracturing fluid should be considered carefully. N_2 is, of course, an inert gas. When N_2 is added in small amounts and without a surfactant, one simply has an additive that is totally inert and relatively immiscible in the fluid, as illustrated in Fig. 7.14. The use of CO_2, however, introduces a reactive component to the fracturing fluid. As Fig. 7.15 illustrates, CO_2 is quite soluble in water. (Note the difference in the scales of the axes in Figs. 7.14 and 7.15.) CO_2 can be converted to carbonic acid, which may be incompatible with fracturing fluids.

The solubility of CO_2 in treating fluids and reservoir fluids can be advantageous when this gas is used in a stimulation treatment. As the solubility curve in Fig. 7.15 shows, a great deal of the CO_2 will go into solution at typical reservoir conditions. Also, dissolved gas does not easily dissipate into the formation, as might be the case with a less-soluble gas, such as N_2. The net result is that the stored energy is maintained at the most advantageous location. When the pressure is subsequently reduced during flowback, the dissolved gas will begin to evolve from the mixture and to impart a solution-gas drive to the treating fluid. This gas-drive phenomenon results in effective removal of the treating fluids from the reservoir.

It is therefore imperative that a fluid commingled with a gas should be flowed back as quickly as possible. Field experience has shown that up to 4 hours can be allowed for shut-in using nitrified treatments without a great loss of the energizing medium into the formation. The obvious advantages and disadvantages to using CO_2 and N_2 should be weighed and the relative cost-effectiveness compared before their use.

7.9 Fracturing-Fluid Additives

The first fracturing treatments contained only gasoline, napalm, and sand. Modern fracturing fluids are more complex. It is not uncommon today to have as many as seven or eight different additives in a typical fracturing fluid. These additives are needed in most cases. One needs, however, to verify the relative compatibility of the various additives. Fig. 7.16 graphically shows chemical incompatibility of two additives. One loses not one but both additives and can endanger the treatment or the well by not having the proper

Fig. 7.16—An example of chemical incompatibility. The flocculant precipitant indicates a reaction between incompatible components.

additives present. Pretreatment tests by the service company should negate potential compatibility problems.

Biocides. Virtually no water-based fracturing fluid should be pumped into a formation without some type of biocide present.[175] Biocides are used to eliminate surface degradation of the polymers in the tanks. A more important purpose is that properly designed biocides will stop the growth of anaerobic bacteria in the formation. Many formations have turned sour because of the growth of *Desulfovibrio* bacteria, which create hydrogen sulfide and turn the formation crude sour. Biocides should be added to fracturing fluids both to maintain gel stability on the surface and to protect the formation from bacterial growth. The proper procedure for adding biocides is to add at least half of the biocide into the tank before it is filled with water. This will give a rapid kill or a concentrated kill of any bacteria in the tank. When the tank is full, the rest of the biocide should be added, and then about 6 to 8 hours should be allowed before gelling up with the selected polymer. Most biocides require some time to achieve a kill.

Breakers. A breaker is an additive that enables a viscous fracturing fluid to be degraded controllably to a thin fluid that can be produced back out of the fracture.[176,177] All breakers used today are internal breakers; i.e., they are incorporated into the fracturing fluid at the surface. In the early days of fracturing, some attempts were made to break a viscous fluid with external means. These attempts were unsuccessful because the viscous fluids themselves act as diverting agents. Currently used breaker systems include enzymes and catalyzed oxidizer breaker systems for low-temperature (70 to 130°F [21 to 54°C]) applications. Conventional oxidizer breaker systems are used for a temperature range of 130 to 200°F [54 to 93°C], and delayed activated oxidizer systems are applicable for temperatures from 180 to 240°F [82 to 116°C]. Weak organic acids

TABLE 7.4—LINEAR GEL BREAK SCHEDULE FOR 24-HOUR SHUT-IN TIME*

BHT	Concentration of Breaker (lbm/1,000 gal fluid gelling agent/1,00 gal)							
	20		40		60		80	
(°F)	E**	O†	E	O	E	O	E	O
60	0.4	—	0.7	—	1.2	—	3.0	—
80	0.1	—	0.2	—	0.4	—	1.0	—
100	0.06	—	0.1	—	0.25	—	0.5	—
120	0.06	—	0.1	—	0.175	—	0.3	—
130	0.06	0.25	—	0.5	0.175	0.8	0.25	1.0
140	0.06	0.175	0.1	0.375	0.15	0.375	0.2	0.55
160	0.06	0.08	0.1	0.2	0.15	0.375	0.2	0.55
180	—	0.075	—	0.15	—	0.275	—	0.425
200	—	0.075	—	0.15	—	0.2	—	0.3

*Break data applicable for guar, HPG, HEC, CMC, and CMHEC.
**E = enzyme.
†O = oxidizer.

are sometimes used as breakers at temperatures above 200°F [93°C].

All the breaker systems are used to degrade the polymers in water-based fracturing gels. Examples of breaker-loading schedules for enzymes, oxidizers, and controlled oxidizer systems are shown in Tables 7.2 through 7.4. Note that one of the most critical factors relating to breaker mechanisms is the pH of the fracturing fluid. Most enzyme breakers will function only between pH 3 and 8, with an optimum at pH 5. Below pH 3 and above 8, the effectiveness of the enzyme breaker is greatly reduced. If enzyme breakers are used in high-pH or very-low-pH fluids, one will have a very serious problem with gel degradation. Enzyme breakers actually break the molecular chains and thereby effectively lower the molecular weight.

Oxidizer breaker systems will function from pH 3 through 14. Oxidizer systems also work by breaking the molecular structure of the polymer. In fluid systems where catalyzed oxidative, conventional, or oxidative systems can be used, the break achieved is superior to that of the enzyme breaker systems. Weak organic acids have also been used, but if the acid contacts carbonate in the reservoir, the breaker will react with the formation rather than with the fluid.

In oil-based gel systems, typical breakers are bicarbonate, lime, and/or water solutions of amines. Weak acids have been used with limited success to degrade the system. Oil gel breaker works by addition of an acid or a base that dissolves slowly in the fluid such that the reaction is forced one way or the other, breaking the gel system. The amine system is a proprietary technique where free-radical generation occurs. Water must be present. It is an extremely critical reaction used only in low-temperature applications.

Pilot tests on breakers should be conducted before they are incorporated into a fracturing treatment. Omission of breakers in the early stages of fracturing treatments, because they lower the stability of the gel and might cause a screenout, may result in an unbroken gel, which will plug off the fracture system. This is particularly the case if the temperature is inadequate to cause gel degradation. It is imperative that breakers be included throughout a treatment in such reservoirs. Breakers can be run at low concentrations in the early stages of a treatment and increased at later stages to enhance breaking and flowback.

Buffers. Common buffering agents are used in fracturing fluids to control the pH for specific crosslinkers and crosslink times.[142] They also speed up or slow down the hydration of certain polymers. Typical products are sodium bicarbonate, fumaric acid, combinations of mono- and disodium phosphate, soda ash, sodium acetate, and combinations of these chemicals. Another and perhaps more important function of a buffer is to ensure that the fracturing fluid is within the operating range of the breakers or degrading agents. As mentioned earlier, some breakers simply do not function outside specific pH ranges. By applying a buffer rather than a strong acid or strong base, one is able to maintain a pH range even though contaminants from formation water or other sources tend to change the pH. Other instances where buffers are useful include counteracting fracture-tank contamination or compensating for water delivered to the location that has a high concentration of carbonate, bicarbonate, or other mineral that affects the pH. By using buffers, one is able to prepare quality fracturing fluids that will hydrate and degrade properly.

Surfactants and Nonemulsifiers. A surfactant (surface-active agent) can be defined as a molecule that seeks out an interface and has the ability to alter the prevailing conditions.[177,178] A surfactant is almost always composed of two parts: a long hydrocarbon chain that is virtually insoluble in water but soluble in oil and a strongly-water-soluble tail. Because there is partial solubility in oil and water, the surfactant will tend to accumulate at the interface of these fluids.

The water-soluble portion of the molecule may be ionically positive (cationic), negative (anionic), or mixed (amphoteric). The ionic charge of the various surfactants used in oilfield stimulation is important in terms of wettability imparted to a given formation. The inherent ionic characteristics of particular formations cause cationic surfactants to leave carbonates water-wet and sandstones oil-wet. Anionic surfactants tend to leave sandstones water-wet and limestones oil-wet. Amphoteric surfactants are organic molecules whose ionic charges depend on the pH of the fluid. Almost all formations are naturally water-wet, which favors oil movement through the rock. Because the water-wet condition is preferred, the ionic nature of the surfactant is an important consideration, and one should be aware of the charge of a surfactant in its selection. It is generally inadvisable to mix cationics with anionics because of the possibility of forming precipitates. Because a large number of formations throughout the world are heterogeneous, limy sands or sandy limes, it is often useful to select a nonionic surfactant, provided that it meets certain nonemulsification criteria.

An emulsion consists of two immiscible fluids, in which one phase exists as fine droplets dispersed throughout the other phase. Oilfield emulsions are either oil in water (where oil droplets exist in the continuous water phase) or water in oil (where oil is the continuous phase). The viscosity of an emulsion can vary from several to several thousand centipoise. If an emulsion is created near the wellbore, severe production blockage may occur.

Because of their surface-active nature, surfactants can act as de-emulsifiers or as emulsifiers. Effectiveness of a surfactant as a de-emulsifier in a particular crude-oil/water system must be determined experimentally. Tests should be run according to specifications set out in API *RP-42*[24] to determine the proper type and concentration of surfactants required to prevent emulsification of a particular crude with a treating fluid. The surfactant should maintain its surface activity at reservoir temperatures and should not be easily stripped out of solution by adsorption from contact with the reservoir rock.

As discussed earlier, some fracturing fluids are composed of hydrocarbon and water that are emulsified to build fluid viscosity. When emulsified fluids are used, it is desirable for the surfactant to adsorb on the formation so that the emulsion will break.

Surfactants are also used to prevent or to treat near-wellbore water blocks. Although not as severe as emulsions, a water block can impair production. Surfactants lower the surface tension of the water and reduce capillary pressure, which results in lower energy required to move the water across boundaries and through the formation matrix.

Another form of well damage that may be treated by surfactants is blockage by fines.[179] Fines can be silts, clay minerals, or drilling-fluids solids. If a surfactant that wets the individual fine particles is used in the fracturing fluid, the particles can be removed from the formation more easily when the broken fracturing fluid is produced back.

Fluorocarbon Surfactants. Fluorocarbon surfactants have been used in oilfield treatments for many years.[22] Fluorocarbon surfactants are very similar to hydrocarbon surfactants, except that in the oil-soluble half of the molecule, hydrogen atoms attached to the carbon chain are replaced by fluorine atoms. The water-soluble portion of the molecule is effectively unchanged. The fluorocarbon products can be classified as cationic, anionic, nonionic, or amphoteric in the same manner as the hydrocarbon surfactants. Fluorocarbon surfactants have certain advantages over hydrocarbon surfactants: they typically are much more surface-active, and therefore tend to yield lower surface tension at equal concentrations and are useful at low concentrations.

In addition to reducing surface tension, fluorocarbon surfactants alter the contact angle at the surface of the pore space. The LaPlace-Young equation can be used to explain the consequence of a contact-angle change on capillary pressure:

$$P_c = \frac{2\sigma \cos \theta_c}{r_p}.$$

This equation states that the pressure difference across an interface in a capillary is a function of two times the surface tension, σ, times the cosine of the contact angle, θ_c, over the radius of the pore, r_p. If one could achieve a 90° [1.6-rad] contact angle, one could produce zero capillary pressure. Fluorocarbon surfactants, in addition to reducing the surface tension, tenaciously adsorb onto the wall of the pore space and cause the contact angle to approach 90° [1.6 rad]. By effectively reducing capillary pressure to near zero, one can produce the wetting-phase fluid, usually water, from the formation and fracture more easily.

Clay Stabilizers. Laboratory studies and field results have indicated that clays and fines present in producing formations may reduce stimulation success. The percentage of clays present may not be as important as the type and location of clays. Kaolinite, illite, and chlorite are the most common types found in sandstone reservoirs. These clays are typically nonswelling, particularly in the presence of potassium chloride water. Quite often, however, they are interspersed with lesser amounts of smectite and mixed-layer clays that are not particularly stable. The introduction of fracturing fluids or a change in temperature, pressure, or ionic environment may cause the clays to become dislodged and to migrate through the pore system of the rock.

As the particles migrate, they may bridge in narrow pore throats and seriously reduce permeability. Once permeability is impaired, specific steps must be taken to repair the damage. Another form of permeability impairment is clay swelling, which reduces the permeability in a formation. Susceptibility of a formation to damage by clay swelling and particle migration appears to depend on the following characteristics: (1) clay content; (2) clay type; (3) clay distribution; (4) pore-size and grain-size distribution; and (5) amount and location of cementing materials, such as calcite, siderite, or silica. Susceptibility to damage may be evaluated through X-ray diffraction, a scanning electron microscope, and thin-section point counting. Damage can be mitigated through the use of clay-stabilizing agents. The common clay-stabilizing agents are as follows.

Potassium chloride (KCl) prevents the dispersion of clay particles by providing sufficient cation concentration to prevent leaching of the exchangeable cations present and keeps individual platelets of the stacked clay particles in a coagulated or condensed state. KCl does very little to prevent migration and provides no residual protection against dispersion by subsequent contact with low-salinity water. KCl is currently the most commonly used antiswelling agent.[8,9] Virtually all treatments in sandstone reservoirs are designed to contain KCl, and it is even used in limestone reservoirs that contain sandstone intervals containing clays.

Ammonium chloride behaves like KCl in preventing clay swelling. It typically is not used in fracturing operations but finds some use in hydrofluoric acid treatments.

Calcium chloride functions like KCl and ammonium chloride. It readily forms precipitates in the presence of high-sulfate or high-alkalinity formation water; however, it appears to be useful in high-methanol/water solutions where KCl and ammonium chloride have limited solubility.

Upon dilution in water, *zirconium salts,* particularly zirconium chloride, form a complex inorganic polymer containing hydroxyl bridging groups.[180,181] The highly charged nature of these polymers causes them to adsorb onto the clay surfaces in an irreversible fashion and may bond the clay particles to the sand grain surfaces. This particular clay stabilizer may be applicable in preflushes ahead of fracturing treatments.

TABLE 7.5—FLUID-LOSS ADDITIVES

Oil Based	Water Based
Adomite Mark II	Silica flour
Silica flour	Adomite aqua
Adomite Aqua	Mixture gum and oil-soluble resin
Lime powder	Mixture gum and talc
Sodium bicarbonate powder	1 to 5% diesel
N_2/CO_2	0.05 to 1% aromatics and surfactant
100-mesh salt or sand	N_2/CO_2
	100-mesh sand, oil-soluble resin, salt, or benzoic acid

Certain *modified polyamines* perform two functions: they enhance the clay-swelling control obtained with KCl and prevent the migration of fines.[182] These products chemically adsorb onto clay particles and thereby keep them in a compact or undispersed state. They may be useful in preventing sloughing and fines generation of the fracture faces during high flow rates in fracturing and flow-by. These products lack the duration of protection of polymeric clay stabilizers, but they do not plug pore spaces the way that high-molecular-weight polymeric clay stabilizers do.

Polymeric clay stabilizers are cationically charged high-molecular-weight polymers that tend to adsorb onto the surface of clays tenaciously, tying them down and negating any fines migration or swelling.[183-186] They need to be applied with care because over-treatment can plug the pore spaces. They are relatively permanent once in place, and some success has been achieved with these products, particularly when they are combined with KCl.

Polymeric solutions of hydroxyaluminum, which is adsorbed tightly onto clay mineral surfaces, may be useful in negating particle migration or clay swelling.[187-191] The requirements of particle overflush and a somewhat lengthy curing time have limited its use somewhat in stimulation applications.

7.10 Fluid-Loss Additives

In the early days, when oil-based fluids were normally used, an excellent fluid-loss additive was developed.[44] This additive, a tall-oil derivative (Adomite Mark II™), gives excellent fluid-loss control when used with soap-type fluids or an oil-based fluid without viscosifier. It cannot be used in the napalm- or aluminum-ester-type gels because of incompatibility problems. For these systems, common non-oil-soluble fluid-loss agents are used.

The most common water-based fluid-loss additive consists of very finely ground silica flour. A related product uses nonswelling clays, silica flour, and guar gum. Another uses an oil-soluble resin and a swellable gum. Mixtures of vegetable compounds, talc, silica flour, and guar gum are used in another product.[191-194] These products tend to plug the face of the fracture, with very little penetration into the formation matrix.

The fluids having the best fluid-loss control were the early guar gums, which had high residues combined with silica flour. Also used successfully was a combination of talc and swellable gums with guar. This mixture provided an excellent fluid efficiency almost independent of permeability. Clean fluids, such as HEC or CMHEC, cannot be treated adequately to give excellent fluid-loss control in high-permeability formations. In fact, many formations of moderate permeability can be severely damaged by use of such clean fluids. To achieve excellent fluid-loss control, one must have not only a bridging material but also a wall-building material. Nonresidue fracturing fluids generally do not have wall-building properties.

More recent efforts to control fluid loss have used either diesel fuel at concentrations up to 5%, or lesser concentrations of aromatic hydrocarbons with surfactants that yield a microemulsion.[54,193] This technique appears to give better fluid-loss efficiency for fracturing fluids used in formations with permeabilities less than 1 md. The fluid-loss control achieved by microemulsions with diesel oil or aromatic hydrocarbons is less efficient in moderate- to high-permeability reservoirs.

Use of very clean fluids (the HPG or cellulose derivatives) often entails adding large concentrations of fluid-loss additives to achieve even moderate fluid-loss control. Because fluid-loss control is a requisite for effective fracturing, use of such clean fluid appears to be somewhat self-defeating.

TABLE 7.6—DIVERTING AGENTS

- Coarse rock salt
- Graded rock salt
- Graded paraformaldehyde
- Flake benzoic acid (fine and coarse)
- Graded oil-soluble resin (low temperature)
- Graded oil-soluble resin (high temperature)
- Ball sealers
- Solution of benzoic acid in alcohol or hydrocarbon*
- Unibeads (was beads)
- Crosslinked polymers
- Slurries of oil-soluble resins*
- Mixture of karaya and oil-soluble resin
- Karaya powder*
- Graded naphthalene
- Oil-external emulsion
- High-concentration linear gel
- Oyster shells
- Polymer-coated sand
- Buoyant particles
- High-quality foams
- Flake boric acid

*Primary function in matrix application.

For many applications, conventional guar gum systems with inexpensive silica flour may be the most cost-effective fluid-loss additive to achieve efficient fracturing fluids. Table 7.5 illustrates many common fluid-loss additives.

Foamers. In the early days of foam fracturing, commercial products that had been used as sudsing agents were used as foam stabilizers and, in fact, performed adequately. As the industry moved toward higher-temperature formations, higher sand concentrations, and larger job sizes, efforts were made to create much more efficient and cost-effective foaming equipment and agents.[152,165] Foamers are now available for virtually any base fluid from fresh water to high-brine fluids contaminated with large amounts of hydrocarbons to water/alcohol mixtures varying from 0 to 100% methanol.

Virtually any base fluid can be foamed with a temperature-stable foaming agent. However, nonionic water-soluble surfactants and fluorocarbon surfactant in hydrocarbons often suffer from cloud-point problems at elevated temperatures. Thus, it is desirable to determine that there is no problem with stability of the foamer during the treatment.

Common stabilizers for foaming treatments include the basic guar, HPG, and xanthan gums. Such materials are added to the fracturing fluid to increase the foam half-life, particularly at elevated temperatures. In this application, the stabilizer must be relatively immune to thermal degradation at BHT's.

In systems that use delayed crosslinkers, it is not uncommon to crosslink the gelling agents used as stabilizers for the foam. Researchers report enhanced stability and viscosity and longer half-lives for such systems under ambient conditions than normally seen when the gelling agents are not crosslinked.[195]

Friction Reducers. Virtually all polymers act as turbulence-suppression agents in the presence of low-viscosity base fluids. When pumped at high rates down small tubular goods, low-viscosity water or hydrocarbon fluids tend to achieve high turbulence, which translates into high friction pressure.[60-74] When high-molecular-weight polymers are added to these fluids, dramatic decreases in pumping friction are seen because of turbulence suppression.

Turbulence suppression is thought to be achieved by an ordering of the fluid through the use of the high-molecular-weight polymer chain and its inherent affinity for water molecules. The long-chain polymer deters turbulence by controlling migration of the individual water molecules, thereby eliminating much of the disorder and turbulence. Its effect is dramatic when the very efficient polyacrylamide materials with friction reduction are used in highly turbulent flow. Low concentrations of guar or HPG (10 to 20 lbm/1,000 gal [1198 to 2397 g/m^3]) are used commonly in fracturing systems today because of their relatively low cost and accessibility. The most efficient and cost-effective friction reducers used for fracturing fluids are low concentrations of polymers and copolymers of acrylamide. These friction reducers are applicable in water and acid systems. Of course, one needs to select the properly charged acrylamide for water- and acid-based systems. Cationically charged species are typically used for acid systems because of the presence of positively charged acid inhibitors. In fresh or 2%-KCl water, negatively charged species function with high efficiency. Slightly anionically charged copolymers of acrylamide are both compatible and functional in acids. The key to selection is to test the friction reducer in the fluid to be pumped. It must have functionality and be compatible with all additives present.

Friction is often reduced in oil-based systems with a high-molecular-weight polyisodecylmethacrylate. This friction reducer for oil-based systems is readily available. Although functional, it will seldom achieve friction reduction higher than 70%. Another way to reduce friction in hydrocarbon fluids is to use low concentrations of an aluminum phosphate ester gel (e.g., 2 gal/1,000 gal [0.002 m^3/m^3]) and a sodium aluminate activator. These systems may be even more effective than the polyisodecylmethacrylate friction reducers.

Friction reducers offer no advantage unless the fluid is to be transported in turbulence. Thus, a fairly viscous oil to be pumped down casing at a low rate offers little opportunity for friction drop reduction. Similarly, an already viscosified fracturing fluid can seldom have its friction reduced further by addition of an acrylamide friction reducer. One has to achieve high turbulence for the friction reducer to be advantageous, and neither the low-rate casing treatment nor the viscous fluid can be assisted by friction reducers.

Temperature Stabilizers. Temperature stability of fracturing fluids is basically a result of the stability of the base chain polymer, the pH of the fracturing fluid, and/or the presence of oxidizing agents. It is known that HPG is more stable than guar and that some of the acrylamides or cellulose derivatives may be more stable than the HPG. None of these products is particularly stable in acid media because of hydrolytic degradation. Therefore, one means of stabilizing a fracturing fluid is to increase the pH into the basic range. Typical pH's for many fracturing fluids are from 8 to 10. The higher pH yields enhanced stability simply by eliminating hydrogen ion in the fluid.

Another basic use for temperature stabilizing is to remove free oxygen from the system. A temperature stabilizer commonly used for this purpose is sodium thiosulfate.[196] It is used as an oxygen scavenger to remove oxidative degradation as a means of breaking down the fracturing fluid. Another temperature stabilizer with the same function is methyl alcohol. Methanol becomes an oxygen scavenger at high temperatures and functions in a solvent-change relationship to give a temperature-stabilizing effect. Methanol is not as cost-effective an oxygen scavenger as sodium thiosulfate, thiourea, and others.

Diverting Agents. Table 7.6, a partial listing of diverting agents, shows that a diverting agent is typically a graded material that is insoluble in fracturing fluids but soluble in formation fluids. Also included are slurries of resins, viscous fluids, and crosslinked fluids. Ball sealers are another type of diverting agent.

The major purpose of a diverting agent is to divert flow of the fracturing fluid to a zone below or above the zone being treated by plugging off either the perforation (if a cased-hole completion) or some part of the formation (if an openhole completion). Some of the products on the list are more applicable to matrix diversion but have been used in fracturing situations.

The most effective diversion is by ball sealers or zone isolation through packers. Particulate diverters, such as flake benzoic acid, rock salt, or other materials, have apparently been used with success in some areas.[197] It should be apparent that both the concentration and type of diverting agent are critical. A major consideration is for the diverting agent to be compatible with the fracturing fluid or for the diverting agent to be run in a spacer fluid between the stages of the fracturing treatment.

References

1. Clark, J.B.: "A Hydraulic Process for Increasing the Productivity of Oil Wells," *Trans.*, AIME (1949) **186,** 1–8.
2. Clark, R.C. *et al.*: "Application of Hydraulic Fracturing to the Stimulation of Oil and Gas Production," *Drill. & Prod. Prac.*, API (1953) 113–22.
3. Wilsey, L.E. and Bearden, W.G.: "Reservoir Fracturing—A Method of Oil Recovery from Extremely Low Permeability Formations," *Trans.*, AIME (1954) **201,** 169–75.
4. Waters, A.B.: "Hydraulic Fracturing—What Is It?" *JPT* (Aug. 1981) 1416.
5. Hubbert, M.K. and Willis, D.G.: "Mechanics of Hydraulic Fracturing," *Trans.*, AIME (1957) **210,** 153–66.
6. Scott, P.P. Jr., Bearden, W.G., and Howard, G.C.: "Rock Rupture as Affected by Fluid Properties," *Trans.*, AIME (1953) **198,** 111–24.
7. Howard, G.C. and Fast, C.R.: "Optimum Fluid Characteristics for Fracture Extension," *Drill & Prod. Prac.*, API (1957) 261–70.
8. Black, H.N. and Hower, W.E.: "Advantageous Use of Potassium Chloride Water for Fracturing Water Sensitive Formations," paper 851-39-F presented at the 1965 Mid-Continent District Meeting, API Div. of Production, Wichita, KS, March 31–April 2.
9. Smith, C.F., Pavlich, J.P., and Solvinsky, R.L.: "Potassium, Calcium Treatments Inhibit Clay Swelling," *Oil & Gas J.* (Nov. 30, 1964) 80.
10. Reed, M.G.: "Stabilization of Formation Clays with Hydroxy-Aluminum Solutions," *JPT* (July 1972) 860–64; *Trans.*, AIME, **253.**
11. Monaghan, P.H. *et al.*: "Laboratory Studies of Formation Damage in Sands Containing Clays," *Trans.*, AIME (1959) **216,** 209–15.
12. Jones, F.O. Jr.: "Influence of Chemical Composition of Water on Clay Blocking of Permeability," *JPT* (April 1964) 441–46; *Trans.*, AIME, **231.**
13. Hewitt, C.H.: "Analytical Techniques for Recognizing Water-Sensitive Reservoir Rocks," *JPT* (Aug. 1963) 813–18.
14. Hower, W.F.: "Influence of Clays on the Production of Hydrocarbons," paper SPE 4785 presented at the 1974 SPE Symposium on Formation Damage Control, New Orleans, Feb. 7–8.
15. Mungan, N.: "Permeability Reduction Through Changes in pH and Salinity," *JPT* (Dec. 1965) 1449–53; *Trans.*, AIME, **234.**
16. Smith, C.F., Pavlich, J.P., and Solvinsky, R.L.: "Potassium Calcium Treatments Inhibit Clay Swelling," *Oil & Gas J.* (Nov. 30, 1964) 80–81.
17. Peters, F.W. and Stout, C.M.: "Clay Stabilization During Fracturing Treatments With Hydrolyzable Zirconium Salts," *JPT* (Feb. 1977) 187–94; *Trans.*, AIME, **263.**
18. Bates, T.G., Gruver, R.M., and Yuster, S.T.: "Influence of Clay Content on Water Conductivity of Oil Sands," *Oil Weekly* (Oct. 21, 1946) 48.
19. Coulter, A.W. Jr., Frick, E.K., and Samuelson, M.L.: "Effect of Fracturing-Fluid pH on Formation Permeability," paper SPE 12150 presented at the 1983 SPE Annual Technical Conference and Exhibition, San Francisco, Sept. 5–8.
20. Hawsey, J.D., Whitesell, L.B., and Kepley, N.A.: "Injection of a Bactericide-Surfactant During Hydraulic Fracturing—A New Method of Corrosion Control," paper SPE 978 presented at the 1964 SPE Annual Meeting, Houston, Oct. 11–14.
21. Tannich, J.D.: "Liquid Removal From Hydraulically Fractured Gas Wells," *JPT* (Nov. 1975) 1309–17.
22. Clark, H.B. *et al.*: "Water Soluble Fluorochemical Surfactant Well Stimulation Additives," *JPT* (July 1982) 1565–69.
23. Allen, T.D. and Roberts, A.P.: "Surfactants for Well Treatments," *Production Operations,* Oil and Gas Consultants Intl. Inc., Tulsa (1978) **2,** 115–16.
24. *RP42, Recommended Practices for Laboratory Testing of Surface-Active Agents for Well Stimulation,* API, Dallas (1977).
25. Graham, J.W., Monaghan, P.H., and Osoba, J.S.: "Influence of Propping Sand Wettability on Productivity of Hydraulically Fractured Oil Wells," *Trans.*, AIME (1959) **216,** 324–29.
26. Penny, S.S. *et al.*: "Enhanced Load Water-Recovery Technique Improves Stimulation Results," paper SPE 12149 presented at the 1983 SPE Annual Technical Conference and Exhibition, San Francisco, Sept. 5–8.
27. Featherston, A.B., Mihram, R.G., and Waters, A.B.: "Minimization of Scale Deposits in Oil Wells by Placement of Phosphates in Producing Zones," *JPT* (March 1959) 29–32.
28. Knox, J.A., Waters, A.B., and Arnold, B.B.: "Checking Paraffin Deposition by Crystal Growth Inhibition," paper SPE 443 presented at the 1962 SPE Annual Meeting, Los Angeles, Oct. 7–10.
29. Sloat, B.: "Controlled Solubility Phosphates—New Application Techniques Simplify Scale Prevention in Waterfloods and Producing Wells," paper SPE 760 presented at the 1963 SPE California Regional Meeting, Santa Barbara, Oct. 24–25.

30. Bauer, K.A. and Bezemer, C.: "Prevention of Carbonate Scale Deposition: A Well-Packing Technique with Controlled Solubility Phosphates," *JPT* (April 1969) 505–14.
31. Tinsley, J.M.: "Design Techniques for Chemical Fracture-Squeeze Treatments," *JPT* (Nov. 1967) 1493–99; *Trans.*, AIME, **240.**
32. McCall, J.M. Jr. and Johnson, R.L.: "Paraffin Treatment in the Well Service Industry," *Proc.*, Southwestern Petroleum Short Course, Lubbock, TX (April 1984) 457–68.
33. Stiff, H.A. Jr. and Davis, L.E.: "A Method For Predicting the Tendency of Oil Field Waters to Deposit Calcium Sulfate," *Trans.*, AIME (1952) **195,** 213–16.
34. Daneshy, A.A.: "Numerical Solution of Sand Transport in Hydraulic Fracturing," *JPT* (Jan. 1978) 132–40.
35. Novotny, E.J.: "Proppant Transport," paper SPE 6813 presented at the 1977 SPE Annual Technical Conference and Exhibition, Denver, Oct. 9–12.
36. Daneshy, A.A.: "Hydraulic Fracture Propagation in Layered Formations," *SPEJ* (Feb. 1978) 33–41.
37. Clark, P.E.: "Design of a Large Vertical Prop Transport Model," paper SPE 6814 presented at the 1977 SPE Annual Technical Conference and Exhibition, Denver, Oct. 9–12.
38. Hannah, R.R. and Harrington, L.J.: "Measurement of Dynamic Proppant Fall Rates in Fracturing Gels Using a Concentric Cylinder Tester," *JPT* (May 1981) 909–13.
39. Shah, S.N.: "Proppant Settling Correlations for Non-Newtonian Fluids Under Static and Dynamic Conditions," *SPEJ* (April 1982) 164–70.
40. Ford, T.F.: "Viscosity Concentration and Fluidity-Concentration Relationships for Suspensions of Spherical Particles in Newtonian Liquids," *J. Phys. Chem.* (Sept. 1960) 1168–74.
41. Clark, P.E. and Guler, N.: "Proppant Transport in Vertical Fractures: Settling Velocity Correlations," paper SPE 11636 presented at the 1983 SPE/DOE Low Permeability Gas Reservoirs Symposium, Denver, March 13–16.
42. Stewart, J.B. and Coulter, A.W.: "Increased Fracturing Efficiency by Fluid Loss Control," *Pet. Eng.* (June 1959) B-43.
43. Hall, C.D. Jr. and Dollarhide, F.E.: "Effects of Fracturing Fluid Velocity on Fluid-Loss Agent Performance," *JPT* (May 1964) 555–57; *Trans.*, AIME, **231.**
44. Hawsey, J.D. and Jacocks, C.L.: "The Use of Fluid-Loss Additives in Hydraulic Fracturing of Oil and Gas Wells," paper SPE 244 presented at the 1961 SPE California Regional Meeting, Bakersfield, Nov. 2–3.
45. Hall, C.D. Jr. and Dollarhide, F.E.: "Performance of Fracturing Fluid Loss Agents Under Dynamic Conditions," *JPT* (July 1968) 763–69.
46. Gatlin, C. and Nemir, C.E.: "Some Effects of Size Distribution on Particle Bridging in Lost Circulation and Filtration Tests," *JPT* (June 1961) 575–78; *Trans.*, AIME, **222.**
47. Pye, D.S. and Smith, W.A.: "Fluid Loss Additive Seriously Reduces Fracture Proppant Conductivity and Formation Permeability," paper SPE 4680 presented at the 1973 SPE Annual Meeting, Las Vegas, Oct. 1–3.
48. King, G.E.: "Factors Affecting Dynamic Fluid Leakoff With Foam Fracturing Fluids," paper SPE 6817 presented at the 1977 SPE Annual Technical Conference and Exhibition, Denver, Oct. 9–12.
49. Shumaker, E.F., Raines, V.B., and Warenbourg, P.A.: "Leak-Off Control Techniques Improve Efficiency of Acid Fracturing Treatments," paper SPE 7168 presented at the 1978 SPE Regional Gas Technology Symposium, Omaha, NE, June 7–9.
50. Settari, A.: "A New General Model of Fluid Loss in Hydraulic Fracturing," *SPEJ* (Aug. 1985) 491–501.
51. Harris, P.C.: "Dynamic Fluid-Loss Characteristics of Nitrogen Foam Fracturing Fluids," *JPT* (Oct. 1985) 1847–52.
52. Harris, P.C.: "Dynamic Fluid-Loss Characteristics of CO_2-Foam Fracturing Fluids," *SPEPE* (May 1987) 89–94.
53. McDaniel, R.R. *et al.*: "An Improved Method For Measuring Fluid Loss at Simulated Fracture Conditions," *SPEJ* (Aug. 1985) 482–90.
54. Zigrye, J.L., Whitfill, D.L., and Sievert, J.A.: "Fluid-Loss Control Differences of Crosslinked and Linear Fracturing Fluids," *JPT* (Feb. 1985) 315–20.
55. Reece, E.T.: "Bacterial Degradation of Cellulose Derivatives," *Ind. & Eng. Chem.* (1957) **49,** 89.
56. Wirick, M.G.: "Study of the Substitution Pattern of Hydroxyethyl Cellulose and Its Relationship to Enzymatic Degradation," *J. Polymer Sci.* (1968) **A-1,** No. 6, 1705.
57. Almond, S.W. and Bland, W.E.: "Effect of Break Mechanism on Gelling Agent Residue and Flow Impairment in 20/40 Mesh Sand," paper SPE 12485 presented at the 1984 SPE Formation Damage Symposium, Bakersfield, Feb. 13–14.
58. Ousterhout, R.S. and Hall, C.D. Jr.: "Reduction of Friction Loss in Fracturing Operations," *JPT* (March 1961) 217–22.
59. Melton, L.L. and Malone, W.T.: "Fluid Mechanics Research and Engineering Application in Non-Newtonian Fluid Systems," *SPEJ* (July 1964) 56–66; *Trans.*, AIME, **231.**
60. Bundrant, C.O. and Matthews, T.A.: "Friction Loss of Fracturing Fluids," paper SPE 531-G presented at the 1955 SPE Annual Meeting, New Orleans, Oct. 2–5.
61. Lord, D.L., Hulsey, B.W., and Melton, L.L.: "General Turbulent Pipe Flow Scale-Up Correlation for Rheologically Complex Fluids," *SPEJ* (Sept. 1967) 252–58; *Trans.*, AIME, **240.**
62. Savins, J.G.: "Drag Reduction Characteristics of Solutions of Macromolecules in Turbulent Pipe Flow," *SPEJ* (Sept. 1964) 203–14; *Trans.*, AIME, **231.**
63. Hoover, M.F. and Padden, J.J.: "Thickened Friction Reducer for Water-Based Oil Well Treating Fluids," U.S. Patent No. 3,442,803 (May 6, 1967).
64. Clark, P.E.: "Stimulation Fluid Rheology—A New Approach," paper SPE 8300 presented at the 1979 SPE Annual Technical Conference and Exhibition, Las Vegas, Sept. 23–26.
65. Rogers, R.E., Veatch, R.W., and Nolte, K.G.: "Pipe Viscometer Study of Fracturing Fluid Rheology," *SPEJ* (Oct. 1984) 575–81.
66. Sinclair, A.R.: "Rheology of Viscous Fracturing Fluids," *JPT* (June 1970) 711–19.
67. Cloud, J.E. and Clark, P.E.: "Alternatives to the Power-Law Fluid Model for Crosslinked Fluids," *SPEJ* (Dec. 1985) 935–42.
68. Govier, G.W. and Aziz, K.: *The Flow of Complex Mixtures in Pipes,* Van Nostrand Reinhold Co., New York City (1972) 184.
69. Dodge, D.W. and Metzner, A.B.: "Turbulent Flow of Non-Newtonian Systems," *AIChE J.* (Feb. 1959) 189–204.
70. Conway, M.W. *et al.*: "Chemical Model for the Rheological Behavior of Crosslinked Fluid Systems," *JPT* (Feb. 1983) 315–20.
71. Buechley, T.C. and Lord, D.L.: "Hydraulic Fracturing Fluid Mechanics—State of the Art," paper 26C presented at the 1973 AIChE National Meeting, New Orleans.
72. Pruitt, G.T. *et al.*: "A Method to Minimize Cost of Pumping Fluids Containing Friction Reducing Additives," *JPT* (June 1965) 641–46.
73. White, G.L.: "Friction Pressure Reducers in Well Stimulations," *JPT* (Aug. 1964) 865–68.
74. Sitaramaiah, G. and Smith, C.L.: "Turbulent Drag Reduction by Polyacrylamide and Other Polymers," *SPEJ* (June 1969) 183–88.
75. Conway, M.W., Pauls, R.W., and Harris, L.E.: "Evaluation of Procedures and Instrumentation Available for Time Temperature Stability Studies of Crosslinked Fluids," paper SPE 9333 presented at the 1980 SPE Annual Technical Conference and Exhibition, Dallas, Sept. 21–24.
76. Holditch, S.A. and Ely, J.W.: "Successful Stimulation of Deep Wells Using High Proppant Concentrations," *JPT* (Aug. 1973) 959–64.
77. Ely, J.W. *et al.*: "Compositions for Fracturing High Temperature Well Formation," U.S. Patent No. 3,898,165 (Aug. 5, 1975).
78. Hsu, C.H. and Conway, M.W.: "Fracturing Fluids for Deep, Hot Formations," *JPT* (Nov. 1981) 2213–18.
79. Seidel, W.R. and Stahl, E.J.: "Gas Well Stimulation with a Viscous Water-Base Fracturing Fluid," *JPT* (Nov. 1972) 1385–90.
80. Williams, J.R., Tinsley, J.M., and Malone, W.T.: "Optimum Fracturing Treatment Design Helps Reach Maximum Profit," paper presented at the 1967 Petroleum Soc. of CIM Annual Technical Meeting.
81. Whistler, R.L.: *Industrial Gums,* Academic Press Inc., New York City (1973).
82. Chrisp, J.D.: "Gelled Compositions Containing Galactomannan Gums," U.S. Patent No. 3,301,723 (1967).
83. Anderson, R.W. and Baker, J.R.: "Use of Guar Gum and Synthetic Cellulose in Oilfield Stimulation Fluids," paper SPE 5005 presented at the 1974 SPE Annual Meeting, Houston, Oct. 6–9.
84. Tiner, R.: "Polymers and Their Use in the Oilfield," *Proc.*, Southwestern Petroleum Short Course, Lubbock, TX (1976).
85. Chatterji, J. and Borchardt, J.K.: "Application of Water-Soluble Polymers in the Oil Field," *JPT* (Nov. 1981) 2042–56.
86. Ely, J.W.: "Fracturing Fluid Systems State of the Art," *Proc.*, Southwestern Petroleum Short Course, Lubbock, TX (April 23–24, 1981).
87. Githens, C.J. and Burnham, J.W.: "Chemically Modified Natural Gum for Use in Well Stimulation," *SPEJ* (Feb. 1977) 5–10.
88. Cooke, C.E. Jr.: "Effect of Fracturing Fluid on Fracture Conductivity," *JPT* (Oct. 1975) 1273–82.
89. Almond, S.W.: "Factors Affecting Gelling Agent Residue Under Low Temperature Conditions," paper SPE 10658 presented at the 1982 SPE Formation Damage Control Symposium, Lafayette, LA, March 24–25.
90. Pober, K.W. *et al.*: "Water-Insoluble Residue Following Acid Hydrolysis of Water-Soluble Polysaccharides," *JPT* (Dec. 1983) 2185–91.

91. Volk, L.J. *et al.*: "A Method for Evaluation of Formation Damage Due to Fracturing Fluids," paper SPE 11638 presented at the 1983 SPE/DOE Symposium on Low Permeability Gas Reservoirs, Denver, March 13–16.
92. White, G.L. and Free, D.L.: "Properties of Various Frac Fluids as Compared to the Ideal Fluid," *Proc.*, AGA Symposium on Stimulation of Low Permeability Reservoirs, Feb. 16–17, 1976, 1–14.
93. Kim, C.M. and Losacano, J.A.: "Fracture Conductivity Damage Due to Crosslinked Gel Residue and Closure Stress on Propped 20/40 Mesh Sand," paper SPE 14436 presented at the 1985 SPE Annual Technical Conference and Exhibition, Sept. 22–26.
94. Pabley, A.S. and Holcomb, D.L.: "A New Stimulation Technique: High Strength Crosslinked Acid," paper SPE 9241 presented at the 1980 SPE Annual Technical Conference and Exhibition, Dallas, Sept. 21–24.
95. Gardner, D.C. and Eikerts, J.V.: "Rheological Characterization of Crosslinked and Delayed Crosslinked Fracturing Fluids by Using a Closed-Loop Pipe Viscometer," paper SPE 12028 presented at the 1983 SPE Annual Technical Conference and Exhibition, San Francisco, Sept. 5–8.
96. Woodroof, R.A. Jr. and Anderson, R.W.: "Synthetic Polymer Friction Reducers Can Cause Formation Damage," paper SPE 6812 presented at the 1977 SPE Annual Technical Conference and Exhibition, Denver, Oct. 9–12.
97. Hendrickson, A.R., Hurst, R.E. and Wieland, D.R.: "Engineered Guide for Planning Acidizing Treatments Based on Specific Reservoir Characteristics," *Trans.*, AIME (1960) **219,** 16–23.
98. Hendrickson, A.R. and Rosene, R.B.: "The Role of Acid Reaction Rates in Planning Acidizing Treatments," paper SPE 71 available from SPE headquarters, Richardson, TX.
99. Barron, A.N., Hendrickson, A.R., and Wieland, D.R.: "The Effect of Flow on Acid Reactivity in a Carbonate Fracture," *JPT* (April 1962) 409–15; *Trans.*, AIME, **225.**
100. Hendrickson, A.R., Rosene, R.B., and Wieland, D.R.: "Acid Reaction Parameters and Reservoir Characteristics Used in the Design of Acidizing Treatments," paper presented at the 1960 ACS Petroleum Section Meeting, Cleveland, OH, April.
101. Nierode, D.E. and Williams, B.B.: "Characteristics of Acid Reaction in Limestone Formations," *SPEJ* (Dec. 1971) 406–18; *Trans.*, AIME, **251.**
102. Smith, C.F., Crowe, C.W., and Wieland, D.R.: "Fracture Acidizing in High Temperature Limestone," paper SPE 3008 presented at the 1970 SPE Annual Meeting, Houston, Oct. 4–7.
103. Williams, B.B. *et al.*: "Characterization of Liquid-Solid Reactions, Hydrochloric Acid-Calcium Carbonate Reaction," *Ind. & Eng. Chem.* (Nov. 1970) **9,** 589.
104. Broaddus, G.C. and Know, J.A.: "Influence of Acid Type and Quantity in Limestone Etching," paper presented at the 1965 API Mid-Continent Meeting, Wichita, KS, March 31–April 2.
105. Nierode, D.E. and Williams, B.B.: "Design of Acid Fracturing Treatments," *JPT* (July 1972) 849–59.
106. Schols, R.S. *et al.*: "An Analysis of the Acidizing Process in Acid Fracturing," *SPEJ* (Aug. 1973) 239–48; *Trans.*, AIME, **258.**
107. Roberts, L.D. and Guin, J.A.: "The Effect of Surface Kinetics in Fracture Acidizing," *SPEJ* (Aug. 1974) 385–95; *Trans.*, AIME, **257.**
108. Coulter, A.W. *et al.*: "Alternate Stages of Pad Fluid and Acid Provide Improved Leak-Off Control for Fracture Acidizing," paper SPE 6124 presented at the 1976 SPE Annual Technical Conference and Exhibition, New Orleans, Oct. 3–6.
109. Knox, J.A., Lasater, R.M., and Dill, W.R.: "A New Concept in Acidizing Utilizing Chemical Retardation," paper SPE 975 presented at the 1964 SPE Annual Meeting, Houston, Oct. 11–14.
110. Carpenter, N.F. and Ernst, E.A.: "Acidizing: With Swellable Polymers," *JPT* (Sept. 1962) 1041–47; *Trans.*, AIME, **225.**
111. Crowe, C.W., Martin, R.C., and Michaelis, A.M.: "Evaluation of Acid-Gelling Agents for Use in Well Stimulation," *SPEJ* (Aug. 1981) 415–24.
112. Norman, L.R., Conway, M.W., and Wilson, J.M.: "Temperature-Stable Acid-Gelling Polymers: Laboratory Evaluation and Field Results," *JPT* (Nov. 1984) 2011–18.
113. Norman, L.R.: "Properties and Early Field Results of a Liquid Gelling Agent for Acid," paper SPE 7834 presented at the 1978 SPE Petroleum Technology Symposium, Hobbs, NM, Oct. 30–31.
114. Pabley, A.S. and Holcomb, D.L.: "A New Method of Acidizing or Acid Fracturing: Crosslinked Acid Gels," *Proc.*, Southwestern Petroleum Short Course, Lubbock, TX (April 17–18, 1980).
115. Smith, M.A., Dawson, J., and Scoggins, D.: "High Temperature, Crosslinked High Strength Acid System," *Proc.*, Southwestern Petroleum Short Course, Lubbock, TX.
116. Deysarkar, A.K. *et al.*: "Crosslinked Fracture Acidizing Acid Gel," paper 82-33-16 presented at the 1982 Petroleum Soc. of CIM Annual Technical Meeting, Calgary, Alta., June 6–9.
117. Williams, B.B., Gidley, J.L., and Schechter, R.S.: *Acidizing Fundamentals,* Monograph Series, SPE, Richardson, TX (1979) **6.**
118. Holtmyer, M.D. and Githens, C.J.: "Field Performance of a New High Viscosity Water Base Fracturing Fluid," paper API-875-24E presented at the 1970 API Spring Meeting, Rocky Mountain Dist., Production Div., Denver, April 27–29.
119. Free, D.L., Frederick, A.F., and Thompson, J.E.: "Fracturing With a High-Viscosity, Crosslinked Gel—Continuous Fracturing Technique," *JPT* (Jan. 1978) 119–22.
120. Holtmeyer, M.D., Githens, C.J., and Tinsley, J.M.: "Fracturing Well Formations," British Patent No. 1,337,651 (1973).
121. Holtmeyer, M.D., Githens, C.J., and Tinsley, J.M.: "Compositions for Fracturing Well Formations," U.S. Patent No. 4,021,355 (1977).
122. Holtmeyer, M.D., Githens, C.J., and Tinsley, J.M.: "Methods for Fracturing Well Formations," U.S. Patent No. 4,033,415 (1977).
123. Tiner, R.L. *et al.*: "Method and Compositions for Fracturing Well Formations," U.S. Patent No. 3,888,312 (1975).
124. Hannah, R.R. and Matson, W.G.: "A Family of Viscous Gelled Water Systems Featuring Moderate Polymer Loadings and Low Residue on Breaking," paper SPE 6380 presented at the 1977 SPE Permian Basin Oil and Gas Recovery Symposium, Midland, March 10–11.
125. Hannah, R.R. and Baker, J.R.: "A New Nondamaging, Aqueous Cross-Linked Gel with Improved Fracturing Properties and Perfect Proppant Support," *Proc.*, Southwestern Petroleum Short Course, Lubbock, TX, 123–28.
126. Lescarboura, J.A., Sifferman, T.R., and Wahl, H.A.: "Evaluation of Fracturing Fluid Stability by Using a Heated, Pressurized Flow Loop," *SPEJ* (June 1984) 249–55.
127. Conway, M.W. and Harris, E.: "A Laboratory and Field Evaluation of a Technique for Hydraulic Fracturing Stimulation of Deep Wells," paper SPE 10964 presented at the 1982 SPE Annual Technical Conference and Exhibition, New Orleans, Sept. 26–29.
128. Gardner, D.C. and Eikerts, J.V.: "The Effects of Shear and Proppant on the Viscosity of Crosslinked Fracturing Fluids," paper SPE 11066 presented at the 1982 SPE Annual Technical Conference and Exhibition, New Orleans, Sept. 26–29.
129. Craigie, L.J.: "A New Method for Determining the Rheology of Crosslinked Fracturing Fluid Using Shear History Simulation," paper SPE 11635 presented at the 1983 SPE/DOE Low-Permeability Gas Reservoirs Symposium, Denver, March 13–16.
130. Harms, S.D., Goss, M.L., and Payne, K.L.: "New Generation Fracturing Fluid for Ultrahigh-Temperature Application," paper SPE 12484 presented at the 1984 SPE Formation Damage Control Symposium, Bakersfield, Feb. 13–14.
131. Baumgartner, S.A. *et al.*: "High Efficiency Fracturing Fluids for High Temperature Reservoirs," paper SPE 11566 presented at the 1983 SPE Production Operations Symposium, Oklahoma City, Feb. 27–March 1.
132. Lagrone, C.C., Baumgartner, S.A., and Woodroof, R.A. Jr.: "Chemical Evolution of a High-Temperature Fracturing Fluid," *SPEJ* (Oct. 1985) 623–28.
133. Kiel, O.M.: "A New Hydraulic Fracturing Process," *JPT* (Jan. 1970) 89–96.
134. Engel, J.D.: "How Superfrac Worked in the Mid-continent," *Oil & Gas J.* (June 15, 1970) **68,** No. 24, 65–68, 70.
135. Payne, K.L. and Harms, S.D.: "Chemical Processes for Controlling Crosslinking Reaction Rates," paper presented at the 1984 AIChE Natl. Meeting, Anaheim, CA (May 20–24).
136. Rummo, G.J.: "Crosslinking Delay Achieved with Gel Additive," *Oil & Gas J.* (Sept. 13, 1982) 84.
137. Harris, P.C.: "Influence of Temperature and Shear History on Fracturing Fluid Efficiency," paper SPE 14258 presented at the 1985 SPE Annual Technical Conference and Exhibition, Sept. 22–26.
138. Kucera, C.H., Smith C.F., and Braunlich, F.H.: "New Oil Gelling Systems Prevent Damage in Water-Sensitive Sands," paper SPE 3503 presented at the 1974 SPE Annual Meeting, New Orleans, Oct. 3–6.
139. Burnham, J.W. *et al.*: "Developments in Hydrocarbon Fluids for High-Temperature Fracturing, " *JPT* (Feb. 1980) 217–20.
140. Hendrickson, A.R., Nesbitt, E.E., and Oaks, B.D.: "Soap-Oil Systems for Formation Fracturing," *Pet. Eng.* (May 1957) B-58.
141. Malone, W.T. and Anderson, T.O.: "Gelled Crude for Formation Fracturing," *Oil & Gas J.* (Feb. 1956) 117.
142. Ely, J.W.: *Stimulation Treatment Handbook, An Engineers Guide to Quality Control,* PennWell Publishing Co., Tulsa, OK (1985).
143. Tindell, W.A., Misak, M.D., and Gras, E.H.: "The Use of Alcohol-Water Mixtures in Fracture Stimulation of Gas Wells," *Proc.*, Southwest Petroleum Short Course, Lubbock, TX (April 18–19, 1974).

144. McLeod, H.O. and Coulter, A.W.: "The Use of Alcohol in Gas Well Stimulation," paper SPE 1633 presented at the 1966 SPE Eastern Regional Meeting, Columbus, OH, Nov. 10–11.
145. Smith, C.F.: "Gas Well Fracturing Using Gelled Non-Aqueous Fluids," paper SPE 4678 presented at the 1973 SPE Annual Meeting, Las Vegas, Sept. 30–Oct. 3.
146. Keeney, B.R. and Frost, J.G.: "Guidelines Regarding the Use of Alcohols in Acidic Stimulation Fluids," *JPT* (May 1975) 552–54.
147. Sinclair, A.R., Terry, W.M., and Kiel, O.M.: "Polymer Emulsion Fracturing," *JPT* (July 1974) 731–38.
148. Bernard, G.G., Holm, L.W., and Jacobs, W.L.: "Effect of Foam on Trapped Gas Saturation and on Permeability of Porous Media to Water," *SPEJ* (Dec. 1965) 295–300; *Trans.*, AIME, **234**.
149. Blauer, R.E. and Kohlhaas, C.A.: "Formation Fracturing with Foam," paper SPE 5003 presented at the 1974 SPE Annual Meeting, Houston, Oct. 6–9.
150. Holcomb, D.L. and Blauer, R.E.: "Foam Fracturing Shows Success in Gas, Oil Formation," *Oil & Gas J.* (Aug. 1975).
151. Plummer, R.S. and Holditch, S.A.: "The Design of Stable Foam Fracturing Treatments," *Proc.*, Southwestern Petroleum Short Course, Lubbock, TX (April 1976).
152. Harris, P.C. and Reidenbach, V.G.: "High-Temperature Rheological Study of Foam Fracturing Fluids," *JPT* (May 1987) 613–19.
153. Holcomb, D.L., Callaway, E., and Curry, L.L.: "Foamed Hydrocarbon Stimulation for Water Sensitive Formations," paper SPE 9033 presented at the 1980 SPE Rocky Mountain Regional Meeting, Casper, WY, May 14–16.
154. Ford, W.G.F.: "Foamed Acid—An Effective Stimulation Fluid," *JPT* (July 1981) 1203–10.
155. Lord, D.L.: "Analysis of Dynamic and Static Foam Behavior," *JPT* (Jan. 1981) 39–45.
156. Holcomb, D.L.: "Foamed Acid as a Means For Providing Extended Retardation," paper SPE 6376 presented at the 1977 SPE Permian Basin Oil and Gas Recovery Conference, Midland, March 10–11.
157. Holcomb, D.L. and Wilson, S.C.: "Foamed Acidizing and Selective Diverting Using Stable Foam for Improving Acid Stimulation," *Proc.*, Southwestern Petroleum Short Course, Lubbock, TX (1978).
158. Scherubel, G.A. and Crowe, C.W.: "Foamed Acid, A New Concept in Fracture Acidizing," paper SPE 7568 presented at the 1978 SPE Annual Technical Conference and Exhibition, Houston, Oct. 1–4.
159. Gaydos, J.S. and Harris, P.C.: "Foam Fracturing: Theories, Procedures and Results," paper SPE 8961 presented at the 1980 SPE/DOE Symposium on Unconventional Gas Recovery, Pittsburgh, PA, May 18–21.
160. Grundmann, S.R. and Lord, D.L.: "Foam Stimulation," *JPT* (March 1983) 597–602.
161. Reidenbach, V.G. *et al.*: "Rheological Study of Foam Fracturing Fluids Using Nitrogen and Carbon Dioxide," *SPEPE* (Jan. 1986) 31–41.
162. David, A. and Marsden, S.S. Jr.: "The Rheology of Foam," paper SPE 2544 presented at the 1969 SPE Annual Meeting, Denver, Sept. 28–Oct. 1.
163. Holcomb, D.L., Callaway, E., and Curry, L.L.: "Chemistry, Physical Nature, and Rheology of an Aqueous Stimulation Foam," *SPEJ* (Aug. 1981) 410–14.
164. Ely, J.: "Recent Mechanical Chemical Improvements in Foam Fracturing," *Proc.*, Southwestern Petroleum Short Course, Lubbock, TX.
165. Harms, S.D. and Payne, K.L.: "Factors Affecting the Selection of Foaming Agents for Foam Stimulation,"
166. Harris, P.C.: "Effects of Texture on Rheology of Foam Fracturing Fluids," paper SPE 14257 presented at the 1985 SPE Annual Technical Conference and Exhibition, Sept. 22–26.
167. Foshee, W.D. and Hurst, R.E.: "Improvement of Well Stimulation Fluids by Including a Gas Phase," *JPT* (July 1965) 768–72.
168. Moran, J.P. and Horton, H.L.: "Use of Inert Gases in Well Completions," paper SPE 584 presented at the 1963 SPE Permian Basin Oil and Gas Recovery Conference, May 9–10.
169. Crawford, H.R. *et al.*: "Carbon Dioxide—A Multipurpose Additive for Effective Well Stimulation," *JPT* (March 1963) 237–42.
170. Hurst, R.E.: "Gas Frac—A New Stimulation Technique Using Liquid Gases," paper SPE 3837 presented at the 1972 SPE Rocky Mountain Regional Meeting, Denver, April 10–12.
171. Boren, R.J. and Johnson, D.L.: "Nitrogen, An Oil Field Look," *World Petroleum* (April 1965) 29–32.
172. Shouldice, P.: "Atomized Acid Improved Well Clean Up," *Canadian Petroleum* (Aug. 1966) 14–15.
173. Black, H.N. and Langsford, R.W.: "Energized Fracturing With 50% CO_2 for Improved Hydrocarbon Recovery," *JPT* (Jan. 1982) 135–40.
174. Justice, W.H. and Nielsen, J.P.: "Improved Techniques Developed for Acidizing Gas Producing and Injection Wells," *Trans.*, AIME (1952) **195**, 285–88.
175. Kalish, P.J. *et al.*: "The Effect of Bacteria on Sandstone Permeability," *JPT* (July 1964) 805–14; *Trans.*, AIME, **231**.
176. Clark, P.E. and Skelton, H.: "A Low-Temperature Breaker for Faster Well Turn-Around and Better Well Clean-Up," *Proc.*, Southwestern Petroleum Short Course, Lubbock, TX (1978) 47–50.
177. Tannich, J.D.: "Liquid Removal From Hydraulically Fractured Gas Wells," *JPT* (Nov. 1975) 1309–17.
178. Boyer, J.P. *et al.*: "Effect of Fracture Fluids in Tight Gas Formations," Report No. GRI-79/0099, Gas Research Inst. (Jan. 1981).
179. Muecke, T.W.: "Formation Fines and Factors Controlling Their Movement in Porous Media," *JPT* (Feb. 1979) 144–50.
180. Veley, C.D.: "How Hydrolyzable Metal Ions React With Clays to Control Formation Water Sensitivity," *JPT* (Sept. 1969) 1111–18.
181. Peters, F.W. and Stout, C.M.: "Clay Stabilization During Fracturing Treatments With Hydrolyzable Zirconium Salts," *JPT* (Feb. 1977) 187–94; *Trans.*, AIME, **263**.
182. Clementz, D.M.: "Clay Stabilization in Sandstones Through Adsorption of Petroleum Heavy Ends," *JPT* (Sept. 1977) 1061–66.
183. McLaughlin, H.C., Alphingstone, E.A., and Hall, B.E.: "Aqueous Organic Polymers for Treating Clays in Oil and Gas Producing Formations," paper SPE 6008 presented at the 1976 SPE Annual Technical Conference and Exhibition, New Orleans, Oct. 3–6.
184. Anderson, R.W. and Kannenberg, B.G.: "Method of Stabilizing Clay Formations," U.S. Patent No. 4,158,521 (June 19, 1979).
185. Williams, L.H. Jr. and Underdown, D.R.: "New Polymer Offers Effective Permanent Clay Stabilization Treatment," *JPT* (July 1981) 1211–17.
186. Young, B.M., McLaughlin, H.C., and Borchardt, J.K.: 'Clay Stabilization Agents—Their Effectiveness in High-Temperature Steam," *JPT* (Dec. 1980) 2121–31.
187. Reed, M.G.: "Formation Permeability Maintenance With Hydroxy-Aluminum Solutions," U.S. Patent No. 3,603,399 (Sept. 7, 1971).
188. Reed, M.G.: "Stabilization of Formation Clays with Hydroxy-Aluminum Solutions," *JPT* (July 1972) 860–64; *Trans.*, AIME, **253**.
189. Blevins, T.R., Cotten, W.R., and Dugas, E.G.: "OH-Al Treatments Sustained Acid Stimulated Production," *World Oil* (Aug. 1, 1973) 26–29.
190. Coppel, C.P., Jennings, H.Y. Jr., and Reed, M.G.: "Field Results from Wells Treated With Hydroxy-Aluminum," *JPT* (Sept. 1973) 1108–12.
191. Haskin, C.A.: "A Review of Hydroxy-Aluminum Treatments," paper SPE 5692 presented at the 1976 SPE Formation Damage Control Symposium, Houston, Jan. 29–30.
192. Penny, G.S., Conway, M.W., and Lee, W.S.: "Control and Modeling of Fluid Leakoff During Hydraulic Fracturing," *JPT* (June 1985) 1071–81.
193. Penny, G.S.: "Nondamaging Fluid Loss Additives for Use in Hydraulic Fracturing of Gas Wells," paper SPE 10659 presented at the 1982 SPE Formation Damage Control Symposium, Lafayette, LA, March 24–25.
194. Williams, B.B.: "Fluid Loss from Hydraulically Induced Fractures," *JPT* (July 1970) 882–88; *Trans.*, AIME, **249**.
195. Watkins, E.K., Wendorff, C.L., and Ainley, B.R.: "A New Crosslinked Foam Fracturing Fluid," paper SPE 12027 presented at the 1983 SPE Annual Technical Conference and Exhibition, San Francisco, Oct. 5–8.
196. Thomas, R.L. and Elbel, J.L.: "The Use of High Temperature Stabilizer in High Temperature Fracturing," paper SPE 8344 presented at the 1979 SPE Annual Technical Conference and Exhibition, Las Vegas, Sept. 23–26.
197. Hannah, R.R., Harrington, L., and Anderson, R.W.: "A Study of Bingham Plastic Flow for Use as a Temporary Diverting Agent in Hydraulic Fracturing," ASME Publication 78-Pet-36.

SI Metric Conversion Factors

bbl	× 1.589 873	E−01	=	m^3
cp	× 1.0*	E−03	=	Pa·s
ft^3	× 2.831 685	E−02	=	m^3
°F	(°F−32)/1.8		=	°C
gal	× 3.785 412	E−03	=	m^3
lbm	× 4.535 924	E−01	=	kg
psi	× 6.894 757	E+00	=	kPa

*Conversion factor is exact.

Chapter 8
Fluid Leakoff

Glenn S. Penny,* SPE, Halliburton Services
Michael W. Conway,* SPE, Halliburton Services

8.1 Overview

This chapter discusses how fluid leakoff during hydraulic fracturing is modeled, calculated, and measured experimentally and examines procedures for converting laboratory data to an estimate of leakoff under actual conditions. It begins by defining the mathematical relationships currently used to calculate leakoff to the formation. This is followed by a discussion of methods of measuring leakoff in the laboratory and the relative effectiveness of various fluid-loss additives as a function of permeability, gel type, additive concentration, and temperature. The effects of such parameters as pressure, temperature, shear, and time at temperature on the leakoff profile of common fracturing fluids are then examined. Methods of converting laboratory data to leakoff coefficients for use in estimating fluid loss in an actual treatment are presented. Finally, laboratory and field methods of obtaining in-situ leakoff coefficients are compared.

8.2 Introduction

The rate of fluid leakoff to the formation during a hydraulic fracturing treatment is one of the most critical factors involved in determining fracture geometry for a given treatment design. The volume of fluid lost during the treatment determines the fracturing-fluid efficiency or the ratio of fracture volume to volume pumped. The efficiency must be estimated accurately to avoid premature job termination because of a sandout or excess pressure caused by gel dehydration. Further, the rate of fluid leakoff influences fracture closure time and may influence the final distribution of proppant within the fracture. The sensitivity of fracture geometry and treatment costs to leakoff is discussed in Sec. 17.9; efficiency vs. fracture closure time is discussed in Sec. 14.3.

The rate of fluid leakoff to the formation is governed by the fracturing-fluid leakoff coefficient, C, which has been defined by a combination of three types of linear flow mechanisms encountered during a fracturing treatment.[1] The three types of linear flow mechanisms are effluent viscosity and relative permeability effects, C_v; reservoir-fluid viscosity/compressibility effects, C_c; and wall-building effects, C_w. The first two coefficients can be calculated from reservoir data and fracturing-fluid viscosity. The third coefficient is computed from leakoff data for fluid-loss additives, which must be determined experimentally. Although each coefficient is derived independently, all act simultaneously in a fracturing treatment and affect the efficiency of the fluid.

Constant-pressure injection of a viscous fracturing fluid into a porous medium under conditions of linear flow is governed by the following relation (C_v is in ft/min$^{1/2}$)[1]:

$$C_v = 0.0469\left(\frac{k_i \Delta p \phi}{\mu_a}\right)^{1/2}, \quad \text{.......................... (8.1)}$$

where

k_i = relative permeability of formation to leakoff effluent,
Δp = difference in pressure between fluid at formation face and initial pore pressure of formation,
μ_a = viscosity of effluent from fracturing fluid at bottomhole fracturing conditions, and
ϕ = fractional porosity of formation.

Both the compressibility and viscosity of the existing reservoir fluid will affect the rate of fracturing-fluid leakoff and is mathematically expressed as[1]

$$C_c = 0.0374\Delta p\left(\frac{k_r c_t \phi}{\mu_f}\right)^{1/2}, \quad \text{...................... (8.2)}$$

where

k_r = formation permeability relative to mobile reservoir fluid,
c_t = total formation compressibility, and
μ_f = viscosity of mobile formation fluid at reservoir conditions.

Again, Δp is expressed in psi, and C_c is in ft/min$^{1/2}$. The actual calculation of the C_c term is quite complex and is shown in detail in the example problem of Sec. 8.3.

Experimentally, the fracturing-fluid coefficient caused by the addition of fluid-loss additives can be determined graphically by plotting the filtrate volume vs. the square root of time, $t^{1/2}$, as in Fig. 8.1. The volume of fluid that is lost before a filter cake is established is known as spurt volume and is the extrapolated y intercept of the plot of volume vs. square root of time. The slope of this line, m, is directly proportional to C_w, which is commonly called the wall-building fluid-loss coefficient.

An approximate value for the rate of spurt leakoff in an actual treatment can be calculated from the properties of the reservoir fluid, treating fluid, and formation. The equations and correlations used to calculate C_v and C_c are dealt with in detail in Sec. 8.3. In many analyses, C_c and C_v are combined in the form of C_{vc}, where

$$C_{vc} = \frac{2C_v C_c}{C_v + (C_v^2 + 4C_c^2)^{1/2}}. \quad \text{...................... (8.3)}$$

When particulate matter and/or polymer in solution is pressurized against a porous medium, the filtration process generates a filter cake. The rate of flow of fluid through the filter cake is governed by C_w, which is defined (in ft/min$^{1/2}$) as

$$C_w = \frac{0.0164m}{A}, \quad \text{.............................. (8.4)}$$

*Now with Stim-Lab Inc.

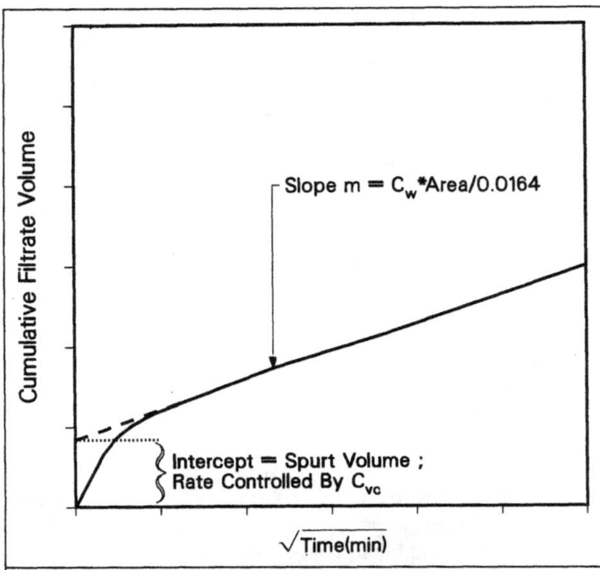

Fig. 8.1—Determination of C_w and spurt from a cumulative volume vs. time½.

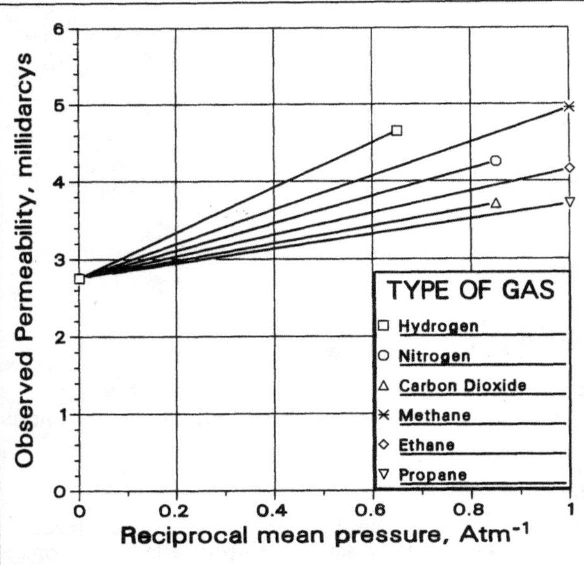

Fig. 8.2—Variation in gas permeability with mean pressure (from Klinkenberg[6] and Calhoun,[7] *Fundamentals of Reservoir Engineering,* revised edition, ©1953, 1960 by the U. of Oklahoma Press).

where m=slope of volume-vs.-$t^{1/2}$ plot (such as Fig. 8.1) and A=area of core used in laboratory test. In a physical sense, m can be expressed (in mL/min½) as

$$m=\frac{k_f A^2 \Delta p'}{\mu_f F_{cV}}, \quad (8.5)$$

where

k_f = filter-cake permeability,
$\Delta p'$ = pressure drop,
μ_f = fracturing fluid filtrate viscosity, and
F_{cV} = ratio of filter cake to filtrate volume=$F_{Vc}/1-f$ [F_{Vc} is the ratio of volume in cake to total fluid-loss volume (cake volume plus filtrate volume)].

The relationship in Eq. 8.5 is seldom solved analytically. It is, however, usually determined experimentally, as in Eq. 8.4, and is shown here only to demonstrate the nature of the wall-building coefficient. The leakoff rate through a filter cake at any time t is governed by the Darcy flow equation. The amount of deposition of the filter cake at a future time t depends on the ratio of the volume of filter-cake-forming components to the filtrate volume.

Because C_w is determined in the laboratory on core samples using the additives in question, it is not an exact engineering quantity, but is highly dependent on additive type, test cell design, measurement technique, and environmental effects (see Secs. 8.5 through 8.7).

The most complete and accurate form of fluid-loss calculations would use a two-phase-flow numerical simulator that could model leakoff into the formation, as well as production before and after a stimulation treatment.[2] In a majority of cases, however, adequate information to use such a simulator does not exist. Therefore, the above single-phase, single-dimension fluid loss equations are commonly used in the design of fracturing treatments.

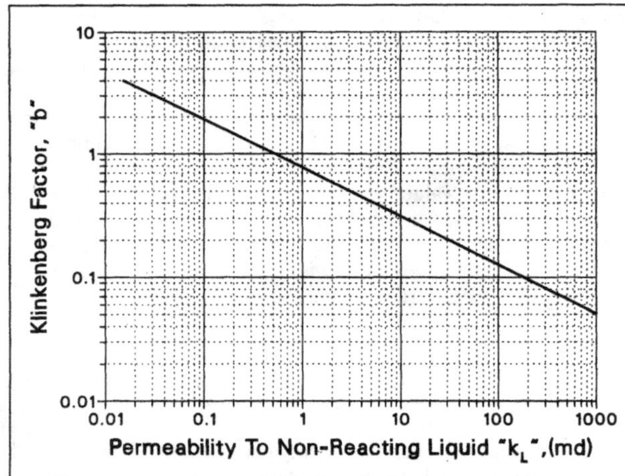

Fig. 8.3—Effect of the permeability of U.S. oilwell cores on the Klinkenberg extrapolation factor (from API[8]).

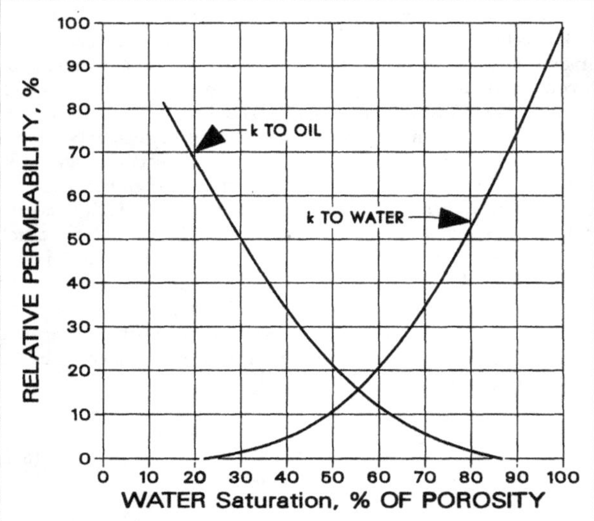

Fig. 8.4—Relative permeability vs. water saturation for a water-wet sandstone.

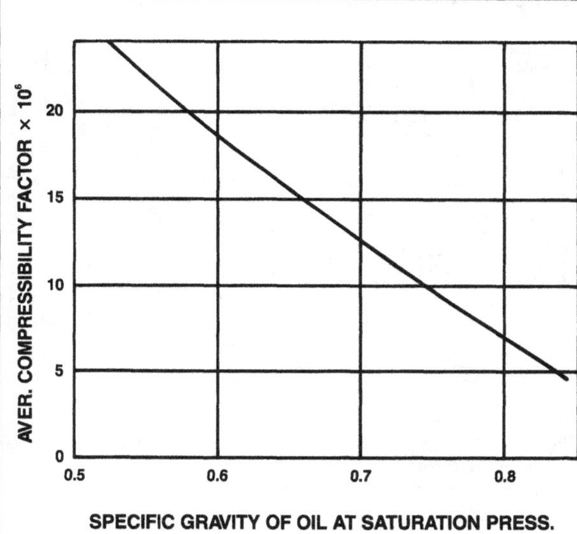

Fig. 8.5—Average compressibility per pound per square inch per barrel of liquid at bubblepoint as function of specific gravity at bubblepoint (from Calhoun,[7] *Fundamentals of Reservoir Engineering*, revised edition, ©1953, 1960 by the U. of Oklahoma Press).

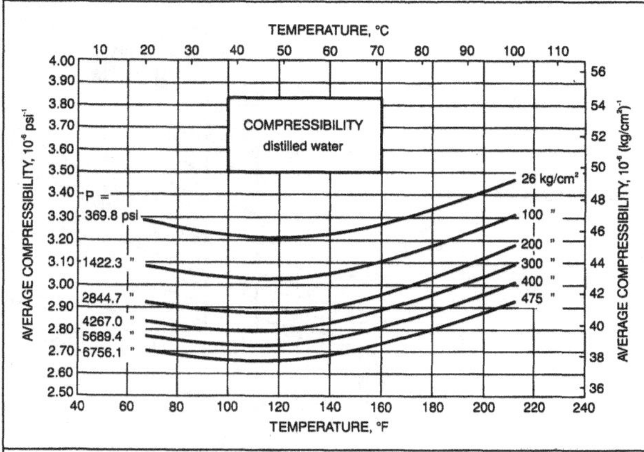

Fig. 8.6—Average compressibility of distilled water (after Long and Chierici[9]).

Recent advances in pressure analysis have made a field measurement of leakoff possible. The basis for the proposed methods is discussed in detail in Chap. 14. Field implementation of the methods and a comparison of results with laboratory-measured data are given in Chap. 13. In this chapter, comparing laboratory data and field leakoff data is discussed in Sec. 8.8.

8.3 Calculation of Combined Fluid-Loss Coefficients

Several methods have been proposed for combining the three separate leakoff coefficients.[1,3-5] The simplest method used by the industry today is to assume that the wall-building coefficient, C_w, will dominate and thus to use that value in the fracture-geometry calculations. Other methods involve combining C_v and C_c, as demonstrated in Eq. 8.3. When combined, they are compared with C_w and the lesser of the two values is used in the fluid-loss calculation.

Another procedure involves separately calculating C_w, C_v, and C_c at the stated Δp and then combining these values to arrive at an overall coefficient, C_t. To derive C_t, the relationship between the effective coefficient and the pressure drop must be determined. The relationship between the magnitude of C_v and C_c and the pressure drop is explicitly stated in each equation (i.e., $C_c \propto \Delta p$ and $C_v \propto \Delta p^{1/2}$). On the other hand, the relationship between the pressure drop and C_w must be implied or measured in the laboratory. The equation for C_t generally assumes a $p^{1/2}$ relationship for the influence of Δp on C_w.

The basic approach of combining C_c, C_v, and C_w (assuming a $p^{1/2}$ dependence) was developed by Williams[4] and Williams *et al.*[5] In this approach, Eqs. 8.6 and 8.7 are used:

$$v = C_t / t^{1/2} \quad \text{(8.6)}$$

and

$$\Delta p_t = \Delta p_v + \Delta p_c + \Delta p_{cw}. \quad \text{(8.7)}$$

New quantities were then defined to partition the pressure drop: $C_c^* = C_c/\Delta p$, $C_v^* = C_v/\Delta p^{1/2}$, and $C_w^* = C_w/\Delta p^{1/2}$. It follows that

$$\Delta p_t = \frac{C_t}{C_c^*} + \frac{C_t^2}{C_v^*} + \frac{C_t^2}{C_w^*}. \quad \text{(8.8)}$$

C_t can now be calculated from the quadratic equation[5] to give

$$C_t = \frac{2C_c C_v C_w}{C_v C_w + [C_w^2 C_v^2 + 4C_c^2 (C_v^2 + C_w^2)]^{1/2}}. \quad \text{(8.9)}$$

When the leakoff rate is dominated by C_{vc}, then $C_t \cong C_{vc}$ and the volume of fluid lost to the formation is simply C_{vc} or $C_t t^{1/2}$. When the equilibrium leakoff rate is dominated by C_w, however, the total amount of fluid lost to the formation may also include a spurt volume, or the amount of fluid that leaks off before reaching C_w. The equation for the volume per unit area of fluid lost to the formation is then

$$V = V_s + 2C_w t^{1/2}. \quad \text{(8.10)}$$

The leakoff rate of the spurt volume, V_s, is controlled by C_{vc}; therefore, the spurt time, t_s, is defined as

$$t_s = \left(\frac{V_s}{2C_{vc}}\right)^2. \quad \text{(8.11)}$$

Example Calculation of Fluid-Loss Coefficients. The following example presents detailed calculations of C_v and C_c and their impact on C_t. The casual reader may find it more expedient to skip the rest of this section and to proceed through the rest of this chapter before reviewing the detailed calculations in this example problem. The reservoir conditions are as follows.

Perforated interval, 8,225 to 8,275 ft;
net interval, 57 ft;
gross interval, 100 ft;
bottomhole pressure (BHP), p_{bh}, 3,600 psi (normally pressured);
bottomhole treating pressure (BHTP), p_{bht}, 4,600 psi (fracture gradient=0.56 psi/ft);
reservoir temperature (BHTP), T_R, 200°F;
porosity, ϕ, 11% (or 0.11);
oil saturation, S_o, 0%;
residual oil saturation, S_{or}, 0%;
water saturation, S_w, 32%;
residual water saturation, S_{wr}, 32%;
chlorides, 36,500 ppm;
permeability, k_i, 0.28 md; and
fracturing fluid, complexed hydroxypropyl guar (HPG) (40 lbm/1,000 gal 2% KCl water).

1. Calculation of C_c.

$$C_c = 0.0374 \Delta p \left(\frac{k_i c_t \phi}{\mu_f}\right)^{1/2}. \quad \text{(8.2)}$$

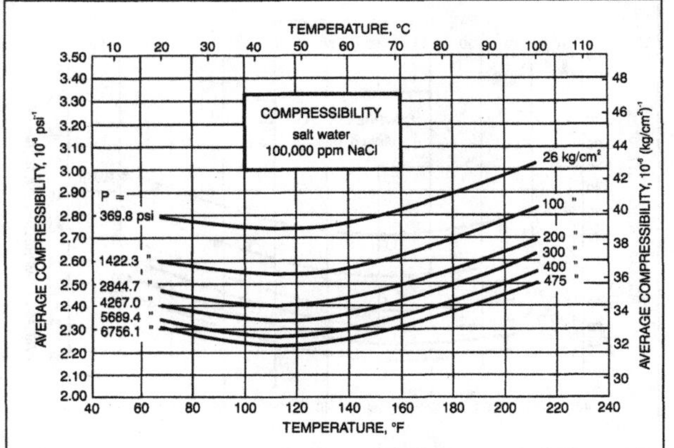

Fig. 8.7—Average compressibility of 100,000-ppm NaCl in distilled water (after Long and Chierici[9]).

The individual terms are obtained from correlations or measured directly.

$\Delta p = p_{bht} - p_{bh} = 1{,}000$ psi.

ϕ = total porosity = 11%.

$$k_i = \frac{(S_o - S_{or})k_o + (S_w - S_{wr})k_w + S_g k_g}{(S_o - S_{or}) + (S_w - S_{wr}) + S_g}, \quad \text{(8.12)}$$

where

k_i = permeability to mobile reservoir fluid,
S_o = S_{or} = 0%,
S_w = S_{wr} = 32%, and
S_g = 68%.

k_i is the measured gas permeability (k_g) = 0.28 md because gas is the only mobile reservoir fluid and the gas permeability was measured at a high differential pressure. If the gas permeability is determined at low pressure, then several different pressures should be used and a plot of measured permeability vs. reciprocal mean

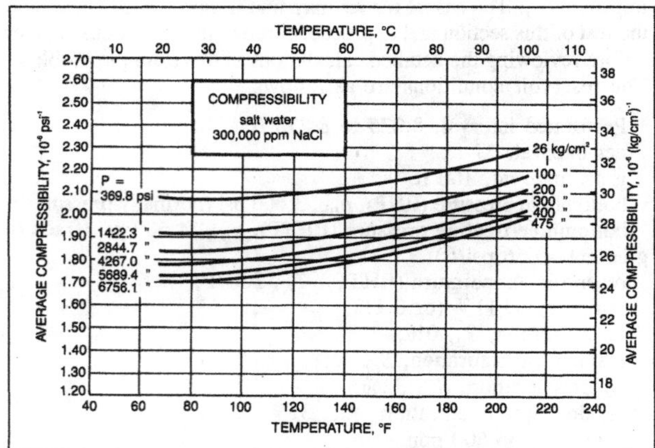

Fig. 8.9—Average compressibility of 300,000-ppm NaCl in distilled water (after Long and Chierici[9]).

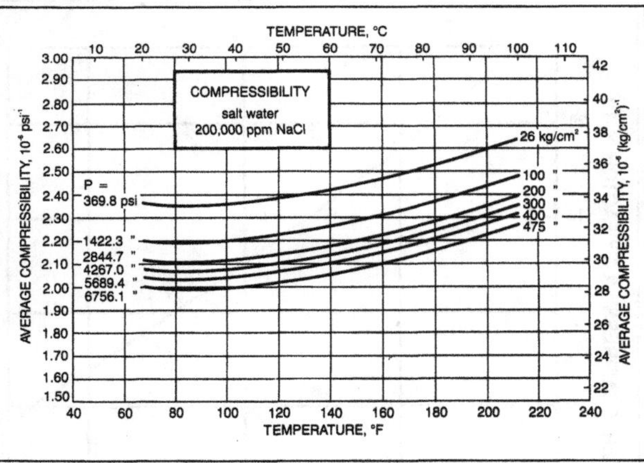

Fig. 8.8—Average compressibility of 200,000-ppm NaCl in distilled water (after Long and Chierici[9]).

pressure is used to extrapolate the permeability to infinite pressure (Fig. 8.2).

The gas permeability will also need to be converted to a fluid-independent permeability.[8] The gas permeability k_i = 0.28 md was measured at 100 psi, and

$$k_L = \frac{k_g}{(1 + b/\bar{p})}, \quad \text{(8.13)}$$

where

k_L = nonreactive liquid permeability,
k_g = measured permeability,
$\bar{p}$ = mean pressure at which gas is flowing, and
b = Klinkenberg constant (Fig. 8.3) for a given gas and porous medium.

$$k_L = \frac{0.28}{(1 + 1.3/50)} = 0.272. \quad \text{(8.13)}$$

If two phases are flowing, then the relative permeability needs to be adjusted on the fluid saturations. If the relative permeability

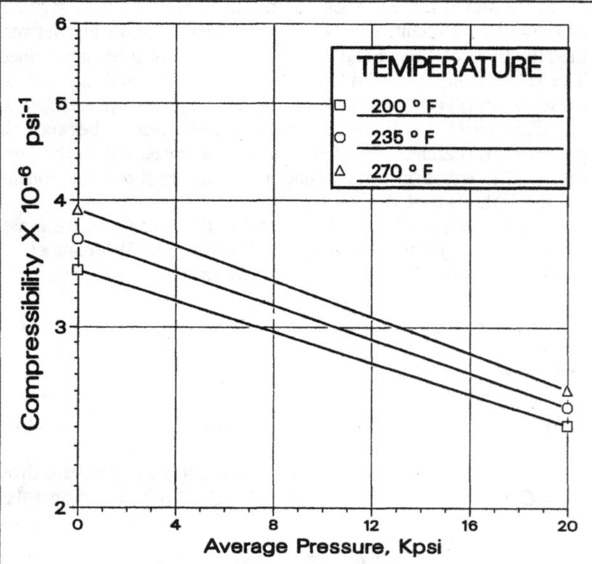

Fig. 8.10—Effect of temperature and pressure on the compressibility of water (from Osif[10]).

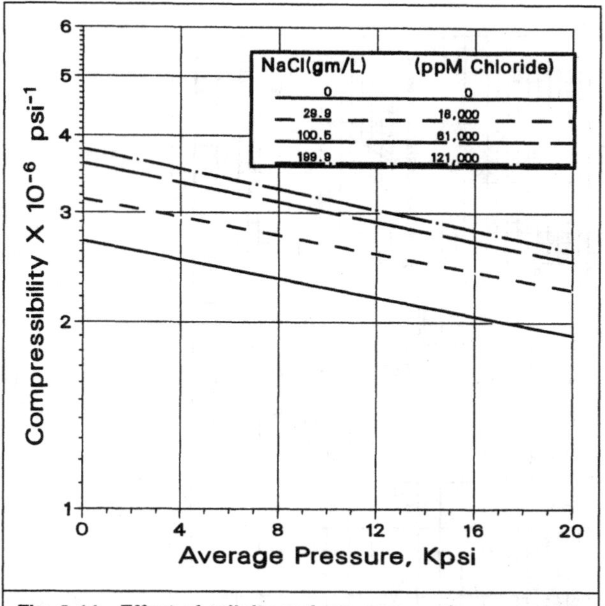

Fig. 8.11—Effect of salinity and pressure on the compressibility of water (from Osif[10]).

information is not available for the specific case in question, then the average values from Fig. 8.4 can be used for oil/water flow.

$$C_t = S_o c_o + S_w c_w + S_g c_g + c_r, \qquad (8.14)$$

where

S_o = oil saturation = 0%,
c_o = compressibility of oil (from Fig. 8.5)—not applicable,
S_w = water saturation = 32%,
c_w = compressibility of water (from Figs. 8.6 through 8.11) = 3×10^{-6},
c_r = compressibility of rock (from Fig. 8.12) = 5×10^{-6},
S_g = gas saturation = 68%, and
c_g = compressibility of gas (see below) = 220×10^{-6}.

The total compressibility is the sum of all the individual compressibilities. The values are obtained in the same manner as for traditional reservoir studies. Gas compressibilities, c_g's, are typically the most difficult to obtain from correlations. The use of pseudocritical pressure and temperature[7] is the most reliable method available. The details of the method are given in the following paragraphs, and example compressibilities under various conditions are given in Table 8.1 to aid in understanding the range of possible values.

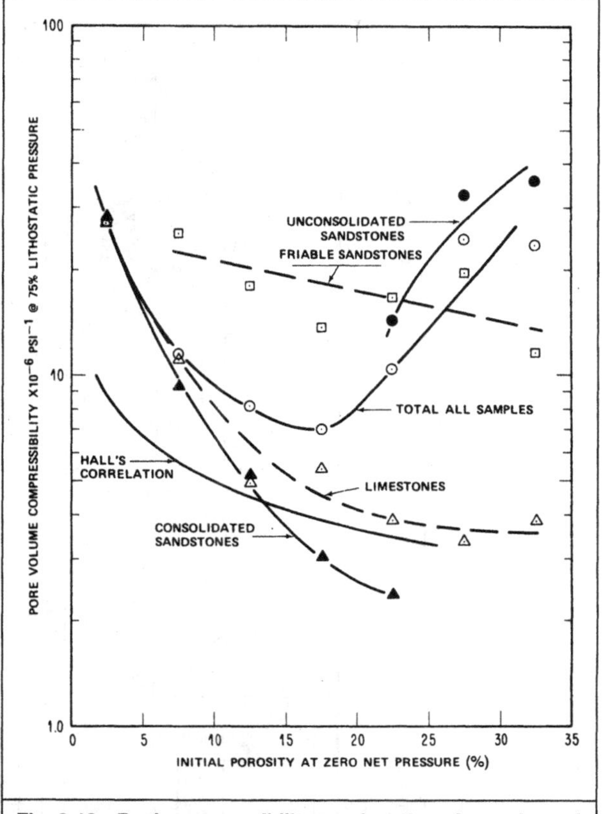

Fig. 8.12—Rock compressibility as a function of porosity and rock type (from Newman[11]).

The pseudocritical temperature can be estimated from Fig. 8.13 by knowing the gas gravity and the amounts of the heavy contaminants N_2, CO_2, and H_2S.

TABLE 8.1—VARIATIONS IN COMPRESSIBILITY AND VISCOSITY FOR TYPICAL GAS WELLS

Depth (ft)	Temperature (°F)	Pressure (psi)	Gas Gravity	Compressibility* (psi^{-1} × 10^6)	Viscosity* (cp)
3,000	100 to 136	2,658 Geopressured**	0.6	351 to 339	0.012 to 0.015
			0.8	260 to 295	
		1,329 Normal†	0.6	856 to 819	
			0.8	1,005 to 964	
		886 Underpressured‡	0.6	1,236 to 1,267	
			0.8	1,471 to 1,424	
8,000	150 to 246	7,088 Geopressured	0.6	65 to 74	0.027 to 0.030
			0.8	42 to 52	0.033 to 0.040
		3,544 Normal	0.6	314 to 248	0.02
			0.8	161 to 200	0.023 to 0.027
		2,364 Underpressured	0.6	445 to 437	0.016
			0.8	420 to 426	0.017 to 0.018
12,000	190 to 334	10,632 Geopressured	0.6	37 to 42	0.032 to 0.035
			0.8	27 to 31	0.038 to 0.045
		5,316 Normal	0.6	114 to 127	0.022 to 0.024
			0.8	92 to 117	0.026 to 0.030
		3,545 Underpressured	0.6	233 to 243	0.02
			0.8	214 to 227	0.021 to 0.027

*The range in compressibility and viscosity is that expected from varying the temperature gradient from 1 to 2.2°F/100-ft depth.
**Geopressured equals $2p_{wh}$.
†Normal pressure equals the hydrostat of a column of water, p_{wh}.
‡Underpressured equals $\frac{2}{3}p_{wh}$.

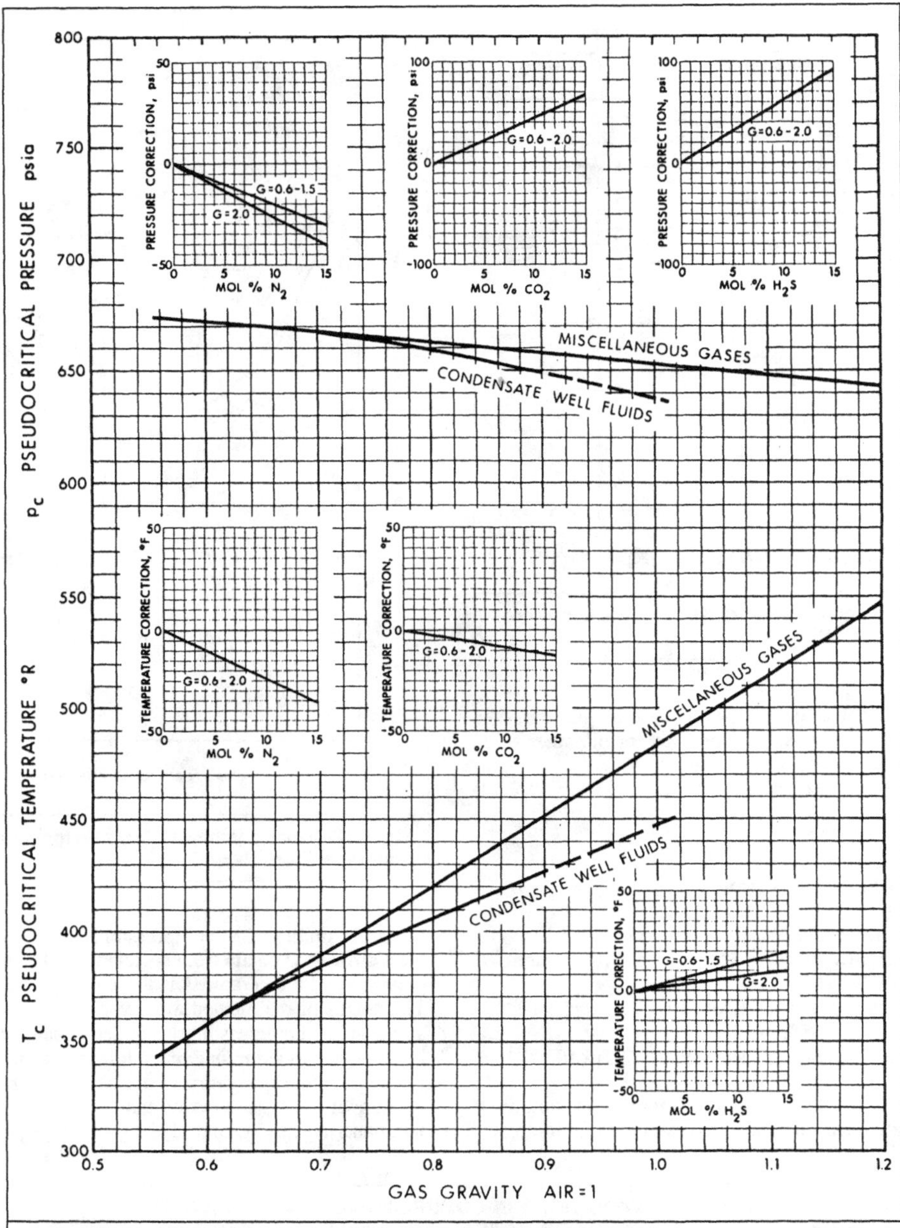

Fig. 8.13—Pseudocritical properties of miscellaneous natural gases (adapted from Brown et al.,[13] inserts adapted from Carr et al.[14]).

The reduced pressure is equal to the reservoir pressure divided by the critical pressure:

$$p_r = p/p_c. \quad (8.15)$$

The reduced temperature is equal to the absolute temperature of the reservoir divided by the critical temperature:

$$T_r = t_R(°F+460)/T_c, \quad (8.16)$$

or

$$T_r = t_R(°C+273.15)/T_c.$$

The reduced compressibility and temperature product, c_rT_r, can be obtained with Figs. 8.14 and 8.15:

$$c_r = \frac{(c_rT_r)}{T_r}. \quad (8.17)$$

The compressibility is then obtained by dividing c_r by p_c.

$$c = c_r/p_c. \quad (8.18)$$

For this example problem, with a 0.65-gravity gas at 200°F [366.5 K] and 3,600 psi [24 821 kPa] from Fig. 8.13, T_c=375°R [208 K] and p_c=668 psi [4610 kPa]:

p_r = 2,600/668=5.38.
T_r = 660/375=1.77.
c_r = 0.26/1.77=0.147.
c_g = 0.147/668
= 220×10^{-6} psi^{-1}.
c_f = 5×10^{-6} psi^{-1} (Fig. 8.12).
c_t = [$0.32(3\times10^{-6})+0.68(220\times10^{-6})+5\times10^{-6}$]
= 155.6×10^{-6}.
μ_f = because gas is the only mobile reservoir fluid,
μ_g = 0.0125 cp (from Fig. 8.16 at g=0.65 and 200°F),

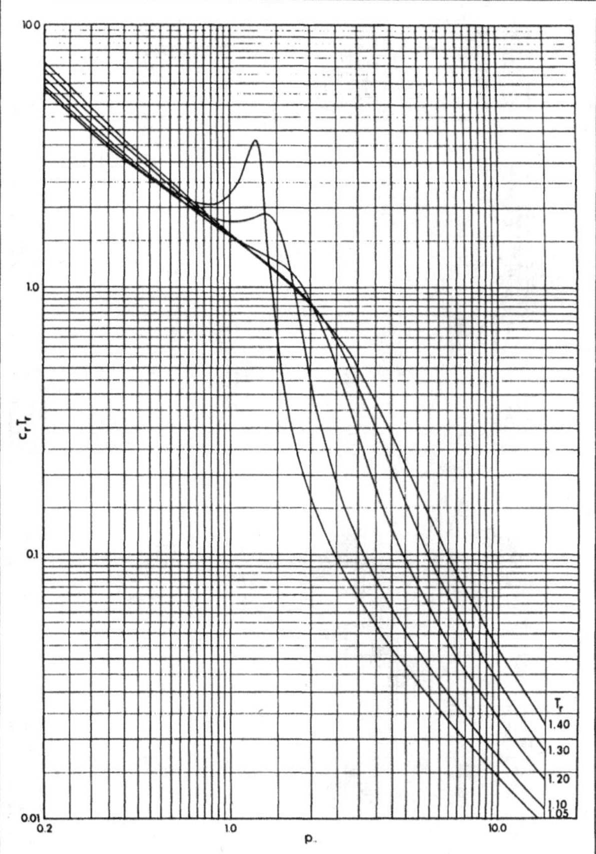

Fig. 8.14—Variation of c_rT_r with reduced temperature and pressure ($1.05 = T_r = 1.4$; $0.2 = p_r = 15.0$).[15] (Previously presented at the 1982 Annual Technical Meeting of the Petroleum Soc. of CIM, Calgary, June; reprinted with permission of the Canadian Inst. of Mining and Metallurgy.)

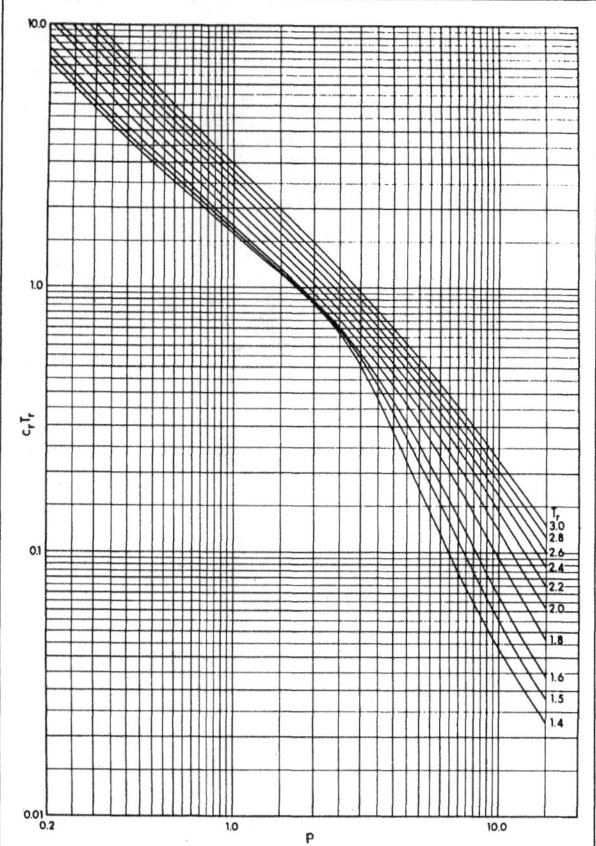

Fig. 8.15—Variation of c_rT_r with reduced temperature and pressure ($1.4 = T_r = 3.0$; $0.2 = p_r = 15.0$).[15] (Previously presented at the 1982 Annual Technical Meeting of the Petroleum Soc. of CIM, Calgary, June; reprinted with permission of the Canadian Inst. of Mining and Metallurgy.)

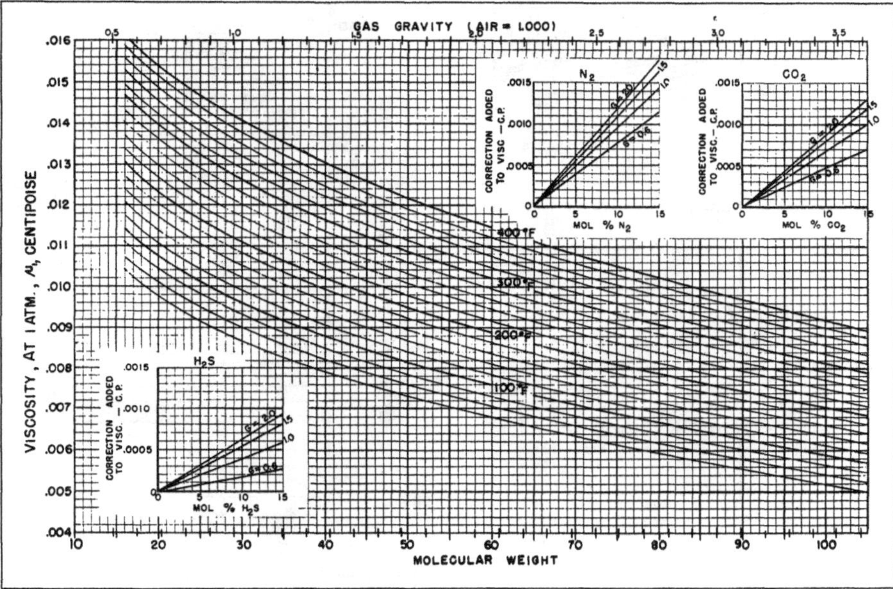

Fig. 8.16—Viscosity of paraffin hydrocarbon gases at 101.325 kPa (adapted from Carr *et al.*[14]).

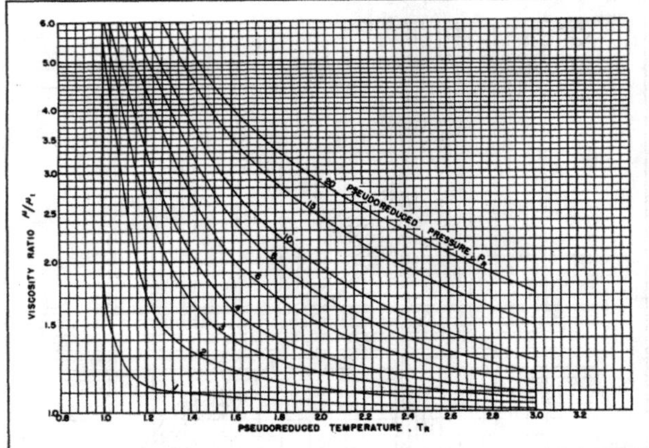

Fig. 8.17—Viscosity ratio vs. pseudoreduced temperature (from Carr *et al.*[14]).

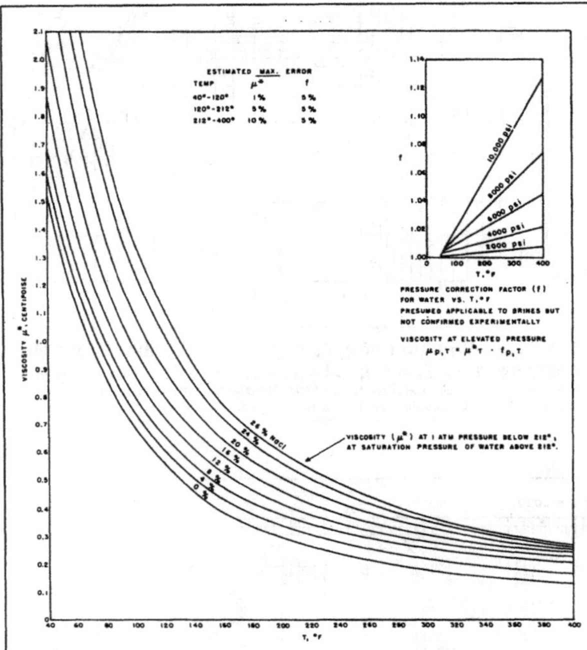

Fig. 8.18—Estimating water viscosity at reservoir temperatures.[50]

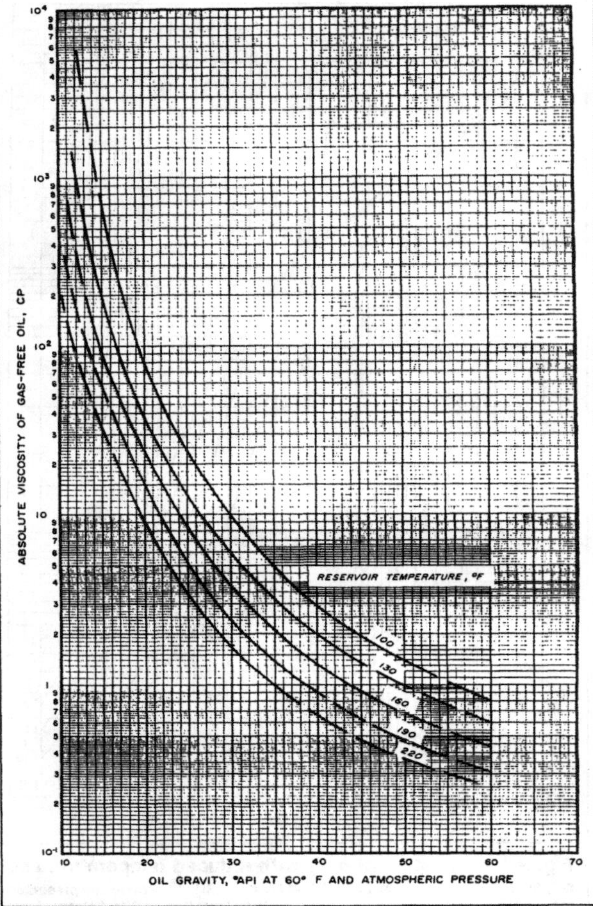

Fig. 8.19—Gas-free oil viscosity at reservoir temperatures.

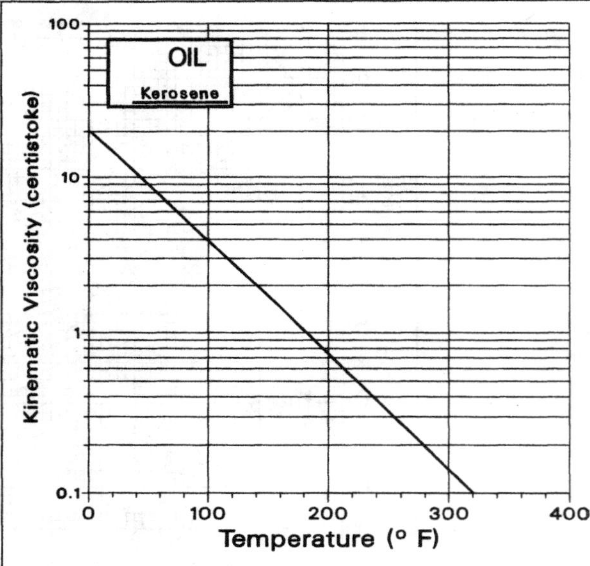

Fig. 8.20—Viscosity of refined oils vs. temperature: kinematic viscosity = absolute viscosity (cp)/specific gravity (from Ref. 16).

μ/μ_1 = 1.6 (from Fig. 8.17 at T_r=1.77 and p_r=5.38),
μ_f = 0.02 cp.
c_c = 0.0374 (1,000 psi)

$$\times\left(\frac{0.28\times10^{-3}\text{darcies}\times155.6\times10^{-6}\ \text{psi}^{-1}\times0.11}{0.02\ \text{cp}}\right)^{1/2}$$

=0.018 ft/min$^{1/2}$.

2. Calculation of C_v.

$$C_v=0.0469\left(\frac{k_i\Delta p\phi}{\mu_a}\right)^{1/2}, \qquad (8.1)$$

where

k_i = 0.272 md (this was the k_L value calculated earlier),
Δp = 1,000 psi,

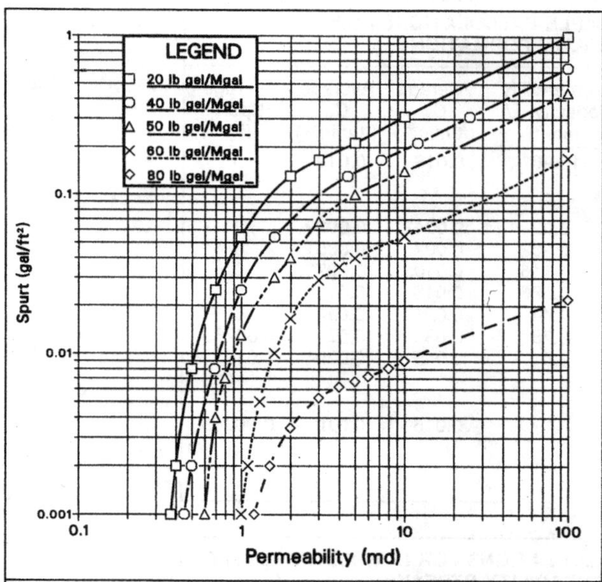

Fig. 8.21—Spurt loss vs. permeability and gel concentration for complexed HPG fluids at 125°F.

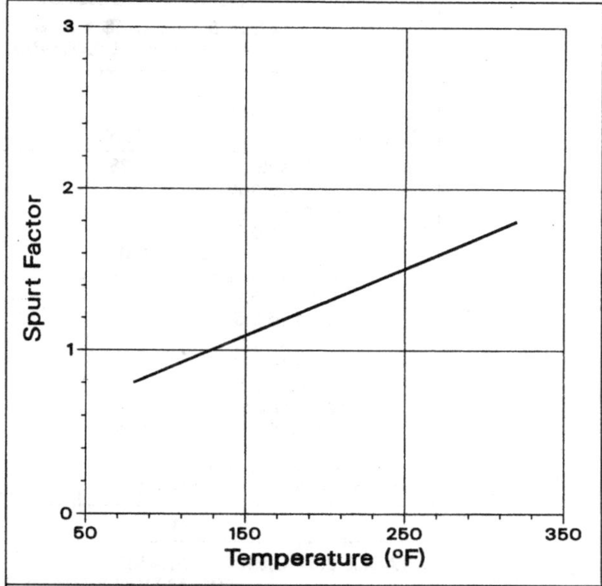

Fig. 8.22—Spurt and C_w correction factor vs. temperature for complexed fluids.

$$\phi = \bar{\phi}(1-S_{or}-S_{wr})$$
$$= 0.11(0.68)$$
$$= 0.075, \text{ and}$$

μ_a = 0.3 cp (see Fig. 8.18 and the discussion of Table 8.2.) Because this was a crosslinked gel and a low-permeability reservoir, it was assumed the effluent will be water.

$$C_v = 0.0469\left(\frac{0.272\times10^{-3}\text{ darcies}\times1{,}000\text{ psi}\times0.075}{0.3\text{ cp}}\right)^{1/2}$$

$$= 0.012 \text{ ft/min}^{1/2}.$$

If crude oil is the stimulation fluid, then the viscosity can be obtained from Fig. 8.19; for refined-hydrocarbon treatments, use Fig. 8.20.

3. Calculation of C_{vc}.

$$C_{vc}=\frac{2C_vC_c}{C_v+(C_v^2+4C_c^2)^{1/2}} \quad \text{(8.3)}$$

$$=\frac{2(0.012\text{ ft/min}^{1/2})(0.018\text{ ft/min}^{1/2})}{(0.012\text{ ft/min}^{1/2})+[(0.012\text{ ft/min}^{1/2})^2+4(0.018\text{ ft/min}^{1/2})^2]^{1/2}}$$

$$=0.00865 \text{ ft/min}^{1/2}.$$

4. Determine Appropriate Spurt and C_w. For calculating spurt, refer to Fig. 8.21. Spurt for 0.28 md at 125°F [52°C] for 40 lbm/1,000 gal [4.8 kg/m³] complexed HPG=0. For temperature correction (discussed in Sec. 8.8), refer to Fig. 8.22.

$$\text{Spurt}=0\times1.8=0 \text{ gal/ft}^2.$$

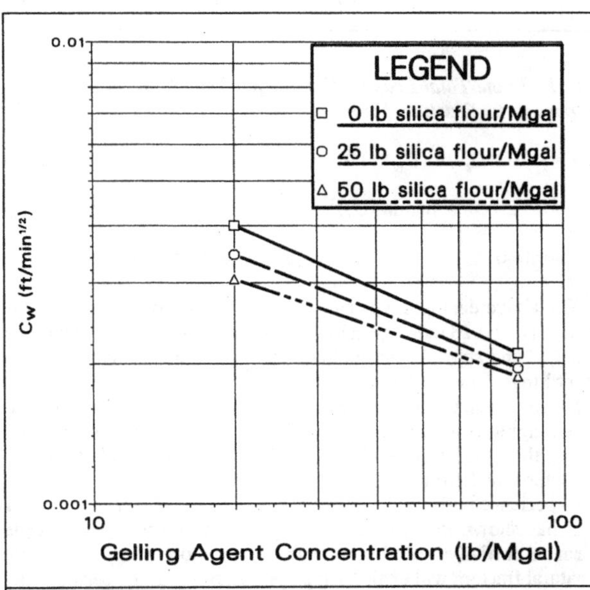

Fig. 8.23—C_w vs. gelling agent and silica concentration for complexed HPG fluids at 125°F.

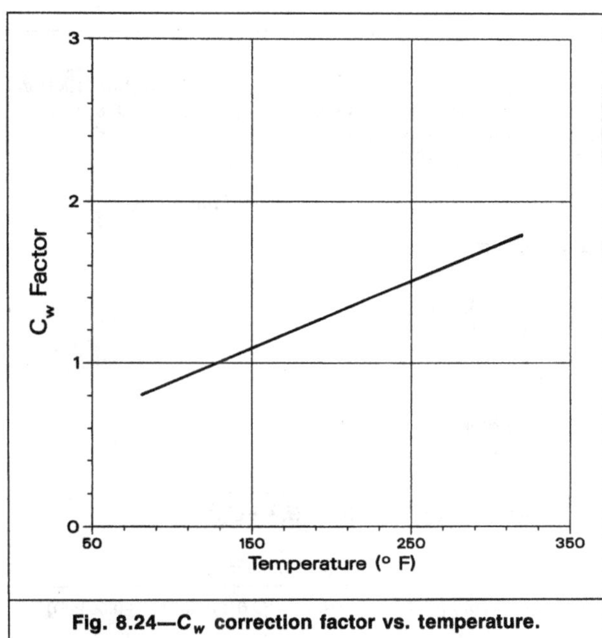

Fig. 8.24—C_w correction factor vs. temperature.

TABLE 8.2—CASES 1 AND 2: EXAMPLE CALCULATIONS FOR A TYPICAL LOW-PERMEABILITY FORMATION

Case	k Range* (md)	$\bar{k}$ (md)	Height (ft)	Permeability Capacity, kh (md-ft)	Average C_c (ft/min$^{1/2}$)	Average C_v (ft/min$^{1/2}$)	Average Spurt (ft)
1	Coefficients from average permeability	0.28	57	15.96	0.018	0.012	0
2	0** to 0.03	0.015	2	0.03	0.0042	0.0029	0
	0.031 to 0.0625	0.047	6	0.282	0.0075	0.0051	0
	0.0626 to 0.125	0.093	12	1.116	0.0105	0.0072	0
	0.126 to 0.25	0.187	18	3.366	0.015	0.01	0
	0.251 to 0.5	0.37	12	4.44	0.021	0.0143	0
	0.51 to 1.0	0.75	5	3.75	0.03	0.02	0.00265
	1.1 to 2.0	1.5	2	3.0	0.042	0.029	0.012
	Coefficients from permeability distribution	0.28			0.015	0.010	0.00065

*Permeability determined at 1,000 psi with dry N_2.
**No flow at 1,000 psi for 1-in. core plugs.

TABLE 8.3—CASES 1, 2, AND 3: EXAMPLE CALCULATIONS FOR LOW-PERMEABILITY FORMATION WITH HIGH-PERMEABILITY STREAKS

Case	k Range (md)	$\bar{k}$ (md)	Height (ft)	Permeability Capacity, kh (md-ft)	Average C_c (ft/min$^{1/2}$)	Average C_v (ft/min$^{1/2}$)	Average Spurt (ft)
1	Coefficients from average of low-permeability component	0.28	57	15.96	0.018	0.012	0
2	Coefficients from total average permeability	22.4	57	1276.8	0.163	0.111	0.072
3	0 to 0.03	0.015	2	0.03	0.0042	0.0029	0
	0.031 to 0.0625	0.047	3	0.141	0.0075	0.0051	0
	0.0626 to 0.125	0.093	8	0.744	0.0105	0.0072	0
	0.126 to 0.25	0.187	14	0.262	0.015	0.01	0
	0.251 to 0.5	0.37	9	0.333	0.021	0.0143	0
	0.51 to 1.0	0.75	5	0.375	0.03	0.02	0.00265
	1.1 to 2.0	1.5	2	3.0	0.042	0.029	0.012
	16 to 32	24.0	2	48.0	0.169	0.115	0.072
	32 to 64	48.0	4	192.0	0.239	0.162	0.13
	64 to 128	96.0	5	480.0	0.338	0.230	0.15
	128 to 240	184.0	3	552.0	0.468	0.318	0.19
	Coefficients from permeability distribution				0.09	0.061	0.0356

To calculate C_w, refer to Fig. 8.23. C_w for 40 lbm/1,000 gal [4.8 kg/m^3] complexed HPG at 125°F [52°C]=0.0028 ft/min$^{1/2}$. Correct for temperature (discussed in Sec. 8.8) by referring to Fig. 8.24. For 200°F [93°C], multiply by 1.3.

$$C_w = 0.0028 \text{ ft/min}^{1/2} \times 1.3 = 0.0036 \text{ ft/min}^{1/2}.$$

Calculated spurt time is

$$t_s = \left(\frac{V_s}{2C_{vc}}\right)^2 \quad \text{(8.11)}$$

$$=0.$$

5. *Calculation of C_t.*

$$C_t = \frac{2C_cC_vC_w}{C_vC_w + [C_w^2C_v^2 + 4C_c^2(C_v^2 + C_w^2)]^{1/2}} \quad \text{(8.9)}$$

$$= \frac{2(0.018)(0.012)(0.0036)}{(0.012)(0.0036) + [(0.0036)^2(0.012)^2 + 4(0.018)^2(0.012^2 + 0.0036^2)]^{1/2}}$$

$$=0.003 \text{ ft/min}^{1/2}.$$

6. *Estimate Total Leakoff Volume per Unit Area for 60-minute Exposure to Fracture Fluid.*

$$V = V_s + 2(C_t \text{ or } C_w)(t)^{1/2} \quad \text{(8.10)}$$

$$= 0 + 2(0.003 \text{ ft/min})(60)^{1/2}$$

$$= 0.046 \text{ ft or } 0.046 \text{ ft} \times 7.48 \text{ gal/ft}^3 = 0.348 \text{ gal/ft}^2.$$

The above example was calculated with the given average permeability of 0.28 md. The calculations can be refined by calculating C_c, C_v, and spurt for individual sections of the formation height exhibiting a range of permeabilities.

Example calculations are presented in Table 8.2 for permeabilities ranging from 0.015 to 2.0 md. The spurt values for the fracturing fluid (40 lbm/1,000 gal [4793 g/m^3] complexed HPG) can be determined from Figs. 8.21 and 8.22.

Experience with cores obtained from tight-gas reservoirs (<0.1 md) has shown the existence of high-permeability streaks in the form of partially healed natural fractures. Core plugs containing a natural fracture will exhibit a high permeability. If the permeability estimated from cores actually represents the true formation permeability, then few problems in selecting the leakoff coefficient will occur. If the natural fractures are poorly communicated in the

TABLE 8.4—COMPARISON OF EXAMPLE CALCULATIONS

Fluid = 40 lbm complexed HPG fluid + 5% diesel
C_w = 0.0036 ft/min$^{1/2}$ at 200°F
Exposure time = 60 minutes

Parameter	Low Permeability	Low Permeability With High-Permeability Streaks
C_c, ft/min$^{1/2}$	0.015	0.09
C_v, ft/min$^{1/2}$	0.010	0.061
C_{vc}, ft/min$^{1/2}$	0.0072	0.044
Spurt, ft	0.00065	0.0356
C_w, ft/min$^{1/2}$	0.0036	0.0036
Spurt time, minutes	0.008	0.6
C_t, ft/min$^{1/2}$	0.003	0.0035
Leakoff volume, $V_s+2C_t(t^{1/2})$, gal/ft^2	0.35	0.94

reservoir, however, then they may not be identified by well tests and the wrong fluid-loss coefficient could be calculated.

The effects of natural fractures are at best difficult to model. It must be stressed that natural fractures may not contribute significantly to production because their permeability is very sensitive to confining stress. However, they can significantly affect fluid loss if they become inflated during a pumping operation. One way to model their influence is to modify the permeability by including a high-permeability component. Because permeability has a log-normal distribution, the distribution of each permeability component is divided accordingly.

In Table 8.3, the high-permeability component is added to the permeability distribution used to construct Table 8.2. Three separate cases for the estimated fluid-loss coefficient are presented. Case 1 considers only the low-permeability component from Table 8.2; Case 2 considers the total average permeability; and Case 3 presents the results of the weighted average of layered reservoir calculations. The data suggest that layered reservoir models will be required to model the leakoff in dual-porosity reservoirs.

A comparison of the leakoff calculations for the example in Table 8.2 to the case illustrated in Table 8.3 is provided in Table 8.4. The effective fluid-loss coefficient, C_t, is only slightly less than C_w in both cases, showing the dominance of the wall-building coefficient on the leakoff rate when a wall-building additive is used. Note that the leakoff volume in the high-permeability case is three times that of the low-permeability case because of spurt loss. This example illustrates the importance of recognizing high-permeability streaks so that spurt loss can be predicted accurately and a spurt-loss-control additive can be selected (Sec. 8.5). These examples also demonstrate that the layered nature of the reservoir provides results different from those obtained from considering only average permeability.

Influence of Filtrate Viscosity. In the above example, the viscosity of the base fluid (water) at bottomhole temperature (BHT) was used to calculate C_v. This approach is valid at low permeabilities where little or no gelled fluid enters the formation; however, at higher permeabilities, the viscosity of the fluid lost during spurt increases as the permeability increases. Only the fluid lost after spurt has the same viscosity as the base fluid.

The influence of filtrate-fluid viscosity was investigated with the data of Roodhart,[17] which shows n' and K' vs. permeability for a 40-lbm/1,000-gal [4793-g/m^3] hydroxyethyl cellulose (HEC) fluid. The viscosity of the filtrate in Table 8.2 was 1.7 cp [1.7 mPa·s] at 0.75 md and 2.4 cp [2.4 mPa·s] at 1.5 md (Table 8.5). These data show that the value of C_v decreases in the high permeability area during spurt.

As illustrated in Table 8.5, the increased filtrate viscosity will result in a lower value of C_{vc} during spurt. The lower value of C_{vc} will produce a longer spurt time. Thereafter, C_{vc} reverts to the value calculated at the viscosity of the base fluid at temperature. The spurt volume and C_t are unaffected. Because the final leakoff volume is the same and spurt times are relatively short compared with pumping times, the base-fluid viscosity is used exclusively in most fracture simulators.

A convenient method for quickly evaluating the influence of the three separate parameters is presented in Fig. 8.25. It is easy to see that C_v and C_c are cumulative in their effect. For example, a value of $C_v=C_c=0.01$ gives $C_{cv}=0.005$. The same nomogram shows that $C_{vc}=C_w=0.01$ gives $C_t=0.005$. On the other hand, when the C_w is significantly less than C_{vc}, then the C_t is only slightly less than C_w.

8.4 Laboratory Measurement of Fluid Leakoff

Fluid leakoff can be measured in the laboratory under static or dynamic conditions. This section presents a review of the equipment and procedures required for running these tests and analyzing fluid-loss data.

Equipment and Procedures for Static Core Testing. Standardized procedures for running static fluid-loss tests with core have been established in API *RP 39*.[9] The type of filter cell or core has not been standardized. The most common cells are the modified Baroid high-pressure, high-temperature cell that accepts a 15/16×1-in. [2.38×2.54-cm] core (Fig. 8.26) and the Halliburton cell that accepts either a 1¾- or 1⅛-in. [4.45- or 2.83-cm] -diameter core up to 4 in. [10 cm] long (Fig. 8.27). API *RP 39* specifies a core length of 1 in. [2.54 cm]. As pointed out later, the core length can be decreased to avoid core interference with the determination of wall-building coefficients.

Test pressure has been standardized to use N_2 to supply a 1,000-psi [6.9-MPa] pressure differential across the core, which represents a realistic pressure drop for most treatments. Variations in the pressure differential, however, will influence the rate of fluid leakoff (see Sec. 8.6).

The general procedure for running a static test is (1) to premix the gelled fluid with additives, (2) to place it into the test cell, (3) to heat to the desired temperature (standard temperature is 125°F [52°C] if none is specified), and then (4) to apply a 1,000-psi [6.9-MPa] pressure differential with N_2. The liquid outlet valve is then opened and the volume of accumulated filtrate is measured vs. time. Standard times of 1, 4, 9, 16, 25, and 36 minutes have been specified to allow easy plotting of volume vs. the square root of time (minutes$^{1/2}$) (Fig. 8.1). The test may need to be run longer if a

TABLE 8.5—FILTRATE-VISCOSITY PARAMETERS

$\bar{k}$ (md)	Length (ft)	n'_{80}	n'_{200}*	K'_{80}	K'_{200}	Viscosity** (10 seconds^{-1})	C_v Capacity Estimated†	C_v Capacity Actual‡
0.015 to 0.093	20	Water at 200°F				0.3	0.122	0.122
0.187	18	0.95	1.0	0.0001	—	0.3	0.1825	0.1825
0.37	12	0.9	1.0	0.0003	—	0.3	0.1712	0.1712
0.75	5	0.8	1.0	0.0088	0.000035	1.7	0.1	0.0425
1.5	2	0.58	0.91	0.0015	0.00006	2.4	0.057	0.0202
						Average	0.0102	0.00945

*The n' and K' of Roodhart *et al.*[7] were corrected to 200°F with the general temperature thinning profiles for HEC gels. The subscript for each power-law parameter represents the temperature at which the parameter was determined.
**The viscosity at 10 seconds^{-1} was used.
†See data in Table 8.1 assuming water was the leakoff fluid.
‡Calculated from actual viscosity data.

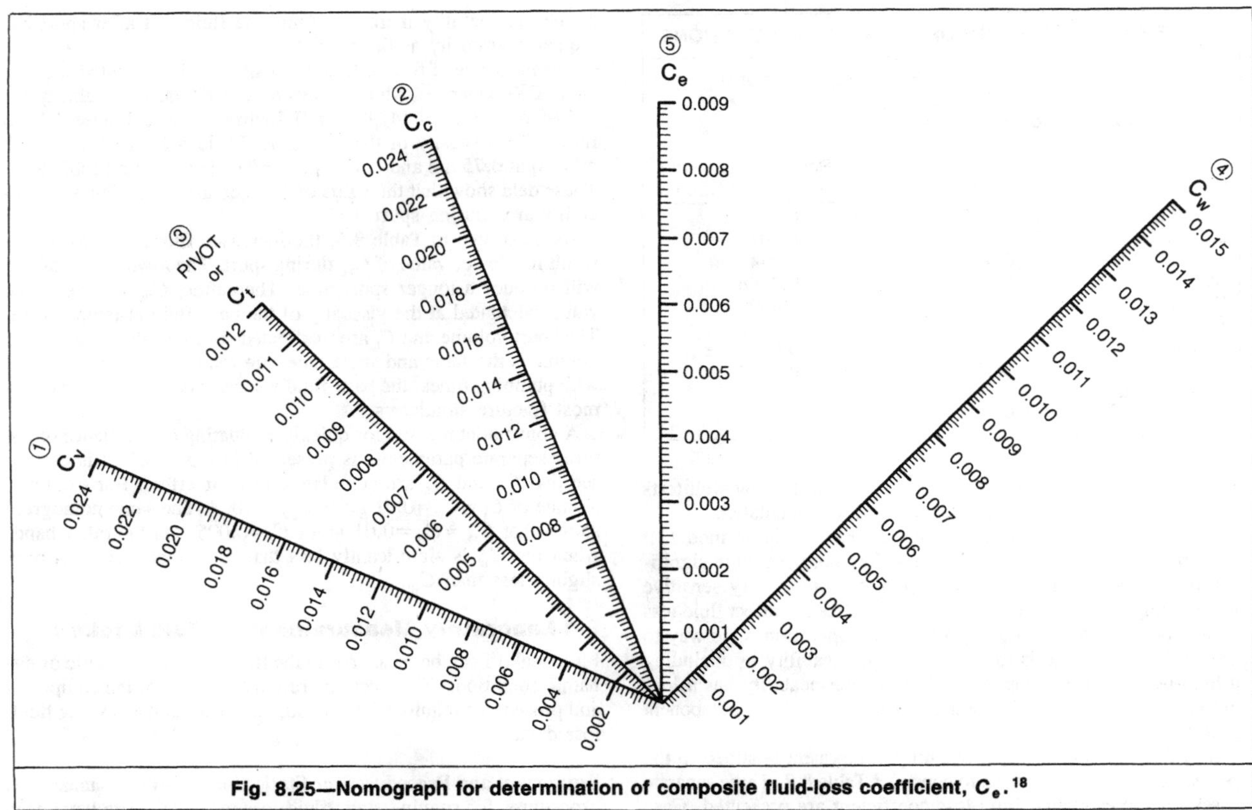

Fig. 8.25—Nomograph for determination of composite fluid-loss coefficient, C_e.[18]

steady-state leakoff rate has not been reached at the 36-minute point or if the data are to be used to design a treatment that will require a 4- to 12-hour (or more) pump time.

Calculation of C_w and Spurt Loss. The static fluid-loss data are plotted as filtrate volume vs. the square root of time. A least-squares fit of the later-time data (16 to 36 minutes) is used to determine the slope, m (in mL/min$^{1/2}$), and the y intercept, b (in mL). Spurt, V_s, is determined in gal/ft^2 from

$$V_s=\frac{b}{A}\times\frac{\text{gal}}{3{,}875}\times\frac{929}{\text{ft}^2} \quad \text{(8.19)}$$

$$=0.246\frac{b}{A}. \quad \text{(8.20)}$$

Spurt and C_w are sometimes evaluated by width or length. They should, however, be interpreted in volume/area. The volume of fluid lost at any point during pumping can be determined from

$$V=V_s+\frac{m}{A}t^{1/2}, \quad \text{(8.21)}$$

where

V = leakoff volume per unit area (or width) at t,
V_s = spurt volume per unit area,
m = slope of volume-vs.-$t^{1/2}$ plot,
A = area, and
t = exposure time for the unit area.

C_w is calculated by taking the derivative of the leakoff velocity, v, with respect to time:

$$\frac{dv}{dt} \text{ of } \frac{m}{A}t^{1/2}=\frac{m}{2At^{1/2}}=\frac{C_w}{t^{1/2}}, \quad \text{(8.22)}$$

$$C_w=\frac{m}{2A},\ \frac{\text{cm}}{\text{minutes}^{1/2}}, \quad \text{(8.23)}$$

or

$$C_w=\frac{m}{2A},\ \frac{\text{cm}}{\text{minutes}^{1/2}}\times\frac{0.0328\ \text{ft}}{\text{cm}}=\frac{0.0164m}{A},\ \text{ft/min}^{1/2}. \quad \text{(8.24)}$$

Core-Preparation Effects. Saturating core wafers before a fluid-loss test is run can have a dramatic effect on the calculated C_w values, particularly in short-term tests. The intent of core preparation is to minimize compressibility and relative permeability effects within the core that result from incomplete liquid saturation.[20] Because the filtrate volume is usually measured after it exits the core, the results of undersaturation are reduced spurt value and C_w. Saturation times of at least 24 hours are recommended, as well as evacuation of the core in a vacuum chamber before the saturating fluid is added. Flowing the core with the saturation fluid before the fluid-loss test also helps ensure complete saturation.

Procedures to Minimize Variations. Variations in C_w values calculated from laboratory fluid-loss tests can result from varying the fluid mixing procedures and the fluid pressurization sequence. It is best to simulate the field mixing sequence as closely as possible. For example, the addition sequence for a 40-lbm/1,000-gal [4793-g/m^3] crosslinked gel containing 5% diesel should be gelling agent and buffers, followed by the surfactant and diesel, with the crosslinker added last. Addition of the diesel earlier in the procedure will interfere with full gel hydration, producing spurious results.[20] Additionally, pressurization of the cell during heatup should be at a minimum. The final 1,000-psi [6.9-MPa] differential should be applied just as the test is beginning. Full pressurization before the fluid exit valve is opened caused some of the largest variations in leakoff volume and C_w in the static test. Even these procedural considerations will not negate early-time discrepancies resulting from core-permeability and -length effects.

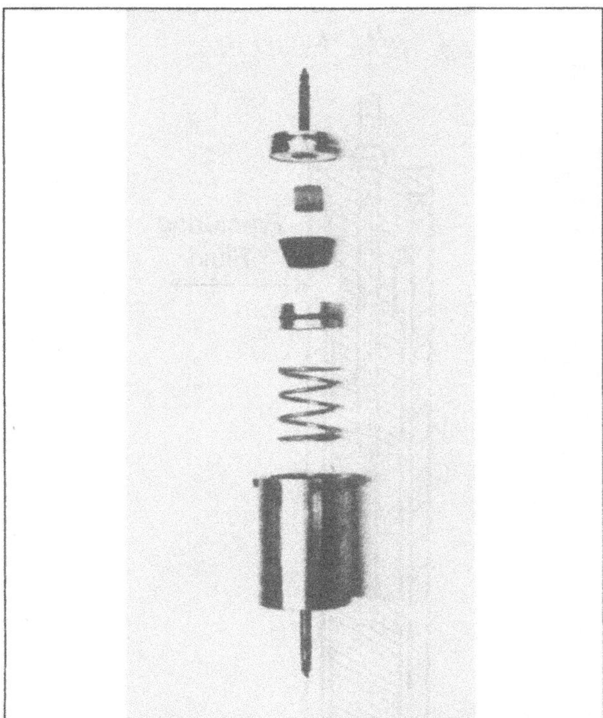

Fig. 8.26—Modified Baroid high-pressure filter cell (from API *RP 39*[19]).

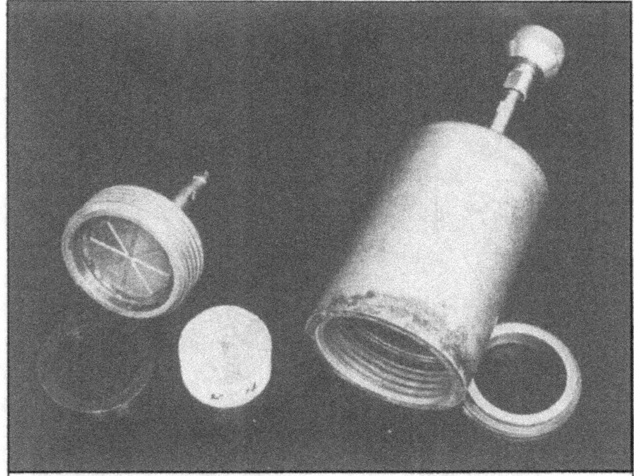

Fig. 8.27—Halliburton high-pressure cell (from API *RP 39*[19]).

Effect of Core Parameters on C_w Measurement. The permeability and length of the selected core can influence the measurement of C_w. For the core to have no effect on the C_w calculation, the pressure differential of the test should be adsorbed by the filter cake, not by the core. In a more practical sense, it is desirable that at least 95% of the pressure drop is across the filter cake. This becomes important at early times in low-permeability core where a significant pressure drop through the core is required to transmit the filter-cake fluid velocity through the core. The result can be a square-root-of-time plot in which the slope increases in the early time or throughout the test (Fig. 8.28).[21] Calculated C_w values will then be lower than the actual value, and a negative spurt value will be calculated.

The permeability and length of the core and/or the length of the test can be selected to achieve accurate C_w values. A plot of C_w effects vs. permeability and core length has been constructed (Fig. 8.29).[22] At 0.01 md, the core length must be 0.22 cm or less to calculate a C_w that is 95% of the ideal value of 0.001 ft/min$^{1/2}$ [0.0003 m/min$^{1/2}$]. Alternatively, it is possible to use a core length of 0.63 in. [1.6 cm] and only the data generated beyond 16 minutes.

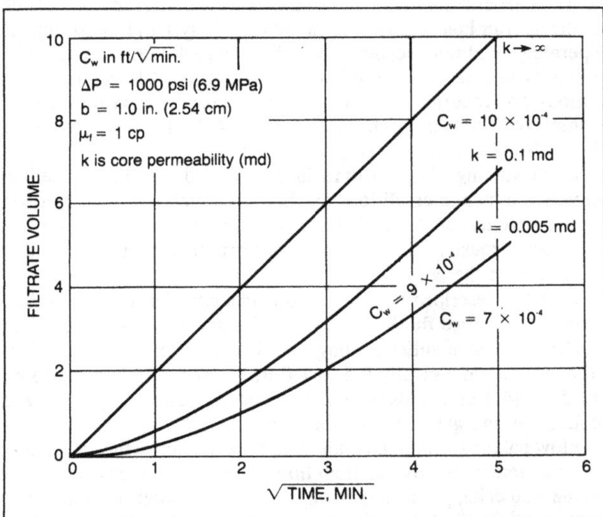

Fig. 8.28—Effect of core permeability on wall building (from Nolte[21]).

Measurement of Fluid Leakoff With Filter Paper. Filter paper is not used as often as core to measure leakoff properties because the filter paper lacks the core/fluid interaction that influences spurt. Filter paper is useful in determining C_w when a fluid/rock interaction is not responsible for reducing the permeability of the filter cake.[23]

When using filter paper, one should use three pieces of Whatman 50 hardened filter paper at a pressure differential of 1,000 psi [6.9 MPa]. Examples of data obtained on filter paper can be found in Ref. 24.

Natural Fracture Leakoff Tests. Leakoff through natural fractures is commonly simulated in a static or dynamic test with a metal disk or a fractured core. The most recent static test, known as a tapered-slot device, is described by Woo and Cramer.[25]

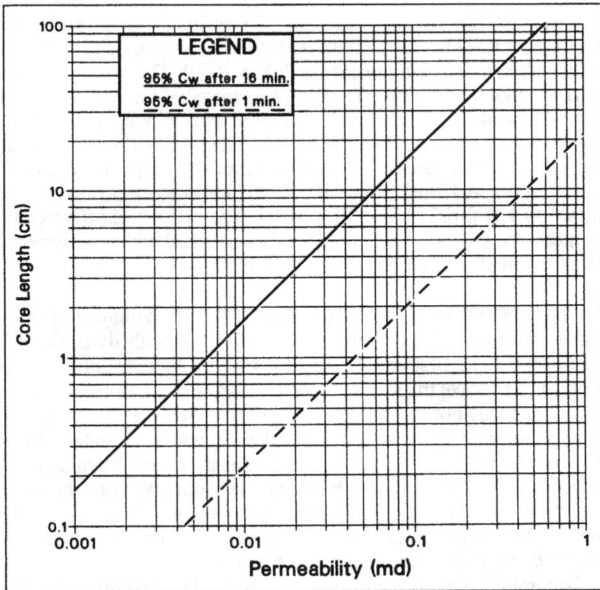

Fig. 8.29—Effect of permeability and core length on calculation of C_w, C_w (ideal) = 0.001 ft/(min)$^{1/2}$; p = 1,000 psi; T = 200°F; m = 0.3 cp.

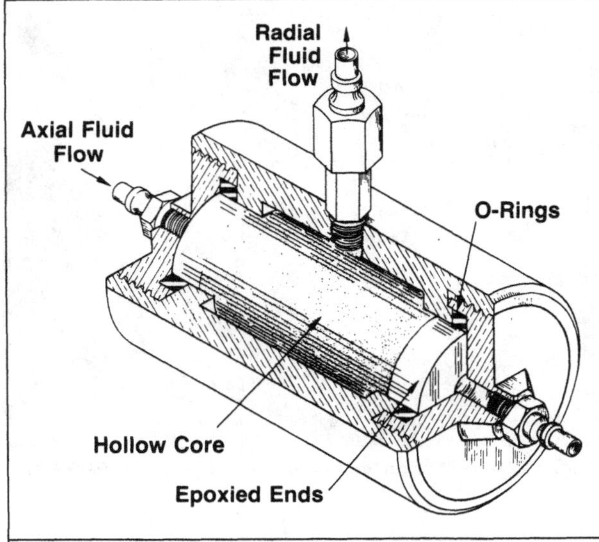

Fig. 8.30—Core and cell assembly for measuring natural fracture leakoff (from Penny *et al.*[20]).

Tapered-Slot Fluid-Loss Test. This test uses a 2.10×0.250-in. [5.33×0.64-cm] -thick stainless-steel disk. A 1.0-in. [2.54-cm] -long slot is cut in the center of the disk. The slot tapers from 0.020 in. [0.05 cm] wide on the inlet face to 0.010 in. [0.025 cm] wide on the exit face. The disk is placed in a 150-mL fluid-loss cell sealed with an O-ring. The cell is then preheated to 200°F [93°C] and loaded with the test fluid, in this case a crosslinked 40-lbm/1,000-gal [4793-g/m^3] HPG fluid containing a fluid-loss additive. The cell cap is secured, and the fluid in the cell is heated to 200°F [93°C]. The stem valves are opened and the test is begun as the 1,000-psi [6.9-MPa] pressure differential is applied.

The tapered-slot fluid-loss tests are performed essentially according to API *RP 39.*[19] Fluid-loss volumes are recorded at 1-, 4-, and 9-minute intervals. Extensive laboratory testing has indicated that in most cases very little additional fluid loss occurs at 16 and 25 minutes. If fluid loss is not controlled, the time is recorded when the cell completely empties (or a "blowout" occurs).[25]

Fractured-Hollow-Core Test. A dynamic test in which fluid is flowed through the center of a fractured hollow core has been reported.[20,26,27] In such tests, a 1.75×4-in. [4.4×10-cm] core containing a ⅛- to ¼-in. [0.32- to 0.64-cm] axial hole is mounted in a double-ended high-pressure cell (Fig. 8.30). The system is pressurized with water until a fracture is propagated through the core-cylinder walls. The flow system is then charged with gelled fluid containing various fluid-loss additives. The fluid is flowed through the core at shear rates of 10 to 100 seconds^{-1}, a pressure differential of 10 to 15 psi [69 to 103 kPa], and a system pressure of 1,000 to 1,500 psi [6.9 to 10.3 MPa]. The radial leakoff velocity is measured vs. time. (Examples of data obtained with both systems can be found at the end of Sec. 8.5.)

Dynamic Fluid-Loss Tests on Matrix. There are a number of principal differences between the results obtained with dynamic and static tests. The first and foremost is the development of an equilibrium filter-cake thickness where the leakoff rate becomes proportional to time, not the square root of time. Second, the length of time a dynamic test should be extended beyond the standard 36 to 90 minutes or longer to detect the nonlinear effects (with square-root-of-time plots) produced by shear. Dynamic tests also show the common fluid-loss additives to be much less effective than static tests. Thus, static tests can cause underestimation of the leakoff that will be experienced during a treatment.

Laboratory models for measuring matrix fluid leakoff under dynamic conditions have revolved around four designs: hollow cores, flow impinging on a core wafer, fluid stirring above a core wafer, and flow through a crack bounded by core wafers.

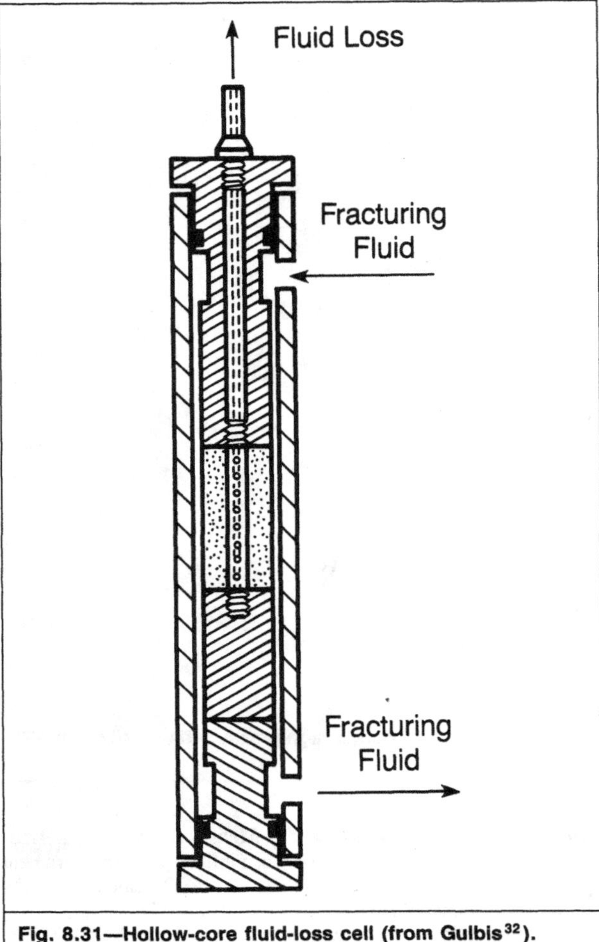

Fig. 8.31—Hollow-core fluid-loss cell (from Gulbis[32]).

The hollow core was introduced by Hall and Dollarhide[28,29] and was later examined by Sinha[30] to evaluate oil-based fluids. Gulbis[31,32] reported the testing of linear and complexed aqueous fluids through a hollow core (Fig. 8.31) after conditioning the fluid in a fracture simulator (Fig. 8.32).

Core samples reported by Gulbis[32] typically are 2 in. [5.1 cm] in length and 1 in. [2.54 cm] in diameter with a ¼-in. [0.64-cm] -diameter axial hole. Fluid is flowed through a ¼-in. [0.64-cm] annulus around the core, and leakoff is measured from the outside to the interior hole and vice versa. When this system is used, typical operating conditions reported are a 500-psi [3.4-MPa] pressure differential and shear rates of 48 to 160 seconds^{-1}. Similar conditions are reported by Penny *et al.*[20] for the measurement of complexed fluids through a 1.75×4-in. [4.4×10-cm] core with a ⅛-in. [0.32-cm] hole.

Fluid stirring above a core in a modified Baroid test cell to simulate dynamic conditions has been examined in a comparison of dynamic test methods.[20] The equipment is simple in design but does not adequately simulate conditions resulting from variable flow velocities at the core face.

Harris[33] described impinging flow on a core for measuring fluid leakoff of foamed fluids. The system features a standard fluid-loss cell (Fig. 8.33) modified to allow flow in and out of the system. Core dimensions are typically 0.5 to 1.0 in. [1.27 to 2.54 cm] long by 1.75 in. [4.4 cm] in diameter. The system features a pressurized collection and gas separation system.

Flow patterns within the impinging-flow cell make the calculation of shear rate very difficult. High flow velocities tend to create a hole in the wall cake, producing ambiguous results if wall-building fluids that create fragile filter cakes (i.e., HPG+5% diesel) are tested.[20]

Flow through a fracture simulator in the laboratory has been examined.[5,17,20,22,34] McDaniel *et al.*'s[34] system is illustrated in

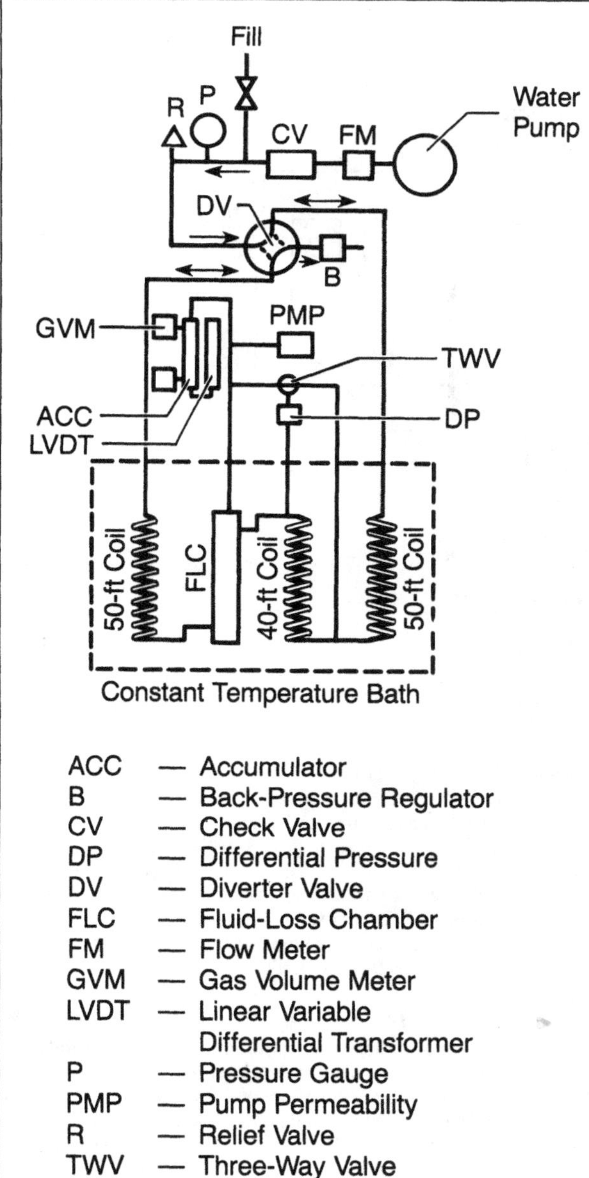

Fig. 8.32—Fracturing-fluid conditioning and fluid-loss system (from Gulbis[32]).

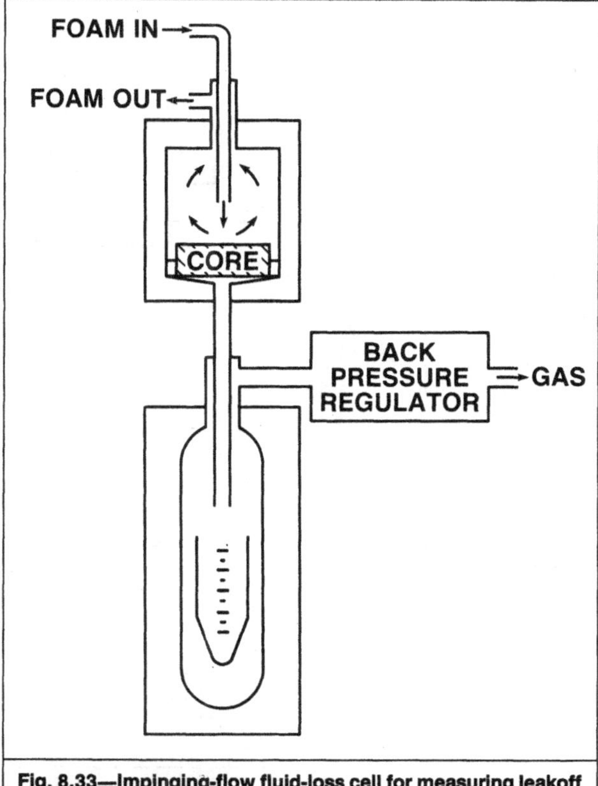

Fig. 8.33—Impinging-flow fluid-loss cell for measuring leakoff of foams (from Harris[33]).

Fig. 8.34. Fluid loss is measured through a 1.5-in. [3.8-cm] -diameter core of variable length (0.79 to 4.7 in. [2 to 12 cm]). The slot in the reported experiments was 0.25 in. [0.64 cm]. Leakoff was measured for several Newtonian and non-Newtonian (complexed) fluids at shear rates of 0 to 123 seconds^{-1} and pressure differentials of 100 to 1,000 psi [0.69 to 6.9 MPa].

A slot device featuring a $^{15}\!/_{16}\times 0.5$-in. [2.4×1.3-cm] core within a ⅛-in. [0.31-cm] slot has been reported by Penny *et al.* (Fig. 8.35).[22] Measurements were reported for linear and complexed fluids at shear rates of 10 to 300 seconds^{-1} and pressure differentials of 10 to 1,000 psi [0.069 to 6.9 MPa].

Roodhart[17] reported the use of a slot device capable of handling six cores simultaneously, each having a diameter of 4.2 cm and a thickness of 1.5 cm (Fig. 8.36). In his work, linear and complexed gels were measured at 0 to 600 seconds^{-1} at pressure differentials of 10 to 2,500 psi [0.069 to 17.2 MPa].

More recently, Penny *et al.*[35] used two 10-in.^{2} [64.5-cm^2] cores mounted in a modified API conductivity cell to measure leakoff. The system was equipped with a shear-history simulator.

Standard Dynamic Conditions. A standard set of conditions has not been established for dynamic testing. Obviously, the core configuration and size will depend on the selected equipment. The factors that can be standardized are the time of the test, the pressure differential, and the shear rates. As discussed above, the minimum time should be 90 minutes. The pressure differential should be 1,000 psi [6.9 MPa], and the shear rates should be 100 seconds^{-1} for mean-tip measurements and 40 seconds^{-1} for the majority of the fracture.[36] The temperature should be selected on the basis of reservoir conditions.

A pressure differential of 1,000 psi [6.9 MPa] is the accepted pressure because it seems to represent most field situations. It becomes important to establish a standard pressure differential when dynamic test results are compared because Δp and shear rate are interrelated in the determination of the rate of buildup of the filter cake. The effects of varying pressure differential are discussed in Sec. 8.5.

The selected shear rate can affect both fluid rheology and the rate of filter-cake deposition (Sec. 8.5); therefore, establishing a shear rate and shear history of the fluid becomes important. To simulate the high shear encountered during pumping down tubing, a shear rate and shear time should be selected that simulate pumping down tubing/casing at a given rate (e.g., a shear rate of 1,000 seconds^{-1} for 4 to 10 minutes). This step is often modified or ignored if the fluid does not contain a crosslinker or if the fluid contains a delayed crosslinker that will not yield during the initial shear period. Shear history has been simulated by pumping 5 to 10 minutes through ¼-in. [0.64-cm] tubing at ⅓ gal/min [1.3×10^{-3} m^3/min], followed by pumping 5 minutes through nominal 1-in. [2.54-cm] tubing.[35]

Heatup is generally accomplished during shearing at the shear rate of the subsequent fluid-loss test. It is recommended that the selected temperature and shear rates during heatup and testing be chosen to represent the conditions encountered throughout the majority of the treatment and near the fracture tip. Fracture shear rates have been published by Conway and Harris[36] and are discussed in Sec. 8.5. The common range of shear rates encountered

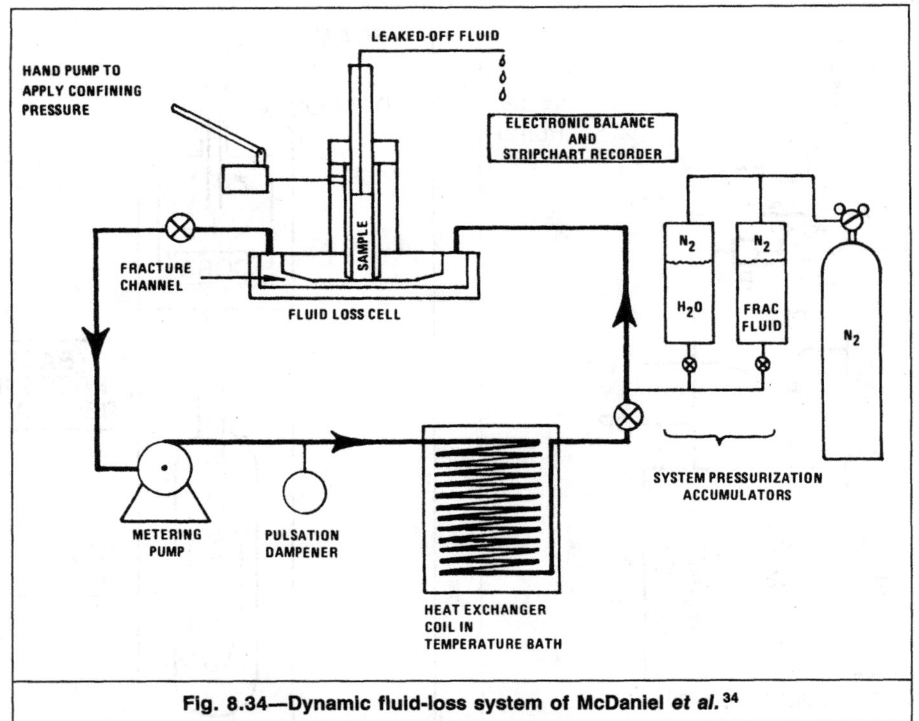

Fig. 8.34—Dynamic fluid-loss system of McDaniel *et al.*[34]

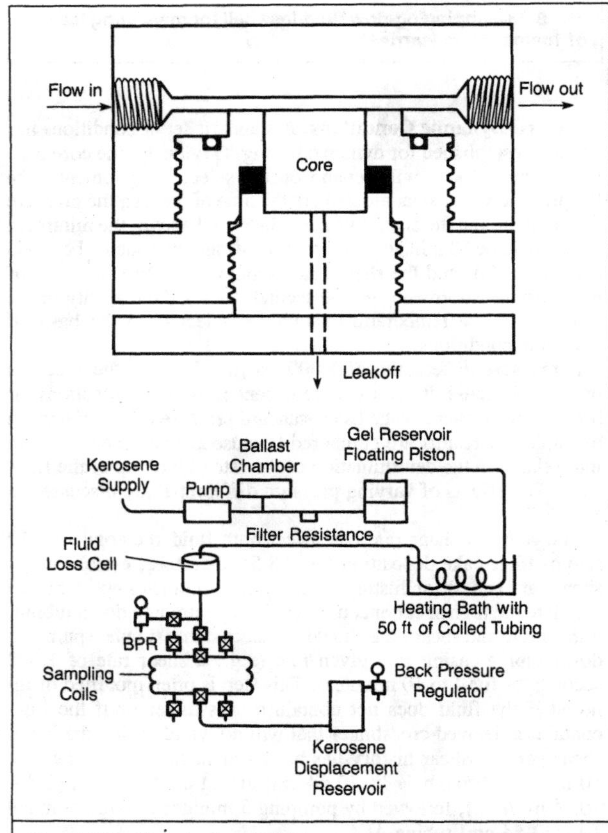

Fig. 8.35—Slot-device fluid-loss cell and loop of Penny *et al.*[22]

during a fracturing treatment has been estimated to be 5 to 200 seconds^{-1}, with 40 to 200 seconds^{-1} representing the shear encountered in high-injection-rate jobs, and 5 to 40 seconds^{-1} at low injection rates. It has been suggested[17,20,32] that standard shear rates should be 40 and 100 seconds^{-1}. The temperature should be selected on the basis of well conditions. No standard temperature has been established. Recent work by Penny[35] suggests modeling the cooldown temperature of a point 50 ft [15 m] into the fracture. Selected test temperatures vs. BHT's are given in Table 8.6.

8.5 Relative Effectiveness of Fluid-Loss Additives

The effectiveness of additives in controlling fluid leakoff is measured in the laboratory in terms of spurt-loss coefficient and wall-building fluid-loss coefficient. Both coefficients are influenced by (1) the type of base fluid, (2) the types of gelling agent and fluid-loss additive and concentration, and (3) the formation permeability and temperature. Data presented in this section come from static fluid-loss tests unless stated otherwise. Variations in the data caused by dynamic conditions are discussed in Sec. 8.6.

For the purposes of this discussion, the relative effectiveness of various additives will be discussed only from the standpoint of water-based, crosslinked fracturing fluids. A discussion of the fluid-loss properties of other fluids, such as water-based linear gels, foams, and oil-based gels, can be found in Sec. 8.9.

Spurt Loss. Spurt-loss control in crosslinked fluids without fluid-loss additives is a function of permeability, gel concentration, and temperature. The effects of permeability and gel concentration are illustrated in Fig. 8.37. In most crosslinked fluids (40 to 50 lbm/1,000 gal [4793 to 5991 g/m^3]), little or no spurt loss is seen until the permeability exceeds 1 md, because $C_{vc} < C_w$ under these circumstances. Above about 2 md, spurt loss is proportional to $k^{1/2}$.

Adding more gelling agent will decrease spurt loss significantly. As Fig. 8.21 shows, spurt loss is decreased from 0.3 to 0.008 gal/ft^2 [0.01 to 0.3×10^{-3} m^3/m^2] by increasing the polymer concentration from 20 to 80 lbm/1,000 gal [2497 to 9586 g/m^3] at a permeability of 10 md. A graph of spurt-loss volume vs. gel concentration with various amounts of silica flour is presented in Fig. 8.37.

Temperature affects spurt loss because the apparent viscosity of the fracturing fluid decreases with temperature. Although more rigorous treatments of this topic are possible, the most expedient

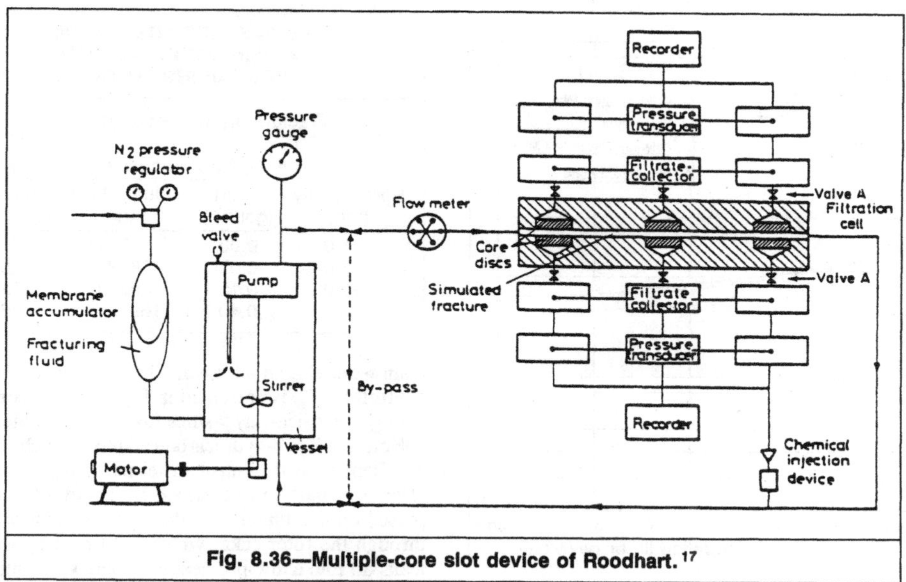

Fig. 8.36—Multiple-core slot device of Roodhart.[17]

TABLE 8.6—TEST TEMPERATURE vs. BHT's

BHT (°F)	Cooldown Temperature at 50 ft Into Formation (°F)
175	120
225	130
275	150
300	165

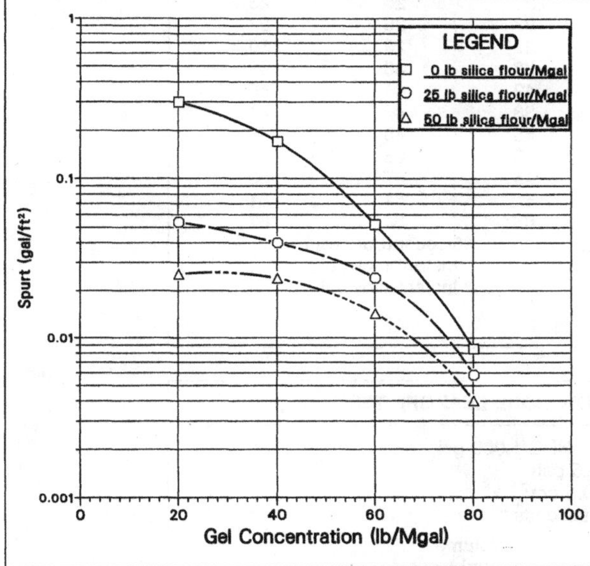

Fig. 8.37—Spurt loss vs. gel and silica-flour concentration for complexed HPG fluids on 10-md cores at 125°F.

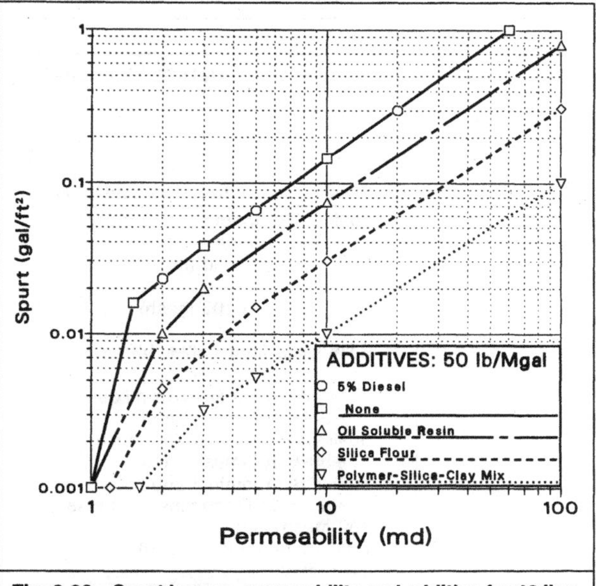

Fig. 8.38—Spurt loss vs. permeability and additive for 40 lbm complexed HPG fluids at 125°F.

manner to account for temperature effects is to apply the empirically derived adjustment factors presented in Fig. 8.22.[20]

Wall-Building Coefficient. The value of C_w for crosslinked fluids without fluid-loss additives depends on gelling-agent concentration and temperature. The value of C_w is independent of permeability, although calculated C_w values can be influenced by core effects, as explained in Sec. 8.4.

Gelling-agent concentration affects the value of C_w as illustrated in Fig. 8.23. C_w is decreased from 0.0040 to 0.0026 ft/min$^{1/2}$ [0.0012 to 0.0008 m/min$^{1/2}$] by increasing the polymer loading from 20 to 50 lbm/1,000 gal [2397 to 5991 g/m^3]. The effects of silica-flour concentration on C_w are also included in Fig. 8.23.

The C_w of complexed fluids increases as temperature increases. The change in C_w is proportional to the change in viscosity of the filtrate because the fluid flowing through the filter cake is water. Temperature correction factors to be applied are shown in Fig. 8.24.

Additives. Different types of additives have characteristic effects on spurt loss and C_w. Additives compared in this discussion are (1) silica flour, (2) oil-soluble-resin/water-soluble-polymer mixtures, (3) polymer/silica-clay mixtures, and (4) diesel.

The effect of silica flour on spurt loss is shown in Fig. 8.37. The trends are seen for all particulate additives; i.e., as the additive concentration is increased, spurt volume decreases. A comparison of various additives in a crosslinked HPG fluid is presented in Fig. 8.38. The order of effectiveness of additives in controlling spurt loss is as follows: polymer/silica-clay > silica flour > oil-soluble

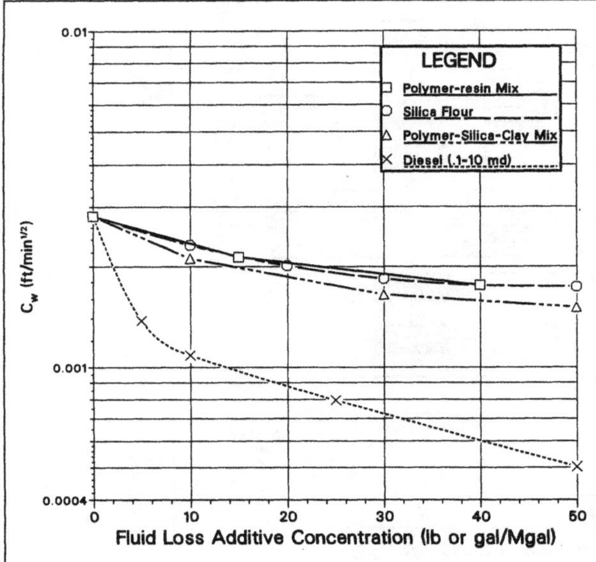

Fig. 8.39—Wall-building coefficient vs. fluid-loss additive type and concentration at 125°F.

TABLE 8.7—SPURT LOSS vs. PERMEABILITY FOR 40 lbm COMPLEXED HPG/1,000 gal CONTAINING 5% DIESEL (125°F)

Fracture area $=4\times500$ ft $\times$ 20 ft $=4\times10^5$ ft^2

	No Particulate		With 50 lbm Silica Flour/1,000 gal	
Permeability (md)	Spurt (gal/ft^2)	Volume Lost (gal)	Spurt (gal/ft^2)	Volume Lost (gal)
1.0	0.00	0	0	0
5.0	0.07	28,000	0.008	3,200
10.0	0.12	48,000	0.025	10,000
100.0	0.40	160,000	0.060	24,000

resin ≫ diesel ≧ no additive. The effect of the various additives in controlling C_w is presented in Fig. 8.39. In controlling C_w, diesel ≫ polymer/silica-clay > silica flour ≧ oil-soluble resin > no additive.

Note that liquid-hydrocarbon additives, such as 5% diesel, provide excellent leakoff control in terms of C_w but have little, if any, effect on spurt. As described by Penny *et al.*,[20] C_w with a dispersed oil is a function of the relative permeability of water (k_{rw}) through the filter cake, which is formed by the gel. As the fluid leaks off into a low-permeability matrix, the mechanism of control is the lowering of k_{rw} because of the trapping of oil within the filter cake. Because oil is not trapped until the cake is formed, oil will have little influence on spurt loss.

TABLE 8.8—TAPERED-SLOT FLUID-LOSS TEST[25] IN 50 lbm COMPLEXED HPG/1,000 gal

	Cumulative Fluid-Loss Volume at 200°F (mL)		
Fluid-Loss Additive, 100 lbm/1,000 gal	1 Minute	4 Minutes	9 Minutes
50% OSR* (nominal 100 mesh)	1	2	2
+ 50% 100-mesh sand	3	5	5
Polymer/inert-solid mixture	12	15	15
	23	25	28
OSR (nominal 100 mesh)	23	26	27
	27	29	30
50% silica flour + 50% 100-mesh sand	38	38	39
	42	44	45
50% polymer/inert-solid mixture	38	40	44
+ 50% 100-mesh sand	45	48	48
50% OSR (nominal 250 mesh)	33	38	43
+ 50% 100-mesh sand	45	47	53
50% OSR (nominal 250 mesh)	51	51	51
+ 50% OSR (nominal 100 mesh)	63	64	67
100-mesh sand		45-second blowout	
OSR (nominal 250 mesh)		Instant blowout	
Silica flour		Instant blowout	

*OSR = oil-soluble resin.

TABLE 8.9—FRACTURED-HOLLOW-CORE LEAKOFF TEST

Fluid = 2% KCl + 60-lbm HPG/1,000 gal
80°F ($\Delta p=25$ psi)
Core area = 40.5 cm^2
Shear rate = 100 seconds^{-1}

	Volume (mL)					
Additive	1 minute	4 minutes	9 minutes	16 minutes	25 minutes	36 minutes
1 lbm/gal 100 mesh*	50.0	200.0	—	—	—	—
50 lbm silica flour*	1.0	5.3	14.3	21.9	—	—
1 lbm/gal 100 mesh + 50 lbm silica flour*	7.0	25.0	42.0	54.0	63	—
0.3 lbm/gal 100-mesh OSR**	10.0	22.0	33.0	—	—	—
0.3 lbm/gal 100-mesh OSR** + 5 gal 325-mesh OSR	7.0	15.0	19.0	—	—	—
0.3 lbm/gal 100-mesh OSR** + 50 lbm silica flour	5.0	10.0	14.0	—	—	—

*Ref. 20.
**Refs. 26 and 27. OSR = oil soluble resin.

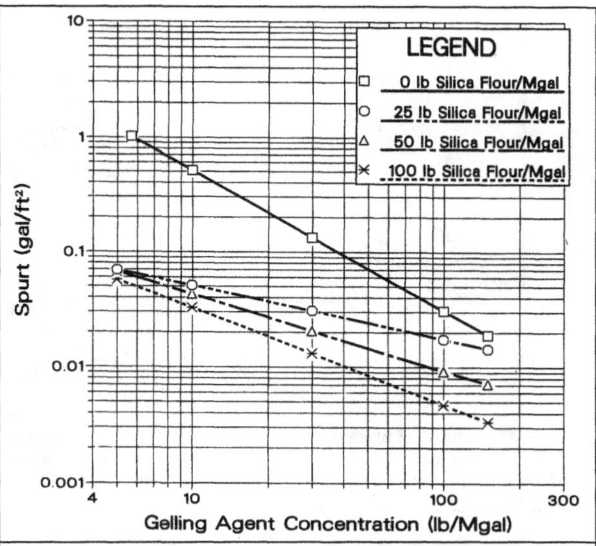

Fig. 8.40—Spurt loss vs. gel and silica-flour concentration for linear gels in water at 125°F through 5- to 10-md cores.

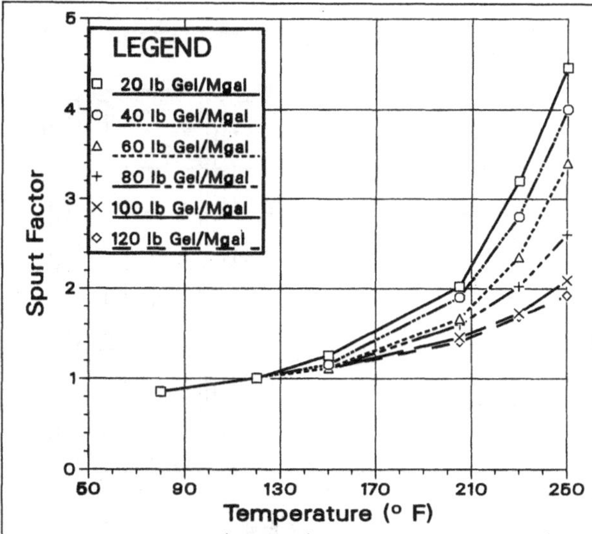

Fig. 8.41—Spurt correction factor vs. temperature with various linear-gel concentrations.

Liquid hydrocarbon offers excellent leakoff control in terms of C_w at all permeabilities, provided that the gel forms a filter cake. However, liquid hydrocarbon does not lower spurt loss below that of the gel alone. Therefore, a particulate additive must be added in combination with hydrocarbon to control spurt loss at formation permeabilities >5 md or in zones containing >5-md streaks. Table 8.7 compares the volume of fluid lost with and without silica flour vs. permeability of a typical fracture. At 10 md, 48,000 gal [182 m^3] would be lost with 5% diesel alone, while only 10,000 gal [38 m^3] will be lost with 5% diesel and 50 lbm/1,000 gal [5991 g/m^3] silica flour.

A logical fluid-loss additive package would be a combination of silica flour and hydrocarbon. The silica flour controls leakoff to high-permeability areas (spurt) and the hydrocarbon follows to reduce C_w.

Additives for Natural Fractures. Various additives have been evaluated for their relative effectiveness in controlling leakoff to natural fractures in both the tapered-slot fluid-loss test[25] and the fractured-hollow-core dynamic tests.[20,26,27]

Data from the tapered-slot fluid-loss test are presented in Table 8.8. From these tests, the three best fluid-loss additive systems in crosslinked fluids, in order of their effectiveness, were (1) a mixture of oil-soluble resin (nominal 100 mesh) and 100-mesh sand, (2) a polymer/inert-solids mixture, and (3) an oil-soluble resin (nominal 100 mesh).

In dynamic fractured-hollow-core tests, where the natural fractures were 10 to 20 μm wide, combinations of 100-mesh sand and silica flour revealed that the silica flour alone reduces leakoff to controllable values, while 100-mesh sand alone has no effect in 60-lbm/1,000-gal [7190-g/m^3] linear gels (Table 8.9).[26,27] In a related study, a combination of oil-soluble resin and silica flour was found to be superior.[20] From these studies, it appears that silica flour is very necessary for control of leakoff to natural fractures.

Sample Problem. Calculate the spurt and C_w for a fracturing fluid to be pumped at these conditions: 5-md permeability, 175°F BHT, a prepad of 20 lbm HPG+25 lbm silica flour per 1,000 gal [2397+2997 g/m^3], and a pad and sand-laden fluid of 40 lbm complexed HPG+25 lbm silica flour per 1,000 gal [4793+2997 g/m^3].

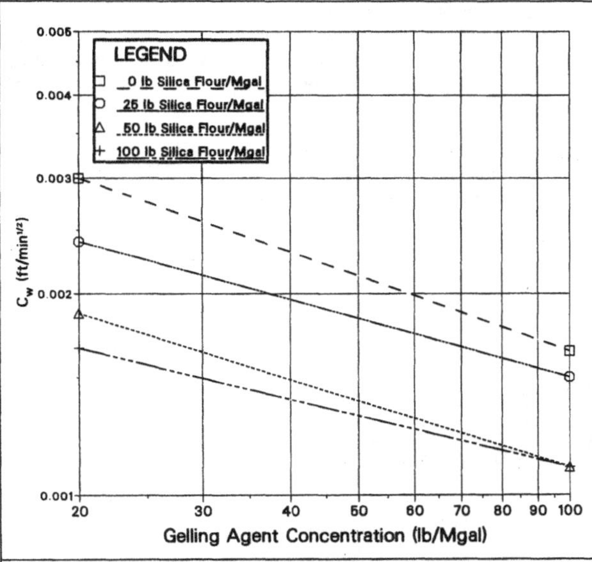

Fig. 8.42—Wall-building coefficient vs. silica-flour concentration for linear gels at 125°F through 0.1- to 100-md cores.

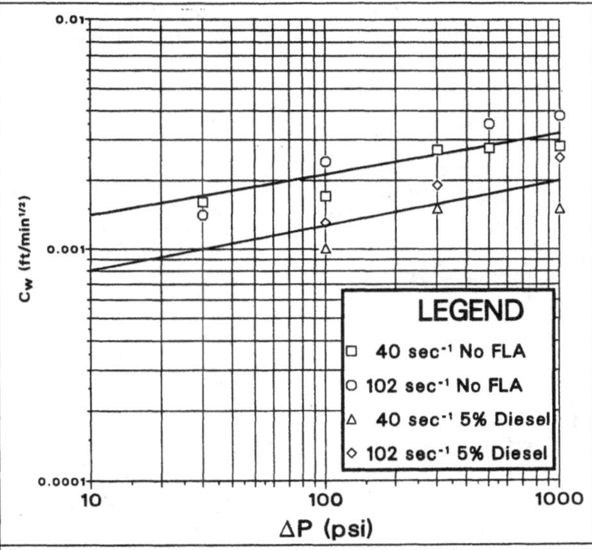

Fig. 8.43—Dynamic leakoff data for 40 lbm complexed HPG gels as a function of pressure drop: permeability=0.1 md; temperature=175°F (Penny *et al.*[22]).

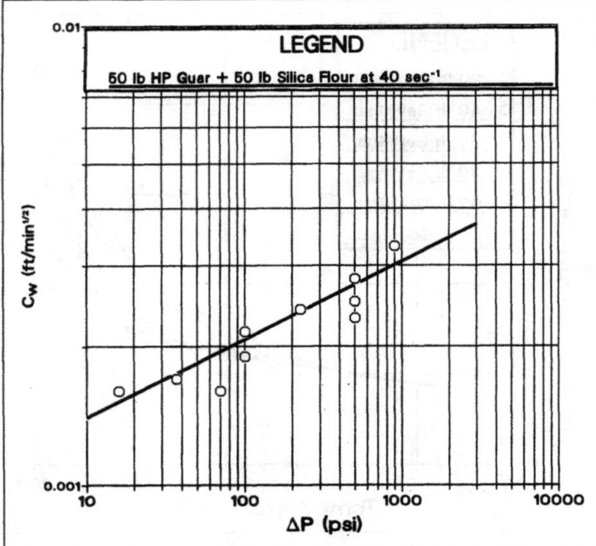

Fig. 8.44—Dynamic leakoff data for linear HPG gels as a function of pressure drop: permeability = 0.1 to 40 md; shear rate 5 to 40 seconds $^{-1}$; temperature = 175°F (Penny *et al.*[22]).

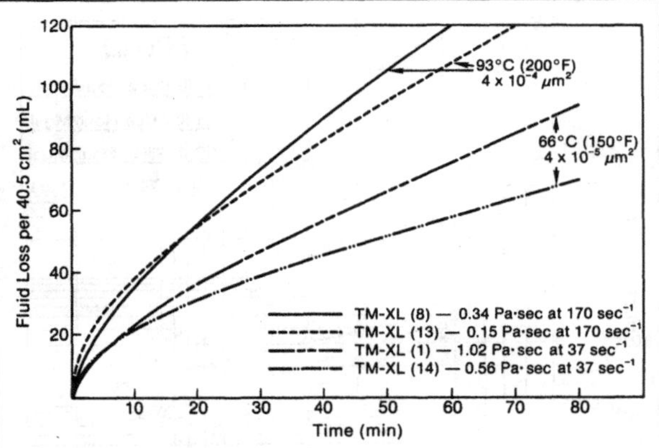

Fig. 8.45—Effect of shear, shear history, and temperature on dynamic fluid loss.[31] **[Previously published in the *J. Cdn. Pet. Tech.* (Oct.–Dec. 1975) 14, No. 4, 77–80; reprinted with permission of the Canadian Inst. of Mining and Metallurgy.]**

Prepad. Refer to Fig. 8.40. At 5 md, spurt for a 20 lbm gel+25 lbm silica flour per 1,000 gal=0.036 gal/ft^2. To correct for temperature, apply a factor of 1.6 from Fig. 8.41.

Prepad spurt=0.036×1.6=0.058 gal/ft^2.

Refer to Fig. 8.42 to determine C_w. C_w for 20 lbm HPG with 25 lbm silica flour equals 0.0024 ft/min^{½}. Apply a temperature correction factor of 1.2 (Fig. 8.24).

Prepad C_w=0.0024×1.2=0.0029 ft/min^{½}.

Pad and Sand-Laden Fluid. For a 40-lbm complexed gel, refer to Fig. 8.37 to determine spurt at 125°F. Spurt is 0.040 gal/ft^2 with 25 lbm silica flour. Apply the temperature correction factor of 1.2 from Fig. 8.22.

Pad spurt=0.048 gal/ft^2.

Refer to Fig. 8.23 to determine C_w. C_w at the desired fluid composition=0.0023 ft/min^{½}. Apply the correction factor of 1.2 (Fig. 8.24).

Pad C_w=0.0023×1.2=0.0028 ft/min^{½}.

8.6 Environmental Effects

Effect of Pressure Drop on C_w. Ideally, C_w should vary as the square root of p. This is, however, true only if the filter-cake permeability is constant. Roodhart's[17] data indicate that gels containing particulate additives and hydrocarbon are incompressible above 300 psi [2.1 MPa] and the permeability of the filter cake is a function of Δp. Tests performed by Penny *et al.*[22] indicate that filter cakes with and without particulate additives are compressible at all values of Δp. Note that the values of C_w vs. pressure in Figs. 8.43 and 8.44 exhibit a slope of 0.17 on a log-log plot of C_w vs. Δp, indicating a dependence on $p^{1/6}$ instead of $p^{½}$. As discussed by Nolte, C_w for a compressible gel is proportional to $(p^{⅓})^{½}=p^{1/6}$.[37,38] To correct a C_w value measured at Δp_1 to an equivalent C_w value at a different pressure drop, Δp_2, use the relationship

$$(C_w)_2=(C_w)_1*\left(\frac{\Delta p_2}{\Delta p_1}\right)^{1/6}, \quad (8.25)$$

where $(C_w)_2$ is the equivalent fluid-loss coefficient at Δp_2 if the fluid-loss coefficient $(C_w)_1$, measured at Δp_1, has a compressible gel filter cake.

The pressure differential during an actual treatment is usually about 1,000 psi [6.9 MPa] (BHTP−BHP) near the wellbore and decreases near the tip. Pressure differentials as high as 3,000 psi [20.7 MPa] have been observed in some reservoirs. In the higher-pressure case, leakoff velocities will increase by a factor of $3^{½}$, or 1.73, over the 1,000-psi [6.9-MPa] laboratory values with incompressible filter cakes and by a factor of $3^{1/6}$, or 1.2, with compressible filter cakes.

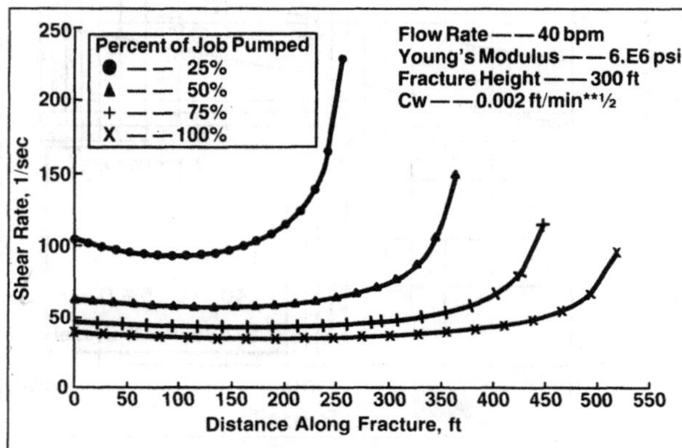

Fig. 8.46—Shear rate vs. distance along a fracture (from Conway and Harris[36]).

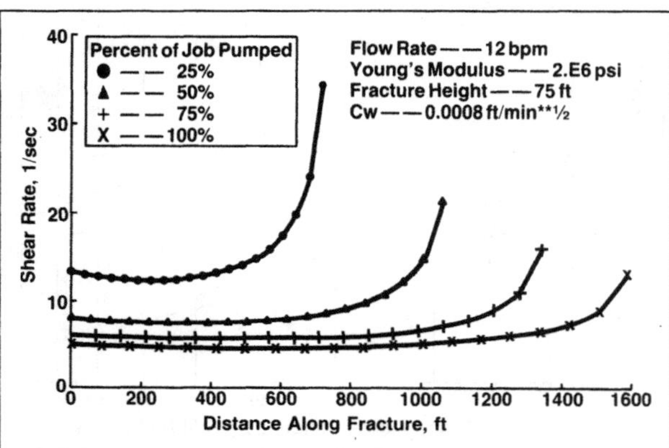

Fig. 8.47—Shear rate vs. distance along a fracture (from Conway and Harris[36]).

TABLE 8.10—DESIGN PARAMETERS FOR FRACTURE-SHEAR-RATE CALCULATIONS[36]

Khristianovich-Zheltov[46] Model

	Minimum to Maximum
Treating rate, bbl/min	12 to 40
Young's modulus, psi $\times 10^6$	2 to 6
Fracture height, ft	75 to 300
C_w, ft/min$^{1/2}$	0.0008 to 0.002
Permeability, md	1.0
Porosity	0.15
Formation compressibility, psi^{-1}	8.3×10^{-5}
Reservoir-fluid viscosity, cp	0.02
n'	0.45
K', (lbf-sec)/ft^2	0.04
Treatment volume, gal	250,000

The shear rate vs. distance along the fracture is plotted in Figs. 8.46 and 8.47 for a broad range of treating conditions. It can be seen that at low flow rates (12 bbl/min), the shear rate even near the tip of the fracture does not exceed 40 seconds^{-1}. At high flow rates (40 bbl/min), the shear rate is $\geq$100 seconds^{-1} in the first 25% of the job and approaches 250 seconds^{-1} near the tip. Throughout most of the job, the shear rate is close to 40 to 60 seconds^{-1}.

Leakoff after a shut-in should decrease proportionally to $\Delta p^{1/2}$ as the pressure bleeds off even with a compressible filter cake. Once the filter cake is compressed, the permeability tends to remain constant as the pressure decreases.

Shear-Rate Effects. The effects of shear rate on fluid leakoff have long been recognized. Several workers have shown that increasing shear results in an increase in leakoff rate over static tests.[5,22,27-32,34,35] Gulbis'[32] data shown in Fig. 8.45 demonstrate the differences in leakoff possible at shear rates of 37 and 170 seconds^{-1}. This being the case, it becomes important to know the shear rate that exists in typical treatments. Average fracture shear rates (Figs. 8.46 and 8.47) were calculated by Conway and Harris[36] using the treating parameters given in Table 8.9.

Shear Effects in Ungelled Fluids. Shear-rate effects are highly dependent on the type of fluid being pumped. Hall and Dollarhide[28,29] observed that kerosene containing 50 lbm/1,000 gal [5991 g/m^3] of a particulate additive follows a $t^{1/2}$ trend for the first 4 to 5 minutes and then becomes linear with time. The leakoff velocity increased linearly from 0.00015 to 0.00035 ft/min [0.046×10^{-3} to 0.11×10^{-3} m/min] as the shear rate was increased from 0 to 1,500 seconds^{-1}.

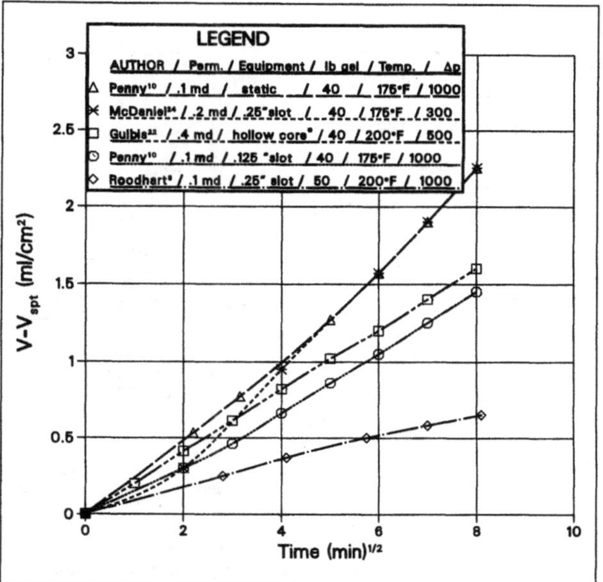

Fig. 8.48—Comparison of dynamic fluid-loss data for 40 or 50 lbm titanate complexed HPG gels at 100 seconds^{-1}.

Variations in the type of oil used were also found to affect the leakoff-vs.-shear profile. The use of crude oil instead of kerosene reportedly resulted in a leakoff rate that approached a square-root-of-time relationship. This was attributed to the asphaltenic and paraffinic insolubles within the tested crude, and possibly the viscosity increase in the oil.[29]

The addition of 50 lbm [22.7 kg] of particulate to water at nearly 1,000-seconds^{-1} shear followed the square root of time for 15 minutes and then approached a linear relationship with time. A calculated leakoff-velocity profile would be 0.0017 ft/min$^{1/2}$ [0.52×10^{-3} m/min$^{1/2}$] for 15 minutes and then increase to 0.0005 ft/min [0.15×10^{-3} m/min].[29]

Shear Effects in Gelled Water. It is generally agreed that base gels or linear gels (noncrosslinked) made from guar, derivatized guar, and derivatized cellulose provide leakoff profiles that are in-

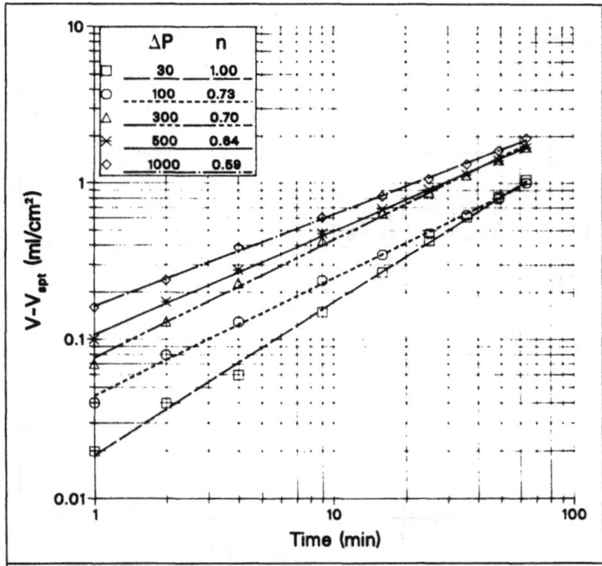

Fig. 8.49—Effect of pressure drop on dynamic leakoff data for 40 lbm HPG gels complexed with delayed titanate at 175°F and a 100-second^{-1} shear rate: permeability = 0.1 md.

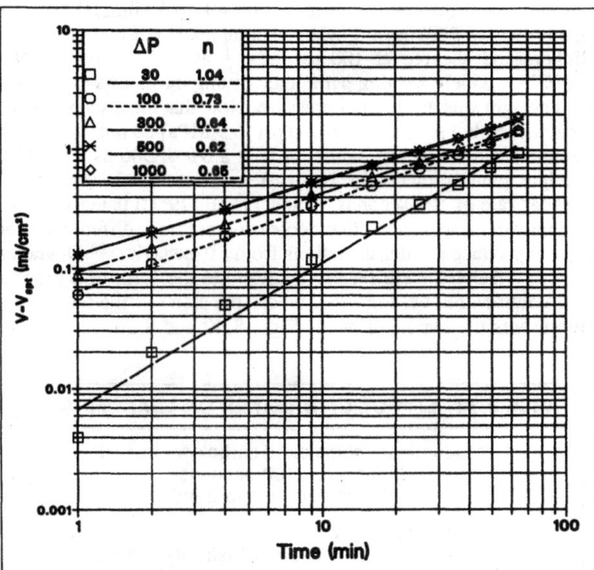

Fig. 8.50—Effect of pressure drop on dynamic leakoff data for 40 lbm HPG gels complexed with delayed titanate at 175°F and a 100-second^{-1} shear rate: permeability = 0.1 md.

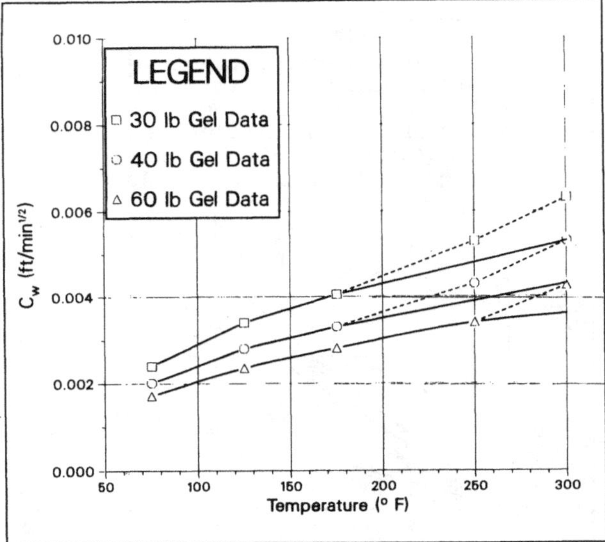

Fig. 8.51—C_w vs. temperature for titanate complexed HPG gels: solid lines = C_w at 80°F × $(\mu_w)^{-1/2}$.

dependent of shear rate. Gulbis[32] and Penny *et al.*[22] have observed that leakoff profiles of a 40-lbm/1,000-gal [4793-g/m^3] HPG are identical at shear rates of 0 to 300 seconds^{-1}.

Several reports indicated that the leakoff profiles of crosslinked fluids *are* dependent on shear rate.[5,22,27-32] Variations in equipment and testing procedures, however, have produced a great deal of uncertainty.

Latent factors that may affect the measured leakoff profile are the complexor chemistry and mixing procedure, the shear history before and during heat-up, the shape of the pathway past the core surface, the core permeability, thickness, and saturation history, the final temperature of the fluid and the cell, the total pressure of the system, and finally the pressure drop across the core. In the work of McDaniel *et al.*,[34] the leakoff profile of a 50-lbm/1,000-gal [5991-g/m^3] complexed HPG fluid increased with increasing shear rates (0 to 123 seconds^{-1}) through a ¼-in. [0.64-cm] slot at a Δp of 300 psi [2.1 MPa] and at 80°F [27°C]; however, the shear-rate dependence disappeared at 175°F [79°C].

Fig. 8.48 shows a comparison of the leakoff vs. $t^{1/2}$ values obtained by four different laboratories for 40- to 50-lbm HPG/1,000-gal [4793- to 5991-g/m^3] systems complexed with titanate crosslinkers and sheared at 100 seconds^{-1} at 175 to 200°F [79 to 93°C]. A widely varying parameter among these tests (apart from cell design) was differential pressure. Note that the tests performed at a differential pressure of 1,000 psi [6.9 MPa] do not vary greatly from a $t^{1/2}$ relationship, while those at a Δp of 300 psi [2.1 MPa] deviate quite dramatically. In a systematic study of varying Δp at a shear rate of 100 seconds^{-1} and 175°F [79°C] using a delayed complexor, it was found that decreasing pressure differential does in fact produce greater deviations from $t^{1/2}$ behavior.[22] Deviations from $t^{1/2}$ behavior are readily analyzed in the form of a graph of log volume/area vs. log time, where the slope of the log-log plot represents the exponent of t^n (Figs. 8.49 and 8.50; see Sec. 8.7 for further discussion). If leakoff is following the square root of time, $n=0.5$. As the pressure differential decreases from 1,000 to 10 psi [6.9 to 0.07 MPa], it can be seen that n increases from 0.5 to 1.0; in other words, it follows time in the low-Δp example.

TABLE 8.12—EFFECT OF BREAKER ON LEAKOFF AT 75°F

Gel = 40 lbm aluminum complexed carboxymethyl HPG/1,000 gal
Shear rate = 100 seconds^{-1}
Permeability = 1 to 2 md

Time at 100 seconds^{-1} and 75°F (hours)	Apparent Viscosity (cp at 100 seconds^{-1})	Spurt (gal/ft^2)	C_w (ft/min$^{1/2}$)
0.5	240	0.014	0.0027
2.0	67	0.024	0.0026
4.0	30	0.093	0.0026

TABLE 8.11—EFFECT OF TIME AT 250°F ON FLUID LOSS[22]

Gel = 40 lbm titanate complexed HPG/1,000 gal
Shear rate = 100 seconds^{-1}
Permeability = 1 to 2 md

Time at 250°F (hours)	Apparent Viscosity (cp at 100 seconds^{-1})	Spurt (gal/ft^2)	C_w (ft/min$^{1/2}$)
0	119	0.011	0.0042
2	4	0.086	0.0036
4	2	0.19	0.0036

Effect of Temperature. Because spurt loss is a function of gel viscosity, spurt volume will increase as gel viscosity decreases (see Fig. 8.21 and the discussion in Sec. 8.5). C_w, on the other hand, is controlled by the viscosity of the base fluid, which is water in most cases, until the base polymer begins to degrade with temperature. The temperature at which a given polymer concentration loses effectiveness can be determined by referring to Fig. 8.51, which presents values of C_w vs. temperature. Calculated points are computed by multiplying the C_w at 80°F [27°C] (where the viscosity of water $\simeq 1.0$) by the square root of the inverse of the viscosity of water at reservoir temperature. Note that the measured values follow the calculated values nicely until ~200°F [~93°C] for a 30-lbm/1,000-gal [3595-g/m^3] gel and up to 250°F [121°C] for a 60-lbm/1,000-gal [7190-g/m^3] gel. These points of departure are considered the thermal degradation point of each polymer loading.[21,22]

Effect of Time at Temperature. When the encountered temperature is greater than the thermal degradation point of a given polymer loading, long-term exposure to temperature will lead to a decrease in gel viscosity and consequently an increase in spurt loss. C_w remains the same, however, even after the viscosity of the gelled fluid approaches 10 cp [10 mPa·s] at 100 seconds^{-1} (Table 8.11).[22] Although the molecular weight of the polymer is reduced by degradation, the absolute amount of polymer available to build a filter cake remains constant. Similar experiments have been run by Zigrye *et al.*[39] at 350°F [177°C].

Effect of Breaker on Leakoff at Low Temperatures. The effect of a breaker on leakoff at low temperatures is similar to the effects of high temperature. The spurt loss of the breaking fluid increases, while the C_w shows little change (Table 8.12).

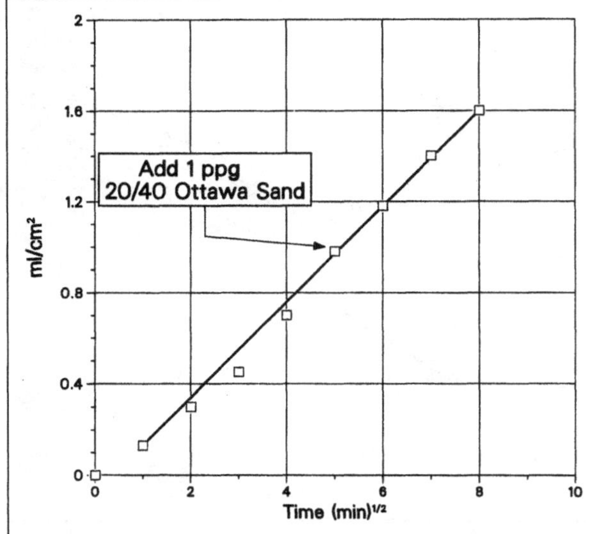

Fig. 8.52—Effect of proppant addition on leakoff observed with 40 lbm complexed HPG gel in a hollow core at 40 seconds^{-1} (from Penny *et al.*[20]).

Effect of Proppant. Experiments conducted by Hall and Dollarhide[29] and Penny[35] and Penny *et al.*[20] have shown that the addition of low concentrations of proppant to a gelled fluid under dynamic conditions has no measurable effect on the leakoff, provided that a filter cake has previously been established (Fig. 8.52). Experimental evidence indicates that the proppant tends to flow in the center of a stream, thus minimizing the possibility of filter-cake scouring. These experiments were conducted with a ⅛- to ⅓-in. [0.32- to 0.85-cm] simulated fracture width and only 1 to 3 lbm/gal [120 to 359 kg/m^3] of 20/40-size sand. Decreased widths, more sand, larger-size sand, or increased shear rates could very well show a difference.

8.7 Modeling of Combined Effects

The equation describing the volume per unit area of fluid lost to the formation (in gal/ft^2) where wall building dominates is[1,4]

$$V=V_s+2C_w(t)^{1/2}. \quad (8.10)$$

In Eq. 8.10, spurt loss is considered to be instantaneous, and t represents the exposure time of a unit area of the fracture. When spurt loss is not instantaneous, the equation can be modified as follows:

$$V=V_s+2C_w(t-t_s)^{1/2}, \quad (8.26)$$

where t_s is generally calculated from the relationship

$$t_s=\left(\frac{V_s}{2C_{vc}}\right)^2. \quad (8.11)$$

The above equations assume a square-root dependence on Δp, filter-cake thickness, and time. Most simulators assume that leakoff is proportional to the square root of time and not a function of pressure. In many cases, however, fluid loss varies as $p^{1/6}$ to $p^{1/2}$, and fluid flow in the fracture (shear) inhibits filter-cake buildup beyond some equilibrium point of filter-cake growth.

Several methods have been proposed to handle shear effects. In one method, leakoff in a dynamic test was defined by the following equations[3,5]:

$$V=V_s+v_n(t-t_s), \quad (8.27)$$

where v_n=the leakoff velocity in ft/min. Alternatively, V can be expressed as

$$V=V_s+C_d(t) \text{ or } (t-t_s), \quad (8.28)$$

where

$$C_d=0.0328\frac{m'}{A}, \quad (8.29)$$

m'=slope of volume vs. time, and
A=area of core.

In this analysis, the spurt volume per unit area is usually fairly large because it encompasses C_{vc} and any wall-building character, C_w, before establishing an equilibrium leakoff velocity, C_d, which approaches proportionality to time.

In a related analysis, Clark and Barkat[40] recently proposed that the leakoff profile of a dynamic test can be modeled by

$$V=V_s(1-e^{-bt})+v_Dt, \quad (8.30)$$

where
V_s = spurt volume per unit area,
b = pack-buildup constant, and
v_D = equilibrium Darcy flow velocity.

Once again, C_{vc} and any wall-building effects are taken into account in the V_s term. The shape of the line is governed by b, the pack-buildup constant, which is a function of gel type and additives. The term b may be able to reflect environmental conditions, such as shear-rate effects and Δp. It is calculated from a curve-fitting simplex algorithm.[40] Examples provided by Clark and Barkat indicate values of b in the range of 0.2 to 0.8 per minute.

The calculated volume of fluid lost to the formation by this method is the same as the volume calculated by the original method, except that it can accurately model the volume of leakoff vs. time before the spurt time is reached. This calculation method requires a knowledge of volume, V_s, and time to be useful.

In a second analysis method proposed by Roodhart,[17] a wall-building phase is made a part of the analysis. The modified equation takes the form

$$V=V_s+2C_wt_A^{1/2}+C_dt_B, \quad (8.31)$$

where
V_s = spurt volume per unit area,
C_w = wall-building coefficient (or A in Ref. 17),
t_A = time at which fluid loss is proportional to $t^{1/2}$ (wall building),
C_d = dynamic leakoff coefficient (or B in Ref. 17), and
t_B = time at which leakoff is proportional to t (or $t-t_A$, where t is total time).

In data generated by Roodhart, V_s and C_w are comparable to static data. The time where leakoff has changed from $t^{1/2}$ proportionally to being proportional to time, t_A, looks to be about 50 minutes at 100 seconds^{-1}, at which point the leakoff approaches a linear time relationship. By this method, it may be possible to estimate dynamic leakoff volumes (i.e., 100 seconds^{-1}) from static test results, *although the actual laboratory data would be preferred.* To do so, the following procedure would be used.

1. Use static spurt value.
2. Apply static C_w value for t_A (~50 minutes).
3. Calculate C_d at t_A, the leakoff velocity during the last minute at C_w.

$$C_d=C_w[t_A^{1/2}-(t_A-1)^{1/2}]. \quad (8.32)$$

4. Apply C_d over the remainder of the treatment time, t_B.

In a third approach, it has been observed that the leakoff profile at 40 to 100 seconds^{-1} can be described by a power of time between 0.5 and 1.0, depending on the shear rate and Δp.[22] Thus, both the shear rate and pressure effects can be accounted for in the exponent of time. The equation takes the form

$$V=V_s+m''t^n, \quad (8.33)$$

where m''=slope of plot of volume vs. t^n and n=0.5 to 1.0, depending on shear rate and pressure drop.

The terms m'' and n are readily determined from a log-log plot of volume vs. time (Figs. 8.46 and 8.50). A least-squares fit of the later-time data is used to calculate m and n. The early-time data are offset by spurt. Spurt can be calculated at an early volume-percent area and time (V_i, t_i) from the relationship

$$V_s=V_i-m''t_i^n. \quad (8.34)$$

Sample Calculations. Examples showing the calculation of volume lost per square foot of formation by the described methods are provided below. Each method will be used to analyze the dynamic laboratory data in Fig. 8.53 for a 40-lbm/1,000-gal [4793-g/m^3] titanate complexed HPG fluid at 100 seconds^{-1} and 200°F [93°C].[32] The volume of fluid lost per square foot will be calculated with a pump time of 4 hours (240 minutes). All four methods shown below are only representative of an element of the fracture area adjacent to the wellbore because the total exposure time referenced is total job time. To model elements of the fracture area any distance away from the wellbore, it is necessary to adjust the times used to be representative of that area element. If a single calculation is made to represent some average value for all the created fracture area, then some time value less than total job time should be used.

1. $V=V_s+C_d(t), \quad (8.28)$

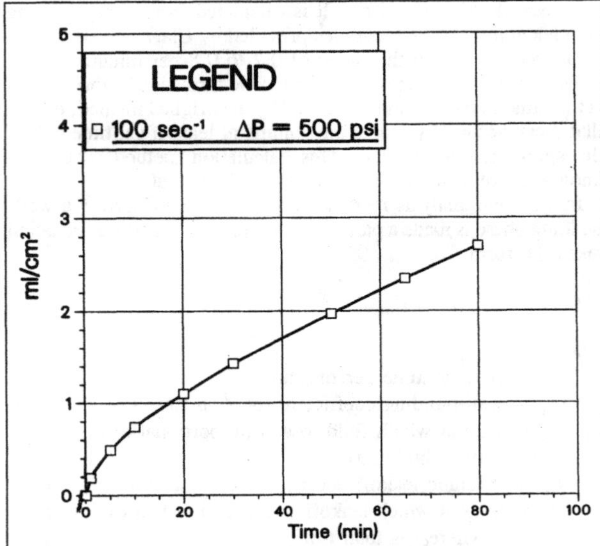

Fig. 8.53—Dynamic leakoff data for sample calculations: fluid = 40 lbm HPG complexed with titanate; permeability = 0.4 md; temperature = 200°F (from Gulbis[32]).

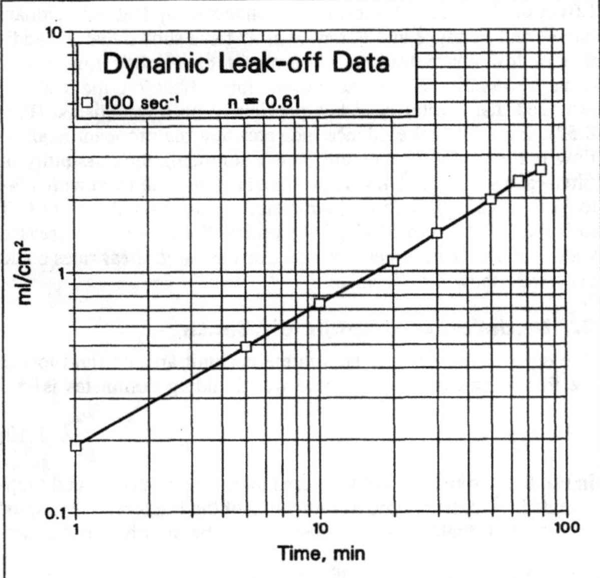

Fig. 8.54—Log-log plot of data in Fig. 8.53: fluid = 40 lbm titanate complexed HPG; permeability = 0.4 md; temperature = 200°F.

V_s = 0.80 mL/cm² = 0.20 gal/ft²,

$$m = \frac{(2.60-1.95)\ \text{mL/cm}^2}{(80-50)\ \text{minutes}} = 0.0216\ \text{cm/min},$$

C_d = 0.0216 cm/min × 0.0328 ft/cm = 0.00071 ft/min,
V = 0.20 gal/ft² + 0.00071 ft/min(240 minutes) × 7.48 gal/ft³
= 0.20 gal/ft² + 1.28 gal/ft²
= 1.48 gal/ft².

Or using a practical application of Eq. 8.30 yields

1a. $V = V_s(1-e^{-bt}) + v_D(t-t_s)$, ... (8.35)

V_s = 1.9 mL/cm² = 0.47 gal/ft² at t_s = 50 minutes,
b = 0.043/min,
v_D = 0.0216 cm/min = 0.00071 ft/min,
V = 0.47 gal/ft² $(1-e^{-0.043\times 240})$ + 0.00071 ft/min × (240 − 50 minutes) × 7.48 gal/ft³
= 0.47 gal/ft² + 1.01 gal/ft² = 1.48 gal/ft².

2. $V = V_s + 2C_w t_A{}^{1/2} + C_d t_B$, ... (8.31)

V_s = 0,
C_w = (1/2)(0.27) cm/min$^{1/2}$ × 0.0164
= 0.0022 ft/min$^{1/2}$ for t_A = 50 minutes,
C_d = 0.00071 ft/min for t_B = 240 − 50 = 190 minutes,
V = 0 + 2(0.0022 ft/min$^{1/2}$)(50 minutes)$^{1/2}$ × 7.48 gal/ft³ + 0.00071 ft/min × (190 minutes) × 7.48 gal/ft³
= 0 + 0.23 gal/ft² + 1.01 gal/ft²
= 1.24 gal/ft².

3. $V = V_s + m''t^n$. .... (8.33)

Plot log mL/cm² vs. log time, weighting later-time data (Fig. 8.54).

n = slope = 0.61,
m'' = y intercept = 0.18 cm/min$^{0.61}$,

V_s = 0,
V = 0 + 0.18 cm/min$^{0.61}$(240 minutes)$^{0.61}$
= 5.1 mL/cm² or 1.25 gal/ft².

Methods 1 and 1a arrive at similar answers (follow t), while Methods 2 and 3 arrive at smaller volumes. Methods 1 and 1a may overestimate the total fluid leakoff.

Conversion of Laboratory Data to a Field Leakoff Coefficient. The final goal of any fluid-loss analysis is to arrive at a leakoff coefficient that will predict the volume of fluid loss during an actual treatment. Three methods of converting laboratory data into a usable field leakoff coefficient are commonly used in most fracture simulators. The first and most widely used method is simply to enter a laboratory spurt and C_w that are characteristic of the fluid additive package for a specific formation permeability and temperature. A second method allows the input of more than one set of fluid parameters. Finally, a third method is the calculation of a coefficient that varies on the basis of a changing fracture environment.

In the first method, static fluid-loss data are normally used in the model. For the most part, dynamic data have been ignored. It is possible, however, to use dynamic data in such a simulator by using an averaging technique that calculates a C_w from the projected volume lost to the formation.

The steps involved in arriving at a C_w value representative of the near-wellbore leak-off conditions are as follows.

1. Determine the projected volume of fluid lost from laboratory dynamic data and pressure correlations by the method of choice (see Modeling of Combined Effects).

2. Calculate the C_w value necessary to arrive at the projected volume and pump time (effective C_w) from the relationship[20,22]

$$C_{we} = \frac{V-V_s}{2(t_p)^{1/2}}. \qquad (8.36)$$

It should be noted that this effective C_{we} is only a rough approximation. If a computerized fracturing simulator can handle more than one C_w value, then more accuracy can be obtained if multiple C_w values are used as a function of the pumping time.

Examples showing the calculation of C_{we} values from laboratory data follow.

Example 1. From the dynamic data in Fig. 8.53, calculate C_{we} (for near-wellbore conditions) for a 4-hour pump time, using

Eq. 8.20 to estimate the volume of leakoff.

$$V = V_s + C_d(t), \quad \ldots\ldots (8.28)$$

V_s = (0.81 mL/cm^2)(0.246)=0.20 gal/ft^2,
C_d = 0.00071 ft/min at 500 psi (calculated from shape of data in Fig. 8.53),
t = 240 minutes, and
V = 1.48 gal/ft^2.

$$C_{we} = \frac{V - V_s}{2(t_p)^{1/2}}$$

$$= \frac{(1.48-0.20)(\text{gal/ft}^2)}{2(240 \text{ minutes})^{1/2}(7.48 \text{ gal/ft}^3)}$$

= 0.0055 ft/min$^{1/2}$, and
V_s = 0.2 gal/ft^2.

To correct C_w for a Δp of 1,000 psi, use Eq. 8.26:

C_{we}=(1,000/500)$^{1/6}$(0.0055)=0.0062 ft/min$^{1/2}$.

Example 2. From the given laboratory data, calculate C_{we} for a 4-hour pump time from

$$V = V_s + 2C_w t_A^{1/2} + C_d t_B, \quad \ldots\ldots (8.31)$$

V_s = 0,
C_w = 0.0022 ft/min$^{1/2}$,
t_A = 50 minutes (estimated from dynamic tests),
C_d = 0.00071 ft/min at 500 psi, and
t_B = 240−50=190 minutes.

Calculate leakoff volume per unit area:

V = 0+2(0.0022 ft/min$^{1/2}$)(50 minutes)$^{1/2}$+0.00071 ft/min(190 minutes)
= 0.031 ft+0.135 ft
= 0.166 ft.

Calculate C_{we}:

$$C_{we} = \frac{0.166 \text{ ft}}{2(240 \text{ minutes})^{1/2}} = 0.0054 \text{ ft/min}^{1/2} \text{ (at 500 psi)},$$

V_s=0

To convert to 1,000 psi using Eq. 8.26, multiply by (1,000/500)$^{1/6}$=1.12:

C_{we}=0.0060 ft/min$^{1/2}$

and

V=0.185 ft or 1.39 gal/ft^2.

Example 3. A simulator is available that allows the input of multiple fluid parameters. Up to three sets of spurt and C_w values can be entered for the same number of fluid volumes. In this example, the shear-rate-vs.-time profile of the job indicates that the fluid will encounter 100 seconds^{-1} for 16 minutes and about 40 seconds^{-1} for the remaining 224 minutes of a 4-hour job. Calculate the leakoff volumes and C_{we} to be entered into the program for each shear rate on the basis of the leakoff data given in Figs. 8.49 and 8.50 for the following set of parameters.

Fluid=40 lbm/1,000 gal titanate complexed HPG,
permeability=0.1 md,
temperature=175°F,
shear rate=100 seconds^{-1} for 16 minutes and 40 seconds^{-1} for 224 minutes, and
Δp=1,000 psi.

Calculate leakoff volume at 100 seconds^{-1}, from Fig. 8.49 or 8.50:

n = 0.65,
m'' = 0.14 cm/min$^{0.65}$,
V_s = 0, and

$$V = V_s + m''t^n \quad \ldots\ldots (8.33)$$

= 0+0.14 cm/min$^{0.65}$(16 minutes)$^{0.65}$
= 0.85 mL/cm^2 or 0.028 ft or 0.21 gal/ft^2.

Calculate leakoff volume while at 40 seconds^{-1}, from Fig. 8.49 or 8.50:

n = 0.59,
m'' = 0.16 cm/min$^{0.59}$, and
V = 0.16 cm/min$^{0.59}$(240$^{0.59}$−16$^{0.59}$)minutes$^{0.59}$
= 3.24 mL/cm^2 or 0.106 ft or 0.79 gal/ft^2.

Calculate input values for C_{we}:

$$C_{we} \text{ at 100 seconds}^{-1} = \frac{0.028 \text{ ft}}{2(16 \text{ minutes})^{1/2}} = 0.0035 \text{ ft/min}^{1/2}$$

and

$$C_{we} \text{ at 40 seconds}^{-1} = \frac{0.106 \text{ ft}}{2(240^{1/2} - 16^{1/2})(\text{minutes}^{1/2})}$$

=0.0046 ft/min$^{1/2}$.

If a single input C_w value were needed, it would be

$$C_{we} = \frac{V_{100} + V_{40}}{2(240 \text{ minutes})^{1/2}} = \frac{0.028 \text{ ft} + 0.106 \text{ ft}}{2(240 \text{ minutes})^{1/2}}$$

=0.0043 ft/min$^{1/2}$.

To convert to 200°F, multiply by 1.3/1.2 (Fig. 8.24). At 200°F,

V=0.134 ft×1.3/1.2=0.145 ft or 1.09 gal/ft^2

and

C_{we}=0.0047 ft/min$^{1/2}$.

Example 4. The conversion of laboratory data to a leakoff coefficient that varies with downhole conditions was proposed by Settari and Price.[2] In their model, the variable C_w is a function of cake erosion caused by shear, cake compaction, and viscosity degradation. Rather than projecting the effects of shear and pressure drop over rather large fluid volumes (as in the above methods), the variable-C_w model calculates the combined effects of each fluid increment.

The following values were put into the variable-C_w program.

Formation.

Permeability=0.4 md,
production=gas,
S_w=40%,
k_{rg}=0.5,
ϕ=0.1,
T_{bh}=200°F, and
p_{bh}=3,000 psi.

TABLE 8.13—CALCULATED LEAKOFF vs. TIME WITH WALL-BUILDING DOMINATING USING THE SETTARI AND PRICE[2] MODEL (see conditions in text)

Time (minutes)	Width (ft)	Leakoff Volume (gal/ft²)
60	0.052	0.39
120	0.099	0.74
180	0.150	1.12
240	0.214	1.60

Treatment.
Fluid=40 lbm/1,000 gal titanate complexed HPG,
fluid-loss additive=none,
pump time=4 hours at 20 bbl/min,
shear rate=100 seconds^{-1} for 16 minutes and 40 seconds^{-1} for 224 minutes,
p_{bht}=4,000 psi,
near-wellbore Δp=1,000 psi, and
rheology at 200°F=

Time (hours)	n'	K'
0	0.62	0.069
1	0.63	0.053
2	0.64	0.041
4	0.66	0.024

Laboratory Fluid-Loss Data. Equilibrium leakoff velocity= 0.00071 ft/min at 100 seconds^{-1} and Δp=500 psi, filtrate viscosity=0.3 cp at 200°F, and point at which cake becomes incompressible=300 psi.

Output of the Program. The program solves for the fluid loss volume while varying (1) the pressure across the filter cake, (2) the apparent viscosity of the fracturing fluid at the time and temperature of each increment, (3) the leakoff velocity (based on shear rate) or wall-building properties, and (4) the viscosity reduction of the fracturing fluid as it flows through the filter cake into the formation.

In an example where wall building was made to dominate leakoff (as in Examples 1 through 3), the calculated leakoff volume vs. time is as given in Table 8.13. When the program is allowed to use C_{vc} in its calculation of leakoff volume, the calculations presented in Table 8.14 result.

This exercise illustrates the variation in leakoff volumes possible when Δp and the filtrate viscosity are altered. The variable-Δp option produces a volume of 1.35 vs. 1.04 gal/ft² [0.055 vs. 0.042 m³/m²] with a constant Δp of 1,000 psi [6.9 MPa]. Likewise, the input of a filtrate viscosity of 1.0 rather than 0.3 cp [1.0 rather than 0.3 mPa·s] (water at 200°F [93°C]) produces a decrease in the estimated volume of 1.35 to 1.01 gal/ft² [0.055 to 0.041 m³/m²].

Comparison of Examples. The calculated leakoff volumes when static and dynamic data are used (Examples 1 through 4) are given in Table 8.15. The volume of fluid lost to the formation in Examples 1 through 4 varies from 0.83 to 1.65 gal/ft² [0.034 to 0.067 m³/m²], depending on the data-generation and analysis methods. In future work, the measurements of leakoff under simulated field conditions will be necessary to validate laboratory procedures and analysis methods.

TABLE 8.15—COMPARISON OF CALCULATED LEAKOFF VOLUMES FOR A 4-HOUR PUMP TIME USING FIVE METHODS (40 lbm complexed HPG/1,000 gal)

Method	Volume Calculation	4-Hour Leakoff Volume (gal/ft²)
Static	$V_s+2C_wt^{1/2}$	0.83
Example 1	V_s+C_dt	1.65
Example 2	$V_s+2C_wt_A{}^{1/2}C_dt_B$	1.65
Example 3	V_s+mt^n	1.09
Example 4	Variable-leakoff simulator w/C_{vc}	1.35

TABLE 8.14—CALCULATED LEAKOFF vs. TIME USING C_{vc} AND C_w IN THE SETTARI AND PRICE[2] MODEL (see conditions in text)

Program Variables Altered	Leakoff Volume (ft³/ft²)	Leakoff Volume (gal/ft²)
None	0.18	1.35
Constant Δp	0.14	1.04
Filtrate viscosity of 1.0 instead of 0.3 cp	0.135	1.01

8.8 Comparison of Laboratory and Field Leakoff Coefficients

It is important to determine whether laboratory tests can accurately predict the true leakoff in a fracture. Two methods, fracturing analysis and in-situ leakoff measurements, have been developed to measure leakoff in the field.

Postfracturing Analysis. Until recently, a method to measure leakoff in the field had not been developed; therefore, operators relied on field experience to tell them whether laboratory data were valid. For example, Crawford[41] reported that experience has shown that multiplying the laboratory C_w value by a factor of 1.5 improved the success rate of jobs being pumped. This is an example of a postfracturing analysis. In this analysis, the common method used to determine field leakoff coefficients in problem areas has been to enter C_{vc} and/or C_w values into a fracture simulator until the program predicts a screenout. The field leakoff coefficient is then identified when the predicted screenout point coincides with observed screenouts in the field. Using this method, Hall and Houk[27] estimated C_t values of 0.002 ft/min$^{1/2}$ for one field and 0.0045 ft/min$^{1/2}$ for another, where the C_t calculated from laboratory data was 0.001 ft/min$^{1/2}$.

In-Situ Measurement of Fluid-Loss Coefficients. Two methods are currently being proposed to evaluate leakoff coefficients during fracturing treatments. Nolte's[37,38] method determines a leakoff coefficient from pressure-decline data by following a pump in treatment, and Nierode's[42] method evaluates leakoff by following the change in BHTP vs. time during the treatment. In either case, changes in pressure are related to generated fracture area in the permeable region. A detailed discussion of these methods can be found in Sec. 13.11 and Chap. 14.

The fluid efficiency calculated from pressure decline is equal to the volume injected minus the volume lost to the permeable region minus the fluid volume lost generating a fracture in nonpermeable formation. If the ratio between the total generated fracture area and the fluid-loss area were known with some certainty, then the in-situ leakoff coefficient could be determined explicitly. This ratio cannot be determined accurately because calculations involve an estimate of net formation height and gross formation height and assumption of a relationship between the generated height, width, and subsequent length of the generated fracture.

Net fracture height is generally estimated from elastic log properties. Gross fracture height either is determined from a log-derived estimate or is based on temperature surveys and/or radioactive surveys after a fluid pump-in test.[43] Note, however, that at least one study[44] has shown that temperature and/or radioactive surveys revealed only the minimum height and may not identify the maximum created height. The relationship between fracture width and/or height or length of the fracture is generally predicted by one of two methods: the Perkins-Kern[45] relationship assumes that the width of the fracture is proportional to the created height, and the Khristianovich and Zheltov[46] approach assumes that the width of the fracture is proportional to the created length. In general, for a given volume of fluid, the Perkins-Kern approach will predict a much greater created fracture area than one would expect from the Khristianovich-Zheltov approach.

In actual practice, it has been demonstrated that for a given set of field-derived data, order-of-magnitude differences in the calculated in-situ leakoff coefficient can arise because of the uncertainty

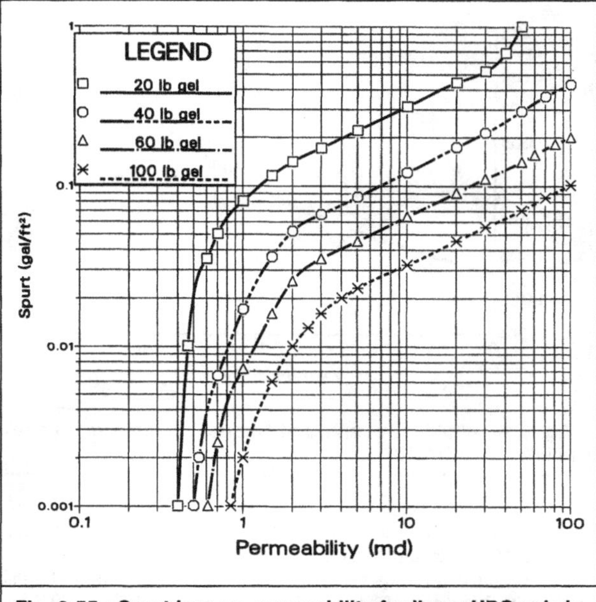

Fig. 8.55—Spurt loss vs. permeability for linear HPG gels in water at 125°F.

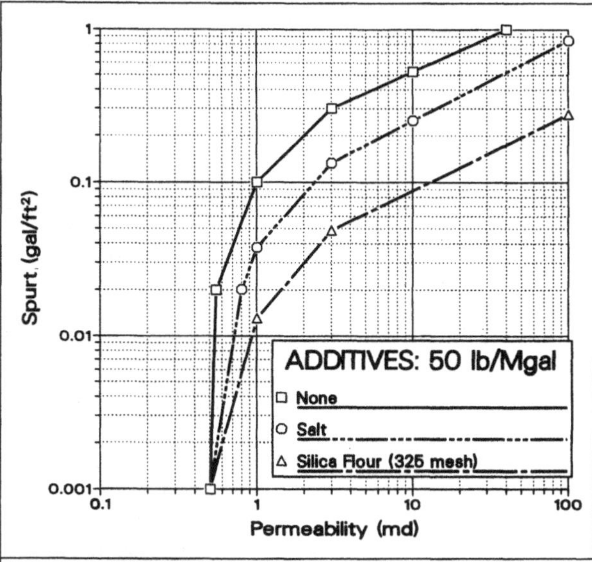

Fig. 8.56—Spurt loss vs. permeability and additive for gelled diesel at 125°F.

in the selection of gross and net heights and the appropriate width equations. We suggest that the most reliable values to be obtained from in-situ testing would be that of fracture area in the permeable region and the apparent fluid efficiency in that zone. A comparison of theoretical leakoff coefficients and those determined with the Nolte and Nierode methods has been presented by Cooper.[47,48] In general, job failures could be related to in-situ values that were much greater than the calculated values.

We propose the following approach to the field determination of in-situ leakoff rates. Previous sections have demonstrated that the laboratory-derived leakoff values are strongly dependent on the method used to produce and to analyze the data. By selecting a simple fluid, such as a 60-lbm/1,000-gal [7190-g/m³] uncrosslinked HPG gel, one can determine in the laboratory, under a variety of conditions, a maximum and a minimum leakoff coefficient. This fluid could then be pumped in the field in a minifracture[21,37] treatment, and the fracture area created in the permeable zone for this fluid can be determined. After suitable time has passed in which the effects of the previously pumped gel have been removed (i.e., flowback and fluid-viscosity degradation), a second fluid containing a maximum fluid-loss additive package could then be pumped. Laboratory-derived values over a variety of test conditions could be used to predict the maximum change in leakoff coefficient expected in the permeable region.

A major change in apparent fluid efficiency while the two previously described fluids are pumped would indicate that the generated fracture area is dominated by a fracture in the permeable region. On the other hand, very little change in the leakoff value would indicate that the majority of the fluid inefficiency occurs because of a generated fracture in a nonpermeable region. This approach, while indicating only a relative effectiveness of a given treatment, would suggest a future direction in that reservoir in providing a more efficient fluid for creating more effective fractures.

8.9 Relative Effectiveness of Fluid-Loss Additives in Fluids Other Than Complexed Water/Sand Fluids

Linear Gels. Above 2.0 to 5.0 md, spurt loss is proportional to $k^{1/2}$. This is illustrated in Fig. 8.40 for a series of linear gels (HPG) in water. Below 2.0 md, spurt quickly approaches zero. Graphs of spurt loss vs. fluid-loss additive and gelling agent that follow are given for the permeability range of 5 to 10 md. The values can easily be adjusted to higher permeabilities using the $k^{1/2}$ relationship, although lower polymer loadings (20 lbm [9 kg]) tend to lose control near 30 to 50 md.

Spurt loss decreases as the gelling-agent concentration is increased. For example, increasing the HPG loading from 10 to 100 lbm/1,000 gal [1198 to 11 980 g/m³] in water lowers the spurt from 0.5 to 0.03 gal/ft² [0.02 to 0.001 m³/m²] in 5- to 10-md cores (Fig. 8.55). Spurt decreases as the inverse of the square root of the apparent viscosity of the gelled fluid, where the apparent viscosity is calculated at a shear rate of 10 seconds^{-1}. An average shear rate of 10 seconds^{-1} was calculated as what may be encountered during the first minute of leakoff for gelled water-based fluids in the pore throat of a 5- to 10-md formation at a pressure differential of 1,000 psi [6.9 MPa].

Temperature increases produce higher spurt values because of gel-viscosity reduction and/or degradation. Data presented in Figs. 8.40 and 8.55 are given for the previously accepted standard temperature of 125°F [52°C]. For other temperatures, a temperature correction factor should be applied. The factor can be determined by referring to Fig. 8.41. In this figure, the sensitivity of lower polymer loadings to temperature becomes obvious.

The wall-building coefficient, C_w, is independent of permeability in most gel systems. (The apparent dependence of C_w on permeability when the gas phase is used will be discussed later.) C_w decreases as the polymer concentration is increased (Fig. 8.42) because of the increased amount of polymer available for deposition in the form of a filter cake. Polymer and additives decrease leakoff by both decreasing the permeability of the filter cake and increasing thickness. The rate of deposition producing thickness can be limited by the shear rate (see Section 8.5).

The C_w data presented in Fig. 8.42 are for water-based gels at 125°F [52°C]. At other temperatures, the C_w correction factors given in Fig. 8.24 should be applied. This correlation considers the effects of temperature on both the viscosity of the base fluid and thermal degradation of the polymer. The separation of these two mechanisms is discussed in Sec. 8.6.

The addition of particulate additives to linear gels produces a decrease in spurt and C_w. Silica flour (325 mesh) is provided as an example in Figs. 8.42 and 8.55. Relatively small amounts of particulate additives (≦25 lbm [≦11 kg]) are highly effective at low polymer loadings (20 lbm [9 kg]). At higher polymer loadings (100 lbm [45 kg]), viscosity inhibits the transport of the fluid-loss additive to the filter cake; therefore, larger amounts of additive are required to obtain equivalent decreases in spurt loss over the polymer alone.

Additives commonly used in leakoff control during fracturing treatments, listed in order of increasing effectiveness in controlling spurt loss, include (1) 5% diesel or oil/surfactant mixtures; (2) high-melting-point ground resins and resin/polymer blends; (3) inhibited

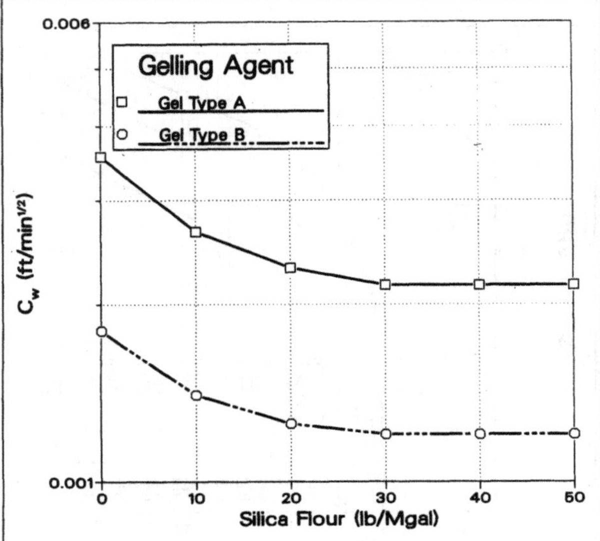

Fig. 8.57—C_w vs. particulate-additive concentration for gelled diesel at 125°F.

polymers, such as guar, cellulose, and karaya; (4) silica flour (325 mesh); and (5) blends of polymers, silica, and clays. A comparison of spurt loss vs. additive type and permeability in a 40-lbm [18-kg] HPG fluid at 125°F [52°C] is provided in Fig. 8.38, while a comparison of the effects of the same additives on wall-building character is given in Fig. 8.39.

Oil-Based Fluids. Spurt and C_w of oils gelled with phosphate/aluminum association polymers depend on the base-fluid viscosity and cleanliness, as well as the gelling-agent composition and concentration. Because oils vary at each field location and gelling agents and fluid-loss additives differ among service companies, no firm numbers can be presented for spurt and C_w of gelled oil. However, common numbers for a gelled diesel at 125°F [52°C] with various additives are presented in Figs. 8.56 and 8.57.

Leakoff of Foam Fracturing Fluids. The dynamic leakoff properties of water-based fluids foamed with N_2 and CO_2 have been measured by Harris.[33,49] When gelling agents like HPG are a part of the base fluid, foams are wall-building fluids. The C_w for a foam is a combination of liquid and gas leakoff velocities. For lack of better data, spurt values are generally assumed to be zero for foams at all permeabilities, while C_w increases with permeability (Fig. 8.58).

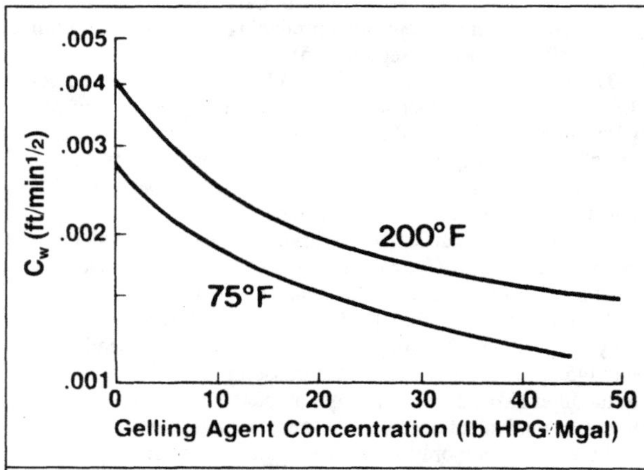

Fig. 8.59—Effect of temperature and gelling-agent concentration on foam C_w in Ohio sandstone (from Harris[49]).

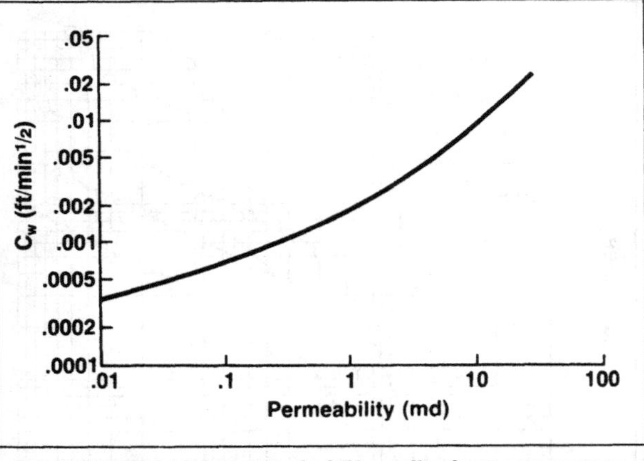

Fig. 8.58—Fluid-loss control of 70-quality foams vs. permeability (from Harris[33]).

The leakoff coefficient decreases with increasing gelling agent (Fig. 8.59). The temperature correlations developed for C_w of water-based gelling agents (Fig. 8.24) are applicable to foams.

Quality has little effect on C_w at low permeabilities (0.1 to 1 md). At higher permeabilities (10 md), C_w decreases with increasing quality (Fig. 8.60).[49]

Nomenclature

A = area
b = Klinkenberg constant
c = compressibility
c_t = total formation compressibility
C = fluid leakoff coefficient
C_c = reservoir-fluid viscosity/compressibility coefficient
C_d = dynamic leakoff coefficient
C_e = composite fluid-loss coefficient
C_t = overall coefficient
C_v = effluent viscosity and relative permeability coefficient
C_{vc} = $C_c + C_v$
C_w = wall-building coefficient
F_{cV} = ratio of filter cake to filtrate volume
F_{Vc} = ratio of volume in cake to total fluid-loss volume

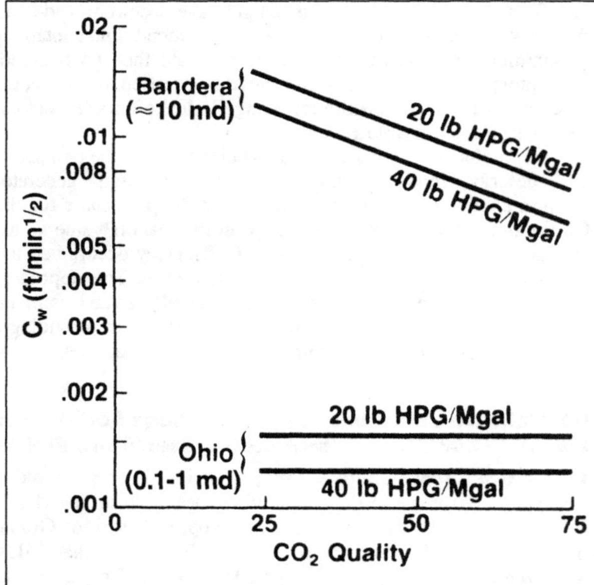

Fig. 8.60—Effect of quality, permeability, and gelling-agent concentration on foam C_w at 75°F (from Harris[49]).

g = gas specific gravity
k_i = relative permeability of formation to leakoff effluent
k_r = formation permeability relative to mobile reservoir fluid
K' = consistency index
m = slope of volume vs. t
m' = slope of volume vs. time
m'' = slope of volume vs. t_n
n' = flow index
p = pressure
p_{wh} = hydrostatic pressure of water per foot
S = saturation
t = time
t_A = time when fluid loss is proportional to $t^{1/2}$
t_B = time when leakoff is proportional to t
T = temperature
v = velocity
v_D = equilibrium Darcy flow velocity
V = volume
μ = viscosity
μ_a = viscosity of effluent
μ_f = viscosity of mobile formation fluid at reservoir conditions
ϕ = fractional porosity of formation

Subscripts

bh = bottomhole
bht = bottomhole treating
e = effective
f = filter cake
g = gas
i = mobile
L = liquid
o = oil
or = residual oil
p = pump
r = reduced
R = reservoir
s = spurt
t = total
w = water
wr = residual water

Superscripts

n = exponent of time on a log-log plot of vol/area vs. time
$-$ = average
$*$ = substitution variable

References

1. Howard, G.C. and Fast, C.R.: *Hydraulic Fracturing,* Monograph Series, SPE, Richardson, TX (1970) **2,** 55.
2. Settari, A. and Price, H.S.: "Simulation of Hydraulic Fracturing in Low-Permeability Reservoirs," *SPEJ* (April 1984), 141–52.
3. Smith, J.E.: "Design of Hydraulic Fracture Treatments," paper SPE 1286 presented at the 1965 SPE Annual Meeting, Denver, Oct. 3–6.
4. Williams, B.B.: "Fluid Loss from Hydraulically Induced Fractures," *JPT* (July 1970) 882–88; *Trans.,* AIME, **249.**
5. Williams, B.B., Gidley, J.L., and Schechter, R.S.: *Acidizing Fundamentals,* Monograph Series, SPE, Richardson, TX (1979) **6.**
6. Klinkenberg, L.J.: "The Permeability of Porous Media to Liquids and Gases," *Drill. and Prod. Prac.,* API (1951) 200
7. Calhoun, J.C.: *Fundamentals of Reservoir Engineering,* U. of Oklahoma Press, Norman (1953).
8. *API RP 27, Recommended Practice for Determining the Permeability of Porous Media,* third edition, API, Dallas (1956).
9. Long, G. and Chierici, G.: "Salt Content Changes Compressibility of Reservoir Brines," *Pet. Eng.* (July 1961) B25–B31.
10. Osif, T.L.: "Effects of Salt, Gas, Temperature, and Pressure on the Compressibility of Water," *SPERE* (Feb. 1988) 175–81.
11. Newman, G.H.: "Pore-Volume Compressibility of Consolidated, Friable, and Unconsolidated Reservoir Rocks Under Hydrostatic Loading," *JPT* (Feb. 1973) 129–34.
12. *Gas Well Testing Theory and Practice,* fourth edition, Energy Resources Conservation Board, Alberta, Canada (1979) Appendix A.
13. Brown, G.G. *et al.*: *Natural Gasoline and Volatile Hydrocarbons,* NGAA, Tulsa, OK (1948).
14. Carr, N.L., Kobayashi, R., and Burrows, D.B.: "Viscosity of Hydrocarbon Gases Under Pressure," *Trans.,* AIME (1954) **201,** 264–72.
15. Mattar, L., Brar, G.S., and Aziz, K.: "Compressibility of Natural Gases," *J. Cdn. Pet. Tech.* (1975).
16. "Viscosity of Liquid Fuels," *Intl. Critical Tables,* first edition, McGraw-Hill Book Co. Inc., New York City (1927) **II,** 146–47.
17. Roodhart, L.P.: "Fracturing Fluids: Fluid-Loss Measurements Under Dynamic Conditions," *SPEJ* (Oct. 1985) 629–36.
18. *The Fracbook Design/Data Manual,* Halliburton Services, Duncan, OK (1971) Fig. 77.
19. *API RP 39, Recommended Practice, Standard Procedure for the Evaluation of Fracturing Fluids,* API, Dallas (1983).
20. Penny, G.S., Conway, M.W., and Lee, W.S.: "Control and Modeling of Fluid Leakoff During Hydraulic Fracturing," *JPT* (June 1985) 1071–81.
21. Nolte, K.G.: "Principles for Fracture Design Based on Pressure Analysis," *SPEPE* (Feb. 1988) 22–30.
22. Harris, P.C. and Penny, G.S.: "Influence of Shear History at Bottomhole Temperature on Fracturing-Fluid Efficiency," *SPEPE* (May 1989) 189–93.
23. Gray, G.R., Darley, H.C.H., and Rogers, W.F.: *Composition and Properties of Oil Well Drilling Fluids,* Gulf Publishing Co., Houston (1980) 270.
24. Schlottman, B.W., Miller, W.K. II, and Lueders, R.K.: "Massive Hydraulic Fracture Design for the East Texas Cotton Valley Sands," paper SPE 10133 presented at 1981 SPE Annual Technical Conference and Exhibition, San Antonio, Oct. 4–7.
25. Woo, G.T. and Cramer, D.D.: "Laboratory and Field Evaluation of Fluid-Loss Additive Systems Used in the Williston Basin," paper SPE 12899 presented at the 1984 SPE Rocky Mountain Regional Meeting, Casper, WY, May 21–23.
26. McMechan, D. and Conway, M.W.: "Recent Developments in Hydraulic Stimulation Treatments in the Fletcher Field of the Deep Anadarko Basin," paper SPE 11604 presented at the 1983 SPE/DOE Symposium on Low Permeability Gas Reservoirs, Denver, March 14–16.
27. Hall, B.E. and Houk, S.G.: "Fluid-Loss Control in the Naturally Fractured Buda Formation," paper SPE 12152 presented at the 1983 SPE Annual Technical Conference and Exhibition, San Francisco, Oct. 5–8.
28. Hall, C.D. Jr. and Dollarhide, F.E.: "Effect of Fracturing Fluid Velocity on Fluid-Loss Agent Performance," *JPT* (May 1964) 555–60; *Trans.,* AIME, **231.**
29. Hall, C.D. Jr. and Dollarhide, F.E.: "Performance of Fracturing Fluid Loss Agents Under Dynamic Conditions," *JPT* (July 1968) 763–69; *Trans.,* AIME, **243.**
30. Sinha, B.K.: "Fluid Leakoff Under Dynamic and Static Conditions Utilizing the Same Equipment," paper SPE 6126 presented at the 1976 SPE Annual Technical Conference and Exhibition, New Orleans, Oct. 3–6.
31. Gulbis, J.: "Dynamic Fluid Loss Study of Fracturing Fluids," paper 82-33-18 presented at the 1982 Annual Technical Meeting of the Petroleum Soc. of CIM, Calgary, Alta., June 6–9.
32. Gulbis, J.: "Dynamic Fluid Loss of Fracturing Fluids," paper SPE 12154 presented at the 1983 SPE Annual Technical Conference and Exhibition, San Francisco, Oct. 5–8.
33. Harris, P.C.: "Dynamic Fluid Loss Characteristics of Nitrogen Foam Fracturing Fluids," *JPT* (Oct. 1985) 1847–52.
34. McDaniel, R.R. *et al.*: "An Improved Method for Measuring Fluid Loss at Simulated Fracture Conditions," *SPEJ* (Aug. 1985) 482–90.
35. Penny, G.S.: "Evaluation of the Effects of Environmental Conditions and Fracturing Fluids on the Long-Term Conductivity of Proppants," paper SPE 16900 presented at the 1987 SPE Annual Technical Conference and Exhibition, Dallas, Sept. 27–30.
36. Conway, M.W. and Harris, E.: "A Laboratory and Field Evaluation of a Technique for Hydraulic Fracturing Stimulation of Deep Wells," paper SPE 10964 presented at the 1982 SPE Annual Technical Conference and Exhibition, New Orleans, Sept. 26–29.
37. Nolte, K.G.: "Determination of Fracture Parameters from Fracturing Pressure Decline," paper SPE 8341 presented at the 1979 SPE Annual Technical Conference and Exhibition, Las Vegas, Sept. 23–26.
38. Nolte, K.G.: "Determination of Proppant and Fluid Schedules From Fracturing-Pressure Decline," *SPEPE* (July 1986) 255–65; *Trans.,* AIME, **283.**
39. Zigrye, J.L, Whitfill, D.L., and Sievert, J.A.: "Fluid-Loss Control Differences of Crosslinked and Linear Fracturing Fluids," *JPT* (Feb. 1985) 315–20.

40. Clark, P.E. and Barkat, O.: "The Analysis of Dynamic Fluid Loss Data," paper SPE 13845 available at SPE headquarters, Richardson, TX.
41. Crawford, H.R.: "Proppant Scheduling and Calculation of Fluid Lost During Fracturing," paper SPE 12064 presented at the 1983 SPE Annual Technical Conference and Exhibition, San Francisco, Oct. 5–8.
42. Nierode, D.E.: "Comparison of Hydraulic Fracture Design Methods to Observed Field Results," *JPT* (Oct. 1985) 1831–39.
43. Dobkins, T.A.: "Improved Methods to Determine Hydraulic Fracture Height," *JPT* (April 1981) 719–26.
44. Smith, M.B., Rosenberg, R.J., and Bowen, J.F.: "Fracture Width: Design vs. Measurement," paper SPE 10965 presented at the 1982 SPE Annual Technical Conference and Exhibition, New Orleans, Sept. 26–29.
45. Perkins, T.K. Jr. and Kern, L.R.: "Widths of Hydraulic Fractures," *JPT* (Sept. 1961) 937–49; *Trans.*, AIME, **222**.
46. Khristianovich, S.A. and Zheltov, Yu.P.: "Formation of Vertical Fractures by Means of Highly Viscous Liquid," *Proc.*, Fourth World Pet. Cong., Rome (1955) **2**, 579–86.
47. Cooper, G.D., Nelson, S.G., and Schopper, M.D.: "Improving Fracturing Design Through the Use of an On-Site Computer System," paper SPE 12063 presented at the 1983 SPE Annual Technical Conference and Exhibition, San Francisco, Oct. 5–8.
48. Cooper, G.D., Nelson, S.G., and Schopper, M.D.: "Comparison of Methods of Determining In-Situ Leakoff Rate Based on Analysis With an On-Site Computer," paper SPE 13223 presented at the 1984 SPE Annual Technical Conference and Exhibition, Houston, Sept. 16–19.
49. Harris, P.C.: "Dynamic Fluid-Loss Characteristics of CO_2-Foam Fracturing Fluids," *SPEPE* (May 1987) 89–94; *Trans.*, AIME, **283**.
50. Matthews, C.S. and Russell, D.G.: *Pressure Buildup and Flow Tests in Wells,* Monograph Series, SPE, Richardson, TX (1967) **1,** 158.

SI Metric Conversion Factors

atm	×	1.013 250*	E+05	=	Pa
bbl	×	1.589 873	E−01	=	m^3
cp	×	1.0*	E−03	=	Pa·s
ft	×	3.048*	E−01	=	m
ft^2	×	9.290 304*	E−02	=	m^2
ft^3	×	2.831 685	E−02	=	m^3
°F		(°F−32)/1.8		=	°C
gal	×	3.785 412	E−03	=	m^3
in.	×	2.54*	E+00	=	cm
(lbf-sec)/ft^2	×	4.788 026	E+01	=	(N·s)/m^2
lbm	×	4.535 924	E−01	=	kg
lbm/gal	×	1.198 264	E+02	=	kg/m^3
psi	×	6.894 757	E+00	=	kPa
psi^{-1}	×	1.450 377	E−01	=	kPa^{-1}

*Conversion factor is exact.

Chapter 9
Fracturing-Fluid Flow Behavior

John R. Cameron, SPE, Amoco Production Co.
Robert K. Prud'homme, Princeton U.

9.1 Overview

This chapter discusses technology pertinent to the flow behavior of fracturing fluids presented according to fracturing-fluid type and flow regime. The discussion addresses experimental techniques used in characterizing fracturing-fluid flow behavior, including steady-shear, dynamic, and turbulent-flow test methods and applications of laboratory data to various aspects of fracturing design. Pressure-drop prediction, gel scheduling, and the design of liquid and foamed fracturing fluids flowing down casing are discussed. The last three sections present observed trends in laminar and turbulent flow for fracturing solutions, gels, and foams, with and without proppant. Environmental, chemical, and flow-deformation effects are examined.

9.2 Introduction

A knowledge of the flow behavior of fracturing fluids is important to fracturing-treatment design. Relations of shear stress as functions of flow are used to predict pressure drops in tubular goods and fractures. Pressure considerations are important with regard to pumping costs, tubular-goods pressure constraints, and fracture geometry. The fluid rheology also influences proppant transport and fluid loss to the rock matrix. The dynamic flow behavior (response to oscillatory testing) has recently been used to define the viscoelastic nature of fracturing fluids. This can indicate the crosslinked structure and expected transient behavior of fracturing fluids.

In fracturing-treatment design, flow behavior is normally specified with suitable shear-stress relations determined under laminar or turbulent flow conditions. Under laminar flow conditions, a fluid moves in "layers" with parallel streamlines. The local velocity at any point is constant in steady laminar flow, and momentum transport occurs on the molecular level. Shear stress can be correlated with a suitable kinematic variable (such as shear rate) by use of an appropriate laboratory viscometer.

As the flow rate increases, a point is reached when minor perturbations to flow are no longer attenuated, causing flow instabilities, such as secondary flows and, in the extreme case, fully developed turbulence. The flow conditions marking the transition to turbulence can be correlated to the Reynolds number (the ratio of the inertia to viscous forces) for some types of fluids. In flow through pipes, Newtonian fluids like water undergo an abrupt transition to turbulence at a Reynolds number of 2,100. For fracturing fluids, the transition to turbulence is often less distinct, depending on the nature of the fluid.

Under turbulent conditions, the local velocity at any point is random, even though the average volumetric flow rate may be constant. Superimposed on the bulk flow are randomly moving eddies, which are fluid-volume elements of macroscopic size. Momentum transport thus proceeds by both molecular and eddy diffusive mechanisms in turbulent flow. Shear stress is correlated with suitable dimensionless variables, including some form of the Reynolds number and the friction factor. Appendix C discusses friction-factor development.

Terminology: Solutions, Gels, Foams, Emulsions, Slurries. Fracturing fluids are classified according to their physical/chemical state. Fluids that comprise solvents containing uncrosslinked polymers will be referred to as "solutions" to be consistent with the current terminology in the fluid and rheological community. Uncrosslinked fracturing fluids (solutions) are often referred to as linear or base gels, even though this is not correct in the colloidal sense.

When the polymer molecules are crosslinked, the colloidal system can be categorized as a "gel."

Fracturing fluids consisting of gaseous or liquid internal phases dispersed in a liquid-external phase are referred to as "foams" and "emulsions," respectively. The objective of fracturing fluids is to transport a suitable proppant into the developing fracture. This introduces a third phase, and the fracturing fluid is then referred to as the proppant-laden fluid or "slurry."

Definition of Rheology. Rheology is the study of deformation and flow—i.e., the relationship between stresses imposed on a material (e.g., a fracturing fluid) and its resulting motion or, inversely, the stresses generated when a material moves under an imposed deformation. For simple Newtonian fluids, the viscosity is the only "material function" required to calculate pressure drop and flow rate. For viscoelastic polymer solutions, gels, and foams, more complex descriptions of the material properties are usually required. In addition to viscous stresses, these systems may demonstrate elastically generated stresses that act perpendicular (normal) to the direction of flow.

One goal of rheology is to represent the material properties responsible for these complex flow phenomena mathematically so that data collected in simple laboratory instruments can be generalized and used to predict performance under process conditions in more complex geometries. The mathematical equation describing the relationship between stress and deformation is called a "constitutive equation" or "rheological equation of state."

Additionally, rheological relationships between gel crosslink density and dynamic moduli can be used to interpret rheological data to aid in the characterization of gels. Rheological measurements are often the most sensitive ways to characterize materials or to compare one class of materials with another.

9.3 Steady Laminar Flows and Methods of Rheological Characterization

Steady Laminar Flows. Practical flow problems can often be modeled by recourse to simpler idealized problems that can be solved analytically, such as certain steady laminar flows. For example, flow in a rock fracture, which must certainly be irregular, is often modeled as flow between parallel plates, and flow in a wellbore is modeled as flow in a tube.

The constitutive equation for a fluid can often be expressed as a simple rheological model having constant experimentally determined parameters. The simplest rheological model is the Newto-

TABLE 9.1—SOLUTIONS FOR STEADY LAMINAR FLOW FOR VARIOUS FLUID MODELS IN VARIOUS GEOMETRIES[1]

Flow in a Tube	Flow Between Parallel Plates
General fluid: $\tau = f(\dot{\gamma})$	
$\tau_w = r_t \Delta p_f / 2L$	$\tau_w = w_f \Delta p_f / 2L$
$\Delta p_f = -\Delta p + \rho g \Delta z$	$\Delta p_f = -\Delta p + \rho g \Delta z$
$\dot{\gamma}_w = \left(\frac{3n+1}{4n}\right)\frac{4q}{\pi(r_t)^3}$,	$\dot{\gamma}_w = \left(\frac{2n+1}{3n}\right)\frac{6q}{Dw_f^2}$,
where	where
$n = \frac{d \ln \tau_w}{d \ln\left[\frac{4q}{\pi(r_t)^3}\right]}$	$n = \frac{d \ln \tau_w}{d \ln\left(\frac{6q}{Dw_f^2}\right)}$
Newtonian fluid: $(\tau = \mu\dot{\gamma})$: μ is constant	
$v_z = \frac{\tau_w r_t}{2\mu}\left[1-\left(\frac{r}{r_t}\right)^2\right]$	$v_y = \frac{\tau_w w_f}{4\mu}\left[1-\left(\frac{2x}{w_f}\right)^2\right]$
$q = \pi(r_t)^4 \Delta p_f / 8\mu L$	$q = w_f^3 D \Delta p_f / 12\mu L$
$\dot{\gamma}_w = 4q/\pi(r_t)^3$	$\dot{\gamma}_w = 6q/Dw_f^2$
Power-law fluid: $(\tau = K'\dot{\gamma}^{n'})$; K' and n' are constant	
$v_z = \left(\frac{\tau_w}{K'}\right)^{1/n'}\frac{r_t}{(1/n'+1)} \times \left[1-\left(\frac{r}{r_t}\right)^{(1/n'+1)}\right]$	$v_y = \left(\frac{\tau_w}{K'}\right)^{1/n'}\frac{w_f}{2(1/n'+1)} \times \left[1-\left(\frac{2x}{w_f}\right)^{(1/n'+1)}\right]$
$q = \frac{\pi r_t^{(3+1/n')}}{(1/n'+3)}\left(\frac{\Delta p_f}{2K'L}\right)^{1/n'}$	$q = \frac{Dw_f^2}{(2/n'+4)}\left(\frac{w_f \Delta p_f}{2LK'}\right)^{1/n'}$
$\dot{\gamma}_w = \left(\frac{3n'+1}{4n'}\right)\frac{4q}{\pi(r_t)^3}$	$\dot{\gamma}_w = \left(\frac{2n'+1}{3n'}\right)\frac{6q}{Dw_f^2}$
$n' = \frac{d \ln \tau_w}{d \ln\left[\frac{4q}{\pi(r_t)^3}\right]}$	$n' = \frac{d \ln \tau_w}{d \ln\left(\frac{6q}{Dw_f^2}\right)}$
Bingham fluid: $(\tau = \tau_o + \mu_p\dot{\gamma})$; τ_o and μ_p are constants	
$q = \frac{\pi r_t^3 \tau_w}{4\mu_p}\left[1 - 4/3\left(\frac{\tau_o}{\tau_w}\right) + 1/3\left(\frac{\tau_o}{\tau_w}\right)^4\right]$	$q = \frac{Dw_f^2 \tau_w}{6\mu_p}\left[1 - 3/2\left(\frac{\tau_o}{\tau_w}\right) + 1/2\left(\frac{\tau_o}{\tau_w}\right)^3\right]$
when $\tau_w \ge \tau_o$	when $\tau_w \ge \tau_o$
Ellis fluid: $[\mu^o/\mu = 1 + (\tau/\tau_{1/2})^{\alpha-1}]$; μ^o, $\tau_{1/2}$, and α are constants	
$q = \frac{\pi r_t^3 \tau_w}{4\mu^o}\left[1 + \frac{4}{\alpha+3}\left(\frac{\tau_w}{\tau_{1/2}}\right)^{\alpha-1}\right]$	$q = \frac{Dw_f^2 \tau_w}{6\mu^o}\left[1 + \frac{3}{\alpha+2}\left(\frac{\tau_w}{\tau_{1/2}}\right)^{\alpha-1}\right]$

nian model that has one constant parameter, a shear-rate-independent viscosity. Another more general model in which the viscosity depends on shear rate is the power-law model. This model has two constant parameters: the power-law index, n', and the consistency index, K'. Appendix C has more information on rheological models.

Table 9.1 shows solutions for the velocity components of Newtonian and power-law fluids flowing in a tube and a slot. Also list-

TABLE 9.1—SOLUTIONS FOR STEADY LAMINAR FLOW FOR VARIOUS FLUID MODELS IN VARIOUS GEOMETRIES (continued)

	Couette Flow*	Falling Sphere	Flow in a Packed Bed**
General fluid: $\tau = f(\dot{\gamma})$	$T_o = \tau_i 2\pi(r_i)^2 L$ $\tau_{r\theta} = (r_i/r)^2 \tau_i$ $\dot{\gamma}_i = \sum_{j=0}^{\infty} 2\left(\frac{\partial \Omega}{\partial \ln \tau}\right)\Big\vert_{\tau=\tau_i \kappa^{-2j}}$		$\Delta p_f = -\Delta p + \rho g \Delta z$
Newtonian fluid: $(\tau = \mu\dot{\gamma})$: μ is constant	$v_\theta = \frac{\Omega r\left[1-\left(\frac{r_i}{r}\right)^2\right]}{(1-\kappa^{-2})}$ $\dot{\gamma}_i = 2\Omega/(1-\kappa^{-2})$	$v_r = v_S\left[3/2\left(\frac{r_{sp}}{r}\right) - 1/2\left(\frac{r_{sp}}{r}\right)^3\right]\cos(\theta)$ $v_\theta = -v_S\left[3/4\left(\frac{r_{sp}}{r}\right) + 1/4\left(\frac{r_{sp}}{r}\right)^3\right]\sin(\theta)$ $v_S = \frac{2r_{sp}^2(\rho_{sp} - \rho_{fl})g}{9\mu}$	Blake-Kozeny Equation: $q = \frac{\pi r_t^2 \Delta p_f d_p^2 \phi^3}{L150\mu(1-\phi)^2}$ for $\frac{d_p \rho q}{\pi r_t^2 \mu(1-\phi)} < 10$
Power-law fluid: $(\tau = K'\dot{\gamma}^{n'})$; K' and n' are constant	$v_\theta = \frac{\Omega r\left[\left(\frac{r_i}{r}\right)^{2/n'} - 1\right]}{[(\kappa)^{-2n'} - 1]}$ $T_o = 2\pi r_i^2 K' L\left[\frac{2\Omega}{n'(1-\kappa^{-2/n'})}\right]^{n'}$ $\dot{\gamma}_i = \frac{2\Omega}{n'(1-\kappa^{-2/n'})}$ $n' = \frac{d \ln T_o}{d \ln \Omega}$		$q = \pi r_t^2\left(\frac{k\Delta p_f}{\mu_{eff} L}\right)^{1/n'}$, where $\mu_{eff} = K'\left[\frac{\left(\frac{25}{12}\right)^{n'}\left(3+\frac{1}{n'}\right)^{n'} 3^{(n'+1)}}{75} \times \frac{d_p^{(1-n')}\phi^{2(1-n')}}{(1-\phi)^{(1-n')}}\right]$ $[=]\,\mathrm{Pa \cdot s}^{n'} \cdot \mathrm{m}^{(1-n')}$
Bingham fluid: $(\tau = \tau_o + \mu_p\dot{\gamma})$; τ_o and μ_p are constants	$T_o = \frac{4\pi(r_i)^2 L\mu_p}{(1-\kappa^{-2})}(\Omega + \tau_o/\mu_p \ln \kappa)$		

*Where only the outside cylinder is rotating.
**From Bird, R.B., Stewart, W.E., and Lightfoot, E.N.: *Transport Phenomena*, John Wiley and Sons Inc., New York City (1960) 196–207.

ed are the velocities for Couette flow and a sphere falling in a Newtonian fluid. One can calculate the shear rate at the boundary, $\dot{\gamma}_w$ or $\dot{\gamma}_i$, from the velocity equation in terms of flow rate, q, or angular velocity, Ω, if the parameters of the rheological model are known. Thus, knowing q or Ω allows the engineer to calculate $\dot{\gamma}_w$ or $\dot{\gamma}_i$ and consequently the shear stress at the wall, τ_w or τ_i, using the appropriate rheological model. From the macroscopic force balance for the general fluid (the first equations for tube, slot, and Couette flows in Table 9.1), the friction pressure drop, Δp_f, or torque, T, for Couette flow can be predicted. The viscosity is predicted by calculating it from its definition ($\mu = \tau/\dot{\gamma}$). For Bingham or Ellis fluids, τ_w or τ_i (and thus Δp_f or T) can be solved implicitly from the listed equations for q if the model constants are known.

For flow of Newtonian and power-law fluids through a packed bed, Table 9.1 gives equations from which Δp_f can be explicitly solved with parameters descriptive of the rheology and the packed bed.

If the rheology is not known, the viscosity can be calculated if the fluid behaves as a general fluid. Here, a general fluid is defined as a homogeneous fluid having a shear stress in steady laminar flow that is a unique function of shear rate. Data of τ_w vs. q or τ_i vs. Ω allow the calculation of the shear rate (see equations in Table 9.1 under General Fluid) and thus the viscosity. Once shear stress and shear rate are known, they can be fit to the appropriate rheological model to determine the model parameters. The next section discusses common laboratory methods for determining viscosity and other "material functions."

TABLE 9.2—RELATIONS FOR DETERMINING THE VISCOMETRIC FUNCTIONS (μ, ψ_1, ψ_2) IN STANDARD EXPERIMENTAL GEOMETRIES[1]

$\mu = \tau_{21}/\dot{\gamma}_{21}$; viscosity

$\psi_1 = -(\tau_{11} - \tau_{22})/(\dot{\gamma}_{21})^2$; primary normal stress coefficient

$\psi_2 = -(\tau_{22} - \tau_{23})/(\dot{\gamma}_{21})^2$; secondary normal stress coefficient

Capillary viscometer:

$$\mu(\dot{\gamma}_w) = \frac{\tau_w}{(q/\pi r_t^3)}\left[3 + \frac{d \ln (q/\pi r_t^3)}{d \ln \tau_w}\right]^{-1}$$

$$\dot{\gamma}_w = \frac{1}{(\tau_w)^2}\frac{d(\tau_w^3 q/\pi r_t^3)}{d\tau_w}$$

$$\tau_w = r_t \Delta p_f/2L$$

Cone-and-plate instrument:

$$\mu(\dot{\gamma}) = \frac{3T_o}{2\pi r_c^3 \dot{\gamma}};\ \dot{\gamma} = \frac{\Omega}{\theta_c}$$

$$\psi_1(\dot{\gamma}) = \frac{2F}{\pi r_c^2 \dot{\gamma}^2}$$

$$\psi_2(\dot{\gamma}) = \{p_a - [\tau_{\theta\theta}(r_c) + p(r_c)]\}/\dot{\gamma}^2$$

$$\psi_1(\dot{\gamma}) + 2\psi_2(\dot{\gamma}) = -\frac{1}{\dot{\gamma}^2}\frac{\partial[\tau_{\theta\theta} + p]}{\partial \ln (r)}$$

Parallel-disk instrument:

$$\mu(\dot{\gamma}_R) = \frac{(T_o/2\pi r_p^3)}{\dot{\gamma}_R}\left[3 + \frac{d \ln (T_o/2\pi r_p^3)}{d \ln \dot{\gamma}_R}\right],$$

where

$$\dot{\gamma}_R = \frac{\Omega r_p}{w_f}$$

$$\psi_1(\dot{\gamma}_R) - \psi_2(\dot{\gamma}_R) = \frac{(F/\pi r_p^2)}{\dot{\gamma}_R^2}\left[2 + \frac{d \ln(F/\pi r_p^2)}{d \ln \dot{\gamma}_R}\right]$$

$$\psi_1(\dot{\gamma}_R) + \psi_2(\dot{\gamma}_R) = \frac{1}{\dot{\gamma}_R^2}\frac{d[\tau_{zz}(0) + p(0)]}{d \ln \dot{\gamma}_R}$$

$$\psi_2(\dot{\gamma}_R) = \frac{p_a - [\tau_{zz}(r_p) + p(r_p)]}{\dot{\gamma}_R^2},$$

where $\tau_{zz}(0) + p(0)$ = total normal stress measured at center and $\tau_{zz}(r_p) + p(r_p)$ = total normal stress measured at rim

Couette viscometer (narrow gap):

$$\mu(\dot{\gamma}) = \frac{T_o(r_o - r_i)}{2\pi r_i^3 L|\Omega_o - \Omega_i|}$$

$$\dot{\gamma} = \frac{|\Omega_o - \Omega_i| r_i}{r_o - r_i}$$

$$\psi_1(\dot{\gamma}) = -[\tau_{rr}(r_o) + p(r_o) - \tau_{rr}(r_i) - p(r_i)] r_i/[\dot{\gamma}^2 (r_o - r_i)] + \frac{\rho r_i^2}{3\dot{\gamma}^2}(\Omega_i^2 + \Omega_o^2 + \Omega_i \Omega_o)]$$

Steady-Shear Rheometry. For steady-shear flow, relationships between stresses and velocity gradients can be specified that define material functions (also called viscometric functions). The shear stress, τ_{12}, is related to the shear rate, $\dot{\gamma}_{21}$, by the viscosity, μ; the normal stresses—τ_{11}, τ_{22}, and τ_{33}—are related to the shear rate squared by the primary and secondary normal stress coefficients, Ψ_1 and Ψ_2. Appendix C provides a more detailed discussion.

The material functions for steady-shear flow can be obtained by a number of experimental geometries. Table 9.2 presents several of these geometries and shows how material functions are obtained from experimentally observable quantities.[1] Several texts are available that describe the details of the theory and practice of rheometry.[1,2] The choice of geometry is usually governed by such considerations as the amount of material available, the shear-rate range to be studied, and available equipment. The following general guidelines may prove helpful.

Capillary Viscometers. Capillary viscometers are generally most useful for obtaining high-shear-rate data, although Liauh and Liu[3] presented data from a capillary viscometer designed to cover both low- and high-shear-rate regimes. Larger volumes of fluid generally are required for capillary viscometers than for rotational viscometers. Data from the capillary viscometer can cover the range of shear rates that characterize process conditions but cannot be used to determine other material functions that might prove useful in polymer characterization. Capillary viscometers also have the advantage of being able to make measurements at elevated pressures and temperatures. The entry in Table 9.2 shows how to obtain the viscosity and shear rate from pressure-drop and flow-rate data with a procedure involving graphical or numerical differentiation of the data. This procedure does not assume that viscosity follows any particular rheological model, although it does assume a homogeneous fluid where shear stress is a unique function of shear rate. Rheologists often fit experimental data to known solutions for flow in tubes, such as those shown in Table 9.1. If the fluid is *assumed* Newtonian, then the "apparent viscosity" is given by

$$\mu_a = \frac{\pi \tau_w r_t^3}{4q} = \frac{\pi(\Delta p - \rho g \Delta z) r_t^4}{8Lq} \quad \text{(9.1)}$$

Or if the fluid is *assumed* to be a power-law fluid, then the power-law parameters can be obtained from a plot of $\ln(q)$ vs. $\ln(\tau_w)$, as suggested by the solution for flow of a power-law fluid in a tube:

$$q = \frac{\pi r_t^3}{(1/n') + 3}\left(\frac{\tau_w}{K'}\right)^{1/n'} \quad \text{(9.2)}$$

Rotational Instruments. Rotational instruments can be used with several different geometries: Couette (concentric cylinders), cone and plate, or parallel plate. The advantage of the narrow-gap Couette and the cone-and-plate geometries is that the shear rate throughout the fluid is nearly uniform. The viscosity can be calculated directly from the measured torque and known geometry. With the parallel-plate geometry, the viscosity can also be determined, but the calculation involves differentiation of the torque vs. angular velocity data, as shown in Table 9.2.[1] In most instances, rotational instruments are limited to lower shear rates for solutions because inertia causes secondary flows and the sample can be thrown out of the gap. However, Kramer *et al.*[4] and Connelly and Greener[5] showed how to obtain data to 50,000 seconds^{-1} using very small gaps between parallel plates. Earlier work[6] showed that high shear rates can also be attained in a conical geometry.

Oscillatory Measurements for Viscoelastic Fluids. To study viscoelastic fluid behavior, such as that exhibited by crosslinked gels, one must use dynamic oscillatory measurements, which can be generated in Couette, cone-and-plate, or parallel-plate geometries. The equations whereby dynamic moduli and viscosity are calculated from experimentally determined torque values, phase lags, and frequency of oscillation are given in Fig. C-1 in Appendix C. For linear viscoelastic measurements, it is necessary to test that the linear regime has been achieved by performing tests at several strain amplitudes. In the linear regime, the calculated moduli are independent of strain amplitude. For polymer solutions, linear behavior is the rule, but for some dispersions and gels, it is the exception. These systems can display nonlinear behavior at strains as low as 0.1%, as shown in Fig. 9.1, where the modulus of a polyacrylamide gel, which shows linear viscoelastic behavior, is contrasted to the behavior of a hydroxypropyl guar (HPG) gel, which shows structural breakdown under even small strains. In the linear regime, the moduli can be determined with the parallel-plate geometry without having to resort to differentiation, as in the case of steady-shear flow.

Blender Vortex Closure Time for Gelation Studies. A common technique to study the kinetics of gelation for polymers crosslinked by metal ions is to introduce the polymer solution into a blender and to observe the closure time of the fluid vortex. The centrifugal forces caused by the circular motion of the fluid driven by the spinning blender blades create a vortex in the center of the blender jar. As the polymer solution gels, its elasticity increases, leading to higher values of the normal stresses and causing the vortex to close. The effect of normal stresses on vortex closure has been modeled by Gordon and Balakrishnan[7] and Armstrong.[8] Although vortex closure tests are rapid, simple, and helpful in screening large numbers of samples, the flow field in the blender affects the measurements. Prud'homme[9] showed that for HPG's crosslinked by titanates, the kinetics are so fast that the rheology of the gel may be determined by the mixing rather than by the chemical kinetics. Knoll[10] recently presented data showing that the rate of increase in the storage modulus with time for several samples did not correspond to the ranking of the samples by vortex closure time.

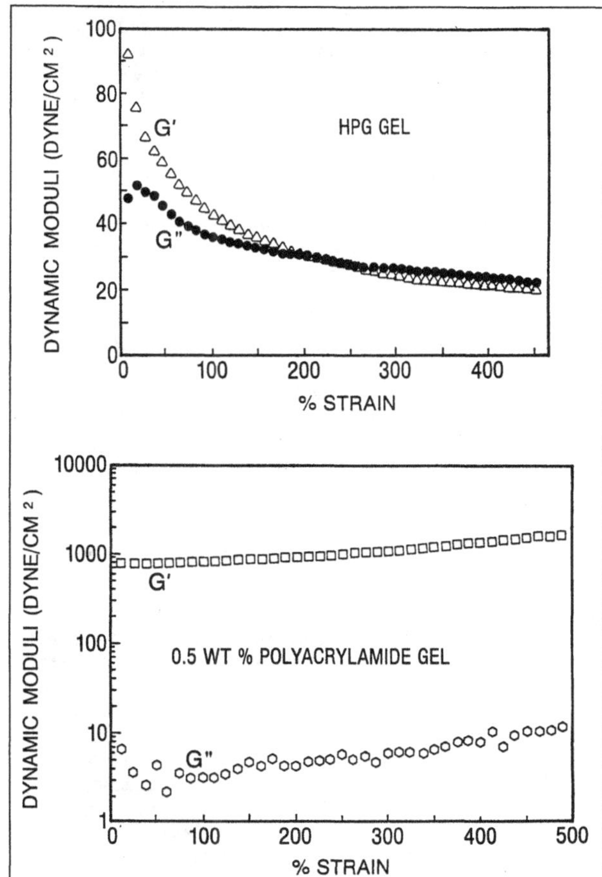

Fig. 9.1—Dynamic moduli of a 40-lbm/1,000-gal HPG gel crosslinked with 0.04 vol% titanium acetyl acetonate solution compared with the moduli of a 0.5 wt% polyacrylamide gel crosslinked with dichromate and bisulfate.[9]

9.4 Application of Laboratory Data in Fracturing Design

Most fracturing design is limited to power-law rheology because model constants for more descriptive rheological models, like the Ellis or Herschel-Bulkley models, are usually not available. The error in using the power-law model is probably not great compared with the uncertainty coming from other factors, such as fluid preparation, thermal and deformation history, and stability phenomena. Design equations based on power-law rheology are presented for pressure drops through various fracturing-related flow geometries, along with a discussion of the application of power-law rheology to gel scheduling.

Pressure-Drop Calculations in Fracturing Design. In hydraulic fracturing design, a knowledge of flowing pressure losses is important for several reasons. High pumping rates are desirable to transport proppant with more efficiency (less fluid loss and less proppant settling) and to generate fracturing pressures large enough to create wide fractures. However, limits on pumping pressures are set by pressure ratings of the pumping equipment and tubular goods and by reservoir characteristics related to maintaining the fracture in the desired zone. The following presents fundamental design equations for calculating pressure drops of incompressible fluids (liquids) in tubular goods, fractures, and perforations.

Tubular Goods. If friction-pressure charts are available for the fracturing-fluid system at the specified conditions, friction pressure (i.e., pressure drop caused by flow resistance, Δp_f) can be read directly. If friction-pressure charts are not available, an estimate of the flow regime may be made by calculating the generalized Reynolds number with power-law constants for the particular fluid system at an estimated average flow temperature in the tubular goods.

The Metzner-Reed generalized Reynolds number as given in Appendix D (Eq. D-3) may be used for tubular flow and its analog for annular flow.

Tubular-flow generalized Reynolds number (oilfield units)[11] is

$$N'_{\text{Re}}=\frac{(547.3)\rho q^{(2-n')}}{(1{,}647.7)^{n'}K_p d^{(4-3n')}}, \quad\text{(9.3)}$$

where $K_p=K'\left(\frac{3n'+1}{4n'}\right)^{n'}$.

Annular-flow generalized Reynolds number (oilfield units)[11] is

$$(N'_{\text{Re}})_{an}=\frac{(820.3)\rho q^{(2-n')}}{(2{,}470)^{n'}K_{an}(d_c-d_t)^{(2-2n')}(d_c+d_t)^{(2-n')}}, \quad\text{(9.4)}$$

where

$$K_{an}=K'\left(\frac{2n'+1}{3n'}\right)^{n'},$$

d_c=casing ID, and

d_t=tubing OD.

If the value of the generalized Reynolds number is <2,000, laminar flow may be assumed (although this has not been well substantiated for annular flow), and the following relations based on power-law rheology may be used to calculate the friction pressures, Δp_f.

Tubular-flow friction pressure (laminar power-law fluid) (oilfield units)[12] is

$$\Delta p_f=(0.333)(1{,}647)^{n'}\frac{LK_p q^{n'}}{d^{(1+3n')}}. \quad\text{(9.5)}$$

Annular-flow friction pressure (laminar power-law fluid) (oilfield units)[12] is

$$\Delta p_f=(0.333)(2{,}741)^{n'}\frac{LK_{an}q^{n'}}{(d_c+d_t)^{n'}(d_c-d_t)^{(1+2n')}}. \quad\text{(9.6)}$$

These relations assume steady laminar flow. The relationship of the geometry-dependent consistency indices, K_p and K_{an}, to the geometry-independent consistency index, K', is a consequence of the particular form of the Newtonian and power-law shear rates.[13,14] K_p and K_{an} are obtained experimentally by plotting wall shear stress, τ_w, vs. Newtonian (nominal) shear rate on logarithmic paper and computing the intercept at a nominal shear rate equal to unity.[12] The equation for annular flow uses an approximation referred to as the narrow-annulus solution. The error in this approximation becomes greater as n' and d_t/d_c decrease. However, as Savins[13] showed, for the majority of fracturing fluids with $n'>0.33$ and $d_t/d_c>0.3$, the maximum expected error is about 5%. K_{an} is equivalent to K_f for flow through a slot or fracture.

If the calculated generalized Reynolds number is >2,000, turbulence is possible, and an appropriate turbulent correlation is necessary to calculate the friction pressure. As discussed in Appendix D, drag-reducing non-Newtonian fluids require correlations involving several experimentally determined parameters or dimensionless groups. A commonly used correlation is that specified in the API *RP 39*,[12] which is based on Bowen's[15] relation:

$$\tau_w=\frac{d\Delta p_f}{4L}=Ad^b\left(\frac{8\bar{v}}{d}\right)^s, \qquad (9.7)$$

with A, b, and s being experimentally determined parameters. This equation is the same as Eq. 9.25 but is written with a parameter, b, which equals $s-j$. This correlation, coupled with the laminar power-law expression for τ_w, predicts a sharp transition from laminar to turbulent flow. For drag-reducing fracturing fluids, a more gradual transition to turbulent flow may be in effect. Correlations predicting gradual transitions will be given in Sec. 9.5.

The critical flow rate for the laminar/turbulent transition, q_c, can be calculated with Bowen's correlation, data for the laminar-flow parameters (n' and K'), and turbulent-flow parameters (A, b, and s).

Tubular-flow critical flow rate (oilfield units)[12] is

$$q_c=(6.071\times10^{-4})d^3\left(\frac{12^bK_p}{Ad^b}\right)^{1/(s-n')} \qquad (9.8)$$

Annular-flow critical flow rate (oilfield units)[12] is

$$q_c=(6.071\times10^{-4})(d_c+d_t)(d_c-d_t)^2\left[\frac{12^b(1.5)^{n'}K_{an}}{A(d_c-d_t)^b}\right]^{1/(s-n')} \qquad (9.9)$$

Note that in the annular case, the expression for the turbulent shear stress is correlated with $8\bar{v}/d_e$, where d_e is the equivalent diameter d_c-d_t:

$$\tau_w=A(d_e)^b\left(\frac{8\bar{v}}{d_e}\right)^s. \qquad (9.10)$$

This same formulation is used in the derivation of the turbulent annular friction pressure. The resulting tubular and annular friction pressures for turbulent flows are as follows.

Tubular flow (turbulent friction pressure) (oilfield units)[12] is

$$\Delta p_f=(0.333)(1{,}647)^s12^{-b}LAd^{(b-3s-1)}q^s. \qquad (9.11)$$

Annular flow (turbulent friction pressure) (oilfield units)[12] is

$$\Delta p_f=(0.333)(1{,}647)^s12^{-b}LA(d_c-d_t)^{(b-2s-1)}\left(\frac{q}{d_c+d_t}\right)^s. \qquad (9.12)$$

Fractures. The pressure drop of a power-law fluid flowing in a fracture is often approximated as laminar flow between infinite parallel plates. The solution[14] to this problem (in oilfield units) is

$$\Delta p_f=(0.167)(80.85)^{n'}L_fK_fw_f^{(-2n'-1)}\left(\frac{q_f}{h_f}\right)^{n'}, \qquad (9.13)$$

where

$$K_f=K_a=K'\left(\frac{2n'+1}{3n'}\right)^{n'}, \qquad (9.14)$$

h_f = fracture height,
w_f = fracture width,
L_f = length down one wing of fracture, and
q_f = injection rate down one wing of fracture.

In fracture-design calculations, this equation is usually modified to account for fluid loss, spatial and temporal changes in geometry, and changes in fluid rheology through changes in K_f and n'. More exact solutions that consider boundary effects and irregular boundaries are solved best by numerical methods. Chaps. 1, 4, 5, 11, and 14 discuss fracture flow in more detail.

Perforation Friction Pressure. A knowledge of pressure losses sustained by fracturing fluids during flow through perforations is important in fracturing-treatment design and analysis. Inadequately perforated zones or plugged perforations can result in unacceptably high pressures in the tubular goods during a fracturing treatment. In limited-entry designs, the number of perforations is intentionally limited to cause high downhole pressures, which can result in simultaneous stimulation of zones with different closure pressures.[16] In fracture-treatment analysis, an accurate estimate of the flowing pressure in the fracture near the wellbore can be made only if the perforation friction pressure is known. Downhole pressure recorders or open-annulus pressure monitoring yields the sum of the perforation and fracture friction pressures. One estimate of perforation friction pressure during an actual treatment using a downhole pressure-monitoring device would come from a measurement of instantaneous shut-in pressure taken early in the fracturing treatment, if negligible pressure losses in the fracture are assumed.

Design equations for perforation friction pressure are necessarily semiempirical and are similar in form to the equation for flow through an orifice. The relative importance of geometry, fluid density, and viscosity to perforation friction pressure can be shown by considering their effect on orifice flow. The pressure drop across an orifice is commonly measured with *vena contracta* taps; one situated one pipe diameter upstream and the other downstream at the *vena contracta* (the point of minimum pressure and minimum cross section of the fluid jet emerging from the orifice). A mechanical energy balance between these two points yields the pressure drop, which can be expressed as[17]

$$\Delta p_o=\frac{\rho(1-\beta^4)\bar{v}_o^2}{2g_cC_o^2}=\frac{8\rho(1-\beta^4)q_o^2}{g_cC_o^2\pi^2d_o^4}\propto\frac{\rho q_o^2}{d_o^4}, \qquad (9.15)$$

where

$\bar{v}_o$ = average velocity through orifice,
q_o = orifice flow rate,
β = ratio of orifice diameter divided by pipe diameter, and
C_o = discharge coefficient.

C_o is empirically determined and is usually expressed as a function of β and the orifice Reynolds number ($d_o\bar{v}_o\rho/\mu$), where d_o is the orifice diameter. Correlations for C_o are also dependent on the sharpness of the orifice edge and the location of the pressure taps. C_o is a correction for deviations from a uniform-velocity profile and for the contraction effect of the *vena contracta*. At low values of orifice Reynolds numbers where laminar flow exists, deviations from a uniform-velocity profile become more pronounced and viscous effects start to affect C_o. At Reynolds numbers <16, C_o is solely dependent on viscous effects; at Reynolds numbers >16, velocity profile and contraction effects become important.[18] As shown in Fig. 9.2 for a square-edged circular orifice, at $\beta<0.4$, C_o is approximately constant for Reynolds numbers >40, varying between 0.7 and 0.6.[19]

Fluid viscosity affects the orifice discharge coefficient mainly through its correlation with the orifice Reynolds number. Because viscous effects become important at lower Reynolds numbers, its impact on perforation friction pressures at typical treatment flow rates is most probably small. For example, water flowing at 0.2 bbl/min [0.53 dm^3/s] through a 0.25-in. [0.635-cm] orifice has an orifice Reynolds number of 106,000, assuming a β of 0.04. The corresponding C_o from Fig. 9.2 would be 0.61. If a fluid 300 times as viscous (300 cp [300 mPa·s]) were used under the same conditions, the orifice Reynolds number would drop to 353, but the value of C_o would only change to 0.68 (an increase of 11%).

The orifice pressure drop, Δp_o, is not necessarily a permanent pressure drop, however. Some of the pressure is recovered downstream from the *vena contracta* as the jet stream expands and slows down. The permanent pressure drop is therefore some fraction of Δp_o. Laboratory tests[20-22] using simulated perforations with little "tunnel"-flow friction (analogous to fracturing treatments where the perforation tunnels are open) have shown the permanent perforation friction pressure to have the form (in oilfield units)

$$(\Delta p_f)_{\text{perf}}=\frac{B\rho q_o^2}{d_o^4}, \qquad (9.16)$$

where B is a proportionality constant that, for design purposes, usually ranges from 0.20 to 0.50.

Perforation friction pressures are often presented in nomograph form, as shown in Fig. 9.3. This nomograph is equivalent to Eq. 9.16, with B =0.3718.

The above correlations were based on data from laboratory devices typically consisting of perforated casing, perhaps with a small nipple or collar on the discharge side that causes higher-than-expected pressure drops. If high perforation friction pressures are undesirable, a perforation program should be designed to maximize the number and size of perforations implemented.

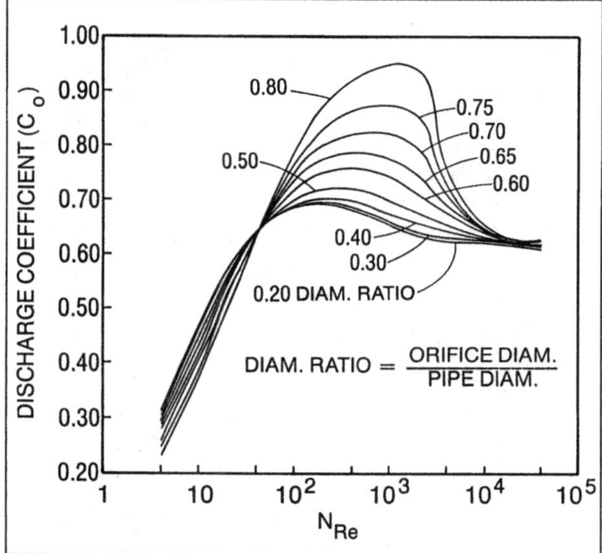

Fig. 9.2—Coefficient of discharge for square-edged circular orifices with corner taps.[19]

Fracturing-Fluid Scheduling. The scheduling of fracturing fluids is a key consideration in fracturing-treatment design. Fluids are scheduled in such a manner as to provide treatment viscosities sufficient to maintain adequate fracture widths and to suspend proppant adequately. Upper limits on fluid viscosities are also set to keep pumping pressures within equipment tolerances and to keep the fracture from breaking out of zone or to prevent the creation of secondary fractures. Fluid loss, which can be influenced by the fluid polymer concentration and fluid exposure time at temperature, is also impacted by fluid scheduling.

Because of the complexity of this problem, models of various degrees of sophistication have emerged. Some of the simplest treatments consider only the fluid material balance with assumed fracture geometry and width and use average effective fluid-loss rates.[23] This is effectively a one-dimensional treatment because only the fracture length is allowed to change. The next level of complexity is the two-dimensional (2D) models, which assume various geometrical shapes and allow the width to change along with the length or radius. Some of the classic models, reviewed by Geertsma and Haafkens,[24] couple fracture mechanics with fluid flow and, in some cases, with mass conservation. Three-dimensional models have recently been developed that require no assumptions as to geometry.[25-27] It is not within the scope of this chapter to discuss fracture simulation; see Chaps. 4 and 5 for more details.

To be useful for fluid scheduling, a model must have provisions to describe the fluid temperature. The simplest approach assumes the fluid to be at reservoir temperature during its entire history in the fracture. At typically high fracturing pump rates, and especially for large fluid volumes, however, the more realistic behavior would be to assume the fluid to increase from ambient temperature at the wellbore to nearly reservoir temperature at some point in the fracture. Poulsen and Lee,[28] using 2D fracture geometry with fluid loss, incorporated temperature effects with a modified fluid-temperature model originally formulated by Whitsitt and Dysart.[29] The predicted fluid-temperature profiles along the length of the fracture make a rather gradual transition from near-ambient temperature to reservoir temperature at a fracture location about one-third the distance from the tip. Other treatments incorporating heat effects in 2D models predict steeper temperature profiles, which can be approximated by two straight lines. The first line, originating at the wellbore at near surface temperature, attains reservoir temperature anywhere from one-half[30] to one-quarter[31] of the fracture length out from the wellbore. Fig. 9.4 shows one such profile.

The design of a fracturing-fluid "gel" schedule requires rheology data as a function of time at temperature and the fluid thermal history through the fracture. For a rigorous calculation of the fluid thermal history, the fluid rheology must be known because calculation of fracture widths and therefore fluid transit times involves the use of viscosity.

A modeling scheme using a specific fluid system with known rheological properties as functions of time at temperature would involve an iterative trial-and-error scheme to specify fluid compositions that give adequate viscosities. A more straightforward approach is not to constrain the model by specifying a particular fluid system beforehand, but rather to assume that fluids will be scheduled appropriately to provide the rheology used by the model in calculating the fluid thermal history in the fracture.

One such method published by Nolte,[31,32] which is an extension of Nordgren's[33] model to accept power-law rheology, assumes that the fracture temperature profile at any time during the fracture treatment is described by a linear temperature increase from surface temperature at the wellbore to reservoir temperature one-fourth the way down the fracture. This method requires a knowledge of only the fracture efficiency, e_f, defined as the fracture volume at the end of pumping divided by the total injected volume. It is presented graphically in Fig. 9.5. Here, the ordinate f_e is the fluid exposure time (at reservoir temperature) normalized to the total pumping time and the abscissa f_{fl} is the fraction of fluid pumped. The parameter x_c is the fraction of fracture length at which the fluid attains reservoir temperature, and f_d is a correction factor to adjust the normalized pad volume, f_p, to be more in agreement with simulated values. Values for x_c and f_d of 0.25 and 0.05 are typically used. The plot of f_e is determined by use of two points, A_o and C, with their respective equations given in Fig. 9.5. Point A_o is the maximum normalized exposure time, $f_{e\max}$, which is assumed to occur at the end of the pad stage, $f_{fl}=f_p$ (the pad leaks

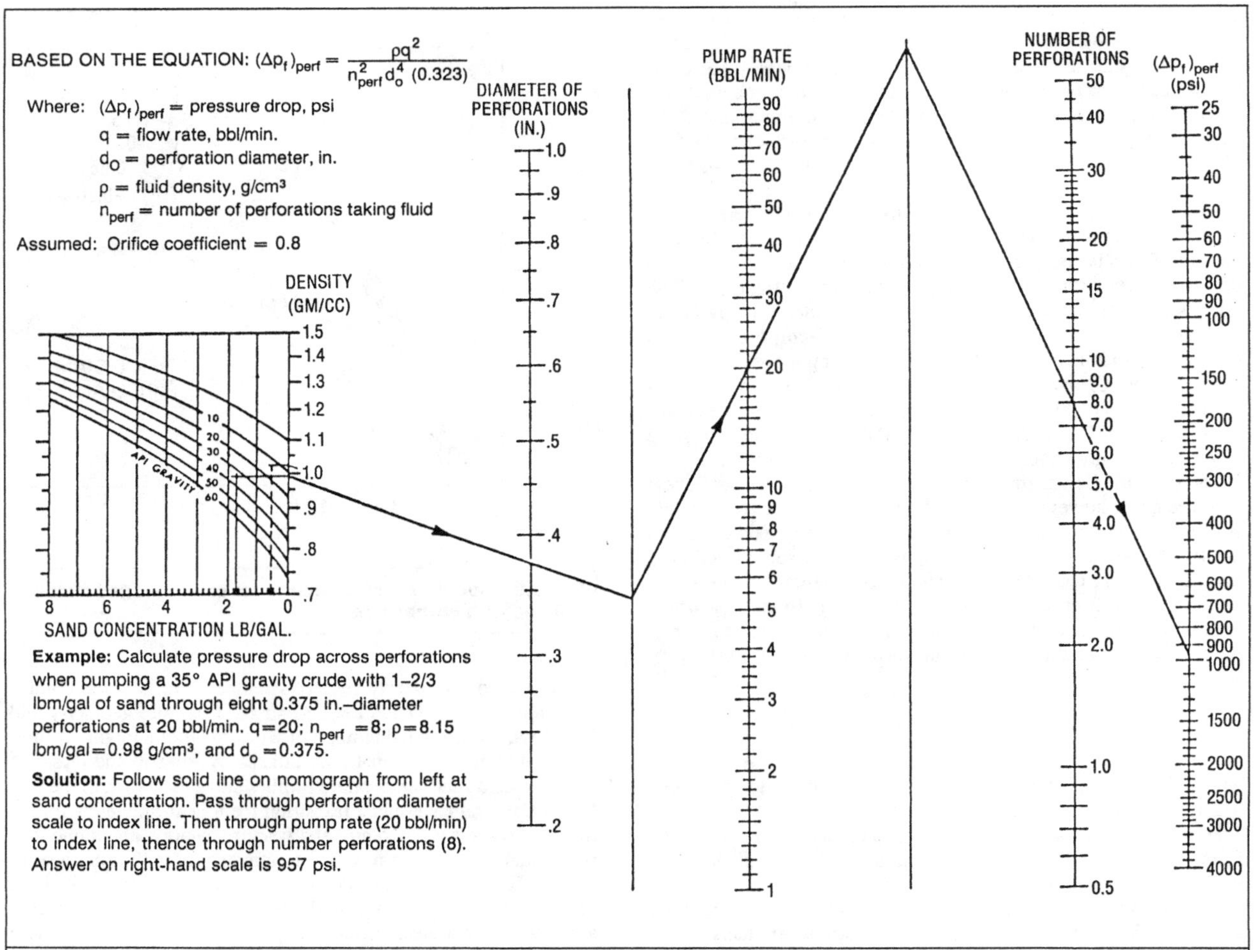

BASED ON THE EQUATION: $(\Delta p_f)_{perf} = \dfrac{\rho q^2}{n_{perf}^2 d_o^4 (0.323)}$

Where: $(\Delta p_f)_{perf}$ = pressure drop, psi
q = flow rate, bbl/min.
d_o = perforation diameter, in.
ρ = fluid density, g/cm³
n_{perf} = number of perforations taking fluid

Assumed: Orifice coefficient = 0.8

DENSITY (GM/CC)
API GRAVITY
SAND CONCENTRATION LB/GAL.
DIAMETER OF PERFORATIONS (IN.)
PUMP RATE (BBL/MIN)
NUMBER OF PERFORATIONS
$(\Delta p_f)_{perf}$ (psi)

Example: Calculate pressure drop across perforations when pumping a 35° API gravity crude with 1-2/3 lbm/gal of sand through eight 0.375 in.-diameter perforations at 20 bbl/min. q=20; n_{perf} =8; ρ=8.15 lbm/gal=0.98 g/cm³, and d_o =0.375.

Solution: Follow solid line on nomograph from left at sand concentration. Pass through perforation diameter scale to index line. Then through pump rate (20 bbl/min) to index line, thence through number perforations (8). Answer on right-hand scale is 957 psi.

Fig. 9.3—Nomograph of perforation pressure drop (courtesy Continental Oil Co., R&D Dept., Production Research Div., Ponca City, OK).

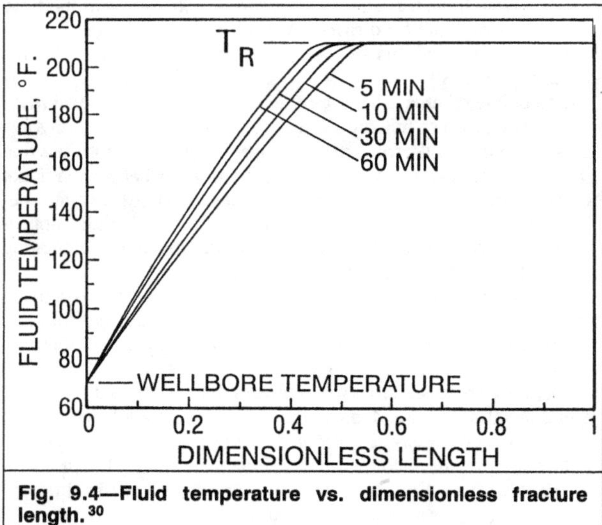

Fig. 9.4—Fluid temperature vs. dimensionless fracture length.[30]

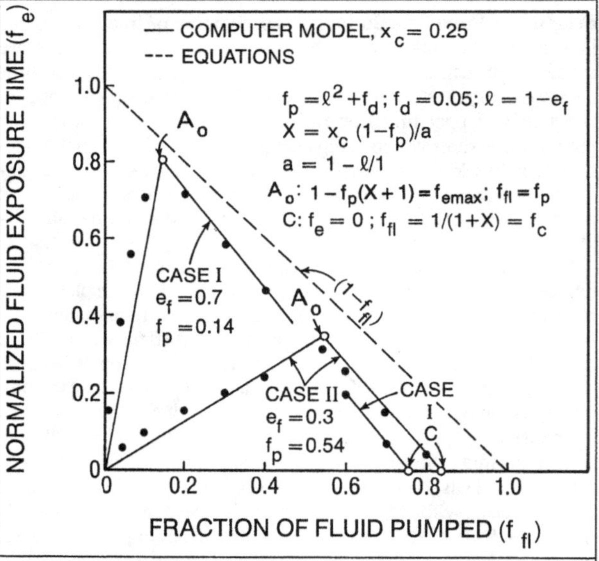

Fig. 9.5—Normalized fluid exposure time to reservoir temperature vs. normalized slurry volume.[31]

off just as the proppant reaches the tip). Point C is the fraction $f_c = 1/(1+X)$, where the fluid front reaches the end of the "cool" region just as injection terminates and thus has an exposure time to reservoir temperature of zero (f_e=0).

With the fluid exposure time known, the fracturing-fluid schedule may be constructed from viscosity data expressed as a function of time at reservoir temperature with composition as a parameter. Fig. 9.6 shows plots containing viscosity data for a hypothetical fluid and simulated exposure times.[32] The first step is to specify a lower range of limiting viscosities. In this particular case, a viscosity range from 50 to 100 cp [50 to 100 mPa·s] at 170-seconds^{-1} shear rate was chosen as the desired range for the viscosity of a fluid element at the end of its exposure time (i.e., the

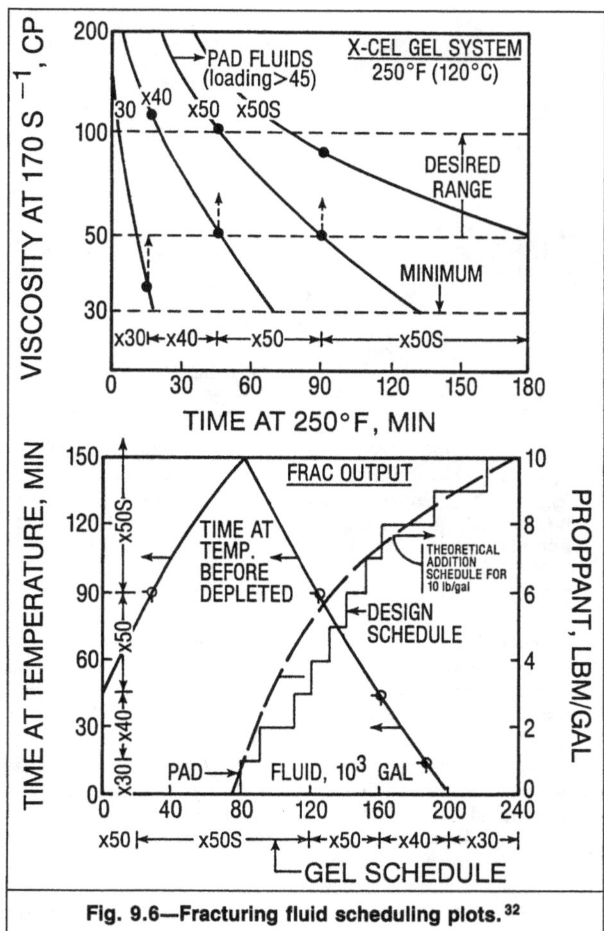

Fig. 9.6—Fracturing fluid scheduling plots.[32]

time at which it completely leaks off or pumping stops, whichever occurs first). The viscosities must be adjusted upward from this lower limit to approximate more closely the viscosities at earlier exposure times or cooler temperatures.

Fig. 9.6 shows the times at which each fluid has viscosities approximately in the desired range. These are indicated above the abscissa on the plot of viscosity vs. time at 250°F [121°C]. Time ranges are then transposed to the time-at-temperature ordinate on the exposure-time plot. Fluid scheduling is achieved by the intersection of each transposed fluid time range with the time-at-temperature curve. The cumulative slurry volumes at each intersection mark the boundaries of each fluid stage. The resulting schedule in this example is typical. The highest polymer concentrations are required at earlier times because exposure times are greatest and proppant concentrations are lowest. Also shown on the bottom plot is the proppant schedule predicted by the simulator. Fluid scheduling for large treatments is usually done in minimum stages of 20,000 gal [76 m^3] because this is the usual fracture-tank size. Proppant scheduling can be done in smaller increments.

After the fracturing-fluid system is scheduled, the fluid-loss coefficients corresponding to the scheduled fluids at the particular composition, temperature, and time range should be checked. If these are not close to the fluid-loss coefficients used in the simulator, a different fluid system may have to be scheduled or appropriate fluid-loss additives specified. The above discussion of fluid scheduling is intended to show the complexity of the problem and one way of approaching it. It is not intended to be a general guide to fluid scheduling. Fluid and proppant scheduling are also discussed in Chap. 11.

9.5 Uncrosslinked Fracturing Fluids

Uncrosslinked fracturing solutions are usually either water- or hydrocarbon-based. Solutions made up of high-molecular-weight water- or hydrocarbon-soluble polymers typically are non-Newtonian and drag-reducing and can be sensitive to temperature and shear rate.

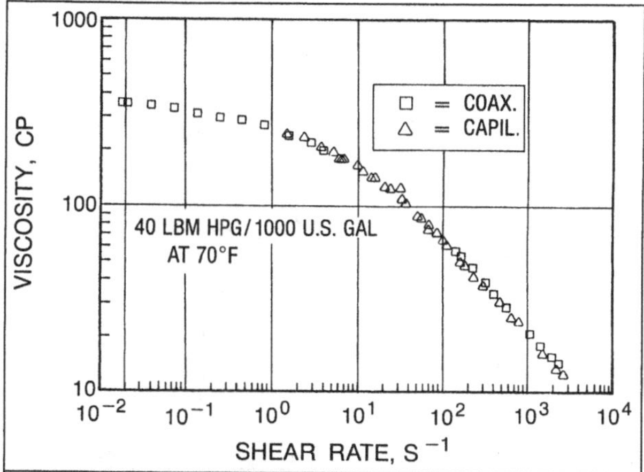

Fig. 9.7—Viscosity vs. shear rate for a 40-lbm/1,000-gal HPG solution measured with capillary and coaxial cylinder viscometers.[36]

Common high-molecular-weight water-soluble polymers used in fracturing solutions include guar and guar derivatives, such as HPG and carboxymethyl hydroxypropyl guar (CMHPG) and cellulose derivatives, such as hydroxyethylcellulose (HEC) and carboxymethyl hydroxyethylcellulose (CMHEC). Although not as extensively used, hydrocarbon solutions like viscous refined oil and lease crude oils have been used to stimulate water-sensitive formations. These hydrocarbon solutions can contain fluid-loss and friction-reducing additives (a high-molecular-weight hydrocarbon polymer at low concentrations).

Although polymer solutions show relatively complex flow behavior, they can readily be characterized by conventional viscometry.[2,34] Both microrheological and macrorheological theories have been developed for polymer solutions,[1,35] but these theories are beyond the scope of this monograph.

Viscosity Behavior. The most common rheological test performed on fracturing fluids is shear-stress/shear-rate testing for apparent viscosity. Fracturing-solution viscosity is a function of shear rate, temperature, polymer concentration, polymer molecular weight, and the chemical environment. Time is also a parameter in that polymers can undergo degradation from chemical reactions and physical deformation.

Effect of Shear Rate. Fracturing solutions are typically pseudoplastic (shear thinning) and exhibit power-law behavior over wall-shear-rate ranges normally expected in fractures (>20 seconds^{-1}). The viscosity of shear-thinning power-law fluids expressed as a function of shear rate becomes (in oilfield units)

$$\mu=47{,}880K'\dot{\gamma}^{(n'-1)}, \quad \text{(9.17)}$$

where

K' = consistency index,
$\dot{\gamma}$ = shear rate, and
n' = power-law index.

Shear-thinning fluids exhibit reductions in viscosity as shear rate increases; therefore, n' is less than one but greater than zero. For fracturing solutions at lower shear rates, a transition to Newtonian behavior with a finite, zero-shear-rate viscosity may become apparent. Newtonian behavior can be described by setting n' equal to unity in Eq. 9.17, in which case the viscosity becomes independent of shear rate and equal to K'. Fig. 9.7[36] shows the viscosity behavior of a 40-lbm/1,000-gal [4.8-kg/m^3] HPG solution at shear rates from 0.02 to 2,000 seconds^{-1}. These data show classic polymer solution behavior with a low-shear-rate Newtonian plateau making a gradual transition to power-law shear rates at higher shear rates.

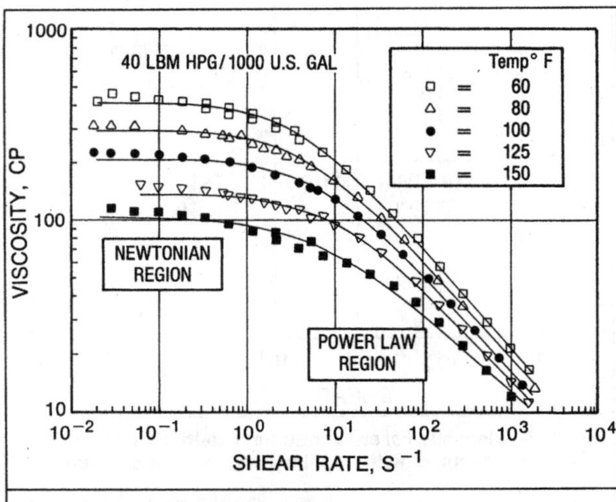

Fig. 9.8—Effect of temperature on the viscosity of a 40-lbm/1,000-gal HPG solution.[36]

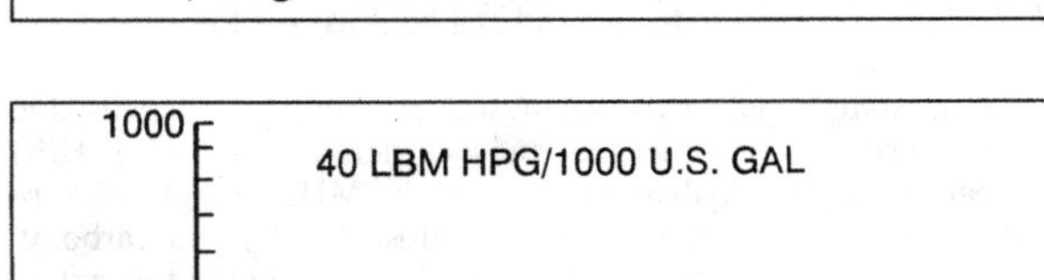

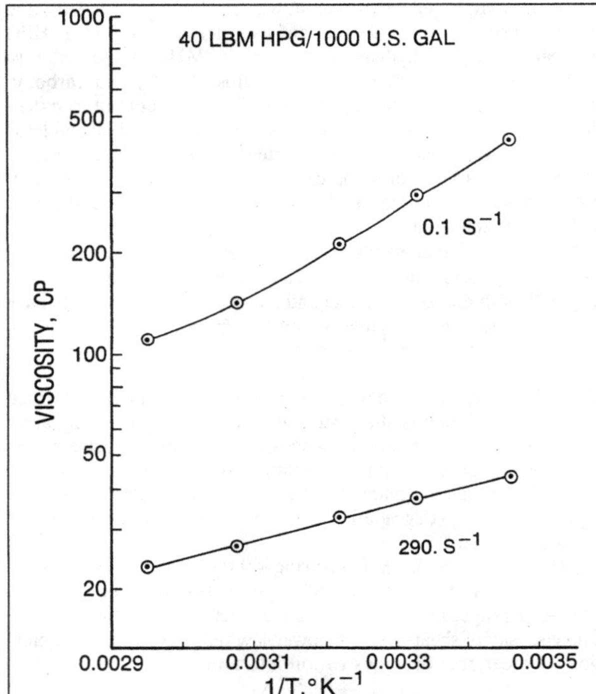

Fig. 9.9—Viscosity vs. 1/*T* at low and high shear rates for a 40-lbm/1,000-gal HPG solution (adapted from data by Guillot and Dunand[36]).

This behavior was modeled by Guillot and Dunand[36] with the three-parameter Ellis model (Table 9.1):

$$\mu=\frac{\mu^o}{1+(\tau/\tau_{1/2})^{\alpha-1}}, \quad \text{(9.18)}$$

where

μ^o = zero shear viscosity,
α = Ellis exponent,
τ = shear stress, and
$\tau_{1/2}$ = shear stress at which $\mu=\mu^o/2$.

The Ellis exponent is obtained from the slope of $(\mu^o/\mu)-1$ vs. $\tau/\tau_{1/2}$ plotted on log-log paper, which is $\alpha-1$. The data in Fig. 9.7 were obtained from both coaxial cylinder and capillary viscometers. The good overlap of data from different viscometers is typical of fracturing solutions and suggests no rheological irregularities, like slip flow, which is discussed later.

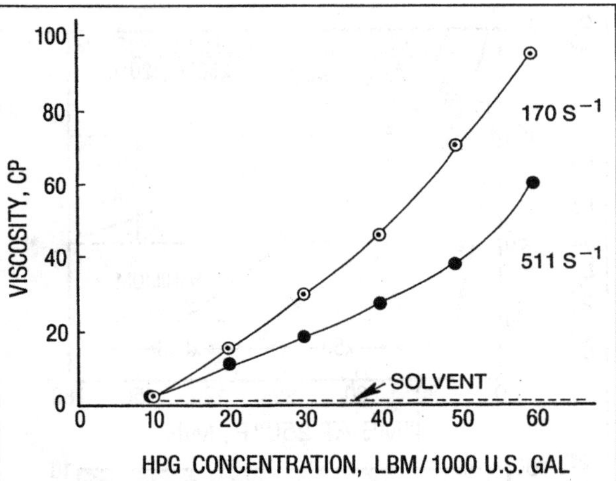

Fig. 9.10—Viscosity vs. polymer concentration at 80°F for a 40-lbm/1,000-gal HPG solution at shear rates of 170 and 511 seconds^{-1} (adapted from data from Dowell Schlumberger.[37]).

Effect of Temperature. Increasing temperature reduces the viscosity of fracturing solutions. Fig. 9.8[36] shows isotherms of viscosity from 60 to 150°F [16 to 66°C] for a 40-lbm/1,000-gal [4.8-kg/m^3] HPG solution over shear rates from 0.02 to 2000 seconds^{-1}. The temperature effect is greater at low shear rates than at high shear rates. Note that in the power-law region at a given value of shear rate, the slopes of the curves decrease as temperature increases. In terms of power-law data, this would be equivalent to an increase in n' with increasing temperature.

Assuming an Arrhenius-like dependence of viscosity on temperature,

$$\mu=A_v \exp\left(\frac{E_f}{RT}\right), \quad \text{(9.19)}$$

where

E_f = flow activation energy,
A_v = constant, and
R = real-gas constant.

The viscosities in Fig. 9.8 at representative low and high shear rates can be replotted vs. $1/T$, as in Fig. 9.9. At 0.1 seconds^{-1}, the slope (E_f/R) is seen to decrease as temperature increases, which may be the result of thermal rupturing of hydrogen bonds. The value of E_f decreases from about 5 kcal/gmol [20.9×10^3 kJ/kmol] at 0.1 seconds^{-1} to 2 kcal/g mol [8.4×10^3 kJ/kmol] at 290 seconds^{-1}, indicating that elevated shearing may reduce intermolecular interactions resulting from molecular entanglements and hydrogen bonding. It is possible for some fracturing solutions to exhibit a maximum in viscosity at some temperature. These, however, are special polymeric blends with a secondary gelling agent designed to hydrate at some intermediate temperature, such as 140°F [60°C]. Also, for some partially gelled hydrocarbon solutions, viscosity can be relatively insensitive to changes in temperature over a considerable range of shear rates and temperatures.

Effect of Polymer Concentration. Increasing the concentration of polymer in fracturing solutions acts to increase the viscosity. Fig. 9.10 shows the result of increasing HPG concentration at shear rates of 170 and 511 seconds^{-1} measured in a Fann Model 50C (Couette) viscometer at 80.6°F [27°C].[37]

For dilute polymer solutions, it is common to express the effect of concentration, C_p, on viscosity in Taylor series form:

$$\mu=\mu_s(1+[\mu]C_p+K^0[\mu]^2C_p^2+\ldots), \quad \text{(9.20)}$$

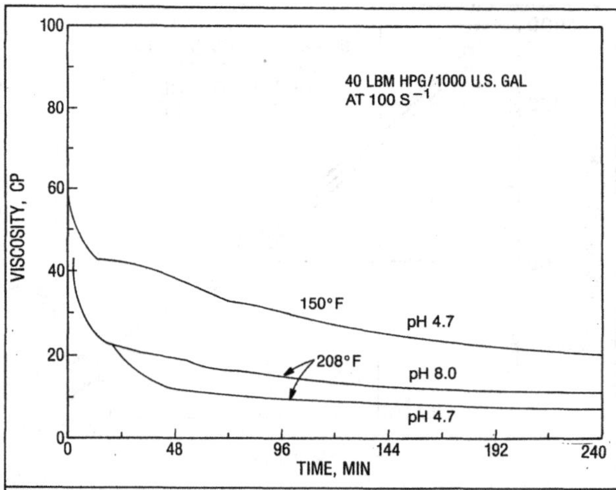

Fig. 9.11—Viscosity reduction at 150°F (pH 4.7) and 208°F (pH 4.7 and 8.0) of a 40-lbm/1,000-gal HPG solution at 100-seconds^{-1} shear rate [data by S.K. Knoll, courtesy Aqualon Co., Houston (1984)].

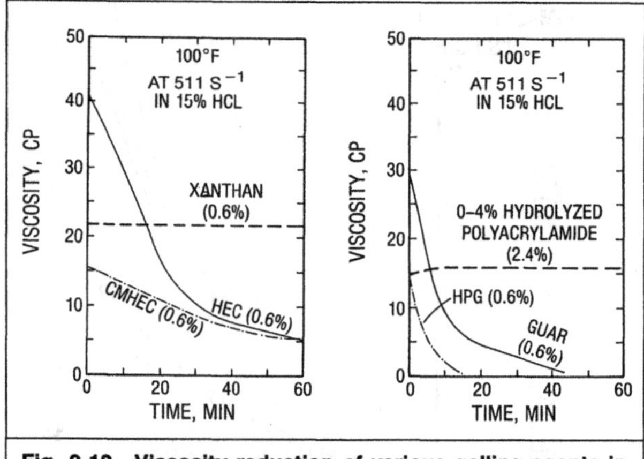

Fig. 9.12—Viscosity reduction of various gelling agents in 15% HCl at 100°F at 511-seconds^{-1} shear rate.[40]

where

$[\mu]$ = intrinsic viscosity,

μ_s = solvent viscosity, and

K^0 = Huggins coefficient.[1]

K^0 has reported values of 0.3 to 0.4 for good solvents and is almost independent of molecular weight.[1] The intrinsic viscosity has dimensions of inverse concentration and is a function of shear rate but not of concentration. It is defined as

$$[\mu] \equiv \lim_{C_p \to 0} \left(\frac{\mu - \mu_s}{\mu_s C_p} \right). \qquad (9.21)$$

Menjivar[38] has shown that there is a critical concentration, C_p^*, at which the hydrodynamic volumes of HPG molecules in solution start to overlap. This was found relatable to intrinsic viscosity as

$$C_p^* = 3.4/[\mu]. \qquad (9.22)$$

Measured values of C_p^* ranged from 16 to 22.5 lbm/1,000 gal [1.92 to 2.70 kg/m^3], corresponding to ranges of $[\mu]$ from 1.81 to 1.25 L/g and of molecular weight, M, from 3×10^6 to 1.8×10^6. A knowledge of C_p^* is important if one wishes to crosslink a polymer solution chemically. At concentrations less than C_p^*, intramolecular crosslinking may dominate intermolecular crosslinking, eventually leading to polymer segregation. Above C_p^*, the polymer solution can no longer be considered dilute and viscosity is no longer correlatable with $[\mu]C_p$. At high concentrations, viscosities of some polymer solutions have been found[1] to correlate with C_pM, but such correlations for fracturing solutions have not yet been published.

Effect of Time at Temperature Under Shear. Fracturing-fluid viscosity typically decreases with time because of physical (e.g., shearing) and chemical degradation. Chemical degradation can result from oxidation, acid hydrolysis, thermal rupture, and enzyme attack. All these mechanisms are accelerated by increasing temperature and can be affected by pH, ionic strength, and trace reducing agents.[39] Above a certain temperature (e.g., 150°F [66°C]) and outside the pH range of 2.5 to 8, most enzymes themselves degrade and are no longer effective breakers.

Fig. 9.11 shows the time dependence of a 40-lbm/1,000-gal [4.8-kg/m^3] HPG solution tested at 149 and 208°F [65 and 98°C] at 100 seconds^{-1}. This system is at pH of 4.7 and 8.0 with no stabilizers.

Fig. 9.12 shows the viscosity reduction with time of HPG and other polymers as a result of acid hydrolysis when used with 15% HCl solution at 100°F [38°C].[40]

Fracturing-fluid polymers can be physically degraded by high shear rates. High-shear tests showed that solutions of guar, HEC, and HPG suffered increasing mechanical degradation in the order listed. Solutions of xanthan gum, on the other hand, were unaffected by shear.[41] Polyacrylamide solutions are even more sensitive than guar-derived polymer solutions.[42,43]

Slip-Flow Phenomena. Slip flow occurs when a fluid has a nonzero velocity at the wall (solid boundary) and thus violates the traditional no-slip boundary condition of fluid mechanics. The no-slip condition at a solid/liquid interface, the usual case for most simple fluids, is the result of attractive molecular forces (van der Waals forces) causing liquid to adhere to the solid surface.

The magnitude of slip flow in uncrosslinked fracturing solutions is small. Fig. 9.7 shows a 40-lbm/1,000-gal [4.8-kg/m^3] HPG solution measured with both capillary and rotating concentric cylinder viscometers. The lack of a viscometer geometry effect supports no-slip. Rogers *et al.*[44] observed similarly good superposition of viscosity data for HPG solutions using different sizes of different-geometry viscometers. There have been some reports of slip flow in polymer solutions[45,46] [e.g., carboxymethyl cellulose (CMC)], but the magnitude of slip is usually small. This is most probably an "apparent" slip phenomenon caused by the formation of a less viscous fluid layer near the wall resulting from polymer migration.[45,47-49]

Uncrosslinked fracturing fluids would not be expected to exhibit slip when tested with viscometers having small stress gradients (concentric cylinder and cone-and-plate viscometers) or tubular viscometers with small length-to-diameter ratios that would not allow sufficient time for polymer migration to establish concentration gradients.

Yield-Stress Phenomena. A finite, zero-shear viscosity, as indicated in Fig. 9.7 for an HPG solution, means that no yield stress is present (finite stress necessary to initiate flow). No yield stresses have also been observed for guar[50] and CMC[51] solutions at concentrations less than 2 wt%. Sometimes a yield point is associated with fracturing solutions, but usually this is the shear stress extrapolated from the power-law portion of the shear-stress/shear-rate curve to zero shear rate. A yield point obtained in this manner can be grossly in error, depending on the low-shear-rate behavior of the shear stress (refer to Appendix C).

Proppant Effect in Steady-Shear Testing. Little has been published regarding steady-shear testing of fracturing solutions containing proppant. Many rotational and capillary viscometers operate with test fluid shear dimensions not much larger than the diameter of 20/40 sand. More important, most viscometers, including large-

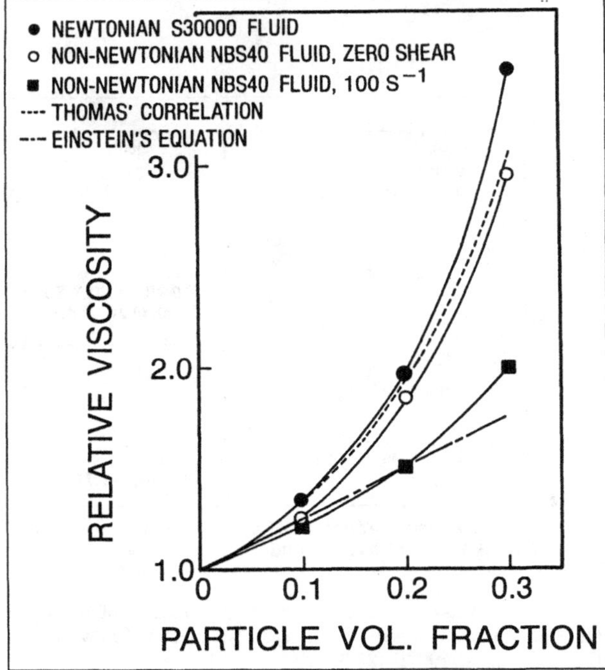

Fig. 9.13—Relative viscosity vs. particle volume fraction for various fluids and correlations.[70]

diameter tubular viscometers that usually are horizontally situated, have the problem of keeping proppant suspended long enough in laminar flow to make a meaningful measurement. Proppant settling is typically quite rapid in fracturing solutions, with reported settling velocities in the range of 0.236 to 94.4 in./min [0.1 to 40 mm/s].[52] For laminar horizontal flow in a tubular viscometer, an estimate of the maximum particle-settling velocity that will still maintain a "uniform" suspension is given by[53]

$$(v_p)_{max} = \frac{4\bar{v}(\bar{\ell}_p)^2}{Ld}, \quad \text{(9.23)}$$

where

$\bar{v}$ = average pipe flow velocity,
$\bar{\ell}_p$ = average interparticle distance,
L = pipe length, and
d = pipe ID.

In essence, $(\bar{v}_p)_{max}$ is the settling velocity required for a particle, at distance ℓ_p from the wall, to reach the wall by the time the particle reaches the end of the pipe.

Recently, Shah and Lord[54] developed a correlation for the velocity required to suspend proppant when flowing in horizontal tubing. They studied proppant transport in both uncrosslinked and crosslinked fracturing fluids and considered the effects of pipe size; polymer-gelling-agent concentration; fluid rheology; proppant size, density, and concentration; and fluid density.

Note that proppant settling is a complex function of fluid rheology, particle Reynolds number, particle-size distribution, particle volume fraction, flow field, and flow geometry. Further discussion of proppant settling is found in Chaps. 10 and 11. Refs. 55 through 63 provide significant background on particle-migration phenomena.

Suspensions in Newtonian Fluids. Extensive theoretical and experimental work has been done on the behavior of suspended particles in Newtonian liquids. Theoretical efforts have looked at the effects of particle character (concentration, shape, and size distribution) and the various forces acting on particles, including hydrodynamic, thermal (Brownian), London-van der Waals, and electrical forces. Discussion of past work is beyond the scope of

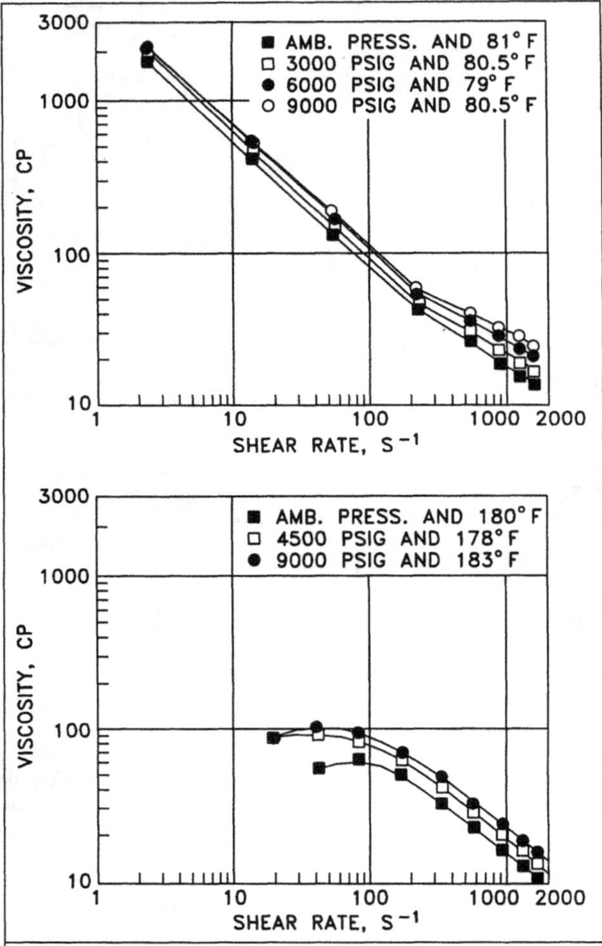

Fig. 9.14—Pressure effect on a partially gelled diesel at ambient temperature and at 180°F [data by J.R. Cameron, courtesy Amoco Production Co. Research, Tulsa, OK (1986)].

this monograph; see Refs. 64 and 65 for good reviews on this subject.

Thomas[66] developed the following semiempirical correlation for suspensions in Newtonian fluids by extrapolating data obtained for a wide variety of particle sizes (0.1 to 435 μm) and shear rates to minimize secondary effects, such as those caused by non-Newtonian behavior, inertial forces, and measuring-instrument wall effects:

$$\mu = \mu_c[1 + 2.5V_p + 10.05V_p^2 + 0.00273\exp(16.6V_p)], \quad \text{(9.24)}$$

where

μ_c = solid-free (continuous phase) fluid viscosity,
V_p = particle volume fraction, and
μ = bulk viscosity of suspension.

Suspensions in Non-Newtonian Fluids. Suspensions in non-Newtonian fluids would be expected to deviate somewhat from Thomas' correlation (Eq. 9.24). Experimental evidence from studies on suspensions[67] suggests that the relative viscosity ($\mu/\mu_c = \mu_r$) decreases with increasing shear rate up to a point, e.g., 20 seconds^{-1}. At higher shear rates, the relative viscosity is constant at a given particle volume fraction. Even at the lowest shear rates, however, the relative viscosity was lower than that for suspensions in Newtonian fluids. Analogous trends were observed for the normal stress behavior, where the ratio of first-normal-stress difference of the suspension to that of the solids-free fluid became constant above 20 seconds^{-1}. At lower shear rates, this ratio increased. Normal stress behaviors of various types of solutions are discussed in more detail in Refs. 1, 42, 51, 68, and 69.

Theoretical studies of suspensions in non-Newtonian shear-thinning fluids support the above experimental trends. Work by Kawase and Ulbrecht[55] predicts decreasing relative viscosities as the power-law index decreases at a given particle volume fraction.

Fig. 9.13 shows relative viscosities for various correlations and theories. Included are the data of Chan and Powell[70] generated using cone-and-plate and parallel-plate viscometry for suspensions of 25- to 38-μm-diameter glass beads in a non-Newtonian fluid (NBS Fluid 40). Their zero-shear-rate relative-viscosity data follow Thomas' data closely, while relative-viscosity data at higher shear rates (up to 100 seconds^{-1}) decrease in keeping with the other studies discussed. At volume fractions <0.2, the high-shear-rate data correspond closely to Einstein's equation for dilute Newtonian fluids—i.e., $\mu_r = 1 + 2.5\ V_p$. For design purposes, the effect of suspended particles on the slope, n', of the power-law region can be considered insignificant.[67] The flow curve of log shear stress vs. log shear rate can be assumed to be shifted vertically a constant amount upon addition of suspended solids even for non-power-law behavior.

Pressure Effect. Increasing the hydrostatic pressure on a liquid generally causes the viscosity to increase. The viscosity of fracturing solutions is normally measured at ambient pressure or at moderate pressures <1,000 psi [6.9 MPa] using the Fann Model 50C viscometer. However, hydraulic fracturing treatments can occur at pressures in excess of 10,000 psi [69 MPa]. Models[71] of liquid flow that relate resistance to flow in terms of fluid voids or the presence of free volume predict viscosity to increase as free volume decreases under conditions of increasing pressure. Therefore, one might expect greater effects of pressure to be observed in more compressible fluids. Because water has very low compressibility, dilute aqueous polymer solutions would probably show little pressure effect. Oil-based fracturing fluids, however, could very well show a significant pressure effect because oils are several times more compressible than water.

Fig. 9.14 shows the effect of pressures up to 9,000 psi [62 MPa] on a partially gelled diesel. Increases of viscosities up to 100% were observed at various temperatures and shear rates. The effect of pressure on an aqueous 60-lbm/1,000-gal [7.2-kg/m^3] HPG solution was negligible, however, over the same 9,000-psi [62-MPa] pressure range.* The curves for the partially gelled diesel show that at ambient temperature, viscosities are similar to those at 180°F [82°C] at shear rates greater than 100 seconds^{-1}. At lower shear rates, however, the ambient-temperature viscosities become much greater. Elevating the temperature apparently shifts the low-shear Newtonian region to higher shear rates. Because proppant transport may depend on low-shear behavior, higher temperatures may affect the proppant-transport capability of gelled oils even though the high-shear viscosities appear adequate. It is evident that the effect of pressure varies with shear rate and temperature for this particular partially gelled diesel.

Turbulent Behavior of Fracturing Solutions. Fracturing solutions can display turbulent behavior that is not correlatable by the usual Newtonian relations (e.g., the Blasius equation; refer to Eq. D-1). Even turbulent correlations for inelastic power-law fluids such as that by Dodge and Metzner (Eq. D-4) may not apply. Viscoelastic fracturing solutions can be drag-reducing as well as shear-thinning. When correlated with a Reynolds number, friction-factor plots can show an additional effect of tubing diameter. See Appendix D for further discussion of drag reduction and turbulence of viscoelastic solutions.

For fracturing solutions, an empirical relation based on Bowen's[15] extension of the Blasius equation has been shown to describe successfully the turbulent behavior of fluids, including CMC and guar solutions. This relation takes the form

$$\frac{d^{(1+j)}\Delta p_f}{4L} = A(8\bar{v})^s, \quad \ldots \quad (9.25)$$

where j, A, and s are empirically determined constants. Values for j in the literature have ranged from 0.0 to 0.266 for a variety of slurries and polymer solutions flowing in tubes of different diameters.[72,73] Values of A and s vary with fluid system and concentration. API has adopted Eq. 9.25 for scaling up the turbulent

*Data from D.C. Gardner, Amoco Production Co. Research Center, Tulsa, OK (1986).

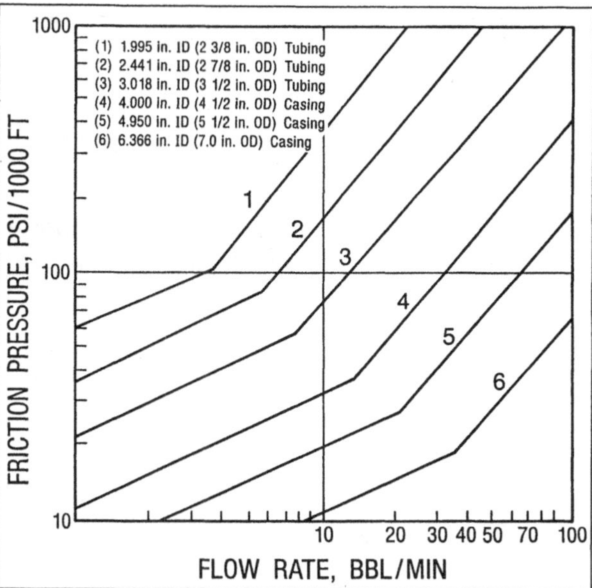

Fig. 9.15—Typical presentation of friction-pressure data for a 60-lbm/1,000-gal polymer solution in tubing and casing (courtesy Dowell Schlumberger).[37]

flow of fracturing fluids through tubing and tubing/casing annuli.[12] API *RP 39* recommends the addition of 0.036 to s (the smooth-pipe flow-curve slope) to compensate for hydraulic roughness in oilfield tubular goods. Also explained is how to calculate the point for the onset of turbulence: equate the expression for the shear stress in the laminar region (in power-law form) to that derived from Bowen's relation. Refer to Sec. 9.4 for presentation of these equations.

Bowen's relations can be cast into friction-factor form by dividing by the kinetic energy and grouping in terms of the solvent Reynolds number, $(N_{Re})_s$.

$$f = \frac{\tau_w g_c}{\frac{1}{2}\rho\bar{v}^2} = \frac{2Ag_c 8^s}{d^{j-1}\bar{v}^{(1-s)}\mu_s (N_{Re})_s}, \quad \ldots \quad (9.26)$$

where $(N_{Re})_s = \rho d\bar{v}/\mu_s$ and μ_s = solvent viscosity.

When $(N_{Re})_s$, ρ, and μ_s are constant, the effect of diameter on f becomes evident when $\bar{v}$ is written in terms of these constants, $\bar{v} = \mu_s (N_{Re})_s/(\rho d)$:

$$f = \frac{2Ag_c(8)^s d^{(2-j-s)}\rho^{(1-s)}}{\mu_s^{(2-s)}(N_{Re})_s^{(2-s)}}. \quad \ldots \quad (9.27)$$

For the empirically determined ranges of j (0.0 to 0.25) and s (1.0 to 1.75), this equation predicts an increase in f with increasing diameter, which is the typical diameter effect for drag-reducing fluids. For values of j and s of 0.25 and 1.75, this would be analogous to the Blasius relation (Eq. D-1) in which the diameter effect vanishes. For other values, drag reduction is predicted. The empirically determined constants j and s apparently embody the viscoelastic non-Newtonian character that can give a diameter effect.

It is also of interest to examine the effect of diameter when friction factor is correlated with the generalized Reynolds number, N'_{Re} (see Eq. 9.3), which is commonly used for drag-reducing fracturing fluids. Writing f in terms of N'_{Re} in Eq. 9.27 and substituting the velocity solved in terms of N'_{Re} (Eq. D-3) shows that

$$f = \frac{2Ag_c(8)^s d^{\left[\frac{n'(n'-s)}{2-n'} + n' - j\right]} \rho^{\left(\frac{n'-s}{2-n'}\right)}}{\left[g_c K_p 8^{(n'-1)} N'_{Re}\right]^{\left(\frac{n'-s}{2-n'} + 1\right)}}. \quad \ldots \quad (9.28)$$

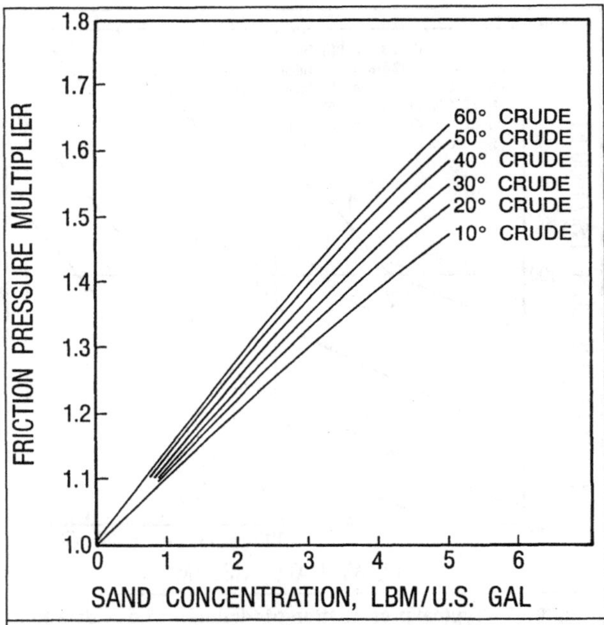

Fig. 9.16—Friction-pressure multiplier vs. pounds of sand per gallon of API gravity crude oil.[77]

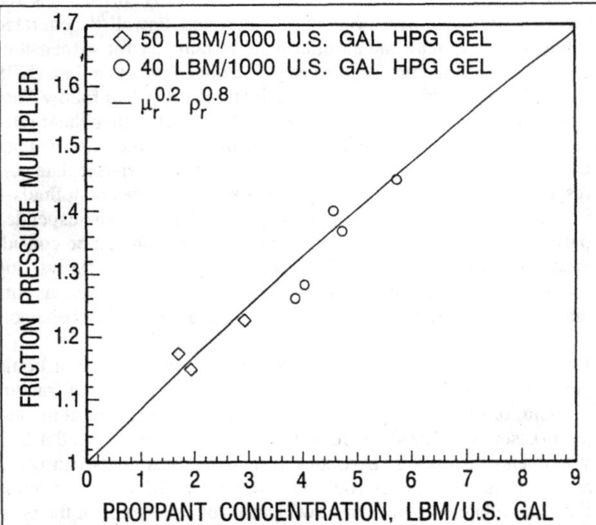

Fig. 9.17—Friction-pressure multiplier vs. proppant concentration in HPG gels flowing at 26 bbl/min through a 5½-in.-casing/2⅜-in.-tubing annulus.[78]

For constant N'_{Re}, ρ, n', and K_p, this relation would predict that

$$f \propto d^{\left[\frac{n'(n'-s)}{2-n'}+n'-j\right]} \qquad (9.29)$$

Thus, depending on the values of n', j, and s, the friction factor could either increase or decrease with diameter. As an example, for $s=1.2$ and $j=0.1$, the exponent of d would be -0.058, -0.011, $+0.041$, $+0.1$, $+0.167$, and $+0.331$ for n' values of 0.1, 0.2, 0.3, 0.4, 0.5, and 0.7, respectively. For typical fracturing fluids with $n'<0.7$ and turbulent values of Δp_f that can be correlated with Bowen's relation (Eq. 9.25), the effect of diameter could be very weak when plotted with N'_{Re} and sometimes may not be observed.[74] If f were correlated with $(N_{\mathrm{Re}})_s$ when $s=1.2$ and $j=0.1$, the exponent of d would always be 0.7, regardless of n' (see Eq. 9.27). The fact that diameter can have a weak effect when f is correlated with N'_{Re} is of practical importance in cases where data from only one pipe size are available. In this event, a fairly good estimate of friction pressures in field-size pipes may be made by assuming no diameter dependence and using the f-vs.-N'_{Re} correlation for the known pipe size.

Bowen's relation is a convenient empirical method for handling fully developed viscoelastic turbulence; however, it does not describe transition to turbulence in a continuous fashion, as has been seen by some observers.[74,75] For gradual transitions to turbulence, semiempirical correlations like those proposed by Seyer and Metzner[76] and Shah[74] have been used successfully.

Any effects of temperature and pressure would be experimentally observable through changes in the coefficients A, j, and s. In their turbulent studies of CMC solutions and titanium dioxide suspensions, Quader and Wilkinson[73] observed no temperature dependence of s but found j and A to vary rather weakly with temperature. Effects of hydrostatic pressure on friction pressure have not been investigated well. The effect of pressure on aqueous fracturing solutions is most likely small but may be significant when hydrocarbon-based solutions are pumped.

Presentation of Turbulent-Design Data. For fracturing solutions, turbulent flow data for pipes or annuli are usually presented as shown in Fig. 9.15. A series of straight lines for different tubing or annulus sizes is plotted on log-log paper with a discontinuity representing transition from laminar to turbulent flow. In the laminar zone, the slope of the line should correspond to n' as measured with a viscometer and the slope in the turbulent zone can vary from about 1.0 to 1.8.

Effect of Proppant on Turbulent Behavior. The effect of proppant on friction pressure has recently received attention because of the need to predict bottomhole treating pressures (BHTP's) for real-time fracturing-treatment analysis. The addition of proppant is generally thought to increase the friction pressure of fracturing solutions. One way to correct for the effect of proppant is simply to multiply the friction pressure of the proppant-free fluid by a suitable correction factor. Crittendon's[77] widely used correlation was originally developed for API-gravity crudes. The origin and generality of this correlation are not clear. Fig. 9.16 shows this correlation.

A recently proposed correction factor, F_C, proposed by Hannah *et al.*[78] represents the friction-pressure ratio for the proppant-laden and -unladen fluids $[(\Delta p_f)_M/(\Delta p_f)_{fl}]$. Their correction factor gives the increase in friction pressure from viscosity and density changes that would be experienced by a Newtonian fluid in turbulent flow upon the addition of proppant at constant flow rate and in a constant-diameter tube:

$$F_C=\mu_r^{0.2}\rho_r^{0.8}=(\Delta p_f)_M/(\Delta p_f)_{fl}. \qquad (9.30)$$

The relative viscosity, μ_r, is given by Eq. 9.24[66] as $\mu_r=\mu/\mu_c$ and thus is a function of only proppant volume fraction, V_p, not fluid rheology. The relative density, ρ_r, is the ratio of the proppant-laden and -unladen fluid densities. F_C is therefore a function of proppant density, proppant concentration, and fluid density (specific gravity) only; it does not vary with fluid rheology, flow rate, proppant size, or flow geometry. The general application of such a correction factor would therefore be suspect. Nevertheless, F_C has been reported[78] to predict the increase in friction pressure with proppant addition accurately. As an example, data for crosslinked HPG gels (see Sec. 9.6) are shown in Fig. 9.17.

Shah and Lee[79] proposed a friction-pressure correlation that considers the effects of fluid rheology, flow rate, proppant size, and pipe size, in addition to densities and proppant concentration. This correlation is in terms of three dimensionless groups and is presented in graphical form in the original paper. At a given proppant concentration, the percent friction-pressure increase over the base gel is reduced by increasing flow rate, increasing HPG concentration, decreasing proppant size, and decreasing proppant density. This correlation was developed with various concentrations of HPG solutions; its generality to other types of solutions has not yet been demonstrated.

Some field observations while delayed crosslinked HPG fracturing fluids were pumped down long, vertical tubing strings have shown friction pressures to be considerably less than those observed in the laboratory, for fluids with and without proppant.[80] Lord and McGowen[80] suggest that this behavior may result from the formation of nonhomogeneous fluids down very long tubing strings that cannot develop under laboratory conditions where fluid is circulated in loops or through relatively short tubing lengths. They developed the following correlation, which applies to fracturing fluids with and without proppant (in oilfield units):

$$\ln\left[\frac{(\Delta p_f)_o}{(\Delta p_f)_{fl}}\right]=2.38-8.024/\bar{v}-0.2365C_{HPG}/\bar{v}$$

$$-0.1639\ln[C_{HPG}]-0.028C_s\exp[1/C_{HPG}], \quad \ldots\ldots (9.31)$$

where

$\bar{v}$ = average velocity,
C_{HPG} = concentration of HPG, and
C_s = proppant concentration.

$(\Delta p_f)_o$ is the friction pressure of the Newtonian water solvent given (in oilfield units) by

$$(\Delta p_f)_o=0.40429d^{-4.8}q^{1.8}L, \quad \ldots\ldots (9.32)$$

where

L = tubing length,
d = diameter, and
q = flow rate.

This correlation predicts no diameter dependence.

Example Problem 1—Calculation of Surface Pumping Criteria for an HPG Solution. Calculate the surface treating pressure when pumping a 40-lbm/1,000-gal [4.8-kg/m^3] HPG base gel at 60 bbl/min [0.16 m^3/s] down 10,000 ft [3050 m] of 5½-in. [13.97-cm] casing with and without 6 lbm sand/gal liquid [720 kg/m^3]. Assume a flowing temperature in the wellbore of 75°F [24°C], a constant BHTP of 7,000 psi [48 MPa], and negligible perforation friction pressure.

Solution Method 1: API Method. Data available for a 40-lbm/1,000-gal [4.8-kg/m^3] HPG base gel are

$$n'=0.45,\ K_p=0.0259(\text{lbf-sec}^{n'})/\text{ft}^2\ [1.24(\text{N}\cdot\text{s}^{n'})/\text{m}^2],$$

$$A=0.001577(\text{lbf-sec}^s)/\text{ft}^{2+b}\left[\frac{0.0755}{3.28^{-b}\text{m}^{2+b}}\cdot\text{Ns}^s\right],$$

$$s=1.164,\ b=1.11,\ \rho_{fl}=8.34\ \text{lbm/gal}\ [999\ \text{kg/m}^3],$$

and

$$\rho_s=22.10\ \text{lbm/gal}\ [2648\ \text{kg/m}^3].$$

Check for turbulence using Eq. 9.3:

$$N'_{Re}=\frac{(547.3)\rho q^{(2-n')}}{(1{,}647.7)^{n'}K_p d^{(4-3n')}} \quad \ldots\ldots (9.3)$$

$$=\frac{(547.3)(8.34)(60)^{(2-0.45)}}{(1{,}647.7)^{0.45}(0.0259)4.95^{[4-3(0.45)]}}$$

$$=51{,}752\ \text{(turbulent)}.$$

Use Eq. 9.11 to calculate the turbulent friction pressure and the suggestion in the API *RP 39*[12] to correct the s parameter for roughness by adding 0.036 to s (i.e., $s=1.164+0.036=1.20$):

$$(\Delta p_f)_{fl}=(0.333)(1{,}647)^s 12^{-b}LAd^{(b-3s-1)}q^s \quad \ldots\ldots (9.11)$$

$$=(0.333)(1{,}647)^{1.2}12^{-1.11}(10{,}000)(0.001577)$$

$$\times(4.95)^{[1.11-3(1.2)-1]}60^{1.2}$$

$$=1{,}236\ \text{psi (HPG solution friction pressure)}.$$

The surface treating pressure is calculated from

$$p_s=(\Delta p_f)_{fl}-\Delta p_H+p_{bh}, \quad \ldots\ldots (9.33)$$

where

$$\Delta p_H=0.05194\rho_f\Delta z\ \text{(psi)}=0.05194(8.34)(10{,}000)$$

$$=4{,}332\ \text{psi [29.9 MPa] (hydrostatic pressure)}$$

and thus

$$p_s=1{,}236-4{,}332+7{,}000=3{,}904\ \text{psi [26.9 MPa]}$$

$$=\text{surface treating pressure without proppant.}$$

For the case of pumping proppant at a concentration of 6 lbm sand/gal liquid [719 kg/m^3] liquid, Hannah *et al.*'s correction (Eq. 9.30) will be used to correct the proppant-free Δp_f for sand:

$$F_C=\mu_r^{0.2}\rho_r^{0.8}=(\Delta p_f)_M/(\Delta p_f)_{fl}. \quad \ldots\ldots (9.30)$$

The slurry density is

$$\rho=\frac{6+8.34}{6/22.1+1}=11.28\ \text{lbm/gal}\ [1352\ \text{kg/m}^3],$$

and thus the relative density is

$$\rho_r=11.278/8.34=1.352.$$

The sand volume fraction is

$$V_p=\frac{6/22.1}{6/22.1+1}=0.2135,$$

and the relative viscosity is obtained from Eq. 9.24:

$$\mu_r=\frac{\mu}{\mu_c}=[1+2.5V_p+10.05V_p^2+0.00273\exp(16.6V_p)] \quad \ldots\ldots (9.24)$$

$$=1+(2.5)(0.2135)+(10.05)(0.2135)^2$$

$$+(0.00273)\exp[(16.6)(0.2135)]=2.086.$$

Thus, from Eq. 9.30, the slurry friction pressure is

$$(\Delta p_f)_M=(2.086)^{0.2}(1.352)^{0.8}(1{,}236)=1{,}822\ \text{psi [12.6 MPa]}.$$

The hydrostatic and surface treating pressures are calculated as

$$\Delta p_H=(0.05194)(11.28)(10{,}000)=5{,}859\ \text{psi [40.4 MPa]}$$

and

$$p_s=1{,}822-5{,}859+7{,}000=2{,}963\ \text{psi [20.4 MPa]}.$$

Solution Method 2: Lord and McGowen's Method. The "field-calibrated" correlation of Lord and McGowen (Eq. 9.31) developed with delayed crosslinked HPG gels will be used assuming no significant crosslinking at the wellbore temperature of 75°F [24°C]:

$$\ln\left[\frac{(\Delta p_f)_o}{(\Delta p_f)_{fl}}\right]=2.38-8.024/\bar{v}-0.2365C_{HPG}/\bar{v}$$

$$-0.1639\ \ln[C_{HPG}]-0.028C_s\ \exp[1/C_{HPG}]. \qquad (9.31)$$

The average fluid velocity is

$\bar{v}=17.156q/d^2$(ft/sec)

$=(17.156)(60)/(4.95)^2=42.0$ ft/sec [12.8 m/s].

The ratio of C_{HPG} to $\bar{v}$ is

$C_{HPG}/\bar{v}=0.0583C_{HPG}d^2/q$

$=(0.0583)(40)(4.95)^2/60$

$=0.952$ (lbm HPG-sec)/(1,000 gal-ft) [0.374 kg·s/m^4].

Eq. 9.32 gives the friction pressure of the water solvent as

$$(\Delta p_f)_o=0.40429d^{-4.8}q^{1.8}L \qquad (9.32)$$

$=(0.40429)(4.95)^{-4.8}(60)^{1.8}(10,000)$

$=2,973$ psi [20.5 MPa].

For the proppant-free case, use of Eq. 9.31 for HPG solution friction pressure gives

$(\Delta p_f)_{fl}=(\Delta p_f)_o\ \exp[-2.38+8.024/42.0+(0.2365)(0.952)$

$+0.1639\ \ln(40)]=(2,973)\ \exp(-1.359)=763.7$ psi [5.3 MPa].

Thus, for surface treating pressure without proppant,

$p_s=(\Delta p_f)_{fl}-\Delta p_H+p_{bh}=763.7-4,332+7,000$

$=3,432$ psi [23.7 MPa].

For 6 lbm sand/gal liquid [719 kg/m^3]$=C_s$, Eq. 9.31 gives

$(\Delta p_f)_M=(\Delta p_f)_o\ \exp[-1.359+(0.028)(6)\ \exp(1/40)]$

$=(\Delta p_f)_o(0.3052)=(2,973)(0.3052)=907.4$ psi [6.26 MPa].

The surface treating pressure with proppant is

$p_s=907.4-5,859+7,000=2,048$ psi [14.1 MPa].

Note that Method 2 gives friction pressures 472 and 915 psi [3.25 and 6.3 MPa] lower than Method 1 for cases without and with proppant, respectively. No recommendation will be made here regarding a preferred method.

Friction-Pressure Data From Field Fracturing Treatments. Acquiring friction-pressure data during actual fracturing treatments is often desirable. Common practice is to stop pumping momentarily during the pad stage to obtain the instantaneous shut-in pressure at the wellhead. By adding the hydrostatic pressure, Δp_H, of the fluid column to the ISIP, the bottomhole pressure, p_{bh}, can be calculated. If friction pressures through the perforations and fracture can be ignored, the friction pressure in the tubular goods, Δp_f, can then be estimated from the surface pressure, p_s, before the ISIP is taken:

$$\Delta p_f=p_s-p_{bh}+\Delta p_H, \qquad (9.33)$$

where (in oilfield units)

$$\Delta p_H=\left(\frac{C_s+\rho_{fl}}{C_s/\rho_s+1}\right)\Delta z(0.05194). \qquad (9.34)$$

9.6 Fracturing Gels (Crosslinked Fluids)

It is common practice to crosslink fracturing solutions to enhance proppant-carrying capabilities and viscosity. Polymer expenses and the risk of potential fracture damage from high amounts of insoluble polymer products are reduced. Aqueous solutions composed of guar, guar derivatives, and cellulose derivatives are commonly crosslinked with metal or organometallic crosslinkers. Metals used in crosslinkers include titanium, zirconium, boron, aluminum, chromium, and antimony. The bonding characters of these metals vary from weakly bonded, as in boron crosslinked solutions, to strongly bonded, in zirconium crosslinked systems. Hydrocarbon fluids, like diesel, kerosene, and crude oil, can be "gelled" with suitable orthophosphate esters complexed with aluminum salts or fatty acids complexed with caustic. The resulting chemical complex can be referred to as an association polymer.

The flow of crosslinked solutions or gels is a rather contradictory proposition. By definition, a gel is a material with a solid continuous (external) phase and a liquid dispersed (internal) phase. Because solids do not flow, a gelled fracturing fluid must not be a gel in the strict sense when it is made to flow. Whether a fracturing "gel" flows as a partially crosslinked solution or as a fractured or degraded gel depends on its chemistry, as well as its thermal and deformation histories. In some cases, fracturing gels may not flow in the conventional sense at all; rather they may flow by some kind of slip mechanism or as nonhomogeneous fluids. The true flow behavior of fracturing gels is currently uncertain, even for controlled conditions in the laboratory.

Viscosity Behavior. As for fracturing solutions, the most commonly performed rheological test for fracturing gels is that for viscosity. Contrary to the situation for fracturing solutions, however, fracturing gels are not easily characterized by conventional viscometry. One reason is that many crosslinked fluids are still reactive during the characterization period, which makes test-result reproducibility difficult. Furthermore, the presence of large normal forces can cause the fluid to crawl away from the viscometer test surfaces, e.g., climbing out of the gap between rotating concentric cylinders. One source of particular consternation is fracturing gels that do not obey conventional scale-up laws for viscometric flows as a result of some kind of slip-flow phenomena or some unknown experimental artifact. When this is the case, the applicability of laboratory data for field use is even more doubtful.

Viscosities of fracturing gels (crosslinked solutions) are usually measured with rotating concentric-cylinder viscometers or tubular viscometers. Rotating concentric-cylinder viscometers have the advantage of maintaining well-defined "shear" histories and nearly uniform test-fluid shear stresses when the ratio of the radii of the inner and outer cylinders is >0.9 (small gaps). Disadvantages include cylinder end effects, mixing of the test fluid with the rheometer pressuring medium (e.g., nitrogen or oil), and the possibility of migration of fluid from or into the annular test region as a result of secondary flows or normal stresses. Moreover, most concentric-cylinder viscometers are designed for testing batch-mixed fluids, whereas fracturing gels are usually made in field operations by continuous addition of crosslinker while the gel is pumped downhole.

Tubular viscometers have the advantage of more closely duplicating the shear-stress profile experienced by the fluid in the field, i.e., varying linearly from zero in the center of the fracture or tubing to a maximum at the wall. Tubular viscometers designed for fracturing fluids usually make provisions for simulating field mixing conditions. The main problem with tubular flow devices is the shear effects introduced by the circulating pump.

For fracturing-design purposes, viscosity data generated with the Fann Model 50C viscometer are specified most commonly. The Fann Model 50C viscometer gained wide use because it was one of the first commercially available rotating viscometers that could make measurements at temperatures to 400°F [204°C] and pressures to 1,000 psi [6.9 MPa]. Elevated pressures are necessary for vapor containment at elevated temperatures.

Effect of Shear Rate. At a given temperature and time, and at wall shear rates normally associated with hydraulic fractures (e.g., 20 to 500 seconds^{-1}), fracturing gels show shear-thinning behavior and are conveniently modeled by the power law (see Eq. 9.17). For design purposes, the rheology is usually given as the

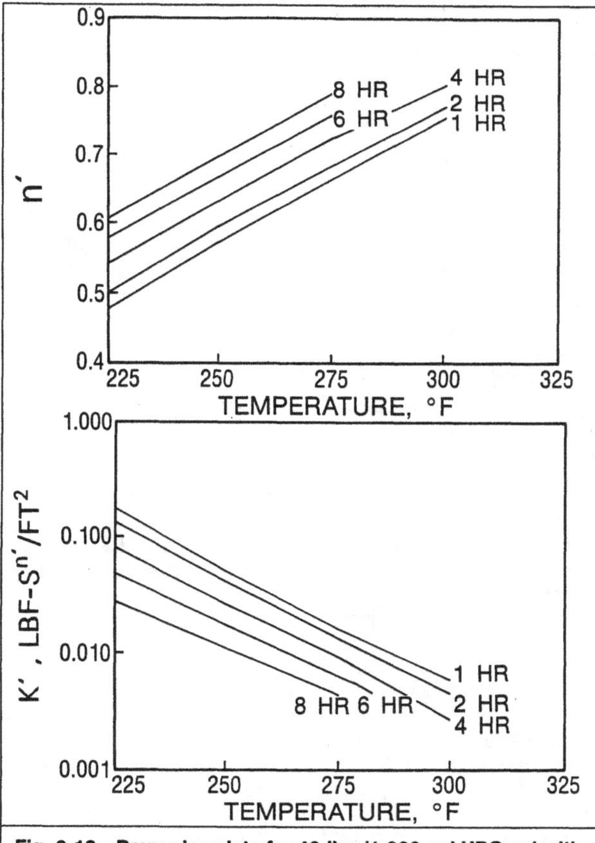

Fig. 9.18—Power-law data for 40-lbm/1,000-gal HPG gel with 0.12 wt% stabilizer.[37]

power-law index, n', and the power-law consistency index, K', as functions of time at a given temperature. Most hydraulic fracturing simulators consider these parameters in the calculation of shear rate and, ultimately, viscosity. For aqueous gels, n' values ranging from 0.2 to 0.9 and K' values ranging from 0.001 to 0.20 (lbf-sec$^{n'}$)/ft^2 [0.0479 to 9.58 (N·s$^{n'}$)/m^2] are common. Values of n' and K' depend on the gel composition, time, temperature, and test procedure. Fig. 9.18 shows n' and K' data for a 40-lbm/1,000-gal [4.8-kg/m^3] HPG solution crosslinked with a titanium crosslinker and with 10 lbm/1,000 U.S. gal [1.2 kg/m^3] solid stabilizer.[37] Testing was done with a Fann Model 50C viscometer. Hydrocarbon gels can show even greater shear thinning with values of n' ranging from 0.1 to 0.9 and typical values of K' ranging from 0.01 to 0.2 (lbf-sec$^{n'}$)/ft^2 [0.479 to 9.58 (N·s$^{n'}$)/m^2]. Fig. 9.14 gives an example of a shear-thinning partially crosslinked diesel gel.

In some gel systems, a certain amount of gel "rehealing" may occur after a period of high shear, with viscosities gradually returning to their prior low-shear-rate values. Tests with the Fann Model 50C viscometer have shown boron or chromium crosslinked gels to have good rehealing properties at moderate temperatures.[81] Hydrocarbon gels also show good rehealing capabilities. Titanium[81] crosslinked gels usually do not show good rehealing qualities. Gel rehealing results from reforming of broken crosslinks rather than reforming of broken polymer.

Low-Shear-Rate Region and Yield-Stress Behavior. Although power-law behavior is usually assumed for the rheology of fracturing gels, deviation from the power law can be expected over broad shear rate ranges. For example, gels with yield stresses (finite stress at zero shear rate) can be represented better by the Herschel-Bulkley model:

$$\tau=\tau_o'+K''\dot{\gamma}^{n''}. \qquad (9.35)$$

Rockefeller[59] described a more general approach in which the flow curve of log shear stress vs. log shear rate is approximated as a series of segments, each with a designated slope, n_i', and intercept, K_i', applicable only to a designated range of shear rates:

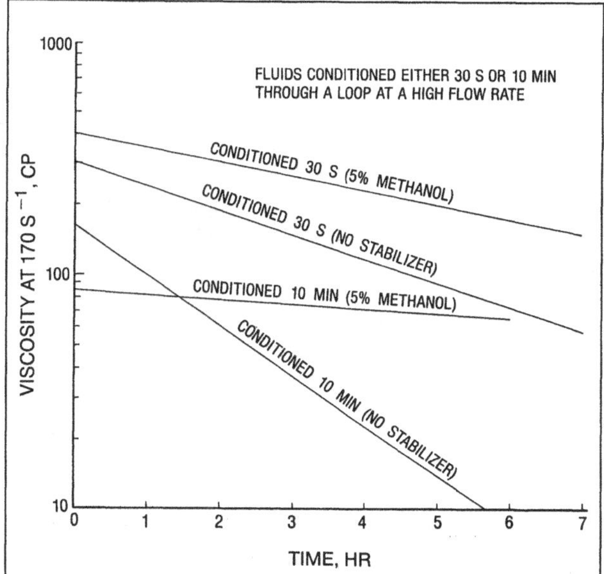

Fig. 9.19—Viscosity as a function of time at 250°F for a 50-lbm/1,000-gal HPG gel with stabilizer and flow conditioning as parameters [data courtesy Amoco Production Research Co. (1980–81)].

$$\tau=K_i'\dot{\gamma}^{n_i'}, \qquad (9.36)$$

where K_i' and n_i' are discrete functions of $\dot{\gamma}$. For the segmented model, a gel with a yield stress could have an n_i' of zero for the zero-shear-rate segment.

A knowledge of low-shear-rate behavior (less than 20 seconds^{-1}) is important with regard to proppant transport because the shear rate varies from a maximum at the fracture wall to zero at the center of the fracture width. The portion of gel subjected to low shear rate is greater for gels with small n' because they have more blunted velocity profiles. Proppant transport is influenced by the prevailing local viscosity and the gel yield stress if present (refer to Chaps. 10 and 11 for more information on proppant transport). Unfortunately, low-shear-rate data are not usually available because of sensitivity limitations in most commercially available high-pressure rheometers. Limited laboratory tests[82,83] have shown yield stresses for crosslinked gels ranging from 0.001 to 0.021 psi [6.9 to 145 Pa].

Effect of Time at Temperature. Design data are generated with the Model 50 Fann viscometer by testing gels at a set temperature and angular velocity over extended times up to 8 hours. Design data reported as functions of time at temperature will generally show viscosity to decrease monotonically with time as the result of mechanical, chemical, and thermal degradation processes. In keeping with this trend, K' will also decrease but n' will increase toward 1.0, the Newtonian case, as shown in Fig. 9.18. For aqueous gels, increasing n' can indicate molecular degradation. Shear-thinning behavior, which becomes more pronounced as n' decreases toward zero, is the result of molecular entanglements.[71] Therefore, as polymer molecules degrade, molecular entanglements become less likely and n' will increase. The viscosity will sometimes be observed to increase before it starts its monotonic decline. Such behavior is most probably indicative of additional crosslinking during testing. At elevated temperatures, the monotonic decline in viscosity over extended periods of time most generally is not thixotropic (i.e., reversible) and may represent polymer degradation. Fig. 9.19 shows typical design data for a 50-lbm/1,000-gal [6.0-kg/m^3] HPG gel crosslinked with an organometallic crosslinker. Fig. 9.19 shows that the decline in viscosity can be moderated by the use of chemical stabilizers that prevent polymer backbone oxidation. Stabilizers like methanol or sodium thiosulfate are commonly used. Chemical degradation of polymer is discussed further in Sec. 9.5.

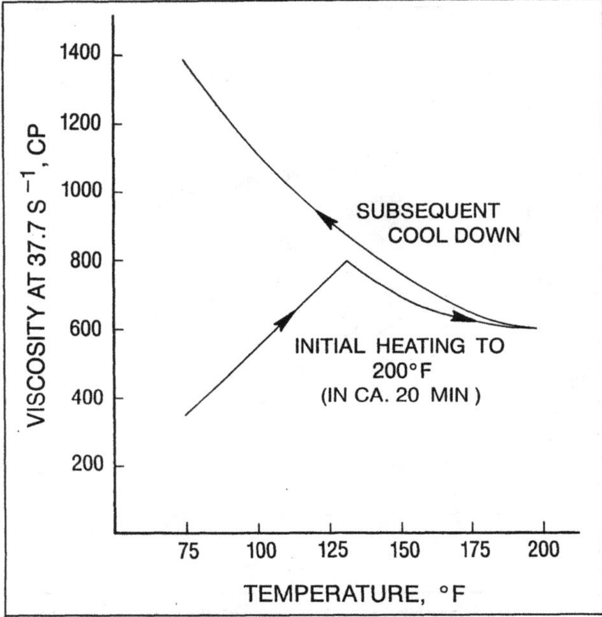

Fig. 9.20—Effect of temperature history on the viscosity of a 40-lbm/1,000-gal HPG gel crosslinked with titanium acetyl acetonate [data by J.R. Cameron, courtesy Amoco Production Research Co., Tulsa, OK (1985)].

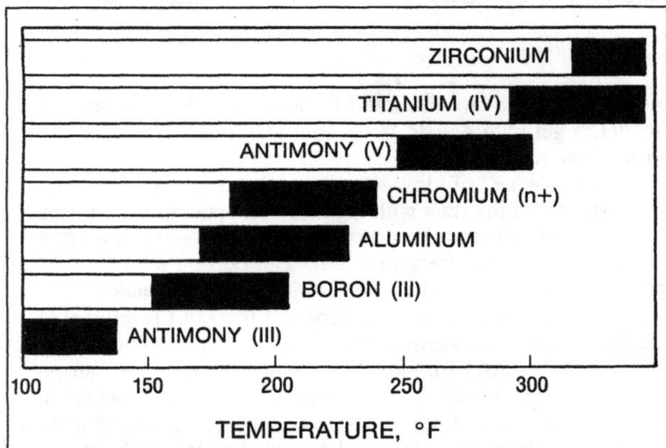

Fig. 9.21—Upper-limit temperature ranges for various aqueous crosslinkers in their usable pH and concentration ranges (data adapted from those of Conway *et al.*[85]).

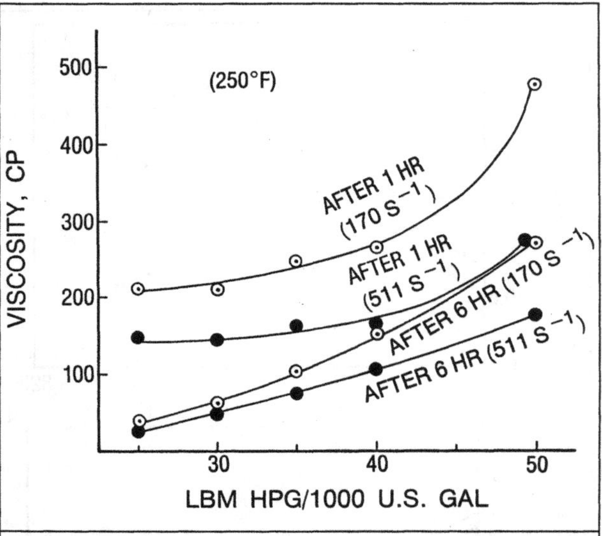

Fig. 9.22—Effect of polymer concentration on the viscosity of an HPG gel with 10-lbm/1,000-gal solid stabilizer at 250°F after 1 and 6 hours.[37]

Also shown in Fig. 9.19 is the effect of flow conditioning, which is discussed later in this section.

The effect of temperature on the viscosity of fracturing gels can be deceptive. For example, a 40-lbm/1,000-gal [4.8-kg/m^3] HPG solution crosslinked with titanium acetyl acetonate can show a viscosity at 75°F [24°C] that is less than what it will attain if it is subsequently heated to 200°F [93°C]. Once equilibrated at 200°F [93°C], however, subsequent cooling of the same sample back to 75°F [24°C] will result in a viscosity greater than that at 200°F [93°C]. Elevated temperature apparently enhances crosslinking kinetics and drives the crosslinking reaction toward completion. Fig. 9.20 illustrates these trends for a gel mixed in a laboratory beaker and tested in a Fann Model 50C viscometer.

Note that during initial heating, a maximum is obtained after which the effects of thermal thinning become dominant. Each type of crosslinker exhibits its own particular temperature behavior, which, as shown later, is also dependent on the shear rate history.

The new generations of delayed-crosslinked water-based gels, which often are temperature-activated, can also be expected to show increasing viscosity during initial heating.

Some gel systems, even when apparently crosslinked equilibrated, have been reported to show unconventional trends with temperature. For example, gelled oils and borate crosslinked gels have been reported[37] to show essentially constant viscosity as temperature increases from 120 to 180°F [49 to 82°C]. Also, n' may actually decrease with increasing temperature with a corresponding increase in K'. Decreasing n' may be an indicator of more intermolecular association resulting from expansion or conformation changes of the polymer molecules as temperature rises. Note, however, that other tests[84] made on borate crosslinked system have shown the conventional temperature trends using generically the same fluid and similar test apparatus.

As temperature is increased, a temperature is eventually attained beyond which viscosity will fall off rapidly and approach uncrosslinked-solution (base gel) values. This behavior may represent both the breaking of the polymer backbone and chemical crosslinks. Crosslinkers that bond to the polymer more strongly will generally provide higher temperature stability. Fig. 9.21 shows the upper-limit temperature ranges for various crosslinker metal ions. The specific temperature at which viscosity becomes unacceptably low depends on shear intensity, pH, and chemical component concentration and type.[85] Menjivar[38] tested the effect of temperature on molecular crosslinking by measuring the storage modulus. He found the storage modulus to break down catastrophically at a given temperature for a particular system (e.g., HPG/titanium acetyl acetonate broke down at about 302°F [150°C]). This breakdown was attributed primarily to crosslink breakage because precautions were taken to minimize polymer degradation from acid hydrolysis and oxidative depolymerization.

Effect of Polymer Concentration. Increasing the concentration of polymer in crosslinked gels is an effective way of increasing viscosity and temperature stability of fracturing gels. Here, temperature stability is used in the design sense as the time at temperature when the gel viscosity is maintained above a certain arbitrary level (e.g., 100 cp [100 mPa·s] at 170 seconds^{-1}). Fig. 9.22 shows the effect of HPG concentration on viscosity at 250°F [121°C] and at various shear rates measured with the Fann Model 50C viscometer.[37] Shown are data for fracturing gels with stabilizer at short and long times.

As mentioned in Sec. 9.5, a minimum polymer concentration is required for effective intermolecular crosslinking. For HPG solutions, this concentration is about 16.7 lbm HPG/1,000 gal [2 kg HPG/m^3].[38] At small concentrations, overlap of molecular hydrodynamic volumes apparently does not occur and intermolecular crosslinking is not favored. Assuming an average molecular weight of 1.5×10^6 for HPG, the above critical concentration implies a

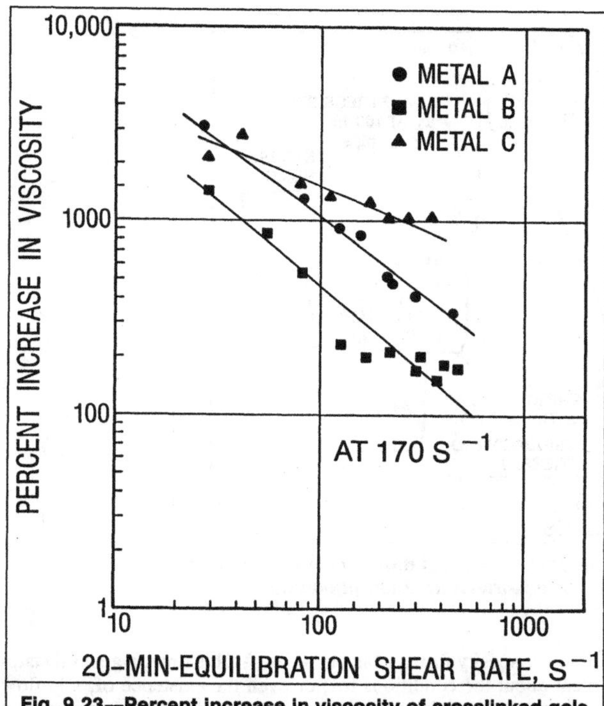

Fig. 9.23—Percent increase in viscosity of crosslinked gels over their base gels as a function of equilibration shear rate.[85]

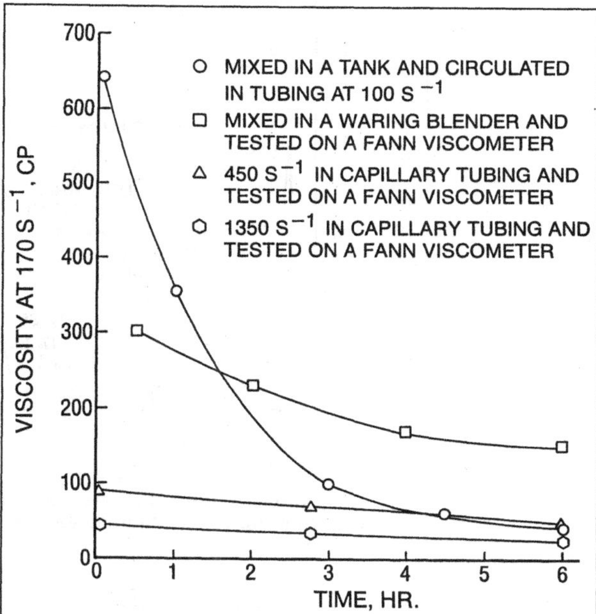

Fig. 9.24—Comparison of 40-lbm/1,000-gal HPG gels conditioned in tubing under laminar flow with batch-mixed HPG gels (tested at 250°F with gel stabilizers).[37,88,89]

diameter of about 0.1 μm for an HPG molecule with a spherical hydrodynamical volume.

Effects of Flow-Deformation and Temperature Histories. Because many times fracturing gels are still crosslinking while being tested, their observed rheology will be sensitive to their flow-deformation and temperature histories. Indeed, this sensitivity is the probable cause for the difficulties encountered by API when trying to formulate a recommended practice for testing crosslinked fracturing fluids. A large experimental variation between and within laboratories was observed in split-sample testing of the "API generic gel" (a 40-lbm/1,000-gal [4.8-kg/m^3] HPG solution crosslinked with titanium acetyl acetonate). Although most design data have been generated with conventional rotating viscometers with batch-mixed gels, it is generally appreciated that these laboratory conditions do not represent field conditions. In addition, because minor variations in laboratory procedures can cause large variations in observed viscosity, it is not easy to justify the use of any particular set of laboratory data for general field use. One may be forced to tailor fracturing-design software around a particular set of rheology data to get a treatment design consistent with field experience.

Research has shown that laminar shear history has a strong effect on the rheology of aqueous gels when fluids are conditioned on rotational viscometers[81,85-87] and in tubular flow devices.[88] This is true particularly for organometallic crosslinked systems (e.g., titanates and zirconates) that can be irreversibly degraded by shear. Boron and chromium gels show greater rehealing tendencies, but they can be restricted to use at lower temperatures (<200°F [<93°C]). Fig. 9.23 shows the percent viscosity increase over that of the base gel at viscosities at 170 seconds^{-1} for gels made with three different metallic crosslinkers plotted vs. the 20-minute equilibration shear rate on the Fann Model 50C viscometer.[85] These data show how viscosity decreases with increasing levels of equilibration shear rate. One explanation is that high shear favors intramolecular crosslinking as opposed to intermolecular crosslinking.[85] Fig. 9.24 shows the effect of 300 seconds of laminar flow conditioning in a coiled capillary tube on the viscosity of a 40-lbm/1,000-gal [4.8-kg/m^3] HPG solution crosslinked with an organometallic crosslinker.[88] After the conditioning period, the gel was pumped into a Fann Model 50C viscometer, where the viscosity at 250°F [121°C] was measured. The viscosity for the fluid conditioned at 1,350 seconds^{-1} is about one-half that of the fluid

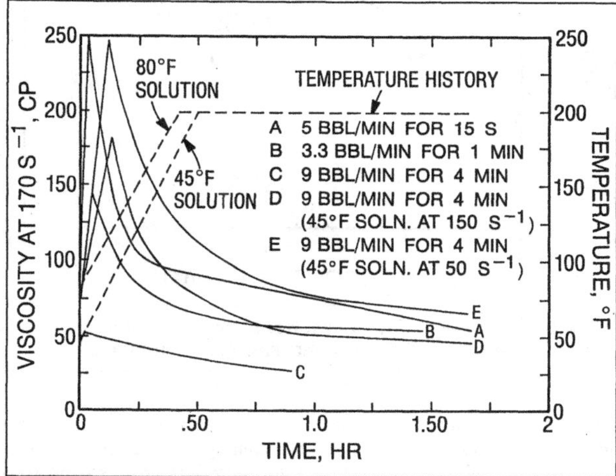

Fig. 9.25—Comparison of 40-lbm/1,000-gal HPG gels crosslinked with titanium acetyl acetonate subjected to various turbulent flow and temperature histories.[90]

conditioned at 450 seconds^{-1}. The data generated with 1,350-seconds^{-1} shear-rate conditioning scaled well to laminar flow through 4,000 ft [1219 m] of 1-in. [2.54-cm] -ID tubing, where friction pressures were only 10% greater than those of the uncrosslinked gel.[88] Also shown for comparison are data for similar gels prepared by batch mixing. One was prepared by mixing in a blender for 3 minutes after vortex closure[37] (the "conventional" API *RP 39* procedure[12]) and the other[89] in a stirred tank followed by circulation in a pipe loop at 100 seconds^{-1}.

Research indicates that turbulent-flow conditioning can also give substantially lower viscosities than those attained with conventional blender conditioning when organometallic crosslinked gels are tested. Fig. 9.19 shows the effect of short and long times in a tubular loop circulated at high rates with a gear pump. Fig. 9.25 shows the effect of various turbulent-flow intensities and temperature histories during conditioning on a 40-lbm/1,000-gal [4.8-kg/m^3] HPG

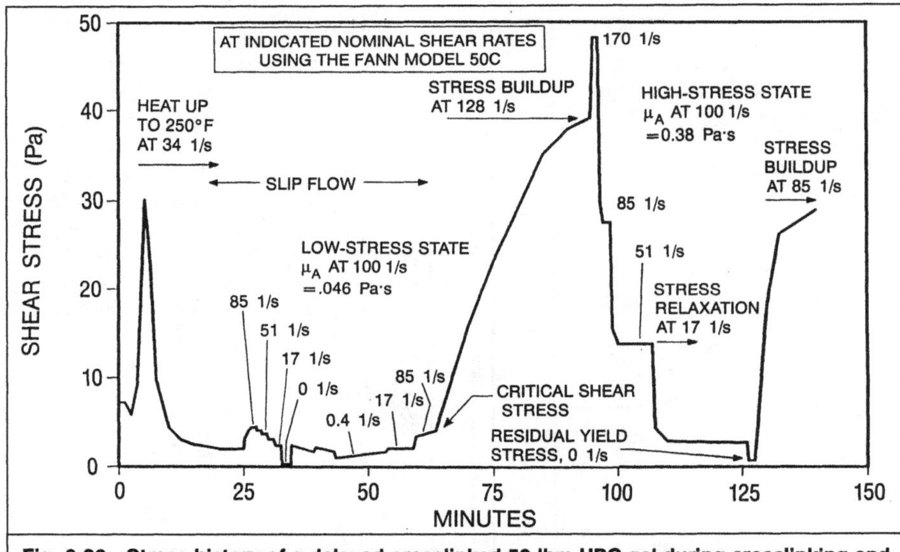

Fig. 9.26—Stress history of a delayed-crosslinked 50-lbm HPG gel during crosslinking and subsequent shear-rate testing demonstrating the stress-transformation phenomenon.[94]

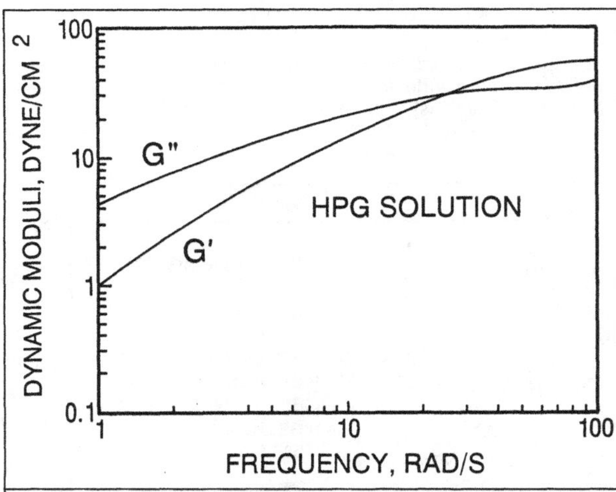

Fig. 9.27—Dynamic moduli vs. frequency for a 40-lbm/1,000-gal HPG solution at 50% strain and room temperature.[9]

solution crosslinked with titanium acetyl acetonate.[90] Conditioning was done to simulate shear stress existing at the designated flow rate through 2.44-in. [6.20-cm] -ID tubing. Viscosities were subsequently measured on the tubular flow loop. Curve A was conditioned under the least severe conditions equivalent to 5 bbl/min [0.013 m^3/s] for 15 seconds. Curves D and E show how cooler temperatures during the conditioning phase result in higher viscosities; cooling most probably retards or delays crosslinking. Curve E, which shows the highest viscosities, was generated after the conditioning period by circulating at a pipe-wall shear rate of 50 $seconds^{-1}$; the other curves were generated at 150 $seconds^{-1}$.

Tubular-flow experiments have shown that both titanium and zirconium crosslinked gels can give better high-temperature performance when crosslinking is delayed so that it does not occur substantially during the conditioning period.[88,90-92] Even when crosslinking occurs under low shear conditions, however, such as those on a Fann Model 50C viscometer, problems with rheological characterization can arise. These are discussed next.

Slip-Flow Behavior. Research both supports[10,93] and refutes[9,10,75,88] slip flow under various conditions. Slip flow is said to occur when flow curves (shear stress vs. shear rate) differ for different viscometer geometries. For fracturing gels, this is probably the result of nonhomogeneous flow, where a thin fluid region of low viscosity forms near the walls. At this time, a general statement about the conditions for, or even the existence of, slip flow in fractures is premature. More carefully conducted and documented experiments are required. If slip (nonhomogeneous flow) is occurring, proper scaling procedures for laboratory data to fracture flow will have to be devised.

Some observers believe that fracturing gels will not slip when crosslinked while flowing. However, recent testing of delayed-crosslinked organometallic HPG gels that were continuously sheared while crosslinking in a Fann Model 50C viscometer has shown that these gels can slip.[94] Slip flow was verified by observing tracer beads through a transparent Lexan viscometer cup. During slip flow, all beads moved at the same angular velocity, usually less than that of the cup, and the apparent viscosity was very low. The bulk fluid was moving as a gelled solid, flowing on thin layers of low-viscosity fluid (probably uncrosslinked HPG solution) at the bob and cup. When the cup angular velocity was raised to a certain level, however, a critical shear stress was reached where the gelled bulk fluid broke down, as indicated by slower bead angular velocities near the bob. At this point, a dramatic and sometimes very slow stress buildup occurred, eventually yielding a relatively large apparent viscosity (see Fig. 9.26). This phase transformation is reversible. Below a certain shear stress, the stress "relaxed" and the bulk fluid gelled once more. Nearly the same stresses (viscosities) resulted during each cycle. That these gels can transform between different phase states may be responsible for some of the data variation and difficulties experienced during testing of these types of fracturing gels. A theory for modeling this type of slip flow that avoids the classic assumption of a finite slip velocity has been published.[94,95]

Viscoelastic Behavior. Fracturing gels are viscoelastic. This is not surprising, because the polymer solutions from which they are formed are also viscoelastic. These fluids exhibit both fluid-like and solid-like properties. Viscoelastic properties can be determined easily by subjecting the test fluid to small-amplitude oscillatory shear strains (see Appendix C). The resultant fluid stress can be factored into an in-phase component, indicative of the fluid elastic nature, and a 90° out-of-phase component, indicative of the fluid's viscous nature. The in-phase component is represented by the storage modulus G' and the out-of-phase component by the loss modulus G''. These moduli are functions of frequency and can also vary with strain amplitude. Curves of G' and G'' vs. frequency are useful in identifying the physical character of a fluid.

Fig. 9.27 shows the dynamic moduli as functions of frequency for a 40-lbm/1,000-gal [4.8-kg/m^3] HPG solution, while Fig. 9.28 shows those for the same HPG solution when crosslinked with 0.04 vol% Tyzor AA™ (titanium acetyl acetonate) crosslinker.[9] At low frequencies, $G''>G'$ but exhibits a crossover point and a minimum

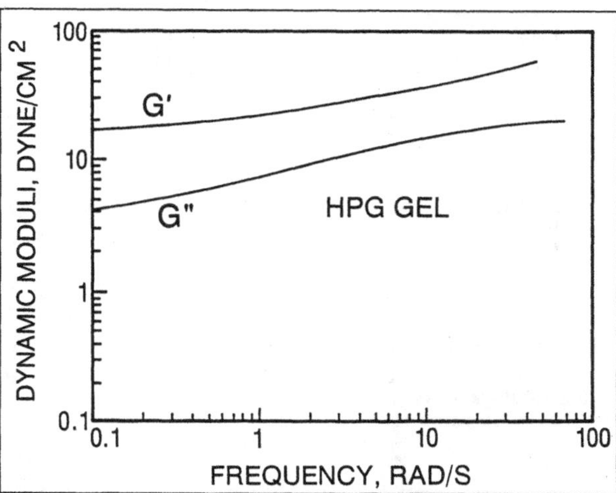

Fig. 9.28—Dynamic moduli vs. frequency for a 40-lbm/1,000-gal HPG crosslinked gel at 100% strain and room temperature.[9]

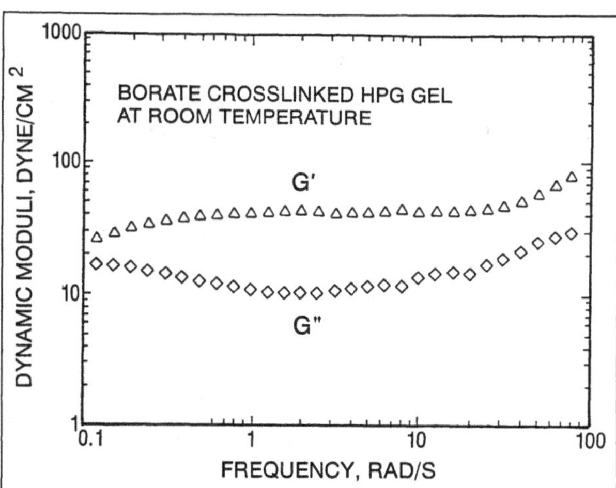

Fig. 9.29—Dynamic moduli vs. frequency for a borate crosslinked 40-lbm/1,000-gal HPG gel at 300% strain and room temperature [data by S.K. Knoll, courtesy Aqualon Co., Houston (1984)].

at high frequency. At concentrations too low for molecular overlap, the crossover of the curves would not be expected to occur and G'' would always be greater than G'.[71] When crosslinked (see Fig. 9.28), the HPG solution shows traditional gel behavior with $G'>G''$ and G' approaching a finite value at low frequencies, representing the equilibrium shear modulus, which is zero in the uncrosslinked state.[71]

Figs. 9.29 and 9.30 show the change in the dynamic moduli for a borate crosslinked HPG solution when the temperature is increased from ambient temperature to 149°F [65°C]. The borate crosslinked gel reverts back to the HPG solution.

Viscoelastic data are usually obtained from nonflowing test samples. However, flowing fluids exhibit altered viscoelastic behavior. Experiments performed on concentrated polymer solutions have shown G' at low frequencies to be shifted to progressively higher frequencies as the magnitude of superimposed shear rate increases.[71] Fracturing gels may exhibit similar viscoelastic response to superimposed shear. Also, during the initial crosslinking period, one can expect superimposed shear rate to affect the developing viscoelasticity, just as it does the developing steady shear stress, by affecting the extent of intermolecular bonding. Fig. 9.28 was obtained from a fracturing gel allowed to crosslink under static conditions. On the other hand, Fig. 9.31[9] shows the viscoelastic behavior of the same fluid after 900 seconds of shear at 1,350 seconds^{-1}. Note that both moduli have been reduced by the shear. Also, at 300% strain, $G'<G''$, indicating disruption of the gel network by larger strains. This viscoelastic behavior may be more like that of an emulsion than a gel or polymer solution.

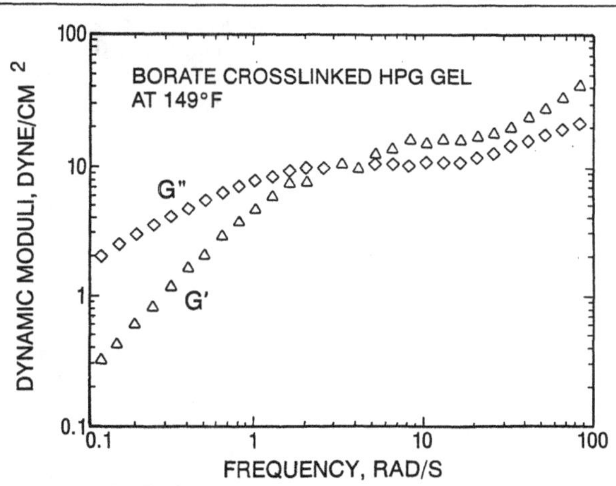

Fig. 9.30—Dynamic moduli vs. frequency for a borate crosslinked 40-lbm/1,000-gal HPG gel at 300% strain and 149°F [data by S.K. Knoll, courtesy of Henkel Corp., Houston (1984)].

Proppant Effect in Steady-Shear Testing. Few data are available on the flow behavior of proppant-laden fracturing gels because of problems inherent with proppant settling and proppant/viscometer incompatibilities.

Hannah *et al.*[83] measured the flow curves of a very viscous fracturing gel (unspecified) with and without 2-lbm/gal [240-kg/m^3] silica flour. They used a Fann Model 50C viscometer and tested at low shear rates to 1.0 seconds^{-1}. They found essentially the same flow curves down to 4 seconds^{-1}, but at <4 seconds^{-1}, the silica-flour flow curve separated above that for the solids-free gel. This may be evidence that the presence of silica flour increases the yield stress and the low-shear viscosity, as would be expected.

Gardner and Eikerts[75] studied the slurry rheology of a CMHEC-Al^{+3} gel with 20/40 proppant using a tubular flow loop. Before the flow curves were taken, the slurries were conditioned at a stress level equivalent to 10 bbl/min [0.0265 m^3/s] through 2.441-in. [6.20-cm] -ID tubing for 1.0 minute. The resulting flow curves, as shown in Fig. 9.32, show a stronger dependence on sand concentration than would be expected for slurries in uncrosslinked polymer solutions or in Newtonian fluids. Whether this is an artifact (perhaps of proppant settling) or a reflection of the effect of proppant on crosslinking during conditioning (e.g., enhanced crosslinking from proppant suppression of turbulence) is not known. Clearly, more experimental evidence is required before the effect of proppant on fracturing gel rheology can be quantified.

Turbulent Behavior. Sec. 9.5 showed that most fracturing solutions exhibited drag-reducing turbulent behavior typical of viscoelastic solutions. Friction-factor correlations with N_{De}, N'_{Re}, and n' can be developed (Appendix D).

For fracturing gels in turbulent flow, the same dimensionless groups will be operable in correlations, assuming that the fluid is only partially crosslinked while flowing. As mentioned earlier in this section, intense shearing or turbulence has been observed to lead to low laminar-flow viscosities for some "crosslinked" systems,[85,88,90,91] implying that intermolecular crosslinking is incomplete during turbulence. This is certainly true for delayed crosslinked gels at low temperatures. Thus, turbulence correlations mentioned for fracturing solutions in Sec. 9.5 may apply in some instances. Shah[74] noted that his correlation involving f, n', and N'_{Re} predicted

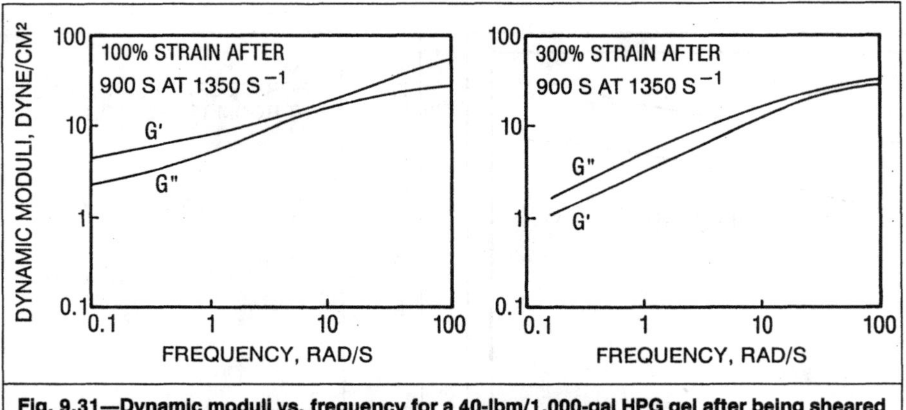

Fig. 9.31—Dynamic moduli vs. frequency for a 40-lbm/1,000-gal HPG gel after being sheared for 900 seconds at 1,350-seconds^{-1} shear rate at room temperature.[9]

friction factors for aqueous and hydrocarbon gels to within 17% when tested in ¾-in. [1.91-cm] tubing. However, whether this would apply when scaling up to field-size tubing is not yet known.

Currently, the only public source of friction pressures for a particular fracturing gel is from service-company technical-data manuals. The presentation of these data is similar to that for fracturing solutions, which was discussed in some detail in Sec. 9.5. Friction pressure ($\Delta p_f/L$) is usually plotted as a function of flow rate (in barrels per minute) on log-log paper. Fig. 9.33 shows representative data for various fracturing gels flowing in 2⅞-in. [7.303-cm] tubing (6.5 lbm/ft [9.7 kg/m]).

These data show that the borate system extends the laminar zone to higher flow rates, indicating that the borate system is achieving higher viscosities in tubular flow. These data are not intended to be used for design purposes and are shown only to illustrate possible qualitative trends.

9.7 Foamed Fracturing Fluids

Foamed fracturing fluids comprise N_2 or CO_2 dispersed as bubbles in water, acid fluids, methanol/water mixtures, or hydrocarbon liquids. A foam is characterized largely according to its quality, Γ, which is defined as the ratio of the dispersed gas volume to the total foam volume at a given temperature and pressure. For fracturing purposes, foam qualities are usually designed to range from 0.65 to 0.85. Aqueous foams are made by mixing a suitable foaming agent or surfactant (e.g., ethoxylated nonylphenol) at concentrations typically <2 vol% into water. This solution is foamed by injection of N_2 or CO_2 into the moving stream of liquid. Sometimes polymer is added to the water to give added viscosity and stability. By crosslinking of the polymer, stability has been improved to the point where foam qualities considerably less than 0.5 can be achieved.[96] Hydrocarbon foams are prepared similarly with a suitable fluorocarbon surfactant in hydrocarbon liquids, such as diesel, kerosene, xylene, and condensate.

Foams gained popularity as fracturing fluids in the early 1970's because of several desirable properties, including low formation damage, energized fluid recovery, low fluid loss (good efficiency), high viscosity, low friction pressure, and good proppant-suspending capabilities.[97] Some disadvantages in using foams include low hydrostatic heads, resulting in high surface pressures, especially when deep wells are fractured; difficulty in achieving high proppant concentrations; usually higher cost; and sometimes difficult field operation. Recent advances in foamers and viscosifiers have made possible the use of foams at temperatures in excess of 300°F [149°C] for periods up to several hours.[98,99] However, fracturing-treatment design suffers from the additional complexity of pumping a very compressible fluid, as well as the additional uncertainties associated with fluid loss, proppant transport, and the rheological character of foamed fracturing fluids.

The rheological character of foams is not easy to quantify because of the many variables to be specified. In addition to the standard parameters of deformation and temperature histories and chemical composition, other variables important to foam rheology include foam quality and texture, viscosity of both phases, interfacial tension, and pressure. Rheological measurements are usually

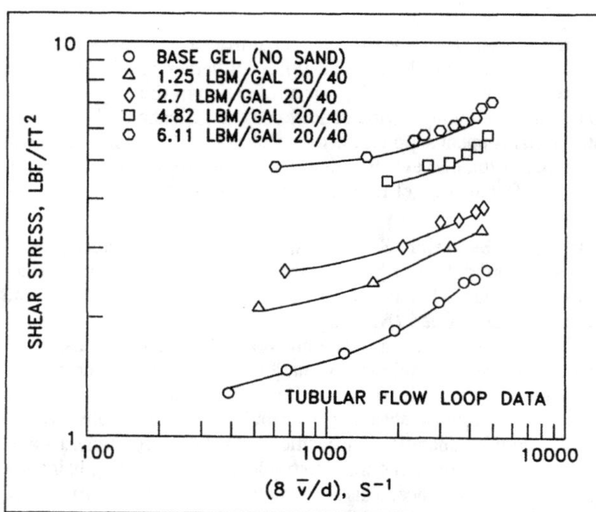

Fig. 9.32—Effect of 20/40 sand concentration on the flow curves of a 60-lbm/1,000-gal CMHEC-Al^{+3} gel at room temperature.[75]

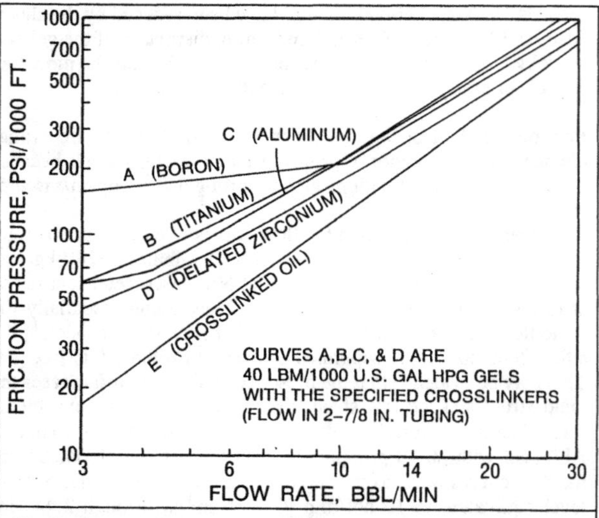

Fig. 9.33—Representative friction-pressure data for various gels flowing through 2⅞-in. tubing.

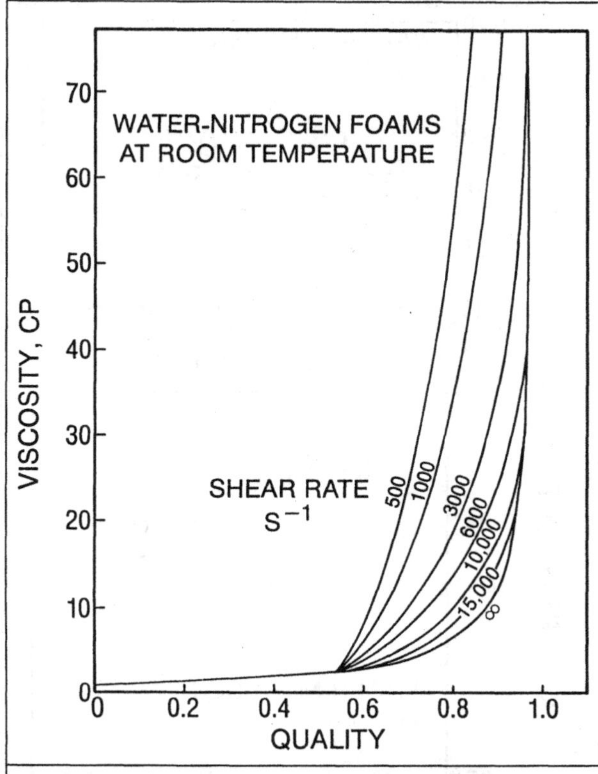

Fig. 9.34—Effect of quality and shear rate on the viscosity of water/N_2 foams at room temperature.[101]

performed in pressurized tubular devices or in the few commercially available high-pressure rheometers. Most testing has been limited to friction pressure and viscosity testing; little has been done regarding the viscoelasticity and normal stress behavior of fracturing foams.

Viscosity Behavior. ***Effect of Quality.*** A key parameter in specifying foam rheology is foam quality. Testing has shown that at foam qualities above a critical value, viscosity increases markedly. Blauer *et al.*[100] divided the range of foam quality into regions on the basis of differences in bubble interactions. Below a quality of 0.52, spherical bubbles may be well dispersed and not in contact with each other. Between qualities of 0.52 and 0.74, bubbles are closely packed and may contact each other during flow, causing bubble interference. Between qualities of 0.74 and 0.95, bubbles must deform to flow, and maximum viscosity is attained in this range. At qualities above 0.95, the foam is no longer stable and inverts to a mist (the liquid becomes the dispersed phase) with a dramatic fall in viscosity. Of course, these designated regions may be shifted somewhat depending on bubble-size distribution.

The functional dependence of viscosity on quality is influenced by other parameters, including shear rate and liquid-phase rheology. Mitchell[101] performed high-shear-rate (>500-seconds^{-1}) tests on water/nitrogen foams flowing through capillary tubes and obtained the data shown in Fig. 9.34. Note the absence of shear-rate dependence below a quality of about 0.55, where the foam displays Newtonian behavior. Above a quality of 0.55, the viscosity becomes shear-thinning and appears to attain a high shear-rate asymptote. At low qualities where bubble interference is not great, there is a rather weak dependence on quality. The behavior at low qualities is in accord with theoretical hydrodynamic predictions, such as that by Taylor,[102] who developed a theory for emulsions with small dispersed-phase concentrations:

$$\mu=\mu_c\left\{1+V_d\left[\frac{2.5(\mu_d/\mu_c)+1}{\mu_d/\mu_c+1}\right]\right\}, \quad (9.37)$$

where μ_c and μ_d are the viscosities of the continuous and dispersed phases, respectively, and V_d is the volume fraction of the dispersed

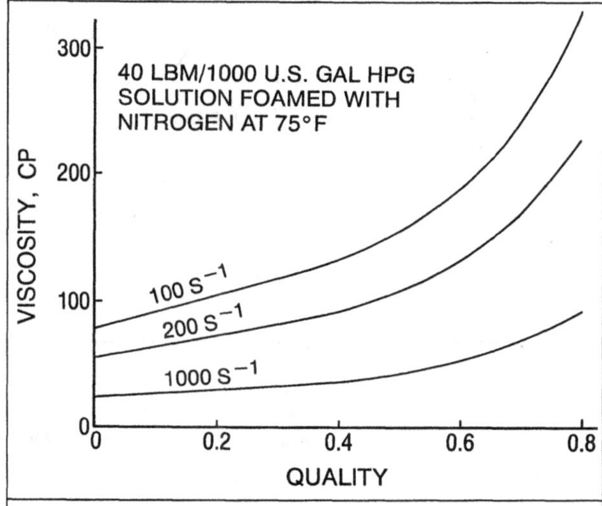

Fig. 9.35—Effect of quality and shear rate on the viscosity of a 40-lbm/1,000-gal HPG solution foamed with N_2 at 75°F.[103]

phase. In the case where (μ_d/μ_c) is small, as for foams, the corresponding expression for foam viscosity, μ_f, becomes

$$\mu_f=\mu_c(1+\Gamma), \quad (9.38)$$

where μ_c is the liquid-phase viscosity. For qualities less than 0.55, Mitchell[101] fit his data with the relation

$$\mu_f=\mu_c(1+3.6\Gamma), \quad (9.39)$$

which shows a stronger dependence on Γ. This is probably the result of hydrodynamic interactions between bubbles, which start to become important at dispersed-phase concentrations greater than 0.10 volume fraction.

For foams made from non-Newtonian liquids, the above relations for the low-quality region would no longer apply. The viscosity for all values of Γ become dependent on shear rate. This is shown in Fig. 9.35, which applies to a 40-lbm/1,000-gal [4.8-kg/m^3] HPG solution at 75°F [24°C] foamed with N_2.[103] It is interesting to note that quantitatively similar data were obtained when CO_2 was used instead of N_2.[103] In this case, the CO_2 was in the liquid state; therefore, the dispersion was an emulsion. Reidenbach *et al.*[103] concluded that this similarity between the N_2 foam and CO_2 emulsion showed the dominance of two-phase structuring over dispersed-phase composition in determining rheological behavior.

Delayed crosslinked gels foamed with N_2 have been observed to have behavior qualitatively similar to foamed uncrosslinked gels but with substantially higher viscosities, as shown in Fig. 9.36.[104] These data compare uncrosslinked and delayed crosslinked 30-lbm-HPG/1,000-gal [3.6-kg-HPG/m^3] N_2 foams at 110°F [43°C] as tested in a tubular viscometer.

At qualities above about 0.50, closely packed bubble interactions result in viscosity varying exponentially with Γ, as shown in Figs. 9.34 through 9.36. Suitable models must reflect this nonlinearity in Γ at higher qualities.

Rheological Models for Foamed Fracturing Fluids. Various rheological models have been proposed in attempts to quantify the observed effects of shear rate. Most commonly, data have been correlated with the Bingham,[105] power-law,[98,106,107] and Herschel-Bulkley[99,103] models:

$$\tau=\tau_o+\mu_p\dot{\gamma}, \quad (9.40)$$

$$\tau=K'\dot{\gamma}^{n'}, \quad (9.41)$$

and

$$\tau=\tau_o'+K''\dot{\gamma}^{n''}, \quad (9.42)$$

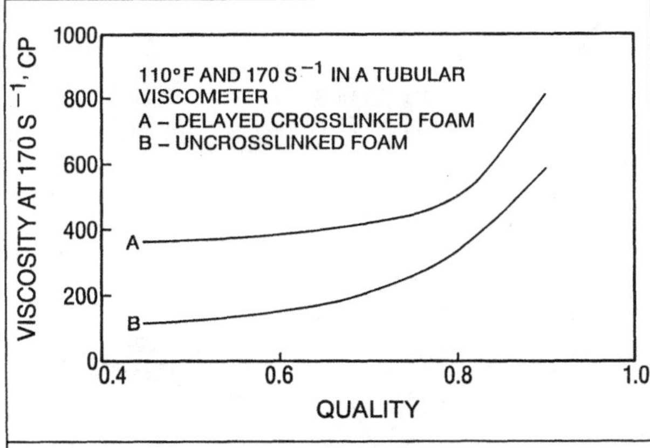

Fig. 9.36—Effect of crosslinking a 30-lbm/1,000-gal HPG/N_2 foam at 110°F.[104]

respectively, where

τ_o, τ'_o = yield stresses,
μ_p = plastic viscosity, and
K'', n'' = consistency and power-law indices for Herschel-Bulkley fluid.

A distinction in yield stress between the Bingham and Herschel-Bulkley models has been made because it may vary for the same set of data if it is determined by extrapolation to zero flow. However, there is actually only one true yield stress for each fluid under a given set of conditions.

Viscometer data can be fit to these models to determine the best fit. As an example, Sanghani and Ikoku[106] used regression analysis to fit their data for an aqueous/air foam at qualities ranging from 0.65 to 0.95. For a 0.80-quality foam, they found the power-law model (which has no yield stress) to correlate their data slightly better than the Herschel-Bulkley model with correlation coefficients of 0.975 and 0.963, respectively. The Bingham model gave the worst fit with a correlation coefficient of 0.919. This demonstrates the difficulty in estimating yield stress by extrapolating from moderate shear rates like those often associated with fracturing-design data.

Although the presence of a yield stress is not easily deduced by extrapolating data, the existence of a yield stress is somewhat intuitive from everyday observations of foam behavior and field performance. Short interruptions during pumping of sand-laden foams have not resulted in problems indicative of sand settling.[97] Krug[108] published yield stresses as a function of foam quality.

Currently, no universal format is available for the presentation of rheology data for foamed fracturing fluids. In the case of power-law formulations, n' and K' data are sometimes presented as functions of liquid-phase composition, quality, and temperature. Figs. 9.37 and 9.38 show data for an unstabilized foam (aqueous/N_2) and a foam stabilized with 40 lbm/1,000 gal [4.8 kg/m^3] of a stabilizing thickener.[37] Note that the stabilized foam gives a greater K' and smaller n' (more shear-thinning) at all temperatures. Note also that perfect stability is presumed in that the effect of time is not considered. Actually, foamed polymer solutions or foamed crosslinked gels are subject to instabilities resulting from polymer degradation at high temperatures by thermal, mechanical, and/or chemical mechanisms.

Harris and Reidenbach[99] developed a correlation for N_2/aqueous foams containing from 0 to 80 lbm/1,000 gal [0 to 9.6 kg/m^3] HPG at 0.0 to 0.80 quality and at temperatures ranging from 75 to 300°F [24 to 149°C]. The liquid phase contained 10 vol% methanol and selected surfactants to maintain thermal and dynamic stability. This correlation is based on the Herschel-Bulkley yield-pseudoplastic model in terms of nominal pipe shear rate ($8\bar{v}/d$). The model constants K''_p and n''_p at a given temperature are functions of temperature, T (°F), the polymer concentration by virtue of the rheology of the liquid phase at 75°F (n'_{75} and K'_{75}), and foam quality:

$$n''_p = n'_{75} \exp[(0.0028 - 0.0019\Gamma)(T-75)] \quad \text{(9.43)}$$

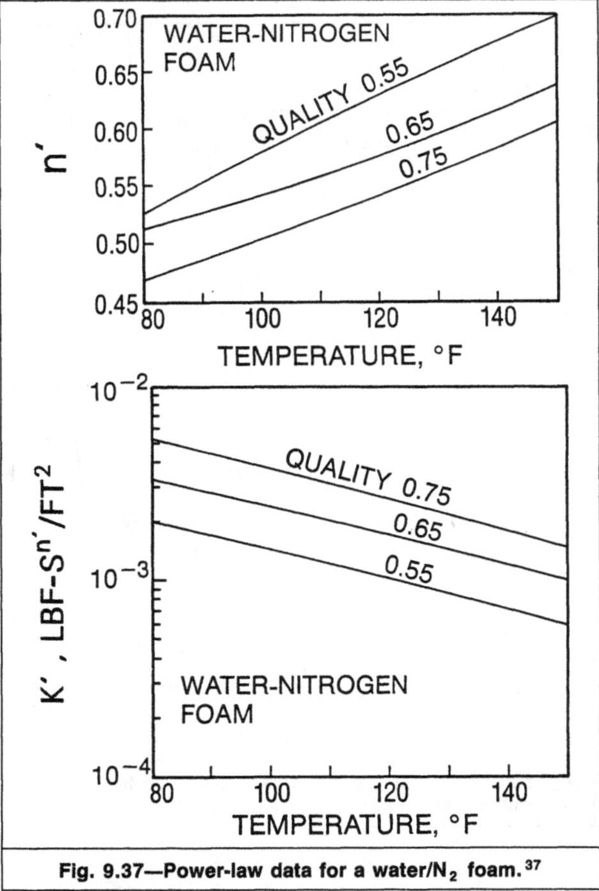

Fig. 9.37—Power-law data for a water/N_2 foam.[37]

and

$$K''_p = K'_{75} \exp[(C_2\Gamma - 0.018)(T-75)] \exp(C_\ell\Gamma + 0.75\Gamma^2), \quad \text{(9.44)}$$

with

$$C_2 = \exp[-(3.1 + 3n'_{75})]$$

and

$$C_\ell = 4(n'_{75})^{1.8}.$$

The yield stress τ^p_o (in lbf/ft^2) was correlated as a function of quality only:

$$\tau^p_o = 0.07\Gamma, \text{ for } \leq 0.6, \quad \text{(9.45)}$$

and

$$\tau^p_o = 0.0002 \exp(9\Gamma), \text{ for } \Gamma > 0.6. \quad \text{(9.46)}$$

These parameters are used with the nominal shear rate in a Herschel-Bulkley-type shear-stress relation:

$$\tau = \tau^p_o + K''_p \left(\frac{8\bar{v}}{d}\right)^{n''_p}. \quad \text{(9.47)}$$

Note that this applies strictly to pipe flow. Modifications of K''_p, n''_p, and τ^p_o may be necessary when this is extended to other geometries.[103] Also, the parameters for this correlation were generated by recirculating foam through a 0.305-in. [0.775-cm] -ID test loop until a given "equilibrium" foam texture was achieved. The pressure differential was observed to increase by a factor of two

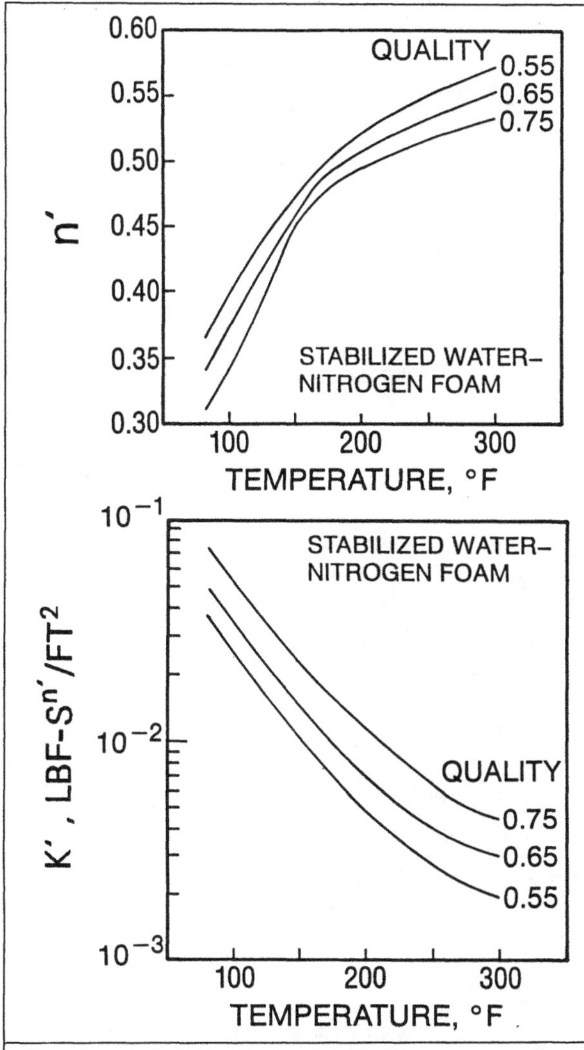

Fig. 9.38—Power-law data for a water/N_2 foam stabilized with 40 lbm thickener/1,000 gal water.[37]

or more during the attainment of this "equilibrium" state, indicating the importance of texture to foam rheology. Thus, application of such correlations to other types of foams generated under different conditions should be done judiciously.

Foam Texture. Foam texture is an important parameter that is usually not specified with foamed-fracturing-fluid design data.[109] Foam texture refers to the bubble-size distribution, bubble shape, and structure of a foam. Fracturing foams normally have mean bubble diameters in the range from 300 to 1,200 μm, with a size distribution varying by a factor of 10.[96,97,110,111] Because of the rather narrow bubble-size distribution and small bubble size relative to most fracturing-related flow passages, foams are treated as homogeneous fluids with conventional rheological models, as already described.

Foam texture depends on a variety of factors: quality, pressure, flow conditions, foam generation technique, and chemical composition. At lower qualities (e.g., <0.5), bubbles are mostly spherical and are not in contact with each other. At higher qualities, different bubble conformations have been observed. For crosslinked foams with qualities up to 0.74, spherical bubbles have been reported.[96] Furthermore, when there is a wide bubble-size distribution, the quality may be appreciably greater than 0.74 without bubble deformation occurring.

Increasing pressure at a constant foam quality has been observed to create finer-texture foams[110,112,113] with more spherical bubbles.[112] Harris[110] observed a small decrease in viscosity of dynamically equilibrated 0.70-quality aqueous/N_2 foams when pressure increases from 200 to 1,000 psi [1.38 to 6.9 MPa].

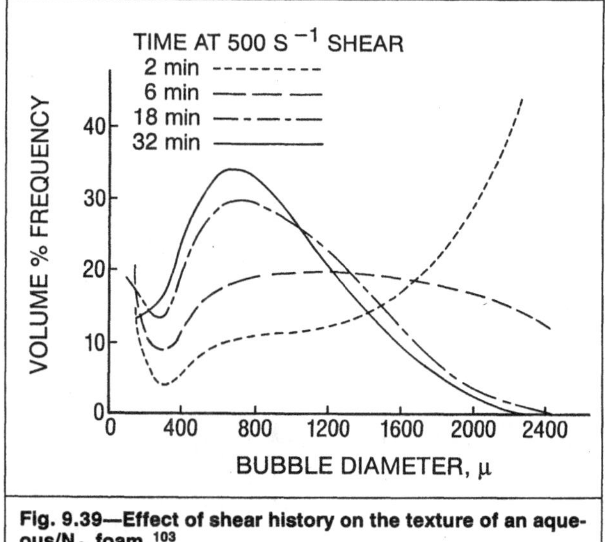

Fig. 9.39—Effect of shear history on the texture of an aqueous/N_2 foam.[103]

The flow conditions experienced by a foam can affect foam texture. Stop-action pictures from a videotaping of a flowing 0.88-quality aqueous/N_2 foam showed a more uniform bubble size while flowing than while static.[111] A similar phenomenon was observed while an aqueous/N_2 foam was recirculated through tubing.[103] In this instance, at relatively high levels of energy input, bubbles degraded to very small uniform sizes with good stability. At low energy levels, the bubbles coalesced and drained, forming large, nonuniform sizes. Fig. 9.39 shows the bubble-size-distribution development when sheared continuously at 500 seconds^{-1} through tubing.[103] Although viscosity increases with time at any given equilibration rate, at any given time, viscosity has been observed to be smaller at higher equilibration rates.[110]

Harris[110] studied the effect of polymer on texture using HPG-solution/N_2 foams equilibrated under various levels of shear rate. No significant change in bubble-size distribution occurred when equilibrated at 1,100 seconds^{-1} for HPG concentrations from 0 to 20 lbm/1,000 gal [0 to 2.4 kg/m^3]. However, when equilibrated at 250 seconds^{-1}, the HPG foams had broader bubble-size distributions.

Harris also studied the effects of nonaqueous liquids on foam textures. Foamed kerosene produces larger bubbles that are more susceptible to shear degradation.

Experiments[110,111] have shown the importance of surfactant type and concentration on texture and therefore on viscosity. It was shown[111] that when an aqueous/N_2 foam of 0.88 quality is made, synergistic blends of fluorocarbon/hydrocarbon surfactants produced the most stable foams (<500-μm bubble size) at the lowest surfactant volume percent concentration. In general, bubble size decreased with increasing surfactant concentration, giving a corresponding increase in viscosity. Surfactant concentrations in excess of 0.6 vol% were necessary to achieve stable foams for all surfactants tried. Because foaming requires energy to create new surfaces, surfactants giving lower surface tensions will promote foaming and give more stable foams. Fig. 9.40 shows surface tensions of classes of surfactants according to their liquid solvents.[114] Smaller bubble sizes require larger surfactant concentrations to give stability.

Foam Stability. Foam stability is important in the design of foam fracturing treatments. Stability in this case refers to the ability of the foam to resist breakdown resulting from bubble collapse or coalescence. Bubble breakdown can be caused by drainage of the liquid from the foam, increasing quality (e.g., from pressure reduction or heating) beyond 0.95, or physical rupture and coalescence. A standard qualitative test for foam stability is the half-life test, which measures the time for one-half of the original liquid phase to drain from a static foam column. Note, however, that half-life time can be strongly affected by the method of foam generation.[110]

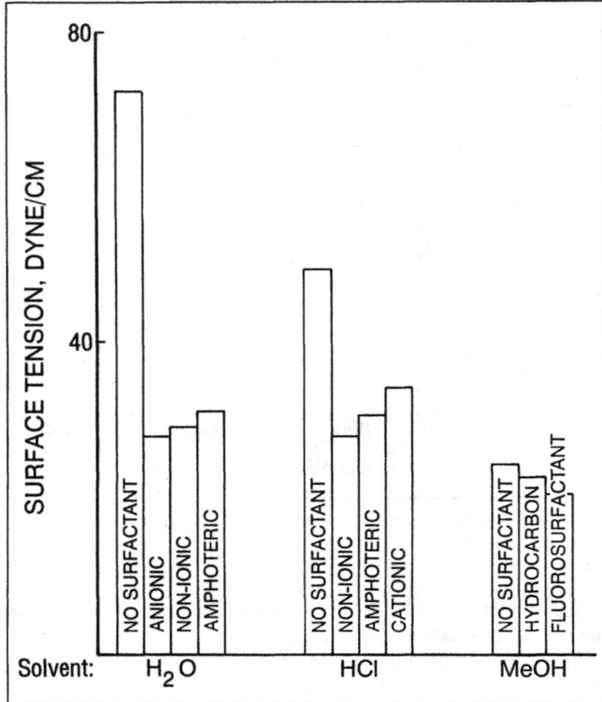

Fig. 9.40—Surface tensions of various surfactant classes in water, HCl, and methanol (MEOH).[114]

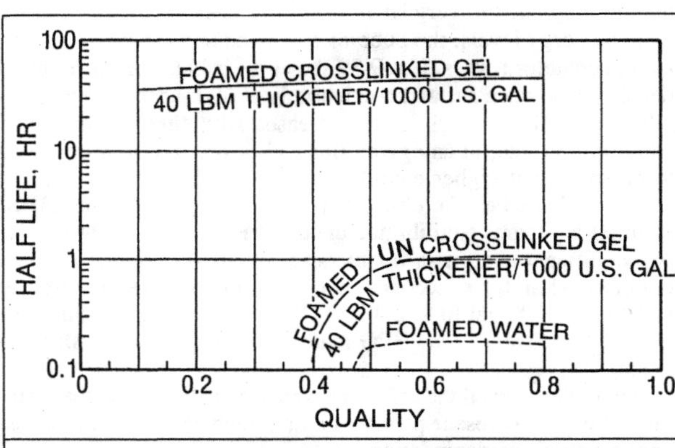

Fig. 9.41—Half-life vs. quality for stabilized and unstabilized foam.[96]

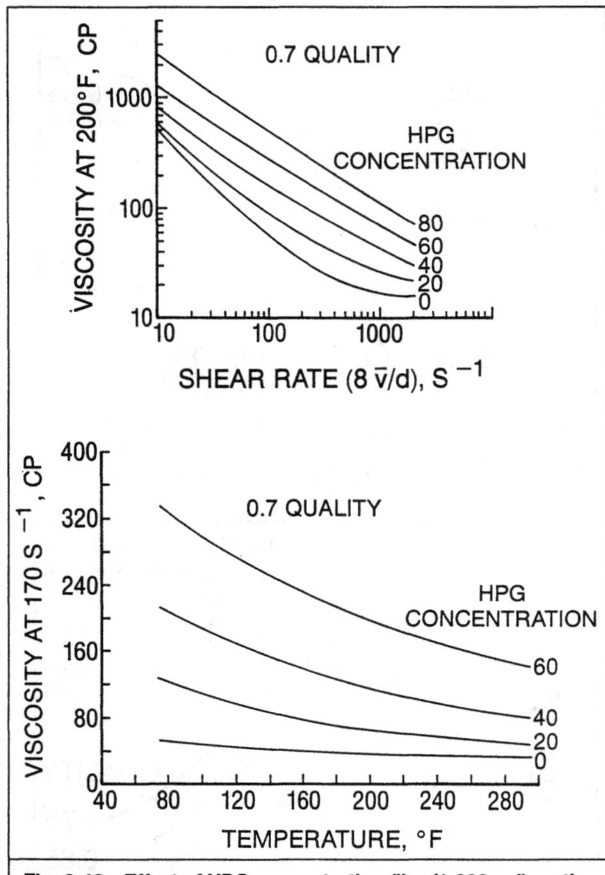

Fig. 9.42—Effect of HPG concentration (lbm/1,000 gal) on the viscosity of a 0.70-quality foam.[99]

It was implied previously that factors producing finer-textured foams (such as surfactant type and concentration, increasing pressure, and mechanical energy input) also promote stability. For high-temperature application, Harris and Reidenbach[99] observed that surfactant type and concentration were more important in preserving dynamic stability (i.e., while flowing) than the addition of thickeners (polymer stabilizers, like HPG). It is not clear, however, what flow intensity is necessary to maintain stability. Whether such conditions typically exist in fractures during proppant transport is not known.

Half-life tests (with static foams) have shown various viscosifiers to increase half-life times significantly.[98] Crosslinked fluids give even longer half-lives. Fig. 9.41 shows the half-lives for different stabilized foams as a function of quality.[96] Note that crosslinking extends the ability to form low-quality foams. In addition to increasing half-life, viscosifiers also increase viscosity, which is implicit in Figs. 9.37 and 9.38, and is shown in Fig. 9.42 for addition of uncrosslinked HPG to a 0.70-quality foam.[99]

Slip Flow. Experimental evidence both supports[115,116] and refutes[99,101] the phenomenon of slip flow in flowing foams. One test for slip flow is whether the flow curve of shear stress as a function of shear rate changes with viscometer geometry. For pipe or capillary flow, this would appear as a diameter effect in laminar flow. The testing of foams in tubing may show diameter effects arising from foam-texture phenomena. For flow in capillary tubes, reports of viscosity dependence on diameter have shown viscosity to be proportional to some power of the diameter.[117]

Turbulent Flow Behavior. Efforts to describe the turbulent flow behavior of fracturing foams have been directed primarily toward the development of suitable empirical correlations for the friction factor as a function of Reynolds number or for the friction pressure as a function of flow rate. A fundamental assumption in these correlations is that foams are pseudohomogeneous fluids with densities that are functions of temperature, pressure, and quality.

A correlation for aqueous/N_2 foams based on the Fanning friction factor was developed by Blauer *et al.*,[100] who correlated the friction factor to a Reynolds number written in terms of the average foam-flowing velocity, $\bar{v}_f$, the foam density, ρ_f, and an effective viscosity, μ_e:

$$(N_{\mathrm{Re}})_f=\frac{\rho_f d \bar{v}_f}{\mu_e}, \qquad (9.48)$$

where μ_e is derived by equating flow-rate expressions for a Bingham plastic fluid and a Newtonian fluid and solving for the Newtonian viscosity in terms of the Bingham plastic parameters. The $(\tau_o/\tau_w)^4$ term in the Bingham-plastic-flow expression (see Eq. 9.51) is presumed negligible, and friction pressures are presumed equal for equal flow rates. The resulting expression for μ_e is

$$\mu_e=\mu_p+\frac{\tau_o d}{6\bar{v}_f}. \qquad (9.49)$$

When data for aqueous/N_2 foams in capillary tubes and in tubing (1.380- and 1.995-in. [3.51- and 5.07-cm] ID) were plotted on the Moody diagram of friction factor vs. $(N_{\mathrm{Re}})_f$, a close correspondence with the classic single-phase Newtonian turbulent curves

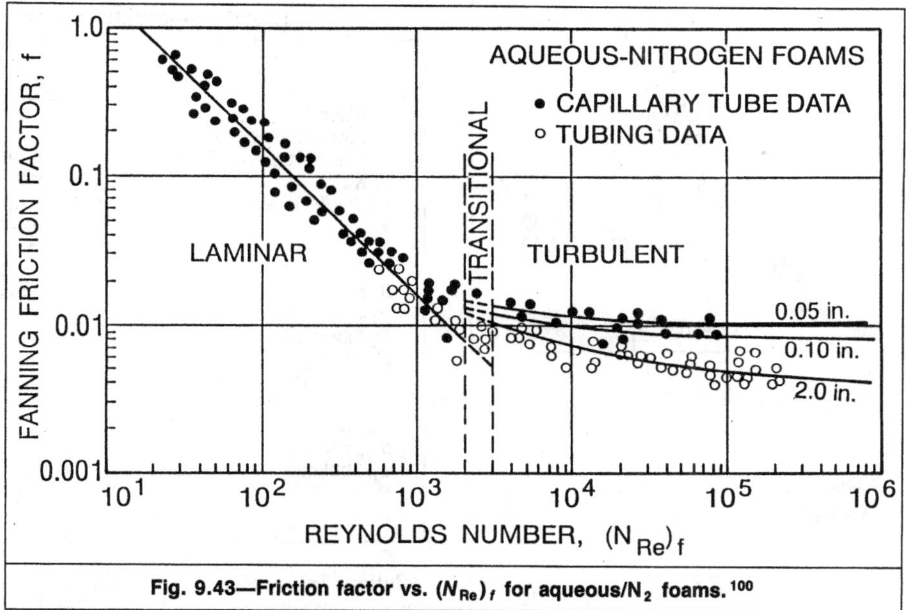

Fig. 9.43—Friction factor vs. $(N_{Re})_f$ for aqueous/N_2 foams.[100]

for smooth pipes resulted.[100] Fig. 9.43 shows these data in a Moody diagram. Blauer *et al.*[100] described the use of μ_e to calculate $(N_{Re})_f$ and thus the friction pressure. If $(N_{Re})_f > 2{,}000$, assume turbulence and read f from the Moody diagram. The friction pressure, $\Delta p_f/L$, is then calculated from the definition of f:

$$f=\frac{g_c d\Delta p_f}{2L\rho_f \bar{v}_f^2}. \quad (9.50)$$

If $f<2{,}000$, assume laminar flow and use the Buckingham-Reiner equation for laminar flow of Bingham plastic fluids through pipes to calculate $(\Delta p_f/L)$:

$$q=\frac{\pi d^2}{4}\bar{v}_f=\frac{\pi(\Delta p_f/L)d^4}{128\mu_p}\left[1-\frac{4}{3}\frac{\tau_o}{\tau_w}+\frac{1}{3}\left(\frac{\tau_o}{\tau_w}\right)^4\right]. \quad (9.51)$$

The above correlation is not appropriate for foams containing polymer because they show drag-reducing properties. A more general relation, proposed by Reidenbach *et al.*,[103] is a modification of the Bowen relation with another parameter, x', included to correlate for changes in density. This correlation takes the form

$$\tau_w=A'\rho_f^{x'}d^b\left(\frac{8\bar{v}_f}{d}\right)^s. \quad (9.52)$$

For Newtonian fluids, the values of x', b, and s are 0.8, 1.6, and 1.8, respectively. Turbulent flow tests were performed in pipes with ID's varying from 0.633 to 1.376 in. [1.61 to 3.50 cm]. Aqueous/N_2 foams with 10 to 40 lbm/1,000 gal [1.2 to 4.8 kg/m^3] HPG produced correlations with values of x, b, and s consistently lower than the Newtonian values. This is typical of drag-reducing fluids. For foams with polymer, the b and s values were identical, implying that shear stress is a function of density and velocity only. The x' exponent was found to be approximately equal to $s-1$ and correlates density changes resulting from changes in quality or pressure. Values for the exponents are shown in Table 9.3. The flow curves of a 0.70-quality foamed water and a 20-lbm/1,000-gal [2.4-kg/m^3] HPG foam are shown in Fig. 9.44 curve-fit with their mathematical models. Note the upward curvature in the laminar region for the 0.70-quality foamed water, which may be indicative of a yield stress. The average laminar/turbulent transition velocity, $\bar{v}_c$, can be obtained by equating the laminar shear stress (e.g., by Eq. 9.47) to the turbulent shear stress (e.g., by Eq. 9.52) and solving for $\bar{v}_c$ by trial and error.

Friction pressure data for foams are offered by some service companies in the form of log-log plots of $\Delta p_f/L$ vs. flow rate that are usually presented as functions of quality and tubular-goods geometry (through tubing, casing, or tubing/casing annuli). Such parameters as pressure, temperature, polymer concentration, surfactant type and concentration, and texture are often unspecified. Because pressure down the tubular goods varies, all the foam properties will vary to some extent. Accordingly, to calculate the total friction pressure down a long treatment string, some kind of incremental numerical scheme is often applied.[97] Many service companies have foam-treatment-design programs. An example of one approach to foam-fluid design is given at the end of this section.

TABLE 9.3—SCALE-UP EQUATION FOR TURBULENT FLOWS OF N_2 FOAMS[103]

$$\frac{\Delta p_f d}{4L}=A'\rho_f^{x'}d^b\left(\frac{8\bar{v}_f}{d}\right)^s$$

External Phase	A'	x'	b	s
Water	1.028×10^{-5}	0.80	1.87	1.80
10 lbm HPG	2.239×10^{-4}	0.37	1.37	1.37
20 lbm HPG	3.873×10^{-4}	0.31	1.31	1.31
30 lbm HPG	6.556×10^{-4}	0.27	1.27	1.27
40 lbm HPG	1.000×10^{-3}	0.23	1.23	1.23

Proppant Effects. Laboratory studies on the effects of proppant on the flow behavior of foamed fracturing fluids have not been published. Until more complete information is available, assuming the following approximations may be useful.

As a first approximation to the effect of proppant on foam viscosity, the proppant can be considered to be a part of the internal phase that effectively increases the foam quality.[118] The resulting foam-slurry viscosity can be estimated as being equal to the viscosity of the proppant-free foam having a quality equal to the sum of the proppant and gas volume fractions in the foam slurry. As an example, consider a water/N_2 foam with an original quality of 0.55. If enough sand were added to this foam to produce a concentration of 6 lbm sand/gal [719 kg sand/m^3] of foam slurry, the volume fractions of the gas and sand would be 0.40 and 0.27, respectively. The sum of these volume fractions would be 0.67, and the viscosity would be estimated as that of the same foam without proppant with a quality of 0.67. Laboratory tests to date have been limited to cases with proppant concentrations <12 lbm proppant/gal [1438 kg proppant/m^3] of foam slurry for N_2 and CO_2 foams and water (external)/diesel (internal) emulsions. Primarily 20/40-mesh

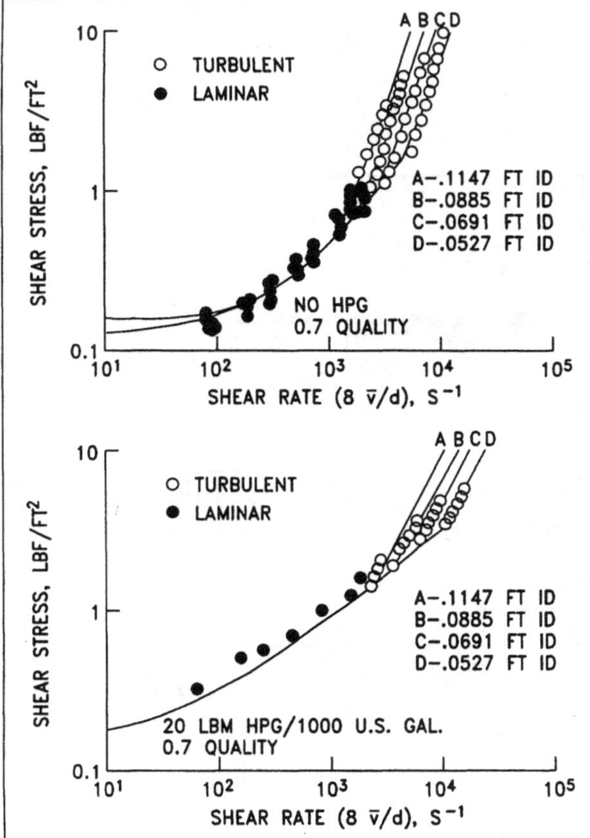

Fig. 9.44—Tubular flow curves (laminar and turbulent) for 0.70-quality water/N_2 foams with and without 20 lbm HPG/1,000 gal.[103]

sands have been tested, but this approximation is expected to hold for coarser grades and for high-strength proppants.

An estimate of the effect of proppant on the turbulent friction pressure can be made by adjusting the density and velocity terms in Eq. 9.52 while holding the correlation parameters A', x', b, and s constant. These parameters apparently are affected mostly by the external-phase rheology and may be determined from turbulent studies of the proppant-free foam. Foam texture may be of secondary importance in this case. Experiments on N_2 foams support this method of friction-pressure estimation, but similar studies for CO_2 have not yet been performed.

Example Problem 2—Calculation of Surface Pumping Criteria for Foam Fracturing. A 50-ft [15.24-m] zone at a 10,000-ft [3048-m] depth will be fractured with stabilized 40-lbm/1,000-gal [4.8-kg/m^3] HPG base gel foamed with N_2. The closure stress is 7,000 psi [48.26 MPa]; the reservoir temperature is 250°F [121°C]. The foam will be pumped down 5½-in. [13.97-cm] casing, and the fracturing-design simulator assumes flow at 60 bbl/min [0.16 m^3/s] in the fracture at reservoir temperature and closure pressure. For the pad stage, 0.7 quality is assumed at bottomhole conditions of closure pressure and temperature. The proppant stage is assumed to have a concentration of 6 lbm sand/gal [719 kg sand/m^3] of foam slurry at bottomhole conditions with a total internal-phase (sand plus N_2) volume fraction of 0.7.

For both the pad and proppant stages, calculate the required treating pressure, foam density and composition, and flow rate at the surface. Also calculate the flow rate and density at the perforations and thereby the perforation treatment necessary with ⅝-in. [1.59-cm] jet perforations to achieve negligible perforation friction pressure (<10 psi [68.95 kPa.]) Assume a flowing-foam temperature in the wellbore of 75°F [24°C], no fluid loss, and a constant fracture pressure of 7,000 psi [48.26 MPa]. Use Eqs. 9.47 and 9.52 in the friction-pressure calculation.

TABLE 9.4—GIVEN PARAMETERS

	Pad ($\Gamma=0.7$)	Slurry ($C_s=6$)
g, ft/sec^2	32.2	32.2
g_c, (lbm-ft)/(sec^2-lbf)	32.2	32.2
R, L-atm/(K-gmol)	0.082	0.082
T, K	297.2	297.2
A', (lbf/ft^2)(gal/lbm)$^{x'}$(1/ft)^b(sec)^s	0.001	0.001
x	0.23	0.23
b	1.23	1.23
s	1.23	1.23
d, ft	0.4125	0.4125
V_{LR}	0.30	0.30
C_s, lbm/gal	0	6.0
ρ_{fl}, lbm/gal	8.34	8.34
ρ_s, lbm/gal	NA	22.10
M_{N_2}, g/gmole	14	14
T_R, K	394.3	394.3
p_R, psi	7,000	7,000
q_R, bbl/min	60	60
Γ_R	0.7	0.4285
z_R	1.31	1.31

Solution. To calculate the surface treating pressure, the relevant force-balance equation must be used. For this problem, the change in casing pressure over a finite interval Δz (with z increasing down the casing) is computed assuming steady-state flow in terms of the change in hydrostatic force and the friction pressure. Because foam is very compressible, rigorously speaking, the change in momentum term should be included in the force balance. As shown later, however, this term is small and can be neglected.

The resulting equation, with the foam density, ρ_f, and the casing wall shear stress, τ_w, evaluated at the midpoint of the interval $(z+\Delta z/2=z_m)$, is written as

$$\Delta p=\frac{g\rho_f(z_m)\Delta z}{g_c}-\frac{4\tau_w(z_m)\Delta z}{d}. \quad (9.53)$$

For the present, it will be assumed that flow in the casing is turbulent (this will be verified later), and Eq. 9.52 with parameters for 40-lbm-HPG/1,000-gal [4.8-kg-HPG/m^3] N_2 foam from Table 9.3 will be used to calculate τ_w—i.e., $A'=0.001$, $x'=0.23$, and $b=s=1.23$ (see Table 9.4 for units of A' and d):

$$\tau_w=A'\rho_f^{x'}d^b\left(\frac{8\bar{v}_f}{d}\right)^s. \quad (9.52)$$

Using Eq. 9.52 and noting that $\bar{v}_f(z_m)=4q_R\rho_R/\pi d^2\rho_f(z_m)$ [by continuity (mass conservation) assuming steady-state flow] yields

$$\Delta p=0.05194\rho_f(z_m)\Delta z-(0.1419)\left(\frac{q_R\rho_R}{d^2}\right)^{1.23}\frac{\Delta z}{\rho_f(z_m)d}. \quad (9.54)$$

To solve this, an expression for $\rho_f(z_m)$ must be developed. One way is to express $\rho_f(z)$ in terms of ρ_R, which has the units of pounds per gallon of foam (slurry if proppant is present) at reservoir temperature and closure pressure—i.e., lbm/gal foam at RC, where RC denotes at reservoir conditions:

$$\rho_f(z)=\frac{\rho_R}{(V_{N_2}+C_s/\rho_s+V_{LR})}, \quad (9.55)$$

where V_{N_2} is the gal N_2 gas at position z/gal foam slurry at RC, C_s is the lbm proppant/gal foam slurry at RC, ρ_s is the lbm proppant/gal proppant, and V_{LR} is the gal liquid/gal foam slurry at RC. The denominator of Eq. 9.55 may be interpreted as being the volume occupied at position z by 1 gal [0.004 m^3] of foam when at reservoir temperature and closure pressure. The quantities ρ_s, C_s, and V_{LR} are specified parameters that will be considered constant because proppants and liquids are relatively incompressible. V_{N_2} is dependent on z and may be expresssed by use of the real-gas law:

$$V_{N_2}(z)=\frac{(1{,}762)ZC_{N_2}RT}{M_{N_2}p(z)}, \quad (9.56)$$

TABLE 9.5—CALCULATION OF PRESSURE DROP FOR FOAM FLOW BY SUCCESSIVE APPROXIMATION
One-Step Interval: Δz = 10,000 ft, $\Delta p = p(10{,}000) - p(0)$, and $p(10{,}000)$ = 7,000 psi

Γ_R	C_s	Estimated $p(z_m)$ (psi)	$Z(z_m)$	C_{N_2}	ρ_R	$V_{N_2}(z_m)$	$\rho_f(z_m)$	q_R (bbl/min)	Δp_H (psi)	Δp_f (psi)	Δp (psi)
					Pad						
0.7	0	7,000	1.360	0.919	3.42	0.548	4.035	60	2,096	970	+1,126
0.7	0	6,437	1.313	0.919	3.42	0.575	3.908	60	2,030	1,002	+1,028
0.7	0	6,486	1.317	0.919	3.42	0.573	3.920	60	2,036	999	+1,039
0.7	0	6,482	1.317	0.919	3.42	0.573	3.919	60	2,035	999	+1,037
					Slurry						
0.4285	6.0	7,000	1.360	0.563	9.066	0.336	9.990	60	5,189	1,230	+3,959
0.4285	6.0	5,021	1.196	0.563	9.066	0.411	9.225	60	4,792	1,407	+3,384
0.4285	6.0	5,308	1.220	0.563	9.066	0.397	9.361	60	4,862	1,387	+3,475
0.4285	6.0	5,262	1.216	0.563	9.066	0.399	9.342	60	4,852	1,390	+3,462
0.4285	6.0	5,269	1.216	0.563	9.066	0.399	9.342	60	4,854	1,389	+3,465

TABLE 9.6—CALCULATED PARAMETERS FOR FOAM FLOW DOWN CASING*

Γ_R	C_s/ρ_s	z (ft)	$p(z)$ (psi)	Z	$V_{N_2}(z)$	$\rho_f(z)$	$q(z)$ (bbl/min)	$\Gamma(z)$	$V_L(z)$	$V_p(z)$	$p^*(z)$	q_c (bbl/min)
						Pad						
0.7	0	0	5,963	1.274	0.602	3.791	54.13	0.667	0.333	0	5,967	42.4
0.7	0	10,000	7,000	1.360	0.548	4.035	50.86	0.646	0.354	0	7,000	39.2
						Slurry						
0.428	0.272	0	3,535	1.072	0.524	8.277	65.72	0.478	0.274	0.248	3,600	41.1
0.428	0.272	10,000	7,000	1.360	0.335	9.990	54.45	0.370	0.331	0.299	7,000	32.5

*q_R = 60 bbl/min—pad and slurry at bottomhole conditions of 250°F and 7,000 psi.
q_c = flow rate at laminar/turbulent transition.
$p^*(z)$ = value calculated using the exact integrated analytical expression with the momentum-change term.
$V_{N_2}(z)$ = gal N_2 at z/gal foam slurry at RC.
$\rho_f(z)$ = lbm foam at z/gal foam slurry at z.
$\Gamma(z)$ = gal N_2 at z/gal foam slurry at z.
$V_L(z)$ = gal liquid phase at z/gal foam slurry at z.
$V_p(z)$ = gal proppant at z/gal foam slurry at z.
C_s/ρ_s = gal proppant/gal foam slurry at RC.
Z = gas compressibility factor.

where T is temperature (K) and Z is the gas compressibility. At 75°F [24°C] and over pressures ranging from 3,500 to 7,000 psi [24.1 to 48.3 MPa], Z may be expressed as[119]

$$Z=[0.779+0.000083p(z)]. \quad (9.57)$$

The term C_{N_2} is the lbm of N_2/gal foam slurry at RC and can also be expressed using the real-gas law as

$$C_{N_2}=\frac{(0.000568)\Gamma_R p_R M_{N_2}}{Z_R R T_R}, \quad (9.58)$$

where

p_R, Z_R,
T_R = pressure, gas compressibility, and temperature at reservoir temperature and closure pressure,
M_{N_2} = molecular weight of N_2,
R = real-gas constant, and
Γ_R = specified foam quality at reservoir conditions.

The values of the given parameters and their units are listed in Table 9.4. The density at bottomhole conditions, ρ_R, may now be calculated as

$$\rho_R=(C_{N_2}+C_s+V_{LR}\rho_{fl}), \quad (9.59)$$

where ρ_{fl} is the density of the liquid phase. The foam flow rate at any position $q(z)$ may now be calculated by use of continuity $[\rho_f(z)q(z)=\rho_R q_R]$ and Eqs. 9.55 through 9.59.

Eq. 9.54 will now be solved for Δp by estimating a pressure at the midpoint $p(z_m)$ and then calculating $\rho_f(z_m)$ from Eqs. 9.55 through 9.59. The resulting Δp will be used to generate the next estimate of $p(z_m)$. This process will be continued until the estimated values of $p(z_m)$ converge. Table 9.5 shows the results of the successive approximation calculation for both the pad and proppant-laden stages. The easiest case of a single interval, Δz, of 10,000 ft [3048 m] $[\Delta p=p(10{,}000)-p(0)]$ was assumed. An initial estimate for $p(z_m)$ of 7,000 psi [48.3 MPa] was used with $p(10{,}000)$ held constant at 7,000 psi [48.3 MPa]. Convergence was obtained for both stages in no more than five steps. For the pad stage, Δp was calculated as +1,037 psi [+7.15 MPa], while that for the slurry stage was +3,465 psi [+23.89 MPa]. The corresponding surface pressures are given in Table 9.6

With a knowledge of $p(z)$, the composition of foam at any position z can be calculated as

$$\Gamma(z)=\frac{V_{N_2}(z)}{V_{N_2}(z)+C_s/\rho_s+V_{LR}}, \quad (9.60)$$

$$V_L(z)=\frac{V_{LR}}{V_{N_2}(z)+C_s/\rho_s+V_{LR}}, \quad (9.61)$$

and

$$V_p(z)=\frac{C_s/\rho_s}{V_{N_2}(z)+C_s/\rho_s+V_{LR}}. \quad (9.62)$$

The desired surface and bottomhole densities, flow rates, and compositions can now be calculated with Eqs. 9.54 through 9.62. Table 9.6 shows the results. Also shown are the surface pressures calculated with the exact integrated solution of the force balance with the momentum-change term. The agreement with the approximated surface pressures is remarkably good. In the exact solution, the momentum-change term, $\Delta(\rho_f \bar{v}_f^2)/g_c$, assuming uniform turbulent velocity at any casing cross section,[17] contributed only +0.53 psi [+3.65 kPa] for the pad stage and +4.68 psi [+32.3

kPa] for the slurry stage. Consequently, neglect of this term is justified.

With the calculated flow rates and densities at the surface and perforations, the original assumption of turbulent flow can be checked by equating the laminar and turbulent expressions (Eqs. 9.47 and 9.52) and solving for the critical velocity, $\bar{v}_c$. In this way, the critical flow rate, q_c, for transition to turbulence can be calculated. In calculation of n_p'' and K_p'' for Eq. 9.47, Eqs. 9.43 and 9.44 can be used, assuming $n'=0.45$ and $K'=0.023(\text{lbf-sec}^{n'})/\text{ft}^2$ [1.10 $\text{Pa}\cdot\text{s}^{n'}$]. Eq. 9.46 is used to compute $\tau_o{}^p$. $\bar{v}_c$ is solved implicitly from the resulting equation:

$$A'\rho_f^{x'}d^b\left(\frac{8\bar{v}_c}{d}\right)^s - K_p''\left[\frac{8\bar{v}}{d}\right]^{n_p''} = \tau_o^p. \quad \text{(9.63)}$$

The resulting critical flow rates are less than the predicted foam flow rates, as shown in Table 9.6, implying that the assumption of turbulence was correct.

With known values of flow rate and density at the perforations, the size and number of perforations required to produce negligible perforation friction pressure can now be calculated. Assuming a perforation friction pressure of <10 psi [<68.95 kPa] to be negligible, Eq. 9.16 can be used to determine the number of ⅝-in. [1.59-cm] -diameter perforations. Letting $q_o=q/n_{perf}$, where q_o is the flow rate per perforation and n_{perf} is the number of perforations for the total flow rate, q, Eq. 9.16 can be solved for n_{perf}:

$$n_{perf} = \left[\frac{B\rho q^2}{(\Delta p_f)_{perf}\, d_o^4}\right]^{1/2} \quad \text{(9.64)}$$

Using a perforation diameter of ⅝ in. [1.59 cm], a flow rate of 54.45 bbl/min [0.144 m^3/s], a density of 9.99 lbm/gal [1.197 kg/m^3], and a B of 0.5 psi-in.^{4}-gal-(no. perf)^2/lbm-(bbl/min)^2 [169{$\text{MPa}\cdot\text{cm}^4\cdot\text{m}^3\cdot(\text{no. perf})^2$}/$\text{kg}\cdot(\text{m}^3/\text{s})^2$] gives 98.5 perforations. Thus, a 50-ft [15-m] perforated interval would require at least 2 shots/ft [7 shots/m]. In most fracturing treatments, however, there are at least 4 shots/ft [13 shots/m] with proper phasing to provide good communication with the fracture. Therefore, a conventional perforation treatment would most probably be sufficient.

Nomenclature

a = defined in Fig. 9.5
A = constant in Bowen's model, Eq. 9.7, $\text{lbf-sec}^s/\text{ft}^{(2+b)}$ [$\text{N}\cdot\text{s}^s/\text{m}^{2+b}$]
A' = constant in Eq. 9.52 for turbulent flow (see Table 9.4 for units)
A_o = point in Fig. 9.5 at (f_{emax}, f_p)
A_v = constant in Eq. 9.19
b = diameter exponent for Bowen's model, Eq. 9.7, dimensionless
B = coefficient for Eq. 9.16 (perforation friction pressure)=0.2 to 0.5
C = point in Fig. 9.5 at (0, f_c)
C_{HPG} = concentration of HPG, lbm/1,000 gal [g/m^3]
C_ℓ = parameter in foam consistency index, Eq. 9.44
C_{N_2} = nitrogen concentration, lbm/gal foam slurry at reservoir conditions [g/m^3]
C_o = orifice discharge coefficient
C_p = polymer concentration, lbm/1,000 gal [g/m^3]
C_p^* = critical polymer concentration for molecular overlap, lbm/1,000 gal [g/m^3]
C_s = proppant concentration, lbm proppant/gal foam slurry at reservoir conditions [g/m^3]
C_2 = parameter in foam consistency index, Eq. 9.44
d = pipe or casing ID, in. [cm]
d_c = casing ID, in. [cm]
d_e = equivalent diameter for annular flow, in. [cm]
d_o = orifice or perforation diameter, in. [cm]
d_p = particle diameter (packed bed), Table 9.1, in. [cm]
d_t = tubing ID, in. [cm]
D = plate depth (Table 9.1)
e_f = fracture efficiency
E_f = flow activation energy
f = friction factor (also polymer solution friction factor)
f_c = fraction of pumped volume that just reaches end of "cool" region in fracture when pumping stops
f_d = correction factor in Fig. 9.5
f_e = fluid exposure time normalized to total pump time
f_{emax} = maximum normalized exposure time
f_{fl} = fraction of fluid pumped
f_p = normalized pad volume
F = total normal force on cone or parallel plate
F_C = correction factor for proppant contribution to friction pressure
g = gravitational acceleration
g_c = gravitational dimensional constant
G' = storage modulus
G'' = loss modulus
h_f = fracture height, ft [m]
j = diameter exponent in Bowen's model, Eq. 9.25
k = permeability to Newtonian fluids (Table 9.1)
K' = power-law consistency index, $\text{lbf-sec}^{n'}/\text{ft}^2$ [$\text{Pa}\cdot\text{s}^{n'}$]
K'' = Herschel-Bulkley consistency index, Eq. 9.35, $\text{lbf-sec}^{n''}/\text{ft}^2$ [$\text{Pa}\cdot\text{s}^{n''}$]
K^0 = Huggins coefficient, dimensionless
K_{an} = power-law consistency index for annular flow, $\text{lbf-sec}^{n'}/\text{ft}^2$ [$\text{Pa}\cdot\text{s}^{n'}$]
K_f = power-law consistency index for slot or fracture flow, $\text{lbf-sec}^{n'}/\text{ft}^2$ [$\text{Pa}\cdot\text{s}^{n'}$]
K_i' = power-law consistency index for segmented model, $\text{lbf-sec}^{n_i'}/\text{ft}^2$ [$\text{Pa}\cdot\text{s}^{n_i'}$]
K_p = power-law consistency index for pipe flow, $\text{lbf-sec}^{n'}/\text{ft}^2$ [$\text{Pa}\cdot\text{s}^{n'}$]
K_p'' = Herschel-Bulkley consistency index for pipe flow, Eq. 9.44, $\text{lbf-sec}^{n_p''}/\text{ft}^2$ [$\text{Pa}\cdot\text{s}^{n_p''}$]
K_{75}' = power-law consistency index at 75°F [24°C], $\text{lbf-sec}^{n_{75}'}/\text{ft}^2$ [$\text{Pa}\cdot\text{s}^{n_{75}'}$]
ℓ = $1-e_f$ (defined in Fig. 9.5)
$\bar{\ell}_p$ = average interparticle distance
L = tube length (Table 9.1) or inner-cylinder length, ft [m]
L_f = length of one wing of fracture, ft [m]
M = molecular weight
M_{N_2} = molecular weight of nitrogen
n = experimental parameter for general fluid (Table 9.1)
n' = power-law index, dimensionless
n'' = Herschel-Bulkley shear-rate exponent
n_i' = power-law index for segmented model
n_p'' = Herschel-Bulkley shear-rate exponent for pipe flow, Eq. 9.43
n_{perf} = number of perforations
n_{75}' = power-law index at 75°F [24°C]
N_{De} = Deborah number
N_{Re} = Reynolds number (Fig. 9.2)
N_{Re}' = generalized Reynolds number
$(N_{Re}')_{an}$ = generalized Reynolds number for annular flow
$(N_{Re})_f$ = Reynolds number for foam flow
$(N_{Re})_s$ = Reynolds number based on solvent properties
$p(z)$ = fluid pressure at depth z
p_a = atmospheric pressure (Table 9.2), psi [kPa]
p_{bh} = bottomhole pressure, psi [kPa]
p_R = reservoir closure pressure
p_s = surface pressure, psi [kPa]
Δp = fluid pressure drop (Table 9.1), psi [kPa]
Δp_f = friction pressure drop (Table 9.1), psi [kPa]
$(\Delta p_f)_{fl}$ = friction pressure drop of unladen fluid, psi [kPa]
$(\Delta p_f)_M$ = friction pressure drop of slurry (mixture), psi [kPa]

$(\Delta p_f)_o$ = friction pressure drop of solvent, psi [kPa]
$(\Delta p_f)_{perf}$ = perforation friction pressure, psi [kPa]
Δp_H = change in hydrostatic pressure, psi [kPa]
Δp_o = pressure drop across an orifice, psi [kPa]
q = flow rate, bbl/min [m^3/s]
q_c = critical flow rate, laminar/turbulent transition, bbl/min [m^3/s]
q_f = one-half pumping flow rate (down one wing of fracture)
q_o = flow rate/perforation, bbl/min [m^3/s]
q_R = foam flow rate at reservoir conditions, bbl/min [m^3/s]
r = radial coordinate (Table 9.1)
r_c = radius of cone-and-plate geometry (Table 9.2)
r_H = hydraulic radius
r_i = radius of inner cylinder (Table 9.1)
r_o = radius of outer cylinder (Table 9.1)
r_p = radius of parallel-plate geometry (Table 9.2)
r_{sp} = radius of sphere (Table 9.1)
r_t = radius of pipe or tubing (Table 9.1)
R = real-gas constant or sphere radius
s = turbulent velocity exponent in Eq. 9.7, dimensionless
T = temperature, °K
T_o = torque on viscometer (Tables 9.1 and 9.2)
T_R = reservoir temperature
$\bar{v}$ = average fluid velocity, ft/sec [m/s]
$\bar{v}_c$ = critical average velocity for laminar/turbulent transition, ft/sec [m/s]
$\bar{v}_f$ = average foam velocity, ft/sec [m/s]
v_i = velocity component in ith coordinate direction $(i=\theta,r,x,z)$
$\bar{v}_o$ = average velocity through an orifice, ft/sec [m/s]
$(v_p)_{max}$ = maximum particle-settling velocity in a horizontal tube that will maintain a suspension
v_S = Stokes settling velocity (Table 9.1)
V_d = volume fraction of dispersed phase
V_L = volume liquid at depth z per gallon of foam slurry at depth z
V_{LR} = volume liquid per gallon of foam slurry at reservoir conditions
V_{N_2} = volume nitrogen at depth z per gallon of foam slurry at reservoir conditions
V_p = particle volume fraction or volume proppant at depth z per gallon of foam slurry at depth z
V_{pn} = pad volume normalized to total pumped volume
w_f = fracture width or plate separation distance, in. [cm] (Table 9.1)
x = rectilinear coordinate (horizontal)
x' = density exponent in Eq. 9.52 for turbulence
x_c = fraction of fracture length at which fluid attains reservoir temperature
X = defined in Fig. 9.5
z = rectilinear coordinate (vertical)
z_m = midpoint of vertical interval Δz, ft [m]
Δz = vertical interval, ft [m]
Z = gas compressibility
Z_R = gas compressibility at reservoir conditions
α = Ellis model exponent (Table 9.1)
β = orifice diameter/pipe diameter
$\dot{\gamma}$ = shear rate, seconds^{-1}
$\dot{\gamma}_i$ = shear rate at inner cylinder in Couette viscometer, seconds^{-1}
$\dot{\gamma}_{ij}$ = component of rate-of-deformation tensor
$\dot{\gamma}_R$ = shear rate at edge of rotating parallel plates, seconds^{-1}
$\dot{\gamma}_w$ = shear rate at wall (e.g., tubular flow), seconds^{-1}
Γ = foam quality (volume gas/volume foam slurry)
Γ_R = foam quality at reservoir conditions
θ = angle (spherical) (Table 9.1)
θ_c = cone angle (Table 9.2)
κ = r_o/r_i (Table 9.1)
μ = viscosity (e.g., for power-law fluids, Newtonian fluids, suspensions), cp [mPa·s]
μ_a = apparent viscosity, cp [mPa·s]
μ_c = continuous-phase or liquid-phase viscosity, cp [mPa·s]
μ_d = dispersed-phase viscosity, cp [mPa·s]
μ_e = effective viscosity for foams, Eq. 9.49, cp [mPa·s]
μ_{efl} = effective "viscosity" for flow of power-law fluid in a packed bed (Table 9.1), Pa·s$^{n'}$·m$^{(1-n')}$
μ_f = foam viscosity, cp [mPa·s]
μ_p = plastic (Bingham) viscosity, cp [mPa·s]
μ_r = relative viscosity (μ/μ_c), dimensionless
μ_s = solvent viscosity, cp [mPa·s]
μ^o = zero-shear viscosity, cp [mPa·s]
$[\mu]$ = intrinsic viscosity (Eq. 9.21), L/g
ρ = density, lbm/gal [kg/m^3]
ρ_f = foam density, lbm/gal [kg/m^3]
ρ_{fl} = fluid density, lbm/gal [kg/m^3]
ρ_r = relative density, dimensionless
ρ_R = foam density at reservoir conditions, lbm/gal [kg/m^3]
ρ_s = density of sand, proppant, or particle, lbm/gal [kg/m^3]
ρ_{sp} = density of sphere (Table 9.1), lbm/gal [kg/m^3]
τ = shear stress, lbf/ft^2 [Pa]
τ_i = shear stress on inner cylinder of Couette viscometer, lbf/ft^2 [Pa]
τ_{ij} = component of extra stress tensor, lbf/ft^2 [Pa]
τ_o = yield stress, Bingham model, lbf/ft^2 [Pa]
τ_o' = yield stress, Herschel-Bulkley model, lbf/ft^2 [Pa]
τ_o^p = yield stress, Herschel-Bulkley model for pipe flow, lbf/ft^2 [Pa]
τ_w = wall shear stress (Table 9.1), lbf/ft^2 [Pa]
$\tau_{1/2}$ = Ellis model shear stress at $\mu^o/2$, lbf/ft^2 [Pa]
ϕ = void fraction of packed bed (Table 9.1)
ψ_1 = primary normal stress coefficient
ψ_2 = secondary normal stress coefficient
Ω = viscometer angular velocity (Table 9.1), rad/sec
Ω_i = angular velocity of inner cylinder (Table 9.2), rad/sec
Ω_o = angular velocity of outer cylinder (Table 9.2), rad/sec

References

1. Bird, R.B., Armstrong, R.C., and Hassager, O.: *Dynamics of Polymeric Liquids: Vol. 1, Fluid Mechanics,* John Wiley and Sons Inc., New York City (1977).
2. Walters, K.: *Rheometry,* Chapman and Hall Ltd., London (1975).
3. Liauh, W.W. and Liu, T.W.: "A Capillary Viscometer for the Study of EOR Polymers," paper SPE 12649 presented at the 1984 SPE/DOE Symposium on Enhanced Oil Recovery, Tulsa, April 15–18.
4. Kramer, J., Uhl, J.T. and Prud'homme, R.K.: "Measurement of the Viscosity of Guar Gum Solutions to 50,000 s^{-1} Using a Parallel Plate Rheometer," *Polym. Eng. Sci.* (1987) **27**, No. 8, 598–602.
5. Connelly, R.W. and Greener, J.J.: "High-Shear Viscometry with a Rotational Parallel-Disk Device," *J. Rheol.* (1985) **29**, 209–26.
6. Merrill, E.W., Ram, A., and Mickley, H.S.: "Effect of Shear on the Intrinsic Viscosity of a Whole Polyisobutylene in Cyclohexane," *J. Polymer Sci.* (1961) **51**, 43–44.
7. Gordon, R.J. and Balakrishnan, C.: *J. Appl. Polym. Sci.* (1972) **16**, 1629–39.
8. Armstrong, R.C.: "Obtaining Constitutive Equations for Macromolecular Fluids from Molecular Theories," PhD dissertation, U. of Wisconsin, Madison (1973) 105–12.
9. Prud'homme, R.K.: "Rheological Characterization of Fracturing Fluids," final reports, PRAC Projects 82–45, 84–45 and 85–45, API, Dallas (1984–85).

10. Knoll, S.K.: "Wall Slip Evaluation in Steady Shear Viscosity Measurements," paper SPE 13904 presented at the 1985 SPE/DOE Low-Permeability Gas Reservoirs Symposium, Denver, May 19–22.
11. Craft, B.C. and Holden, W.R.: *Well Design, Drilling and Production*, Prentice-Hall Inc., Englewood Cliffs, NJ (1962) 64–74.
12. *RP 39, Recommended Practice for Standard Procedures for Evaluation of Hydraulic Fracturing Fluids*, second edition, API, Dallas (1983).
13. Savins, J.G.: "Generalized Newtonian (Pseudoplastic) Flow in Stationary Pipes and Annuli," *Trans.*, AIME (1958) **213**, 325–32.
14. Skelland, A.H.P.: *Non-Newtonian Flow and Heat Transfer*, John Wiley & Sons Inc., New York City (1967) 110–11.
15. Bowen, R.L. Jr.: "Designing Turbulent-Flow Systems," *Chem. Eng.* (July 24, 1961) 143–50.
16. Lagrone, K.W. and Rasmussen, J.W.: "A New Development in Completions Methods—The Limited Entry Technique," *JPT* (July 1963) 695–702.
17. Bennett, C.O. and Meyers, J.E.: *Momentum, Heat, and Mass Transfer*, second edition, McGraw-Hill Book Co. Inc., New York City (1974) 75–79.
18. Grose, R.D.: "Orifice Flow at Low Reynolds Number," *J. Pipelines* (1983) **3**, 207–14.
19. *Chemical Engineer's Handbook*, fifth edition, R.H. Perry and C.H. Chilton (eds.), McGraw-Hill Book Co. Inc., New York City (1973) Sec. 5, 12–14.
20. Stekoll, M.H.: "New Light on Fracturing Through Perforations," *Oil & Gas J.* (Oct. 29, 1956) 95–97.
21. Brown, R.W. and Gilbert, B.: "Pressure Drop Across Perforations Can Be Computed," *Pet. Eng.* (Sept. 1957) B-82–B-88.
22. Kraemer, J.W.: "Fluid Flow Rate Through Perforations," *Pet. Eng.* (Jan. 1959) B-44–B-46.
23. Harrington, L.J., Whitsitt, N.F., and Hannah, R.R.: "Prediction of the Location and Movement of Fluid Interfaces in a Fracture," *Proc.*, Southwestern Petroleum Short Course, Lubbock, TX (April 26–27, 1973).
24. Geertsma, J. and Haafkens, R.: "A Comparison of the Theories for Predicting Width and Extent of Vertical Hydraulically Induced Fractures," *J. Energy Res. Technol.* (March 1979) **101**, 8–19.
25. Settari, A. and Cleary, M.P.: "Three-Dimensional Simulation of Hydraulic Fracturing," *J. Pet. Tech.* (July 1984) 1177–90.
26. Dougherty, R.L. and Abou-Sayed, A.S.: "Evaluation of the Influence of In-Situ Reservoir Conditions on the Geometry of Hydraulic Fractures Using a 3D Simulator," paper SPE 13275 presented at the 1984 SPE Annual Technical Conference and Exhibition, Houston, Sept. 16–19.
27. Bouteca, M.J.: "3D Analytical Model for Hydraulic Fracturing: Theory and Field Test," paper SPE 13276 presented at the 1984 SPE Annual Technical Conference and Exhibition, Houston, Sept. 16–19.
28. Poulsen, D.K. and Lee, W.S.: "Fracture Design with Time- and Temperature-Dependent Fluid Properties," paper SPE 12483 presented at the 1984 SPE Formation Damage Control Symposium, Bakersfield, CA, Feb. 13–14.
29. Whitsitt, N.F. and Dysart, G.R.: "The Effect of Temperature On Stimulation Design," *JPT* (April 1970) 493–502.
30. Biot, M.A., Massé, L., and Medlin, W.L.: "Temperature Analysis in Hydraulic Fracturing," *JPT* (Nov. 1987) 1389–97; *Trans.*, AIME, **283**.
31. Nolte, K.G.: "Determination of Proppant and Fluid Schedules from Fracturing-Pressure Decline," *SPEPE* (June 1986) 255–65; *Trans.*, AIME, **281**.
32. Nolte, K.G.: "Application of Fracture Design Based on Pressure Analysis," *SPEPE* (Feb. 1988) 31–42.
33. Nordgren, R.P.: "Propagation of a Vertical Hydraulic Fracture," *SPEJ* (Aug. 1972) 306–14; *Trans.*, AIME, **253**.
34. Ferry, J.D.: *Viscoelastic Properties of Polymers*, third edition, John Wiley and Sons Inc., New York City (1980).
35. Bird, R.B. *et al.*: *Dynamics of Polymeric Liquids: Volume 2, Kinetic Theory*, John Wiley and Sons Inc., New York City (1977).
36. Guillot, D. and Dunand, A.: "Rheological Characterization of Fracturing Fluids by Using Laser Anemometry," *SPEJ* (Feb. 1985) 39–45.
37. *Fracturing Fluids: Engineering Data*, Dowell Schlumberger, Houston (1984).
38. Menjivar, J.A.: "On the Use of Gelation Theory to Characterize Metal Cross-linked Polymer Gels," *Proc.*, ACS Div. of Polymeric Materials, Science and Engineering, Philadelphia, PA (1984) **51**, 88–95.
39. Anderson, R.W. and Baker, J.R.: "Use of Guar Gum and Synthetic Cellulose in Oilfield Fluids," paper SPE 5005 presented at the 1974 SPE Annual Meeting, Houston, Oct. 6–9.
40. Crowe, C.W., Martin, R.C., and Michaelis, A.M.: "Evaluation of Acid-Gelling Agents for Use in Well Stimulation," *SPEJ* (Aug. 1981) 415–24.
41. Liption, D. and Burnett, D.B.: "Comparisons of Polymers Used in Workover and Completion Fluids," paper SPE 5872 presented at the 1976 SPE California Regional Meeting, Long Beach, April 8–9.
42. Chang, H.D. and Darby, R.: "Effect of Shear Degradation on the Rheological Properties of Dilute Drag-Reducing Polymer Solutions," *J. Rheol.* (1983) **27**, No. 1, 77–88.
43. Maerker, J.M.: "Shear Degradation of Partially Hydrolyzed Polyacrylamide Solutions," *SPEJ* (Aug. 1975) 311–22; *Trans.*, AIME, **259**.
44. Rogers, R.E., Veatch, R.W. Jr., and Nolte, K.G.: "Pipe Viscometer Study of Fracturing Fluid Rheology," *SPEJ* (Oct. 1984) 575–81.
45. Astarita, G., Marrucci, G., and Palumbo, G.: "Non-Newtonian Gravity Flow Along Inclined Surfaces," *I&EC Fund.* (1964) **3**, No. 4, 333–39.
46. Metzner, A.B., Cohen, Y., and Rangel-Nafaile, C.: "Inhomogeneous Flows of Non-Newtonian Fluids: Generation of Spatial Concentration Gradients," *J. Non-Newtonian Fluid Mech.* (1979) **5**, 449–62.
47. Tirrell, M. and Malone, M.F.: "Stress-Induced Diffusion of Macromolecules," *J. Polym. Sci.* (1977) **15**, 1569–83.
48. Dutta, A. and Mashelkar, R.A.: "Hydrodynamics in Media with Migrating Macromolecules: Development of FDCF Asymptote," *J. Non-Newtonian Fluid Mech.* (1984) **16**, 279–302.
49. Brunn, P.O. and Chi, S.: "Macromolecules in Nonhomogeneous Flow Fields: A General Study for Dumbbell Model Macromolecules," *Rheol. Acta* (1984) **23**, 163–71.
50. Naik, S.C.: "Apparent Viscosity Characteristics of Guar Gum Sols," *Rheology, Vol. 2: Fluids*, G. Astarita, G. Marrucci, and L. Nicolais (eds.), Plenum Press, New York City (1980) 341–45.
51. Elson, T.P., Solomon, J., and Nienow, A.W.: "The Interaction of Yield Stress and Viscoelasticity on the Weissenberg Effect," *J. Non-Newtonian Fluid Mech.* (1982) **11**, 1–22.
52. Clark, P.E. and Quadir, J.A.: "Prop Transport in Vertical Fractures," paper SPE 10261 presented at the 1981 SPE Annual Technical Conference and Exhibition, San Antonio, Oct. 5–7.
53. Govier, G.W. and Aziz, K.: *The Flow of Complex Mixtures in Pipes*, Robert E. Krieger Publishing Co. Inc., Malabar, FL (1982) 19–20.
54. Shah, S.N. and Lord, D.L.: "Hydraulic Fracturing Slurry Transport in Horizontal Pipes," paper SPE 18994 presented at the 1989 SPE Rocky Mountain Regional Meeting/Low Permeability Reservoirs Symposium, Denver, March 6–8.
55. Kawase, Y. and Ulbrecht, J.J.: "Rheological Properties of Suspensions of Solid Spheres in Non-Newtonian Fluids," *Chem. Eng. Commun.* (1983) **20**, 127–36.
56. Ho, B.P. and Leal, L.G.: "Inertial Migration of Rigid Spheres in Two-Dimensional Unidirectional Flow," *J. Fluid Mech.* (1974) **65**, 365–400.
57. Gauthier, F., Goldsmith, H.L., and Mason, S.G.: "Particle Motions in Non-Newtonian Media, I: Couette Flow," *Rheol. Acta* (1971) **10**, 344–64.
58. Gauthier, F., Goldsmith, H.L., and Mason, S.G.: "Particle Motions in Non-Newtonian Media, II: Poiseuille Flow," *Trans. Soc. Rheol.* (1971) **15**, No. 2, 297–330.
59. Rockefeller, H.A.: "The Segmented 'Power-Law' Model...A Generalized Method to Describe Fracturing Fluid Rheology," paper presented at the 1984 Natl. Spring AIChE Conference, Anaheim, CA, May 23.
60. Bannister, C.E.: "Rheological Evaluation of Cement Slurries: Methods and Models," paper SPE 9284 presented at the 1980 SPE Annual Technical Conference and Exhibition, Dallas, Sept. 21–24.
61. Karis, T.E., Prieve, D.C., and Rosen, S.L.: "Anomalous Lateral Migration of a Rigid Sphere in Torsional Flow of a Viscoelastic Fluid," *J. Rheol.* (1984) **28**, No. 4, 381–92.
62. Ho, B.P. and Leal, L.G.: "Migration of Rigid Spheres in a Two-Dimensional Unidirectional Shear Flow of a Second-Order Fluid," *J. Fluid Mech.* (1976) **76**, Part 4, 783–99.
63. Brunn, P.: "The Hydrodynamic Wall Effect for a Disperse System," *Intl. J. Multiphase Flow* (1981) **7**, 221–34.
64. Jinescu, V.V.: "The Rheology of Suspensions," *Intl. Chem. Eng.* (July 1974) **14**, No. 3, 397–420.
65. Jeffrey, D.J. and Acrivos, A.: "The Rheological Properties of Suspensions of Rigid Particles," *AIChE J.* (May 1976) **22**, No. 3, 417–32.
66. Thomas, D.G.: "Transport Characteristics of Suspension: VIII. A Note on the Viscosity of Newtonian Suspensions of Uniform Spherical Particles," *J. Colloid Sci.* (1965) **20**, 267–77.
67. Highgate, D.J. and Whorlow, R.W.: "Rheological Properties of Suspensions of Spheres in Non-Newtonian Media," *Rheol. Acta* (1970) **9**, 569–76.
68. Leppard, W.R.: "Viscoelasticity: Stress Measurements and Constitutive Theory," PhD dissertation, U. of Utah, Salt Lake City (1975).
69. Cameron, J.R.: "A Rheological Characterization of Coal-Derived Liquids," PhD dissertation, U. of Michigan, Ann Arbor (1981).
70. Chan, D. and Powell, R.L.: "Rheology of Suspensions of Spherical Particles in a Newtonian and a Non-Newtonian Fluid," *J. Non-Newtonian Fluid Mech.* (1984) **15**, 165–79.
71. Ferry, J.D.: *Viscoelastic Properties of Polymers*, second edition, John Wiley and Sons Inc., New York City (1970) 320–26.

72. Lord, D.L., Hulsey, B.W., and Melton, L.L.: "General Turbulent Pipe Flow Scale-Up Correction for Rheologically Complex Fluids," *SPEJ* (Sept. 1967) 252–58; *Trans.*, AIME, **240**.
73. Quader, A.K.M.A. and Wilkinson, W.L.: "Correlation of Turbulent Flow Rate—Pressure Drop Data for Non-Newtonian Solutions and Slurries in Pipes," *Intl. J. Multiphase Flow* (1980) **6**, 553–61.
74. Shah, S.N.: "Correlations Predict Friction Pressures of Fracturing Gels," *Oil & Gas J.* (Jan. 16, 1984) 92–98.
75. Gardner, D.C. and Eikerts, J.V.: "The Effects of Shear and Proppant on the Viscosity of Cross-Linked Fracturing Fluids," paper SPE 11066 presented at the 1982 SPE Annual Technical Conference and Exhibition, New Orleans, Sept. 26–29.
76. Seyer, F.A. and Metzner, A.B.: "Turbulence Phenomena in Drag Reducing Systems," *AIChE J.* (May 1969) **15**, No. 3, 426–34.
77. Crittendon, B.C.: "The Mechanics of Design and Interpretation of Hydraulic Fracture Treatments," *JPT* (Oct. 1959) 21–39.
78. Hannah, R.R., Harrington, L.J., and Lance, L.C.: "The Real-Time Calculation of Accurate Bottomhole Fracturing Pressure From Surface Measurements Using Measured Pressures as a Base," paper SPE 12062 presented at the 1983 SPE Annual Technical Conference and Exhibition, San Francisco, Oct. 5–8.
79. Shah, S.N. and Lee, Y.N.: "Friction Pressures of Proppant-Laden Hydraulic Fracturing Fluids," *SPEPE* (Nov. 1986) 437–45; *Trans.*, AIME, **281**.
80. Lord, D.L. and McGowen, J.M.: "Real-Time Treating Pressure Analysis Aided by New Correlation," paper SPE 15367 presented at the 1986 SPE Annual Technical Conference and Exhibition, New Orleans, Oct. 5–8.
81. Conway, M.W., Pauls, R.W., and Harris, L.E.: "Evaluation of Procedures and Instrumentation Available for Time-Temperature Stability Studies of Cross-Linked Fluids," paper SPE 9333 presented at the 1980 SPE Annual Technical Conference and Exhibition, Dallas, Sept. 21–24.
82. Cloud, J.E. and Clark, P.E.: "Alternatives to the Power-Law Fluid Model for Crosslinked Fluids," *SPEJ* (Dec. 1985) 935–42.
83. Hannah, R.R., Harrington, L., and Anderson, R.W.: "A Study of Bingham Plastic Flow for Use as a Temporary Diverting Agent in Hydraulic Fracturing," paper 78-PET-36 presented at the 1978 Petroleum Div. of the ASME Energy Technology Conference and Exhibition, Houston, Nov. 5–9.
84. *Pacesetter Rheology Handbook*, The Western Co. of North America, Fort Worth, TX (1982).
85. Conway, M.W. *et al.*: "Chemical Model for the Rheological Behavior of Crosslinked Fluid Systems," *JPT* (Feb. 1983) 315–20.
86. Wilson, M.J.: "Determination of Several Variables Affecting Laboratory Measurements of Cross-linked Fracture Fluids," MS thesis, Texas A&M U., College Station (1982).
87. Prud'homme, R.K.: "Rheological Characterization of Fracturing Fluids," final report, PRAC Project 86-45, API, Dallas (1986).
88. Craigie, L.J.: "A New Method for Determining the Rheology of Cross-linked Fracturing Fluid Using Shear History Simulation," paper SPE 11635 presented at the 1983 SPE/DOE Low-Permeability Gas Reservoirs Symposium, Denver, March 14–16.
89. Lescarboura, J.A., Sifferman, T.R., and Wahl, H.A.: "Evaluation of Fracturing Fluid Stability by Using a Heated Pressurized Flow Loop," *SPEJ* (June 1984) 249–55.
90. Gardner, D.C. and Eikerts, J.V.: "Rheological Characterization of Crosslinked and Delayed Crosslinked Fracturing Fluids Using a Closed-Loop Pipe Viscometer," paper SPE 12028 presented at the 1983 SPE Annual Technical Conference and Exhibition, San Francisco, Oct. 5–8.
91. Shah, S.N. and Watters, L.T.: "Time and Shear Effects On Rheological Properties of Crosslinked Fluids—An Evaluation Method," *SPEPE* (Jan. 1986) 55–61.
92. Royce, T.N., Beck, L.M., and Rickards, A.R.: "Rheological Characteristics of Adjustable Crosslinked Fracturing Fluids," paper SPE 13178 presented at the 1984 SPE Annual Technical Conference and Exhibition, Houston, Sept. 16–19.
93. Pilehvari, A. and Clark, P.E.: "Rheology of Hydraulic Fracturing Fluids: Wall Slip During Viscosity Measurement," *JPT* (Oct. 1985) 1840–46.
94. Cameron, J.R., Gardner, D.C., and Veatch, R.W. Jr.: "New Insights on the Rheological Behavior of Delayed Crosslinked Fracturing Fluids," paper SPE 18209 presented at the 1988 SPE Annual Technical Conference and Exhibition, Houston, Oct. 2–5.
95. Cameron, J.R.: "Viscometry of Nonhomogeneous Flows and the Behavior of a Titanium-Crosslinked Hydroxypropyl Guar Gel in Couette Flow," *J. Rheol.* (1989) **33**, 15–46.
96. Watkins, E.K., Wendorff, C.L., and Ainley, B.R.: "A New Crosslinked Foamed Fracturing Fluid," paper SPE 12027 presented at the 1983 SPE Annual Technical Conference and Exhibition, San Francisco, Oct. 5–8.
97. Blauer, R.E. and Kohlhaas, C.A.: "Formation Fracturing with Foam," paper SPE 5003 presented at the 1974 SPE Annual Meeting, Houston, Oct. 6–9.
98. Wendorff, C.L. and Ainley, B.R.: "Massive Hydraulic Fracturing of High-Temperature Wells with Stable Frac Foams," paper SPE 10257 presented at the 1981 Annual SPE Technical Conference and Exhibition, San Antonio, Oct. 5–7.
99. Harris, P.C. and Reidenbach, V.G.: "High-Temperature Rheological Study of Foam Fracturing Fluids," *JPT* (May 1987) 613–19; *Trans.*, AIME, **283**.
100. Blauer, R.E., Mitchell, B.J., and Kohlhaas, C.A.: "Determination of Laminar, Turbulent, and Transitional Foam Flow Losses in Pipes," paper SPE 4885 presented at the 1974 SPE California Regional Meeting, San Francisco, April 4–5.
101. Mitchell, B.J.: "Viscosity of Foam," PhD dissertation, U. of Oklahoma, Norman (1970).
102. Taylor, G.I.: "The Viscosity of a Fluid Containing Small Drops of Another Fluid," *Proc. Roy. Soc. A* (1932) **138**, 41.
103. Reidenbach, V.G. *et al.*: "Rheological Study of Foam Fracturing Fluids Using Nitrogen and Carbon Dioxide," *SPEPE* (Jan. 1986) 31–41; *Trans.*, AIME, 281.
104. Craighead, M.S., Hossaini, M., and Freeman, E.R.: "Foam Fracturing Utilizing Delayed Cross-Linked Gels," paper SPE 14437 presented at the 1985 SPE Annual Technical Conference and Exhibition, Las Vegas, Sept. 22–25.
105. Mitchell, B.J.: "Test Data Fill Theory Gap on Using Foam as a Drilling Fluid," *Oil & Gas J.* (Sept. 6, 1971) 96–100.
106. Sanghani, V. and Ikoku, C.U.: "Rheology of Foam and its Implications in Drilling and Cleanout Operations," topical report, Contract No. DE-AC19-79BC10079, U.S. DOE (June 1982).
107. Patton, J.T., Holbrook, S.T., and Hsu, W.: "Rheology of Mobility-Control Foams," *SPEJ* (June 1983) 456–60.
108. Krug, J.A.: "Air and Water Requirements for Foam Drilling Operations," MS thesis, Colorado School of Mines, Golden (1971).
109. Wendorff, C.L. and Earl, R.B.: "Foam Fracturing Laboratory," paper SPE 12025 presented at the 1983 SPE Annual Technical Conference and Exhibition, San Francisco, Oct. 5–8.
110. Harris, P.C.: "Effects of Texture on Rheology of Foam Fracturing Fluids," paper SPE 14257 presented at the 1985 SPE Annual Technical Conference and Exhibition, Las Vegas, Sept. 22–25.
111. Holcomb, D.L., Callaway, E., and Curry, L.L.: "Chemistry, Physical Nature, and Rheology of Aqueous Stimulation Foams," *SPEJ* (Aug. 1981) 410–14.
112. Holcomb, D.L., Ewing, B.C., and Scott, M.A.: "The Structure of Flowing Aqueous Stimulation Foams at High Pressure," paper 8155 68C presented at the 1983 AIChE Natl. Meeting, Denver, Aug. 28–31.
113. Rand, P.B. and Kraynik, A.M.: "Drainage of Aqueous Foams: Generation-Pressure and Cell-Size Effects," *SPEJ* (Feb. 1983) 152–54.
114. Harms, S.D. and Payne, K.L.: "Factors Affecting the Selection of Foaming Agents for Foam Stimulation," *Proc.*, 30th Annual Southwestern Petroleum Short Course, Lubbock, TX (1983) 66–74.
115. Wenzel, H.G. Jr., Stelson, T.E., and Brungraber, R.J.: "Flow of High Expansion Foam in Pipes," *Proc.*, American Soc. of Civil Engineers, *J. Eng. Mechanics Div.* (Dec. 1967) 153–56.
116. Beyer, A.H., Millhone, R.S., and Foote, R.W.: "Flow Behavior of Foam as a Well Circulating Fluid," paper SPE 3986 presented at the 1972 SPE Annual Meeting, San Antonio, Oct. 8–11.
117. Hirasaki, G.J. and Lawson, J.B.: "Mechanisms of Foam Flow in Porous Media: Apparent Viscosity in Smooth Capillaries," *SPEJ* (April 1985) 176–90.
118. Harris, P.C., Klebenow, D.E., and Kundert, D.P.: "Constant Internal Phase Design Improves Stimulation Results," paper SPE 17532 presented at the 1988 SPE Rocky Mountain Regional Meeting, Casper, WY, May 11–13.
119. Glasstone, S.: *Thermodynamics for Chemists*, D. Van Nostrand Company Inc., New York City (1947) 27–30.

SI Metric Conversion Factors

atm	× 1.013 250*	E+05	=	Pa
°API	141.5/(131.5+°API)		=	g/cm^3
bbl	× 1.589 873	E−01	=	m^3
cp	× 1.0*	E+00	=	mPa·s
dynes/cm^2	× 1.0*	E−01	=	Pa
ft	× 3.048*	E−01	=	m
ft^2	× 9.290 304*	E−02	=	m^2
°F	(°F−32)/1.8		=	°C
gal	× 3.785 412	E−03	=	m^3
in.	× 2.54*	E+00	=	cm
lbf	× 4.448 222	E+00	=	N
(lbf-sec)/ft^2	× 4.788 026	E+01	=	Pa·s
lbm	× 4.535 924	E−01	=	kg
lbf-ft	× 1.355 818	E+00	=	N·m
lbm/gal	× 1.198 264	E+02	=	kg/m^3
psi	× 6.894 757	E+00	=	kPa

*Conversion factor is exact.

Chapter 10
Proppant Transport

Ali Daneshy, SPE, Halliburton Services

10.1 Overview

The physical events affecting proppant transport are described as they occur during a fracturing treatment. The laws governing proppant settling inside the fracture are presented, together with a brief description of supporting laboratory data. The influence of fluid viscosity on particle transport and the resulting proppant distribution inside the fracture are discussed. This chapter shows how heating of the fluid inside the fracture may accelerate proppant settling. The differences between proppant distribution at the end of the pumping and after fracture closure are shown graphically. Finally, the causes of screenout and proppant flowback are discussed.

10.2 Introduction

The fluid pressure applied inside a hydraulic fracture keeps it open during fluid injection. The fluid is usually mixed with proppant particles, which keep the fracture open when fluid pressure has declined after the treatment.

The effectiveness of a hydraulic fracturing treatment depends on several factors: (1) the propped length of the fracture, (2) the conductivity of the propped fracture (defined as the product of fracture width and permeability), and (3) the propped height of the fracture, especially the position of the proppant relative to the producing formation.

These factors in turn depend on fluid and proppant characteristics and amount and the manner of their injection into the fracture.

To understand proppant transport inside the fracture, one first needs to examine the sequence of events in a fracturing treatment.

1. The first fluid increment injected inside the fracture is called "prepad." This fluid has low viscosity and does not contain any proppant.

2. The next fluid increment is called "pad." This fluid has higher viscosity than prepad fluid, but still does not contain any proppant.

3. The propping agent is introduced next by mixing it with the fracturing fluid and forming a slurry. The concentration of the propping agent in the slurry is usually defined in pounds per gallon or kilograms per cubic meter. In fracturing treatments, the proppant concentration usually starts at lower values and increases as the treatment progresses.

Early in the treatment, the prepad fluid occupies the leading edge of the fracture. Behind it is the pad, followed by the slurry, which extends to the wellbore.

Because the fracturing fluid is at a pressure higher than reservoir and the formation is permeable, the fracturing fluid penetrates the formation as leakoff. The highest leakoff rate into the formation occurs at the fracture tip, which is also occupied by the lower-viscosity prepad fluid. As a result of this leakoff, the prepad and pad volumes remaining inside the fracture decrease with time. This results in pad fluid and slurry getting closer to the fracture tip with job progress, which consequently increases the rate of fluid loss from them.

The injected slurry also undergoes leakoff, but because only the fluid part can usually penetrate the formation, the proppant concentration in the fracture gradually increases. As the fracturing treatment progresses and the early stages of the slurry get closer to the fracture tip, the rate of leakoff and consequently the rate of increase of propping-agent concentration in the slurry gradually increase.

As the slurry moves inside the fracture, the propping agent gradually moves downward inside the fracture as a result of gravity. As we will see later, the rate of proppant settling depends on many factors, including fluid viscosity and proppant size and density. For lower-viscosity fluids, the proppant settling rate is relatively high; consequently, the proppant settles into a bed at the bottom of the fracture. In these instances, the concentration of suspended proppant does not rise significantly.

For high-viscosity fluids, the proppant settling rate is very slow. Therefore, for long-duration treatments, the proppant can travel a considerable distance in the fracture.

In the next section we examine the physical laws governing particle settling inside a fluid and then use them to study proppant distribution inside the fracture.

10.3 Fundamentals of Particle Settlement

Stokes' Law. The distribution of proppant inside the fracture depends on its settling velocity in the fracturing fluid. For a spherical particle, the drag coefficient, C_D, is given by Stokes' law as

$$C_D=\tfrac{4}{3}\frac{\rho_p-\rho_f}{\rho_f}\frac{gd_p}{{v_t}^2}, \quad \text{(10.1)}$$

where

g = gravity acceleration,
d_p = particle diameter,
v_t = terminal particle-settling velocity,
ρ_p = proppant density, and
ρ_f = fluid density.

Terminal settling velocity is the velocity of a single particle in an infinite medium.

The Reynolds number, N_{Rep}, is defined as

$$N_{Rep}=\frac{d_p v_t \rho_f}{\mu}. \quad \text{(10.2)}$$

Depending on the value of the Reynolds number, three relationships can be identified between drag coefficient and Reynolds number.[1]

Stokes region:

$$C_D=\frac{24}{N_{Rep}}, \; N_{Rep}<0.0. \quad \text{(10.3)}$$

Substituting Eqs. 10.1 and 10.2 into Eq. 10.3 yields

$$v_t=\frac{gd_p^2(\rho_p-\rho_f)}{18\mu}. \quad \text{(10.4)}$$

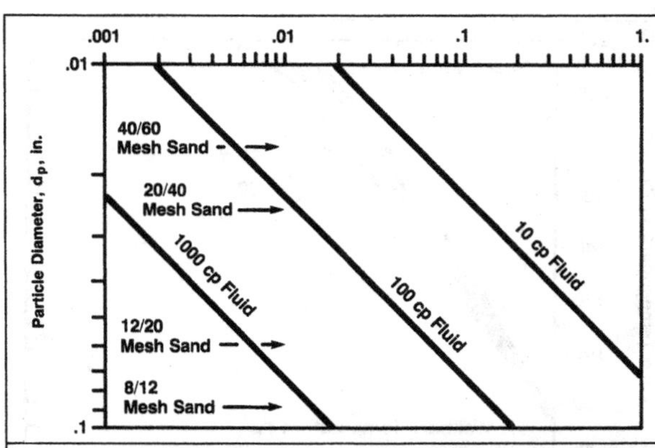

Fig. 10.1—Terminal settling velocity as a function of particle diameter and fluid viscosity.

Intermediate region:

$$C_D=\frac{18.5}{N_{Rep}^{0.6}},\ 2<N_{Rep}<500. \quad (10.5)$$

Making the same substitutions as above yields

$$v_t=\left[\frac{0.072(\rho_p-\rho_f)}{\rho_f^{0.4}}\frac{gd_p^{1.6}}{\mu^{0.6}}\right]^{0.71}. \quad (10.6)$$

Newton's region:

$$C_D\simeq 0.44,\ 500<N_{Rep}<200{,}000. \quad (10.7)$$

Substituting Eq. 10.1 into 10.7 will yield the equation for terminal settling velocity for very high values of Reynolds number. Such cases will very seldom be encountered inside hydraulic fractures.

Eqs. 10.4 and 10.6 show that particle-settling velocity increases with increasing particle diameter or density and decreases with increasing fluid viscosity or density.

Non-Newtonian Fluids. Most fracturing fluids have a non-Newtonian behavior. For engineering computations, it is usually assumed that these fluids behave according to the power-law model, where rheological behavior of the fluid is characterized by

$$\tau=K\dot{\gamma}^{n}, \quad (10.8)$$

where

- τ = shear stress,
- $\dot{\gamma}$ = shear rate,
- n = flow-behavior index, and
- K = fluid-consistency index.

For power-law fluids one can define an apparent viscosity, μ_a, corresponding to viscosity for Newtonian fluids. The apparent viscosity can be computed from

$$\mu_a=K(\dot{\gamma})^{n-1}. \quad (10.9)$$

The apparent viscosity of power-law fluids is a function of the prevailing shear rate.

For a particle settling in a fluid, the shear rate, $\dot{\gamma}_p$, is

$$\dot{\gamma}_p=3v_t/d_p. \quad (10.10)$$

The Reynolds number can be computed by substituting $\dot{\gamma}_p$ from Eq. 10.10 into Eq. 10.9 to get apparent viscosity and then substituting the resultant μ_a for μ in Eq. 10.2. This will give

$$N'_{Rep}=\frac{d_p{}^n v_t{}^{2-n}\rho_f}{3^{n-1}K}. \quad (10.11)$$

The expression for settling velocity can be derived in the same manner as for Newtonian fluids. It has the form

$$v_t=\left[\frac{gd_p^{n+1}(\rho_p-\rho_f)}{18K(3)^{n-1}}\right]^{1/n},\ N_{Rep}<0.1. \quad (10.12)$$

A similar expression can be derived for higher values of Reynolds number. Fig. 10.1 shows the variations of v_t with fluid viscosity and particle diameter. Note that 40/60 sand in 10-cp [10-mPa·s] fluid settles at a velocity of approximately 0.4 ft/sec [0.12 m/s], and in 100-cp [100-mPa·s] fluid at 0.004 ft/sec [0.0012 m/s], which is a much slower velocity.

Experimental Results. Most of the available experimental results are related to the application of Stokes' law to describing observed particle-settling rates. Hannah and Harrington[2] measured particle-settling rates in crosslinked water-based fracturing fluid. Their test cell consisted of two concentric cylinders, with the outer cylinder rotating. The settling rate was measured in the gap between the two cylinders. They concluded that their results are best described by Stokes' law.

Novotny[3] also used rotating concentric cylinders for his experimental work. He derived the Stokes law for non-Newtonian fluids in a form slightly different from Eq. 10.12. The difference comes from Novotny's expression for shear rate around the particle, which he described by

$$\dot{\gamma}_p=v_t/d_p.$$

Novotny recommended using an effective shear rate on the particle that is the vector sum resulting from proppant settling (the expression above) and the shear rate imposed by fluid motion, $\dot{\gamma}_I$:

$$\dot{\gamma}_e=[(v_t/d_p)^2+\dot{\gamma}_I{}^2]^{1/2}. \quad (10.13)$$

The apparent viscosity of the fluid in Eq. 10.2 is calculated for the above shear rate, and resulting N_{Rep} is substituted into Eq. 10.3 to give the settling velocity. Novotny also considered the hindering effect of fracture walls on particle settling and presented correlations that depend on particle diameter, fracture width, and Reynolds number. His experimental data also showed that in concentrated slurries the particles settle more slowly because of the increase in slurry viscosity and density. He suggested using

$$v_{ts}/v_t=\phi^{5.5},\ N'_{Rep}<2 \quad (10.14)$$

or

$$v_{ts}/v_t=\phi^{3.5},\ 2<N'_{Rep}<500, \quad (10.15)$$

where v_{ts} = settling velocity in the slurry and ϕ = slurry porosity.

Roodhart[4] used parallel plates 0.2 to 0.8 in. [0.5 to 2 cm] apart in his experimental work. From these data, he recommended adding a term corresponding to zero shear viscosity to the computed settling velocity. The corresponding settling velocity is given by

$$v_t=\frac{gd_p{}^2(\rho_p-\rho_f)}{18\mu_0}+\frac{gd_p{}^2(\rho_p-\rho_f)}{18K(\dot{\gamma}_z)^{n-1}}, \quad (10.16)$$

where $\dot{\gamma}_z$ = shear rate corresponding to lowest effective viscosity location and μ_0 = effective viscosity at zero shear.

Eq. 10.16 is applicable to shear rates below 25 seconds^{-1}. At higher shear rates, Roodhart recommended using a three-parameter model to describe fluid behavior.

Clark and Guler[5] and Clark and Quadir[6] studied proppant transport in crosslinked and uncrosslinked fracturing fluids. They reported that their experimental work fit a modified form of the Stokes' law.

$$v_e=av_t, \quad (10.17)$$

where

v_t = settling velocity calculated from Stokes' law,
v_e = experimental settling velocity, and
a = correlation coefficient.

The coefficient a depends on the type of gel and crosslinker. Clark and Guler gave three expressions, two for hydroxypropyl guar (HPG) crosslinked with borate ion and one for uncrosslinked gels. Acharya[7] suggested the following modifications to Stokes' law:

$$v_t=\left[\frac{(\rho_p-\rho_f)g d_p^{n+1}}{18Kf(n)}\right]^{1/n},\ N'_{\mathrm{Rep}}<2, \quad (10.18)$$

where

$$f(n)=3^{(3n-3)/2}\left[\frac{33n^5-63n^4-11n^3+97n^2+16n}{4n^2(n+1)(n+2)(2n+1)}\right], \quad (10.19)$$

and

$$v_t=\frac{3\rho_f}{4(\rho_p-\rho_f)g d_p}\left[\frac{24f(n)}{N'_{\mathrm{Rep}}}+\frac{f_2(n)}{N'_{\mathrm{Rep}}f_3(n)}\right]^{-1/2}, \quad (10.20)$$

$$2<N'_{\mathrm{Rep}}<500,$$

$$f_2(n)-10.5n=3.5,$$

and

$$f_3(n)-0.32n=0.13.$$

Shah[8] conducted extensive experiments on settling of various particles in stagnant and in moving columns of fluid. He found that his experimental results deviated from Stokes' law, but could be expressed by the empirical expression

$$Y=AX^B+C, \quad (10.21)$$

where

$Y = (C_D^{2-n}N'^{2}_{\mathrm{Rep}})^{1/2}$,
$X = N'_{\mathrm{Rep}}$, and
A,B,C = constants.

He gave the values of A, B, and C for different fluids. A plot of Y vs. X would consist of a series of converging lines, each line corresponding to a different n. Because Y, as defined by Eq. 10.21, is independent of v_t and depends only on fluid and particle properties, the plot, or Eq. 10.21, becomes a convenient way of determining the settling velocity without trial and error. Fluid parameters n and K in Shah's correlations are those measured with a Fann Model 35™ viscometer.

Dunand and Soucemarianadin[9] used a cylindrical model for particle-settling experiments. They observed good agreement between their data and an Ellis model (three-parameter) representation of the fracturing fluid. They also observed that concentrated slurries settled two to three times faster than single particles. Kirby and Rockefeller[10] measured static settling of slurries from 0 to 12 lbm/gal [0 to 1438 kg/m^3] in concentration in crosslinked and uncrosslinked fracturing fluids. They also observed a clustering effect that resulted in faster settling velocities. Gottschling *et al.*[11] experimentally observed that nitrogen gas will transport 20/40-mesh sand if the fluid is moving at a high velocity.

Effect of Fluid Viscoelasticity. In the discussion so far, it has been implicitly assumed that fracturing fluids behave like a power-law model. For most uncrosslinked gels, or some highly sheared crosslinked gels, this is a reasonable assumption.

Most crosslinked fracturing fluids are viscoelastic; i.e., their rheological properties are time- and/or shear-dependent. During a fracturing treatment, these fluids may lose part of their viscosity while they are exposed to high shear rates during flow inside the wellbore. Inside the fracture, the shear rate is reduced greatly and fluid returns to laminar flow. Some crosslinked fluids will rebuild some of their viscosity inside the fracture. The rheological properties of such fluids are therefore a function of time and shear.

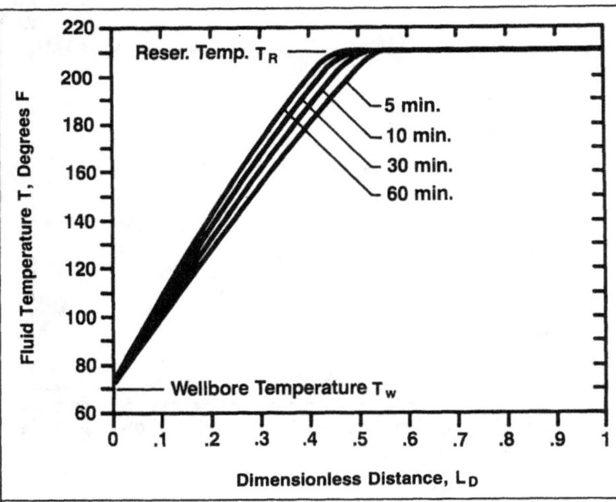

Fig. 10.2—Temperature distribution inside the fracture as a function of distance and time.[13]

The study of viscoelasticity of crosslinked fracturing fluids is very complex. Although the industry recognizes the need for viscoelastic characterization of fluids, a definitive solution to the problem has not yet been formulated. However, some general observations have been made regarding particle settling in viscoelastic fluids.

1. For low Reynolds numbers ($N'_{\mathrm{Rep}}<2$), the fluid elasticity does not affect settling rates.[5] A power-law model is adequate for describing fluid rheology.
2. For intermediate Reynolds numbers ($2<N'_{\mathrm{Rep}}<500$), fluid viscoelasticity will increase the settling rate of the proppant.[7]

Until fluid viscoelasticity is better understood, experimental results offer a reasonable tool for computing particle settling inside the fracture. Emphasis should be placed on adequate modeling of the shear history of the fluid before its rheological characteristics are measured.

10.4 Temperature Distribution Inside the Fracture

In most fracturing treatments, producing-formation temperature is higher than ambient conditions of fracturing fluid. Thus, fluid heats up as it moves inside the fracture. The heating of fluid may change its rheological behavior and consequently the distribution of proppant.

Fluid temperature inside the fracture can be computed from several correlations.[12-14] These indicate that fluid temperature is a function of formation and fluid thermal conductivities, specific heats, static bottomhole temperature (BHT), and surface earth and fluid temperatures. The exact formulations can be found in Refs. 12 through 14 and will not be repeated here. In general, by the time fluid has traveled approximately halfway through the length of fracture, it is essentially at BHT (Fig. 10.2).

BHT is one of the key parameters considered in the selection of a reasonably stable fluid for the duration of the treatment such that, to the extent possible, fluid has sufficient viscosity to transport proppant away from the wellbore. Figs. 10.3 and 10.4 show variations of n and K for a crosslinked HPG gel at 200°F [93°C]. Because K has a stronger influence on fluid apparent viscosity, these figures demonstrate the reduction of viscosity with both time and high temperature.

At high BHT's, fluid moving inside the fracture gradually heats up. If an increase in temperature reduces fluid viscosity, proppant may settle faster to the bottom of the fracture.

Once fluid injection stops, the well is usually shut in. During this time, the cooler fluid near the wellbore will also get to BHT. The mathematical expressions for fluid heat-up during the shut-in period are given by Poulsen and Lee.[14] Heating of fluid causes a viscosity

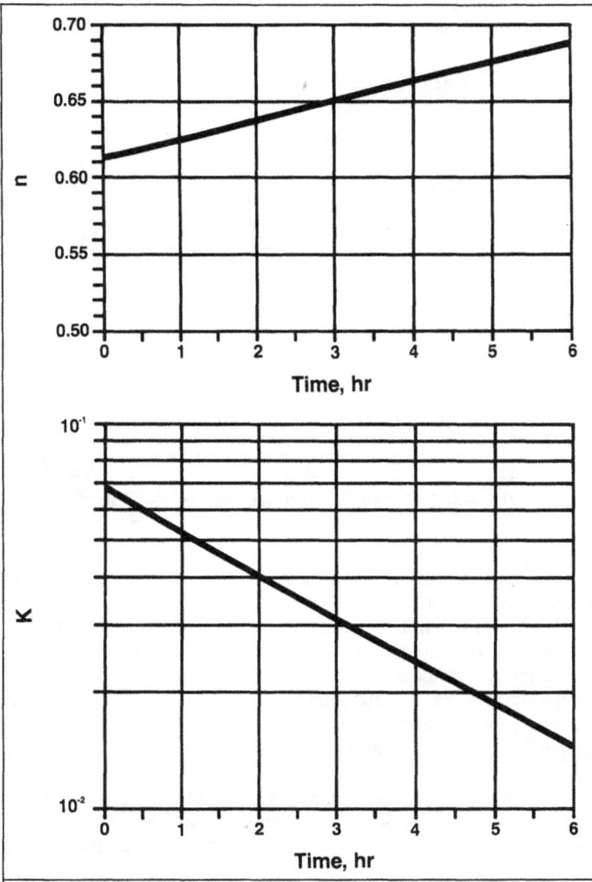

Fig. 10.3—Example variations of *n* and *K* of fracturing fluid with time at a given temperature.[14]

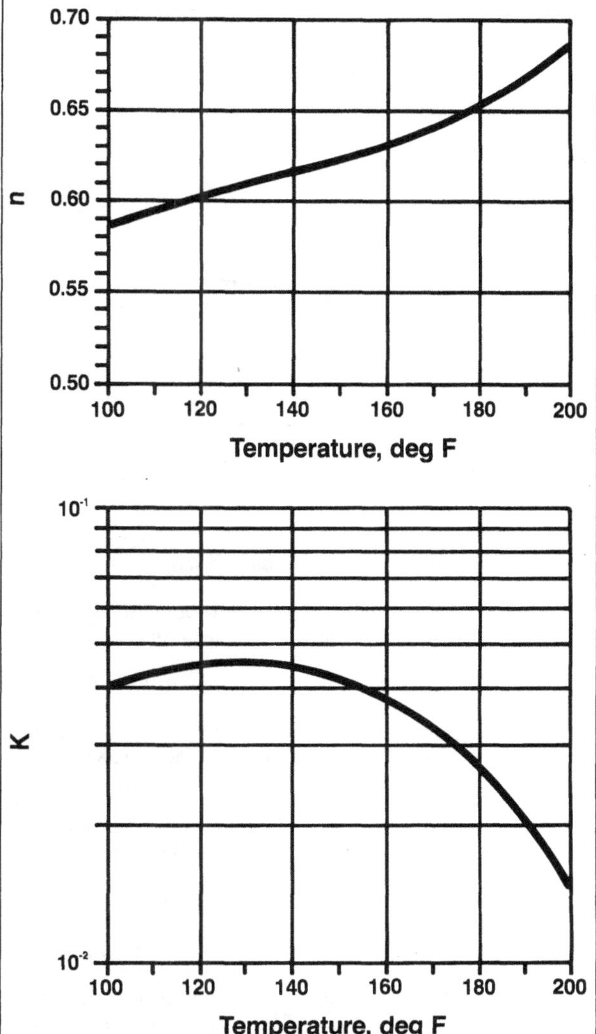

Fig. 10.4—Example variations of *n* and *K* of fracturing fluids with temperature.[14]

reduction and faster proppant settling. Proppant will continue its fall until it either becomes trapped between closing fracture walls or settles to the bottom.

Effect of Proppant on Fluid Temperature. The correlations presented in Refs. 12 through 14 ignore the effect of proppant on fluid temperature. Actually, the influence of proppant can easily be included in the computations by replacing thermal properties and mass of fluid in the fracture with those of the slurry.

These computations show that the amount of heat absorbed by proppant is slightly different from that of the fluid it replaces. Except for very high proppant concentrations, one can generally assume the temperature of slurry to be the same as an equal-volume fluid.

10.5 Variations of Shear Rate Inside the Fracture

Shear rate of the fluid inside the fracture is influenced by the following factors.

1. The position of the fluid perpendicular to the fracture walls. Fracturing-fluid velocity is zero at the walls (ignoring leakoff) and is maximum halfway between the walls. Velocity profile for a power-law fluid is[3]

$$v=\bar{v}\frac{(2n+1)}{(n+1)}\left[1-\left(\frac{2y}{w}\right)^{(n+1)/n}\right], \quad \text{(10.22)}$$

where

v = fluid velocity in the fracture,
$\bar{v}$ = average fluid velocity in the fracture,
y = position perpendicular to fracture walls ($y=0$ describes fracture centerline halfway between the walls), and
w = fracture width.

Shear rate is expressed by

$$\dot{\gamma}_f=\frac{dv}{dy}=\frac{2\bar{v}}{w}\left[\frac{(2n+1)}{n}\right]\left(\frac{2y}{w}\right)^{1/n}. \quad \text{(10.23)}$$

Eq. 10.23 shows shear rate to be zero at $y=0$, which is halfway between fracture walls, and maximum at $y=w/2$, which is at the fracture wall.

If, at the wellbore, fracture height is h, its average width is $\bar{w}$, and injection rate is i, then

$$\bar{v}=i/2h\bar{w} \quad \text{(10.24)}$$

and

$$\dot{\gamma}=\frac{i}{h\bar{w}^2}\left(\frac{2n+1}{n}\right)\left(\frac{2y}{\bar{w}}\right)^{1/n}. \quad \text{(10.25)}$$

Eq. 10.25 shows that at the fracture intersection with the wellbore, shear rate is most sensitive to fracture width. As width increases, shear rate decreases. Because large widths are generated with higher-

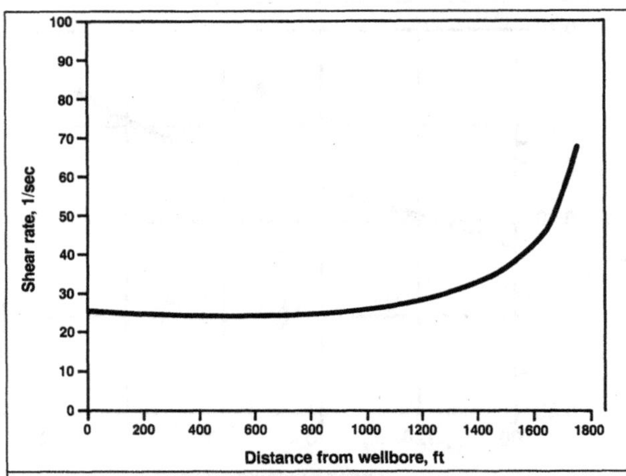

Fig. 10.5—Example variations of shear rate of fracturing fluid in the fracture with distance.

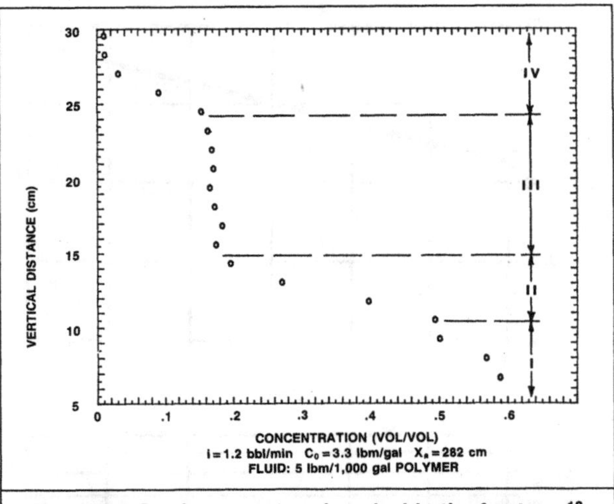

Fig. 10.6—Sand-transport regions inside the fracture.[18]

viscosity fluids, higher-viscosity fluids undergo less shear in the fracture.

Shear rate is also influenced by the flow-behavior index, n. As n increases to approach unity (Newtonian fluids), shear rate decreases.

2. The position of fluid along the fracture. Eq. 10.25 can be used for any point along the fracture, provided that i, h, and $\bar{w}$ are replaced by their local values. Most design computations assume h to be constant. The localized values of i and $\bar{w}$ decrease slowly along the first half of fracture near the wellbore. Eq. 10.25 shows that $\dot{\gamma}$ is more sensitive to $\bar{w}$. Near the fracture tip, $\bar{w}$ becomes very small. In these regions, shear rate increases rapidly, despite small values of flow rate. Fig. 10.5 shows a typical example of shear-rate variations along the fracture length.

3. Fluid leakoff. The direction of fluid movement for leakoff is perpendicular to the fracture walls. This would result in a higher fluid velocity at the fracture wall but will leave centerline velocity unchanged.

At any given time, fluid-leakoff rate is highest near the fracture tip and lowest at the wellbore. Therefore, leakoff effect is most pronounced near the fracture tip.

4. Temperature. For most fracturing fluids, n approaches unity as fluid temperature is increased. Eq. 10.25 shows that as n increases, shear rate will decrease. Because the highest fluid temperatures are encountered near the fracture tip, increases in temperature will have a small decreasing effect on shear rate.

Effect of Fluid Shear Rate on Particle Settling. For Newtonian fluids ($n=1$), viscosity is independent of shear rate. Therefore, proppant encounters the same viscosity regardless of position. In laboratory experiments, fracture walls were modeled with plexiglass sheets to allow visual inspection of particles during their fall in the fracture. In these experiments, plexiglass sheets were not scratched by sand particles during their frequent use. This observation indicates that in Newtonian fluids, proppants move to where local velocity is at a maximum, which is at fracture centerline. In these tests, the horizontal velocity of particles was found equal to the average fluid velocity in the fracture (i.e., the particles and the fluid travel at the same velocity).

In non-Newtonian fluids, experiments indicate that particles also tend to stay away from the center of the model where shear rate would be zero. As each particle encounters a lower viscosity (as a result of higher shear rate), its falling velocity increases. For example, when two particles are released at the same point just seconds apart, one would usually observe that the second particle follows the path of the first particle and gradually catches up to it.

In fracture treatments, width is usually too narrow to allow significant migration of the proppant toward the wall or centerline, although minor movements might occur. In many treatments where low proppant concentrations were used, the introduction of proppant did not increase fluid bottomhole pressure (BHP). This would imply that particles did not significantly touch fracture walls. In other treatments when higher proppant concentrations were used, one could see an immediate gradual increase in BHP, indicating wall interference to particle movement. These observations imply that fracture geometry and dimensions are such that shear-rate variations perpendicular to the fracture walls will not cause significant proppant migration.

In laboratory measurements, it is important for the non-Newtonian fluids to be in motion. Settling velocities measured in stagnant fluids have been observed to be much lower than when fluid was in motion.[3]

10.6 Proppant Deposition Inside the Fracture

Numerical techniques are usually used for computing proppant deposition inside the fracture. As a final step, one usually computes fracture geometry (width, length, and height) as a function of time for the specific fracture parameters.

Two basic methods of computing fracture geometry are in widespread use by the industry. One is Perkins and Kern's[15] method, which assumes constant height, elliptical cross section in a vertical plane, and a polynomial profile in a horizontal plane. The other is Khristianovich and Zheltov's[16] technique, which assumes constant height, almost elliptical fracture cross section in a horizontal plane, and rectangular cross section in a vertical plane. If the fracture geometry is computed by the Perkins and Kern method, then in proppant-transport computations, the elliptical fracture profile in the vertical plane is replaced by an equivalent rectangular shape.

In numerical solution of proppant transport, the hydraulic fracture is divided into small elements (Appendix E). In each element, the local fracture width, fracture volume, and leakoff volume are computed at different times. The injected fluid is also divided into small increments. The position of each fluid increment with regard to length elements is established through mass-balance principles.

As each fluid increment moves along the fracture length, proppant is assumed to move horizontally with the fluid containing it. In the vertical direction, proppant is assumed to move downward as a result of gravity. Vertical settling of the particle can be computed by any of the equations presented earlier in this chapter. If proppant falls enough to reach an existing bed or fracture bottom, it is assumed to stop moving and to settle into the bed. For particles still in suspension, one can compute the increase in concentration resulting from leakoff, and correct the settling velocity accordingly for the next timestep.

Proppant deposition inside hydraulic fractures can be classified into three broad categories, depending on whether proppant forms into a bed (high settling velocity), stays mostly in suspension (low settling velocity), or a combination of both.

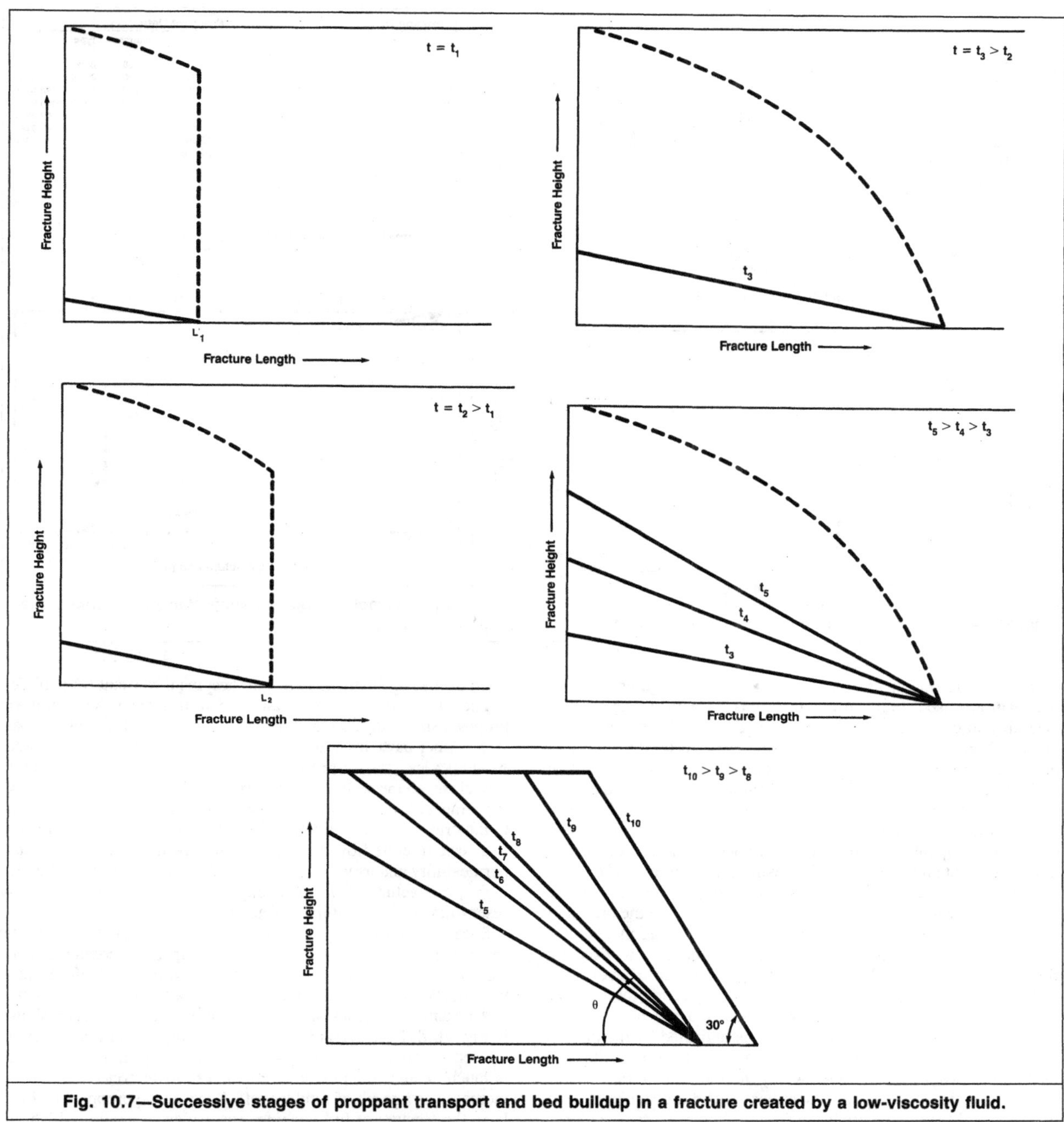

Fig. 10.7—Successive stages of proppant transport and bed buildup in a fracture created by a low-viscosity fluid.

Proppant Deposition With Low-Viscosity Fluids. In low-viscosity fluids, proppant-settling rate is high. Consequently, proppant soon falls to the fracture bottom to form a bed. Examples of low-viscosity fluids include linear gels (at low gel concentrations) or very-shear-sensitive crosslinked fluids.

Most of the sand-transport computations in low-viscosity fluids are based on the experimental and theoretical work of Babcock *et al.*[17] From their experimental work, they identified four regions in the fracture (Fig. 10.6). Region 1 is a stationary bed of proppant that has settled to the fracture bottom. This bed is loosely packed and has a proppant volumetric concentration of 0.5 to 0.6. Region 2 is a fluidized bed (Babcock *et al.* call it a saltation layer) where proppant particles roll over the existing bed before settling on it. Region 3 consists of proppant partices that settle because of gravitional forces and are resisted by fluid viscosity. The average proppant concentration in this region is approximately the same as that of the injected slurry. Region 4 does not contain any proppant. A paper by Medlin *et al.*[18] recently confirmed that observation. In this model, the effect of natural or stimulated turbulence at the fracture entrance is taken into consideration. Consequently, Region 4 contains particles carried by fluid turbulence.

Proppant-transport computations in hydraulic fractures usually ignore Regions 2 and 4 during early computations. Region 2 is usually very narrow (a few inches), and fluid turbulence created by perforations (Region 2) dies very rapidly in the fracture. As a result, these computations consider only the suspended and the deposited proppant.

Babcock *et al.* observed that as the proppant bed builds up inside the fracture, the cross-sectional area of fracture open to slurry flow decreases, resulting in higher fluid velocity. They postulated that a situation may arise where the fluid velocity may become so large that the particles may be perfectly transported by the fluid. In this situation, the height of the proppant bed will not increase.

Medlin *et al.*[18] and Biot and Medlin,[19] using their experimental and theoretical work, argued that an equilibrium bank height is very seldom established inside the fracture. They note that, aside from viscous drag, neither fluidized bed nor turbulence can create a perfect transport environment inside a large fracture.

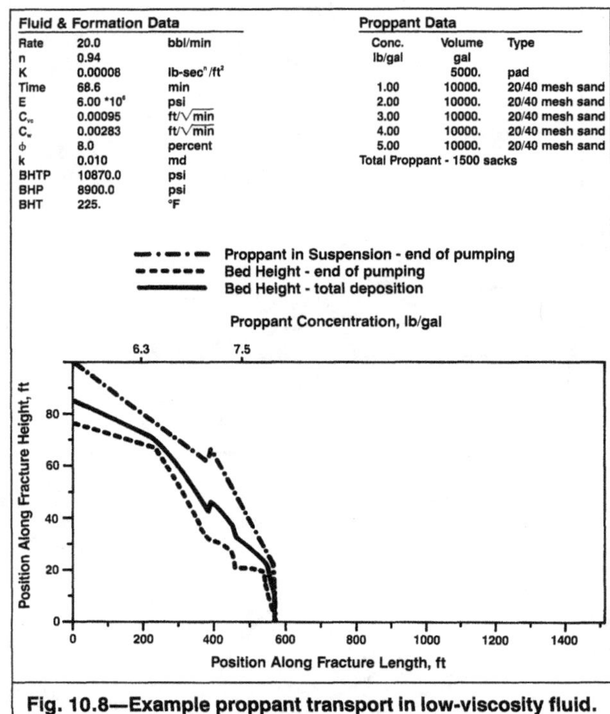

Fig. 10.8—Example proppant transport in low-viscosity fluid.

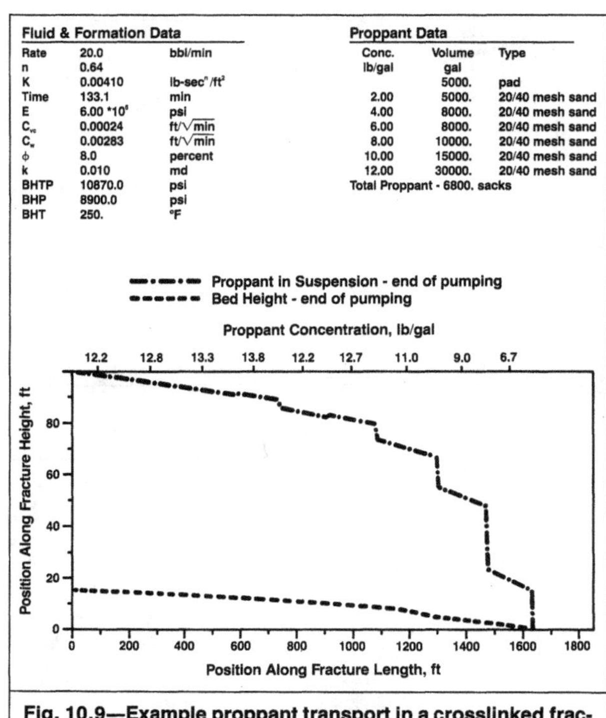

Fig. 10.9—Example proppant transport in a crosslinked fracturing fluid.

In hydraulic fracturing, especially when very large amounts of proppant are used, one can obtain propped fracture lengths considerably in excess of what viscous-drag computations would indicate. In addition, without a mechanism to account for perfect proppant transport, one would compute a screenout (caused by filling of the fracture near the wellbore) to occur at much smaller amounts of proppants than have actually been pumped inside the fracture. As a result, a mechanism (other than viscous drag) must exist to carry the proppant away from the wellbore. Medlin *et al.* recognized this point and offered possible explanations; however, no firm resolution of this point exists in the literature. For the time being, it is usually assumed that turbulence can occur above the proppant bed and consequently can explain the field observations.

Numerical computations for particle transport in low-viscosity fluids are usually based on the following sequence of events.

At time t, slurry has moved to a distance d_1 inside the fracture (Fig. 10.7). During this time, proppant contained in the upper portion of the slurry has settled down, as indicated by the dashed line. Proppant in the lower portion of the fracture has settled to its bottom, forming a bed, as shown by the solid line in Fig. 10.7a.

At a later time, $t_2 > t_1$, proppant has moved a distance d_2 into the fracture, has settled more in the fracture, and has formed a higher and longer bed (Fig. 10.7b).

The proppant settling and bed growth continue until those particles entering the top of the fracture at early time fall to its bottom (Fig. 10.7c). Continued pumping of slurry after this time will not substantially change the proppant travel distance in the fracture, but it will increase proppant bed height (Fig. 10.7d).

As the proppant bed continues to grow, it will progressively reduce the cross section open to flow, resulting in higher slurry velocities. If the proppant bed grows high enough, it can increase fluid velocity to a point where it can carry the proppant in perfect transport. The height corresponding to this condition is called "equilibrium bed height." The mathematical relationships between fluid, fracture, and proppant properties and equilibrium bed height are given in Appendix F.

As more proppant is introduced into the fracture, the bed continues to grow in length at the equilibrium height (Fig. 10.7e). Proppant starts to roll over the existing bed and increase the bed angle, θ, as shown in Fig. 10.7e.

At this point, bed growth is mainly in the form of increasing θ, until θ reaches the angle of repose (approximately 30° [0.5 rad]). Then the bed again grows in length, keeping θ constant (Fig. 10.7e).

The above discussion pertains to a treatment with uniform proppant size. Industrial treatments sometimes use different proppant sizes—finer particles at the beginning and coarser ones at the end. In such cases, coarser proppant will settle faster and therefore will stay closer to the wellbore. The proppant bed will have a step-like appearance. Proppant bed thickness formed in this case will be equal to the created fracture width, which will be large. This will result in a large fracture flow capacity, defined as the product of fracture permeability and its width, kw. Large kw values are best suited for high postfracture flow rates expected from high-permeability reservoirs or from naturally fractured formations.

Because fluid usually does not carry the proppant close to the fracture tip, there is not much danger of proppant becoming bridged close to the tip because of narrow fracture width. Therefore, these treatments usually do not require large pad volumes.

An example of an actual design with low-viscosity fluids is shown in Fig. 10.8. Note that at the end of pumping, most of the proppant has settled into a bed at the fracture bottom. Propped fracture length is much smaller than created length. Proppant concentration has increased as a result of fluid leakoff. Because the proppant bed will keep the fracture at full opening, suspended proppant, which has relatively high settling velocity, will soon drop on top of the existing bed and add to its height.

Proppant Deposition With High-Viscosity Fluids. High-viscosity fluids allow little proppant settling, keeping it in suspension for a long time. Proppant covers a large area of the fracture and, consequently, of the producing formation. One can usually inject larger amounts of proppant inside the fracture and carry it farther away from the wellbore than in a similar equilibrium bank treatment. Crosslinked gels, stable foams, and emulsions are examples of high-viscosity fluids.

The high-viscosity slurry moving inside the fracture does not change much during its early life. While proppant concentration may increase slightly because of leakoff, this increase is usually minor.

The proppant weight, W_p, contained in an area of the fracture, A, is given by

$$W_p = A\bar{w}C_p,$$

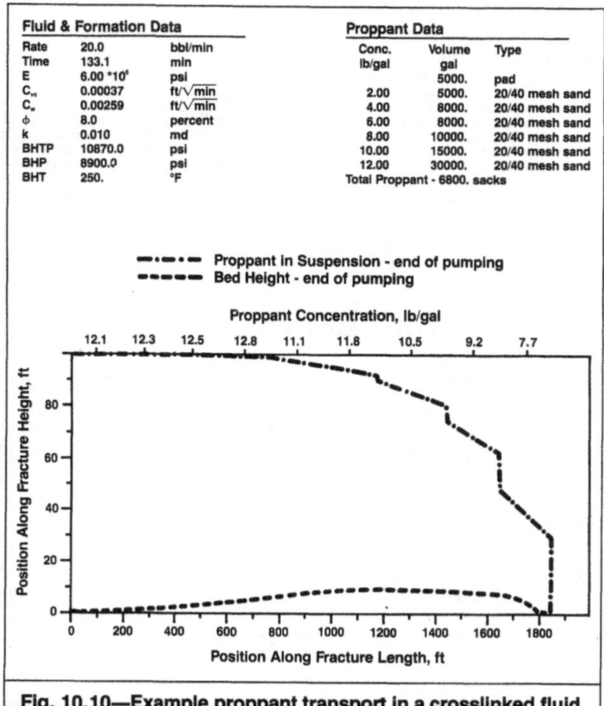

Fig. 10.10—Example proppant transport in a crosslinked fluid, including temperature effect on fluid viscosity.

where $\bar{w}$=average fracture width across area A and C_p=proppant concentration in the slurry. Thus, weight of proppant per unit surface area of fracture would be $\bar{w}C_p$. This value usually decreases as slurry moves away from the wellbore because $\bar{w}$ decreases faster with distance from wellbore than C_p increases because of leakoff. This phenomenon, coupled with the usual increase in proppant concentration during treatment, yields a fracture with successively smaller amounts of proppant along its length and thus lower flow capacity. This could result in a fracture tip that, although propped, does not make a significant contribution to production. Thus, selection of propping-agent concentrations used with high-viscosity fluids is important. In fact, one can determine a desirable optimum flow capacity inside the fracture on the basis of reservoir engineering and then determine the pumping schedule that would yield it.[20]

In computing proppant transport with high-viscosity fluids, one needs to consider other physical aspects of the treatment. Leakoff rate is higher close to the fracture tip than anywhere else. Therefore, proppant concentration can get high if the slurry travels to the tip area and stays there. The remedy is to increase pad volume and to prevent long residence of slurry at the fracture tip. Another potential hazard is for sand to get trapped near the tip because of narrow fracture width. The remedy again is to increase pad volume. Fig. 10.9 shows an example fracture design with a crosslinked fracturing fluid. There is a small proppant bed built at the fracture bottom at the end of pumping. Most of the proppant is in suspension. It has traveled far inside the fracture, and its concentration has increased slightly as a result of leakoff.

Static BHT in this example is 250°F [120°C]. The n and K used in these calculations correspond to an average temperature for the entire treatment.

As discussed earlier, fluid temperature inside the fracture increases as the fluid travels away from the wellbore. At high BHT's, fluid viscosity may gradually decrease with time, thus reducing its ability to transport proppant.

When the example in Fig. 10.9 is recomputed, allowing fluid rheology to change with time and temperature, deposition profiles change to what is shown in Fig. 10.10. At an assumed BHT of 250°F [120°C], the selected fluid loses viscosity with time according to relationships depicted in Fig. 10.11. Deposition profiles in Fig. 10.10 show a proppant bed at the end of pumping that reaches its maximum midway along the fracture length. This results from loss of fluid viscosity with time and temperature.

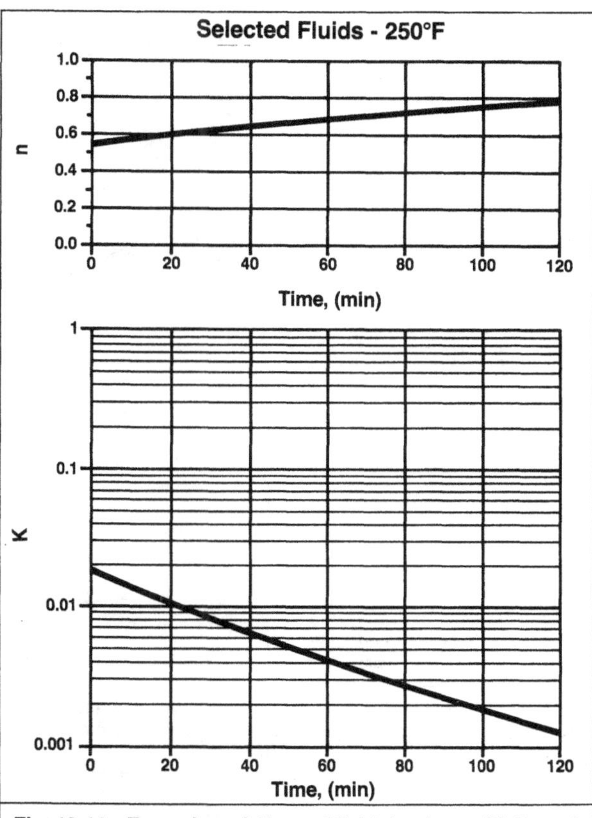

Fig. 10.11—Example variations of fluid rheology with time at a fixed temperature.

When the effect of proppant is considered in temperature computations, the results change very little (Fig. 10.12).

Effect of High Proppant Concentrations on Transport. Ever since the introduction of hydraulic fracturing, attempts have been made to increase the maximum concentration and the amount of proppant pumped into a fracture. Proppant concentrations as high as 20 lbm/gal [2397 kg/m^3] are known to have been pumped during fracturing treatments. Operators continue to see better results with higher concentrations.[21,22]

There are several logical reasons for the use of high proppant concentrations. Conductivity of a propped fracture can deteriorate with time because of the release and migration of formation fines, chemical depositions inside the fracture, rearrangement of proppant, embedment, crushing, or a combination of any of these. In fact, field examples indicate a reduction of effective conductivity even during early fracture life. High concentrations of proppant provide sufficient excess conductivity to compensate for later losses.

Another advantage of high concentrations is related to fracture closure time. Because more of the fracture volume is filled with proppant, the fracture closes earlier. This would result in less proppant settling during fracture closure. Because most perforations are adjacent to producing formations where the proppant is introduced, less settling will leave more proppant in the producing zone.

Walls of hydraulic fractures are usually not smooth; they contain roughnesses and offsets that are likely spots for proppant bridging and trapping. These rough spots increase production because they tend to keep the fracture open and give it much higher conductivity. As proppant concentration increases, the probability of the proppant's getting trapped also increases. As shown later, however, if these traps occur very near the wellbore, they will also increase the probability of proppant flowback or a screenout.

A negative aspect of high proppant concentration may be higher settling velocity of aggregate vs. individual particles. While it is true that a slurry with more proppant has higher viscosity and therefore proppant should fall more slowly in it, laboratory experiments show "lumping," where several particles stay so close

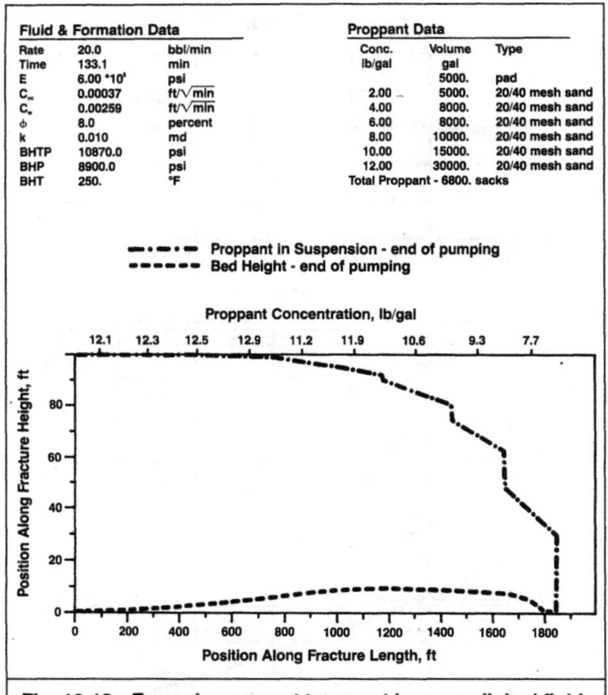

Fig. 10.12—Example proppant transport in a crosslinked fluid, including temperature effect on fluid viscosity and effect of proppant on fracture temperature.

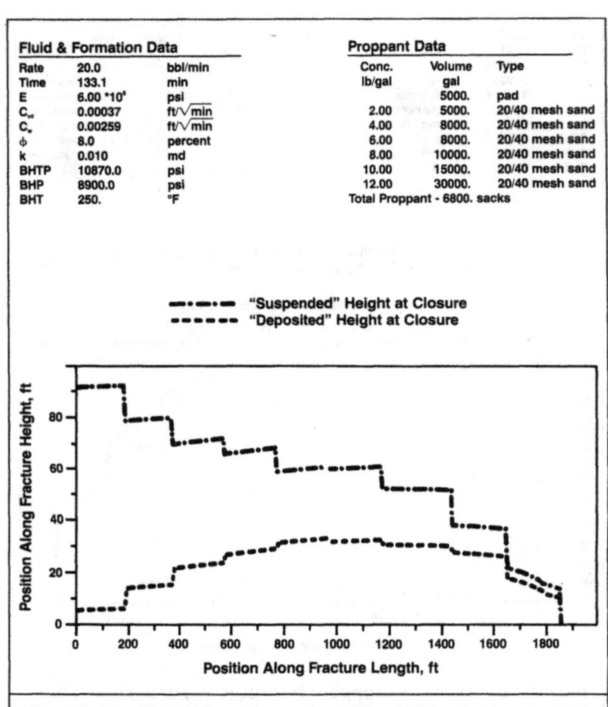

Fig. 10.13—Example proppant profiles in the fracture at the time of fracture closure.

to each other that they behave like a larger particle and therefore settle faster. The mathematics of lumping is not yet well defined, but it does occur in laboratory experiments.

Generally, the field trend has been toward higher proppant concentrations. These treatments generally yield better production increases with a longer life.

Effect of Fracture Closure Time on Proppant Transport. The deposition of the propping agent inside the fracture continues beyond the end of pumping until either all the proppant settles into a bed or the fracture closes and traps the suspended proppant.

Nolte[23] presented methods and equations for the calculation of closure time. Briefly, the technique involves performing a minifracture without a propping agent and flowing the well back after the treatment. By performing the computations presented by Nolte, one obtains a fluid-loss coefficient, closure time, and other fracturing parameters. These numbers are then used to design the main treatment.

The closure time depends on the fluid-loss coefficient. Small leakoff results in long closure times. In fact, in minifractures, closure occurs when the fluid in the fracture leaks into the surrounding rock. In the presence of proppant, the fracture will close earlier than computed through the minifracture process. Although the closure time still depends on fluid leakoff, it also depends on the fraction of fracture volume occupied by proppant. The closure time gets shorter as the volume of proppant increases.

In hydraulic fracturing, it is desirable to distribute the proppant throughout the fracture in the producing formation. This can often be accomplished by keeping the proppant in suspension. But regardless of how well fluid suspends the proppant during injection, if the fracture takes a long time to close, proppant will settle mostly into a bed at the fracture bottom, which may even be in a nonproducing zone.

A rough estimate of fracture closure time with proppant may be obtained by the following method. Note the fluid efficiency (percent of injected volume still in the fracture at the end of pumping) and closure time computed from a minifracture process. Suppose the computed efficiency is 80% and closure time is 4 hours. Compute the bulk volume of the proppant and the percentage of the injected volume it constitutes. Suppose the proppant constitutes 20% of the injected volume. At the end of pumping, approximately 80% of injected volume is still in the fracture, of which 60% is fluid. A rough estimate of closure time would be $(0.6/0.8)\times4=3$ hours. This rough computation points out that two methods are available to reduce closure time: increase the proppant concentration or decrease fluid efficiency.

Higher proppant concentrations are becoming increasingly popular in the oil industry because of the above implications, as well as for other reasons which were discussed earlier.

To decrease fluid efficiency, one needs to increase fluid leakoff. This is somewhat contrary to the early trend in the industry, which was to reduce leakoff for increasing fracture length and decreasing chances of screenout. However, one can strike a reasonable balance between the two. Because long closure times usually occur with viscous fluids, one may be able to let the viscosity of these fluids control leakoff and use only small amounts of fluid-loss additive, enough to create sufficient fracture width.

One exception is foams, which have very low fluid-loss coefficients and consequently high fluid efficiencies. This results in very long computed closure times (on the order of several days) that would be undesirable considering the previous discussion. Yet foams are known to be effective fracturing fluids often used by the industry, sometimes because of their very low leakoff. In foam treatments, however, the well is often flowed back shortly after pumping ends. Although small volumes of proppant may occasionally flow back, the advantages of foam outweigh its disadvantages.

Fig. 10.13 shows the deposition profiles of the earlier example at the time of fracture closure.

Location of Proppant With Respect to Pay Zone. Proppant-transport profiles provide two critical types of data: location of proppant, and its conductivity. Because the usual intent in hydraulic fracturing is to place the proppant in the pay zone, it is important to mark the location of the pay with respect to the fracture (Fig. 10.14). In this example, the bottom 20 ft [6 m] of the fracture is almost completely propped but contributes little to production. The pay zone has trapped suspended proppant at closure. Flow capacity of the fracture in the pay is less than its average flow capacity in the propped area. In the computation of production increase, it is im-

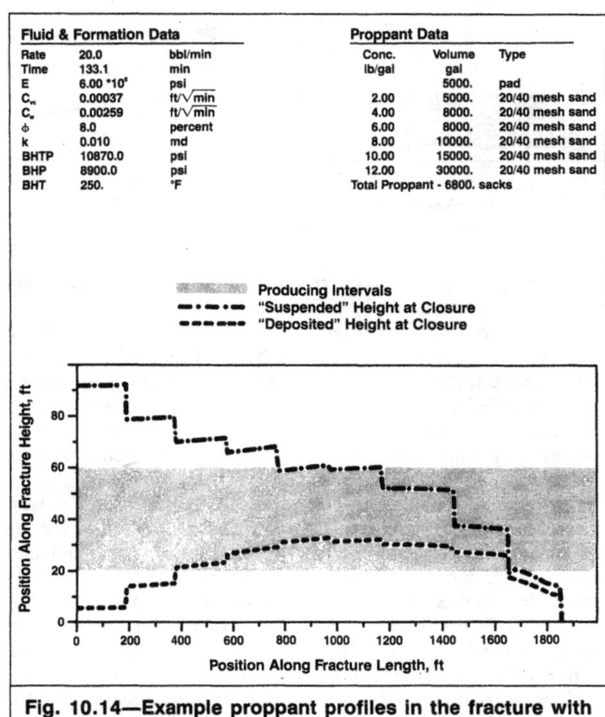

Fig. 10.14—Example proppant profiles in the fracture with respect to pay zone.

portant to use the value corresponding to the pay zone. The flow capacity in the pay zone may be adjusted by varying the proppant concentration in the injected fluid.

Effect of Barriers and Width. Barriers are formations that slow the growth of the fracture, either because they are subject to higher stresses or because of their mechanical properties. If a formation acts as a barrier to fracture growth, then even if the fracture extends into it, the fracture width in it is likely to be narrower than the adjacent nonbarrier zone. The narrower width will create lower flow capacity compared with the wider parts of the fracture. Because the barrier by definition slows the growth of the fracture, it will contain less fracturing fluid and consequently less proppant.[24] In fact, if the barrier is impermeable (as is usually assumed by the petroleum industry) because prepad and pad do not leak into it, it may contain only minor fractions of the total proppant. A more drastic effect will be created if the fracture width in the barrier is so small that it either prevents proppant entry into it or causes proppant bridging.

10.7 Screenout

In the course of a fracturing treatment, fluid pressure needed for further fluid injection occasionally exceeds the limitations of the injection well conduit (tubing or casing), wellhead equipment, or pumping units. This necessitates premature termination of the treatment. This condition is referred to as screenout or sandout. The word draws its origin from its frequent connection with sand as a propping agent. In more technical language, sandout is that condition where pressure needed for further fluid injection is beyond wellbore or equipment tolerance.

Before examining causes of sandouts, one would find it useful to compute the amount of pressure needed for fracture extension. Suppose a created fracture has a surface area A_f and the least in-situ principal stress is σ_1. The force resisting fracturing, F_f, is

$$F_f = A_f \sigma_1 . \quad (10.26)$$

The force needed for fracture extension has to be larger than F_f to provide fracture opening, leakoff, and surface energy.

Suppose the fluid inside the fracture pressurizes area A_1 to a pressure p_1, area A_2 to pressure p_2, etc. Then the force exerted

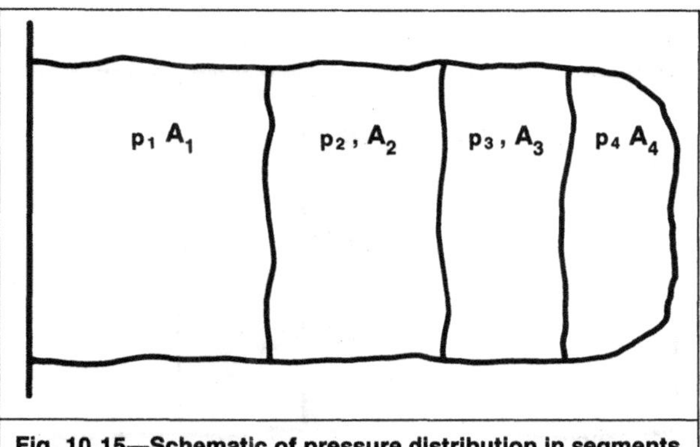

Fig. 10.15—Schematic of pressure distribution in segments of a hydraulic fracture.

by pressurized fluid, F_p, is (Fig. 10.15)

$$F_p = A_1 p_1 + A_2 p_2 + \ldots . \quad (10.27)$$

Because F_p needs to be larger than F_f to cause fracture extension,

$$A_1 p_1 + A_2 p_2 + \ldots > A_f \sigma_1 . \quad (10.28)$$

Obviously,

$$A_f > A_1 + A_2 + \ldots$$

because fluid does not penetrate the entire length of the fracture.[11,20,21]

The inequality (Eq. 10.28) states that the average fluid pressure inside the fracture has to be larger than the least in-situ principal stress.

Fluid pressure inside the fracture has its highest value at the wellbore and decreases in the direction of flow because of frictional losses.

Suppose that for some reason the pressure in region A_2 drops below σ_1. The inequality states that the fluid pressure between A_2 and wellbore has to become large enough to compensate for the drop. Eq. 10.28 can be written as

$$\frac{A_1}{A_f}\frac{p_1}{\sigma_1} + \frac{A_2}{A_f}\frac{p_2}{\sigma_1} + \ldots > 1 . \quad (10.29)$$

If $p_2/\sigma_1 < 1$, then $p_1/\sigma_1 > 1$. The degree of increase is related to how low p_2 drops and how large A_2 is. The more p_2 drops, or the larger A_2 gets, the more p_1 has to increase.

Screenouts usually occur because the fracturing fluid cannot be pumped into the fracture because of some restriction to fluid flow. This restriction will increase the frictional-pressure drop. As a result, the pressure increases behind the restriction. For example, suppose that a vertical 1-ft [0.3-m] -long segment of fracture that is 50 ft [15 m] high and 0.5 in. [1.3 cm] wide is filled with proppant and as a result has a permeability of 150 darcies. Assuming the flow rate of a fluid that has 40-cp [40-mPa·s] viscosity to be 5 bbl/min [0.8 m³/min] through this section, one can calculate frictional-pressure drop resulting from flow to be

$$\Delta p = \frac{Q\mu\Delta X}{1.12kA}$$

$$= \frac{5 \text{ bbl/min} \times 60 \text{ minutes} \times 24 \text{ hours} \times 40 \text{ cp} \times 1 \text{ ft}}{1.12 \times 150 \text{ darcies} \times 50 \text{ ft} \times (0.5/12) \text{ ft}}$$

$$= 823 \text{ psi } [5675 \text{ kPa}].$$

Thus, fluid pressure behind this restriction (e.g., p_1) will be 823 psi [5675 kPa] above the value in front of it (p_2). Fig. 10.16 gives a graphic representation of the problem.

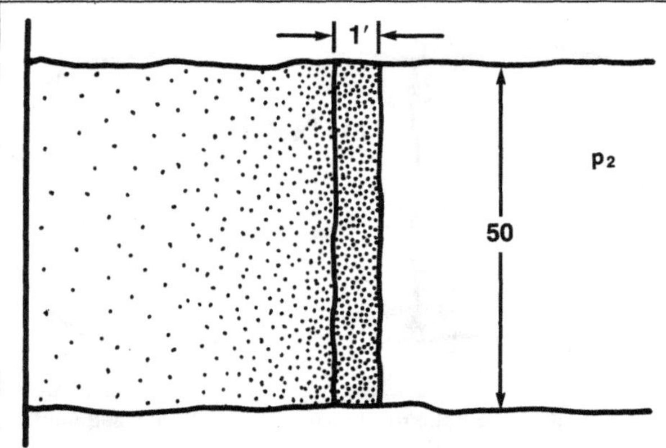

Fig. 10.16—Schematic of proppant blockage inside a hydraulic fracture.

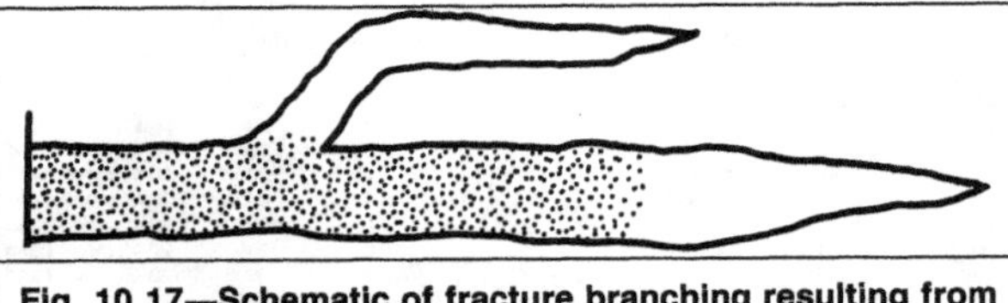

Fig. 10.17—Schematic of fracture branching resulting from proppant blockage.

The maximum value of p_1 is usually fixed and depends on the maximum pressure limitation for the treatment. Thus, Eq. 10.28 states that the closer the restriction to the wellbore, the lower A_1, and therefore the more likely that the restriction could cause a screenout. In fact, if the restriction occurs close to the fracture tip, it is not likely to cause a screenout because A_1 will be so large that it will provide enough force for fracture propagation.

The area near each perforation is a very likely place for proppant to become trapped and packed because the fracture that has initiated from the perforation may have to reorient itself to become perpendicular to the least in-situ principal stress.[25] Suppose a perforation is ½ in. [1.3 cm] in diameter, the fracture is ½ in. [1.3 cm] wide, the fluid has 40-cp [40-mPa·s] viscosity, the flow rate through this perforation is ¼ bbl/min [0.04 m^3/min], and the area 1 ft [0.3 m] in radius around this perforation is filled with proppant and has 150-darcy permeability. One can calculate pressure drop caused by radial flow through this 1-ft [0.3-m] -radius bed to be

$$\Delta p=\frac{i\mu \ln\dfrac{r_e}{r_w}}{7.07kh}$$

$$=\frac{0.25 \text{ bbl/min}\times 60 \text{ minutes}\times 24 \text{ hours}\times 40 \text{ cp}\times \ln(1/0.25)}{7.07\times 150 \text{ darcies}\times (0.5/12) \text{ ft}}$$

$$=452 \text{ psi.}$$

Obviously, such a perforation is not likely to accept any fluid after the packed bed is formed around it. The flow rate through it will be distributed through other open perforations that could be targets for the same event. The packed bed of proppant with a 1-ft [0.3-m] radius around a perforation has a volume of 0.07 ft^3 [0.002 m^3] and at a concentration of 4 lbm/gal [479 kg/m^3] would take less than 10 seconds to form. The severe influence of perforations is obvious.

Screenouts in the field can be divided into two main categories, sudden and gradual. In sudden screenouts, the pressure rise happens very quickly and without prior warnings. These cases can usually be traced to perforations or areas very near the wellbore. At the time of screenout, pressure at the wellbore is much higher than the previous extension pressure. The inequality (Eq. 10.28) states that if the force exerted by the fluid is insufficient to cause fracture extension, then A_1 must be very small compared with A_f and that pressure in most of the fracture (p_2, p_3, etc.) must be reduced to $<\sigma_1$ in magnitude.

In sudden screenouts, the wellbore is sometimes filled with proppant to a point where most or all of the perforations are covered. In these cases, a usual remedy for subsequent treatments is to increase fluid viscosity or rate to increase its ability to carry proppant. In some other cases, the well is opened quickly and fluid is reversed out. There are instances where reverse flow removes the restriction and allows resumption of pumping.

In low-viscosity fluids, Medlin *et al.*[18] reported screenouts in their laboratory experiments caused by the filling of the fracture with proppant. As mentioned earlier (in Sec. 10.6), they argued that both turbulence and fluidized beds in fractures have negligible effect on proppant transport. Therefore, it is possible to screen out a fracture by simply filling it with proppant near the wellbore. Such screenouts will usually occur suddenly.

In gradual screenouts, the rate of pressure increases slowly rises. Restriction to flow could be near the perforations or inside the fracture. If the source is near the perforations, this rise indicates that the number of perforations accepting fluid is gradually decreasing, thus creating high perforation friction pressure. A restriction at the perforation is not likely to be removed with further pumping; therefore, pressure is likely to stay high throughout the remainder of treatment with a high chance of eventual screenout.

If the restriction to flow is inside the fracture, then the fluid is not flowing freely throughout the fracture area. As long as pumping continues, fracture length and width will grow. This could cause removal of some of the restrictions to flow. The rate of pressure increase caused by a given restriction is lower if it is away from the wellbore (larger A_1 in Eq. 10.28).

An increase in fluid pressure inside the fracture makes it possible for a secondary fracture to extend. This can be the opening and extension of an intersected natural fracture. The secondary fracture will soon reorient itself and become parallel to the original fracture (Fig. 10.17). But bypassing of the flow restriction could cause a decrease in fluid pressure. The farther the original restriction is from the wellbore, the greater the chances are of a secondary fracture initiation and a subsequent pressure relief. Therefore, once again, the influence of flow restrictions is stronger near the wellbore.

If pad volumes, fluid viscosities, or injection rates are low or leakoff is high, these could also create a pressure increase at the wellbore. In these cases, fracture width is not large enough to allow free proppant movement. The collision of proppant with fracture walls creates a friction-pressure increase. In addition, any slowing of proppant movement with respect to carrying fluid creates a condition similar to reducing the fracture area open to flow and consequently increases fluid pressure. In these situations, the restriction to flow occurs wherever proppant moves. The creation of secondary fractures does not remove the obstructions because fluid has to travel through some of the original fracture; besides, any secondary fracture will be equally narrow and subject to the same problem.

Thus, in these cases, pressure continues to rise and becomes further aggravated as proppant concentration is increased, until it creates a screenout.

Pressure-Out. In most industrial screenouts, proppant is considered responsible for creating high fluid pressure in one way or another. Although rare, there are cases where the increase in pressure is not associated with a proppant. Here they are called pressure-outs.

Laboratory experiments[26,27] have shown that if one restricts vertical growth of the fracture, a pressure increase will result. The magnitude of this increase depends on the length/height ratio and how severely the vertical growth of the fracture has been reduced. For example, if the hydraulic fracture has stopped completely at an interface, once its length/height ratio exceeds approximately 3/1 to 5/1, the pressure begins to increase at the wellbore. At a given injection rate, the lower the fracture height is, the faster the length/height ratio and the wellbore pressure will increase. Thus, it is possible to advance into a situation where pressure needed to extend

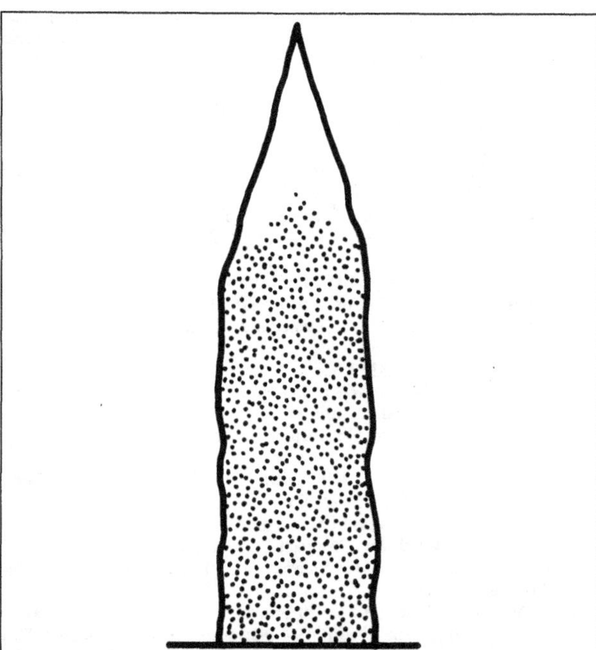

Fig. 10.18—Schematic of loose proppant on top of a proppant bed.

the fracture exceeds allowable wellbore pressure. This can sometimes happen before any proppant is introduced into the fracture.

As fluid pressure inside the fracture increases, its chances of breaking into adjacent zones also increases. If this happens, pressure increase would be relieved.

If fracture height is high enough, the influence of a barrier may not show until after a proppant is introduced into the fracture. In such cases, one would observe a steady pressure rise independent of proppant concentration or amount. Pressure increases that are caused by proppant would be additional.

10.8 Proppant Flowback

When wells are flowed back after fracturing treatments, the fluid occasionally carries small volumes of proppant back to the surface, which would erode surface chokes and valves. Sometimes the fluid cannot carry the proppant to the surface; it settles into the wellbore and occasionally can cover enough of the perforations to restrict production, thus necessitating a cleanup operation. Other potential damage caused by proppant flowback is reduction in propped fracture length, height, or flow capacity, any of which can reduce postfracture production rate. In fact, if enough of the proppant in the first few feet of the fracture near the wellbore flows back to cause complete fracture closure with very little flow capacity, most of the fracture's beneficial effect on production may be lost.

Two conditions can cause sand flowback: a fracture that has not completely closed yet or a fracture that is partially packed with proppant, leaving some of it free to move. If any portion of the proppant is not trapped between fracture walls and is therefore freely in suspension, fluid backflow can bring it back to the wellbore. This may happen if the well is flowed back before the fracture has had enough time to close while the fluid still maintains sufficient viscosity. A common practice to reduce this problem is to give the fracture time to close and to flow it back slowly. Most of the existing practices are based on experience and include leaving the well shut in overnight and flowing it back through a choke to reduce return velocity. In some cases, such as in some foam treatments, the return of a small amount of proppant is considered less damaging to well productivity than leaving the well shut in for a long time. Therefore, proppant flowback is accepted as a necessary risk.

Proppant may also flow back if part of the fracture is packed with it. This type of proppant flowback is more likely to happen during production. Fig. 10.18 graphically describes what happens. Above the packed bed of proppant, the fracture does not close sharply and stays open for a short distance. The upper layers of proppant are thus free to move. As the reservoir fluid (or fracturing fluid during cleanup) moves through the bed of proppant, it can, with sufficient velocity, pick up the upper layer of proppant and move it closer to the wellbore. If the fluid velocity is high enough (high producing oil or gas wells), the proppant can be carried all the way to the surface. With lower production rates, the proppant may move toward the wellbore and settle in other existing cavities in the fracture. After a while, such cavities will fill up, causing the proppant to flow into the borehole. In these cases, proppant flowback occurs later during production and mostly through the bottom perforations, which are also fed by gravitational settling of loose proppant.

As discussed earlier, the simplest method for prevention of proppant flowback is to ensure fracture closure before the fluid is flowed back. But there are cases when proppant flows back anyway or it occurs later during production when the fracture is clearly closed. There are no sure methods known to the industry to prevent proppant flowback completely. Existing methods range from using resin-coated proppants (which are supposed to adhere in the fracture and form a dense aggregate around perforations) to pumping a resin inside the fracture (which is supposed to adhere to existing proppant) to intentionally causing a screenout at the end of the treatment to form a packed bed of proppant around perforations. None of these methods work all the time, and no criteria are available to know when to do what. The successful cases are usually results of experimentation and a little bit of old-fashioned good luck.

Nomenclature

a = correlation coefficient
A = fracture area, ft^2 [m^2]
A_f = created fracture surface area, ft^2 [m^2]
A,B,C = constants in Eq. 10.21
C_D = drag coefficient
C_p = proppant concentration in slurry, lbm/gal [kg/m^3]
C_{vc} = fluid-loss coefficient
C_w = effective fluid-loss coefficient
d_p = particle diameter, in. [cm]
E = Young's modulus of formation
F_f = force resisting fracturing, lbf [N]
F_p = force exerted by pressurized fluid, lbf [N]
g = acceleration of gravity
h = fracture height, ft [m]
i = injection rate, bbl/min [m^3/min]
k = fracture permeability, darcies
K = fluid-consistency index
L = distance slurry moves inside fracture, ft [m]
L_D = dimensionless distance
n = flow-behavior index
N_{Rep} = proppant Reynolds number
N'_{Rep} = proppant Reynolds number for non-Newtonian fluids
p = pressure, psi [kPa]
r_e = external radius, ft [m]
r_w = wellbore radius, ft [m]
t = time
T = fluid temperature, °F [°C]
T_R = reservoir temperature, °F [°C]
T_w = wellbore temperature, °F [°C]
v = fluid velocity in fracture, ft/sec [m/s]
$\bar{v}$ = average fluid velocity in fracture, ft/sec [m/s]
v_e = experimental settling velocity, ft/sec [m/s]
v_t = particle terminal settling velocity, ft/sec [m/s]
v_{ts} = settling velocity in slurry, ft/sec [m/s]
w = fracture width, ft [m]
$\bar{w}$ = average fracture width, ft [m]
W_p = weight of proppant, lbm [kg]
y = position perpendicular to fracture walls
$\dot{\gamma}$ = shear rate
$\dot{\gamma}_e$ = effective shear rate

$\dot{\gamma}_I$ = shear rate imposed by fluid motion
$\dot{\gamma}_p$ = shear rate of particle settling in fluid
$\dot{\gamma}_z$ = shear rate corresponding to lowest effective viscosity location
θ = bed angle, degrees [rad]
μ = viscosity
μ_0 = effective viscosity at zero shear
ρ_f = fluid density
ρ_p = proppant density
σ_1 = least in-situ principal stress
τ = shear stress
ϕ = slurry porosity

References

1. McCabe, W.L. and Smith, J.C.: *Unit Operations of Chemical Engineering,* McGraw-Hill Book Co. Inc., New York City (1956) Chap. 7.
2. Hannah, R.R. and Harrington, L.J.: "Measurement of Dynamic Proppant Fall Rates in Fracturing Gels Using a Concentric Cylinder Tester," *JPT* (May 1981) 909-13.
3. Novotny, E.J.: "Proppant Transport," paper SPE 6813 presented at the 1977 SPE Annual Technical Conference and Exhibition, Denver, Oct. 9-12.
4. Roodhart, L.P.: "Proppant Settling in Non-Newtonian Fracturing Fluids," paper SPE 13905 presented at the 1985 SPE/DOE Low Permeability Gas Reservoirs Symposium, Denver, May 19-22.
5. Clark, P.E. and Guler, N.: "Proppant Transport in Vertical Fractures: Settling Velocity Correlations," paper SPE 11636 presented at the 1983 SPE/DOE Low Permeability Gas Reservoirs Symposium, Denver, March 14-16.
6. Clark, P.E. and Quadir, J.A.: "Proppant Transport in Hydraulic Fractures: A Critical Review of Particle Settling Velocity Equations," paper SPE 9866 presented at the 1981 SPE/DOE Low Permeability Gas Reservoirs Symposium, Denver, May 27-29.
7. Acharya, A.: "Particle Transport in Viscous and Viscoelastic Fracturing Fluids," *SPEPE* (March 1986) 104-10.
8. Shah, S.N.: "Proppant Settling Correlations for Non-Newtonian Fluids Under Static and Dynamic Conditions," *SPEJ* (April 1982) 164-70.
9. Dunand, A. and Soucemarianadin, A.: "Concentration Effects of the Settling Velocities of Proppant Slurries," paper SPE 14259 presented at the 1985 SPE Annual Technical Conference and Exhibition, Las Vegas, Sept. 22-25.
10. Kirkby, L.L. and Rockefeller, H.A.: "Proppant Settling Velocities in Nonflowing Slurries," paper SPE 13906 presented at the 1985 SPE/DOE Low Permeability Gas Reservoirs Symposium, Denver, May 19-22.
11. Gottschling, J.C., Royce, T.N., and Shuck, L.Z.: "Nitrogen Gas and Sand: A New Technique for Stimulation of the Devonian Shale," *JPT* (May 1985) 901-07.
12. Whitsitt, N.F. and Dysart, G.R.: "The Effect of Temperature On Stimulation Design," *JPT* (April 1970) 493-502.
13. Biot, M.A., Massé, L., and Medlin, W.L.: "Temperature Analysis in Hydraulic Fracturing," *JPT* (Nov. 1987) 1389-97; *Trans.,* AIME, **283**.
14. Poulsen, D.K. and Lee, W.S.: "Fracture Design With Time- and Temperature-Dependent Fluid Properties," paper SPE 12483 presented at the 1984 SPE Formation Damage Control Symposium, Bakersfield, CA, Feb. 13-14.
15. Perkins, T.K. Jr. and Kern, L.R.: "Widths of Hydraulic Fractures," *JPT* (Sept. 1961) 937-49; *Trans.,* AIME, **222**.
16. Khristianovich, S.A. and Zheltov, Y.P.: "Formation of Vertical Fractures by Means of Highly Viscous Liquid," *Proc.,* Fourth World Pet. Cong. (1955) **II,** 579-86.
17. Babcock, R.E., Prokop, C.L., and Kehle, R.O.: "Distribution of Propping Agents in Vertical Fractures," *Prod. Monthly* (Nov. 1967) 11-18.
18. Medlin, W.L., Sexton, J.H., and Zumwalt, G.L.: "Sand Transport Experiments in Thin Fluids," paper SPE 14469 presented at the 1985 SPE Annual Technical Conference and Exhibition, Las Vegas, Sept. 22-25.
19. Biot, M.A. and Medlin, W.L.: "Theory of Sand Transport in Thin Fluids," paper SPE 14468 presented at the 1985 SPE Annual Technical Conference and Exhibition, Las Vegas, Sept. 22-25.
20. Soliman, M.: "Fracture Conductivity Distribution Studied," *Oil & Gas J.* (Feb. 10, 1986) 89-93.
21. Shah, S.N. and Smith, D.R.: "Ultra High Sand Concentration Boosts Oklahoma Production," *Pet. Eng. Intl.* (Jan. 1985) 50-60.
22. Pauls, R. *et al.*: "Successful Stimulation of the Olmos Formation Using Oil-Base Fluids and High Proppant Concentrations," paper SPE 13817 presented at the 1985 SPE Production Operations Symposium, Oklahoma City, March 10-12.
23. Nolte, K.G.: "Determination of Fracture Parameters From Fracturing Pressure Decline," paper SPE 8341 presented at 1979 SPE Annual Technical Conference and Exhibition, Las Vegas, Sept. 23-27.
24. Daneshy, A.A.: "Propagation of Hydraulic Fracture and Its Conductivity in Layered Media," *Proc.,* First Japan-United States Joint Seminar on Hydraulic Fracturing and Geothermal Energy, Tokyo (Nov. 2-5, 1982) 45-60.
25. Daneshy, A.A.: "Experimental Investigation of Hydraulic Fracturing Through Perforations," *JPT* (Oct. 1973) 1201-06; *Trans.,* AIME, **255**.
26. Daneshy, A.A. and Conrad, N.: "Fluid Pressure Variations During Hydraulic Fracturing," *Proc.,* 19th U.S. Symposium on Rock Mechanics, Stateline, NV (May 1-3, 1978) **11,** 8-17.
27. Daneshy, A.A.: "Hydraulic Fracture Propagation in Layered Formations," *SPEJ* (Feb. 1978) 33-41.

SI Metric Conversion Factors

bbl	× 1.589 873	E−01	= m^3
cp	× 1.0*	E−03	= Pa·s
degrees	× 1.745 329	E−02	= rad
ft	× 3.048*	E−01	= m
°F	(°F−32)/1.8		= °C
gal	× 3.785 412	E−03	= m^3
in.	× 2.54*	E+00	= cm
lbm	× 4.535 924	E−01	= kg
psi	× 6.894 757	E+00	= kPa

*Conversion factor is exact.

Chapter 11
Fracture Treatment Design

D.E. Nierode, SPE, Exxon Production Research Co.

11.1 Overview

Chapter Objectives. The objectives of this chapter are to discuss the procedure for designing fracturing treatments, to emphasize the critical design factors that determine design effectiveness, and to discuss optimal treatment selection. After studying the material and example cases in this chapter, the reader should be able to design a fracturing treatment.

This chapter puts all the technology discussed in other chapters together into one cohesive unit to model the entire fracturing process. The prefracture well test analysis methods are used to determine an effective reservoir permeability. The fluid mechanics of non-Newtonian fluid flow in pipe is combined with linear elastic rock-mechanics models for fracture growth estimation. Proppant placement in the fracture is modeled, as is the closing of the fracture at the end of pumping. Finally, the production response of the well is estimated with techniques discussed in Chap. 15.

This chapter follows the chronological steps of an actual design. First, the input data needed to conduct a treatment design are discussed. The prefracture production behavior is then described, followed by the evaluation of well productivity response to a fracture treatment relative to the decisions that must be made about treatment size and materials. Finally, the economic implications of optimal treatment selection are presented.

The results of calculations for an example design case are presented at crucial steps in the design process. It is anticipated that the reader will learn the details of the design process as much from study of the example cases as from study of the text material. Calculations from a complex computer simulation program and from a simple procedure are presented. Finally, the iterative process whereby the optimal materials are selected in the optimal amount and distribution is discussed. Throughout this chapter, discussions are restricted to vertical plane fractures because almost all the hydrocarbon-bearing reservoirs in need of stimulation fracture in the vertical plane.[1] Acid fracturing stimulation will not be a part of this chapter because it has already been covered elsewhere.[2,3]

Fracture Treatment Design Philosophies. The design methodology presented here can be used with two quite different philosophies in mind. One can either believe that the calculations represent actual, quantitative, fracture behavior, or use the calculations in a relative way to select directionally "better" treatments.[4] Recent evidence[5,6] indicates that current fracture-dimension models may be quantitative representations of actual fracture behavior in support of the first concept. In either event, when designs are done with these models as the basis, jobs are usually pumped away successfully.

Fortunately, belief that fracturing is or is not quantitatively modeled by today's design methods does not materially affect one's ability to design a treatment. It affects only the expectations from the well once it has been treated and begins to produce reservoir fluids. In addition, it is not necessary to use a computer program to design a treatment. Hand calculations and graphical design methods can be used to design fracture treatments effectively.[7-10]

It is worthwhile to summarize here the design method that will be discussed in more detail in the remainder of the chapter. The design process consists of the following steps.

1. Gather all the required well data, including reservoir, completion, and approximate treatment parameters.
2. Select a few suitable fluids, or specify a viscosity value that is thought to be needed for the particular treatment. If a viscosity value is specified, a final design step will be needed to specify service-company fluids that will provide that viscosity level.
3. Estimate the production response that would be obtained from small and large treatments for both sand and high-strength proppant.
4. Do detailed modeling of sand and high-strength proppant treatments to determine the effect of overall job size on productivity response to treatment.
5. Select the treatment size and type that will give the best return on investment, augmenting design calculations as needed to reach the economic limiting size.
6. Refine the final treatment design to place the desired fracture into the well at minimum risk and expense.
7. If a viscosity value was specified, convert this value into specific service-company fluids.

Step 5 will not be emphasized here. For simplicity, only payout time will be considered as the economic limitation on treatment design. More detailed economics will be covered in Chap. 17. Once the design sequence in this chapter is understood, one can apply it to optimize any economic model by simply repeating the design iteratively until the economic function is maximized.

11.2 Data Requirements

General Data Requirements. Information about the reservoir characteristics and the mechanical completion configuration is needed to design a successful fracturing treatment. The more complete and consistent the information is, the more reliable the design will be.

Fracturing information can be broken down into uncontrollable and controllable parameters. Uncontrollable parameters, reservoir characteristics that cannot be modified, include the following.

1. Reservoir permeability and porosity.
2. Reservoir net sand thickness and areal extent.
3. Reservoir stress levels.
4. Reservoir temperature and pressure.
5. Reservoir fluid properties and saturations.
6. Adjacent barrier thicknesses and areal extent.
7. Adjacent barrier stress levels.

Controllable parameters are completion characteristics that can be varied to optimize treatment effectiveness, even though many of them may be determined by other constraints. For example, maximum tubing size may be constrained by production casing size limitations imposed during drilling. These are typical controllable parameters.

1. Wellbore casing, tubing, and wellhead configurations.
2. Wellbore downhole equipment.
3. Perforation location and quantity.

4. Fracturing fluid and proppant.
5. Fracturing treatment rate and materials schedules.

From the uncontrollable parameters, reservoir data are needed to estimate well productivity both before and after stimulation. If there are no data on well flow capacity other than log-determined porosities, then the response to a fracturing treatment will not be predictable. Similarly, treatment size and selection of materials cannot be made intelligently. If good prefracture production data are available, then the optimal treatment can be determined. A complete set of prefracture production information consists of drawdown production data with sufficient duration to exhibit decline, good pressure-buildup data, and in-situ-corrected core-permeability data.

From the controllable parameters, completion data are needed so that the mechanical aspects of fracturing treatment design can be addressed. The well must be capable of safely resisting the high pressure and injection rates during the treatment. Completion components sometimes need to be changed in a well before it can be fractured. For example, if the wellhead has a low pressure rating, it must be either changed or bypassed with a pressure-isolation tool. If there are several downhole restrictions or other undesirable devices in the completions string, the tubing may need to be changed to enable high-pressure injection of large amounts of fluid and proppant.

Fortunately, not all the many pieces of information needed to design a fracturing treatment need be known with great accuracy. Some parameters, such as Poisson's ratio of the rock, need be known only roughly. Other parameters, such as fracture gradient, must be known rather accurately in situations where required surface pressures are at the limits of equipment. If the fracture gradient is much higher than expected, it is possible to have all the equipment assembled for a job and not be able to inject into the well at fracturing rates.

The next two sections discuss the specific data needed for treatment design and indicate the relative importance of each piece.

Uncontrollable Reservoir Parameters. Reservoir parameters are of critical importance for the determination of expected well response to fracturing. If the permeability is hundreds of millidarcies, then fracturing will not increase well productivity by alteration of reservoir flow pattern. Any increase in production will be caused by either damage removal or vertical capture of additional productive zones. If permeability is extremely low, very long propped fractures will be needed to drain the sands effectively in a reasonable time span. Similar statements can be made about the influence of reservoir pressure and fluid viscosity on pre- and post-stimulation well productivity.

The following items of reservoir data are discussed in order of relative importance in effective treatment design. A later section will show the influence of some of these parameters on stimulated production rate.

Effective Permeability. The increase of the well productivity index (PI) that will result from a specific fracturing treatment depends most directly on reservoir permeability. The abscissas on both the McGuire-Sikora[11] and Tannich-Nierode[12] curves for predicting stimulation ratio depend inversely on reservoir permeability. The higher the permeability, the lower the potential stimulation, with everything else constant. Reservoir permeability is probably the most important parameter determining postfracture production rates.

The best source of reservoir permeability is a prefracture production test analyzed in conjunction with a pressure-buildup test. These data can be analyzed with either a complex reservoir simulator computer program or the various type curves for pressure-buildup analysis. The result should be a value of effective reservoir permeability that, along with the other reservoir parameters, quantifies the observed prefracture production rate and the pressure-buildup behavior.

Core data can also be used to estimate prefracture well flow capacity with a somewhat lower level of confidence than actual production data. Routine core-permeability and -porosity data need to be suitably corrected[13] to in-situ conditions for this purpose. If the cores cover the entire interval to be fractured, then the core-permeability values can be used directly to estimate well productivity. If the cores cover only part of the fractured interval, then the data can be used to calibrate porosity logs. If a suitable correlation can be found between porosity and permeability, porosity logs can be used to estimate the total net productive footage and effective permeability.

Reservoir Thicknesses. The reservoir thicknesses that contribute to production, that accept leakoff of fracturing fluid, and that are ultimately penetrated by fracture height need to be known. These three thicknesses are often different. For example, fracture height of 150 ft [46 m] could be achieved during a treatment where 75 ft [23 m] of interval accept the fluid loss and only 50 ft [15 m] contribute to final production.

Net productive thickness is important because reservoir kh product determined from a pressure-buildup test must be separated into permeability and thickness for fracture calculations. If the h value is overestimated, the design will be overly conservative and use too much fluid. It may also not use a sufficiently conductive proppant. If the net footage is underestimated, permeability will be too high, and high-strength proppant might be used unnecessarily.

Long-term recovery estimates will also depend heavily on the choice of the correct k and h values. For example, if h is overestimated, reserves will be correspondingly overestimated.

Net productive footage may be different before and after fracturing because of vertical fracture growth away from the perforated intervals. A major contribution to the increase of production after fracturing can come from the capture of additional net productive footage[14,15] through vertical fracture height growth.

The interval thickness that accepts fluid loss during fracturing can be greater than the net productive sand for several reasons. In some instances, fluid loss occurs in zones that are not ultimately held open with proppant as a result of proppant settling. In other cases, some zones accept fluid loss at the high pressure levels during fracturing but contribute little or no flow in the production flow direction. Some natural fractures can behave this way with high fluid loss in the injection direction and essentially no contribution to production.

The best source for net productive and fluid-loss footages is a combined analysis of core data and porosity logs. Gross fracture-height determination is discussed in detail in Chap. 16.

Fracture Gradient. Fracture gradient is the bottomhole pressure (BHP) needed to propagate a fracture divided by the depth of the reservoir. This value is of extreme importance in the determination of permissible rates and pressures for the injection of fracturing materials. The value also has an influence on fracture geometry because the net fluid pressure level in the fracture influences both the fluid loss and the width-to-length distribution.

Fracture gradient is determined most easily from pressure records of fracture rate injections into nearby wells in the reservoir. When pumping is abruptly ceased, the instantaneous bottomhole shut-in pressure divided by depth is the fracture gradient.[16-19] Measurements in nearby wells completed in the same reservoir are often a close enough estimate of the behavior of a new well. However, some operators now do an intentional minifracture treatment in the new well ahead of the normal fracture treatment to estimate fracture gradient and other fracture parameters. Pressures during and after treatment are similarly monitored to provide fracture treatment information.[20-24] This is discussed in more detail in Chap. 14.

Static Temperature and Pressure. Static reservoir temperature is needed so that a stable fluid can be chosen to withstand potential long-term exposure to the temperature. Underestimation of the reservoir temperature can cause screenouts as a result of degradation of proppant-carrying capacity as the fluid becomes too hot. Overestimation of temperature can cause the undesirable slow return of fracture fluid as the well is put back on production after stimulation.

The best source for reservoir temperature is a wellbore measurement suitably corrected to undisturbed conditions. For example, routine maximum logging temperatures can be corrected for thermal effects caused by circulation and other recent drilling activities. It is often best to determine reservoir temperature on a fieldwide basis to arrive at a geothermal gradient for the reservoir at that location. Measured values taken in quiescent wells that have been shut in for months or years should be given the most weight in defining true static reservoir temperature. Once a localized static gradient is determined for the field or area studied, that value can be used for estimation of bottomhole temperature in new wells.

Initial reservoir pressure is a parameter of great importance that is often not directly measured, even though it is simple to determine as a part of the perforation process. Immediately after perforating, the pressure level that occurs at the perforations in an underbalanced condition is the reservoir pressure. Gross estimates of the pressure can be obtained from the mud weight used to drill the section, but this is usually an overestimate of actual pressure. In those instances when a significant pressure response is not obtained after perforating, a small acid treatment, called a hole-opening treatment, can be performed. A reasonably good estimate of reservoir pressure can then be obtained after acid cleanup.

A good number for reservoir pressure is necessary for the correct analysis of well productivity before and after fracturing. It is of critical importance in long-term recovery estimates. It is of near-term importance because it determines the required fracturing pressure. To fracture a well, one must overcome the total stress in the reservoir, which is the sum of the rock frame stress and the pore pressure. An unusually high pore pressure can make injection into the well impossible if pressure limits on the well equipment are not high enough.

Reservoir pressure is the driving force for the recovery of fracturing fluid after the well has been put back into production. Normal hydrostatic reservoir pressure levels are usually sufficient to drive most of the load fluids back to the surface. If reservoir pressure is significantly below hydrostatic, consideration should be given to the use of low-density foam fluids to promote cleanup.

Reservoir Fluid Density, Viscosity, and Compressibility. Reservoir fluid properties are important in the evaluation of well productivity and the design of the fracture treatment. Density, viscosity, and compressibility have obvious implications for the long-term recovery of reserves in the stimulated well. They also influence the fluid-loss behavior of fracture fluid during the treatment.[25-28] If viscosity is low and compressibility is high, the reservoir will provide little restriction to flow during the spurt-loss part of fluid loss to the reservoir.

These fluid parameters ideally come from laboratory tests on reservoir fluid samples taken at downhole conditions. More commonly, the only data available come from routine well test fluid samples where density at ambient conditions yields a fluid-gravity value. Viscosity and compressibility are then estimated from correlation curves. These parameters are not of great importance for the design of the fracture treatment itself but are important for the evaluation of well productivity in the short and long term.

Rock Mechanical Properties. The rock properties, Young's modulus and Poisson's ratio, are needed for fracture-propagation calculations with fracture models. Fortunately, they are not needed with great accuracy for this purpose. Ideally, core material could be tested at realistic in-situ conditions to yield the values. This is done infrequently because of both a lack of core and the high cost of such testing. Fortunately, the parameters can be estimated with sufficient accuracy by the methods discussed below or can be directly measured from the interpretation of the full-wave train of long-spaced sonic logs.

Poisson's ratio does not vary greatly for hydrocarbon-bearing rocks. Table 11.1 shows reasonable values that are dependent on gross hardness and rock type. The gross measures of hardness are soft, medium, and hard (S, M, and H), as indicated in the first entry of the formation code. The rock types are sandstone, dolomite, and limestone (S, D, and L), as in the second entry of the formation code.

For computational purposes, selection of rock type and a subjective measure of soft, medium, or hard strength are sufficiently accurate to determine fracture dimensions. A more in-depth knowledge will be necessary when three-dimensional (3D) models are routinely used for fracture-height estimation.[29-33] Poisson's ratio will then be needed for all rock layers to be fractured, including shales.

Young's modulus can be approximated from sonic travel time and other reservoir parameters:

$$E=\frac{2.16\times10^8(1-2\nu)(1+\nu)[165(1-\phi)+\phi\rho_f]}{(1-\nu)(\Delta t_c)^2} \quad \ldots\ldots (11.1)$$

TABLE 11.1—APPROXIMATE POISSON'S RATIOS

Formation code	SS	MS	HS	SD	MD	HD	SL	ML	HL
Poisson's ratio	0.20	0.17	0.15	0.30	0.27	0.25	0.28	0.25	0.23

Vertical Stress Distribution. The distribution of vertical stresses is currently believed[34,35] to control fracture-height growth. The weight of the earth's overburden generates horizontal stresses in the reservoir rock that may be augmented by tectonic contributions. The minimum principal horizontal stress may vary vertically as rock lithology changes with depth. The height to which a fracture will grow depends directly on the vertical distribution of the minimum horizontal stress. We would like to know this distribution so that we can either determine the gross fracture height that goes into a constant-height model or use the distribution as input to a fully 3D model.

Unfortunately, we often do not know the vertical stress distribution, and it is difficult and expensive to measure directly in a wellbore. In constant-height models, a pure guess is often made for gross fracture height, with errors being on the high side. The guess can be become educated if data for height are available from either temperature or radioactive tracer surveys in similar wells that have already been fractured.[36-38] Three-dimensional models will overcome the need to guess fracture height but are only truly useful if the vertical stress and rock-property distribution are known. Gross fracture height is not usually an extremely important parameter for design purposes as long as it is known within a factor of two. It can become of overriding importance if excessive height growth can cause production of unwanted fluids. In surveillance work, the height should be determined from postfracture survey data.

Barrier Characteristics. The thickness, areal extent, and stress level of barriers above and below the pay zone directly influence the success of a fracture treatment. If the barriers are thick and highly stressed, then fracture height will be confined mostly to the pay zone, and two-dimensional (2D) fracture models can be used effectively to model fracture growth. If the barriers are thin and of stress level similar to the pay zone, then the fracture will grow through the barriers toward other permeable zones. In the latter situation, a radial, penny-shaped-crack fracture model would best decribe fracture growth until effective barriers are eventually encountered. When effective barriers control fracture-height growth, 2D models can again describe fracture length and width growth. Unfortunately, it is difficult to predict when a particular barrier will be effective at stopping fracture-height growth unless extensive stress and rock-property data are available. The fully 3D models will have to be used in situations when stabilized fracture height does not occur.

Controllable Completion Parameters. The mechanical condition of the well completion must be evaluated to ensure that the fracture treatment can be injected into the well safely and at acceptable rates to form a good fracture. Many completion decisions made during the lifetime of a well do not take into account the potential for future fracture stimulation. At the time of fracture design, the well condition must be evaluated and any necessary changes made to allow for successful treatment.

Tubing and Wellhead Characteristics. Fracture treatment must usually be injected through tubing in the well. The tubing must be of large enough diameter and without significant restrictions so that the fluid can be injected at high rate without large frictional pressure losses. If selected on deliverability considerations alone, the tubing in a 15,000-ft [4570-m] gas well is sized to be 2⅜ in. [6 cm] in diameter; it would then be impossible to inject a fracture treatment at rates above 5 bbl/min [0.8 m^3/min] at a surface pressure limit of 15,000 psi [103 MPa]. In this case, it would be necessary to go to higher pressures or to change the tubing to a larger diameter to reduce the frictional losses.

The wellhead assembly may similarly have been selected from production rate considerations only and may be of too low a pressure rating to permit high-pressure fracture injection. In this circumstance, the wellhead can be changed for one of higher rating, temporarily replaced with a fracture valve, or most commonly, can be isolated from treating pressures with a special "wellhead isolation" tool.

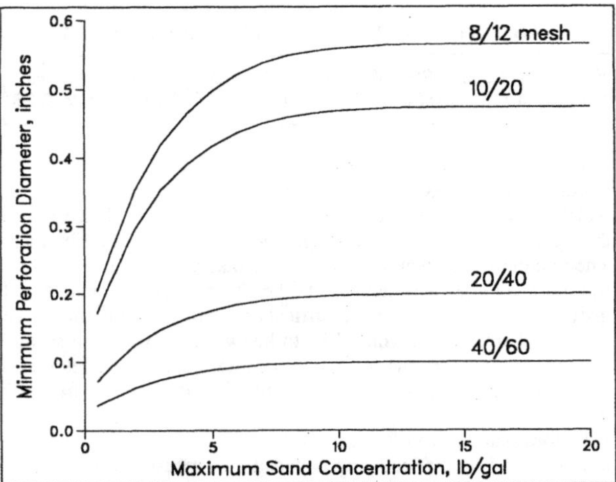

Fig. 11.1—Minimum perforation diameter vs. proppant size and concentration.

Casing Characteristics. The production casing string must be of sufficient integrity to withstand the high pressures that occur during fracturing. In some special instances, entire treatments are pumped down the production casing to achieve high treating rates. In this instance, the casing must be able to withstand fracturing pressures throughout its entire length. When injection is down a tubing/packer assembly, the casing below the packer must be able to withstand full treating pressures. Of particular concern in this case is worn or corroded casing that may divert the entire treatment into some undesirable zone. Squeezed perforations in this interval may also be especially vulnerable to breakdown and subsequent loss of treating materials.

In deep wells, the production casing/tubing annulus above the packer must often be pressurized to some moderate value to minimize the stresses on the tubing string. A 15,000-psi [103-MPa] surface pressure injection into 3½-in. [9-cm] tubing in a 15,000-ft [4570-m] -deep well will often require 3,000 to 5,000 psi [20.7 to 34.5 MPa] of backup pressure maintained during the treatment on the production tubing/casing annulus. Another somewhat specialized application involves simultaneous injection of fracturing materials down the tubing and production casing/tubing annulus. In terms of pressure, this situation is little different from straight casing injection.

Packer Type and Location. Special packers are needed in many wells fractured through tubing. The packer must be able to withstand the large pressure forces that tend to drive the packer body uphole during the fracture treatment. The required packers usually grip the casing in both upward and downward directions and are generally permanent. Less secure packers usually cannot be used during fracturing operations unless special situations pertain. For example, injection down both tubing and the casing/tubing annulus can be accomplished with an inflatable packer on the end of the tubing. When the treatment is over, the packer can be inflated to isolate flowback to the tubing.

In very deep wells, an unlatched packer assembly is usually used. If the tubing were firmly latched to the packer body, the thermal stresses induced by cool fluid injection could part the tubing or connections. Instead, a sliding seal assembly is used so that the tubing can float in the packer's polished-bore receptacle at an equilibrium position. The multiple seal locations for production and injection directions must be carefully selected to maintain annular seal during all operating conditions.

The packer should be located sufficiently above the perforated interval to allow a smoothing out of the streamlines of injection fluids before they enter the perforations. This is especially important to promote efficient seating of ball sealers used for diverting treatment stages. The packer should also be spaced sufficiently above the perforation interval to ensure that sand fill occurs inside the casing only after the final sand stage has been underdisplaced. Packers are typically placed 100 to 200 ft [30 to 60 m] above the shallowest perforation, and several hundred feet of rathole are usually provided below the deepest perforation.

Perforation Distribution. The number and location of perforations in a casing completion are important to the extent that the perforations help or hinder the injection of fracturing materials. With an extremely high number of perforations, the flow rate into any one perforation may be so low that a moderate- to low-viscosity fluid may not be able to carry proppant into the perforation tunnel. If there are very few perforations, the large pressure drop across them may restrict overall injectivity. A moderate number of perforations (from about 50 to several hundred) is usually recommended for wells to be fractured. The inclination to shoot many more perforations can be countered by relying on vertical fracture-height growth to link unperforated sands with nearby perforations.

The distribution of perforations becomes important if diversion is planned during the treatment. For ball sealers to divert fracture stages successfully over 500- to 1,000-ft [150- to 300-m] -thick sections, one would need to place limited numbers of perforations at isolated locations. For example, shooting 10 perforations per interval with 100-ft [30-m] separations between intervals has given good stage separation with ball sealers.[16] Closer spacing or the use of more perforations would lead to ineffective staging, with several or all stages entering the same intervals.[15]

To divert by limited entry,[39,40] a very limited number of perforations are placed across the completion interval so that the pressure drop in each perforation is hundreds of psi. The perforation pressure drop causes increased wellbore pressure that can force fluids into higher-fracture-gradient zones. For example, during pumping at 25 bbl/min [4 m^3/min] into 15 perforations, the pressure drop would be about 500 psi [3.4 MPa] across each perforation. During injection, the fluids will break down the zone with the lowest fracture gradient first. Because the pressure drop across the perforations is 500 psi [3.4 MPa], however, the wellbore pressure can still be maintained 500 psi [3.4 MPa] higher than the lowest fracture gradient, enabling fluid to be distributed to other less easily fractured zones. More zones can be fractured with this technique than with high perforation density. An increase in perforation diameter as a result of erosion during pumping of the treatment can lead to a reduction in the amount of excess wellbore pressure maintained. Limited-entry perforating is discussed further in Chap. 12.

Perforation Size. Perforation size directly affects proppant sizes and concentrations that can be pumped during a treatment. A perforation must be large enough relative to maximum proppant diameter to prevent bridging across the perforation diameter.[41] Fig. 11.1 illustrates the minimum perforation size necessary to inject various sizes of proppants at various concentrations successfully. For example, to pump 20/40-mesh sand at 10 lbm/gal [1198 kg/m^3], the perforation needs to be at least 0.17 in. [0.43 cm] in diameter. If maximum sand concentration were only 1.5 lbm/gal [180 kg/m^3], the sand could be pumped through a perforation as small as 0.10 in. [0.25 cm] in diameter. The general rule of thumb is that perforation diameter must be at least six times the maximum particle diameter to allow injection of concentrations as high as 20 lbm/gal [2397 kg/m^3] without perforation bridging. At concentrations less than about 6 lbm/gal [719 kg/m^3], the diameter requirement is somewhat reduced. In the limit at 0.5 lbm/gal [60 kg/m^3], perforation diameter needs to be only twice the maximum particle size.

Another consideration in perforation size is fracture-fluid degradation. If perforation diameter is too small, the high shear rates that occur in the perforation tunnel can irreversibly destroy gel structure. Good high-viscosity fluid can lose its ability to carry proppant, and a screenout can ensue. Shear degradation is covered in more detail in Chap. 9.

Fracturing Fluid and Proppant Selection. The selection of the specific fracturing fluid and proppant provides some control over the ultimate response of the well to the fracturing treatment. Particular fluids and proppants can be selected to provide advantages over other combinations. If a very viscous fluid is selected, fracture-height growth will be promoted and little proppant settling will occur. This is the preferred treatment if it is desired that additional net permeable footage be put into communication with the wellbore by fracture-height growth and maximum propped fracture length be attained. It must be emphasized that fracture-height growth is controlled predominantly by rock stresses and rock properties and that the effect of fracture fluid viscosity is, at best, of secondary importance.

If a very-low-viscosity fluid is selected, fracture-height growth would tend to be minimized and proppant-carry distance would be short. Treatment with a low-viscosity fluid should be considered if capture of additional productive interval is not desired and if short, well-packed fractures are wanted. Very low viscosity would tend to promote the formation of an equilibrium bank of proppant to give maximum propped fracture width and minimum propped length.

Another advantageous combination of fluid and proppant is a moderate-viscosity fluid and a maximum-density proppant. This combination could provide moderate fracture-height growth at the same time as some proppant banking occurs to increase propped width. High-density proppant could be ceramic proppants with a gravity of about 3.6 relative to water or even steel shot with a gravity of about 7.8. Any fluid/proppant system can best be selected to exploit particular characteristics of viscosity or density contrast with the aid of a detailed fracture-growth simulator that takes into account the specific shear rates, temperatures, and velocities that occur at various locations along the fracture length.

Injection-Rate and -Pressure Limitations. When all the mechanical aspects of the well completion are taken into account, the basic requirement is that the well be capable of accepting fracturing fluids at sufficient rates and pressures to perform the designed fracture treatment. The engineer has considerable latitude in decisions to arrive at a completion that will allow this condition. He must evaluate the pressure and rate limitations of the well soundly and modify the limitations as needed.

The best way to obtain acceptable injection conditions is to plan for them in the original well development plan. Anticipation of fracturing at these initial stages can lead to a completion that will not be limited by pressures or treating rates.

Logistical Limitations. The remaining information about a well that can have some bearing on fracture treatment design concerns logistical and site-specific factors. For example, the well might be located in a remote part of the world where it would be difficult to obtain high-strength proppant at a reasonable cost in a reasonable time. In this instance, it would probably be the best engineering decision to select the local sand as proppant even though sand quality may be not be optimal. In remote U.S. locations, such as in the Rocky Mountains, it may be most economical to fracture with foams to minimize the expense involved in trucking water for the fracture fluid to and away from the wellsite.

Safety considerations often dictate permissible treatments. Some wellsites may be located in urban areas where 20,000-psi [138-MPa] injection pressures may involve undue hazard to the public. In this case, it would be prudent to run a special fracture treatment string of larger-diameter tubing to reduce surface injection pressures. Foam fluids may also be avoided in favor of conventional gelled waters for safety reasons. The transport of the gaseous component of the foam is more hazardous than the transport of water, and the flowback of the well with foam is at higher pressures and rates than a water treatment. It would be safer in this situation to use a gelled water and to swab the well to recover the load water.

Another factor commonly encountered in distant locations is limited selection of fracturing-fluid materials. If the only polymer available in the country is a natural guar gum, the job is designed with this type of gelled water. Job size is also often determined by the quantity of polymer or proppant currently available.

It is these types of site-specific considerations that can dramatically alter the fracture design process. It is up to the design engineer to work within these constraints to design the best treatment.

Example Case Data. The data contained in Table 11.2 represent parameters for a moderate-permeability gas well. These data will be used to design a fracture treatment throughout the remainder of the text. A gas well was selected because it represents the most complicated design case because of gas well turbulence. An oilwell treatment is designed with the same method without turbulence considerations.

TABLE 11.2—DATA FOR EXAMPLE CASE

Reservoir Parameters	
Static reservoir pressure, psi	4,500
Static reservoir temperature, °F	250
Gas-filled porosity, fraction	0.05
Net sand thickness, ft	50
Permeability, md	0.1
Sonic travel time, μsec/ft	70
Fluid density, lbm/ft³	10
Compressibility, psi⁻¹	0.00021
Poisson's ratio, dimensionless	0.17
Young's modulus, psi	6.44×10^6
Gas gravity, dimensionless	0.6
Initial fracture gradient, psi/ft	0.75
Prefracture production rate, Mcf/D	725
Prefracture surface flowing pressure, psi	100
Prefracture flowing BHP, psi	250
Postfracture surface flowing pressure, psi	500
Postfracture flowing BHP, psi	875
Economic Limitation	
Payout time for $k=0.1$ md, days	60
Payout time for $k=1.0$ md, days	15
Completion Parameters	
Borehole diameter, in.	6.75
Production casing ID, in.	4.89
Production tubing ID, in.	2.44
Midpoint depth of perforations, ft	9,200
Depth to packer, ft	9,000
Perforation diameter, in.	0.20
Well spacing, acres	320
Total feet of perforations, ft	50
Perforation density (0° phasing), shots/ft	2
Treatment Variables	
Estimated gross fracture height, ft	250
Initial fracture treating rate, bbl/min	25
Maximum treating pressure, psi	10,000
Crosslinked gel fluid-loss coefficient, ft/min$^{1/2}$	0.002
Crosslinked gel power-law n value	0.658
Crosslinked gel power-law K value	0.00438
Small Job	
20/40-mesh sand, lbm	100,000
Crosslinked water-based gel, gal	80,000
Fluid/Sand Schedule	
42,000 gal pad volume	
3,600 gal with 1 lbm/gal sand or bauxite	
3,600 gal with 2 lbm/gal sand or bauxite	
30,000 gal with 3 lbm/gal sand or bauxite	
Large Job	
20/40-mesh sand, lbm	1,000,000
Crosslinked water-based gel, gal	310,000
Fluid/Sand Schedule	
84,000 gal pad volume	
3,600 gal with 1 lbm/gal sand or bauxite	
10,000 gal with 2 lbm/gal sand or bauxite	
50,000 gal with 3 lbm/gal sand or bauxite	
50,000 gal with 4 lbm/gal sand or bauxite	
50,000 gal with 5 lbm/gal sand or bauxite	
63,000 gal with 6 lbm/gal sand or bauxite	

11.3 Fractured-Well Productivity Estimation

To select the size of treatment and types of materials during the design process, the effect of a given fracture treatment on productivity must be estimated. This section will discuss the methods for predicting the increase in productivity that comes from alteration of the reservoir flow pattern. This productivity increase is called "reservoir stimulation" to distinguish it from the productivity gain resulting from damage removal. Throughout this chapter, when a well is successfully fractured, the gain in productivity is considered to have two possible components, reservoir stimulation and damage removal. The main methods to predict productivity increase from reservoir stimulation are discussed in this section.

Productivity-Increase Estimation Methods. ***Pseudosteady-State Curves.*** The McGuire-Sikora[11] stimulation curves shown in Fig. 11.2 estimate the gain in PI that can be obtained for an oil well

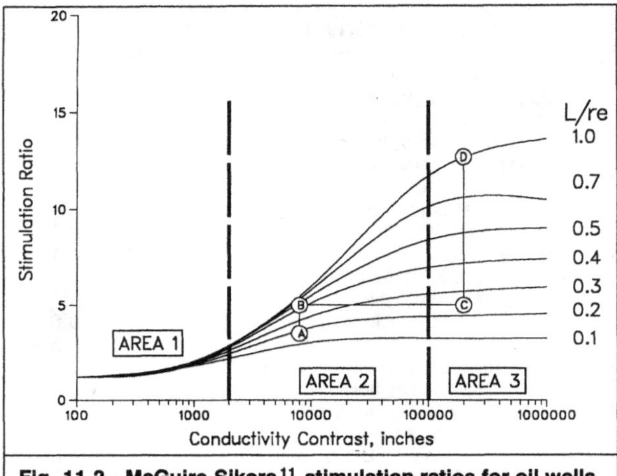

Fig. 11.2—McGuire-Sikora[11] stimulation ratios for oil wells.

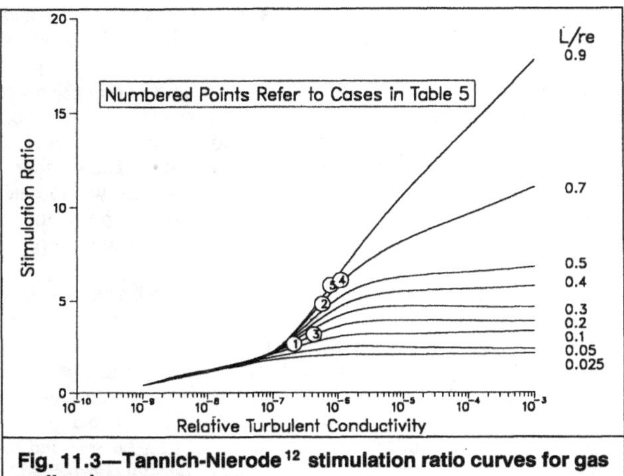

Fig. 11.3—Tannich-Nierode[12] stimulation ratio curves for gas wells—base case.

from alteration of reservoir flow pattern by a fracture. Increases in PI, commonly called the stimulation ratio, are plotted vs. dimensionless fracture length (L/r_e, where L is propped fracture length and r_e is the well drainage radius) and fracture permeability contrast. The curves were generated from an electric analog laboratory model by measuring electrical potential differences, as detailed in Chap. 15. The assumptions behind these curves restrict their use to pseudosteady-state conditions for slightly compressible reservoir fluids, as found in undersaturated oil wells. It is important to note that these curves give the increases of pseudosteady-state PI that would be realized at the same average reservoir pressure before and after fracturing. The unfractured well would obviously take a much longer time to arrive at the datum reservoir pressure. The curves in Fig. 11.2 apply directly for a well on 40-acre [16-ha] spacing, with extension to other spacings achieved by appropriate use of the spacing parameters on both axes. The parametric groups that control stimulation ratio response will be discussed later.

A second pseudosteady-state set of curves developed by Tannich and Nierode[12] for gas wells is shown in Fig. 11.3. The shapes of these curves are similar to the shapes of the McGuire-Sikora curves in Fig. 11.2, but the correlating parameters are a little different. PI ratio is plotted vs. fractional fracture length and a conductivity group called relative turbulent conductivity, C_{rB}. The effect of turbulence on fracture conductivity is entered in terms of the fracture turbulence coefficient:

$$C_{rB}=\frac{w_p\mu}{k}\sqrt{\frac{z(T+460)}{\beta_f\gamma(p_e^2-p_w^2)}}\sqrt{\frac{2,640}{r_e}}. \qquad (11.2)$$

The curves in Fig. 11.3 apply directly to a well on 640-acre [260-ha] spacing. Extension to other spacings is achieved by appropriate use of the spacing parameter on each axis. The PI ratio is given specifically in Eq. 11.3. When Fig. 11.3 is entered with a calculated relative turbulent conductivity taking spacing into account, the PI-ratio-axis parametric value must be divided by the spacing term to arrive at the actual PI ratio, J'/J_o:

$$X=\left(\frac{J'}{J_o}\right)\left(\frac{7.50}{\ln\dfrac{0.220r_e}{r_w}}\right). \qquad (11.3)$$

Fig. 11.3 was generated from many computer calculations for a fractured gas well with a finite-difference reservoir-flow simulator and correlation of the results. Because gases are of low viscosity and high compressibility, turbulence has a major effect on stimulation ratio.

It is important to note that the Tannich-Nierode stimulation ratios are at the same time of production for pseudosteady-state flow. In other words, the PI ratio could be for the point in time of 30 days as long as the before- and after-fracture PI's were both at pseudosteady state. Reservoir pressure for the fractured well would be lower than for the unfractured well because of greater postfracture withdrawals.

The curves in Figs. 11.2 and 11.3 can be used to estimate the effect of fracture stimulation on the pseudosteady-state productivity of oil and gas wells, respectively. They are used during the design process to evaluate whether additional fracture length or proppant permeability will improve stimulated well response.

Finite-Difference Simulators. Another way to assess the response of a well to fracture stimulation is to model that specific well directly with a finite-difference reservoir simulator.[6,42-45] This method does not need to rely on pseudosteady-state assumptions and can be used to assess long-term recovery potential for a complex depletion strategy in a multiphase reservoir. As computers become faster and cheaper, direct simulation in this manner will increasingly become the norm. The base-case example design discussed in the remainder of this chapter will present results of calculations by use of one such simulator, as well as calculations based on pseudosteady-state curves.

Controllable Factors Affecting Recovery and Production Rate. The three factors that can be at least partially controlled in a fracture treatment are fracture length, fracture permeability/width product, and fracture height. All three factors directly affect the stimulation response of a well. The effect of each parameter will be discussed in the next sections by reference to the McGuire-Sikora and Tannich-Nierode curves.

Effect of Propped Fracture Length. The McGuire-Sikora chart can be broken into three areas, as indicated in Fig. 11.2. If the fractional length and permeability contrast parameters of a fractured well fall in Area 1, then stimulation response is quite low unless the well is severely damaged. This generally occurs when reservoir permeability is higher than about 10 md. At this permeability level, proppants cannot provide a fracture flow channel that is sufficiently more permeable than the reservoir rock to change the reservoir flow pattern greatly. Note that in Area 1, increased length, L, provides little additional stimulation beyond a dimensionless fracture length, L/r_e, of 0.10 or 0.20. In Area 2, higher stimulations are possible and length begins to provide increased stimulation. Fracture proppants are able to provide good flow channels to the wellbore in competition with radial flow through the reservoir. In Area 3, proppant permeability is sufficiently high that length becomes the dominant variable constraining stimulation. Here, the longer the fracture, the larger the stimulation.

It is important to note that longer propped fracture length can give increased stimulation when permeability contrast is high, but when that contrast is low, larger treatments do not always increase stimulation. In other words, larger treatments will not always give significantly greater stimulation.

Effect of Fracture Permeability/Width Product. As Fig. 11.2 indicates, a treatment that will result in a stimulation within Area 1 can be improved only if permeability contrast is increased. Longer propped fracture lengths can help only after the contrast has also been increased.

TABLE 11.3—FRACTURE DIMENSIONS FROM ANALYTICAL MODELS

	Simplified Models		Computer Models	
	PKN	GDK	PKN	KZ*
Dynamic created length, ft	1,155	831	1,200	708
Dynamic wellbore width, in.	0.212	0.239	0.195	0.280
Average width, in.	0.131	0.188	0.120	0.220
Total suspension propped length, ft	449	468	565	428
Total suspension propped average width, in.	0.054	0.051	0.043	0.058
Equilibrium bank propped length, ft	184	128		
Equilibrium bank propped average width, in.	0.131	0.187		
Sand-carry distance, ft	20,013	14,347		
Stimulation ratio (total suspension)	2.25	2.38		
Stimulation ratio (equilibrium bank)	2.61	2.74		
Final estimated stimulation ratio	2.25	2.38	2.43	2.59
Final estimated production rate, MMcf/D	1.58	1.67	1.70	1.81

*KZ refers to Khristianovich and Zheltov for model calculations based on their fracture model, as in Ref. 46.

Because the permeability contrast parameter in both Figs. 11.2 and 11.3 is on a log scale, order-of-magnitude changes are necessary to affect the obtained stimulation greatly. The only permeability contrast factor we can change by that much is proppant-pack permeability. As discussed in Chap. 6, the two proppants that provide extremes in permeability are sand and high-strength bauxite. Pack permeabilities differ by one to two orders of magnitude, depending on the closure-stress level. Thus, a treatment with sand that is in Area 1 of Fig. 11.2 could move to Area 2 or 3 if high-strength bauxite were used instead. The effect on permeability contrast of width increase caused by higher proppant concentration is limited to about a factor of two. The width effect will be demonstrated in the example case.

Effect of Fracture Height. The third somewhat controllable parameter is gross fracture height. Height has at least two effects on stimulation response. A height increase can increase response by capturing new net footage of sand that had not previously been in communication with the wellbore. However, stimulation can also be decreased by excessive height growth into unproductive sections, with subsequent reduction in useful propped length. Undesired gas or water production can also result from uncontrolled fracture-height growth.

The ability to control fracture height is marginal at best. In most cases, it cannot really be controlled. Instead, we predict height as accurately as possible and base our design on that value. In some instances, height growth can be reduced somewhat by a reduction in treating injection rate and treatment size. In other cases, height growth can be reduced somewhat by lowering downhole treating pressures through the use of less viscous fluids. Neither method has good field data to support its worth.

Selection of the Best Treatment. In keeping with the preceding discussions of McGuire-Sikora and Tannich-Nierode stimulation curves, the treatment that will give the most stimulation is that treatment with the greatest permeability contrast and the longest useful propped length. Two descriptive terms distinguish the direction for improvement of a given treatment: length-limited and conductivity-limited.

Length Limitation. Length limitation means that a given fractured well can make effective use of increased fracture length. A length-limited fracture also can make only limited use of a more permeable fracture until the fracture first becomes longer. In other words, a well with a length-limited fracture would benefit more from a larger treatment than from a change to a more conductive fracture material. The length-limited areas in Figs. 11.2 and 11.3 can be spotted easily. Look for a place where the fractional-length lines are separated, such as Locations A and C in Fig. 11.2.

In the design process (discussed later), when a treatment is length limited, an effort is made to increase treatment size or to adjust the sand schedule to maximize fracture length to take advantage of increased-length.

Conductivity Limitation. Conductivity limitation means that a given fractured well can make effective use of a more conductive fracture. Conductivity limitation implies that longer fracture length would not greatly benefit well productivity. In Figs. 11.2 and 11.3, the conductivity-limited areas occur when the fractional-length lines are close together. If the proppant is changed for one of higher permeability, or if much wider propped fractures are formed by alterations in the proppant schedule, the treatment may no longer be conductivity-limited.

The path shown in Fig. 11.2 shows a well that initially is length-limited at Point A. As treatment size is increased to Point B, it becomes conductivity-limited. When proppant type and proppant bank geometry are changed, it again becomes length-limited at Point C. Larger treatments are then done on Path CD until the treatment has reached the drainage radius and cannot be much more effective.

11.4 Fracture-Dimension Simulation

To estimate the production response of a well to the presence of a fracture, we need to know the fracture width, length, height, and permeability. This information comes from simulator models of the fracturing process that range from simple hand calculation procedures to complex computer programs that must be run on powerful computers.

General Fracture Parameter Needs. The fracture-dimension information needed for a specific treatment is fracture height, propped fracture length, and average propped width or propped-width distribution. As discussed, fracture height is a variable that can be only grossly estimated with today's technology. Educated guesses can come from field experience in nearby wells, from more sophisticated vertical-stress profile information, or from analysis of lithological changes in rock properties. However, all these methods are approximate at this time. Total fracture length and width distributions can currently be predicted for fixed height by the methods discussed in Chap. 4, which detailed the theory behind the results presented later in this section.

Final proppant distribution as it affects length and width can be modeled by methods ranging from simple to highly complex. These methods have already been discussed in Chap. 10; results of these calculations for the example case in this chapter will be discussed later.

Analytical Constant-Height Models. To calculate dynamic fracture dimensions for a treatment without the assistance of a main-frame computer, one can use either the approximate solution to the governing rock mechanics equations known as the Geertsma-deKlerk[7] (GDK) solution or the approximate solution known as the Perkins-Kern-Nordgren[8,9] (PKN) solution. Final proppant distribution can be bounded for each by the limiting cases of total proppant suspension and equilibrium bank formation. These limiting cases have been discussed in Chap. 10. The likelihood of either limit can be assessed by a simple calculation of sand carry distance with only Stokes'-law proppant settling. Resulting width and length estimates for the small fracturing treatment in Table 11.2 are shown in Table 11.3 for these simple analytic models. The numbers in Table 11.3 were calculated for sand proppant by following the

```
TIME SINCE TREATMENT BEGAN        63.67 MIN                    PAD VOLUME REMAINING AHEAD OF PROPPANT =      8578. GALS
FLUID VOLUME PUMPED =      64289. GAL                                      (FOR EACH WING)
TOTAL PROPPANT INJECTED =      56728. LBS
DYNAMIC FRACTURE LENGTH =        618.4 FT            PERCENT PAD VOLUME REMAINING AHEAD OF PROPPANT =      40.8 %
DYNAMIC FRACTURE WIDTH
          AT THE WELLBORE =      0.2580 INCH
          AVERAGE =      0.2026 INCH
FLUID TEMPERATURE ENTERING PERFORATIONS =          81.5 DEG F
```

FRACTURE CHARACTERISTICS AT THE RESPECTIVE LENGTHS ARE:

FRACTURE LENGTH (FEET)	FRACTURE HEIGHT (FEET) 0. 50. 100. 150. 200. 250.	SUSPENDED PROPPANT	WIDTH,IN	CONCENTRATIONS LB/GAL	CONCENTRATIONS LB/SQFT	LAYERS
0.	X---------X---------X---------X---------X---------X					
10.3	X..X	20/ 40 MESH SAND	0.258	3.0	0.43	1.99
20.6	X..X	20/ 40 MESH SAND	0.258	3.1	0.43	2.01
30.8	X..X	20/ 40 MESH SAND	0.258	3.1	0.44	2.02
40.9	X..X	20/ 40 MESH SAND	0.257	3.1	0.44	2.04
50.9	X..X	20/ 40 MESH SAND	0.257	3.2	0.44	2.05
60.8	X..X	20/ 40 MESH SAND	0.257	3.2	0.45	2.07
70.6	X..X	20/ 40 MESH SAND	0.256	3.2	0.45	2.08
80.4	X..X	20/ 40 MESH SAND	0.256	3.2	0.45	2.10
90.1	X..X	20/ 40 MESH SAND	0.255	3.3	0.45	2.11
99.8	X..X	20/ 40 MESH SAND	0.255	3.3	0.46	2.12
118.8	X... X	20/ 40 MESH SAND	0.253	3.4	0.46	2.15
128.3	X... X	20/ 40 MESH SAND	0.252	3.4	0.47	2.16
137.7	X... X	20/ 40 MESH SAND	0.251	3.4	0.47	2.17
147.0	X... X	20/ 40 MESH SAND	0.251	3.5	0.47	2.18
156.3	X... X	20/ 40 MESH SAND	0.250	3.5	0.47	2.19
165.6	X... X	20/ 40 MESH SAND	0.249	3.6	0.47	2.20
174.8	X... X	20/ 40 MESH SAND	0.247	3.6	0.48	2.21
183.9	X... X	20/ 40 MESH SAND	0.246	3.6	0.48	2.22
193.0	X... X	20/ 40 MESH SAND	0.245	3.7	0.48	2.23
202.1	X... X	20/ 40 MESH SAND	0.244	3.7	0.48	2.24
211.1	X.. X	20/ 40 MESH SAND	0.242	3.2	0.42	1.95
220.1	X.. X	20/ 40 MESH SAND	0.241	2.8	0.37	1.73
229.1	X.. X	20/ 40 MESH SAND	0.240	2.6	0.35	1.62
238.0	X.. X	20/ 40 MESH SAND	0.238	2.6	0.34	1.58
246.9	X.. X	20/ 40 MESH SAND	0.237	2.6	0.34	1.58
255.8	X... X	20/ 40 MESH SAND	0.235	2.2	0.30	1.38
264.7	X... X	20/ 40 MESH SAND	0.233	1.9	0.25	1.17
273.5	X.. X	20/ 40 MESH SAND	0.231	1.6	0.21	0.99
282.4	X.. X	20/ 40 MESH SAND	0.230	1.4	0.19	0.89
	X---------X---------X---------X---------X---------X					

Fig. 11.4—Simulated fracture dimensions for example case—pumping time is 64 minutes.

method discussed in Appendix G. It is seen that for either a GDK or a PKN fracture shape, the fracture treatment was nearly at total-suspension conditions.

Computer-Implemented Constant-Height Models. Many oil and service companies have modeled fracture growth and proppant transport on high-speed computers where it is possible to solve the governing equations[46-48] more completely than the solutions discussed in the preceding section. The most complete of these models simulate the entire process and include the following steps.

1. Prefracture data, involving either estimates of prefracture productivity from specified permeability or calculation of effective permeability from prefracture production data, are analyzed. Multiphase-flow simulators are used to model wellbore hydraulics for these calculations. Reservoir flow range from simple radial-flow-equation calculations to complex computer simulations is simulated.

2. Fluid friction loss and density effects in tubing during fracture treatment pumping are modeled to estimate treating rates and pressures during the treatment.

3. The dynamic fracture dimensions are modeled by means of a finite-difference solution to the governing rock-mechanics equations. Fluid loss is considered in great detail.

4. Proppant transport is modeled in great detail by considering many phenomena that affect fall rate and by a massive bookkeeping effort to track the current locations of various sand stages in the dynamically growing fracture.

5. Fracture closure is modeled to evaluate the extent to which proppant settles before the fracture closes on it and to estimate how much time should elapse before the well is returned to production.

6. Postfracture production response is calculated for the anticipated postfracture producing conditions. This can involve anything from a relatively simple steady-state calculation to a complete reservoir-simulator estimate.

Partial results of such a detailed computer simulation[6,49] for the small treatment example case data in Table 11.2 is shown in Figs. 11.4 through 11.7. Each figure is a representation of one wing of a vertical fracture at a particular time during the treatment. These figures can be visualized most easily if turned sideways with the wellbore in a vertical orientation. At the time shown in Fig. 11.4, the 3-lbm/gal [359-kg/m^3] stage of proppant has been partially pumped. The dynamic fracture length is about 620 ft [190 m], and the wellbore and average widths are 0.26 and 0.20 in. [0.66 and 0.51 cm], respectively. In the pictorial representation of the fracture dimensions, the dots signify that proppant is in suspension. Any bank that has formed at the bottom is signified by B (bank). Note that the proppant front has progressed to only the first 282 ft [86 m] of the total length and that the proppant concentration is increasing above its inlet values because of leakoff. Also note that at this time, nearly half of the pad volume has been lost to leakoff, with only about 40% of the original volume left inside the fracture ahead of the proppant front.

Fig. 11.5 shows similar conditions just at the point that the entire treatment has been pumped into the well. Note the slight amount of settling of proppant at the top of the fracture near the tip. Beyond this point in time, the fracture will close, as shown in Figs. 11.6 and 11.7. In Fig. 11.6, the fracture has only partially closed on proppant, while in Fig. 11.7, every part of the fracture has closed.

The final propped fracture dimensions from Fig. 11.7 have been included in Table 11.3 for comparison. Note that the results are similar to the hand calculations.

Height-Growth Simulation Models. At this writing, fracture models that do not need to assume constant fracture height are being developed.[36-38] With additional information about rock properties and stress distributions in the reservoir to be fractured, these programs can predict the height, length, and width growth rates. At some future date, they will be refined to include a complete simulation of the fracturing process, as discussed previously, and will be able to model the overall fracturing process better. Sample calculations of fracture dimensions are presented in Chap. 5.

11.5 Selection of Materials

The final decisions to be made before designing a fracture treatment are the selection of fracture fluid with additives and the selection

```
 TIME SINCE TREATMENT BEGAN       79.45 MIN                 PAD VOLUME REMAINING AHEAD OF PROPPANT =     6795. GALS
FLUID VOLUME PUMPED =      78880. GAL                                  (FOR EACH WING)
TOTAL PROPPANT INJECTED =      100499. LBS
DYNAMIC FRACTURE LENGTH =         708.0 FT           PERCENT PAD VOLUME REMAINING AHEAD OF PROPPANT =     32.4 %
DYNAMIC FRACTURE WIDTH
          AT THE WELLBORE =      0.2798 INCH
          AVERAGE =      0.2197 INCH
FLUID TEMPERATURE ENTERING PERFORATIONS =        81.1 DEG F

FRACTURE                FRACTURE HEIGHT  (FEET)                 FRACTURE CHARACTERISTICS AT THE RESPECTIVE LENGTHS ARE:
 LENGTH
 (FEET)                                                                                          CONCENTRATIONS
          0.        50.       100.       150.      200.      250.  SUSPENDED PROPPANT    WIDTH,IN  LB/GAL  LB/SQFT  LAYERS
    0.    X---------X---------X---------X---------X---------X
    9.6   X.................................................X  20/ 40 MESH SAND      0.280     3.0      0.46     2.16
   19.2   X.................................................X  20/ 40 MESH SAND      0.280     3.0      0.47     2.17
   28.6   X.................................................X  20/ 40 MESH SAND      0.280     3.1      0.47     2.18
   47.4   X.................................................X  20/ 40 MESH SAND      0.279     3.1      0.48     2.21
   56.7   X.................................................X  20/ 40 MESH SAND      0.279     3.2      0.48     2.23
   75.1   X.................................................X  20/ 40 MESH SAND      0.278     3.2      0.49     2.25
   93.2   X.................................................X  20/ 40 MESH SAND      0.277     3.3      0.49     2.28
  120.1   X................................................ X  20/ 40 MESH SAND      0.276     3.3      0.50     2.31
  146.5   X................................................ X  20/ 40 MESH SAND      0.274     3.4      0.51     2.35
  155.2   X................................................ X  20/ 40 MESH SAND      0.273     3.4      0.51     2.36
  163.8   X................................................ X  20/ 40 MESH SAND      0.272     3.5      0.51     2.37
  198.1   X................................................ X  20/ 40 MESH SAND      0.269     3.6      0.52     2.40
  215.0   X................................................ X  20/ 40 MESH SAND      0.267     3.7      0.52     2.42
  248.4   X...............................................  X  20/ 40 MESH SAND      0.262     3.8      0.53     2.45
  273.2   X...............................................  X  20/ 40 MESH SAND      0.258     3.9      0.53     2.46
  297.8   X...............................................  X  20/ 40 MESH SAND      0.254     4.0      0.53     2.48
  322.2   X...............................................  X  20/ 40 MESH SAND      0.249     4.1      0.53     2.48
  338.5   X...............................................  X  20/ 40 MESH SAND      0.246     4.1      0.53     2.48
  354.7   X..............................................   X  20/ 40 MESH SAND      0.242     4.2      0.53     2.48
  362.8   X..............................................   X  20/ 40 MESH SAND      0.240     4.0      0.51     2.35
  370.9   X..............................................   X  20/ 40 MESH SAND      0.238     3.9      0.49     2.27
  379.0   X..............................................   X  20/ 40 MESH SAND      0.236     3.6      0.46     2.14
  387.1   X.............................................    X  20/ 40 MESH SAND      0.234     3.4      0.43     1.99
  395.2   X.............................................    X  20/ 40 MESH SAND      0.232     3.2      0.40     1.87
  403.3   X.............................................    X  20/ 40 MESH SAND      0.230     2.9      0.36     1.69
  411.5   X.............................................    X  20/ 40 MESH SAND      0.228     2.7      0.35     1.61
  419.6   X............................................     X  20/ 40 MESH SAND      0.225     2.6      0.32     1.50
  427.8   X............................................     X  20/ 40 MESH SAND      0.223     2.3      0.29     1.36
          X---------X---------X---------X---------X---------X
```

Fig. 11.5—Simulated fracture dimensions for example case—pumping time is 80 minutes.

```
 TIME SINCE TREATMENT BEGAN      105.43 MIN                 TOTAL FLUID TO BE LEAKED OFF TO CLOSURE=     20469. GALS
 FLUID VOLUME PUMPED =      78880. GAL                                 (FOR EACH WING)
 TOTAL PROPPANT INJECTED =      100499. LBS
 PROPPED FRACTURE LENGTH =        427.8 FT                         FLUID REMAINING TO LEAK OFF =      9903. GALS
 PROPPED FRACTURE WIDTH                                                (FOR EACH WING)

          AT THE WELLBORE =      0.2193 INCH
          AVERAGE =      0.2051 INCH
 FLUID TEMPERATURE ENTERING PERFORATIONS =        236.8 DEG F
 TIME ELASPED SINCE PUMPING STOPPED =         25.98 MIN

FRACTURE                FRACTURE HEIGHT  (FEET)                 FRACTURE CHARACTERISTICS AT THE RESPECTIVE LENGTHS ARE:
LENGTH
(FEET)                                                                                           CONCENTRATIONS
          0.        50.       100.       150.      200.      250.  SUSPENDED PROPPANT    WIDTH,IN  LB/GAL  LB/SQFT  LAYERS
    0.    X---------X---------X---------X---------X---------X
    9.6   XB................................................X  20/ 40 MESH SAND      0.219     4.0      0.46     2.16
   28.6   XB................................................X  20/ 40 MESH SAND      0.219     4.1      0.47     2.18
   47.4   XB................................................X  20/ 40 MESH SAND      0.219     4.1      0.48     2.21
   65.9   XB................................................X  20/ 40 MESH SAND      0.218     4.2      0.48     2.24
   84.2   XB................................................X  20/ 40 MESH SAND      0.218     4.3      0.49     2.27
  102.2   XB................................................X  20/ 40 MESH SAND      0.217     4.4      0.49     2.29
  120.1   XB................................................X  20/ 40 MESH SAND      0.216     4.4      0.50     2.31
  137.7   XB................................................X  20/ 40 MESH SAND      0.215     4.5      0.50     2.34
  146.5   XB............................................... X  20/ 40 MESH SAND      0.215     4.6      0.51     2.35
  181.0   X................................................ X  20/ 40 MESH SAND      0.212     4.7      0.51     2.39
  206.6   X................................................ X  20/ 40 MESH SAND      0.210     4.8      0.52     2.41
  231.8   X................................................ X  20/ 40 MESH SAND      0.207     5.0      0.52     2.43
  256.7   X................................................ X  20/ 40 MESH SAND      0.204     5.1      0.53     2.45
  281.4   X...............................................  X  20/ 40 MESH SAND      0.201     5.2      0.53     2.47
  289.6   X...............................................  X  20/ 40 MESH SAND      0.200     5.3      0.53     2.47
  305.9   X...............................................  X  20/ 40 MESH SAND      0.198     5.4      0.53     2.48
  314.1   X...............................................  X  20/ 40 MESH SAND      0.197     5.4      0.53     2.48
  330.4   X...............................................  X  20/ 40 MESH SAND      0.194     5.5      0.53     2.48
  338.5   X...............................................  X  20/ 40 MESH SAND      0.193     5.6      0.53     2.48
  354.7   X...............................................  X  20/ 40 MESH SAND      0.190     5.7      0.53     2.48
  379.0   X..............................................   X  20/ 40 MESH SAND      0.185     4.9      0.46     2.14
  387.1   X..............................................   X  20/ 40 MESH SAND      0.184     4.5      0.43     1.99
  403.3   X.............................................    X  20/ 40 MESH SAND      0.180     3.8      0.36     1.69
  411.5   X.............................................    X  20/ 40 MESH SAND      0.178     3.6      0.35     1.61
  427.8   X...........................................      X  20/ 40 MESH SAND      0.175     3.1      0.29     1.36
          X---------X---------X---------X---------X---------X
```

Fig. 11.6—Simulated fracture dimensions for example case during fracture closure—time since pumping stopped is 26 minutes.

```
 TIME SINCE TREATMENT BEGAN      195.43 MIN                   TOTAL FLUID TO BE LEAKED OFF TO CLOSURE=     20469. GALS
FLUID VOLUME PUMPED =       78880. GAL                                  (FOR EACH WING)
TOTAL PROPPANT INJECTED =     100499. LBS
PROPPED FRACTURE LENGTH =         427.8 FT                              FLUID REMAINING TO LEAK OFF =           0. GALS
PROPPED FRACTURE WIDTH                                                     (FOR EACH WING)

          AT THE WELLBORE =     0.0548 INCH
          AVERAGE =     0.0581 INCH
FLUID TEMPERATURE ENTERING PERFORATIONS =        243.7 DEG F
TIME ELASPED SINCE PUMPING STOPPED =         115.98 MIN

FRACTURE                 FRACTURE HEIGHT  (FEET)                    FRACTURE CHARACTERISTICS AT THE RESPECTIVE LENGTHS ARE:
 LENGTH
 (FEET)                                                                                         CONCENTRATIONS
          0.        50.       100.       150.       200.       250.  SUSPENDED PROPPANT   WIDTH,IN  LB/GAL  LB/SQFT  LAYERS
    0.    X---------X---------X---------X---------X---------X
    9.6   XBBBBBBBBBBBBBBBBBBBBBBBBBBBBBBBBBBBBBBBBBBBBBBBB X   20/ 40 MESH SAND     0.055     35.2     0.46     2.16
   28.6   XBBBBBBBBBBBBBBBBBBBBBBBBBBBBBBBBBBBBBBBBBBBBBBBB X   20/ 40 MESH SAND     0.056     35.2     0.47     2.18
   47.4   XBBBBBBBBBBBBBBBBBBBBBBBBBBBBBBBBBBBBBBBBBBBBBBBB X   20/ 40 MESH SAND     0.056     35.2     0.48     2.21
   65.9   XBBBBBBBBBBBBBBBBBBBBBBBBBBBBBBBBBBBBBBBBBBBBBBB  X   20/ 40 MESH SAND     0.057     35.2     0.48     2.24
   75.1   XBBBBBBBBBBBBBBBBBBBBBBBBBBBBBBBBBBBBBBBBBBBBBBB  X   20/ 40 MESH SAND     0.057     35.2     0.49     2.25
   93.2   XBBBBBBBBBBBBBBBBBBBBBBBBBBBBBBBBBBBBBBBBBBBBBBB  X   20/ 40 MESH SAND     0.058     35.2     0.49     2.28
  102.2   XBBBBBBBBBBBBBBBBBBBBBBBBBBBBBBBBBBBBBBBBBBBBBBB  X   20/ 40 MESH SAND     0.058     35.2     0.49     2.29
  128.9   XBBBBBBBBBBBBBBBBBBBBBBBBBBBBBBBBBBBBBBBBBBBBBBB  X   20/ 40 MESH SAND     0.059     35.2     0.50     2.33
  163.8   XBBBBBBBBBBBBBBBBBBBBBBBBBBBBBBBBBBBBBBBBBBBBBBB  X   20/ 40 MESH SAND     0.060     35.2     0.51     2.37
  215.0   XBBBBBBBBBBBBBBBBBBBBBBBBBBBBBBBBBBBBBBBBBBBBBB   X   20/ 40 MESH SAND     0.062     35.2     0.52     2.42
  240.1   XBBBBBBBBBBBBBBBBBBBBBBBBBBBBBBBBBBBBBBBBBBBBBB   X   20/ 40 MESH SAND     0.062     35.2     0.53     2.44
  265.0   XBBBBBBBBBBBBBBBBBBBBBBBBBBBBBBBBBBBBBBBBBBBBBB   X   20/ 40 MESH SAND     0.063     35.2     0.53     2.46
  273.2   XBBBBBBBBBBBBBBBBBBBBBBBBBBBBBBBBBBBBBBBBBBBBBB   X   20/ 40 MESH SAND     0.063     35.2     0.53     2.46
  289.6   XBBBBBBBBBBBBBBBBBBBBBBBBBBBBBBBBBBBBBBBBBBBBBB   X   20/ 40 MESH SAND     0.063     35.2     0.53     2.47
  297.8   XBBBBBBBBBBBBBBBBBBBBBBBBBBBBBBBBBBBBBBBBBBBBBB   X   20/ 40 MESH SAND     0.063     35.2     0.53     2.48
  314.1   XBBBBBBBBBBBBBBBBBBBBBBBBBBBBBBBBBBBBBBBBBBBBBB   X   20/ 40 MESH SAND     0.063     35.2     0.53     2.48
  322.2   XBBBBBBBBBBBBBBBBBBBBBBBBBBBBBBBBBBBBBBBBBBBBBB   X   20/ 40 MESH SAND     0.063     35.2     0.53     2.48
  338.5   XBBBBBBBBBBBBBBBBBBBBBBBBBBBBBBBBBBBBBBBBBBBBB    X   20/ 40 MESH SAND     0.063     35.2     0.53     2.48
  346.6   XBBBBBBBBBBBBBBBBBBBBBBBBBBBBBBBBBBBBBBBBBBBBB    X   20/ 40 MESH SAND     0.063     35.2     0.53     2.48
  362.8   XBBBBBBBBBBBBBBBBBBBBBBBBBBBBBBBBBBBBBBBBBBBBB    X   20/ 40 MESH SAND     0.060     35.2     0.51     2.35
  379.0   XBBBBBBBBBBBBBBBBBBBBBBBBBBBBBBBBBBBBBBBBBBBB     X   20/ 40 MESH SAND     0.054     35.2     0.46     2.14
  395.2   XBBBBBBBBBBBBBBBBBBBBBBBBBBBBBBBBBBBBBBBBBBBB     X   20/ 40 MESH SAND     0.048     35.2     0.40     1.87
  411.5   XBBBBBBBBBBBBBBBBBBBBBBBBBBBBBBBBBBBBBBBBBBB      X   20/ 40 MESH SAND     0.041     35.2     0.35     1.61
  419.6   XBBBBBBBBBBBBBBBBBBBBBBBBBBBBBBBBBBBBBBBBBB       X   20/ 40 MESH SAND     0.038     35.2     0.32     1.50
  427.8   XBBBBBBBBBBBBBBBBBBBBBBBBBBBBBBBBBBBBBBBBB        X   20/ 40 MESH SAND     0.035     35.2     0.29     1.36
          X---------X---------X---------X---------X---------X
```

Fig. 11.7—Simulated fracture dimensions for example case after fracturing closure—time since pumping stopped is 116 minutes.

of proppant. The characteristics and properties of a wide range of these materials have already been discussed in previous chapters. The aspects discussed in the remainder of this section relate to the thought process whereby the list of potential materials is reduced to a manageable number.

Fluid Considerations. The five technical aspects upon which a fracturing fluid is selected or rejected are viscosity, fluid friction loss, fluid loss, cleanup, and compatibility with reservoir rock and fluids.

Cost and availability are certainly two other factors that receive consideration once a fluid has met the technical requirements.

Viscosity. The level of viscosity provided by a fracturing fluid will affect the width and length distributions within the created fracture volume, as well as the final distribution of proppant. Most treatments designed today use moderate- to high-viscosity fluids that will generate sufficient fracture width for easy injection of high proppant concentrations with nearly total suspension of the proppant during pumping. The preference for relatively high viscosities is based on the desire to achieve wide propped fractures and to avoid screenouts that would occur if proppant dropped out of the fluid in either the casing or the fracture. During the design process, numerous combinations of fluids and proppants can result in fractures that are at times too narrow or that allow too much proppant settling. The goal of the design engineer should be to consider all the plausible fluid/proppant combinations, even though some of the fluid systems are not highly viscous. These systems should be studied with a computer in the office where screenouts do not cause large financial losses.

At fracture flow conditions, if average fluid viscosity along the fracture is at least 50 to 100 cp [50 to 100 mPa·s], proppants will be almost totally suspended during pumping. After pumping, excessive proppant settling can occur even though viscosity is high if fluid loss is very low. The fluid can break and lose its viscosity before the fracture closes on the proppant adjacent to the pay zone. Current fluid systems that are totally suspending for the reservoir parameters in the example well in this chapter are 40-lbm/1,000-gal [4793-g/m^3] crosslinked gels, brine-external polymer emulsion, most foams with gelled aqueous phases, and some gelled oils.

If average fluid viscosity in the fracture is less than about 10 cp [10 mPa·s], significant proppant bank formation can occur during treatment injection, especially if proppant is more dense than sand. For the reservoir conditions in Table 11.2, a 20-lbm/1,000-gal [2397-g/m^3] uncrosslinked guar gel would give significant bank buildup, and a 10-lbm/1,000-gal [1198-g/m^3] gel or slick water would give a complete equilibrium bank.

Fig. 11.4 shows a proppant-pack profile for a 40-lbm/1,000-gal [4793-g/m^3] crosslinked gel demonstrating nearly perfect proppant suspension. Figs. 11.6 and 11.7 illustrate closing behavior where little settling occurs. The formation of an equilibrium bank with a thin fracturing fluid is illustrated in Figs. 11.8 and 11.9. A 10-lbm/1,000-gal [1198-g/m^3] linear gelled water gives the proppant distribution shown in Fig. 11.8 early in the treatment. By the end of the treatment, an equilibrium bank has formed over the first 434 ft [132 m], as seen in Fig. 11.9.

The fluid selection process from the viscosity viewpoint should involve the following steps.

1. Select a fluid that is viscous enough to give nearly total suspension for initial scoping calculations and gross treatment sizing.
2. Lower the viscosity of this fluid by decreasing gel concentration to achieve less complete suspension and yet to place the treatment successfully.
3. Try much-lower-viscosity fluids to form equilibrium banks only if wider propped fractures are needed, such as in soft formations.

These calculations should not be pursued in earnest until the other aspects of fluid and proppant selection have been decided.

Fluid Friction Loss. The ease with which a fluid can be pumped through long tubing strings has a direct relationship to fluid viscosity. However, traditional fracturing thought places pumpability as a consideration separate from the viscosity considerations discussed earlier. The history of this approach comes from past experience where laboratory measurements of fluid viscosity in a fracture usually did not agree with the way that it pumped in tubing. For

```
 TIME SINCE TREATMENT BEGAN        143.58 MIN
FLUID VOLUME PUMPED =    136872. GAL
TOTAL PROPPANT INJECTED =    306875. LBS
DYNAMIC FRACTURE LENGTH =       1206.1 FT
DYNAMIC FRACTURE WIDTH
          AT THE WELLBORE =     0.2883 INCH
          AVERAGE =     0.2264 INCH
FLUID TEMPERATURE ENTERING PERFORATIONS =        79.2 DEG F

FRACTURE                 FRACTURE HEIGHT  (FEET)                FRACTURE CHARACTERISTICS AT THE RESPECTIVE LENGTHS ARE:
 LENGTH
 (FEET)                                                                                        CONCENTRATIONS
          0.        50.       100.      150.      200.      250.   SUSPENDED PROPPANT   WIDTH,IN   LB/GAL   LB/SQFT   LAYERS
      0.  X---------X---------X---------X---------X---------X
          |BBBBBBBBBBBBBBBBBBBBB.........................   |
          |BBBBBBBBBBBBBBBBBBBBB.........................   |
          |BBBBBBBBBBBBBBBBBBBBB.........................   |
    99.1  XBBBBBBBBBBBBBBBBBBBBB.........................   X   20/ 40 MESH SAND     0.287      4.5      0.67      3.10
          |BBBBBBBBBBBBBBBBBBBB.......................      |
          |BBBBBBBBBBBBBBBBBBBB.......................      |
          |BBBBBBBBBBBBBBBBBBBB.......................      |
   182.5  XBBBBBBBBBBBBBBBBBBBB.......................      X   20/ 40 MESH SAND     0.285      5.0      0.72      3.35
          |BBBBBBBBBBBBBBBBBB.....................          |
   253.9  XBBBBBBBBBBBBBBBBBB.....................          X   20/ 40 MESH SAND     0.282      5.5      0.77      3.59
          |BBBBBBBBBBBBBBBB....................             |
   315.6  XBBBBBBBBBBBBBBBB....................             X   20/ 40 MESH SAND     0.278      6.0      0.82      3.82
          |BBBBBBBBBBBBBBB..................                |
   369.6  XBBBBBBBBBBBBBBB..................                X   20/ 40 MESH SAND     0.274      6.6      0.87      4.03
          |BBBBBBBBBBBBB.................                   |
   417.5  XBBBBBBBBBBBBB.................                   X   20/ 40 MESH SAND     0.270      7.0      0.90      4.18
          |BBBBBBBBBBBB............                         |
   460.3  XBBBBBBBBBBBB............                         X   20/ 40 MESH SAND     0.266      7.0      0.88      4.11
          |BBBBBBBBBB.....                                  |
   498.4  XBBBBBBBBBB.....                                  X   20/ 40 MESH SAND     0.262      6.0      0.78      3.61
   532.6  XBBBBBBBBB...                                     X   20/ 40 MESH SAND     0.259      6.5      0.81      3.75
          |BBBBBBB..                                        |
   563.5  XBBBBBBB..                                        X   20/ 40 MESH SAND     0.255      6.9      0.83      3.88
   591.7  XBBBBB.                                           X   20/ 40 MESH SAND     0.251      7.3      0.86      4.01
          |BBBB                                             |
   617.6  XBBBB                                             X
   641.4  XBB                                               X
          X---------X---------X---------X---------X---------X
```

Fig. 11.8—Simulated fracture dimensions for example case—10-lbm/1,000-gal linear gel fluid. Significant banking of proppant.

```
 TIME SINCE TREATMENT BEGAN        298.95 MIN
FLUID VOLUME PUMPED =    268053. GAL
TOTAL PROPPANT INJECTED =  1013116. LBS
DYNAMIC FRACTURE LENGTH =       1795.4 FT
DYNAMIC FRACTURE WIDTH
          AT THE WELLBORE =     0.3517 INCH
          AVERAGE =     0.2762 INCH
FLUID TEMPERATURE ENTERING PERFORATIONS =        77.8 DEG F

FRACTURE                 FRACTURE HEIGHT  (FEET)                FRACTURE CHARACTERISTICS AT THE RESPECTIVE LENGTHS ARE:
 LENGTH
 (FEET)                                                                                        CONCENTRATIONS
          0.        50.       100.      150.      200.      250.   SUSPENDED PROPPANT   WIDTH,IN   LB/GAL   LB/SQFT   LAYERS
      0.  X---------X---------X---------X---------X---------X
          |BBBBBBBBBBBBBBBBBBBBBBBBBBBBBBBBBBBBBBBBBBBBBB.. |
          |BBBBBBBBBBBBBBBBBBBBBBBBBBBBBBBBBBBBBBBBBBBBBB.. |
          |BBBBBBBBBBBBBBBBBBBBBBBBBBBBBBBBBBBBBBBBBBBBBB.. |
          |BBBBBBBBBBBBBBBBBBBBBBBBBBBBBBBBBBBBBBBBBBBBBB.. |
          |BBBBBBBBBBBBBBBBBBBBBBBBBBBBBBBBBBBBBBBBBBBBBB.. |
          |BBBBBBBBBBBBBBBBBBBBBBBBBBBBBBBBBBBBBBBBBBBBBB.. |
          |BBBBBBBBBBBBBBBBBBBBBBBBBBBBBBBBBBBBBBBBBBBBBB.. |
   434.2  XBBBBBBBBBBBBBBBBBBBBBBBBBBBBBBBBBBBBBBBBBBBBBB.. X   20/ 40 MESH SAND     0.341      9.1      1.37      6.36
          |BBBBBBBBBBBBBBBBBBBBBBBBBBBBBBBBBB               |
          |BBBBBBBBBBBBBBBBBBBBBBBBBBBBBBBBBB               |
   518.4  XBBBBBBBBBBBBBBBBBBBBBBBBBBBBBBBBBB               X
          |BBBBBBBBBBBBBBBBBBBBBBBBBBBBBB..                 |
   577.8  XBBBBBBBBBBBBBBBBBBBBBBBBBBBBBB..                 X   20/ 40 MESH SAND     0.333     10.9      1.52      7.06
          |BBBBBBBBBBBBBBBBBBBBBBBBBB....                   |
   624.4  XBBBBBBBBBBBBBBBBBBBBBBBBBB....                   X   20/ 40 MESH SAND     0.330     11.7      1.57      7.30
          |BBBBBBBBBBBBBBBBBBBBBB.....                      |
   662.4  XBBBBBBBBBBBBBBBBBBBBBB.....                      X   20/ 40 MESH SAND     0.327     12.4      1.62      7.51
          |BBBBBBBBBBBBBBBBBB.......                        |
   694.6  XBBBBBBBBBBBBBBBBBB.......                        X   20/ 40 MESH SAND     0.324     13.0      1.66      7.69
   722.7  XBBBBBBBBBBBBBBB........                          X   20/ 40 MESH SAND     0.322     13.6      1.69      7.85
   747.8  XBBBBBBBBBBBB.........                            X   20/ 40 MESH SAND     0.320     14.2      1.72      8.00
   770.7  XBBBBBBBBBB........                               X   20/ 40 MESH SAND     0.318     14.7      1.75      8.11
   791.9  XBBBBBBBB........                                 X   20/ 40 MESH SAND     0.316     15.0      1.76      8.17
   811.5  XBBBBBBB....                                      X   20/ 40 MESH SAND     0.314     15.1      1.75      8.15
   829.6  XBBBB..                                           X   20/ 40 MESH SAND     0.312     15.0      1.74      8.09
   846.2  XBB                                               X
   862.0  X                                                 X
          X---------X---------X---------X---------X---------X
```

Fig. 11.9—Simulated fracture dimensions for example case—10-lbm/1,000-gal linear gel fluid. Equilibrium bank formed during pumping.

TABLE 11.4—ESTIMATED TREATING RATES AND PRESSURES FOR EXAMPLE CASE

Treating Fluid Type	Treating Rate (bbl/min)	Surface Injection Pressure (psi)	Fluid Head Pressure (psi)
40 lbm/1,000 gal crosslinked gel	25	9,075	4,025
	23.3	8,500	4,025
Brine-external emulsion	25	21,900	3,500
	12.7	8,500	3,500
Slick water, 20 lbm/1,000 gal linear gel	25	7,950	4,025
Nitrogen foam (70% quality with	25	9,710	2,625
40 lbm/1,000 gal linear gel)	20	8,660	2,575
	15	8,035	2,545
Crosslinked gelled oil	25	7,785	3,220

example, the Fann Model 50™ method of studying crosslinked aqueous gels led to reasonable fluid-viscosity estimates in flow along the fracture but predicted that friction loss in the tubing would be much greater than it actually was in the field. For this fluid and most others, it was found best to treat pumpability as a fluid property separate from viscosity for flow in a fracture.

The various service-company fluid systems have been studied in the field and laboratory for their friction loss behavior in flow through tubing, each with a set of curves for estimation of friction loss. The use of these data has already been discussed in Chap. 9. Some fluids can be eliminated from consideration quickly because of low injection rates and high treating pressures from excessive pipe friction. Potential fluids will be compared in terms of the various injection rates that can be achieved at a specific surface treating pressure or in terms of treating pressures at a specific rate.

Table 11.4 gives the estimates of surface treating pressures and rates that could be achieved for various fluid systems. If surface pressure is limited to 8,500 psi [58.6 MPa], then brine-external emulsion, nitrogen foam gel, and crosslinked gel could not be pumped at the desired rate of 25 bbl/min [4 m^3/min]. Because crosslinked gel is just over that limit, surface injection pressure could be reduced to 8,500 psi [58.6 MPa] by lowering the rate to 23.3 bbl/min [3.7 m^3/min].

Additional pressure losses come from flow through the perforations and along the fracture. Both of these losses are usually on the order of a few hundred psi and are neglected. Further discussion of them is contained in Chap. 12 concerning limited entry fracture staging.

Cleanup. From the design viewpoint, the three most important aspects of the fluid for acceptable cleanup behavior are (1) rapid breaking of the fluid after the treatment is over to a very-low-viscosity system that is easily produced back; (2) low residue content of the broken fluid so that the proppant pack is not irrevocably damaged; and (3) fluid density low enough that the static reservoir pressure will adequately flow the fracture liquids back to the surface.

The breaking behavior of fracture fluids was discussed in Chap. 7. With the addition of appropriate chemical breakers in the correct amounts, most fluids will break as desired. Those fluids that use either thermal or adsorption-type breaks will function appropriately if used according to their guidelines.

Excessive residue can be eliminated by selection of refined fluids. Either manmade polymers or chemically processed natural substances can be made with very low residue content.

The main area of concern in cleanup is recovery of treatment fluids. Because an estimate of static reservoir pressure is needed for a complete design, it is a simple matter to evaluate whether this pressure is sufficient to allow fracture fluid to be produced by natural flow. If the reservoir has subnormal pressure, gelled oils and foams should be considered. Gelled oils could be flowed back if the reservoir is only slightly underpressured. If pressure is very low, a foam should be considered as a viable choice to recover fracture liquids rapidly without the lengthy process of swabbing.

Table 11.4 shows the fluid head pressure for the fluids considered in the example case. Because the example well is normally pressured, fracture fluid density is not a problem during cleanup. Had the formation pressure been 3,000 psi [20.7 MPa], only a foam would have cleaned up with drive from reservoir pressure.

Compatibility With Reservoir Fluid and Rock. The petrographic characteristics of a reservoir rock and the fluid contained in its pores can occasionally cause a sensitivity to fracturing fluids. The best way to evaluate this potential problem is through laboratory coreflow studies, though all too often the first discovery of the problem comes from the surveillance of field treatment data. This has already been discussed in Chap. 7.

The design engineer should be concerned about any information that comes from geological or chemical analyses of the reservoir about to be fractured. For example, a clay analysis may reveal that the reservoir rock contains a significant amount of swellable clay like montmorillonite. With this information, the fracture fluid should be selected to minimize clay swelling potential. This can be achieved by the addition of special swelling-preventive chemicals or through the use of 2% KCl in a water-based fluid.

Another factor to consider is the potential precipitation of various compounds mainly involving iron ions. When this is known to pose a problem, chemical additives are available to prevent it. Some reservoirs exhibit a general sensitivity to any water composition for a variety of possible reasons. When this is suspected, oil-based fluids should be considered.

A final potential reservoir incompatibility with fracture fluid is the adverse thermal interaction between fracture fluid and reservoir fluids. Precipitation of asphaltenes and paraffins can occur in an otherwise undersaturated reservoir fluid if it is cooled by fracturing fluid. Unfortunately, redissolution of the precipitate can be quite slow, even though the fluid surrounding the precipitate is undersaturated. Precipitated asphaltenes can plug some pores totally and never be in significant contact with undersaturated reservoir fluids.

To prevent asphaltene or paraffin deposition, fracture fluids can be heated to a high enough temperature to prevent precipitation, or a solubilizing fracture fluid can be used. If special PVT analyses on bottomhole reservoir fluid samples are available, one would know the specific temperature at which precipitation occurs. If the precipitation temperature is less than about 150°F [66°C], a moderate-sized fracturing treatment could be heated above this temperature to prevent precipitation. A more practical solution may be to use a gelled oil fluid that contains enough aromatic components to redissolve precipitate where it occurs.

Final Fluid Selection. After the technical aspects have been considered, the list of potential fracture fluids is likely narrowed to a few generic types. At this point, such other factors as cost and availability are considered to arrive at a few fluids (three or four at the most) that will be considered in a complete design. This choice can often be narrowed to a single fluid that has been found to perform acceptably in similar treatments.

It is desirable to limit the number of fluids to avoid excessive design calculations. Other factors, such as injection rate and sand schedule, are more important for study. Of course, all parametric variations are multiplied together to arrive at the total number of cases that must be considered.

In the base design example case, no water sensitivity is specified to limit water-based fluids, which are the front runners based on cost and availability. Emulsions and foam are eliminated because of excessive treating pressure. The costs of both of these fluids are also higher than that of gelled water. Gelled oil is a viable fluid in terms of technical requirements, but it also is more expensive than gelled water. In summary, crosslinked gelled water is the fluid ultimately selected because of its adequate technical standards and lowest cost.

Proppant Considerations. The main aspects of proppant selection from a design viewpoint are[50-53] proppant permeability at stress compared with reservoir permeability and proppant transportability through perforations and along the fracture.

Proppant Permeability. Fortunately, there are fewer choices for proppants than for fluids. Chap. 6 showed that there are no more than five or six different proppants in use today. The basic difference between these proppants is proppant-pack permeability under high-stress conditions. Sand and intermediate-density and high-strength proppants are the three main types of proppants. If a computer program is available to do the design calculations, it is often simplest to consider all three as the propping agent. At the very least, calculations should be done with both sand and high-strength proppant to span the range of proppant characteristics.

As already seen during the discussion of the McGuire-Sikora and Tannich-Nierode curves, it is possible in certain cases to go from a conductivity-limited condition to a length-limited condition by changing from sand to high-strength proppant. The design method presented in this chapter automatically looks at this span in potential stimulation by looking at the effects of proppant type and the size of treatment. For the example well, designs for small and large treatments with both sand and bauxite proppant are done as a starting point, and the results are plotted in Fig. 11.3. Depending on the relative location of these points in Fig. 11.3, a decision is made to consider further designs with more or less proppant of one or both types. This procedure is less complicated than other gas well design methods presented in the literature because the Tannich-Nierode curves presented in Fig. 11.3 represent detailed reservoir simulator calculations. The use of these curves is similar to the design of an oilwell treatment with McGuire-Sikora curves (Fig. 11.2). The only difference is that the conductivity contrast parameter in Fig. 11.3 includes turbulence.

Proppant Transportability. As long as a fracturing fluid has moderate viscosity and treating rate, typical treatments will have no difficulty transporting the proppant from the blender tub to the perforation entrance. The only potential for trouble in the tubing is proppant settling if an extended shut-in should occur during pumping of the proppant stages. Beyond the wellbore, the first high-potential location for screenout is the perforation tunnel itself. The tunnel must have a cross-sectional area that allows proppant passage without bridging. As mentioned previously, Fig. 11.1 indicates the minimum perforation diameters needed for pumping various proppant sizes at various concentrations. The required perforation opening is smallest for low concentrations and approaches an asymptote beyond about 6 lbm/gal [719 kg/m^3]. Behind each of these curves is the laboratory finding that perforation diameter must be at least six times greater than the maximum proppant particle diameter to prevent bridging of high-concentration slurries. Ultra-low-density slurries could eliminate bridging with a perforation-to-particle ratio as low as 2. One proppant selection criterion is for the proppant to be small enough to pass through the perforations without bridging.

Fracture width at the wellbore must also be large enough to accept the initial stages of proppant. It is generally required that the ratio of fracture width to maximum proppant diameter be at least 2½ to 3. These values come from bridging experiments on flow of proppant between parallel plates. It is usually true that once the fracture is wide enough to accept the first proppant stages, subsequent stages will not have bridging problems in the fracture at the wellbore.

However, screenouts can occur along the fracture length with proppant bridging between the fracture walls if concentration becomes too great.[49,54-60] About the only way to assess this possibility is by use of a detailed computer simulator to track the concentrations relative to widths along the fracture length. The potential for bridging depends on the specific fluid/proppant combination chosen. A fluid with high fluid loss or low viscosity may not generate enough width to prevent bridging. A proppant that is too large may similarly bridge even in a viscous, low-fluid-loss fluid.

The final location for screenouts is at the fracture tip.[61] If all the pad fluid has leaked off at some point in the treatment, the proppant front is then at the tip, and bridging there can prevent further fracture growth. This is called a "tip pack" if done intentionally. Once the tip becomes encumbered with proppant, the fracture volume may not grow fast enough to accommodate the incoming proppant and a screenout will occur because the fracture volume has filled up. The example problem will look closely at the behavior of the pad volume during the treatment to arrive at a pad volume that is sufficiently large to prevent bridging at the fracture tip. Detailed discussion of proppant transport is presented in Chap. 10.

11.6 Optimal Treatment Selection

This section discusses the essence of fracture treatment design. As stated in the Overview, the main objective of this chapter is to present a design method that will lead to a economically optimal proppant fracture treatment. For simplicity, the controlling economic function used will be treatment payout time. This economic measure does not account for the time value of money but has the advantage of simplicity so that economic matters will not cloud the technical design process itself. Once the details of the design method are understood, it is a simple matter to incorporate other more realistic economic functions. Many of these better yardsticks are discussed in Chap. 17.

Meaning of Optimization. The two levels of optimization that can be done with a fracture treatment are economic optimization of accelerated recovery over the lifetime of the well and refinement of treatment procedures and schedules to place the optimal treatment into the well at minimum cost.

The optimization process is broken into two parts because of the complexity of the operation being optimized. The first step is a gross optimization leading to the selection of the best treatment size; the second is a specific refinement of the job procedure to place this job at minimum cost. These two steps together are what is normally defined as the optimization of a less complex process. The two points will be discussed in the remainder of this section.

Design Process To Achieve Optimal Economic Stimulation. The practical goal of most fracturing treatments is to obtain the highest production rates possible immediately after the treatment. In the past, long-term recovery implications were taken into account only to the extent that excessive rates might damage overall reservoir performance. As computers become more accessible, it is expected that long-range recovery implications will become the main focus of fracture treatment design. Today, long-range recovery implications are given careful scrutiny only in expensive or uncertain production areas where fracturing is an integral part of reservoir development.

The iterative trial-and-error process to arrive at maximum stimulation consists of the following steps.

1. Obtain and analyze thoroughly the design input information indicated in Sec. 11.2.
2. Simulate treatments within the bounds imposed by the design input data investigating different materials and different amounts of these materials. Do four initial designs, considering sand and bauxite as proppant with job sizes of 100,000 and 1,000,000 lbm [45.4 and 453.6 Mg] total.
3. Select that combination of materials that gives the most stimulation at the economic limit and refine the fluid and proppant volumes.
4. Assess the sensitivity of the design to critical variables that are somewhat uncertain by repeating Steps 1 through 3.

It must be emphasized that the most critical piece of design input information is undamaged/unstimulated reservoir productivity. As discussed in Sec. 11.3, increased productivity after a successful fracture treatment can come from reservoir stimulation and/or damage removal. Reservoir stimulation is a term reserved for that gain in productivity that results from alteration of reservoir flow pattern in the vicinity of the fracture. Reservoir stimulation gains will be very high if effective reservoir permeability is very low, and insignificant reservoir stimulation will result if permeability is too high. For most reservoirs, if permeability is more than 5 to

TABLE 11.5—DESIGN RESULTS SUMMARY FOR EXAMPLE CASE

Case	Total Fluid Volume (gal)	Pad Volume (gal)	Pad Left Ahead of Proppant (%)	Total Proppant (lbm)	Proppant Schedule (lbm/gal)	Proppant Type	Propped Length (ft)	Propped Width (in.)	Stimulation Ratio	Stimulated Rate (MMscf/D)
1	79,000	42,000	32.4	100,000	1,2,3	Sand	430	0.058	2.59	1.81
2	310,000	84,000	4.9	1,000,000	1,2,3,4,5,6	Sand	1,580	0.152	4.72	3.31
3	79,000	42,000	32.9	100,000	1,2,3	Bauxite	420	0.043	3.09	2.16
4	310,000	84,000	5.3	1,000,000	1,2,3,4,5,6	Bauxite	1,330	0.115	5.98	4.19
5	560,000	168,000	5.0	2,000,000	1,2,3,4,5,6	Sand	2,250	0.214	5.71	4.00
6	490,000	200,000	13.4	1,400,000	1,2,3,4,5,6	Sand	1,805	0.191	5.38	3.77
7	390,000	100,000	3.8	1,400,000	1,2,3,4,5,6	Sand	1,850	0.177	5.23	3.66
8	340,000	50,000	0.8	1,400,000	1,2,3,4,5,6	Sand	1,800	0.170	5.07	3.55
9	315,000	25,000	0.1*	1,400,000	1,2,3,4,5,6	Sand	1,750	0.167	4.97	3.48
10	187,000	100,000	30.5	300,000	1,2,3,4,5	Bauxite	725	0.075	4.16	2.91
11	137,000	50,000	15.2	300,000	1,2,3,4,5	Bauxite	815	0.068	4.25	2.97
12	112,000	25,000	4.4	300,000	1,2,3,4,5	Bauxite	840	0.062	4.18	2.93
13	97,000	10,000	0.6*	300,000	1,2,3,4,5	Bauxite	820	0.059	4.05	2.84
14	374,000	84,000	2.7	1,400,000	1,2,3,4,5,6	Sand	1,840	0.175	5.19	3.63
15	129,000	42,000	11.4	300,000	1,2,3,4,5	Bauxite	830	0.066	4.25	2.98
16	584,000	84,000	0.8	1,000,000	2	Sand	1,140**	0.085	3.14	2.20
17	334,000	84,000	4.1	1,000,000	4	Sand	1,660	0.133	4.32	3.02
18	250,000	84,000	8.4	1,000,000	6	Sand	1,350	0.171	4.72	3.30
19	209,000	84,000	12.8	1,000,000	8	Sand	1,165	0.203	4.85	3.39
20	187,000	84,000	16.3	1,000,000	10	Sand	1,050	0.232	4.91	3.44
21	154,000	84,000	24.8	1,000,000	15	Sand	850	0.285	4.65	3.25
22†	310,000	84,000	24.6	1,000,000	1,2,3,4,5,6	Sand	2,315	0.106	3.85	2.69
23†	584,000	84,000	17.1	1,000,000	2	Sand	2,250**	0.060	2.55	1.79
24†	334,000	84,000	23.1	1,000,000	4	Sand	2,450	0.096	3.39	2.37
25†	310,000	84,000	29.2	1,000,000	6	Sand	1,935	0.122	3.93	2.75
26†	209,000	84,000	32.8	1,000,000	8	Sand	1,650	0.144	4.26	2.98
27†	187,000	84,000	43.2	1,000,000	10	Sand	1,380	0.160	4.37	3.06
28†	154,000	84,000	67.4	1,000,000	15	Sand	860	0.183	4.10	2.87
29†	61,000	42,000	64.2	100,000	2,4,6,8	Sand	355	0.069	2.67	1.87
30†	225,000	84,000	31.8	1,000,000	2,4,6,8,10	Sand	1,750	0.140	4.47	3.13
31†	61,000	42,000	64.7	100,000	2,4,6,8	Bauxite	340	0.052	3.09	2.17
32†	225,000	84,000	32.9	1,000,000	2,4,6,8,10	Bauxite	1,660	0.108	6.34	4.44
33†	315,000	84,000	23.3	2,000,000	2,4,6,8,10,12	Sand	2,460	0.144	7.29	5.10
34†	95,000	42,000	51.4	400,000	2,4,6,8,10	Bauxite	820	0.087	4.52	3.17
35†	305,000	75,000	17.3	2,000,000	2,4,6,8,10,12	Sand	2,677	0.179	6.00	4.20
36†	330,000	100,000	19.6	2,000,000	2,4,6,8,10,12	Sand	2,650	0.188	5.20	3.64
37†	380,000	150,000	23.4	2,000,000	2,4,6,8,10,12	Sand	2,588	0.189	5.27	3.69
38‡	79,000	42,000	32.4	100,000	1,2,3	Sand	430	0.058	1.41	9.86
39‡	310,000	84,000	4.9	1,000,000	1,2,3,4,5,6	Sand	1,580	0.152	1.84	12.86
40‡	79,000	42,000	32.9	100,000	1,2,3	Bauxite	420	0.043	1.41	9.86
41‡	310,000	84,000	5.3	1,000,000	1,2,3,4,5,6	Bauxite	1,540	0.115	2.34	16.38
42‡	41,250	21,000	31.4	50,000	1,2,3	Sand	300	0.042	1.36	9.50
43‡	16,800	10,000	38.2	10,000	1,2	Sand	135	0.018	1.19	8.32
44‡	41,250	21,000	31.4	50,000	1,2,3	Bauxite	290	0.031	1.60	11.22

*Proppant bridging occurred near the fracture tip.
**Effective propped fracture length reduced by excessive proppant settling.
†Using the PKN model for fracture cross section.
‡Reservoir permeability increased to 1 from 0.1 md for base case.

10 md, fracturing will provide very little reservoir stimulation. Total stimulation in this case could still be quite high if the well were severely damaged during drilling or completion. The damage in this case could also have been removed in other ways without the risk of a fracture treatment.

Initial Scoping Calculations. The objective of doing the four scoping designs is to be able to put four data points on either the McGuire-Sikora or Tannich-Nierode curves so that the working location on these curves is defined. After the economics is computed for each of these cases, another one or two cases are often required to complete the selection of gross treatment size. For the base example case being carried through this chapter, an initial treatment simulation has already been presented in Figs. 11.4 through 11.7 for 100,000 lbm [45.4 Mg] of sand. The only auxiliary calculation done before this case could be run was the determination of effective reservoir permeability from the data presented in Table 11.2. The well can produce 725 Mcf/D [20.5×10^3 m^3/d] of gas if effective reservoir permeability is about 0.10 md. Reservoir pressure, flowing BHP, net sand thickness, fluid viscosity and compressibility, and drainage radius were used in the pseudosteady-state radial-flow equation to arrive at this permeability value. If buildup or more detailed data were available, reservoir permeability could be determined more confidently. The radial-flow calculation implicitly assumed that while producing 725 Mcf/D [20.5×10^3 m^3/d] the well was undamaged and unstimulated. Time spent in determination of this parameter is well spent because it is one of the main factors governing stimulation ratio on both the McGuire-Sikora and Tannich-Nierode curves.

Table 11.5 contains the details of the parameters and results of the first four scoping calculations as Cases 1 through 4. All subsequent references to design cases will refer to the case numbers in this table. The first four cases are plotted in Fig. 11.3. It is seen that Treatment 1 is somewhat length-limited in that a larger treatment could give increased stimulation. The larger treatment, Treatment 2, mostly relieves the length limitation. At Point 2, maximum stimulation is almost achieved within the constraints of sand as a proppant.

To achieve significantly greater stimulation, permeability contrast must be increased by either significantly increasing propped width or changing to a higher-permeability proppant. Points 3 and 4 in Fig. 11.3 show that small and large bauxite jobs do give significantly greater stimulation than sand. The 1,000,000-lbm [453.6-Mg] bauxite treatment gives the most stimulation of all four treatments considered.

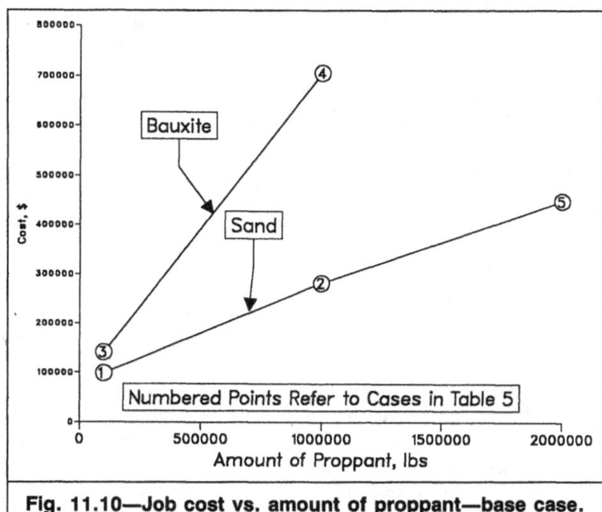

Fig. 11.10—Job cost vs. amount of proppant—base case.

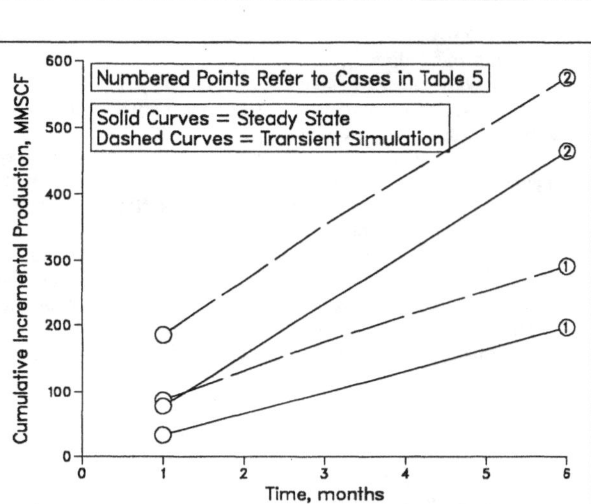

Fig. 11.11—Incremental production from fracturing vs. time—sand base case.

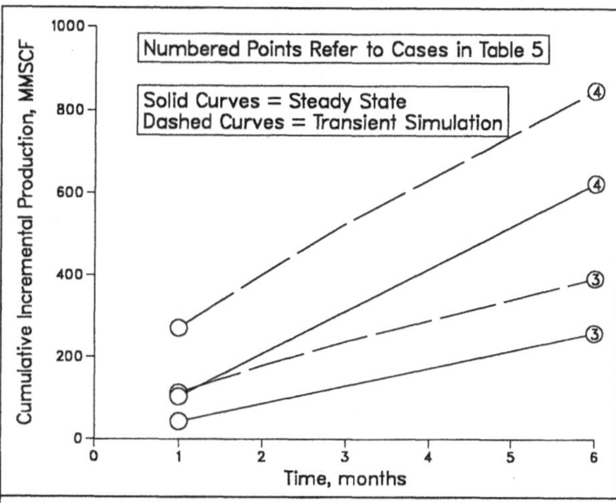

Fig. 11.12—Incremental production from fracturing vs. time—bauxite base case.

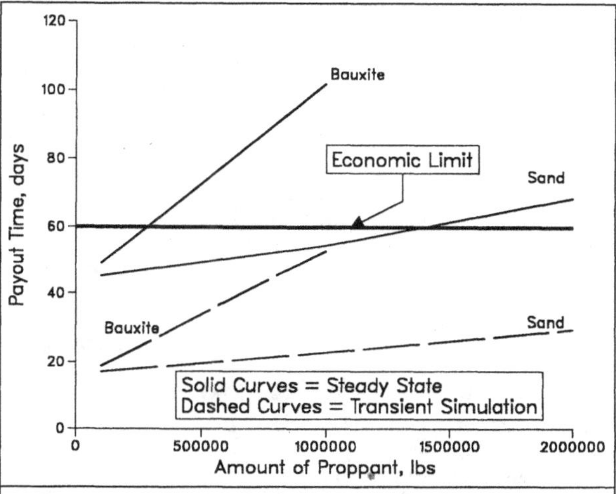

Fig. 11.13—Payout time vs. amount and type of proppant—base case.

All four design cases shown in Fig. 11.3 were done without particular regard to proppant schedule. Stage sizes of moderate volume were selected, with the final stage containing most of the proppant. A later section addresses proppant-schedule refinement once a gross treatment size has been selected.

Initial Cost and Payout-Time Estimation. Highest stimulation does not automatically make Treatment 4 in Fig. 11.3 the best treatment for the well. If treatment design were that simple, the best job would always be that which forms a fracture to the drainage radius with the most permeable proppant. Instead, the economic function will be the final discriminator to make the selection of the best treatment size for the well.

Each of the four initial treatments was evaluated for approximate cost according to the algorithm in Appendix H. The resulting costs are plotted in Fig. 11.10 vs. proppant quantity. It is seen that treatment costs increase with amount of proppant because of increased costs of proppant, fluid, pumping time, and equipment. A fifth case with 2,000,000 lbm [907.2 Mg] of sand was added to Fig. 11.10 after the first pass through the design process because the 1,000,000-lbm [453.6-Mg] sand job paid out sooner than the economic limit of 60 days. This is one place where experience comes into the design process. Experienced engineers would probably have eliminated the need for the fifth calculation because they would have adjusted their initial job sizes to encompass the economic bounds from prior experience.

Incremental Recovery Estimates. To calculate almost any economic function, including payout time, the accelerated recovery resulting from the fracture must first be estimated. Two ways to determine this are either simple pseudosteady-state estimates from the stimulation ratios given by Figs. 11.2 and 11.3 or from the more detailed estimates of a reservoir-flow simulator. Once obtained, incremental production will be used to calculate incremental revenue directly.

Figs. 11.11 and 11.12 show the incremental production calculated from both pseudosteady-state considerations and transient reservoir simulator runs. Fig. 11.11 is for sand cases and Fig. 11.12 for the bauxite case. The pseudosteady-state estimates were calculated by multiplying the prefracture production rate by the appropriate stimulation ratio with adjustment for change in drawdown. The reservoir simulator values are significantly higher because of transient flow that occurs before pseudosteady state is reached.

With an incremental revenue gain of \$2/Mcf [\$0.07/m^3] produced, the payout times of the jobs are plotted in Fig. 11.13. Several conclusions can be drawn from this figure relative to optimal treatment size.

1. The optimal bauxite treatment can use only about one-fourth as much proppant as the sand job.

2. A larger treatment can be justified if transient gas recovery is accounted for rather than pseudosteady-state recovery estimates.

3. With the conservative pseudosteady-state estimates, the permissible bauxite treatment can use about 300,000 lbm [136 Mg] of proppant vs. about 1,400,000 lbm [635 Mg] of sand.

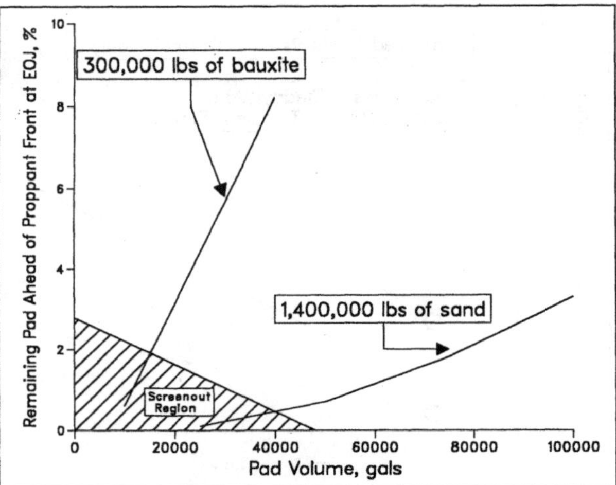

Fig. 11.14—Pad volume remaining ahead of proppant at the end of the job—base case.

At this point it is strongly suspected that the 1,400,000-lbm [635-Mg] sand treatment is the economically optimal treatment. As seen in Table 11.5, Case 14 has a greater stimulation ratio than Case 15. The 300,000-lbm [136-Mg] bauxite job, Case 15, gives significantly less stimulation at the same payout time. The sand job has a pseudosteady-state rate of 3.6 vs. 3.0 MMcf/D [102×10^3 vs. 85×10^3 m^3/d] for bauxite, and the sand job recovers 7.0 Bcf [198×10^6 m^3] in 30 years vs. 6.8 Bcf [193×10^6 m^3] for the bauxite job. For illustrative purposes, both cases will be carried through the remainder of the text. In daily practice, the bauxite job would be dropped at this point in favor of further design work on the sand job. The section below will pursue the refinement of each one of these two treatments to use minimum fluid and optimal proppant schedule to place the desired amount of proppant.

11.7 Optimal Proppant and Fluid Scheduling

The selection of the optimal treatment in Sec. 11.6 led to a final judgment on fluid and proppant that would be pumped during the job, including approximate volumes of both proppant and fluid. The economic viability of the treatment was also determined in that section. This section presents the final refinement of the optimal treatment to arrive at the final optimal treatment design. The specific fluid schedule is first optimized in terms of the minimum pad volume, and then the sand schedule is optimized to make use of minimal fluid to place the desired fracture length.

The bauxite treatment tentatively selected from the initial scoping treatment design used 42,000 gal [160 m^3] of pad volume and a 1-, 2-, 3-lbm/gal [120-, 240-, 360-kg/m^3] sand schedule that involved arbitrary amounts of fluid for each proppant stage. The sand case used 84,000 gal [320 m^3] of pad with sand stages from 1 to 6 lbm/gal [120 to 719 kg/m^3]. Each treatment will be varied and the design repeated to assess the effect of pad volume and proppant schedule.

Pad Volume Effect. Fig. 11.14 shows the results of simulations done with decreasing amounts of pad volume for both a 300,000-lbm [136-Mg] bauxite job and a 1,400,000-lbm [635-Mg] sand treatment. It is seen that the limiting case where all the pad volume has leaked off at the end of the treatment would require a pad volume of about 20,000 gal [76 m^3] for the 1,400,000-lbm [635-Mg] job and about 5,000 gal [19 m^3] for the 300,000-lbm [136-Mg] job. It is also seen that these values need to be increased to about 42,000 and 20,000 gal [159 and 76 m^3], respectively, to avoid proppant bridging along the fracture length while the treatment is being pumped. The shaded region was determined by analysis of a detailed computer simulation of the proppant placement. As fluid leaks off along the fracture length, the concentration increase is monitored until it becomes high enough to cause bridging. This type of analysis would be nearly impossible by hand calculation.

These latter values are the bare minimum volumes that would place the job away if all other design parameters are well known. Because the other parameters are generally not all well known, a safety factor on pad volume is usually applied. The factor is usually in the range of two to four times the minimum requirement. For this case, let us use twice the amount and select 84,000 and 42,000 gal [318 and 159 m^3] as pad volumes.

At this point, the amount of pad volume has been specified with the job still proceeding successfully. Notice that the final volumes were identical to the initial rough estimates used for scoping calculations. This was merely a coincidence because a somewhat arbitrary safety factor of two was chosen to reach the final design volumes. It will be seen in a later design for the PKN fracture model applied to this same example case that final pad volume will need to be increased. It generally will be found that the final pad volume selected will be somewhat different from the initial value selected.

Next, a design step to find the minimum gel requirement would be conducted, with repeated designs done at lower gel loadings. The details of this analysis will not be presented here because the subject was discussed in Chap. 7. For the base example case, it was found that gel loadings as low as 30 lbm/1,000 gal [3595 g/m^3] would provide a very acceptable treatment.

Proppant Schedule Effect. The specific proppant schedule is chosen with the following factors in mind.

1. The lowest one or two stages should be pumped with a volume of 1½ to 2 wellbore volumes to ensure that the well will take the treatment without early screenout.
2. The middle stages should be pumped at sufficient volume and proppant concentrations to provide the half of the fracture extending from midlength to tip with a propped width sufficient to make it very conductive, but not at so high a concentration as to cause bridging problems or to waste proppant on excessive width.
3. The final stages are pumped with largest volume and highest concentration to provide good propped fracture width throughout the half of the fracture extending from wellbore to midpoint.
4. The final stage is sometimes pumped at very high concentration to ensure that the very-near-wellbore region is extremely well propped.

The overall goal is to prop the fracture open with the right amount of sand located in the right places. The ideal final shape is triangular, with the widest dimension at the wellbore and monotonically decreasing width toward the tip. Various rules of thumb have developed in terms of a minimum of 2 lbm/ft^2 [9.8 kg/m^2] of fracture area or four proppant layers across the width. These rules can be useful in overall treatment-size scoping but are not hard and fast requirements in design. The most important consequences of specific fracture width profile are well production rates and recovery over time.

Ideally, the optimal proppant schedule could be found by repeatedly simulating treatments that generate the desired fracture length with various proppant schedules. In practice, this is a difficult task because it takes several attempts with each proppant schedule to arrive at the specified propped fracture length and because there are many potential proppant schedules.

A less exact but simpler approach is often used to evaluate the adequacy of the stair-step proppant schedule assumed earlier in the design process. Instead of searching for the optimal proppant schedule, this less exact method determines only whether average proppant concentration is too high or too low. The simple method consists of simulating the placement of the optimal treatment size with several single-proppant-stage treatments. For example, a 100,000-lbm [45.4-Mg] stimulation would be simulated with a treatment that consisted of an appropriate pad volume with all the proppant pumped at 2 lbm/gal [240 kg/m^3]. It would be simulated further by a treatment with all proppant pumped at 4 lbm/gal [480 kg/m^3]. Additional simulations at 6, 8, 10, and 15 lbm/gal [720, 960, 1200, and 1800 kg/m^3] would complete the evaluation of proppant schedule. The number of proppant schedule variations is thus restricted in this case to six.

For the convenience of simulating only a few proppant schedules, one certainly gives up accuracy in determining the optimal proppant schedule because each of the six treatments uses different amounts

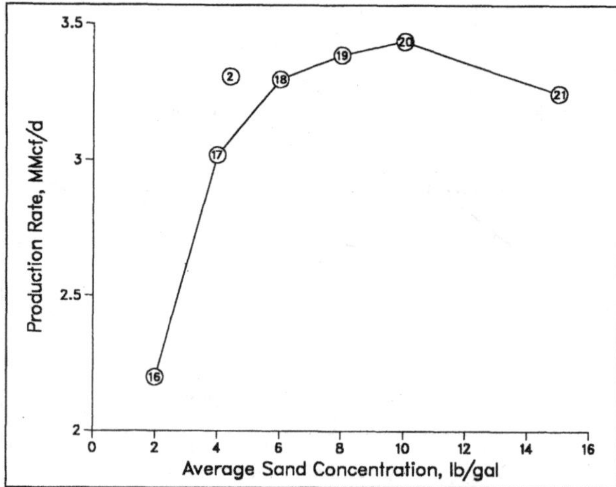

Fig. 11.15—Effect of sand schedule on steady-state production rate—base case.

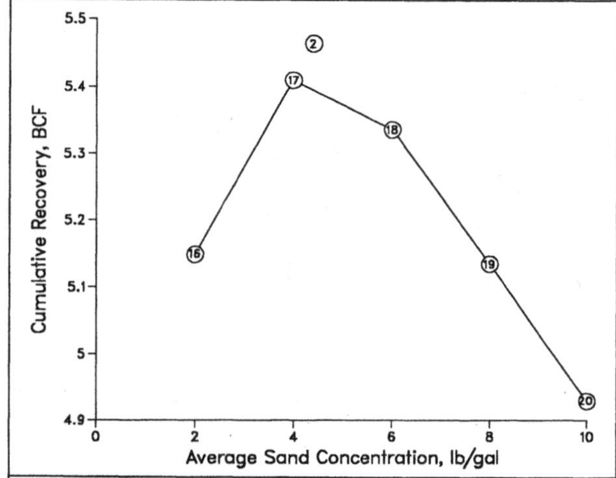

Fig. 11.16—Effect of sand schedule on recovery at 10 years—base case.

of fluid and has a different fracture length. Thus, the different amount of stimulation from each treatment is affected by both proppant schedule and propped fracture length. The inaccuracies introduced are greatest as the single proppant concentration value deviates the most from the average concentration of the stair-step schedule. To minimize the influence of fracture length, one could evaluate a narrower range of single-concentration treatments. For example, if the stair-step treatment averaged 5 lbm/gal [599 kg/m^3], one could compare its stimulation to that obtained from 4- and 6-lbm/gal [480- and 720-kg/m^3] treatments.

The following example illustrates the use of the simple method for evaluating proppant schedule for the base example case being carried through this chapter. Figs. 11.15 and 11.16 illustrate the results of multiple simulations of a 1,000,000-lbm [453.6-Mg] sand treatment with treatments at constant sand concentrations of 2, 4, 6, 8, 10, and 15 lbm/gal [240, 480, 720, 960, 1200, and 1800 kg/m^3], and the initial stair-step concentration with an average of 4.4 lbm/gal [527 kg/m^3]. The pseudosteady-state stimulation ratios from the Tannich-Nierode curves are plotted in Fig. 11.15 for the specific treatments numbered in Table 11.5. A similar plot in Fig. 11.16 was determined by putting the specific fracture lengths and widths in a finite-difference simulator and comparing the estimated recovery at 10 years of production.

Figs. 11.15 and 11.16 indicate that the stair-step proppant schedule is about as good as the constant-concentration treatment at 4 lbm/gal [480 kg/m^3]. Fig. 11.15 indicates that higher concentration would give a very slight benefit, while Fig. 11.16 shows that the stair-step treatment is optimal. In other words, the optimal or best proppant schedule for the example case in this chapter is basically the stair-step schedule initially assumed. Increasing or decreasing the average concentration would provide less stimulation. The required fracture length could be propped with less fluid if the entire treatment were done at about 4 lbm/gal [480 kg/m^3]. However, the safety factor involved with stair-step increases would be sacrificed.

At this point, small changes in stage volumes can be made to make tankage requirements more convenient. This was already done when minimum pad volume was doubled from 20,000 to 42,000 gal [76 to 159 m^3] for the bauxite treatment. Depending on the specific field configuration, allowances should be made for the undrainable volumes of each tank. Typically, in a 500-bbl [79-m^3] tank, as much as 50 bbl [8 m^3] cannot be drained.

As the following PKN example shows, when the proposed proppant schedule deviates significantly from the optimum of the curves like Figs. 11.15 or 11.16, then the schedule should be modified.

A detailed analysis of the 300,000-lbm [136-Mg] bauxite job similar to Figs. 11.15 and 11.16 will not be presented here because the analysis also showed that the 1-, 2-, 3-lbm/gal [120-, 240-, 360-kg/m^3] staging was close to the optimal schedule.

Long-Term Recovery Estimates. Now that the optimal treatment size and specific pumping schedule have been selected, long-term recovery estimates can be made for reservoir engineering study.

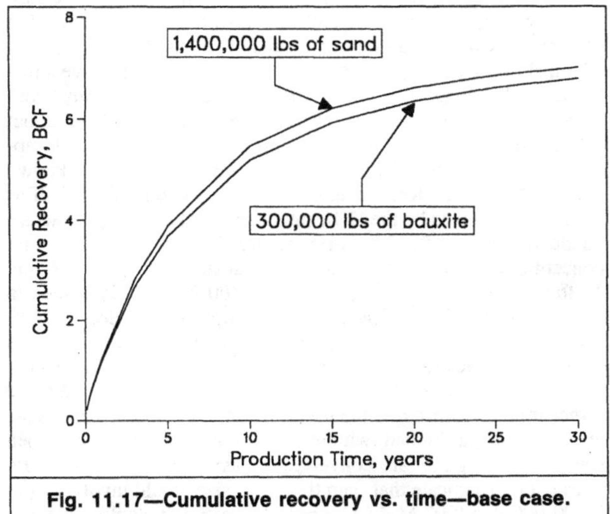

Fig. 11.17—Cumulative recovery vs. time—base case.

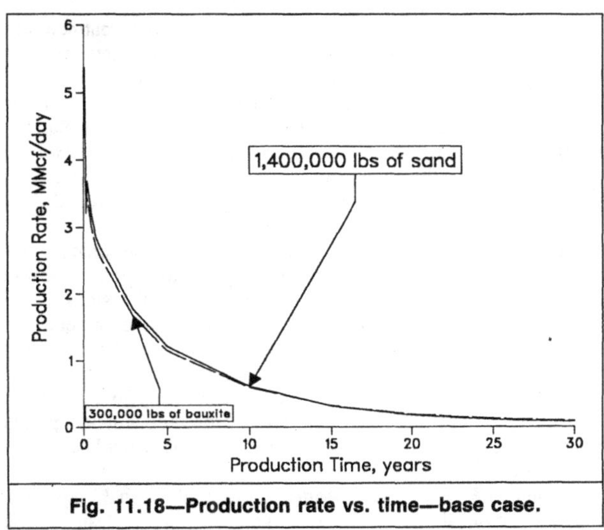

Fig. 11.18—Production rate vs. time—base case.

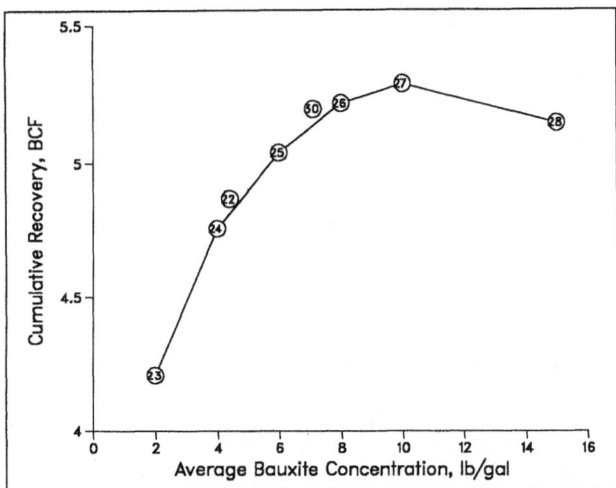

Fig. 11.19—Effect of sand schedule on recovery at 10 years—PKN case.

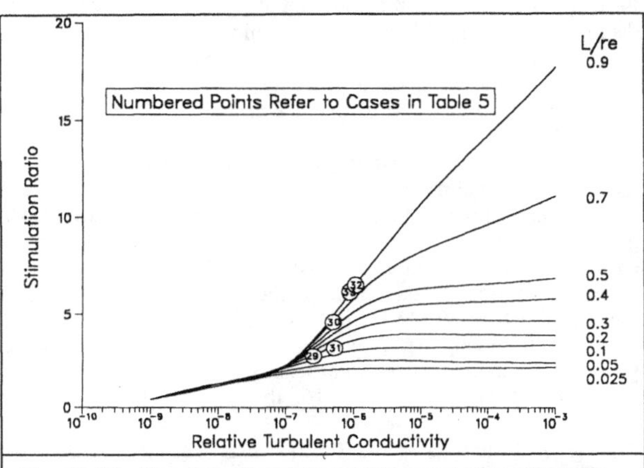

Fig. 11.20—Tannich stimulation ratio curves for gas wells—PKN case.

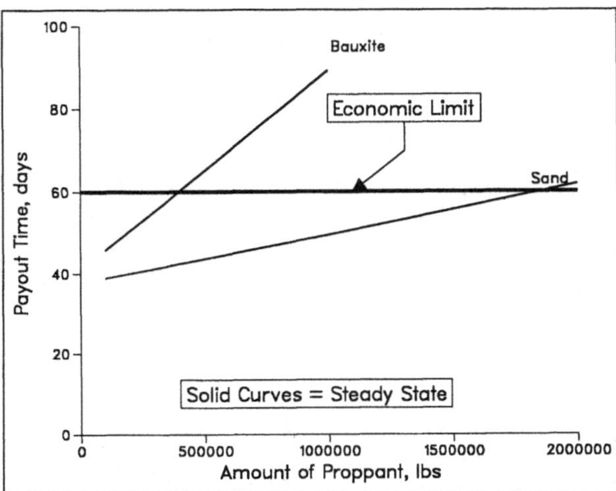

Fig. 11.21—Payout time vs. amount and type of proppant—PKN case.

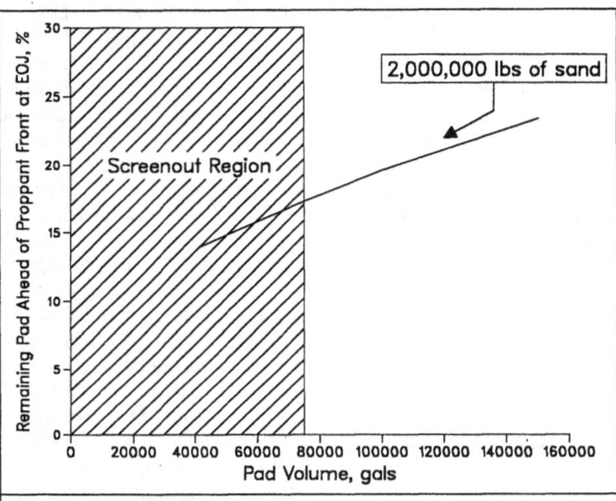

Fig. 11.22—Pad volume remaining ahead of proppant at the end of the job—PKN case.

Figs. 11.17 and 11.18 display the estimated recovery response of the example well to a realistic production schedule. The first year would be at 875-psi [6-MPa] flowing BHP. From Years 1 through 5, pressure would be reduced to 500 psi [3.4 MPa]. Subsequent years up to 30 would be at 100 psi [6.9 MPa].

Estimates for the 300,000-lbm [136-Mg] bauxite treatment are included for comparison. The sand treatment would have greatest recovery, but other long-term factors may alter this conclusion. In a particular well, it may be that silicate dissolution by reservoir water would degrade the proppant pack over the long term. Bauxite would withstand this better than sand. However, an appropriate place to take something like this into account would have been during the proppant selection when sand would have been disqualified.

Alternative Designs/Uncertainty Assessment. Often some parameters used in an optimal design or some of the specific design methods are questionable. In this situation, additional designs should be done to evaluate the influence of changes of parameters or methods. The remainder of this section will consider two such changes: change of the model to PKN from the base-case model of GDK and an increase in reservoir permeability to 1 md from the base-case value of 0.1 md. The design procedure will be repeated to assess the changes in the optimal treatment that would be caused by the altered information.

Change to PKN Fracture Model. Cases 22 through 37 represent the design cases that were run to arrive at the optimal design, Case 36. The PKN model was used in all these cases to model the dimensions of the fracture during pumping. As a preliminary look at the gross behavior of the well for this situation, Case 2 was run for a 1,000,000-lbm [453.6-Mg] sand job with the expectation that this job would be near the optimum judging from the results of the base case design. Cases 23 through 28 were also run to assess the effect of average sand concentration; results are plotted in Fig. 11.19, which shows that the proposed job, Case 22, does not have sufficiently high average sand concentration to be near the optimal condition. If average sand concentration were increased from about 4 to about 8 lbm/gal [480 to 960 kg/m^3], the design would approach the optimum. This is a demonstration of the well-known facts[62,63] that the PKN model gives fracture widths about half as wide as those given by the GDK model and that equivalent fracture conductivities in a PKN fracture require about twice the proppant concentration. Case 30 was run with sand stages of 2, 4, 6, 8, and 10 lbm/gal [240, 480, 720, 960, and 1200 kg/m^3]. It is seen in Fig. 11.19 that Case 30 is close to the optimum region.

Cases 29 through 33 were run as the initial scoping cases equivalent to the previous scoping, Cases 1 through 5. These data, plotted on the Tannich-Nierode curves in Fig. 11.20, show that the larger treatments provide the greatest stimulation. The treatments are mostly length-limited, with lesser conductivity limitation. When Cases 29 through 33 are converted to payouts, the results shown in Fig. 11.21 indicate that sand treatment size would involve about 2,000,000 lbm [907 Mg] and a bauxite treatment would use about

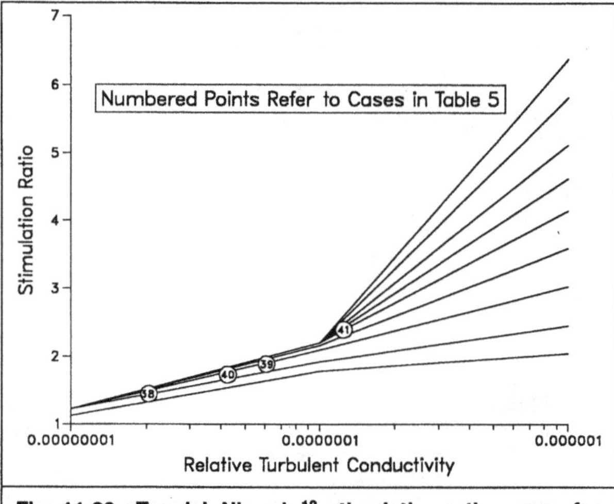

Fig. 11.23—Tannich-Nierode[12] stimulation ratio curves for gas wells—k=1 md.

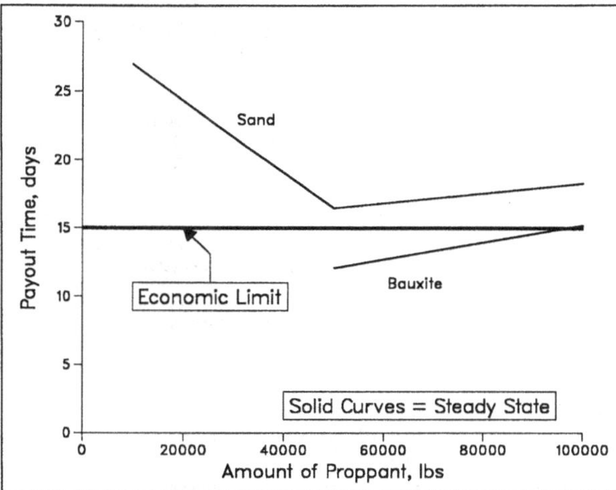

Fig. 11.24—Payout time vs. amount and type of proppant—k=1 md.

400,000 lbm [181 Mg]. It is immediately seen in Table 11.5 that the sand treatment gives the greatest stimulation and that sand is the preferred proppant.

Cases 35 through 37 were run to determine pad volume. These data, plotted in Fig. 11.22, show that the minimum pad volume would be around 75,000 gal [284 m^3]. A safety factor of two places the desired pad volume at around 150,000 gal [568 m^3], so Case 37 is the optimal case for the PKN model.

The main difference between the optimal PKN design and the base GDK case is the near doubling of average sand concentration. The increase in sand concentration and narrower predicted fracture width also lead to a significant increase in pad volume.

Change to Permeability of 1 md. If reservoir permeability is 1 md instead of the 0.1 md assumed in the base-case design, the optimal design will be Case 40 in Table 11.5. Because of the higher permeability, the proppant will need to be switched from sand to the more permeable bauxite material.

Cases 38 through 41 are the four general scoping cases with stimulation ratios plotted in Fig. 11.23. Note that the stimulated state is in the part of the Tannich-Nierode curves where the sand treatments are conductivity-limited. Only when proppant is changed to bauxite to gain higher contrast does increased length become useful.

Note that the example problem parameters in Table 11.2 specify that when reservoir kh product increases to 50 md-ft [15 md·m] from the base case value of 5 md-ft [1.5 md·m], the economic payout time decreases from 60 to 15 days. Allowable payout time decreases because of a more detailed economic analysis where the time value of money is taken into account. This is discussed further in Chap. 17.

When payout times were calculated for the initial scoping, Cases 38 through 41, all these cases were found to have a payout time above the economic limit. Thus, it was necessary to simulate the additional smaller jobs shown as Cases 42 through 44. The payout times of all of these cases are plotted in Fig. 11.24, which shows that a sand proppant case cannot be found to satisfy the economic requirement. When the job size is large, payout time is too long because minimal stimulation results from a large treatment expense. When job size is very small, job cost is small, but the amount of stimulation is still too small to pay out the treatment.

When a high-strength proppant, such as sintered bauxite, is used, a treatment of slightly less than 100,000 lbm [45.4 Mg] can be justified. Thus, Treatment 40 is the optimal treatment size with a payout time of about 15.2 days. The design procedure would proceed beyond this point to evaluate pad volumes and proppant schedules in a manner similar to that discussed for the prior examples.

11.8 Special Design Considerations

The discussions in the preceding sections presented a simple design method that one could use as a basis to design a treatment for almost any well. Unfortunately, the number of special constraints that can occur in a specific case are numerous. The design tools presented are therefore just that, tools. The common sense, thought, and experience that the engineer brings to the design are at least as important as the tools with which he designs the treatment. The remainder of the chapter will focus on three often-encountered situations: naturally fractured reservoirs, undesired height growth, and unconsolidated reservoir fracturing.

Naturally Fractured Reservoirs. Luckily, many naturally fractured reservoirs have such high productivity that hydraulic fracturing is not necessary to enhance recovery. Some reservoirs, however, like parts of the Cotton Valley of east Texas, have moderate natural-fracture density where hydraulic fracturing is still needed for economic development. The natural fractures may have such a small capacity for gas storage that they are not major factors in well productivity, but at the same time they can have a pronounced effect on the behavior of a proppant fracturing treatment. The fractures can greatly increase fluid-loss rate and cause a screenout.

An analogy of what happens during any fracturing treatment is blowing up a balloon with pinholes in it. The inflation of the balloon is the equivalent of the opening of the fracture, and the loss of air from the pinholes in the balloon is equivalent to the leakoff of fluid from the fracture faces into the rock pores. In natural fractures, the analogy can be extended to include tears or gashes to simulate the natural fractures. Fluid loss is greatly increased by these fractures.

To pump treatments away successfully, one needs to pump enough pad volume to create a fracture volume that is greater than the proppant volume and/or to pump at fast enough rates to keep up with the fluid loss.

Example of Natural Fractures. Table 11.6 summarizes the history of a naturally fractured reservoir where early screenouts were not cured either by increasing pad volume or by adding 100-mesh sand to restrict fluid loss into the fractures. The small treatments on Wells A and B were successful with small pad volumes and low treating rates only because the planned job size was also small. When increased sand volumes were attempted in Wells C and D, both treatments screened out. Treatment D attempted to improve chances by increasing the total pad volume and adding 100-mesh sand in the middle of the pad to plug the fractures. The treatment screened out anyway.

A surveillance of the treatments done on Wells A through D was done by modeling the actual treatment pumped with a fracturing simulator using effective fluid-loss coefficient as the fit parameter. As Table 11.6 shows, all the treatments were successfully modeled by a fluid-loss coefficient of around 0.008 ft/min$^{1/2}$ [0.0024 m/min$^{1/2}$]. The treatments on Wells C and D were simulated to have sand arrive at the fracture tip when the fluid loss coefficient was this value. The small treatments in Wells A and B were cor-

TABLE 11.6—NATURALLY FRACTURED RESERVOIR SURVEILLANCE

Well	Treating Rate (bbl/min)	Pad Volume (bbl)	Total Volume (bbl)	Design Sand Amount (lbm)	Actual Sand Amount (lbm)	Effective Fluid Loss (ft/min $^{1/2}$)
A	15	600	1,836	155,000	155,000	0.0080*
B	15	667	1,738	135,000	135,000	0.0080*
C	20	1,386	5,060	450,000	140,000	0.0083
D	20	3,900**	8,372	675,000	110,000	0.0078
E	50*	10,000	24,000	2,100,000	1,700,000	0.0080

*Treatment predicted to pump successfully with this high a fluid-loss coefficient.
**Pad volume includes 23,000 lbm of 100-mesh sand.

rectly predicted to be successful even if the fluid-loss coefficient was as high as 0.008 ft/min $^{1/2}$ [0.0024 m/min $^{1/2}$]. The fact that both successful and unsuccessful treatments were modeled by the same value of coefficient gave credence to that value as the correct fluid-loss coefficient for the reservoir.

The treatment on Well E was designed with the 0.008-ft/min $^{1/2}$ [0.0024-m/min $^{1/2}$] coefficient value. To simulate the successful pumping of 2,100,000 lbm [953 Mg] of sand, it was necessary to pump the job at 50 bbl/min [8 m^3/min] and to use an extremely large pad of 10,000 bbl [1590 m^3]. As Table 11.6 shows, the treatment was able to pump away only 1,700,000 lbm [771 Mg] of the planned treatment with the treatment screening out at this point. Subsequent analysis showed that the screenout was caused by reduced injection rate near the end of the treatment. Several pumps and part of the blender operation were lost for the last 300,000 lbm [136 Mg] of sand pumped and treating rate dropped to 35 bbl/min [5.6 m^3/min]. Calculations later showed that if the rate had been at 35 bbl/min [5.6 m^3/min] for the entire treatment, only slightly more than 1,000,000 lbm [453.6 Mg] would have been pumped before screenout. The analysis therefore reinforces the correctness of the surveillance conclusion that fluid-loss coefficient is about 0.008 ft/min $^{1/2}$ [0.0024 m/min $^{1/2}$]. This field example illustrates the use of two methods for keeping up with the high rate of fluid loss in naturally fractured reservoirs: increasing the treating rate and increasing the pad volume. The only other approach would be to use special fluid-loss additives, such as 100-mesh sand, in the pad fluid to restrict leakoff. The effectiveness of this approach must be determined on a field-by-field basis because plugging of natural fractures with particles will depend on the ratio of particle size to fracture width. In the previous example, 100-mesh sand did not provide a discernible benefit.

Undesired Height Growth. Fracture height growth is sometimes undesirable. In some instances, gas or water is present in close proximity to the zone about to be fractured and would be produced if the fracture extended into them. In other situations, fracture length is sacrificed at the expense of height growth to make the stimulation dollar ineffectively spent. In these cases, a few things can be done to minimize undesirable fracture height growth. Treating rate and downhole treating pressure are kept as low as possible to minimize height growth.

One way to keep treating pressure low is to use the thinnest possible fracturing fluid. Treating pressure at the wellbore is greater than that pressure required at the fracture tip because of the friction pressure loss along the fracture length. A thick fluid may have 300- to 500-psi [2.1- to 3.4-MPa] pressure loss, while a thin fluid may have less than 100-psi [690-kPa] loss. If confinement depends on stress differences as low as a few hundred psi, then the use of a thin fluid can make a difference in height growth.

There are no good field examples of successful reduction of height growth by either rate or fluid changes, but there are many examples of the occurrence of undesirable height growth. McDonald[64] presented field data for acidized fractures showing height growth that exceeded design expectations. Others[65,66] have shown similar results for proppant fractures in chalks to sandstones.

Unconsolidated Reservoir Fracturing. Some reservoirs that have only loosely consolidated reservoir rock can still be fractured.[67-69] This is possible because under in-situ confining stress, even unconsolidated material can behave like brittle rock. The core that can be disintegrated by hand on a desk can still behave like a solid rock when in a highly compressed state underground.

Potential problems that can occur in such soft or unconsolidated rock involve loss of propped fracture integrity either through embedment or physical mixing of proppant with reservoir material. To diminish these possibilities, one can use very high proppant concentration or place an equilibrium bank treatment to yield maximum propped width. Another approach is to use epoxy-coated sand to retain fracture integrity by forming a literally solid proppant bank. An example of unconsolidated reservoir fracturing is given by Strubhar *et al.*[67] Similar results and concepts are discussed in several papers on fracturing of soft chalk formations.[64,65]

Example of Unconsolidated Reservoir Fracturing. The particular problem discussed by Strubhar *et al.* in diatomaceous earth fracturing was the need to place a wide fracture that had a vertical fracture height nearly equal to fracture length. The wide propped fracture was required because of the potentially severe embedment that might occur in the soft earth if only one or two layers of proppant pack were placed. Several authors[64,65] have measured embedment in various rock strengths and agreed that embedment occurs only to the extent of about one-half grain width in even the softest formations. It appears that once a one-half grain width enters the rock, the embedment of succeeding layers is impeded by the first layer's entry. Thus, the goal in soft rock fracturing is to place at least four or five layers of proppant pack so that three or four effective layers remain after embedment.

Strubhar *et al.* needed a nearly circular fracture with equal height and length growth because the fracture interval was as thick as 2,000 ft [610 m] and log interpretation of productive sands was very difficult. In essence, the fracturing treatment was to be used to search for productive zones by growing vertically away from the perforated intervals and connecting all zones with the wellbore. It was found that the more viscous the fluid used in treatment of these wells was, the better both needs were satisfied. A thick fluid like a cross-linked gel was able to carry high sand concentrations to satisfy the need for multiple propped layers, and the high viscosity also promoted vertical height growth because wellbore treating pressures were increased by friction loss along the fracture length.

Equilibrium Bank Placement. If the need for vertical fracture height growth were not present, another way to place a wide propped fracture would be to use a very-low-viscosity fluid to form an equilibrium bank of proppant. This would result in a greater width of propped fracture than attainable by any other means. An example of such a placement was in the fifth treatment placed in the Rio Blanco MHF-3 well in the western U.S. Fig. 11.10 shows an example of equilibrium bank formations. Such treatments are not often done because of the increased risk of screenout compared with totally suspending jobs.

Nomenclature

E = Young's modulus of elasticity, psi [kPa]
J' = productivity index of fractured well in turbulent flow, Mscf/D-psi^2 [std m^3/d·kPa^2]
J_o = productivity index of undamaged, unfractured well if flow were laminar, Mscf/D-psi^2 [std m^3/d·kPa^2]
k = effective reservoir permeability, md

K = power-law constant, (lbf-sec^n)/ft^2 [Pa·s]
L = fracture length, ft [m]
p_e = reservoir pressure at drainage radius, psi [kPa]
p_w = flowing BHP, psi [kPa]
r_e = drainage radius, ft [m]
r_w = wellbore radius, ft [m]
Δt_c = sonic travel time, μsec/ft [μs/m]
T = static reservoir temperature, °F [°C]
w_P = propped fracture width, in. [cm]
X = chart value in Eq. 11.3
z = gas compressibility factor, dimensionless
β_f = fracture turbulence coefficient, ft^{-1} [m^{-1}]
γ = gas specific gravity, dimensionless
μ = gas viscosity, cp [mPa·s]
ν = Poisson's ratio, dimensionless
ρ_f = reservoir fluid density, lbm/ft^3 [kg/m^3]
ϕ = bulk reservoir porosity, fraction

References

1. Veatch, R.W. Jr.: "Overview of Current Hydraulic Fracturing Treatment and Design Technology—Part 1," *JPT* (April 1983) 677–87.
2. Williams, B.B., Gidley, J.L., and Schechter, R.S.: *Acidizing Fundamentals*, SPE Monograph Series (1979) **6**, 53–67.
3. Williams, B.B. and Nierode, D.E.: "Design of Acid Fracturing Treatment," *JPT* (July 1972) 849–59.
4. Kiel, O.M.: "A New Hydraulic Fracturing Process," *JPT* (Jan. 1970) 89–96.
5. Gidley, J.L. *et al.*: "Stimulation of Low-Permeability Gas Formations by Massive Hydraulic Fracturing—A Study of Well Performance," *JPT* (April 1979) 525–31.
6. Nierode, D.E.: "Comparison of Hydraulic Fracture Design Methods to Observed Field Results," *JPT* (Oct. 1985) 1831–39.
7. Geertsma, J. and deKlerk, F.: "A Rapid Method of Predicting Width and Extent of Hydraulically Induced Fractures," *JPT* (Dec. 1969) 1571–81; *Trans.*, AIME, **246**.
8. Perkins, T.K. and Kern, L.R.: "Widths of Hydraulic Fractures," *JPT* (Sept. 1961) 937–49; *Trans.*, AIME, **222**.
9. Nordgren, R.P.: "Propagation of a Vertical Hydraulic Fracture," *SPEJ* (Aug. 1972) 306–14; *Trans.*, AIME, **253**.
10. Geertsma, J. and Haafkens, R.: "A Comparison of the Theories for Predicting Width and Extent of Vertical Hydraulically Induced Fractures," *Trans.*, ASME (1979) **101**, 8–19.
11. McGuire, W.J. and Sikora, V.J.: "The Effect of Vertical Fractures on Well Productivity," *Trans.*, AIME (1960) **219**, 401–03.
12. Tannich, J.D. and Nierode, D.E.: "The Effect of Vertical Fractures on Gas Well Productivity," paper SPE 15902 available from SPE, Richardson, TX.
13. Rosepiler, M.J.: "Determination of Principal Stresses and Confinement of Hydraulic Fractures in Cotton Valley," paper SPE 8405 presented at the 1979 SPE Annual Technical Conference and Exhibition, Las Vegas, Sept. 23–26.
14. Williams, B.B. *et al.*: "A Staged Fracturing Treatment for Multisand Intervals," *JPT* (Aug. 1973) 897–904.
15. Von Albrecht, C. *et al.*: "Stimulation of Asphaltic Deep Wells and Shallow Wells in Lake Maracaibo, Venezuela," *Proc.*, 10th World Pet. Cong., Bucharest, Romania (1979) **3**, 55–62.
16. Hubbert, M.K. and Willis, D.G.: "Mechanics of Hydraulic Fracturing," *Trans.*, AIME (1957) **210**, 153–66.
17. Eaton, B.A.: "Fracture Gradient Prediction and Its Application in Oilfield Operations," *JPT* (Oct. 1969) 1353–60; *Trans.*, AIME, **246**.
18. Eaton, B.A.: "Fracture Gradient Prediction and Application," *Proc.*, Fourth Intl. Symposium on Salt, Northern Ohio Geological Society (1973) 153–75.
19. Salz, L.B.: "Relationship Between Fracture Propagation Pressure and Pore Pressure," paper SPE 6870 presented at the 1977 SPE Annual Technical Conference and Exhibition, Denver, Oct. 9–12.
20. Nolte, K.G.: "Determination of Fracture Parameters From Fracturing Pressure Decline," paper SPE 8341 presented at the 1979 SPE Annual Technical Conference and Exhibition, Las Vegas, Sept. 23–26.
21. Nolte, K.G.: "Principles for Fracture Design Based on Pressure Analysis," *SPEPE* (Feb. 1988) 22–30; "Application of Fracture Design Based on Pressure Analysis," *SPEPE* (Feb. 1988) 31–42.
22. Nolte, K.G. and Smith, M.B.: "Interpretation of Fracturing Pressures," *JPT* (Sept. 1981) 1767–75.
23. Nolte, K.G.: "Determination of Proppant and Fluid Schedules From Fracturing Pressure Decline," *SPEPE* (July 1986) 255–65.
24. Allam, A.M. *et al.*: "The Use of Bottomhole Pressure Analysis in Optimizing Frac Design—Anadarko Basin of Western Oklahoma," paper SPE 9262 presented at the 1980 SPE Annual Technical Conference and Exhibition, Dallas, Sept. 21–24.
25. Williams, B.B.: "Fluid Loss From Hydraulically Induced Fractures," *JPT* (July 1970) 882–88; *Trans.*, AIME, **249**.
26. Crawford, H.R.: "Proppant Scheduling and Calculation of Fluid Lost During Fracturing," paper SPE 12064 presented at the 1983 SPE Annual Technical Conference and Exhibition, San Francisco, Oct. 5–8.
27. Rubin, M.B.: "On Fluid Leak-Off During Propagation of a Vertical Hydraulic Fracture," paper SPE 10556 available from SPE, Richardson, TX.
28. Cooper, G.D., Nelson, S.G., and Schopper, M.D.: "Comparison of Methods for Determining In-Situ Leakoff Rate Based on Analysis With an On-Site Computer," paper SPE 13223 presented at the 1984 SPE Annual Technical Conference and Exhibition, Houston, Sept. 16–19.
29. Clifton, R.J. and Abou-Sayed, A.S.: "On the Computation of the Three-Dimensional Geometry of Hydraulic Fractures," paper SPE 7943 presented at the 1979 SPE Low-Permeability Gas Reservoirs Symposium, Denver, May 20–22.
30. Ahmed, U. *et al.*: "A Step-By-Step Approach to Hydraulic Fracture Treatment Design, Implementation, and Analysis for Tight Gas Sands," paper SPE 10829 presented at the 1982 SPE/DOE/GRI Unconventional Gas Recovery Symposium, Pittsburgh, PA, May 16–18.
31. Ahmed, U.: "A Practical Hydraulic Fracturing Model Simulating Necessary Fracture Geometry, Fluid Flow and Leakoff, and Proppant Transport," paper SPE 12880 presented at the 1984 SPE/DOE/GRI Unconventional Gas Recovery Symposium, Pittsburgh, PA, May 13–15.
32. Frohne, K.H.: "Third Annual Western Gas Sands Program Review," Report DOE/METC/85-7(DE85001961), U.S. DOE (1984).
33. Cleary, M.P.: "Comprehensive Design Formulae For Hydraulic Fracturing," paper SPE 9259 presented at the 1980 SPE Annual Technical Conference and Exhibition, Dallas, Sept. 21–24.
34. Warpinski, N.R., Schmidt, R.A., and Northrop, D.A.: "In-Situ Stresses: The Predominant Influence on Hydraulic Fracture Containment," *JPT* (March 1982) 653–64.
35. Simonson, E.R., Abou-Sayed, A.S., and Clifton, R.J.: "Containment of Massive Hydraulic Fractures," *SPEJ* (Feb. 1978) 27–32.
36. Dobkins, T.A.: "Methods to Better Determine Hydraulic Fracture Height," paper SPE 8403 presented at the 1979 SPE Annual Technical Conference and Exhibition, Las Vegas, Sept. 23–26.
37. Dobkins, T.A.: "Improved Methods to Determine Hydraulic Fracture Height," *JPT* (April 1981) 719–26.
38. Bundy, T.E.: "Prefracture Injection Surveys: A Necessity for Successful Fracture Treatments," *JPT* (May 1982) 995–1001.
39. Lagrone, K.W. and Rasmussen, J.W.: "A New Development in Completion Methods—The Limited Entry Technique," *JPT* (July 1963) 695–702.
40. Mathias, J.P.: "Successful Stimulation of Thick, Low-Pressure, Water-Sensitive Gas Reservoir By Pseudolimited Entry," *JPT* (Feb. 1971) 185–90.
41. Gruesbeck, C. and Collins, R.E.: "Particle Transport Through Perforations," *SPEJ* (Dec. 1982) 857–65.
42. Bennett, E.N. and Forgerson, C.D.: "Predicting Reserves and Forecasting Flow Rates of Relatively Tight Gas Wells Using Limited Performance Data," *JPT* (May 1975) 585–91.
43. Eilerts, C.K.: "Methods for Estimating Deliverability After Massive Fracture Completions in Tight Formations," paper SPE 5112 presented at the 1974 SPE Deep Drilling and Production Symposium, Amarillo, TX, Sept. 8–10.
44. Holdtich, S.A. and Morse, R.A.: "The Effects of Non-Darcy Flow on the Behavior of Hydraulically Fractured Gas Wells," *JPT* (Oct. 1976) 1169–78.
45. Holditch, S.A. *et al.*: "An Automated Method of Matching Production Performance Using Dimensionless Solutions," paper SPE 12846 presented at the 1984 SPE/DOE/GRI Unconventional Gas Recovery Symposium, Pittsburgh, PA, May 13–15.
46. Khristianovich, S.A. and Zheltov, Y.P.: "Formation of Vertical Fractures by Means of Highly Viscous Fluids," *Proc.*, Fourth World Pet. Cong., Rome (1955) **2**, 579–86.
47. Zheltov, Y.P.: "An Approximate Method of Calculating the Size of Fissures Produced by the Hydraulic Fracturing of a Formation," *Isvest. Akad. Nauk SSSR, Otdel. Tekh. Nauk* (1957) No. 3, 180–82.
48. Sneddon, I.N.: "The Distribution of Stress in the Neighborhood of a Crack in an Elastic Solid," *Proc.*, Royal Soc. London (1946) **A**, 187, 229.
49. Novotny, E.J.: "Proppant Transport," paper SPE 6813 presented at the 1977 SPE Annual Technical Conference and Exhibition, Denver, Oct. 9–12.

50. Steanson, R.E., Elbel, J.L., and Wendorff, C.: "How to Select A Proppant Agent for Hydraulic Fracturing Treatments," *Pet. Eng. Intl.* (July 1979) 82–92.
51. Holditch, S.A.: "Criteria for Propping Agent Selection," third edition, Norton Alcoa Proppants, Dallas (1988).
52. Clark, H.C.: "A Method for Selection of Optimum Proppant in the Cotton Valley," paper SPE 11617 presented at the 1983 SPE/DOE Low-Permeability Gas Reservoirs Symposium, Denver, March 14–16.
53. Montgomery, C.T. and Steanson, R.E.: "Proppant Selection: The Key to Successful Fracture Stimulation," *JPT* (Dec. 1985) 2163–72.
54. Harrington, L.J., Hannah, R.R., and Williams, D.: "Dynamic Experiments on Proppant Settling in Crosslinked Fracturing Fluids," paper SPE 8342 presented at the 1979 SPE Annual Technical Conference and Exhibition, Las Vegas, Sept. 23–26.
55. Shah, S.N.: "Proppant Settling Correlations for Non-Newtonian Fluids Under Static and Dynamic Conditions," *SPEJ* (April 1982) 164–70.
56. Clark, P.E. *et al.*: "Prop Transport in Vertical Fractures," paper SPE 10261 presented at the 1983 SPE/DOE Low-Permeability Reservoirs Symposium, Denver, March 14–16.
57. Kern, L.R., Perkins, T.K., and Wyant, R.E.: "The Mechanics of Sand Movement in Fracturing," *Trans.*, AIME (1959) **216,** 403–05.
58. Schols, R.S. and Visser, W.: "Proppant Bank Buildup in a Vertical Fracture Without Fluid Loss," paper SPE 4834 presented at the 1974 SPE European Meeting, Amsterdam, May 28–29.
59. Daneshy, A.A.: "Numerical Solution of Sand Transport in Hydraulic Fracturing," *JPT* (Jan. 1978) 132–40.
60. van Domselaar, H.R. and Visser, W.: "Proppant Concentration in and Final Shape of Fractures Generated by Viscous Gels," *SPEJ* (Dec. 1974) 531–36.
61. Smith, M.B., Miller, W.K. II, and Haga, J.: "Tip Screenout Fracturing: A Technique for Soft, Unstable Formations," *SPEPE* (May 1987) 95–103.
62. Coulter, G.R. and Wells, R.D.: "The Advantages of High Proppant Concentrations in Fracture Stimulation," *JPT* (June 1972) 643–50.
63. Shah, S.N., Smith, D.R., and Donaldson, D.E.: "Ultra-High Sand Concentration in Hydraulic Fracturing—A Case History," paper SPE 11578 presented at the 1983 SPE Production Operations Symposium, Oklahoma City, Feb. 27–March 1.
64. McDonald, S.W.: "Evaluation of Production Tests in Oil Wells Stimulated by Massive Acid Fracturing Offshore Qatar," *JPT* (March 1983) 496–506.
65. Hartley, R. and Bosma, M.G.R.: "Fracturing in Chalk Completions," *JPT* (Jan. 1985) 73–79.
66. Warnock, W.E., Harris, P.C., and King, D.S.: "Successful Field Applications of CO_2-Foam Fracturing Fluids in the Arkansas-Louisiana-Texas Region," *JPT* (Jan. 1985) 80–88.
67. Strubhar, M.K. *et al.*: "Fracturing Results in Diatomaceous Earth Formations, South Belridge Field, California," *JPT* (March 1984) 495–502.
68. Strickland, F.G.: "Reasons for Production Decline in the Diatomite, Belridge Oil Field: A Rock Mechanics View," *JPT* (March 1985) 521–26.
69. Lambert, S.A., Dolan, R.T., and Gallus, J.P.: "Fracturing Poorly Consolidated Sandstone Formations," *Proc.*, 30th Annual Southwestern Petroleum Short Course, Lubbock, TX (1983).

SI Metric Conversion Factors

acres	× 4.046 873	E+03	=	m^2
bbl	× 1.589 873	E−01	=	m^3
degrees	× 1.745 329	E−02	=	rad
ft	× 3.048*	E−01	=	m
ft^2	× 9.290 304*	E−02	=	m^2
ft^3	× 2.831 685	E−02	=	m^3
°F	(°F−32)/1.8		=	°C
gal	× 3.785 412	E−03	=	m^3
in.	× 2.54*	E+00	=	cm
lbm	× 4.535 924	E−01	=	kg
psi	× 6.894 757	E+00	=	kPa
psi^{-1}	× 1.450 377	E−01	=	kPa^{-1}

*Conversion factor is exact.

Chapter 12
Well Completions

Stephen A. Holditch, SPE, Texas A&M U.

12.1 Overview

This chapter presents information concerning both the completion strategy required to fracture-treat a well successfully and the completion techniques necessary to implement the chosen strategy safely. The chapter begins by briefly reviewing several basic subjects, such as drilling, cementing, perforating, and completion fluids. These brief discussions are followed by a more detailed discussion of completion strategy. The chapter continues with thorough discussions concerning tubular design and treatment diversion and concludes by presenting information and ideas concerning several special well-completion problems.

12.2 Drilling Aspects

Drilling personnel are sometimes judged by how fast they can drill a well to a specified depth. However, when one is drilling a well that will be completed and fracture-treated in a low-permeability reservoir, the speed of the rig personnel is of secondary importance. Of primary importance is drilling a gauge borehole. If the borehole is in gauge, it substantially improves the chances of obtaining both an accurate suite of openhole geophysical logs and a satisfactory primary cement job. When an accurate set of logs is obtained so that the potential producing zones can be easily identified and when the primary cementing operation successfully isolates the potential producing zones, most completion problems will have been eliminated.

To design a casing program for the drilling operation, an engineer needs to estimate several completion parameters before spudding the well. In consultation with a geologist and a reservoir engineer, the engineer designing the fracture treatment should estimate values for depth to the pay zone, initial reservoir pressure, fracture gradient, created fracture height, Young's modulus of the pay zone, and optimum fracturing-fluid viscosity. With estimates of these parameters, an engineer can calculate values of surface injection pressure during a fracture treatment for a variety of tubular configurations. From these calculations, a preliminary completion strategy can be chosen, and from this strategy, the casing string and bit program can be designed. It is imperative to design a casing string large enough and strong enough to allow the proper size of fracture treatment to be performed at the optimum injection rate.

Because virtually all low-permeability wells must be fracture-treated to provide economic flow rates and ultimate recoveries, the entire drilling program must be centered around the fracture treatment. If a few dollars are saved by running small casing or casing that is not strong enough to withstand fracturing pressures and the ability to fracture-treat the well successfully is restricted, then the overall profit from the well may be greatly reduced or completely eliminated.

Perhaps the most important contribution to formation evaluation from drilling operations is a gauge borehole. An accurate log evaluation is a necessity in deciding how to complete a low-permeability reservoir. Because the logs are affected by hole size and mud-filtrate invasion and because the hole size is often a function of the mud system and bit hydraulics, the drilling-mud system should be monitored closely. In most low-permeability reservoirs, formation damage near the wellbore caused by mud-filtrate invasion is not of major importance. Shallow damage around the wellbore can easily be removed or bypassed with a stimulation treatment. Therefore, the mud system and bit hydraulics should be designed to minimize hole washout problems and to maximize the chances of obtaining a gauge borehole and thus an accurate log suite.

Another important aspect of successful stimulation is the ability to pump the fracture fluid into specified intervals behind the casing. This can be accomplished only if the primary cement job isolates the multiple intervals. It is well understood by the industry that to maximize the chances of obtaining an acceptable primary cement job, the borehole must be in gauge. If the borehole is in gauge and the casing can be both rotated and reciprocated during the cementing operation, the primary cementing operation is usually successful.

In summary, it is always of concern that a well be drilled for the least feasible cost while safety regulations are observed. The term "least cost" is usually synonymous with "fast drilling," but, when all costs, including both evaluation and completion costs, are considered, fast drilling times may not necessarily mean that the total well costs are minimized. Of primary importance to the success of a completion are (1) an accurate log suite and (2) an effective primary cement job. Both of these factors can be accomplished with a high degree of certainty if the drilling personnel drill a gauge borehole while minimizing mud-filtrate invasion into the potential producing intervals.

12.3 Casing and Wellbore Configuration

Fracture treatment conditions should be considered in the casing design. Many times, the fracture treatment will be pumped down the casing or a live annulus will be used to monitor the bottomhole pressure (BHP) during the fracture treatment. In those situations, the casing must be designed to withstand the full fracturing BHP. In other situations, the fracture treatment will be pumped down tubing, and to minimize the burst pressure on the top joint of tubing, backup pressure will be held on the casing/tubing annulus. In virtually every situation, pressure will be exerted on the casing at the surface; therefore, the burst pressure of the casing string should be designed to withstand the maximum pressure expected during the fracture treatment.

As discussed in Sec. 12.7, it is common to fracture-treat a well down casing and then to run small tubular goods into the well after the fracture treatment. In some reservoirs, however, it may be necessary to perform the fracture treatment down tubing. In a few instances, multistage fracturing treatments are performed and the well is completed as a dual completion. Figs. 12.1 through 12.5 illustrate several common wellbore configurations.

Fig. 12.1 presents a wellbore diagram for the simple case where the fracture treatment will be pumped down the production casing. The conductor pipe is used to spud the well, and the surface casing is set to protect the freshwater sands and to provide a base for mounting the blowout preventors and, later, the wellhead. It is very important that cement be circulated back to the surface when the surface casing is set. For a well drilled to a moderate depth in a

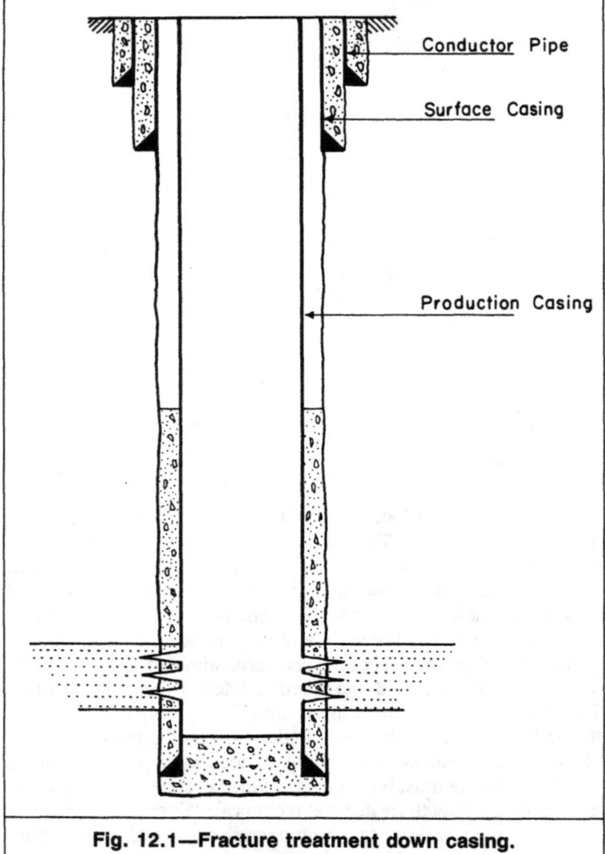

Fig. 12.1—Fracture treatment down casing.

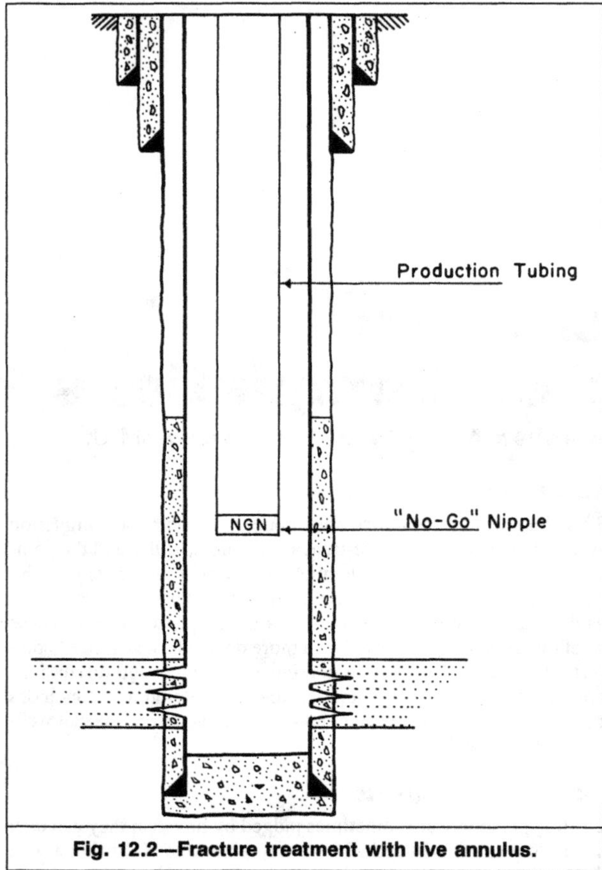

Fig. 12.2—Fracture treatment with live annulus.

normally pressured area, where no unusual or hazardous drilling problems are encountered, the well will usually be drilled to total depth, logged, and evaluated; then casing would be run from total depth back to the surface.

When fracture-treating a well down casing, one must design the casing string for maximum burst conditions. The pressure on the outside of the casing will be hydrostatic pressure in the earth. Inside the casing, the pressure distribution can be calculated by

$$p_{(D)}=p_{si}+Dg_{ph}-Dg_{pfr}. \quad \text{(12.1)}$$

The worst conditions will occur if the fracture treatment screens out and the surface pressure, p_{si}, reaches a predetermined maximum allowable value. If the local hydrostatic gradient in the formation outside the casing is equal to or greater than the hydrostatic gradient of the fracture fluid, then the maximum burst pressure will be at the top of the casing string. However, if the fracture fluid slurry exerts a hydrostatic pressure gradient that is higher than the hydrostatic pressure gradient in the earth, then the maximum burst pressure will be at the bottom of the hole if the fracture treatment screens out.

While the treatment is being pumped, the pressure inside the casing can be constant with depth or can either increase or decrease with depth. As can be illustrated with Eq. 12.1, when the hydrostatic-pressure gradient, g_{ph}, exactly equals the friction-pressure gradient, g_{fr}, the pressure of the fracturing fluid in the casing, $p_{(D)}$, will remain constant with depth, D. If g_{ph} of the fracture fluid slurry is greater than g_{fr} of the fracture fluid slurry, the pressure in the casing will increase with depth. If $g_{ph}<g_{fr}$, the pressure in the casing will decrease with depth. The value of g_{ph} will be a function of the base fluid density, proppant type, and proppant concentration. If nitrogen (N_2) or carbon dioxide (CO_2) is used, g_{ph} will also be a function of pressure and temperature. g_{fr} will be a function of pipe diameter, injection rate, proppant concentration, and crosslink time (if appropriate), as well as pressure and temperature if N_2 or CO_2 is used in the treatment.

It is apparent that it may be difficult to predict where the maximum burst pressure will occur. To determine maximum burst pressure, it is recommended that the pressure profile inside the casing be calculated for the two following situations: (1) pumping pad with no problems and (2) a screenout where the surface pressure reaches the maximum allowable value while carrying proppant at maximum concentration.

To compute the pressure in the casing during pumping, one should assume that the pressure at the bottom of the wellbore is equal to the treating BHP, which is simply the pressure required to propagate the fracture. Then the pressure profile up the hole can be computed by subtracting the hydrostatic head of the slurry and adding the pipe friction-pressure drop.

To compute the pressure in the casing if a screenout occurs, one should assume that the surface pressure increases to the maximum allowable pressure for the treatment. The pressure in the casing will then be the maximum allowable surface pressure plus the hydrostatic head of the fracture fluid slurry.

After determining the pressure distribution inside the casing for these two conditions, the design engineer can then subtract the pressure outside the casing to determine the expected burst pressures during the treatment. One can then either *design* a casing string to withstand the expected maximum burst pressure or *limit* the maximum allowable surface pressure if the casing string has already been set and cemented. It is hoped that the design engineer will consider these burst limitations before running the production casing.

When a casing (or tubing) string is designed for burst, it is recommended that the engineer design for the worst possible conditions. Even though the fracture gradient in a local area may be known, the BHP will normally increase during a fracture treatment. This pressure increase is usually caused either by increases in backstress or by friction-pressure drop down the fracture. Therefore, using the maximum estimate of fracture gradient and adding a safety factor to determine the maximum expected surface treating pressure is always recommended. In many cases, 2,000+ psi [13.8+ MPa] should be added as a safety factor to account for the excess pressure needed to propagate a long fracture. Table 12.1 illustrates the process.

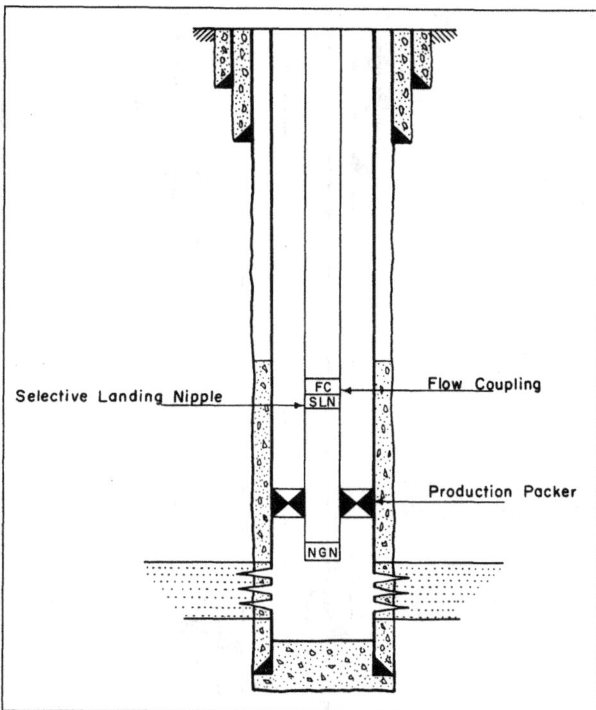

Fig. 12.3—Fracture treatment down tubing with packer.

When designing casing, one should use a safety factor for burst of about 1.1; therefore, the casing string for the example problem should be designed for 8,450×1.1 or 9,295-psi [64-MPa] burst. If no screenout occurs, the maximum burst pressure will be at the top of the casing and will be equal to the maximum injection pressure. If a screenout does occur and the fluid gradient in the well is greater than the fluid gradient outside the casing, the maximum burst will be at the bottom of the casing.

The design engineer should also consider the effects of temperature cooling during the treatment on the tension force on the casing. When the casing is hung in the casing spool, a tension force will be left on the casing. The maximum tension force will be on the top joint of the casing. As cool fracture fluids are pumped down the casing, the steel casing will try to shrink and the tension force in the casing will be increased. Eq. 12.2 can be used to relate a change in temperature to a change in length for a steel tube.

$$\Delta L_4 = 0.0000069 L \Delta T. \qquad (12.2)$$

The relationship between the change in length caused by cooling, ΔL_4, and the resulting tension force is a function of whether or not the casing is buckled. This topic is discussed in more detail in Sec. 12.8. For now, it is adequate simply to remind the design engineer that the tension force on the casing hanger can be substantially increased during a fracture treatment as the casing temperature is reduced.

The wellbore configuration illustrated in Fig. 12.1 provides the most basic approach to well completion for stimulation. If the casing string is strong enough to withstand maximum burst pressure, few problems should be encountered.

Fig. 12.2 presents an illustration of a fracture treatment down tubing with a live annulus. This treatment configuration is used so that the fracturing BHP can be measured directly while the treatment is being pumped. Normally, the treatment is pumped down the tubing and the annulus is used to monitor the BHP. The annulus must be kept full of a fluid of known density and the surface pressure on the annulus must be measured.

In certain situations, the fracture treatment can be pumped down the casing/tubing annulus and the tubing string can be used to measure the fracturing BHP. If desired, a surface-readout wireline pressure gauge can be run in the tubing to record the pressure near the bottom of the well directly. If the fracture treatment is pumped

TABLE 12.1—EXAMPLE BURST-PRESSURE CALCULATION

Well depth, ft	10,000
Maximum estimate of fracture gradient, psi/ft	0.8
Estimated treating BHP, psi	8,000
Fracture-fluid hydrostatic gradient, psi/ft	0.44
Fluid gradient with 5-lbm/gal sand, psi/ft	0.57
Fluid friction-pressure gradient, psi/ft	0.20
Earth's hydrostatic gradient, psi/ft	0.465

Under Normal Pumping of Pad

$p_{si} = 8{,}000 + (10{,}000)(0.2) - (10{,}000)(0.44)$
$= 5{,}400$ psi

Maximum Allowable Surface Pressure Should Be No Less Than

$(p_s)_{max} = p_{si} + 2{,}000$ psi
$= 7{,}400$ psi

Maximum BHP During a Screenout

$(p_{bh})_{max} = (p_s)_{max} + Dg_{ph}$ (gradient inside casing)
$= 7{,}400 + (10{,}000)(0.57)$
$= 13{,}100$

Burst Under Worst Conditions

$(\Delta p_{burst} = (p_{bh})_{max} - g_{ph} D$ (gradient outside casing)
$= 13{,}100 - (0.465)(10{,}000)$
$= 8{,}450$ psi

down the casing/tubing annulus, a blast joint should be used as the top tubing joint to protect the tubing from erosion where the fracture fluid enters the annulus.

When designing the casing configuration for a treatment that will be pumped down a well with a live annulus, one must compute both the burst and the tension forces on the casing string for the maximum expected conditions. The engineer should also determine the maximum burst and/or collapse forces that will be applied to the tubing string.

In many situations, it is not desirable to expose the casing to large fracturing pressures. This could be the case in older wells or in wells where the formation is highly geopressured and the casing is protected to minimize the potential of casing damage. For these and other reasons, many fracture treatments are pumped down tubing with the annulus isolated from the fracturing BHP by a packer. Fig. 12.3 illustrates this typical wellbore configuration. It is recommended that selective landing nipples be run both above and below the packer. These landing nipples can be used to hang pressure bombs and to install blanking plugs or for any number of other applications. Flow couplings should be used above each landing nipple to minimize tubing failures resulting from erosion.

When performing a fracture treatment down tubing, one must determine the maximum allowable surface treating pressure from the burst pressure of the tubing string. Using the same logic previously presented in the discussion of Fig. 12.1, the design engineer should calculate the maximum surface injection pressure during the pad and add about 2,000 psi [13.8 MPa] to set a value for maximum allowable surface pressure. If the treatment is pumped without any problems, the maximum burst pressure will occur at the surface. If the treatment screens out during the last part of the treatment, however, and the slurry density in the tubing is greater than the fluid density in the casing/tubing annulus, then the maximum burst pressure will be in the tubing at the packer.

The tubing pressure will cause an upward-acting force below the packer, and the annular pressure will cause a downward-acting force above the packer. This differential force across the packer should be calculated and the packer selected for the treatment should obviously be strong enough to withstand this differential force.

Another design consideration is that forces are generated not only by the actual pressures during a fracture treatment but also by the changes in pressure and temperature applied to the tubing string from the time it is landed to the point when maximum conditions are encountered during the fracture treatment. These changes in temperature and pressure can increase the tension force on the top joint of tubing and increase the stress on the packer. This topic will be discussed in more detail in Sec. 12.8.

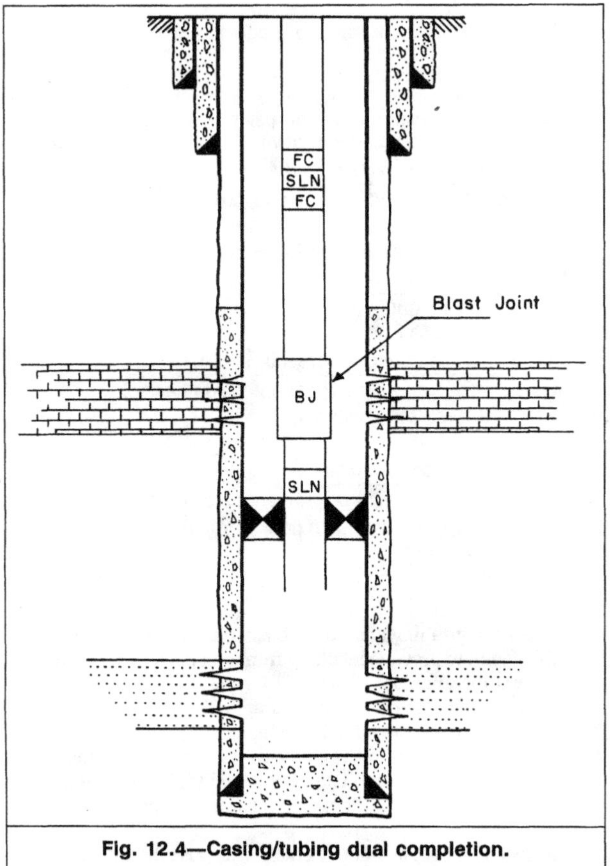

Fig. 12.4—Casing/tubing dual completion.

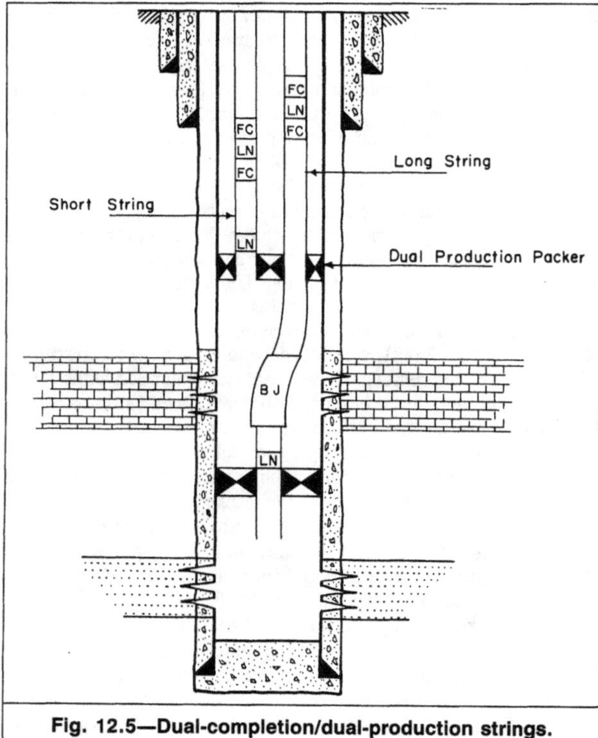

Fig. 12.5—Dual-completion/dual-production strings.

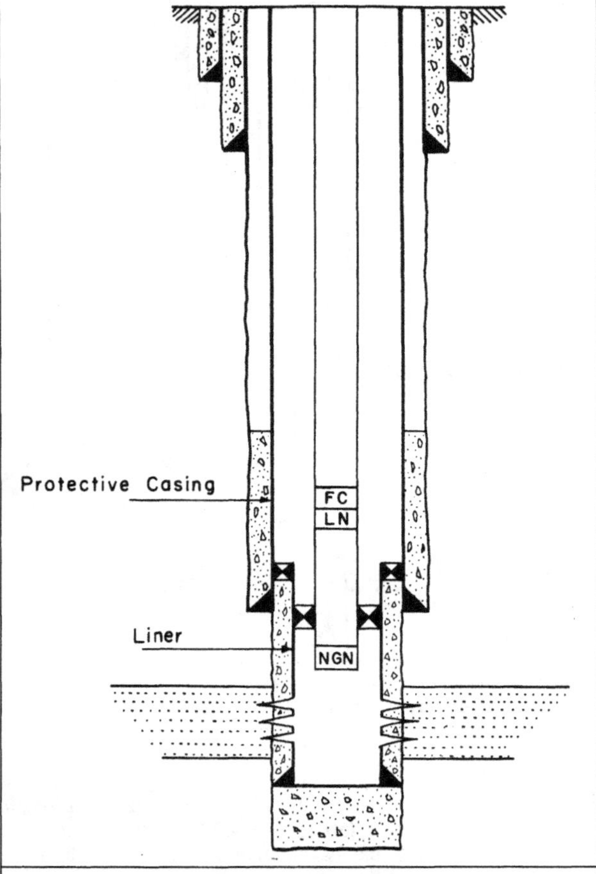

Fig. 12.6—Single completion—protective casing and liner.

It is always important to recognize the weak points in any well-completion plan and to minimize problems. For that reason, such devices as sliding sleeves in the tubing and differential valve tools in the casing should not be used in a well where a large, high-pressure fracture treatment is required to obtain commercial flow rates and recoveries.

Figs. 12.4 and 12.5 present typical wellbore diagrams for dual completions. In certain situations, it may be necessary to complete and fracture-treat multiple reservoirs that cannot be commingled. In such situations, each reservoir can be fracture-treated individually with an appropriate diverting technique; then the downhole hardware can be installed. Figs. 12.4 and 12.5 illustrate the results, while Sec. 12.9 presents information concerning methods of treatment diversion.

Fig. 12.6 presents a diagram of a single completion, where protective casing had to be set and a liner was hung and cemented to complete the casing string. It is well known that obtaining an adequate cement bond around a liner is very difficult. In many situations, the liner top has to be squeezed to obtain a seal. Even when a seal is obtained, the quality of the cement bond of the liner is usually questionable.

It is never recommended that a fracture treatment be performed down casing when a liner top is directly exposed to the treatment fluid. Even if properly tested to the maximum expected fracture pressure before the treatment, the liner top could still break down during the fracture treatment under prolonged exposure to high pressure. Also, as the temperature is reduced during the treatment, the liner will try to contract, and this extra tensional force caused by cooling could result in a liner-top failure.

If a fracture treatment is to be pumped in a well where a liner has been set, the treatment should be pumped down tubing. The liner top should be isolated from the fracturing BHP's with either a packer in the liner or a polished-bore receptacle in the liner top.

Finally, the design engineer should always be cognizant of potential weak links in the downhole tubular goods. Liner tops, squeeze perforations, and differential valve tools all pose potential problems during a high-pressure pumping operation. It is best to isolate these potential problems behind a tubing string and a packer; however, even when a packer and tubing are used, one should remember that backup pressure is normally held on the casing/tubing annulus to minimize the burst pressure on the top joint of tubing.

Any weak link in the casing must be tested to a pressure greater than it will be exposed to during a fracture treatment. If a failure is going to occur, it is best that it occur during a test before the fracture treatment rather than halfway through the stimulation treatment.

12.4 Cementing

To maximize the chances of success, each well should be completed so that the hydraulic fracture can be initiated in the desired reservoir. The fracture may eventually propagate out of zone; however, the design engineer should minimize the chances of out-of-zone migration at or near the wellbore by obtaining a primary cement seal around the production casing or liner.

To determine whether the primary cementing operation was successful, a cement bond log can be run to estimate the quality of the cement bond near the zone of interest. Many papers are available to explain the use and interpretation of cement evaluation logs.[1-5] Special care should be taken to ensure that one does not misinterpret the log data and conclude that a poor primary cement bond exists when the problem is really a microannulus or poor log data.[6]

If the cement bond log or other data from production logs and production tests definitely indicates that a channel exists behind the casing that will allow the fracturing fluids to migrate to another location, then the completion engineer will have to repair the primary cement by squeezing additional cement behind the casing at critical locations. A thorough discussion of squeeze cementing is presented in Ref. 7.

The key information used in the decision to squeeze a well before a fracture treatment is the estimation of where the fracture fluids will go if the squeeze operation is not performed. For example, in many deep, low-permeability reservoirs, the formations of interest consist of interbedded sandstones, siltstones, and mudstones bounded by massive marine shales. There are no permeable water zones, and even if a poor cement bond exists, the fracture fluid will almost assuredly enter the porous and permeable gas intervals in preference to the shales. In this situation, a squeeze-cementing operation would not usually be advisable.

On the other hand, some reservoirs are composed of several permeable layers containing different fluids. To prevent the fracture treatment from migrating behind pipe from an oil zone to a water or gas zone, or from a gas zone to a water zone, etc., the well should be squeeze-cemented if the primary cementing operation did not adequately isolate the various layers in the formation.

It is sometimes difficult to decide whether a well should be squeeze-cemented, but when the decision has been made, the following comments should be considered. First, the completion engineer should design and perform a low-pressure, low-water-loss cement squeeze. The objective of a squeeze-cementing operation is to place cement in the annulus between the casing and the formation. Pumping cement at pressures above the fracture gradient and subsequently pumping most of the cement into a hydraulic fracture will not be of much benefit.

Second, the completion engineer must be selective about the location of perforations to be used during the squeeze operation. If the well is to be squeezed below the zone of interest, it is best to perforate as deeply as feasible and to use a cement retainer to squeeze the well. The retainer can be left in the well; thus, the lower set of squeeze perforations can be isolated from the fracture-treatment pressures. With the retainer set as deep as possible, the length of the rathole can be maximized. A deep rathole is often advantageous when the engineer is trying to run production logs.

If the well has to be squeezed above the zone to be fracture-treated, the placement of the squeeze perforations will be a function of how the well is to be completed. If the well is to be fracture-treated down casing or down tubing with no packer in the well (see Figs. 12.1 and 12.2), the squeeze perforations should be placed very near the top of the completion interval. Then, if the squeeze perforations happen to break down during the fracture treatment, minimal harm should result.

If the well is to be fracture-treated down tubing with a packer in the hole (see Fig. 12.3), the squeeze perforations should be placed up the hole so that they will be above the packer and isolated from the fracturing BHP during the treatment. It must be remembered that pressure is usually held on the casing/tubing annulus during a fracture treatment; therefore, the squeeze perforations should be tested to a pressure greater than the pressure expected during the fracture treatment. If the squeeze perforations are going to break down, one would prefer the failure to occur during a test with a rig on the well so that the failure can be corrected without risk to the well. If the squeeze perforations break down during the fracture treatment and sufficient backup pressure cannot be held on the casing/tubing annulus, then the treatment may have to be prematurely halted.

In summary, a well should not be squeeze-cemented unless absolutely necessary. Many times, the fracture fluids will enter the correct intervals even if the primary cementing operation failed to achieve complete zonal isolation. Prefracture logging and injection tests can be run to determine whether the primary cement bond is going to allow fluid flow behind the casing. If a squeeze-cementing operation is necessary, the engineer should design a low-pressure squeeze with low-water-loss cement, and the locations of the squeeze perforations should be chosen on the basis of the methods used to complete and to fracture-treat the well.

12.5 Perforating

The main concerns involving perforations are simply safety and obtaining the correct number of perforations in the proper formations. The safety aspect refers mainly to well control. In highly geopressured areas or highly depleted formations, perforating may best be accomplished by use of a through-tubing gun with the wellhead installed. By using a proper size of lubricator and perforating the well underbalanced, one should maximize the chances of obtaining penetration through the casing and cement and into the formation. A well would normally be perforated through tubing in an underbalanced condition when the fracture treatment is going to be pumped down the tubing.

In a normally pressured reservoir that will be fracture-treated down casing, it is usually better to perforate the formation overbalanced with a casing gun. Clean fluid should be circulated into the wellbore before perforating to minimize the amount of debris left in the perforation tunnel.

More detailed information concerning perforating can be found elsewhere[8,9]; however, most of that information deals with the situation where a formation is not going to be hydraulically fracture-treated. McLeod[10] presented information concerning the effects of the perforation tunnel on the performance of gas flow and non-Darcy effects around the borehole. Another paper by Saucier and Lands[11] presented useful information concerning a laboratory study of perforations in stressed formation rocks. The information in these publications can cause concern about the status of the productivity index of a typical formation as a result of poor flow performance through perforation tunnels. However, these concerns are not meaningful if one performs a perforation breakdown treatment before testing or fracture-treating the formation.

It is widely understood that most perforations, even when they are shot under ideal conditions, do not result in effective communication between the reservoir and the wellbore. Breaking down the perforations with a nondamaging fluid and a diverting agent is usually necessary before one can be assured that the production rates and flowing pressures measured in a well are representative of the true influx from a formation. Therefore, it is normally advisable to perform a perforation breakdown treatment before a prefracture well test is run and the hydraulic fracture treatment is performed.

The best method to open perforations is usually to pump a nondamaging fluid, such as a mixture of KCl water and surfactant or an acid solution, and to drop ball sealers interspersed in the fluid. Other methods to open perforations are also available. In certain situations, one may choose to use a selective injection packer. Regardless of the method chosen, a perforation breakdown treatment should be performed on virtually every well to be fracture-treated.

TABLE 12.2—EXAMPLE CALCULATION OF OPEN PERFORATIONS

d_{pf}	i_{pf}	n_{pf}
0.325	3.0	4
0.300	2.55	5
0.275	2.14	6
0.250	1.77	7

Normally, knowing that all the perforations are open and accepting fluid is important before the fracture treatment is pumped. If a perforation ballout treatment is performed and the data indicate that only a few of the perforations are open, it is advisable to run production logs and to reperforate any interval that is not open. It is extremely important to be certain that the proper communication has been achieved with the formation before the hydraulic fracture treatment is begun.

The choice of a fluid to be used in the ballout treatment is quite important. Acid can be used in limestone, dolomite, or competent sandstone reservoirs. In carbonate reservoirs, the acid will react with the formation materials and open up the flow channels between the wellbore and the formation quite successfully.

In some sandstone formations, acid should not be used during the perforation-breakdown treatment. KCl solution mixed with surfactants or methanol can normally be used to perform the ballout treatments in gas reservoirs. In oil reservoirs, diesel oil can be used to ball out the perforations. Such fluid mixtures should be nondamaging to most sandstones and can be easily produced back. If the formation is slightly underpressured, N_2 or CO_2 can be used to help produce the fluids from the reservoir.

Special care should be taken to prevent fluid damage to the formation during the ballout treatment. Specifically, some sandstone reservoirs are cemented with calcium carbonate cement. If acid is pumped into these reservoirs, it is possible that the formation can become unconsolidated, collapse, and begin flowing particles of formation into the wellbore. Therefore, a nondamaging fluid should be chosen for the perforation-breakdown treatment. Always remember that the purpose of the treatment is simply to open the perforations so that a valid prefracture well test can be performed and the hydraulic fracture treatment can be successfully pumped.

It is therefore necessary to calculate the number of holes that are taking fluid during an injection treatment. Eqs. 12.3 through 12.6 can be used to determine the number of holes that are open and accepting fluid:

$$p_{si}=p_{bhf}+p_{fr}+p_{pfr}-p_h, \quad\text{(12.3)}$$

$$p_{bhf}=p_{ss}+p_h, \quad\text{(12.4)}$$

$$g_f=p_{bhf}/D, \quad\text{(12.5)}$$

and

$$p_{pfr}=\frac{0.2369 i_{pf}^2 \rho}{d_{pf}^4 \alpha^2}. \quad\text{(12.6)}$$

Eq. 12.3 can be used to determine the surface injection pressure, p_{si}, if values are known for treating BHP, p_{bhf}; the friction pressure drop in the tubular goods, p_{fr}; the friction pressure drop through the perforations, p_{pfr}; and the hydrostatic head, p_h, in the borehole. Eq. 12.4 can be used to determine the treating BHP if one measures the instantaneous shut-in pressure (ISIP), p_{ss}, during the early stages of an injection treatment.

The ISIP is measured by first establishing injection into a set of perforations with a fluid that is completely filling the wellbore at a rate just barely high enough to create a hydraulic fracture. After a few barrels of fluid are pumped, the pumps should be shut down so that the friction pressure terms in Eq. 12.3 go to zero. The resulting surface pressure is the value for ISIP. By adding the value of hydrostatic head of the fluid in the wellbore to the value of ISIP determined during the shut-down period, one can compute the value for treating BHP. Eq. 12.5 can be used to determine a value for fracture gradient. The fracture gradient is computed by dividing the value of treating BHP by the depth to the midpoint of the perforations.

Eq. 12.6 can be used to determine the friction-pressure drop through the perforations. Because fluid flow through multiple perforations is a parallel-flow problem, the average injection rate through each perforation should be similar. The average injection rate used to compute friction-pressure drop is i_{pf}, in units of barrels per minute per perforation. Note in Eq. 12.6 that the diameter of the perforation is raised to the fourth power and that the perforation coefficient, α, is squared. These variables are quite important in the analysis of the condition of the borehole perforations. The perforation coefficient is an efficiency number that corrects for the fact that all perforations are not perfectly circular and smooth orifices. It is well known in the industry that after some proppant has been pumped through perforations, the abrasive nature of the propping agents will erode and/or smooth the perforation opening such that the hole diameter is enlarged and the perforation coefficient increases. When this occurs, the perforation friction will decrease.

Eqs. 12.3 through 12.6, along with an adequate estimate of pipe friction, can be used to predict the surface injection pressure during a fracture treatment. However, to calculate the number of perforations open after injection into the well and measurement of the surface injection pressure, Eqs. 12.3 through 12.6 should be rearranged:

$$p_{pfr}=p_{si}+p_h-p_{bhf}-p_{fr}, \quad\text{(12.7)}$$

$$i_{pf}=\sqrt{\frac{p_{pfr} d_{pf}^4 \alpha^2}{0.2369\rho}}, \quad\text{(12.8)}$$

and

$$n_{pf}=i/i_{pf}. \quad\text{(12.9)}$$

Eq. 12.7 can be used to determine the value of perforation friction after the surface injection pressure and the treating BHP at a given injection rate are measured. Once the value of perforation friction is known, Eq. 12.8 can be used to calculate the injection rate through each perforation. Finally, Eq. 12.9 can be used to determine the number of perforations actually accepting fluid during any injection test.

The following example illustrates how Eqs. 12.7 through 12.9 can be used to determine the number of open perforations.

Example.

Given

i = 12 bbl/min,
D = 6,000 ft,
p_{si} = 5,350 psi,
p_{fr} = 2,000 psi,
p_{bhf} = 4,200 psi,
d_{pf} = 0.375 in.,
ρ = 8.5 lbm/gal,

$$p_{pfr} = 5{,}350+6{,}000(0.4412)-4{,}200-2{,}000 = 1{,}797 \text{ psi},$$

$$i_{pf} = \sqrt{\frac{(1{,}797)(0.375)^4(0.95)^2}{(0.2369)(8.5)}} = 4 \text{ bbl/min-perforation, and}$$

$$n_{pf} = 12/4=3 \text{ holes.}$$

For this example, an injection treatment was being performed at 12 bbl/min [2 m³/min] in a formation that was 6,000 ft [1830 m] deep with a surface injection pressure of 5,350 psi [36.9 MPa]. The friction pressure in the tubular goods was 2,000 psi [13.8 MPa], and the treating BHP for this particular formation was 4,200 psi [29 MPa]. The injection treatment was being performed with an

8.5-lbm/gal [1019-kg/m^3] fluid, and the shaped charge used in this particular gun should have given a perforation diameter of 0.375 in. [0.95 cm]. From Eq. 12.7, the perforation friction was determined to be 1,797 psi [12.39 MPa]. With this value in Eq. 12.8, the specific injection rate was calculated to be 4 bbl/min [0.64 m^3/min] per perforation. Finally, the number of perforations that are accepting fluid can be determined with Eq. 12.9; in this example, it was three holes.

Under these conditions, a perforation breakdown treatment should be pumped in an attempt to open up more of the perforations. If a perforation breakdown treatment has already been attempted and only three holes are open, then it is recommended that the well be reperforated before a fracture treatment is performed.

It must be stressed that the perforation diameter is very critical to the preceding calculation. Table 12.2 illustrates how the calculation of the number of open perforations changes for the data presented above if the perforation diameter is not well known.

Engineers will always debate the correct type and number of perforations required to complete a particular reservoir properly. Remember that there is a significant difference between producing oil and gas through a perforation tunnel vs. producing oil and gas through an orifice in the casing after the perforation tunnel and possibly the cement around the casing have been destroyed by the hydraulic fracture treatment. Normally, even when small-diameter perforations and limited entry are used to help initiate fluid injection into specific intervals, the limited number of holes available for production will not inhibit the flow of oil and gas into the wellbore. It is reasonable to expect that the perforation tunnels and the integrity of the cement sheath around the casing will be destroyed during a high-volume, high-pressure fracture treatment. Therefore, oil and gas can flow down the fracture to the wellbore and through the holes in the casing with very little restriction.

The most important problem concerns shear degradation of the polymer fluids through the perforation during the fracture treatment. Many of the fracturing fluids used in deep, high-temperature applications are composed of long-chain polymers that are degraded under high-shear conditions. If these polymers are pumped through small-diameter holes at very high shear rates, the possibility will exist that the polymers will be degraded to the point where a significant loss of viscosity occurs. For this reason, if only a single interval is to be fracture-treated and limited-entry diversion techniques are not required, the zone should be perforated with a casing gun or a tubing-conveyed gun with large-diameter perforations to minimize the shear-degradation effects on the polymer fracturing fluids. Also, the communication between the wellbore and the fracture should be better with 90 or 120° [1.6- or 2.1-rad] phasing than with 0° [0-rad] phasing.

In some reservoirs where limited-entry perforating has to be used, a delayed-crosslinker system should be pumped so that the majority of the crosslinking reaction occurs downstream of the perforations. If the crosslinking occurs in the tubular goods, the high-shear conditions through the perforations could significantly decrease the apparent viscosity of the gel fluid. If the polymer is allowed to crosslink fully in the fracture, however, the shear-degradation effects are minimized in both the tubular goods and the perforations and the abilities to create wide fractures and to carry proppant long distances from the wellbore are maximized.

12.6 Completion and Breakdown Fluids

One of the most important decisions for a completion engineer concerns the type of fluids to be used during the completion and stimulation operations on a particular well.[12,13] Most of the emphasis concerning completion fluids has been directed toward gravel-pack operations. Other considerations have also been discussed concerning the use of excessive pipe dope, dirty fluids, and dirty tubular goods, and how these factors are detrimental to the completion attempt on any well. All these concerns are valid, and a prudent engineer will take the steps necessary to use clean fluids and clean tubular goods in any completion attempt. Minimizing the amount of pipe dope used when tubular goods are being tripped in and out of a borehole is widely accepted as a method of minimizing formation damage. It is assumed that most completion engineers are aware of these problems and will adhere to the well-known solutions by using clean fluids and acceptable practices.

The most important part of choosing fluids in a stimulation treatment is to consider the compatibility of the fluid with both the formation minerals and the formation fluids. The obvious problem that can occur when a water-based fluid is used to stimulate an oil reservoir is the creation of an emulsion. An emulsion will form whenever oil and water are mixed together. In many situations, the use of a surfactant will cause the emulsions to break and not affect well performance. Therefore, in the stimulation of an oil reservoir, it is important to obtain representative samples of the fluids in the reservoir and to run fluid compatibility tests to ensure that the proper fluids and surfactants are used to minimize emulsion problems. Another consideration in oil reservoirs is the amount of paraffins and asphaltenes found in the oil. It is possible that the use of some surfactants or some fluids, such as CO_2, can cause the oil to emulsify as a result of fairly unpredictable chemical reactions. Again, the best method for determining potential problems is to run compatibility tests in the laboratory between any injection fluid and the reservoir fluid before injection.

In dry-gas reservoirs, the creation of emulsions is not usually a problem. The most critical aspect in these reservoirs is normally the relative permeability and capillary pressure effects of injecting a liquid into a dry-gas reservoir. It is important that the wettability effects and capillary effects of the fluid be considered before injection. Many times, a 2% KCl solution with the proper surfactant or a small amount of methanol can be used to complete a well and to perform a breakdown treatment in a dry-gas reservoir. However, some reservoirs that contain small pore throats lined with clay particles may be water-sensitive, and fluids other than water should be used during the completion and perforation breakdown. One common solution to such problems is the use of foamed fluids. Foamed fluids may contain between 60 and 80% gas, such as N_2 or CO_2, and the remaining fluid can be water, a water/methanol mixture, or even an oil-based fluid.

In certain situations, a weak acid, such as 7½% HCl, can be used to break down a set of perforations before a well test or stimulation treatment. In carbonate reservoirs, the use of acid is a widely accepted and usually successful practice.

To justify the use of an acid in a breakdown treatment, one must identify a material in the formation that will be dissolved by the acid, and with that material dissolved, the communication between the formation and the wellbore will be improved. In a few reservoirs, such as the Cotton Valley sandstones, acid does not react adversely with the formation materials and can be used successfully to perform a breakdown treatment. In other reservoirs, however, such as the deep Wilcox or Vicksburg formations in south Texas, the use of acid during the prefracture breakdown treatments can weaken the formation to the point where the formation will collapse if high flowing pressure gradients are applied after the acid treatment. In such reservoirs, the use of any form of acid during the well completion is not recommended.

In summary, one should be concerned with the types of fluids used during completion and stimulation of oil and gas reservoirs. In oil reservoirs, the most important consideration is usually the prevention of emulsions if a water-based fluid is used during the completion of that formation. In gas or gas-condensate reservoirs, the most important consideration is usually the minimization of the relative permeability and capillary pressure effects so that the injected fluids are cleaned up in a reasonable amount of time. A weak acid can be used to break down the perforations in carbonate reservoirs and in some sandstone reservoirs.

In most sandstone reservoirs, however, a mixture of 2% KCl water with surfactants and methanol can be used successfully to break down the perforations. In very-low-pressure or water-sensitive reservoirs, foamed fluids would be recommended as both completion and breakdown fluids. Formation damage is normally not an overriding concern if a massive hydraulic fracture treatment is going to be performed on a particular well completion. It is good practice, however, to minimize formation damage so that prefracture well tests can be performed and representative values of formation permeability, reservoir pressure, and skin effect can be computed from such well tests.

12.7 Completion Strategy

To complete a well successfully, the engineer should consider the following 10 items.

1. Investment required.
2. Number of pay intervals encountered by the borehole.
3. Desired producing rates.
4. Reserves in the various intervals.
5. Reservoir rock and fluid properties.
6. Stimulation requirements.
7. Sand control requirements.
8. Future workover requirements.
9. Artificial-lift considerations.
10. Future secondary-recovery needs.

The ideal completion is the one that can be obtained for the lowest cost—considering both the initial completion cost and the subsequent operating costs—that meets or most nearly meets the demands placed on the well for most of its life. Although this definition is somewhat academic, it correctly implies that a prudent engineer will attempt to provide a functional completion for many years to come at the lowest possible cost to the operator.

Of concern in the design of a completion is always the number of producing zones that are separated vertically in the reservoir. To determine whether different producing intervals should actually be treated as a single reservoir, one must first determine whether these intervals will all be connected by a single hydraulic fracture. If a particular zone is separated from a second zone by a thin shale or silt layer with little stress contrast among the various intervals, then one can usually assume that a single hydraulic fracture will connect both intervals. In such a situation and when these different reservoir layers can be commingled with no damage to either layer, the well should be completed as if all the different layers were actually a single reservoir.

If two productive intervals are separated by a shale member 50 to 100 ft [15 to 300 m] thick that will clearly be a barrier to fracture growth, then the decision to commingle these intervals must be based on the potential for reservoir damage and the governing rules from the state regulatory agency in charge of such matters. In essence, it is necessary for an engineer to analyze the logs and core to determine the number of distinctly different producing horizons that have been contacted by the borehole. Then, depending on the net hydrocarbon pay, hydrocarbon content, presence or lack of barriers to fracture growth among the various intervals, and the state regulatory rules, a decision must be made concerning the number of pay zones to be completed and produced in the well.

Another important consideration is the estimation of the desired producing rates after successful stimulation. In most tight reservoirs, early flow rates can be quite large; after a short amount of time, however, the expected flow rates will decline to rather modest values. In these instances, it is usually best to have 2⅜- or 2⅞-in. [6- or 7.3-cm] tubing in the well to maximize the lifting of liquids by natural gas. In other cases, however, hydraulic fracturing can be used to increase flow rates from several hundred to several thousand barrels of oil per day or from several million to tens of millions of cubic feet per day. In those instances, one may prefer to complete the well with 3½-in. [9-cm] tubing. In summary, the completion engineer should be cognizant of the expected flow rates, the expected GOR's and gas/water ratios, and how these factors will change with time.

A reserve analysis should be performed on each potential interval to help decide which zones should be completed. The hydrocarbons in place, the primary reserves if no stimulation occurs, and the secondary recovery of reserves because of fracture stimulation should be computed. The engineer should also determine how reserves will be affected by various stimulation treatment sizes, the use of compression or artificial lift, and the size of the tubular goods to be used in the subject well. The results of all these calculations can be quite meaningful when one must decide exactly which zones to complete and the size of the stimulation treatment necessary for each of the zones.

The determination of certain reservoir properties can also be quite important to the completion engineer (see Chap. 2). Information concerning reservoir heterogeneity, reservoir pressure, and reservoir drive mechanism, and properties concerning permeability, porosity, and rock strength all play an important part in how the well should be completed.

The completion engineer must determine the expected surface injection pressures for a variety of wellbore configurations. Also of importance is an estimate of the volume of fluid required to stimulate each layer of the formation successfully. Pumping these fluids will affect the stresses on the tubular goods, and the changes in stress resulting from the stimulation treatment must be computed before the final tubular-goods configuration is designed.

Sand-control treatments, such as the standard gravel-pack completion used in many unconsolidated formations, are not usually applied in situations where wells are hydraulically fracture-treated. The production of sand after a stimulation treatment is a common occurrence, however, that can cause well-completion problems. The sand being produced back often consists of crushed fracture sand or fluid-loss additives, such as 100-mesh sand. In certain situations, high pressure gradients across a weak formation can lead to formation collapse. Weeks[14] indicated that sand-flowback problems could be minimized by reperforating a well with 4 shots/ft [13 shots/m] after the stimulation treatment but before well cleanup. In other areas, resin-coated proppants have been tried to prevent flowback of sand from the fractures. The completion engineer should be aware of any such problems and design the completion to minimize sand flowback. In many situations, sand flowback can be minimized by flowing the well under controlled conditions with small chokes.

When designing a completion, the engineer should be aware that most wells must eventually be worked over. For an onshore well in a normally pressured area, most workovers can be performed using conventional techniques with a workover rig. Therefore, the well should be completed with a retrievable packer so that the packer and tubular goods can be removed when desired. In other cases, such as deep, geopressured reservoirs or in remote locations where a workover rig can be difficult to obtain or expensive to move in, wells should be completed so that most workovers can be performed through tubing.

Permanent well completions and concentric workover techniques have been used successfully for many years.[15-17] When such techniques are applied, it is very important that all equipment run into the well be calipered to ensure that all ID's and OD's of the systems, such as the packer seals, the nipples in the tubing, blast joints, and flow couplings, have been measured and recorded. Knowing these diameters can be quite useful when choosing the wireline tool size to run into a borehole for performing such tasks as paraffin cutting, corrosion inhibition, scale removal, and running BHP surveys. In summary, the well completion should be organized so that when and if the time comes to work the well over, all necessary information has been recorded in the well file and the workover can proceed with minimal problems.

An example of a future workover requirement is a fracture-stimulated oil well where a rod-pumping system must be installed within a few days or a few weeks after a stimulation treatment. In this situation, if a packer is used to stimulate the well, it should be retrievable because the packer will be removed in a very short time. Also, the use of 100-mesh sand to seal natural fractures during a stimulation treatment is not good practice when a rod pump must be run soon after the stimulation treatment. In certain circumstances, some of the 100-mesh sand flows back into the wellbore and tends to plug up the pump. In those instances, the use of 100-mesh oil-soluble resins may be a viable alternative. In other cases, wells are placed on gas lift after stimulation. If such a system is to be used, one must consider the tubing and gas-lift mandrel size to ensure that the correct size of casing is chosen for the well.

After carefully considering the various reservoirs encountered by the wellbore, the engineer must make a series of decisions concerning the completion configuration. The first is the basic completion type: openhole, liner, or a perforated-casing completion. Then the number of completions within the wellbore must be decided and the final tubing and casing configurations designed.

Openhole completions have been used successfully in many formations. They can be effective when only one pay zone is present, the formation is competent, and there are no problems with selective stimulation or selective production of specific intervals. Openhole completions are usually successful in fractured carbonate reservoirs

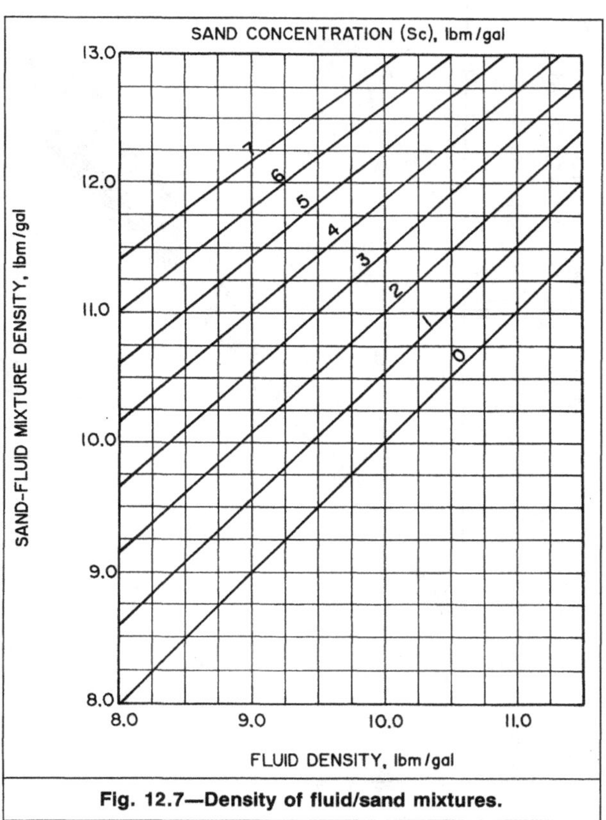

Fig. 12.7—Density of fluid/sand mixtures.

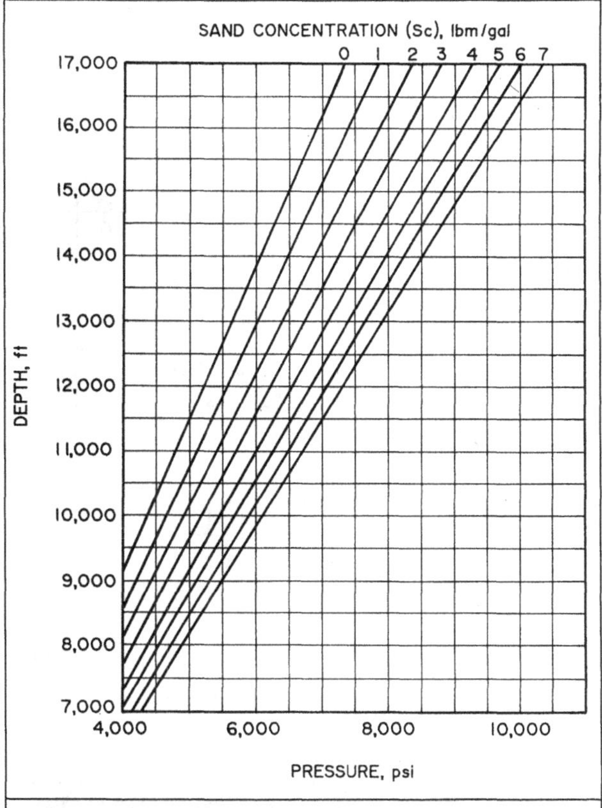

Fig. 12.8—Hydrostatic pressure for various fluid/sand concentrations.

where there is no problem with excessive formation-water production.

Liner completions are used in many situations when intermediate pipe has been set before total depth is reached. An intermediate casing string can be run to isolate problem formations, such as salt layers or sloughing shales. Intermediate casing is also run when the wellbore is going to encounter a geopressured interval. Fig. 12.6 illustrates a typical liner completion. To maximize the chances of stimulating a formation behind a liner successfully, it is usually best to run a packer inside the liner or to use a polished-bore receptacle in the top of the liner so that excessive pressure is never applied to the liner top.

The most common completion is the perforated-casing completion. This system is most successful because it gives the best control over testing and stimulating individual zones and adapts easily to multiple completions.

Once the basic completion method has been chosen, the engineer must determine the number of completions to be attempted within the wellbore and the size and type of tubular goods that must be used to stimulate the well successfully. It is very important that, when the tubing and casing configurations are chosen, both the stimulation requirements and the producing requirements of the formation after it has been stimulated are considered. When packers, bridge plugs, or tubing strings are run into a wellbore, all equipment should be calipered and recorded by the field personnel. OD's of packers and ID's of sealbores, selective landing nipples, and all other equipment should be calipered, recorded, and placed in the well file for future reference. It is also important to record the data concerning the pickup or set-down weight actually applied to the packer, as well as the weight, size, and grade of tubular goods run into the wellbore. With detailed planning and careful field execution, the engineer can maximize the completion success.

Fracturing Down Casing. Perhaps the best technique for stimulating and completing most wells is to perform the fracture treatment down casing, to flush the treatment with a clean, solids-free fluid, and then to run in with the packer and tubing before the fracture fluids are produced back. In some situations where clean brine fluids are not dense enough to overbalance the BHP after the treatment, a lubricator can be used to run in a retrievable packer on wireline with either a pumpout or a pushout plug in the base of the packer. After the packer has been set, the wellbore pressure can be bled off and the tubing run. The main advantage of fracturing down casing is that a viscous fluid can be pumped at high injection rates with low injection pressures (compared with similar treatments performed down smaller tubular goods).

To maximize the proppant placement in the fracture and to minimize the effects of fluid leakoff, high injection rates should be used. In earlier studies on fracturing when most fracture fluids exhibited Newtonian behavior, there appeared to be a direct correlation in many situations between fracture-height growth and injection rate. This direct correlation to fracture-height growth was obviously caused by the increase in friction pressure down the fracture as injection rate was increased. With the pseudoplastic, non-Newtonian fluids currently in use, however, injection rate is not usually a critical parameter to fracture-height growth. Most fracture-design equations indicate that the apparent viscosity of the fluid in the fracture will decrease as injection rate is increased and that the two parameters virtually offset each other. Therefore, the high injection rates obtained down casing will usually improve proppant distribution in the fracture and minimize the time allowed for fracture-fluid leakoff, while not dramatically affecting the friction-pressure drop down the fracture.

In summary, performing a fracture treatment down casing can be quite beneficial when it allows the use of high injection rates at low surface injection pressures. The higher injection rates, e.g., 25 to 50 bbl/min [4 to 8 m^3/min], that can normally be achieved down casing can be important to the success of a stimulation treatment. After the fracture treatment has been performed, small-diameter tubing can be run into the wellbore to minimize liquid-loading problems. For example, many wells can be fracture-treated down 5½-in. [14-cm] casing, and then 2⅜-in. [6-cm] tubing can be run to produce the well.

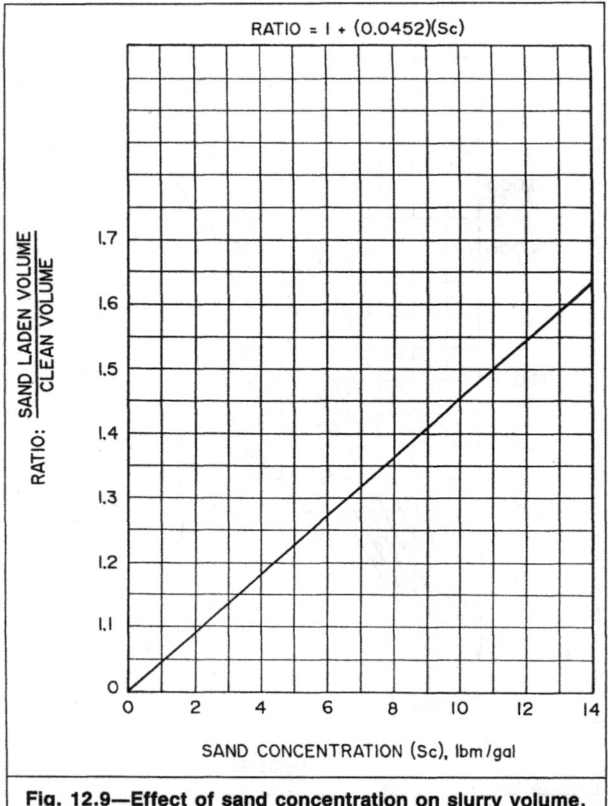

Fig. 12.9—Effect of sand concentration on slurry volume.

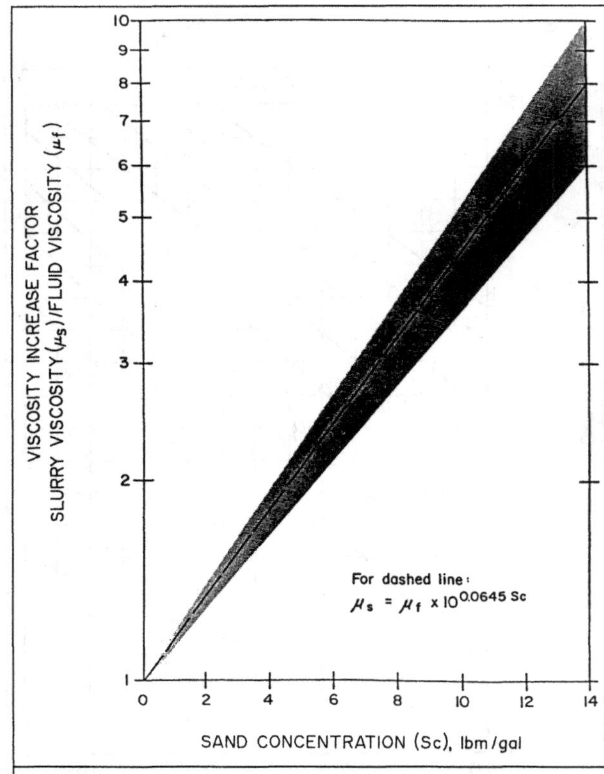

Fig. 12.10—Increase in slurry viscosity vs. sand concentration.

Fracturing Down Tubing. In certain situations, the fracture treatment should be pumped down tubing with a packer to isolate the annulus. When any weak spots exist in the casing because of corrosion, erosion, or some mechanical problem, such as a weak liner top or a differential valve tool, it is best not to pump the fracture treatment down casing. In other instances, such as in highly geopressured formations or in extremely-low-pressure formations, it may be best to stimulate a well down tubing so that well control can be maintained at all times. After the stimulation treatment and a brief shut-in period, the well can be produced back and thus can minimize the amount of time the fracture fluid stays in the formation.

The size of the tubing string necessary for successful stimulation will be determined by the desired injection rate during the fracture treatment. In shallow reservoirs, it is sometimes possible to fracture-treat down 2⅜-in. [6-cm] tubing at 12 to 20 bbl/min [2 to 3 m^3/min]. As depth and fracturing pressures increase, 2⅞- or 3½-in. [7.3- or 9-cm] tubing must be run to obtain the desired injection rates, which may be from 15 to 25 bbl/min [2.4 to 4 m^3/min]. Normally, 2⅜- and 2⅞-in. [6- and 7.3-cm] tubing can be run inside 5½-in. [14-cm] casing. To run 3½-in. [9-cm] tubing, it is best to have 7-in. [17.8-cm] casing. It is always good practice to maintain sufficient clearance between the casing and the tubing so that wash-over pipe can be used if the tubing ever has to be fished from the wellbore.

To understand the fracturing process, it is helpful if one can compute the fracturing BHP during a fracture treatment. By plotting the increase in fracturing BHP vs. the increase in time during the treatment, one can determine whether the fracture was contained in a particular formation or whether the fracture tended to grow vertically out of the desired interval. The difficulty in calculating fracturing BHP's is magnified when pseudoplastic, non-Newtonian fluids are used during stimulation. As long as clean fluids are run with no large changes in injection rate, it is relatively easy to calculate fracturing BHP's. When sand is added to the pseudoplastic fluids, however, both the hydrostatic head and the friction pressure are increased.

The increase in friction is currently not well understood, and the best technique for correcting the friction-pressure drop for the presence of a granular propping material is at best empirical. When the proppant concentration is about 1 to 3 lbm/gal [119 to 359 kg/m^3], the proppant will usually affect hydrostatic pressure more than friction; therefore, the surface injection pressure will tend to decrease as proppant is added at low concentrations. However, when the proppant concentration increases to 4 to 6 lbm/gal [479 to 719 kg/m^3], or even to 8 to 12 lbm/gal [959 to 1438 kg/m^3], the friction pressure will increase more than the hydrostatic pressure of the slurry. In such situations, the surface injection pressure will begin to increase more than the downhole pressure. Figs. 12.7 and 12.8 present information concerning the effects of sand on the density and hydrostatic pressure of the various fluids. Fig. 12.9 illustrates the increase in slurry volume caused by the addition of sand at various proppant concentrations. Fig. 12.10 shows an approximate effect of proppant concentration on the slurry viscosity vs. the base-fluid viscosity. Even though this relationship is not a direct measure of friction-pressure gradient, it illustrates that the slurry viscosity is affected dramatically by proppant concentration, and an increase in the friction-pressure drop of the slurry will follow. These data can be reproduced with the two basic relationships that the weights are additive and the volumes are additive. Therefore, data similar to those presented in Figs. 12.7 through 12.9 can be computed for other proppants that may not have the same density as sand.

In summary, many formations are fracture-treated down tubing successfully. Because the tubing diameter is smaller than that of the casing, small changes in friction properties of the fluid can make large changes in the surface injection pressures. A design engineer who supervises fracture treatments in the field must be aware of the basic relation between surface injection pressure, hydrostatic gradient, and friction-pressure changes as the proppant is added to the system during the treatment.

Fracturing With Live Annulus. Perhaps the best method of completing most wells would be to have a tubing string in the well without a packer so that the annulus has not been packed off. Such a completion is illustrated in Fig. 12.2. If the fracture treatment is pumped down the tubing, then the annulus can be used to measure the fracturing BHP. In some situations, the fracture treatment can be pumped down the casing/tubing annulus and the tubing can be used to measure fracturing BHP. If one decides to pump down the annulus, a blast joint must be used at the top to prevent erosion of the tubing where sand enters the annulus. If the correct sizes of casing and tubing are run, then the benefits of fracturing down casing at high injection rates with low injection pressures can be obtained with this completion method. Also of benefit to the success of the treatment is the direct measurement of fracturing BHP's. Knowledge of the BHP during the fracture treatment can be used to determine whether fracture containment is being maintained or to foresee possible screenouts before they actually occur (see Chap. 14 for details). Another benefit of this completion type is that if a screenout does occur, the well can be reverse-circulated to remove most of the proppant material that might tend to bridge in the casing or tubing. Yet another advantage of this system is that the well can be produced immediately after the fracture treatment, thus minimizing the amount of time the formation is exposed to the fracture fluid.

In summary, a fracture treatment down a live annulus or down tubing using the live annulus to measure fracturing BHP has many advantages over fracturing the well either down casing or down tubing with a packer in the well. It is recommended that this live-annulus system be used whenever possible. The main concern is to ensure that the casing has been designed to withstand the maximum allowable pressure.

12.8 Tubular Design

It is important to determine the maximum stresses that will be applied to the tubular goods during the fracture treatment. The first comprehensive treatment of this problem was published by Lubinski *et al.*[18] Other papers[19,20] have since been published for computing the stresses that occur during stimulation for special cases, such as combination strings or dual-completion strings. The well-understood concepts of tubing movement and tubing stresses should be computed for every fracture treatment when tubing is in the wellbore.

Tubing Strength. When designing the tubulars for a well, the engineer should determine the maximum burst pressures, collapse pressures, and tension forces that will be applied to both the casing string and the tubing. Two common situations are usually faced by the well-completion engineer. First, in many instances, the casing and tubing have already been run and the engineer must choose a value for maximum allowable surface injection pressure judging from existing conditions. The second, and most desirable, situation is the case where the well has not yet been completed so that the casing and tubing can be designed and run on the basis of the fracture-treatment requirements.

Safety factors should be used for all tubular designs. Although each operator may have different criteria, the following safety factors are usually appropriate: tension, 1.6 to 1.8; burst, 1.3 to 1.33; and collapse, 1.0 to 1.2. All tubulars run into the well should be able to withstand the maximum stress conditions expected during the fracture treatment after being derated by the appropriate safety factor.

Remember that all tubulars will be handled several times before being run into a well and will be exposed to the weather for several weeks or months; therefore, corrosion and/or erosion can weaken the tubular goods before they are ever run into the wellbore. It is best to inspect all tubular goods and to send a rabbit through each joint of tubing to ensure that only high-quality pipe that meets or exceeds all minimal strength requirements is obtained and used in the well completion.

When a well has to be fracture-treated at high pressures, it is imperative that the casing and tubing be designed to withstand the high pressures. It is usually not wise to run a combination string of N-80 and J-55 casing in an attempt to reduce the cost of the casing slightly if, by doing so, the maximum allowable injection pressure is restricted during the stimulation treatment. In most cases, the stronger-grade casing should be used from top to bottom so that the fracture treatment can be pumped at the desired injection rates. One of the worst mistakes made during the completion is running in a string of pipe that is not strong enough to withstand the maximum pressures expected during the stimulation treatment.

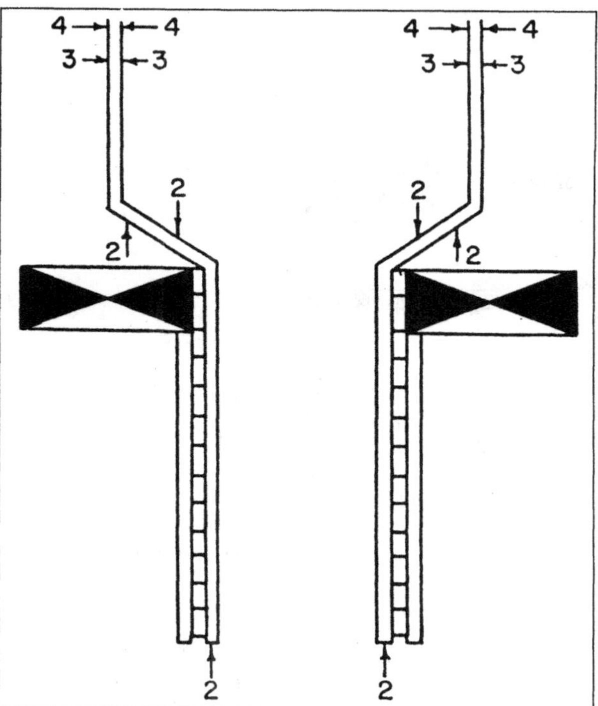

Fig. 12.11—Forces trying to change tubing length: 1—helical buckling, based on force changes at packer; 2—piston force, based on force changes at packer; 3—ballooning force, based on average force changes; and 4—temperature force, based on average temperature changes.

Helical Buckling. Lubinski *et al.*[18] thoroughly discussed the problem of tubing stresses. This type of stress can be divided into two major parts: the stresses caused by the actual pressures acting on actual areas at the packer and the stresses caused by changes in pressure and temperature in the tubulars from the time the tubing is landed until the maximum conditions are encountered during the fracture treatment. If the tubing is anchored in the packer so that tubing movement is disallowed, the stresses must be computed. If a locator seal assembly is used, tubing movement is allowed if the tubing tries to shorten, but no movement is allowed if the tubing tries to elongate.

The most logical method of solving for the forces induced in the tubing string during stimulation is first to solve for the forces and/or length changes introduced by changes in pressure and temperature. The four different effects that create these length and force changes are (1) piston effect, (2) buckling effect, (3) ballooning effect, and (4) temperature effect. Fig. 12.11 illustrates where these forces act on the tubing. The piston, buckling, and ballooning effects are caused by changes in pressure in the wellbore. The temperature effect is related only to changes in temperature and is not affected by changes in pressure.

Eqs. 12.10 through 12.23 can be used to compute the forces and length changes caused by changes in pressure and temperature.

Piston Effect.

$$F_1=[(A_p-A_o)\Delta p_o]-[(A_p-A_i)\Delta p_i] \quad\text{(12.10)}$$

and

$$\Delta L_1=\frac{F_1 L}{EA_s}. \quad\text{(12.11)}$$

Helical Buckling.

$$F_2=0$$

and

$$\Delta L_2=-\frac{\Delta r^2 A_p^2(\Delta p_i-\Delta p_o)^2}{8EI(W_s+W_i-W_o)}. \quad (12.12)$$

Ballooning Effect.

$$F_3=0.6(\overline{\Delta p_o}A_o-\overline{\Delta p_i}A_i) \quad (12.13)$$

and

$$\Delta L_3=\left(\frac{0.2L}{1.0\times 10^7}\right)\left[\frac{R^2(\overline{\Delta p_o}-\overline{\Delta p_i})}{R^2-1}\right]. \quad (12.14)$$

Temperature Effect.

$$F_4=207A_s\Delta T \quad (12.15)$$

and

$$\Delta L_4=0.0000069L\Delta T. \quad (12.16)$$

Slackoff Effect.

$$\Delta L_m=\left(\frac{F_m L}{EA_s}\right)+\left[\frac{\Delta r^2 F_m{}^2}{8EI(W_s+W_i-W_o)}\right]. \quad (12.17)$$

Total effect.

$$F_p=F_1+F_3+F_4+F_m \quad (12.18)$$

and

$$\Delta L_t=\Delta L_1+\Delta L_2+\Delta L_3+\Delta L_4+\Delta L_m. \quad (12.19)$$

Actual force.

$$F_a=[(A_p-A_o)p_o]-[(A_p-A_i)p_i]. \quad (12.20)$$

Top-joint tension.

$$F_{tgs}=F_{ta}+F_a-F_p. \quad (12.21)$$

Effective tubular weight.

$$W_i=0.0034d_i^2\rho_i \quad (12.22)$$

and

$$W_o=0.0034d_o^2\rho_o. \quad (12.23)$$

The signs must be correct when Eqs. 12.10 through 12.23 are solved. As Lubinski *et al.*[18] suggested, the following sign convention has been adopted: negative=shortening or tension and positive=lengthening or compression.

To determine values for Δp_o, Δp_i, $\overline{\Delta p_o}$, $\overline{\Delta p_i}$, and ΔT, one must always subtract the landing conditions from the current or expected conditions during the fracture treatment.

The following example problem illustrates how one can solve a typical tubing-movement problem.

Tubing-Movement Problem. *Data.*

Depth to packer=10,000 ft.

Tubing=2⅞ in., 6.5 lbm/ft, N-80.

Casing=7 in., 35 lbm/ft, N-80.

Packer bore=1.95 in.

Ambient temperature=70°F.

Temperature gradient=1.3°F/100 ft.

Landing Conditions:

Static temperature.

11-lbm/gal mud in hole.

Set-down weight=8,000 lbm.

Fracturing Conditions:

Maximum surface pressure=10,000 psi.

Casing pressure at surface=4,000 psi.

Annulus fluid=10 lbm/gal brine.

Fracture-fluid density=8.33 lbm/gal.

Average tubing temperature=75°F.

Calculate.

1. Total slack-off required when landing the tubing.
2. How many inches the tubing will actually move during the fracture treatment if the well screens out while pumping sand at 5 lbm/gal with a packer that allows limited motion.
3. The stress on the tubing during a screenout if the packer allows no motion.

Solution.

Landing Conditions:

$T_R=70+(0.013)(10{,}000)=200°\text{F}$.

$T_a=(70+200)/2=135°\text{F}$.

$p_i=p_o=(0.052)(11)(10{,}000)=5{,}720$ psi.

$\overline{p_i}=\overline{p_o}=5{,}720/2=2{,}860$ psi.

$W_s=0.542$ lbm/in.

$W_i=0.0034(2.441)^2\times 11=0.223$ lbm/in.

$W_o=0.0034(2.875)^2\times 11=0.309$ lbm/in.

Fracture Conditions:

$T_a=75°\text{F}$.

$p_i=10{,}000 \text{ psi}+(10{,}000)(0.052)(10.87)=15{,}652$ psi.

$\overline{p_i}=(10{,}000+15{,}652)/2=12{,}826$ psi.

$p_o=4{,}000+(0.052)(10)(10{,}000)=9{,}200$ psi.

$\overline{p_o}=(4{,}000+9{,}200)/2=6{,}600$ psi.

$$\rho_i=\frac{(1\text{ gal})8.33\text{ lbm/gal}+5\text{ lbm}}{1\text{ gal}+(5\text{ lbm})\left(\dfrac{\text{gal}}{8.33\text{ lbm}\times 2.65}\right)}$$

$$=10.87\text{ lbm/gal (slurry density).}$$

$W_s=0.542$ lbm/in.

$W_i=0.0034(2.441)^2\times 10.87=0.220$ lbm/min.

$W_o=0.0034(2.875)^2\times 10=0.281$ lbm/in.

Changes in Conditions:

$\Delta T=75-135=-60°\text{F}$.

$\Delta p_i=15{,}652-5{,}720=9{,}932$ psi.

$\Delta p_o=9{,}200-5{,}720=3{,}480$ psi.

$\overline{\Delta p_i}=12{,}826-2{,}860=9{,}966$ psi.

$\overline{\Delta p_o}=6{,}600-2{,}860=3{,}740$ psi.

1. Slack-off.

$$\Delta L_m=\frac{LF_m}{EA_s}+\frac{\Delta r^2 F_m{}^2}{8EI(W_s+W_i+W_o)},$$

where

L = 120,000 in.,

F = 8,000 lbf,

A_s = 1.812 in.^{2},
W_s = 0.542 lbm/in.,
Δr = 1.564 in.,
E = 30×10^6 psi,
I = 1.611 in.^{4},
W_i = 0.223 lbm/in.,
W_o = 0.309 lbm/in.

$$\Delta L_m=\frac{(120{,}000)(8{,}000)}{(30\times10^6)(1.812)}+\frac{(1.564)^2(8{,}000)^2}{(8)(30\times10^6)(1.611)(0.542+0.223\text{-}0.309)}$$

$$=17.66+0.89$$

$$=18.55 \text{ in., lengthening.}$$

Piston effect.

$$F_1=[(A_p-A_o)\Delta p_o]-[(A_p-A_i)\Delta p_i],$$

$$A_p=\frac{\pi}{4}(1.95)^2=2.986 \text{ in.}^2,$$

$$A_i=\frac{\pi}{4}(2.441)^2=4.68 \text{ in.}^2,$$

$$A_o=\frac{\pi}{4}(2.875)^2=6.49 \text{ in.}^2,$$

$$A_s=A_o-A_i=6.49-4.68=1.81 \text{ in.}^2,$$

$$F_1=[(2.986-6.49)(3{,}480)]-[(2.986-4.68)(9{,}932)]$$

$$=4{,}631 \text{ lbf, and}$$

$$\Delta L_1=\frac{F_1L}{EA_s}$$

$$=\frac{(4{,}631)(10{,}000)(12)}{(30\times10^6)(1.81)}$$

$$=+10.23 \text{ in.}$$

Helical Buckling.

$$\Delta L_2=\frac{-\Delta r^2A_p{}^2(\Delta p_i-\Delta p_o)^2}{8EI(W_s+W_i-W_o)},$$

$$\Delta r=(d_i-d_o)/2$$

$$=(6.004-2.875)/2$$

$$=1.564 \text{ in.}$$

$$I=\frac{\pi}{64}(d_o{}^4-d_i{}^4)$$

$$=\frac{\pi}{64}[(2.875)^4-(2.441)^4]$$

$$=1.61 \text{ in.}^4$$

$$\Delta L_2=\frac{-(1.564)^2(2.986)^2(9{,}932-3{,}480)^2}{(8)(30\times10^6)(1.61)(0.542+0.22-0.281)}$$

$$=-4.88 \text{ ft.}$$

Ballooning effect.

$$F_3=0.6(\overline{\Delta p_o}A_o-\overline{\Delta p_i}A_i)$$

$$=0.6[(3{,}740)(6.49)-(9{,}966)(4.68)]$$

$$=-13{,}421 \text{ lbf.}$$

$$\Delta L_3=\left[\frac{(0.2)(120{,}000)}{10^7}\right]\left[\frac{(1.18)^2(3{,}740)-9{,}960}{(1.18)^2-1}\right]$$

$$=-29.07 \text{ in.}$$

Temperature effect.

$$F_4=207A_s\overline{\Delta T}$$

$$=(207)(1.81)(-60)$$

$$=-22{,}480 \text{ lbf.}$$

$$\Delta L_4=0.0000069L\Delta T$$

$$=(0.0000069)(120{,}000)(-60)$$

$$=-49.68 \text{ in.}$$

2. Total length change for packer with limited motion.

$$\Delta L_t=\Delta L_1+\Delta L_2+\Delta L_3+\Delta L_4+\Delta L_m$$

$$=10.23-4.88-29.07-49.68+18.55$$

$$=-54.85 \text{ in.}$$

3. $F_p=F_1+F_3+F_4+F_m$

$$=4{,}631-13{,}421-22{,}480+8{,}000$$

$$=-23{,}270 \text{ lbf.}$$

$$F_a=[(A_p-A_o)p_o]-[(A_p-A_i)p_i]$$

$$=[(2.986-6.49)9{,}200]-[(2.986-4.68)15{,}652]$$

$$=-5{,}722 \text{ psi.}$$

$$F_{tgs}=F_{ta}+F_a-F_p.$$

$$F_{tgs}=(10{,}000)(6.5)-5{,}722-(-23{,}270)$$

$$=82{,}548 \text{ lbf.}$$

For 2⅞-in., 6.5-lbm/ft [7.3-cm, 9.67-kg/m], N-80 tubing, the maximum allowable tensile force for the pipe body would be 80,000 psi$\times$1.81 in.^{2} or 144,800 lbf [551.6 MPa$\times$11.7 cm^2 or 644$\times$10^3 N] (where 1.81 in.^{2} [11.7 cm^2] is the cross-sectional area of steel). If nonupset couplings are used, the maximum allowable tensile force will be 105,570 lbf [469.6$\times$10^3 N]. Therefore, to achieve a tension safety factor of 1.8, one would need to use upset couplings in this well.

To minimize tensile forces on the tubing during screenout conditions, however, one could use a packer that allows limited motion and install an 8-ft [2.4-m] sealbore extension with 10 ft [3 m] of seals. The seals will actually move only 4.56 ft [1.4 m] if the well screens out, but the excess seals should be run to provide a safety factor.

Transferring weight from the tubing hanger to the packer by slacking off the tubing will also minimize tension on the top joint during a fracture treatment. If too much weight is slacked off, however, the tubing may severely buckle and cause problems when concentric workovers are performed with either wireline tools or coil tubing. If only 10,000 to 15,000 lbf [44.5$\times$10^3 to 66.7$\times$10^3

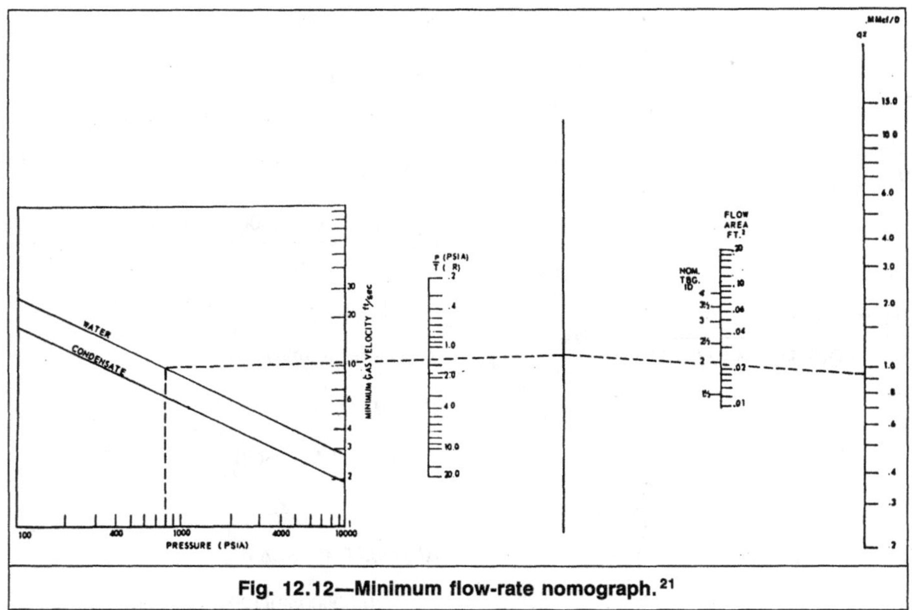

Fig. 12.12—Minimum flow-rate nomograph.[21]

N] is applied, few problems usually will occur. Problems will result, however, if a slackoff force greater than 25,000 lbf [111×10^3 N] is applied when the tubing is landed.

Effect of Tubular Size on Producing Characteristics. In most low-permeability reservoirs, execution of the fracture treatment is the most critical stage in reservoir development. All aspects of the drilling, completion and production of the reservoir can be successful, but if the fracture treatment is not successful, the well will probably not pay out. Therefore, it is extremely important that the tubular size be chosen to maximize the chances of successfully fracture-treating the low-permeability reservoir. The tubular size and strength are determined by the required injection rates and injection pressures for a particular reservoir.

The engineer should also be concerned with the effect of tubular size on the producing characteristics of the well. Turner *et al.*[21] presented a study designed to determine the minimum flow rate required for continuous removal of liquids from gas wells and examined two mechanisms of liquid removal in wellbores. They developed both a continuous-film model that hypothesized that the fluid being lifted in a gas well would ride the walls of the tubular goods and an entrained-drop model that hypothesized that all the liquid is removed in small droplets. For the drop model, the authors used a relationship between surface tension and droplet size; they also used Stokes' law to determine the terminal settling velocity of a liquid droplet in a gas medium. After developing these equations, Turner *et al.* compared their theoretical predictions with field data and determined that the entrained-drop model predicted the actual field performance better than the continuous-film model. The entrained-drop model was corrected empirically to fit the field data better:

$$v_t = 20.4 \frac{\sigma^{1/4}(\rho_L-\rho_g)^{1/4}}{\rho_g^{1/2}} . \quad \ldots\ldots (12.24)$$

A graphical solution to this equation is given in Fig. 12.12. The following example illustrates how the Turner *et al.* technique can be used to predict liquid loading in gas wells.

Example of Liquid Loading. Given: A well is producing gas and water up 2⅜-in. [6-cm] tubing against a backpressure of 800 psi [5.5 MPa]. What gas flow rate is required to keep the liquids continuously unloaded? The flowing tubing temperature is 100°F [38°C].

Solution: change temperature to Rankin.

$°R = 100 + 460 = 560°R.$

Compute p/T:

$$\frac{800 \text{ psi}}{560°R} = 1.43.$$

Following the dashed curve in Fig. 12.12, one can determine that the gas flow rate must be at least 950 Mcf/D [26.9×10^3 m^3/d] up 2⅜-in. [6-cm] tubing against 800-psi [5.5-MPa] backpressure to keep the well continuously unloaded.

Several points should be made about this example to clarify the technique. First, the Turner *et al.* technique does not predict that the well is going to load up and die when the minimum flow rate is reached. The minimum flow rate is that required for continuous removal of liquids. If the actual flow rate drops below this calculated minimum value, the well will begin heading and slugging fluid periodically. This slug flow may continue for months or years before a workover would be required to improve the flow performance of the well. It is often found in decline-curve analyses that the decline rate increases when the minimum gas flow rate has been reached.

Note also that the data gathered by Turner *et al.* ranged from wells producing only a few barrels of water per million cubic feet to wells that produced as much as 130 bbl water/MMcf gas [0.73×10^{-3} m^3 water/m^3 gas]. Therefore, the correlation developed is applicable for a wide range of producing conditions. The correlation is accurate regardless of whether a well produces only 2 or as much as 130 bbl/MMcf [11.2×10^{-6} to 0.73×10^{-3} m^3/m^3] of liquid. The important point is that a minimum gas velocity is required to lift the liquid to the surface. If a well produces a large volume of water, it will simply load up faster than a well producing only a small volume of water. Of practical significance to the liquid-loading problem is that the gas well performance can usually be improved with small-diameter tubular goods and compression. Both aspects will increase the gas velocity and improve the liquid-lifting capacity of the natural gas.

Other papers[22-27] written on the liquid-loading problem illustrate the fact that small tubular goods are needed for efficient removal of liquids in low-productivity gas formations. Therefore, the design engineer must face the problem of using large tubular goods during the fracture treatment to maximize injection rates and minimize injection pressures but needing small tubular goods after the fracture treatment to produce these low-productivity formations effectively. In many situations, 2⅞-in. [7.3-cm] tubing can be used to stimulate the well successfully. In other cases, one may run 2⅜-in. [6.0-cm] tubing without a packer, fracture treat the well down the annulus, and then produce the well through the 2⅜-in. [6.0-cm] tubing string.

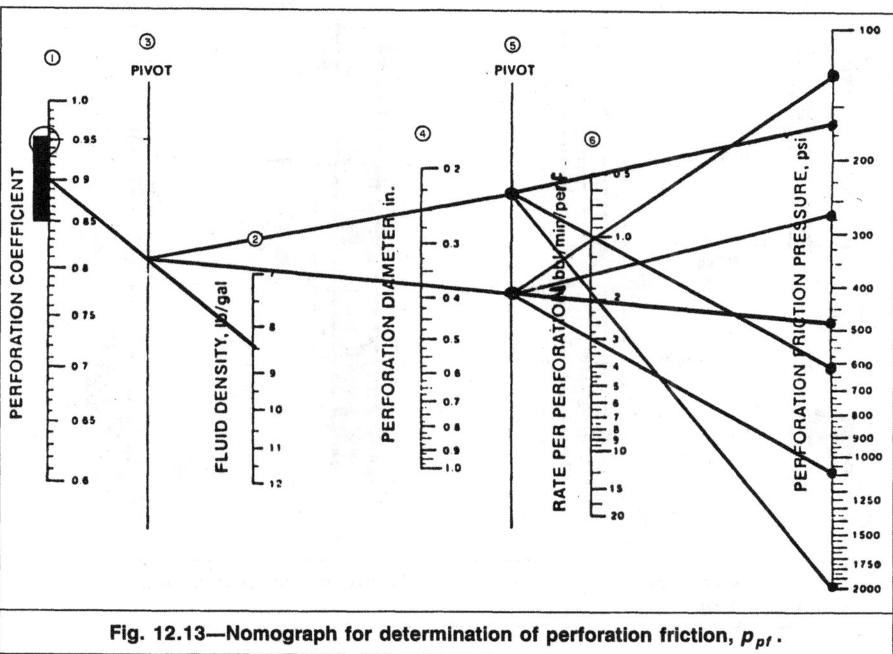

Fig. 12.13—Nomograph for determination of perforation friction, p_{pf}.

There is no easy or straightforward solution to the liquid-loading problem in many cases. The design engineer must balance the fracture-treatment injection requirements with the well producing requirements as dictated by the individual well conditions and economics.

12.9 Treatment Diversion

The efficient, simultaneous treatment of multiple sets of perforations spread over a long vertical section has been a problem in well stimulation for many years. Numerous treatment diversion methods have been used to treat multiple zones with various degrees of effectiveness. Harrison[28] presented histories of and example applications for most diverting agents. He explained many of the techniques used since 1936 and included information on diverting both acid and fracture treatments. The following sections briefly discuss most of the currently applied diverting techniques.

Limited Entry. One method for diverting a fracture treatment is the limited-entry technique. This diverting method was first reported by Lagrone and Rasmussen,[29] who reported that the limited-entry technique was used in 363 wells in Texas and New Mexico. In the limited-entry technique, the number of perforations in a well is limited and sufficient injection rate is provided for the restricted flow capacity of the perforations to divert the treatment to a greater portion of the perforated interval.

The key to limited-entry diversion is to use perforation friction to increase the pressure inside the wellbore to a pressure greater than that necessary to fracture-treat all perforated intervals. Fig. 12.13 can be used to illustrate how perforation size and the number of perforations can be used to control perforation friction. The numerical values down Bar 1 represent a perforation coefficient that ranges from 0.6 to 1.0. The perforation coefficient is an efficiency factor that takes into consideration the smoothness and roundness of the perforation. A typical value for the perforation coefficient is 0.9. If an 8.5-lbm/gal [1018-kg/m^3] fracturing fluid were being pumped, the pivot point on Bar 3 is established by connecting Bars 1 and 2. Once the pivot point is established, two lines can be drawn: for a perforation diameter of 0.25 in. [0.64 cm], which is typical of a through-tubing gun, and for a perforation diameter of 0.375 in. [0.95 cm], which is typical of a casing gun.

These two cases can then be expanded for various injection rates in terms of barrels per minute per perforation through Bar 6. The small-perforation case, 0.25 in. [0.64 cm], shows that if the flow rate through the perforation is 0.5 bbl/min [0.08 m^3/min], the perforation friction will be about 170 psi [1172 kPa]. If the injection rate were increased to 1 bbl/min-perforation [0.16 m^3/min·perforation], the friction-pressure drop through each perforation would increase to 600 psi [4137 kPa]. Finally, if the injection rate were 2 bbl/min-perforation [0.32 m^3/min·perforation], the perforation friction would be about 2,000 psi [13 790 kPa].

For the larger-diameter perforation, an injection rate of 1 bbl/min-perforation [0.16 m^3/min·perforation] will result in only 125-psi [862-kPa] perforation friction. Increasing the injection rate to 2 bbl/min-perforation [0.32 m^3/min·perforation] will give about 480-psi [3310-kPa] perforation friction. Finally, at 3 bbl/min-perforation [0.48 m^3/min·perforation] through the 0.375-in. [0.95-cm] hole, the friction-pressure drop will be about 1,100 psi [7585 kPa].

It is obvious from these examples in Fig. 12.13 that small-diameter perforations and injection rates of 1 to 3 bbl/min-perforation [0.16 to 0.48 m^3/min·perforation] will be required to generate enough perforation friction to ensure that the pressure in the wellbore is large enough to overcome the stresses in the various zones and to cause injection through each set of perforations. The limited-entry technique usually works much better if the perforations are broken down with ball sealers before the fracture treatment. If the different sets of perforations are not effectively broken down before the treatment, it is conceivable that even the high injection pressures experienced as a result of the limited number of holes may not adequately open up each set of perforations.

One of the problems with limited-entry perforating as a diverting agent is that the perforations will erode as proppant is pumped through each hole. If a through-tubing gun has been used to perforate the well and a 0.25-in. [0.64-cm] perforation has resulted, it is conceivable that the perforation can erode considerably during the treatment. When this occurs, the perforation friction can be decreased quite dramatically. It is often suspected that in many situations perforation erosion is a major cause of inadequate treatment coverage. The pad fluid in the early portions of a treatment can be successfully diverted with limited entry, but it is doubtful in many situations that the entire treatment can be pumped successfully because of the loss of perforation friction as the treatment continues. Lagrone and Rasmussen[29] suggested that fracture treatments be pumped at a maximum rate and pressure to maximize the perforation friction to offset any losses resulting from perforation erosion.

The limited-entry technique is very useful in deep wells where other forms of diversion become very expensive. For the limited-entry technique to work, a good cement bond is needed around the casing, the proper number and size of perforations must be placed in each porous interval, and barriers to fracture growth must exist between each porous interval. The limited-entry technique should

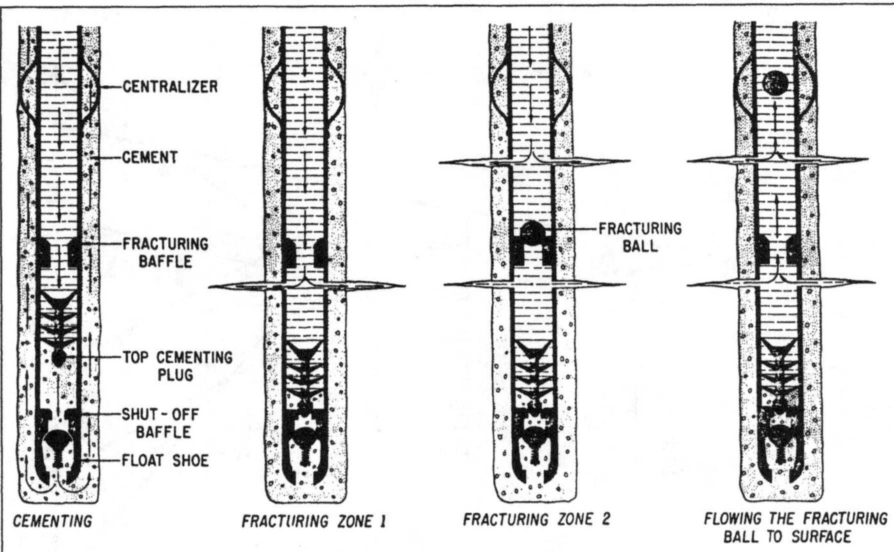

Fig. 12.14—Two-stage fracturing procedure for 3½- and 4½- or 5½-in. casing (courtesy Halliburton Services).

be used to pump into multiple zones where multiple fractures will be created. If no barriers to fracture growth exist, and it is expected that the fracture created by the treatment will be a single, continuous vertical fracture, then it does not matter where the fluid leaves the wellbore and enters the hydraulic fracture. The limited-entry technique is to be used only where multiple vertical fractures are going to be created.

Multistage Treatment With Ball Sealers. The limited-entry technique has been used successfully in many situations to divert fracture treatments; however, some intervals are best treated with a pseudo-limited-entry technique where the treatment stages are diverted with ball sealers. This technique requires a prescribed number of perforations to be placed in two or more intervals. A multistage fracture treatment is then pumped and a prescribed number of ball sealers are dropped between each stage. Such a technique has been described by Webster *et al.*[30] and by Stipp and Williford.[31] These authors demonstrated that with an effective cement bond around the pipe and proper design of the perforating and treatment diversion system, it is possible to stimulate multiple sets of perforations effectively.

Remember, however, that this treatment technique, although successfully applied in many situations, can be extremely difficult to control. The multistage fracturing technique with ball sealers for diversion works well when some stimulation fluid needs to be pumped into all existing perforations; i.e., when a short, highly conductive hydraulic fracture is needed to stimulate a fairly-high-permeability formation where fracture barriers exist between the various sets of perforations. If deeply penetrating hydraulic fractures are needed to stimulate very-low-permeability reservoirs, this particular diverting technique may not be successful because of the tendency of the ball sealers to be eroded off perforations or to fall off perforations if the pumps are shut down for some mechanical reason.

A recent innovation in ball sealer diversion was the introduction of lightweight ball sealers[32] designed to be slightly less dense than the fluid carrying them. Therefore, if a ball bypasses a perforation, it will float back up and seat on the perforation. Such a mechanism virtually ensures 100% efficiency for the ball sealers. When the ball sealers used are slightly more dense than the fluid being pumped, the ball-seating efficiency is never 100%. In most situations, about 50% excess ball sealers are used to offset this efficiency problem. The lightweight ball sealers, however, have reduced the need for dropping excess ball sealers.

One problem with the lightweight ball sealers, however, is that they will be returned to the surface as the well is produced. Therefore, special equipment must be arranged to handle these balls as they are produced up the tubing and through the flowlines and choke. The dense ball sealers are normally allowed to sink into the rathole and do not usually cause problems with the surface equipment.

In summary, the pseudo-limited-entry technique, which uses multistage treatments separated by ball sealers, has been applied for many years. This technique is generally believed to be adequate and has application in certain situations; however, it is clearly recognized that the lack of control involved with both the limited-entry and the pseudo-limited-entry techniques can result in inadequate stimulation in portions of a formation.

Packer and Bridge Plug. The most reliable technique for diverting fracture treatments into multiple intervals is the packer-and-bridge-plug method. A typical treatment would involve perforating a lower zone, performing a fracture treatment of that zone, setting a bridge plug above that particular interval, and then perforating and fracture-treating the next zone. This process can be repeated several times before the well is cleaned out and all the zones previously fracture-stimulated are commingled. In certain situations, the well may be completed as a dual completion. The main advantage of the packer-and-bridge-plug method is that it is the most positive way to divert a fracture treatment. When the method is performed correctly, each zone should be optimally stimulated and income from the well should be maximized.

The disadvantages with this method are that a workover rig must be required to move the packer and bridge plugs in the well, the method is time-consuming, and mechanical problems can occur with the retrievable bridge plugs normally used to isolate intervals.

If the pressure differential across a retrievable bridge plug becomes too large, then a drillable plug must be set between the fracture treatment stages. Under most conditions, the drillable plugs can be removed quite easily, but situations could exist where the drillable plugs are difficult to remove from the well. It is very undesirable to have to drill a plug and to circulate in a well with an open fracture above the plug. Many times, lost circulation or drilling fines can damage the fracture.

Hinn[33] illustrated, however, that the packer-and-bridge-plug method has proved to be the most economical diverting technique in a particular Cotton Valley area in east Texas. He used decline-curve analyses to project future recoveries and reserves for a number of different wells that had been fracture-treated with several different treatment diversion techniques. From the results of that study, Hinn concluded that the packer-and-bridge-plug technique was the most economical method of treatment diversion because the reserves from those wells were much greater than those from similar wells diverted with the Pine Island technique.

The cost of treatment diversion should not be the sole indicator of the most economical diversion technique. Field data and projected well performance as a function of the treatment diversion technique should be collected. It is quite likely, as was the case in the Cotton Valley, that the most expensive diverting technique is also the most economical diverting technique because it will improve cumulative recovery and cash flow.

Baffles. Baffles installed in the casing string have been used for many years to divert stimulation treatments. Fig. 12.14 illustrates a two-stage fracturing procedure that uses a casing baffle and a fracturing ball. The baffle is placed between the two intervals to be fracture-stimulated. The lower zone can be fractured, a ball dropped to seat in the baffle, and the upper zone perforated and fracture-treated. Flowing the well will remove the fracturing ball, and the baffle can be drilled up with a normal rock bit. This technique has been widely used in shallow areas in Texas and the Appalachian basin.

Pine Island Technique. Another widely applied treatment diversion method is the Pine Island fracturing technique. Fig. 12.15 illustrates how this method is applied. A typical well may have three zones to be stimulated. The lower interval can be perforated and fracture-treated, and then a sand plug is placed over the first zone. After the sand plug has been pressure-tested, the second interval can then be perforated and fracture-treated. The second zone would also be covered with a sand plug, and the third zone would be perforated and stimulated. After the third fracture treatment, the wellbore can be cleaned out with coil tubing or a conventional workover rig.

Many different versions of the Pine Island technique can be applied. For instance, one could fracture-treat and clean up each zone before setting the next sand plug. If all the different stages can be pumped in 1 or 2 days, all zones could conceivably be cleaned up simultaneously after all treatments have been pumped.

As with the other treatment diversion techniques, problems can exist in the field. With the Pine Island technique, it is sometimes very difficult to place the sand plug precisely where it is needed. If too much sand is left in the borehole, it could cover up part of the next zone to be perforated and stimulated; therefore, the excess sand would have to be washed from the borehole. In other cases, the sand plug tends to move downhole or through the perforations to void places behind the pipe, and extra time is required to dump additional sand and to allow it to settle at the proper position. To maximize the effectiveness of the Pine Island technique, a sand mixture of 20/40- and 100-mesh sand is sometimes used to minimize the permeability of the sand in the wellbore. Even though some fluid may initially leak through the plug, the addition of fluid-loss additives should quickly seal the plug and prevent fluid from migrating through the plug.

Other Diverting Techniques. Other techniques have been tried to stimulate multiple intervals economically. In fracture acidizing treatments, wax beads, benzoic acid flakes, or other granular additives can sometimes be used to plug perforations and/or the fracture and to divert stimulation treatments successfully. Such diverting methods, however, do not work efficiently when high-viscosity fracture fluids are used to carry granular propping agents. The high-viscosity fluids will create extremely wide fractures. The granular diverting agents will not successfully seal these fractures near the wellbore and therefore cannot be successfully used to divert a fracture treatment.

Other techniques to divert treatments, such as the use of viscous slurries or viscous pills, have also been applied. As with the granular techniques, these viscous pills are difficult to control and cannot be continuously and effectively used as a stimulation method.

The best way to ensure treatment diversion is to apply the packer-and-retrievable-bridge-plug method. This method is usually the most expensive, but is also the most effective way of stimulating each zone. The limited-entry technique, pseudo-limited-entry technique, the use of baffles and fracturing balls, and the Pine Island technique can all be used in certain situations to provide effective fracture-treatment diversion. The fracture-design engineer must evaluate each method for a particular reservoir and the resulting reservoir performance as a function of treatment diversion techniques to determine the optimum method for any particular situation.

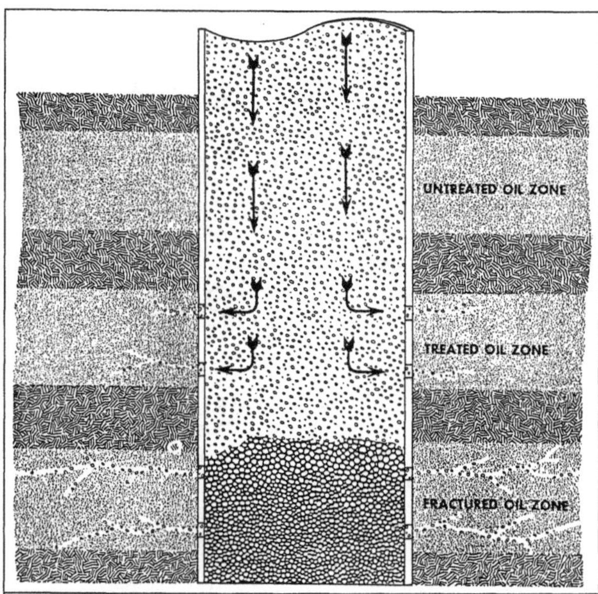

Fig. 12.15—Pine Island fracturing technique (courtesy Halliburton Services).

Nomenclature

A_i = area based on tubing ID, in.^{2} [cm^2]
A_o = area based on OD, in.^{2} [cm^2]
A_p = area of packer bore, in.^{2} [cm^2]
A_s = area of steel in pipe body, in.^{2} [cm^2]
d_i = casing ID, in. [cm]
d_o = tubing OD, in. [cm]
d_{pf} = perforation diameter, in. [cm]
D = depth, ft [m]
E = Young's modulus for steel, 30×10^6 psi [207×10^6 kPa]
F_a = actual force, lbf [N]
F_m = mechanical force, lbf [N]
F_p = total force at packer, lbf [N]
F_{tgs} = top-joint force, lbf [N]
F_1 = piston force, lbf [N]
F_2 = helical force, lbf [N]
F_3 = ballooning force, lbf [N]
F_4 = temperature force, lbf [N]
g_f = fracture gradient, psi/ft [kPa/m]
g_{fr} = friction-pressure gradient, psi/ft [kPa/m]
g_{ph} = hydrostatic pressure gradient, psi/ft [kPa/m]
i = total injection rate, bbl/min [m^3/min]
i_{pf} = specific injection rate, bbl/min-perforation [m^3/min·perforation]
I = moment of inertia, in.^{4} [cm^4]
L = length of tubing or casing, in. [cm]
ΔL_m = length change caused by mechanical force, in. [cm]
ΔL_t = total length change caused by changes in temperature and pressure, in. [cm]
ΔL_1 = length change caused by piston force, in. [cm]
ΔL_2 = length change caused by buckling, in. [cm]
ΔL_3 = length change caused by ballooning force, in. [cm]
ΔL_4 = length change caused by temperature force, in. [cm]
n_{pf} = number of open perforations

p_{bhf} = fracturing BHP, psi [kPa]
$p_{(D)}$ = pressure at depth D, psi [kPa]
p_{fr} = pipe friction, psi [kPa]
p_h = hydrostatic pressure, psi [kPa]
p_{pfr} = perforation friction, psi [kPa]
p_{si} = surface injection pressure, psi [kPa]
p_{ss} = instantaneous shut-in pressure, psi [kPa]
Δp_i = change in pressure in tubing at packer, psi [kPa]
$\overline{\Delta p_i}$ = change in average pressure in tubing, psi [kPa]
Δp_o = change in pressure in annulus at packer, psi [kPa]
$\overline{\Delta p_o}$ = change in average pressure in annulus, psi [kPa]
Δr = clearance between casing ID and tubing OD, in. [cm]
T_R = reservoir temperature, °F [°C]
ΔT = change in average temperature, °F [°C]
v_t = terminal settling velocity, ft/sec [m/s]
W_i = weight of fluid inside tubing, lbm/in. [kg/cm]
W_o = weight of fluid in annulus displaced by tubing, lbm/in. [kg/cm]
W_s = weight of steel, lbm/in. [kg/cm]
α = discharge coefficient, usually 0.9
ρ = fracturing-fluid density, lbm/gal [kg/m^3]
ρ_g = gas density, lbm/ft^3 [kg/m^3]
ρ_L = liquid density, lbm/ft^3 [kg/m^3]
σ = interfacial tension, dynes/cm^2 [Pa]

References

1. Grosmangin, M., Kokesh, F.P., and Majani, P.: "A Sonic Method for Analyzing the Quality of Cementation of Borehole Casings," *JPT* (Feb. 1961) 165–71; *Trans.*, AIME, **222**.
2. Pickett, G.R.: "Prediction of Interzone Fluid Communication Behind Casing by Use of Cement Bond Log," *Proc.*, 1966 SPWLA Annual Logging Symposium, May 8–11.
3. Brown, H.D., Grijalva, V.E., and Raymer, L.L.: "New Developments in Sonic Wave Train Display and Analysis in Cased Holes," paper presented at the 1970 SPWLA Annual Logging Symposium, May 3–6.
4. Cooke, C.E. Jr.: "Radial Differential Temperature (RDT) Logging—A New Tool for Detecting and Treating Flow Behind Casing," *JPT* (June 1979) 676–82.
5. Froelich, B. *et al.*: "Cement Evaluation Tool: A New Approach to Cement Evaluation," *JPT* (Aug. 1982) 1835–41.
6. Fitzgerald, D.D., McGhee, B.F., and McGuire, J.A.: "Guidelines for 90% Accuracy in Zone-Isolation Decisions," *JPT* (Nov. 1985) 2013–22.
7. Smith, D.K.: *Cementing*, Monograph Series, SPE, Richardson, TX (1987) **4**.
8. Buzarde, L.E. Jr. *et al.*: *Production Operations Course I—Well Completions*, Video Tape Series, SPE, Richardson, TX (1972).
9. Allen, T.O. and Roberts, A.P.: *Production Operations 1*, Oil and Gas Consultants, Inc., Tulsa, OK (1978).
10. McLeod, H.O. Jr.: "The Effect of Perforating Conditions on Well Performance," *JPT* (Jan. 1983) 31–39.
11. Saucier, R.J. and Lands, J.F.: "A Laboratory Study of Perforations in Stressed Formation Rocks," *JPT* (Sept. 1978) 1347–53; *Trans.*, AIME, **265**.
12. Tuttle, R.N. and Barkman, J.H.: "New Nondamaging and Acid-Degradable Drilling and Completion Fluids," *JPT* (Nov. 1974) 1221–26.
13. Millhone, R.S.: "Completion Fluids for Maximizing Productivity—State of The Art," *JPT* (Jan. 1983) 47–55.
14. Weeks, S.G.: "Revised Completion Procedures for S.W. Texas Low-Strength, Low-Permeability, Geopressured Deep Wilcox Sands," paper SPE 7913 presented at the 1979 SPE Symposium on Low Permeability Gas Reservoirs, Denver, May 20–22.
15. Huber, T.A. and Tausch, G.H.: "Permanent-Type Well Completion," *Trans.*, AIME (1953) **198**, 11–16.
16. King, T.G.: "The Application of Permanent Completion Techniques to Offshore Operations," *JPT* (Sept. 1966) 1031–40.
17. Frank, W.J. Jr.: "Improved Concentric Workover Techniques," *JPT* (April 1969) 401–08.
18. Lubinski, A., Althouse, W.S., and Logan, J.L.: "Helical Buckling of Tubing Sealed in Packers," *JPT* (June 1962) 655–70; *Trans.*, AIME, **225**.
19. Hammerlindl, D.J.: "Movement, Forces, and Stresses Associated with Combination Tubing Strings Sealed in Packers," *JPT* (Feb. 1977) 195–208.
20. Hammerlindl, D.J.: "Packer-to-Tubing Forces for Intermediate Packers," *JPT* (March 1980) 515–27.
21. Turner, R.G., Hubbard, M.G., and Dukler, A.E.: "Analysis and Production of Minimum Flow Rate for the Continuous Removal of Liquids from Gas Wells," *JPT* (Nov. 1969) 1475–82.
22. MacDonald, R.M.: "Fluid Loading in Low Permeability Gas Wells in the Cotton Valley Sands of East Texas," paper SPE 9855 presented at the 1981 SPE Symposium on Low-Permeability Gas Reservoirs, Denver, May 27–29.
23. Vosika, J.L.: "Use of Foaming Agents to Alleviate Liquid Loading in Greater Green River TFG Wells," paper SPE 11644 presented at the 1983 SPE Symposium on Low Permeability Gas Reservoirs, Denver, March 14–16.
24. Lea, J.F.: "Avoid Premature Liquid Loading in Tight Gas Wells by Using Prefrac and Postfrac Test Data," *Oil & Gas J.* (Sept. 20, 1982) 123–28.
25. Libson, T.N. and Henry, J.R.: "Case Histories: Identification of and Remedial Action for Liquid Loading in Gas Wells—Intermediate Shelf Gas Play," *JPT* (April 1980) 685–93.
26. Ikoku, C.U. and Ilobi, M.I.: "Minimum Gas Flow Rate for Continuous Liquid Removal in Gas Well," paper SPE 10170 presented at the 1981 SPE Annual Technical Conference and Exhibition, San Antonio, Oct. 5–7.
27. Henderson, L.J.: "Deep Sucker-Rod Pumping for Gas Well Unloading," paper SPE 13199 presented at the 1984 SPE Annual Technical Conference and Exhibition, Houston, Sept. 16–19.
28. Harrison, N.W.: "Diverting Agents—History and Application," *JPT* (May 1972) 593–98.
29. LaGrone, K.W. and Rasmussen, J.W.: "A New Development in Completion Methods—The Limited Entry Technique," *JPT* (July 1963) 695–702.
30. Webster, K.R., Goins, W.C. Jr., and Berry, S.C.: "A Continuous Multistage Fracturing Technique," *JPT* (June 1965) 619–25.
31. Stipp, L.C. and Williford, R.A.: "Pseudolimited Entry: A Sand Fracturing Technique for Simultaneous Treatment of Multiple Pays," *JPT* (May 1968) 457–62.
32. Erbstoesser, S.R.: "Improved Ball Sealer Diversion," *JPT* (Nov. 1980) 1903–10.
33. Hinn, R.L. Jr.: "Case History: A Multizone Tight Gas Completion Technique for the Blocker Cotton Valley Field, East Texas," paper SPE 11934 presented at the 1983 SPE Annual Technical Conference and Exhibition, San Francisco, Oct. 5–8.

SI Metric Conversion Factors

bbl	× 1.589 873	E−01	=	m^3
°F	(°F−32)/1.8		=	°C
ft	× 3.048*	E−01	=	m
ft^2	× 9.290 304*	E−02	=	m^2
ft^3	× 2.831 685	E−02	=	m^3
in.	× 2.54*	E+00	=	cm
in.^{2}	× 6.451 6*	E+00	=	cm^2
lbm	× 4.535 924	E−01	=	kg
lbm/gal	× 1.198 264	E+02	=	kg/m^3
psi	× 6.894 757	E+00	=	kPa
°R	× 5/9		=	K

*Conversion factor is exact.

Chapter 13
Field Implementation of Hydraulic Fracturing

H.O. McLeod, SPE, Conoco Inc.
G.D. Cooper, SPE, Guydon Software Services

13.1 Overview

This chapter presents the pretreatment arrangements and communications necessary before treatment execution. The use of checklists for inspection and quality control before treatment initiation is important. We provide an overview of fracturing execution with an emphasis on quality control, safety, communication, and computerized process monitoring and control.

13.2 Pretreatment Planning

The details of pretreatment planning are given in Chaps. 2 and 11. It is important for the operations supervisor to have on hand a well file, data from the prefracture breakdown test, and the summary of the fracturing design (including the data used in making the design calculations). The purpose of this file is to make available the information needed to make decisions on site when problems are encountered either before or during the treatment.

Included in the well file should be the completion sketch showing sizes of tubular goods and pressure ratings on the wellhead equipment. Copies of the well logs, or at least reproductions of the log sections showing the formation and barrier sections above and below, should be available. One of the well logs is the cement bond log, which indicates the formation isolation behind the casing. One should also be familiar with the fracture height barriers and the location of nearby water zones or thief zones in case the fracture breaks through barriers or establishes communication behind pipe during the treatment. It is important to have sufficient rathole below the perforated interval to contain the volume of proppant left in the wellbore after shutdown.

Reservoir or formation data compiled for the fracturing-treatment design should be on hand. These data are pertinent to special problems with fluid-loss control that affect decisions regarding pressure-outs or premature shutdowns. Water saturations and water analyses are important in determining whether fracturing fluid or formation fluid is being returned after a treatment.

Perforating data are required for the main fracturing treatment and for the pretreatment breakdown test, particularly where limited perforating is used. The number and size of open perforations are needed in case proppants bridge in perforations during the treatment and cause unexpected pressure increases.

The prefracture breakdown test (see Chap. 14) data and interpretation should be available to the operations supervisor. The fracturing treatment may have been redesigned according to the prefracture breakdown tests, especially if the estimated fracture height has changed or a better estimate of fluid-loss coefficient has been determined from the pressure falloff. The latest revised design figures—including the schedules of rates and wellhead pressures expected, fluid volumes, appropriate additives, and proppant concentrations—should also be available. Any concerns about the safe proppant concentration limits (pounds of proppant per gallon of gelled fluid) should have been discussed by the engineering and operations personnel, especially if proppant size increases in the final stages.

13.3 Fracturing Equipment

This section briefly describes the different equipment units used in storing, blending, and pumping fracturing fluids and slurries. Fig. 13.1A shows a schematic of equipment on a well site for conventional pumping units where pump pressures are normally less than 10,000 psi [69 MPa]. Pressure intensifiers are frequently used where surface pressures exceed 10,000 psi [69 MPa], or for long pumping times of 4 hours or more at pressures greater than 8,000 psi [55 MPa]. Such a typical hookup is shown in Fig. 13.1B. Fig. 13.2 is a photograph of fracturing equipment on location for a large treatment.

Before the fracturing treatment, the base gel of polymer and brine may have been mixed and stored in fracturing-fluid tanks. Water alone is stored in the tanks when a liquid gel concentrate is to be used. Mixing of the gel and quality-control testing are described later. During pumping, the fluid moves from the tanks to a common header and is then pulled to the blender by the centrifugal pump mounted on the blender. Fluid-loss additives may be added at the blender or may have been premixed in the initial gel preparation in the fluid tanks the day before. The proppant is added at the blender in a prescribed schedule, such as that illustrated in Table 13.1. The photograph in Fig. 13.3 shows bulk sand handlers feeding sand to a conveyor belt, which carries it to the blender and mixes it with fluid from the fracture tanks shown. The gelled fluid pads and the fluid/proppant mixtures (slurries) are pumped from the blender to the manifold for intake to high-pressure pumps or pressure intensifiers. Fig. 13.4 shows a manifold hookup to pump trucks. The crosslinking agents may be added at the blender or downstream of the blender at the intake manifold to the high-pressure pumps. If diesel oil is used as a fluid-loss additive, mixing it in the blender allows better dispersion of diesel-oil droplets in the viscous gel. The required number of pumps selected for the design rate are hooked up to the intake manifold along with the backup pumps in case any of the on-line pressure pumps shut down. Centrifugal pumps prime the intake manifold for good fluid supply to the high-pressure pumps. The high-pressure pumps are tied into a high-pressure manifold leading to the high-pressure line to the wellhead. Wellhead isolation tools are often used to isolate and protect the wellhead from the high pumping pressures involved in fracturing treatments. Such a tool is shown in Fig. 13.5.

Densimeters are needed at the blender to enable the blender operator to control the proppant concentrations. A densimeter is also needed near the wellhead to monitor the final slurry concentration going down the treating string. This final densimeter is used to time the final overflush or displacement to prevent overflushing the proppant in the fracture and to ensure that the final design pumped concentration is placed in the fracture next to the wellbore.

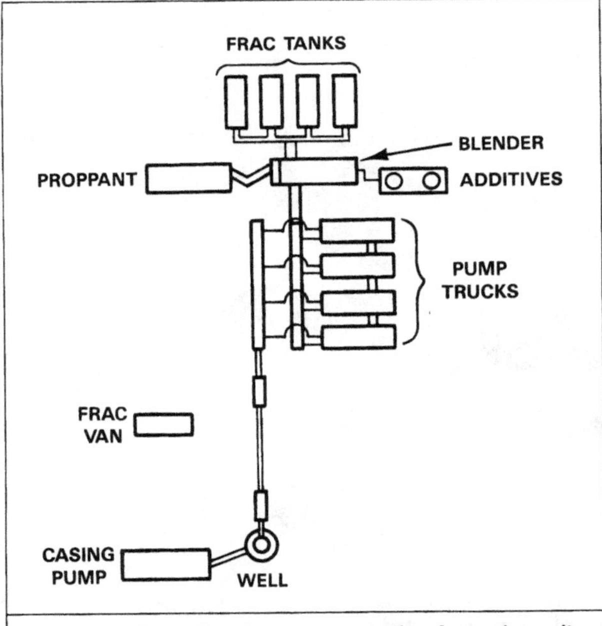

Fig. 13.1A—Equipment layout—conventional pumping units.

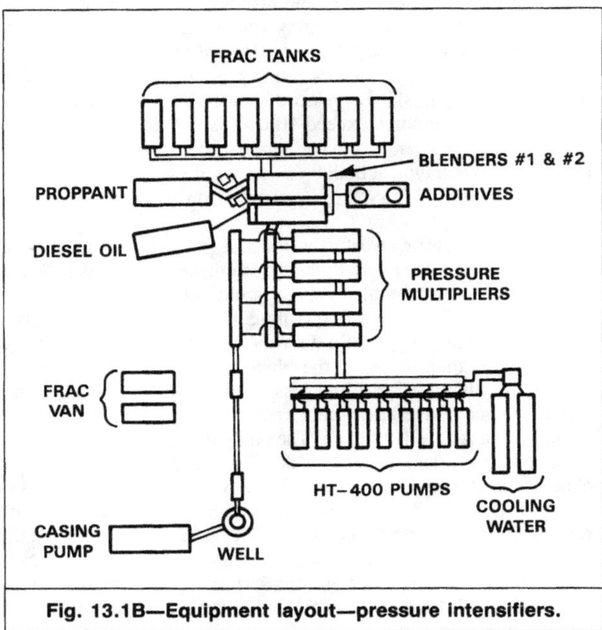

Fig. 13.1B—Equipment layout—pressure intensifiers.

Fig. 13.2—Site layout (courtesy Halliburton Services).

Fig. 13.3—Bulk sand handler, conveyor belt, tanks, and blender (courtesy BJ Titan).

For critical fracturing treatments, especially for exploratory wells, there will be both a treatment command center, or fracture van, and a monitoring van, or on-site computer. The monitoring equipment and monitoring van are described in Secs. 13.10 and 13.11, where plots of significant fracturing data are generated in real time by the on-site computer for review by personnel.

The service company supervisor, hereafter called the service supervisor, is located in the treatment command center where he is in communication with the operators of the blender, sand transporter, and pumps. The operating company representative is often located in this treatment command center to follow the treatment progress and to make any decisions about premature shutdowns, rate changes, or proppant concentration changes. In this center, the pump volumes and proppant concentration are checked against the prearranged treatment schedule.

Fracturing-Fluid Tanks. Fracturing-fluid tanks hold the treatment fluids and segregate the various types of fluids that may be used. Several sizes and shapes of fracturing-fluid tanks are available. The rectangular, 500-bbl [80-m^3] -capacity mobile tank has rear wheels, so it can be pulled like a trailer, thereby facilitating rig-up time and use of available space. Cylindrical tanks, either upright or horizontal, are used in some areas, but they are more difficult to handle. Ideally, tanks should be lined to prevent iron from contaminating water and interfering with proper gelation and cross-linking of fracturing fluids. Linings should be inspected to certify their integrity. Splits or holes can lead to spaces under the lining that harbor bacteria or incompatible chemicals from prior use.

These tanks are connected to the fluid/proppant proportioning unit (blender) by 4-in. [10-cm] flexible hoses. The hose connection to the tanks may be on the face of the tank itself or, more commonly, on an 8- to 10-in. [20- to 25-cm] manifold extending from the tank. The extended manifold allows several tanks to be hooked up in series by simply connecting each manifold. A recirculating line is built into the tanks to mix the fluid and additives properly.

Tanks also may have sloped, U-shaped bottoms so that as much treating fluid as possible is used.

Bulk Handling Equipment for Proppants. Several bulk handling systems are available to deliver the proppant to the fluid/proppant proportioning unit (blender) at the desired rate in the simplest manner.

Small jobs may require only dump trucks positioned over the blender hopper to gravity feed the proppant into the blender at a rate controlled manually by a chute opening. These units hold up to 400 ft^3 [10 m^3] of proppant.

Other treatments may require large quantities of proppant delivered at high rates. To accommodate high-demand situations, field storage bins are used. Their capacities vary from 2,000 to 4,000 ft^3 [55 to 115 m^3]. The bins are divided into several compartments

TABLE 13.1—PUMPING SCHEDULE

Step	Fluid	Injection Rate (bbl/min)	Liquid Volume (gal)	Cumulative Liquid Volume (gal)	Proppant lbm/gal	Proppant Mesh	Proppant Type	Stage Volume (gal)	Cumulative Slurry Volume gal	Cumulative Slurry Volume bbl	Tank	Remarks
1	WF-50	20	40,000	40,000	Pre-pad			40,000	40,000	952	1,2	Fluid to contain 2% KCl and Freflow F plus the following per 1,000 gal: 50-lbm WF + 0.5 gal D-47 + 25% MEOH + 60 lbm Adomite Aqua
2	YF650	20	60,000	100,000	Pad			60,000	100,000	1,429	3,4,5	Fluid to contain 2% KCl plus the following per 1,000 gal: YF650 + J444 + 1 gal F-75 + 0.5 gal D-47 + 10 lbm J353 + 30 lbm Adomite Aqua + 5% D.O.
3	YF650	20	25,000	125,000	Pad			25,000	125,000	595	6,7	Delete Adomite Aqua
4	YF650	20	9,804	137,804	½	20/40	Ottawa	10,000	135,000	238	7	
5	YF650	20	15,238	150,042	1	20/40	Ottawa	16,000	151,000	381	7,8	
6	YF650	20	11,009	161,051	2	20/40	Ottawa	12,000	163,000	286	8,9	
7	YF650	20	8,929	169,980	3	16/20	I-PROP	10,000	173,000	238	9	
8	YF650	20	8,621	178,601	4	16/20	I-PROP	10,000	183,000	238	9,10	
9	YF650	20	8,333	186,934	5	16/20	I-PROP	10,000	193,000	238	10	
10	YF650	20	4,878	191,812	6	16/20	I-PROP	6,000	199,000	143	10,11	Add breaker
11	YF650	20	2,100	193,912	Flush			2,100	201,100	50	11	Cut crosslinker + diesel

Fig. 13.4—Manifold (courtesy Halliburton Services).

Fig. 13.5—Wellhead isolation tool (courtesy Halliburton Services).

of various sizes, each with one or more adjustable gates that are controlled hydraulically to regulate proppant rate. The proppant drops onto a conveyor belt that runs underneath the bin and carries the proppant to the blender. Each proppant bin is capable of delivering 10,000 to 20,000 lbm/min [4535 to 9070 kg/min], depending on the power rating of the hydraulic system. Each unit has its own power package.

If several proppant-storage bins are required and/or rates exceed a single unit capacity, the bins may be positioned along a conveyor belt that delivers the proppant to the blender. The conveyers can deliver up to 30,000 lbm/min [13.6 Mg/min] of proppant and may be single- or double-belt units with a completely redundant hydraulic system. A setup of bulk proppant handlers is shown in Fig. 13.6.

Fluid/Proppant Proportioners. Fluid/proppant proportioners are commonly called blenders because their primary function is to take all the premixed fluid, various liquid and dry additives, and proppant and "blend" them uniformly at the desired proportion and rates. The blender simultaneously discharges the fluid/slurry to supercharge the high-pressure pumps that inject it downhole. Fig. 13.7 is a schematic of a blender.

The fluid from the fracturing tanks, the proppant, and other liquids or dry additives are brought together and mixed thoroughly in the blender "tub." Agitation by one or a combination of paddles,

Fig. 13.6—Bulk handlers (courtesy BJ Titan).

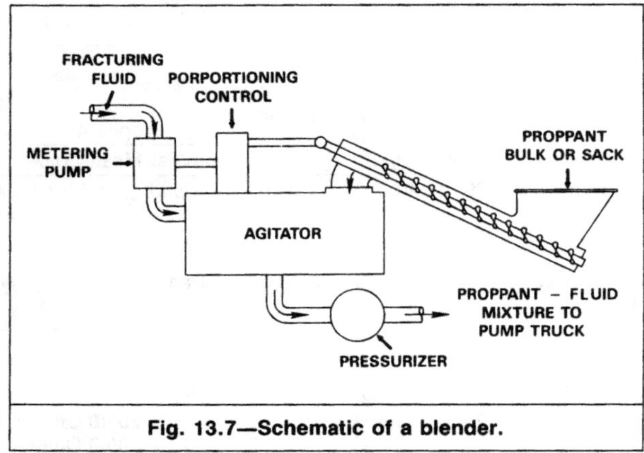

Fig. 13.7—Schematic of a blender.

augers, and fluid jets is required to mix the components uniformly. Blender capacities vary from 6 to 20 bbl [0.96 to 3.2 m^3]. Suction pumps draw fluid from the fracturing tanks and discharge it into the tub. Units have one or two centrifugal pumps, depending on liquid injection rates; in most cases, only one pump is used at a time. The size of the centrifugal pump and the power package determine the blender rate capacity.

Because many blender tubs are located at the same height as the deck, about 6 to 8 ft [1.8 to 2.4 m] above ground level, an auger-type system is usually required to transport the proppant to the tub level. Other blenders lower their tub to ground level and are able to gravity feed proppant into it. With the elevated tub, proppant rate is controlled by adjusting the speed of the augers (usually three). The capacity depends on auger size and available hydraulics to power the system. Belt-type systems are also used to move proppants to the tub. Most units can deliver from 10,000 to 20,000 lbm/min [4535 to 9070 kg/min]; however, some units approach 30,000 lbm/min [13.6 Mg/min]. Where the tub is lowered to ground level, proppant rate is controlled by gate openings from a vertical sand bin. From a practical standpoint, blender proppant capacities are about 25,000 lbm/min [11.3 Mg/min] at about 22 lbm/gal [2636 kg/m^3]. Above these limits, control is very difficult.

Dry additives, such as gellant and fluid-loss material, are added to the treating fluid by an eductor or venturi-type mixer. Dry additives may be premixed in the fracturing fluid before the job or added to the fluid during the treatment. The rate of additives is controlled by an adjustable vane feeder. The fluid from the eductor goes directly into the tub.

Many blenders are equipped with several pumps to add liquids at selected times during mixing and treatment. These can be used for surfactants, liquid polymers, crosslinking agents, and hydrocarbon liquids for fluid-loss control.

After the fluid, proppant, and other additives are mixed in the tub, the discharge centrifugal pump removes fluid from the tub and supercharges the high-pressure pump units through 4-in. [10-cm] hoses on a special manifold arrangement that is discussed later.

Fluid/proppant proportioning units (blenders) are divided into two groups according to discharge rate, 0 to 50 bbl/min [0 to 8 m^3/min] or 0 to 100 bbl/min [0 to 16 m^3/min]. The rating depends on the size of the centrifugal pumps and the available power package to drive them. A new type of blender has been developed on the basis of computer control of blending proppant and fluids in a vortex mixer, as described in Sec. 13.17.

Pumping Units. Pumping provides the necessary horsepower to create and to propagate the desired fracture with the designed fluids, pump rates, and injection pressure. These units are classified into two large groups: conventional pump units and intensifiers, or multipliers. An important rating is the hydraulic horsepower the unit is capable of producing.

Conventional Pump Units. These units are designed to operate at pressures from 0 to 10,000 psi [0 to 69 MPa]. With special preparations, they can be used at higher pressures, but their reliability is greatly reduced. The units have three main components: power package, power end, and fluid end. The power package provides the horsepower to drive the pumps and actually determines the hydraulic horsepower the unit can provide. The most common are diesel engines and turbines, but electric motors are sometimes used. The brake horsepower of the engine determines the hydraulic horsepower of the pump unit. Power from the engine is transferred to the power end or gear end of the pump, which activates the plungers on the pump. The fluid end is the portion of the pump through which all the treatment fluid passes going to the well.

Plunger size in the fluid end determines the working pressure of the unit. As the treating pressure increases, plunger diameter decreases. Plunger diameters vary from 3 to 6.75 in. [7.6 to 17.1 cm], depending on the injection pressure. As the plunger size decreases, so does pump rate. The most common plunger size is 5 in. [12.7 cm] in diameter. Plunger stroke varies with the type of pump used; however, most pumps use a 6- to 8-in. [15- to 20-cm] stroke to reduce the number of valve cycles and to increase pump life.

Pumping units may be truck- or skid-mounted, depending on the location and use. Truck-mounted versions can be body- or trailer-mounted with one or two pumps per unit.

The hydraulic horsepower rating of each pump unit is dependent on the size of the power package used, the plunger diameter, which controls the rate, and the number of pumps on the unit. The rating ranges from 700 to 1,600 hhp [520 to 1195 kW].

Intensifiers. Intensifiers, or multipliers, are designed to operate at treating pressures from 10,000 to 20,000 psi [69 to 138 MPa] for long periods of time. They are more reliable than conventional pumps for long pump times (more than 2 hours).

The intensifier pump has a high-pressure fluid end through which the treating fluid is pumped to the well. The power for the intensifier pump is provided by separate conventional or specially designed pump units that pump fluid to drive the intensifier plungers. The fluid used by the power units may be either oil or water. As many as four power units may be used to drive a single intensifier pump. Of course, the number of units determines the hydraulic-horsepower output of the intensifier unit. Each intensifier can produce 1,000 to 5,000 hhp [745 to 3730 kW], depending on the model and power units used.

The power units apply hydraulic pressure to the plunger on the intensifier pump. The plunger has a large diameter at the power end and reduces to a smaller diameter at the fluid end where the treating fluid is pumped to the well. The diameter change allows the power units to pump at about one-third the treating pressure of the intensifier unit; i.e., the intensifier output of 10,000 to 20,000 psi [69 to 138 MPa] will be powered by conventional pumps operating at 3,500 to 7,000 psi [24 to 48 MPa].

Another feature of the intensifier/multiplier units is the long pump stroke, about 65 to 70 in. [165 to 178 cm], which allows fewer valve cycles and pump strokes per volume of fluid pumped and greatly prolongs pump life. Such a unit with accompanying power units is shown in Fig. 13.8.

Injection Manifolds/Headers. When several pumping units are required to perform a treatment, hooking each unit to the blender and to the high-pressure treating line to the wellhead can take several hours. It not only takes a great deal of time but also results in a

Fig. 13.8—Pressure intensifiers (courtesy The Western Co. of North America).

Fig. 13.9—Nitrogen tanks (courtesy BJ Titan).

tangle of high-pressure treating connections and supercharge hoses. This complicates rigging up and is a safety hazard because leaks are hard to detect and difficult to fix.

This condition was remedied by developing a treatment manifold. It is usually mounted on a trailer that can be positioned between the pumping units. The trailer is outfitted with a top suction manifold for connection to the blender to carry treating fluid to the pump units. Underneath is a high-pressure discharge manifold for connecting the pump units to the well treating line. The pumping units are connected to the suction and to the discharge, saving several hours of job preparation time, uncluttering the area between the pump units, and creating a safer working area. Such a manifold is shown in Fig. 13.4.

Wellhead Isolation Tool. Many times an operator will want to complete a well with the tubing and wellhead equipment in place. The fracturing treatment can damage the wellhead because of excessive pressure and the abrasive action of the proppant. The wellhead isolation tool isolates the wellhead from these potentially damaging conditions (see Fig. 13.5). The tool is attached to the wellhead. A tubular mandrel made of high-strength tubular steel is extended through the wellhead and sealed in the tubing. The tool, which can be installed under pressure without killing the well, uses a high-pressure rubber seal and can accommodate most tubing sizes and wellhead arrangements. Wellhead isolation tools are available to handle treating pressures up to 20,000 psi [138 MPa].

Liquefied Gas Transports, Pumps, and Heat Exchangers. N_2 and CO_2 are commonly used in stimulation treatments to make stable foams for the primary treating fluid or for gas assist in well cleanup.

Nitrogen. Transported and pumped by specialized equipment, N_2 is delivered to location in a large vacuum bottle at −320°F [−196°C]. It must be vaporized before being injected downhole or mixed with other treating fluids. The liquid nitrogen is fed to a high-pressure triplex pump that pumps it through a diesel-fired vaporizer (heat exchanger). The gaseous N_2 is heated to 80 to 100°F [27 to 38°C] before it is injected into the treating stream.

The triplex pumps that inject N_2 have different pressure and rate classifications from other pumps. The unit or units selected will depend on the anticipated rates and pressures required to conduct the treatment. Pressure ratings go up to 15,000 psi [103 MPa]. Rate capabilities are usually designed in standard cubic feet per hour, which varies with injection pressure. Liquid N_2 storage normally accompanies the pump unit. N_2 units come mounted on the truck body or on a trailer. The trailer units usually have more storage capacity. Additional N_2 comes in transports that have a capacity of ±7,200 gal [±27 m^3] (670,000 scf [18 970 m^3]). For very large jobs, field storage tanks are used that contain several thousand gallons. Fig. 13.9 shows a typical hookup of N_2 tank trucks.

Carbon Dioxide. Pumping CO_2 differs from pumping N_2. CO_2 is delivered in a liquid state at approximately 0°F [−18°C] and is pumped as a liquid into the treating fluid. It is not preheated. Conventional triplex fracturing pumps are used to inject the liquid CO_2. A special booster system receives liquid CO_2 from transports or field storage units and supercharges the suction of the conventional triplex pump. These booster systems are designed specifically to handle liquid CO_2 safely and efficiently.

13.4 Treatment Preparation

This section reviews the work done before a treatment. The materials brought to location should be examined to ensure that they were selected for the treatment and that the necessary amounts are present on site to do the job. The required treating equipment to complete the treatment, including sufficient backup, should be on site. The pumping schedule must be reviewed to confirm the schedule and the changes in additives for various stages during the treatment. Table 13.1 is an example of a pumping schedule. Parker[1] and others[2-7] provided excellent guidelines for treatment preparation and execution.

Wellsite Preparation. Wellsite preparation often receives too little attention. The key considerations are adequate size and accessibility. A location should be large enough to accommodate the required equipment and fracture tanks easily. Crowded conditions slow the preparation process by allowing only one phase of the job to be performed at a time. When time is lost, the natural tendency is to hurry and take short cuts to get back on schedule, which usually leads to additional problems, time delays, and often a sacrifice in job quality. The best method is to resist the time pressure and to perform the preparation correctly. The investment is substantial and the consequences too important to be jeopardized by trying to save 1 or 2 days.

Sometimes equipment placement and material handling are made awkward or impractical by items left on location from the drilling operation or the premature installation of production equipment, such as tank batteries, separators, and flare pits. These obstacles often result in improper placement of equipment and materials and/or limit the size of the job. They may also create an unsafe condition by crowding the pumping equipment too close to the wellhead. An effort should be made to keep the location as unobstructed as possible until operations are complete.

To avoid untimely delays, unsafe conditions, and other restrictions, prepare a scaled drawing of the location, including such things as mud pits, roads, lines, etc. Also make scale models representing equipment, sand storage units, and fracture tanks. After the job requirements have been determined, fit these into the scale of the wellsite. This is helpful in determining the most efficient arrangement. It will also show whether the location can handle the proposed treatment adequately or may point out modifications that must be made beforehand (Fig. 13.1). Be sure to consider the following points.

1. For safety, plan for at least 50 ft [15 m], and preferably 100 ft [30 m], from the nearest manned piece of equipment to the wellhead to make the well accessible if problems develop that require special equipment for well control or workover.

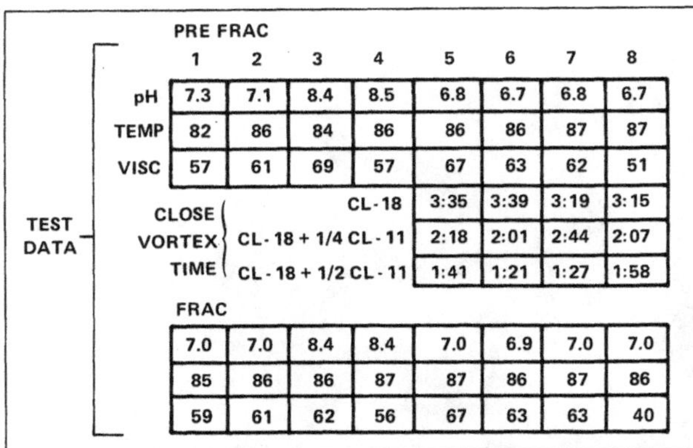

TEST DATA		1	2	3	4	5	6	7	8
PRE FRAC									
pH		7.3	7.1	8.4	8.5	6.8	6.7	6.8	6.7
TEMP		82	86	84	86	86	86	87	87
VISC		57	61	69	57	67	63	62	51
CLOSE VORTEX TIME	CL-18					3:35	3:39	3:19	3:15
	CL-18 + 1/4 CL-11					2:18	2:01	2:44	2:07
	CL-18 + 1/2 CL-11					1:41	1:21	1:27	1:58
FRAC									
		7.0	7.0	8.4	8.4	7.0	6.9	7.0	7.0
		85	86	86	87	87	86	87	86
		59	61	62	56	67	63	63	40

Fig. 13.10—Fluid property measurement chart for quality control.

2. Allow adequate space behind the fracture tanks to allow refilling between stages or emptying contaminated tanks without the removal of equipment in place. Make sure the ground is competent enough to support water-hauling equipment.

3. If possible, make the location large enough to accommodate all fracture tanks without locating some remotely.

4. Have an unobstructed route open from the service road to the wellhead for emergency service and transportation of injured persons in case of an accident.

5. Have available unobstructed, open escape routes for personnel at every manned position on wellsite.

6. Remove all objects from location because they hinder equipment and fracture-tank placement and create a safety hazard.

7. Do not install or construct production equipment or facilities prematurely; doing so will probably interfere with the operation and will be a possible safety hazard.

8. If possible, slope the location for drainage to avoid excessive accumulation of water from rain. This will help prevent unwanted delays and help avoid problems with moving heavy equipment.

9. Prepare ditches to contain potential leaks of flammable materials, such as diesel oil, on site and to divert any flow safely away from potential fire hazards.

After the final arrangements have been decided, take several photographs and distribute them and other plans and drawings to the people involved in conducting the operation.

Everyone must operate from the same plan, thus minimizing the confusion that arises from miscommunication.

Fluid Preparation. First, all tanks used to prepare the gel should be extremely clean. Most gel systems are sensitive to pH and chemical contaminants that can prevent gel hydration and/or interfere with the crosslinking mechanisms. Steam cleaning and flushing the tank with fresh water yields the best results.

One major source of gel problems, particularly in the summer, is bacterial contamination. Gelling agents actually serve as a food source; the gel structure is destroyed by bacterial enzymes. Sulfate-reducing bacteria are some of the most common, yielding a characteristic black color to the fluid and a strong H_2S odor. However, bacterial contamination can manifest itself in other ways. In the early stages of growth, bacterial byproducts can interfere with the crosslinking/complexing capability of the gel system. At this point, base-gel viscosity is not reduced, as in advanced stages, nor is there any other abnormal indication of severe contamination. Therefore, it is imperative that pilot crosslinking/complexing tests of each tank be made periodically until just before pumping. To control the bacterial growth, a bactericide should be placed in the fracture tank before it is filled. This treatment is more effective and less expensive than trying to eliminate the bacteria after they have already contaminated the fracture tank. The polymer degradation caused by bacterial growth in a fracturing fluid can be detected by rheology measurements, as discussed in Sec. 13.9.

Water sources should be tested to ensure compatibility with the processed gel system. This will avoid costly losses of time and money as a result of placing incompatible water on location. After the tanks are filled with the approved water, another check of each tank should be made to verify water quality. Iron content and pH are checked. pH can be adjusted, but water should not contain more than 50 ppm dissolved iron. Tanks containing unsuitable water can be identified and corrected before gelling.

When the gel in the tanks has been mixed, check for proper gel hydration and uniform mixing by measuring viscosity and crosslinking time. Fig. 13.10 shows such data measured on site. Again, tanks containing unsuitable gel can be identified properly and adjusted before pumping. If gel quality is verified, problems during pumping can be avoided. If adverse situations occur, look for other problems. Ensuring gel quality also helps in assessing the effectiveness of the treatment in postfracture analysis and performance.

A word of caution should be mentioned here: gel preparation should begin only after the well is ready for treatment. Numerous unexpected delays, such as packer failures, perforating problems, and casing/tubing leaks, can postpone a job indefinitely. During this period, the gel can deteriorate to a point where it is unusable, resulting in a loss of thousands of dollars.

An important option is the use of liquid gel concentrate, which can be mixed during the job. One problem with this system is the accurate metering of the liquid gel concentrate to achieve a uniform viscosity. The use of one or two work tanks as residence tanks dampens fluid-property variations and delivers a more uniform fluid to the blenders. More detailed quality-control guidance is provided by Ely.[2]

13.5 Propping Agents

Check the condition of the proppant bulk haulers brought to location. The top hatches can be inspected with a flashlight to see the condition and cleanliness of the proppant delivered. Samples can be collected through the hatches with a sample-can-and-pole arrangement to inspect the quality of the proppant. Check to see that the sand haulers were clean before the proppant was loaded. Inspection of bins on site has disclosed contaminants from prior hauling of cement and other trash. Trash or fines mixed in with the proppant may result in poorer performance of the fractured well after the treatment. Consult Chap. 6 on fracturing-proppant specifications and testing.

First, proppant samples should be inspected visually to ensure cleanliness and the proper type of proppant. Second, testing sieves on location can be used to inspect the proper grading of the proppant visually to determine whether too many fines are present (a special problem with some sources of mined quartz sand). A lapse in quality control by proppant suppliers can lead to out-of-size proppant, as described in Chap. 6.

13.6 Job Execution

Successful job execution depends on good planning, preparation, supervision, attention to details, quality control, and communication. Nowhere else in the industry is so much money spent in so short a time. It is well worth the cost to select well-trained, dedicated personnel, well-maintained equipment, and good-quality materials. A lack of one of these prerequisites will often result in unsatisfactory stimulation and therefore a lower return on investment, even though the costs of a poorly executed treatment may be substantially lower.

Equipment Operation Checks. While the fluids are being mixed (usually the day before or the morning of the treatment, depending on treatment size), review and check the operation and location of all equipment to ensure that all equipment operators are familiar with their tasks and that all equipment is working before the treatment is begun. A checklist for this work is included in Table 13.2.

Pressure Testing. The service company crew must pressure test all lines to the maximum pressure limit with the wellhead plug valve closed. This may take an hour or more as leaks are detected and connections are tightened or replaced. Once the entire pressure system has been made leak-tight and pressure holds steady for at least 1 minute with no leaks spotted, the equipment and injection system are ready for the treatment.

Safety Meeting. The safety meeting is the first and most important part of the "tailgate" meeting (pretreatment meeting where everyone is assembled at a protected site away from the noise of motors on location). Safety standards should be reviewed, including those discussed in Sec. 13.9. For example, a safe meeting area is designated in case of emergency. Smoking and nonsmoking areas are also designated. Personnel are assigned to operate critical valves. In this meeting, review the job details and point out any critical procedures that affect safety.

Operating Meeting. After the safety meeting, it is important to establish an understanding between all involved personnel of the importance of communication during the treatment. For example, if a pump is not working, if crosslinker is not feeding properly, of if proppant is not moving to the blender, the malfunction should be communicated to the fracturing van or treatment command center. Early communication of malfunctions minimizes the problems that may occur hours later during the treatment. Problems with the blender (especially sand-concentration fluctuations) are particularly important in crosslinked-fracturing-fluid treatments. Surprises are dangerous, and good communication prevents surprises.

After reviewing the treatment, explain the reasons for any special mixing or procedures that will take place. Allow time for questions from equipment operators and other personnel.

Contingency Plans. The most common problems encountered during the treatment, pump breakdowns and fluid pressure leaks, often cause unplanned shutdowns and startups. Corrective actions should be decided beforehand. Prior arrangements are made to coordinate pump shutdown and repair and to bring standby pumps on line.

The slurry blender is the heart of the fracturing operation. The critical part of the fracturing operation is to have the standby blender located so that the proppant delivery can switch from one blender to the other in a short time and so that fluid pumping will not be stopped; or if it is stopped, the standby blender will be operating in less than 30 seconds. A pretreatment switching exercise should be performed while the pad is pumped to ensure that both blenders are operable and that the switch can take place without the treatment being shut down. Two proper blender setups are shown in Fig. 13.11. The third important contingency is for a premature "pressure-out" (pump pressure increases rapidly to the maximum allowable pressure limit). Usually a pressure-out in crosslinked fluid fracturing is signaled by a gradual rise in pressure during injection (see Chap. 14). This pressure rise is usually exponential. A real-time linear plot of the pump pressure and calculated fracturing bottomhole pressure (BHP) in the fracturing van is a big help in predicting a pressure-out. Such a plot is described in Sec. 13.14. Faster pressure-outs (screenouts) are also possible when proppant bridges the perforations or the fracture near the perforations. Pumps are shut down quickly when the screenout occurs. A pressure-relief valve and blow line can be connected to the high-pressure injection line at the wellhead to protect the tubing and wellhead from catastrophic pressures.

Communication. The importance of communication must be emphasized during the "tailgate" meeting when all personnel are gathered to discuss safety and operating procedures.

The first discussion is on the potential treatment of a well. Free exchange of information, compilation of data and well information, explanation of fluids, additives, and equipment needed, and significant or potential well problems should be discussed. Pretreatment planning sessions are often held for critical wells or for new or different treatment procedures. These sessions are very helpful when new or improved on-site quality-control procedures are initiated. In this meeting, the design and treatment schedule, wellsite and well preparation, treating equipment and materials, and general logistics of the operation are reviewed. Designation of authority for various tasks will minimize confusion later on site.

On-site review of the treatment after the equipment is in place and before fluids are mixed is important to ensure that any late charges are known and understood. During the treatment, hand-held radios facilitate good communication and treatment control.

Fig. 13.11—Two properly aligned blenders (courtesy BJ Titan).

On critical or large treatments, it is not uncommon to have several personnel observe or monitor the blender operations, fluid quality-control testing, and treatment data. This rapid accessibility to current events is essential to minimize problems and to make necessary corrections. Decisions on rate changes, proppant or schedule changes, and early shutdowns can be made if warranted. This communication helps prevent surprises during treatment and allows safer decisions to be made.

Post-job communication is important. Personnel should be gathered after the job and before the equipment is rigged down. Acknowledgment of good performance can contribute to future job performance. Practices that need improving can also be discussed at this time while the events are fresh. Any lapse in safety or operating procedures should be reviewed and discussed to emphasize the importance of these procedures. Treatment execution is a complicated and intensive operation.

Treatment Monitoring. All fracturing treatments need good quality control and effective monitoring. The practices discussed here are as important for volumes of 50,000 gal [190 m^3] as for 1,000,000 gal [3785 m^3] or more. An error in large treatments can affect the final treatment control more because any percentage error in measurements will involve more volume.

Massive-hydraulic-fracturing treatments require large volumes of fluid (usually more than 200,000 gal [760 m^3]), sand, and extended pump time. It is not uncommon for treatments to last 6 to 12 hours. Consequently, even relatively small variations in the fluid and sand rates become significant when allowed to continue unmonitored for the duration of the job. Failure to compensate for a rate deviation of 1 bbl/min [0.03 m^3/s] can result in an error of hundreds of barrels of fluid. This error can result in a major compromise of the proposed treatment. It is possible to run out of gelled fluid before the final highest proppant concentration is pumped to achieve maximum stimulation. Most instruments used to monitor fluid and sand rates are usually accurate to only $\pm 5\%$. Verification and ap-

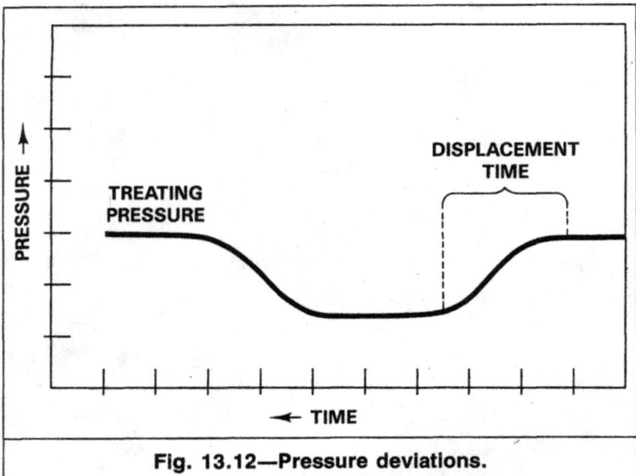

Fig. 13.12—Pressure deviations.

propriate adjustments should be made at frequent intervals to keep variations from the proposed treatment to a minimum.

Injection rate can be verified three ways; all three should be used as a countercheck for each other. The turbine flowmeter provides the visual record in the fracture monitor; it can be calibrated to correspond to the actual rate. Flowmeter accuracy is affected by changes in viscosity of the treatment fluid, however, and is accurate by a factor of $\pm 5\%$. As a result, it should not be the only source of rate verification.

Counting pump strokes and knowing their displacement capacity can be used to check rate. Variables in this method can be sources of error; it is useful, however, and approximates the injection rate.

A third, more accurate method of determining injection rate is by measuring the volume of fluid pumped from the fracture tank and the time required to do it. The time intervals should be long enough to minimize lost time in reading the tank strap. Usually a 5- to 6-minute interval is sufficient. Additionally, not all the tank valves may be holding properly, causing an erroneous reading. Simply check the level of the other tanks periodically to ensure that all values are holding.

The volumes of the different stages should be measured primarily by tank strapping rather than using the totalizer in the fracture monitor. All three methods should be used to crosscheck each other in determining the actual rate as closely as possible. However, the tank strap vs. time, which is the most reliable, requires that accurate data on tank volume strap be available.

Measuring the rate of sand is an equally difficult task. Small errors throughout the job can result in a substantial error at the conclusion of the treatment. Sand rate can be verified several ways; all methods should be used in conjunction to minimize the error.

The most reliable method of monitoring proppant use is by gauging the proppant storage units. A schedule should exist for emptying the individual compartments at predetermined stages during the treatment. The increments should be frequent so that errors can be corrected without a major compromise in job design.

Most blenders are equipped with proppant augers to control the rate at which proppant is added to treatment fluid. Each revolution of the auger delivers a certain quantity of proppant. However, that quantity can differ with each blender. Each blender must be calibrated to determine the rotation rate required to deliver the correct amount of proppant at various injection rates. The rates are subject to change with age because wear decreases the auger efficiency, requiring frequent recalibration. Calibration must also include corrections for proppant density and size.

Radioactive densimeters are very effective in monitoring proppant concentrations, especially any sudden changes in concentration. They need to be calibrated against total sand pumped during the treatment to monitor the entire treatment more effectively.

Continual monitoring of the fluid and proppant volumes enables problem areas to be detected early and allows reasonable changes in the planned schedule. Some of the earlier sand stage volumes may be reduced to allow the more critical, heavier sand concentrations to be pumped as scheduled.

Pressure Deviations. One of the most critical areas to evaluate during a treatment is the fluctuation of treating pressures. Correctly identifying the source of the pressure change is necessary for effective remedial action. Four sources of pressure deviations are mechanical problems, changes in gel properties, variations in proppant concentration, and formation response.

The most common mechanical problem causing abnormally high treating pressures is flow restriction through perforations. A pressure drop is created when fluid flows through a perforation. When some of the perforations are not accepting fluid, the flow rate per perforation is increased, causing a higher pressure drop. This results in a higher surface treating pressure than anticipated and may result in an undesirable distribution of the treatment. If the surface treating pressure is higher than anticipated, take an instantaneous shutdown pressure to verify the BHP to determine whether the fracturing BHP was estimated correctly. Knowing the injection rate and friction pressure down the treatment conductor allows the number of open perforations to be calculated. This number may be less than the number of actual perforations when some perforations are plugged or are not connected to the fracture or when perforation diameters are less than expected. Actual perforated holes in the casing may have smaller diameters than expected from perforating-charge data sheets. It may be necessary to reacidize and/or to reperforate before the fracture treatment is continued. These problems are usually taken care of in the minifracture and ballout before the main treatment.

When pressures rise and fall, it is difficult to determine whether the cause is the fracturing behavior of the formation or gel properties. One key factor is the time over which the pressure change occurs (Fig. 13.12). If that time corresponds to the time it takes to displace the tubing or casing, it probably reflects a change in the treatment-fluid properties. Other pressure changes usually are related to actual changes in fracture geometry and dynamics.

If the pressure changes are determined to be fluid-related, the following causes should be investigated. Difficulty in maintaining proppant concentration will cause pressure variations as a result of the gain or loss in hydrostatic pressure. A change in the polymer (gel) concentration can also cause the pressure to vary. Poor polymer mixing can cause layering of different concentrations in the tank, sometimes causing rising pressures near the end of pumping a tank and decreasing pressure when starting a new tank. Uniform mixing procedures can minimize the effect. Switching to a different-density proppant will affect pressure the same way.

A pressure drop may occur when the crosslinking mechanism of the treating fluid fails. Again, the pressure drop will normally take place over the time required to displace the fluid through the tubing. This should be corrected as quickly as possible. It often results in a screenout because proppant transport capabilities are diminished, fracture width is reduced with the decreased viscosity, and fluid loss may increase. Pressure increases occur when crosslinker is added too rapidly, resulting in a crosslinker concentration that is too high.

Pressure fluctuations become greater in magnitude as the injection rate increases, primarily because they are related to friction pressure. At lower injection rates, the problems associated with gel properties may go undetected but are no less detrimental.

Significant pressure changes caused by the fracturing behavior of the formation are usually a result of out-of-zone fracturing or screenout. These pressure variations are discussed in Chap. 14.

Pressure increases preceding a screenout usually have characteristics that serve as an early warning; however, they are not always present. When the pressure increase is sudden, screenouts usually occur near the wellbore. Causes can be unusually high sand concentrations pumped by mistake, loss of viscosity owing to crosslinking failure, or perforation restrictions, as discussed earlier. There is little warning and minimal remedial action that can be taken. When a screenout occurs early in a treatment as a result of fluid and/or proppant mixing problems and at a point where the design will be compromised critically, flow the well back to open the plugged perforations. Resume pumping if possible. This technique may be successful if the block is superficial and not tightly packed.

The other type of screenout occurs some distance from the wellbore. The pressure increases gradually at first, then more rapidly

TABLE 13.2—CHECKLIST FOR BALLOUTS AND FRACTURING JOBS

1. Fracture Tanks
   A. Make sure they are clean, inside and outside
   B. Make sure they are internally coated and that the coating is not damaged
   C. Make sure that the valves operate and do not leak
   D. Make sure that the top is skid proofed and the hatches are fully opening
   E. Make sure that the tanks are placed on level ground and leaning slightly forward
   F. Check to ensure that the ladders are secure
   G. Measure the tank depths and verify gauging procedures
   H. Check the condition of the suction lines and determine the height of the suction above the bottom
   I. Add biocide with the first load of water
2. Fluid
   A. Dictate the source of the fluids
   B. Make sure that clean dedicated transports are used
   C. Verify fluid volume hauled vs. fluid volume charged
   D. Check fluid cleanliness by smell, color, and freshness
   E. Measure pH and specific gravity and look for any signs of bacterial growth
   F. Measure the fluid temperature and the ambient temperature
   G. Run a pilot test to make sure the fluid will gel
   H. Once the fluid has been gelled, have the service company measure pH, temperature, and apparent viscosity
   I. Repeat these measurements just before pumping to ensure no gel deterioration
3. Blenders
   A. Always have a primary blender and a backup
   B. Make sure they are located in a position where either can receive sand without shutdown
   C. Inspect the hoppers for size, cleanliness, and exposed bearings
   D. Check the operation of all augers, tub paddles, and pumps
   E. Inspect the mixing tubs—look for leakage, additive injection points, leveler operation, and paddle operation
   F. Check the condition of hydraulic hoses—make sure the hydraulic oil supply is sufficient
   G. Inspect suction and discharge hoses for proper connections, proper routing, and good condition
   H. Inspect and diagram the flow path of fluids through the blender
   I. Check the diversion valves to ensure operation; practice switching valves during prepad stage to check operator competence and to measure pressure and rate changes
   J. Ascertain who will be the primary blender operator and who will be the backup operator
   K. Verify that the densiometers operate and know their location.
   L. Operate both blenders at all times during the fracture treatment; maintain a state of constant readiness
4. Proppant Transport
   A. Check the operation of the conveyor system
   B. Check the condition of all hydraulic hoses
   C. Make sure that a backup hydraulic system is available with quick-connect couplings
   D. Determine who will be the primary and backup operators and their levels of experience with the equipment
   E. Swing the conveyor system to ensure that it will reach either hopper
   F. Check the condition of all conveyor belts
   G. Check the number of compartments, capacities, and present volumes; determine location of each proppant and the sequence of emptying the compartments
   H. Determine that the proper type and amount of proppant is on location and check the quality for cleanliness and size
   I. Check the inside of the proppant transport when the job is finished to verify usage.
5. Additives
   A. Make sure that the specified additives are on location
   B. Verify that the quantity is adequate
   C. Determine how they will be injected and who will do it
   D. Determine how the additives will be measured
   E. Make sure that the additive pumps operate
   F. Make sure that the diesel tanks are full and determine how the diesel will be measured and by whom
   G. Be sure to gauge the diesel tanks during the job in addition to monitoring the flowmeter
   H. Check for fire extinguishers
   I. Check tanks at the end of the job
6. Manifold
   A. Understand the flow path from the blender and pump trucks through the manifold and to the well
   B. Know what is injected and where
   C. Check the condition of all hoses
   D. Review the valving system and make sure it is operable in case of a failure
   E. Check for any internal restrictions
   F. Make sure that the sample catcher is double-valved
   G. Determine who will catch the samples
   H. Determine the measurements to be made on the sample
   I. Determine what changes will be made if the sample is not as expected

as the fracture fills with sand. There is a noticeable general upward trend in the surface treating pressure. At one point, there may be one or several sharp breaks in the pressure. This usually precedes the steep pressure increase, however, and screenout is almost a certainty. Experience with repeated treatments in an area provides guidance on screenouts. There is often enough warning to switch to flush in time to displace proppant from the tubing string.

Proppant Densimeters. Multiple proppant densimeters are needed on site for critical treatments that involve high-temperature formations, difficult fluid-loss control (as in naturally fractured formations), and/or long pump times (more than 2 hours), where continuous proppant concentration monitoring is essential to prevent premature screenouts and to ensure consistent blending for high proppant concentrations (8 lbm/gal [960 kg/m^3] or more). Densimeters are used at the blender outlet and near the wellhead on the high-pressure treating line. If diesel oil is added at the blender, the densimeter at the blender and the wellhead should give about the same reading. If the diesel oil is added downstream of the blender at the intake manifold, the densimeters at the blender and at the

TABLE 13.2—CHECKLIST FOR BALLOUTS AND FRACTURING JOBS (courtesy Conoco Inc.) (continued)

7. Pump Trucks
   A. Verify that the horsepower required is on location; determine how it is distributed
   B. Know the horsepower available per truck
   C. Always have sufficient backup horsepower, truck for truck
   D. Make sure that fire extinguishers are by each truck
   E. Understand the drive mechanism of each pump truck
   F. Know the pump capacity and rating for each truck
   G. Determine who the operators are for each set of units
   H. Keep all engines running
   I. Know at all times how many pumps are on line
   J. If pumps go down, determine if it is temporary or permanent and re-examine the backup
   K. Pressure test all lines and check for leaks; repair all leaks
   L. Pressure test casing pump truck and line; use double pop-off bleed line and visual pressure gauge
   M. Monitor the amount of water pumped into the casing; this could indicate problems
8. Injection Lines
   A. Verify that lines are of sufficient size
   B. Ensure that they are staked down.
   C. Determine the location of the densiometer and the backup densiometer
   D. Pressure test and repair all leaks
   E. Use double valves above the tree and double-wing valves
   F. Do not use swages into the tree or screwed connections
   G. Check each valve for the proper number of turns to open and close fully
   H. Determine who will close valves in case of emergency
   I. Locate the closing device
   J. Check to be sure that there are visual pressure gauges
   K. Determine the location and number of pressure-monitoring devices
   L. Check discharge lines frequently and make sure they are staked down
9. Tree
   A. Determine if the space between the tubing wraparound and the blowout-preventer coupling to the tree is pressurized; if not, pressure up and monitor the pressure during the job to uncover a tree leak, packoff leak, seal assembly problem, or tubing leak
   B. Make sure that casing hanger holddown screws are in and tight; if not, and if pressure was experienced after the primary cement job, it probably means that the production string has dropped
   C. Personally open and close all valves; double check to ensure that the proper valves are open and closed
   D. Make sure that the tree is rated for the anticipated surface pressure
10. Tree Saver
   A. Be sure that the packoff is new
   B. Check to make sure that the tube is in good condition
   C. Carefully measure the tube and the tree to ensure that it fits perfectly in the tubing
   D. Stab into tree very carefully to avoid valve damage
   E. Install choke-in-tree choke assembly
   F. The tree row has two valves; know how they operate
   G. In the event of an emergency, make sure you know exactly who will operate the valves and be sure he has everything needed at the tree
11. Safety Meeting
   A. Assemble everyone away from the noisy area
   B. Keep a current head count
   C. Ask anyone who is there but doesn't belong to leave
   D. Select a safe meeting area in case of emergency
   E. Designate smoking and nonsmoking areas
   F. Check to determine that safety equipment is available and used (hard hats, eye protection, dust masks, safety shoes, and ear plugs)
   G. Determine who the foam-fire-fighting operator is
   H. Tell the entire group who will operate the valves in case of emergency
   I. Have at least three company people actively involved in the job
12. Operating Meeting
   A. Determine whether all the service company employees are from the same district
   B. If they are not all from the same district, make it clear that you demand cooperation
   C. Compliment the group members if they did a good job the last time out; review the problems if the job was not smooth
   D. Spell out exactly what you expect; know who will gauge the tanks and make sure that is his only job; make sure that everyone in the meeting knows what everyone else is expected to do
   E. Let the group know the maximum pressures permitted
   F. Spell out the flush procedure under conditions of both a successful job and a screenout; do not overflush

wellhead will be slightly different. The high-pressure treating line should have a working backup densimeter. The densimeters should be calibrated before the treatment and should be monitored during the treatment to ensure that all are working.

Control and monitoring of the blender operation are critical. Sudden changes in proppant adding rate can cause spikes—i.e., too high a concentration at the beginning of each stage. It is better to increase the sand concentration between each sand concentration level gradually.

Fluctuations in blender liquid level can change pumped proppant concentration. In long pumping jobs, it may be easy for sudden changes in blender operations to be overlooked. Long hours by service company personnel preparing for the job and long work weeks in periods of high activity can lessen alertness and motivation.

Sampling During Treatment. Special sampling manifolds can be inserted in the line from the blender to the pump intake manifold. Two sampling manifolds are needed (one for backup). Each sampler has two valves in series so that one can be shut off if the outer valve shuts off with sand during the treatment. A sample should be caught periodically (every 5 minutes) to check the mixed slurry concentrations and to detect any deterioration in the gelled water. Erratic

TABLE 13.3—CHECKLIST FOR PROJECT QUALITY CONTROL

Well name ____________

Field ____________

Date ____________

Service Company

Perforaters ____________

Loggers ____________

Acid Company ____________

Fracturing Company ____________

Other ____________

Before the job

1. Engineers
   - ☐ A. Specify system
     - ☐ 1. Sand type (e.g., 20/40 Ottawa)
     - ☐ 2. Gel system (40-lbm/gal HPG system or 30-lbm/gal HPG titanate crosslink system)
     - ☐ 3. Fluid additives; amount/1,000 gal, type
       - ☐ a. Breaker—specify where it is to be added, breaker time
       - ☐ b. Fluid loss
       - ☐ c. Surfactant
       - ☐ d. Biocide—specify where and when it is to be added
       - ☐ e. Miscellaneous
   - ☐ B. Rate
   - ☐ C. Maximum treating pressure, expected treating pressure (make sure wellhead is adequately sized)
   - ☐ D. Tubing size (example 2⅞-in. workstring)
   - ☐ E. Quality-control equipment (if available and necessary)
     - ☐ 1. Calibrated recording densiometer in same van with pressure recorder
     - ☐ 2. Fann viscometer
   - ☐ F. Special safety precautions
   - ☐ G. Meet with foreman, technician and service company representative to discuss job
2. Field personnel
   - ☐ A. Make sure the location is clean and correct size (have a picture of how everything will be set up before all equipment arrives)
   - ☐ B. Have pit dug
   - ☐ C. Order clean fracture tanks; check fracture tanks, blender, and transports for cleanliness
   - ☐ D. Watch biocide go in first load of tank
   - ☐ E. Test water quality (pH, iron content) before gelling up
   - ☐ F. Make sure all chemicals are on location before gelling up
   - ☐ G. Coordinate with service company personnel as needed
   - ☐ H. Collect sand samples, perform sieve analysis
   - ☐ I. Make sure all equipment is in working order
   - ☐ J. Have a prejob safety inspection by foreman

During the job

1. Engineers
   - ☐ A. Monitor pressure, rate, and recording densiometer in fracturing van
   - ☐ B. Keep foreman informed
   - ☐ C. Know when you are going to flush (if problems arise—equipment parameters, fracture gradient of reservoir, etc.)
   - ☐ D. Monitor barrel counter; know when each stage begins and ends
2. Field personnel
   - ☐ A. Watch blender operations; check screw rotation; ensure that all additives are being used in correct proportions
   - ☐ B. Collect samples of gel and sand-laden fluid
   - ☐ C. Ensure that safety precautions are followed
   - ☐ D. Keep engineer informed
   - ☐ E. Monitor all equipment to ensure that it is working

After the job

- ☐ 1. Ensure that all chemicals were used
- ☐ 2. Weigh sand trucks to ensure that the correct amount of sand was pumped
- ☐ 3. Check location to make sure that sand or other additives have not been dumped on the ground
- ☐ 4. Meet with field personnel concerning job quality and critique job

TABLE 13.3—CHECKLIST FOR PROJECT QUALITY CONTROL[5] (continued)

Overall Critique

Perforators

General remarks (Did the guns fire? Hole shape and consistency? Gun swelling?)

Overall service (circle one): Excellent Good Fair Poor

Loggers

General remarks (Log quality: are they on depth? Tool failure?)

Overall service (circle one): Excellent Good Fair Poor

Acid Company

General remarks (pH? Equipment failure?)

Overall service (circle one): Excellent Good Fair Poor

Fracture job

1. Design
   Was the job designed correctly? Explain.
   Was it redesigned if necessary?
   What could be done to improve the design?
2. Service Company
   Were there any failures of the service company equipment?
   Safety attitude (circle one): Excellent Good Fair Poor
   Were all additives on location?
   Overall performance of service company (circle one): Excellent Good Fair Poor
3. Our Quality Control
   Comments
4. Areas to be improved

Signatures

Foreman ______ Date ______

Engineer ______ Date ______

Technician ______ Date ______

Technician ______ Date ______

Initial each box.

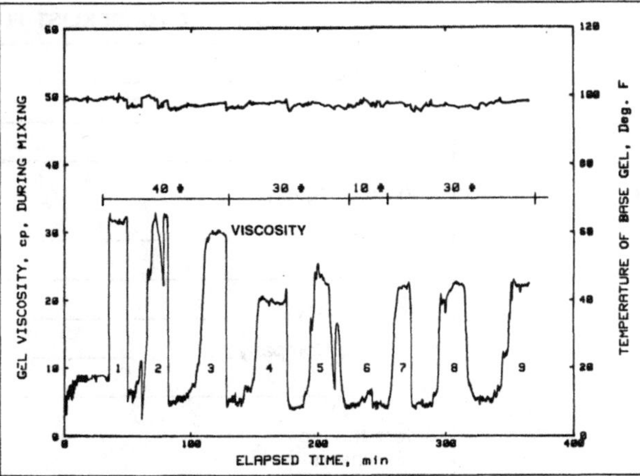

Fig. 13.13—Computer-monitored tank mixing before proppant stimulation shown in Fig. 13.14; note 40-lbm/1,000 gal gel system (courtesy BJ Titan).

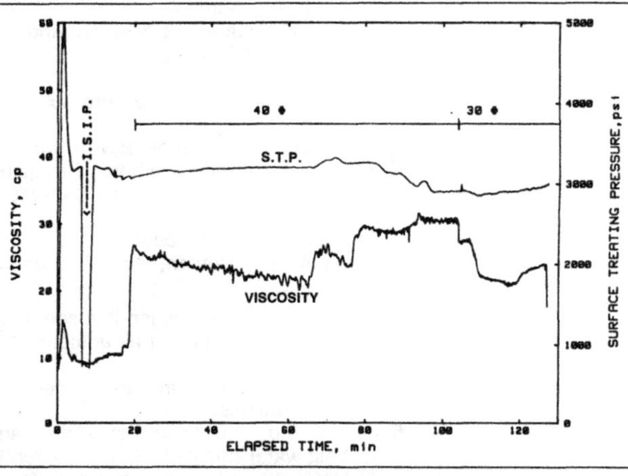

Fig. 13.14—Effects of biodegradation on 40-lbm system after 24 hours as monitored by a computer (courtesy BJ Titan).

variations in sampled proppant concentration will show problems in proppant transfer or blending. The crosslinking time should be checked on each tank as it is pumped, about every 5 to 10 minutes. The minimum is twice on each tank: once at the beginning and once near the end of the tank evacuation.

Final Flush Coordination. Most service company blenders have "clean/dirty" bypass lines. It is important to cut cleanly from the final maximum proppant concentration to clean fluid, and to pump the proper displacement volume. Proppant should be displaced from the surface lines and tubing to a point above the top perforation (preferably about 50 to 100 ft [15 to 30 m] above the top perforation to allow for variations in pipe diameter and error in flowmetering). Sufficient rathole should be in place for the volume of proppant in the wellbore to settle below the bottom perforation after the fluid has broken.

If it is not possible to bypass the blender, the densimeter is used to control the final displacement or flush. When the proppant is shut off to the blender, the sand concentration from the blender will decline gradually. The volume displacement should start as soon as the sand densimeter at the wellhead begins to fall. This ensures that the final designed high slurry proppant concentration will not be overflushed from the wellbore. Some dispersion and mixing with flush can occur in surface lines. The wellhead densimeter should always be used to monitor the final concentration to ensure that final concentration is maintained and placed in the fracture at the wellbore.

Fracture Closure and Fluid Recovery. Well shut-in times depend on the break time of the final gel pumped and fracture closure time; both should be accomplished before flowback. Break tests should be conducted at conditions approximating the downhole temperature profile determined by a fracture temperature heat-up model. Fluid samples can be stored in a heating bath at the service company field laboratory. When the samples break at formation temperature, it is safe to flow the well. Prematurely flowing back an unbroken gel can flush proppant out of the fracture at the wellbore and results in poor conductivity near the wellbore where it is most needed. Field experience with fluid systems and flowback after treatments in a given formation, depth, and field area offer valuable guidance in this process.

The well should be flowed back at a preselected rate to prevent production from the fracture. Dual chokes and valves in place at the wellhead allow one choke to be shut and exchanged while the other choke controls flow from the well. If poorly broken gel carries proppant out of the wellbore, severe erosion results. All lines leading to the waste pit must be chained to two iron stakes to prevent whipping of the pit line during well flow. The valves to the tubing/casing annulus on the wellheads should be opened slightly so that pressure can be bled off as the returning fluids heat the fluid in the annulus. If annulus valves are closed, the rising temperature will increase pressure in the annulus, which can cause casing failure or tubing collapse.

13.7 Quality-Control Worksheets and Checklist

Tables 13.2 and 13.3 are examples of a checklist and a quality-control worksheet.[5,6] The first checklist concerns the operation of

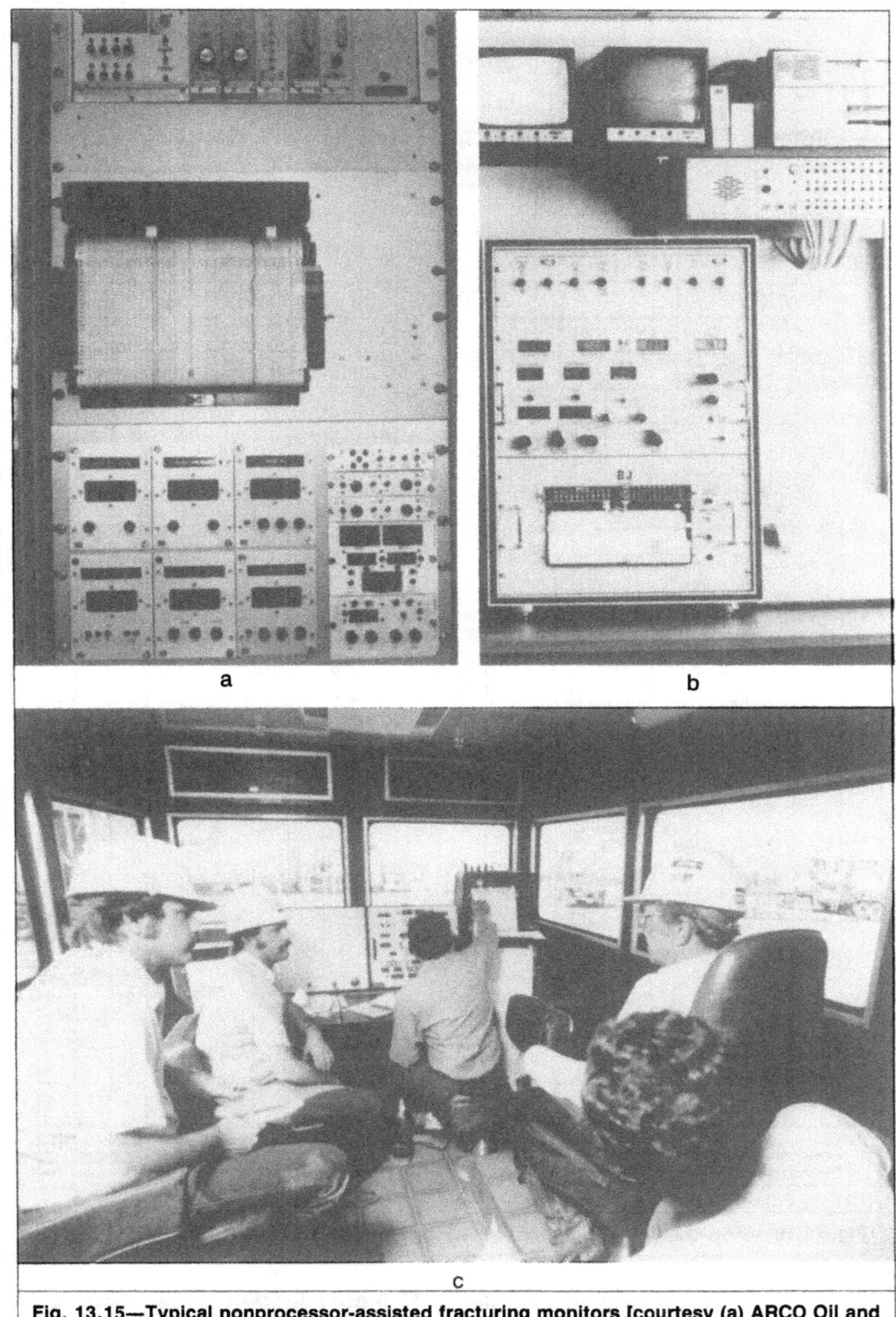

Fig. 13.15—Typical nonprocessor-assisted fracturing monitors [courtesy (a) ARCO Oil and Gas Co., (b) BJ Titan, and (c) The Western Co. of North America].

the different pieces of equipment on location. The other quality-control worksheet is used in monitoring fluid and proppant properties and general treatment progress.

13.8 Safety Standards

A safe working environment for personnel is the most important aspect of all fracturing operations. All fittings and lines should be pressure rated to above the fracturing pressure limit on location to prevent any catastrophic failures. The high-pressure pumping lines up to the valve on the wellhead should be tested to this pressure limit to ensure that all fittings and valves do not leak and can withstand the maximum pressure expected. Make sure that the wellhead is rated to the pressure expected. If the wellhead will not take the expected pressure, then a tree saver should be used to isolate the wellhead from the pressure. The pressure limit is set by the burst rating of the tubing in the well or the casing if the treatment is down casing. If a sealbore packer is used, tubing movement is calculated to ensure that the tubing will not come out of the packer during treatment.

Access to the wellsite should be controlled. Only authorized vehicles should be allowed in, and the road(s) must be clear during the pumping operation.

During the tailgate safety meeting before the treatment, designate and show all personnel on site the escape routes from the wellsite and the emergency regrouping area. This procedure is most important when energized fluids, such as CO_2 or N_2, and especially gelled oil fluids are used.

Contingency plans must be made for any emergency that may arise during pumping. Actions to be taken should be written and understood by all persons directly involved with the contingency procedure. Emergency tasks and responsibilities should be defined

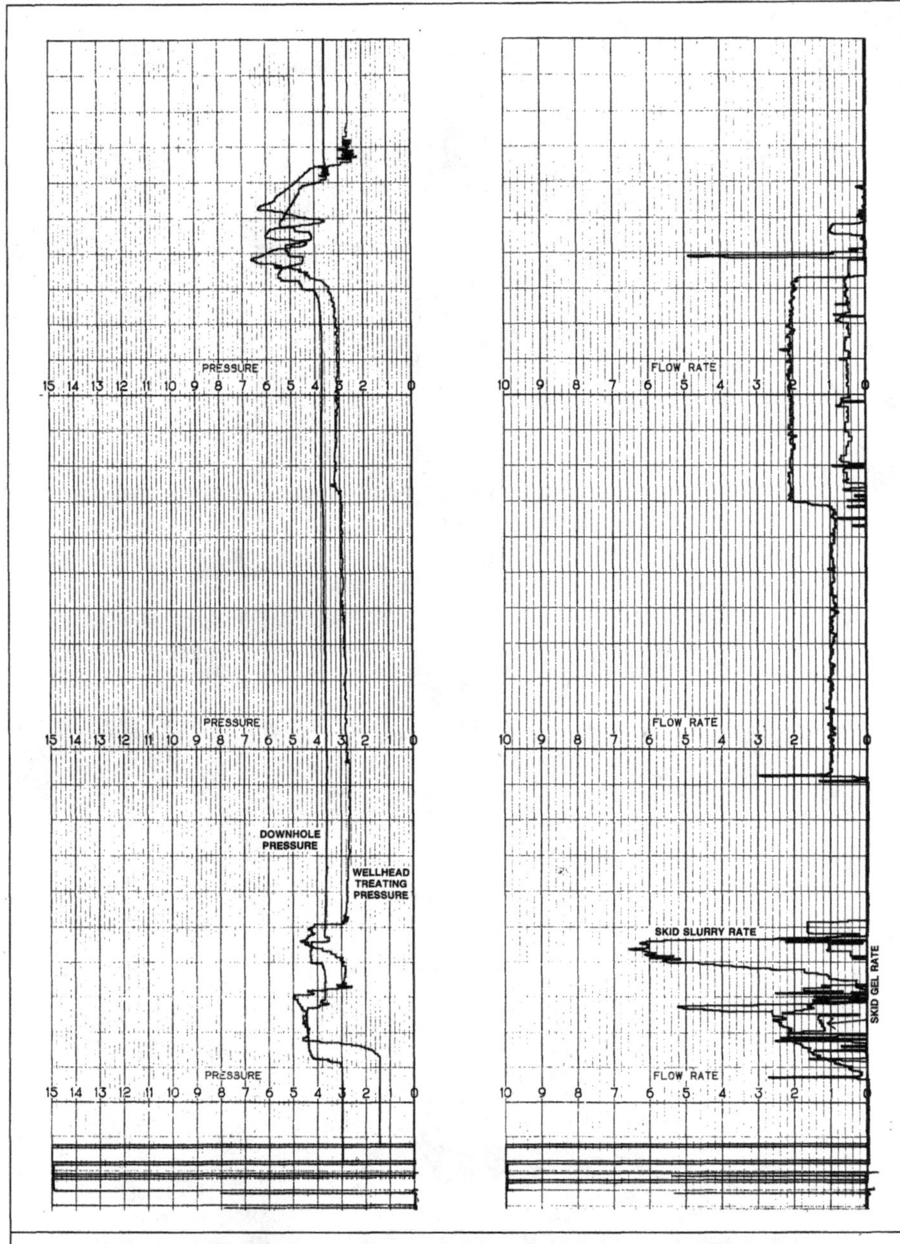

Fig. 13.16—Strip-chart recording from fracturing monitor (courtesy ARCO Oil and Gas Co.).

before the treatment begins. There must be no smoking around the wellhead or around any hydrocarbon storage on site. Safe smoking areas should be designated. Personnel relief and backup should be coordinated to ensure that critical personnel are relieved occasionally so that operating personnel are alert at all times.

All service company personnel should wear fire-retardant safety clothing when pumping flammable fracturing fluids (e.g., diesel oils or alcohol). Fire-control equipment should be on site with a designated operator. During fracturing treatment, headphones or ear protectors, standard safety shoes, and hard hats should be worn by all on-site personnel.

When pressurized fluids (N_2 or CO_2) are pumped, all on-site personnel must be familiar with the dangers of these high-pressure gases. People untrained in the use of these fluids must avoid the high-pressure lines during and after the treatment until the lines are taken down and disconnected. Accidents with these fluids can occur during shutdown, when valves are closed and opened. Personnel must be trained in the proper bleeding of fluid lines. All untrained observers should leave the site during the shutdown and pressure-bleedoff operations.

13.9 Introduction to Field Computer Measurements

The person overseeing a hydraulic fracturing stimulation is concerned not only with developing an effective stimulation design, but also with its successful implementation. The effectiveness can be enhanced greatly by monitoring devices and on-site computer presentation and evaluation of information during and after the stimulation process.

The microprocessor on location currently functions in three modes: (1) data collection and presentation, with creation of permanent libraries of data that can be reanalyzed as new techniques are developed; (2) on-location process analysis, process control of blending and pumping operations, and quality control during and after hydraulic stimulation; and (3) fracture design and research modeling generally based on information discerned during the stimulation process and from on-site testing techniques.

Monitoring equipment provides individual surface and sometimes wellbore parameters to be assimilated and viewed real-time on a central display. This is the electronic window through which one can observe an overall view of the ongoing stimulation process. Monitors convert the signals from an array of devices—such as

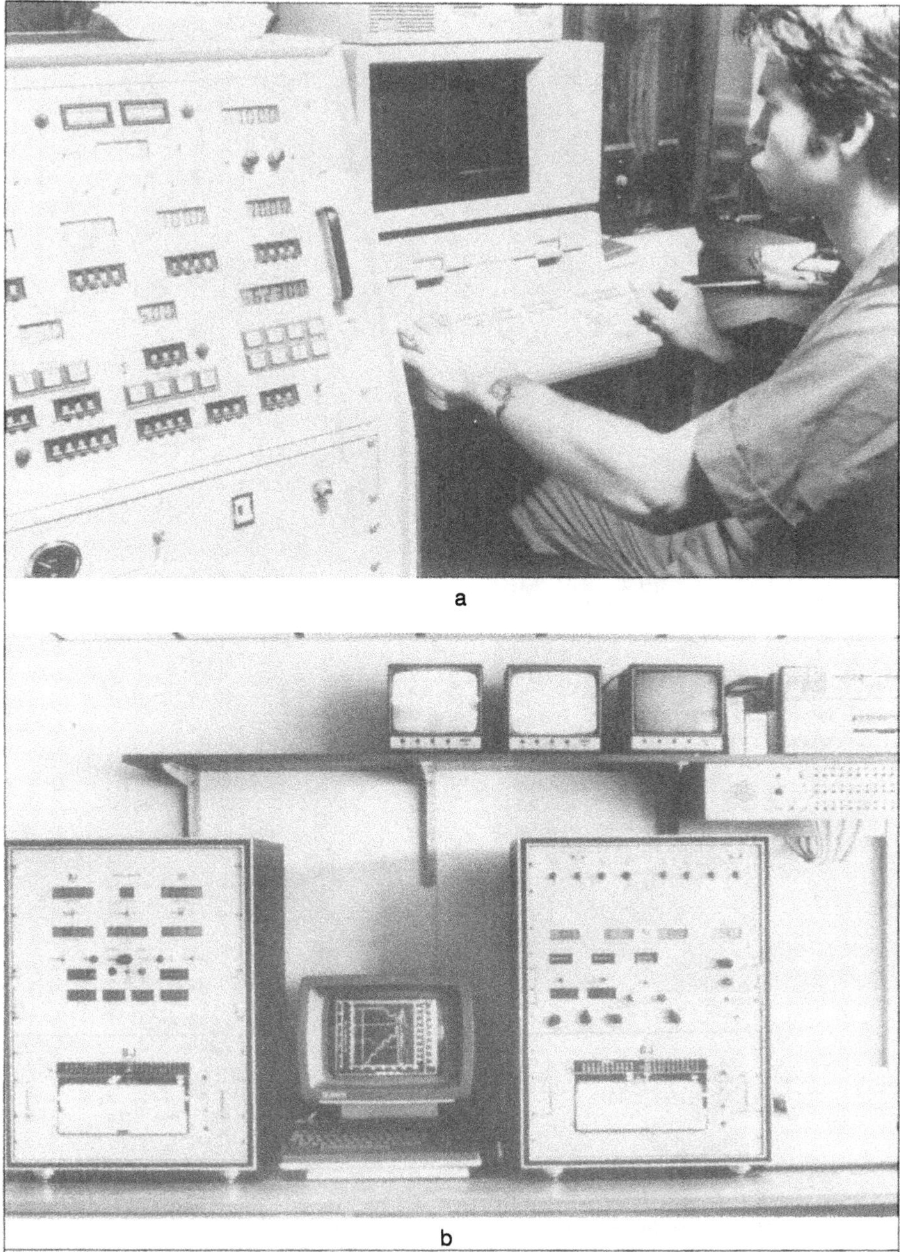

Fig. 13.17—Microprocessor-assisted fracturing monitors with internal analog-to-digital converter. An IEEE-488 interface exists in the back of the monitor allowing digital information flow between the monitor and external computers and peripherals [courtesy (a) The Western Co. of North America and (b) BJ Titan].

pressure transducers, proppant densimeters, flowmeters, and rheological apparatus—to digital formats that aid in staging the job by totaling pumping volumes and proppant, all within a predetermined pressure limit. Such instrumentation has become a necessity in accurately performing complicated proppant stimulation designs.

On-site computers and their analyses can greatly aid in evaluating whether premature job termination, before design objectives were obtained, resulted from misjudgment in design, poor execution, and/or quality control. As an example, harmful degradation of a water-based gel can occur in a 24-hour period. The viscosities in Figs. 13.13 and 13.14 were measured by an in-line viscometer and recorded by computer during tank mixing and proppant stimulation, respectively, a day later. Comparison of the 40-lbm/1,000 gal [4793-g/m^3] gel viscosity in Fig. 13.13 to that of Fig. 13.14 indicates that gel degradation had occurred. This degradation was later attributed to bacterial contamination. The 40-lbm/1,000 gal [4793-g/m^3] gel was used during the pad, the stage in which the majority of total fracture length was intended to occur. This quality-control failure was the probable cause of the premature screenout one-third of the way through the proppant stages.

13.10 Fracturing Monitors

The fracturing monitor is the central unit from which the service company treater and well operator can view pertinent information important to decision making during the stimulation process. In its most basic form, instrumentation is connected to the monitor by cables. The monitor then supplies voltage to and receives input from the sensors. The input received is converted internally to digital displays of the various parameters and totals of fluid volumes and proppant weights. Normally, strip-chart recorders are used to provide a permanent record of rates, pressures, and proppant con-

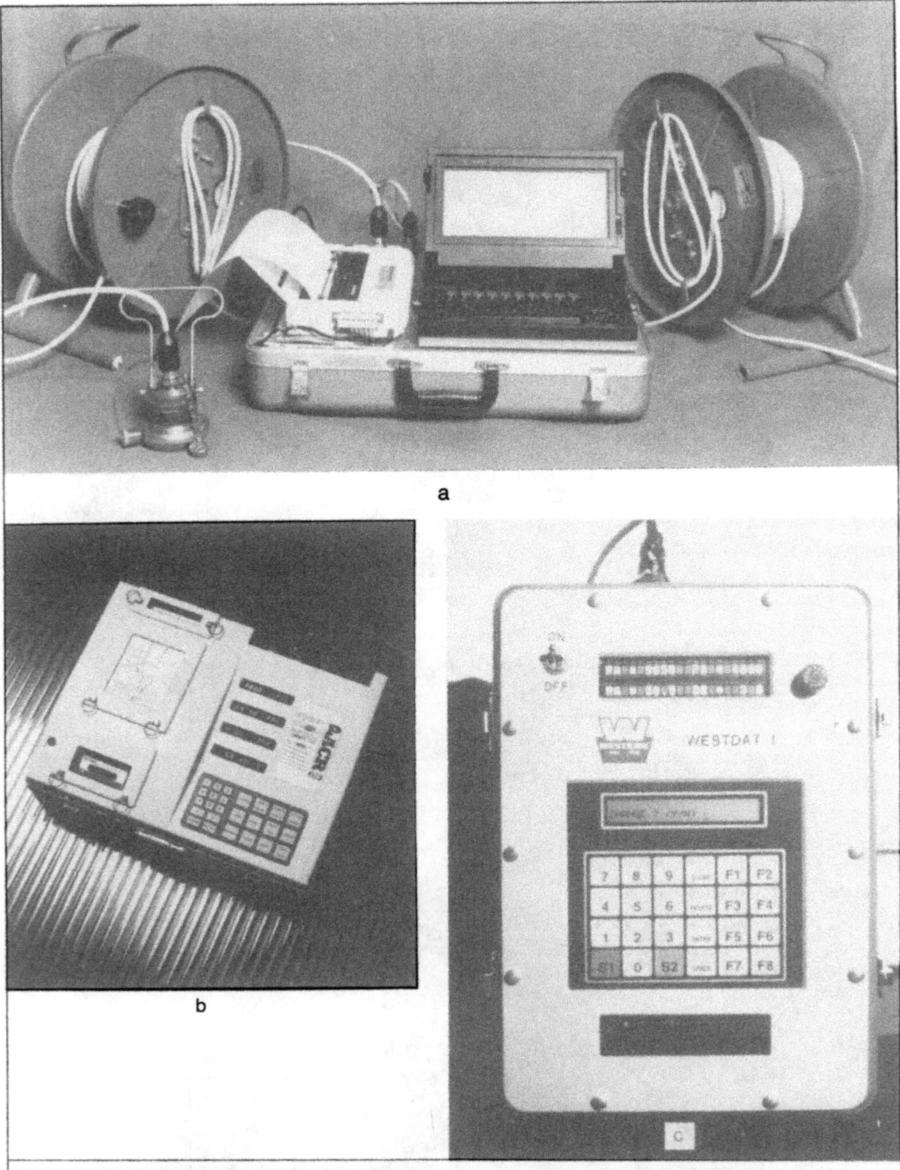

Fig. 13.18—Examples of battery-powered, portable, processor-assisted fracturing monitors and data-recording units [courtesy (a) Halliburton Services, (b) Dowell Schlumberger, and (c) The Western Co. of North America].

centrations. Such a monitoring system is shown in Fig. 13.15; a strip chart from a proppant stimulation is illustrated in Fig. 13.16.

More sophisticated monitors use internal microprocessors with programmable firmware. These systems can instantly provide many of the needed calculations and some modeling capabilities during the stimulation process, such as staging information, hydrostatic-head calculations, position of fluid fronts in the treating string, surface and downhole foam qualities, and fracturing pressure. These internal processors also convert all vital information to a digital format, allowing information to be transferred directly to and from interfaces, and to other computer systems performing various analyses. The information can also be sent to peripherals, like graphic displays, mass storage devices for permanent storage and later analysis, printers, and even satellite links. Examples of such systems are shown in Fig. 13.17.

Some process-assisted monitors are compact and battery powered and have built-in printers and mass storage systems. These features allow them to act as data loggers before, during, and after stimulation without tying up a large amount of equipment. Later, the information can be played back from the magnetic storage device for analysis by other computers. Such systems are in a gray area between being a fracturing monitor and an on-site computer monitoring and analysis system. They can be a very useful, cost-effective tool when analysis of the stimulation data in detail, on a real-time basis, is not necessary. Several of these systems are pictured in Fig. 13.18.

13.11 Computer Measurements

Historically, the fracturing monitor was the link between the treater, field engineers, and stimulation process. Decisions were based on digital readouts and strip-chart recordings of surface parameters. Since the introduction of on-site computers in the late 1970's, they have gradually become a base for engineering decision making during and immediately after the stimulation process. When CRT monitors are made available to the treater, the graphical presentation of parameters assists service company personnel in spotting problems more readily than when the information is simply displayed digitally.

The on-site computer not only enhances the quality of data gathering and presentation, but also allows rapid analysis of the

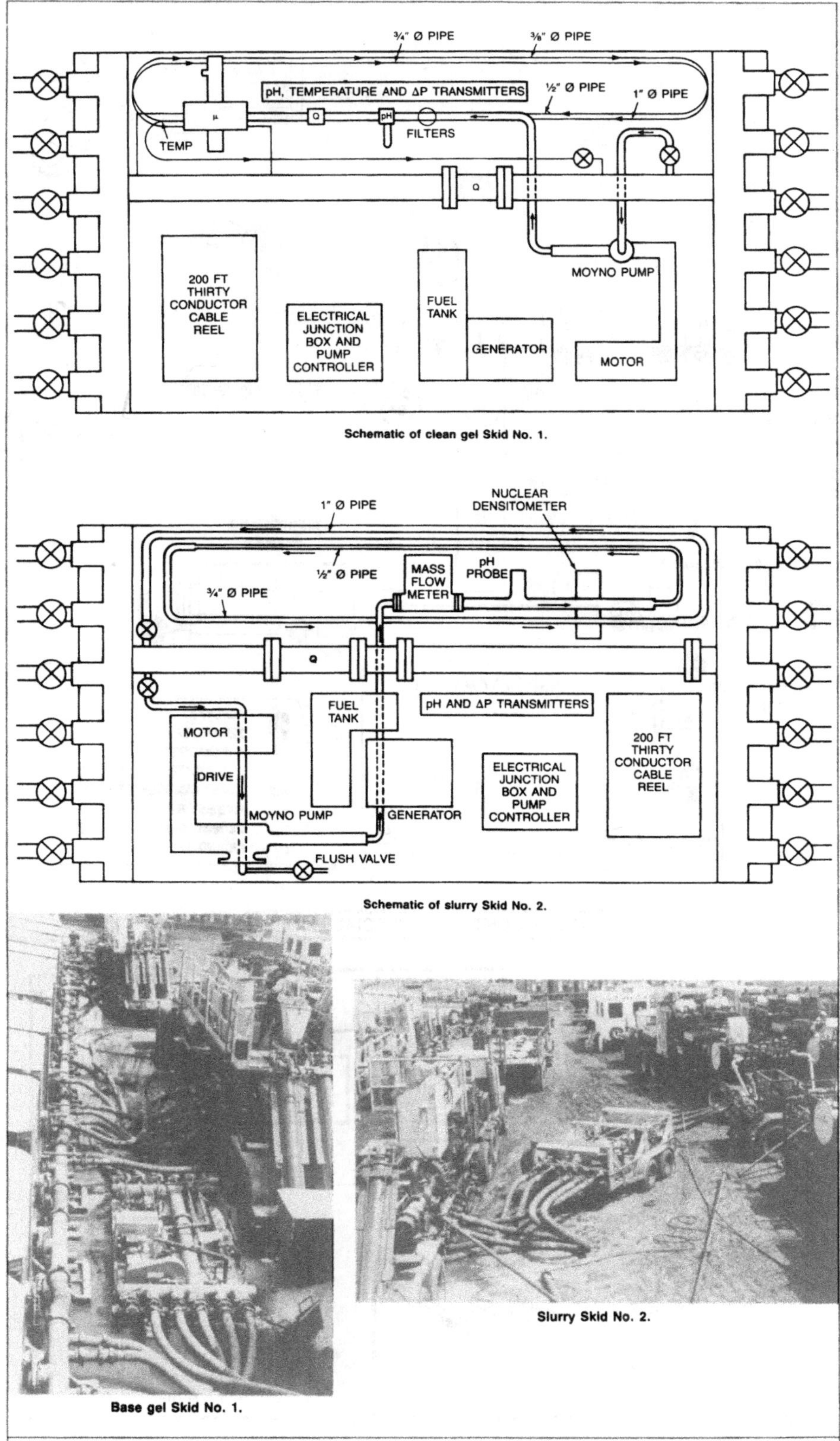

Schematic of clean gel Skid No. 1.

Schematic of slurry Skid No. 2.

Slurry Skid No. 2.

Base gel Skid No. 1.

Fig. 13.19—Skid units: Skid 1 is for gel side; Skid 2 is for slurry side.[12] (μ = viscometer; Q = turbine or magnetic flowmeter.)

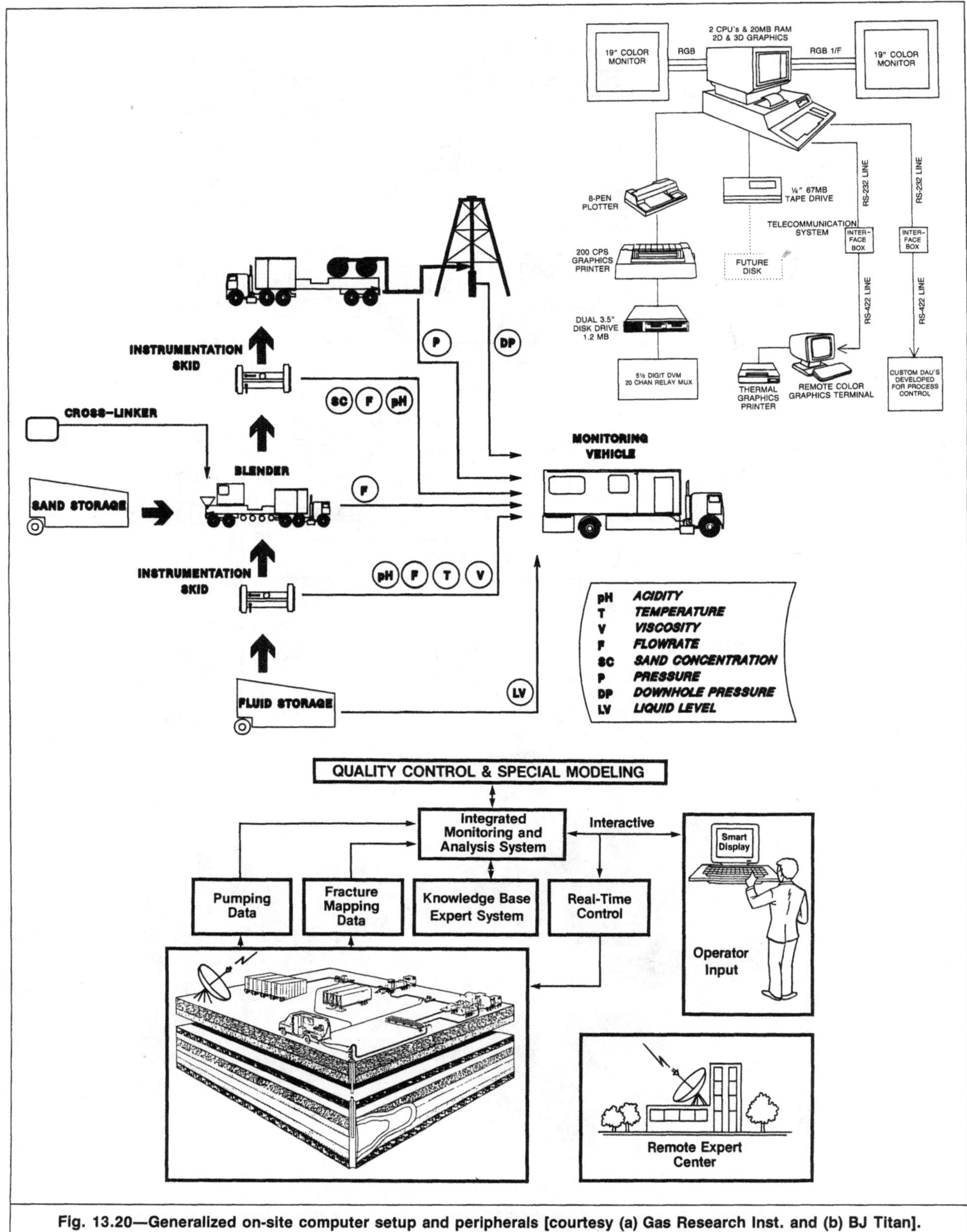

Fig. 13.20—Generalized on-site computer setup and peripherals [courtesy (a) Gas Research Inst. and (b) BJ Titan].

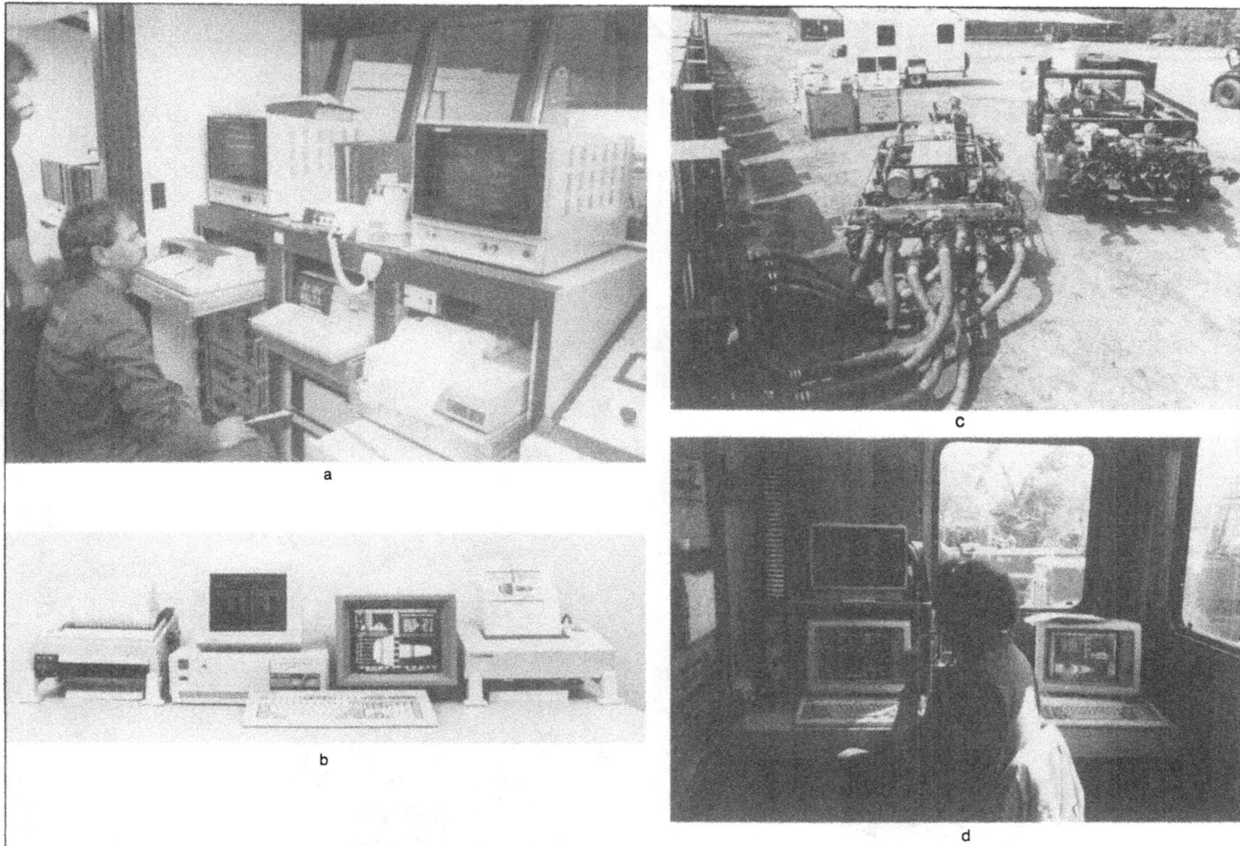

Fig. 13.21—Various on-site computer systems and monitoring van with skids [courtesy (a) Halliburton Services, (b) The Western Co. of North America, and (c) and (d) ARCO Oil and Gas Co.].

information.[8-11] This ability is generally referred to as process monitoring. A data-acquisition unit usually is used to convert selected analog voltage signals from transducers and instrumentation to a digital format that the computer can assimilate.

In some cases, the transducers are centralized on mobile skid units (Fig. 13.19). Skid units can be set up rapidly, allow independence from standard oilpatch instrumentation, and allow modification of the instrumentation for particular research needs.[12] Generally, there are two skids: one for the clean side (gel measurement) and one for the dirty side (slurry measurement). Although early skid designs were limited to low rates (not more than 30 bbl/min [0.08 m^3/s]), newer designs are capable of working at 100 bbl/min [0.26 m^3/s].

The measured and modeled parameters can then be recorded on permanent storage systems, graphically displayed, and printed. The information generally is processed at rates dictated by the type of analysis being performed. During prefracturing analysis, the pressure may be monitored in fractions of a second, whereas proppant stimulations can be monitored in fractions of a minute, although much faster collection rates can be achieved if necessary. These systems rely on software and firmware. Therefore, many different formats of analyses can be performed during stimulation, including sophisticated fracture-propagation modeling. Fig. 13.20 is a generalized view of an on-site computer setup. Various types of infield systems and configurations are pictured in Figs. 13.21 and 13.22.

Note that not all on-site computer monitoring systems require a large amount of space. On-site systems range in size from lap-top computers that can be carried in a briefcase to mobile trailer-mounted mainframes. The peripherals (large-screen monitors and printers) usually require larger work areas and in turn require a separate space or van from that of the fracturing monitor and treater. Newer-generation monitoring systems are merging the fracturing monitor and on-site computer into a single unit (Fig. 13.23).

13.12 Determination of Downhole Treating Pressure

Of particular value is the application of the on-site computer for fracturing analyses either during or immediately after stimulation. Most of these analysis techniques are determined from the downhole or wellbore treating pressure. During and after prefracturing tests (non-proppant-laden fluid tests) and proppant stimulation jobs, the real-time downhole treating pressure can be determined directly using a quartz crystal or indirectly through a static string, or modeled from surface parameters.

Some situations limit the use of each of these measurement techniques. Direct measurement of downhole treating pressure is not generally used during proppant stimulations. The sensor and wireline could become stuck in the event of a screenout. However, side-mount systems for tubing jobs can be used at extra expense to protect the sensors and wireline from such occurrences. The indirect method may not be applicable because of the critical pressures set for the tubing or casing. Neither the indirect nor computer modeling technique can be used if the formation will not support the hydrostatic head of the fluid column.

Because the downhole pressures measured by all three configurations include perforation friction pressure, steps should be taken before analysis to ensure that this friction effect is minimized.

Non-real-time downhole-pressure-bomb devices, retrieved after testing, can provide data to be analyzed later. Some of these devices have internal data loggers that record information and then allow direct data transfer to computers after retrieval. Accurate systems with fast response times and high resolutions must be used.

Of the above-mentioned techniques for determining downhole treating pressure, the least accurate is modeling from surface parameters. Reasons for this are explained in Modeling of Downhole Treating Pressure.

The primary goal of any on-site computer system is to gather information that can be applied toward improving job design and

Fig. 13.22—Various on-site computer systems [courtesy (a) BJ Titan, (b) ARCO Oil and Gas Co., and (c) The Western Co. of North America].

quality control. By using in-situ tests, such as those introduced by Nolte[13] and Nolte and Smith,[14] we can determine key formation properties before, during, and after proppant stimulation on the basis of pressure analyses. These series of tests are generally referred to as prefracturing tests and fracturing pressure slope analysis. Application of such tests has provided useful information for improving design and determining effectiveness of design execution.[8,9,11,15-19]

13.13 Prefracturing Tests

Three tests are commonly performed by computers before proppant stimulation: the step-rate injection test, pump-in flowback (in-situ-stress test), and the minifracturing pressure-decline analysis.[13,15,18] Examples of on-site computer-generated analysis for each is given in Figs. 13.24 through 13.27.

The minifracturing/pressure-decline analysis has become an important test for design information. The rate and fluid type used are the same as those intended for the pad of the main proppant stimulation. After the desired volume has been pumped, the well is shut in and the pressure decline monitored and recorded by the computer (Fig. 13.26). With the techniques developed by Nolte,[13] the data collected during the pressure-decline test can be used to derive in-situ leakoff. This technique is based on a series of calculations and incorporates curve matching performed by the computer (Fig. 13.27). Well parameters—such as height or radius, formation stresses, pump time, and fluid rheology—are then related to the curve match point and applied to the calculation of the in-situ leakoff.[13,18] Nolte provided other equations for calculation of fluid efficiency, length, average width, time until closure, and estimated pad and proppant scheduling.[13,16] A typical on-site computer-generated output sheet resulting from flowback and pressure-decline tests appears in Tables 13.4 and 13.5. A proppant-fluid scheduling design based on minifracture information appears in Fig. 13.28. (A more detailed discussion of prefracturing tests is given in Chap. 14.) As can be seen, these tests provide specific information that can be used in fracturing design.

13.14 Computerized Nolte-Smith Slope Analysis—Fracturing Pressure

During nonproppant fracturing testing or proppant stimulation, the net pressure in the fracture or fracturing pressure can be used to determine downhole fracture conditions.[14,15,17,19] The slopes of the log of fracturing pressures vs. the log of elapsed time or volume can indicate the effectiveness of leakoff control, the point of critical pressure, efficient fracture extension, unconfined height growth, and flow restrictions in the fracture during the stimulation process. By computerized monitoring with slope analysis and display of the log-log plot, the information can be used instantly to identify the downhole causes of many problems that can occur during a stimulation treatment and to improve fracture design of offset wells. The slope trends may also be useful in determining the type of geometry being created.[16,18] Examples of these fracturing-pressure slopes are presented in Figs. 13.29 and 13.30, along with their surface parameter plots as recorded by an on-site computer. A more in-depth discussion of this analysis technique appears in Chap. 14.

13.15 On-Site Modeling With Computers

Modeling of Downhole Treating Pressure. An attempt can be made to calculate downhole treating pressure from surface parameters with a computer model.[10,11] Although this technique is qualitative, a degree of accuracy can be achieved in the slope-trend information, particularly in determination of flow restriction and unconfined height growth during stimulation (Figs. 13.31 and 13.32, respectively). Because this technique does not require a downhole

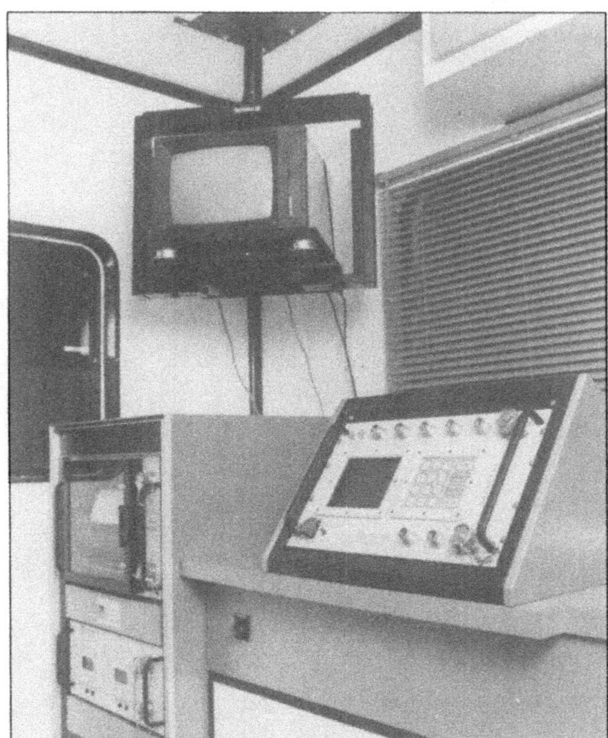

Fig. 13.23—Centralized system in which a computerized fracturing monitor with CRT display (right) and an on-site computer monitoring system (top) exist in the same work area (courtesy Dowell Schlumberger).

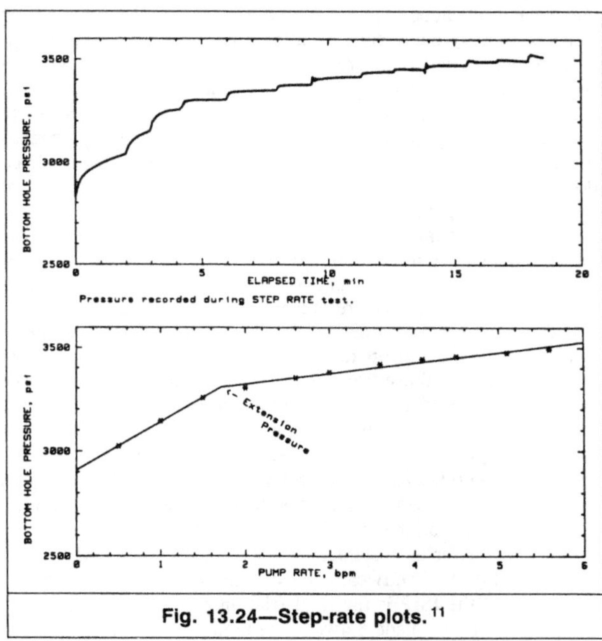

Fig. 13.24—Step-rate plots.[11]

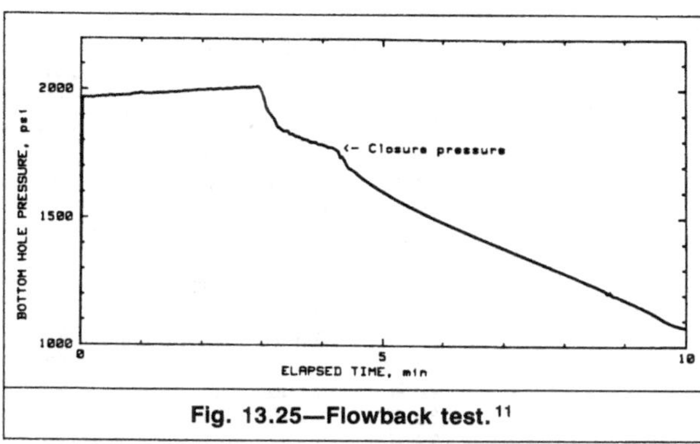

Fig. 13.25—Flowback test.[11]

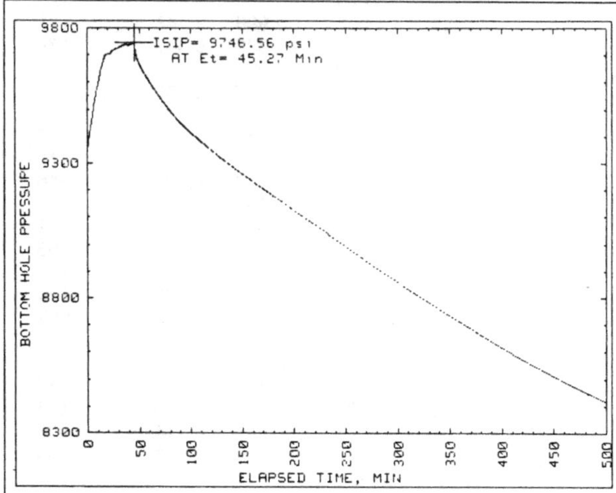

Fig. 13.26—Pressure decline following minifracturing test; direct quartz crystal measurement (courtesy BJ Titan).

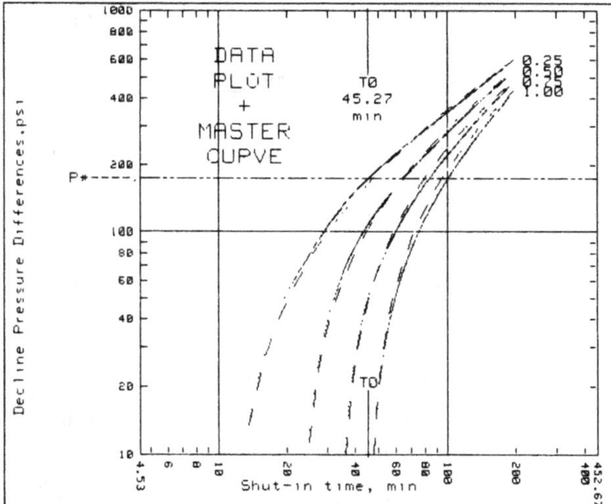

Fig. 13.27—Curve matching for Fig. 13.26 pressure data, establishing p^* for in-situ leakoff calculation (courtesy BJ Titan).

TABLE 13.4—PRESSURE DIFFERENCES, Δp, AS A FUNCTION OF DIMENSIONLESS TIME, t_D (courtesy BJ Titan)

Shut-in Time	t_D	Pressure	Δp (0.25,t_D)	Δp (0.50,t_D)	Δp(0.75,t_D)	Δp (1.0,t_D)
0.07	0.001	9,739	0	0	0	0
0.23	0.005	9,733	0	0	0	0
11.07	0.244	9,622	0	0	0	0
11.23	0.248	9,621	0	0	0	0
11.40	0.252	9,620	1	0	0	0
11.57	0.255	9,619	2	0	0	0
22.40	0.495	9,556	65	0	0	0
22.57	0.498	9,555	66	0	0	0
22.73	0.502	9,554	67	1	0	0
22.90	0.506	9,553	68	2	0	0
33.73	0.745	9,497	124	58	0	0
33.90	0.749	9,496	125	59	0	0
34.07	0.753	9,496	126	59	1	0
34.23	0.756	9,495	127	60	2	0
45.07	0.996	9,448	174	107	49	0
45.23	0.999	9,447	174	108	49	0
45.40	1.003	9,446	175	109	50	1
45.57	1.007	9,446	176	109	51	1
56.40	1.246	9,407	214	148	89	40
56.57	1.250	9,407	215	148	90	40
56.73	1.253	9,406	215	149	90	41
56.90	1.257	9,405	216	150	91	41
67.73	1.496	9,370	252	185	127	77
67.90	1.500	9,369	252	186	127	78
68.07	1.504	9,369	253	186	128	78
79.23	1.750	9,334	287	221	162	113
90.73	2.004	9,300	321	255	196	147
101.73	2.247	9,269	352	286	227	178
113.23	2.501	9,237	385	318	260	210
124.23	2.744	9,207	415	348	290	240
124.73	2.755	9,205	416	350	291	242
135.73	2.998	9,176	446	379	321	271
158.73	3.507	9,114	507	441	382	333
169.73	3.750	9,085	536	470	411	362

TABLE 13.5—MINIFRACTURE TEST ANALYSIS (courtesy BJ Titan)

Customer: demonstration	Date: May 18, 1982
Well name: 10,000-ft well	Analyzed by: SGN
Well number: in Wyoming	
Supplied Information	
Data file	7 MG #9
Formation information	
Young's modulus, psi	4.00×10^6
Poisson's ratio	0.200
Effective Young's modulus, psi	4.17×10^6
Gross fracture height, ft	100.0
Net pay height, ft	78.0
Minifracture information	
Fluid pumped, gal	10,000
Pumping time, minutes	45.3
n'	0.7500
Viscosity exponent	1.00
Information obtained from pump-in/flowback tests	
ISIP, psi	9,746.6
Closure stress, psi	9,000.0
Calculated Properties	
p^*, psi	175.50
Combined leak-off coefficient, ft/min$^{1/2}$	0.000511
Decline-pressure ratio	1.87172
Closure time (after shutdown), minutes	93.95
Fluid efficiency, %	65.1776
Fracture length (wellbore to tip), ft	276.43
Average fracture width, in.	0.18913

n' = power-law exponent; p^* = match pressure point.

pressure sensor or tubing/casing configuration without a packer, and because it can be used down tubing or casing, it is widely used as a way to derive downhole treating pressure. Of particular concern in such models is accurately accounting for fluid friction pressure with changing rates, proppant concentrations, treating string sizes, perforation friction, and fluid rheology. Because fluids can vary greatly from surface to wellbore owing to viscosity changes in base gel and crosslinking time and rate fluctuations, the modeled downhole pressure and its slope trends can be off by a significant amount. If proppant and fluid parameters are being measured correctly, however, and input from instrumentation is correct, the model can achieve a qualitative determination of downhole treating pressure. As a case in point, comparisons of modeled and measured downhole pressures are presented in Figs. 13.29 and 13.30.

Differential pressure transducers are sometimes used on site to establish a pressure drop for the clean and slurry fluid. Scaling models similar to those used to derive theoretical friction curves based on laboratory flow-loop systems are then used to relate the pressure drop to the actual treating string ID and length. Such systems are pictured in Figs. 13.19, 13.33, and 13.34. However, such systems do not as yet adequately account for crosslinking as the fluid moves down the treating string and into the fracture.

A useful check on the modeled friction pressure can be made by shutting down for an instantaneous shut-in pressure (ISIP) early in the pad (Fig. 13.35). Because the difference between the pumping pressure and the ISIP is the total friction pressure at that rate, a comparison can be made between the modeled and actual friction pressure. When making such comparisons, however, one must consider that the shut-in total friction pressure includes the effects of the fracture (wellbore-to-tip pressure drop); the modeled friction pressure does not. Furthermore, friction-pressure effects of proppant will not be accounted for because the shut-in occurs on non-proppant-laden fluid.

Modeling of Fracture Geometry. Many of the two-dimensional (2D) and planar-three-dimensional (3D) fracture-design-modeling

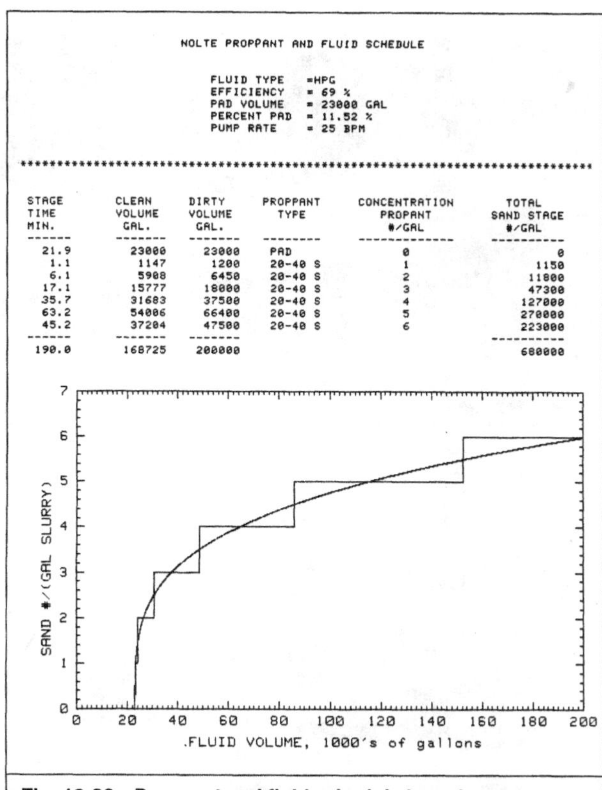

NOLTE PROPPANT AND FLUID SCHEDULE

FLUID TYPE =HPG
EFFICIENCY = 69 %
PAD VOLUME = 23000 GAL
PERCENT PAD = 11.52 %
PUMP RATE = 25 BPM

STAGE TIME MIN.	CLEAN VOLUME GAL.	DIRTY VOLUME GAL.	PROPPANT TYPE	CONCENTRATION PROPANT #/GAL	TOTAL SAND STAGE #/GAL
21.9	23000	23000	PAD	0	0
1.1	1147	1200	20-40 S	1	1150
6.1	5908	6450	20-40 S	2	11800
17.1	15777	18000	20-40 S	3	47300
35.7	31683	37500	20-40 S	4	127000
63.2	54006	66400	20-40 S	5	270000
45.2	37204	47500	20-40 S	6	223000
190.0	168725	200000			680000

Fig. 13.28—Proppant and fluid schedule based on pressure-decline fluid efficiency (courtesy BJ Titan).

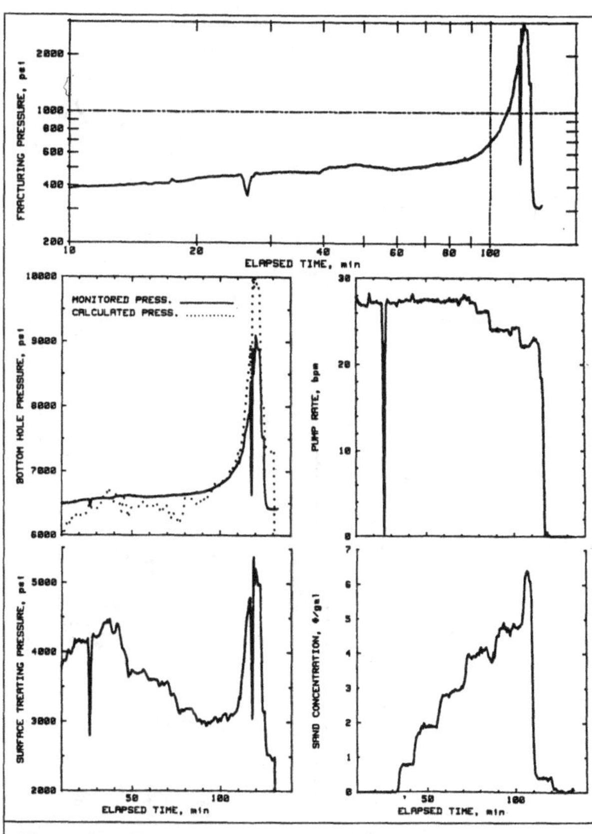

Fig. 13.29—Field results from fracturing down annulus in east Texas. Pressure monitored in static tubing.[11]

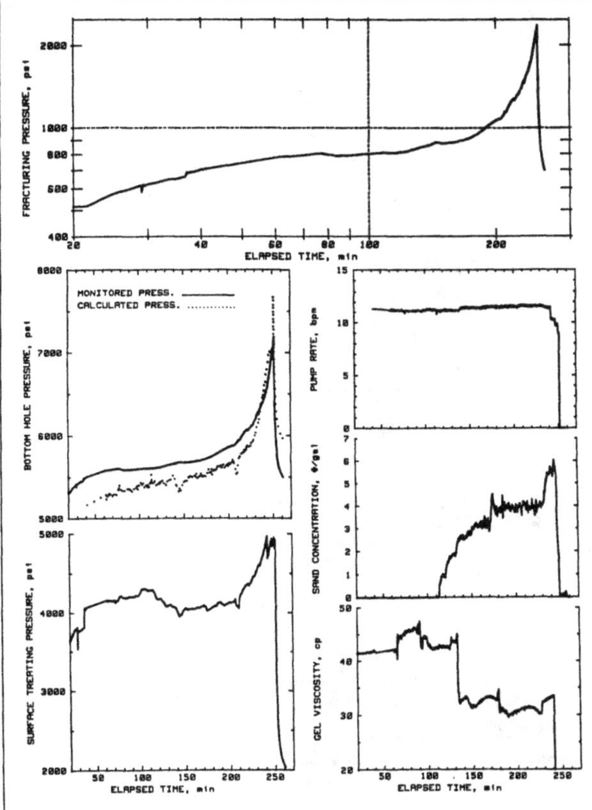

Fig. 13.30—Field results from fracturing down tubing in east Texas. Pressure monitored in static annulus.[11]

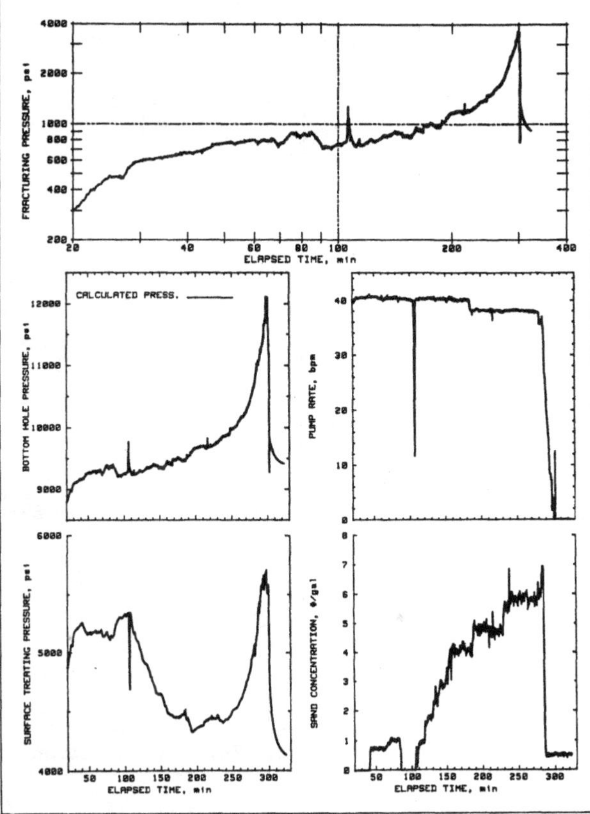

Fig. 13.31—Field results from fracturing down tubing in east Texas. BHP calculated with computer model.[11]

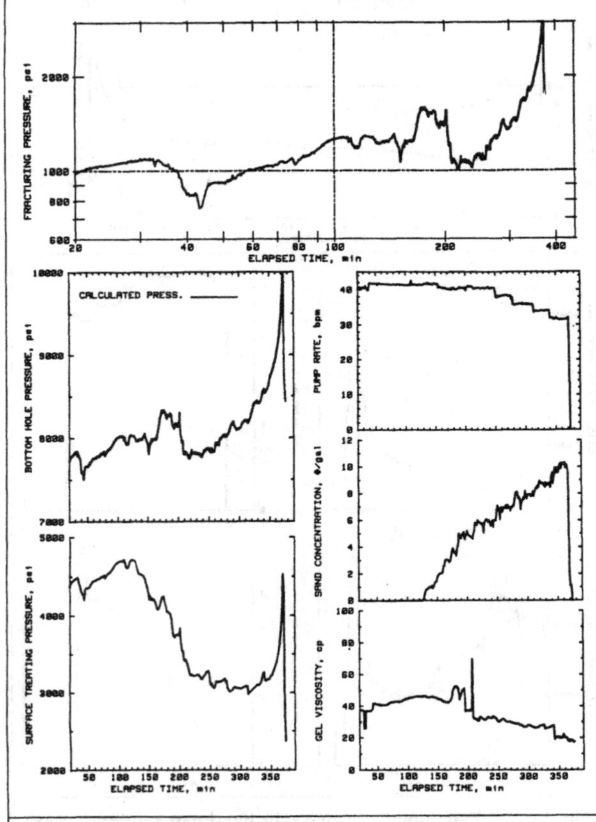

Fig. 13.32—Field results from fracturing down casing in east Texas. BHP calculated with computer model.[11]

Fig. 13.33—On-site rheological clean-side flow-loop system (courtesy Halliburton Services).

Fig. 13.34—On-site rheological clean-side flow-loop system (courtesy The Western Co. of North America).

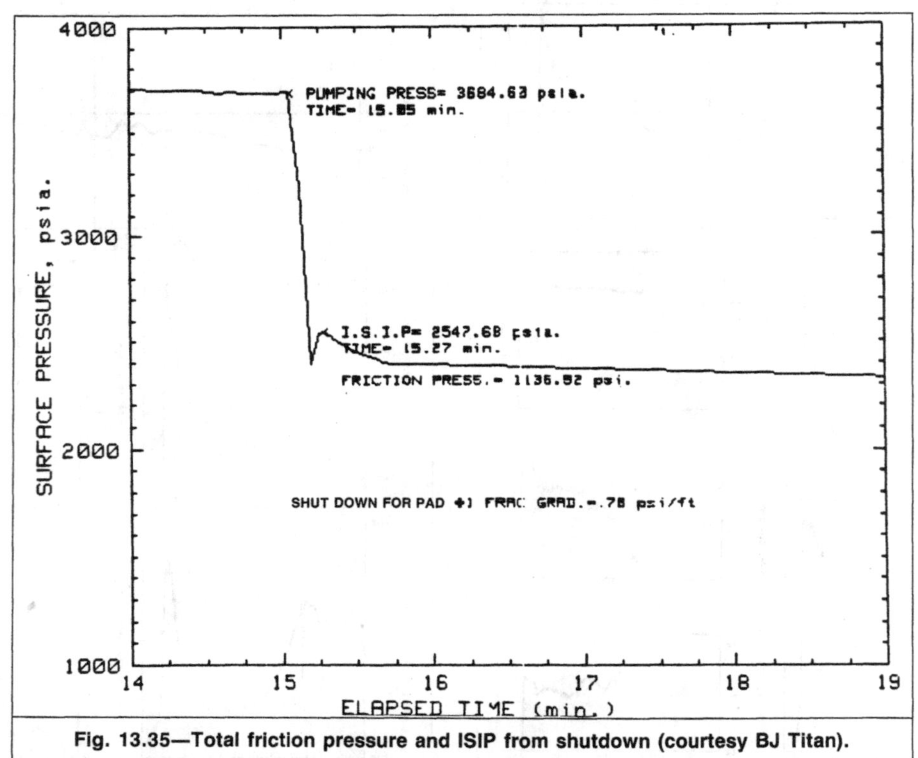

Fig. 13.35—Total friction pressure and ISIP from shutdown (courtesy BJ Titan).

TABLE 13.6—FRACTURE DESIGN OUTPUT (Courtesy BJ Titan)

Perfect Support Transport Model

Fluid	T40
Pump rate, bbl/min	22
Fluid volume, gal	70,000
Total proppant weight, lbm	225,000

Proppant Schedule

Stage	Fluid Volume (gal)	Proppant Description	Proppant Concentration (lbm/gal)
1	15,000	None	0.0
2	5,000	20/40 sand	1.0
3	5,000	20/40 sand	2.0
4	10,000	20/40 sand	3.0
5	10,000	20/40 sand	4.0
6	10,000	20/40 sand	5.0
7	15,000	20/40 sand	6.0

Proppant Distribution

Volume of Fluid Pumped (gal)	Time (minutes)	Distance From Wellbore (ft)	Proppant Concentration (lbm/gal) At Wellbore	Proppant Concentration (lbm/gal) At Distance From Wellbore	Height Slurry Top (ft)	Height Bank Top (ft)	Final Sand Concentration (lbm/ft^2)	Propped Width (in.)
5,831	6.3	1,175.3	0.0	0.0	0.0	0.0	0.00	0.00
10,000	10.8	1,126.8	0.0	0.0	0.0	0.0	0.00	0.00
15,000	16.5	1,064.1	1.0	1.9	80.0	0.0	0.52	0.06
20,000	22.4	996.3	2.0	3.8	80.0	0.0	1.04	0.12
25,000	28.5	922.6	3.0	5.6	80.0	0.0	1.52	0.17
30,000	34.6	845.5	3.0	5.3	80.0	0.0	1.43	0.16
35,000	41.0	761.3	4.0	6.8	80.0	0.0	1.84	0.21
40,000	47.4	672.6	4.0	6.3	80.0	0.0	1.71	0.19
45,000	54.0	575.1	5.0	7.5	80.0	0.0	2.02	0.23
50,000	60.7	471.8	5.0	6.9	80.0	0.0	1.87	0.21
55,000	67.5	357.9	6.0	7.7	80.0	0.0	2.08	0.23
60,000	74.4	236.7	6.0	7.1	80.0	0.0	1.91	0.21
65,000	81.3	107.7	6.0	6.5	80.0	0.0	1.76	0.20
70,000	86.7	0.0	6.0	6.0	80.0	0.0	1.62	0.20

and production-increase programs used in fracturing stimulation design have been incorporated into the on-site computer capabilities. The less CPU-intensive fracturing stimulation models that run rapidly[20-23] and planar-3D models can be used on location to generate a complete stimulation design, as seen in Tables 13.6 and 13.7. Generally, such models are run after the minifracturing pressure-decline analysis, using the information derived from on-site testing. These design models are also used during proppant-stimulation monitoring to modify proppant and fluid-volume scheduling if needed.

The introduction of true 32- and 64-bit desktop microcomputers makes even more sophisticated modeling available on location. Many of these systems have multiple CPU's and multitasking operating systems; i.e., the systems can run more than one program simultaneously. This allows a single computer to run one program dedicated to the input of surface and downhole parameters from various instruments. A second program, running at the same time, deals with general analyses, output to mass storage devices, graphic terminals, and printers. A third program, using instrumentation input, can be dedicated to generating graphical and numerical representation of geometry during treatment, based on 2D or planar-3D models.[24] Such a multiprocessor multitasking system is pictured in Fig. 13.36.

Some systems even allow a look-forward or look-back capability that can be generated in real time without disrupting the current information flow. For example, one might wish to see what effect cutting crosslinker rate or extending a proppant stage time will have on fracture geometry toward the end of the job. This can be done without disrupting the real-time displays and other modeling. Another possibility would be to recall information from previous jobs for comparison with current information. An example of such a system is shown in Fig. 13.37.

Another approach is to have one computer perform the tasks of gathering the transducer information, displaying the information and calculated downhole parameters, and then passing the needed information through an interface to another computer or computers dedicated to running a real-time 2D or planar-3D analysis (Fig. 13.38). Example graphical outputs for 2D and planar-3D models currently being run in real time are presented in Figs. 13.39 through 13.41.

Advancements in the speed of desktop computers are allowing researchers to investigate the feasibility of using fully 3D and production models on site. This is being attempted in both a design/redesign mode and in real time. These models are extremely CPU-time-intensive.[25,26] Even on mainframe systems, designs

Fig. 13.36—Parallel-processing, multitasking system for process monitoring and control. This system also has a two-dimensional real-time fracture-geometry model (courtesy BJ Titan).

TABLE 13.7—PROPPANT PROFILE STUDY—PERFECT SUPPORT FLUIDS (Courtesy The Western Co. of North America)

Operator: SPE 1987	Representative:—
Field: —	Formation: —
County: Tarrant	State: Texas
Well: —	Date: Dec. 1, 1988

Fluid studied: Saturn 50H
Total volume: 188,971 gal
Fluid penetration: 2,828 ft

Permeability to stimulation fluid, md	4.20	Fracturing pressure, psi	16,900
Permeability to reservoir fluid, md	7.00	Reservoir-pressure, psi	13,000
Leakoff-fluid viscosity, cp	1.00	n'	0.6600
Reservoir-fluid viscosity, cp	0.02	K', (lbf-sec)$^{n'}$/ft^2	0.0770
Reservoir fluid compressibility, psi^{-1}	7.20×10^{-5}	Young's modulus, psi	5.00×10^6
		Width, in.	0.624
Stimulation fluid C-III, ft/min$^{1/2}$	4.00×10^{-4}	Combined fluid-loss coefficient, ft/min$^{1/2}$	6.22×10^{-4}
Total fracture height, ft	50		
Net fracture height, ft	40	Porosity, %	12.0
Injection rate, bbl/min	10	Fluid spurt loss, cm^3	0.00

Fluid Volume (gal)	Proppant Type	Surface Proppant Concentration (lbm/gal)	Location in Fracture (ft)	Fracture Proppant Concentration (lbm/ft^2)	Cumulative Proppant (lbm)
35,000	—	0.00	2,498 to 2,828	0.000	0
35,000	—	0.00	2,106 to 2,498	0.000	0
20,000	Bauxite	0.50	1,845 to 2,106	0.382	10,000
20,000	Bauxite	0.50	1,552 to 1,845	0.341	20,000
10,000	Sand	1.00	1,387 to 1,552	0.607	30,000
10,000	Sand	1.00	1,212 to 1,387	0.570	40,000
5,000	WPROP1	2.00	1,117 to 1,212	1.056	50,000
5,000	WPROP1	2.00	1,019 to 1,117	1.021	60,000
5,000	WPROP2	3.00	913 to 1,019	1.423	75,000
5,000	WPROP2	3.00	804 to 913	1.372	90,000
5,000	WPROP3	3.50	686 to 804	1.477	107,500
5,000	WPROP3	3.50	562 to 686	1.419	125,000
5,000	WPROP4	4.00	431 to 562	1.526	145,000
5,000	WPROP4	4.00	295 to 431	1.463	165,000
5,000	WPROPLT	4.50	151 to 295	1.563	187,500
5,000	WPROPLT	4.50	0 to 151	1.495	210,000

Total fracturing-fluid volume, gal 180,000

K' = flow-consistency index.

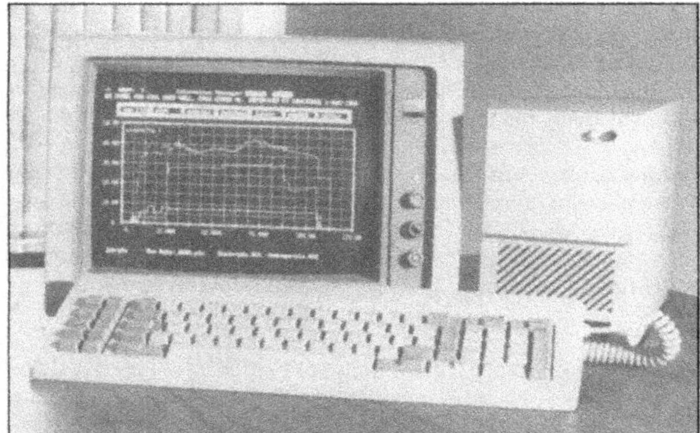

Fig. 13.37—PC-based system with add-on CPU board. This system allows real-time planar-3D modeling with history matching and look-forward/look-back capabilities (courtesy Gas Research Inst.).

Fig. 13.38—Two interfaced computers: one (right) for process monitoring and providing information to the computer on the left, which is for 2D or planar-3D fracture-geometry modeling in real time (courtesy ARCO Oil and Gas Co.)

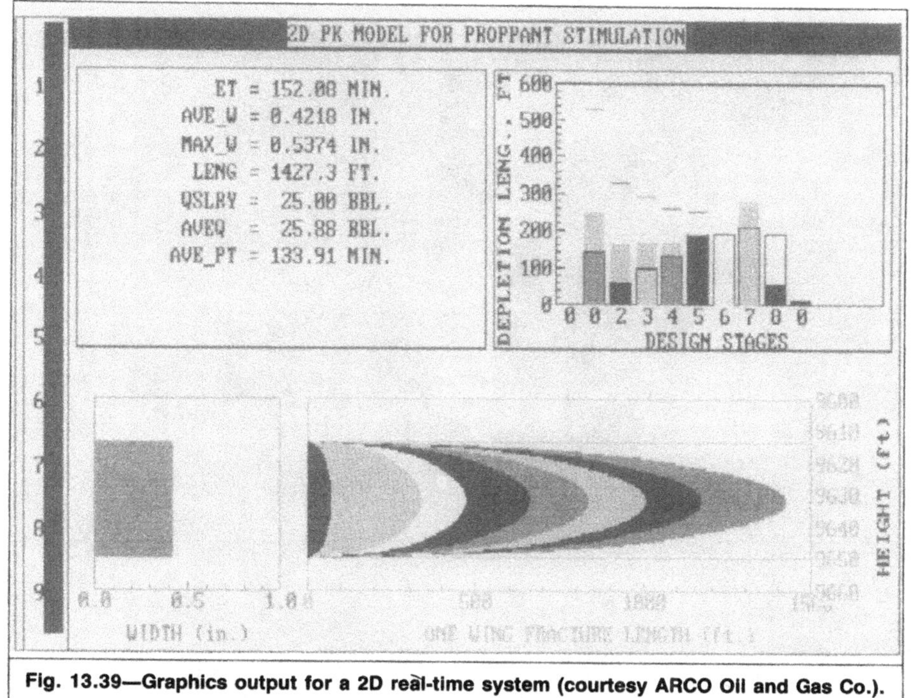

Fig. 13.39—Graphics output for a 2D real-time system (courtesy ARCO Oil and Gas Co.).

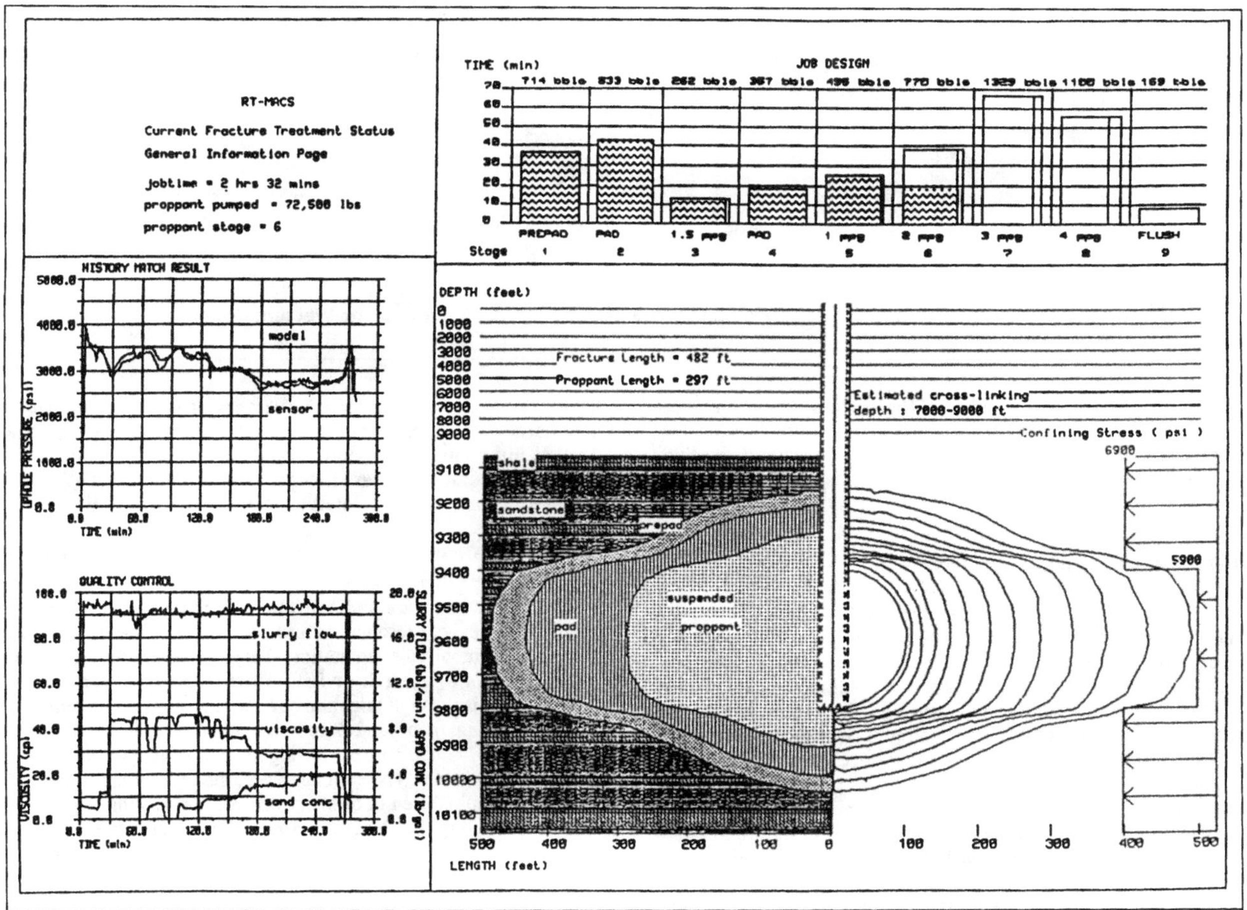

Fig. 13.40—Planar-3D real-time model with history matching (courtesy Gas Research Inst.).

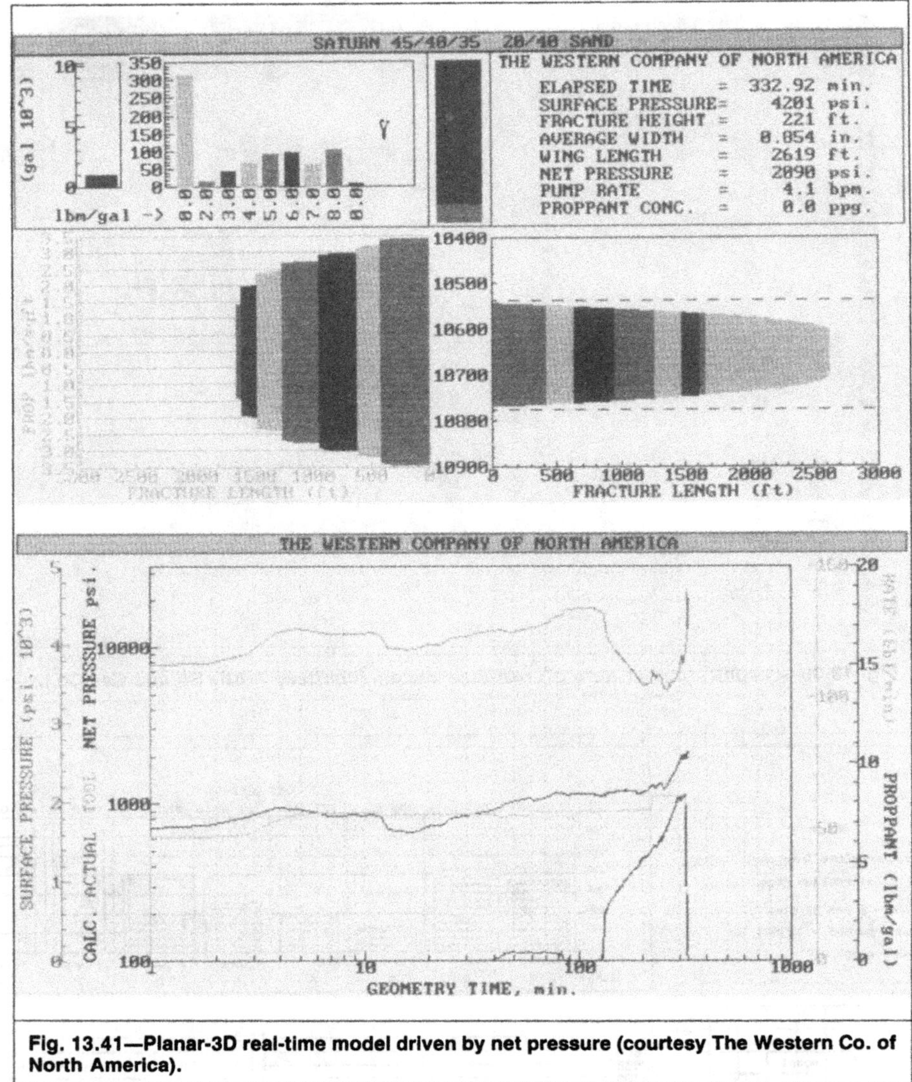

Fig. 13.41—Planar-3D real-time model driven by net pressure (courtesy The Western Co. of North America).

may take hours to run to completion. Researchers are currently seeking an acceptable balance between model complexity, iteration techniques, and available hardware. The hardware necessary to achieve acceptable computing speeds for such models is discussed in Sec. 13.18.

Data Bases. The on-site computers generate a tremendous amount of information. Some systems have taken advantage of this information flow by incorporating interactive data bases into the real-time system, which allows previous job information to be accessed in any combination of ways during and after treatment. Some data bases even have intelligent assistants that allow the user to relate previous job histories to the current job. For example, one could ask the data base to retrieve, from all jobs present in the data base, pressure information for those jobs that screened out in a particular formation during the 6-lbm/gal [720-kg/m^3] proppant stage. Such data bases allow operators to accumulate and centralize years of information on stimulation design, results of design, and quality control. Data bases generally include sections on production history that are kept current on office computers. An example short form output from an onsite computer is given in Table 13.8. This output is generated immediately after the stimulation and is given to the operator for his or her reports. The data base from which this short form is generated is filled in by the computer during monitoring and is actually 10 pages long with information about pressures, rates, proppant, totals, and end-of-job averages of specific parameters. Furthermore, job performance and quality-control information, as well as information on fracture design and modeling, is included. The last part of the data base is concerned with poststimulation production history and is updated throughout the life of the well.

13.16 Remote Viewing With On-Site Computer Satellite Links

With current hardware and software technology, it is possible to transmit the exact presentation and evaluation of the on-site computer to a remote viewing station. Such a station could be set up in an operator's research or field office. This has been done on a limited basis through telephone or satellite links. A satellite link system for support of land-based on-site monitoring is pictured in Fig. 13.42. Although remote viewing in land-based fracturing stimulation has not yet become commonplace, the technology is available and is used quite often in offshore drilling, cementing, and stimulation operations. Engineers in areas like Canada and the North Slope will appreciate the introduction of such systems for fracturing stimulation.

13.17 Process Control With Computers

Until now, only field process monitoring and modeling have been discussed. Another use of microprocessors is in process control of blending and pumping operations during hydraulic fracturing stimu-

Fig. 13.42—Land-based mobile satellite communication system for support of on-site monitoring of fracturing stimulation (courtesy Halliburton Services).

lation. This may involve the use of small dedicated processors or preprogrammed microchips. Such systems are commonly referred to as "black boxes" and are analogous to the processors currently used to control engine processes in modern automobiles. A single processor may be dedicated to controlling tub fluid levels of blenders and rate of proppant addition; another holds pumps at a constant preselected rate. Examples of microprocessor-controlled blenders and their proppant ramping capabilities are shown in Fig. 13.43.

Although such control systems are just being introduced to land-based stimulation, more sophisticated systems are in current use on offshore stimulation boats (North Sea) that have combined the benefits of process monitoring and control. Various computer systems are used in the stimulation control center, depending on the fracturing boat. Some use several linked, small mainframe computers, each dedicated to a specific task, while others use a single mainframe with multitasking capabilities, with each program dedicated to a specific function. In either case, the computer or computers perform the functions of process monitoring (job monitoring) and process control (job execution). One system will process real-time data and present the information. The second and sometimes third system(s), dedicated to the execution of the treatment, may in turn be the overseer and controller of other smaller processors controlling the blending and pumping units.

Most process control systems will operate in three modes: (1) automatic, where the computer controls all the stimulation processes (blending and pumping) from a stimulation design prerecorded on a storage medium and/or entered by an operator during job execution; (2) semiautomatic, where selected parameters are manually controlled while others are computer controlled; and (3) manual, where all functions are manually controlled and the computer(s) are left in the process-monitoring mode. Any of these modes can be activated during job execution. A fracturing-stimulation ship and computer control and monitoring system is pictured in Fig. 13.44; examples of land-based process control systems are pictured in Figs. 13.45 and 13.46.

a

b

Fig. 13.43—Process-controlled blender with proppant ramping capabilities. Process controller box appears in lower left corner. An actual ramping plot generated by a process control computer ramping a blender shows the relationship between target concentrations (asterisks) and actual [courtesy (a) Dowell Schlumberger and (b) The Western Co. of North America].

13.18 The Microprocessor Revolution

The stand-alone desktop microprocessor has revolutionized on-site data collection, analysis, and modeling. Because of their compactness, they are easily transported to location, bringing with them the computing power that was, until recently, available only with stationary large mainframe systems. A rapid desktop evolution is occurring in the MS- or PC-DOS-compatible machines. These computers are making quantum leaps in CPU clock speeds, add-on boards for creating a multiple-CPU environment, and graphics-support technology on almost a monthly basis.

Some of the recent hardware and software developments allowing researchers to investigate CPU-time-intensive models, such as those

Fig. 13.44—(a) Fracturing stimulation ship with (b) computerized process monitoring control center (courtesy BJ Titan).

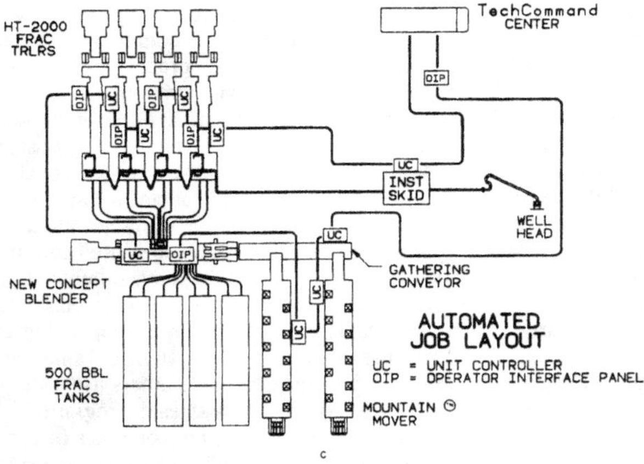

Fig. 13.45—Process monitoring and control center for land-based fracturing stimulation. (a) The CRT on the left is for monitoring and process control of blender operations, the middle for process monitoring, and the right for monitoring and process control of pressure pump controllers. (b) The computer on the left is for design and redesign of the job using fracturing-geometry design models (courtesy Halliburton Services).

required to generate a fully 3D fracture simulation on a desktop level, are as follows.

1. Parallel-processor capabilities, where two or more CPU's work together to achieve a common goal or solution.

2. Mathematics accelerator boards, which are specialized separate CPU's for performing floating-point mathematics and many other functions, freeing the main CPU('s) to perform other tasks simultaneously.

3. Graphics boards, with separate CPU's for creating graphics and even allowing 3D rotation of screen images without slowing down the main CPU('s).[27]

4. Reduced-instruction-set processors and Array processors, which allow desktops to achieve speeds to within 1/24 of a Cray X-MP-2 (one processor) computer.

5. Multitasking operating systems, which allow the programmer to have multiple applications running simultaneously and share results through a common memory.

6. Large random-access memory (RAM) boards, allowing data transfer to occur at RAM speed.

Mainframe terminology is now applicable for desktop systems. The CPU speeds are presented in millions of instructions per second (MIP's) and RAM storage is in megabytes. As can be seen in Table 13.9,[28*] the 80386, 68000 series chips, and add-on boards for PC-compatible processors are approaching the number-crunching capability of many popular microworkstations and mainframes. A key point, however, is the language speed. New BASIC, FORTRAN, and C COMPILERS, a few of the languages used, support the above chips and add-on boards. They alone are resulting in faster execution times because of management techniques incorporated into the compilers and the ability to generate 32-bit code.

*Personal correspondence with G. White and J. Schmit, Arco Oil & Gas Co., Plano, TX (1987).

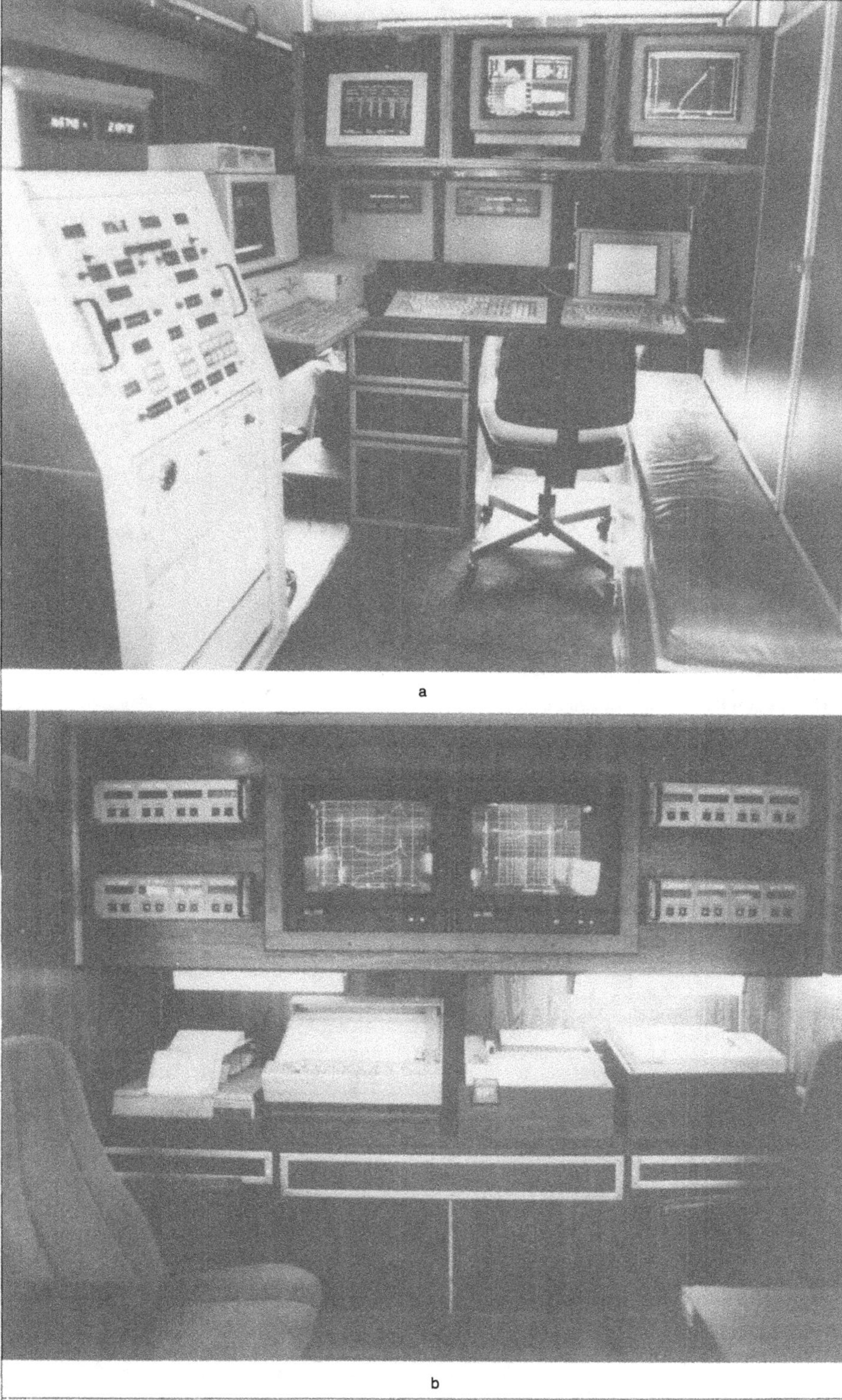

a

b

Fig. 13.46—Process monitoring and control center. (a) The upper level right monitor shows computer-controlled proppant ramping, the middle monitor displays real-time planar-3D model, and the left monitor displays planar-3D model information, job totals, and pressure and rate averages. The CRT in the lower left is another computer for process monitoring and drives the monitors, printers, and plotters shown in (b) (courtesy The Western Co. of North America).

TABLE 13.8—STIMULATION DATA BASE XIII (SHORT FORM, courtesy ARCO Oil and Gas Co.)

A proppant on Sept. 3, 1987, pumped by Well Service using All American Frac Gel and 20/40 mesh. The operator was John Doe.

Well-Completion Information

	Casing	Tubing
Size, in.	5.5	
Weight, lbm	80	
Type	N-80	
Measured depth, ft	8,950	

Treatment Information

		Design	Computer	Service Company
Gel total, bbl		14,242	14,129.2	14,265.2
Fluid total, bbl		14,632.0	14,560.3	14,625.1
Slurry total, bbl		15,893.4	15,903.4	16,097.3
Proppant total, lbm		1,168,500	1,243,747	1,259,321
Final proppant concentration, lbm/gal		7.00	7.93	8.1
Average job rate, bbl/min				53.90
Average job pressure, psi				4,245.88
Average job proppant concentration, lbm/gal				2.27
Net pay height, ft		97		
Pad, %		28.8		
Perforated interval, ft				8,497 to 8,724
Holes shot		71		
Holes open		47		
ISIP's, psi				
Breakdown	3,240			
Pad	3,176			
End of job	3,994			
Average fracture gradient, psi/ft				0.74
Perforation friction at 60 bbl/min, psi				193
Breakdown comments				good ball action
ISIP comments				none
Start of screenout				bbl minutes
Premature shutdown comments				none
Stages cut short				none
Stage problem comments				none
Mechanical problems comments				downhole pump problems in maintaining consistent rate
Fluid problem comments				Tanks 2 and 3 viscosities were 10 cp low; see plots during 8-lbm stage
Communication problem comments				none
Tracer information for job				0.3 mCi/1,000 lbm 6-lbm stage; 0.4 mCi thereafter
Tracer company used				Trace Services
Measured gross fracture height, ft				
Logging company				
Fracturing treatment overall comments				prefracture bacteria containment problems caused delay in gelling of fluids; uncertain hydration of fluids the night before the job resulted in uncertain gel viscosities because of new LGC diesel-based system

Postfracture production

Initial rate

Gas, Mscf/D

Oil, B/D

Water, B/D

	30 Days	6 Months
	Rate(Mscf/D, B/D)/Cumulative(MMscf/bbl)	Rate(Mscf/D, B/D)/Cumulative(MMscf/bbl)
Gas		
Oil		
Water		

TABLE 13.9—RESULTS FOR THE DOUBLE-PRECISION LINPACK[27] BENCHMARK ON MAINFRAMES, MICROWORKSTATIONS, APPLE, AND MS/DOS-COMPATIBLE COMPUTERS (courtesy ARCO Oil and Gas Co.)

Machine	Hardware	Software	MFlops
	Mainframes		
IBM 3090	IBM	VS-FORTRAN	
		Level 3	6.4
		Level 2	6.4
		Level 1	3.17
		Level 0	2.26
IBM 4381	IBM	VS-FORTRAN	
		Level 3	0.95
VAX 11/785	DEC	VAX FORTRAN	0.21
	Microworkstations		
SUN SPARC*,**	RISC CPU Weitek FPA	Sun FORTRAN	1.6
SUN 3/75*	16.6-MHz 68020 Weitek FPA	Sun FORTRAN	0.40
uVAXII	DEC	VAX FORTRAN	0.13
Mac IIx	16-MHz 68031/68882	Lang Systems	0.088
		Max IIx	0.077
Mac II	16-MHz 68020 68881	ABSOFT FORTRAN	0.05
	PC Compatible		
Everex Step	33-MHz 80386	Microsoft FORTRAN	0.71
Compaq Deskpro	25-MHz 80386 80387-25 FPA	Microsoft FORTRAN	0.26
Definicon-780**	16.6-MHz 68020 68881	SVS FORTRAN	0.092

*As reported by Sun; all other values are test results.
**Available for MSDOS PC's as an add-on coprocessor.

Even the size of the computers themselves has been reduced drastically. PC laptop systems have as powerful a number-crunching capability as the larger PC desktops. Some even have add-on slots that allow the addition of large memory and other CPU boards, yet the weight is below 15 lbm [6.8 kg] and the dimensions allow the computer to be placed in a briefcase or easily carried in a car or airplane (Fig. 13.47).

The use of computers on site by the fracturing stimulation industry is not limited by hardware or even software capabilities, but mainly by the number of analysis techniques currently available. As new instrumentation and analyses are introduced, the on-site computer is available to play a vital role in allowing researchers to evaluate, modify, and incorporate new techniques into the daily activities of hydraulic fracturing stimulation.

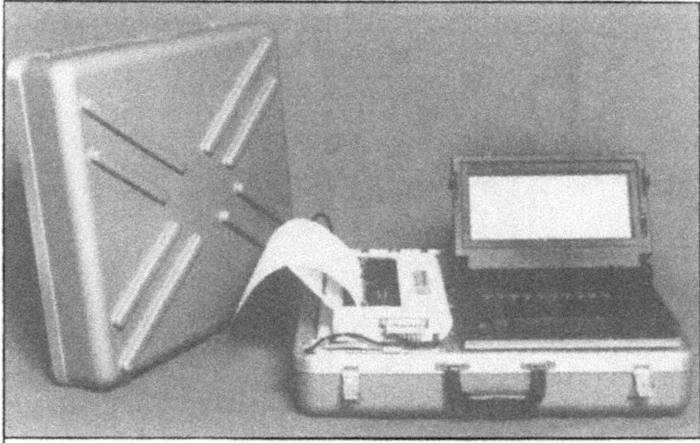

Fig. 13.47—Battery-powered laptop monitoring system and printer with carrying case (courtesy Halliburton Services).

Acknowledgments

We express our appreciation to the managements of Conoco Inc. and Guydon Software Services for permission to publish this chapter and to personnel of these companies for assistance in the preparation of this manuscript, especially Cecil D. Parker, Theresa Burton Yocom, Steve Ferda, and Woody Bernard of Conoco Inc., and Steve Nelson of Guydon Software Services.

References

1. Parker, C.D.: "Logistical and Operational Considerations for Massive Hydraulic Fracturing," *JPT* (July 1981) 1189–95.
2. Ely, J.W.: *Stimulation Treatment Handbook,* PennWell Publishing Co., Tulsa, OK (1985).
3. Osborne, M.W., McLeod, H.O., and Schroeder, H.D.: "The Analysis and Control of Hydraulic Fracturing Problems," paper SPE 9868 presented at the 1981 SPE/DOE Low Permeability Gas Reservoirs Symposium, Denver, May 27–29.
4. Robinson, J.C.: "Practical On-Site Inspection and Testing Improves Fracture Treatment Quality," paper SPE 14435 presented at the 1985 SPE Annual Technical Conference and Exhibition, Las Vegas, Sept. 22–25.
5. Burton, T.L. and Ferda, S.R.: "Fracture Stimulation Quality Control Program," *Proc.*, 33rd Annual Southwestern Petroleum Short Course, Lubbock, TX (1986) 69–79.
6. Bernard, W.: "Checklist—Ballout and Frac Jobs," Conoco Inc., Houston (1984).
7. Howard, G.C. and Fast, C.R.: *Hydraulic Fracturing,* Monograph Series, SPE, Richardson, TX (1970) **2,** 125.
8. Erdle, J.C., Bell, J., and Bezier, C.: "Results of Hydraulic Fracturing Treatment BHP Analysis in Peru," paper SPE 10310 presented at the 1981 SPE Annual Technical Conference and Exhibition, San Antonio, Oct. 4–7.
9. Elbel, J.L. *et al.*: "Stimulation Study of Cottage Grove Formation," *JPT* (July 1984) 1199–1205.
10. Hannah, R.R., Harrington, L.J., and Lance, L.C.: "Real-Time Calculation of Accurate Bottomhole Fracturing Pressure from Surface Measurements Using Measured Pressures as a Base," paper SPE 12062 presented at the 1983 SPE Annual Technical Conference and Exhibition, San Francisco, Oct. 5–8.
11. Cooper, G.D., Nelson, S.G., and Schopper, M.D.: "Improving Fracturing Design Through the Use of an On-Site Computer System," paper SPE 12063 presented at the 1983 SPE Annual Technical Conference and Exhibition, San Francisco, Oct. 5–8.
12. Pearson, C.M.: "Development and Application of an Operator's Stimulation Monitoring System," paper SPE 16903 presented at the 1987 SPE Annual Technical Conference and Exhibition, Dallas, Sept. 27–30.
13. Nolte, K.G.: "Determination of Fracture Parameters from Fracturing Pressure Decline," paper SPE 8341 presented at the 1979 SPE Annual Technical Conference and Exhibition, Las Vegas, Sept. 23–26.

14. Nolte, K.G. and Smith, M.B.: "Interpretation of Fracturing Pressures," *JPT* (Sept. 1981) 1767–75.
15. Nolte, K.G.: "Principles for Fracture Design Based on Pressure Analysis," *SPEPE* (Feb. 1988) 22–30.
16. Nolte, K.G.: "Determination of Proppant and Fluid Schedules from Fracturing-Pressure Decline," *SPEPE* (July 1986) 255–65; *Trans.*, AIME, **283**.
17. Nolte, K.G.: "Application of Fracture Design Based on Pressure Analysis," *SPEPE* (Feb. 1988) 31–42.
18. Nolte, K.G.: "A General Analysis of Fracturing Pressure Decline With Application to Three Models," *SPEFE* (Dec. 1986) 571–83; *Trans.*, AIME, **284**.
19. Smith, M.B., Miller, W.K. II, and Haga, J.: "Tip Screenout Fracturing: A Technique for Soft, Unstable Formations," *SPEPE* (May 1987) 95–103; *Trans.*, AIME, **283**.
20. Perkins, T.K. and Kern, L.R.: "Widths of Hydraulic Fractures," *JPT* (Sept. 1961) 937–49; *Trans.*, AIME, **222**.
21. Nordgren, R.P.: "Propagation of a Vertical Hydraulic Fracture," *SPEJ* (Aug. 1972) 306–14; *Trans.*, AIME, **253**.
22. Khristianovich, S.A. and Zheltov, Y.P.: "Formation of Vertical Fractures by Means of Highly Viscous Liquid," *Proc.*, Fourth World Pet. Cong., Rome (1955).
23. Geertsma, J. and de Klerk, F.: "A Rapid Method of Predicting Width and Extent of Hydraulically Induced Fractures," *JPT* (Dec. 1969) 1571–81; *Trans.*, AIME, **246**.
24. Cleary, M.P., Barr, D.T., and Willis, R.M.: "Enhancement of Real-Time Hydraulic Fracturing Models With Full 3D Simulation," paper SPE 17713 presented at the 1988 SPE Gas Technology Symposium, Dallas, June 13–15.
25. Clifton, R.J. and Abou-Sayed, A.S.: "On the Computation of the Three-Dimensional Geometry of Hydraulic Fractures," paper SPE 7943 presented at the 1979 SPE Symposium on Low-Permeability Gas Reservoirs, Denver, May 20–22.
26. Settari, A. and Cleary, M.P.: "Three-Dimensional Stimulation of Hydraulic Fracturing," *JPT* (July 1984) 1177–90.
27. Boone, J.J., Ingraffea, A.R., and Roegiers, J.-C.: "Visualization of Hydraulically Driven Fracture Propagation in Poroelastic Media Using a Superworkstation," *JPT* (June 1989) 574–80.
28. Dongarra, J.J. *et al.*: *LINPACK Users' Guide*, SIAM Publications, Philadelphia, PA (1979).

SI Metric Conversion Factors

bbl	×	1.589 873	E−01	=	m^3
cp	×	1.0*	E−03	=	Pa·s
cycles/sec	×	1.0*	E+00	=	Hz
ft	×	3.048*	E−01	=	m
ft^2	×	9.290 304*	E−02	=	m^2
ft^3	×	2.831 685	E−02	=	m^3
gal	×	3.785 412	E−03	=	m^3
in.	×	2.54*	E+00	=	cm
(lbf-sec)/ft^2	×	4.788 026	E+01	=	(N·s)/m^2
lbm	×	4.535 924	E−01	=	kg
psi	×	6.894 757	E+00	=	kPa
psi^{-1}	×	1.450 377	E−01	=	kPa^{-1}

*Conversion factor is exact.

Chapter 14
Fracturing-Pressure Analysis

K.G. Nolte, SPE, Dowell Schlumberger Inc.

14.1 Overview

The change in fluid pressure during and after a fracture treatment can be used to infer the manner in which the fracture is propagating and design parameters for subsequent treatments. The theoretical framework and example applications of fracturing pressure analysis are outlined in this chapter.

Fracturing-pressure analysis is similar to the pressure-transient analysis applied in reservoir engineering with respect to both the physical principles and application. Both provide a means to interpret complex phenomena occurring deep in the earth by the pressure resulting from fluid movement in rock formations. An important difference, however, is that reservoir-pressure analysis has developed into a mature discipline during the past 50 years, while fracturing-pressure analysis is in a relative state of infancy.

Historical Development. The relatively slow development of fracturing-pressure analysis did not result because of a late start. Injection pressures have been measured as a safety precaution since the first treatment in 1947, and the relationship between pressure and fluid volume in a fracture was reported[1] in 1953.* Also, the important coupling of the pressures resulting from fluid flow in a fracture and the elastic deformation of the fracture was studied[2] in 1955, and the most applicable conclusion for the subsequent sections was presented[3] in 1958: "The observation of... bottomhole pressures during fracturing operations is necessary to a complete understanding and possible improvement of this process." In 1961, laboratory experiments of pressure vs. injected volume were reported[4] with an analysis of fracture widths for both horizontal and vertical fractures. An additional analysis of pressure resulting from flow in a fracture was not reported until 1968.[5] Further development of fracturing-pressure analysis did not begin again until the latter part of the 1970's, when "massive" treatments[6] of tight-gas formations became common. These large treatments changed the character of fracturing from a relatively low-cost, high-success-ratio operation to a relatively expensive operation with costs comparable to drilling and other completion expenses. Consequently, a renewed interest in fracturing research began during the late 1970's.

An analysis[7] of the combined effect of fluid pressure, rock properties, and in-situ stresses on vertical-fracture growth was presented in 1976. This study quantified the important contribution that higher in-situ stresses, for the bounding formations relative to the pay formation, have in limiting vertical height growth. The role of stress differences was indicated[1] in 1953. In 1977, the use of a numerical simulator was reported[8] for predicting the decline in pressure as a fracture closed after fluid injection was terminated. A theoretical analysis of this pressure decline (i.e., "falloff" in reservoir terminology) was presented[9] in 1979 that permitted the inference of fracture design parameters and dimensions. An analysis was also presented[10] for interpreting the fracturing process during injection through changes in the bottomhole fluid pressure.

These analyses of fracturing pressure have been performed routinely in the field since the advent of specially equipped computer vans provided by the service companies.

**Transaction* publication for many cited references occurred later.

Also, the use[11,12] of numerical simulations to calibrate field-recorded pressure during and after injection for determining design parameters, such as height growth and fluid loss, began in 1979. Calibrated simulations were found to be an effective means for increasing the effectiveness and reducing costs of subsequent massive treatments in tight-gas fields. The analysis of fracturing pressures has been applied for the design of fractures in other types of applications; e.g., injection wells,[13] primary production of oil in hard[14] and soft[15] formations, and geothermal wells.[16]

14.2 A Porous-Balloon Analogy For a Fracture

The purpose of this section is to provide insight into the analyses of fracturing pressures through the simplified mathematics and intuition that result by considering the fracture as a porous balloon surrounded by a confining pressure.

Fig. 14.1 illustrates a balloon that is assumed to inflate with a constant cross-sectional area, A (side view), and expanding width (top view) as fluid is injected. The assumption of a constant area is made to simplify the analysis; in reality, however, the area of a fracture would generally increase with injection. The balloon's surface is porous, which permits fluid to leak through the surface and provides the analogy of fluid loss to the formation being fractured. The rate of fluid loss, q_ℓ, through the surface is assumed to be equal to the product of the constant C and the area A and to be independent of time or pressure. In general, the fluid loss from an actual fracture would depend on time and pressure. It is assumed that the balloon is elastic and that the average width, $\bar{w}$, increases in proportion to the net pressure, p_n, at the inlet. The net pressure is the inlet pressure minus the value of the confining pressure, p_c, which acts to deflate or to close the balloon and provides the analogy to the in-situ stress in the formation. In addition, the balloon or fracture does not begin to inflate until the fluid pressure exceeds p_c—i.e., $p_n>0$. The average width is proportional to the net pressure multiplied by the elastic compliance, c, of the balloon. The balloon's compliance is analogous to the elasticity of the formation, which is deformed during fracturing. The fracture's compliance is proportional to a characteristic dimension of the fracture and inversely proportional to the elastic modulus of the formation.

The two parameters, p_c and c, define a straight-line relationship for average width in terms of pressure, as illustrated in Fig. 14.1.

Relationships in Terms of Time. The balloon's response to a constant injection rate of fluid, q_i, is illustrated in Fig. 14.2. The center illustration shows the fraction of stored volume, V_f, on top of the axis and the fraction of lost volume, V_ℓ, on the bottom of the axis. At the time that injection stops, t_i, the sum of these volumes is equal to the total volume injected, V_i. After injection stops, for this analogy the stored volume decreases at the same rate as fluid was lost during injection until the fracture closes with zero width at a time that is t_c greater than the injection time. From similar right triangles, with the equal angle, a, in Fig. 14.2, the

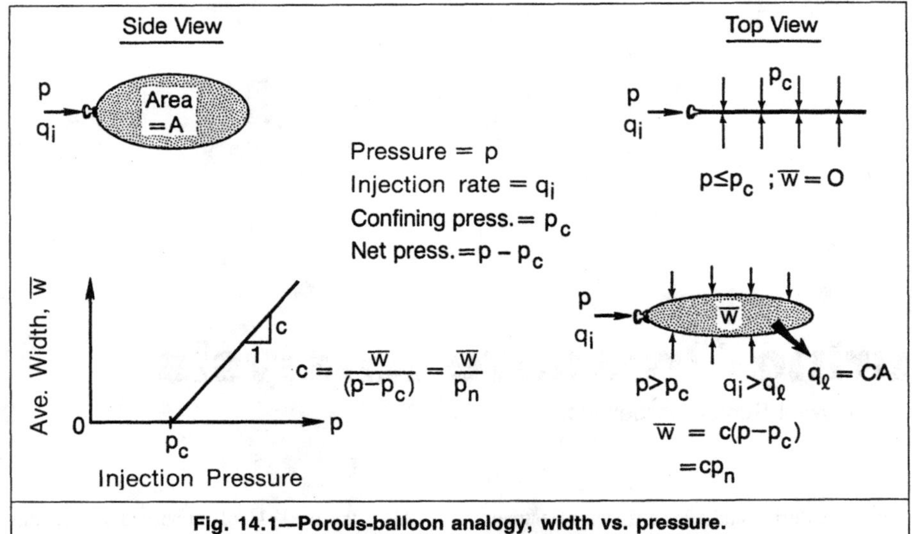

Fig. 14.1—Porous-balloon analogy, width vs. pressure.

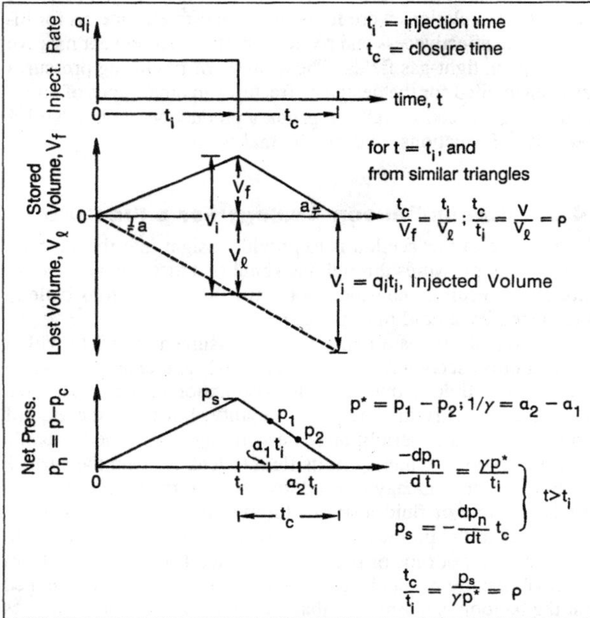

Fig. 14.2—Material balance for balloon analogy, injection and closure.

injection and closure times are related to the stored and lost volumes when injection stops. This relationship is

$$\frac{t_c}{t_i}=\frac{V_f}{V_\ell}=\rho. \quad (14.1)$$

The variable ρ, defined as the ratio of the stored to loss volumes in Eq. 14.1 can be shown to be inversely proportional to the dimensionless variable of the classic Carter area equation,[17] and is related[9] to the fluid efficiency, η, defined as the ratio of fracture to injected volumes, by the expression

$$\eta=\frac{V_f}{V_i}=\frac{\rho}{1+\rho}. \quad (14.2)$$

The correspondence of ρ to the Carter variable is given by Eq. 14.4:

$$V_f=\overline{w}A, \quad (14.3)$$

$$V_\ell=q_\ell t_i=CAt_i,$$

and

$$\rho=\frac{V_f}{V_\ell}=\frac{\overline{w}}{Ct_i}, \quad (14.4a)$$

whereas the inverse of the Carter variable is

$$1/f_C=\frac{\overline{w}}{2C\sqrt{\pi t_i}} \quad (14.4b)$$

and reflects the assumed square-root-of-time dependence for fluid loss, which is ignored for the balloon analogy.

Thus, determining the ratio of closure time to injection time defines the variable ρ, which also depends on the ratio of fracture width and fluid-loss coefficient, C. Hence, if C were also known, Eq. 14.4 provides the means of determining the average width as

$$\overline{w}=Ct_i\rho \quad (14.5)$$

and the fracture area from Eqs. 14.2 and 14.3 as

$$A\overline{w}=V_f=\frac{V_i\rho}{1+\rho}, \quad (14.6)$$

and using Eq. 14.5,

$$A=\frac{V_i}{Ct_i(1+\rho)}. \quad (14.7)$$

For these and following expressions, it is assumed that injection time and injected volume are known.

Relationships in Terms of Pressure. The lower illustration in Fig. 14.2 is in terms of the net pressure, which is related to the width, $\overline{w}=V_f/A$, by the elastic compliance, c; i.e., $p_n=\overline{w}/c=V_f/Ac$. From this relation, the pressure can be used to find relationships similar to Eqs. 14.1, 14.5, and 14.7 and to find the loss coefficient, C. The rate of pressure decline, after injection stops and before closure, is

$$\frac{dp_n}{dt}=\frac{1}{Ac}\frac{dV_f}{dt}, \quad (14.8)$$

assuming that A and c are constant. Also, for this time period,

$$-\frac{dV_f}{dt}=\frac{dV_\ell}{dt}=CA, \quad (14.9)$$

and combining and rearranging Eqs. 14.8 and 14.9 yields

$$C=-c\frac{dp_n}{dt}. \quad (14.10)$$

Thus, the fluid-loss coefficient can be determined from the rate of pressure decline if the fracture's compliance is known.

Determining rate of changes from actual data can be complicated by signal noise. Eq. 14.10 can be defined in terms of pressure differences between times that are long enough generally to eliminate signal noise. As shown in the lower illustration in Fig. 14.2, the pressures used for the difference, p^*, are selected at two times

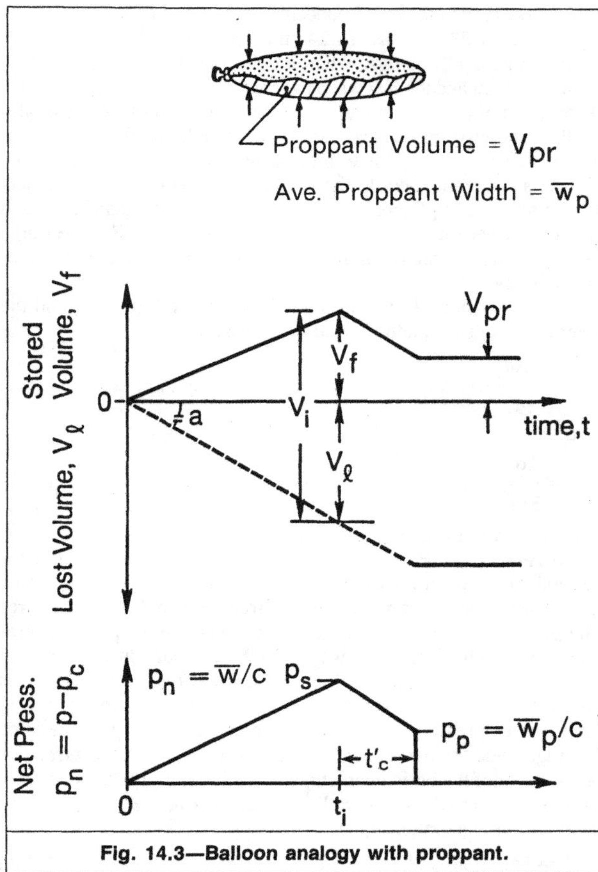

Fig. 14.3—Balloon analogy with proppant.

that are multiples (α_1 and α_2) of the injection time. As a result, Eq. 14.10 can be expressed as

$$-\frac{dp_n}{dt}=\frac{p_1-p_2}{(\alpha_2-\alpha_1)t_i}=\frac{\gamma p^*}{t_i} \qquad (14.11)$$

and from Eq. 14.10

$$C=-c\frac{dp_n}{dt}=c\frac{\gamma p^*}{t_i}, \qquad (14.12)$$

where $\gamma=1/(\alpha_2-\alpha_1)$ and t_i define the rate of pressure decline in terms of the characteristic decline pressure, p^*.

The net pressure when injection stops, p_s, is related to the time of closure, t_c, and rate of pressure decline by

$$p_s=-\frac{dp_n}{dt}t_c, \qquad (14.13)$$

and combining Eqs. 14.1, 14.11, and 14.13 results after rearranging in

$$\rho=\frac{t_c}{t_i}=\frac{p_s}{\gamma p^*}. \qquad (14.14)$$

This expression indicates that ρ in Eqs. 14.1 through 14.7 can equivalently be defined in terms of pressure or the ratio of closure to injection times.

From the definition of compliance, the average width at time t_i can be defined in terms of the corresponding pressure,

$$\bar{w}=cp_s,$$

and from Eq. 14.14,

$$\bar{w}=c\rho\gamma p^*. \qquad (14.15)$$

Also, substituting Eq. 14.12 into Eq. 14.7 yields

$$A=\frac{V_i}{c\gamma p^*(1+\rho)} \qquad (14.16)$$

for the area in terms of the pressure difference, p^*, and ρ.

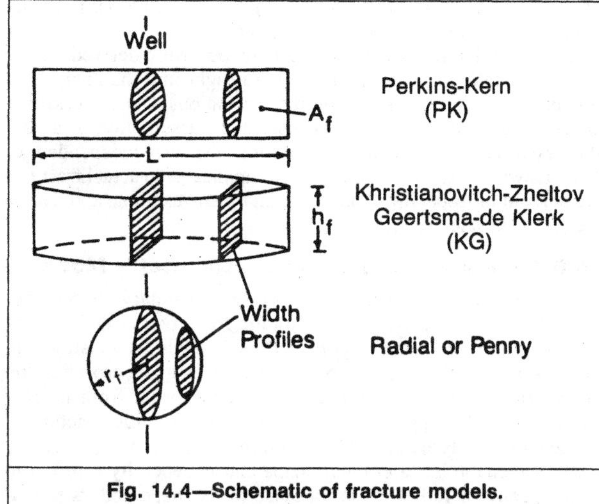

Fig. 14.4—Schematic of fracture models.

Effect of Proppant. The prior discussion assumed that after injection, the balloon, or analogously the fracture, closed completely without any remaining width. This would be the case for "calibration treatments"[9] performed to determine the fracturing parameters discussed above; however, the purpose of fracturing generally is to create a propped fracture with a finite propped width, $\bar{w}_p$, after the fracture closes.

Fig. 14.3 illustrates the effect of proppant for the balloon analogy. The volume of proppant, V_{pr}, including its porosity, ϕ, is related to the solid volume, V_s, as

$$V_{pr}=\frac{V_s}{1-\phi}. \qquad (14.17)$$

The volume fraction of proppant, relative to the total injected slurry volume, V_i, will be denoted as

$$f_{Vp}=\frac{V_{pr}}{V_i}=\frac{V_s}{V_i(1-\phi)}, \qquad (14.18)$$

the time to close with proppant as t_c' at a net pressure, p_p, average width, $\bar{w}_p$, and fracture volume $V_f=V_{pr}$.

The average propped width can be inferred from the pressure when closure occurs by the compliance

$$\bar{w}_p=cp_p. \qquad (14.19)$$

Also, in terms of average values, the ratio of propped width to the maximum achieved width, $\bar{w}_s$, at time t_i is

$$\bar{w}_p/\bar{w}_s=p_p/p_s, \qquad (14.20)$$

which provides, in terms of the indicated net pressure ratio, a measure of the effectiveness of the treatment in providing propped width.

The actual value of the parameter ρ (used in Eqs. 14.1 through 14.16) can be found in terms of that inferred from the time to close on the proppant, t_c', as

$$\rho'=\frac{t_c'}{t_i} \qquad (14.21a)$$

and

$$\frac{\rho}{\rho'}=\frac{t_c}{t_c'}=\frac{V_f}{V_f-V_{pr}}, \qquad (14.21b)$$

where the relationship between times and volumes results from the similar triangles seen in Figs. 14.2 and 14.3. Substituting Eqs. 14.2 and 14.18 into Eq. 14.21b and simplifying leads to

$$\rho=\frac{\rho'+f_{Vp}}{1-f_{Vp}}. \qquad (14.22)$$

Thus, if proppant is injected and the volume fraction of proppant is known (Eq. 14.18), information about the propped width (Eqs.

14.19 and 14.20) and the parameter ρ can be found (Eq. 14.22) from the time to close on the proppant.

It is noted that the equations in this section were derived for the simplified balloon analogy to provide insight into the analysis of fracturing pressures. As a result, these equations *do not necessarily apply* for an actual fracture because of the assumption of a constant area during injection and fluid loss that was independent of time. However, analogous expressions applicable to actual fractures are derived in Appendix J, and a summary and example application are given in later sections.

14.3 Dependence on Fracture-Geometry Models

This section discusses the three basic or elementary shapes considered for the geometry of a hydraulic fracture and their consequence on the fracturing response. These models are illustrated in Fig. 14.4, with the upper two having a rectangular cross section in the fracture plane. The associated names result from investigators[2,4,5,] who applied these models to hydraulic fracturing. These two models are applied to fractures in the vertical plane. The lower schematic has a circular shape and is generally referred to as the penny or radial model because of its shape and its propagation in the radial direction. This model is applied either in the vertical plane for massive, uniform formations where the horizontal penetration (i.e., $2r_f$) is less than the formation thickness or for fractures in the horizontal plane that result when the fluid pressure exceeds the overburden. Each of the three models has a different relationship between pressure and width, or equivalently a different elastic compliance, c in Fig. 14.1 for the balloon analogy. As a result, an analysis of fracturing pressures generally requires an assumption of a basic model and the use of the corresponding compliance.

Compliance. The elastic-compliance relationships for the models come from the work of Sneddon[18] and Sneddon and Elliott.[19] The rectangular models result for an infinitely extending elastic solid with a pressurized crack having a dimension, d, in one direction and extending through the infinite solid in the other direction. For this geometry, the pressurized crack has the shape of an ellipse with the major axis d and the minor axis the maximum width, w. For an ellipse, $\bar{w}=\pi w/4$ and the results[19] can be expressed in terms of the net pressure as

$$\bar{w}=\frac{\pi d}{2E'}p_n$$

or

$$\bar{w}=cp_n;\ c=\frac{\pi d}{2E'} \qquad (14.23)$$

and

$$E'=\frac{E}{1-\nu^2},$$

where

E' = plane-strain modulus of elastic solid,
E = Young's modulus, and
ν = Poisson's ratio.

Eq. 14.23 is applied to the rectangular fractures with the infinite direction corresponding to the length, L, and the finite dimension, d, to the height h_f for the Perkins and Kern[4] (PK) model; and the infinite direction corresponding to h_f and d to L for the Khristianovich-Zheltov[2] and Geertsma-deKlerk[5] (KG) model. Thus, the compliance for these models from Eq. 14.23 is

$$c=\frac{\pi h_f}{2E'} \qquad (14.24a)$$

for the PK model and

$$c=\frac{\pi L}{2E'} \qquad (14.24b)$$

for the KG model.

The applicability of these rectangular models generally assumes that the vertical height, h_f, is constant during the extension of the length. Because of the correspondence of the finite direction in going from Eq. 14.23 to Eqs. 14.24, the applicable[20,21] rectangular model is generally the PK when the fracture has a length greater than the height and the KG when the height is greater than the length. This range of model application is shown in Fig. 14.5[22] as a ratio of the predicted maximum width (i.e., $4/\pi$ times Eq. 14.23) for the models to the actual maximum width for a range of L/h_f. This figure is also equivalent to the ratio of compliance for the models to the actual compliance. A width ratio near one indicates good agreement between the model and actual behavior. For the range of L/h_f near one, the radial model is more applicable than the rectangular models.

The radial model[18] also has an elliptical width profile, and the average width is related to the net pressure by

$$\bar{w}=\frac{16r_f}{3\pi E'}p_n \qquad (14.25a)$$

and

$$c=\frac{16r_f}{3\pi E'}, \qquad (14.25b)$$

where r_f is the radius of the fracture.

Correction for Pressure Gradients. The previous relation for compliance was for the case in which the pressure within the crack or fracture was constant across the area. For hydraulic fractures the pressure varies[2] within the fracture because of pressure gradients caused by fluid flow away from the wellbore during and after injection.[4] As a result, the fluid pressure at the wellbore, p_w, is greater than the average pressure in the fracture. An equivalent uniform net pressure, $\hat{p}_n$, can be defined that would give the same average width as the actual varying pressure in the fracture. For application of the basic geometry models, a multiplying parameter, β, can be defined for the wellbore net pressure, $p_n=p_w-p_c$, to provide the equivalent $\hat{p}_n$:

$$\hat{p}_n=\beta p_n \qquad (14.26a)$$

or

$$\beta=\hat{p}_n/p_n. \qquad (14.26b)$$

This multiplying correction for the pressure gradient can be applied to the prior expressions for compliance, c (Eqs. 14.24 and 14.25), to yield

$$c_f=\beta c, \qquad (14.27)$$

where c_f is the fracture compliance that relates the average width to the wellbore value of net pressure, p_n. For the ideal case of constant pressure, or equivalently, zero gradient, $\beta=1$. The values of β during and after injection will differ.[9] Specific values for β are given in Eqs. 14.36 and 14.37.

Vertical Variations of Modulus and Stress. The compliance is inversely proportional to the elastic modulus of the formation. As a result, a representative value must be obtained for a valid determination of the compliance. The most representative values of modulus are obtained from laboratory tests on samples obtained from the formation being fractured with the test conditions representative of the in-situ pressure and fluid-saturation conditions. Data can also be obtained by sonic measurements in the laboratory or in situ by logging; however, values obtained by sonic (dynamic vibrations) measurements are generally greater than the applicable modulus value for relatively small loading rates. Because fractures will generally cross formations with different moduli, the combined modulus of the various formations must be considered. An average value, weighted by the thickness of the formations, can provide an approximate value; however, an exact analysis of the combined effect requires an elastic finite-element analysis. Fig. 14.5 shows the width profile for a finite-element analysis of a uniformly pressured fracture crossing two formations with different moduli. The upper 0.6 of the fracture is in a formation with a modulus of seven units and the lower 0.4 with a modulus of three units. This figure also shows the profile for a uniform formation with a modulus of six units with the same average width as the case of two formations. This example indicates that the weighted average ($0.6\times7+0.4\times3=5.4$) would underestimate the actual equivalent modulus by 10%.

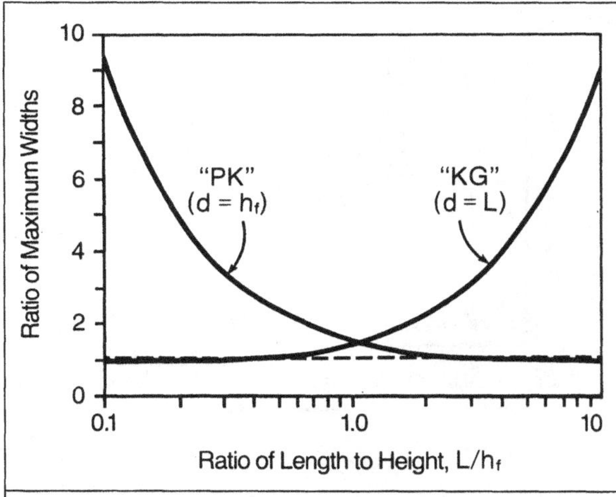

Fig. 14.5—Ratio of model to actual width for constant pressure.

In addition to variations of modulus, in-situ rock stress variations within a formation, fracture migration through formations with different stresses, or interbed slippage can affect[10] the width compliance. The closure pressure, p_c, reflects a global average of the minimum in-situ stress over the formation. The effect of interbed slippage was found[10] from an analysis of a specific example to produce less than a 10% increase in compliance for slip displacements up to one-half the maximum fracture width. The vertical migration of a fracture will tend to be arrested by slippage at an interface.

The PK model in Fig. 14.4 is generally the applicable rectangular model for vertical growth into higher-stress formations because this model predicts increasing pressure with time and tendency for growth, while the KG model predicts decreasing pressure (as discussed later). For the PK model, the effect on the width compliance by vertical growth into higher in-situ-stress formations is small[10] if the value of height used in Eq. 14.24 is the original height. Equivalently, the compliance experiences a change approximately inversely proportional to the increase in height if the h_f value in Eq. 14.24 corresponds to the changing height. The latter dependence is shown in Fig. 14.7, where the change in average width with pressure and height is expressed by the stiffness multiplier, s.[23] The multiplier would be applied to the denominator in Eq. 14.24 to give

$$c=\frac{\pi h_f}{2sE'}, \qquad (14.28)$$

from which the product sE' could be considered as a pseudomodulus of the formation, which would give the same average width as a uniformly stressed formation. Fig. 14.7 shows that s increases approximately equal to h_f/h_i, or that c is relatively constant if defined in terms of h_i for this idealized example.

The effect of an interbedded high-stress zone is illustrated by the example[23] shown in Fig. 14.8. The figure shows that for the initial height, $h_f/h_i=1$, the value of s is about 2. Equivalently, the compliance in Eq. 14.28 would be about half that for a uniformly stressed formation.

Figs. 14.7 and 14.8 indicate that vertical growth into higher-stress formations or interbedded zones of higher stress can have a significant effect on the width compliance of a fracture; these considerations must be included for an accurate determination of compliance. The effect of growth on compliance can be minimized by defining compliance in terms of the initial height, h_i.

Fluid Flow. The power-law model is generally assumed to represent the flow relationship between shear stress, τ, and shear rate, $\dot\gamma$, for fracturing fluids. This model can be expressed as

$$\tau_s=K\dot\gamma^n \qquad (14.29)$$

($n=1$ for the case of a Newtonian fluid), and leads[4] to a pressure gradient in the fracture that can be expressed in the form of the

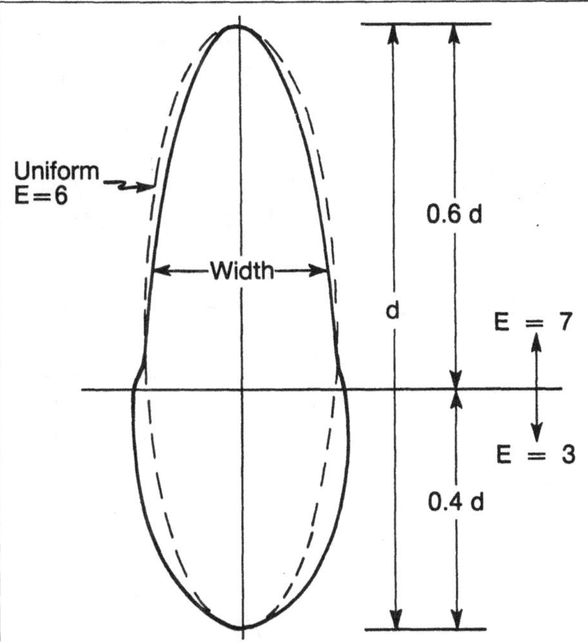

Fig. 14.6—Effect of moduli variation and definition of equivalent uniform modulus.

proportionality (i.e., $\propto$) for simplicity:

$$\frac{dp}{dz}\propto\frac{Kv^n}{w^{n+1}};\ v=\frac{q}{wh} \qquad (14.30a)$$

and

$$\frac{dp}{dz}\propto\frac{K}{w^{2n+1}}\left(\frac{q}{h}\right)^n. \qquad (14.30b)$$

This expression relates the gradient down the fracture length to the fluid velocity, v, or flow rate, q. Introducing the compliance ($w=cp_n$) leads to

$$\frac{dp_n}{dz}\propto\frac{K}{(cp_n)^{2n+1}}\left(\frac{q}{h}\right)^n \qquad (14.31a)$$

or

$$p_n^{2n+1}dp_n\propto\frac{K}{c^{2n+1}}\left(\frac{q}{h}\right)^n dz. \qquad (14.31b)$$

Integrating along the fracture length and assuming that p_n is negligible at the tip of the fracture leads to

$$p_n^{2n+2}\propto\frac{K}{c^{2n+1}}\left(\frac{q}{h}\right)^n L \qquad (14.32a)$$

or

$$p_n\propto\left[\frac{K}{c^{2n+1}}\left(\frac{q}{h}\right)^n L\right]^{1/(2n+2)}. \qquad (14.32b)$$

Introducing the dependence of the compliance on the appropriate dimension (h_f, L, or r_f in Eqs. 14.24 and 14.25) and noting for the radial model that $h\propto L\propto r_f$ leads to

$$p_n\propto(E^{2n+1}Kq^n)^{1/(2n+2)}\left(\frac{L}{h_f^{3n+1}}\right)^{1/(2n+2)} \qquad (14.33a)$$

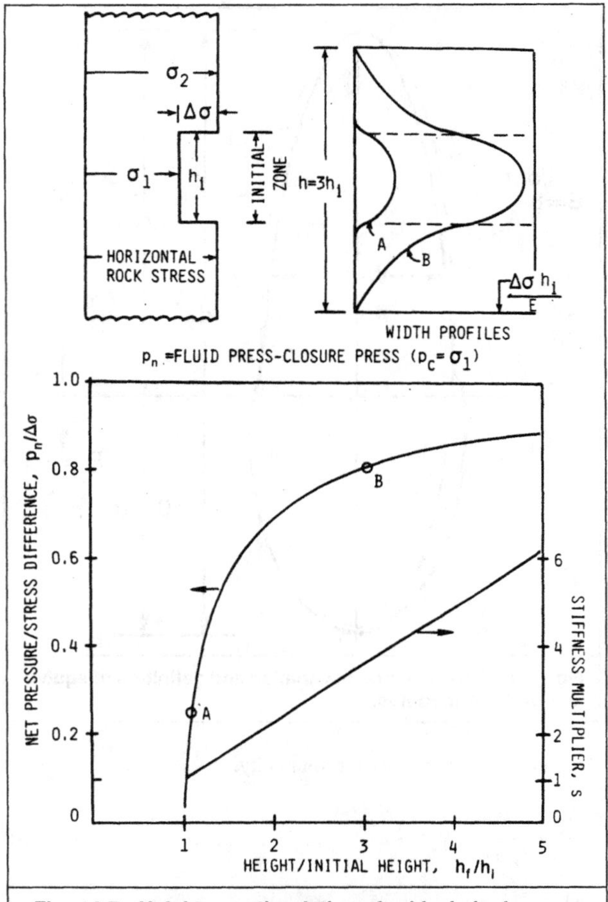

Fig. 14.7—Height-growth relations for ideal single zone.

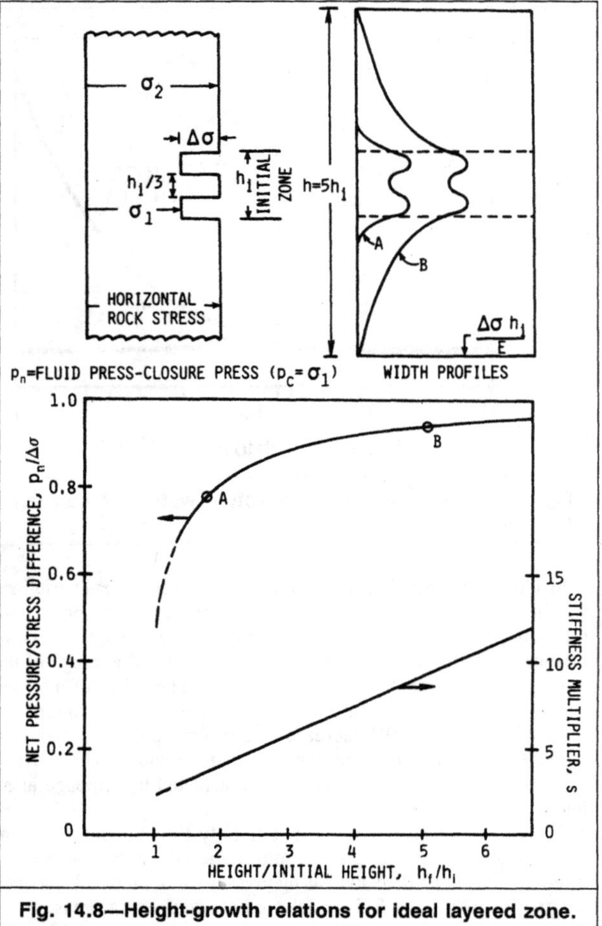

Fig. 14.8—Height-growth relations for ideal layered zone.

for the PK model,

$$p_n \propto (E^{2n+1}Kq^n)^{1/(2n+2)}\left(\frac{1}{h_f^n L^{2n}}\right)^{1/(2n+2)} \quad \text{........ (14.33b)}$$

for the KG model, and

$$p_n \propto (E^{2n+1}Kq^n)^{1/(2n+2)}\left(\frac{1}{r_f^{3n}}\right)^{1/(2n+2)} \quad \text{........... (14.33c)}$$

for the radial model.

These proportionalities for the designated models indicate the effect on pressure from variations of the fluid viscosity, K; injection rate, q; modulus, E; fracture extension, L or r_f; and a constant height, h. In particular, the equation indicates that p_n will increase with extension for the PK model and decrease with extension for the KG and radial models and that p_n is strongly dependent on E ($E^{3/4}$ for $n=1$) for all models and inversely proportional to h for the PK model, $n=1$.

Multiplying Eqs. 14.32b by the compliance provides the expression for fracture width

$$\bar{w}=cp_n \propto \left[cKL\left(\frac{q}{h}\right)^n\right]^{1/(2n+2)} \quad \text{................... (14.34)}$$

Again introducing the appropriate compliances for each model leads to

$$\bar{w} \propto \left(\frac{Kq^n}{E}\right)^{1/(2n+2)} (h_f^{1-n}L)^{1/(2n+2)} \quad \text{............. (14.35a)}$$

for the PK model,

$$\bar{w} \propto \left(\frac{Kq^n}{E}\right)^{1/(2n+2)} (L^2/h_f^n)^{1/(2n+2)} \quad \text{............. (14.35b)}$$

for the KG model, and

$$\bar{w} \propto \left(\frac{Kq^n}{E}\right)^{1/(2n+2)} (r_f^{2-n})^{1/(2n+2)} \quad \text{................ (14.35c)}$$

for the radial model. These correspond to the classic width equations[4,5] (noting that $n=1$ and K=viscosity for a Newtonian fluid).

The direct integration of Eqs. 14.30 for the pressure gradient is applicable[4] only for the PK model since, for the other models, the width at any point in the fracture depends on the pressure at all locations. As a result, the equivalent uniform net pressure, $\hat{p}_n$, is equal to the average net pressure for determining the value of β in Eq. 14.26 only for the PK model. Therefore, for the PK model, approximations for β during pumping, β_p, and shut-in, β_s, can be determined from the average net pressure following a method[9] similar to that used between Eqs. 14.30 and 14.31. Assuming that velocity, Eq. 14.30a, is essentially constant along the fracture[24] during pumping and that flow rate is essentially constant along the fracture during shut-in,

$$\beta_p=(\hat{p}_n/p_n)_p \approx (n+2)/(n+3+a) \quad \text{................ (14.36a)}$$

and

$$\beta_s=(\hat{p}_n/p_n)_s \approx (2n+2)/(2n+3+a), \quad \text{.............. (14.36b)}$$

where n is the exponent of the power-law fluid model and a is the exponent for the variation of power-law coefficient (equivalent to viscosity) along the fracture,

$$K(z)=K_w(2z/L)^a,$$

and K_w is the value of K at the well. For $a=1$, the variation is linear from K_w to zero at the tip; for $a=0$, K is constant along the fracture. As a result, for constant viscosity ($a=0$) and a Newtonian fluid ($n=1$), Eqs. 14.36 give for the PK model

$$\beta_p \approx 3/4$$

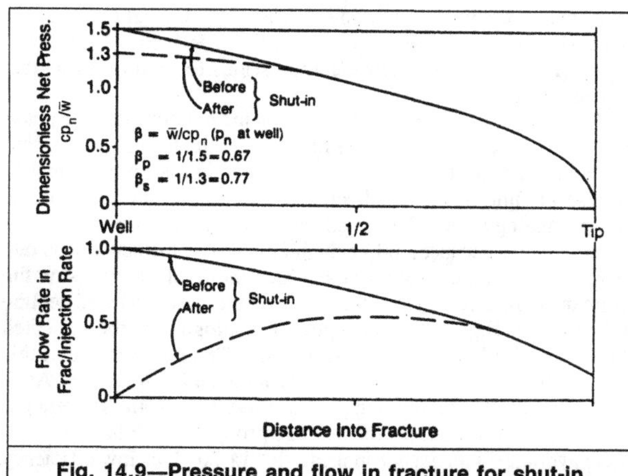

Fig. 14.9—Pressure and flow in fracture for shut-in.

and

$\beta_s \approx 4/5$.

This value of β_p is equal to the value inferred from a more complete analysis[25] for the same conditions and no fluid loss ($\eta \to 1$) and approximately equal to the value inferred for predominant fluid loss ($\eta \to 0$). Fig. 14.9 shows values of $\beta_p = 0.67$ and $\beta_s = 0.77$ derived from a numerical simulator[23] for $n = 0.5$ and an increasing viscosity near the well because of the cooled entrance.

Equivalent derivations for the values of β have not been reported for the KG or radial models. Because the value of β_s is required for application of a pressure-decline analysis, an estimate of

$$\beta_s \approx 0.9 - 0.95 \quad \text{.......} (14.37)$$

can be used for the KG or radial models. This range of values is based on the observation that the pressure gradient is concentrated near the fracture tip for these models,[2,5] and for small flow rates near the well during shut-in, a nearly constant pressure distribution should result.

Bounds for Time Dependence of Penetration and Pressure. Fluid loss and pressure changes during injection depend on the manner in which the fracture extension increases with time. For a constant injection rate, the fracture growth can be bounded by two extremes: the case of no fluid loss (i.e., $V_f = V_i = q_i t_i$) for which the efficiency $\eta = V_f/V_i = 1$ and the case of almost total fluid loss (i.e., $V_\ell \to V_i = q_i t_i$ and $V_f \to 0$) for which efficiency $\eta = V_f/V_i \to 0$. For the latter case and the general assumed dependence of V_ℓ on the square root of time t, the fracture area increases[5,25] in proportion to $t^{1/2}$; i.e.,

$$V \propto t;\ \eta \to 1$$

and

$$A \propto t^{1/2};\ \eta \to 0.$$

Lower Bound. The bound of $\eta \to 0$ implies that for the three basic models in Fig. 14.4,

$$t^{1/2} \propto A = h_f L$$

for the PK and KG models and

$$t^{1/2} \propto A = \pi r_f^2$$

for the radial model, or for constant h_f that

$$L \propto t^{1/2} \quad \text{.......} (14.38a)$$

for the PK and KG models and

$$r_f \propto t^{1/4} \quad \text{.......} (14.38b)$$

for the radial model for the fracture penetration with time.

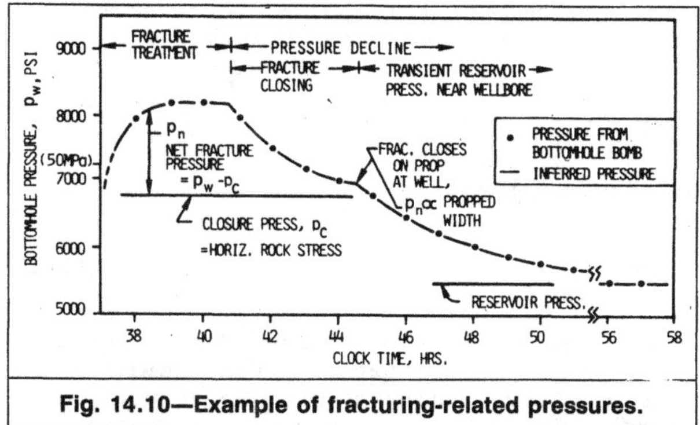

Fig. 14.10—Example of fracturing-related pressures.

Combining Eqs. 14.38 and 14.33 for constant-injection conditions (K, n, q, and h_f) leads to

$$p_n \propto t^{**}\left[\frac{1}{4(n+1)}\right],\ \eta \to 0 \quad \text{.......} (14.39a)$$

for the PK model,

$$p_n \propto t^{**}\left[\frac{-n}{2(n+1)}\right],\ \eta \to 0 \quad \text{.......} (14.39b)$$

for the KG model, and

$$p_n \propto t^{**}\left[\frac{-3n}{8(n+1)}\right],\ \eta \to 0 \quad \text{.......} (14.39c)$$

for the radial model for the time dependence of the net pressure. The notation ** implies the exponent operation.

Upper Bound. The bound of $\eta \to 1$ implies for the models that

$$t \propto V = \bar{w}A = \bar{w}(h_f L) \quad \text{.......} (14.40a)$$

for the PK and KG models and

$$t \propto V = \bar{w}A = \bar{w}(\pi r_f^2) \quad \text{.......} (14.40b)$$

for the radial model for constant-injection conditions (K, q, n, h_f). Combining Eqs. 14.40 and 14.35 leads to

$$t \propto L^{(2n+3)/(2n+2)}$$

for the PK model,

$$t \propto L^{(n+2)/(n+1)}$$

for the KG model, and

$$t \propto r_f^{(3n+6)/(2n+2)}$$

for the radial model, or

$$L \propto t^{**}\left(\frac{2n+2}{2n+3}\right) \quad \text{.......} (14.41a)$$

for the PK model,

$$L \propto t^{**}\left(\frac{n+1}{n+2}\right) \quad \text{.......} (14.41b)$$

for the KG model, and

$$r_f \propto t^{**}\left(\frac{2n+2}{3n+6}\right) \quad \text{.......} (14.41c)$$

for the radial model for the increase in fracture penetration with time. Combining Eqs. 14.41 and 14.33 gives

$$p_n \propto t^{**}\left(\frac{1}{2n+3}\right),\ \eta \to 1 \qquad (14.42a)$$

for the PK model and

$$p_n \propto t^{**}\left(\frac{-n}{n+2}\right),\ \eta \to 1 \qquad (14.42b)$$

for the KG and radial models for the time dependence of p_n.

Assumptions. The above expressions for pressure were based on an assumption of the fluid viscosity dominating the pressure distribution (i.e., Eqs. 14.30) and ignoring the fracture toughness of the formation. This assumption is generally valid for commercial applications of fracturing that use high-viscosity fluids and create fractures with dimensions in excess of 50 ft [15 m]. For the case of small-scale fractures created with low-viscosity fluids, fracture toughness can dominate and result in different exponents for time—e.g., $-\frac{1}{5}$ for the radial model[4] in Eqs. 14.42. Also, for the radial model, the use of Eq. 14.30 assumes that fluid enters the fracture over a significant portion of the complete height ($2r_f$) in Fig. 14.4. If fluid entry was limited to a section that was very small compared with $2r_f$, such as a limited perforated section or a horizontal fracture from a vertical wellbore, the relatively higher velocity of the fluid entering the fracture would result in a different pressure response than indicated in Eqs. 14.39 and 14.42. For this case,[5] the pressure decrease with time would be less than indicated by Eqs. 14.39 or 14.42.

Log-Log Plots. The expressions for penetration (Eqs. 14.38 and 14.41) and net pressure (Eqs. 14.39 and 14.42) are in terms of time raised to an exponent. As a result, a log-log plot of these variables vs. time would yield a straight line with a slope equal to the respective exponent. Similarly, the time expressions for penetration can be combined with the width expressions in Eqs. 14.35 to yield similar relationships for width. As presented in a following section, the log-log plot of net pressure vs. time provides a basis for interpreting pressures during fracturing.

14.4 Example of Fracturing Pressures

The following sections integrate the concepts from the balloon analogy and the specific relations developed for the fracture models into a basis for interpreting and analyzing fracturing pressures during and after injection. The different phases of the fracturing-pressure response are illustrated in the following example.

Fig. 14.10 shows a bottomhole recording[23] of the fluid pressure during and after a fracture treatment of a tight-gas sand formation. The recording contains all phases of the fracturing-pressure response—the injection period, the pressure decline as the fracture closes, the time when the fracture closed on the proppant (Time 44), and the recovery of the local increase in reservoir pressure caused by fluid loss during the treatment. The closure pressure, p_c, for the formation is indicated in the figure and is analogous to the confining pressure for the balloon shown in Fig. 14.1. This is the pressure that must be exceeded by the injected fluid to maintain an open unpropped fracture and that is the reference pressure for the net pressure. The net pressure governs the fracture's propagation and resulting dimensions. Appendix I discusses considerations for measuring fracturing pressures and defining the closure pressure.

A qualitative interpretation of the pressure shown in Fig. 14.10 can be made from information presented in the prior sections. The increasing net pressure during the first 2 hours of the treatment indicates that the PK model (Fig. 14.4), for a confined-height fracture in the vertical plane, is the most applicable basic model. This is the only model that predicts increasing pressure with increasing penetration; i.e., Eqs. 14.33. The nearly constant pressure during the last hour of the treatment is consistent with substantial vertical height growth into higher-stressed bounding formation(s), as shown in Fig. 14.7 for conditions beyond Point B. For this assumption, Fig. 14.7 shows that the pressure is being governed by the height growth and is being regulated to a value somewhat less than the horizontal stress in the bounding bed. Because the net pressure for this period is on the order of 1,400 psi [10 MPa], Fig. 14.7 would imply stress differences between the various formation layers that exceed 1,400 psi [10 MPa].

After injection stops, Fig. 14.10 shows the pressure-decline phase as the fracture closes because of fluid loss to the formation. During the closing phase, the rate of pressure decline is proportional to the rate of fluid loss, as indicated in Eqs. 14.8 and 14.9 for the balloon analogy. For the actual case in Fig. 14.10, the rate of decrease slows because, unlike the balloon case, the actual loss rate decreases with time. The point at which the fracture closes on the proppant is inferred[8] from the point of increased rate of decline at Time 44 in Fig. 14.10. This point is analogous to the indicated closure time in Fig. 14.3 for the balloon. The net pressure at this point is about 0.17 of the net pressure at the end of injection. After accounting for the decrease in pressure resulting from different β's (i.e., Fig. 14.9) at shut-in, the average propped width is about 0.20 of the average width at shut-in using Eq. 14.20. This low efficiency of propped width is consistent with the relatively low proppant concentrations used for this 1978 treatment (i.e., a maximum of 3 lbm sand/gal fluid [359 kg /m^3]).

After the fracture's closure is restricted by proppant, the pressure decline is no longer governed by the fracture compliance and fluid loss (i.e., Eq. 14.10) but by the decay in the increased reservoir pressure near the fracture. This phase of the pressure decline is governed by reservoir parameters and results because of communication between the fluid in the wellbore and the formation. The transient reservoir pressure after closure was not considered in the balloon analogy; hence the actual fracturing pressure differs from the sudden drop in pressure shown in Fig. 14.3. The pressure response for the closure of a fracture without proppant (e.g., calibration treatment) generally does not show any apparent change of character on a plot similar to Fig. 14.10. As discussed in Appendix I, a plot of pressure vs. the square root of time or vs. G (defined in Eq. 14.63) generally produces an apparent slope change when an unpropped fracture closes.

This example indicates the insights into the fracturing process that can be obtained from a recording of the bottomhole fracturing pressure. Additional interpretations and inferences of fracturing parameters from pressure are outlined in the following sections.

14.5 Interpretation and Simulation of Pressure While Fracturing

The manner in which the bottomhole pressure (BHP) changes with injection time provides a basis for interpreting the fracturing process.[10] Fig. 14.11 illustrates the characteristic changes in pressure for idealized fracture geometries using the previously developed information. The following discussion provides a basis for interpreting fracturing pressure and assumes conditions of nearly constant injection rate and fluid properties.

Magnitude of Pressure. For the normal state of rock stress, the overburden or vertical stress exceeds[26] the horizontal stress. Because fractures follow the path of least resistance or travel in a plane perpendicular to the lowest stress, they will normally be in the vertical plane. Also, by definition, the BHP must exceed the closure pressure, p_c, which reflects a global average of the rock stress perpendicular to the fracture plane. Therefore, if the BHP is less than the overburden stress (approximately 1 psi/ft [22.6 kPa/m] of depth), the fracture is only in the vertical plane, but if the pressure exceeds the overburden or vertical stress, a fracture can develop in the horizontal plane. This condition can occur (1) near the surface where tectonic or erosional activity produced more horizontal stress than the vertical, (2) in formations with low in-situ shear strength that produce nearly equal vertical and horizontal stresses, or (3) under vertical fracturing conditions that produce pressures that exceed the overburden stress and produce both a horizontal and vertical fracture. This condition of companion fractures has been observed in shallow coalbed[27] and limestone[28] formations.

Thus, as indicated by Fig. 14.11, the magnitude of the pressure provides an interpretation[26] of whether the fracture plane is vertical (pressure less than overburden) or is horizontal (pressure greater than overburden).

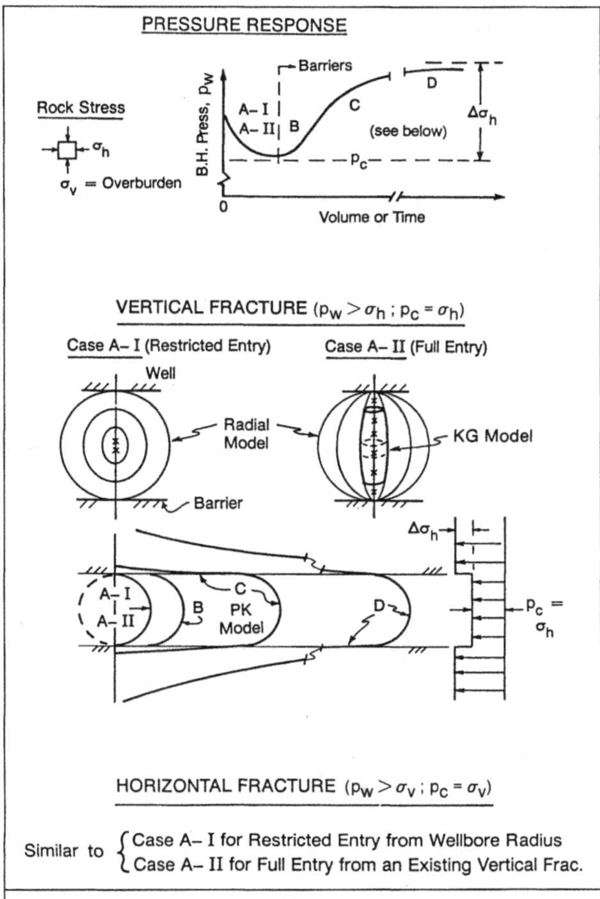

Fig. 14.11—Evolution of fracture geometry and pressure.

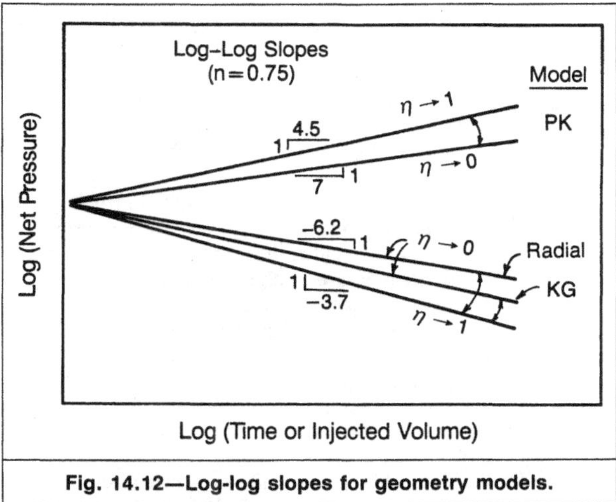

Fig. 14.12—Log-log slopes for geometry models.

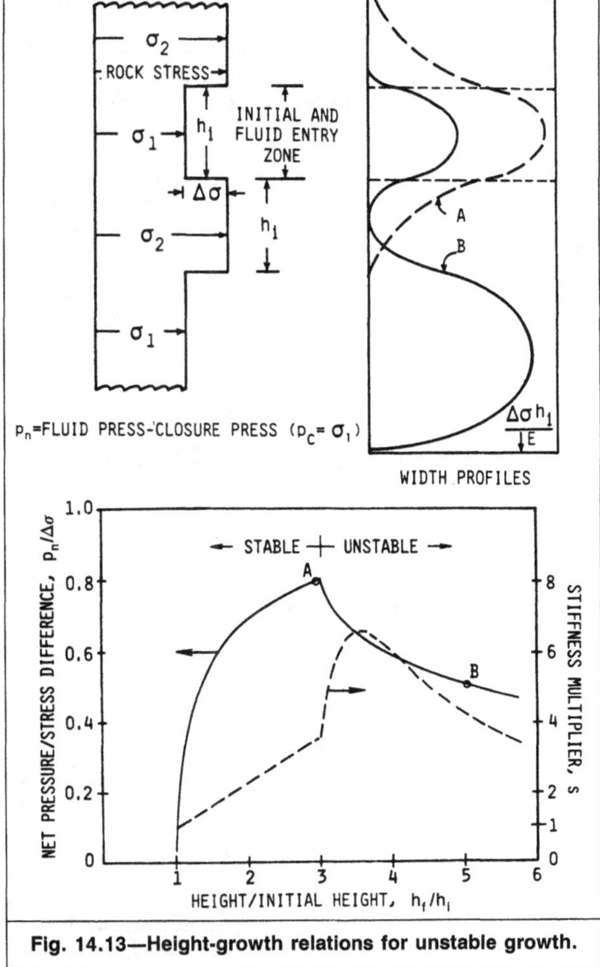

Fig. 14.13—Height-growth relations for unstable growth.

Increasing or Decreasing Pressure. The combination of fluid flow and fracture compliance for the three basic models provided a power-law relation between net pressure and injection time (i.e., Eqs. 14.39 and 14.42), or equivalently injected volume, with the exponent depending on the fluid property, n, and efficiency, η. Fig. 14.12 shows the log-log straight lines that result for this type of relationship. The figure uses a value of $n=0.75$ and includes the bounds of efficiency approaching both unity and zero.

Fig. 14.12 indicates that a small and negative log-log slope would indicate a fracture propagating in a manner similar to the KG or radial models illustrated in Fig. 14.4. The fracture evolving with decreasing pressure in accordance with these two models is shown as Cases A-I and A-II in Fig. 14.11. For both cases, the fracture is permitted to evolve into its preferred shape of a circle and is not influenced by any restriction, such as a higher-stress barrier. Also, both cases would apply for fractures in the vertical or horizontal planes. Case A-I represents the condition for which fluid entry is restricted essentially to a point source. Such a condition for a vertical fracture could result either from a limited perforated section or a wellbore that is inclined to the plane of minimum stress. For a horizontal fracture, restricted fluid entry would result for the case without a companion vertical fracture (i.e., from Fig. 14.11 $\sigma_h > p_w > \sigma_v$). Case A-II represents the condition of fluid entry from a line source through the complete section of the fracture. This could occur for a vertical fracture from an uncased wellbore or a long perforated section and for a horizontal fracture with a companion vertical fracture (i.e., $p_w > \sigma_v$ and $p_w > \sigma_h$). For the full-entry case, the early stages of the evolution would be consistent with the assumption for the KG model, as indicated by Case A-II, Fig. 14.11.

Fig. 14.12 indicates that a small and positive log-log slope would indicate a fracture propagating in a manner similar to the PK model illustrated in Fig. 14.4. As shown in Fig. 14.11, the applicability of this increasing-pressure model for vertical fractures comes after the radial model encounters barriers above and below. The barriers force the fracture from its preferred circular shape and result in increasing pressure as the fracture becomes long relative to its vertical height, as shown by Case C. If the upper or lower barrier results from higher stresses, Fig. 14.7 indicates that the restricted vertical growth and the log-log slope in Fig. 14.12 would be valid until p_n approached a value of about half the stress difference between the initial zone and the barrier with the smaller stress. As the net pressure continued to increase and exceeded 0.8 of the stress difference (Point B, Fig. 14.7), significant height growth would occur with small pressure increases and ultimately approach a limiting value of $p_n \approx 0.9$ of the stress difference, or equivalently a total pressure slightly below the stress of the barrier being penetrated. This condition is shown as D in Fig. 14.11. If the injection continues with significant vertical growth, the fracture can encounter a formation of lower stress, as shown in Fig. 14.13 for heights greater than indicated by Point A. For this condition of unrestricted vertical growth, the pressure would decrease in a manner similar to the unrestricted growth of a radial fracture, Case A in Fig. 14.11.

Thus, decreasing net pressures during the initial growth period can be interpreted as an indication of the unrestricted growth of a radially evolving fracture in either the horizontal or vertical plane, whereas increasing net pressures with a small log-log slope (e.g., $<1/5$) after the initial growth period is indicative of a vertical fracture extending primarily in length with restricted height growth. If after this period pressures begin to approach a period of nearly a constant value, one possible cause is significant height growth through a stress barrier. Subsequently decreasing pressure is indicative of unrestricted (unstable) vertical growth into a lower-stress zone.[10]

Constant Pressure. In addition to significant growth through a stress barrier, another interpretation for a period of constant pressure is the opening of secondary fractures.[10] This secondary fracturing occurs in natural fissures or cracks within the formation that are crossed by the primary fracture. These cracks generally will have relatively higher permeability than the formation matrix, and fluid will penetrate relatively deeply into the cracks with a pressure nearly equal to the pressure in the fracture. The cracks will open and begin to act as a propagating fracture when the fluid pressure exceeds the rock stress acting across them. When this pressure magnitude is reached, the naturally occurring cracks open, act as a pressure regulator for this pressure magnitude, and can thieve a significant portion of the injected fluid. The net pressure that will open the pressure-regulating cracks[10] is $\Delta\sigma/(1-2\nu)$, where $\Delta\sigma$ is the difference between the principal stresses in the plane perpendicular to the primary fracture (i.e., difference in horizontal stresses, $\sigma_H-\sigma_h$, for a vertical fracture) and ν is Poisson's ratio of the rock.

Formation Pressure Capacity. A period of constant net pressure for a vertical fracture can occur either because of significant height growth into a stress barrier, from secondary fracturing of cracks or fissures in the formation, or the formation of a companion horizontal fracture when the pressure for a vertical fracture exceeds overburden pressure. For these cases, a significant amount of the injected fluid becomes unavailable for the horizontal extension of the fracture in the formation to be stimulated. Also, this loss of fluid and the resulting inefficient extension is created when the fracturing pressure exceeds a level of pressure determined by the stresses in the formation or bounding formations. As a result, this limiting pressure can be called the formation pressure capacity, and a fracture in a formation becomes analogous to any other pressure vessel that has a pressure capacity. Exceeding the pressure capacity leads to loss of integrity and permits undesirable fluid loss. In the case of a vertical fracture, the fluid is lost through the top or bottom when the capacity of the stress barrier or overburden is exceeded and through the side when the formation's capacity to maintain fissures closed is exceeded. Fig. 14.14 shows the rock stresses that define the formation pressure capacity and their requirements to confine a net pressure of p_n efficiently.[10] Eq. 14.33a for the PK model indicates (1) that the net pressure is approximately proportional to rock modulus and inversely proportional to height and (2) that the viscosity, i.e., K, and extension, L, appear as a product with the same exponent. Consequently, the pressure capacity of a formation will be reached with less penetration for a relative increase in modulus, decrease in formation height, and proportional increase in viscosity (i.e., twice the viscosity results in half the penetration). Because modulus and height cannot be changed, the only controllable variable for an efficient injection rate is viscosity. After viscosity is decreased to the minimum required value,[23] the fracture penetration that can be achieved without exceeding the capacity becomes limited. Therefore a limiting penetration for efficient fracture extension from conventional practices[11] is defined by the in-situ rock stresses, which normally will be the same for a large area of a field development.

Thus, the formation capacity should be defined and appropriate treatment-design changes made as early as possible in the development program.

Increased Rate of Pressure. An increasing rate of pressure change that approaches a unit (1 to 1) or greater log-log slope of net pressure vs. time will result for a fracture that has restricted growth in all directions of its plane. This condition can occur from proppant bridging, commonly called a screenout. For this case, continued injection results primarily in expanding width, as was the case for the balloon analogy. The equivalent uniform net pressure, $\hat{p}_n$, which defines the average width, $\bar{w}$, in terms of the fracture compliance, c, is related to the wellbore net pressure by

$$\hat{p}_n=\beta p_n \quad \text{(14.43a)}$$

and

$$\hat{p}_n=\bar{w}/c, \quad \text{(14.43b)}$$

where β accounts for the pressure gradient caused by fluid flow. Also, $\bar{w}=V_f/A_f$, where V_f and A_f are the fracture volume and area. In addition, $V_f=\eta V_i$ and $V_i=q_i t$, where η, V_i, and q_i are the fracture's fluid efficiency, total injected fluid volume, and injection rate. Thus,

$$\hat{p}_n=\frac{\bar{w}}{c}=\frac{\eta V_i}{A_f c}, \quad \text{(14.44a)}$$

$$\hat{p}_n=\frac{\eta q_i}{A_f c}t=Mt, \quad \text{(14.44b)}$$

and

$$M\equiv\frac{\eta q_i}{A_f c}. \quad \text{(14.44c)}$$

From Eqs. 14.24 and 14.25, for the compliance, c, and the relationships for area above Eqs. 14.38,

$$M=\frac{2E'\eta q_i}{\pi}\left(\frac{1}{h_f^2 L}\right) \quad \text{(14.45a)}$$

for the PK model,

$$M=\frac{2E'\eta q_i}{\pi}\left(\frac{1}{h_f L^2}\right) \quad \text{(14.45b)}$$

for the KG model, and

$$M=\frac{2E'\eta q_i}{\pi}\left(\frac{3\pi}{32r_f^3}\right) \quad \text{(14.45c)}$$

for the radial model.

Fig. 14.15 illustrates on a log-log plot the relationships in Eqs. 14.43 and 14.44. The upper illustration indicates the relationship between the net pressure at the wellbore, p_n, and the equivalent uniform pressure, $\hat{p}_n$, which is a measure of the average width. For increasing fracture extension, Eqs. 14.45 indicate that M would be decreasing. If the fracture extension were restricted at time t_o shown in Fig. 14.15, the value of M would become a constant, denoted as M_o. As a result, Eqs. 14.44 indicate that $\hat{p}_n$ would then be linearly dependent on injection time and result in a unit log-log slope indicated on the center illustration for the restriction occurring along the tip of the fracture's area. As the fracture width increased in a ballooning fashion (i.e., constant area and increasing width), the pressure gradient from fluid flow would significantly decrease

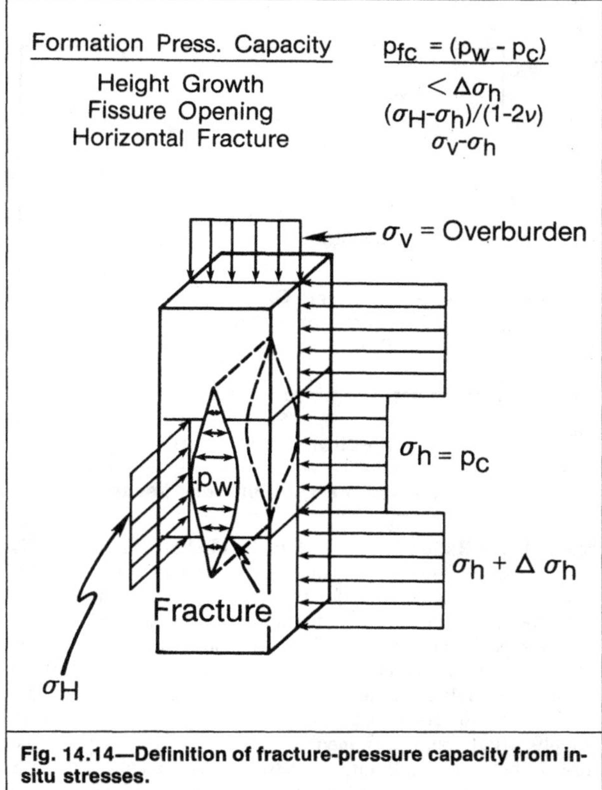

Fig. 14.14—Definition of fracture-pressure capacity from in-situ stresses.

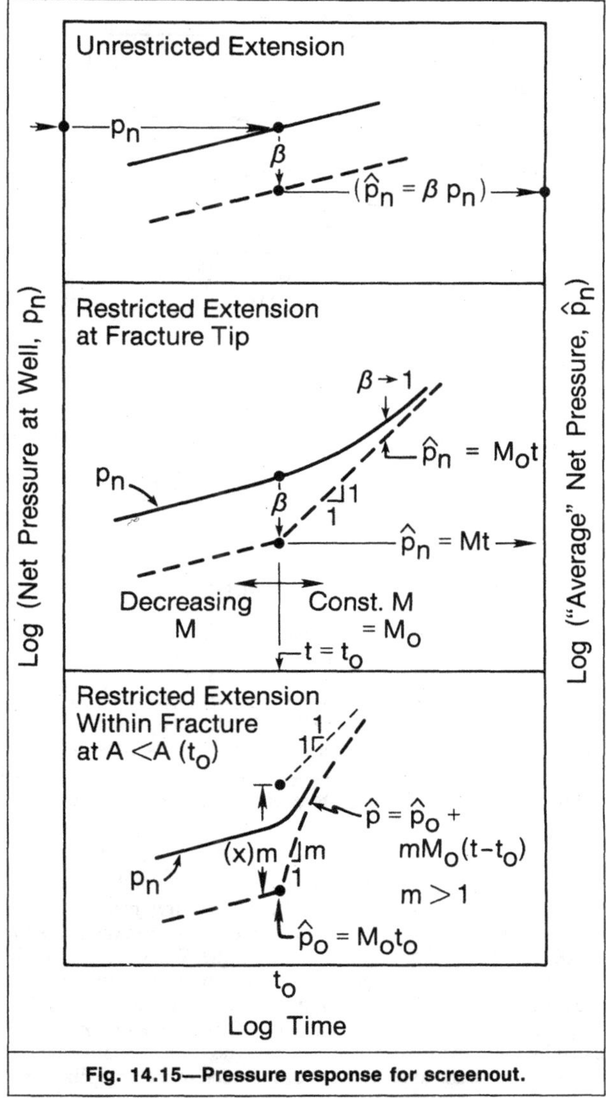

Fig. 14.15—Pressure response for screenout.

(Eqs. 14.30) and the value of p_n would approach $\hat{p}_n$ with β approaching unity. Thus, some period of time after the restricted growth occurred at t_o, p_n would begin to increase at a greater rate and approach the unit slope. The period of time before the wellbore pressure responds to the restriction can be significant and, as seen from the illustration, is on the order of $1/\beta$. This value for injection (i.e., $1/\beta_p$) is indicated to be 1.5 in Fig. 14.9, or equivalently on the order of half the value of t_o, the time of injection before the restriction occurred.

The lower illustration in Fig. 14.15 shows the condition for the restriction occurring between the wellbore and the fracture's extremities. If M_r represents the value of M in Eqs. 14.45 for the case where the restriction occurs within the fracture, the ratio of M_r to M_o is

$$m=\frac{M_r}{M_o}=L/L_r \quad \text{(14.46a)}$$

for the PK model,

$$m=\frac{M_r}{M_o}=L^2/L_r^2 \quad \text{(14.46b)}$$

for the KG model, and

$$m=\frac{M_r}{M_o}=r_f^3/r_{fr}^3 \quad \text{(14.46c)}$$

for the radial model, where L and r_f represent the fracture extremities and L_r and r_{fr} represent the distance to the restriction. The lower illustration in Fig. 14.15 indicates that the rate of increase in $\hat{p}_n$ is m times greater than for the case of a restriction at the extremities (i.e., $d\hat{p}_n/dt = mM_0$) and that the initial log-log slope is m from Eqs. 14.46. Because of the nature of the log-log plot, $\hat{p}_n$ will approach a unit slope that is a multiple of m times greater [denoted by arrow and "(x)m" in Fig. 14.15] than the unit slope for a restriction at the extremities. Because the pressure gradient for flow is the greatest near the extremities and this region is not active for a restriction within the fracture, the pressure will be nearly uniform in the active portion and will result in a value of β nearly equal to unity. Consequently, p_n will approach $\hat{p}_n$ in a much shorter period of time than for the case of a restriction at the extremities. This illustration and discussion indicate that a log-log slope greater than unity is a diagnostic for a restriction between the wellbore and fracture extremities, with the slope increasing as the restriction approaches the wellbore.

The dimension of the unrestricted fracture can be estimated[10] from the rate of pressure increase, $dp_w/dt = dp_n/dt = mM$, by use of Eqs. 14.45 and 14.46 with $\eta=1$ and the appropriate values of q_i, E', and h_f. This value of η will provide an upper bound on the dimension L or r_f. Eqs. 14.46 indicate larger values for the rate of pressure increase corresponding to restrictions nearer the wellbore. In particular, the r_f^3 relationship in Eqs. 14.46 for the radial model will give very large log-log slopes for restrictions within the fracture ($m=8$ for $r_{fr}/r_f=½$), whereas if one complete wing of a constant-height fracture becomes restricted at the wellbore ($L/L_r=2$), Eqs. 14.46 indicate that $m=2$ for the PK model and that $m=4$ for the KG model.

Thus, a log-log slope approaching one can indicate restricted fracture extension at the fracture's extremities, whereas a slope greater than one can indicate a restriction within the fracture.

Examples for Restricted-Height Vertical Fractures. Fig. 14.16,[23] a summary of the prior sections, shows on log-log plots

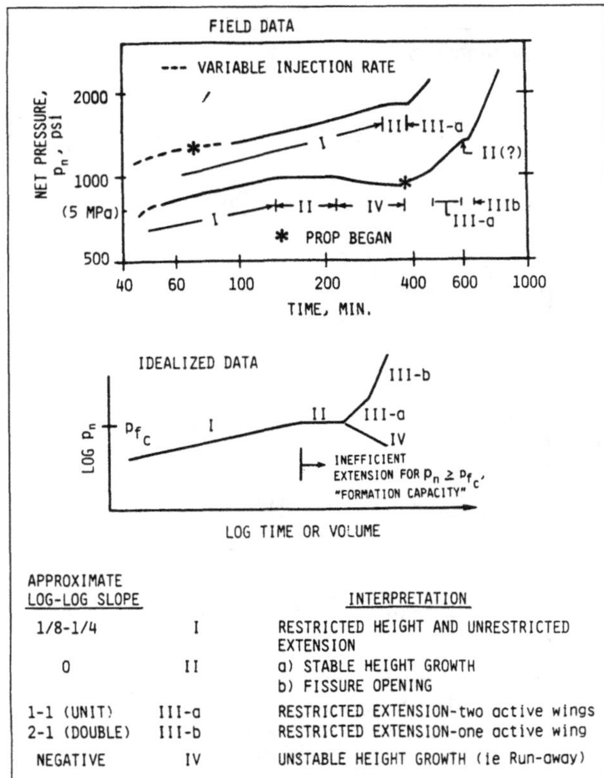

Fig. 14.16—Log-log slope interpretation for field and idealized data.

actual field and idealized data and an interpretation of the data based on the slopes. The field data are for massive treatments in tight-gas fields using crosslinked polymer fluids. The top example is for a 300,000-gal [1135-m^3] treatment of the Wattenberg field in Colorado. This formation is unusual[11] in that very high net pressure, 1,700 psi [11.7 MPa], can be confined with only modest height growth. The pressure capacity of the formation, p_{fc}, indicated by the Type II zero slope, is governed[11] by opening natural fissures. This leads to accelerated dehydration of the proppant slurry and eventually proppant bridging, causing the restricted extension indicated by the unit slope denoted as III-a. The bottom field example is for a 900,000-gal [3410-m^3] treatment in the Cotton Valley field of east Texas. This example illustrates all five types of interpretive slopes. The interpretation of the data is that the initial period of Type I slope indicates extension with restricted height, followed by the Type II slope, which defines the formation capacity from significant growth through a stress barrier (i.e., Fig. 14.7), followed by the Type IV slope, which indicates unstable vertical growth through a lower-stressed formation (i.e., Fig. 14.13) until proppant was introduced. The proppant bridged in the nearly-zero-width higher-stress barrier shown by the width profile in Fig. 14.13 and restricted further vertical growth.[11] Also, during the prior 250-minute period (i.e., 150 to 400 minutes) of significant vertical growth, the horizontal growth was retarded. As a result, the very high polymer concentration (>1%) of the fluid formed a thick polymer filter cake at the fracture tip that restricted further horizontal extension. Thus, the complete extremities of the fracture were restricted either by proppant or polymer cake, and continued injection was stored in ballooning width indicated by the Type III-a slope for the next 3 hours. After this period, the pressure was constant for a relatively short period of time (at 600 minutes, potentially from fissure opening) and subsequently increased on a 2 : 1 slope indicated by III-b. This later slope would indicate that half of the prior fracture area became restricted to flow and could have resulted from one wing of the fracture becoming blocked to flow.

The indication of the formation pressure capacity, p_{fc}, from the data for these two examples led to design[11] changes that resulted in more effective and efficient treatments for subsequent wells.

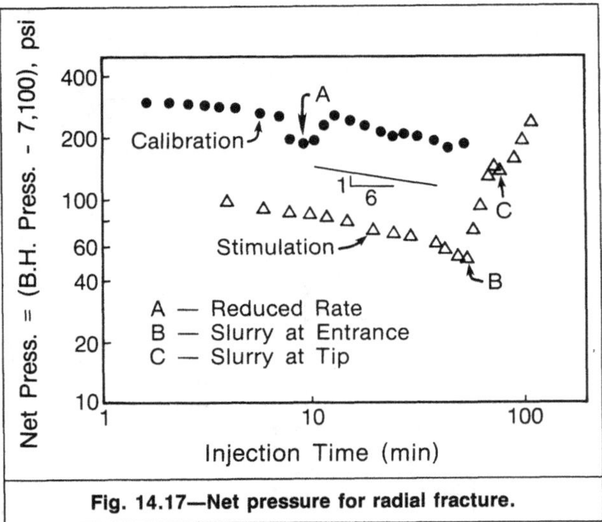

Fig. 14.17—Net pressure for radial fracture.

Example for Radial Fracture. Fig. 14.17 shows the interpretive log-log plot for radially propagating fractures in a massive chalk formation of the Valhall field in the Norwegian sector of the North Sea. The data[15] are for a calibration treatment without proppant that was performed to determine the fluid-loss coefficient for a gelled-oil fluid in the chalk formation, and the lower set of data is for the stimulation treatment of the same well with sand using a similar but lower-viscosity fluid. The closure pressure of 7,100 psi [49 MPa] was determined from a series of tests performed directly before the calibration treatment. As a result of these injections into the relatively-high-permeability oil reservoir, the closure pressure was greater[15] for the calibration treatment than the stimulation treatment performed sometime later. The higher closure pressure for the calibration treatment and relatively lower-viscosity fluid of the nonproppant portion of the stimulation produced the relative difference in net pressure values shown for the two treatments. Both curves, before Point B, show a negative slope of about ⅙ expected for a radial fracture with the low efficiencies obtained. The stimulation shows a relatively significant increase in pressure as the higher-viscosity fluid with proppant entered the fracture (Point B in Fig. 14.17) and progressed through the fracture until the slurry reached the fracture tip (Point C). A design objective[15] for the treatment was to restrict the fracture extension (i.e., tip screenout) at Point C and to continue injecting proppant with increasing width to obtain a wide proppant pack in the soft formation. The subsequent pressure increase after Point C was an indication that this objective was achieved.[15]

Simulation of Pressures. The Introduction noted the similarity of fracturing-pressure and pressure-transient analyses of reservoirs. Numerical simulation is a commonly applied and powerful tool for reservoir-pressure analysis and prediction; however, the reported application of simulation for fracturing pressures is relatively limited. The initial application reported[8] was for the pressure decline after injection for predicting fracture closure time. The reported applications of pressure simulation during injection include fracturing of the Cottage Grove formation in Oklahoma[14] (an oil sand shown in Fig. 14.18). Other applications[11,12,23] are for gas sands shown in Figs. 14.19 through 14.21. Except for the calibration treatment shown in Fig. 14.21, all three of these examples show a near-constant-pressure period indicative of the formation capacity illustrated in Fig. 14.16. The simulated causes for the capacity were vertical height growth for Figs. 14.18 and 14.20 and the opening of natural fissures for Fig. 14.19. In the latter case, the pressure capacity of about 1,700 psi [11.7 MPa] was the same value shown by the upper curve in Fig. 14.16 for a stimulation of the same formation and indicates the similarity of the rock stresses surrounding these wells. The Poisson's ratio log shown with Fig. 14.20 was found[11] from the compressional and shear velocities from a sonic log and defines the "elastic" in-situ stress log shown in Fig. 14.20 for the horizontal stress resulting from the overburden and pore pressure (see Chap. 4). Because of other effects,[29] the actual rock

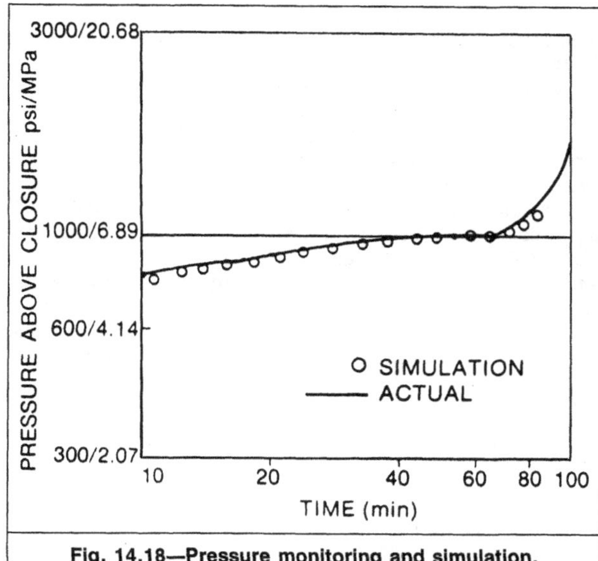

Fig. 14.18—Pressure monitoring and simulation.

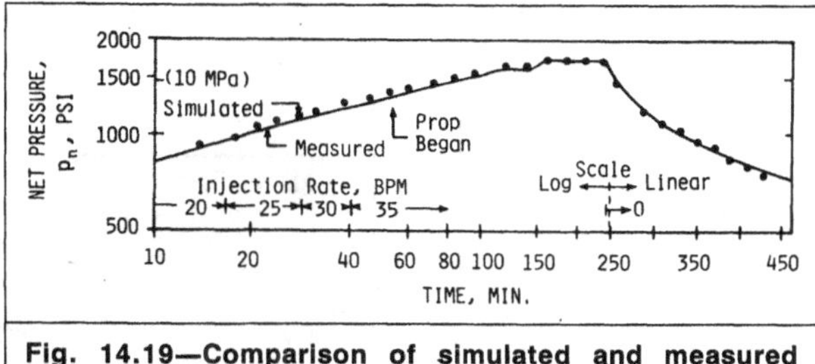

Fig. 14.19—Comparison of simulated and measured pressures.

stresses differ from the elastic stresses and must be found by small fracturing treatments in selected zones. These actual stresses in selected zones can be correlated[10] with the elastic stresses to provide a relationship for deriving the "calibrated" stress log from the elastic log. The pressure/height/ stiffness correlation shown in Fig. 14.20 was derived from the calibrated stress log and used for the simulation of the treatment pressure shown in the figure. The pressure for this example indicates a period of one-half rate in Fig. 14.20. The reduction in net pressure, and hence the tendency for height growth, is shown to be very minor for the reduction of rate[23] by one-half, as predicted by Eqs. 14.33. The dependence on rate is the $n/(2n+2)$ power, or $\frac{1}{6}$ power for $n=0.5$, and indicates only an 11% decrease in net pressure for a 50% reduction in rate. The simulation of the decreasing pressure, during the last hour of the treatment, indicated height growth into the lower-stress zones above Point C in Fig. 14.20.

The same numerical simulation program[23] was used for these reported examples. The program incorporates the combined elasticity/fluid-flow equation (Eqs. 14.31); a prescribed relationship between pressure, height, and stiffness multiplier as shown in Fig. 14.20; the fracture compliance as modified for height growth (i.e., Eq. 14.28); and the continuity equation for flow[25] that incorporates fluid loss[17] to the formation. For these examples, the numerical simulation was used to calibrate or to confirm values for the parameters that govern the pressure response (e.g., Eqs. 14.32 and 14.28). The calibrated parameters were then used to make design changes in subsequent treatments. Even though there is not a unique set of simulation parameters that match a pressure response, a calibrated set provides the rational basis for investigating the effect of changing the controllable parameters and producing more effective stimulations[11,14] in the future.

14.6 Calibration Treatments

While the above simulations were for actual treatments, a special case is a calibration treatment, which is performed without proppant before the actual stimulation treatment. The calibration treatment permits a more effective and efficient fracture stimulation to be performed later on the same well. Fig. 14.21 illustrates an example of a calibration treatment of a gas well in the Dakota basin of New Mexico. This figure was reported[12] as part of a comprehensive field and analytical study for determining the information required for an effective fieldwide fracturing program. Fig. 14.21 includes a numerical simulation of the injection pressures, the calibrated parameters, a curve-matching technique[9] for determining the fluid-loss coefficient from the pressure decline after injection, and the resulting design schedule for the subsequent stimulation treatment. The decline analysis, outlined in the following section, is similar to Eq. 14.12 for the balloon analogy. Fig. 14.17 also includes an example of a calibration treatment from which the injection-pressure response indicated a radial fracture and the subsequent decline was used for calibrating the fluid-loss coefficient.

Calibration treatments are generally performed for critical stimulations or at the beginning of a fieldwide program[12] to increase the effectiveness of fracturing. These treatments should be performed without proppant because the presence of proppant can affect the pressures (i.e., screenout), obstruct the free closure of the fracture (which would invalidate the decline analysis), and jeopardize the subsequent stimulation because of a screenout by the proppant remaining from the calibration treatment. The size of a calibration treatment can generally be small compared with the stimulation treatment, which is an economic advantage. As Figs. 14.16 through 14.20 indicate, the nature of the pressure response (radial or confined height) is indicated during the first 30% of the treatment. In addition, the fluid-loss control for a treatment is established by the initial nonproppant portion (pad) of the treatment. As a result, calibration treatments should be performed with the intended fluid system and additives, injection rate, and generally a volume approximately equal to the pad for the actual stimulation.

14.7 Analysis of Decline Pressure and Closure Time

The section on the balloon analogy provided relationships for width, penetration, fluid efficiency, and loss coefficient in terms of the rate of pressure decline and closure time after injection. Similar relationships,[9,30] in terms of an actual fracture instead of the idealized balloon, are presented here with an example. The mathematical details and considerations for application are given in Appendix J.

Assumptions. The fluid-loss coefficient, C, will be assumed to equate the rate of fluid loss through a unit area of one side of the fracture by

$$q_\ell(t)=\frac{C}{\sqrt{t-\tau}}, \quad \text{(14.47)}$$

where t=current time and τ=time at which the unit area was first exposed to fluid. Hence $t-\tau$ is the time that the area has been exposed to fluid. This expression for fluid loss is known as the Carter relation,[17] with the coefficient C generally depending on pressure. To achieve closed-form expressions for the analysis, the assumption that C is independent of pressure is required. Corrections for this assumption are given in Appendix J (Eq. J-35). Note that the pressure-independent assumption is generally made for wall-building fluids that constitute most fluids used currently. In addition, Eq. 14.47 does not include the effect of spurt loss, which may be important while the fracture area is increasing during injection but would generally not be significant after injection stops. The effect of spurt is accounted for by a multiple to C, denoted by κ. An expression for κ is given in Appendix J (Eq. J-37) and can be estimated from laboratory data. The time of first fluid exposure, τ, can be found from the upper and lower bounds given in Eqs. 14.38 and 14.41 if the injection rate and fluid properties remain essentially constant. In addition, the permeable area, A_p, that is subject to fluid loss is assumed to be a constant fraction, f_p, of the total fracture area, A_f; A_f is assumed to remain constant after injection stops.

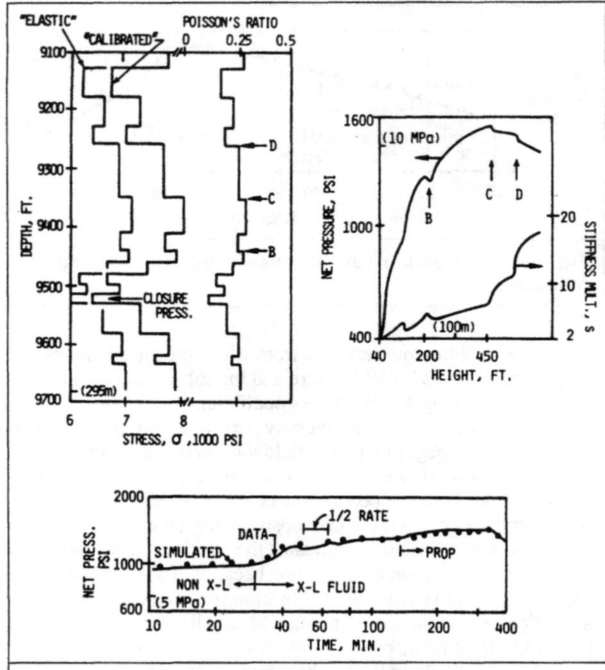

Fig. 14.20—Example application with calibrated stress from log.

The fracture's compliance, c_f, is assumed to be independent of pressure for the analyses based on pressure. This is the case (Eqs. 14.24 and 14.25) for the idealized models shown in Fig. 14.4. When height growth occurs, as shown in Fig. 14.7, the decline analysis can be applied to the latter stages of closing if no proppant is present. This is also discussed in more detail in Appendix J (see Fig. J-2). The direct applications of the pressure analyses also require that the closure pressure of the fracture remain constant. As discussed in Appendix J, meaningful changes in closure pressure generally will not occur for highly compressible (i.e., gas) or low-permeability reservoirs. For moderate- and high-permeability-liquid reservoirs, however, meaningful closure pressure changes can occur and the changing closure pressure must be considered for the analyses based on pressure.[13]

The assumption is also made that the fluid's volume changes resulting from temperature and pressure are negligible compared with the fracture's volume. This will be the case for water-based fluids without an added gas phase and will generally be a good approximation for fluids with an added gas phase if the volume is based on bottomhole conditions. If the pressure-decline analysis is performed without a downhole shut-in or complete displacement of the wellbore with water, volume changes in the borehole caused by temperature and pressure changes can produce invalid results if hydrocarbons or gases are present in the wellbore. Although the rate of volume expansion may be calculated, the resulting injection into the closing fracture cannot be directly incorporated into the analysis based on pressure. Fig. I-2 indicated that, for deeper and higher-temperature formations, the density changes for water in the wellbore can have a significant effect on surface inferences of BHP.

Basic Fluid-Loss Relationships. The total rate of fluid loss after injection for time t_i is given by Eq. J-8 in Appendix J as

$$q_\ell(t)=\frac{2Cf_pA_f}{\sqrt{t_i}}f(t_D) \quad \text{(14.48a)}$$

and

$$t_D=(t-t_i)/t_i=t/t_i-1, \quad \text{(14.48b)}$$

where A_f is one face of the fracture area and f_p is the ratio of fluid loss to fracture areas.

The dimensionless loss-rate function, $f(t_D)$, in terms of the dimensionless shut-in time, t_D, is shown in the top of Fig. 14.22. The function is given by the upper ($\eta\rightarrow1$) and lower ($\eta\rightarrow0$) bounds. Eq. 14.48a differs from the balloon analogy (Fig. 14.1) by the time dependence, $2f(t_D)/\sqrt{t_i}$, for the rate of loss.

The volume of fluid lost during a period of time after shut-in is found by integrating Eqs. 14.48a:

$$V_{\ell s}=2Cf_pA_f\sqrt{t_i}\,[g(t_D)-g(0)] \quad \text{(14.49a)}$$

and

$$g(0)=\begin{Bmatrix}4/3\\ \\ \pi/2\end{Bmatrix}, \quad \text{(14.49b)}$$

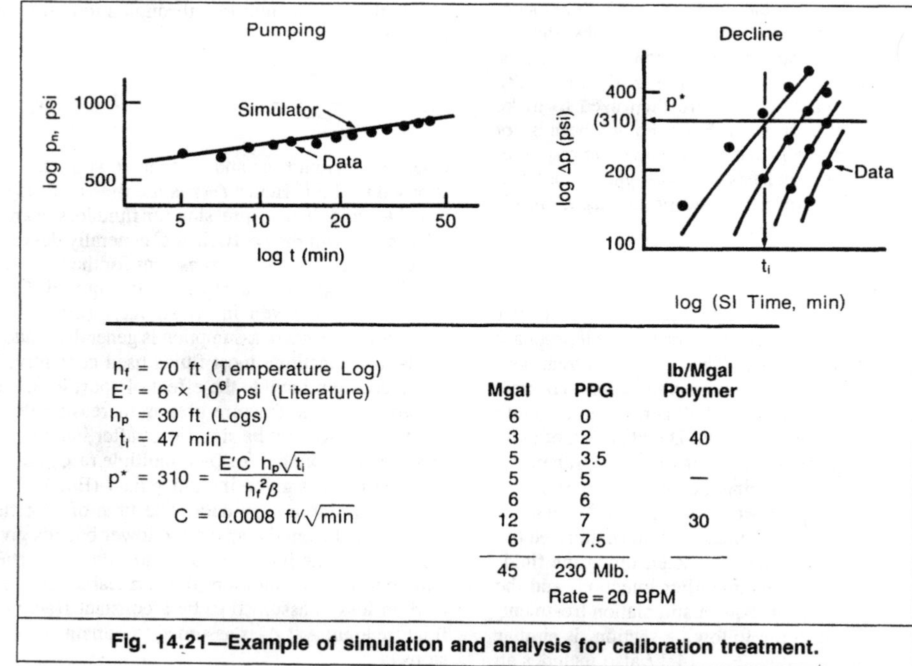

Mgal	PPG	lb/Mgal Polymer
6	0	
3	2	40
5	3.5	
5	5	—
6	6	
12	7	30
6	7.5	
45	230 Mlb.	

Rate = 20 BPM

Fig. 14.21—Example of simulation and analysis for calibration treatment.

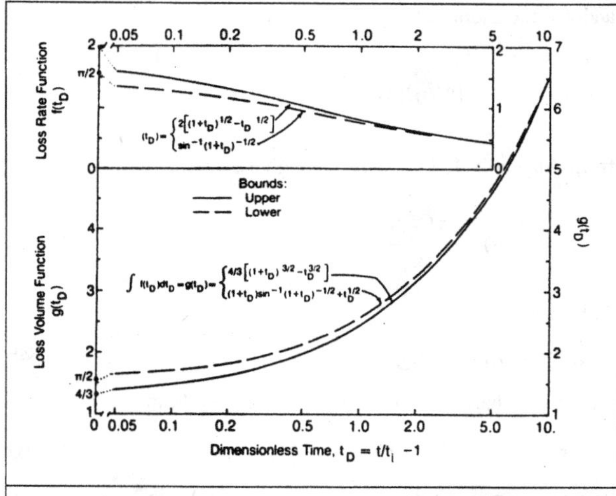

Fig. 14.22—Dimensionless functions for loss rate and volume.

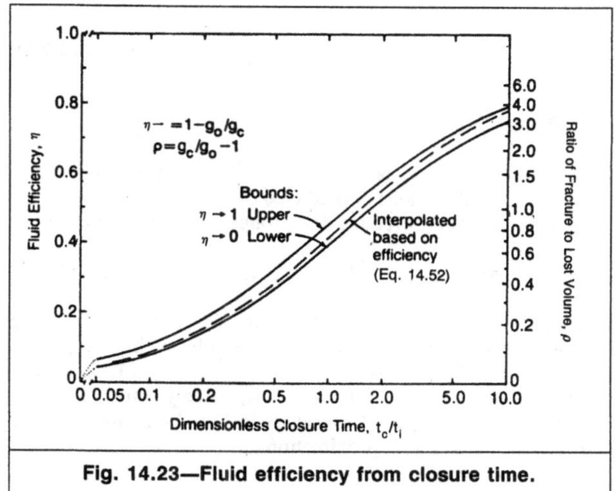

Fig. 14.23—Fluid efficiency from closure time.

for which $g(t_D)$ is the dimensionless loss-volume function, $g(0)$ is its value at $t_D=0$, and the values in the braces indicate the upper and lower bounds, respectively. The bounds for this function are shown on the lower portion of Fig. 14.22. When an unpropped fracture closes at a time t_c after injection stops, or in terms of dimensionless time, $t_{cD}=t_c/t_i$, the total fluid loss after shut-in equals the fracture's volume at shut-in, V_f. Therefore, from Eq. 14.49,

$$V_f=2Cf_pA_f\sqrt{t_i}\,(g_c-g_0), \quad (14.50)$$

$$g_c=g(t_{cD});\ g_0=g(0).$$

The volume of fluid lost during injection for time t_i is given by Eq. J-13 of Appendix J as

$$V_\ell=2\kappa Cf_pA_f\sqrt{t_i}\,g_0, \quad (14.51)$$

and using Eq. 14.49b,

$$\pi>\frac{V_\ell}{\kappa Cf_pA_f\sqrt{t_i}}>8/3,$$

with the spurt-loss correction κ included. Eq. 14.51 is analogous to Eq. 14.3 for the balloon.

The above expressions provide the basic relationships for the analysis of the fracture's pressure decline and closing time. These expressions are in terms of bounds, which indicates that the actual case falls between the bounds. Fortunately, the bounds are very close and their average would never differ by more than ±10% from the actual value. In addition, a more accurate result can be found by weighting the bounds in terms of their corresponding efficiencies; e.g.,

$$g=\eta g_u+(1-\eta)g_\ell, \quad (14.52)$$

where g_u and g_ℓ represent the upper and lower bounds for g, respectively.

Relations Based on Loss Coefficient and Closure Time. Most fluid systems currently used build an effective wall cake on low-permeability formations as they are lost into the formation. For these systems, the fluid-loss coefficient depends on the properties of the cake—i.e., the fluid system, not the formation properties. Therefore, for these systems, the loss coefficient will not vary significantly from one application to the next for a particular reservoir condition, and a value of the coefficient determined from an in-situ calibration (discussed later) can be considered a known value. Expressions presented in this section are applicable if the loss coefficient, C, is known and the fracture's closure time is found after a treatment.

If proppant is placed in the fracture, a correction must be applied for the time of closure, as illustrated for the balloon analogy in Fig. 14.3 and given by Eqs. 14.17 through 14.22. The *correction for proppant* in an actual fracture is *not included* in the following but is given in Appendix J by Eqs. J-20 and J-22.

The average fracture width, $\bar{w}$, can be found from Eq. 14.50 because $V_f=A_f\bar{w}$:

$$\bar{w}=2Cf_p\sqrt{t_i}\,(g_c-g_0), \quad (14.53)$$

which is in terms of the known loss coefficient and g evaluated at the closure time without proppant.

The parameter ρ, as used with the balloon analogy, is the ratio of the stored fracture volume to loss volume at the end of injection, $t=t_i$,

$$\rho\equiv V_f/V_\ell, \quad (14.54a)$$

which, from $V_f=A_f\bar{w}$ and Eq. 14.51, can be expressed as

$$\rho=\frac{\bar{w}}{2\kappa Cf_p\sqrt{t_i}\,g_0}. \quad (14.54b)$$

For $\kappa=f_p=1$,

$$\rho=\frac{\bar{w}}{2g_0C\sqrt{t_i}}\approx\frac{\bar{w}}{3C\sqrt{t_i}}, \quad (14.54c)$$

where the approximation of $2g_0\approx3$ follows from Eq. 14.49b. The last expression indicates that ρ is proportional to the inverse of the variable for the classic Carter area equation,[17] which is given in Eq. 14.4b. Also, ρ is related to the fluid efficiency, η, which is based on the total injected volume, V_i,

$$\eta=V_f/V_i, \quad (14.55)$$

by

$$\eta=\rho/(1+\rho) \quad (14.56a)$$

and

$$\rho=\eta/(1-\eta). \quad (14.56b)$$

As derived in Appendix J, η and ρ are related to closure time through g_c by

$$\rho=(g_c/g_0-1)/\kappa \quad (14.57a)$$

and

$$\eta=1-\frac{\kappa}{(\kappa-1+g_c/g_0)} \quad \text{(14.57b)}$$

$$=1-g_0/g_c;\ \kappa=1,$$

where $\kappa=1$ if spurt loss is negligible. Likewise, if η or ρ is known from other considerations or predicted by a design model, the time to close (using the correction for proppant if appropriate) is

$$t_{cD}=t_c/t_i=g^{-1}[g_0(\kappa\rho+1)], \quad \text{(14.58)}$$

where g^{-1} implies the inverse function of g. Because Eq. 14.58 cannot be solved analytically, a numerical solution is given in Fig. 14.23 showing the interpolated value based on Eq. 14.52. For the balloon analogy, $t_{cD}=\rho$, Eq. 14.1. Surprisingly, this result from the analogy falls between the bounds in Fig. 14.23 for $t_{cD}<0.5$ and provides a useful approximation:

$$t_{cD}=t_c/t_i\approx\rho, \quad \text{(14.59)}$$

$$t_{cD}<0.5, \text{ or } \rho<0.5,\ \eta<0.33.$$

In terms of ρ (i.e., t_{cD}) and C, the average width (from Eq. 14.54b) is

$$\bar{w}=2\kappa Cf_p\sqrt{t_i}\,g_0\rho \quad \text{(14.60a)}$$

and fracture area derived in Appendix J (Eq. J-24) is

$$A_f=\frac{V_i}{2\kappa Cf_pg_0\sqrt{t_i}\,(1+\rho)}=(h_fL) \quad \text{(14.60b)}$$

for the PK and KG models, and

$$A_f=\frac{V_i}{2\kappa Cf_pg_0\sqrt{t_i}\,(1+\rho)}=(\pi r_f^2) \quad \text{(14.60c)}$$

for the radial model, where the areas for each of the basic models of Fig. 14.4 are given in parentheses. Eqs. 14.60 are the analog of Eq. 14.7 for the balloon.

Relations Based on Pressure and Modulus. As indicated by the balloon analogy, an analysis of the pressure decline after shut-in can be used to determine the basic parameter ρ and, if the fracture compliance is known, the fluid-loss coefficient, C. These values can then be used with Eqs. 14.60 for the fracture dimensions. The compliance, from Eqs. 14.24, 14.25, and 14.27, is

$$c_f=\frac{\pi\beta}{2E'}(h_f) \quad \text{(14.61a)}$$

for the PK model,

$$c_f=\frac{\pi\beta}{2E'}(L) \quad \text{(14.61b)}$$

for the KG model, and

$$c_f=\frac{\pi\beta}{2E'}\left(\frac{32}{3\pi^2}r_f\right) \quad \text{(14.61c)}$$

for the radial model. These values indicate that the modulus and one dimension of the fracture must be specified for determining the loss coefficient.

Defining the difference in net pressure between the value at a specified time t_D^* and subsequent times t_D as

$$\Delta p_n(t_D^*,t_D)=p_n(t_D^*)-p_n(t_D) \quad \text{(14.62)}$$

and a dimensionless difference function in terms of g as

$$G(t_D,t_D^*)=\frac{4}{\pi}[g(t_D)-g(t_D^*)], \quad \text{(14.63)}$$

from Appendix J, Eqs. J-26 through J-28,

$$\Delta p_n(t_D^*,t_D)=\frac{\pi Cf_p\sqrt{t_i}}{2c_f}G(t_D,t_D^*) \quad \text{(14.64a)}$$

and

$$\Delta p_n(t_D^*,t_D)=p^*G(t_D,t_D^*), \quad \text{(14.64b)}$$

where the characteristic decline pressure is defined by

$$p^*\equiv\Delta p_n(t_D^*,t_D) \text{ for } G(t_D,t_D^*)=1. \quad \text{(14.65)}$$

Combining Eqs. 14.64a and 14.64b gives the analog of Eq. 14.12 for the balloon,

$$C=\frac{2p^*}{\pi f_p\sqrt{t_i}}c_f, \quad \text{(14.66)}$$

and substituting Eqs. 14.61,

$$C=\frac{\beta p^*}{f_p\sqrt{t_i}\,E'}(h_f) \quad \text{(14.67a)}$$

for the PK model,

$$C=\frac{\beta p^*}{f_p\sqrt{t_i}\,E'}(L) \quad \text{(14.67b)}$$

for the KG model, and

$$C=\frac{\beta p^*}{f_p\sqrt{t_i}\,E'}\left(\frac{32}{3\pi^2}r_f\right) \quad \text{(14.67c)}$$

for the radial model using Eqs. 14.61.

Fig. 14.24 shows the dimensionless difference function, G, for the expected range of t_D^* values in terms of the efficiency bounds. This figure can be used as a master curve along with actual pressure differences (defined by Eq. 14.62) for a pressure-difference matching technique to define p^*. Similarly, from Eq. 14.64a, p^* can be defined by the slope of a p_n (p_w if p_c is constant) vs. $G(t_D)=G(t_D,\ 0)$ plot. The inferred value of p^* from either method can then be used with Eqs. 14.67 for defining C. Examples of determining p^* from decline data are given in a following section.

The value of p^* and the net pressure after shut-in, p_s, can be used to define the basic parameter ρ as in the balloon analogy (Eq. 14.14). From Appendix J, Eq. J-30,

$$\rho=\frac{\eta}{1-\eta}=\frac{p_s}{\kappa p^*}\begin{Bmatrix}3\pi/16\\1/2\end{Bmatrix}, \quad \text{(14.68)}$$

where the expressions in the braces are for the upper and lower bounds, respectively, for efficiency. Combining Eqs. 14.67 and Eq. 14.60 for A_f gives the fracture penetration—i.e., L or r_f—in terms of p^*, ρ, and h_f as

$$\frac{V_iE'}{2\kappa g_0p^*(1+\rho)}=\beta(h_f^2L) \quad \text{(14.69a)}$$

for the PK model,

$$\frac{V_iE'}{2\kappa g_0p^*(1+\rho)}=\beta(h_fL^2) \quad \text{(14.69b)}$$

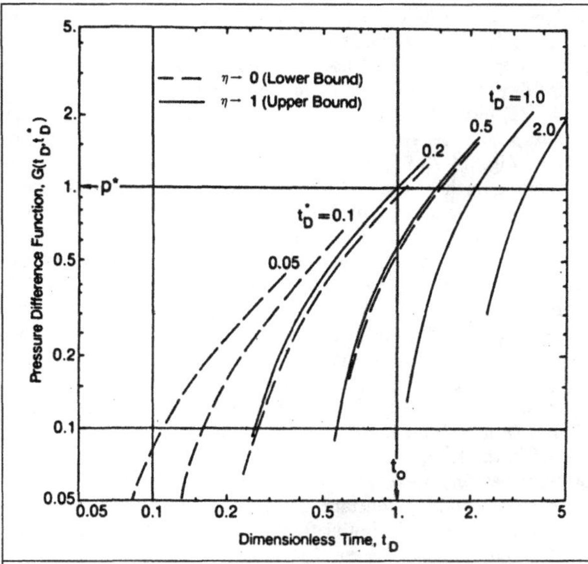

Fig. 14.24—Dimensionless-difference function—master for curve matching.

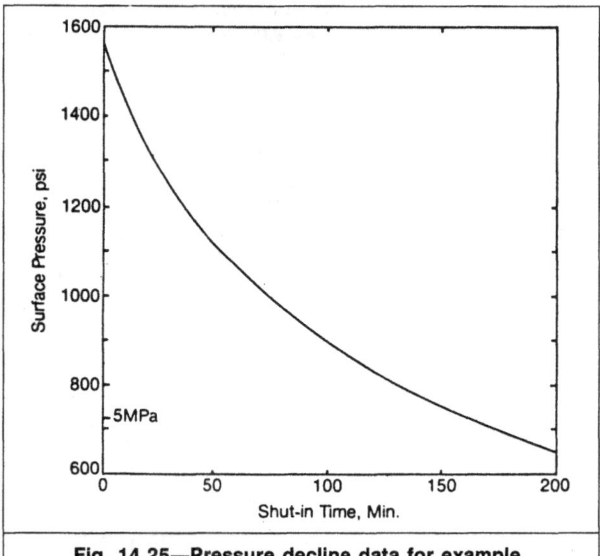

Fig. 14.25—Pressure decline data for example.

for the KG model, and

$$\frac{V_i E'}{2\kappa g_0 p^*(1+\rho)} = \beta\left(\frac{32}{3\pi} r_f^3\right) \quad \text{(14.69c)}$$

for the radial model. Also, the same combination, using Eq. 14.60a for $\bar{w}$, gives

$$\bar{w} = \frac{2 g_0 \kappa \rho p^*}{E'} \beta(h_f) \quad \text{(14.70a)}$$

for the PK model,

$$\bar{w} = \frac{2 g_0 \kappa \rho p^*}{E'} \beta(L) \quad \text{(14.70b)}$$

for the KG model, and

$$\bar{w} = \frac{2 g_0 \kappa \rho p^*}{E'} \beta\left(\frac{32}{3\pi^2} r_f\right) \quad \text{(14.70c)}$$

for the radial model. Eqs. 14.69 and 14.70 provide the analogs for Eqs. 14.15 and 14.16 for the balloon.

Expressions similar to Eqs. 14.69 and 14.70 in terms of the shut-in pressure, p_s, are given in Appendix J, Eqs. J-31b and J-32b.

Approximate Proppant Schedule. The efficiency, or equivalently ρ, defines[31] an approximate proppant addition schedule, as summarized in Appendix J, Eqs. J-33 and J-34. In particular, for the first proppant slurry to reach the fracture tip at the end of injection, the pad fraction (initial fraction of total injected volume without proppant) is

$$f_{pad} \approx \frac{1-\eta}{1+\eta} = \frac{1}{1+2\rho}. \quad \text{(14.71)}$$

The pad fraction can also be expressed in terms of the closure time without proppant by using $G_c = G(t_{cD}, 0)$ from Eq. 14.63 and the approximations (Eqs. J-29)

$$\rho \approx G_c/2 \quad \text{(14.72a)}$$

and

$$\eta \approx G_c/(2+G_c). \quad \text{(14.72b)}$$

Substitution of Eqs. 14.72 into 14.71 gives

$$f_{pad} \approx 1/(1+G_c). \quad \text{(14.73)}$$

Example. The pressure-decline expressions outlined above will be applied to the data from the Wattenberg tight-gas field in the Denver basin.[30] The formation is generally about 8,200 ft [2500 m] deep and has a permeability of $<5\times10^{-6}$ darcies and a temperature of 265°F [130°C]. The data were collected from a calibration treatment without proppant to determine the fluid-loss coefficient before a massive hydraulic fracture treatment on the well. The calibration treatment consisted of 500 bbl=2,800 ft³ [80 m³] at a rate of 5 bbl/min [13.2×10^{-3} m³/s], with a pump time, t_i, of 100 minutes. The fluid was an emulsion consisting of two-thirds condensate and one-third water-based polymer. For this fluid and formation, spurt loss is negligible. Also, the pressures during pumping were less than the natural fissure-opening pressure of the formation, interpreted to be p_n=1,700 psi [11.7 MPa] (Figs. 14.16 and 14.19). Thus, there is no correction for these effects and κ is unity.

The pressure-decline curve is shown in Fig. 14.25 with a surface shut-in pressure of about 1,550 psi [10.7 MPa]. Because of the formation's low permeability and the relatively high compressibility of gas, significant changes in closure pressure would not be expected. This was the case; Fig. 14.26 shows that both the pump-in/flowback test before the calibration treatment and a plot of decline pressure vs. square root of shut-in time after the treatment give essentially the same surface reference value of 750 psi [5.2 MPa] for the closure pressure. These methods for determining closure pressure are discussed in Appendix I. Based on this value of closure pressure, the time for the fracture to close, t_c, was 150 minutes after shut-in (Fig. 14.25 or 14.26) and the net shut-in pressure, p_s, was 1,550−750=800 psi [5.5 MPa].

Fig. 14.27 shows a plot of the log of pressure difference (Eqs. 14.62) vs. log of shut-in time with the difference taken from the dimensionless times 0.2, 0.5, and 1.0. These reference values are appropriate on the basis of the dimensionless closure time of 150/100=1.5. This figure also shows the match with the theoretical-difference curve from Fig. 14.24 and indicates a match pressure, p^*, of 370 psi [2.6 MPa]. The theoretical curves at t_D=1 are aligned with the pressure-difference data for a shut-in time equal to the injection time, t_i, and are moved vertically until most of these curves overlie the data. The indicated p^* value corresponds to the theoretical-difference value of unity. The data before closure (t_D=1.5) were matched by the later two theoretical curves (t_D^*=0.5 and 1). For times greater than the closure, the data deviate from the curves, which can be inferred as an indication of a closed fracture and deviation from the assumption of a closing

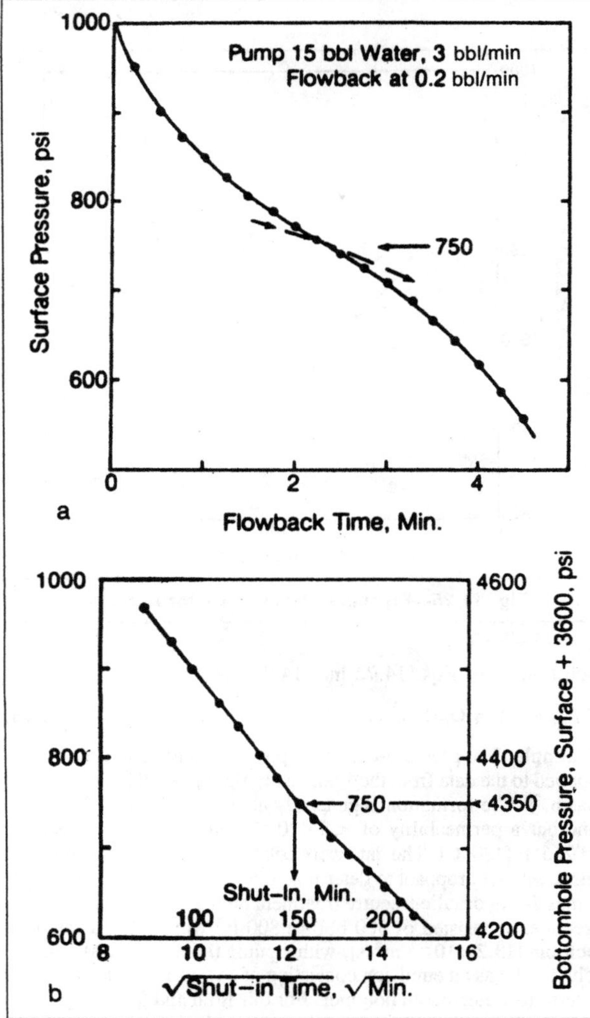

Fig. 14.26—Interpretations of closure pressure for example: (a) pump-in/flowback before treatment; (b) pressure decline after treatment.

fracture. Fig. 14.28 shows the same decline data on a plot of surface pressure (which can be used for Δp_n if closure pressure is constant) vs. $G(t_D)=G(t_D,0)$. As indicated by Eq. 14.64b, the slope of this plot is p^*. The plot shows the indication of closure in a manner similar to the square-root plot in Fig. 14.26, as discussed in Appendix I. The *closure time or pressure must be determined* to ensure that the matching technique (Fig. 14.27) or determination of the slope (Fig. 14.28) uses data before closure.

The deviation of the data from the $t_D^*=0.2$ curve (about 70 psi [483 kPa] higher) in Fig. 14.27 or for $G<0.5$ in Fig. 14.28 can be interpreted[9] as additional fracture extension[4] during this time period. Fig. 14.28 indicates that the actual shut-in pressure, p_s, is greater than the "ideal" value extrapolated to $G=0$. The "ideal" value has no physical significance. The "ideal" value of p_s, however, is consistent with the ideal assumptions for the analysis—e.g., no extension after shut-in—and, when used with Eq. 14.68, provides the same estimate of ρ as the closure-time analysis of Eq. 14.57b. When the deviation results from extension, Eq. 14.68 with the actual value of p_s provides the better estimate of ρ at the time of shut-in and hence is the better estimate for conditions during pumping. For cases with height growth into stress barriers, the actual value of p_s would be lower than the ideal because of the increase in compliance (Fig. J-2) during the closing of the height growth. As discussed in the Height Growth section in Appendix J, after the net pressure has decreased to less than half its value at shut-in, the compliance is essentially constant, and a valid decline analysis can be obtained by use of the initial fracture height for the compliance.

Other required parameters for an analysis of the example data are the height and elastic modulus. The gross sand section, which is the appropriate fracture height, h_f, of 60 ft [18.3 m], was inferred from the spontaneous potential (SP) log. The SP log also indicated the primary permeable section height, h_p, to be 32 ft [9.8 m]. These values imply that $f_p=h_p/h_f=0.53$. From core tests, the appropriate modulus for the composite section of the fracture height is about 4×10^6 psi [28 GPa]. Thus, the relevant parameters are V_i=2,800 ft^3 [80 m^3], $E'=4\times10^6$ psi [28 GPa], h_f=60 ft [18.3 m], f_p=0.53, κ=1.0, p_s=800 psi [5.5 MPa], p^*=370 psi [2.6 MPa], t_i=100 minutes, t_c=150 minutes, and t_{cD}=1.5.

From Fig. 14.23 for time to close, η=0.50 is indicated; whereas Eq. 14.68, based on the ratio of p_s/p^*, indicates a value of ρ=1.18 for which the interpolation formula (Eq. 14.52 with η=0.5) was used for the two bound values. This value of ρ from the pressure ratio indicates an η=0.54 (Eqs. 14.56). Thus, the two different bases of determining fluid efficiency differ by less than 10%, and an average value of 0.52 will subsequently be used. The different values result from the difference in the actual and ideal values of p_s. Also, the value of G at closure (i.e., G_c) of 1.75 from Fig. 14.28 can be used to approximate the efficiency from Eq. 14.73 as 0.47. As previously indicated, if the variance for η is a result of fracture extension after shut-in, the larger value of η based on the pressure ratio is more representative of the conditions during pumping. From the value of η=0.52 selected for the following analysis, the corresponding values of ρ, Eqs. 14.56, and the interpolated value of g_0, Eqs. 14.50 and Eq. 14.52, are ρ=1.08 and g_0=1.44.

The following values of β (Eqs. 14.36 and 14.37) will be used: the value of 0.77 from Fig. 14.9 for the PK model, the assumed

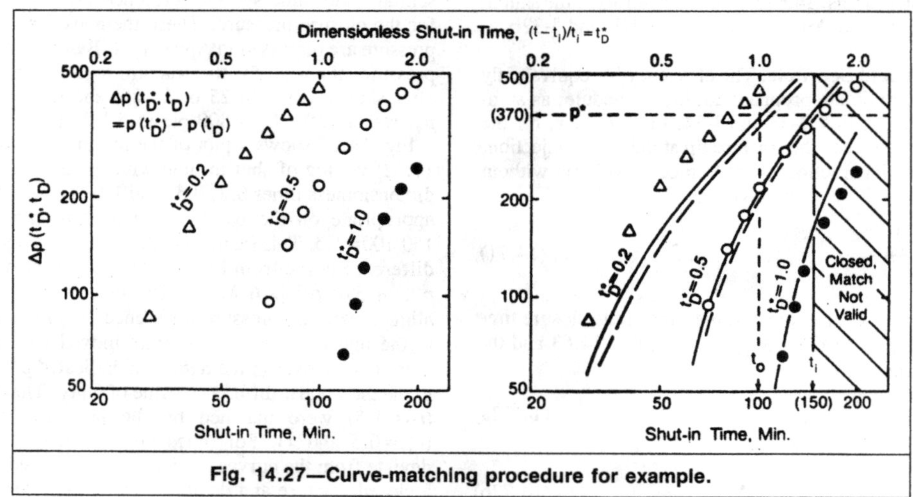

Fig. 14.27—Curve-matching procedure for example.

value of 0.9 for the KG model, and a value of $0.93=3\pi^2/32$ selected for convenience with the radial model.

For this example, it will be assumed that the loss coefficient, C, is unknown and is to be found from the analysis. Thus, the pressure relations, Eqs. 14.67 through 14.70, must be used. Also, the surface pressure can be used without corrections for closure-pressure changes because the closure pressure was found to be essentially constant.

Substituting the values for the appropriate parameters given above into Eqs. 14.69 indicates the inferred tip-to-tip penetration for the PK, KG, and radial models, respectively:

$$L = 1{,}700 \text{ ft [520 m]},$$

$$L = 290 \text{ ft [88 m], and}$$

$$2r_f = 230 \text{ ft [70 m]}.$$

For the above and subsequent values of inferred parameters, only two significant figures will be indicated to be consistent with the relative accuracy of the values of E' and h_f. Also, for purposes of illustration, the value of $f_p=0.53$ will be used for the radial model, whereas in practice this model is generally applicable for massive formations with fluid loss over the complete fracture area, $f_p=1.0$.

Using these values of penetration for the KG and radial models in Eqs. 14.67 for the loss coefficient and other appropriate parameters indicated above yields

$$C = 8.1\times10^{-4} \text{ ft}/\sqrt{\text{min}} = 3.2\times10^{-5} \text{ m}/\sqrt{\text{s}},$$

$$C = 4.6\times10^{-3} \text{ ft}/\sqrt{\text{min}} = 1.8\times10^{-4} \text{ m}/\sqrt{\text{s}}, \text{ and}$$

$$C = 1.8\times10^{-3} \text{ ft}/\sqrt{\text{min}} = 7.2\times10^{-5} \text{ m}/\sqrt{\text{s}}$$

for the PK, KG, and radial models, respectively.

For the emulsion fluid used and the gas-filled low-permeability formation, fluid-loss resistance is likely dominated by a compressible wall cake; i.e., $e=1/6$ for Eq. J-35 of Appendix J. Therefore, for a reservoir BHP of 3,200 psi [22.1 MPa], a closure BHP of 4,350 psi [30 MPa], and $p_s=800$ psi [5.5 MPa], the pressure correction to C would be less than $(1+800/1{,}350)^{1/6}=1.08$. This maximum correction is within the uncertainty of the other parameters and will subsequently be ignored. If the loss resistance were dominated by the reservoir (e.g., a liquid-filled reservoir), however, the maximum correction would be with $e=1$ in Eq. J-35. For this case, the maximum correction would be 1.6, which would be very significant. If this were the case, the pressure dependence for C should be considered in subsequent applications.

The average fracture width at shut-in, based on the pressure relations, is found from Eqs. 14.70 to be

$$\bar{w}=0.16 \text{ in.}=4.1\times10^{-3} \text{ m},$$

$$\bar{w}=0.91 \text{ in.}=2.3\times10^{-3} \text{ m},$$

and

$$\bar{w}=0.36 \text{ in.}=9.1\times10^{-3} \text{ m}$$

for the PK, KG, and radial models, respectively.

The above-inferred values for penetration, loss coefficient, and width indicate significant variations in each of the parameters on the basis of the fracture model assumed to be applicable. The increasing net pressures shown in Figs. 14.16 and 14.19 for treatments of other wells in the Wattenberg field indicate that the PK model is most applicable.

Based on $\eta=0.52$ for this example, Eq. 14.71 would give an approximate pad fraction of 0.32 for a propped stimulation, whereas based on $G_c=1.75$, Eq. 14.73 would indicate an approximate pad fraction of 0.36. The actual required pad fraction would generally be larger for a stimulation treatment than indicated by the calibration treatment because of the reduced efficiency[31] for the generally larger stimulation treatment.

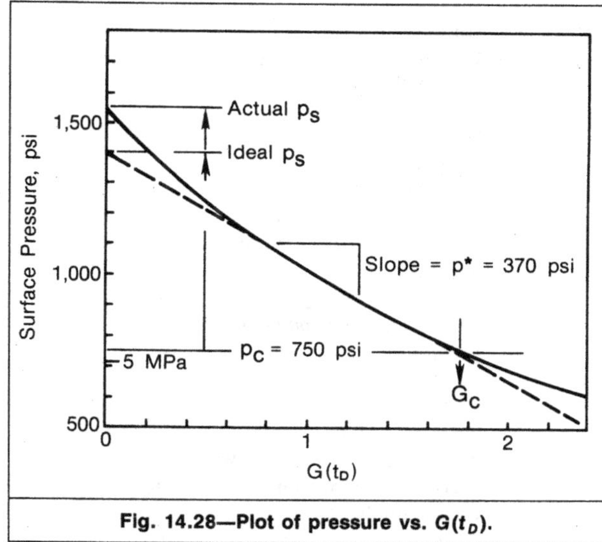

Fig. 14.28—Plot of pressure vs. $G(t_D)$.

Nomenclature

a = exponent for variation of K
A = general variables for area
A_f = fracture area
A_p = permeable, or loss, area
c = compliance
c_f = fracture-width compliance
C = fluid-loss coefficient
d = characteristic dimension for model
e = exponent
E = Young's modulus
E' = plane-strain elastic modulus
$f(t_D)$ = dimensionless loss-rate function
f_C = variable of Carter area equation, Eq. 14.4b
f_p = ratio of permeable to fracture areas, $A_p/A_f=h_p/h_f$
f_{pad} = pad volume fraction based on V_i
f_{Vp} = volume fraction of proppant, V_{pr}/V_i
g_c = $g(t_c/t_0)$
g_ℓ = lower bound for g
$g(t_D)$ = dimensionless loss-volume function
g_u = upper bound for g
g_0 = $g(t_D=0)$
$G(t_D,t_D^*)$ = dimensionless difference function, Eq. 14.63
$G(t_D)$ = $G(t_D,0)$, variable for decline plot, Fig. 14.28
h_f = vertical fracture height
h_p = permeable section height
K = coefficient for power-law fluid
L = fracture length, tip to tip
m = M_r/M_o
M = rate of pressure change
M_o = M for tip screenout
M_r = M for restriction less than A_f
n = power-law exponent for fluid
p = pressure
p^* = characteristic decline pressure, Eq. 14.65
p_c = closure pressure
p_{fc} = formation pressure capacity
p_n = net pressure, p_w-p_c, at wellbore
$\hat{p}_n$ = equivalent uniform p_n to define $\bar{w}$
p_p = net pressure for propped width
p_s = net pressure at shut-in
p_w = pressure in wellbore
q = flow rate in fracture
q_i = constant injection rate of fluid
q_ℓ = fluid-loss rate

r_f = fracture radius
s = stiffness multiplier
t = time since start of pumping
t_c = closure time after shut-in
t_{cD} = dimensionless closure time, t_c/t_i
t_D = dimensionless time, $t/t_i - 1$
t_D^* = reference value of t_D
t_i = injection or pumping time
v = fluid velocity in fracture
V_f = fracture volume, end of pumping
V_i = total volume injected
V_ℓ = loss volume during pumping
$V_{\ell s}$ = loss volume after pumping ends
V_{pr} = proppant volume with voids
V_s = solid volume of proppant
w = fracture width
$\bar{w}$ = average fracture width, end of pumping
$\bar{w}_p$ = average propped fracture width
z = variable distance along fracture length
α = time multiple
β = $\hat{p}_n/p_n$
γ = characteristic time variable for analogy
$\dot{\gamma}$ = shear rate
Δ = difference
η = fluid efficiency, V_f/V_i
κ = correction for C while pumping, e.g., spurt loss
ν = Poisson's ratio
ρ = V_f/V_ℓ, ratio of stored to loss volumes
σ = stress
τ = time fracture area was created
τ_s = shear stress
ω_{pr} = specific weight of proppant

Subscripts

h = minimum horizontal
H = maximum horizontal
i = injection
s = shut-in
v = vertical
w = well

Superscripts

$^-$ = average
$'$ = for closure on proppant

References

1. Harrison, E., Kieschnich, W.F., and McGuire, W.J.: "The Mechanics of Fracture Induction and Extension," *Trans.*, AIME (1954) **201**, 252–63.
2. Khristianovich, S.A. and Zheltov, Y.P.: "Formation of Vertical Fractures by Means of Highly Viscous Liquid," *Proc.*, Fourth World Pet. Cong., Rome (1955) Sec. II, 579–86.
3. Godbey, J.K. and Hodges, H.D.: "Pressure Measurements During Fracturing Operations," *Trans.*, AIME (1958) **213**, 65–69.
4. Perkins, T.K. and Kern, L.R.: "Widths of Hydraulic Fractures," *JPT* (Sept. 1961) 937–49; *Trans.*, AIME, **222**.
5. Geertsma, J. and deKlerk, F.: "A Rapid Method of Predicting Width and Extent of Hydraulic Induced Fractures," *JPT* (Dec. 1969) 1571–81; *Trans.*, AIME, **246**.
6. Fast, C.R., Holman, G.B., and Covlin, R.J.: "The Application of Massive Hydraulic Fracturing to the Tight Muddy 'J' Formation, Wattenberg Field, Colorado," *JPT* (Jan. 1977) 10–16.
7. Simonson, E.R., Abou-Sayed, A.S., and Clifton, R.J.: "Containment of Massive Hydraulic Fractures," *SPEJ* (Feb. 1978) 27–32.
8. Novotny, E.J.: "Proppant Transport," paper SPE 6813 presented at the 1977 SPE Annual Technical Conference and Exhibition, Denver, Oct. 9–12.
9. Nolte, K.G.: "Determination of Fracture Parameters from Fracturing Pressure Decline," paper SPE 8341 presented at the 1979 SPE Annual Technical Conference and Exhibition, Las Vegas, Sept. 23–26.
10. Nolte, K.G. and Smith, M.B.: "Interpretation of Fracturing Pressures," *JPT* (Sept. 1981) 1767–75.
11. Nolte, K.G.: "The Application of Fracture Design Based on Pressure Analysis," *SPEPE* (Feb. 1988) 31–42.
12. Veatch, R.W. and Crowell, R.F.: "Joint Research/Operations Programs Accelerate Massive Hydraulic Fraturing Technology," *JPT* (Dec. 1982) 2763–75.
13. Smith, M.B.: "Stimulation Design for Short, Precise Hydraulic Fractures," *SPEJ* (June 1985) 371–79.
14. Elbel, J.L. *et al.*: "Stimulation Study of Cottage Grove Formation," *JPT* (July 1984) 1199–1205.
15. Smith, M.B., Miller, W.K., and Haga, J.: "Tip Screenout Fracturing: A Technique for Soft, Unstable Formations," *SPEPE* (May 1987) 95–103.
16. Morris, C.W. and Sinclair, A.R.: "Evaluation of Bottomhole Treatment Pressure for Geothermal Well Hydraulic Fracture Stimulation," *JPT* (May 1984) 829–36.
17. Howard, G.C. and Fast, C.R.: "Optimum Fluid Characteristics for Fracture Extension," Appendix by R.D. Carter, presented at Mid-Cont. Dist. Spring Meeting, Tulsa, 1957; also *Hydraulic Fracturing*, Monograph Series, SPE, Richardson, TX, **2**.
18. Sneddon, I.N.: "The Distribution of Stress in the Neighborhood of a Crack in an Elastic Solid," *Proc.*, Royal Soc. London (1946) **A**, 187, 227.
19. Sneddon, I.N. and Elliott, A.A.: "The Opening of a Griffith Crack Under Internal Pressure," *Quart. Appl. Math.* (1946) **11**, 262.
20. Perkins, T.K.: "Discussion of On the Design of Vertical Hydraulic Fractures," *JPT* (Jan. 1973) 93–95.
21. Geertsma, J. and Haafkens, R.: "A Comparison of the Theories for Predicting Width and Extent of Vertical Hydraulically Induced Fractures," *J. Energy Resources Tech.* (March 1979) 8–19; *Trans.*, ASME, **101**.
22. Barree, R.D.: "A Practical Numerical Simulator for Three-Dimensional Fracture Propagation in Heterogeneous Media," paper SPE 12273 presented at the 1983 SPE Symposium on Reservoir Simulation, San Francisco, Nov. 15–18.
23. Nolte, K.G.: "Principles of Fracture Design Based on Pressure Analysis," *SPEPE* (Feb. 1988) 22–30.
24. Nolte, K.G.: "Fluid Flow Considerations in Hydraulic Fracturing," paper SPE 18537 presented at the 1988 SPE Eastern Regional Meeting, Charleston, WV, Nov. 1-4.
25. Nordgren, R.P.: "Propagation of a Vertical Hydraulic Fracture," *SPEJ* (Aug. 1972) 306–14; *Trans.*, AIME, **253**.
26. Hubbert, M.K. and Willis, D.G.: "Mechanics of Hydraulic Fracturing," *Trans.*, AIME (1957) **210**, 153–66.
27. Mahoney, J.W. *et al.*: "Effects of a No-Proppant Foam Stimulation Treatment on a Coal-Seam Degasification Borehole," *JPT* (Nov. 1981) 2227–35.
28. Wood, M.D. *et al.*: "Fracture Proppant Mapping by Use of Surface Superconducting Magnetometers," paper SPE 11612 presented at the 1983 SPE/DOE Low Permeability Gas Reservoirs Symposium, Denver, March 14–16.
29. Prats, M.: "Effect of Burial History on the Subsurface Horizontal Stresses of Formations Having Different Material Properties," *SPEJ* (Dec. 1981) 658–62.
30. Nolte, K.G.: "A General Analysis of Fracturing Pressure Decline With Application to Three Models," *SPEFE* (Dec. 1986) 571–83.
31. Nolte, K.G.: "Determination of Proppant and Fluid Schedules From Fracturing-Pressure Decline," *SPEPE* (July 1986) 255–65; *Trans.*, AIME, **283**.

Chapter 15
Postfracture Formation Evaluation

W. John Lee, SPE, Texas A&M U.

15.1 Overview

This chapter discusses techniques used to analyze wells after a fracture treatment. Discussed first are methods of analyzing productivity-index increase expected in fracturing, applicable for stabilized flow in both damaged and undamaged wells and for unsteady-state flow. Then methods of analyzing postfracture production are explained. In this section, the influence of fracture length and conductivity, drainage area size and shape, and formation permeability is detailed. Also, modeling techniques for postfracture production are analyzed, including dimensionless solutions to flow equations and finite-difference reservoir simulations. To allow comparison of unfractured- and fractured-well production forecasts, prefracture production forecasts are also discussed. Flow regimes in hydraulically fractured formations are treated next, followed by a discussion of postfracture transient test analysis techniques. This chapter deals briefly with the effects of nonideal conditions, including fracture-face damage and water blocking, and concludes with a discussion of postfracture transient test design. Examples illustrating major points in this chapter have been collected in Appendix K.

15.2 Analysis of Productivity-Index Increase

Productivity Index for Oil and Gas Wells. Comparison of productivity indices of a well after and before fracturing is a simple and convenient measure of treatment success. Comparison of productivity indices is more meaningful than comparison of rates because rate is related to drawdown imposed, and the pressure drawdown in a well before and after fracturing may change dramatically. In contrast, productivity index, the ratio of rate to pressure drawdown, is influenced more directly by formation and completion properties.

For an oil well, productivity index, J, is defined as

$$J=\frac{q_o}{(\bar{p}-p_{wf})}. \qquad (15.1)$$

From the pseudosteady-state-flow equation for an oil well,

$$\bar{p}-p_{wf}=\frac{141.2q_oB_o\mu_o}{kh}\left[\ln\left(\frac{r_e}{r_w}\right)-0.75+s\right]. \qquad (15.2)$$

Thus, during pseudosteady-state flow of oil,

$$J=\frac{q_o}{(\bar{p}-p_{wf})}=\frac{kh}{141.2B_o\mu_o\left[\ln\left(\frac{r_e}{r_w}\right)-0.75+s\right]}. \qquad (15.3)$$

For a gas well, the corresponding form of the pseudosteady-state-flow equation is

$$J=\frac{q_g}{\bar{p}_a-p_{a,wf}}=\frac{kh}{141.2\bar{B}_g\bar{\mu}\left[\ln\left(\frac{r_e}{r_w}\right)-0.75+s'\right]}, \qquad (15.4)$$

where p_a is adjusted pressure, defined in Chap. 2. In terms of pressure itself, an adequate approximation to the pseudosteady-state-flow equation is

$$\bar{p}-p_{wf}=\frac{141.2q_g\bar{B}_g\bar{\mu}}{kh}\left[\ln\left(\frac{r_e}{r_w}\right)-0.75+s'\right], \qquad (15.5)$$

where

$$\bar{B}_g=\frac{5.04T\bar{z}}{p_{av}}=\frac{10.08T\bar{z}}{(\bar{p}+p_{wf})}. \qquad (15.6)$$

Thus, the flow equation may be written as

$$\bar{p}^2-p_{wf}{}^2=\frac{1{,}422q_gT\bar{z}\bar{\mu}}{kh}\left[\ln\left(\frac{r_e}{r_w}\right)-0.75+s'\right]. \qquad (15.7)$$

If we place rate and *all* the terms that are functions of pressure on the left side of the equation and the constant terms on the right side, the result is

$$\frac{q_g\bar{\mu}\bar{z}}{(\bar{p}^2-p_{wf}{}^2)}=\frac{kh}{1{,}422T\left[\ln\left(\frac{r_e}{r_w}\right)-0.75+s'\right]}. \qquad (15.8)$$

Thus, for a gas well at pseudosteady state, the group

$$\frac{\bar{q}\bar{\mu}\bar{z}}{(\bar{p}^2-p_{wf}{}^2)}$$

should be constant; we can consider this group to be a sort of productivity index, just as $q_o/(\bar{p}-p_{wf})$ is for an oil well. For highly simplified work with gas wells, we can define J as $q_g/(\bar{p}-p_{wf})$, but for more accurate calculations, we must remember that this group is not constant and independent of pressure as it is for an oil well. In fact, we have just seen that

$$\frac{q_g}{(\bar{p}-p_{wf})}=\text{constant}\frac{(\bar{p}+p_{wf})}{\bar{\mu}\bar{z}}. \qquad (15.9)$$

For more accurate work, then, the constant group

$$\frac{q_g\bar{\mu}\bar{z}}{(\bar{p}^2-p_{wf}{}^2)}$$

should be used to characterize the ability of a gas well to produce with a given pressure drawdown.

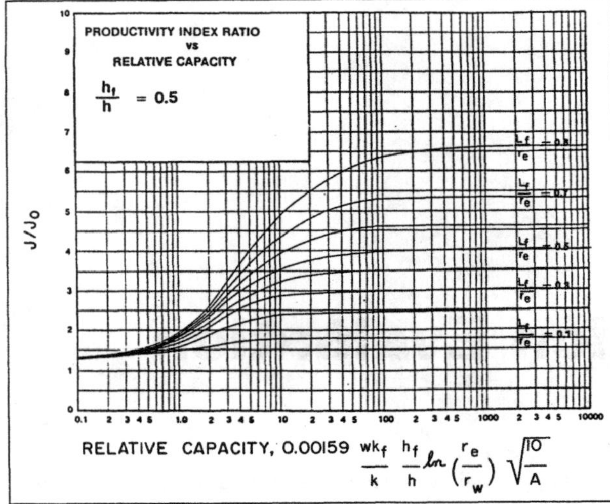

Fig. 15.1—Tinsley *et al.*'s[2] correlation of increase in productivity from fracturing, $h_f/h=1.0$.

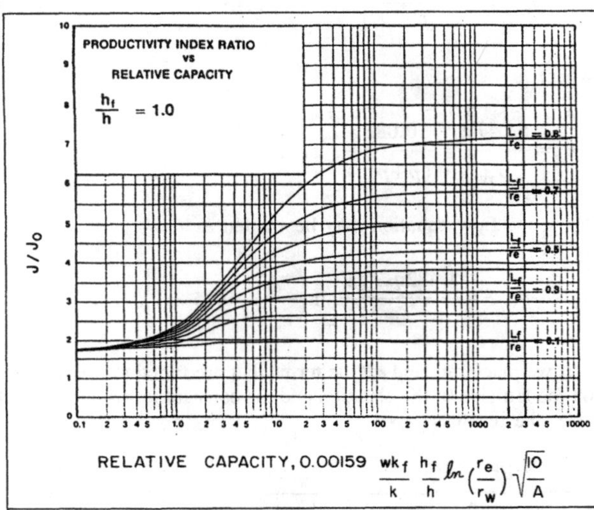

Fig. 15.2—Tinsley *et al.*'s[2] correlation of increase in productivity from fracturing, $h_f/h=0.5$.

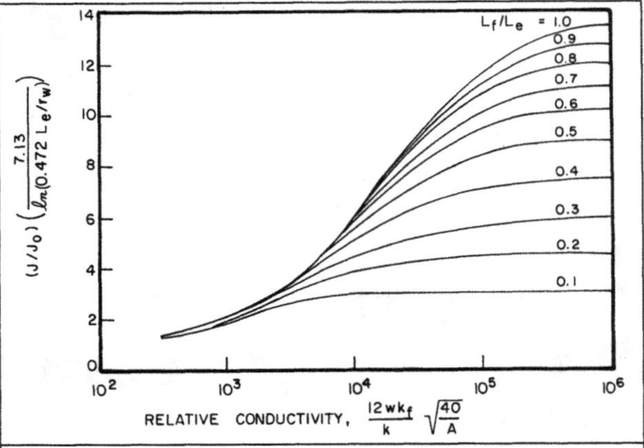

Fig. 15.3—McGuire and Sikora's[3] correlation of increase in productivity from fracturing.

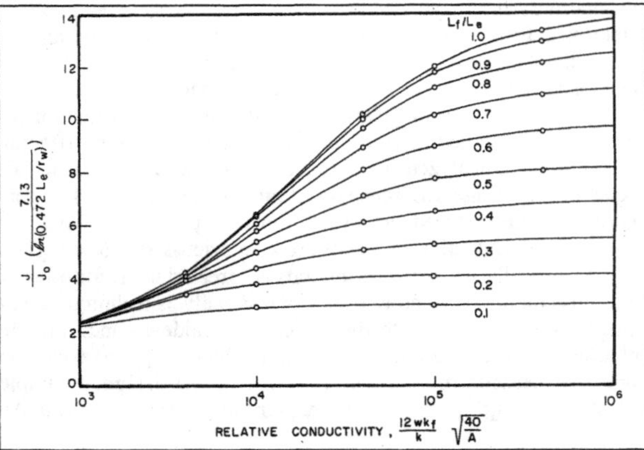

Fig. 15.4—Holditch's[4] modification of McGuire and Sikora's correlation of increase in productivity from fracturing.

Productivity-Index Ratio. For both oil and gas wells, the effect of fracturing can be represented conveniently as the ratio of productivity index after and before fracturing, J/J_o. Widely used methods of calculating J/J_o were developed by Prats,[1] Tinsley *et al.*,[2] and McGuire and Sikora[3] for wells at *steady state* or *pseudosteady state*. Low-permeability formations with long fractures require months or years of constant-rate production to achieve pseudosteady state. Steady state is achieved only after similar long periods of constant-rate production. Additionally, constant pressure is required at the drainage boundary of a well, which is a convenient approximation for solving equations, but one unlikely to be achieved in practice. Also, many wells are produced at conditions more closely approximating production at constant bottomhole pressure (BHP) than at constant rate, and methods of calculating J/J_o based on constant-rate techniques become even more approximate.

Prats' Method. Prats'[1] method is the simplest, most easily applicable technique for determining productivity-index ratio; its weakness is the highly idealized conditions of applicability. Prats found that

$$J/J_o=\frac{\ln(r_e/r_w)}{\ln(r_e/0.5L_f)}. \quad \text{(15.10)}$$

The assumptions on which Prats' analytic solution is based include steady-state flow (constant rate and constant pressure at the drainage radius), cylindrical drainage area, incompressible fluid flow, infinite fracture conductivity, and propped fracture height equal to formation height. Application of Prats' method is illustrated in Example 15.1 in Appendix K.

Tinsley* et al. *Chart. Tinsley *et al.*[2] also predicted postfracture-to-prefracture productivity-index ratio, J/J_o, but with methods and assumptions significantly different from those Prats used. Tinsley *et al.* used electrolytic model studies to generate J/J_o and, significantly, provided results for various fractions of formation height propped. Tinsley *et al.*'s results are in the form of figures (Figs. 15.1 and 15.2) rather than a simple equation as provided by Prats. The assumptions on which these figures are based include steady-state flow (constant-rate production and constant pressure at the drainage radius), cylindrical reservoir geometry, incompressible fluid flow, and an r_e/r_w in the field equal to that used in the model (or, more realistically, that the effect of the difference is negligible). Tinsley *et al.* examined the influence of the fraction of formation thickness with a propped fracture, whereas the other two methods assumed that the entire net pay thickness contains proppants. Application of Tinsley *et al.*'s charts is illustrated in Example 15.1 in Appendix K.

McGuire-Sikora Chart. McGuire and Sikora[3] also developed a chart (Fig. 15.3) for estimating productivity-index-ratio change as a result of fracturing. In some ways, the assumptions on which this chart is based (using an electric analog study) more closely model actual field conditions. This chart is based on the assumptions of pseudosteady-state flow (constant-rate production with no flow across the outer boundary), square drainage area, compressible fluid flow, and a fracture propped throughout the entire productive in-

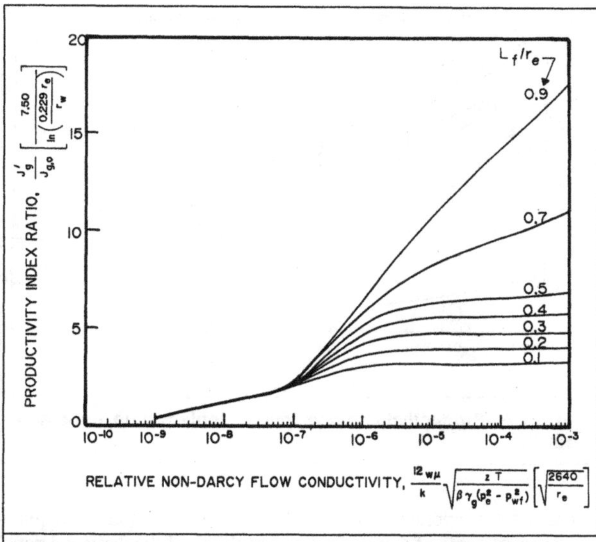

Fig. 15.5A—Stimulation ratio from reservoir-flow-pattern alteration for gas wells.

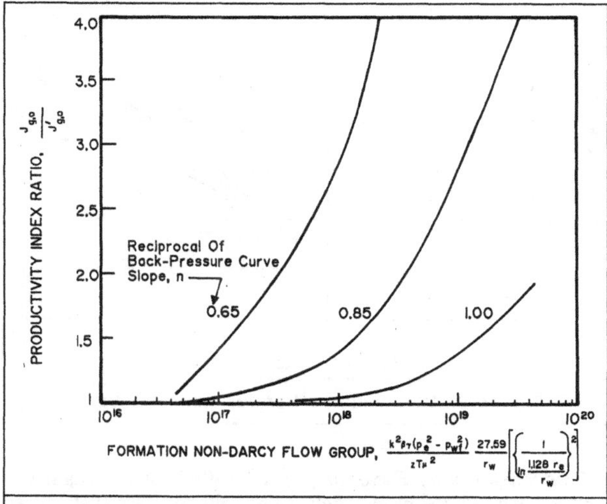

Fig. 15.5B—Productivity increase from removal of near-wellbore non-Darcy flow restriction.

terval. The ratio of productivity indices for stabilized constant-BHP production is essentially the same as for pseudosteady-state flow.

The abscissa on the McGuire-Sikora chart is relative conductivity,

$$\frac{12wk_f}{k}\sqrt{\frac{40}{A}},$$

and the ordinate is a scaled productivity increase,

$$(J/J_o)\left[\frac{7.13}{\ln\left(0.472\frac{L_e}{r_w}\right)}\right].$$

Scaling factors are required to transform the results of the McGuire and Sikora experiments from the 40-acre [16-ha] drainage area and L_e/r_w of 2,640 modeled to the drainage area and L_e/r_w for the field situation being analyzed.

Several conclusions about postfracture well performance can be reached from information given in the McGuire-Sikora chart.

1. In low-permeability reservoirs (with high relative conductivity —e.g., 10^5), greater productivity increases will result from increasing fracture length than from increasing fracture conductivity.
2. For a given fracture length, there is an optimal fracture conductivity. Increasing fracture conductivity beyond this optimal value will be unprofitable. As an example, for $L_f/L_e=0.5$, there is essentially no increase in J/J_o for relative conductivity greater than about 10^5.
3. The theoretical maximum increase in productivity-index ratio for an undamaged well is about 13.6 ($L_f/L_e=1$, relative conductivity$=10^6$).

Holditch[4] simulated fracture well performance with a modern finite-difference reservoir simulator and correlated his results as McGuire and Sikora did. He made the same assumptions: a well centered in a square reservoir with no-flow outer boundaries, slightly compressible liquid flow, fracture propped throughout the production interval, and pseudosteady-state flow. Holditch's results (Fig. 15.4) broadly confirm those of McGuire and Sikora, although there is some difference in detail. Where the results differ, those from the finite-difference simulator are more accurate. Application of Holditch's modification of the McGuire-Sikora chart is illustrated in Example 15.1 of Appendix K.

Tannich and Nierode's Chart. Tannich and Nierode[5] developed a graphic method similar to the McGuire-Sikora chart but specifically for gas wells and including the effects of non-Darcy flow. The method consists of two charts: the first describes the part of the stimulation caused by reservoir-flow-pattern modification in direct analogy to the McGuire-Sikora chart, and the second gives the amount of stimulation attributable to the reduction of radial flow, near-wellbore, non-Darcy-flow pressure loss. The two components of the overall stimulation are shown on the right side of Eq. 15.11.

$$\frac{J_g'}{J_{g,o}'}=\left(\frac{J_g'}{J_{g,o}}\right)\left(\frac{J_{g,o}}{J_{g,o}'}\right). \quad (15.11)$$

The stimulation component resulting from alteration of the reservoir flow pattern is shown in Fig. 15.5A, where productivity-index ratio multiplied by a scale factor is shown as a function of dimensionless fracture length and relative non-Darcy flow conductivity, C_{nr}. Eq. 15.12 shows that C_{nr} includes non-Darcy flow effects in the fracture through the fracture turbulence coefficient, β_f. Values for this coefficient can be obtained from laboratory data correlations, such as that reported by Cooke.[6]

$$C_{nr}=\frac{12w\mu}{k}\sqrt{\frac{zT}{\beta_f\gamma_g(p_e^2-p_{wf}^2)}}\sqrt{\frac{2,640}{r_e}}. \quad (15.12)$$

The stimulation contribution from the reduction of non-Darcy flow restriction in the reservoir rock itself is represented by the second chart (Fig. 15.5B). The productivity-index-ratio increase is shown as a function of the reciprocal slope, n, of the backpressure curve and the formation non-Darcy flow group, G_{nf}. As shown in Eq. 15.13, G_{nf} contains the formation non-Darcy flow factor, β, to quantify the influence of non-Darcy flow in the rock matrix. Specific values can be estimated from such correlations as that reported by Katz *et al.*[7]:

$$G_{nf}=\frac{k^2\beta\gamma_g(p_e^2-p_{wf}^2)}{zT\mu^2}\left[\frac{27.59}{r_w}\left(\frac{1}{\ln\frac{1.128r_e}{r_w}}\right)^2\right]. \quad (15.13)$$

The stimulation-ratio component from Fig. 15.5B is included or excluded from the overall stimulation predicted for a fracture treatment depending on well-completion history or pressure-buildup-test analysis results. Often the reservoir non-Darcy flow restriction has already been eliminated by an effective matrix acid treatment or a fracture-rate hole-opening treatment. In these cases, only the component resulting from reservoir flow-pattern modification would be achieved from a subsequent proppant fracture treatment.

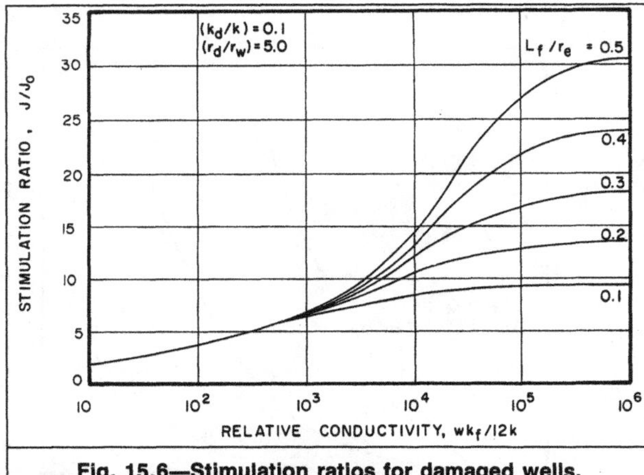

Fig. 15.6—Stimulation ratios for damaged wells.

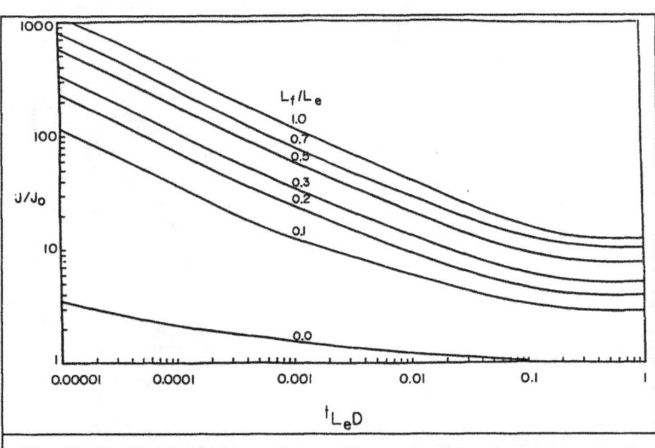

Fig. 15.7—Productivity ratio vs. dimensionless time, constant-rate flow.

Productivity-Index Ratio for Damaged Wells. When a damaged well is fractured, the productivity-index-ratio increase at stabilized conditions can be far larger than predicted by such techniques as the McGuire-Sikora chart. This chart assumes no formation damage at the time of fracturing; in fact, severe damage exists in many cases and may even be a fundamental reason for the fracture treatment.

Raymond and Binder[8] developed an approximate method of predicting productivity-ratio increase for finite-conductivity fractures in pseudosteady-state flow in cylindrical drainage areas. They showed that

$$\frac{J}{J_o}=\frac{\dfrac{k}{k_d}\ln\left(\dfrac{r_d}{r_w}\right)+\ln\left(\dfrac{r_e}{r_d}\right)}{\dfrac{k}{k_d}\ln\dfrac{r_d+\dfrac{w}{\pi}\left(\dfrac{k_f}{k_d}-1\right)}{r_w+\dfrac{w}{\pi}\left(\dfrac{k_f}{k_d}-1\right)}+\ln\dfrac{L_f+\dfrac{w}{\pi}\left(\dfrac{k_f}{k}-1\right)}{r_d+\dfrac{w}{\pi}\left(\dfrac{k_f}{k}-1\right)}+\ln\left(\dfrac{r_e}{L_f}\right)}. \quad (15.14)$$

This equation is a reasonable approximation for $L_f/L_e \leq 0.5$, as Raymond and Binder showed in a comparison with the McGuire-Sikora chart.

Fig. 15.6 is an example of calculations made with the Raymond-Binder equation for the specific case of $r_w=0.25$ ft [0.076 m], $r_d=1.25$ ft [0.38 m], and $k_d/k=0.1$; i.e., damage extends five times the wellbore radius into the formation, and the permeability in the damaged zone is 10% of the native formation permeability. Fig. 15.6 shows that for $L_f/L_e=0.5$, J/J_o can be >30 for relative conductivity, $wk_f/12k \cong 1\times10^6$.

Quantitative application of the Raymond-Binder equation requires estimates of depth of damage and permeability of the damaged zone. While it is not possible to determine both these quantities for the usual stimulation candidate, it is usually possible to determine skin factor, s, from a prefracture pressure-buildup test. Skin factor and damaged-zone properties are related by the equation

$$s=\left(\frac{k}{k_d}-1\right)\ln\left(\frac{r_d}{r_w}\right). \quad (15.15)$$

One can then estimate "reasonable" values of r_d and determine the corresponding value of k_d from

$$k_d=\frac{k}{\left[\dfrac{s}{\ln\left(\dfrac{r_d}{r_w}\right)}+1\right]}. \quad (15.16)$$

Alternatively, if one has laboratory data providing estimates of k_d/k, r_d can be estimated with a known skin factor and the equation

$$r_d=r_w\exp\left(\frac{s}{\dfrac{k}{k_d}-1}\right). \quad (15.17)$$

Productivity-Index Ratio During Unsteady-State Flow. In lower-permeability reservoirs with long fractures, steady state or pseudosteady-state is not achieved until considerable time has passed, often weeks or months. Until this time (which we can call time to stabilized flow), methods like the McGuire-Sikora chart are not applicable for analysis of productivity-index increase. Instead, the

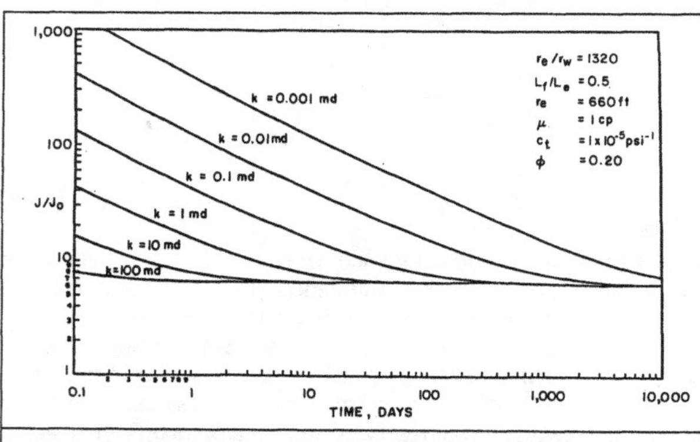

Fig. 15.8—Productivity ratio vs. real time, constant-rate flow.

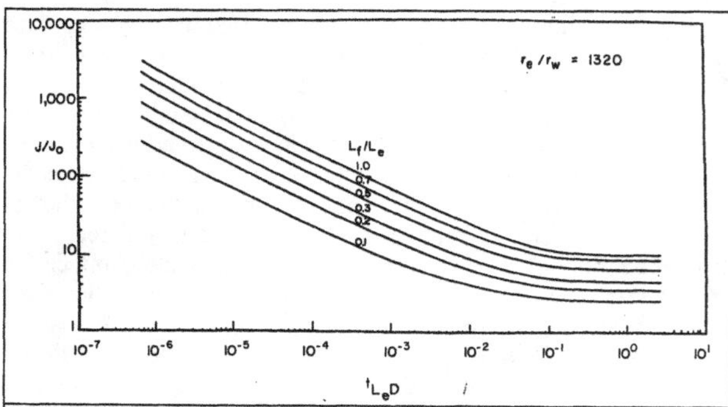

Fig. 15.9—Productivity ratio vs. dimensionless time, constant-BHP flow.

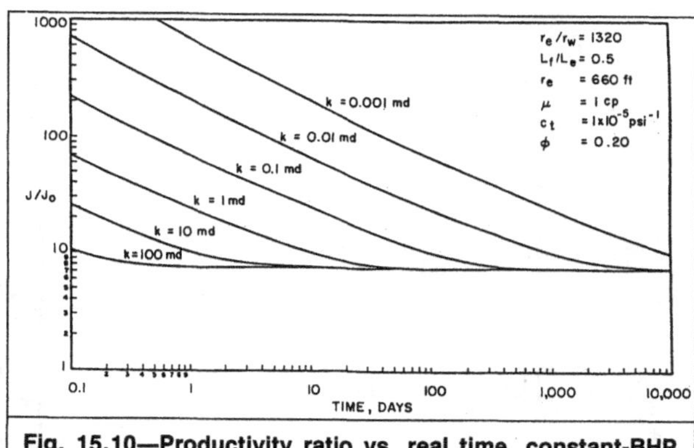

Fig. 15.10—Productivity ratio vs. real time, constant-BHP flow.

unsteady-state flow before stabilization must be taken into account. This situation was studied by Morse and Von Gonten[9] for slightly compressible liquids for both constant-rate and constant-BHP production.

Fig. 15.7 shows the ratio of instantaneous productivity index (during unsteady-state constant-rate flow), J, to *stabilized* prefracture productivity index, J_o. This ratio is plotted as a function of dimensionless time,

$$t_{L_eD}=\frac{0.0002637kt}{\phi\mu c_T L_e^2}, \qquad (15.18)$$

with the parameter L_f/L_e (ratio of fracture length to drainage radius) varying from 0 to 1.0. Results were generated for square drainage areas and for infinitely conductive fractures.

Note in Fig. 15.7 that J/J_o reaches a stabilized value at $t_{L_eD}\cong$ 0.25 for all fracture lengths. Note the large values of J/J_o at early times; in general, the longer the fracture, the greater the ratio of J/J_o at early times.

These results are made less abstract in Fig. 15.8, which gives J/J_o vs. time for fixed fracture length (L_f/L_e=0.5), several representative permeabilities, and other properties fixed. Note that J stabilizes in less than 1 day for k=100 md but requires almost 10,000 days (27.4 years) to stabilize for k=0.01 md.

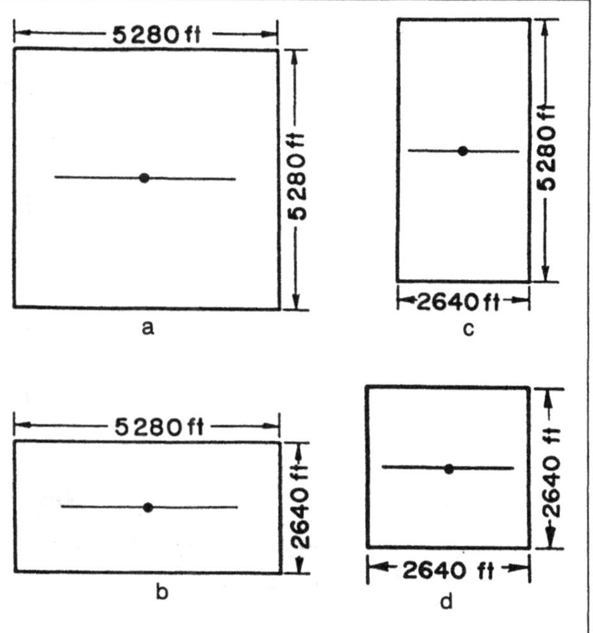

Fig. 15.11—Plane view of fractured well for (a) 640-acre spacing, (b) 320-1 spacing, (c) 320-2 spacing, and (d) 160-acre spacing.

Fig. 15.9 is similar to Fig. 15.7 but presents results of constant-BHP production simulations. Again, J/J_o reaches a stable value at $t_{L_eD}\cong$0.25. Duration and magnitude of the unsteady-state effects are illustrated in Fig. 15.10 for a specific set of reservoir properties and for a wide range of permeabilities.

Figs. 15.7 and 15.9 can be used to construct J/J_o charts for any fractured-well properties of interest (limited, of course, to infinitely conductive fractures).

Productivity-index ratio during unsteady-state flow is illustrated in Example 15.1, Appendix K.

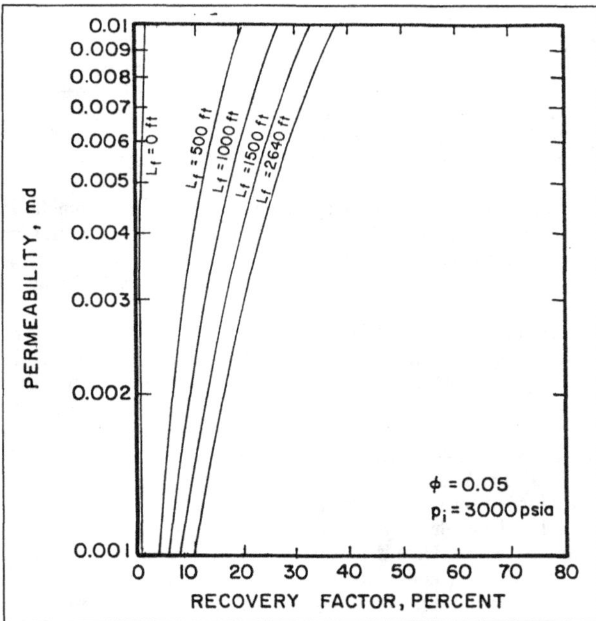

Fig. 15.12—Permeability vs. percent recovery of gas at the end of a 30-year period for 640-acre well spacing.

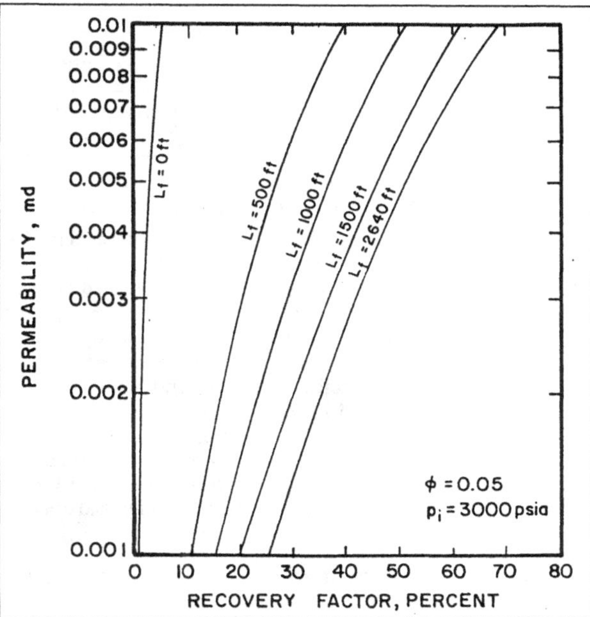

Fig. 15.13—Permeability vs. percent recovery of gas at the end of a 30-year period for 320-1 well spacing.

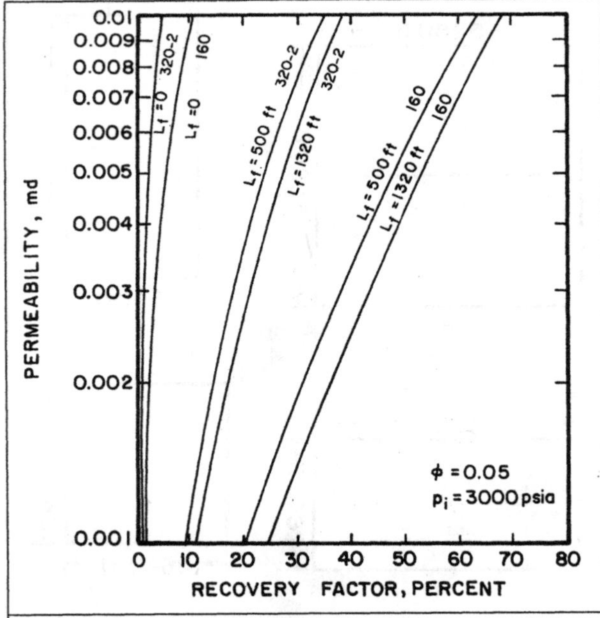

Fig. 15.14—Permeability vs. percent recovery of gas at the end of a 30-year period for 320-2 and 160-acre well spacing.

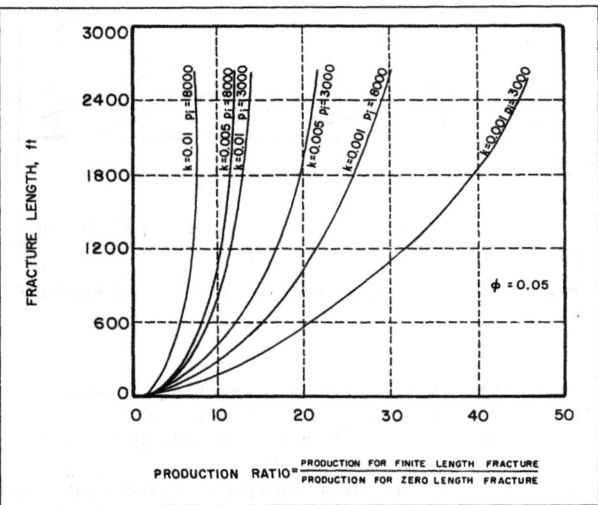

Fig. 15.16—Fracture length vs. production ratio for 320-1 well spacing, production for 30 years.

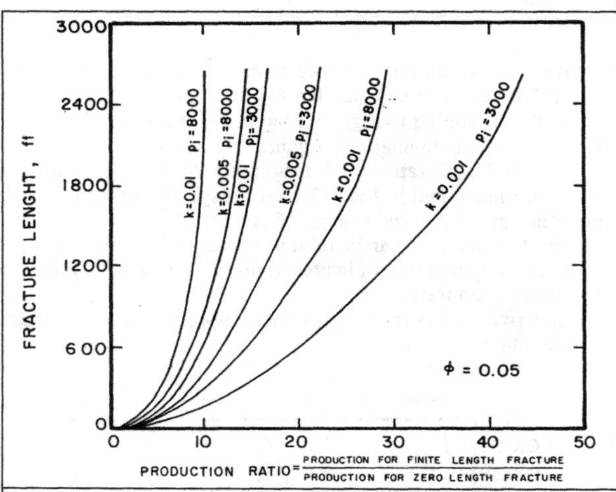

Fig. 15.15—Fracture length vs. production ratio for 640-acre well spacing, production for 30 years.

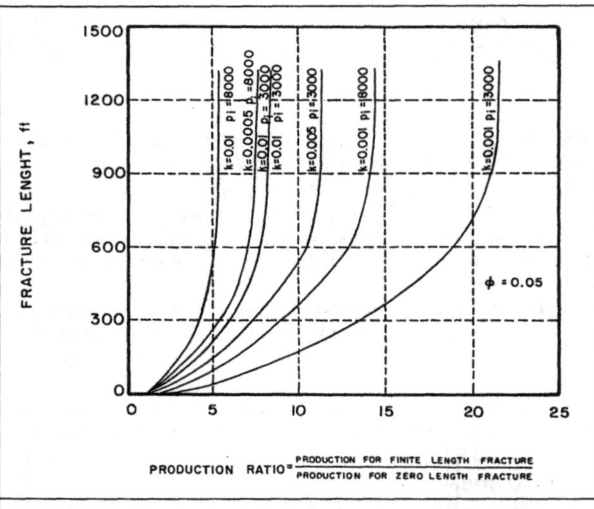

Fig. 15.17—Fracture length vs. production ratio for 320-2 well spacing, production for 30 years.

15.3 Ultimate Recovery for Fractured Wells

Introduction. The first fracture treatment was performed in 1947. During the first 25 years of fracturing, most reservoirs produced at economic rates, and individual wells were stimulated primarily to remove damage or to accelerate production and increase present-value profit. The emphasis in fracture design was placed on productivity increase, not on increased recovery. However, decline curves from many of these wells indicated that, in fact, fracture treatments frequently led to improved recovery.

In the low-permeability reservoirs currently being developed, the importance of increasing ultimate recovery with fracturing has been magnified. Factors that affect ultimate recovery include fracture length, fracture conductivity, formation permeability, and drainage-area size and shape.

Lemon *et al.* Study. Lemon *et al.*[10] systematically studied the effects of fracture and reservoir properties on recovery from low-permeability gas reservoirs. They simulated production at constant surface pressure and calculated production *in a 30-year well life.* This condition is extremely important: given enough time (and without an economic limit rate), any gas well would eventually recover all the gas from the pore space with which it is in pressure communication. Further, *one* well could drain an entire reservoir. The issue, then, is gas produced in a reasonable well life and at a profitable rate.

Lemon *et al.* studied four drainage areas: a 640-acre [259-ha] square (Fig. 15.11a); a 320-acre [130-ha] rectangle with width half the length and (1) a fracture oriented parallel to the long dimension (Fig. 15.11b; spacing called "320-1") and (2) the fracture oriented parallel to the short dimension (Fig. 15.11c, called "320-2"); and a 160-acre [65-ha] square (Fig. 15.11d).

Figs. 15.12 and 15.13 show calculated recovery factors as functions of formation permeability for 640- and 320-1-acre [259- and 130-1-ha] drainage areas. The parameter on each graph is fracture length. Note the general trend to improved recovery as formation permeability and fracture length increase; note also that the recovery as a percentage of gas in place in the drainage area is higher at a given permeability and fracture length for the 320-1-acre [130-1-ha] well spacing pattern than for the 640-acre [259-ha] pattern. Fig. 15.14 shows recovery factor as a function of per-

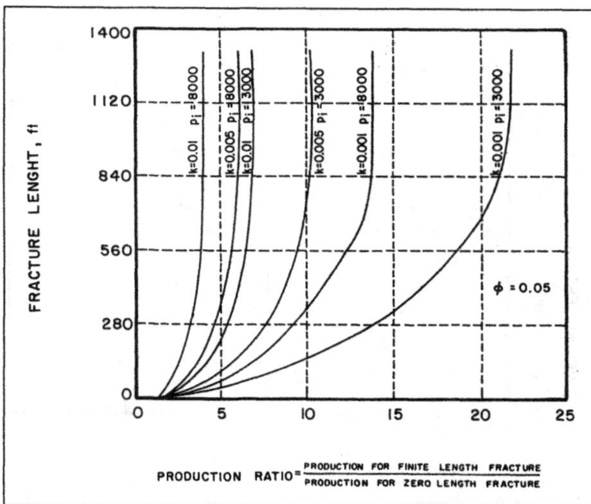

Fig. 15.18—Fracture length vs. production ratio for 160-acre well spacing, production for 30 years.

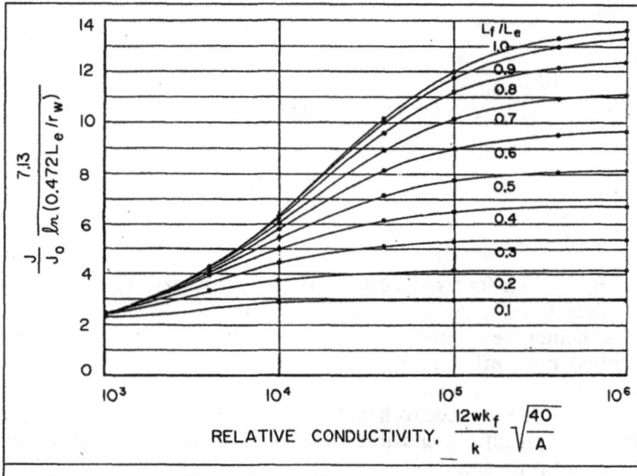

Fig. 15.19—Stimulation ratio for square drainage area.

meability with a fracture-length parameter for 160- and 320-2-acre [65- and 130-2-ha] drainage patterns. Figs. 15.15 through 15.18 show the effect of fracture length on 30-year production for the various drainage-area patterns, with selected permeabilities and discovery pressures, p_i, as parameters. Note that in many cases (particularly for higher permeabilities), fracture lengths greater than some minimum have negligible effect on recovery.

Holditch[4] extended Lemon *et al.*'s study and simulated fractured-well performance under the following conditions:

$$
\begin{aligned}
p_i &= 7{,}500 \text{ psia [51.7 MPa]},\\
p_{wf} &= 1{,}500 \text{ psia [10.3 MPa]},\\
h &= 100 \text{ ft [30 m]},\\
\phi_g &= 0.10,\\
\text{economic limit} &= 150 \text{ Mcf/D [4248 m}^3\text{/d]},\\
A &= 160 \text{ acres [65 ha]},\\
w &= 0.0167 \text{ ft [0.005 m]},\\
k_f &= 300{,}000 \text{ md},\\
\gamma_g &= 0.65, \text{ and}\\
\text{well life} &= 20 \text{ years.}
\end{aligned}
$$

As in the Lemon *et al.* study, specific numerical results depend on fracture and formation properties assumed in the simulation. Nevertheless, general trends identified in these sorts of studies are valid over wide ranges of property values.

Holditch studied wells with either 40- or 160-acre [16- or 65-ha] square drainage areas (with fracture geometry similar to that in Fig. 15.11a), and 2×1 and 5×1 rectangles with the fracture oriented along the long axis (similar to the geometry shown in Fig. 15.11b).

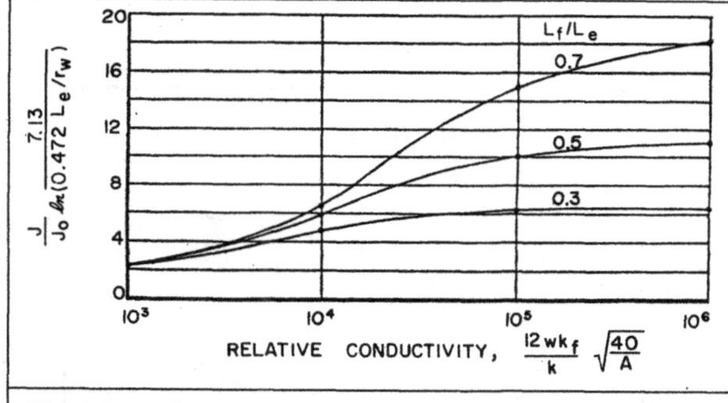

Fig. 15.20—Stimulation ratio for 2×1 rectangular drainage area.

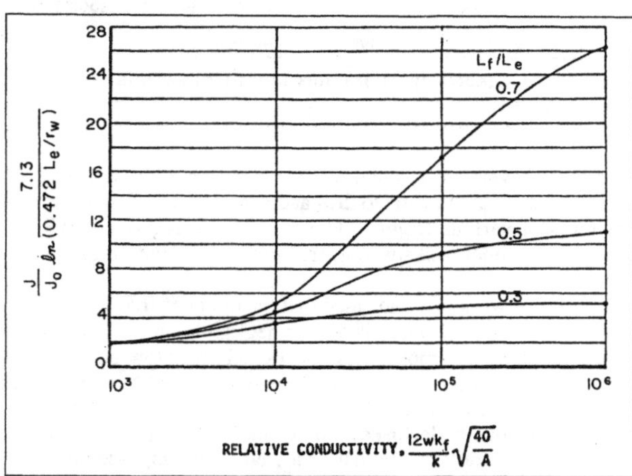

Fig. 15.21—Stimulation ratio for 5×1 rectangular drainage area.

Figs. 15.19 through 15.21 show stabilized productivity-index increases, plotted in a way similar to the McGuire-Sikora chart. Fig. 15.19, for a square drainage area, is the simulator verification of the McGuire-Sikora chart mentioned in Sec. 15.2. Note the 11-fold increase in J/J_o at a relative conductivity of 1×10^6 for L_f/L_e=0.7.

Fig. 15.20 is a similar chart for a 2×1 rectangle, and Fig. 15.21 is for a 5×1 rectangle. Note the 18-fold increase in J/J_o for a 2×1 rectangle and a 26-fold increase for a 5×1 rectangle at a relative conductivity of 1×10^6 for L_f/L_e=0.7.

The first reaction to these results is that rectangular drainage areas have a significant advantage (at least, if one can determine fracture orientation and create drainage areas so that the fractures are favorably oriented in the drainage area). Further examination of these results reveals that much longer fracture lengths are required to achieve the large production increases in the rectangular drainage areas. These longer fractures will, of course, cost a great deal more to create. A more accurate display of the effects of all parameters is found in Fig. 15.22, which is a graph of ultimate recovery vs. fracture length for different permeabilities and fracture lengths. The following conclusions can be drawn from this graph.

1. For a given fracture length, ultimate recovery is greater in a square drainage area at higher permeabilities (0.1 md).
2. A slightly rectangular drainage area (e.g., 2×1) may provide better results in lower-permeability reservoirs (0.01 md).
3. In extremely-low-permeability reservoirs (0.001 md), a highly rectangular drainage area (5×1) may provide the best results.
4. Ultimate recovery is highly dependent on reservoir permeability.

These results indicate that, to plan the development of a reservoir, the engineer should prepare graphs of ultimate recovery vs. fracture

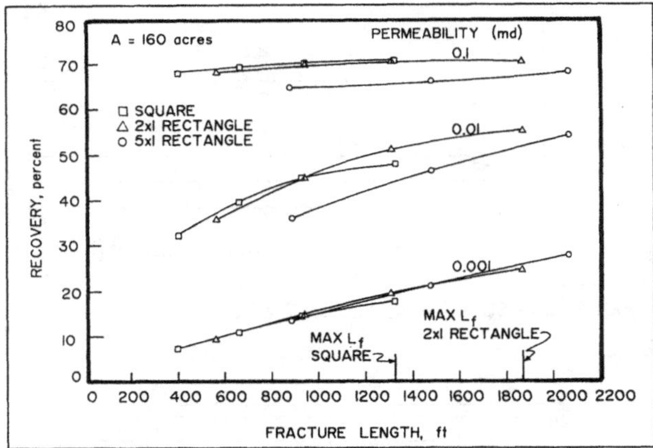

Fig. 15.22—Recovery vs. fracture length for various drainage-area shapes.

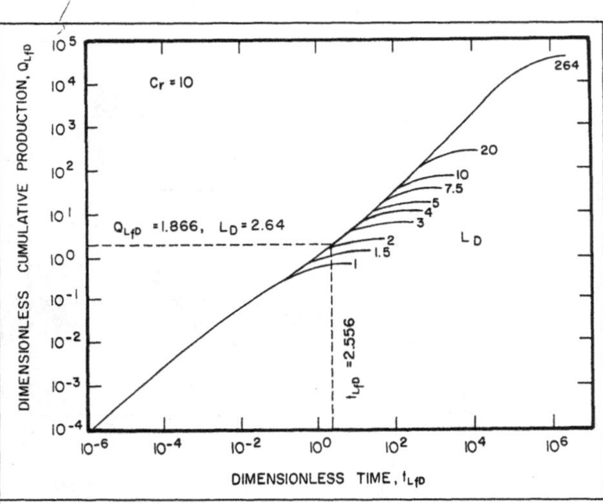

Fig. 15.23—Example type-curve graph, square reservoir, C_r = 10.

length for several different well spacings. An economic analysis can be made with the costs to drill and complete a well and to create fractures of various lengths. The results of this analysis will indicate the optimal well spacing, fracture length, and drainage-area shape.

Holditch's work also indicated that the effect of fracture conductivity on ultimate recovery is almost negligible unless the fracture heals completely. In most cases, and especially in low-permeability formations, fracture conductivity affects only the life of a well. A low-conductivity fracture will decrease the initial productivity index, but the well will not decline as rapidly and will level off at a productivity index slightly below that of a well with a high-conductivity fracture. Thus, as an example, if a well with a high-conductivity fracture produces 5 Bcf [142×10^6 m^3] in 10 years, a well with a low-conductivity fracture may require 15 years to produce the 5 Bcf [142×10^6 m^3]. As producing time approaches 20 to 30 years, the ultimate recovery of the two wells may be similar.

15.4 Predicting Performance With Type Curves

Introduction. In the determination of optimal fracture properties, a critical part of the economic analysis is projection of future performance for the unfractured well and for the fractured well with various fracture lengths and conductivities. While these projections can be made with a finite-difference fractured-well simulator, they can also be made by hand with type curves.

Type curves are simply graphs of solutions to the equations describing fluid flow in a porous medium. Use of type curves is most often associated with pressure-transient data analysis. When key reservoir parameters have been determined, however, type curves can also be used to predict future performance. If a well is assumed to produce at a constant flowing BHP (a reasonable assumption for tight gas wells beyond the first months of production), type curves that assume constant-BHP production can be used to predict performance.

Type Curves for Fractured-Well-Performance Prediction. Holditch *et al.*[11] used a two-dimensional, one-phase reservoir simulator to generate more than 200 fractured-well type curves for a variety of reservoir conditions. Among these is a set of constant-pressure type curves that includes the effects of a finite reservoir. One of these curves is shown in Fig. 15.23; the others may be obtained from the authors of Ref. 11.

These curves are log-log graphs of dimensionless cumulative production, Q_{LfD}, vs. dimensionless time, t_{LfD}, for various values of dimensionless fracture conductivity, C_r, and dimensionless fracture penetration, L_D. These dimensionless quantities are defined below in terms of real pressure and real time:

$$Q_{LfD}=\frac{0.8936QB}{\phi c_T h L_f^2(p_i-p_{wf})}, \quad (15.19)$$

$$t_{LfD}=\frac{0.0002637kt}{\phi\mu c_T L_f^2}, \quad (15.20)$$

$$C_r=\frac{wk_f}{\pi L_f k}, \quad (15.21)$$

and

$$L_D=L_e/L_f. \quad (15.22)$$

Fig. 15.23 is for a well centered in a square reservoir, with dimensionless fracture conductivity, C_r, of 10.

Although the dimensionless solutions plotted on type curves are applied most efficiently with a computer, hand analysis is not difficult. Also, hand analysis best illustrates the procedures used to predict performance with these dimensionless solutions. Example 15.2 in Appendix K illustrates the basic method in terms of real time and pressure.

Type Curves for Unfractured-Well-Performance Prediction. Another important type curve in fracture evaluation is for radial flow into a well centered in its drainage area and with a skin factor. The well is assumed to be produced at constant BHP. This type curve is important because it allows us to model future performance of a well before a fracture treatment; thus, we can model the benefit resulting from fracturing.

This curve, shown in Fig. 15.24, is a log-log graph of dimensionless cumulative production, Q_D, vs. dimensionless time, t_D, with the parameter r_{eaD}. The definitions of these quantities are given below:

$$Q_D=\frac{0.8936QB}{\phi c_T h r_{wa}^2(p_i-p_{wf})}, \quad (15.23)$$

$$t_D=\frac{0.0002637kt}{\phi\mu c_T r_{wa}^2}, \quad (15.24)$$

$$r_{eaD}=r_{ea}/r_{wa}, \quad (15.25)$$

$$r_{ea}=\sqrt{A/\pi}, \quad (15.26)$$

and

$$r_{wa}=r_w e^{-s}, \quad (15.27)$$

where s=skin factor and A=drainage area in which the well is centered.

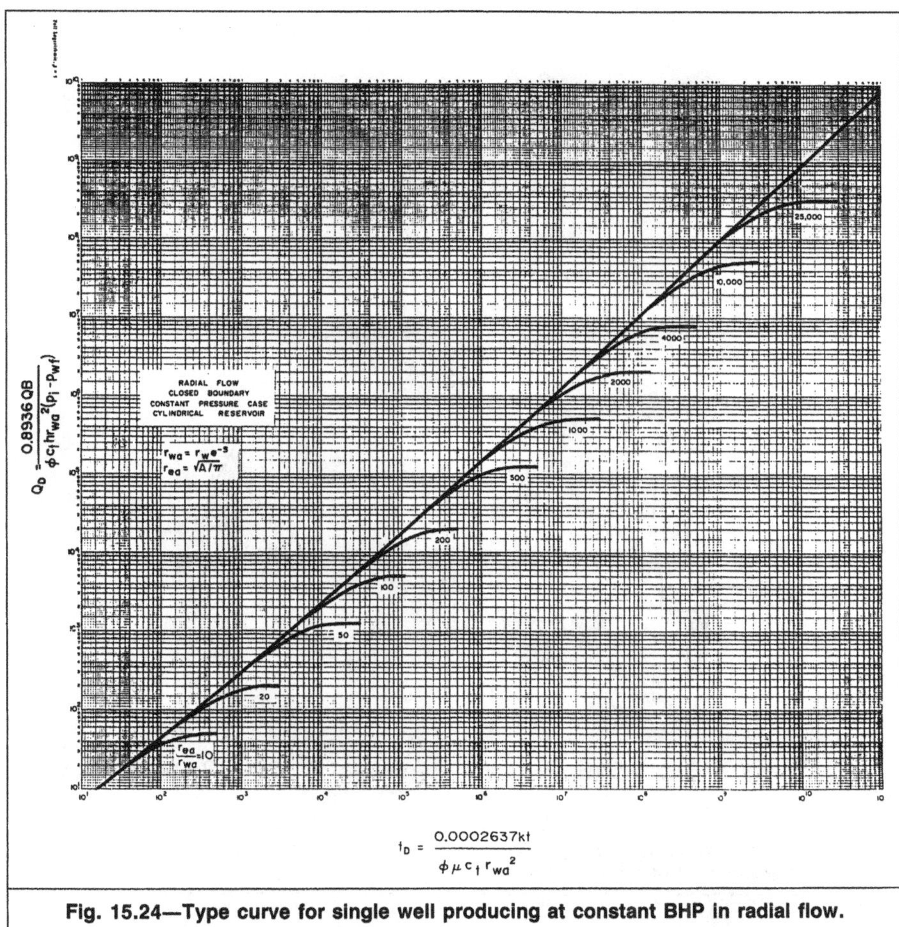

Fig. 15.24—Type curve for single well producing at constant BHP in radial flow.

Fig. 15.24 is based on the familiar solution of van Everdingen and Hurst[12] for these boundary conditions. Example 15.3 in Appendix K illustrates use of this type curve.

15.5 Flow Regimes in Hydraulically Fractured Formations

Cinco-Ley and Samaniego-V.[13] described four distinct flow periods for hydraulically fractured wells, with successive periods separated by transition periods. These periods, with flow patterns, are illustrated in Fig. 15.25. The successive periods are fracture linear, bilinear, formation linear, and pseudoradial flow.

Fracture linear flow (Fig. 15.25a) occurs at very small values of dimensionless time. During this flow period, most of the fluid entering the wellbore comes from expansion of the fluid contained in the fracture, and the flow is essentially linear. This flow period is of no practical use because it ends at an extremely early time; its duration is given by

$$t_{L_fD} \cong \frac{(0.1)(C_r)^2}{\eta_{fD}^2}. \qquad (15.28)$$

Bilinear flow (Fig. 15.25b) occurs when fluid flows linearly into the fracture from the formation and when, in the fracture, the fracture-tip effects have not yet influenced well behavior. The bilinear flow period can last for a substantial period of time; e.g., for $C_r \geq 1$, it continues to

$$t_{L_fD} \cong \frac{0.01}{C_r^2}. \qquad (15.29)$$

During this flow period, p_{wf} is a linear function of $t^{1/4}$ on Cartesian coordinate paper, and a line with a slope of ¼ occurs in this time region on log-log paper.

Formation linear flow (Fig. 15.25c) occurs only in fractures with high conductivity ($C_r \geq 100$). It continues to $t_{L_fD} \cong 0.016$ for these fractures and is characterized on a Cartesian coördinate plot by the fact that p_{wf} is a linear function of $t^{1/2}$ and on a log-log plot by a line with slope of ½. Quantitative analysis of these plots is given in Sec. 15.6.

Pseudoradial flow (Fig. 15.25d) occurs with fractures of all conductivities—the higher the conductivity, the later the drainage pattern can be characterized as essentially radial. Cinco-Ley and Samaniego-V. found that pseudoradial flow begins at $t_{L_fD} \cong 5$ for fractures with $C_r = 100$ and at smaller t_{L_fD} values for lower C_r's—e.g., at $t_{L_fD} = 2$ for $C_r = 0.1$. This start of pseudoradial flow is characterized by the start of a straight line on a graph of p_{wf} vs. log t. Of course, the time at which the straight line starts is somewhat subjective; personal experience shows that the semilog straight line is reached for most practical purposes at t_{L_fD} values varying from 3 for $C_r = 100$ to 1.5 for $C_r = 0.1$. The start of the semilog straight line for various values of C_r is shown in Fig. 15.26.

The same flow regimes that appear in flow tests also appear in pressure-buildup tests at about the same dimensionless times as in flow tests. The physical interpretation is that the pressure has built up to essentially the same value throughout the region in a buildup test. For example, when the pseudoradial regime is reached, pressure has built up to a uniform level in a circular pattern centered at the wellbore. The radius of the circle and the pressure level within the circle increase with increasing shut-in time.

An example calculation of flow-regime duration is given in Appendix K, Example 15.4.

15.6 Postfracture Pressure-Transient Test Analysis

Introduction. The objective of postfracture transient test analysis is to provide estimates of fracture length, fracture conductivity, and formation permeability. Lower-than-expected values for any one

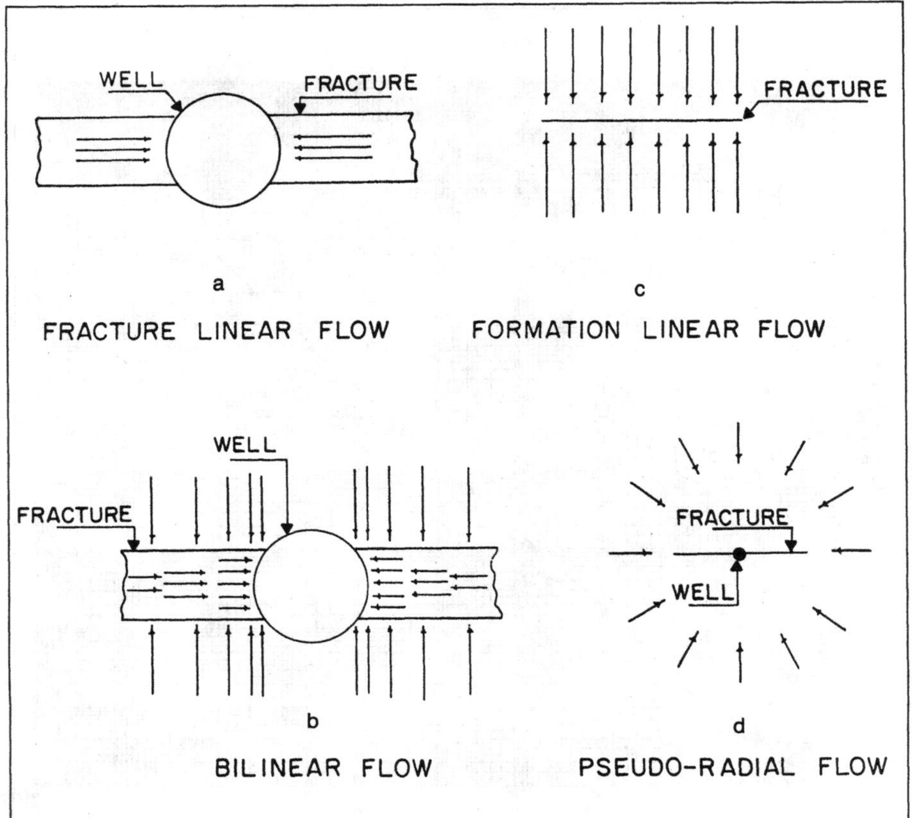

Fig. 15.25—Flow patterns for fractured wells: (a) fracture linear flow; (b) bilinear flow; (c) formation linear flow; (d) pseudoradial flow.

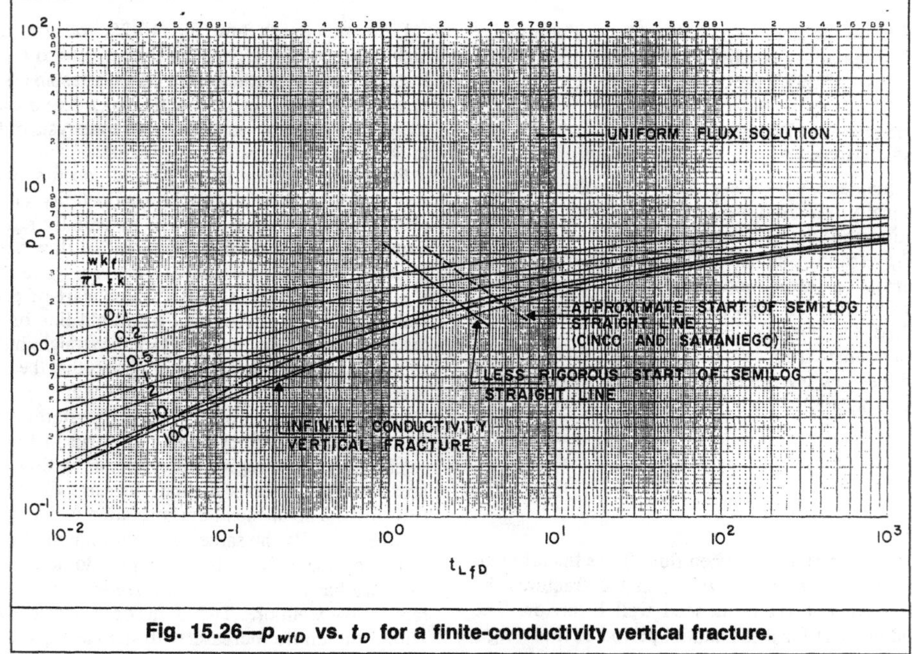

Fig. 15.26—p_{wfD} vs. t_D for a finite-conductivity vertical fracture.

TABLE 15.1—SUMMARY OF PLOTTING AND ANALYSIS TECHNIQUES—RADIAL OR PSEUDORADIAL FLOW

	Case			
Flow Test	Oil	Gas, With Adjusted Variables*	Gas, With Pressure and Time	Gas, With Pressure Squared and Time
Semilog graph variables	p_{wf} vs. t	$p_{a,wf}$ vs. t_a	p_{wf} vs. t	$p_{wf}{}^2$ vs. t
Permeability from slope m of semilog straight line	$k=\dfrac{162.6q_oB_o\mu_o}{mh}$	$k=\dfrac{162.6q_g\bar{B}_g\bar{\mu}}{mh}$	$k=\dfrac{162.6q_g\bar{B}_g\bar{\mu}}{mh}$	$k=\dfrac{1{,}637q_gT\bar{z}\bar{\mu}}{mh}$
Skin-factor calculation	$s=1.151 \times\left(\dfrac{\bar{p}-p_{1hr}}{m}-\log\dfrac{k}{\phi\mu_oc_Tr_w{}^2}+3.23\right)$	$s'=1.151\left(\dfrac{\bar{p}-p_{1hr}}{m}-\log\dfrac{k}{\phi\bar{\mu}\bar{c}_Tr_w{}^2}+3.23\right)$		$s'=1.151 \times\left[\dfrac{\bar{p}^2-p_{1hr}{}^2}{m}-\log\dfrac{k}{\phi\bar{\mu}\bar{c}_Tr_w{}^2}+3.23\right]$
Buildup Test				
Semilog graph variables	p_{ws} vs. $(t_p+\Delta t)/\Delta t$	$p_{a,ws}$ vs. $(t_{ap}+\Delta t_a)/\Delta t_a$	p_{ws} vs. $(t_p+\Delta t)/\Delta t$	$p_{ws}{}^2$ vs. $\left(\dfrac{t_p+\Delta t}{\Delta t}\right)$
Permeability from slope m of semilog straight line	$k=\dfrac{162.6q_gB_o\mu_o}{mh}$	$k=\dfrac{162.6q_g\bar{B}_g\bar{\mu}}{mh}$	$k=\dfrac{162.6q_g\bar{B}_g\bar{\mu}}{mh}$	$k=\dfrac{1{,}637q_gT\bar{z}\bar{\mu}}{mh}$
Skin-factor calculation	$s=1.151 \times\left(\dfrac{p_{1hr}-p_{wf}}{m}-\log\dfrac{k}{\phi\mu c_Tr_w{}^2}+3.23\right)$	$s'=1.151\left(\dfrac{p_{1hr}-p_{wf}}{m}-\log\dfrac{k}{\phi\bar{\mu}\bar{c}_Tr_w{}^2}+3.23\right)$		$s'=1.151 \times\left[\dfrac{p_{1hr}{}^2-p_{wf}{}^2}{m}-\log\dfrac{k}{\phi\bar{\mu}\bar{c}_Tr_w{}^2}+3.23\right]$
Definition of $p_{D,MBH}$	$\dfrac{kh(p^*-\bar{p})}{70.6q_oB_o\mu_o}$	$\dfrac{kh(p_a{}^*-\bar{p}_a)}{70.6q_g\bar{B}_g\bar{\mu}}$		$\dfrac{kh(p^{*2}-\bar{p}^2)}{711q_gT\bar{z}\bar{\mu}}$

*For flow tests of gas wells, a semilog graph of $p_{a,wf}$ vs. t (not t_a) will lead to more accurate results.

of these variables could be the explanation for an unsuccessful fracture treatment.

Four main methods are used to analyze postfracture pressure-transient tests: a method based on the assumption that the pseudoradial flow regime has been achieved, a method based on analysis of the linear flow regime if present, use of type curves, and use of reservoir simulators in a history-matching mode.

Pseudoradial Flow Methods. In a flow test, we consider the flow to be pseudoradial around a fractured well when the drainage pattern has completed a transformation from rectangular (linear flow) through elliptic (transitional flow), to almost radial. (We use the term "pseudo" or "almost" radial because the drainage area never becomes completely circular in a fractured well; it simply becomes close enough to a circle to be considered circular for practical purposes.) More specifically, equations derived for strictly radial flow can be applied with negligible error when pseudoradial flow is achieved. Time required to achieve pseudoradial flow for an infinitely conductive fracture is given by

$$t_{L_fD}=\frac{0.0002637kt}{\phi\mu c_TL_f{}^2}\cong 3. \quad\text{(15.30)}$$

In a pressure-buildup test, the same time is required for the buildup pattern to be considered pseudoradial. The physical interpretation is that the pressure has built up to essentially a uniform level first in a narrow rectangle (linear regime), then an ellipse (transitional regime), and finally a near circle.

When the pseudoradial regime is reached, conventional semilog analysis can be used to allow calculation of permeability and skin factor. Skin factor is useful in postfracture test analysis because Prats[1] showed that for a highly conductive fracture, the relationship between skin factor and fracture half-length is

$$L_f\cong 2r_we^{-s}. \quad\text{(15.31)}$$

Thus, to use this method, we plot p_{wf} vs. log t (or equivalent) for a flow test, determine the position and slope of the semilog straight line, calculate k and s, and then determine fracture half-length. Four cases are of major interest: slightly compressible liquids (usually oil); gases with time and adjusted pressure used as plotting functions; gases with ordinary pressure and time used as plotting functions; and gas with pressure squared and time used as plotting functions. Working equations for each case are summarized in Table 15.1.

The simplicity of this method makes it highly appealing. Unfortunately, it is useful only rarely in practice—for short fractures in relatively-high-permeability formations. The fundamental problem is that the semilog straight line must be reached (and, in fact, about a ⅓-cycle of semilog straight line is desired for accuracy). This requires $t_{L_fD}\geq 3$ or

TABLE 15.2—TIME TO REACH SEMILOG STRAIGHT LINE FOR FRACTURED WELLS

k (md)	Time to Reach Semilog Straight Line (hours)		
	$L_f=100$ ft	$L_f=500$ ft	$L_f=1{,}000$ ft
0.01	9,120	228,000	912,000
0.1	912	22,800	91,200
1.0	91.2	2,280	9,120

$$t\geq 11{,}400\phi\mu c_TL_f/k^2, \quad\text{(15.32)}$$

where t is in hours. As examples, consider a gas well with $\phi=0.2$, $\mu=0.02$ cp [0.02 mPa·s], $c_T=2\times10^{-4}$ psi^{-1} [0.29×10^{-4} kPa^{-1}], and the values of L_f and k given in Table 15.2. The semilog method will not be applicable for the most common well fractured (low permeability, long fracture).

In addition, for gas wells, the apparent skin factor, s', calculated from transient test analysis is distorted by non-Darcy flow; in all cases, there is a possibility of distortion because of an incompletely perforated interval. Further, the method is valid for only high-conductivity fractures; i.e., for

$$C_r=\frac{wk_f}{\pi L_fk}\geq 100. \quad\text{(15.33)}$$

For lower fracture conductivities, calculated fracture lengths from skin factor will be pessimistic. Cinco-Ley and Samaniego-V.[13] provided a chart (Fig. 15.27) relating r_{wa} and L_f for finite-conductivity fractures; this can be useful if one can estimate s and C_r because $r_{wa}=r_we^{-2s}$.

In summary, the pseudoradial flow analysis technique for fractured wells is of *some* value because short, highly conductive fractures are occasionally created in higher-permeability formations. Their occurrence is rare enough, however, that other techniques are also needed.

Linear Flow Methods. The assumption of linear flow (or in Cinco-Ley and Samaniego-V.'s classification scheme, formation linear flow) is excellent up to a dimensionless time, t_{L_fD}, of 0.016, for production into a highly conductive vertical fracture with half-length L_f, once wellbore-storage effects have diminished. In terms of dimensionless variables, linear flow is modeled by

$$p_D=(\pi t_{L_fD})^{½} \quad\text{(15.34)}$$

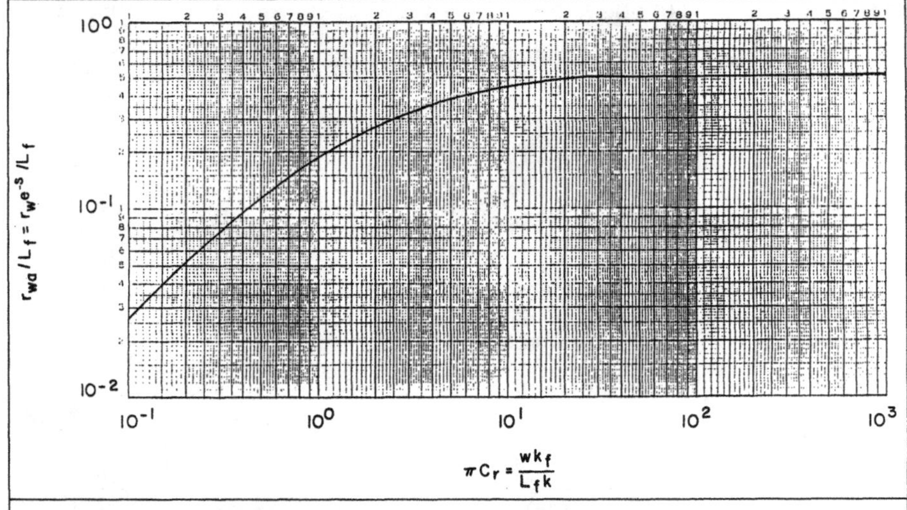

Fig. 15.27—Effective wellbore radius vs. dimensionless fracture conductivity for a vertical fracture.

TABLE 15.3—SUMMARY OF PLOTTING AND ANALYSIS TECHNIQUES—LINEAR FLOW

	Case			
Flow Test	Oil	Gas, With Adjusted Variables*	Gas, With Pressure and Time	Gas, With Pressure Squared and Time
Cartesian-coordinate graph variables	p_{wf} vs. $\sqrt{t}$	$p_{a,wf}$ vs. $\sqrt{t_a}$	p_{wf} vs. $\sqrt{t}$	$p_{wf}{}^2$ vs. $\sqrt{t}$
$\sqrt{k}L_f$ from slope m_L of straight line	$\sqrt{k}L_f=\frac{4.064qB}{m_Lh}\left(\frac{\mu}{\phi c_T}\right)^{1/2}$	$\sqrt{k}L_f=\frac{4.064q_g\bar{B}_g}{m_Lh}\left(\frac{\bar{\mu}}{\phi\bar{c}_T}\right)^{1/2}$		$\sqrt{k}L_f=\frac{40.93q_gT\bar{z}}{m_Lh}\left(\frac{\bar{\mu}}{\phi\bar{c}_T}\right)^{1/2}$
Buildup Test				
Cartesian-coordinate graph variables	p_{ws} vs. $\sqrt{\Delta t_e}$	$p_{a,ws}$ vs. $\sqrt{\Delta t_{ae}}$	p_{ws} vs. $\sqrt{\Delta t_e}$	$p_{ws}{}^2$ vs. $\sqrt{\Delta t_e}$
$\sqrt{k}L_f$ from slope m_L of straight line	$\sqrt{k}L_f=\frac{4.064qB}{m_Lh}\left(\frac{\mu}{\phi c_T}\right)^{1/2}$	$\sqrt{k}L_f=\frac{4.064q_g\bar{B}_g}{m_Lh}\left(\frac{\bar{\mu}}{\phi\bar{c}_T}\right)^{1/2}$		$\sqrt{k}L_f=\frac{40.93q_gT\bar{z}}{m_Lh}\left(\frac{\bar{\mu}}{\phi\bar{c}_T}\right)^{1/2}$

Note: $\Delta t_e=\frac{1}{\left(1+\frac{\Delta t}{t_p}\right)}$

*For flow tests of gas wells, a graph of $p_{a,wf}$ vs. $\sqrt{t}$ (not $\sqrt{t_a}$) will lead to more accurate results.

and is present for fractures with $C_r \geq 100$. On a log-log graph of p_{wf} vs. t, linear flow is characterized by a straight line with a slope of one-half.

For a pressure-buildup test, Agarwal[14] stated that Eq. 15.34 also describes a test during the linear flow region when we use equivalent time $\Delta t_e = \Delta t/(1+\Delta t/t_p)$ in the definition of dimensionless time.

Working equations for analysis of linear-flow-region data obtained in buildup and flow tests are summarized in Table 15.3. Three cases of interest are treated: slightly compressible liquids (usually oil); gases with adjusted time and adjusted pressure used as plotting functions; and gases with ordinary pressure and time used as plotting functions.

The procedure for analyzing *constant-rate* flow-test data with a linear flow regime (suggested by Millheim and Cichowicz[15]) is explained below for a slightly compressible liquid.

1. Plot p_{wf} vs. $\sqrt{t}$ on ordinary Cartesian coordinate paper.
2. Determine the slope, m_L, of the linear region.
3. From independent knowledge of k, estimate the fracture half-length, L_f, from

$$L_f=\frac{4.064qB}{m_Lh}\left(\frac{\mu}{\phi c_T k}\right)^{1/2}. \qquad (15.35)$$

The procedure for a buildup test with a linear flow regime is outlined below for the case of a slightly compressible liquid.

1. Plot p_{ws} vs. $\sqrt{t_e}$, where $\Delta t_e = \Delta t/(1+\Delta t/t_p)$ on ordinary Cartesian coordinate paper.
2. Determine the slope, m_L, of the linear region.
3. From independent knowledge of k, estimate the fracture half-length, L_f, from

$$L_f=\frac{4.064qB}{m_Lh}\left(\frac{\mu}{\phi c_T k}\right)^{1/2}. \qquad (15.36)$$

The linear flow analysis method has important limitations.

1. It is valid only for fractures with high C_r; in fact, strictly speaking, linear flow occurs when there is uniform flux into a fracture rather than infinite fracture conductivity. Uniform flux means the same rate of flow from the formation to the fracture per unit cross-sectional area at all points along the fracture.
2. Only the earliest data in a test exhibit linear flow in a high-conductivity fracture:

$$t_{L_fD} \leq 0.016 \qquad (15.37)$$

or

$$t \leq 60\phi\mu c_T L_f{}^2/k. \qquad (15.38)$$

Some of these same early data may be distorted by wellbore storage. The duration of wellbore storage can be estimated from the Barker-Ramey[16] type curve, which is discussed later.

3. Using the method to estimate fracture half-length requires independent knowledge of formation permeability, k. Millheim and Cichowicz suggested determining k from the data in the pseudoradial flow region; however, this region is usually not reached in low-permeability formations with long fractures. Accordingly, one needs a prefracture test.

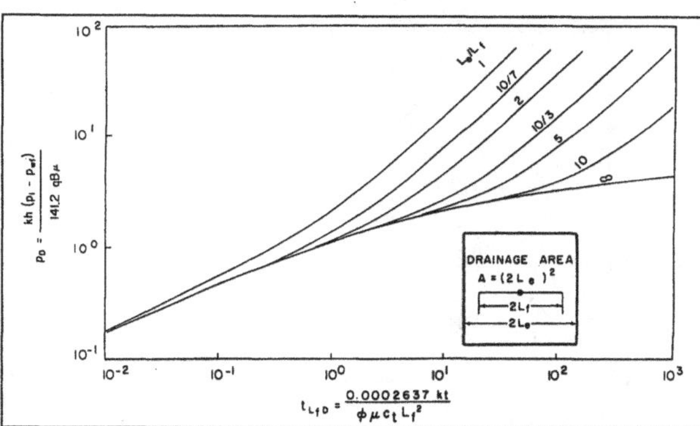

Fig. 15.28—Dimensionless pressure for vertically fractured well in the center of a closed square—no wellbore storage, infinite-conductivity fracture.

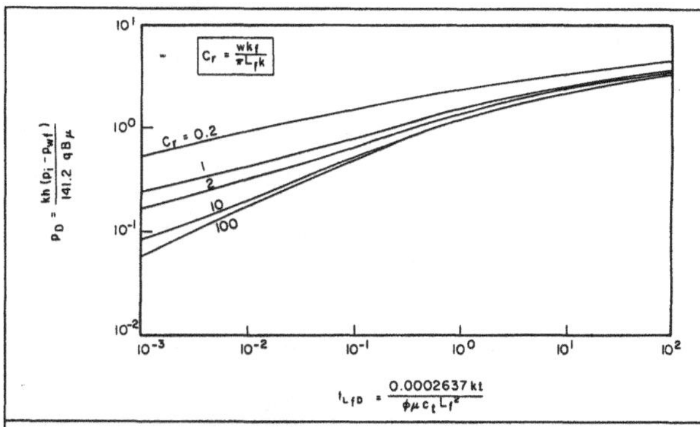

Fig. 15.29—Type curves—finite-conductivity, vertical fracture, constant-rate production.

Type-Curve Methods. Type-curve methods provide considerably more generality than pseudoradial or linear flow analysis techniques because they span the entire time range from linear (or bilinear) to pseudoradial flow and include all the transition regions between the various flow regimes. This section discusses four type curves with different and useful features that should allow the interpretation of most postfracture tests. A systematic procedure is suggested for selecting the particular type curve that is potentially most helpful in the analysis of a specific test. The type curves discussed here include the Gringarten *et al.* [17] curve for infinite-conductivity fracture, finite reservoir; the Cinco-L. and Samaniego-V. [18] curve for finite-conductivity fractures and constant-rate tests; the Agarwal *et al.* [19] curve for finite-conductivity fractures and constant-BHP tests; and the Barker-Ramey curve [16] for fractured wells with wellbore storage.

Gringarten* et al. *Type Curve. This type curve, [17] shown in Fig. 15.28, is a graph of solutions to flow equations modeling vertical hydraulic fractures in a finite reservoir with the following major assumptions: (1) the fracture has infinite conductivity; (2) the well is centered in a square drainage area with no-flow boundaries, and the drainage area is $(2L_e)^2$, where L_e is the perpendicular distance from the well to the reservoir boundary; (3) the fracture has two equal-length wings of length L_f; and (4) wellbore-storage effects are ignored. This curve is potentially useful for analyzing *constant-rate* flow tests and, more important, pressure-buildup tests. To use the curve, we plot pressure change, Δp, since the start of a test vs. elapsed test time, Δt, on tracing paper or on log-log paper with the same size of log cycles as the type curve. We then find the curve that best matches the actual test data. If boundary effects have been felt during the test, the test data will deviate from an earlier fit along the L_e/L_f curve and will be matched by an L_e/L_f "stem" for a finite reservoir. The value of L_e/L_f for the matching stem reflects the reservoir size.

The pressure match point provides the formation-permeability estimate, just as with the Gringarten *et al.* [20] radial-flow type curve discussed in Chap. 3. From the value of p_D and Δp at a match point, we can calculate formation permeability:

$$k=\frac{141.2qB\mu}{h}\left(\frac{p_D}{\Delta p}\right)_{MP} \quad \text{. . . . . (15.39)}$$

The time match point provides an estimate of fracture half-length:

$$L_f=\left[\frac{0.0002637k}{\phi\mu c_T}\left(\frac{\Delta t}{t_{L_fD}}\right)_{MP}\right]^{1/2} \quad \text{. . . . . (15.40)}$$

The linear flow region extends up to $t_{L_fD}\cong 0.016$ on this type curve; note that for $L_e/L_f=\infty$, the curve has a slope of one-half. The pseudoradial flow region begins at $t_{L_fD}\cong 3$ (and, perhaps, as small as 1).

If, based on the type curve match of a test, a number of data points with $t_{L_fD}<0.016$ appear, it will be useful to prepare a square-root-of-time plot (p_{wf} vs. $\sqrt{t}$ or p_{ws} vs. $\sqrt{\Delta t_e}$). A straight line should appear, and its slope will be inversely proportional to $\sqrt{k}L_f$. This can serve as a check on the results of the type-curve analysis, and it may suggest the need for a more consistent type-curve fit (given a limited range of data, type-curve fits are notoriously nonunique).

If, in the type-curve match of a test, a number of data points with $t_{L_fD}>1$ to 3 appear, it will be useful to prepare a semilog plot of the data [p_{wf} vs. log t or p_{ws} vs. log $(t_p+\Delta t)/\Delta t$] and to determine whether a semilog straight line exists. When it does, it provides a unique formation-permeability estimate that can then be used to refine the type-curve match and to provide a permeability estimate for a possible linear-flow analysis. In type-curve matching, when k is known, a pressure match point can be precalculated:

$$(\Delta p)_{MP}=\frac{141.2qB\mu}{kh}(p_D)_{MP}. \quad \text{. . . . . (15.41)}$$

We can choose any convenient dimensionless pressure on the type curve and, using the known permeability, calculate the corresponding Δp at the match point. Then, in the matching procedure, we can align the matching values of p_D and Δp and slide the test data over the type curve *horizontally only*, leading to a unique match.

Cinco-L.* et al. *Type Curve. This type curve, [18] shown in Fig. 15.29, is a graph of flow-equation solutions modeling finite-conductivity vertical hydraulic fractures in an infinite-acting reservoir. Some of the major assumptions on which this solution is based are outlined below.

1. The fracture has finite conductivity that is uniform throughout the fracture and is characterized by the dimensionless fracture conductivity

$$C_r=\frac{wk_f}{\pi kL_f}. \quad \text{. . . . . (15.42)}$$

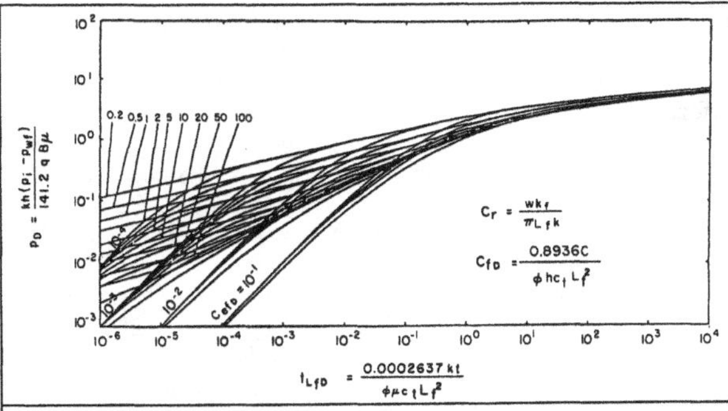

Fig. 15.30—Dimensionless pressure vs. dimensionless time for a well with a fully penetrating vertical fracture (zero fracture skin damage).

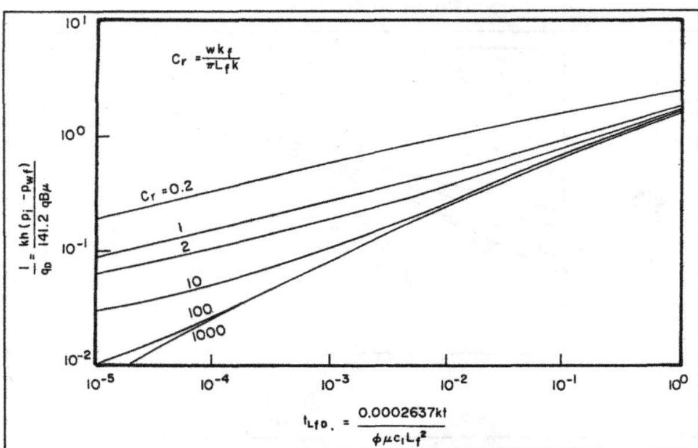

Fig. 15.31—Type curves—finite-conductivity vertical fractures, constant-BHP production.

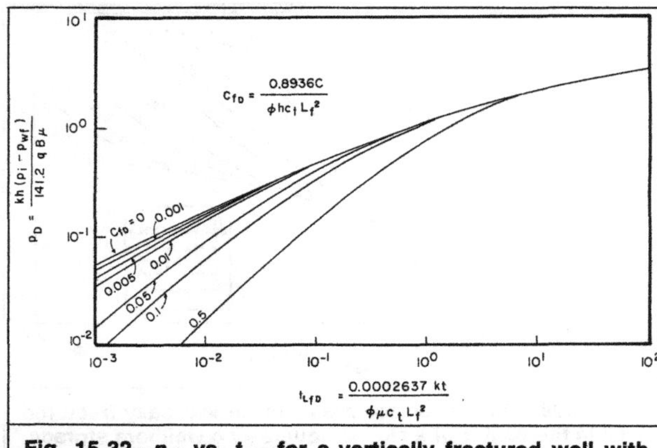

Fig. 15.32—p_D vs. t_{LD} for a vertically fractured well with wellbore storage.

2. The fracture has two equal-length wings of length L_f.
3. Wellbore-storage effects are ignored.
4. The reservoir is infinite-acting; i.e., no boundary effects are felt during the test.

This type curve is potentially useful for analyzing *constant-rate* flow tests and pressure-buildup tests. To use the curve, we plot pressure change, Δp, since the start of a test vs. elapsed test time, Δt, on tracing paper or on log-log paper with the same size of log cycles as the type curve. In principle, we then attempt to find the curve that best matches the test data. In many cases, though, the fit will be nonunique because for lower-conductivity fractures, many curves have the same shape. To use this curve, it is virtually essential that an independent estimate of formation permeability be available, as from a prefracture pressure-buildup test.

When a match is attempted without a prefracture permeability estimate, we estimate k from the pressure match point and L_f from the time match point, just as with the Gringarten *et al.*[17] type curve. The parameter on the best-fitting type curve provides an estimate of C_r and thus, once k and L_f are known, of fracture conductivity, w_{kf}:

$$w_{kf} = C_r(\pi k L_f). \qquad (15.43)$$

The uniqueness of the match is improved significantly when a prefracture permeability estimate is available and the pressure match point is precalculated.

The linear flow region exists before $t_{LfD} = 0.016$ and then only for higher-conductivity fractures ($C_r \geq 100$). At values of $t_{LfD} < 10^{-4}$, a transition from fracture linear flow to formation linear flow has been completed. Pseudoradial flow occurs for $t_{LfD} \geq 1$ to 3 for higher-conductivity fractures and at slightly earlier times for lower-conductivity fractures (Fig. 15.26).

Wellbore storage frequently distorts transient test data at very early times in wells with finite-conductivity fractures, further complicating attempts to analyze early-time data. Although too complicated for use in type-curve analysis in an already complicated situation, Fig. 15.30 (developed by Cinco-Ley and Samaniego-V.[21]) can be helpful in providing qualitative understanding of the effects of wellbore-storage distortion of transient tests in wells with finite-conductivity fractures. In this figure, the wellbore-storage coefficient for a fractured well, C_{fD}, is defined as $0.8936\, C/\phi h c_T L_f^2$. One important conclusion from Fig. 15.30 is that, the lower the C_r, the longer wellbore-storage distortion lasts for a given value of the wellbore-storage coefficient, C_{fD}.

Agarwal* et al. *Type Curve. A type curve similar to the one proposed by Agarwal *et al.*[19] is shown in Fig. 15.31. Like Cinco-L. *et al.*'s[18] type curve, this is a graph of flow-equation solutions modeling finite-conductivity vertical fractures in an infinite-acting reservoir, but it models flow tests conducted at constant BHP. Some of the major assumptions on which this solution is based are outlined below.

1. The fracture has finite conductivity that is uniform throughout the fracture and is characterized by the dimensionless fracture conductivity

$$C_r = \frac{wk_f}{\pi k L_f}. \qquad (15.42)$$

2. The fracture has two equal-length wings of length L_f.
3. The reservoir is infinite-acting; i.e., no boundary effects are felt during the test.

We should note that when a well produces at constant BHP, there *is* no wellbore storage distortion of the test data; thus, to assume constant-BHP production implies a lack of wellbore-storage problems.

This type curve is a plot of the reciprocal of dimensionless rate, q_D, vs. dimensionless time, t_{LfD}, with parameters of dimensionless fracture conductivity, C_r. The curve is potentially useful for analyzing constant-BHP flow tests and, importantly, long-term production data in wells produced at essentially constant pressure as long as the reservoir is infinite-acting.

To use the curve, we plot the reciprocal of rate, $1/q$, vs. elapsed test time, t, on tracing paper or log-log paper with the same size of log cycles as the type curve. In principle, we then attempt to find the curve that best matches the test data. The fit frequently is nonunique because for lower values of C_r, the curves have essentially the same shape. To use this curve, it is virtually essential that an independent estimate of formation permeability be available, as from a prefracture pressure-buildup test.

When a match is attempted without a prefracture permeability estimate, we determine k from the rate match point:

$$k = \frac{141.2qB\mu}{h(p_i - p_{wf})}\left(\frac{1/q_D}{1/q}\right)_{MP}. \qquad (15.44)$$

Fracture length is estimated from the time match point after permeability has been established.

$$L_f = \left[\frac{0.0002637k}{\phi\mu c_T}\left(\frac{t}{t_{LfD}}\right)_{MP}\right]^{1/2}. \qquad (15.45)$$

The parameter on the best-fitting type curve provides an estimate of C_r and thus, once k and L_f are known, fracture conductivity, wk_f:

$$wk_f = C_r(\pi k L_f). \qquad (15.43)$$

When a permeability estimate is available, a unique type-curve match can be established by first precalculating a rate match point:

$$(1/q)_{MP} = \frac{141.2B\mu}{kh(p_i - p_{wf})}(1/q_D)_{MP}. \qquad (15.46)$$

TABLE 15.4—SUMMARY OF ANALYSIS TECHNIQUES—TYPE-CURVE PLOTS

Case	Interpretation of Unit-Slope Line (USL)	Interpretation of Pressure Match Point	Interpretation of Time Match Point in Ramey and Gringarten[15] Type Curve	Interpretation of Time Match Point in Gringarten *et al.*[17] Type Curve	Interpretation of Time Match Point in Fractured-Well Type Curves
Δp vs. Δt, or Δp_a vs. Δt_a	$C_D=\frac{0.0372q_g\bar{B}_g}{\phi h\bar{c}_T r_w^2}\left(\frac{\Delta t}{\Delta p}\right)_{USL}$	$k=\frac{141.2q_g\bar{\mu}\bar{B}_g}{h}\left(\frac{p_D}{\Delta p}\right)_{MP}$	$\phi\bar{c}_T=\frac{0.0002637k}{\bar{\mu}r_w^2}\left(\frac{\Delta t}{t_D}\right)_{MP}$	$C_D=\frac{0.0002637k}{\phi\bar{\mu}\bar{c}_T r_w^2}\left(\frac{\Delta t}{t_D/C_D}\right)_{MP}$	$L_f=\left[\left(\frac{0.0002637k}{\phi\bar{\mu}\bar{c}_T}\right)\left(\frac{\Delta t}{t_{L_fD}}\right)_{MP}\right]^{1/2}$
$\Delta(p^2)$ vs. Δt	$C_D=\frac{0.375q_gT\bar{z}}{\phi h\bar{c}_T r_w^2}\left[\frac{\Delta t}{\Delta(p^2)}\right]_{USL}$	$k=\frac{1{,}422q_gT\bar{z}\bar{\mu}}{h}\left(\frac{p_D}{\Delta p^2}\right)_{MP}$	$\phi\bar{c}_T=\frac{0.0002637k}{\bar{\mu}r_w^2}\left(\frac{\Delta t}{t_D}\right)_{MP}$	$C_D=\frac{0.0002637k}{\phi\bar{\mu}\bar{c}_T r_w^2}\left(\frac{\Delta t}{t_D/C_D}\right)_{MP}$	$L_f=\left[\left(\frac{0.0002637k}{\phi\bar{\mu}\bar{c}_T}\right)\left(\frac{\Delta t}{t_{L_fD}}\right)_{MP}\right]^{1/2}$

Note: For drawdown test, $\Delta p=p_i-p_{wf}$ (or equivalent); $\Delta t=t=$ flow time (or equivalent). For buildup test, $\Delta p=p_{ws}-p_{wf}(\Delta t=0)$ (or equivalent); $\Delta t=\Delta t_e=\Delta t/(1+\Delta t/t_p)$ (or equivalent). For slightly compressible liquid, equations given for Δp vs. Δt are applicable, with q_oB_o replacing $q_g\bar{B}_g$ and μ_o replacing $\bar{\mu}$.

We can choose any convenient dimensionless rate on the type curve and, using the known permeability, calculate the corresponding value of $(1/q)$ at the match point. Then, in the matching procedure, we can align the matching values of $(1/q_D)$ and $(1/q)$ and slide the test data over the type curve *horizontally only*, leading to a unique match.

Barker-Ramey Type Curve. These authors[16] developed a type curve (Fig. 15.32) that includes the effect of wellbore storage on a constant-rate flow test or buildup test. The curve is for an infinite-conductivity fracture in an infinite-acting reservoir.

The formation permeability is estimated from the pressure match point and fracture half-length from the time match point:

$$k=\frac{141.2qB\mu}{h}\left(\frac{p_D}{\Delta p}\right)_{MP} \quad \text{.......................... (15.39)}$$

and

$$L_f=\left[\frac{0.0002637k}{\phi\mu c_T}\left(\frac{\Delta t}{t_D}\right)_{MP}\right]^{1/2} \quad \text{.................... (15.40)}$$

The matching parameter is a dimensionless wellbore-storage coefficient:

$$C_{fD}=\frac{0.8936C}{\phi hc_TL_f^2}, \quad \text{.............................. (15.47)}$$

where C, the dimensional wellbore-storage coefficient, for a gas well is

$$C\cong c_{wb}V_{wb}, \quad \text{.................................. (15.48)}$$

and for an oil well with a moving liquid/gas interface in the wellbore and unchanging tubing-head pressure is

$$C\cong 25.65A_{wb}/\rho_{wb}. \quad \text{.......................... (15.49)}$$

Recommended Procedure for Type-Curve Analysis. The procedure suggested below has proved effective for postfracture buildup and constant-rate flow-test analysis with type curves. This procedure is intended to clarify which of the curves presented earlier can be most helpful.

Regardless of the curve ultimately used, test data should be plotted on log-log coordinates as Δp vs. Δt. A match should first be attempted on the Cinco-Ley and Samaniego-V. finite-conductivity type curve (Fig. 15.29). For long fractures in low-permeability formations, this curve will usually be required and frequently will be the only curve needed. Early-time data with slope significantly steeper than any plotted curve probably indicate wellbore-storage distortion of the test data. If a significant fraction of the data have these early steep slopes, the Barker-Ramey curve (Fig. 15.32) may provide a superior test interpretation. Note, however, that this curve is limited to fractures with infinite conductivity (effectively, $C_r\geq 100$).

If late-time data have slopes that steepen *and* if fracture conductivity is high ($C_r\geq 10$ and, preferably, $C_r\geq 100$), the Gringarten *et al.*[17] curve (Fig. 15.28) may be used and may provide some proof that boundary effects are the cause of the steepened slopes.

The basic tool for matching constant-BHP flow tests or long-term production performance is the Agarwal *et al.*[19] type curve (Fig. 15.31). This curve should be adequate in most cases until boundary effects are felt.

In type-curve matching, the analyst should always check for possibilities of the pseudoradial flow regime at later times in the test ($t_{L_fD}\geq 1$ to 3 and all C_r values) and linear flow at earlier times ($t_{L_fD}\leq 0.016$ *and* $C_r\geq 100$). If the pseudoradial regime is present, permeability can be estimated in a much less ambiguous way than from a type curve and can be used to force a better type-curve fit. Further, even if a prefracture permeability estimate is available, the postfracture estimate is likely to be better (if different) because more formation may be in communication with the wellbore after fracturing. Linear-flow analysis, given a permeability estimate, can provide an excellent fracture-length estimate for high-conductivity fractures.

A summary of match-point interpretation and other useful type-curve analysis techniques is given in Table 15.4. Details are provided for buildup and constant-rate flow tests and for oil- and gas-well tests.

Example applications of type curves (and other techniques) to postfracture test analysis are given in Appendix K, Examples 15.5 through 15.8.

Limitation of Type-Curve Analysis. Although type curves provide a more general method of test analysis than do pseudoradial and linear-flow analyses, they still have significant limitations in some applications. Major limitations are itemized below.

1. There is frequently a significant uniqueness problem with existing type curves; i.e., test data can fit equally well in several different positions on existing type curves, particularly for low-conductivity fractures ($C_r<10$). *If* a formation-permeability estimate is available, this problem essentially vanishes, but formation-permeability estimates require achievement of pseudoradial flow (highly unlikely in low-permeability formations with long fractures) or prefracture tests in the same interval open to the well after fracturing.
2. Type curves necessarily focus on key features (e.g., finite conductivity) and ignore other complications (e.g., wellbore storage, boundaries). Thus, they may be vastly oversimplified models of the tested formation.
3. Fracture closure, leading to continuously variable fracture conductivity along the length of a fracture or with continued test time, is almost certainly a factor in many wells in practice but is not included in existing type curves.
4. In most gas-well buildup tests, non-Darcy flow in the fracture after shut-in is a virtual certainty. This non-Darcy flow causes the "apparent" fracture conductivity to increase continuously throughout a test. Its existence can render type curves useless because even finite-conductivity type curves have fixed (constant) values of fracture conductivity as parameters.

Because of these limitations, many operators choose to use finite-difference reservoir simulators to analyze important tests.

Use of Finite-Difference Reservoir Simulators. Finite-difference reservoir simulators, used in a history-matching mode, provide the most generally applicable method of postfracture pressure-transient test analysis. Lee and Holditch[22] summarized this method of test analysis and observed that uniqueness problems frequently associated with history matching are minimized if one matches not only the data from a given test, but also all test data obtained on the well and the long-term production history of the well.

TABLE 15.5—PRESSURE-BUILDUP DATA FOR WELL 12

t_p, hours	680
p_i, psi	2,200
q_g, Mcf/D	325
h, ft	45
ϕ_g	0.045
μ_{gi}, cp	0.015
c_{gi}, psi	$1{,}000 \times 10^{-6}$
B_{gi}, RB/Mcf	0.267
γ_g	0.70
T, °F	180

Δt (hours)	p_{ws} (psia)
0	103
0.5	190
1	270
1.5	319
2	365
3	438
4	500
5	551
6	597
7	636
8	670
10	726
12	774
14	815
16	848
18	880
20	906
24	955
28	992
32	1,027
36	1,057
40	1,084
48	1,131
56	1,171
64	1,206
72	1,236
88	1,287
104	1,330
120	1,366
136	1,399
156	1,433
180	1,467
204	1,498
228	1,523
252	1,544
276	1,575
300	1,590
324	1,605
348	1,624
624	1,761
960	1,813
1,750	1,933

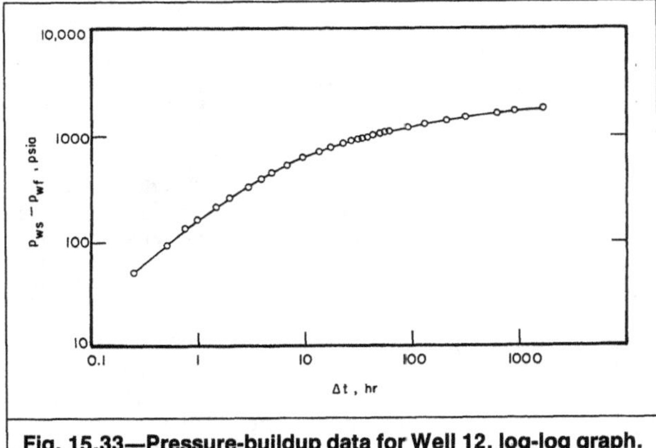

Fig. 15.33—Pressure-buildup data for Well 12, log-log graph.

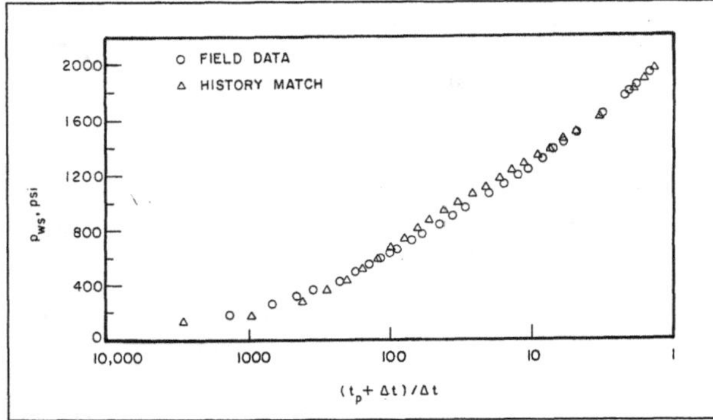

Fig. 15.34—Pressure-buildup data for Well 12, Horner graph with results from history match.

TABLE 15.6—ANALYSIS RESULTS FOR WELL 12

Method	k (md)	L_f (ft)	wk_f (md-ft)
Pseudoradial	0.012	100	
Gringarten *et al.*[16] type curve	0.045	45	
Cinco-Ley-Samaniego-V. type curve	0.052	150	5
History match	0.001	700	20

Simulator history matching is the only practical tool available for handling the simultaneous effects of wellbore storage, reservoir boundaries, finite fracture conductivity, fracture closure, and non-Darcy flow. It also offers the prospect of taking into account possible effects of fracture and reservoir heterogeneity, although the chances of obtaining a unique reservoir description diminish when heterogeneities become important.

An example of the difference in results from the most reliable type curves and other traditional analysis techniques and a finite-difference simulator history match is provided by Well 12 in Ref. 22. A full suite of analysis techniques was attempted for Well 12. The buildup data are presented in Table 15.5 and in Figs. 15.33 and 15.34. The data in Fig. 15.33 indicate that wellbore storage dominated the buildup test for the first 10 hours. After about 10 hours, a half-slope is seen on the log-log graph. The match of these data resulted in estimates of 0.045 md and a 45-ft [14-m] fracture with Gringarten *et al.*'s[16] type curve and 0.052 md and a 150-ft [46-m] fracture with Cinco-Ley and Samaniego-V.'s type curve.

Pseudoradial flow analysis, with the maximum slope in Fig. 15.34, resulted in a permeability estimate of 0.012 md and a fracture length of about 100 ft [30 m]. Note that the slope of the buildup curve still has to increase considerably to reach 2,200 psi [15.2 MPa] at a Horner time of 1.0.

The results from the history-matching computer run are also presented in Fig. 15.34. This match is excellent because the data between 72 and 1,750 hours are matched almost perfectly. The early-time data do not provide a perfect match, probably because of small variations in fracture conductivity and wellbore storage that were not modeled precisely.

For Well 12, the properties estimated with various techniques are summarized in Table 15.6. The history-match result is almost certainly superior (Fig. 15.34); note the dramatic difference in re-

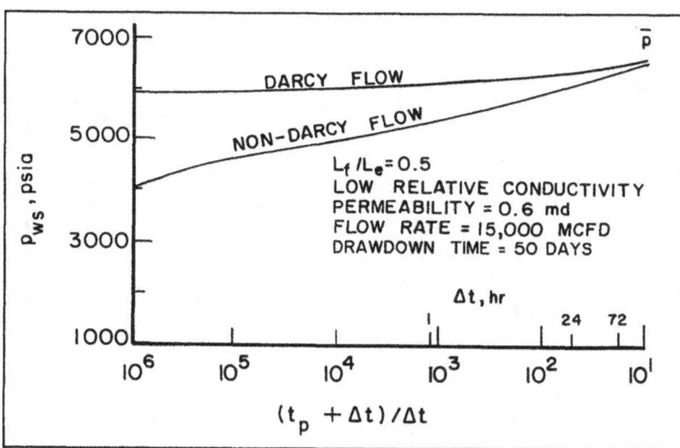

Fig. 15.35—Effect of non-Darcy flow in the fracture on the shape of a buildup curve for small relative conductivity.

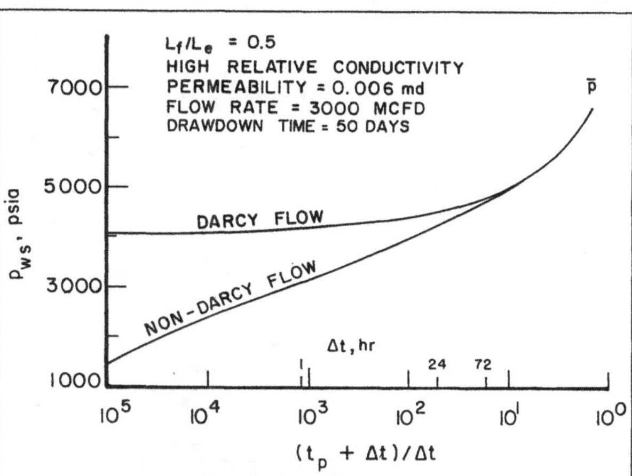

Fig. 15.36—Effect of non-Darcy flow in the fracture on the shape of a buildup curve for large relative conductivity.

sults compared with those derived from type curves. It is not fair to criticize the results from the pseudoradial method in this case because conditions for its applicability were not reached (pseudoradial flow was not achieved); note, however, that this was a 1,750-hour pressure-buildup test.

15.7 Effect of Nonideal Conditions

Introduction. This section briefly discusses complications that offer special difficulties in interpreting postfracture well performance, including non-Darcy flow, damage to the fracture face, water blocking, unequal-length fracture wings, and variable fracture conductivity.

Non-Darcy Flow. Holditch and Morse[23] studied the effect of non-Darcy flow on the behavior of hydraulically fractured gas wells. Notable in their study was the observation that in a buildup test, non-Darcy flow can continue for a substantial time after the well is shut in, and the apparent fracture conductivity can continue to increase throughout much of the test. The non-Darcy flow continues as pressure gradients dissipate during the buildup test because the flow velocity in the narrow fracture remains quite high, and they diminish in magnitude. Figs. 15.35 and 15.36 compare simulated buildup-curve shapes for low and high fracture conductivities, respectively. Not shown are log-log graphs of the test data. In these graphs, compared with the Cinco-Ley and Samaniego-V. finite-conductivity type curve, the buildup-test data continue to move toward increasingly higher C_r values throughout the test. A single value of C_r simply does not describe the entire test. A practical way to analyze tests with non-Darcy flow is to use a reservoir simulator for history matching. Obviously, the simulator must model non-Darcy flow in the fracture.

Damage to Fracture Face. Cinco-Ley and Samaniego-V.[21] presented a study of pressure-transient test behavior for hydraulically fractured wells with damage around the fracture face. Their major finding was that on log-log plots of Δp vs. Δt from a transient test, the graph is a flat, almost horizontal line that approaches the curve for undamaged fractures at late times. Fig. 15.37 shows such a graph (with dimensionless variables) for a very-low-conductivity fracture (C_r=0.2); Fig. 15.38 shows a similar graph for a high-conductivity fracture (C_r=100). The parameter on both graphs is the fracture skin factor, s_{fs}, defined as

$$s_{fs}=\frac{\pi}{2}\frac{w_s}{L_f}\left(\frac{k}{k_s}-1\right), \quad \text{.........................(15.50)}$$

where w_s=width of damaged zone and k_s=permeability of damaged zone.

It is difficult to characterize the formation and the fracture when there is fracture damage. Numerical-simulator history matching is perhaps the best current technique that is generally available. There is a significant uniqueness problem in this history matching, however. The transient behavior of a well with $1 \leq C_r \leq 50$ is similar to that of a well with a damaged infinite-conductivity fracture. Data spanning a wide time range (particularly long-time data) are required to remove this ambiguity.

Water Blocking. Holditch[24] presented a study of the combined effects of formation damage and fracture-fluid invasion that, in extreme cases, can lead to water blocking in a fractured well. This study, for gas reservoirs, was based on the observation that *some* reduction in gas permeability along the fracture face occurs when fracturing fluid enters any formation. This reduction is caused by formation damage (e.g., clay swelling) and movement of fines. In addition, the presence of liquid in the formation reduces the permeability to gas.

The study was performed with a single-phase (gas) and a two-phase (gas/water) reservoir simulator. Some of the questions addressed in the study included the following.

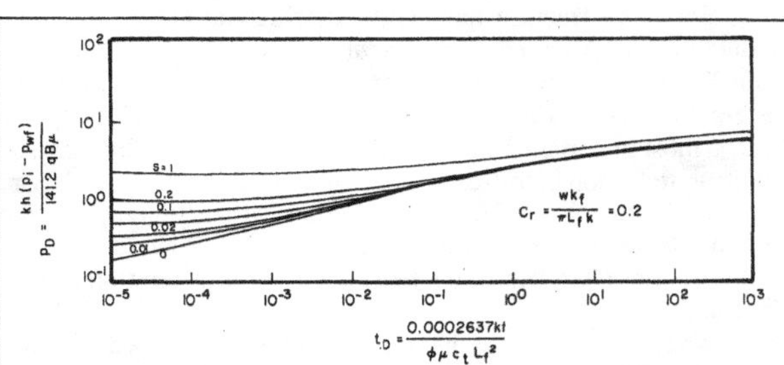

Fig. 15.37—Dimensionless pressure drop vs. dimensionless time for a fractured well with fracture skin damage ($wk_f/\pi L_f k$=100).

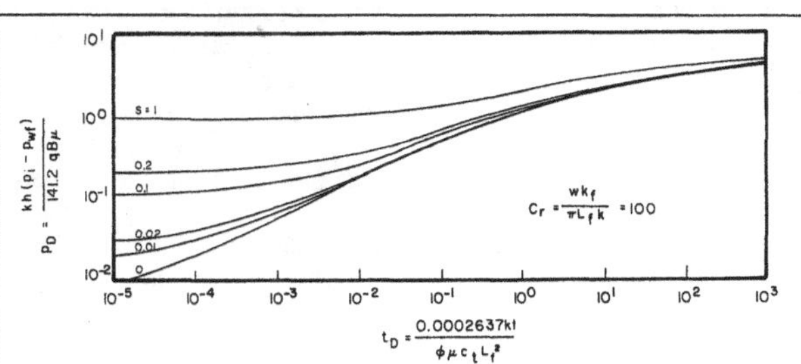

Fig. 15.38—Dimensionless pressure drop vs. dimensionless time for a fractured well with fracture skin damage ($wk_f/\pi L_f$=100).

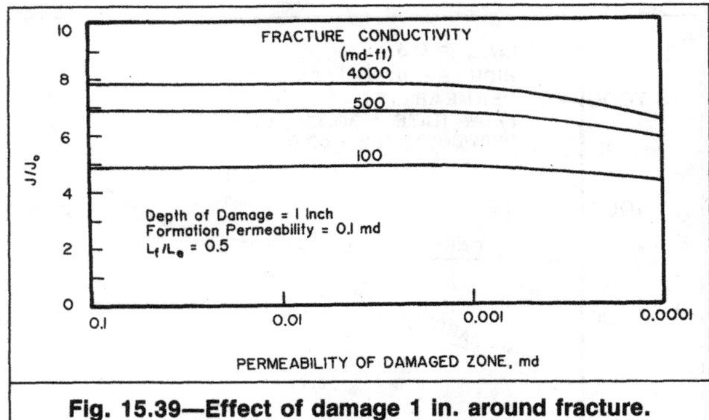

Fig. 15.39—Effect of damage 1 in. around fracture.

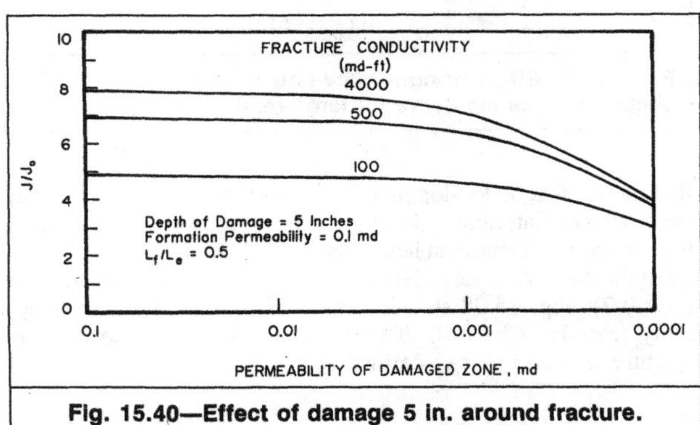

Fig. 15.40—Effect of damage 5 in. around fracture.

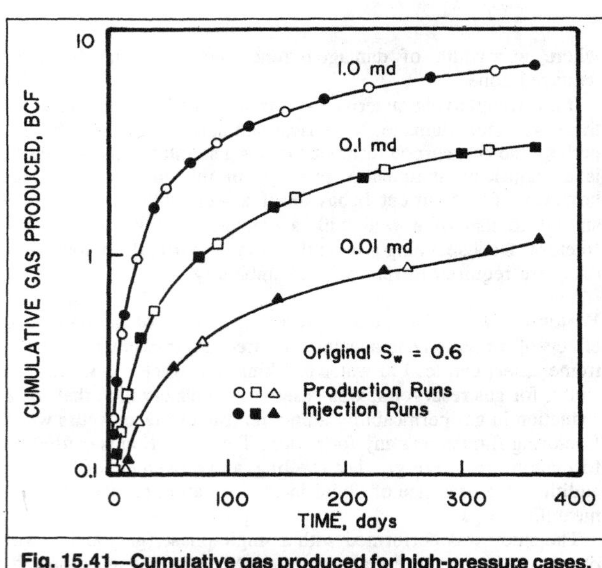

Fig. 15.41—Cumulative gas produced for high-pressure cases.

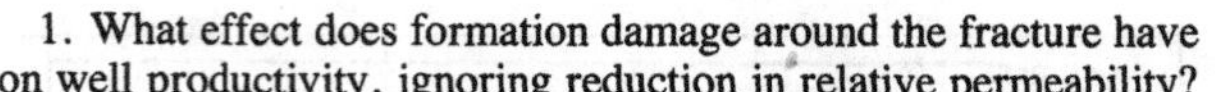

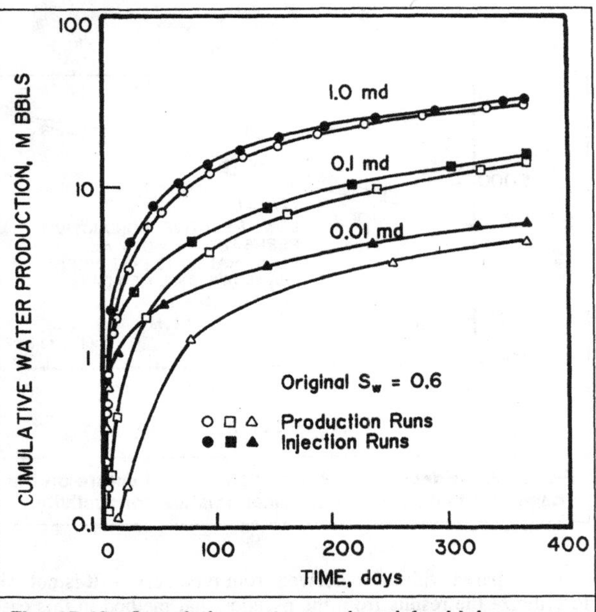

Fig. 15.42—Cumulative water produced for high-pressure cases.

TABLE 15.7—INPUT DATA FOR TWO-PHASE RUNS WITH NO DAMAGE

Parameter	High-Pressure Case	Low-Pressure Case
Reservoir pressure, psi	7,500	2,325
Net gas pay, ft	100	100
Total porosity, %	20	12
Fracture length, ft	660	660
Initial fracture conductivity, md-ft	2,000	2,000
Reservoir temperature, °F	300	150
Gas gravity	0.65	0.7
Well spacing, acres	160	160

TABLE 15.8—LOW-PRESSURE (2,325 psi) RESERVOIR NO FORMATION DAMAGE

Case	Remarks	Permeability (md)	Water Saturation (%)	Flowing BHP (psi)
1	Production run	0.17	60	2,000
2	Injection run	0.17	60	2,000
3	Production run	0.17	60	1,000
4	Injection run	0.17	60	1,000
5	Production run	0.0017	60	2,000
6	Injection run	0.0017	60	2,000
7	Production run	0.0017	60	1,000
8	Injection run	0.0017	60	1,000

1. What effect does formation damage around the fracture have on well productivity, ignoring reduction in relative permeability?
2. What effect does fracturing fluid have on productivity, considering only relative permeability changes with no formation damage?
3. What are the combined effects of fluid invasion and formation damage?
4. Under what conditions can complete water blocking occur?

The effects of formation damage alone are summarized in Figs. 15.39 and 15.40, showing results for 1- and 5-in. [2.5- and 12.7-cm] damaged zones. Conclusions that can be drawn from these figures include the following.

1. Unless damage is deep and severe, formation damage along the fracture face will not significantly affect the well's productivity index.
2. For fixed fracture length, the detrimental effect of damage reduction (J/J_o) increases as fracture conductivity increases.
3. For fixed fracture conductivity, the reduction in J/J_o increases as fracture length decreases.

The most important findings in the study of fracture-fluid invasion (with no reduction in formation permeability) included the following.

1. With the driving force provided by capillary pressure, fluid imbibes from the region of high water saturation toward the region of lower water saturation, decreasing the water saturation near the fracture.
2. When production begins, fracture water is produced, also lowering the water saturation near the fracture and increasing the permeability to gas.
3. Simulated results indicated little, if any, harm caused by filtrate invasion in situations in which no formation damage occurred.

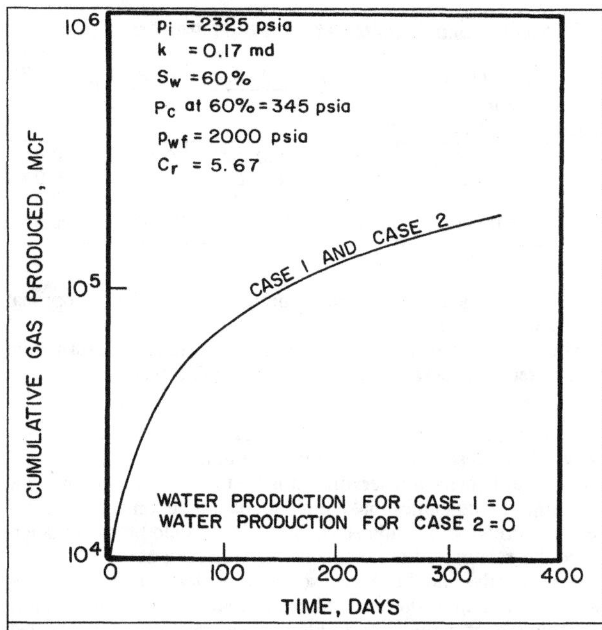

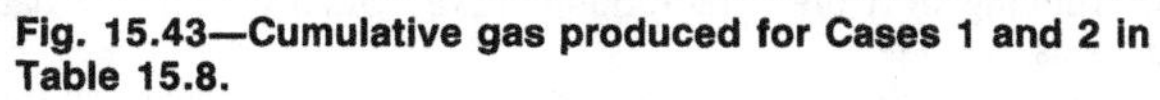

Fig. 15.43—Cumulative gas produced for Cases 1 and 2 in Table 15.8.

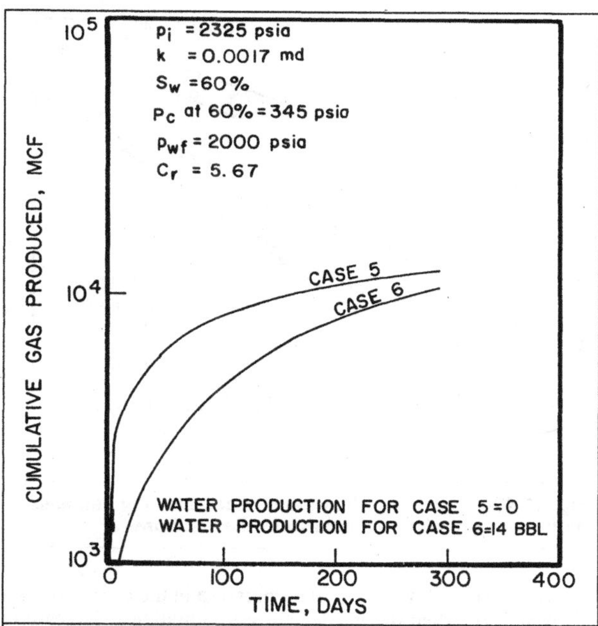

Fig. 15.45—Cumulative gas produced for Cases 5 and 6 in Table 15.8.

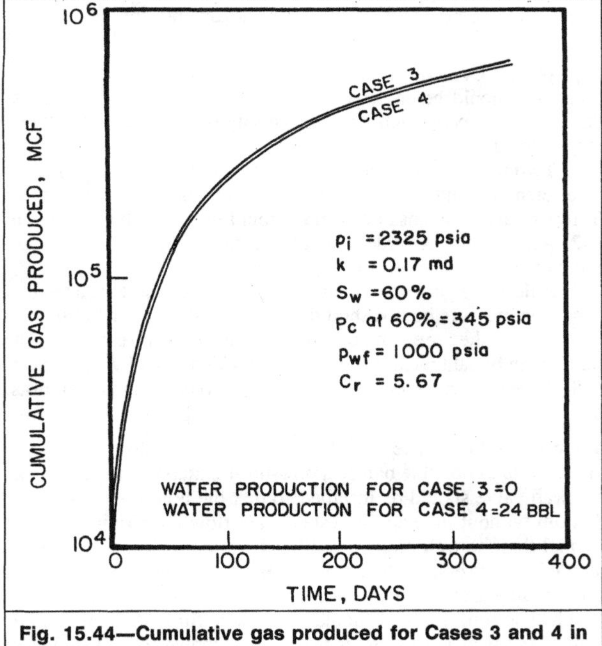

Fig. 15.44—Cumulative gas produced for Cases 3 and 4 in Table 15.8.

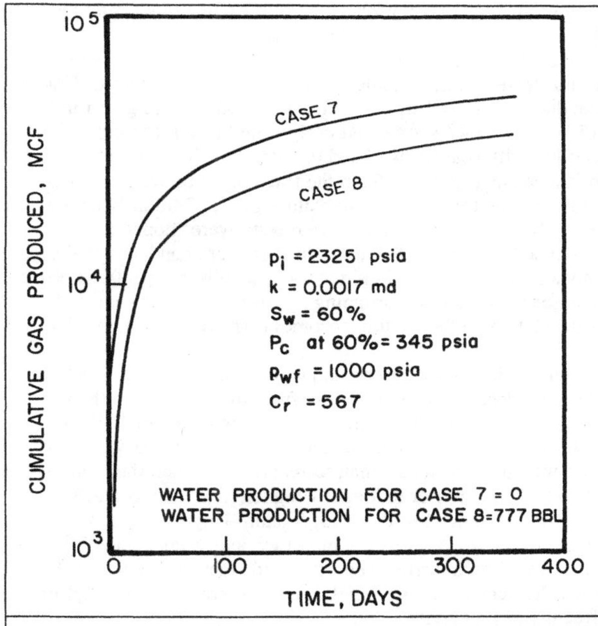

Fig. 15.46—Cumulative gas produced for Cases 7 and 8 in Table 15.8.

Figs. 15.41 and 15.42 compare cumulative water and gas production from formations with and without fracture-fluid invasion and with high initial pressure. Properties of the formation are summarized in Table 15.7. In the figures, "production runs" means that no fracture fluid was injected (formation water had some mobility); "injection runs" were identical except that fracture fluid (water) was injected and subsequently produced back. Fig. 15.41 shows that there was virtually no difference in cumulative gas recovery in the two cases for any of the formation permeability levels modeled.

With lower pressure in the reservoir, the fracture-fluid cleanup was less rapid, but it still occurred. Formation properties for the low-pressure reservoir are summarized in Table 15.7. Eight cases were studied for low-pressure reservoirs; their features are summarized in Table 15.8. Results are shown in Figs. 15.43 to 15.46.

The final issue addressed in Holditch's study was the combined effect of fluid invasion and formation damage. He observed that reduced permeability near the fracture increases capillary pressure at a given water saturation. Water can be trapped in this zone if the pressure drawdown is not great enough to overcome the capillary end effect—i.e., capillary forces tending to hold the water in the formation rather than letting it flow into the fracture and thus to the wellbore. Complete water blocking can occur in this case.

Even when the drawdown is large enough to overcome the capillary end effect, significant reduction in gas permeability (and increased cleanup problems) can occur when permeability reduction

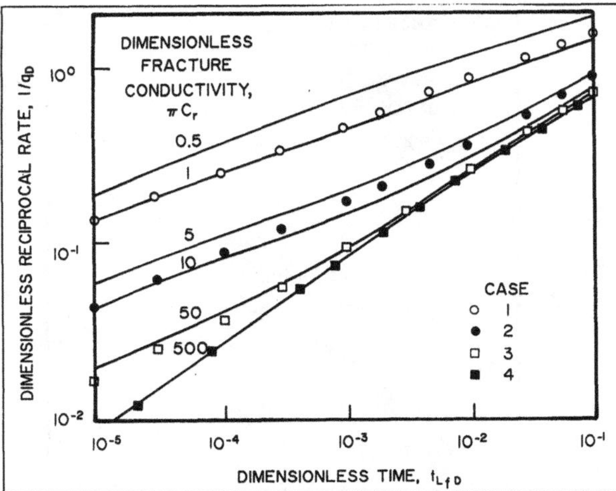

Fig. 15.47—Response when fracture conductivity decreases monotonically with distance from the wellbore.

(damage) is severe, when capillary pressure in the damaged zone is elevated, or when the irreducible gas saturation is increased.

Unequal-Length Fracture Wings and Variable Fracture Conductivity. Bennett *et al.*[25] studied the influence of variable fracture conductivity and unequal wing length on the response of vertically fractured wells. They used a finite-difference model and modeled both constant-rate and constant-BHP production.

Fig. 15.47 illustrates the typical response when fracture conductivity decreases monotonically away from the wellbore. Fracture conductivities used in the four cases modeled are given in Table 15.9. Fig. 15.47 shows that early-time behavior is controlled by conductivity near the well and that late-time behavior is controlled by the average value of C_r in the fracture. In between, the test data cross curves for intermediate values of C_r. Similar behavior was also observed when constant-rate tests were modeled.

Bennett *et al.* noted that the kind of fracture-conductivity decrease modeled and presented in Fig. 15.47 can arise frequently in practice if higher-permeability propping agents are injected near the end of a fracture and if the fracture becomes narrower as it extends laterally into the formation.

Another important observation was that if the fracture conductivity does not decrease monotonically from the wellbore to the fracture tip, the late transient behavior cannot be correlated simply.

In the study of unequal-length fracture wings, it was observed that well performance began to be affected when the wing-length ratio (ratio of shorter to longer wing length) was less than 0.3. For wing-length ratios less than 0.3, fracture-length estimates using the Agarwal *et al.*[19] or the Cinco-Ley and Samaniego-V.[18] type curves will be in error. The estimated fracture length will be less than the actual fracture length, and the error will depend on the fracture conductivity.

15.8 Gas-Well Test Design

Because a large fraction of fractured wells are gas wells, this section deals with the design problems peculiar to these wells. Earlougher[26] treats oilwell test design extensively, so there is no need to duplicate that discussion here.

Objective. The principal objective of gas-well testing is to learn as much as possible about the reservoir so that long-term producing characteristics can be predicted and well stimulation treatments can be optimized. Therefore, it is necessary that the following items be considered for each well.

1. The well completion should be designed so that swab valves, nipples, packers, etc., are adequate for running the pressure-testing equipment.
2. The well should be cleaned up before a test is run.
3. Pressure-measurement devices (including surface gauges) should be properly chosen and calibrated.

TABLE 15.9—SUMMARY OF INPUT PARAMETERS

	Section A: $L_{D1}=0.25$, $L_{D2}=0.25$, $L_{D3}=0.25$, $L_{D4}=0.25$				
Case	Zone 1 πC_{r1}	Zone 2 πC_{r2}	Zone 3 πC_{r3}	Zone 4 πC_{r4}	πC_r
1	1.0	0.5	0.375	0.125	0.5
2	10.0	5.0	3.750	1.250	5.0
3	100.0	50.0	37.500	12.500	50.0
4	1,000.0	500.0	375.000	125.000	500.0

4. Gas and liquid should be accurately measured with optimal separator conditions.

The remainder of this discussion presents detailed information to be used as general guidelines for designing well tests.

Well Completion. In most cases, a multizone well test in a low-permeability reservoir is virtually impossible to analyze correctly. Therefore, in critical reservoirs, it is best to test only one interval at a time. If economics make multizone completions necessary, a temperature/gradiomanometer/spinner log should be run to determine the relative inflow characteristics of each zone.

A swab valve should be used so that a lubricator can be installed without shutting in the well. Shut-in periods immediately before the running of a pressure-buildup survey often complicate data analysis.

Nipples should be run in the tubing string at the packer so that the BHP bombs can be hung as near the perforations as possible. The ID's of all nipples, packers, seals, etc., should be calipered at the wellsite before this equipment is run into the well.

Equipment Considerations. The best available pressure-survey company should be selected on the basis of the reliability of its equipment and personnel. After a company is selected, the following steps are suggested.

1. Two pressure recorders should be run in tandem for all tests. The element and clock serial numbers should be recorded. Temperature-compensated bombs should be used when available.
2. All pressure recorders should be calibrated in the shop immediately before being sent to the field.
3. A dead-weight tester should be taken to the field for all testing.
4. The pressure bombs should be calibrated in the field immediately before they are run and after they have been retrieved. The bombs can be calibrated in the lubricator with the dead-weight tester.
5. A maximum recording thermometer should be run on all tests.

Field Procedures. The well should be flowing as close to a "stabilized" rate as possible before a pressure-buildup test is performed. A swab valve should be installed so that the bombs can be run to bottom without interrupting established flow conditions. It is essential that the well remain stabilized while the bombs are rigged up and run. Pressure should never be bled off before or during a survey unless unsafe well conditions exist. After the bombs have reached bottom, the well should continue to flow for about 2 hours before the well is shut in. If the well is known to be heading or slugging, the flow period should be 10 times as long as the heading cycle. The gas and liquid rates should be measured carefully during this period. Records of estimated gas volume produced during cleanup should be kept so that cumulative gas production before test unit measurements can be calculated. The exact time when the well is shut in should be recorded. After the pressure bombs are pulled, a gradient survey should be run over the entire section tested to below the perforations. The well should not be reopened to flow until the chart has been inspected and, in many cases, preliminary data analysis is completed. Leaving the well shut in allows additional buildup points to be taken if needed for proper reservoir evaluation.

It is important to produce the well at a constant rate before the buildup survey is run. The choke size should not be altered during the 48 hours before the buildup test.

For low-permeability or damaged wells, the flow rate may be less than 100 Mcf/D [2830 m^3/d] and stabilized rates will not be attained. If this is the case, it may be necessary to begin the buildup

TABLE 15.10—PRODUCING-TIME REQUIREMENTS FOR GAS WELLS

Production Rate (Mcf/D)	Permeability (md)	Producing Time (hours)
<50	0.005	500
50 to 100	0.01	250
500 to 1,000	0.1	25
>1,000	1.0	2.5

TABLE 15.11—MINIMUM SHUT-IN TIME REQUIREMENTS FOR GAS WELLS

$\bar{p}$, psia 4,000
p_{wf}, psia 1,000
T, °R (°F) 660 (200)
c_{wb}, psi^{-1} 0.001

Flow Rate (Mcf/D)	V_{wb} (bbl)	Minimum Shut-in Time (hours)
100	30 (tubing)	136
100	200 (casing)	909
250	30	54
250	200	364
500	30	27
500	200	182
1,000	30	14
1,000	200	91

TABLE 15.12—REPRESENTATIVE SHUT-IN TIME REQUIRED FOR SEMILOG ANALYSIS OF GAS-WELL BUILDUP TESTS

Flow Rate (Mcf/D)	Buildup Time (hours)
100	400
100 to 250	160
250 to 500	90
500 to 1,000	48
1,000	24

TABLE 15.13—SHUT-IN TIME REQUIRED TO REACH SEMILOG STRAIGHT LINE IN FRACTURED GAS WELLS

Permeability (md)	Fracture Length (ft)	Shut-in Time To Reach Straight Line (days)
0.01	100	197
0.01	500	4,940
0.01	1,000	19,700
0.1	100	20
0.1	500	490
0.1	1,000	1,970
1.0	100	2
1.0	500	49
1.0	1,000	197

period if the flowing tubing pressure remains fairly constant in the 50- to 100-psi [345- to 690-kPa] range for 8 to 12 hours. For these low-rate conditions, accurate rate measurement is difficult, so it is often advantageous to begin the buildup test before the well begins to head or before measurable flow ceases.

Office Procedures. All data obtained during the well test should be reported by the service company. For fractured reservoirs, the early-time data can be critical to the analysis. It is suggested that pressures be read using the following schedule.

Every 2 minutes for the first 20 minutes.
Every 5 minutes from 20 to 60 minutes.
Every 15 minutes from 1 to 6 hours.
Every 30 minutes from 6 to 24 hours.
Every 60 minutes from 24 to 72 hours.
Every 180 minutes after 72 hours.

If possible, all charts for a specific well, field, etc., should be read by the same person using the same chart reader to provide consistency in the data for the specific area and to improve the quality of conclusions based on analysis of the data. The same pressure bomb should be run back into the well during future surveys to maintain data consistency.

Flow-Rate Measurements. A portable well test separator should be used to test the gas and liquid rates for a minimum of 12 to 24 hours before the pressure-buildup test. Samples of the gas, water, and condensate should be collected and sent to a laboratory for analysis. The analyses should consist of (1) gas, condensate, and water gravity; (2) Btu content of the well gas and dry gas; (3) composition of gas so that the potential corrosion problems can be anticipated and economic potential of gasoline byproducts can be assessed; and (4) a complete water analysis to include resistivity and concentration (ppm) of Na, K, Ca, Mg, Fe, OH, total solids, Cl, Br, HCO_3, SO_4, CO_3, and H_2S.

Producing-Time Requirements. The well should be produced as required to unload the wellbore and to clean up any extraneous fluids that may have invaded the formation. These fluids could consist of mud filtrate, cement filtrate, acid, workover fluids, and fracturing fluid. The choke size, flowing tubing pressure, and estimated gas rates should be recorded during this initial cleanup period. The recovered volumes of acid, water, and fracturing fluids should be measured. After the well has cleaned up, a production test should be run before the buildup test.

Various criteria can be used to determine when the well has cleaned up sufficiently to begin the production test.

1. The amount of workover, acid, or fracturing fluid being produced has decreased to a small percentage of the flow stream. This may require 2 to 4 weeks for wells treated with massive stimulation treatments.

2. The gas/water ratio (assuming the water is formation water) has leveled off.

3. The flow rate and flowing tubing pressure have "stabilized" as well as can be expected. A rule of thumb is that the flow rate and tubing pressure are changing less than 5 to 10% per day.

If a well has not been fractured, the duration of the well test can be less than is required for a fractured well. After the well has cleaned up, a well test separator should be used to measure the "stabilized" flow rate. If a swab valve is not used and the well has to be shut in to install the lubricator and/or to run the bombs to the bottom, the well should be produced at the same "stabilized" rate for 10 times the duration of the *shut-in* time required to install the lubricator.

The entire production period should be long enough to ensure that a significant portion of the reservoir has been affected. This requirement means that both wellbore storage and skin (damage) effects have been overcome and that the radius of investigation in the reservoir is large enough to obtain a true, in-situ reservoir permeability.

The testing times presented in Table 15.10 are required to achieve a radius of investigation of 100 ft [30 m]. To measure the in-situ permeability in a well producing at rates less than 200 to 500 Mcf/D [5665 to 14 160 m^3/d], the well should be produced for about 10 days before shut-in for a pressure-buildup test. For wells producing at rates greater than 500 Mcf/D [14 160 m^3/d], a 1- to 2-day production period should be sufficient.

In general, the production period for a fractured gas well should be longer than for a well with a radial flow pattern (assuming the same permeability). It is, of course, impossible to conduct such a flow test on a new well if immediate results are required. In such a case, the flow period for a fractured well should be chosen from the criteria presented previously for radial flow, but the operator must realize that a longer flow period would have provided considerably more accurate data for analysis. When possible, a fractured well should be produced for 3 to 6 months before the pressure-buildup test is run.

Shut-In Time Requirements. After the well is cleaned up and produced long enough to draw down the pressure in a significant portion of the reservoir, the pressure-buildup data can be recorded. Of primary importance to the proper analysis of the buildup test is the test duration.

The test should be run long enough to ensure that the wellbore storage (afterflow) effects have been eliminated and the straight-line portion of the Horner graph can be determined. In low-permeability gas reservoirs, wellbore storage can last for several days. The minimum shut-in time (in hours) can be estimated for a radial flow case by

$$t_{wbs} \cong \frac{20(\bar{p}^2 - p_{wf}^2)V_{wb}c_{wb}}{Tq_g}, \quad \text{(15.51)}$$

where

$\bar{p}$ = average reservoir pressure,
p_{wf} = flowing BHP at instant of shut-in,
V_{wb} = wellbore volume,
c_{wb} = gas compressibility in wellbore,
T = reservoir temperature,
q_g = gas flow rate before shut-in, and
t_{wbs} = time required for wellbore-storage effects to diminish.

The "typical" data in Table 15.11 were used to calculate for various flow rates the shut-in times required for wellbore-storage distortion of test data to cease. These shut-in times are the minimum values required to get past the distorting effects of wellbore storage. To identify the semilog straight line correctly on a pressure-buildup-test graph, the duration of the buildup survey must be two to three times longer than the value calculated from the above equation. Therefore, for radial flow, assuming that tubing is in the well, the duration of the buildup test should be as shown in Table 15.12. The shut-in time requirements for fractured gas wells are much longer than the times required for wells having radial flow. To achieve a minimum of analyzable data on the semilog straight line for a fractured well in an infinite-acting reservoir, the well must be shut in until the value of dimensionless time exceeds 5.0. Dimensionless time is given by

$$t_{L_fD} = \frac{0.0002637kt}{\phi\mu c_T L_f^2}. \quad \text{(15.20)}$$

Assuming the following properties,

ϕ = 0.05,
μ = 0.02 cp, and
c_g = 0.00025 psi^{-1},

the shut-in time requirements as a function of permeability and fracture length would be as shown in Table 15.13. It is quite apparent that in the majority of cases, the semilog straight line is not reached in a test of a fractured well. Type curves and simulators will most often be required to analyze the test data.

The results of the test analysis become more reliable as the duration of the pressure-buildup test is increased. Therefore, for low-permeability, fractured gas wells, buildup tests should be run for a minimum of 7 days, preferably for 14 days or more. If the formation permeability is 1.0 md or greater and a small fracture treatment was performed, a 3- to 7-day buildup test will be adequate.

Nomenclature

A = drainage area, ft^2 [m^2]
A_{wb} = wellbore area, ft^2 [m^2]
B = FVF, reservoir vol/surface vol
B_g = 5.04 Tz/p, gas FVF, RB/Mscf [res m^3/std m^3]
$\bar{B}_g$ = gas FVF evaluated at p_{av}, RB/Mscf [res m^3/std m^3]
B_{gi} = gas FVF evaluated at p_i, RB/Mscf [res m^3/std m^3]
B_o = oil FVF, RB/STB [res m^3/stock-tank m^3]
c_f = pore-space compressibility, psi^{-1} [kPa^{-1}]
c_{fT} = total compressibility in fracture, psi^{-1} [kPa^{-1}]
c_g = gas compressibility, psi^{-1} [kPa^{-1}]
c_{gi} = gas compressibility evaluated at p_i, psi^{-1} [kPa^{-1}]
c_o = oil compressibility, psi^{-1} [kPa^{-1}]
c_T = $c_oS_o+c_gS_g+c_wS_w+c_f$, total compressibility, psi^{-1} [kPa^{-1}]
$\bar{c}_T$ = total compressibility evaluated at p_{av}, psi^{-1} [kPa^{-1}]
c_w = water compressibility, psi^{-1} [kPa^{-1}]
c_{wb} = compressibility evaluated at average pressure and temperature in wellbore, psi^{-1} [kPa^{-1}]
C = wellbore-storage coefficient, bbl/psi [m^3/kPa]
C_D = 0.8936 $C/\phi c_T h r_w^2$, dimensionless wellbore-storage coefficient (slightly compressible liquid)
C_{fD} = 0.8936 $C/\phi c_T h L_f^2$, dimensionless wellbore-storage coefficient for fractured wells (slightly compressible liquid)
C_{nr} = relative non-Darcy flow conductivity
C_r = $w_{kf}/\pi L_f k$, dimensionless fracture conductivity
D = non-Darcy flow coefficient, D/Mscf [d/std m^3]
G_{nf} = formation non-Darcy flow group
h = net pay thickness, ft [m]
h_f = propped interval thickness, ft [m]
J = $q/(\bar{p}-p_{wf})$, productivity index, STB/(D-psi) [stock-tank m^3/(d·kPa)]
J_g' = $q_g/(\bar{p}^2-p_{wf}^2)$, productivity index of fractured gas well with non-Darcy flow, Mscf/(D-psi^2) [std m^3/(d·kPa^2)]
$J_{g,o}$ = productivity index of undamaged, unfractured gas well with Darcy flow, Mscf/(D-psi^2) [std m^3/(d·kPa^2]
$J_{g,o}'$ = productivity index of undamaged, unfractured gas well with non-Darcy flow, Mscf/(D-psi^2) [std m^3/(d·kPa^2)]
J_o = productivity index before stimulation, STB/(D-psi) [stock-tank m^3/(d·kPa)]
k = effective formation permeability, md
k_d = damaged-zone permeability, md
k_f = propped fracture permeability, md
k_s = permeability of damaged fracture face, md
L_D = L_x/L_e, dimensionless distance
L_e = square drainage-area half-width, ft [m]
L_f = fracture half-length, ft [m]
m = slope of semilog straight line, psi/log$_{10}$ cycle [kPa/log$_{10}$ cycle]
m_L = slope of square-root-of-time-plot straight line, psi/hr$^{1/2}$ [kPa/h$^{1/2}$]
n = reciprocal of backpressure curve slope, dimensionless
p = pressure, psi [kPa]
p^* = pressure obtained when linear portion of Horner graph is extrapolated to Horner time ratio of 1, psi [kPa]
$\bar{p}$ = average static drainage-area pressure, psi [kPa]
p_a = $(\bar{\mu}\bar{c}_T/2p_{av})p_p$, adjusted pressure, psi [kPa]
$\bar{p}_a$ = adjusted pressure evaluated at $\bar{p}$, psia [kPa]
p_{av} = ½$(\bar{p}+p_{wf})$, arithmetic average pressure, psi [kPa]
$p_{a,wf}$ = adjusted pressure evaluated at p_{wf}, psia [kPa]
p_D = $kh\Delta p/141.2qB\mu$, dimensionless pressure (slightly compressible liquid)
p_e = pressure at external radius of drainage, psi [kPa]
p_i = original reservoir pressure, psi [kPa]

$p_p = 2\int_0^p pdp/\mu z$, pseudopressure, psia^2/cp [kPa^2/Pa·s]
p_{wf} = flowing BHP at shut-in in buildup test; also flowing BHP in drawdown test, psi [kPa]
p_{ws} = shut-in BHP, psi [kPa]
p_{1hr} = pressure on semilog straight line at test time of 1 hour, psia [kPa]
Δp = pressure change from start of transient test, psi [kPa]
Δp_a = change in adjusted pressure from start of transient test, psi [kPa]
P_c = capillary pressure, psia [kPa]
q = liquid flow rate at surface, STB/D [stock-tank m^3/d]
q_D = $141.2qB\mu/kh\Delta p$, dimensionless flow rate (slightly compressible liquid)
q_g = gas flow rate at surface, Mscf/D [std m^3/d]
q_o = oil flow rate at surface, STB/D [stock-tank m^3/d]
Q = cumulative production, STB [stock-tank m^3]
Q_D = $0.8936QB/\phi c_T h r_{wa}^2 (p_i - p_{wf})$, dimensionless cumulative production (slightly compressible liquid)
Q_{L_fD} = $0.8936QB/\phi c_T h L_f^2 (p_i - p_{wf})$, dimensionless cumulative production referred to fracture half-length (slightly compressible liquid)
r_d = radius of damaged zone, ft [m]
r_e = drainage radius, ft [m]
r_{ea} = $\sqrt{A/\pi}$, drainage radius of circle with same area as reservoir, ft [m]
r_{eaD} = r_{ea}/r_{wa}, dimensionless apparent drainage radius
r_w = wellbore radius, ft [m]
r_{wa} = $r_w e^{-s}$, apparent wellbore radius, ft [m]
s = skin factor, dimensionless
s' = $s + Dq_g$, effective skin factor for gas well, dimensionless
s_{fs} = $(\pi/2)(w_s/L_f)[(k/k_s)-1]$, fracture-face skin factor, dimensionless
S_g = gas saturation, fraction
S_o = oil saturation, fraction
S_w = water saturation, fraction
t = time, hours
t_a = $\overline{\mu c_T} t_{ap}$, adjusted time, hours
t_{ap} = $\int_0^t dt/\mu c_T$, pseudotime, hr-psia/cp [h·kPa/Pa·s]
t_D = $0.0002637kt/\phi\mu c_T r_w{}^2$, dimensionless time (slightly compressible liquid)
t_{L_eD} = $0.0002637kt/\phi\mu c_T L_e{}^2$, dimensionless time referred to square drainage-area half-width (slightly compressible liquid)
t_{L_fD} = $0.0002637\ kt/\phi\mu c_T L_f{}^2$, dimensionless time referred to fracture half-length (slightly compressible liquid)
t_p = effective producing time, hours
t_{wbs} = duration of wellbore-storage distortion, hours
Δt = time elapsed from start of transient test, hours
Δt_a = adjusted time elapsed from start of transient test, hours
Δt_{ae} = $\Delta t_a/(1+\Delta t_a/t_{pa})$, adjusted equivalent time, hours
Δt_e = $\Delta t/(1+\Delta t/t_p)$, equivalent time, hours
T = reservoir temperature, °R [K]
V_{wb} = wellbore volume, bbl [m^3]
w = fracture width, ft [m]
w_s = width of damaged zone near fracture face, ft [m]
z = gas-law deviation factor, dimensionless
$\bar{z}$ = gas-law deviation factor evaluated at p_{av}, dimensionless
β = reservoir matrix non-Darcy flow coefficient, 1/ft [1/m]
β_f = fracture non-Darcy flow coefficient, 1/ft [1/m]
γ_g = gas specific gravity (air=1.0)
η_{fD} = $k_f \phi c_T / k \phi_f c_{fT}$, dimensionless hydraulic diffusivity in fractured well
μ = viscosity, cp [Pa·s]
$\bar{\mu}$ = viscosity evaluated at p_{av}, cp [Pa·s]
μ_{gi} = gas viscosity at p_i, cp [Pa·s]
μ_o = oil viscosity, cp [Pa·s]
ρ_{wb} = density of liquid in wellbore, lbm/ft^3 [kg/m^3]
ϕ = total porosity, fraction
ϕ_f = fracture porosity, fraction
ϕ_g = ϕS_g, gas porosity, fraction

References

1. Prats, M.: "Effect of Vertical Fractures on Reservoir Behavior—Incompressible Fluid Case," *SPEJ* (June 1961) 105–18; *Trans.*, AIME, **222**.
2. Tinsley, J.M. *et al.*: "Vertical Fracture Height—Its Effect on Steady-State Production Increase," *JPT* (May 1969) 633–38; *Trans.*, AIME, **246**.
3. McGuire, W.J. and Sikora, V.J.: "The Effect of Vertical Fractures on Well Productivity," *Trans.*, AIME (1960) **219**, 401–03.
4. Holditch, S.A.: *Quarterly Low-Permeability Gas Well Research Report for Fall, 1975*, quarterly report, Petroleum Engineering Dept., Texas A&M U., College Station.
5. Tannich, J.D. and Nierode, D.E.: "The Effect of Vertical Fractures on Gas Well Productivity," paper SPE 15902 available at SPE headquarters, Richardson, TX.
6. Cooke, C.E.: "Conductivity of Fracture Proppants in Multiple Layers," *JPT* (Sept. 1973) 1101–07; *Trans.*, AIME, **255**.
7. Katz, D.L. *et al.*: *Handbook of Natural Gas Engineering*, McGraw-Hill Book Co. Inc., New York City (1969) 31.
8. Raymond, L.R. and Binder, G.G. Jr.: "Productivity of Wells in Vertically Fractured, Damaged Formations," *JPT* (Jan. 1967) 120–30; *Trans.*, AIME, **240**.
9. Morse, R.A. and Von Gonten, W.D.: "Productivity of Vertically Fractured Wells Prior to Stabilized Flow," *JPT* (July 1972) 807–11.
10. Lemon, R.F., Patel, H.J., and Dempsey, J.R.: "Effects of Fracture and Reservoir Parameters on Recovery From Low Permeability Gas Reservoirs," paper SPE 5111 presented at the 1974 SPE Annual Meeting, Houston, Oct. 6–9.
11. Holditch, S.A. *et al.*: "An Automated Method of Matching Production Performance With Dimensionless Solutions," paper SPE 12846 presented at the 1984 SPE/DOE/GRI Unconventional Gas Recovery Symposium, Pittsburgh, PA, May 13–15.
12. van Everdingen, A.F. and Hurst, W.F.: "The Application of the Laplace Transformation to Flow Problems in Reservoirs," *Trans.*, AIME (1949) **186**, 305–24.
13. Cinco-Ley, H. and Samaniego-V., F.: "Transient Pressure Analysis for Fractured Wells," *JPT* (Sept. 1981) 1749–66.
14. Agarwal, R.G.: "A New Method to Account for Producing Time Effects When Drawdown Type Curves are Used to Analyze Pressure Buildup and Other Test Data," paper SPE 9289 presented at the 1980 SPE Annual Technical Conference and Exhibition, Dallas, Sept. 21–24.
15. Millheim, K.K. and Cichowicz, L.: "Testing and Analyzing Low-Permeability Gas Wells," *JPT* (Feb. 1968) 193–98; *Trans.*, AIME, **243**.
16. Barker, B.J. and Ramey, H.J. Jr.: "Transient Flow to Finite Conductivity Vertical Fractures," paper SPE 7489 presented at the 1978 SPE Annual Technical Conference and Exhibition, Houston, Oct. 1–3.
17. Gringarten, A.C., Ramey, H.J. Jr., and Raghavan, R.: "Unsteady-State Pressure Distributions Created by a Well with a Single Infinite-Conductivity Vertical Fracture," *SPEJ* (Aug. 1974) 347–60; *Trans.*, AIME, **257**.
18. Cinco-L., H., Samaniego-V., F., and Dominguez, N.: "Transient Pressure Behavior for a Well With a Finite-Conductivity Vertical Fracture," *SPEJ* (Aug. 1978) 253–64.
19. Agarwal, R.G., Carter, R.D., and Pollock, C.B.: "Evaluation and Performance Prediction of Low-Permeability Gas Wells Stimulated by Massive Hydraulic Fracturing," *JPT* (March 1979) 362–72; *Trans.*, AIME, **267**.
20. Gringarten, A.C. *et al.*: "A Comparison Between Different Skin and Wellbore Storage Type-Curves for Early-Time Transient Analysis," paper SPE 8205 presented at the 1979 SPE Annual Technical Conference and Exhibition, Las Vegas, Sept. 23–26.

21. Cinco-Ley, H. and Samaniego-V., F.: "Effect of Wellbore Storage and Damage on the Transient Pressure Behavior of Vertically Fractured Wells," paper SPE 6752 presented at the 1977 SPE Annual Technical Conference and Exhibition, Denver, Oct. 9–12.
22. Lee, W.J. and Holditch, S.A.: "Fracture Evaluation With Pressure Transient Testing in Low-Permeability Gas Reservoirs," *JPT* (Sept. 1981) 1776–92.
23. Holditch, S.A. and Morse, R.A.: "The Effects of Non-Darcy Flow on the Behavior of Hydraulically Fractured Wells," *JPT* (Oct. 1976) 1169–79.
24. Holditch, S.A.: "Factors Affecting Water Blocking and Gas Flow from Hydraulically Fractured Gas Wells," *JPT* (Dec. 1979) 1515–24.
25. Bennett, C.O. *et al.*: "Influence of Fracture Heterogeneity and Wing Length on the Response of Vertically Fractured Wells," *SPEJ* (April 1983) 219–30.
26. Earlougher, R.C. Jr.: *Advances in Well Test Analysis,* Monograph Series, SPE, Richardson, TX (1977) **5,** Chap. 13.

SI Metric Conversion Factors

acres	× 4.046 873	E−01	=	ha
bbl	× 1.589 873	E−01	=	m^3
cp	× 1.0*	E−03	=	Pa·s
ft	× 3.048*	E−01	=	m
ft^3	× 2.831 685	E−02	=	m^3
°F	(°F−32)/1.8		=	°C
in.	× 2.54*	E+00	=	cm
psi	× 6.894 757	E+00	=	kPa
psi^{-1}	× 1.450 377	E−01	=	kPa^{-1}
°R	× 5/9		=	K

*Conversion factor is exact.

Chapter 16
Fracture Azimuth and Geometry Determination

Lewis L. Lacy, Exxon Production Research Co.
Michael Berry Smith, NSI Technologies Inc.

16.1 Overview

Knowledge of hydraulic fracture azimuth can be important in determining the optimum well location for tight-gas sands, waterflooding, EOR, thermal recovery from geothermal wells, and thermal stimulation of tar sands. For the same capital-investment cost, infill well locations can be selected to optimize drainage in low-permeability reservoirs and to avoid early water breakthrough during water injection. This chapter reviews current techniques to map hydraulic fractures away from the wellbore or to predict fracture orientation from near-wellbore data. Active geophysical techniques used to monitor hydraulic fracture growth include tiltmeter arrays and triaxial borehole seismic tools.

The techniques used to predict fracture orientation consist of logging methods and oriented-core analyses. In openhole completions, the initial orientation of fractures at the wellbore can be determined by sonic televiewers, closed-circuit television cameras, and impression packers. The orientation of elongated boreholes has also been used to predict fracture orientation. Analysis of anisotropy in the mechanical or elastic properties of oriented cores has likewise been used to predict the direction of the maximum principal horizontal stress and consequently to determine the fracture orientation. Many of these methods yield hydraulic fracture orientations with an accuracy of 5 to 15°, which is adequate for most reservoir engineering applications.

16.2 Introduction

A knowledge of the orientation of propped fractures is important to the long-term recovery from tight gas sands (formations with permeabilities of 0.01 md or less) and from low-permeability oil reservoirs. As Smith[1] discussed, the fracture orientation becomes important when the spacing between wells approaches the propped fracture length, which can be 1,000 to 2,500 ft [300 to 760 m] long for one wing of the fracture. Fig. 16.1 shows how incomplete reservoir drainage can result if fracture orientation is ignored when the well-spacing pattern is designed. With the same number of wells deployed in a favorable pattern, long-term recovery efficiency can be improved from 20 to 50% by preventing drainage patterns from overlapping.

Knowledge of fracture orientation can be useful in selecting the optimum well locations for waterflooding or EOR applications. As Fig. 16.2 illustrates, hydraulic fractures can develop, or natural fractures can open to channel injected water over large distances in the reservoir. Producing wells located perpendicular to the fracture direction will achieve a better areal sweep efficiency than wells sited parallel to the fracture direction. When fracturing occurs, areal sweep efficiency will be reduced. The sweep efficiency depends on the fracture orientation and size. In several field cases, water breakthrough was observed to occur after several weeks of injection. Water breakthrough caused by fractures often requires expensive remedial actions. Calculations indicate that fracture length may grow in excess of 1,000 ft [300 m] during several years of injection. Even in short injection tests lasting 10 to 20 hours in high-permeability (100- to 300-md) reservoirs, hydraulic fractures 50 to 150 ft [15 to 45 m] long have been estimated to occur from injectivity index data. Because long-term production loss could be from 10 to 80% if fractures occur, it is important that reservoir engineers consider the possibility of fractures, both natural and hydraulic, when designing the injection pattern.

Techniques for mapping or predicting fracture orientation are listed in Table 16.1. The techniques may be classified in terms of active fracturing techniques (those using tiltmeter arrays or borehole seismic tools), logging techniques (those using televiewers), and the passive or predictive techniques (those using oriented-core analyses). Active techniques remotely monitor a geophysical property of the fracture away from the wellbore. Therefore, the active techniques typically have a large radius of investigation that may range from 10 to 1,500 ft [3 to 455 m] from the wellbore. The logging and predictive core techniques are characterized by a radius of investigation limited to the wellbore or the near-wellbore region. Most of the predictive techniques are based on the concept that the maximum principal horizontal stress, σ_{H1}, determines the orientation of vertical fractures. The orientation of σ_{H1} can be known through study of the anisotropy of elastic or mechanical properties of cores. The logging techniques may be either passive, such as the study of borehole elongation data, or active fracturing techniques, such as the use of television to monitor an openhole hydraulic fracture.

Formation Stresses and Fracture Orientation. Determination of fracture azimuth by all techniques depends on interpretation of geophysical data with models and theory. While the models reviewed here are usually adequate for this purpose, field experience indicates that idealized models may not work for all formations and field conditions. A brief discussion of hydraulic fracture models for various formation stresses will help explain why fracture azimuth determination may be more complex in some formations.

The geometry of hydraulic fractures is usually controlled by the principal formation stresses. For hydraulic fractures to occur, the bottomhole injection pressure must overcome both the pore pressure and the minimum in-situ compressive stress in the formation. Fig. 16.3 illustrates four hydraulic-fracture geometries that can occur, depending on the relative magnitude of the in-situ principal stresses in the formation. The vertical principal stress, σ_V, is produced by the overburden, or weight of rock, above the point of interest. The two principal horizontal stresses, σ_{H1} and σ_{H2}, also depend on depth, where σ_{H1} is the maximum principal horizontal stress and σ_{H2} is the minimum principal horizontal stress. In general, σ_{H1} and σ_{H2} are not equal because the horizontal stresses are often caused by continental, regional, and local tectonic forces. When σ_{H1} and σ_{H2} are greater than σ_V, the hydraulic fracture is horizontal, or pancake-shaped, as shown in Fig. 16.3a. Horizontal fractures often occur in shallow rock formations because the overburden is normally less than the horizontal stresses. If the prin-

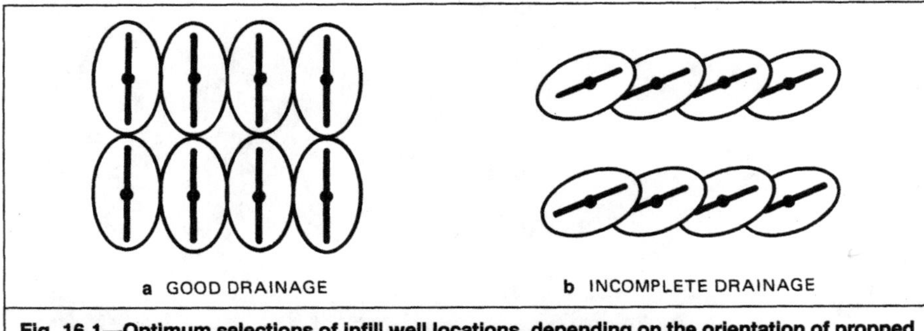

Fig. 16.1—Optimum selections of infill well locations, depending on the orientation of propped fractures in low-permeability reservoirs: (a) optimum recovery; (b) inefficient recovery.

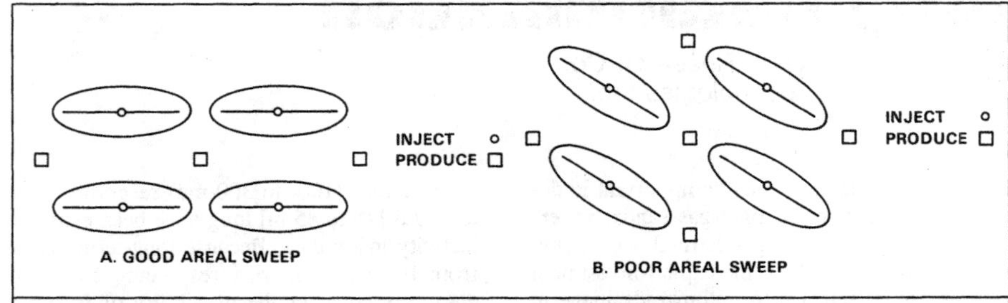

Fig. 16.2—Fracture orientation can be important for waterflooding and EOR operations in reservoirs when fractures channel the injected water in such a manner as to reduce the areal sweep efficiency: (a) good areal coverage; (b) poor areal coverage.

TABLE 16.1—TECHNIQUES USED TO MAP OR TO PREDICT THE HYDRAULIC-FRACTURE ORIENTATION

Active Fracturing Techniques
Tiltmeter arrays
Triaxial borehole seismic
Openhole Logging Techniques
Borehole-elongation orientation
Television camera
Sonic televiewers
Impression packers
Predictive Oriented-Core Techniques
Strain relaxation
Compressional-wave velocity
Thermal expansion
Differential-strain curve
Fracture (point-load) test
Residual stress (overcoring and blind-hole drilling)

cipal stresses are approximately equal, the hydraulic fracture may have a dipping angle, β (Fig. 16.3c). Dipping fractures can also occur in relatively shallow wells and may follow a preferred dip to the formation.

After the vertical stress becomes larger than the horizontal stresses for sufficiently deep wells, the hydraulic fracture is vertical and oriented toward the maximum horizontal stress, σ_{H1} (Fig. 16.3b). Differences in the horizontal stresses, $\sigma_{H1}-\sigma_{H2}$, of 300 to 1,000 psi [2.1 to 6.9 MPa] are probably sufficient to provide a definite trend of the vertical fracture toward σ_{H1}.

As Figs. 16.3c and d illustrate, intermediate fracture geometries between purely horizontal and simple biwing vertical fractures may sometimes occur. Branching vertical fractures (Fig. 16.3d) are probably not common but may occur when the two horizontal stresses are approximately equal and less than σ_V. When $\sigma_{H1}-\sigma_{H2}$ is less than the tensile strength of formation rock ($T_s \approx 0$ to 500 psi [0 to 3.4 MPa]), the fracture orientation will depend on the anisotropy in the rock strength. Because many formations often contain unhealed natural fractures and faults, the primary anisotropy in the formation strength will correspond to the orientation of these natural fractures, as indicated in Fig. 16.3d.

The fracture geometries illustrated in Fig. 16.3 demonstrate three important considerations in the determination of hydraulic fracture orientation: (1) relative magnitudes of σ_V, σ_{H1}, σ_{H2}, and T_s are important; (2) techniques that work well for simple biwing vertical fractures may fail for branching or dendritic fractures; and (3) fracture orientation can be predicted from logging or core data if the orientation of σ_{H1} is known and $(\sigma_{H1}-\sigma_{H2}) \gg T_s$.

16.3 Active Fracture Mapping Methods

Tiltmeter Arrays. ***Principle of Operation.*** When subterranean fractures are created by the injection of fluid into a formation, the earth's surface will be displaced by small amounts.[2] Several models[3-5] exist for calculating the amount of surface displacement. For example, Davis[4] calculated the surface deformation associated with hydraulic fractures for fracture geometries shown in Figs. 16.3a, b, and c. A three-dimensional presentation of the displacement for vertical and horizontal fractures is shown in Figs. 16.4 and 16.5, respectively. The actual vertical displacements are very small. Typical vertical displacements depend on the size and depth of the fracture, with typical displacements of 0.001 to 0.05 in. [0.003 to 0.13 cm].

The small surface displacement associated with hydraulic-fracture treatments can be measured by sensitive bubble-level indicators that can detect shifts or tilts in the earth's surface. Sensitivity of currently available tiltmeters is about 10 nrad, which is equivalent to a sensitivity of 2×10^{-3} seconds of an arc. This sensitivity corresponds to detecting a 0.08-in. [2-mm] uplift over a distance of 140 miles [225 km] or a 3×10^{-8}-in. [0.76-nm] uplift over a distance of 1,000 ft [300 m].

The tiltmeters are placed in shallow boreholes in a circular array around the wellbore and packed in dry sand. The borehole casing may be PVC pipe or steel. In deep-fracture tests, steel casing is preferred because the increased stiffness of steel increases the moment arm of the tiltmeter. Because the tilt signals associated with hydraulic fracture treatments are two-dimensional vectors, the downhole meter orientation must be carefully controlled to within ±5°.

A typical tiltmeter array[2,6,7] will consist of 12 to 18 meters arranged at surveyed positions around the well. One possible array

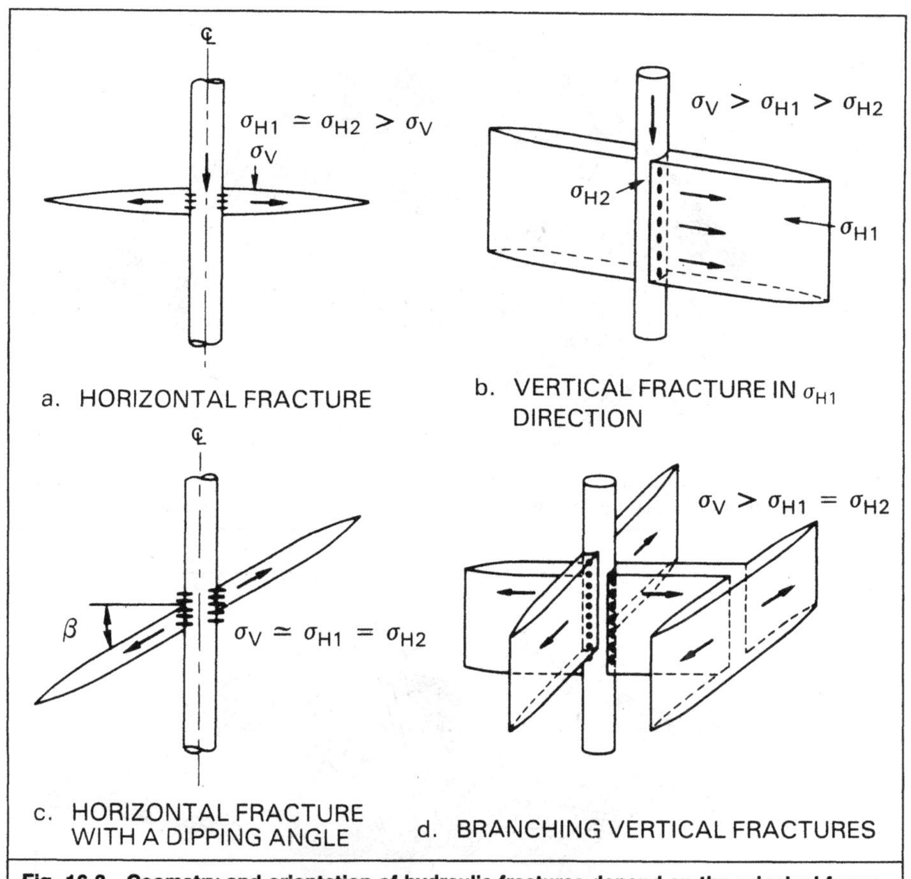

Fig. 16.3—Geometry and orientation of hydraulic fractures depend on the principal formation stresses.

design is shown in Fig. 16.6, where the radius of the array is about 40% of the fracture depth. Theoretical calculations indicate that the maximum tilt response should occur at this radius.

Because tiltmeters can be made with high sensitivity to detect small surface strains associated with massive hydraulic fracture treatments, the meters are also sensitive to other environmental influences: solar or lunar tides induced in the earth's crust; direct thermoelastic and gravitational surface strains associated with the daily heating of the earth; local effects, such as the influence of wind on nearby trees; rain and subsequent changes in subsurface water; and surface traffic, such as cars, trucks, and livestock. These environmental influences can generate a tilt signal equal to or usually much greater than the signals associated with deeper fracture treatments. Consequently, efforts are made to eliminate the environmental effects or to document their influence at each site.

Each meter site records the tilt signal at equal time intervals in two orthogonal directions (x and y directions). The $+y$ axis is usually pointed toward the well. A typical signal for one axis of the tiltmeter is shown in Fig. 16.7. In Fig. 16.7, the meter was located about 840 ft [255 m] from the treatment well, which received 5,400 bbl [860 m^3] of fracturing fluid pumped at an average rate of 46 bbl/min [7.3 m^3/d]. In hydraulic fracture treatments with depths less than about 4,000 ft [1220 m], tiltmeter data can be used to determine the type of fracture (vertical or horizontal), fracture size as a function of time, fracture orientation, and dip.

The amplitude of tiltmeter signals declines rapidly with fracture depth. In the deeper fracture jobs, up to 11,000 ft [3350 m] deep, the tiltmeter signals associated with the fracture treatment are smaller than environmental influences. For example, a typical tilt signal can be seen in Fig. 16.8a for 17 days of data, Fig. 16.8b for 3 days of data, and Fig. 16.8c for 10 hours of data during a massive hydraulic fracture treatment. The daily combined earth tides and thermoelastic data are typically about 500 to 1,000 nrad, whereas the fracture-generated signal may be only 10 to 40 nrad for the deeper fractures. Extracting the fracture-generated signals from the background data requires special care and analyses. Some of the earlier field tests conducted in the Cotton Valley formation of east Texas gave inconclusive results because of the difficulty of extracting small signals. Subsequent field tests were successful, however, when improvements were made in the array design and analytical techniques.

Field Experience. As Evans[7] and Evans and Holzhausen[8] recently reviewed, tiltmeter arrays can determine hydraulic-fracture geometry, orientation, and dip for shallow fracture treatments up to 4,000 ft [1220 m] deep. When the depth of the fracture is less than about 2,000 ft [610 m], tiltmeter arrays have also been successfully used to map the time-dependent fracture radius or length. For shallow fractures, tiltmeters can monitor fractures with lengths as small as 50 ft [15 m]. For deeper fracture treatments, the fracture orientation has been determined for depths up to 11,000 ft [3350 m]. As Lacy[6] indicated, the fracture orientation can be determined to an accuracy of 5 to 15° for a fracture radius of 500 to 1,200 ft [150 to 360 m].

Fig. 16.9 compares the hydraulic fracture orientation mapped by tiltmeter arrays with the fracture orientation mapped by the triaxial borehole seismic tool. Both data sets have been rotated by the same angle and do not represent the true compass orientation of the hydraulic fracture (No. 1). The hydraulic treatment depth was 8,500 ft [2590 m]. In addition to the hydraulic fracture, both data sets indicated secondary peaks or directions (Nos. 2 through 4) that correspond to the orientation of natural fractures in the formation. In this well, the radius of investigation for the triaxial borehole seismic tool was 360 ft [110 m] for a fracture 50 ft [15 m] high and was generated by pumping 420 bbl [67 m^3] of slick water into the formation. The tiltmeter array monitored a fracture treatment of the same well and zone for a fracture length of about 1,200 ft [360 m]. The orientation of the hydraulic fracture agreed within 4° in the two data sets. The borehole seismic data indicated that the natural fractures were only about 100 ft [30 m] long.

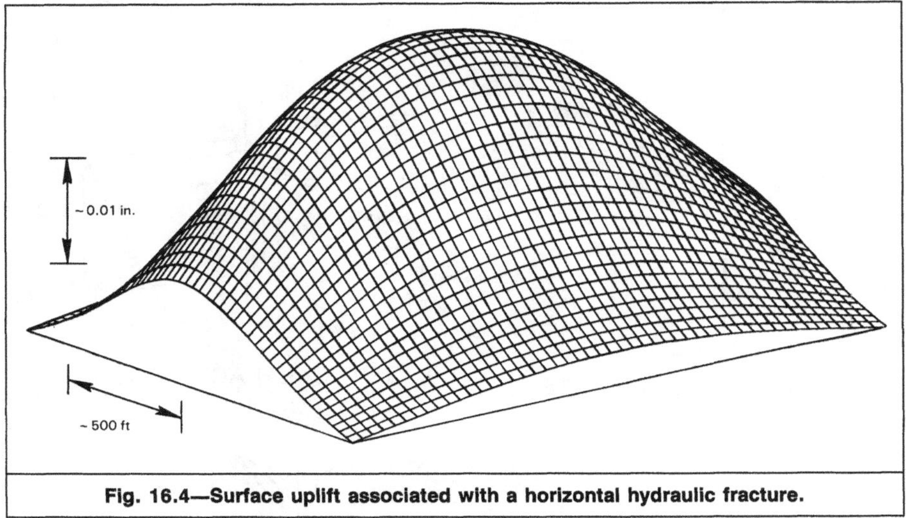

Fig. 16.4—Surface uplift associated with a horizontal hydraulic fracture.

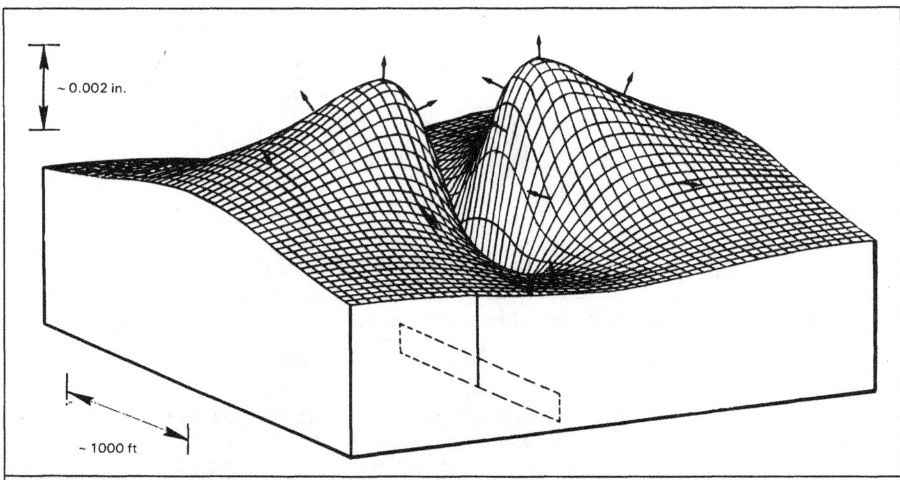

Fig. 16.5—Three-dimensional representation of the earth's surface displacement caused by a vertical fracture (dotted lines). The arrows correspond to the expected tilt directions.

Application and Limitations. As mentioned, tilt signals associated with hydraulic fracture treatments decrease rapidly as a function of depth. For hydraulic-fracture depth >7,000 ft [>2130 m], the tilt signals are often smaller than the background tilt data. Tiltmeters are suitable, however, for fracture orientation tests for fracture depths up to 11,000 ft [3350 m]. Previous field tests[6] indicate that the tilt signals calculated from uplift models tend to underestimate the tilt signal by a factor of 5 or more. The reason for this lack of agreement between theory and experiment for the deeper-fracture test is not clear. The experimental data suggest that the meters may be detecting a surface strain caused by long-wavelength seismic signals.

Triaxial Borehole Seismic. ***Principle of Operation.*** When a subterranean formation is hydraulically fractured, the formation gives off acoustic and seismic sounds associated with the fracturing process.[6,9-14] The sounds originate from formation rock near the fracture as well as from the turbulent flow of fracturing fluid. After the pumping has ceased, the background noise associated with fluid flow decays to a low level within a few minutes. The fractured formation will continue to emit both audible (20 to 1,000 cycles/sec [20 to 1000 Hz]) and inaudible (<20 cycles/sec [<20 Hz]) sounds as the fracture length changes and fluid leaks off into the formation. The postpumping fracturing sounds are believed to originate at the extreme ends, top, bottom, and interior of the fractures.[10-12]

The postpumping sounds produced by the fracture are not continuous but consist of discrete bursts of energy referred to as microseismic events. Most of the microseismic events are believed to be caused by cracking sounds associated with tensile or shear failure of rock near the fracture zone. Each microseismic event lasts from 0.1 to 0.5 second and has a typical frequency in the range of 15 to 600 cycles/sec [15 to 600 Hz]. For hard sandstone formations, each event will release 5 to 20 W of acoustic power. Fig. 16.10b shows a typical microseismic event.

Microseismic signals can be detected with a triaxial borehole seismic tool clamped below the perforations[6,9,12] in the fractured well or clamped in a nearby offset well.[10,11,14] A schematic of a typical borehole seismic test is shown in Fig. 16.11. The three geophones are arranged perpendicular to each other so that three components of motion can be resolved for each event. In a vertical well, one geophone will give a vertical component of the event; the other geophones give the two horizontal components of each detected signal.

Field Technique and Data. In a typical field application, the borehole seismic tool is clamped in an unfractured well[9] a few feet below the bottom perforation. The orientation of the clamped tool in the borehole is usually determined by recording four surface seismic shots located at surveyed positions around the well.[12] Fig. 16.10 shows the downhole detection of a surface seismic shot for three orthogonally mounted seismic detectors. An explosive charge of 5 to 20 lbm [2.3 to 9.1 kg] of Cordite or other explosive placed at depths of 200 to 300 ft [60 to 90 m] is adequate to determine the orientation of the downhole tool to within ±5° of true north.

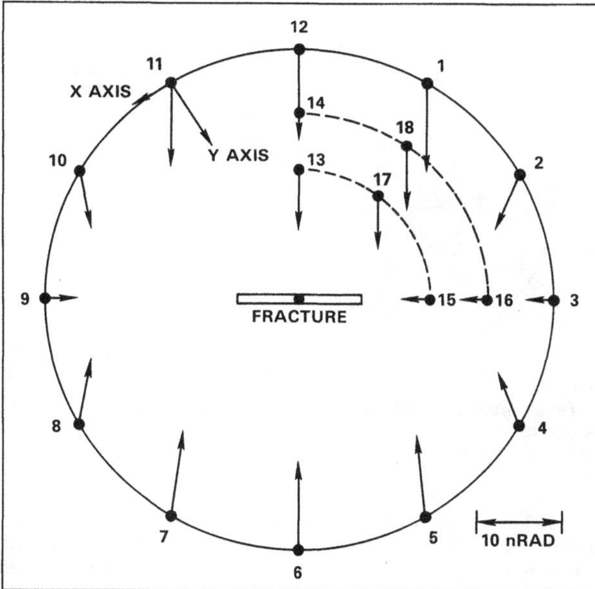

Fig. 16.6—Tiltmeter array design and calculated tilt vectors for a vertical hydraulic fracture.

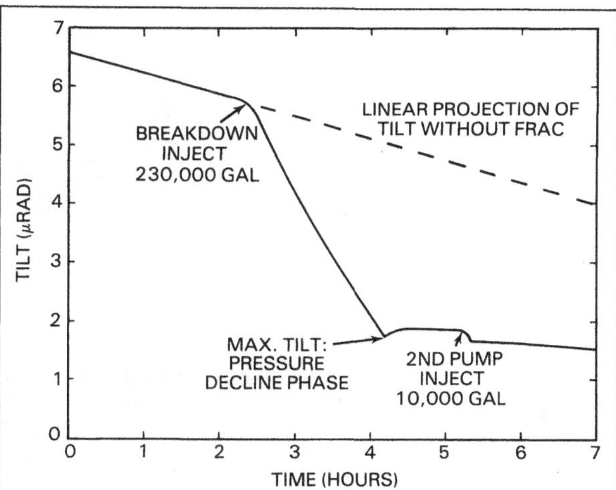

Fig. 16.7—Typical time-dependent signal for one axis of a tiltmeter monitoring the surface displacement associated with a fracture treatment at a depth of about 1,700 ft. The tiltmeter was installed in a shallow borehole located 840 ft northwest of the treatment well.

In offshore applications or if the surface rights cannot be obtained, the tool orientation can be determined with a gyroscope package mounted coaxially to the tool. The gyroscope technique of determining tool orientation is less desirable because the orientation accuracy may be degraded, and the probability of mechanical failure is higher. Offshore applications may also use normal offshore seismic sources for tool orientation.

After downhole orientation of the tool is determined, the formation is fractured with about 500 bbl [80 m^3] of slick water or cross-linked gel. The fluid is usually pumped in two or three different stages with a 20-minute to 2-hour listening period between stages. The hydraulic-fracture length for one wing of the fracture will typically be 200 to 600 ft [60 to 180 m]. The hydraulic-fracture height for this minifracture may be between 50 and 350 ft [15 and 105 m], depending on field conditions.

In a normal fracturing test, 50 to 100 microseismic events will be recorded. To determine the angular direction of the events, the two horizontal channels are plotted together after the data have been digitized.[12] A hodograph of a typical signal is shown in Fig. 16.12. The hodograph will trace out an elliptic pattern whose major axis defines the azimuth angle, Θ, between the two horizontal detectors. If the frequency and phase of the incoming wave are constant, the major axis for the ellipse will maintain a constant angle. These events are referred to as single-phase events. For approximately one-half of the events, the axis of the ellipse will suddenly shift as the phase of the incoming wave changes. These events are referred to as microseismic events.

The seismic events contain both p- and s-wave components. These events are believed to be associated with rock breaking or shear slippage along a fracture plane.[12] By determining the time delay, Δt, between the first arrival of the p-wave and the subsequent arrival of the s-wave, we can determine the radial distance, Δr, to the seismic source:

$$\Delta r = \Delta t v_p \left(\frac{v_s}{v_p - v_s} \right), \qquad (16.1)$$

where v_p is the p-wave velocity and v_s is the s-wave velocity. A study of field test data from three fields in east Texas indicates that Δr can vary between 10 and 720 ft [3 and 220 m]. The lower value is established by our ability to resolve phase shifts that are close together. The upper value corresponds to the seismic detector's sensitivity to resolve weak seismic signals absorbed and attenuated in reservoir rock. Because the attenuation coefficient varies linearly

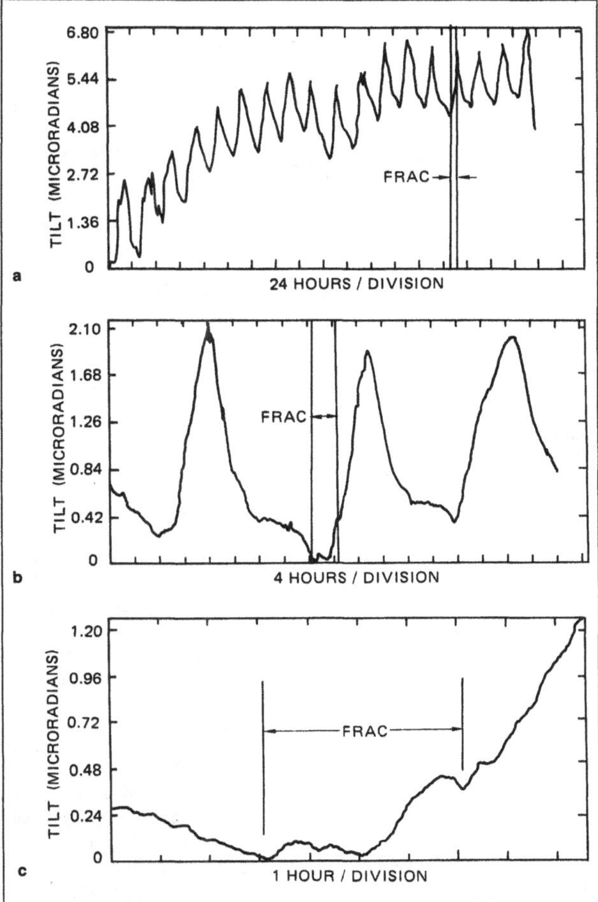

Fig. 16.8—Typical signals from one axis of a tiltmeter for a fracture-treatment depth of about 8,500 ft: (a) 24 hours per division; (b) 4 hours per division; (c) 1 hour per division.

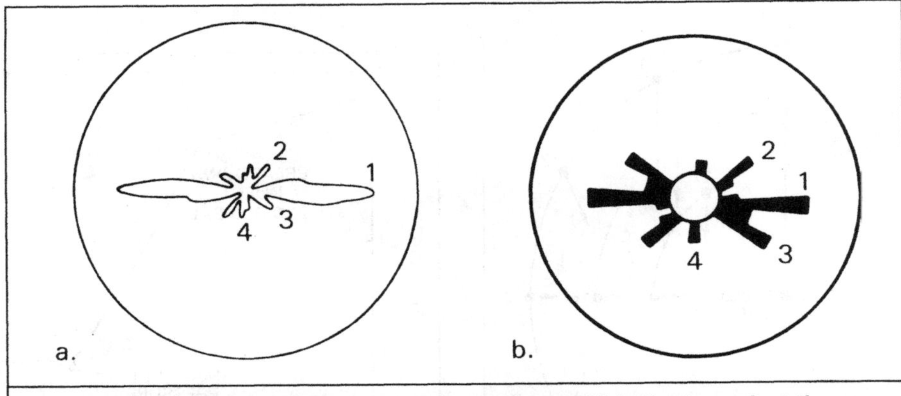

Fig. 16.9—Comparison of field data from an east Texas well for fracture-orientation mapping with (a) tiltmeters and (b) triaxial borehole seismic tool.

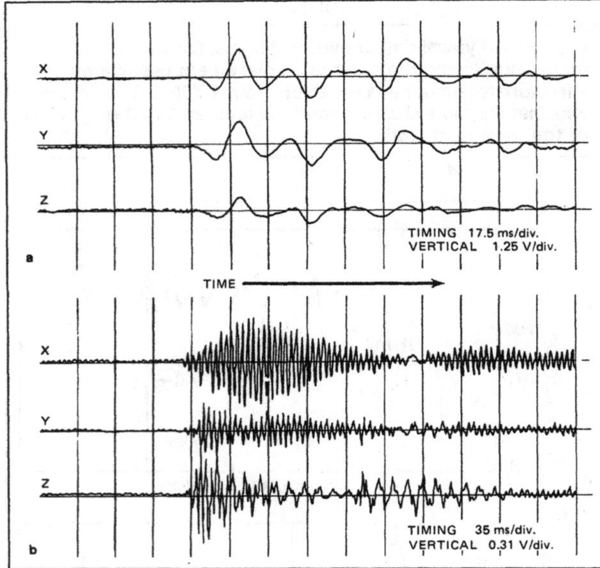

Fig. 16.10—(a) Surface seismic shot detected by the *X, Y, Z* seismometers; (b) microseismic event detected by the *X, Y, Z* seismometers.

with the sound frequency, all the higher-frequency waves will be absorbed within 200 ft [60 m] of the source.

Fig. 16.13 shows a plot of the number of seismic events vs. Δr for an east Texas Cotton Valley sandstone well. Note that most of the events came from the first 160 ft [50 m] around the wellbore. The solid curve corresponds to an exponential decay curve with a decay length of 100 ft [30 m]. No events were recorded for $\Delta r \geq 410$ ft [≥ 125 m] because this was the maximum value calculated for the hydraulic fracture length.

Influence of Natural Fractures. Natural formation fractures give off seismic events when stimulated with fracturing fluid. This has been demonstrated by several field tests conducted in the Cotton Valley formation of east Texas. Even when the injection pressure is below the fracture pressure, natural formation fractures channel the injected slick water along preferred directions in the formation. As illustrated schematically in Fig. 16.14, the seismic signals from natural formation fractures can be confused with similar signals generated by the hydraulic fracture. Thus, the key to applying the borehole seismic technique in naturally fractured reservoirs is to resolve and to classify the events into those occurring from the hydraulic fracture vs. those occurring from open, natural fractures.

Fig. 16.15 shows a polar plot of the number of single-phase and seismic events recorded in an east Texas well. The polar plot has been rotated so that the maximum values point to the left or right and do not correspond to the true compass direction. A polar plot of only the number of microseismic events for the same data set is shown in Fig. 16.15b. Four possible hydraulic fracture directions, corresponding to the maximum number of events, are discernible in the data.

A further classification of events to remove those associated with natural fractures that do not intersect the wellbore helps identify events associated with the hydraulic fracture. A polar plot of the number of events recorded in the well after the data from wellbore nonintersecting natural fractures are removed is shown in Fig. 16.15c. As plotted, Fig. 16.15c is a true radial representation of the orientation of both natural and hydraulic fractures in the formation. The main peak corresponds to the hydraulic fracture. The hydraulic fracture is about 50 ft [15 m] high and 360 ft [110 m] long measured from the wellbore. The size of the fracture was estimated through study of the radial location of the microseismic signal by use of Eq. 16.1. Hydraulic fracture orientation can usually be determined to within $\pm 5°$.

The two secondary peaks on each side of the hydraulic fracture correspond to directions of shear fractures in the formation. The microseismic signals from the shear fractures occur within about 100 ft [30 m] of the wellbore. Thus, the hydraulic fracture is about three times longer than the natural shear fractures. A confirmation that the secondary peaks correspond to open, natural fractures has been obtained by study of the orientation of natural fractures in oriented cores from one of the same wells and zones tested with the borehole seismic tool. The data from this study are shown in Fig. 16.16, where the hydraulic fracture orientation was investigated by five independent techniques, and the natural fracture orientation was determined by borehole seismic measurement and oriented-core studies. The orientation of natural formation fractures determined by borehole seismic measurement has also been correlated with the orientation of surface lineaments identified by a computer study of LANDSAT photographs.

One important conclusion derived from the data in Fig. 16.16 is that the hydraulic fracture orientation does not correspond to the natural fracture orientation, as illustrated in Fig. 16.3b. The hydraulic and natural fracture orientations may coincide when both the principal horizontal stresses are approximately equal, as illustrated in Fig. 16.3d.

Field Experience. The usefulness of the triaxial borehole seismic tool for mapping the orientation of hydraulic fractures is well established in the petroleum literature.[6,10-15] These applications include recovery from tight-gas sands, geothermal wells, and waterflood reservoirs. Previous field experience indicates that borehole seismic tests are usually successful when conducted in the hydraulically fractured well or in offset wells within 200 to 400 ft [60 to 120 m] of the fractured well. These tests will fail to give reliable data when there is too much electrical noise caused by electronic failure in the tool or electrical connections. The borehole seismic tool may give only single-phase events in shallow[15] or highly overpressured reservoirs. However, results usually agree with other techniques such as tiltmeter arrays, various oriented-

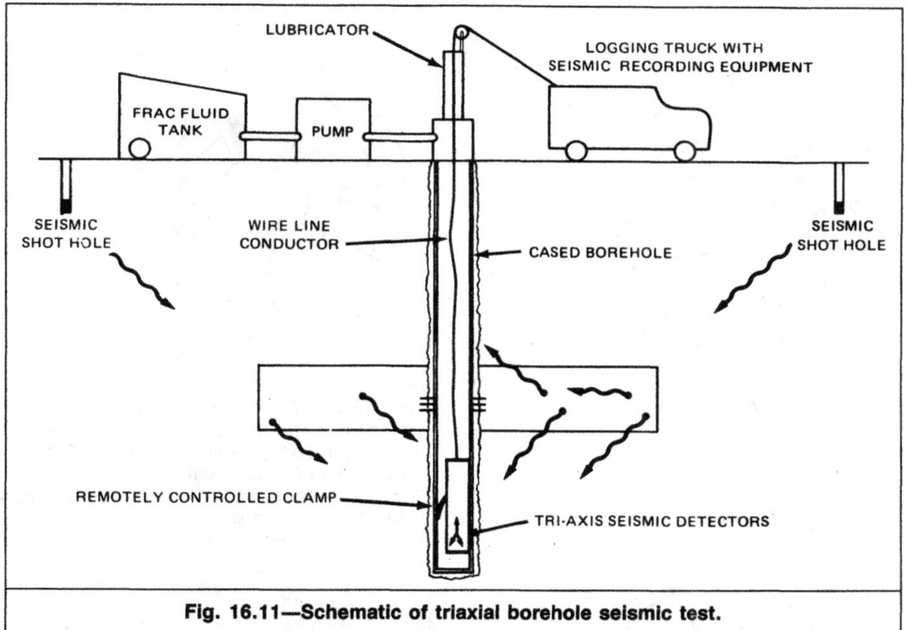

Fig. 16.11—Schematic of triaxial borehole seismic test.

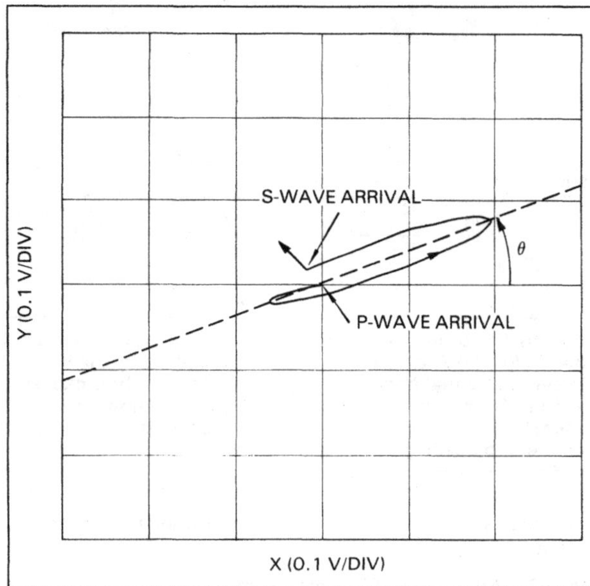

Fig. 16.12—Hodograph of the *X* and *Y* channels used to determine the horizontal azimuth angle and the delay time between the arrival of the *p*- and *s*-waves.

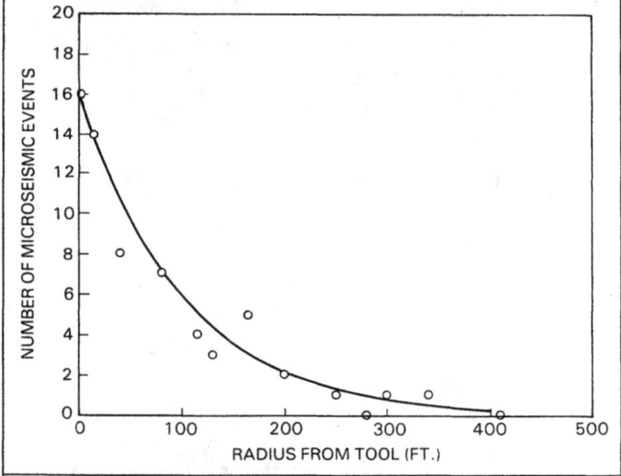

Fig. 16.13—Number of microseismic events recorded vs. the radius of excitation for the events.

core tests, and data from dipmeter logs.[6,16] The resonant/acoustic noise associated with a previously existing propped fracture has also been successfully detected in a borehole seismic test to determine fracture azimuth.

The triaxial borehole seismic method has several advantages over other techniques.

1. It is an active technique with a large radius of investigation (650 ft [200 m]).
2. It can be used for mapping both natural and hydraulic fractures.
3. This technique works in both open- and cased-hole completions.
4. Accuracy does not depend on relative magnitudes of horizontal stresses or hydraulic-fracture depth.

The normal limits for the use of the borehole seismic tool are bottomhole temperatures of 390°F [200°C] and bottomhole pressures of 20,000 psi [138 MPa]. The primary disadvantages of

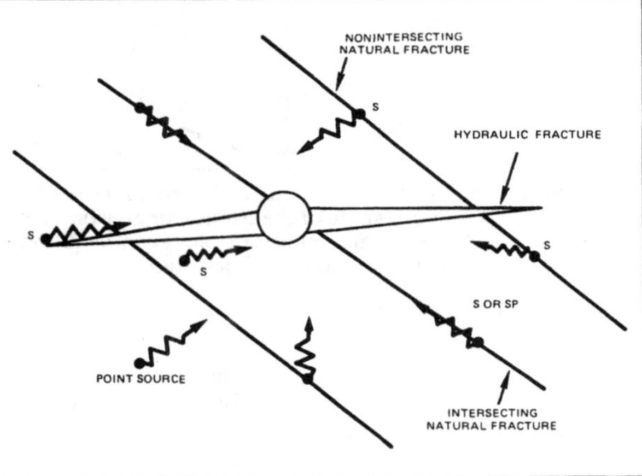

Fig. 16.14—Classification of seismic (S) and single-phase (SP) sources in terms of the primary hydraulic fracture and both wellbore-intersecting and -nonintersecting natural fractures.

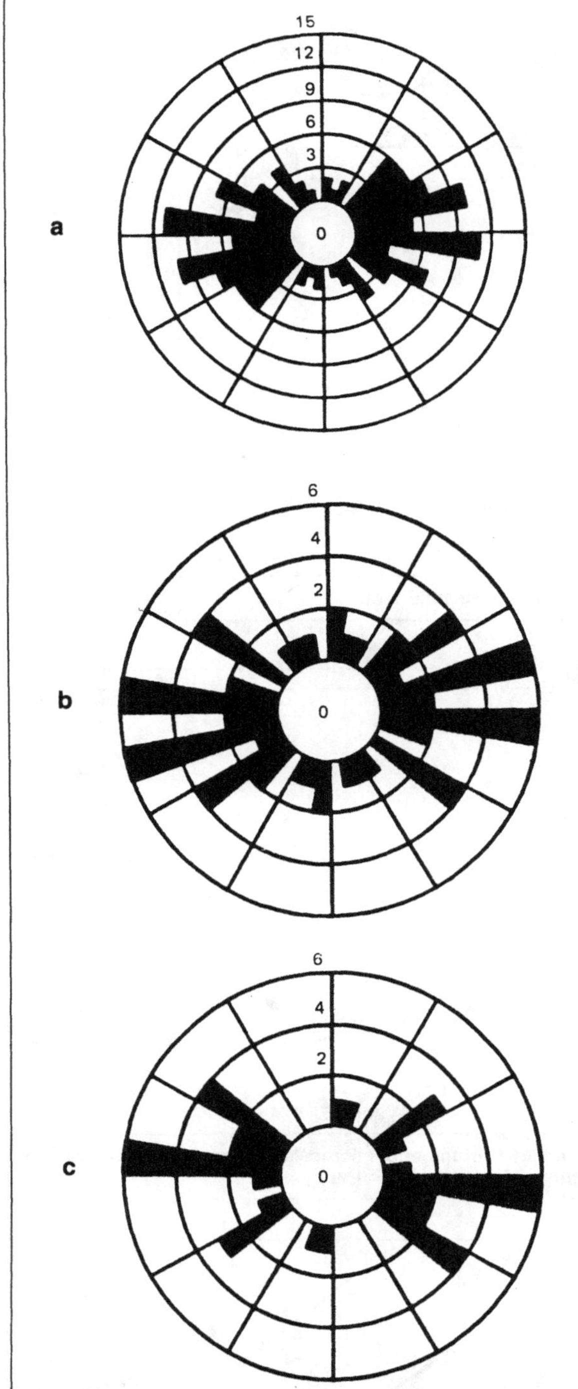

Fig. 16.15—Polar plot of the number of events recorded vs. the horizontal azimuth angle of the events for an east Texas well: (a) single-phase and microseismic events; (b) only microseismic events; (c) data set corrected for nonintersecting natural fractures.

its use are the possibility of a stuck tool in openhole completions and the requirement of surface rights and access to several seismic shotholes in normal applications.

16.4 Logging Techniques

This section covers several logging techniques that directly measure the borehole azimuth of a fracture, as well as an indirect log that predicts the in-situ stress orientation from measuring wellbore deformation.

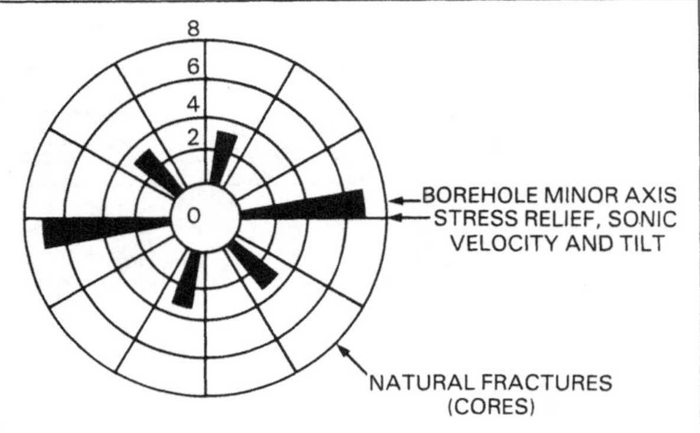

Fig. 16.16—Comparison of polar-plotted triaxial borehole seismic data with average stress relief, minimum sonic velocity, and natural fracture orientation from oriented cores. The tiltmeter data for a massive hydraulic fracture treatment and borehole-elongation data (minor axis) are also shown for the same east Texas well.

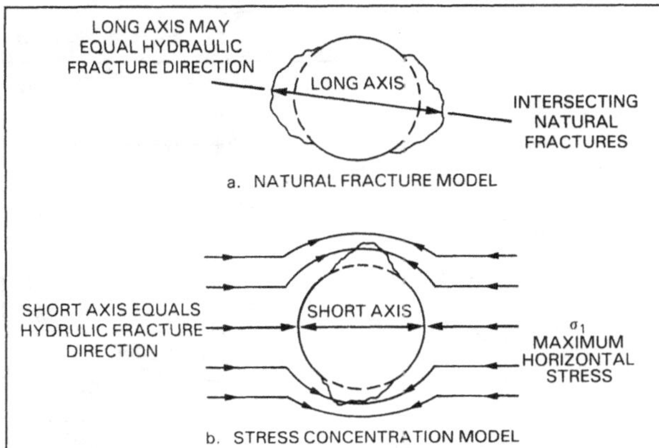

Fig. 16.17—Different models, or interpretations, of elliptic boreholes. (a) The borehole ellipticity is caused by the bit intersecting natural fractures or joints. (b) The borehole causes a stress concentration along a line perpendicular to the maximum horizontal stress, and this stress concentration causes failure and spalling of the wellbore wall.

Direct Measurements. Three procedures are available to measure azimuth directly from the borehole trace of the fracture: (1) impression packers, (2) sonic borehole televiewer, and (3) downhole television. The advantages of these procedures are their ease of analysis and interpretation and their ability to generate reliable data. One disadvantage is the need for an openhole completion or for special tests before casing is run. Also, for the fracture azimuth determination to be reliable, the procedures are subject to some restrictions.

The wellbore should be circular to prevent wellbore geometry from affecting the fracture initiation point, and the fracture must be created hydraulically; i.e., the wells must be intact before hydraulic breakdown. If the wellbore is fractured before the hydraulic fracture, then the azimuth determined from the wellbore trace is possibly associated with existing natural fractures, shear failure around the borehole, or other phenomena, and the hydraulic fracture could reorient itself to the in-situ stress field after leaving the wellbore. Indications of this type of behavior were noted in Colorado.[17]

Impression Packers. Of these three procedures, impression packers are the oldest. Early work by Fraser and Pettitt[18] and Anderson and Stahl[19] documented their use to study hydraulic-fracture orientation and azimuth. The procedure consists of running

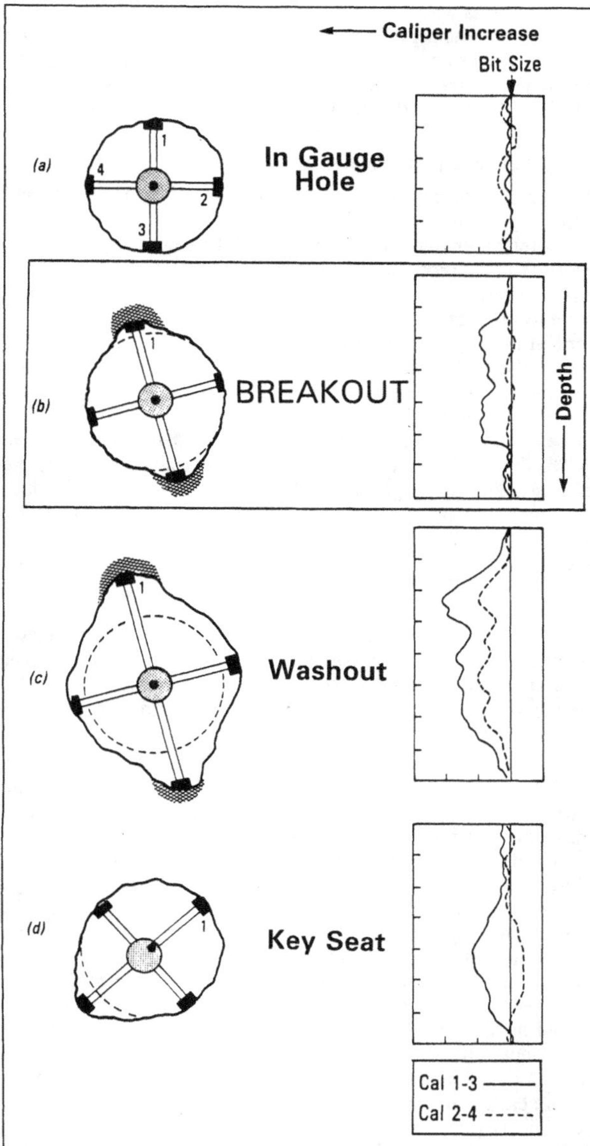

Fig. 16.18—Typical four-arm-caliper log behavior for various types of wellbore eccentricity (after Plumb and Hickman[30]).

inflatable packers on tubing and setting the packers across the desired openhole interval. The exterior of the packer is covered with deformable rubber, so when the packer is inflated, the outer covering deforms into and retains an impression of wellbore features, including fractures. Fracture azimuth is then determined from running single- or multishot survey tools below the packer.

Some advantages of impression packers are obvious, including ease of interpretation and insensitivity to wellbore fluids. One additional advantage is the possibility that the inflation of the packer might open up small, unpropped fractures that might be missed with other logging tools. However, the requirement for rig time and limited ability to sample long, vertical intervals are severe limitations. Also, as with any procedure, results are not 100% assured, with spalling along the fracture edge being one source of problems.[17] More recent work involved the running of two impression packers,[16] with one good impression resulting; no reasons were given for the failure of the other attempt. For the successful run in Ref. 17, the fracture azimuth agreed relatively well with other measuring techniques, and in general, when wellbore conditions and operations permit, impression packers still offer a reliable method of determining fracture azimuth.

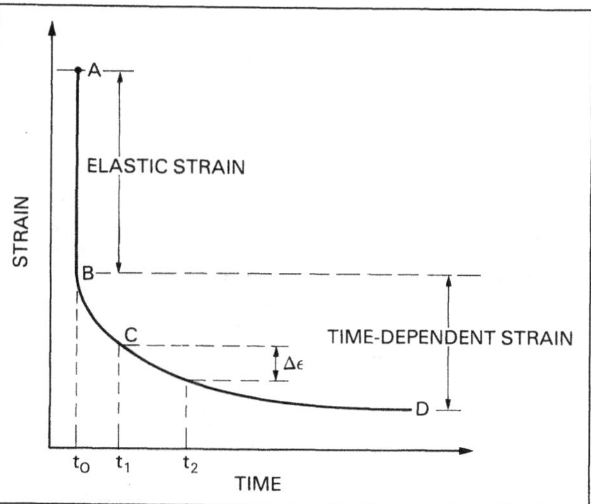

Fig. 16.19—Ideal strain-vs.-time behavior of a core. Elastic strain relaxation (A-B) when the core is cut, followed by time-dependent relaxation from (B-C), where the core is recovered. Measurements are made from C to D.

Borehole Televiewer. The second technique, the sonic borehole televiewer, is a logging tool introduced by Zemanek *et al.*[20] It is a "sonar" tool, which, in principle, should be an excellent fracture-identification tool. Unfortunately, the tool has not always performed up to its potential.

The tool consists of a crystal that emits high-frequency (about 1.3×10^6-cycles/sec [1.3-MHz]) sonic pulses, then receives and records the reflection of these pulses from the borehole wall. The lack of any reflection indicates the possible existence of a fracture. The crystal is then rotated to create circumferential scans of the wellbore during vertical logging. Typical operations are to log vertically at 5 ft/min [1.5 m/min], while the crystal rotates at 3 rev/sec, giving 36 circumferential scans/ft [118 scans/m]. Because the tool operates at sonic frequencies, it is insensitive to wellbore fluids, although heavy muds or gas-cut fluids can cause problems by attenuation of the signal and creation of false reflections. Because the sonic borehole televiewer operates on normal logging equipment, it is relatively simple to use. One problem is that the crystal must be "centered" with the borehole wall for good reflections to occur; any borehole ellipticity creates blind areas.[21] Wellbore deviation greater than a few degrees can decentralize the tool, creating the same effect.

A final unknown is the fracture width that can be discerned. Because the sonic pulse has a finite wavelength (about 1 mm), it must be assumed that fracture width must be a significant fraction of this wavelength to prevent a reflection. It may be that the inability to "see" very narrow fractures may account for some of the published failures. For example, Bredehoeft *et al.*[21] reported a 50% success rate in logging zones known to have been hydraulically fractured, with Hirsch *et al.*[22] reporting similar problems while trying to identify natural fractures. In a successful application, Smith *et al.*[23] reported good comparison with downhole television results; however, the borehole televiewer log was most interpretable over intervals where spalling had occurred along the fracture edges. In general, if spalling is necessary, then successful application of the borehole televiewer to indicate hydraulic fracture azimuth may depend on the mechanical properties of a particular formation. In that case, it may not be possible to predict in advance where the tool can be used successfully.

Downhole Television. The final logging technique discussed here is the use of downhole closed-circuit television. This is clearly the best logging technique as far as results are concerned, and because the tool directly views the borehole wall, interpretation is very simple. There are operational difficulties associated with delivering visibly clean fluid to the bottom of the wellbore, but it can be done.[23] While television logging cannot be done routinely, if such a tool is available, it offers a reliable method of determining fracture azimuth at the wellbore and supplies additional data about fracture width and height as part of the process.

Borehole Elongation. It has been noted for several years that many wellbores wash out, or spall, to create elliptic cross sections, with the long axis of these noncircular sections sharing a common azimuth. Cases where the minimum hole diameter corresponds to bit diameter, called "breakouts," have been reported from many different areas.[24-26] Obviously, many factors affect the stability and cross-sectional shape of a wellbore, including the drilling program, bit condition, and physical/chemical interactions between drilling fluids and the formation. Rock/mud interactions could be particularly important because, as discussed below, the mechanical properties and the "failure mode" of the formation can affect the orientation of wellbore breakouts.

A circular borehole intersecting a formation with unequal horizontal stresses will create a stress concentration around the borehole, as illustrated in Fig. 16.17. Bell and Gough[27] theorized that breakouts are caused by shear failure induced by this stress concentration as a result of two conditions: unequal horizontal stress and appreciable shear strength of the rock. As shown in Fig. 16.17, unequal stresses cause a preferential stress concentration on the side of the wellbore perpendicular to the maximum stress direction, and if the shear strength is high enough, failure of the borehole and spalling will be limited to this region. In such a case, breakouts will develop with the long axis of the elliptical borehole perpendicular to the expected azimuth of hydraulic fractures. This is the usually accepted interpretation of breakouts, and Teufel *et al.*[16] showed this to be consistent with other measurements of fracture azimuth in a Mesaverde well in Colorado. Such an interpretation is also consistent with waterflood behavior noted in the Pembina oil field in Alberta.[24] Fig. 16.16 compares the orientation of the minor borehole axis with other techniques for an east Texas well.

Other interpretations of wellbore breakouts include the possibility (Fig. 16.17) that noncircular cross sections are related to the bit intersecting pre-existing joints or natural fractures in the formation.[26,28] If natural fractures or joints are responsible for borehole elongation, the borehole major axis will be approximately parallel to the natural fracture azimuth. Still another possibility exists for weaker rocks that fail in a "ductile" or "plastic" manner. Depending on the material properties and in-situ stresses, boreholes in such cases may have long axes parallel to either the maximum or the minimum horizontal stress.[29]

That alternative interpretations exist for wellbore breakout data does not detract from this technique, but simply emphasizes, as with other measurements, that care must be taken with the data. The advantage of measuring wellbore breakouts is the operational ease of obtaining the data, leading to such measurements being made in many wells over long, vertical intervals. Thus, even where it may not be a good technique as a primary indicator of azimuth, it could serve as a powerful tool for extrapolating data both laterally and vertically, detecting changes, and defining areas where more comprehensive tests are needed.

Two procedures are available for measuring breakouts, possibly the better of which is the use of the sonic borehole televiewer. Though normally used to determine directly the wellbore trace of fractures, the borehole televiewer is also capable of generating very comprehensive and detailed oriented-caliper logs. The more common procedure, however, is the use of four-arm, oriented-caliper logs available on commercial dipmeter tools. Such logs are routinely available and are easily included in a normal logging program. A good review of the interpretation procedure is presented in Ref. 30. Fig. 16.18 reviews the interpretation of four-arm calipers for various types of borehole elongation.

Summary. Several logging techniques exist for determining hydraulic fracture azimuth, and where wellbore conditions permit, logs and techniques that directly map the wellbore trace of the fracture can produce reliable results. These techniques require prefracture tests to show that an essentially circular, intact wellbore exists before the hydraulic fracture. The use of oriented-caliper logs to measure wellbore breakouts is more indirect, but the ease of implementation makes this a useful procedure. Such logs should be run in any area where fracture azimuth might be of interest, at least to determine whether breakouts exist. Even in areas where interpretation might not be straightforward, if breakouts exist, the logs may be useful to extrapolate data from other techniques both vertically and laterally and to define where additional tests are needed.

16.5 Oriented-Core Analysis

The use of oriented cores to predict fracture azimuth has been suggested for many years,[31,32] and recent consistent results from strain-relaxation measurements[6,33,34] have begun to generate acceptance of the process. This section describes several types of core analyses and, where possible, compares core analysis results with other techniques. First, however, the oriented-coring process is described briefly. This is important because several instances have shown the analyses to be more reproducible within a given core section than from core to core in the same well. This suggests that a portion of the uncertainty in the data comes from the coring process, making quality control a must for successful data collection.

The chief advantage of core analysis for fracture-azimuth prediction is its ease of application. If coring operations are scheduled for other reasons, the additional work required to orient and to analyze the core is small compared with other fracture-azimuth measuring procedures. Also, because most coring is done early in the life of a field, the data collection is very timely. The biggest disadvantage is related to the type of measurement; because it is an indirect measurement, it is difficult to ensure that the answer is correct. Of the core analyses discussed below, direct measurement of strain relaxation has been the most successful, with no published failure when used on whole, intact sandstone core samples.

Oriented Coring. While several oriented-coring processes are available, the most common process uses a special shoe on the core barrel with triangular knives to cut three grooves in the core. One of these, the reference groove, is at a known orientation to an azimuth lug attached to the inner core barrel. An orientation tool is mounted above the core barrel in such a manner that the orientation lug is visible when the tool takes a picture of its compass. The correction between the reference knife and the orientation lug can be preset in the shop, but a preferred technique is to hoist the barrel in the derrick and to use an optical aligning device to determine their relative orientation; this is then recorded for future calculations.[35] While quality control over the coring process is critical, equally important is treating the core with care after recovery. The multishot tool is preset to take pictures at set time intervals, and shortly before each time point, drilling is interrupted to allow a stabilized reading. Depth is recorded at each point, later allowing the core to be oriented with depth. Because this tying of orientation to depth is indirect, the biggest source of error comes from incomplete core recovery, breaks in the core, or a spiraling reference groove.

The problem with incomplete recovery is obvious and often irreversible because there will be no absolute depth scale. Breaks allow the core barrel to rotate abruptly, causing a shift in the azimuth of the reference groove, so that a small depth error can create a large error in orienting the core. In any case, several feet of core should be reassembled so that the change in reference-groove orientation over an interval can be correlated with the recorded change in groove orientation to determine whether the depth markings are correct or to quantify the possible error resulting from uncertainty in depth.

For all coring tests, it is recommended that the test be performed on a minimum of six coring pieces or a minimum of two coring pieces per coring run. Averaging the data from 6 to 12 coring pieces will improve the accuracy and reliability of the data. As discussed recently,[6] the mapped fracture will usually occur within one standard deviation of the predicted fracture orientation.

Finding the orientation of the in-situ stress field by measuring time-dependent strain relaxation of cores has been discussed for several years. The idea of the process is simple and is based on laboratory observations that the stress/strain behavior of rocks is not purely elastic but is a function of loading rate and time.[36] In such a case, strains stored in the rock by the in-situ stresses will not be released instantly when the core is cut but will relax over many hours. If the core can be recovered and instrumented during

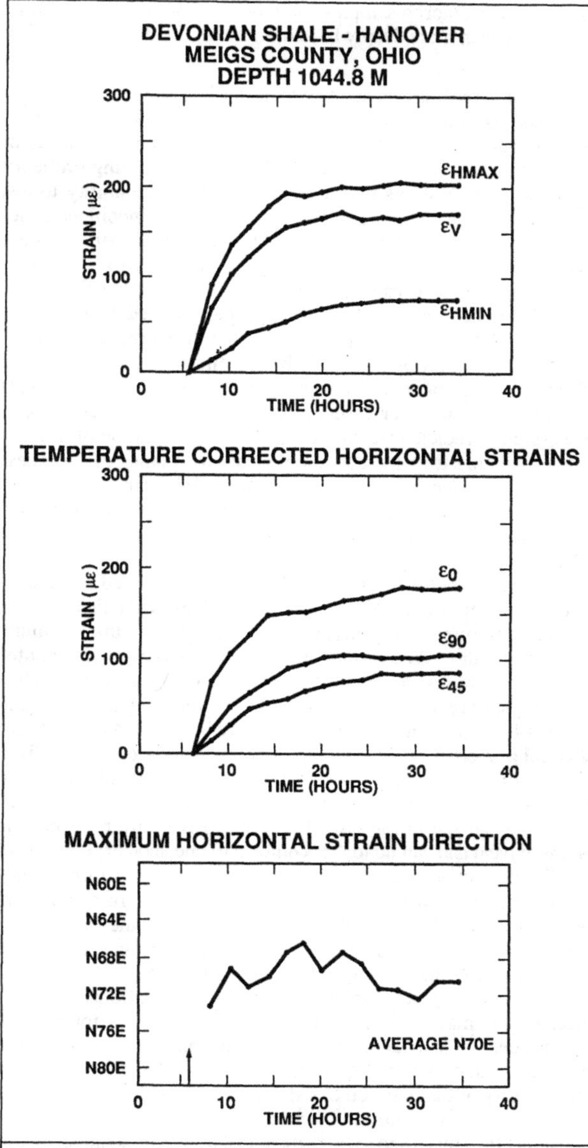

Fig. 16.20—Actual strain-vs.-time behavior of a core. (a) Raw data—strain vs. time recorded starting hours after the core was out. (b) Data from (a) corrected for temperature fluctuations that occurred during the measurement period. (c) Principal-stress direction calculated at different times during the strain-recovery measurements (after Blanton and Teufel[43]).

this time, then the orientation of the stresses can be determined by measuring relaxation in different directions.

When a core is cut, an instantaneous elastic strain relaxation is followed by a period of time-dependent relaxation, as shown ideally in Fig. 16.19. Voight[37] suggested that for an isotropic material, the magnitudes of the time-dependent relaxation in different directions would be proportional to the elastic strains, so measuring time-dependent relaxation would allow a direct measurement of the direction and magnitude of the in-situ stresses. Of course, Fig. 16.19 shows strain recovery in only one direction, but in general, three horizontal measurements and one vertical measurement are made. More complex data can be taken for deviated wells where one principal stress does not parallel the core.

Early work in recording strain relaxation (see Swolfs[38]) produced inconsistent results, possibly because of the direct use of strain gauges on the samples. Strain gauges sample the relaxation of the small volume of rock directly beneath the gauge, so inhomogeneities may significantly affect the results. More consistent results have been generated with clip-on gauges introduced by Teufel[33] that measure expansion of a large core sample.

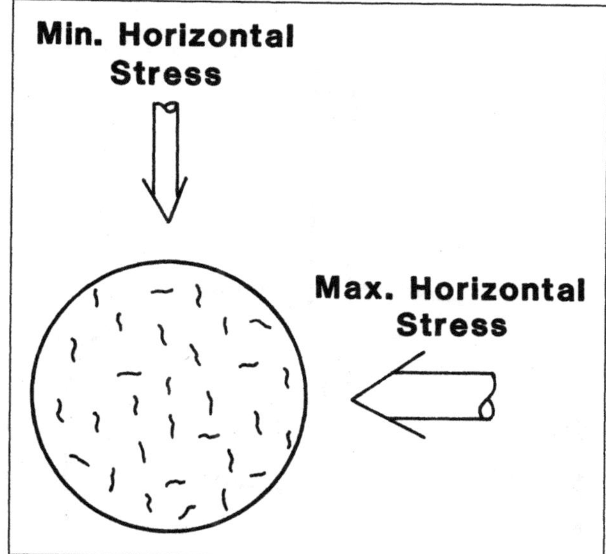

Fig. 16.21—Conceptual model of microcrack formation. Cracks will form in all directions, but will be statistically oriented by the in-situ stresses, with more cracks perpendicular to the maximum in-situ-stress direction.

The process involves selecting several core samples as soon as possible after the core reaches the surface. The selected samples should come from intact core sections to ensure good orientation data. The samples are removed to a reasonably-constant-temperature environment after the cores are sealed to prevent moisture evaporation. Thermocouples are usually attached to core pieces to determine the core temperature. If the core temperature is 6°F [3.3°C] or more above the constant-temperature environment, the core pieces should be cooled by ice for several minutes until they reach the environmental temperature. After isothermal conditions are reached, the clip-on deformation gauges are attached to the samples. Care should be taken to avoid samples from intervals where macrofractures are present, because such samples typically evidence little time-dependent strain relaxation.[39,40] These measurements are then used to calculate the orientation of the in-situ stresses.[41]

Fig. 16.20a presents typical data taken from strain-relaxation measurements on a shale sample.[43] Fig. 16.20b gives the measured strains for three horizontal directions after correction for temperature fluctuations that occurred during the monitoring period. The measurements are used to calculate the maximum and minimum horizontal strains seen in Fig. 16.20b, and Fig. 16.20c depicts typical uncertainty in determining the maximum-horizontal-strain direction (this is the hydraulic fracture azimuth). The biggest uncertainty is the assumption that the principal axes of anelastic (or time-dependent) strain relaxation coincide with the principal-stress axes.

It has been noted that the mechanical properties of rocks change when samples are removed from their in-situ state.[44] One explanation is the opening of microcracks caused by the release of in-situ stresses.[33,45] Evidence for this is circumstantial, however, and other factors, such as relaxation of residual strains related to past stress, may also play a role.[46] However, the opening of microcracks seems to be a dominant mechanism for many rock types, particularly for samples recovered from all but near-surface locations, and such a mechanism offers an explanation for the increased anisotropy observed in rocks after removal from in-situ conditions. In fact, this increase in anisotropy forms the basis for several laboratory techniques discussed below. A better understanding of the internal mechanisms will be needed in the future to predict where strain relaxation is applicable; for example, tests in a shallow limestone[42] showed no anisotropies, and, as discussed, samples from intervals containing visible macrofractures typically display no anelastic strain recovery.

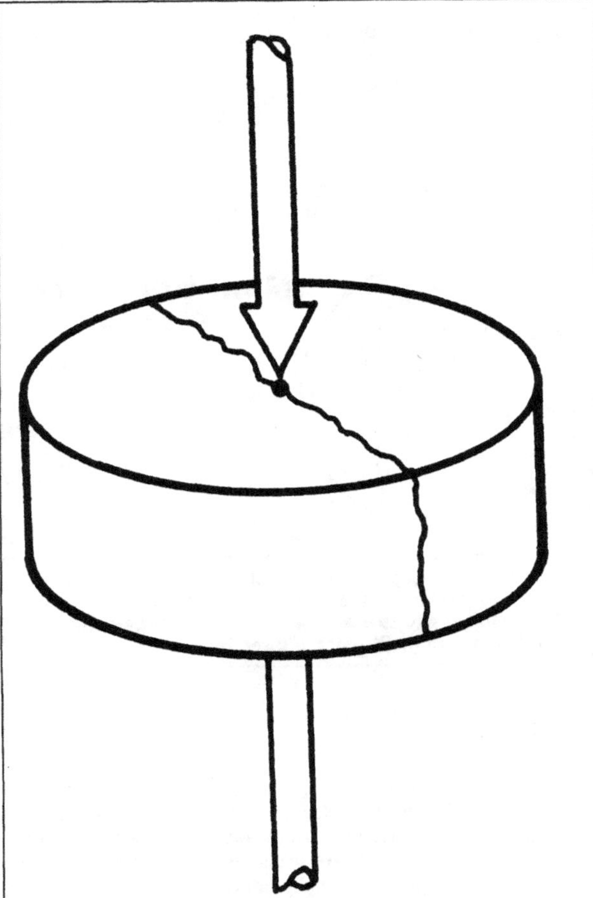

Fig. 16.22—Point-load tests to measure tensile-strength anisotropy. Wafer samples are loaded, and, typically, one tensile fracture forms as shown. Several samples are tested, and if the azimuth of the fracture groups around a common direction, a plane of weakness exists in the rock.

Possibly more important to the accuracy of strain-relaxation techniques is the question of sample isotropy. If release of the in-situ stresses creates anisotropies within the fabric of the rock, then past stress states, if different from the current state of stress, may have imprinted anisotropies into the rocks. In such cases, the measured strain relaxation may relate to the rock anisotropy, not the current in-situ stress orientation. One safeguard against such misinterpretation is to conduct laboratory velocity measurements as discussed, and strain-relaxation measurements generally should always be followed by laboratory tests to determine whether anisotropies exist in the samples and whether these anisotropies remain when the samples are restressed to in-situ conditions.

While these unknowns are real and must be considered in any analysis, the technique has proved reliable in several tests where azimuth was also measured with other procedures. These include tests in a volcanic tuff in Nevada[33]; a low-permeability Mesaverde sandstone[13]; a low-permeability gas sand in the Cotton Valley formation[6,34]; and a high-porosity, high-permeability sandstone in Oklahoma.[15] Cotton Valley strain-relaxation data[6] are compared with other techniques in Fig. 16.16. In all these cases, measurement of anelastic strain recovery correctly predicted hydraulic fracture azimuth. In another published test on granite,[40] however, the principal directions of the time-dependent strains differed from the elastic strain relief and thus presumably would yield an incorrect prediction for fracture azimuth. In this case, the anelastic strains were apparently related to existing microcracks along the grain boundaries of the rock. No discussion was given concerning whether laboratory tests could have identified this discrepancy. This test was conducted with very shallow samples, and the possibility exists that such shallow, near-outcrop samples could be more sensitive to existing rock fabric than more deeply buried, more highly stressed samples.

Laboratory Analyses. In addition to anelastic-strain-recovery measurements, several laboratory techniques have been developed to determine fracture azimuth from oriented cores. The advantages of these are clear: (1) elimination of delays and scheduling problems attendant on operations tied to drilling and (2) the ability to test more samples, giving more opportunity to observe inhomogeneities and vertical variations in behavior. However, laboratory tests are one step further removed from the data of interest—orientation of the in-situ stresses—because the tests depend on mechanical property changes in the rock to create anisotropies related to the stress removal. This may not always be true, and direct measurement of strain recovery may be related to the current in-situ stresses, while laboratory tests may reveal anisotropies that are the sum of all past stress states.[15] In other cases, laboratory analyses of cores have successfully predicted fracture azimuth.[6,16,17,32] Some of the currently available tests include simple tests for planes of weakness in the rock; overcoring to measure residual strains; and mechanical, wave-velocity, and thermal-expansion tests to measure anisotropy in the elastic moduli of the rock.

Basis for Laboratory Tests. It is generally recognized that the in-situ stress controls the orientation and azimuth of hydraulic fractures, with the fracture usually oriented perpendicular to the minimum stress. How, then, do core measurements relate to fracture azimuth? While there are several possible answers, the most appealing is that anisotropies are induced in the rocks by removal of the in-situ stresses from the sample. It has generally been noted that the mechanical properties of rocks change when samples are removed from their in-situ conditions.[44] Strickland and Ren[45] hypothesized that a major factor was the opening of microcracks along grain boundaries, as idealized in Fig. 16.21. As the stresses are relieved, the grains expand and can no longer conform to one another; thus, small cracks open up between grains. Because the grains are irregularly shaped, cracks will open up with all orientations. In a statistical sense, however, they are preferentially oriented by the stress directions; i.e., more cracks open up perpendicular to the maximum-stress direction than in any other orientation.

Such a microcrack model explains many observations, but other mechanisms may also be important. For example, a microcrack model does not satisfactorily explain why some samples exhibit a plane of weakness parallel to the fracture azimuth.[17,32] Because the majority of the microcracks should be perpendicular to the expected fracture azimuth, one would expect the plane of weakness to parallel the cracks and to be perpendicular to the fracture azimuth. Another possible internal mechanism is residual strain,[46] where a portion of the strain in the rock grains is transferred to the cementation material and is thus locked into the rock as residual strain. Because tensile tests used to define a plane of weakness have to overcome both the rock strength and these residual stresses, such a mechanism would predict a plane of weakness perpendicular to the minimum in-situ stress (parallel to the expected hydraulic fracture azimuth). Probably both of these, together with other mechanisms, play roles, but until the internal mechanisms are more clearly defined, testing in a new area should begin with a comprehensive suite of tests. This will define the easiest and best procedures for that particular formation and will help avoid misinterpretations.

Strength Tests. The simplest laboratory procedure is tensile-strength tests to determine whether a plane of weakness exists in the rock; the most common procedure is a point-load test. For this test (discussed in detail in Ref. 47), a vertical core plug is cut into horizontal wafers typically 1 in. [2.5 cm] in diameter by ¼ in. [0.64 cm] thick. These wafers are loaded (Fig. 16.22), and such a point load creates a radially symmetric tensile stress in the sample, typically creating a single fracture, as shown in Fig. 16.22. Several wafers (typically 10 to 20) are tested, and if the fracture planes share a common direction, then a plane of weakness exists in the rock.

Simple tests like these successfully predicted fracture azimuth in a low-permeability gas sand in Colorado,[17,32] where the results were compared with electric-potential surveys. In another area (east Texas Cotton Valley formation), the tests produced very inconsistent results. This procedure does offer a good method of testing unoriented cores to give a probability of their yielding good data.

Differential-Strain-Curve Analysis. Differential-strain-curve analysis was introduced by Strickland *et al.*[48] Its use to determine fracture azimuth is based on the microcrack model. Because the distribution of microcracks is preferentially oriented by the in-situ stresses, the cracks will create anisotropic elastic moduli, so by measuring the stress/strain response of the rock in different directions as the sample is restressed, one can determine the in-situ stress orientations and thus fracture azimuth.

For this test, a cubic sample is cut from an oriented core, and several strain gauges are attached to it. The sample is sealed with a coating, placed in a pressure vessel, and restressed with a hydrostatic confining pressure. Pressure vs. strain is recorded for each strain gauge and is used to find the magnitude and directions of the principal strains with normal Mohr's circle equations.[44] When pressure is first applied to the sample, it produces large strains because of the low stiffness of closing the microcracks. Because a greater number of the cracks are perpendicular to the direction of maximum in-situ stress, the strains are initially larger in this direction. Thus, the axis with the largest strains is assumed to coincide with the axis of maximum in-situ stress.

While differential-strain-curve analysis has been applied to a wide variety of rocks, few occasions where other techniques were also used have been reported. Teufel *et al.*[16] reported good success in a low-permeability Mesaverde sandstone. In a higher-porosity sandstone,[15] differential-strain-curve analysis did not predict the fracture azimuth, with the indicated azimuth apparently related to internal fabric in the rock from past stress states.

Residual-Strain Overcoring. As the in-situ stresses are released by the coring, some of the compressive strain energy in the rock grains may be transferred to the cementation material, thus locking in a memory of the in-situ stress state.[32] Measuring this can give an indication of hydraulic fracture azimuth, and this is typically done by strain-gauge overcoring.

Three strain gauges (mounted on a rosette at 45° to one another) are bonded to a horizontal surface of a core sample, a zero reading is recorded, and the gauges are overcored. If residual strains are locked into the core, then overcoring, by destroying the cementation bonds, will release more strain energy, allowing further expansion of the core. After the overcore barrel is removed, the strain gauges are recorded for several hours, and the direction of maximum expansion is interpreted to be the maximum in-situ-stress direction. In past work, the direction of maximum residual-strain release agreed with other fracture-azimuth measurements in the Muddy-J formation in Colorado[32] and the Cotton Valley formation of east Texas.

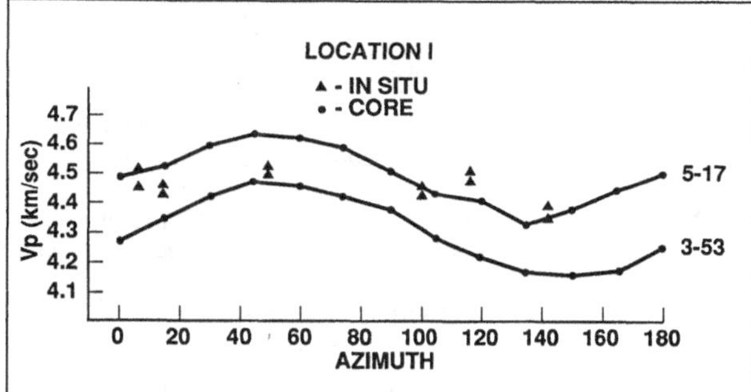

Fig. 16.23—Compressional velocity vs. azimuth for Machias sandstone. Triangles represent velocities measured in situ. Connected dots represent laboratory-measured velocities after removal of core from the outcrop (after Engelder and Plumb[49]).

Along with being a potential process for measuring fracture azimuth, residual-strain techniques can aid in defining the internal mechanisms in the rock and thus in interpreting other oriented-core analyses.

Wave-Velocity Anisotropy. Another laboratory process is to measure compressional velocities of the rock in different directions, either by direct measurements across a full-diameter core or by cutting small horizontal plugs in different directions. The interpretation of such data generally is related to the microcrack model of rock deformation. If the release of the in-situ stresses creates cracks that are preferentially oriented, then the direction of lowest velocity will be perpendicular to the majority of the cracks. The azimuth with the lowest velocity is then parallel to the maximum-stress direction and thus parallel to the azimuth of the hydraulic fracture.

Typical data (from the Machias sandstone in this case) are shown in Fig. 16.23.[49] In this example, the direction of lowest velocity parallels the direction of maximum expansion measured when this shallow sample was cored. Thus, the velocity anisotropy gave a true measurement of the direction of maximum stress. However, Engelder and Plumb[49] also discussed several examples, including other locations in the same sandstone formation, where such agreement was not the case. Lacy[6] reported good agreement between the minimum-velocity direction and fracture azimuth as determined by other procedures in the Cotton Valley formation in Texas.

Directional-velocity tests are also valuable in studying the internal mechanisms causing the anisotropy in the rock, thus aiding in the interpretation of other tests. This is done by measuring the velocity anisotropy of the rock in bench-top tests, then restressing the samples

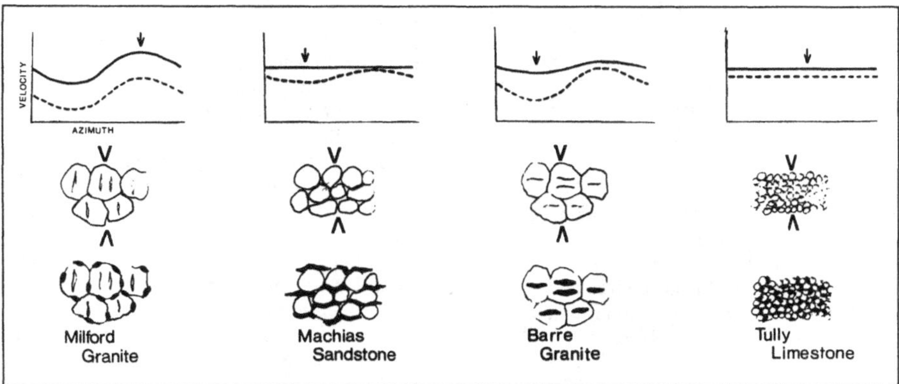

Fig. 16.24—Conceptual model of different types of behavior of velocity anisotropy vs. in-situ stress. The top row consists of plots of velocity vs. azimuth: the solid line is the in-situ velocity and the dashed line is core laboratory data (orientation of the maximum compressive stress is shown by the arrow above each plot). The next row shows four rocks under load, where the maximum-stress direction is indicated by the arrows; pre-existing microcracks are shown where appropriate. The last row shows the growth or opening of microcracks upon removal of the load (after Engelder and Plumb[49]).

to approximately in-situ-stress levels and repeating the measurements. Disappearance or a substantial reduction of the anisotropy is a reasonable indication that the rock behavior is related to the release of the in-situ stresses.[34] On the other hand, if the anisotropy persists, the rock behavior may be related to other fabric properties of the rock, and oriented-core analyses could be misinterpreted or yield erroneous results for fracture azimuth.

Ref. 49 gives a good review of some different types of velocity anisotropy behavior summarized in Fig. 16.24. While the particular examples referred to are for different rock types, the general behavior trends should apply to any rock type.

Thermal-Expansion Anisotropy. Lacy[6] investigated a third procedure for measuring the elastic anisotropy created by microcracks—measuring the anisotropic expansion while heating a sample. Fig. 16.21 indicates that the density of microcracks perpendicular to the maximum-stress axis will cause a lower modulus in this direction, so the axis of maximum expansion during heating will correspond to the direction of maximum in-situ stress. Measurements of this type correspond well with other core data, borehole seismic data, and tiltmeter data for a low-permeability gas sandstone in east Texas[6] and with stress-relief and borehole-elongation data from the Prudhoe Bay field in Alaska.

One advantage of this procedure is the ease of directly combining the technique with anelastic-strain-recovery measurements, particularly because it is generally necessary to make thermal-strain measurements to correct the time-dependent measurements. The same measuring equipment used to record the time-dependent strains can be used, and the tests can be done either on site or later in the laboratory. For thermal-expansion tests, the cores are usually cooled in a refrigerator or by ice to about 40°F [4°C]. Core expansion is then measured as the cores warm to room temperature.

Twinned-Calcite-Strain Analysis. It is interesting to note that the discussion and examples of oriented-core analyses relate to sandstone formations. In fact, the small amount of data available for limestone formations indicates that many of the analysis techniques, which are based on induced anisotropy, may not be applicable. Core analysis done as part of a fracture-azimuth test in a shallow limestone[42] (Oolagah limestone) and recent laboratory velocity tests on the Tully limestone[49] showed no measurable anisotropies in the rock. For calcite rocks, however, an additional laboratory technique is available.

When subjected to shear stress, calcite will deform by a process known as twin gliding on certain crystal planes. This deformation process creates a visually discernible disturbance in the crystal, called twin lamellae,[50] and measuring the orientation of these allows determination of the directions of the stress field that caused the deformation.[51] While this is not an attractive procedure because of the extensive work required (up to 50 hours of universal stage microscope observations per sample[16]), it does exist for limestone and may yield reliable data where other techniques are not applicable.

Ref. 16 discusses calcite-strain measurements on a calcite-cemented sandstone, and the results compare well with these from other procedures.

16.6 Summary

This chapter discusses several methods of measuring fracture azimuth, and work comparing these methods[6,15,16] has shown that each procedure can, under proper conditions, give accurate and reliable results for fracture azimuth. Also, depth is not a factor, because some procedures are essentially depth-independent. However, each procedure has limitations. Research has defined these limitations, and they must be recognized and honored when fracture azimuth is measured in a new area or formation. Just as multiple logs are run to evaluate an unknown formation, several fracture-azimuth measuring procedures should be used in a new area to guard against misinterpretations and to determine the best technique for that area.

This section briefly reviews the procedures and discusses considerations for setting up a fracture-azimuth measuring program in a new area. Some considerations are obvious, such as deciding which procedures are possible. For example, if the need to determine fracture azimuth arises very late, perhaps when a field is to be waterflooded, and no drilling is scheduled, then oriented-core analysis would be difficult. This section ignores such considerations.

"Preparatory Testing." When the need for fracture-azimuth information on a new field is determined, the first step should be to review existing data, including the structural geology of the region (do tectonic features exist that indicate fracture azimuth?), multiarm-caliper logs, and laboratory tests on existing unoriented cores. If oriented-caliper logs exist, they should be analyzed to determine whether breakouts are present and, if so, whether well-to-well and formation-to-formation trends are evident.

If coring is planned for the area for geologic or other reasons, oriented cores should definitely be included in the azimuth program. If coring is not already planned, however, simple laboratory work can aid in the determination of whether cores should be cut strictly for azimuth determination. At a minimum, these tests include tensile-strength (point-load) tests and velocity-anisotropy measurements. These tests should give a positive indication of results before coring strictly for fracture azimuth is considered.

Finally, if some idea of the area involved is available, plans should be made for determining fracture azimuth at several locations. In general, it cannot be assumed that azimuth will be constant. Such a lateral spread of data should be collected as early as possible to use the information to develop the best well pattern.

Direct Measurements. As early as possible, some form of direct technique should always be included in a program for measuring fracture azimuth. Such measurement is needed to confirm the analysis of indirect methods. These indirect methods, such as oriented-core analysis or oriented-caliper logs, can then, if justified, be used to extrapolate the data laterally and to define possible areas where additional measurements are required. The primary considerations involved in selecting a direct procedure include (1) depth, (2) terrain, (3) reservoir pressure and temperature, and (4) rock type.

Borehole Trace of Fractures. If openhole completions are possible, then directly measuring fracture azimuth at the borehole with impression packers, a sonic borehole televiewer, or (if a tool is available) downhole television logging should be considered. For competent formations where the wellbore fracture is initiated hydraulically from a reasonably circular, intact wellbore, such direct wellbore measurement offers a reliable procedure for determining azimuth.

Advantages of these procedures are their relative simplicity because all operations are limited to the wellsite and the use of usual oilfield equipment. Disadvantages include the need for prefracture tests to ensure that the fracture is hydraulically created and the intrusion of such measurements on normal well-completion procedures. Additional considerations would include possible temperature (or pressure) limits on logging tools or packers and the difficulty in using impression packers in an overpressured reservoir because of the need to "kill" the well.

Triaxial Borehole Seismic. This is one of the easiest techniques to apply in cased holes. Normal field operations require only about 6 hours to perform the tests, which are usually done 1 or 2 days before a hydraulic fracture treatment. With proper precautions and good field practices, this technique offers high reliability and good accuracy (5°).

Tiltmeters. Tiltmeters have been commercially available for several years and have the advantage of not interfering with rig or completion schedules. Disadvantages include difficulties in obtaining surface rights for instrument sites and in extracting weak signals from the background data for deeper (7,000- to 11,000-ft [2130- to 3350-m]) fracture jobs. The technique is also sensitive to environmental conditions (e.g., rain).

Indirect Measurements. The advantage of indirect measurements is that they are sometimes easier to use and do not intrude on normal drilling and completion operations, making it possible to extend the data collected from direct measurements laterally and vertically.

The main disadvantages of the indirect oriented-core techniques are total cost (including rig time) and the difficulty of acquiring good oriented core with reliable survey data.

Borehole Ellipticity. Measuring wellbore breakouts emphasizes these advantages because the only operation involved is an oriented-caliper log (four-arm dipmeter or sonic borehole televiewer) in the openhole logging program. Such logs should be included in any fracture-azimuth determination procedure because this is clearly the best procedure for producing a spread of data across a field.

Oriented-Core Analysis. If coring operations are scheduled for other reasons, then the additional complication of orienting the core should always be considered. Particularly early in the life of a field, when several cores may be cut for geologic reasons, core analysis can yield a spread of data across an area in a very timely manner with minimal intrusion into normal operations.

For a new area, the analyses should include, at a minimum, on-site strain-relaxation and thermal-expansion measurements, laboratory tests for tensile strength, and laboratory tests for velocity anisotropies (under both bench-top and simulated in-situ conditions). If facilities are available, such additional tests such as strain-curve analysis and strain-gauge overcoring for residual strains are also desirable. In general, a test suite as complete as possible should be obtained for the first core. This is necessary to check the analysis (because all core-analysis procedures are based on assumptions about the internal deformation mechanisms in the rock), to determine quality control of one procedure against another, and to determine the best procedure for future cores. For each coring test, it is recommended that the test be performed on a minimum of six coring pieces or a minimum of two coring pieces per coring run. Averaging the data from 6 to 12 coring pieces will improve the accuracy and reliability of the data.

References

1. Smith, M.B.: "Effect of Fracture Azimuth on Production With Application to the Wattenberg Gas Field," paper SPE 8298 presented at the 1979 SPE Annual Technical Conference and Exhibition, Las Vegas, Sept. 23–26.
2. Wood, W.D.: "Method of Determining Change in the Subsurface Structure due to Application of Fluid Pressure to the Earth," U.S. Patent No. 4,272,696 (1981).
3. Maruyama, T.: "Statistical Elastic Dislocations in an Infinite and Semi-infinite Medium," *Bull.*, Earthquake Research Inst., Tokyo U. (1964) **289**.
4. Davis, P.M.: "Surface Deformation Associated with Dipping Hydrofracture," *J. Geophysical Res.* (1983) **881**, No. 87, 5826.
5. Pollard, P.O. and Holzhausen, G.: "On the Mechanical Interaction Between a Fluid-Filled Fracture and the Earth Surface," *Tectonophysics* (1979) **53I**, 27.
6. Lacy, L.L.: "Comparison of Hydraulic Fracture Orientation Techniques," *SPEFE* (March 1987) 66–76; *Trans.*, AIME, **283**.
7. Evans, K.: "On the Development of Shallow Hydraulic Fractures as Viewed Through the Surface Deformation Field: Part 1—Principles," *JPT* (Feb. 1983) 406–10.
8. Evans, K. and Holzhausen, G.: "On the Development of Shallow Hydraulic Fractures as Viewed Through the Surface Deformation Field: Part 2—Case Histories," *JPT* (Feb. 1983) 411–20.
9. Schuster, C.L.: "Detection Within the Wellbore of Seismic Signals Created by Hydraulic Fracturing," paper SPE 7448 presented at the 1978 SPE Annual Technical Conference and Exhibition, Houston, Oct. 1–4.
10. Pearson, C.: "The Relationship Between Microseismicity and High Pore Pressuring During Hydraulic Stimulation Experiments in Low Permeability Granite Rock," *J. Geophysical Res.* (Sept. 1981) **86,** 7855–64.
11. Albright, J.N. and Pearson, C.F.: "Acoustic Emissions as a Tool for Hydraulic Fracture Location: Experience at the Fenton Hill Hot Dry Rock Site," *SPEJ* (Aug. 1982) 523–30.
12. Dobecki, T.L.: "Hydraulic Fracture Orientation by Use of Passive Borehole Seismics," paper SPE 12110 presented at the 1983 SPE Annual Technical Conference and Exhibition, San Francisco, Oct. 5–8.
13. Batchelor, A.S., Baria, R., and Hearn, K.: "Monitoring the Effects of Hydraulic Stimulation by Microseismic Event Location: A Case Study," paper SPE 12109 presented at the 1983 SPE Annual Technical Conference and Exhibition, San Francisco, Oct. 5–8.
14. Hart, C.M. *et al.*: "Fracture Diagnostics Results for the First Multiwell Experiment's Paludal Zone Stimulation," *SPEFE* (Sept. 1987) 320–26; *Trans.*, AIME, **283**.
15. Smith, M.B. *et al.*: "A Comprehensive Fracture Diagnostics Experiment: Part 2—Comparison of Fracture Azimuth Measurements," *SPEPE* (Nov. 1986) 423–31; *Trans.*, AIME, **281**.
16. Teufel, L.W. *et al.*: "Determination of Hydraulic Fracture Azimuth by Geophysical, Geological, and Oriented-Core Methods at the Multiwell Experiment Site, Rifle, Colorado," paper SPE 13226 presented at the 1984 SPE Annual Technical Conference and Exhibition, Houston, Sept. 16–19.
17. Smith, M.B. *et al.*: "The Azimuth of Deep, Penetrating Fractures in the Wattenberg Field," *JPT* (Feb. 1978) 185–93.
18. Fraser, C.D. and Pettitt, B.E.: "Results of a Field Test To Determine the Type and Orientation of a Hydraulically Induced Formation Fracture," *JPT* (May 1962) 463–66.
19. Anderson, T.O. and Stahl, E.J.: "A Study of Induced Fracturing Using an Instrumental Approach," *JPT* (Feb. 1967) 261–67; *Trans.*, AIME, **240**.
20. Zemanek, J. *et al.*: "The Borehole Televiewer—A New Logging Concept for Fracture Location and Other Types of Borehole Inspection," *JPT* (June 1969) 762–74; *Trans.*, AIME, **246**.
21. Bredehoeft, J.D. *et al.*: "Hydraulic Fracturing to Determine the Regional In Situ Stress Field, Piceance Basin, Colorado," *Bull.*, GSA (Feb. 1976) **87,** 250–58.
22. Hirsch, J.M. *et al.*: "Recent Experience With Wireline Fracture Detection Logs," paper SPE 10333 presented at the 1981 SPE Annual Technical Conference and Exhibition, San Antonio, Oct. 4–7.
23. Smith, M.B., Rosenberg, R.J., and Bowen, J.F.: "Fracture Width: Design vs. Measurement," paper SPE 10965 presented at the 1982 SPE Annual Technical Conference and Exhibition, New Orleans, Sept. 26–29.
24. Gough, D.I. and Bell, J.S.: "Stress Orientations from Oil Well Fractures in Alberta and Texas," *Cdn. J. Earth Sci.* (1981) **18,** 638.
25. Thorpe, R. and Springer, J.: "Relationship Between Borehole Elongation and In Situ Stress Orientation at the Nevada Test Site," paper presented at the 1982 U.S. Rock Mechanics Symposium, Berkeley, CA, Aug. 25–27.
26. Babcock, E.A.: "Measurement of Subsurface Fractures from Dipmeter Logs," *AAPG Bull.* (July 1978) **62,** 1111.
27. Gough, D.I. and Bell, J.S.: "Stress Orientation from Borehole Well Fractures with Examples from Colorado, East Texas, and Northern Canada," *Cdn. J. Earth Sciences* (1982) **19,** 1358.
28. Brown, R.O., Forgotson, J.M., and Forgotson, J.M. Jr.: "Predicting the Orientation of Hydraulically Created Fractures in the Cotton Valley Formation of East Texas," paper SPE 9269 presented at the 1980 SPE Annual Technical Conference and Exhibition, Dallas, Sept. 21–24.
29. Detournay, E. and Fairhurst, C.: "Generalization of the Ground Reaction Curve Concept," *Issues in Rock Mechanics, Proc.*, 23rd U.S. Rock Mechanics Symposium, R.E. Goodman and F.E. Heuze (eds.), Berkeley, CA (Aug. 25–27, 1982).
30. Plumb, R.A. and Hickman, S.H.: "Stress-Induced Borehole Elongation: A Comparison Between the Four-Arm Dipmeter and the Borehole Televiewer," *J. Geophys. Res.* (1985) **90,** 5513–21.
31. Komar, C.A., Overbey, W.K. Jr., and Watts, R.J.: "Prediction of Fracture Orientation from Oriented Cores and Aerial Photos, West Poison Spider Field, Casper, Wyoming," Morgantown Energy Research Center Report MERC/RI-76/1, Morgantown, WV (March 1976).
32. Logan, J.M. and Teufel, L.W.: "The Prediction of Massive Hydraulic Fracturing from Analyses of Oriented Cores," *Proc.*, 19th U.S. Rock Mechanics Symposium, Reno, NV (May 1–3, 1978) 340–44.
33. Teufel, L.W.: "Prediction of Hydraulic Fracture Azimuth From Anelastic Strain Recovery Measurements of Oriented Core," *Proc.*, 23rd U.S. Rock Mechanics Symposium (Aug. 25–27, 1982) 238–46.
34. Teufel, L.W.: "Determination of the Principal Horizontal In Situ Stress Directions from Anelastic Strain Recovery Measurements of Oriented Cores from Deep Wells: Application to the Cotton Valley Formation of East Texas," *Geomechanics AMD* (1983) **57,** 56–63.
35. Rowley, D.S., Burk, C.A., and Manuel, T.: "Oriented Cores," Christensen Technical Report, Christensen Diamond Products (Feb. 1981).
36. Robertson, E.C.: *Viscoelasticity of Rocks in State of Stress in the Earth's Crust*, W. Judd (ed.), (1964) 181–224.
37. Voight, B.: "Determination of the Virgin State of Stress in the Vicinity of a Borehole from Measurements of Partial Anelastic Strain Tensor in Drill Cores," *Rock Mech. Eng. Geol.* (1968) **6,** 210–15.
38. Swolfs, H.S.: "Determination of In Situ Stress Orientation in a Deep Gas Well by Strain Relief Techniques," Technical Report 75-43, Terra Tek, Salt Lake City, UT (1975).
39. Swolfs, H.S., Lingle, R., and Thomas, J.M.: "Strain Relaxation Tests on Selected Cores from El Paso Natural Gas Well Largo Canyon No. 288," Technical Report 76-50, Terra Tek, Salt Lake City, UT (1976).
40. Engelder, T.: "The Time-Dependent Strain Relaxation of Algerie Granite," *Intl. J. Rock Mech. Mining Sci.* (1984) **21,** No. 2, 63–73.

41. Blanton, T.L.: "The Relation Between Recovery Deformation and In-Situ Stress Magnitudes," paper SPE 11624 presented at the 1983 SPE/DOE Low Permeability Gas Reservoirs Symposium, Denver, March 14-16.
42. Smith, M.B., Logan, J.M., and Wood, M.D.: "Fracture Azimuth—A Shallow Experiment," *J. Energy Res. Tech.* (June 1980) **102.**
43. Blanton, T.L. and Teufel, L.W.: "A Field Test of the Strain Recovery Method of Stress Determination in Devonian Shales," paper SPE 12304 presented at the 1983 SPE Eastern Regional Meeting, Champion, PA, Nov. 9-11.
44. Engelder, T.: "General Characteristics of Strain Relaxation: A Note on Sample Preparation for Large-Scale Tests," *Geophysical Res. Letters* (July 1981) **8,** No. 7, 687-89.
45. Strickland, F.G. and Ren, N.K.: "Use of Differential Strain Curve Analysis in Predicting In Situ Stress State for Deep Wells," *Rock Mechanics, A State of the Art, Proc.*, 21st Rock Mechanics Symposium, D.A. Summers (ed.), Rolla, MO (1980).
46. Friedman, M.: "Residual Elastic Strain in Rocks," *Tectonophysics* (1972) **15,** 297-330.
47. McWilliams, J.R.: "The Role of Microstructure in the Physical Properties of Rock," *Testing Techniques for Rock Mechanics—STP 402,* ASTM (1966) 175-89.
48. Strickland, F.G., Feves, M.L., and Sorrells, D.: "Microstructural Damage in Cotton Valley Cores," paper SPE 8303 presented at the 1979 SPE Annual Technical Conference and Exhibition, Las Vegas, Sept. 23-26.
49. Engelder, T. and Plumb, R.: "Changes in In Situ Ultrasonic Properties of Rock on Strain Relaxation," *Intl. J. Rock Mech. Mining Sci.* (1984) **21,** No. 2, 75-82.
50. Handin, J.W. and Griggs, D.: "Deformations on Yule Marble. Part II. Predicted Fabric Changes," *Bull.*, GSA (1951) **62,** 863-86.
51. Groshong, R.H., Teufel, L.W., and Gasteiger, C.: "Optimum Strategy of Calcite Strain Gauge Technique," *Bull.*, GSA (1984) **95,** 357-63.

SI Metric Conversion Factors

degrees	× 1.745 329	E−02	=	rad
ft	× 3.048*	E−01	=	m
gal	× 3.785 412	E−03	=	m^3
in.	× 2.54*	E+00	=	cm
psi	× 6.894 757	E+00	=	kPa

*Conversion factor is exact.

Chapter 17
Economics of Fracturing

Ralph W. Veatch Jr., SPE, Amoco Production Co.

17.1 Overview

This chapter addresses methods for investigating economic aspects of hydraulic fracturing. Discussions focus on maximizing potential profits derived from accelerated production, reduced operating costs, and possibly increased ultimate recovery. The pertinent economic criteria are given, and examples are presented to demonstrate analysis methods and the sensitivity of economics to certain fracture treatment design and formation parameters. The most important parameters have been included, and procedures for evaluating their effects are shown. Results of studies by other investigators are also discussed.

17.2 Introduction

This chapter attempts to provide an overall approach to fracturing economics. Currently, there is somewhat of a paucity of literature available on the subject as it relates directly to fracture treatment design. Howard and Fast[1] presented a relatively broad coverage. Frick[2] discusses petroleum economic applications in general but does not address fracturing specifically. Most of the field case studies available show only results specific to a given field and do not address the overall economics of fracture design. A comprehensive study by the Natl. Petroleum Council[3] on tight-gas formations covered some fracturing but was generally directed to full-cycle economics for developing tight-gas formations. Warembourg *et al.*[4] recently presented a very good discussion that outlined general procedures and approaches to optimizing fracturing treatment economics.

The intent of this chapter is to focus on the economics pertinent to the specifics of fracturing treatment design. The discussion addresses approaches directly related to fracturing and methodologies for conducting economic investigations and presents examples of various parametric studies. Other examples can be found in Chap. 11, Sec. 11.6, Optimal Treatment Selection.

Investigating the economics associated with fracturing, either for a single well or for a total field program, can range from a relatively simple to a very complex task. In some cases, the economics may not be highly sensitive to treatment design parameters—for example, with relatively small treatments where a large part of the treatment costs result from mobilization, transportation, and minimum rental charges. For these situations, design parameters, such as hydraulic-horsepower pumping costs, may dominate the economic picture. Here it may be relatively easy to arrive at an optimum economic treatment design.

In other cases where the economics is highly dependent on both treatment design and formation parameters, the number of significant factors could be very large. In very deep formations where in-situ closure stresses are high and where long tubulars impose high friction losses, the costs of high-strength propping agents and the hydraulic horsepower may control the economics of the treatment. In very tight formations that require massive hydraulic fracturing (MHF) treatments to achieve deeply penetrating fractures, the well spacing and the material costs may govern treatment economics. Optimal economic design is particularly important for MHF treatments, which can make up a large portion of the total well costs.

An example of the relative treatment cost as a percentage of total well cost in three major U.S. tight-gas basins is given in Fig. 17.1.[5] As can be seen for treatments of 500,000 gal [1890 m^3] or larger, fracturing costs can approach half the total well cost (including fracturing). In areas where MHF treatments account for a large fraction of the total well costs, the importance of fracturing is equal to or greater than that of development drilling for increasing reserves.

The fracturing requirements imposed by different formation properties (e.g., depths, in-situ stresses, permeabilities, thicknesses, and temperature) present situations where the economics could be sensitive to a wide variety of different treatment design parameters.

This chapter covers some general concepts of fracturing economics and integrates net present value (NPV) with commonly observed producing-performance-decline profiles (e.g., constant percentage, hyperbolic, and harmonic). Example cases are given that investigate the effects of well spacing, formation permeability, fracture conductivity, net pay, fracture height, and fluid loss on NPV to show some of the prediction methods used for selecting economically optimized fracture treatment designs.

17.3 General Concepts of Fracturing Economics

Economic design of fracture treatments generally has three basic requirements: (1) to evaluate what oil and/or gas production rates and recoveries might be expected from various fracture lengths and fracture conductivities for a given reservoir and relate these to cash flow income, (2) to determine the fracture treatment requirements to achieve the desired fracture lengths and conductivities and relate these to costs, and (3) to select the fracture lengths and conductivities where the income and costs combine to maximize economic returns. Fig. 17.2 illustrates these concepts. Ideally, a reservoir-performance simulator will provide predictions of the production rates and recoveries for various fracture lengths and conductivities. From these data, a revenue estimate can be developed for various fracture lengths. As can be seen, the revenue estimate as a function of fracture length is usually not a linear relationship. The rate of revenue growth diminishes with increasing fracture length and eventually reaches a relatively flat slope.

A hydraulic fracturing simulator usually is required to compute treatment volumes, types of materials, and pumping schedules necessary to achieve various fracture lengths and conductivities. With these data, a relationship between fracture length (and conductivity) and treatment cost can be generated. This is depicted in the lower portion of Fig. 17.2, where treatment costs accelerate with increasing fracture length.

The final step of investigating the net profit is shown on the right side of Fig. 17.2. This represents the combining of income and costs to relate net revenue to fracture length. The net-revenue curve generally exhibits some optimal point beyond which the cost to achieve longer fractures exceeds the revenue generated by production from the additional length. Thus, treatment designs that maximize economics (i.e., optimal treatments) are identified.

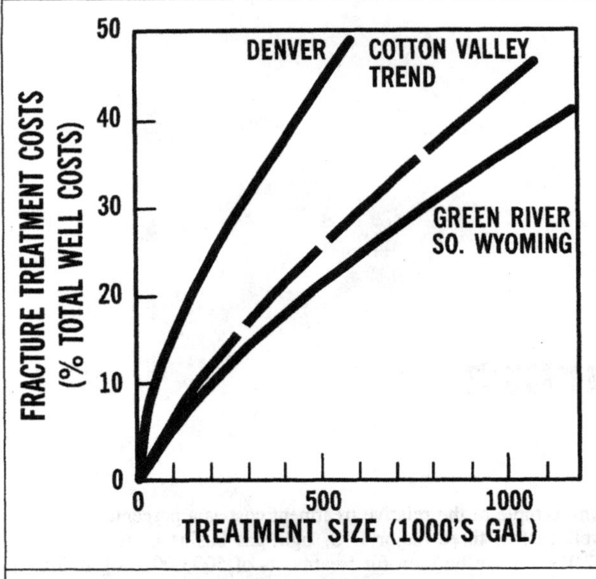

Fig. 17.1—Relative cost of MHF vs. treatment volumes (after Veatch[5]).

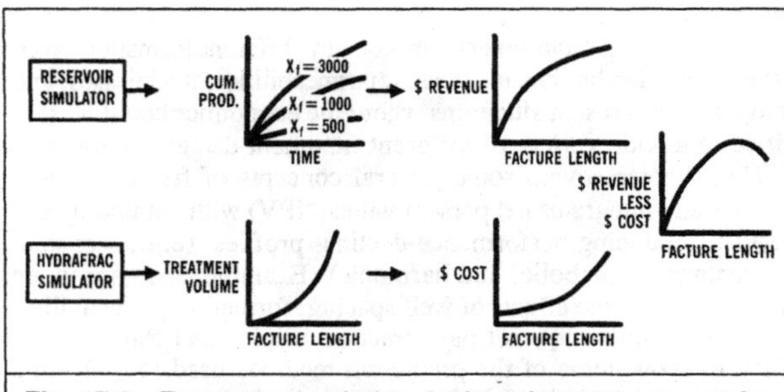

Fig. 17.2—Fracture stimulation design: the total concept for optimization (after Veatch[5]).

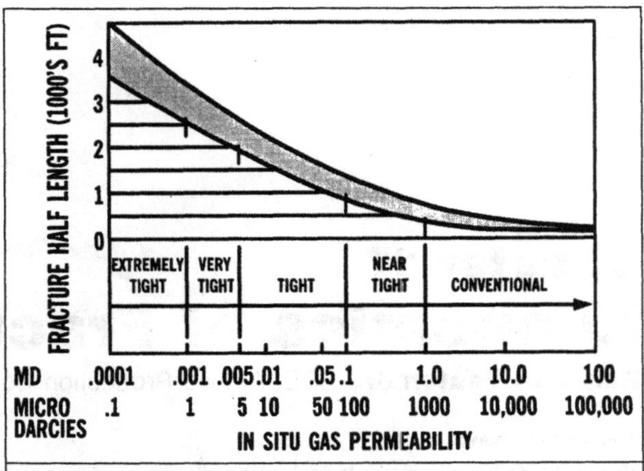

Fig. 17.3—Desired fracture half-lengths for different formation permeabilities (after Elkins[6]).

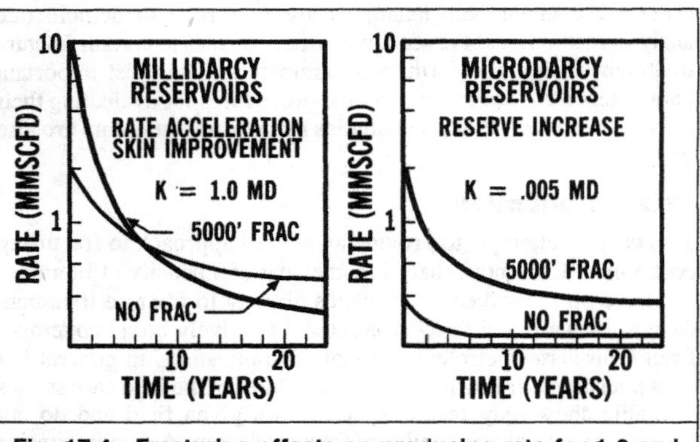

Fig. 17.4—Fracturing effects on producing rate for 1.0 and 0.005 md.

The specific procedures for determining the best treatment design (for deriving maximum profit) from hydraulically fracturing a given well may not always conform precisely to the conceptual steps, but they will always involve some type of balance between treatment costs and revenues generated from the production responses associated with a treatment. The specific examples in this chapter and those in Chap. 11 provide a few of the many approaches applicable to economically optimizing treatment designs.

17.4 Reservoir Response to Fracture Penetration and Conductivity

A major factor in optimization involves achieving the appropriate balance between the fracture characteristics and the formation properties that govern reservoir performance. High-permeability reservoirs require high fracture conductivities but usually do not need deeply penetrating fractures. Low-permeability formations require deeply penetrating fractures, and some can tolerate lower fracture conductivities. Some typical length requirements are illustrated in Fig. 17.3, an example by Elkins.[6] As shown, fracture half-length (i.e., wellbore to tip) requirements are often less than 1,000 ft [300 m] for "conventional" permeability reservoirs ($k>1.0$ md). But extremely low-permeability formations ($k=0.0001$ md) may require half-lengths as long as 3,500 to 4,500 ft [1070 to 1370 m].

Generalized examples of how fracturing affects producing rates and cumulative production for different permeabilities are shown in Figs. 17.4 and 17.5. Fig. 17.4 shows a comparison of gas producing rates between a radial flow case and a deeply penetrating fracture case in 0.005- and 1.0-md formations. Fig. 17.5 shows cumulative recovery data for the same two cases. These examples were obtained from computer simulation runs where all conditions were identical except for formation permeability and fracture length. For the higher-permeability case, fracturing has a significant effect on accelerating producing rate but a relatively small impact on ultimate cumulative recovery. In the lower-permeability case, however, fracturing significantly accelerates producing rates and increases cumulative recovery over the economic life of a well.

From an economic standpoint, the main benefits of rate acceleration include an increase in present value from accelerated income and a potential reduction in the time required for a well to reach its abandonment rate, which can result in reduced costs associated with producing a well or field, including reductions in operating, lifting, overhead, interest, and other miscellaneous costs.

In cases where both rate acceleration and reserve increases are achieved, there may be little or no effective reduction in expenses associated with production costs (in fact, expenses may increase by longer producing times). Here the economics may be significantly impacted by the increased cumulative production.

17.5 Producing-Performance Profiles

To investigate the economics associated with fracturing adequately, one must determine a profile of estimated producing-rate behavior, cumulative production performance, and operating expenses for both the fractured and unfractured cases. These are necessary to compare estimates of the cash flow, present worth, return on investment, and rates of return that might be expected for both cases. Production performance can be predicted in many ways. It is beyond the scope of this work even to try to cover all the possibilities. A wide and varied assortment of analytical, graphical, and computerized methods are available to estimate effects of fracture length and fracture conductivity on well productivity for a particular formation.

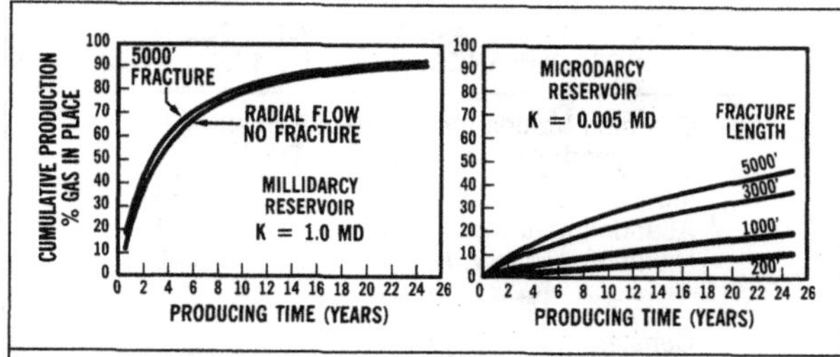

Fig. 17.5—Fracturing effects on cumulative recovery for 1.0 and 0.005 md.

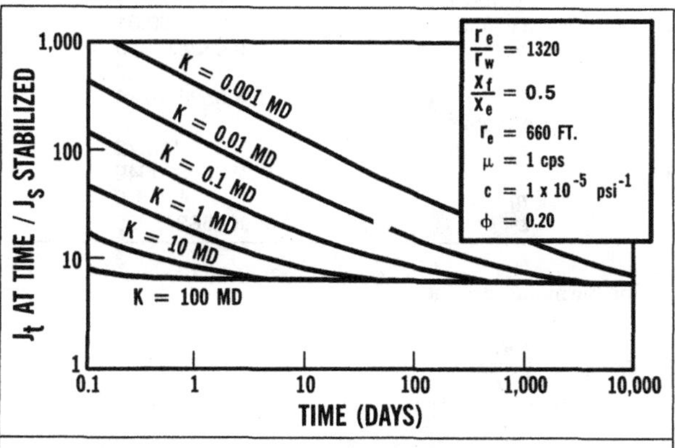

Fig. 17.6—"Relative" productivity ratio vs. real time for different permeabilities (after Morse and Von Gonten[7]).

The approaches used to perform economic studies will depend on whether the performance behavior is following steady-state, semisteady-state, or unsteady-state flow. Guidelines, such as those presented by Morse and Von Gonten[7] in Fig. 17.6, can be used to determine when flow behavior is approaching semisteady-state conditions. For the particular case shown, the "relative" productivity ratio J_f/J_s data converge to a common value for stabilized flow. For the 10-md case, this occurs in 3 or 4 days; for the 0.01-md case, it takes about 10 years. In general, steady-state approaches are usually amenable to permeabilities greater than 10 md. For levels less than 1 md, one should consider using approaches developed for unsteady-state (transient) performance behavior. In the 1-to-10-md range, the method that provides reasonably suitable results for the particular economic study should be selected.

Steady-State or Semisteady-State Behavior. These conditions occur when the reservoir has a relatively high permeability and the steady- or semisteady-state performance regime is established in a relatively short time. Here it is possible to determine production-rate increases with the Prats,[8] McGuire and Sikora,[9] or Tinsley *et al.*[10] approaches. Performance-decline behavior can be estimated by methods developed for constant-percentage, hyperbolic, or harmonic decline trends. The choice of approach to use for determining increases achieved from fracturing and performance-decline behavior is important and can be critical in economic optimization studies. Refer to the above references for guidance in selecting a method.

Production Increase. Prats' method incorporates the curve shown in Fig. 17.7; McGuire and Sikora's approach uses the curves in Fig. 17.8; and Tinsley *et al.*'s method uses Fig. 17.9. Computations for rates of increase are discussed in Chap. 15. An example case is also presented in Chap. 11 that includes the Tannich-Nierode curves for productivity-index ratio for gas wells. This accounts for non-Darcy gas flow in the fracture.

Performance Decline. Experience and previous studies[2,11] have shown that semi-steady-state decline-rate behavior often falls into one of three categories: (1) constant percentage, (2) hyperbolic, or (3) harmonic. This can hold true for both hydraulically fractured

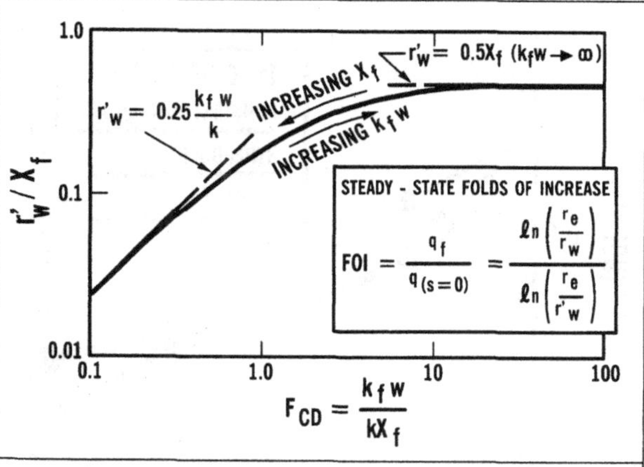

Fig. 17.7—Producing-rate increase curve (after Prats[8]).

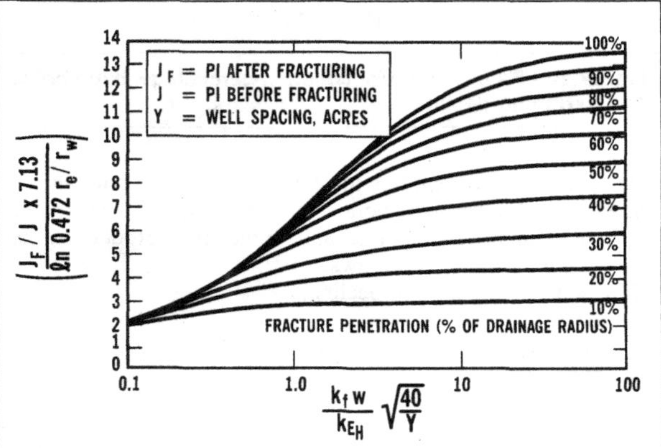

Fig. 17.8—Producing-rate increase curves (after McGuire and Sikora[9]).

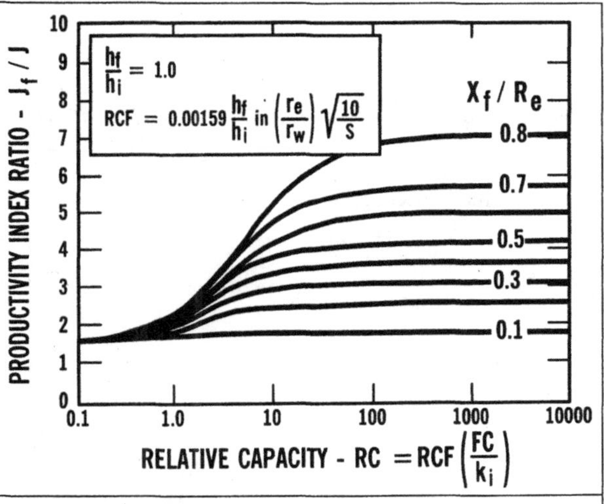

Fig. 17.9—Producing-rate increase curves (after Tinsley *et al.*[10]).

and unfractured wells. The decline rate equation basic to all three has the general form

$$a \equiv -\frac{d}{dt}\ln q = -\frac{1}{q}\frac{dq}{dt} = bq^m, \quad \text{(17.1)}$$

where a is decline rate, t is time, q is producing rate, and b and m are constants.

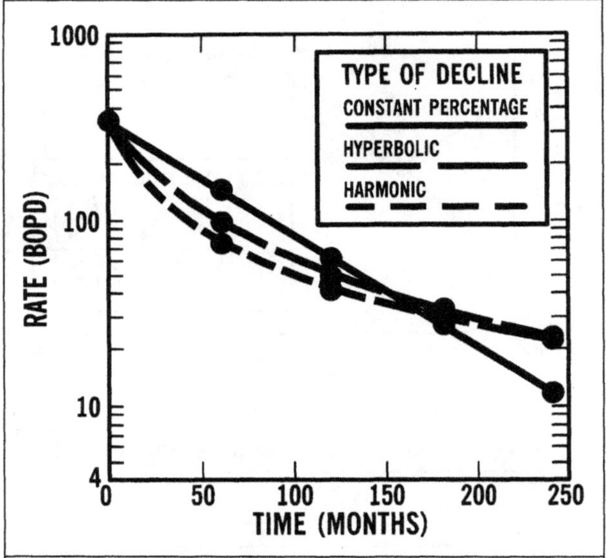

Fig. 17.10—Rate vs. time for constant-percentage, hyperbolic, and harmonic decline.

The exponent m identifies the category (i.e., for constant percentage, $m=0.0$; for hyperbolic, $0.0<m<1.0$; and for harmonic, $m=1.0$). The producing rate and cumulative production performance for each is as follows.

For constant percentage decline,

$$a=\frac{\ln q_i-\ln q}{t}=\frac{\ln(q_i/q)}{t}=\text{constant}, \quad (17.2)$$

$$q=q_i e^{-at}, \quad (17.3)$$

$$N_p=\frac{q_i-q}{a}=\frac{q_i(1-e^{-at})}{a}, \quad (17.4)$$

and

$$t=\frac{1}{a}\ln(q_i/q)=\frac{\ln\left(1-\dfrac{aN_p}{q_i}\right)}{-a}. \quad (17.5)$$

For hyperbolic decline,

$$a_i=bq_i{}^m=\frac{(q_i/q)^m-1}{mt}, \quad (17.6)$$

$$q=q_i(1+ma_it)^{-1/m}, \quad (17.7)$$

$$N_p=\frac{q_i}{(1-m)a_i}\left[q_i^{(1-m)}-q^{(1-m)}\right]$$

$$=\frac{q_i}{(1-m)a_i}\left[1-(1+ma_it)^{(m-1)/m}\right], \quad (17.8)$$

and

$$t=\frac{1}{ma_i}\left[\left(\frac{q_i}{q}\right)^m-1\right]=\frac{1}{ma_i}\left\{\left[1-\frac{N_p(1-m)a_i}{q_i}\right]^{m/(m-1)}-1\right\}. \quad (17.9)$$

TABLE 17.1—PARAMETERS FOR DECLINE PERFORMANCE, EXAMPLE 17.5

Common Values for Constant-Percentage, Hyperbolic, and Harmonic Decline	
Initial rate, q_i, BOPD	336
Abandonment rate, q_a, BOPD	5
Ultimate recovery, N_p, bbl	722,000
Decline Parameters	
Constant percentage	
t_a, months	301.54
a	0.013954032
m	0
b	1.0
Hyperbolic	
t_a, months	760.31
a_i	0.033853855
m	0.7
b	0.000052818286
Harmonic	
t_a, months	1,110.72
a_i	0.059600949
m	1.0
b	0.0000058278038

For harmonic decline,

$$a_i=bq_i=\frac{\left(\dfrac{q_i}{q}-1\right)}{t}, \quad (17.10)$$

$$q=\frac{q_i}{1+a_it}, \quad (17.11)$$

$$N_p=\frac{q_i}{a_i}\ln(q_i/q)=\frac{q_i}{a_i}\ln(1+a_it), \quad (17.12)$$

and

$$t=\frac{1}{a_i}\left(\frac{q_i}{q}-1\right)=\frac{1}{a_i}\left[e^{(N_pa_i/q_i)}-1\right], \quad (17.13)$$

where

a_i = initial decline rate (cycles/unit time),
q_i = initial producing rate, and
N_p = cumulative oil production.

In Eqs. 17.1 through 17.13, the units of time, cumulative production, producing rate, and decline rate must be consistent; e.g., if time is expressed in months and cumulative production in barrels, then producing rate must be in barrels per month and decline rate in cycles per month.

Comparisons of rate behavior are shown in Fig. 17.10 for the three types of decline. The pertinent data are given in Table 17.1. All three cases have the same values for initial rate, abandonment rate, and cumulative production at abandonment. However, the different decline profiles can yield significantly different economic results. Therefore, it is important during economic studies to use the appropriate decline-performance model. Merely using increase or cumulative recovery may not provide the appropriate data to evaluate economics adequately.

Also, it may be that even though steady- or semisteady-state conditions prevail, none of the three common decline rate patterns will predict decline performance acceptably. For these situations, more sophisticated approaches, such as computer simulation models, must be used.

Unsteady-State (Transient) Performance Behavior. If the reservoir has low permeability and transient flow dominates much of a well's early life, either two- or three-dimensional (2D or 3D) numerical computerized reservoir simulators or dimensionless type curves, such as those presented by Holditch *et al.*,[12] Agarwal *et*

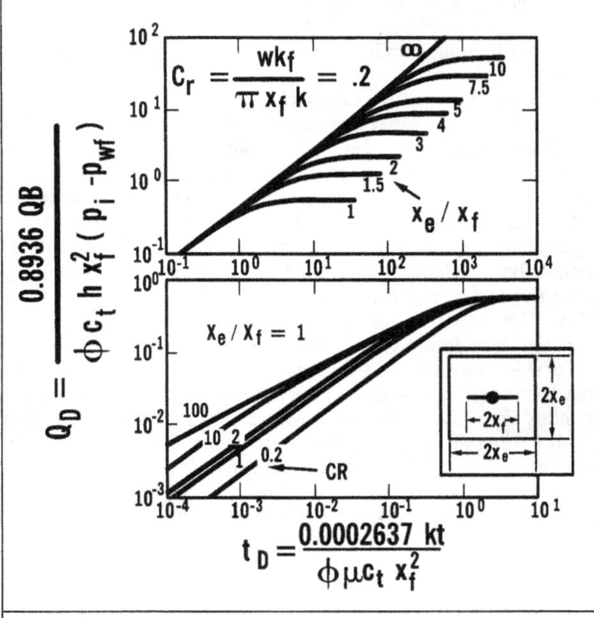

Fig. 17.11—Dimensionless cumulative-production type curves (after Holditch *et al.*[12]).

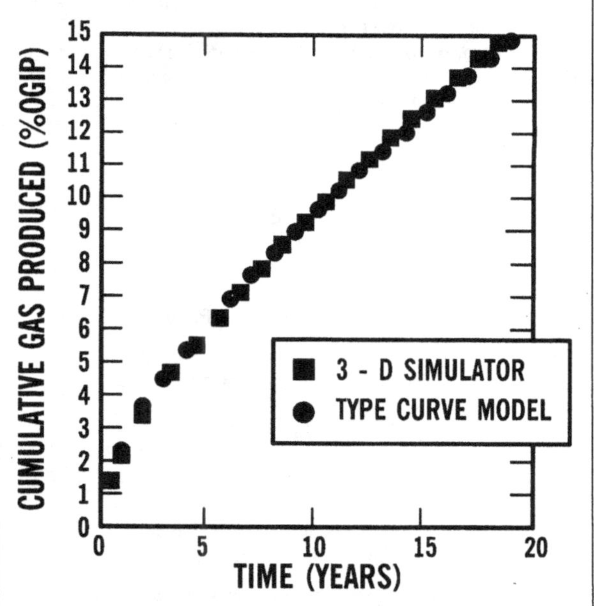

Fig. 17.12—Comparison of type-curve model to a 3D, single-phase reservoir model (after Holditch *et al.*[12]).

al.,[13] Gringarten *et al.*,[14] and Cinco-L. *et al.*[15] are necessary to predict performance before and after fracturing. For many cases, type curves are relatively faster and less expensive than iterative finite-difference computerized reservoir simulators. Dimensionless cumulative-production type curves such as those developed by Holditch *et al.*[12] and shown in Fig. 17.11, are very amenable to economic optimization studies. With these kinds of curves, one can easily determine the incremental production between two given time points and compute revenues for the period. With automated interpolative schemes, one can quickly compute cumulative production values for a given set of time points and the incremental production for each period without a large computer.

Fig. 17.12 shows a comparison made by Holditch[12] between the results from type-curve data and those obtained with a 3D reservoir simulator. As was the case with this example, type-curve methods can provide very fast, relatively accurate predictions (e.g., $\pm 10\%$) for many types of reservoirs. In addition, the interpolative schemes can be automated on much smaller computers than required for iterative reservoir simulation programs.

17.6 General Economic Criteria

To investigate the economics of fracturing, one should consider the following.

1. Present value (PV) (also called present worth and denoted as V_P in the equations) of the cash flow resulting from the well production and expense streams (for both the unfractured and fractured cases). The PV of each case is the sum of all cash flow discounted to some specific point in time at a stated discount rate. It is the value today of some amount in the future given an investment opportunity at a specific discount rate.
2. NPV or V_{nPf} of the net cash flow resulting from the fracturing treatment. This is computed as the V_{Pf} (PV after treatment) minus V_{Puf} (PV before treatment) minus the present value of all costs or investments associated with the treatment.
3. Payout (PO) time or discounted payout (DPO) time for the treatment. PO (t_{po} in the equations), the length of time required for the cumulative cash position to reach zero, is on an after-tax basis unless otherwise designated. DPO (t_{Dp} in the equations) is the length of time for the NPV of the cumulative cash position to reach zero.
4. Return on investment (ROI or R_i) or discounted return on investment (DROI or R_{Di}). ROI is the ratio of cumulative net cash flow over the project life to the maximum cash outlay. DROI for fracturing economics is the ratio of the NPV of the cumulative net cash flow over the project life to the PV of the cost or investment for the treatment.
5. Rate of return (ROR or i_R) (also called profitability index, PI or I_p). This is the compound interest rate whose discount factors will make the present worth of a project's net cash flow equal to zero. It is equal to the constant effective annual percent earnings on the unrecovered capital remaining in a project. Synonyms include marginal efficiency of capital, internal rate of return, true yield, and discounted-cash-flow method.

These terms, along with other economics-related terms pertinent to fracturing, are defined in the Glossary at the end of this chapter. The application of the specifics to fracturing are discussed in following sections. PV, NPV, DPO, DROI, and PI take into account the time value of money; PO and ROI do not. Accordingly, PV, NPV, DPO, and DROI are associated with a given interest rate. Values calculated for these terms have meaning only when the interest rate, i, used in the calculations is specified. The terms are often written as PV_i, DPO_i, or $DROI_i$. For example, PV_{15} implies PV calculated for a 15% annual interest rate.

The terms presented here represent the criteria generally used throughout the industry for evaluating the economic viability of a project. There is no intent to suggest that one given criterion (or set) is more appropriate than any other. Many of the case studies used as examples in subsequent sections will use NPV as the basic criterion for economic comparisons; however, there is no implication that NPV is the more important criterion. The choice of criteria for making economic decisions will depend heavily on the financial goals of the investor.

PV. PV, or V_P, is related to future value, V_F, by

$$V_P = V_F D^n, \quad \text{(17.14)}$$

where

n = total number of compounding periods over which the interest rate applies,

$$D = \frac{1}{1+(i_A/n_A)} = \text{discount factor}, \quad \text{(17.15)}$$

i_A = annual interest rate (fractional), and
n_A = number of compounding periods per annum.

In Eq. 17.15, the annual discount rate is synonymous with i_A and the periodic discount rate with i_A/n_A.

PV Calculations for Discrete Increments. For some future time period where revenues and expenses are accumulated incrementally, such as monthly payments, the PV for L incremental time periods (i.e., $t=t_1; t_2 \ldots t_L$) would be the sum of the PV's for each increment:

$$V_P\Big]_1^L=\sum_{n=1}^{L}\Delta V_P\Big]_{n-1}^{n}=\sum_{n=1}^{L}\Delta P\Big]_{n-1}^{n}D^n, \qquad (17.16)$$

where

$\Delta V_P\Big]_{n-1}^{n}$ = incremental present value for the time increment (t_{n-1},t_n) and

$\Delta P\Big]_{n-1}^{n}$ = incremental net cash flow for period (t_{n-1},t_n),

$$\Delta P\Big]_{n-1}^{n}=(I\Delta H)\Big]_{n-1}^{n}-E\Big]_{n-1}^{n}, \qquad (17.17)$$

where

$\Delta H\Big]_{n-1}^{n}$ = incremental hydrocarbon production (i.e., ΔN_p or ΔG_p) for the period (t_{n-1},t_n),

$I\Big]_{n-1}^{n}$ = average net hydrocarbon value per unit produced for the period (t_{n-1},t_n) after taxes and royalty; i.e.,

I = $(S-E_T)(1-i_R)$, where

S = unit sales price of hydrocarbon (\$/bbl or \$/Mcf),

E_T = taxes per unit hydrocarbon produced (\$/bbl or \$/Mcf), including local, state, and national, that will reduce the revenue received by the operator,

i_R = royalty interest (fractional) including all overriding royalties that will reduce the net income received by an operator, and

$E\Big]_{n-1}^{n}$ = total expenses (excluding unit taxes on hydrocarbon sales) for the period (t_{n-1},t_n). This includes operator, lifting, overhead, and interest costs, E_i, plus all other miscellaneous expenses attributed to the well or the treatment for the period (t_{n-1},t_n). Expenses for interest costs, E_i, on monies expended for the treatment—i.e., for the period (t_{n-1},t_n)—are

$$E_i\Big]_{n-1}^{n}=\frac{i_A}{n_A}\left(C_T-\sum_{j=1}^{n}\Delta P\Big]_{j-1}^{j}\right), \qquad (17.18)$$

where C_T=total cost of the fracturing treatment.

The above discussion presents a general concept of the factors that should be considered for computing PV's. Determining values for P_{n-1}^{n} may require expressions that differ somewhat from those given. For example, it may be more appropriate to consider lifting cost expenses, E_ℓ, on a per-unit-hydrocarbon-production basis rather than for over a given time increment. In such a case, one may have lifting expenses expressed in terms of an average dollars per barrel instead of in dollars per month. For these cases, lifting costs would then be omitted from the E_{n-1}^{n} term, and I would be expressed as

$$I_\ell=(S-E_T)(1-i_R)-E_\ell. \qquad (17.19)$$

Thus, when the components necessary for determining present worth are considered, the expressions should be structured appropriately to take into account the type of information available and whether certain income and/or expenses are expressed in terms of income or expense per unit of hydrocarbon production (e.g., dollars per barrel, dollars per thousand cubic feet) or in terms of incremental time periods (dollars per month).

Continuous PV. If one considers that funds are continuously flowing in and out of business and that earnings are immediately used to earn additional funds, it may be more appropriate to use a continuous, rather than a discretized (or incremental), discount factor to compute PV. Here, the time increments become relatively small and thus the number of compounding units becomes very large.

Simplified Methods for Computing PV's. Computation of PV's for cash flow streams that change with time (which is the usual case for oil and gas wells) are generally not amenable to using standard discount tables or curves. They often require direct use of Eq. 17.16—summing the incremental values. To do this, it is necessary to have the net cash flow for each future time increment. For certain types of decline behavior, however, such as constant-percentage, hyperbolic, and harmonic, these procedures can be simplified by using the decline-rate expressions for cumulative production (Eq. 17.4, 17.8, or 17.12) for specific time increments to develop PV equations that do not require the incremental net-cash-flow data.

Consider the situation for constant net hydrocarbon values per unit volume (I=constant), constant monthly expense (E=constant), monthly time increments, and producing rates expressed in barrels of oil per month. Coupling the decline equations with Eq. 17.16 resulted in Eqs. 17.20 through 17.23 for each of the three decline cases.

For the constant-percentage-decline discrete case,

$$V_p\Big]_1^L=\frac{A(e^a-1)}{a}\left[\frac{1-B^{(L+1)}}{1-B}-1\right]-D\Big]_1^L E \qquad (17.20)$$

Here, the terms that involved summations were in the forms of geometric series that could be reduced to the simpler forms shown.

For the constant-percentage-decline continuous case,

$$V_P\Big]_0^t=\frac{I_\ell q_i(B^t-1)}{\ln B}. \qquad (17.21)$$

For the hyperbolic-decline, discrete case,

$$V_P\Big]_1^L=\frac{A}{(1-m)a_i}\left[D-D^L(1+ma_iL)^{(m-1)/m}+(D-1)\sum_{n=1}^{L-1}D^n(1+ma_in)^{(m-1)/m}\right]-D\Big]_1^L E. \qquad (17.22)$$

For the harmonic-decline, discrete case,

$$V_P\Big]_1^L=\frac{A}{a_i}\left[D^L\ln(1+a_iL)+(1-D)\sum_{n=1}^{L-1}D^n\ln(1+a_in)\right]-D\Big]_1^L E. \qquad (17.23)$$

In Eqs. 17.20 through 17.23,

$$A=q_iI, \qquad (17.24)$$

$B=De^{-a}$,

L=number of monthly increments, and

$$D\Big]_1^L=\left[\frac{1-D^{(L+1)}}{1-D}-1\right]. \qquad (17.25)$$

Expressions like these can be programmed easily on small computers or programmable calculators to compute PV.

The data contained in Table 17.2 show the first 6-month increments from the three decline examples presented in Sec. 17.5 (Fig. 17.10). It is important to note the significance of early-time cash flow on PV's and the differences resulting from different production-decline performance. Figs. 17.13 and 17.14 show accumulated cash

TABLE 17.2—CASH FLOW AND PV's (FIRST 6 MONTHS) FOR CONSTANT PERCENTAGE, HYPERBOLIC, AND HARMONIC DECLINE

Initial production rate, q_i = 336 BOPD, abandonment rate, q_a = 5 BOPD, ultimate recovery, N_p = 722,000 bbl, oil price = $25/bbl, taxes = $5/bbl, royalty interest = 0.875, operating expenses = $3,000/month, annual discount rate = 15.0%.

Time *t* (months)	Instantaneous Rate at *t* (BOPD)	Incremental Production (bbl/month)	Incremental Future Value ($)	Incremental PV ($)	Cumulative Production (bbl)	Cumulative Future Value ($)	Cumulative PV ($)
Constant Percentage: a = 0.0139540322 cycles/month							
1	331	10,156	174,729	172,572	10,156	174,729	172,572
2	327	10,015	172,266	168,039	20,171	346,995	340,611
3	322	9,876	169,838	163,625	30,047	516,834	504,236
4	318	9,740	167,443	159,326	39,787	684,277	663,563
5	313	9,605	165,081	155,140	49,391	849,358	818,702
6	309	9,471	162,751	151,061	58,863	1,012,109	969,764
Hyperbolic Decline: a_i = 0.033853855, m = 0.7, b = 0.0000528182863							
1	325	10,057	172,994	170,858	10,057	172,994	170,858
2	314	9,730	167,276	163,171	19,786	340,269	334,028
3	305	9,421	161,859	155,938	29,207	502,127	489,966
4	295	9,127	156,729	149,131	38,334	658,856	639,097
5	286	8,849	151,865	142,719	47,184	810,721	781,816
6	278	8,586	147,247	136,671	55,769	957,968	918,487
Harmonic Decline: a_i = 0.059600949, b = 0.0000058278038							
1	317	9,934	170,841	168,732	9,934	170,841	168,732
2	300	9,390	161,326	157,367	19,323	332,167	326,099
3	285	8,903	152,798	147,209	28,226	484,965	473,308
4	271	8,464	145,112	138,077	36,690	630,077	611,385
5	259	8,066	138,149	129,829	44,755	768,225	741,214
6	247	7,703	131,811	122,343	52,459	900,036	863,557

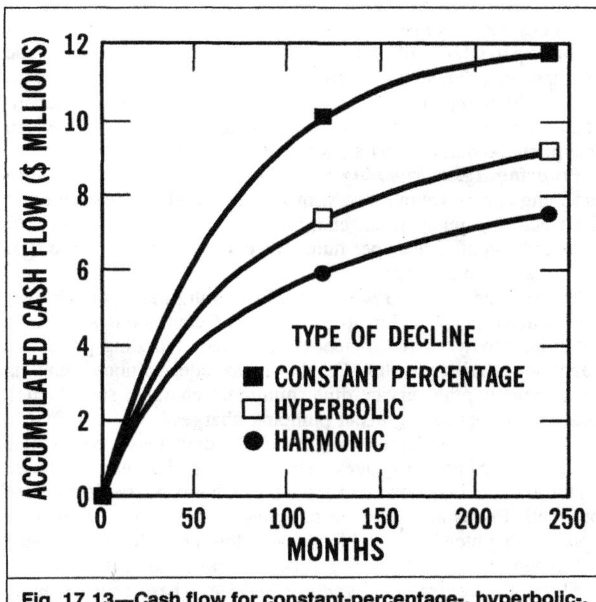

Fig. 17.13—Cash flow for constant-percentage-, hyperbolic-, and harmonic-decline examples.

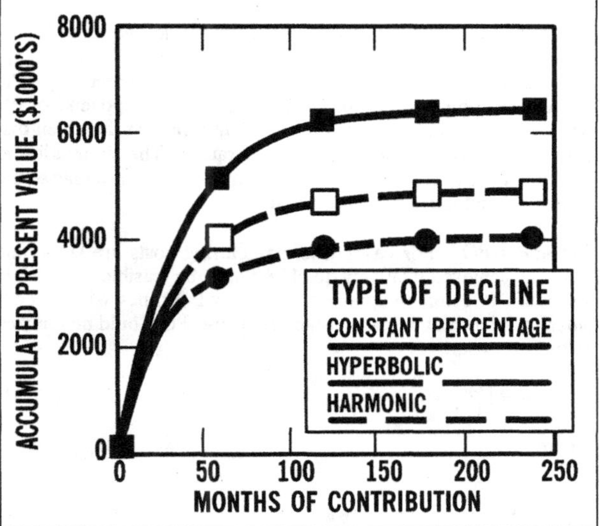

Fig. 17.14—Cumulative PV for constant-percentage-, hyperbolic-, and harmonic-decline examples.

flow and PV's for the first 240 months of the Sec. 17.5 example cases. Fig. 17.13 shows that cash flow is still increasing significantly after 250 months for both the hyperbolic- and the harmonic-decline cases. But as Fig. 17.14 shows, the accumulated PV's for all cases practically reached full value in approximately 120 to 150 months.

Also, there is a significant difference in the PV's realized for the three cases, even though initial and abandonment rates and ultimate recovery are identical for each case.

NPV. This is the term ultimately used to investigate PV economics resulting from a treatment:

$$V_{nP}=V_{Pf}-V_{Puf}-C_T, \qquad (17.26)$$

where C_T is the PV of the treatment cost and the subscripts f and uf denote the fractured and unfractured cases, respectively. The compounding period (i.e., number of time increments) for computing V_{Pf} usually will be different from that for V_{Puf}. NPV includes the entire net cash flow between initial, q_i, and abandonment, q_a, producing rates.

ROI and DROI$_i$. The term ROI does not account for the time value of money. For a fracturing treatment, the ROI is computed by

$$P_c=\frac{P\Big|_1^{L_a}}{C_T}, \qquad (17.27)$$

TABLE 17.3—PI OR ROR, EXAMPLE 17.6

Time (years)	Cash Flow ($1,000)	PV_{30} ($1,000)	PV_{35} ($1,000)	PV_{40} ($1,000)
0	-120	-120	-120	-120
1	75	57.7	55.5	53.5
2	60	35.5	32.9	30.6
3	45	20.5	18.3	16.4
4	30	10.5	9.0	7.8
5	15	4.0	3.3	2.8
		+8.2	-1.0	-8.9

where

$$P\Big]_1^{L_a}=\sum_{n=1}^{L_a} P_{n-1}^n \quad\text{.......(17.28)}$$

and L_a corresponds to the number of time increments at the time of abandonment, t_a. Interest expenses, E_i, are not included in the P terms.

DROI_i is somewhat a measure of capital efficiency and in general approximates the present worth realized per unit of discounted dollar invested. Thus, it takes into account the time value of money.

For a fracturing treatment,

$$R_{Di}=\frac{V_P}{C_T}, \quad\text{.......(17.29)}$$

where

$$V_P=\sum_{n=1}^{L_{af}} (V_{Pf})_{n-1}^n-\sum_{n=1}^{L_{uf}} (V_{Puf})_{n-1}^n. \quad\text{.......(17.30)}$$

Note that here the total time increments are not the same for the fractured- and unfractured-case PV terms. Subscripts f and uf imply the fractured- and unfractured-case increments, respectively. The interest cost on C_T should be included in the E_i components of PV when V_P is computed. The PV's are determined at the given discount factor corresponding to interest rate, i. The summation to increment L corresponds to the total number of increments to the time of abandonment for each case.

PO and DPO$_i$. For cases where fracturing costs are small and PO's are achieved relatively quickly, it may be feasible, practically speaking, to compute approximations for PO times without considering interest and discount factors. Here, PO would be realized at time t_p when

$$\left|C_T-P\Big]_1^{L_P}\right|=0, \quad\text{.......(17.31)}$$

where

$$P\Big]_1^{L_P}=\sum_{n=1}^{L_P} (P_{f,n-1}^n-P_{uf,n-1}^n). \quad\text{.......(17.32)}$$

In the above equations, the interest expense component, E_i, was not included in the P term. Also, PO does not account for the time value of money.

In many cases, it may be more appropriate to consider discount and interest factors to develop a more realistic discounted-cash-flow profile. Here, PO time, t_p, can be approximated by considering the PV's of the cash flow. DPOI_i would occur at a time t_{pi} when

$$\left|C_T-V_{nP}\Big]_1^{L_P}\right|=0. \quad\text{.......(17.33)}$$

Here, the interest cost on C_T should be included in the E_i component of PV when computing

$$V_{nP}\Big]_1^{L_P}.$$

ROR or PI. ROR or PI may be thought of as the discount rate that sets the sum of the discounted cash inflows equal to the sum of the discounted cash outlays.

For fracturing projects, it could be approximated by determining the interest rate, i, at which

$$C_T=V_{nP}\Big]_1^{L_P}. \quad\text{.......(17.34)}$$

To determine PI values, it may be necessary to iterate through several sets of computations and then use an interpolative scheme to arrive at an estimate for PI. For example, presume that a fracturing treatment that cost $120,000 yielded the net cash flow shown in Table 17.3. In this example, we will assume that all discounting is done annually at the end of each year throughout the 5-year life of the well. The total present worth for three discount factors shows that the PI is very near 35%. Linear interpolation would yield an estimate of PI≈34.5%.

17.7 Elements of Fracturing-Treatment Costs

Fracturing treatment costs involve primarily pumping and blending charges and material costs for fracturing fluids, fluid additives, and propping agents. In some cases, associated activities, such as well pulling costs and tubular rentals, contribute significantly to the total treatment costs. Some of the types of costs associated with fracturing treatments both from stimulation service companies and other associated contractors and suppliers are discussed in this section.

Stimulation Service Company Costs. Even though the structures of service company price schedules may vary somewhat between companies, treatment costs usually include the components discussed below.

Fracturing Pumping Equipment. Pump truck charges occur for all trucks, except pressure-multiplier pumps, per well for a period usually up to 4 hours of continuous on-location service per hydraulic horsepower ordered. Hydraulic-horsepower pumping charges increase with pumping-pressure-increment increases. Other costs include additional pumping time, nonpumping service time, minimum pump truck charges, and standby pumping equipment.

Propping-Agent Pumping Charge. These charges apply when propping agents are pumped with any fluid and are in addition to the fracturing pump truck charges, per one hundred pounds or fraction thereof. Prices per hundredweight are based on the type and size of the proppant.

Pressure-Multiplier Pumps. Pressure-multiplier pump charges are incurred per well for up to 4 hours of continuous service on location, per hydraulic horsepower ordered. Pumping prices are based on pumping pressure. Other costs include additional pumping time, nonpumping service time, minimum charges, standby unit charges, and propping-agent pumping charges.

Blender Services. These charges include continuous proportioning and mixing of propping agent and fracturing fluid and are based on the average injection rate for the first 4 hours or fraction thereof per well. Prices are based on pumping rate. Other costs include nonpumping blender time; blender standby; and other blender and equipment charges, such as paddle mixers and densitometers.

Slurry-Concentration Handling Service. These charges apply when propping agents are pumped with any fluid and are in addition to blender and propping-agent pumping charges. Prices depend on propping-agent concentration.

Auxiliary Stimulation Equipment. These items include sand-handling equipment, radioactive material for tagging sand, wellhead protective injection equipment, manifolds, nitrogen, CO_2 equipment, flowmeters, fracturing support units, special equipment (tanks, transfer pumps, valves, and wellheads), ball-sealer equipment, treating connections left on location, and sand concentrators.

Stimulation Technical Support. This includes field support, such as treatment monitoring services and mobile laboratories. It also includes laboratory analyses, consulting, and design support.

Transportation Charges. These charges include mileage, proppant delivery charges, transport units, and delivery charges for product sales without service.

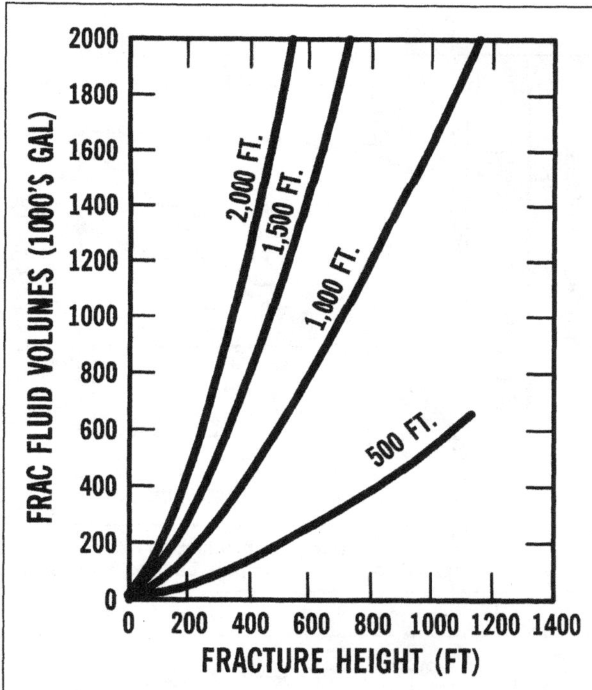

Fig. 17.15—Typical base-case fracture treatment design curves used in Natl. Petroleum Council 1980 tight-formation-gas study.[3]

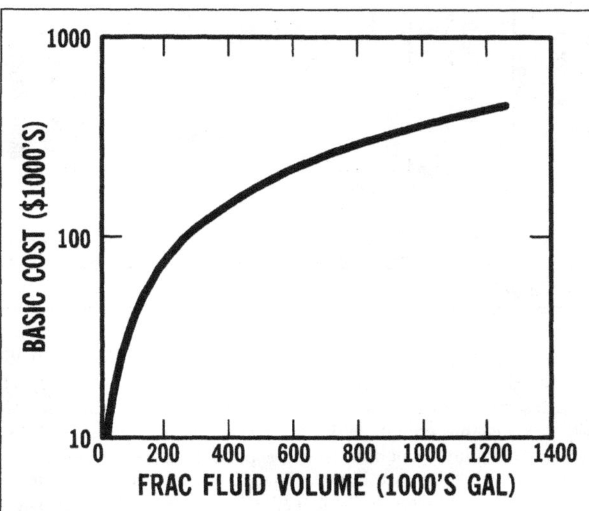

Fig. 17.16—Typical base-case fracture treatment cost curves used in Natl. Petroleum Council 1980 tight-formation-gas study.[3]

Miscellaneous Service Charges. These include handling customer-furnished materials, incomplete services, personnel expenses, group stimulation treatments, combination acidizing and fracturing jobs, multizone fracturing, license fees, and return of materials.

Fracturing Material and Material Services. These involve gelling agents and all additives for the fracturing fluid, nitrogen, CO_2, services for special fluid systems, propping agents, diverting agents, ball sealers, and license fees.

Other Associated Treatment Costs. These include site-preparation costs, tank rentals, fluid hauling, costs for materials not supplied by the service companies, workover rig and equipment costs, tubular rentals, perforating, and special logs and tests. They also include materials not furnished by the stimulation service company.

TABLE 17.4—FORMATION, RESERVOIR, DECLINE PERFORMANCE, AND WELL DATA FOR BASE-CASE EXAMPLE 17.8

Depth, ft	6,100
Net pay, ft	10
Permeability, md	5
Porosity, %	20
Water saturation, %	20
Oil viscosity at reservoir conditions, cp	1
Oil FVF	1.1
Oil recovery factor, %OIP*	40
Initial reservoir pressure, psi	2,500
Producing wellbore bottomhole pressure, psi	500
Spacing, acres/well	160
Wellbore diameter, in.	10.25
Completion mode	Openhole
Tubular length, ft	6,000
Tubular ID, in.	4.670
Prefracture initial producing rate, BOPD	$2 \times kh$
Post/prefracture production increase	Per Prats[8]
Abandonment rate	
(both pre- and postfracture), BOPD	5
Decline performance	Constant-percentage
(both pre- and postfracture)	decline
Recoverable oil at abandonment	
(both pre- and postfracture), bbl	722,000

*OIP = oil in place.

TABLE 17.5—FRACTURE DESIGN DATA FOR BASE-CASE EXAMPLE 17.8

Fracture orientation	Vertical
Fracture closure pressure, psi	4,000
Elastic modulus, psi	4×10^6
Fracture height, ft	180
Fracture-propagation model	Per Perkins and Kern[16]
Pumping rate, bbl/min	40
Tubulars, ft	6,000
Casing ID, in.	4.670
Tubular volume, gal	5,340
Maximum allowable treating pressure, psi	8,000
Fracturing fluid properties	
Density (without proppant), lbm/gal	8.4
Apparent viscosity in the fracture, cp	100
Fluid-loss coefficient, ft/min$^{1/2}$	0.002
Spurt loss, gal/ft^2	0.01
Pipe friction (in 4.620-in. ID), psi/1,000 ft	
At 1 bbl/min	20
At 40 bbl/min	95
At 100 bbl/min	270
Propping agent data	
Average concentration in the fluid, lbm/gal	3.0
Average fracture conductivity, md-ft	3,000

Price Discounts, Competitive Bids. Price discounting or competitive bidding is practiced to accommodate the prevailing economic climate. Under certain conditions, treatment costs can be reduced significantly below computed price schedule rates. When making economic forecasts, one should take these situations into account.

Methods for Estimating Fracturing Treatment Costs. For economic optimization studies where a large number of fracture designs must be investigated, it is necessary to develop treatment-cost-estimation methods to accommodate this need. One method is to develop a job-pricing structure similar to that used by stimulation service companies for computer or programmable-calculator execution. Another is to develop "typical" treatment design and cost curves for a specific formation, such as those shown in Figs. 17.15 and 17.16. These data were used in the 1980 Natl. Petroleum Council study on tight-gas reservoirs.[3] They can also be programmed for automatic interpolation and provide a method for rapidly generating massive sets of estimates at minimal cost. However, they do require curves that are applicable to the particular range of formations pertinent to any specific study.

TABLE 17.6—ECONOMIC DATA FOR BASE-CASE EXAMPLE 17.8

Company working interest, %	100
Company net interest, %	87.5
Crude oil price, $/bbl	25
Crude oil taxes, $/bbl	5
Well operating costs, $/month	3,000
Annual discount rate, %	15.0
Well drilling costs, $/well	145,000
Fracturing treatment costs	
Fracturing fluid, $/gal	0.50
Propping agents, $/100 lbm	
For 3,000-md-ft conductivity	8
For 6,000-md-ft conductivity	60
For 7,000-md-ft conductivity	70
Associated fixed costs, $/well	5,000

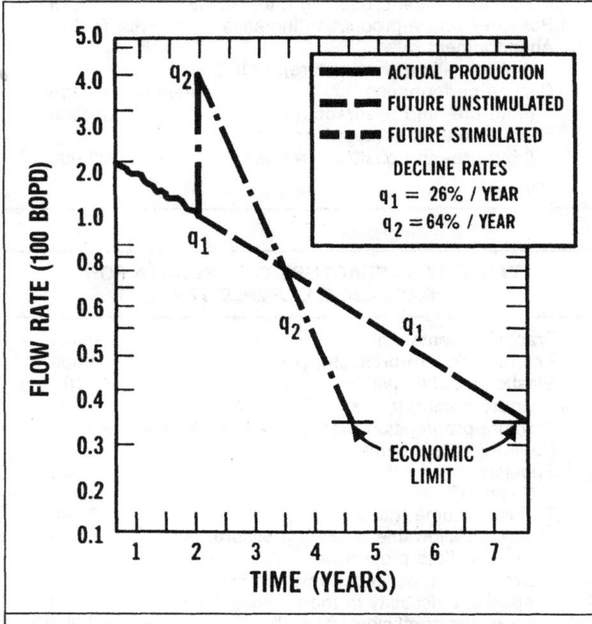

Fig. 17.17—Effect of fracture stimulation constant-percentage decline (after Holditch[17]).

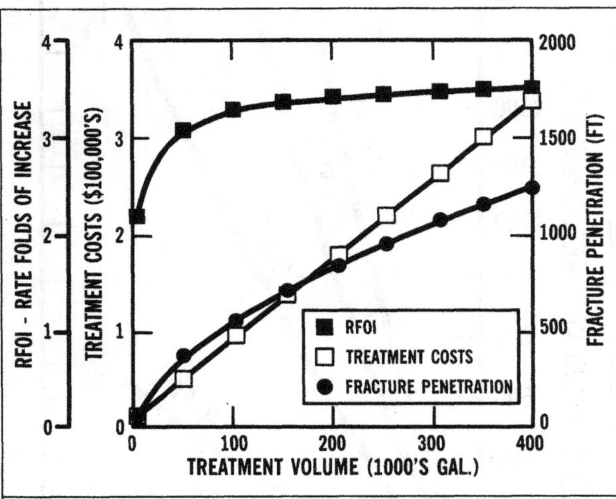

Fig. 17.18—Fracture penetration, production rate increase, and treatment cost vs. volume pumped, base-case example.

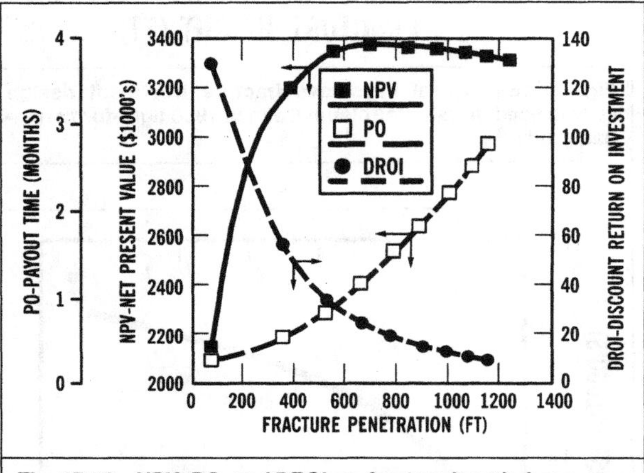

Fig. 17.19—NPV, PO, and DROI vs. fracture length, base-case example.

17.8 Fracturing-Treatment Economics—Base-Case Example

This section discusses the application of the general concepts of economic studies presented in Sec. 17.3. Consider a well with the formation, reservoir, performance behavior, and wellbore properties given in Table 17.4. The fracturing design data are shown in Table 17.5, and the economic criteria are listed in Table 17.6. Fracturing prices from 1985 price schedules were used to estimate treatment costs. A Perkins-Kern[16] fracture-propagation model with constant height was used to compute fracture lengths for various treatment volumes and rates. In the computations, proppant concentrations yielded the design fracture conductivity for the entire fracture length. Prats'[8] method (Fig. 17.7) was used to estimate producing-rate increase results from fracturing. Both pre- and post-fracturing production behavior followed constant-percentage decline, such as the example shown in Fig. 17.17, as presented by Holditch.[17]

The results are depicted in Figs. 17.18 and 17.19. Fig. 17.18 shows the fracture penetration, increase, and total treatment costs associated with the different treatment volumes. Fig. 17.19 shows the NPV's, PO's, and DROI achieved for various fracture lengths. For these conditions, the maximum PV ($3.387 million) occurs for a fracture penetration of 730 ft [220 m]. Note that the shorter PO times and the higher DROI values occur for small treatment volumes and short fractures. This prevailed in practically all the subsequent example cases presented in this chapter.

Another item of interest is the comparison of PV's with that achieved from hyperbolic or harmonic decline performance. The postfracture PV profiles for the three cases at the optimum point (e.g., constant-percentage decline PV of $6.513 million at 730-ft [220-m] fracture penetration) correspond to those previously discussed in Sec. 17.6 and shown in Fig. 17.14. The prefracture values and NPV's are listed in Table 17.7. The significant difference in these values emphasizes the need to apply the appropriate performance-prediction models to investigate economics.

Another factor that may play a significant role in economic studies is the choice of fracture-propagation models. Geertsma and Haafkens[18] demonstrated that different models will yield different fracture length and width results for similar treatment design conditions. One should be aware of how these differences may affect the evaluation of results yielded from economic studies and use the hydraulic fracturing design simulator that applies to the particular cases.

17.9 Economic Analysis Methods and Example Studies

The example in Sec. 17.8 shows the types of results obtained when economics is applied to fracture treatments. This section discusses examples that show how one might address some of the significant questions pertinent to fracture treatment design from an economics standpoint. In all the examples, PV was chosen as the economic criterion of interest. The data in Tables 17.4 through 17.6 serve as the base-case data. Particular parameters were varied to conduct

TABLE 17.7—NPV RESULTS FROM FRACTURING FOR CONSTANT-PERCENTAGE, HYPERBOLIC, AND HARMONIC DECLINE, BASE-CASE EXAMPLE 17.8

	Constant Percentage ($1,000)	Hyperbolic ($1,000)	Harmonic ($1,000)
V_{Pf}	6,513.9	5,018.7	4,134.3
V_{Puf}	2,983.5	2,546.5	2,289.4
Treatment cost	142.9	142.9	142.9
NPV	3,387.4	2,329.3	1,702.0

Production data

	Postfracturing	Prefracturing
Initial rate, q_i, BOPD	336	100
Abandonment rate, q_a, BOPD	5	5
Ultimate recovery, N_p, bbl	722,000	722,000
Producing life, months		
Constant percentage	302	750
Hyperbolic	761	1,227
Harmonic	1,112	1,508

the different investigations. The parameters that were changed are mentioned in the discussions for each study; unless stated otherwise, the remaining parameters conform to the data in the tables.

Scope of Studies. Examples are shown that address optimizing well spacing for hydraulically fractured formations, the effect of permeability on fracture penetration requirements, the design of propping-agent programs to maximize NPV, the effects of net pay on incremental economics, the impact of fracture vertical growth on NPV, and the design of fluid-loss-additive programs. Discussion is also given to previous studies[1] on optimizing pumping rate. The results of other investigators are included for the various examples.

Well Spacing. During the life of almost every field, the question of optimum spacing arises, either during the initial development of a field, soon after discovery, or later in a field's life, when it appears that the initial spacing may be too sparse and infill drilling prospects start to look attractive. In reservoirs where fracturing is an integral part of the development program, the questions of well density and appropriate fracture-length requirements are paramount: should a field be developed on a low-well-density spacing with long, deeply penetrating fractures, or is development more economical with a higher density and shorter fractures? And what is the economically optimum combination of well spacing, fracture length, and conductivity?

Optimization can range from simple approaches to those requiring the use of highly sophisticated economic, fracturing, and reservoir-performance simulators. The approach depends on the complexity of the reservoir, the fracturing requirements, and the economic parameters associated with a given field. No single procedure is considered universally applicable to investigating field-development optimization other than comparing the results that evolve from a number of scenarios and selecting the set that provides the most attractive economic returns. The following example is a very basic approach, but it shows some of the necessary elements involved.

Consider a field with these average reservoir properties: $h=10$ ft [3 m], $k=10$ md, and $q_i=200$ BOPD [32 m^3/d oil]. The other parameters are given in Tables 17.4 through 17.6. Steady-state flow dominates the producing life, and performance follows a constant-percentage decline both before and after fracturing.

Studies were made with a Perkins-Kern[16]-type model with constant fracture height of 180 ft [55 m] for well spacings of 20, 40, 80, 160, 320, and 640 acres/well [8, 16, 32, 65, 130, and 260 ha/well]. The data in Fig. 17.20 show the results in terms of NPW vs. fracture length for each spacing from 20 to 640 acres/well [8 to 260 ha/well]. The NPV's represent the contribution from all wells in a 640-acre [260-ha] section. For example, with drilling and development costs of $145,000/well (excluding fracturing), the total drilling cost normalized to a 640-acre [260-ha] section for 320-, 160-, and 80-acre [130-, 65-, and 32-ha] spacings would be $290,000, $580,000, and $1,160,000, respectively.

Fig. 17.20 shows that the maximum increase in present worth per section is achieved with a spacing of 160 acres/well [65 ha/well].

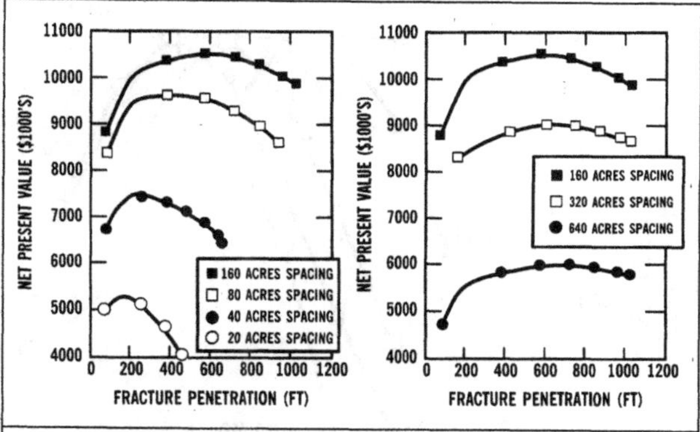

Fig. 17.20—Total NPV per section vs. fracture length for various well spacing.

TABLE 17.8—OPTIMUM PV RESULTS FOR 10-md PERMEABILITY, 10-ft SPACING STUDY, EXAMPLE CASE 17.9

Spacing (acres/well)	Optimum Fracture Penetration (ft)	Optimum* NPV ($1,000)	Maximum Optimum NPV (%)
20	160	5,319.1	50.5
40	200	7,513.2	71.4
80	400	9,627.2	91.5
160	550	10,525.8	100.0
320	610	9,054.3	86.0
640	730	6,026.1	57.2

*Normalized to 640 acres.

TABLE 17.9—OPTIMUM WELL SPACING WITH FRACTURING, EXAMPLE CASE 17.9

Permeability (md)	Net Pay (ft)	Optimum Spacing (acres/well)
1	10	20 to 40
	50	40
5	10	80
	50	80
10	10	160
	50	160

Notice the trend of the values of optimum fracture penetration associated with each drainage area. They are approximately ⅓ to ½ the drainage radius for spacings up to 160 acres/well [65 ha/well]. For this particular study (i.e., permeability of 10 md, etc.), fracture lengths beyond 600 to 700 ft [185 to 215 m] are not economical, regardless of the spacing size.

To study the effect of permeability and thickness variations, the above process was repeated for different sets of permeability (1, 5, and 10 md) and net pay thickness (10 and 50 ft [3 and 15 m]) values. The results are given in Tables 17.8 and 17.9 and Fig. 17.21. In Fig. 17.21 the percent of maximum achievable NPV is the optimum NPV from each spacing as a percent of the maximum for the set. This is the case for each permeability/net-pay set. For example, Table 17.8 lists the optimum values from Fig. 17.20 and the ratio of the value for each spacing as a percent of the maximum value for the set. The data from the table correspond to the 10-md/10-ft [3-m] net-pay curve in Fig. 17.21. The results for all cases are summarized in Table 17.9. For this study, net pay had very little effect except for the 1-md case.

The above example demonstrates one of many ways to investigate the optimum well density and fracture lengths for various combinations of permeability and net pay. It must be emphasized that the results shown are applicable only for cases that fall within the range of the data and assumptions prescribed for the examples

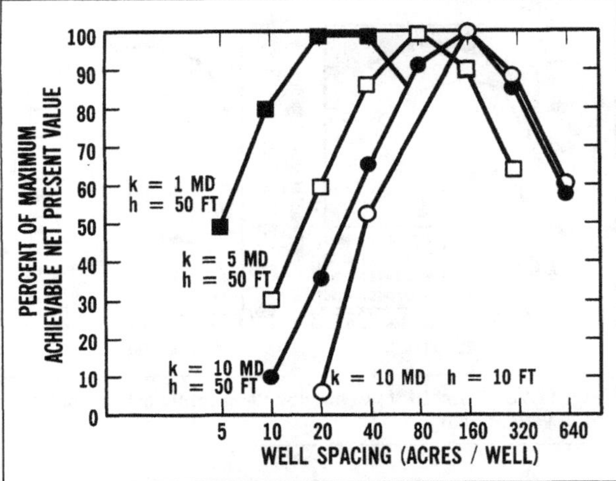

Fig. 17.21—Optimum NPV vs. well spacing for 1-, 5- and 10-md example data.

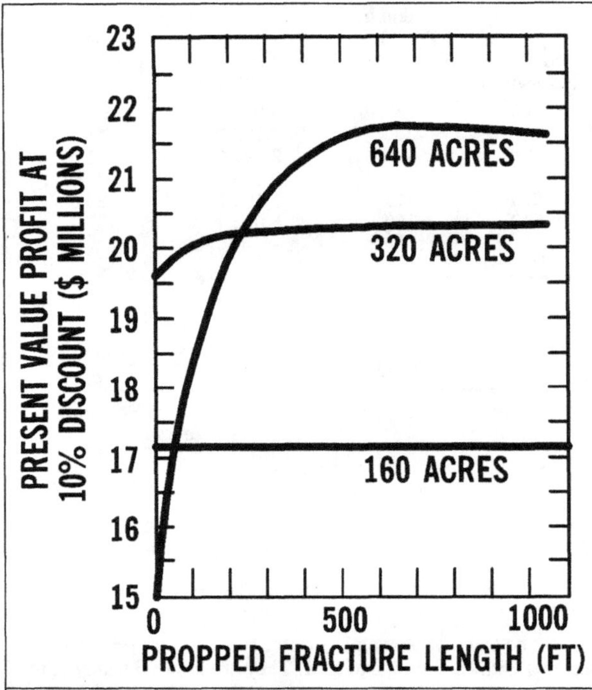

Fig. 17.22—Optimization graph for a 0.1-md-permeability Vicksburg sandstone (after Holditch *et al.*[12]).

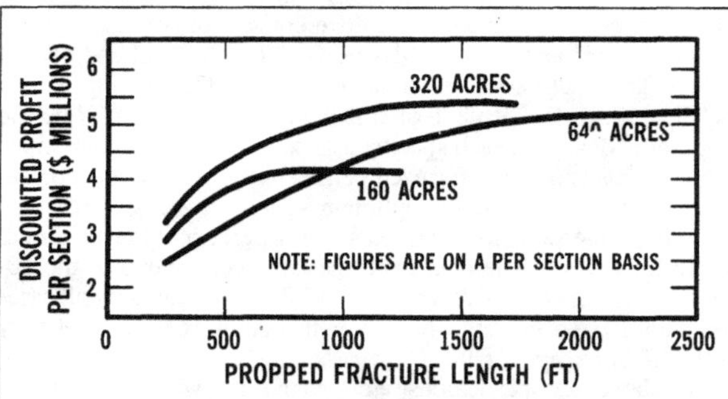

Fig. 17.23—Discounted PV profit vs. fracture length: 0.04-md permeability, 10% discounted Cotton Valley lime (after Kozik and Holditch[19]).

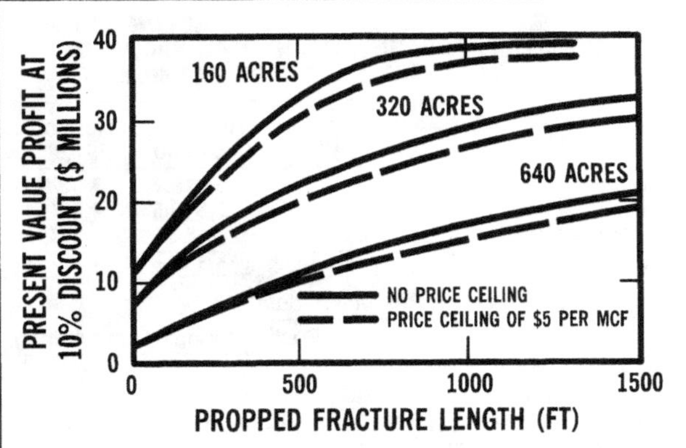

Fig. 17.24—Optimization graph for a 0.02-md-permeability Edwards Reef reservoir (after Holditch *et al.*[12]).

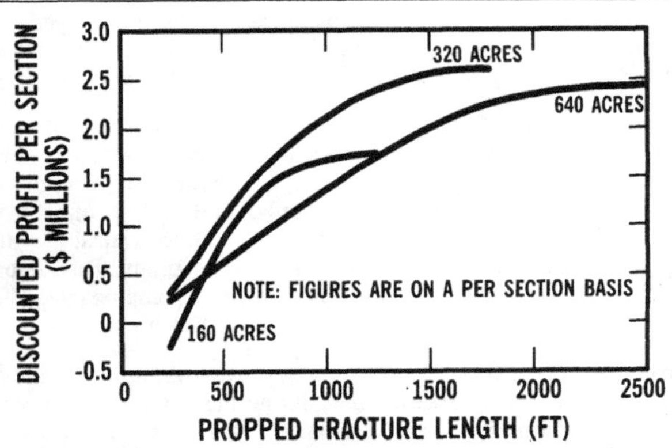

Fig. 17.25—Discounted PV profit vs. fracture length: 0.01-md permeability, 10% discounted Cotton Valley lime (after Kozik and Holditch[19]).

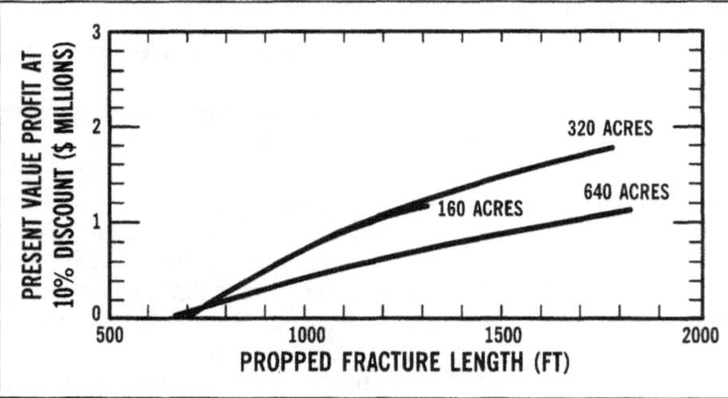

Fig. 17.26—Optimization graph for a 0.005-md-permeability Fahler sandstone reservoir (after Holditch *et al.*[12]).

shown. Other conditions and reservoirs may yield significantly different results. For example, results of studies by Holditch *et al.*[12] and Kozik and Holditch[19] in low-permeability gas formations are shown in Figs. 17.22 through 17.26. For these cases, transient flow dominated the performance-decline behavior. And for these cases (but not for the examples), increases in recoverable reserves by fracturing contributed significantly to the economics. Other studies by the Natl. Petroleum Council,[3] Rosenberg *et al.*,[20] and Brashear *et al.*[21] show more complex approaches to economical spacing investigations in tight-gas formations.

Another factor in well-spacing economics for tight-gas formations involves fracture azimuth, as alluded to in the NPC study and by Jeu[22] and Smith.[23] In azimuth studies, as opposed to many situations where income acceleration dominates the picture, the impact

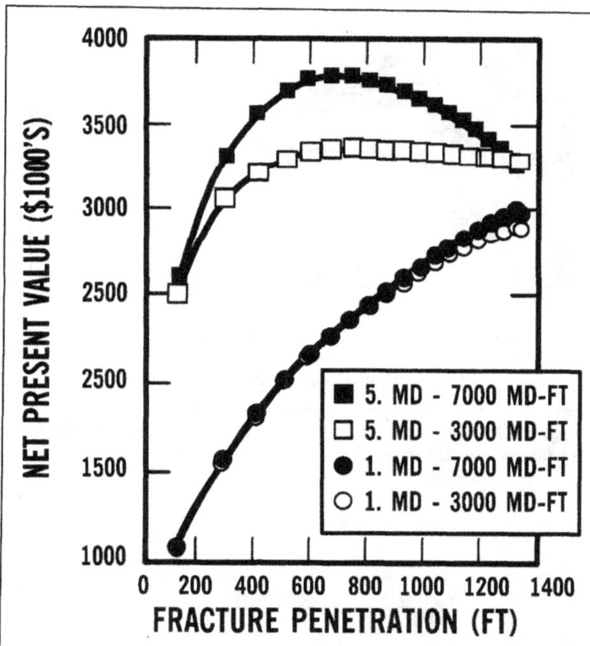

Fig. 17.27—NPV vs. fracture length: 1- and 5-md permeabilities and 3,000- and 7,000-md-ft fracture conductivities.

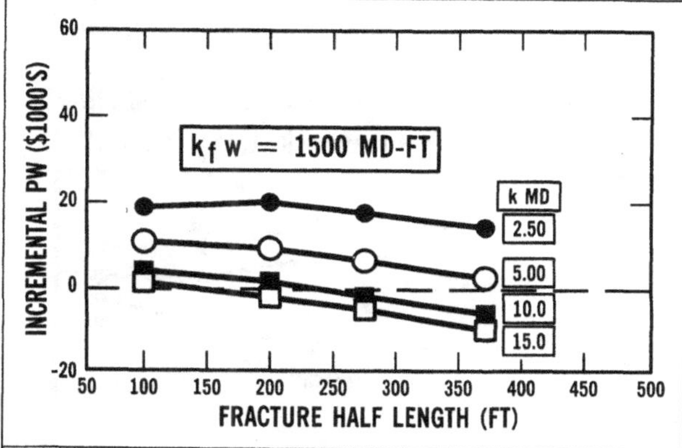

Fig. 17.28—Incremental PV vs. fracture half-length: $k_f w$ = 1,500 md-ft, 2.5- to 15.0-md permeabilities (after Britt[24]).

of late-time well interference from overlapping drainage areas of inefficient well patterns becomes important. In such cases, the economics may not be dominated as strongly by early-time producing rates as emphasized previously.

Formation Permeability, Fracture Penetration, and Fracture Conductivity. As stated previously, one of the most important considerations in fracturing economics is formation permeability. As a prime factor affecting producing rate, permeability has a major impact on cash flow and economics. Many investigators have alluded to relationships between formation permeability, fracture conductivity, and fracture penetration as they relate to producing-rate increase and recoverable reserve increases. The examples given here show that even without reserve increases, the cash flow from rate acceleration has a significant effect on economics. Cases were examined for 1-, 5-, and 10-md permeabilities; 10-ft [3-m] net pay intervals with initial producing rates of 20, 100, and 200 BOPD [3.2, 16, and 32 m^3/d oil], respectively; and fracture conductivities of 3,000 and 7,000 md-ft [915 and 2135 md·m]. Average proppant prices for fracture conductivities of 3,000 md-ft [915 md·m] were $8.00/CWT and for conductivities of 7,000 md-ft [2135 md·m] were $70.00/CWT. Other pertinent data are given in Tables 17.4 through 17.6. The results of NPV vs. fracture penetration for the 1- and 5-md cases are shown in Fig. 17.27.

For the 1-md case, optimum NPV's are achieved with fracture penetrations approaching the 1,320-ft [400-m] drainage radius. Note that very little economic benefit is realized from the higher fracture conductivity for the 1-md case. The higher cost of the proppant offset the difference in production-rate increase (i.e., 8.61 for 7,000 md-ft [2135 md·m] vs. 6.76 for 3,000 md-ft [915 md·m]). The higher-permeability cases showed maximum NPV's for much shorter fracture penetrations (e.g., on the order of 550 to 700 ft [170 to 215 m]). Here there were significant differences between the results for the 3,000- and 7,000-md-ft [915- and 2135-md·m] fracture conductivities.

An extension of this study to the 10-md case also revealed that the NPV's for 5-md permeability were significantly higher than those achieved for the 10-md case because the rate of increase for 5 md at the optimum fracture lengths were higher than for the 10-md case, even with 7,000-md-ft [2135-md·m] conductivity. These results show that conductivities much greater than 7,000 md-ft [2135 md·m] are required for the 10-md formation, even if the costs for doing so may initially appear to be prohibitive.

The studies by Holditch *et al.*[12] and Kozik and Holditch[19] implied the significant effect that permeability has on economic fracture-length requirements for tight reservoirs where transient flow dominates the performance. By inspecting Figs. 17.22 through 17.26, one can observe that optimum economic fracture-length requirements can exceed 1,500 to 2,500 ft [460 to 760 m] as permeabilities approach the 1- to 10-μd range. This is consistent with the Natl. Petroleum Council study and the data by Elkins[6] (Fig. 17.3). Britt[24] investigated the effects of higher permeability on fracture-length requirements. Fig. 17.28 shows that for his study, with a fracture conductivity of 1,500 md-ft [460 md·m], fracture half-lengths of 200 ft [60 m] provide optimum economics for permeabilities of 2.5 md. Also, length requirements decrease as permeabilities increase, as is consistent with other studies.

Propping-Agent Selection. Achieving the appropriate fracture conductivity is an important part of fracture design. The decision to use sand instead of higher-strength manufactured synthetic proppants and the choice of proppant size and concentration play a significant role in economics. The manufactured proppants (e.g., intermediate or high strength and resin coated) are considerably more costly than sand, and their use often requires economic justification. Several authors[24-29] discussed the impact that fracture conductivity has on postfracturing performance. However, many of these deal with production performance rather than economics *per se*. The examples presented here demonstrate how one might investigate the comparative economics for selecting an optimum proppant design. Examples in Chap. 11 depict alternative methods for economically optimizing proppant selection.

Given a 5-md formation with 10-ft [3-m] net pay, comparisons were made for three proppant programs: sand with an average fracture conductivity of 3,000 md-ft [915 md·m] and cost of $8.00/CWT and two manufactured proppants, one with a conductivity of 6,000 md-ft [1830 md·m] that costs $60.00/CWT and one that provides 7,000 md-ft [2135 md·m] at a cost of $70.00/CWT. The remaining data are given in Tables 17.4 through 17.6. The NPV vs. fracture-penetration results (similar to those shown in Fig. 17.27) were examined to determine optimum NPV points for each case. The results showed that the optimum points for all three conductivity cases occurred at approximately the same fracture penetration

TABLE 17.10—FRACTURING-CONDUCTIVITY-OPTIMIZATION RESULTS FOR 5-md, 10-ft NET PAY, EXAMPLE CASE 17.9

Fracture Conductivity (md-ft)	Optimum Fracture Penetration (ft)	Proppant Cost ($1,000)	Maximum NPV ($1,000)	Increases Over the 3,000-md-ft Case (%)
3,000	700	40	3,387	0
6,000	660	316	3,725	10.0
7,000	680	396	3,804	12.3

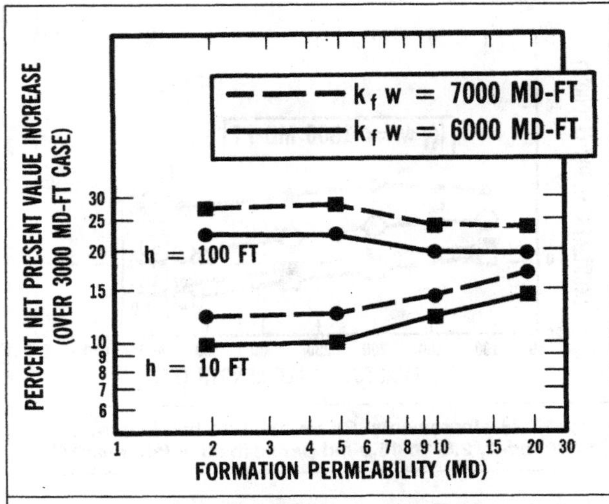

Fig. 17.29—Percent NPV increase over results with 3,000-md-ft vs. formation permeability.

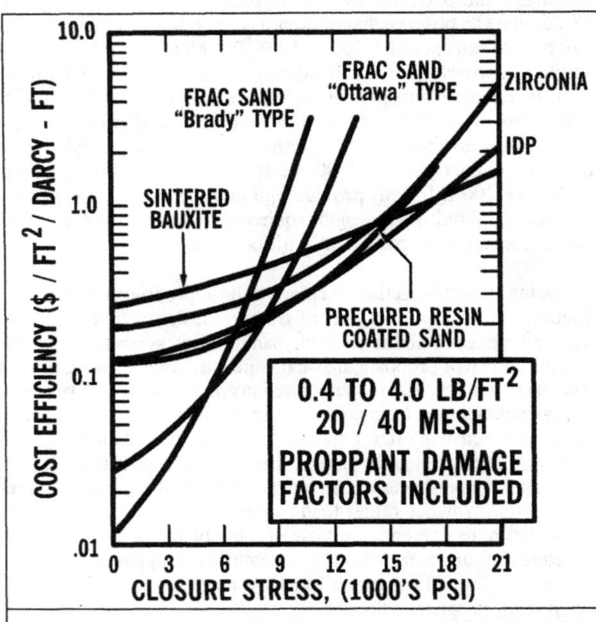

Fig. 17.30—Effect of proppant type on cost efficiency (after Phillips and Anderson[25]).

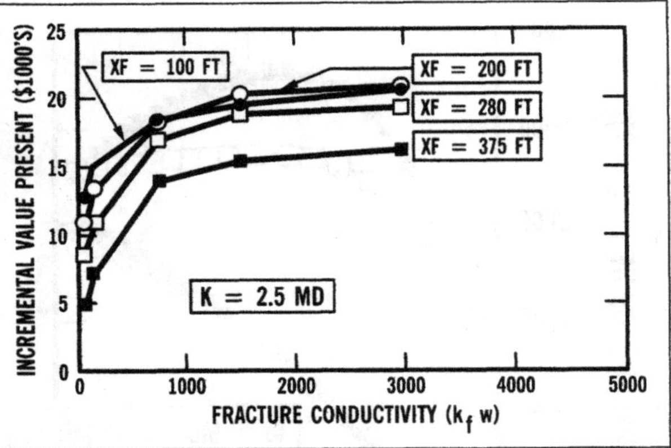

Fig. 17.31—Incremental present worth ($1,000) vs. fracture conductivity for 2.5-md permeability (after Britt[24]).

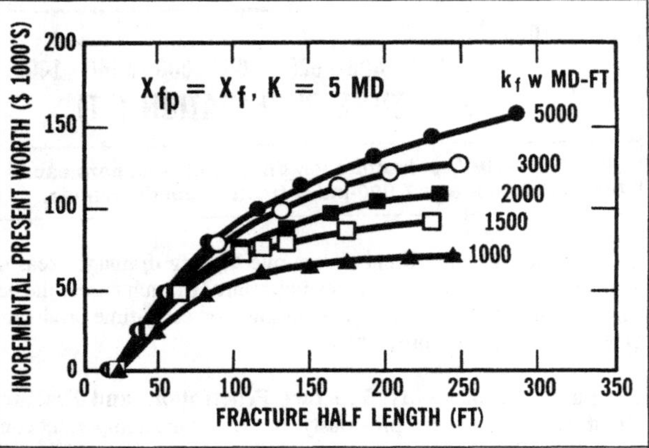

Fig. 17.32—Incremental present worth vs. fracture half-length: $k_f w_f$ = 1,000 to 5,000 md-ft (after Britt[24]).

(≈700 ft [215 m]). The data at the maximum points are summarized in Table 17.10. The total treatment cost differences were essentially comparable to the proppant cost differences. As can be seen, the relatively large expenditure for the higher conductivity yielded significant increases in NPV's.

The study was extended to investigate other permeability levels and the effect of increased net pay. Cases were run for permeabilities of 2, 5, 10, and 20 md and net pays of 10 and 100 ft [3 and 30 m]. The results presented in Fig. 17.29 show the percent increase in optimum NPV achieved over the 3,000-md-ft [915-md·m] case for the different permeabilities and net pays. These values were computed for each of the cases with the optimum (maximum) NPV's from data similar to those shown in Fig. 17.27 and Table 17.10. It is interesting to note that both permeability and net pay have a significant impact on NPV. For the 10-ft [3-m] -net-pay case, NPV increases on the order of 10 to 15%. At each permeability level, there is a significant difference between the 10- and 100-ft [3- and 30-m] -net-pay case, with the 100-ft [30-m] case increasing from 20 to 30%.

Examples of other studies show the significance of fracture conductivity on economics. Data by Phillips and Anderson[25] on studies of the effect of proppant type and closure stress on cost efficiency are shown in Fig. 17.30. Investigations by Britt[24] shown in Figs. 17.31 and 17.32 depict how conductivity affects the overall economics for his studies. Fig. 17.31 indicates that for his conditions, fracture conductivities on the order of 2,000 md-ft [610 md·m] or greater are required to maximize NPV's for the 2.5-md case; Fig. 17.32 shows that for a 5-md case, conductivities in excess of 5,000 md-ft [1525 md·m] may be required. Again, it must be emphasized that the results of an economic study should be construed to apply only to the conditions specified for that study.

Net Pay. In some of the cases discussed, it became apparent that net pay is an important consideration in fracturing economics. It is usually not highlighted in parametric fracturing studies, where the focus is on formation permeability, fracture penetration, and conductivity requirements, so the net pay is normalized out of performance trends. Therefore, its impact is sometimes overlooked. However, the incremental economics contributed by net pay can be a very significant factor in economic studies.

It is common in tight-gas reservoirs to develop curves like Fig. 17.33 that show gas recovery as a function of permeability and propped fracture penetration. Here, for each permeability level there is some fracture penetration beyond which there is little additional recovery. These studies do not often show significantly different values for larger or smaller net-pay values. In high-permeability formations, studies like this usually show little effect of fracture penetration on recovery and very little effect, if any, from variations in net pay. However, this is not the case where economics is concerned. An example case for a 5-md formation shows a significant effect of net pay on fracture penetration requirements to optimize NPV's. The results shown in Fig. 17.34 depict the percent increase in NPV (i.e., postfracturing values expressed as a percent of prefrac-

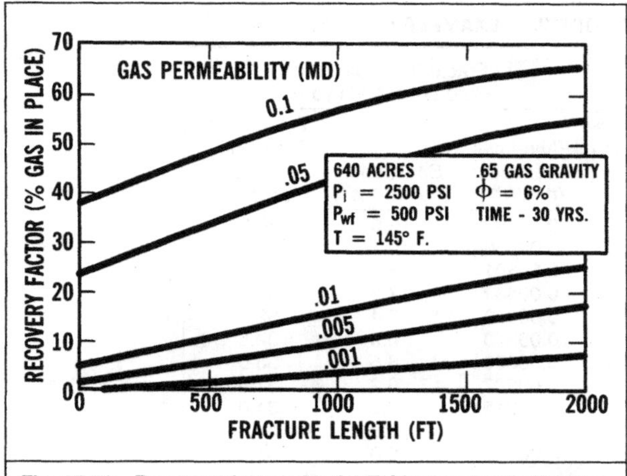

Fig. 17.33—Recovery factors for the Fahler sand with 640-acre well spacing and a 30-year life (after Holditch *et al.*[12]).

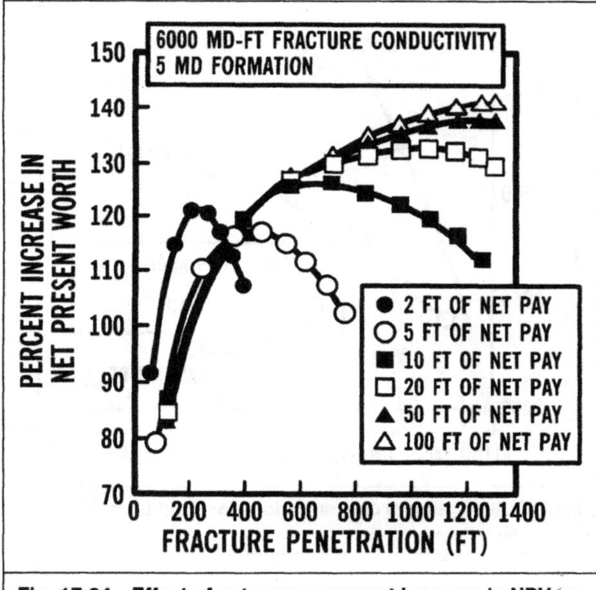

Fig. 17.34—Effect of net pay on percent increase in NPV vs. fracture penetration.

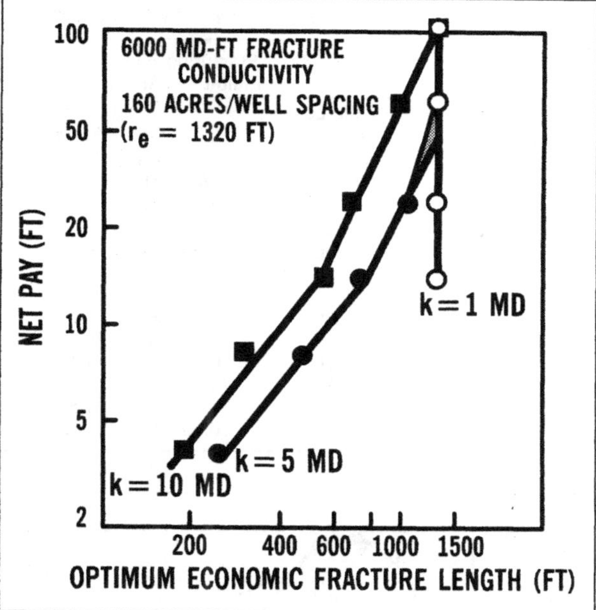

Fig. 17.35—Effect of net pay and permeability on optimum economic fracture penetration.

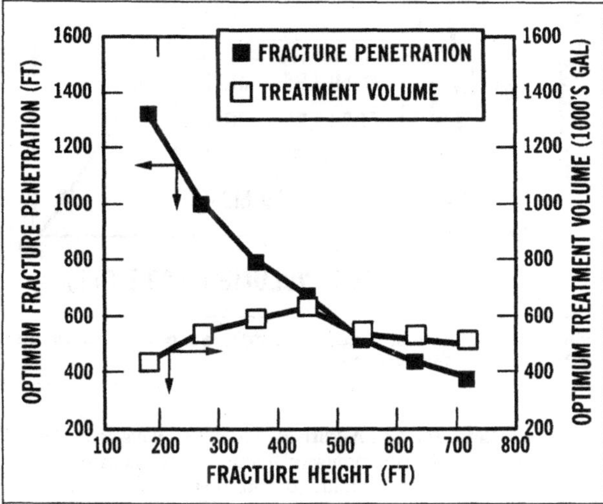

Fig. 17.36—Effects of fracture height on optimum fracture penetration and treatment volume.

turing PV's) vs. fracture penetration for net pays of 2, 5, 10, 20, 50, and 100 ft [0.6, 1.5, 3, 6, 15, and 30 m] and a fracture conductivity of 6,000 md-ft [1830 md·m]. The remainder of the data are given in Tables 17.4 through 17.6. Fig. 17.34 shows how the optimum fracture penetration (i.e., the penetration at which the maximum NPV's occur) increases as net pay increases.

Data such as these were also computed for 1- and 10-md cases. The combined results for the three permeability levels are summarized in Fig. 17.35, which shows the fracture penetration at the maximum NPV points vs. net pay (i.e., the optimum fracture penetration vs. net pay). The data here reflect how incremental economics can be affected by incremental producing rates. It is interesting to note that fracture penetration requirements approach values on the order of the drainage radius to optimize economics, even at relatively high permeability levels for sufficiently large values of net pay.

Fracture Height. Discussions in previous chapters provide a perspective of the importance of fracture height as it relates to fracturing-fluid volume requirements: as vertical growth increases, the volumes, and obviously the costs, required to achieve a given length increases significantly. Addressing fracture height from an economic standpoint brings forth other important aspects that reinforce the need for reliable fracture-height data when treatments are designed. In addition to the obvious increase in costs, fracture height can have a significant impact on optimum economical penetration, which in turn could affect well-spacing requirements.

As an example, cases were run for a 1-md formation with 10 ft [3 m] of net pay. Fracture heights of 180, 270, 360, 450, 540, 630, and 720 ft [55, 80, 110, 135, 165, 190, and 220 m] were investigated. The remaining data are given in Tables 17.4 through 17.6. The resulting optimum fracture lengths and treatment volume requirements are shown in Fig. 17.36. The optimum points were determined on the basis of maximum NPV for each given height. As can be seen, there was a significant effect on optimum length, but the optimum treatment volumes did not change dramatically over the wide range of fracture heights. At a height of 180 ft [55 m], the optimum fracture penetration approaches the 160-acre [65-ha] -spacing drainage boundary, whereas at heights on the order of 600 to 700 ft [183 to 215 m], the optimum lengths of 300 to 400 ft [90 to 120 m] suggest that one may need to investigate the economics for closer well spacing for these particular cases.

TABLE 17.11—DATA FOR FLUID-LOSS ADDITIVE, EXAMPLE CASE 17.9

	Fracturing Fluid with Fluid-Loss Additive A			Fracturing Fluid with Fluid-Loss Additive B		
Fluid-Loss Additive Concentration (lbm/1,000 gal)	Fluid-Loss Coefficient (ft/min$^{1/2}$)	Spurt Loss (gal/ft^2)	Total Fluid Cost with Additive (cents/gal)	Fluid-Loss Coefficient (ft/min$^{1/2}$)	Spurt Loss (gal/100 ft^2)	Total Fluid Cost With Additive (cents/gal)
0	0.00700	12.0	30.0	0.00700	12.0	30.0
10	0.00450	7.5	31.8	0.00560	10.1	30.5
20	0.00295	4.9	33.6	0.00501	8.8	31.0
30	0.00220	3.3	35.4	0.00432	7.9	31.5
40	0.00180	2.4	37.2	0.00400	7.4	32.0
50	0.00155	1.7	39.0	0.00373	6.9	32.5
60	0.00140	1.3	40.8	0.00350	6.5	33.0
70	0.00132	0.94	42.6	0.00327	6.2	33.5
80	0.00130	0.70	44.4	0.00315	6.0	34.0
90	0.00129	0.55	46.2	0.00303	5.8	34.5
100	0.00128	0.45	48.0	0.00299	5.6	35.0

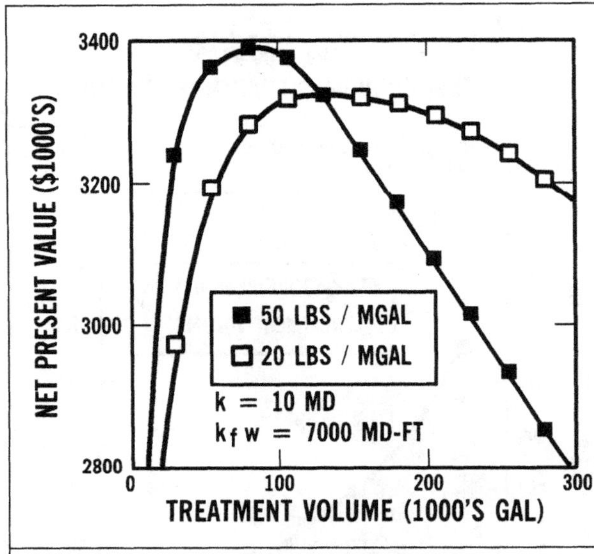

Fig. 17.37—Effects of Fluid-Loss Additive A concentration on NPV vs. treatment volume.

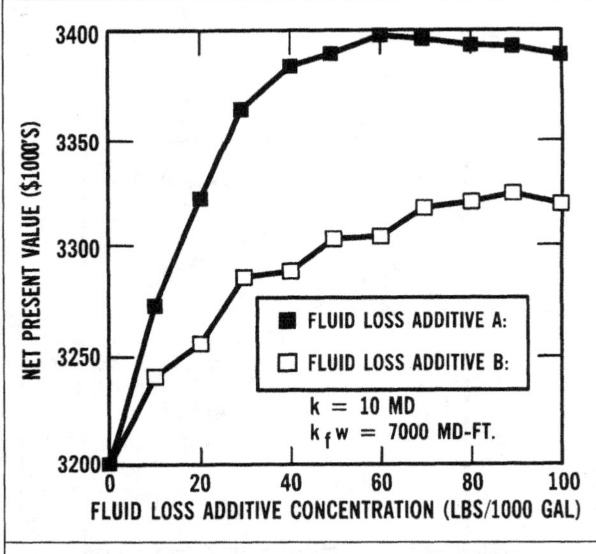

Fig. 17.38—Optimum concentrations for Fluid-Loss Additives A and B.

Fluid-Loss Behavior and Additives. Fluid loss plays an extremely important role in the economics of fracturing. It can affect design on the same order of magnitude (or more) as fracture height, formation permeability, or net pay. It can be controlled to some degree, however, through the use of additives. In some cases, the selection of the type and quantities of these additives is not intuitively obvious. Some additives are expensive, but others, while cheaper, do not effect sufficient fluid-loss reduction. What concentrations are needed? Economic investigations can provide insight for determining the appropriate fluid-loss-additive design.

Consider the example of a 10-md formation with 10 ft [3 m] of net pay. With this relatively high permeability, one would expect to need some type of fluid-loss additive. To investigate the type and amount requires data such as those shown in Table 17.11, which provides fluid-loss coefficient, spurt loss, and cost data for two different systems. These data are typical of what is available from laboratory (or field) tests. Additive A is more expensive, costing \$1.80/lbm [\$3.96/kg], compared with Additive B at \$0.50/lbm [\$1.10/kg]. Additive A, however, yields better fluid-loss and spurt-loss control.

A series of studies was made to find the maximum NPV's at various additive concentrations ranging (in equal increments) from 10 to 100 lbm/1,000 gal [1200 to 12 000 g/m^3] fracturing fluid. Fig. 17.37 shows the results for Additive A concentrations of 20 and 50 lbm/1,000 gal [2400 and 5990 g/m^3]. Here, maximum NPV's were achieved for volumes of 130,000 and 85,000 gal [492 and 322 m^3]. Fracture lengths achieved for these volumes at the maximum NPV points were 440 and 570 ft [135 and 174 m], respectively.

Additional results are shown in Fig. 17.38. The maximum NPV's vs. additive concentration show that Additive A at 60-lbm/1,000-gal [7190-g/m^3] concentration yields the optimum economics. Even though the lower fluid loss achieved at the higher additive concentrations (i.e., 70 to 100 lbm/1,000 gal [8390 to 12 000 g/m^3]) yielded deeper fracture penetrations per unit volume of material (fluid and proppant), the higher cost of the material volume requirements offset the incremental dollar returns. The total treatment costs associated with larger material requirements for the higher-fluid-loss, lower-priced Additive B, offset any savings attributable to the lower price of the material.

Again, one should be extremely careful about generalizing the results of this type of study. Economics can be very sensitive to fluid-loss behavior. It is essential to have reliable quantification of the fluid-loss-control performance of each fluid-loss-additive system when economic comparisons are made. For example, applying studies to 1-md formations shows that the higher-priced Additive A did not yield sufficient NPV's to justify its use.

Pumping Rate. The primary economic factors associated with pumping rate are hydraulic-horsepower costs and fluid leakoff to the formation. Hydraulic horsepower costs are usually attributed predominantly to pipe and perforation friction, although in some

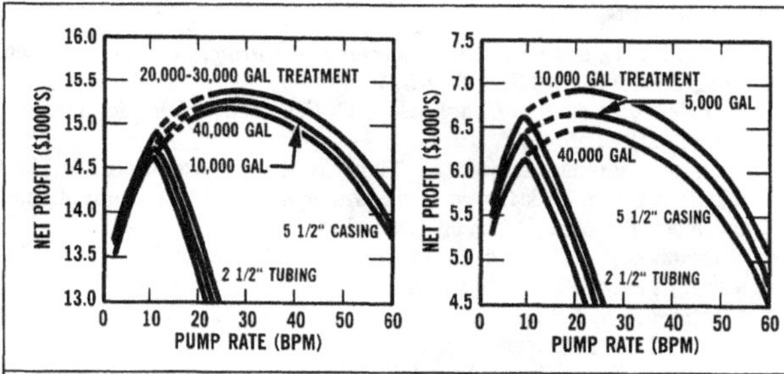

Fig. 17.39—Effect of pump rate and treatment volume on net profit (after Howard and Fast[1]).

cases friction in the fracture may also be a contributor. These are usually investigated by computing friction and hydrostatic-head effects from published data to determine surface wellhead pumping pressure requirements. Examples of this are discussed in detail in Chap. 11 and by Howard and Fast.[1]

Usually, such studies are important in the comparison of treatment costs for various tubular-goods configurations (e.g., 2½- vs. 5½-in. [6.4- vs. 14-cm] casing) to see which configurations can be tolerated from an economic standpoint. They are also important in some instances to investigate whether the added costs for friction-reducing agents are justifiable. This is usually more critical with fluids that inherently exhibit relatively high friction behavior, such as oil/water emulsions and gelled oils. In addition to the higher friction, these fluids have lower densities and consequently less hydrostatic head to counteract the friction loss.

Pumping rate obviously compensates for fluid leakoff to the formation. It would be very unusual, however, for pumping rate to compete economically with fluid-loss additives to overcome leakoff. In formations where leakoff is low, optimum rates will usually be dominated by friction-loss behavior and hydraulic-horsepower costs. In the higher-leakoff cases where the volume requirements are affected by pumping rates, fracturing material costs will enter the picture. Such studies as those presented by Howard and Fast can be used to investigate optimum rates. One of their examples shown in Fig. 17.39 depicts the effects of rate on economics. They have shown that the 2½-in. [6.4-cm] tubing case has a very narrow optimum rate range, whereas it is broader for 5½-in. [14-cm] casing. Also, the lower friction costs for the casing result in higher net profits.

17.10 Risk, Advanced Technology, and Oil or Gas Price

The previous examples have shown how in some cases changes in the physical parameters associated with a formation or fracturing treatment can impact economics. Changes in the economic climate, such as those created by increases or decreases in hydrocarbon prices or variations in discounting rates, can also have a significant effect. And as experienced with physical parameter variations, price or discount rate changes may affect certain aspects of design more critically than other areas. These effects can play a major role in many economic studies and the sensitivity associated with them is often investigated.

Methods similar to these discussed previously can be used to conduct such studies. Results of an example of one such study, conducted by Wyman *et al.*[30] and shown in Fig. 17.40, depict the relationship between PV profit and gas price for various fracture penetration in the Canadian Elmworth field. From this, one could infer that sufficiently high gas prices could justify large investments for achieving longer fracture penetrations. Studies[3,20,21] have discussed the effects of price, discount rates, and risk on economics. The example shown in Fig. 17.41 depicts how the advanced technology of fracturing can increase the total gas resource for various price and risk scenarios.

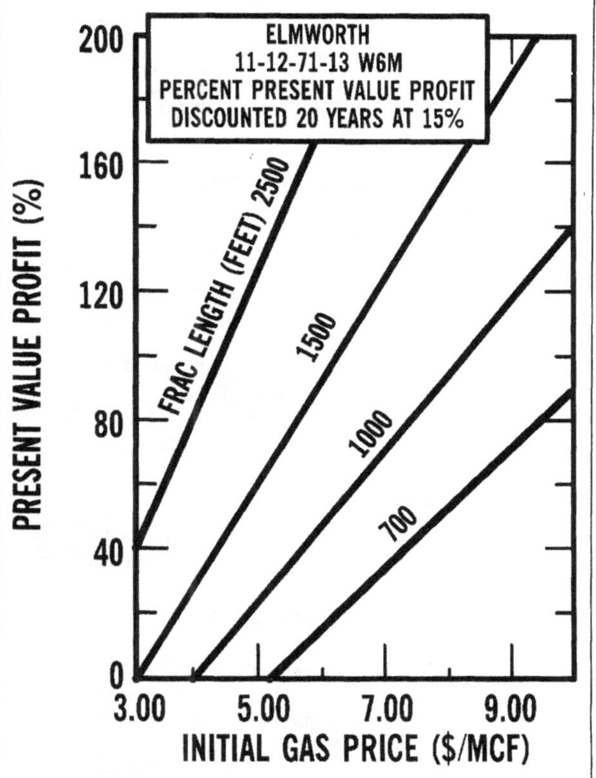

Fig. 17.40—Percent NPV profit as a function of initial gas price and fracture penetration (after Wyman *et al.*[30]).

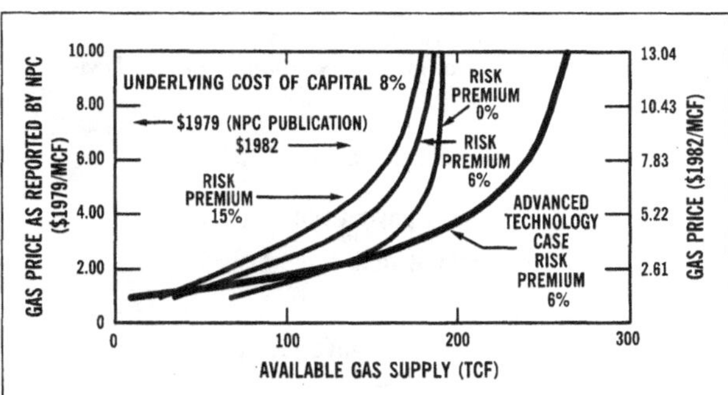

Fig. 17.41—Effect of rate of return on cost-supply curves, Natl. Petroleum Council base-case technology (blanket and lenticular formations combined) (after Rosenberg *et al.*[20]).

Nomenclature

a = constant-percentage decline rate
a_i = initial decline rate
A = $q_i I$
b = constant
B = De^{-a}
C_T = total cost of fracturing treatment

$$D = \frac{1}{1+(i_A/n_A)} = \text{discount factor}$$

$$D_c = \frac{1}{e^{n(i_A/n_A)}} = \text{continuous discount factor for } n \text{ compounding periods}$$

$$D\Big|_1^L = \frac{1-D^{(L+1)}}{1-D} - 1 = \text{discrete discount factor for } L \text{ months}$$

E_i = interest expense
E_ℓ = lifting cost expense
E_{n-1}^n = total expenses

E_T = taxes per unit hydrocarbon produced, \$/bbl or \$/Mcf
G_P = cumulative gas production
h = height, net pay, ft [m]
H_{n-1}^n = total hydrocarbon production (ΔN_p, or ΔG_P) for period (t_{n-1}, t_n)
i = monthly interest rate
i_A = annual interest rate (fractional)
i_r = rate of return
i_R = royalty interest (fractional)
I_{n-1}^n = average net hydrocarbon value per unit produced for period after taxes and royalty, $I=(S-E_T)(1-i_R)$, and $I_\ell=(S-E_T)(1-i_R)-E_\ell$
I_p = profitability index
J = productivity index, unfractured case
J_f = productivity index, fractured case
J_s = productivity index, stabilized flow
J_t = productivity index, flowing at time t before stabilized flow
k = permeability, md
L_a = number of time increments to abandonment
L_{af} = number of time increments to abandonment after fracturing
L_{uf} = number of time increments to abandonment, unfractured case
L_P = number of time increments to payout
m = constant
n = total number of compounding periods over which interest rate applies
n_A = number of compounding periods per annum
n_i = number of monthly increments
N_P = cumulative production at time t
P_c = undiscounted return on investment
P_{cD} = discounted return on investment
P_f = cash flow after fracturing
P_{n-1}^n = total net cash flow for period (t_{n-1}, t_n)
P_s = unit sales price of hydrocarbon, \$/bbl or \$/Mcf
P_{uf} = cash flow before fracturing
$P\Big|_1^{L_a}$ = cash flow to abandonment
$P\Big|_1^{L_P}$ = cash flow up to payout time, t_p
q = producing rate at time t
q_a = abandonment rate
q_i = initial producing rate
R_{Di} = discounted return on investment
R_i = return on investment
t = time
t_a = time to well abandonment
t_{Dpi} = discounted payout at interest rate i
t_L = Lth time increment
t_n = nth time increment $=n\Delta t$
t_p = payout time
t_{pi} = payout time at interest rate i
t_{po} = payout
V_F = future value
V_{nP} = net present value
$V_{nP}\Big|_1^{L_P}$ = net present value of cash flow up to payout time, t_p
V_P = present value
V_{Paf} = present value after fracturing
V_{Pbf} = present value before fracturing
V_{pf} = net present value resulting from fracturing $=V_{Paf}-V_{Pbf}$
V_{Puf} = present value, unfractured case
x_f = fracture length

Superscript

L = number of time increments

References

1. Howard, C.G. and Fast, C.R.: *Hydraulic Fracturing*, Monograph Series, SPE, Richardson, TX (1970) **2**, Chap. 9.
2. *Petroleum Production Handbook*, T.C. Frick (ed.), SPE, Richardson, TX (1962) Chap. 38.
3. "Unconventional Gas Sources: Tight Gas Reservoirs—Part I," Tight Gas Reservoirs Task Group of the Unconventional Gas Committee of the Natl. Petroleum Council (Dec. 1980) **V**.
4. Warembourg, P.A. *et al.*: "Fracture Stimulation Design and Evaluation," paper SPE 14379 presented at the 1985 SPE Annual Technical Conference and Exhibition, Las Vegas, Sept. 22-25.
5. Veatch, R.W.: "Overview of Current Hydraulic Fracturing Design and Treatment Technology—Part 1," *JPT* (April 1983) 677-87.
6. Elkins, L.E.: "Western Tight Sands Major Research Requirements," *Proc.*, GRI/AGA/U.S. DOE Intl. Gas Research Conference, Chicago, IL (June 9-12, 1980) 394-408.
7. Morse, R.A. and Von Gonten, W.D.: "Productivity of Vertically Fractured Wells Prior to Stabilized Flow," *JPT* (July 1972) 807-10.
8. Prats, M.: "Effect of Vertical Fractures on Reservoir Behavior—Incompressible Fluid Case," *SPEJ* (June 1961) 105-16; *Trans.*, AIME, **222**.
9. McGuire, W.J. and Sikora, V.J.: "The Effect of Vertical Fractures on Well Productivity," *Trans.*, AIME (1960) **219**, 401-03.
10. Tinsley, J.M. *et al.*: "Vertical Fracture Height—Its Effect on Steady-State Production Increase," *JPT* (May 1969) 633-38; *Trans.*, AIME, **246**.
11. Guerrero, E.T.: *Practical Reservoir Engineering*, The Petroleum Publishing Co., Tulsa, OK (1968) 72-75.
12. Holditch, S.A. *et al.*: "The Optimization of Well Spacing and Fracture Length in Low Permeability Gas Reservoirs," paper SPE 7496 presented at the 1978 SPE Annual Technical Conference and Exhibition, Houston, Oct. 1-4.
13. Agarwal, R.G., Carter, R.D., and Pollock, C.B.: "Evaluation and Performance Prediction of Low-Permeability Gas Wells Stimulated by Massive Hydraulic Fracturing," *JPT* (March 1979) 362-72; *Trans.*, AIME, **267**.
14. Gringarten, A.C., Ramey, H.J. Jr., and Raghavan, R.: "Unsteady-State Pressure Distribution Created by a Well With a Single Infinite-Conductivity Vertical Fracture," *SPEJ* (Aug. 1974) 347-60; *Trans.*, AIME, **257**.
15. Cinco-L., H., Samaniego-V., F., and Dominguez-A., N.: "Transient Pressure Behavior for a Well With a Finite-Conductivity Vertical Fracture," *SPEJ* (Aug. 1978) 253-64.
16. Perkins, T.K. Jr. and Kern, L.R.: "Widths of Hydraulic Fractures," *JPT* (Sept. 1961) 937-49; *Trans.*, AIME, **222**.
17. Holditch, S.A.: "Economic Production of Tight Gas Reservoirs Looks Better," *Oil & Gas J.* (Feb. 4, 1974) 99-102.
18. Geertsma, J. and Haafkens, R.: "A Comparison of Theories for Predicting Width and Extent of Vertical Hydraulically Induced Fractures," *Trans.*, ASME (1979) **101**, 8-19.
19. Kozik, H.G. and Holditch, S.A.: "A Case History for Massive Hydraulic Fracturing the Cotton Valley Lime Matrix, Fallon and Personville Fields," *JPT* (Feb. 1981) 229-44.
20. Rosenberg, J.I. *et al.*: "A Sensitivity Analysis of the Natl. Petroleum Council Study of Tight Gas," paper SPE 11645 presented at the 1983 SPE/DOE Symposium on Low-Permeability Gas Reservoirs, Denver, March 14-16.
21. Brashear, J.P., Rosenberg, J.I., and Mercer, J.: "Tight Gas Resource and Technology Appraisal: Sensitivity Analyses of the National Petroleum Council Estimates," paper SPE 12862 presented at the 1984 SPE/DOE/GRI Unconventional Gas Recovery Symposium, Pittsburgh, May 13-15.
22. Jeu, S.J.: "A Different Approach to Optimizing MHF Using Fracture Azimuth and Reservoir Data," paper SPE 12190 presented at the 1983 SPE Annual Technical Conference and Exhibition, San Francisco, Oct. 5-8.
23. Smith, M.B.: "Effect of Fracture Azimuth on Production With Application to the Wattenberg Gas Field," paper SPE 8298 presented at the 1979 SPE Annual Technical Conference and Exhibition, Las Vegas, Sept. 23-26.
24. Britt, L.K.: "Optimized Oilwell Fracturing of Moderate-Permeability Reservoirs," paper SPE 14371 presented at the 1985 SPE Annual Technical Conference and Exhibition, Las Vegas, Sept. 22-25.
25. Phillips, A.M. and Anderson, R.W.: "Use of Proppant Selection Models to Optimize Fracturing Treatment Designs in Low Permeability Reservoirs," paper SPE 13855 presented at the 1985 SPE/DOE Symposium on Low-Permeability Gas Reservoirs, Denver, May 19-22.
26. Montgomery, C.T. and Steanson, R.E.: "Proppant Selection—The Key to Successful Fracture Stimulation," *JPT* (Dec. 1985) 2163-72.
27. Norman, M.E., Cipolla, C.L., and Webb, M.L.: "The Application of Manufactured Proppants in Moderately Permeable Oil Reservoirs,"

paper SPE 12357 presented at the 1983 SPE Production Technology Symposium, Lubbock, TX, Nov. 14–15.
28. Cooke, C.E. Jr.: "Fracturing With a High-Strength Proppant," *JPT* (Oct. 1977) 1222–26.
29. Holditch, S.A.: *Criteria for Propping Agent Selection,* second edition, Norton Alcoa Proppants, Dallas (1980).
30. Wyman, R.E., Holditch, S.A., and Randolph, P.L.: "Analyses of an Elmworth Hydraulic Fracture in Alberta," *JPT* (Sept. 1980) 1621–30.
31. Roberts, C.N.: "Fracture Optimization in a Tight Gas Play: Muddy 'J' Formation, Wattenberg Field, Colorado," paper SPE 9851 presented at the 1981 SPE/DOE Low-Permeability Gas Reservoirs Symposium, Denver, May 27–29.
32. Zahner, R.L. and Crafton, J.W.: "A Review of the Long-Term Performance of Hydraulic Fracture Treatments in the Muddy 'J' Sand, Wattenberg Field," paper SPE 13873 presented at the 1985 SPE/DOE Symposium on Low-Permeability Gas Reservoirs, Denver, May 19–22.
33. Meng, H.Z. and Brown, K.E.: "Coupling of Production Forecasting, Fracture Geometry Requirements and Treatment Scheduling in the Optimum Hydraulic Fracture Design," paper SPE 16435 presented at the 1987 SPE/DOE Low Permeability Reservoirs Symposium, Denver, May 18–19.

Glossary

Net cash flow: A schedule or other type of statement of all cash transactions—i.e., outflow (expenses, capital expenditures, and taxes) and inflow in an accounting period. The summation of these several outflows and inflows in a single year is referred to as the annual net cash flow. In equation form, net cash flow = +revenue − investment − operating cost − overhead − taxes.

Cumulative net cash flow: The successive addition of annual net cash flows.

Discounting: The inverse of compounding. The process of determining the value today of some amount in the future.

Discount rate: The interest rate assumed in discounting future amounts to their present value.

Discount factor: The inverse of the compound factor—i.e., $1/(1+i)$.

Expense: An accounting term that indicates that part or all of past or current expenditures are to be deducted from revenue during the current period. Included in expense are operating costs, noncash charges, and overhead.

Full-cycle economics: Measures of performance that encompass the entire time span of the project regardless of the portion of the project already under way or completed.

Incremental economics: Measures of project performance that compare a project to its next best alternative.

Investment: An expenditure intended to acquire or to improve property—real or personal, tangible or intangible—yielding revenue or services. Such expenditures are cash outlays when incurred and may or may not be capitalized on the financial books.

Operating costs: Costs directly attributable to conducting operations designed to provide revenues—e.g., raw materials, fuel, operating labor, maintenance, and repairs.

Overhead (OH): All costs of materials and services that are not directly adding to or readily identifiable with specific operations but that are necessary costs of doing business—e.g., engineering, accounting, research, and general office expense.

Payout (PO): The length of time required for the cumulative cash position to reach zero. Payout is on an after-tax basis unless otherwise designated.

Present value (PV), present worth (PW): The sum of all cash effects discounted to some specific point in time at a stated discount rate. The value today of some amount in the future given an investment opportunity at a specific discount rate.

Present worth profile: A collection of present worths at a range of discount rates.

Profitability index (PI), rate of return (ROR): The compound interest rate whose discount factors will make the present worth of a project's net cash flow equal to zero. The profitability index is equal to the constant effective annual percentage earnings on the unrecovered capital remaining in a project. Synonyms include marginal efficiency of capital, internal rate of return, true yield, discounted-cash-flow method, and others.

Rate of return (ROR): The maximum interest rate one could pay on capital tied up over the life of an investment and still break even.

Revenue: Funds received or derived from the sale of products and services.

Return on investment (ROI): A project measure of performance calculated as follows: the ratio of cumulative net cash flow over the project life to maximum cash outlay.

Royalty: The percentage of gross production given to the land or mineral rights owner as compensation for use of the property.

Sensitivity analysis: A technique for testing, through multiple case comparison, the degree to which a project is affected by the change in one or more components of the investment analysis—e.g., changes in price, production, royalty, and investment costs.

Ultimate net cash flow: The sum of the yearly net cash flows over the life of the project.

Working interest: The obligation to pay a given percentage of all costs and investments in return for that same percentage of production and revenue. A 40% working interest results in 40% of the total net cash flows in each year.

SI Metric Conversion Factors

acres	× 4.046 873	E−01	=	ha
bbl	× 1.589 873	E−01	=	m^3
cp	× 1.0*	E−03	=	Pa·s
ft	× 3.048*	E−01	=	m
ft^2	× 9.290 304*	E−02	=	m^2
ft^3	× 2.831 685	E−02	=	m^3
°F	(°F−32)/1.8		=	°C
gal	× 3.785 412	E−03	=	m^3
in.	× 2.54*	E+00	=	cm
lbm/ft^2	× 4.882 428	E+00	=	kg/m^2
lbm/gal	× 1.198 264	E+02	=	kg/m^3
psi	× 6.894 757	E+00	=	kPa

*Conversion factor is exact.

Appendix A
Typical Products Available From Service Companies

John W. Ely, SPE, S.A. Holditch & Assocs.

Water-Based Polymers

Guar Gum.

1. Powdered guar gum polymer; delayed hydration, designed for batch-mix applications; typically no internal breaker present; high residue, 10 to 14% after complete breakdown.
2. Powdered guar gum polymer; rapid hydrating, designed for continuous-mix application; typically contains internal breaker; high residue, 10 to 14% after complete breakdown.
3. Powdered guar gum polymer; delayed hydration, designed for batch-mix applications; typically no internal breaker; new formulation; low residue, 6 to 8% after complete breakdown.
4. Powdered guar gum polymer; rapid hydrating, designed for continuous-mix applications; typically contains internal breakers; new formulation; 6 to 8% residue after complete breakdown.

Hydroxypropyl Guar (HPG) Gum.

1. Powdered HPG gum; delayed hydration, designed for batch-mix applications: typically contains no internal breaker; 1 to 3% residue after complete breakdown.
2. Powdered HPG viscosifier; rapid hydrating, designed for continuous-mix applications; typically contains internal breaker; low residue, 1 to 3% after complete breakdown.

Hydroxyethylcellulose (HEC).

1. Powdered HEC viscosifier; delayed hydration polymer, designed for use as a secondary gel or can be used in batch-mix applications; essentially no residue after degradation.
2. Powdered HEC viscosifier; rapid hydration, designed for continuous-mix applications.

Carboxymethylcellulose (CMC). Powdered CMC viscosifier; rapid hydration, designed for both batch- and continuous-mix applications.*

Carboxymethylhydroxyethylcellulose (CMHEC). Powdered CMHEC viscosifier, designed for rapid hydration; can be used in both batch and continuous applications.*

Xanthate Gum. Powdered xanthate polymer; designed for viscosifying 15% or lower-strength HCl; can be used as a viscosifier for water; can be batch mixed or mixed continuously.*

Carboxymethylhydroxypropyl Guar (CMHPG). Powdered CMHPG viscosifier; delayed hydration, designed for batch-mix applications; typically no internal breaker; 1 to 3% residue after breaking.

Supplementary to Chap. 7.

*Products designed for continuous-mix applications and stated to be useful in batch-mix applications typically are very difficult to mix because of their rapid viscosification. The service companies have a difficult time mixing these products without lumps.

Specially Modified HPG.

1. Moderately high MS HPG (fairly high substitution of moles propyleneoxide per molecule allowing viscosification of up to 80% methanol); powdered HPG viscosifier; delayed hydration, designed for batch-mix applications; typically no internal breaker.
2. Very high MS substituted HPG (a high enough substitution of moles propyleneoxide that this product is soluble in 100% methanol); powdered HPG viscosifier; delayed hydration, designed for batch-mix applications; typically no internal breaker.

Liquid Viscosifiers for Acid.

1. Acrylate emulsion copolymer of acrylate and other proprietary materials used for viscosifying up to and including 28% acid; relatively good temperature stability and, in most cases, crosslinkable.
2. Liquid viscosifier for acid up to 15%; surfactant formulation designed to viscosify 15% acid, but looses viscosity above and below 15% range.

Powdered Viscosifiers for Acid.

1. As mentioned earlier, CMHEC is used by many service companies as a low-temperature gel for HCl up to and including 28%, especially when the product is crosslinked.
2. Xanthan gum is also used as a viscosifier for acid up to 15% strength.
3. Powdered acrylate polymers are offered by the service companies as viscosifers similar to the liquid viscosifiers previously mentioned; some are crosslinkable.

Friction Reducers

Friction Reducers for Fresh Water Only.

1. Liquid, highly anionic polyacrylamide friction reducer for water; typical loadings from ¼ to 1 gal/1,000 gal [0.25 to 1 m^3/1000 m^3].
2. Powdered, anionic polyacrylamide friction reducer for water and acid; loadings from ¼ to 2 lbm/1,000 gal [30 to 240 g/1000 m^3].

Friction Reducers for Water and Acid.

1. Liquid, anionic copolymer friction reducer for water and acid; typical loadings of ½ to 2 gal/1,000 gal [0.5 to 2 m^3/1000 m^3].
2. Liquid, cationic polyacrylamide friction reducer for acids, brines, and fresh water; typical loadings of ¼ to 2 gal/1,000 gal [0.25 to 2 m^3/1000 m^3].
3. Powdered, anionic friction reducer for acid, brines, and fresh water; typical loadings of 2 to 5 lbm/1,000 gal [240 to 600 g/m^3].
4. Powdered, cationic friction reducer for acid, brines, and fresh water; typical loadings of 2 to 5 lbm/1,000 gal [240 to 600 g/m^3].

Friction Reducer for Hydrocarbons.

1. Liquid concentrate friction reducer for hydrocarbons; concentrate of polyisodecylmethacrylate product, an emulsion concentrate, typically requires an activator.

2. Liquid solution of friction reducer for hydrocarbons; typical loadings of 7 to 10 gal/1,000 gal [7 to 10 $m^3/1000$ m^3]; same product as No. 1 but more dilute.

Fluid-Loss Additives.

1. Selectively graded fine-mesh silica flour used in water, oil, and acid; typical grading through 200 on 325 mesh; some service companies sell two products, one as above and a 200-mesh product.
2. Combination of graded oil-soluble resin and degradable high-molecular-weight polymer; nondamaging fluid-loss additive used in water and acid.
3. 100-mesh benzoic acid used in water, acid, or foam fracturing treatments.
4. 100-mesh sand used in water, oil, and acid.
5. 100-mesh oil-soluble resin used in water and acid; although called 100 mesh, it typically approximates a wider range, more in the 200-mesh range.
6. 100-mesh salt used in saturated brine and hydrocarbon fracturing.
7. Catchall products that are mixtures of nonswelling clays, talc, and starches; good fluid-loss additive for water and oil.
8. Liquid fluid-loss additive consisting of aromatics plus surfactant; typically used at 5 gal/1,000 gal [5 $m^3/1000$ m^3] as an alternative to 5% diesel.
9. Adomite Aqua™; proprietary fluid-loss additive supplied by Nalco used by most service companies; quite similar to No. 7.
10. Adomite Mark II™; proprietary fluid-loss additive supplied by Nalco used by most oil service companies; not compatible with the standard oil gels from the service companies.
11. Diesel or other hydrocarbons used with emulsifying surfactant.

Breakers

Breakers for Water-Based Fluid Systems.

1. Enzyme breakers for guar, guar derivatives, and cellulose derivatives; temperature range of 70 to 130°F [21 to 54°C].
2. Oxidizer breaker for guar, guar derivatives, and cellulose derivatives; typically sodium or ammonium persulfate; temperature range 120 to 200°F [49 to 93°C].
3. High-temperature oxidizer breaker for guar, guar derivatives, and cellulose; temperature-activated breaker typically used in the temperature range of 160 to 230°F [71 to 110°C].
4. Acid breaker for guar, guar derivatives, and cellulose derivatives; typically weak, carboxylic acids used in temperature ranges above 200°F [93°C].
5. Low-temperature breaker activator for high-pH fracturing fluids; typically used with oxidizer breakers in temperature range of 70 to 120°F [21 to 49°C] to allow them to degrade high-pH cross-linked fracturing fluids.

Breakers for Oil Gel Systems.

1. Low-temperature oil breaker for aluminum octoate gels; typically organic compound that hydrolizes to form an acid.
2. Granular breaker for aluminum phosphate ester oil gels; typically weak organic acid.
3. Alternative breaker for aluminum phosphate ester for oil gels; typically weak base products, such as sodium bicarbonate or lime; functionality is from 100°F [38°C] and above.
4. Low-temperature breaker for soaps created from the reaction of caustic and fatty acids; typically is weak organic acid.
5. High-temperature breaker for above-mentioned soaps; typically ammoniacal compound.
6. Liquid breaker for aluminum phosphate ester systems; typically amine-type compound; requires presence of water for functionality.

Diverting Agents

1. Oil-soluble resin in aqueous solution.
2. Graded rock salt.
3. Flake benzoic acid.
4. Alcohol solution of benzoic acid.
5. Wide range graded oil-soluble resin used up to temperatures approaching 350°F [177°C].
6. Unibeads.
7. Polymer-coated sand that swells upon contact with water.
8. Oil-soluble graded naphthalene.

Acid Inhibitors

1. Special inhibitor for organic acids, such as formic and acetic.
2. Low-temperature HCl inhibitor for use up to 150°F [66°C]; not functional above these temperatures; relatively inexpensive.
3. Inhibitor for HCl and HF for bottomhole temperatures (BHT's) up to 250°F [121°C].
4. Inhibitor for HCl and HF up to BHT of 350°F [177°C].
5. Powdered synergistic additive for extending inhibition times at elevated temperatures.
6. Liquid synergistic additive for extending inhibition times at elevated temperatures.
7. HCl inhibitor for water wells up to BHT's of 120°F [49°C].

Acid Retarders

1. Oil-wetting surfactant for limestone reservoirs for moderate temperature applications up to 200°F [93°C].
2. Oil-wetting surfactant for limestone reservoirs for temperatures up to 350°F [177°C].

Emulsifiers

1. Oil-external emulsifier for HCl and HCl/organic-acid mixtures.
2. Cationic emulsifier for preparing acid or water-external polyemulsion fluids.
3. Anionic emulsifier for preparing acid or water-external polyemulsion fluids.

Emulsifiers 1 and 2 have been used in the preparation of CO_2 emulsion, which converts into CO_2 foams in CO_2-foam fracturing.

Clay Stabilizers

1. Cationic polymeric clay stabilizer; typically with acrylic backbone chain.
2. Cationic clay stabilizer of a surfactant nature, not polymeric.
3. Claylock stabilizer developed by Chevron (hydroxyaluminum).

Surfactants

Fluorosurfactants.

1. Nonionic fluorosurfactant for water and acid systems; product yields excellent surface-tension reduction; not effective as nonemulsifier.
2. Cationic fluorosurfactant for water and acid systems; product used primarily as an absorbing product in sandstone formations to yield a "nonwet" condition.
3. Blends of nonemulsifiers and nonionic fluorosurfactants; separate products for acid and brine.

Nonemulsifiers.

1. Nonionic nonemulsifier for water.
2. Nonionic nonemulsifier for acid.
3. Specifically formulated nonionic nonemulsifier product for use in both water and acids.
4. Anionic nonemulsifier for water.
5. Anionic nonemulsifier for acid.
6. Anionic nonemulsifier for oil.
7. Nonionic nonemulsifier for oil.
8. Cationic nonemulsifier for acid.
9. Cationic nonemulsifier for water.
10. Cationic nonemulsifier for oil.

Fines Suspender

1. Nonionic fines suspender for water and acid.
2. Cationic fines suspender for water and acid.
3. Anionic fines suspender for water and acid.
4. Amphoteric fines suspender for water and acid.

Antisludge Agent

1. Antisludge agent for acid.

pH-Control Agents

Most pH-control agents are buffers, but some may be strong acids or strong bases; listed below are typical buffers, strong acids, and bases.

1. Sodium hydroxide; used as a pH-control agent in borate systems.
2. Mixture of mono- and disodium phosphate; used as a low-pH buffer for linear gels.
3. Fumaric acid; used as a low-pH buffer for both linear and crosslinked fracturing fluids.
4. Formic acid; used as a low-pH buffer for some crosslinked gels and some linear gels.
5. Sodium carbonate; used as a strong-base buffer for crosslinked gel systems.
6. Sodium bicarbonate; used as a buffer for crosslinked fracturing fluids and as a breaker for some oil gels.
7. Ammonium hydroxide; strong base used as a buffer for some crosslinked fracturing fluids.
8. Sodium acetate; sold as a liquid and as a solid; used as a low-pH buffer for crosslinked and linear gels.
9. Sulfamic acid; weak acid sometimes used as a buffer.

Crosslinkers

1. Potassium pyroantimonate; low-pH crosslinker.
2. Titanium triethylnolamine; very common crosslinking agent with a pH of 6 to 9.
3. Titanium acetylacetinate moderately-low-pH crosslinking agent used as a delay system with a pH range of 5 to 6.
4. Sodium tetraborate; powdered crosslinker sometimes used in making borate crosslink gels.
5. Boric acid; alternative borate crosslinker frequently used in creating borate crosslink gels.
6. Zirconium oxychloride crosslinking agent used in preparing some acid crosslink gels of CMHPG or CMHEC.
7. Zirconium lactate crosslinker used in a pH range of 9 to 10 for HPG.
8. Zirconium triethylnolamine crosslinker for high-pH fluids.
9. Aluminum acetate crosslinking agent used in the pH range of 3 to 5 with CMHPG, or CMHEC.

Crosslinker Control Agents

1. Proprietary products to slow the rate of crosslinking.
2. Dilute acetinate form used to slow crosslink time, particularly in titanium crosslinks.
3. Proprietary products supplied by some service companies to accelerate crosslink time, particularly in cold weather.
4. Rapid-crosslink-time products blended with delayed systems to control crosslink time.
5. Blends of delayed and rapid crosslinking agents to yield early viscosity were required.

Foamers

Water-Based Foamers.
1. Anionically charged foaming agent for use in fresh water and brines.
2. Zwitterion foaming agent used in water and acid.
3. Cationic foaming agent used in water and acid.

Foamers for Hydrocarbons.
1. Fluorocarbon foamers used for foaming crude oils.
2. Aliphatic foamers used for foaming hydrocarbons.

Foaming Agents for Water and Methanol. Foaming agent with capability of foaming water containing up to 80% methanol.

Foaming Agent for 100% Methanol. Primarily fluorocarbon-type foamers.

Scale Inhibitors.
1. Phosphonate-type scale inhibitors.
2. Acrylate-type scale inhibitors.

Gel Stabilizers.
1. Liquid stabilizer for high-temperature, methanol or isopropanol, is typically used at 5 to 10% concentration.
2. Powdered stabilizer for high temperatures; typically sodium thiosulfate, in an anhydrous form or pentahydrate containing five waters.

Defoamers.
1. Defoamers for aqueous fluids may be silicate material but in many cases simply tributylphosphate.
2. Defoamer for oil-based fluids, typically silicone-type product.

Radioactive Materials.
1. Radioactive sand used for approximating fracture height near wellbore.
2. Radioactive interprop or bauxite; used for the same purpose as radioactive sand.
3. Liquid radioactive iodine.
4. Liquid radioactive scandium.
5. Liquid radioactive iridium.

Nos. 3 through 5 are all used for tracing fluids and their entrance into perforations or fracture systems.

Oil Gelling Additives.
1. Fatty acids; either toll-oil or fish-oil fatty acids for creating soap in combination with caustic or other products; in limited use today.
2. Aluminum octoate; powdered viscosifier in limited use as a viscosifier for high-gravity oils.
3. Standard aluminum phosphate ester; gelling agent widely used by the service companies.
4. Sodium aluminate activator for the aluminum phosphate ester.
5. High-temperature stable version of aluminum phosphate ester; used for treating deeper, higher-temperature wells.
6. Continuous-mix version of aluminum phosphate ester.

Iron-Control Additives

1. Citric acid; widely used sequestrant buffering agent.
2. Acetic acid buffering agent often used in conjunction with citric acid; used for many years in the oil field.
3. Ethylenediametetraacetic acid; widely used iron and calcium sequestrant favored by some service companies.
4. Proprietary and nonproprietary blends containing eryothorbic acid under various names; very popular iron-control system.

Biocides

1. Gluteraldehyde; widely used biocide.
2. Carbamates; widely used low-pH biocide for control of bacterial growth in fracturing fluids.
3. Isothiazolines; biocide, often used for bacteria control.
4. Combination of aldehyde and isothiazoline biocide.
5. Proprietary triazine containing biocide; widely used for bacteria control.
6. Proprietary biocide called Adocide™ supplied by Nalco.
7. Proprietary biocide Adomal™ supplied by Nalco.

Mutual Solvents

1. Ethyleneglycolmonobutylether (EGMBE).
2. Asol™ mutual solvent prepared by Amoco.
3. Special mutual solvent containing EGMBE with surfactants; used at approximately one-half the concentration of standard EGMBE.
4. Asol A-28™ mutual solvent from Amoco.
5. Super Asol™ mutual solvent from Amoco.

Corrosion Inhibitors

1. Multipurpose completion-fluid inhibitor consisting of oxygen scavengers and other proprietary ingredients.
2. H_2S-corrosion inhibitor; proprietary ingredients to protect against corrosion of H_2S supplied both in the form of a completion-fluid product and as a coating product for tools put into the hole.

Weighting Agents

1. Liquid calcium chloride.
2. Liquid calcium bromine.
3. Liquid calcium zinc chloride.

Paraffin Control

1. Paraffin dispersant used primarily in combination with water or heated water to disperse and to remove paraffin from tubing.

2. Liquid paraffin inhibitors that function as crystal poisoning agents and disallow paraffin growth.
3. Combination of paraffin dispersants and inhibitors.
4. Solid paraffin inhibitors in low- and high-temperature versions.

Miscellaneous Products

1. Ammonium bifluoride.
2. High-strength HCl.
3. Calcium chloride.
4. Ammonium chloride.
5. Potassium chloride.
6. Polyvinylalcohol.
7. All types of aromatic and aliphatic solvents.
8. Sodium chloride.
9. Special blends of surfactants, mutual solvents, and paraffin dispersants.
10. Oil-based mud-dispersant products.
11. Scale converters.

Processes

Acid Systems.

1. Nonemulsifying acid—any acid plus a surfactant to improve clean-up, to lower surface tension, and to prevent emulsions.
2. Nonemulsifying acid plus iron control; the same acid as above plus iron sequestrants.
3. Non-HF mud-removal acids; all service companies have some formulation of surfactants to disperse and to suspend mud and fines; typically high concentration of surfactants; used in situations where HF acid is not recommended or needed.
4. Non-HF mud-removal acids plus iron-stabilizing agents; a non-HF mud-removal acid containing iron sequestrants.
5. Aromatic acid dispersion; listed in various terminologies; a mixture of aromatics and acid with surfactant to disperse the aromatic in the acid.
6. High-strength HCl; HCl strengths above 20%.
7. Intensified acid; some service companies call low concentrations of HF, less than 2%, intensified HCl.
8. Fines-removal or fines-suspension acid; HCl plus a fines suspender.
9. Fines-removal acid plus iron control; fines suspender and iron-chelating agents.
10. Iron-control acid and iron-stabilized acid; HCl or hydrochloric HF acid contains iron sequestrant.
11. Perforating acid; acetic acid typically in the 10% range.
12. Formic acid; used in high-temperature situations where HCl is overly corrosive.
13. Hydrocarbon- or oil-soluble acid; nonaqueous acid; acetic acid is mixed with hydrocarbon.
14. Alcoholic acid; mixtures of alcohol and HCl.
15. Retarded acids.
    a. Mixture of HCl and acetic.
    b. Mixture of HCl and formic.
    c. Mixture of HCl and a chemical retarder.
    d. Mixture of HCl and acetic plus chemical retarder.
    e. Mixture of HCl and formic plus chemical retarder.
    f. Oil-external HCl emulsion.
    g. Oil-external HCl emulsions and retarder.
    h. Oil-external HCl emulsions plus formic acid and retarder.
    i. HCl-external emulsion.
    j. Linear gel viscosified acid.
    k. Crosslinked high-viscosity acid system.
    l. Alternating stages of viscous gel and acid to produce differential etching.
16. HF acids.
    a. HCl and HF combination; typically 12% HCl, 3% HF or 12% HCl, 6% HF, etc.
    b. HCl/HF acid plus a surfactant.
    c. HCl/HF plus a surfactant and fines suspender.
    d. Self-generating mud-acid proprietary systems developed and licensed by Shell Development Co.; uses methylformate to give a retarded HF system.
    e. Chemically retarded HF acid system; may be fluoroboric acid or ABF reacted with aluminum chloride.
    f. HF/formic-acid mixture.

Linear Water-Based Gel Systems.

1. Water and friction reducer.
2. Viscosified water: 10 to 60 lbm/1,000 gal [1.2 to 7.2 kg/m^3] of guar derivatives, cellulose derivative, etc.
3. Viscosified water; 10 to 60 lbm/1,000 gal [1.2 to 7.2 kg/m^3] of guar derivatives, cellulose derivatives containing fluid-loss additives.

Crosslinked Gel System.

1. Crosslinked guar gum systems.
2. Crosslinked HPG systems.
3. Crosslinked HPG with 3 to 5% hydrocarbon fluid loss.
4. Crosslinked HPG with high-temperature stabilizers.
5. Crosslinked CMHEC.
6. Crosslinked CMHEC for high temperature with stabilizers.
7. Crosslinked CMHPG systems.
8. Crosslinked CMHPG with temperature stabilizers.
9. Crosslinked systems as above with delayed crosslinked mechanisms.

Alcohol/Water Systems.

1. Crosslinked methanol systems; 20 to 100% methanol.
2. Linear gel systems containing from 20 to 100% alcohol.

Oil Gel Systems.

1. Oil without viscosifier.
2. Gelled oil using fatty-acid viscosifiers.
3. Gelled oil with aluminum octoate.
4. Gelled oil systems with aluminum phosphate esters.
5. High-temperature gelled oil systems with aluminum phosphate esters.
6. Continuous-mix gelled oil systems with aluminum phosphate esters.
7. Water-external emulsion system developed by Exxon with two-thirds oil, one-third water.

Foam Fracturing Systems

1. Water and N_2 foam with or without polymers.
2. Acid and N_2 foam.
3. Hydrocarbon and N_2 foam.
4. Methanol/water and pure methanol N_2 foam.
5. Water and CO_2 foam.
6. Water/methanol and CO_2 foam.
7. Crosslinked gelled water, N_2 or CO_2 foam.

Polymer Temporary Plugs

1. Guar or HPG system; linear gel plugs; typically high concentrations.
2. Cellulose-derivative high-concentration gel for temporary plugs.
3. Crosslinked guar or HPG systems for temporary plugs.
4. Crosslinked guar or HPG with secondary-gelling-agent plugs.

Appendix B
Selection of Water-, Non-Water-, or Acid-Based Fracturing Fluids

John W. Ely, SPE, S.A. Holditch & Assocs.

Sandstone Formations

The vast majority of all hydraulic fracturing treatments with proppant are conducted using water-based fluids. Water-based fluids are normally cost-efficient and safe to pump and have physical properties that can be controlled with precision through recent technological advancements. Occasions arise, however, when, because of formation incompatibility, one has to use some form of hydrocarbon fluid, such as alcohol- or oil-based systems. The decision on the particular fluid to use should be based on safety considerations and fluid compatibility with the formation.

Certain producing horizons cannot accept any water-based-fluid invasion without severe permeability damage or, in some cases, total productivity reduction. Some formations become deconsolidated if water is placed on them. If the cementing materials holding the sand grains together are water-soluble, then the formation should not be treated with water-based fluids.

Another common reason for not using water-based fluids is the presence of water-swellable or migratory clays. A great deal of advancement has been made within the industry in the control of clay swelling and migration by use of various salts—such as potassium chloride—and amine and polymeric amine compounds.

The particular producing horizon should be evaluated on the basis of core flow tests and previous experience in the area before a fracturing fluid is selected. The first fracturing fluid used in a particular area commonly is oil- or methanol-based and has successful results. Because of this early success, many engineers are hesitant to use a water-based fluid. Examples of this scenario have recently occurred in the Olmos formation in south Texas and the Canyon sand in southwest Texas. After only nonaqueous fluids were used on these formations, it was ascertained that current technology would make water-based fluids usable in treating these formations, particularly in selected areas.

In many cases, core flow tests indicate some potential damage caused by fracturing fluid. Some damage can be tolerated without seriously detracting from the overall well productivity. The cost-effectiveness and superior technology available in water-based fluids should always be a consideration in the selection of the available fracturing fluids.

Criteria for the Selection of a Base Fluid in Sandstone Reservoirs

1. Have water-based fracture treatments been conducted successfully in this formation?

Yes. Aqueous fluids can, of course, be used.

No. Consider their use on the basis of the formation properties.

2. If only oil- or methanol-based treatments have been conducted, were definite engineering criteria developed to support their selection?

No. An attempt at water-based treatments should be considered.

Yes. If the treatment failed, consider foam treatments, quick-fluid-recovery treatments, or possibly the incorporation of various clay-stabilizing agents or surfactants to enhance the success of water-based fluids.

3. Does the formation contain 10% or more of clays that can migrate or swell to reduce permeability?

Yes. Consider hydrocarbon- or methanol-based treatments, but only after studying the various possibilities in reference to foam, quick turnaround, or clay-stabilizing materials.

No. A strong possibility exists that water-based fluids can be used to treat this formation.

4. Are the cementing materials that hold sand grains together water-soluble?

Yes. Use nonaqueous fluids for stimulation.

No. Water-based fluids can be considered.

5. Is the formation susceptible to water-blocking problems?

Yes. Consider foamed fluids or rapid turnaround of the fracturing fluid after the treatment before selecting nonaqueous fluids.

No. Strongly consider an aqueous fluid.

6. Would rapid recovery of the water-based fluid negate the need for nonaqueous fluid?

Yes. Use a water-based fracturing fluid, of course.

No. Consider nonaqueous fluids.

7. Would use of foam (N_2 or CO_2) eliminate the need for non-aqueous fluids?

Yes. An aqueous fluid can be used.

No. Strongly consider a hydrocarbon- or alcohol-based fluid.

As can be seen from previous discussion and careful evaluation of the base-fluid checklist, from the standpoints of economics and superior rheological properties, water-based fluids should be used wherever possible.

Carbonate Formations

For formations reactive to HCl or other acids, there is an alternative to proppant fracturing. Acid fracturing has been a successful stimulation technique for more than 50 years. Techniques have been developed to improve the success of fracture acidizing in the last 2 decades that make the process even more viable. Obviously, the formation must have high solubility in acid to be a good candidate for acid fracturing. A good rule of thumb is that the formation should be at least 75% soluble in acid. Stimulation treatments in lower-solubility formations have been successful, but the vast majority of successful treatments have been in formations with limestone or dolomite concentrations exceeding the 75% solubility criteria. The goal is a heterogeneous etched flow pattern across the face of the rock.

It should also be mentioned that many carbonate formations have been successfully stimulated with proppant fracturing. Here we will attempt to delineate a few specific criteria that should be established to ascertain whether one should use proppant- or acid-based fracturing techniques.

Supplementary to Chap. 7.

TABLE B-1—POTENTIAL FRACTURING FLUIDS

Nonaqueous Fluids
Refined oil (no friction reducer)
Gelled oil (phosphate ester, low temperature)
Water-external emulsion (two-thirds oil, one-third H_2O)*
Oil-based foam
Gelled oil (phosphate ester, high temperature)
Gelled methanol/water
Gelled methanol (linear gel)
Foamed methanol
Crosslinked methanol/water (20 to 80% methanol)
Foamed methanol/water (20 to 80%)
Aqueous Fluids
Linear gel (guar, HPG, or cellulose derivative)
Low-temperature crosslinked gel
Low, neutral, high pH
Guar, HPG, CMHPG, CMHEC
High-temperature crosslinked gel (delayed crosslink)
HPG/cationic guar
Linear gel, secondary gel system
Foamed water
Crosslinked foam

*Although two-thirds oil, this fluid cannot normally be used in extremely-water-sensitive formations.

Criteria for the Selection of Acid-Based Fracturing Fluid

1. Is acid solubility 75% or higher?

No. Chances of successful acid fracture stimulation are relatively low.

2. When exposed to acid, does the formation rock release large quantities of fines?

Yes. Do not consider acid fracturing as a first attempt on stimulation. Use low-strength acids if lesser amounts of fines are generated.

3. Is the formation rock relatively soft?

Yes. Anticipate fracture closure in a relatively short period of time and consider proppant fracturing using high proppant concentrations.

4. Do you anticipate excessive closure pressures combined with minimum rock strength?

Yes. Expect relatively quick closure of the fracture matrix and low conductivity.

5. Does the formation BHT exceed 250°F [121°C]?

Yes. Even the best techniques available today make acid fracturing or deep penetration of the acid-etched fractures a dim possibility.

6. Have successful proppant fracturing treatments been conducted?

Yes. If proppant fracturing works successfully in a formation and an excellent production increase is achieved, stay with that technique.

7. If proppant fracturing and acid fracturing have been conducted, which of the two is most cost-effective?

Typically, proppant fracturing, where applicable, is somewhat more cost-effective.

Acid fracturing has been widely used and is a very successful stimulation technique in certain areas. The major advantage of acid fracturing is that a heterogeneously etched fracture has essentially infinite conductivity. The advantages of an infinitely conductive flow path for oil wells where fluids of much higher viscosity need to be conducted are readily apparent. Acidizing or acid fracturing techniques can also be used under low-injection-rate conditions without fear of screenout in areas close to water.

Terminology and Definitions for Fracture Fluid Selection Flowcharts (Table B-1)

Low permeability: <0.1 md.

Moderate permeability: >0.1 md<2 md.

High permeability: >2 md.

Damage removal: ≤50-ft [≤15-m] penetration from wellbore breaking through damage or skin; can be final treatment or small fracture in tight well to evaluate possible need for larger treatment; normally means very small, short pump time, so temperature stability of fluids is not a factor.

Low pressure: requires externally supplied energy to unload well.

High pressure: Sufficient bottomhole pressure to unload fracture fluid after fracturing.

Small created fracture height: <40 ft [<12 m].

Medium created fracture height: >40<100 ft [>12<30 m].

Large created fracture height: <100 ft [<30 m].

Low temperature: 70 to 150°F [21 to 66°C].

Moderate temperature: 150 to 250°F [66 to 121°C].

High temperature: 250 to 375°F [121 to 191°C].

Deep penetrating fractures: Propped fractures extending to 60% of the drainage radius.

Non-Water-Based Fluids Used in Water-Sensitive Formations—Deeply Penetrating Fracture (Fig. B-1)

1-A. 65- to 70-quality N_2 oil foam; 65- to 70-quality CO_2 or N_2 methanol/water foams; energized refined oil; energized (N_2) (low-viscosity) phosphate ester; or energized methanol/water linear gel. (Notes 3 through 5 and 7.)

2-A. 70- to 80-quality N_2 oil foam; 70- to 80-quality CO_2 or N_2 methanol/water foam; energized (moderate-viscosity) phosphate ester oil gel; or energized methanol/water linear gel. (Notes 4, 5, and 7.)

3-A. Energized (N_2) high-temperature-stable (moderate-viscosity) phosphate ester oil gel or energized N_2 or CO_2 40 to 60 lbm [18 to 27 kg] crosslinked methanol/water gel. (Notes 4, 5, and 7.)

4-A. 65- to 70-quality N_2 oil foam; 65- to 70-quality CO_2 or N_2 methanol/water foam; 30 to 50 lbm [14 to 23 kg] energized crosslinked methanol/water gel; or energized N_2 (moderate-viscosity) phosphate ester oil gel. (Notes 4, 5, and 7.)

5-A. 70- to 80-quality N_2 oil foam; 70- to 80-quality CO_2 or N_2 methanol/water foam; or energized N_2 or CO_2 30 to 50 lbm [14 to 23 kg] crosslinked methanol/water gel. (Notes 4 and 7.)

6-A. Energized N_2 high-viscosity, high-temperature-stable phosphate ester oil gel or energized CO_2 or N_2 40 to 60 lbm [18 to 27 kg] crosslinked methanol/water gel. (Notes 4, 5, and 7.)

7-A. ≥70-quality oil foam; ≥70-quality CO_2 or N_2 methanol/water foams; energized (N_2) (moderate-viscosity) phosphate ester oil gel; or energized N_2 or CO_2 30 lbm [14 kg] crosslinked methanol/water gel. (Notes 4, 5, and 7.)

8-A. 70- to 80-quality N_2 oil foam; 70- to 80-quality CO_2 or N_2 methanol/water foam; energized N_2 (moderate to high-viscosity) phosphate ester oil gel; or energized N_2 or CO_2 30 to 50 lbm [14 to 23 kg] crosslinked methanol/water gel. (Notes 4, 5, and 7.)

9-A. Energized N_2 high-temperature-stable (high-viscosity), phosphate ester oil gel; energized N_2 or CO_2 crosslinked methanol/water gel; or energized (N_2) high-temperature-stable, linear gel plus secondary gel. (Notes 4, 5, and 7.)

10-A. Refined oil plus bridging fluid loss; low-viscosity phosphate ester gelled oil plus bridging fluid loss; or linear methanol/water gel plus bridging fluid loss. (Notes 3 through 6.)

11-A. Moderate-viscosity phosphate ester gelled oil plus bridging fluid loss; 30 to 50 lbm [14 to 23 kg] crosslinked methanol/water plus bridging fluid loss; or linear methanol/water gel plus bridging fluid-loss agents. (Notes 4 through 6.)

12-A. High-temperature-stable (moderate-viscosity) phosphate ester oil gel plus bridging fluid loss or 30 to 50 lbm [14 to 23 kg] crosslinked methanol/water gel plus bridging fluid loss. (Notes 4 through 6.)

13-A. Moderate-viscosity phosphate ester oil gel plus bridging fluid loss or 30 to 40 lbm [14 to 18 kg] crosslinked methanol/water gel plus bridging fluid loss. (Notes 4 through 6.)

14-A. Moderate- to high-viscosity phosphate ester oil gel plus bridging fluid loss or 30 to 50 lbm [14 to 23 kg] crosslinked methanol/water gel plus bridging fluid loss. (Notes 4 through 6.)

15-A. High-temperature-stable, high-viscosity phosphate ester oil gel plus bridging fluid loss of 40 to 60 lbm [18 to 27 kg] crosslinked methanol/water gel plus bridging fluid loss. (Notes 4 through 6.)

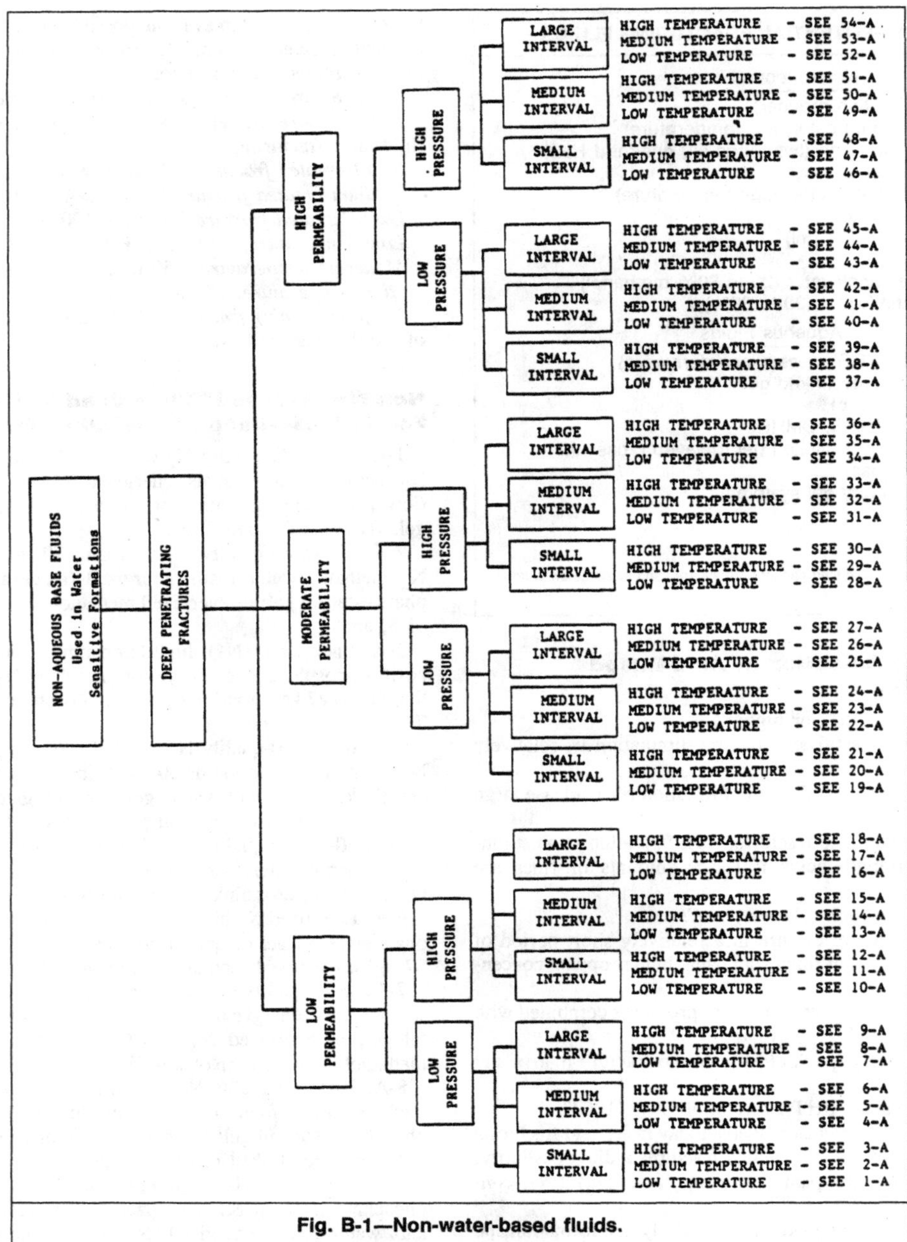

Fig. B-1—Non-water-based fluids.

16-A. Moderate-viscosity phosphate ester oil gel plus bridging fluid loss or 30 to 40 lbm [14 to 18 kg] crosslinked methanol/water gel plus bridging fluid loss. (Notes 4 through 6.)

17-A. Moderate- to high-viscosity phosphate ester oil gel plus bridging fluid loss or 30 to 50 lbm [14 to 23 kg] crosslinked methanol gel plus bridging fluid loss. (Notes 4 through 6.)

18-A. High-temperature-stable, high-viscosity phosphate ester oil gel plus bridging fluid loss; 40 to 60 lbm [18 to 27 kg] crosslinked methanol/water gel plus bridging fluid loss; or high-temperature-stable methanol/water linear gel plus secondary gel plus bridging fluid loss. (Notes 4 through 6.)

19-A. 65- to 70-quality N_2 oil foam plus bridging fluid loss, energized refined oil plus bridging fluid loss; 65- to 70-quality foamed methanol/water plus bridging fluid loss; energized methanol/water linear gel plus bridging fluid loss; or energized (low-viscosity) phosphate ester oil gel plus bridging fluid loss. (Notes 3 through 7.)

20-A. 70- to 80-quality N_2 oil foam plus bridging fluid loss; 70- to 80-quality foamed N_2 or CO_2 methanol/water gel plus bridging fluid loss; energized (N_2) (moderate-viscosity) phosphate ester oil gel plus bridging fluid loss; or 30 to 40 lbm [14 to 18 kg] crosslinked energized methanol/water plus bridging fluid loss. (Notes 4 through 7.)

21-A. 40 to 50 lbm [18 to 23 kg] energized N_2 or CO_2 crosslinked methanol/water gel plus bridging fluid loss; or energized (N_2) high-temperature-stable (high-viscosity) phosphate ester oil gel plus bridging fluid loss. (Notes 4 through 7.)

22-A. 65- to 75-quality N_2 oil foam plus bridging fluid loss; 65- to 75-quality CO_2 or N_2 methanol/water foam plus bridging fluid loss; 30 to 50 lbm [14 to 23 kg] energized methanol/water crosslinked plus bridging fluid loss; or energized (N_2) phosphate ester oil gel plus bridging fluid loss. (Notes 4 through 7.)

23-A. 70- to 80-quality N_2 oil foam plus bridging fluid loss; 70- to 80-quality CO_2 or N_2 foam plus bridging fluid loss; or energized (N_2) (moderate-viscosity) phosphate ester oil gel plus bridging fluid loss; or energized N_2 or CO_2 30 to 50 lbm [14 to 23 kg] crosslinked methanol/water gel plus bridging fluid loss. (Notes 4 through 7.)

24-A. Energized (N_2) high-temperature-stable, high-viscosity phosphate ester oil gel plus bridging fluid loss or energized CO_2 or N_2 40 to 60 lbm [18 to 27 kg] crosslinked methanol/water gel plus bridging fluid loss. (Notes 4 through 7.)

25-A. Energized (N_2) phosphate ester (moderate-viscosity) oil gel plus bridging fluid loss; $\geq$70-quality N_2 oil foam plus bridging fluid loss; energized CO_2 or N_2 30 to 40 lbm [14 to 18 kg] crosslinked methanol/water gel plus bridging fluid loss; or ≥ 70-quality CO_2 or N_2 foamed methanol/water gel plus bridging fluid loss. (Notes 4 through 7.)

26-A. 70- to 80-quality N_2 oil foam plus bridging fluid loss; 70- to 80-quality CO_2 or N_2 methanol/water foam plus bridging fluid loss; energized (N_2) phosphate ester oil gel (moderate-viscosity) plus bridging fluid loss; or energized N_2 or CO_2 30 to 50 lbm [14 to 23 kg] crosslinked methanol/water gel plus bridging fluid loss. (Notes 4 through 7.)

27-A. Energized (N_2) high-temperature-stable (high-viscosity) phosphate ester oil gel plus bridging fluid loss; energized CO_2 or N_2 crosslinked methanol/water gel plus bridging fluid loss; or high-temperature-stable energized (N_2) linear gel plus secondary gel in methanol/water plus bridging fluid loss. (Notes 4 through 7.)

28-A. Refined oil plus bridging fluid loss; low-viscosity phosphate ester oil gel plus bridging fluid loss; or methanol/water linear gel plus bridging fluid loss. (Notes 3 through 6.)

29-A. Moderate-viscosity phosphate ester oil gel plus bridging fluid loss or 30 to 50 lbm [14 to 23 kg] methanol/water crosslink plus bridging fluid loss. (Notes 4 through 6.)

30-A. High-temperature-stable (moderate-viscosity) phosphate ester oil gel plus bridging fluid loss or 30 to 60 lbm [14 to 27 kg] methanol/water crosslink plus bridging fluid loss. (Notes 4 through 6.)

31-A. Moderate-viscosity phosphate ester oil gel plus bridging fluid loss or 30 to 40 lbm [14 to 18 kg] methanol/water crosslink plus bridging fluid loss. (Notes 4 through 6.)

32-A. Moderate- to high-viscosity phosphate ester oil gel plus bridging fluid loss or 30 to 50 lbm [14 to 23 kg] methanol/water crosslink plus bridging fluid loss. (Notes 4 through 6.)

33-A. High-temperature-stable, high-viscosity phosphate ester oil gel plus bridging fluid loss or 40 to 60 lbm [18 to 27 kg] methanol/water crosslink plus bridging fluid loss. (Notes 4 through 6.)

34-A. Moderate-viscosity phosphate ester oil gel plus bridging fluid loss or 30 to 50 lbm [14 to 23 kg] methanol/water crosslink plus bridging fluid loss. (Notes 4 through 6.)

35-A. High-viscosity phosphate ester oil gel plus bridging fluid loss or 40 to 60 lbm [18 to 27 kg] methanol/water crosslink plus bridging fluid loss. (Notes 4 through 6.)

36-A. High-temperature-stable (high-viscosity) phosphate ester oil gel plus bridging fluid loss; 50 to 60 lbm [23 to 27 kg] methanol/water crosslink plus bridging fluid loss; or high-temperature-stable methanol/water linear plus secondary gel plus bridging fluid loss. (Notes 4 through 6.)

37-A. 65- to 70-quality N_2 oil foam plus bridging fluid loss; 65- to 70-quality methanol/water foam plus bridging fluid loss; energized 30 lbm [14 kg] crosslinked methanol/water gel plus bridging fluid loss; or energized (N_2) phosphate ester (low-viscosity) oil gel plus bridging fluid loss. (Notes 4 through 7.)

38-A. 70- to 80-quality N_2 oil foam plus bridging fluid loss; 70- to 80-quality CO_2 or N_2 methanol/water foam plus bridging fluid loss; energized phosphate ester (moderate-viscosity) oil gel plus bridging fluid loss; or energized 30 to 50 lbm [14 to 23 kg] crosslinked methanol/water gel plus bridging fluid loss. (Notes 4 through 7.)

39-A. Energized high-temperature-stable (high-viscosity) phosphate ester oil gel plus bridging fluid loss or 40 to 60 lbm [18 to 27 kg] energized crosslinked methanol/water plus bridging fluid loss. (Notes 4 through 7.)

40-A. 65- to 75-quality N_2 oil foam plus bridging fluid loss; 65- to 75-quality CO_2 or N_2 methanol/water foam plus bridging fluid loss; energized phosphate ester (low-viscosity) oil gel plus bridging fluid loss; or energized 30 lbm [14 kg] crosslinked methanol/water plus bridging fluid loss. (Notes 4 through 7.)

41-A. 70- to 80-quality N_2 oil foam plus bridging fluid loss; 70- to 80-quality CO_2 or N_2 methanol/water foam plus bridging fluid loss; energized 30 to 50 lbm [14 to 23 kg] crosslinked methanol/water foam plus bridging fluid loss; or energized phosphate ester (moderate-viscosity) oil gel plus bridging fluid loss. (Notes 4 through 7.)

42-A. Energized (N_2) (high-viscosity) high-temperature-stable phosphate ester oil gel plus bridging fluid loss or 40 to 60 lbm [18 to 27 kg] energized crosslinked methanol/water gel plus bridging fluid loss. (Notes 4 through 7.)

43-A. 65- to 70-quality N_2 oil foam plus bridging fluid loss; 65- to 70-quality N_2 or CO_2 methanol/water foam plus bridging fluid loss; energized 30 lbm [14 kg] crosslinked methanol/water gel plus bridging fluid loss; or energized (N_2) phosphate ester (moderate-viscosity) oil gel plus bridging fluid loss. (Notes 4 through 7.)

44-A. 70- to 80-quality N_2 oil foam plus bridging fluid loss; 70- to 80-quality N_2 or CO_2 methanol/water foam plus bridging fluid loss; energized 30 to 40 lbm [14 to 18 kg] crosslinked methanol/water plus bridging fluid loss; or energized (N_2) phosphate ester (moderate-viscosity) oil gel plus bridging fluid loss. (Notes 4 through 7.)

45-A. Energized CO_2 or N_2 methanol/water 40 to 60 lbm [18 to 27 kg] crosslink plus bridging fluid loss or N_2 energized high-temperature-stable (high-viscosity) phosphate ester oil gel plus bridging fluid loss. (Notes 4 through 7.)

46-A. Refined oil plus bridging fluid loss; low-viscosity phosphate ester oil gel plus bridging fluid loss; or methanol/water linear gel plus bridging fluid loss. (Notes 3 through 6.)

47-A. Moderate-viscosity phosphate ester oil gel plus bridging fluid loss or 30 to 50 lbm [14 to 23 kg] methanol/water crosslink plus bridging fluid loss. (Notes 4 through 6.)

48-A. High-temperature-stable (moderate-viscosity) oil gel plus bridging fluid loss or 30 to 60 lbm [14 to 27 kg] methanol/water crosslink plus bridging fluid loss. (Notes 4 through 6.)

49-A. Moderate-viscosity phosphate ester oil gel plus bridging fluid loss or 30 to 40 lbm [14 to 18 kg] methanol/water crosslink plus bridging fluid loss. (Notes 4 through 6.)

50-A. Moderate- to high-viscosity ester oil gel plus bridging fluid loss or 30 to 50 lbm [14 to 23 kg] methanol/water crosslink plus bridging fluid loss. (Notes 4 through 6.)

51-A. High-temperature-stable (high-viscosity) phosphate ester oil gel plus bridging fluid loss or 40 to 60 lbm [18 to 27 kg] methanol crosslink plus bridging fluid loss. (Notes 4 through 6.)

52-A. Moderate-viscosity phosphate ester oil gel plus bridging fluid loss or 30 to 50 lbm [14 to 23 kg] methanol/water crosslink plus bridging fluid loss. (Notes 4 through 6.)

53-A. High-viscosity phosphate ester oil gel plus bridging fluid loss or 40 to 60 lbm [18 to 27 kg] methanol/water crosslink plus bridging fluid loss. (Notes 4 through 6.)

54-A. High-temperature-stable (high-viscosity) phosphate ester oil gel plus bridging fluid loss; 50 to 60 lbm [23 to 27 kg] crosslinked methanol/water plus bridging fluid loss; or high-temperature-stable primary methanol/water linear gel plus secondary gel plus bridging fluid loss. (Notes 4 through 6.)

Water-Based Fluids—Deeply Penetrating Fracture (Fig. B-2)

1-B. ≤65-quality CO_2 or N_2 foam, energized polyemulsion or energized linear guar gel plus bridging fluid loss. (Notes 2, 6, and 7.)

2-B. 65- to 75-quality N_2 or CO_2 foam; 30 to 40 lbm [14 to 18 kg] energized crosslinked guar; or energized polyemulsion fluid. (Notes 2 and 7.)

3-B. ≥75-quality N_2 or CO_2 foam or 40 to 60 lbm [18 to 27 kg] energized crosslinked HPG. (Note 7.)

4-B. ≤65-quality CO_2 or N_2 foam or ≤30 lbm [≤14 kg] energized crosslinked guar or energized polyemulsion fluid. (Notes 2 and 7.)

5-B. 65- to 75-quality CO_2 or N_2 foam or 30 to 60 lbm [14 to 27 kg] energized crosslinked guar or energized polyemulsion fluid. (Notes 2 and 7.)

6-B. ≥75-quality CO_2 or N_2 foam; 40 to 60 lbm [18 to 27 kg] energized crosslinked HPG; or energized high-temperature-stable linear gel plus secondary gel plus bridging fluid loss. (Notes 2, 6, and 7.)

7-B. ≤70-quality CO_2 or N_2 foam; energized ≤30 lbm [≤14 kg] crosslinked guar; or energized polyemulsion. (Notes 2 and 7.)

8-B. 70- to 80-quality CO_2 or N_2 foam or 40 to 50 lbm [18 to 23 kg] energized crosslinked guar. (Note 7.)

9-B. ≥75-quality CO_2 or N_2 foam or 40 to 60 lbm [18 to 27 kg] energized crosslinked HPG plus hydrocarbon fluid loss. (Notes 6 and 7.)

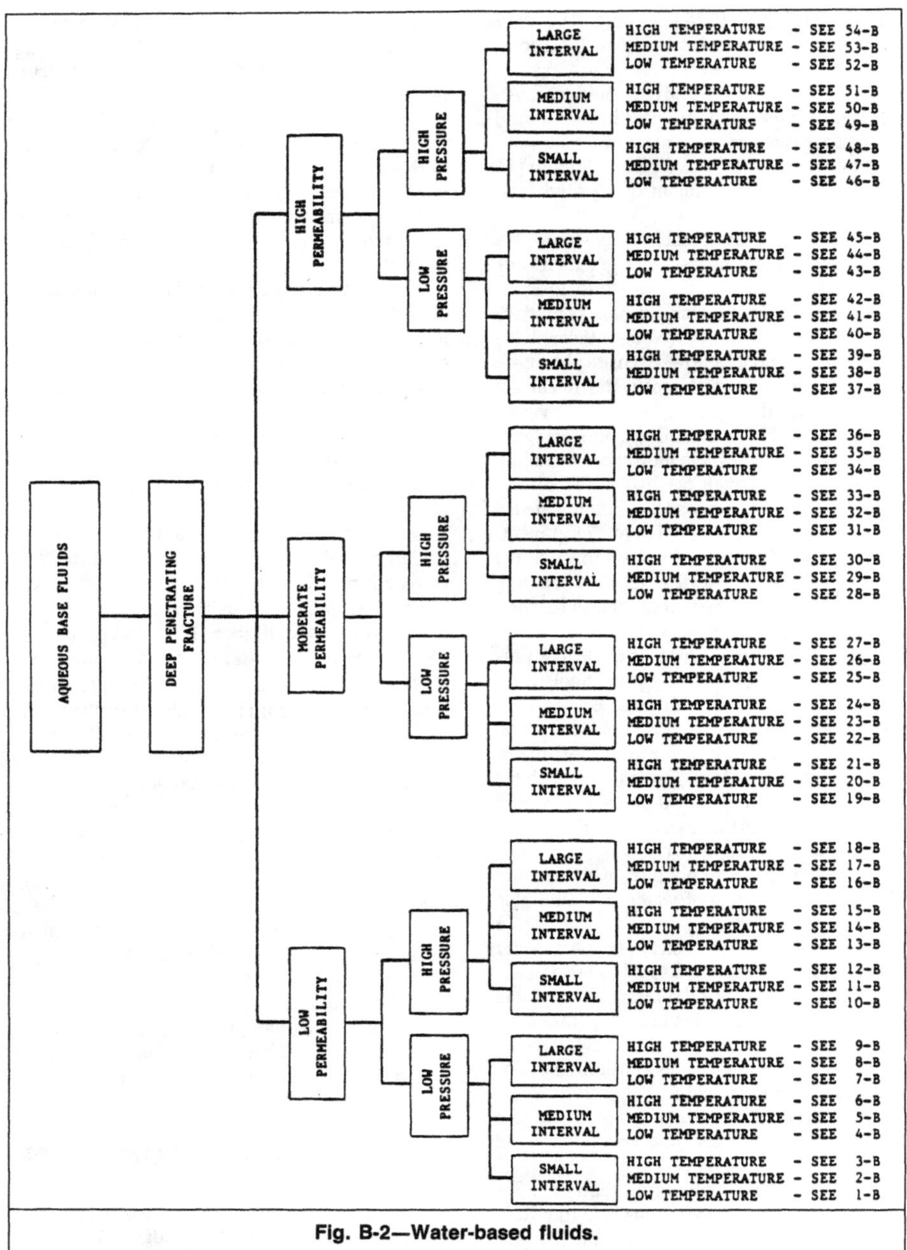

Fig. B-2—Water-based fluids.

10-B. Linear guar gel plus bridging fluid loss or polyemulsion fluid or ≤30 lbm [≤14 kg] guar crosslink plus hydrocarbon fluid loss. (Notes 2 and 6.)

11-B. 30 to 50 lbm [14 to 23 kg] guar crosslink plus hydrocarbon fluid loss or polyemulsion fluid. (Notes 2 and 6).

12-B. 40 to 60 lbm [18 to 27 kg] HPG crosslink plus hydrocarbon fluid loss or high-temperature-stable linear gel plus secondary gel plus bridging fluid loss. (Note 6.)

13-B. 30 lbm [14 kg] guar crosslink plus hydrocarbon fluid loss or polyemulsion fluid (Notes 2 and 6.)

14-B. 40 to 60 lbm [18 to 27 kg] guar crosslink plus hydrocarbon fluid loss or polyemulsion fluid. (Notes 2 and 6.)

15-B. 40 to 60 lbm [18 to 27 kg] HPG crosslink plus hydrocarbon fluid loss or high-temperature-stable linear gel plus secondary gel plus bridging fluid loss. (Note 6.)

16-B. Polyemulsion fluid or 30 to 40 lbm [14 to 18 kg] guar crosslink plus hydrocarbon fluid loss. (Notes 2 and 6.)

17-B. 40 to 60 lbm [18 to 27 kg] guar crosslink plus hydrocarbon fluid loss. (Note 6.)

18-B. 40 to 60 lbm [18 to 27 kg] HPG crosslink plus hydrocarbon fluid loss. (Note 6.)

19-B. ≤65-quality CO_2 or N_2 foam plus bridging fluid loss; energized polyemulsion plus bridging fluid loss; or energized linear guar gel plus bridging fluid loss. (Notes 2, 6, and 7.)

20-B. 65- to 75-quality CO_2 or N_2 foam plus bridging fluid loss; 30 to 40 lbm [14 to 18 kg] energized crosslinked guar plus bridging fluid loss; or energized polyemulsion plus bridging fluid loss. (Notes 2, 6, and 7.)

21-B. ≥75-quality CO_2 or N_2 foam plus bridging fluid loss or 40 to 60 lbm [18 to 27 kg] energized HPG crosslink plus bridging fluid loss. (Notes 6 and 7.)

22-B. ≤65-quality CO_2 or N_2 foam plus bridging fluid loss; ≤30 lbm [≤14 kg] energized crosslinked guar plus bridging fluid loss; or energized polyemulsion plus bridging fluid loss. (Notes 2, 6, and 7.)

23-B. 65- to 75-quality CO_2 or N_2 foam plus bridging fluid loss or 30 to 50 lbm [14 to 23 kg] energized crosslinked guar plus bridging fluid loss. (Notes 6 and 7.)

24-B. ≥75-quality CO_2 or N_2 foam plus bridging fluid loss or 40 to 60 lbm [18 to 27 kg] energized crosslinked HPG plus bridging fluid loss. (Notes 6 and 7.)

25-B. ≤70-quality CO_2 or N_2 foam plus bridging fluid loss; 30

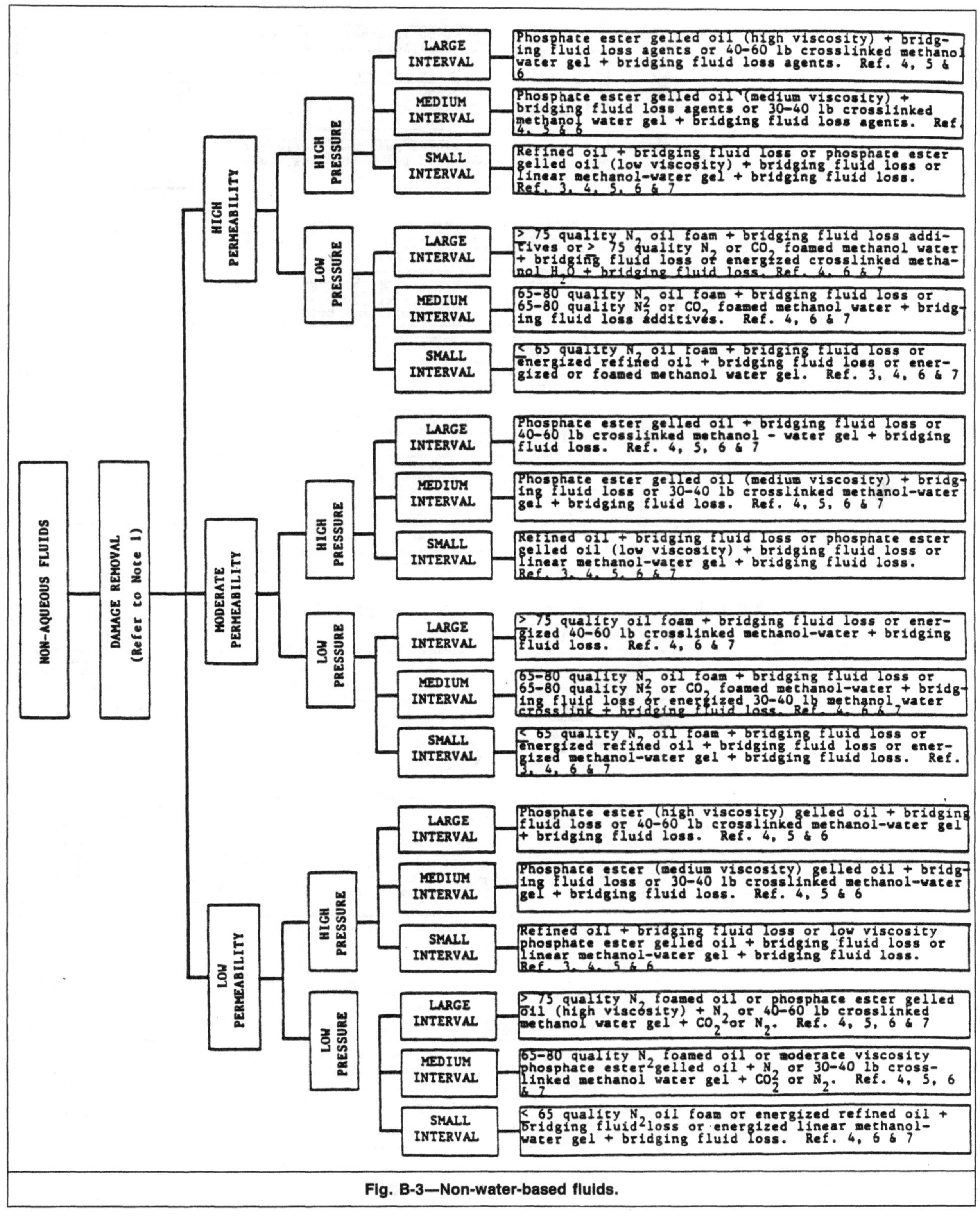

Fig. B-3—Non-water-based fluids.

lbm [14 kg] energized crosslinked guar plus bridging fluid loss; or energized polyemulsion plus bridging fluid loss. (Notes 2, 6, and 7.)

26-B. 70- to 80-quality CO_2 or N_2 foam plus bridging fluid loss or 40 to 50 lbm [18 to 23 kg] energized crosslinked guar plus bridging fluid loss. (Notes 6 and 7.)

27-B. ≥75-quality CO_2 or N_2 foam plus bridging fluid loss or 40 to 60 lbm [18 to 27 kg] energized crosslinked HPG plus bridging fluid loss. (Notes 6 and 7.)

28-B. Linear guar gel plus bridging fluid loss; polyemulsion fluid plus bridging fluid loss; or ≤30 lbm [≤14 kg] guar crosslink plus hydrocarbon fluid loss. (Notes 2 and 6.)

29-B. 30 to 50 lbm [14 to 23 kg] guar crosslink plus hydrocarbon fluid loss or polyemulsion fluid plus bridging fluid loss. (Notes 2 and 6).

30-B. 40 to 50 lbm [18 to 23 kg] HPG crosslink plus hydrocarbon fluid loss or high-temperature-stable linear gel plus secondary gel plus bridging fluid loss. (Note 6.)

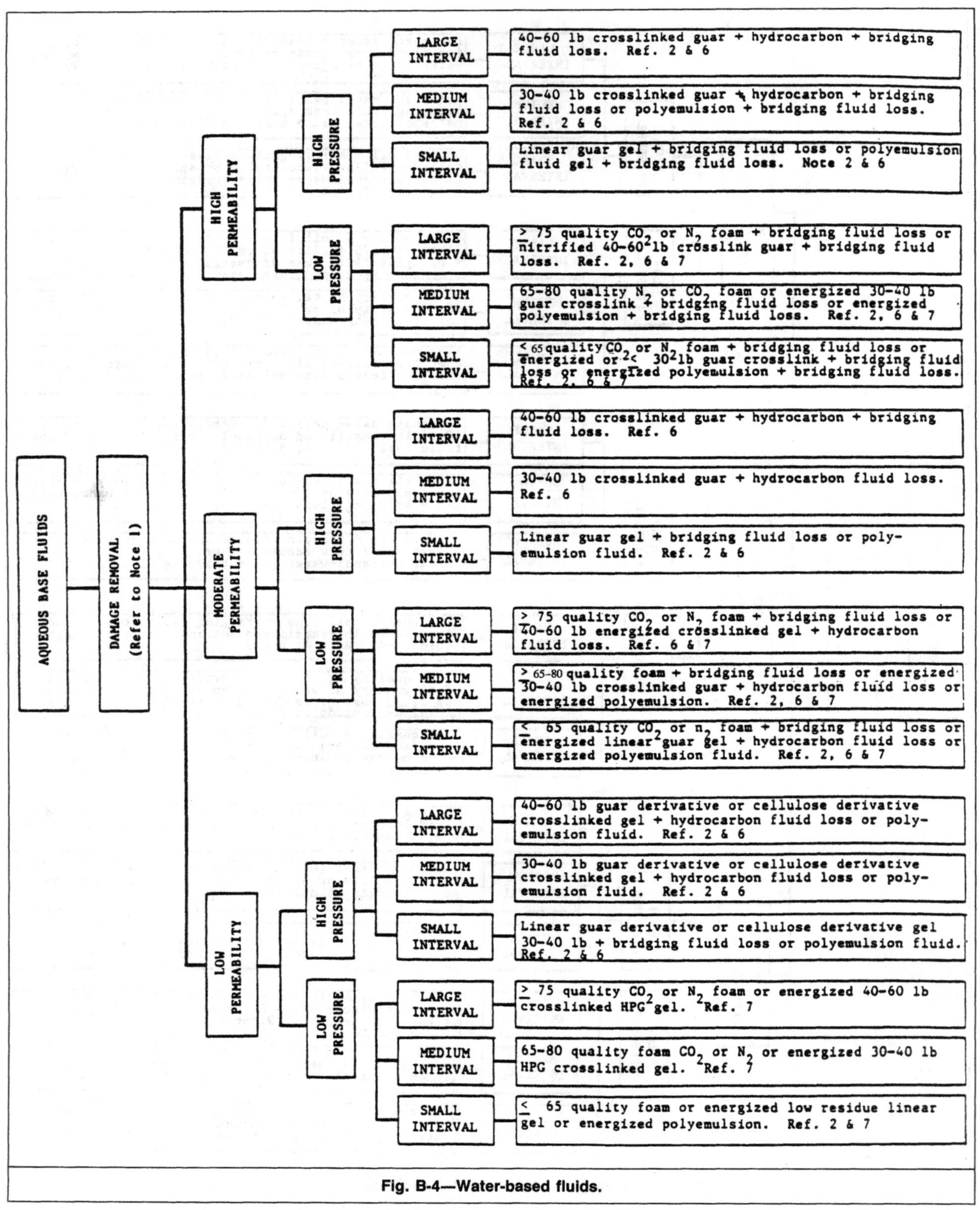

Fig. B-4—Water-based fluids.

31-B. Polyemulsion plus bridging fluid loss or ≤30 lbm [≤14 kg] guar crosslink plus hydrocarbon fluid loss. (Notes 2 and 6.)

32-B. 40 to 60 lbm [18 to 27 kg] guar crosslink plus hydrocarbon fluid loss or polyemulsion plus bridging fluid loss. (Notes 2 and 6.)

33-B. 40 to 60 lbm [18 to 27 kg] HPG crosslink plus hydrocarbon fluid loss or high-temperature-stable linear gel plus secondary gel plus bridging fluid loss. (Notes 2 and 6.)

34-B. Polyemulsion plus bridging fluid loss or 30 to 40 lbm [14 to 18 kg] guar crosslink plus hydrocarbon fluid loss. (Notes 2 and 6.)

35-B. 40 to 60 lbm [18 to 27 kg] guar crosslink plus hydrocarbon fluid loss. (Note 6.)

36-B. 40 to 60 lbm [18 to 27 kg] HPG crosslink plus hydrocarbon fluid loss. (Note 6.)

37-B. ≤65-quality CO_2 or N_2 foam plus bridging fluid loss; ≤30 lbm [≤14 kg] energized guar crosslinked gel plus bridging fluid loss; or energized polyemulsion plus bridging fluid loss. (Notes 2, 6, and 7.)

38-B. 65- to 75-quality N_2 or CO_2 foam plus bridging fluid loss;

30 to 40 lbm [14 to 18 kg] energized crosslinked guar plus bridging fluid loss; or energized polyemulsion plus bridging fluid loss. (Notes 2, 6, and 7.)

39-B. ≥75-quality N_2 or CO_2 foam plus bridging fluid loss or 40 to 60 lbm [18 to 27 kg] energized crosslinked HPG plus bridging fluid loss. (Notes 6 and 7.)

40-B. ≤65-quality CO_2 or N_2 foam plus bridging fluid loss; ≤30 lbm [≤14 kg] energized crosslinked guar plus hydrocarbon plus bridging fluid loss; or energized polyemulsion plus bridging fluid loss. (Notes 2, 6, and 7.)

41-B. 65- to 75-quality CO_2 or N_2 foam plus bridging fluid loss; 30 to 60 lbm [14 to 27 kg] energized crosslinked guar plus bridging fluid loss; or energized polyemulsion plus bridging fluid loss. (Notes 2, 6, and 7.)

42-B. ≥75-quality CO_2 or N_2 foam plus bridging fluid loss or 40 to 60 lbm [18 to 27 kg] crosslinked HPG plus bridging fluid loss. (Notes 6 and 7.)

43-B. ≤70-quality CO_2 or N_2 foam plus bridging fluid loss; ≤30 lbm [≤14 kg] energized crosslinked guar plus bridging fluid loss; or energized polyemulsion plus bridging fluid loss. (Notes 2, 6, and 7.)

44-B. 70- to 80-quality CO_2 or N_2 foam plus bridging fluid loss or 40 to 50 lbm [18 to 23 kg] energized crosslinked guar plus bridging fluid loss. (Notes 6 and 7.)

45-B. ≥75-quality CO_2 or N_2 foam plus bridging fluid loss or 40 to 60 lbm [18 to 27 kg] energized crosslinked HPG plus bridging fluid loss. (Notes 6 and 7.)

46-B. Linear guar gel plus bridging fluid loss; polyemulsion plus bridging fluid loss; or ≤30 lbm [≤14 kg] guar crosslink plus hydrocarbon plus bridging fluid loss. (Notes 2, 6, and 7.)

47-B. 30 to 40 lbm [14 to 18 kg] guar crosslink plus hydrocarbon plus bridging fluid loss or polyemulsion plus bridging fluid loss. (Notes 2, 6, and 7).

48-B. 40 to 50 lbm [18 to 23 kg] HPG crosslink plus hydrocarbon plus bridging fluid loss or high-temperature-stable linear gel plus secondary gel plus bridging fluid loss. (Notes 6 and 7.)

49-B. Polyemulsion plus bridging fluid loss or ≤30 lbm [≤14 kg] guar crosslink plus hydrocarbon and bridging fluid loss. (Notes 2 and 6.)

50-B. 40 to 60 lbm [18 to 27 kg] guar crosslink plus hydrocarbon plus bridging fluid loss or polyemulsion plus bridging fluid loss. (Notes 2 and 6.)

51-B. 40 to 60 lbm [18 to 27 kg] HPG crosslink plus hydrocarbon fluid loss plus bridging fluid loss or high-temperature-stable linear gel plus secondary gel plus bridging fluid loss. (Note 6.)

52-B. Polyemulsion plus bridging fluid loss or 30 to 40 lbm [14 to 18 kg] guar crosslink plus hydrocarbon and bridging fluid loss. (Notes 2 and 6.)

53-B. 40 to 60 lbm [18 to 27 kg] guar crosslink plus hydrocarbon plus bridging fluid loss. (Note 6.)

54-B. 40 to 60 lbm [18 to 27 kg] HPG crosslink plus hydrocarbon and bridging fluid loss. (Note 6.)

Note 1

Because of its very nature, a typical damage-removal treatment is a comparatively small treatment. For this reason, we do not break down in Figs. B-3 and B-4 the low, moderate, and high BHT's. We do suggest that a fairly large cooldown prepad be run ahead of the actual fracture treatment in high-temperature zones. This may also be a good idea in moderate- and low-temperature zones. One must always consider that, when conducting very small fracturing treatments, the actual fracture width, particularly if the zone of interest is quite large, may be quite small even after a fairly high percentage pad volume. We suggest running a fracture design and having at least a 0.2-in. [0.5-cm] fracture width for the entrance of 20/40 sand before sand is initiated. To achieve the needed width, one may need to run an economical cooldown fluid prepad that will also control fluid loss and help obtain the needed fracture width. Obviously, if the damage-removal treatment becomes of moderate size and pump time becomes fairly lengthy, then one certainly would want to move toward the more stable polymers, temperature stabilizers, delayed crosslinkers, etc. Of course, it is quite unusual to run small damage-removal treatments in deep, high-temperature reservoirs.

Note 2

Polyemulsion fluid is recommended as the fluid of choice if the well in question is an oil well and if there is crude oil available to prepare the emulsion. This consideration, of course, is based on economics.

Note 3

Refined oil, which typically means viscous oils for carrying proppant, can be used only if the tubulars or casing treated down can withstand the high friction pressure of these viscous refined oils.

Note 4

The selection of energized methanol/water gel or methanol/water crosslink system over an oil system is based on whether the formation to be treated is a gas well or an oil well. The percentage water one can use in a linear methanol/water gel or a crosslink methanol/water gel has to be based on the particular sensitivity of the formation to be treated. Linear and crosslinked alcohol/water systems up to and including 100% alcohol are available. One needs to confer with the service company about temperature limitations in and investigate potential gel degradation problems with these various polymers and crosslink polymers.

Note 5

The phosphate ester gelled oils have an apparent viscosity based on loading of a liquid additive. By varying the loading of the liquid additive, one can have a fairly-low-viscosity, moderate-viscosity, or high-viscosity system. Low viscosity typically means a loading of some 6 gal/1,000 gal [6 m^3/1000 m^3] or lower; a moderate viscosity means 8 gal/1,000 gal [8 m^3/1000 m^3]; and high viscosity means 10 gal/1,000 gal [10 m^3/1000 m^3] or more.

Note 6

The quantity of bridging fluid-loss additive used in a low-permeability formation will, of course, be lower than that used in a moderate- or high-permeability formation. Because we are discussing permeability ranges, it should be apparent that the very low permeability may not require additives, and formations with 0.2 md may require for instance 20 lbm/1,000 gal [2.4 kg/m^3]. Concentrations of hydrocarbon fluid loss also will vary, depending on permeability. Refer to Chap. 8 for concentration recommendations.

Note 7

Many stimulated wells do not contain energizing gases and cannot unload fluids. These wells are jetted back, swabbed, or pumped after treatment. Fluid sensitivity, tendency to imbibe treatment fluids, or simply relative cost of unloading techniques will dictate whether an energizing medium should be used.

Note that the energizing medium, whether foam or simply added gas, functions as a fluid-loss-control additive and should be replaced if the energizing medium is negated.

Appendix C
Rheological Models and Friction Factors

J.R. Cameron, SPE, Amoco Production Co.
R.K. Prud'homme, Princeton U.

Material Functions for Model Laminar Flows

This section considers two of the most commonly encountered and useful laminar flow fields: steady shear and dynamic oscillatory shear. These velocity fields are shown in Fig. C-1. For these flows, relationships between the stresses and velocity gradients that define "material functions" can be specified. For example, the coefficient relating shear stress in steady flow and velocity gradient is the viscosity. The viscosity, a constant, is sufficient to describe the flow behavior of a Newtonian liquid in any flow field. For a polymeric fluid, however, a different viscosity must be defined for each shear rate; viscosity is a function of shear rate, not a constant coefficient. Furthermore, for polymeric fluids in different flow fields, different material functions must be used. The steady-shear viscosity is not sufficient because there is no direct relationship between the steady-shear viscosity and the elongational viscosity.

Steady-Shear Flows. Consider the flow of a fluid between two parallel plates produced by moving the top plate in the x direction, as shown in Fig. C-1. The stresses generated by the flow act parallel (i.e., shear stresses) and perpendicular to the direction of shear (normal stresses). The experimentally observable stresses perpendicular to the direction of flow include the stress arising from fluid motion and the isotropic hydrostatic pressure. It is customary to eliminate the isotropic pressure by taking the difference between normal stresses; in fact, these are the differences that are experimentally measured, as shown in Sec. 9.3.

Shear stress, τ_{yx}, primary normal stress difference, $\tau_{xx}-\tau_{yy}=N_1$, and secondary normal stress difference, $\tau_{yy}-\tau_{zz}=N_2$, are related to the velocity gradient, $\dot{\gamma}_{yx}$, thereby defining the material functions for steady-shear flow.

Eqs. C-1 through C-3 define the viscosity, primary normal stress coefficient, and secondary normal stress coefficient, respectively:

$$\tau_{yx}\equiv-\mu\dot{\gamma}_{yx}, \quad\text{(C-1)}$$

$$\tau_{xx}-\tau_{yy}\equiv-\psi_1\dot{\gamma}_{yx}^2, \quad\text{(C-2)}$$

and

$$\tau_{yy}-\tau_{zz}\equiv-\psi_2\dot{\gamma}_{yx}^2. \quad\text{(C-3)}$$

These material functions are generally functions of shear rate. The normal stress coefficients are defined in terms of the square of the velocity gradient because the stress difference must be an even power of shear rate; i.e., changing the direction of the shear (making $\dot{\gamma}_{yx}$ negative) does not change the direction or sign of the normal stress, whereas changing the direction of the velocity gradient does change the direction of the shear stress.

Supplementary to Chap. 9.

An example of the viscosity of a hydroxypropyl guar (HPG) solution as a function of shear rate is shown in Fig. 9.8. The normal stresses are generally small for dilute polymer solutions of low molecular weight or polymers having relatively rigid backbones, such as polysaccharides and cellulosics. Fig. C-2 shows the primary normal stress difference of a 40-lbm/1,000-gal [4.8-kg/m^3] HPG solution and a gelled-diesel-based gel. The secondary normal stress difference is exceedingly difficult to measure and its usefulness in practical problems has not been demonstrated. The best estimate is that the secondary normal stress coefficient is $1/10$ to $1/4$ the size of the primary normal stress difference and opposite in sign.

Dynamic Oscillatory Shear Flow. It is possible to impose a sinusoidally varying shear field on a fluid and to measure the amplitude of the torque and the phase angle between the imposed shear and the resulting torque. These measurements are called dynamic oscillatory measurements or dynamic mechanical measurements. For experiments in which the stress is proportional to the imposed strain (i.e., linear viscoelasticity), a large body of literature exists on the relationship between dynamic oscillatory material functions and polymer molecular structure and interactions.[1]

Consider the oscillating velocity field shown in Fig. C-1:

$$v_x=[\dot{\gamma}_{max}\cos(\omega t)]y, \quad\text{(C-4)}$$

where ω is the frequency, $\dot{\gamma}_{max}$ is the maximum velocity gradient, and the maximum value of the strain is given by $\gamma_{max}=\dot{\gamma}_{max}/\omega$. The stress will also be oscillatory and will have some maximum value, τ_{max}, and some phase shift, ϕ, from the imposed shear:

$$\tau_{yx}=\tau_{max}\cos(\omega t-\phi). \quad\text{(C-5)}$$

The stress can be decomposed into two terms: one in phase with the velocity and one 90° out of phase. These can be written in terms of the maximum velocity gradient,

$$\tau_{yx}=-\eta'\dot{\gamma}_{max}\cos(\omega t)-\eta''\dot{\gamma}_{max}\sin(\omega t), \quad\text{(C-6)}$$

where

$$\tau_{max}\cos\phi\equiv-\eta'\dot{\gamma}_{max} \quad\text{(C-7a)}$$

and

$$\tau_{max}\sin\phi\equiv-\eta''\dot{\gamma}_{max}. \quad\text{(C-7b)}$$

This defines the two dynamic viscosity coefficients, η' and η''. η' approaches the zero-shear-rate viscosity in steady shear at low frequencies and is the viscous contribution to τ_{yx} associated with energy dissipation. η'' is the elastic contribution to τ_{yx} associated

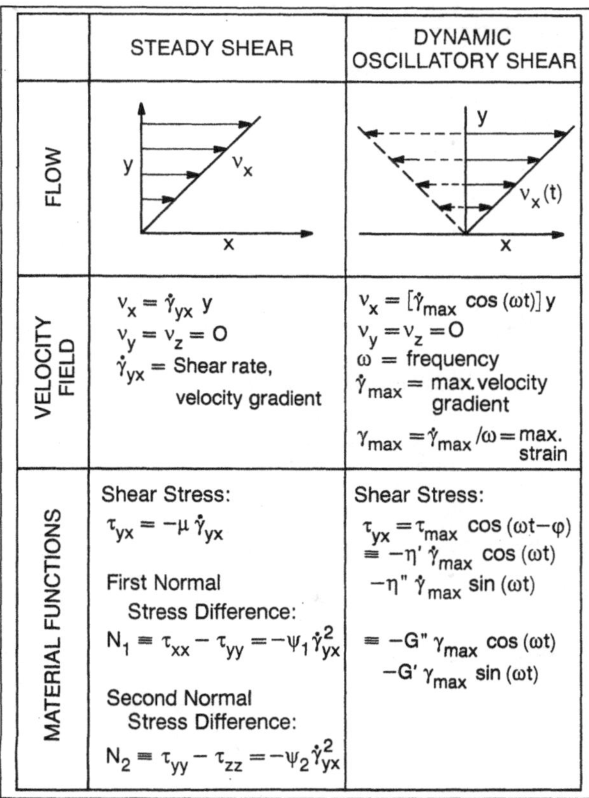

Fig. C-1—Velocity fields and material functions for steady shear and dynamic oscillatory shear between parallel plates.

with energy storage. Alternatively, coefficients can be defined in terms of the maximum strain instead of the strain rate:

$$\tau_{yx} = -G''\gamma_{max} \cos(\omega t) - G'\gamma_{max} \sin(\omega t), \qquad \text{(C-8)}$$

where

$$\tau_{max} \cos\phi \equiv -G''\gamma_{max} \qquad \text{(C-9a)}$$

and

$$\tau_{max} \sin\phi \equiv -G'\gamma_{max}. \qquad \text{(C-9b)}$$

This defines the two functions G' and G'', the storage and loss moduli, respectively. G', related to the stress in phase with strain, qualitatively provides information about the elasticity of a material. For example, an ideal elastic rubber band would have all its stress in phase with strain or displacement. G'', the loss modulus, is related to stress out of phase with displacement and therefore in phase with rate of displacement or shear rate. For a purely viscous liquid, all the stress would be out of phase with displacement.

Keep in mind that linear viscoelasticity assumes that the stress is linearly proportional to strain and that the stress response involves only the first harmonic and not higher harmonics in frequency (i.e., the stress is a sinusoidal). Both of these conditions should be verified experimentally for each material being studied.

These linear viscoelastic dynamic moduli are functions of frequency. They can be related to the structure of polymer solutions and gels. Fig. C-3 shows the dynamic moduli for a polymer solution during gelation.[2] The material begins as a solution in Fig. C-3a and ends as a solid gel in Fig. C-3d. For a polymer solution at low frequency, elastic stresses relax and viscous stresses dominate with the result that the loss modulus, G'', is higher than the storage modulus, G'. Both decrease with decreasing frequency, but G' decreases more quickly. For a gel, the stress cannot relax and therefore is independent of frequency. Also, because the gel is highly elastic, G' is higher than G''.

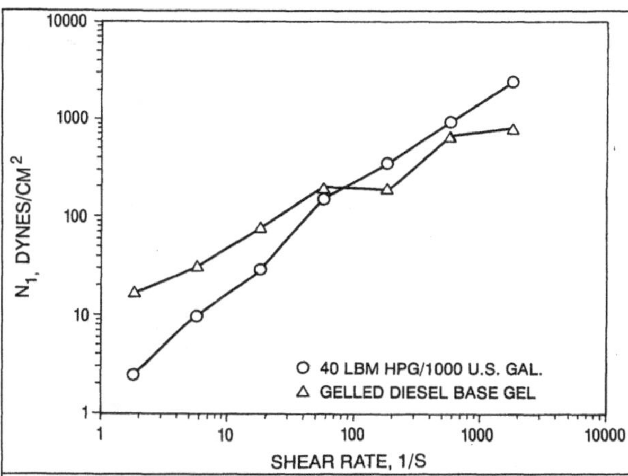

Fig. C-2—First normal stress difference for a 40-lbm/1,000-gal HPG solution and a gelled-diesel-based gel at 75°F and ambient pressure.

Linear viscoelastic measurements can also be used in conjunction with classic polymer kinetic theory to relate the storage modulus of a gel with the number density of crosslinks. By following the storage modulus with time, one can measure the chemical kinetics of gel formation.[3,4] Polymer kinetic theory has shown that the frequency-independent, low-frequency limit of the storage modulus for a gel is equal to

$$G' = \nu k T_a + G_{en}, \qquad \text{(C-10)}$$

where

ν = number density of crosslinks,
k = Boltzmann's constant,
T_a = absolute temperature, and
G_{en} = contribution to modulus from entanglements.

Experiments have shown that G_{en} is very small for dilute polyacrylamide gels.[4]

Rheological Models

Generalized Newtonian Fluid. The simplest rheological equation of state is the Newtonian fluid where only one material function—viscosity—is needed to characterize the fluid in any flow. The constitutive equation for a Newtonian fluid is

$$\vec{\vec{\tau}} = -\mu \vec{\vec{\dot{\gamma}}}, \qquad \text{(C-11)}$$

where $\vec{\vec{\dot{\gamma}}}$, the rate-of-deformation tensor, is defined by

$$\vec{\vec{\dot{\gamma}}} = \nabla\vec{v} + (\nabla\vec{v})^T, \qquad \text{(C-12)}$$

where the superscript T implies the transpose of the tensor. In component form, this tensor becomes

$$\dot{\gamma}_{ij} = \frac{\partial v_j}{\partial x_i} + \frac{\partial v_i}{\partial x_j}, \qquad \text{(C-13)}$$

where x_i are the components of the position vector $\vec{x}$. For a shear flow given in Fig. C-1, the shear stress is given by

$$\tau_{yx} = -\mu\dot{\gamma}_{yx}. \qquad \text{(C-14)}$$

For non-Newtonian fluids, the Newtonian fluid model may be generalized by allowing the viscosity to be a function of shear rate leading to the *generalized Newtonian fluid* model. The viscosity is made a function of the square root of the second invariant of the rate-of-deformation tensor. Continuum-mechanics arguments show that the viscosity can depend only on certain combinations

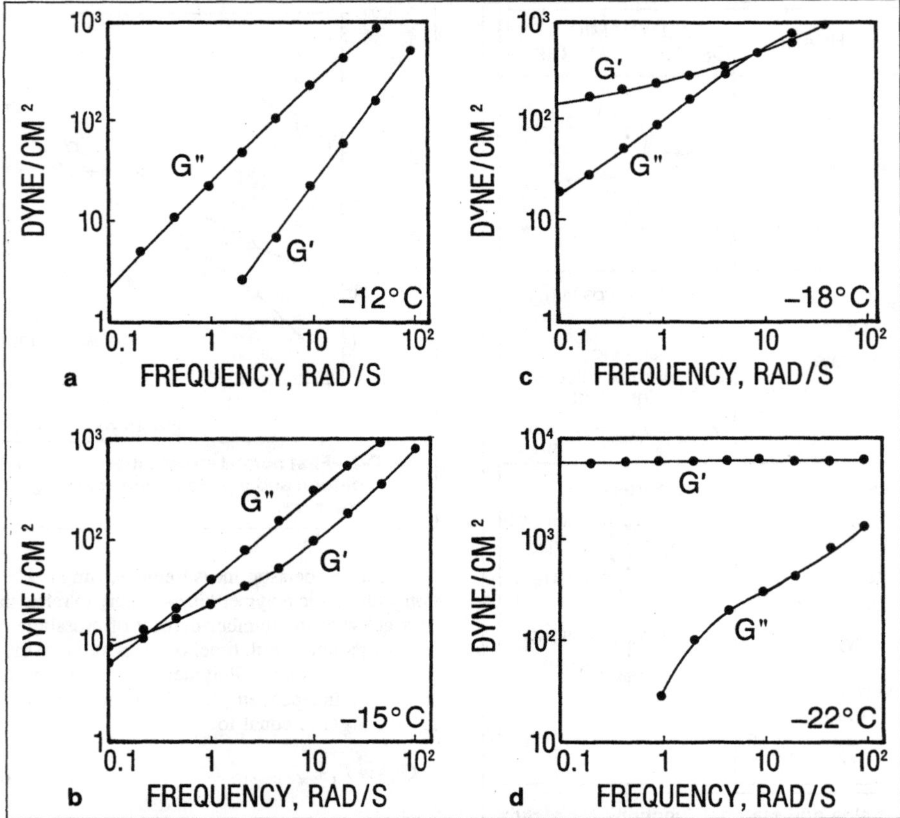

Fig. C-3—Storage, G', and loss, G'', moduli as functions of frequency, ω, for a 9.5 wt% solution of polystyrene in carbon disulfide.[2]

TABLE C-1—EXAMPLES OF RHEOLOGICAL MODELS HAVING SHEAR-RATE-DEPENDENT VISCOSITIES

Model	Parameters	Equation
Power law	n', power-law index K', consistency index	$\mu = K'\dot{\gamma}^{(n'-1)}$
Ellis	μ^o, zero-shear viscosity $\tau_{1/2}$, shear stress, where $\mu = \mu^o/2$ $(\alpha - 1)$, slope of log $(\mu^o/\mu - 1)$ vs. log $(\tau/\tau_{1/2})$	$\dfrac{\mu^o}{\mu} = 1 + \left(\dfrac{\tau}{\tau_{1/2}}\right)^{\alpha-1}$
Carreau	μ^o, zero-shear viscosity μ_∞, infinite-shear viscosity λ, time constant n', power-law index	$\dfrac{\mu - \mu_\infty}{\mu^o - \mu_\infty} = [1 + (\lambda\dot{\gamma})^2]^{(n'-1)/2}$

of the rate-of-deformation tensor called invariants. Of the three independent combinations that can be formed, the second invariant is chosen because the first invariant is equal to zero for an incompressible fluid and the third invariant is equal to zero in a shear flow. The second invariant is defined by

$$\text{II} = \sum_i \sum_j \dot{\gamma}_{ij}\dot{\gamma}_{ji}. \quad \text{(C-15)}$$

Instead of II directly, the viscosity is made a function of the square root of ½II, which is the magnitude of the rate-of-deformation tensor and is symbolized by

$$\dot{\gamma} = \sqrt{\tfrac{1}{2}\sum_i \sum_j \dot{\gamma}_{ij}\dot{\gamma}_{ji}} = \sqrt{\tfrac{1}{2}\text{II}} \quad \text{(C-16)}$$

For shear flow, the magnitude of the rate-of-deformation tensor is just equal to the velocity gradient or shear rate.

The generalized Newtonian fluid model works well for modeling steady-shear flows but not for transient or elongational flows.

One of the most popular generalized Newtonian fluid models is the power-law model. This two-parameter model describes the curve

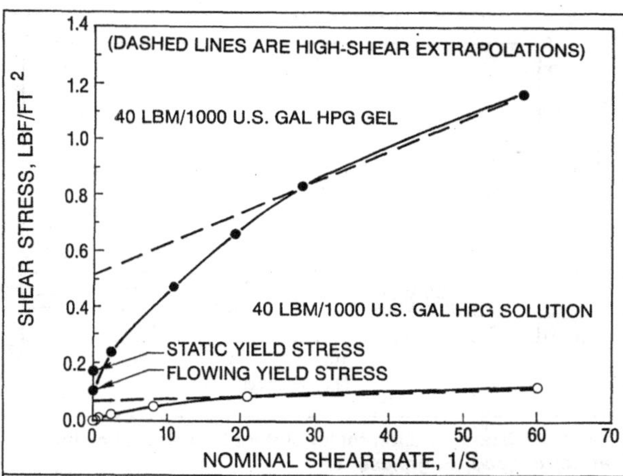

Fig. C-4—Flow curves of shear stress vs. shear rate for an HPG solution and a batch-mixed HPG/titanium gel at 74°F. Tests were done in a capillary tube by flowing under constant gas pressure.

of viscosity vs. shear rate on a log-log plot as a straight line. The viscosity in terms of the shear rate and the magnitude of the shear stress, τ, is

$$\mu=K'\dot{\gamma}^{(n'-1)}=\tau/\dot{\gamma}, \qquad \text{(C-17)}$$

where n' and K' are the power-law parameters. This model is attractive because it can be used conveniently in the analytical solution of flow problems, it describes the high-shear-rate viscosity behavior of polymer solutions under process conditions, and it requires only two parameters. However, it cannot describe the constant low-shear-rate viscosity observed at low flow rates, as shown in Fig. 9.8. Therefore, it fails for slow flows where the characteristic shear rates are in the Newtonian or transition regime. Other three-parameter models must be chosen to model these flows. Several models are shown in Table C-1 in terms of the shear rate and the magnitude of the shear stress.

For reacting systems, like crosslinked fracturing fluids, it is often desirable to express the time dependence of the fluid rheology. This has been done by making the parameters in the power-law model functions of time:

$$n'=n'(t) \qquad \text{(C-18a)}$$

and

$$K'=K'(t). \qquad \text{(C-18b)}$$

This method of modeling crosslinking fluids has serious shortcomings because it cannot model the effect of shear history on gel rheology.[5] Turian,[6] who made the power-law parameters a function of temperature, used a similar concept for nonisothermal problems:

$$n'=n^0+\frac{B'(T_a-T_0)}{T_0} \qquad \text{(C-19a)}$$

and

$$K'=K_0'e^{-A''(T_a-T_0)/T_0}. \qquad \text{(C-19b)}$$

The parameters in Eqs. C-19 were established for a number of water-soluble parameters.

Models With Yield Stresses. Very concentrated emulsions, foams, dispersions, and gels display yield stresses. A true yield stress implies that the material will not flow until some minimum stress is achieved. In many systems, there may be a very slow flow at low shear stresses; however, over the time scale of practical importance, the material can be treated as if it had not yielded. This can be called an "apparent" yield stress.

Fig. C-4 shows a plot of shear stress vs. nominal shear rate for an HPG solution and a batch-mixed HPG gel at room temperature. These data were taken by forcing the fluids through capillary tubing using air at regulated pressures. For the HPG gel, flow was initiated by increasing gas pressure in steps until flow was observed. The corresponding wall shear stress at which flow was initiated is shown in Fig. C-4 as the "static" yield stress and has a value of 0.173 lbf/ft^2 [8.28 Pa]. When the gas pressure was reduced, flow stopped at a nonzero pressure corresponding to a wall shear stress of 0.115 lbf/ft^2 [5.51 Pa] and is referred to as the "flowing" yield stress in Fig. C-4. These yield stresses are the apparent or actual yield stresses.

When the HPG solution was tested, however, flow was observed at all pressures tested. At the lowest pressure tested, the corresponding wall shear stress was 0.000907 lbf/ft^2 [0.043 Pa], and the nominal shear rate was about 0.1 second^{-1}, giving a nominal viscosity of about 430 cp [430 mPa·s]. Extrapolation of the lowest-shear-rate data to zero shear rate resulted in a zero yield stress, as expected for this rather dilute HPG solution.

Fig. C-4 also shows what would happen if the yield stress were estimated by extrapolation from high-shear-rate data. The extrapolated yield stresses are considerably higher than the actual yield stresses, giving a finite yield stress for the HPG solution that is clearly erroneous.

Models describing the shear stress vs. shear rate for materials with yield stress have been reviewed[6] and are presented in Table C-2 in terms of shear rate and the magnitude of the shear stress.

Models With Elastic Effects. Polymer rheologists have constructed an impressive array of models meant to represent viscoelastic fluids. A survey of these models can be found in Table 9.4-1 of Ref. 6. Most of these models were developed by researchers involved in polymer melt processing where elastic effects often are quite large. In fracturing and oil-recovery fluids, viscoelastic effects generally are secondary rather than primary effects.

TABLE C-2—RHEOLOGICAL MODELS FOR MATERIALS WITH YIELD STRESS

Model	Parameters	Equation
Bingham	τ_o, yield stress	$\tau=\tau_o+\mu_p\dot{\gamma}$ for $\tau>\tau_o$
	μ_p, plastic viscosity	$\tau\le\tau_o$ for $\dot{\gamma}=0$
Casson	τ_o, yield stress	$\sqrt{\tau_{yx}}=\sqrt{\tau_o}+\sqrt{\mu_p}\sqrt{\dot{\gamma}}$ for $\tau>\tau_o$
	μ_p, plastic viscosity	$\tau\le\tau_o$ for $\dot{\gamma}=0$
Herschel-Bulkley	τ_o', yield stress	$\tau=\tau_o'+K''\dot{\gamma}^{n''}$ for $\tau>\tau_o'$
	K'' consistency index	
	n'' shear rate exponent	$\tau\le\tau_o'$ for $\dot{\gamma}=0$

Turbulent-Flow Friction Factors

This section will describe the steps involved in going from the equation of motion that is most useful in laminar flow to the macroscopic balance equation used for the solution of turbulent flow problems. The goal is to show the reader the difference between a "material function," which describes the properties of a fluid, and the "friction factor," which involves information on the fluid and the flow geometry.

No analytical solutions are available for turbulent flow because the fluid velocities fluctuate randomly. Consequently, engineering design must be done by referring to correlations determined from experimental data. The experimental correlations are presented in terms of friction factor, f, vs. Reynolds number, N_{Re}. Below we briefly show the relationship between the equation of motion and friction factor.

Following Bird *et al.*,[7] an equation of mechanical energy can be derived by forming the scalar product of $\vec{v}$ times the equations of motion to obtain the equation of mechanical energy. For steady, turbulent flow, the fluid velocity vectors are replaced by an average fluid velocity, $\bar{v}$, assumed constant across the flow direction. The mechanical energy equation is integrated over the volume of the system to yield[7]

$$\left(\Delta \tfrac{1}{2}\bar{v}^2 + g\Delta z + \int_{p_1}^{p_2} \frac{1}{\rho}\,dp\right)w + \frac{dW}{dt} + E_v = 0, \quad \text{(C-20)}$$

which is sometimes called the Bernoulli equation. The term involving the integral over pressure can be simplified for an incompressible fluid[7]:

$$\int_{p_1}^{p_2} \frac{1}{\rho}\,dp = \frac{1}{\rho}(p_2 - p_1). \quad \text{(C-21)}$$

For a foam, the compressibility of the gas phase must be considered and the integral evaluated. The term dW/dt is the rate at which the system performs work on the surroundings; e.g., the energy required to pump a fluid would be $-dW/dt$. w is the mass flow rate, and E_v is the rate at which energy is irreversibly converted to thermal energy by dissipation. Empirical relations for energy dissipation, E_v, for various fittings, elbows, and straight pipe sections are available in the form

$$E_v = \tfrac{1}{2} w \bar{v}^2 \frac{L}{r_H} f \quad \text{(C-22)}$$

for straight pipes and

$$E_v = \tfrac{1}{2} w \bar{v}^2 e_v \quad \text{(C-23)}$$

for bends, elbows, etc., where

- L = length of pipe,
- r_H = hydraulic radius defined as cross-sectional area divided by wetted perimeter (r_H=one-half the pipe radius),
- $\bar{v}$ = average velocity,
- e_v = friction loss factor, and
- f = friction factor.

The friction factor depends on the Reynolds number and the roughness, ϵ, of the pipe, as shown in Fig. C-5.[8] The friction factor therefore is not a fundamental quantity and cannot predict flow behavior. It is only a convenient way of presenting experimental data. It applies only to the specific geometry used to obtain the experimental data and only to fluids with similar rheology. For example, the addition of small amounts of polymer suppresses the onset of turbulence in pipe flow and changes the friction-factor/Reynolds-number relationship markedly.

As an example of the use of friction factors, calculate the pressure drop at a given flow rate, q, in a long section of vertical pipe of diameter d.

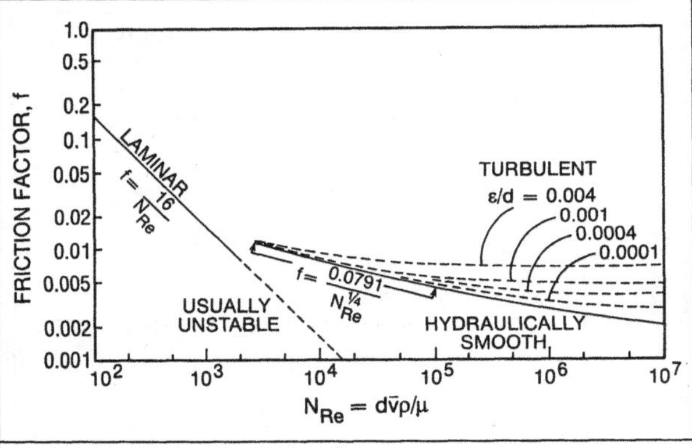

Fig. C-5—Moody-friction-factor plot for tube flow showing the effect of tube roughness, ϵ/d.[8]

1. Calculate the Reynolds number to check for turbulence; i.e., $N_{Re} \geqq 2{,}100$ implies turbulence.
2. For flow in a pipe, the average velocity is constant across any cross section; therefore, $\Delta\bar{v}^2=0$. No external work is done on the fluid in the pipe; therefore, $dW/dt=0$.
3. Eq. C-20, the Bernoulli equation, is written with Eqs. C-21 and C-22 as

$$g\Delta z + 1/\rho(\Delta p) + \tfrac{1}{2}\bar{v}^2 \frac{L}{r_H} f = 0.$$

The pressure drop is found by rearranging this equation:

$$\Delta p = -\rho g \Delta z - \frac{\rho}{2}\bar{v}^2 \frac{L}{r_H} f,$$

where Δz=height of the pipe, L. The average velocity, $\bar{v}$, is calculated from the flow rate ($\bar{v}=q/\pi r_t^2$); the hydraulic radius is $r_t/2$; and the friction factor is determined from an appropriate correlation with N_{Re} such as shown in Fig. C-5.

Nomenclature

$A'', B', K_0',$
n^0, T_0 = parameters in Turian's model, Eqs. C-19
d = pipe or casing ID, in. [cm]
e_v = friction-loss factor
E_v = rate of energy dissipation to heat
f = friction factor
g = gravitational acceleration
G' = storage modulus
G'' = loss modulus
G_{en} = contribution to modulus from molecular entanglements
k = Boltzmann's constant
K' = power-law index, lbf-sec^n/ft^2
K'' = Herschel-Bulkley consistency index
L = length, ft [m]
n' = power-law index
n'' = Herschel-Bulkley shear-rate exponent
N_{Re} = Reynolds number
N_1 = primary normal stress difference
N_2 = secondary normal stress difference
p = pressure, psi [kPa]
p_1 = lower integral limit, Eq. C-20
p_2 = upper integral limit, Eq. C-20
Δp = fluid pressure drop
q = flow rate, bbl/min [m^3/min]
r_H = hydraulic radius, in. [cm]
r_t = radius of pipe or tubing

t = time
T_a = absolute temperature
$\vec{v}$ = velocity vector
$\bar{v}$ = average fluid velocity, ft/sec [m/s]
v_i = velocity component, $i=x, y, z$
w = mass flow rate
dW/dt = rate of work performed on surroundings by system
$\vec{x}$ = position vector
x_i = component of position vector
y = rectilinear coordinate (horizontal)
Δz = vertical interval
α = Ellis model exponent
$\dot{\gamma}$ = shear rate
$\dot{\gamma}_{i,j}$ = component of rate-of-deformation tensor; $i, j=x, y,$ or z
γ_{max} = maximum dynamic strain
$\dot{\gamma}_{max}$ = maximum dynamic shear rate
$\vec{\vec{\dot{\gamma}}}$ = rate-of-deformation tensor
ϵ = roughness
η' = dynamic viscosity
η'' = imaginary part of complex viscosity
λ = time constant in Carreau model
μ = viscosity
μ_p = plastic (Bingham) viscosity
μ^o = zero-shear viscosity
μ_∞ = infinite-shear viscosity
ν = number density of crosslinks
ρ = density, lbm/gal [kg/m^3]
τ = shear stress
$\tau_{i,j}$ = component of extra stress tensor, $i, j=x, y$ or z
τ_{max} = maximum dynamic shear stress
τ_o = yield stress, Bingham model
$\vec{\vec{\tau}}$ = extra stress tensor
$\tau_{½}$ = Ellis model shear stress at $\mu^o/2$
ϕ = phase shift
ψ_1 = primary normal stress coefficient
ψ_2 = secondary normal stress coefficient
ω = dynamic frequency

References

1. Ferry, J.O.: *Viscoelastic Properties of Polymers,* third edition, John Wiley and Sons Inc., New York City (1980).
2. Clark *et al.*: "Rheological Properties of Polystyrene-Carbon Disulfide Gels," *Polymer Preprints* (1983) **24,** No. 2, 86–87.
3. Prud'homme, R.K. *et al.*: "Rheological Monitoring of the Formation of Polyacrylamide/Cr^{+3} Gels," *SPEJ* (Oct. 1983) 804–08.
4. Uhl, J.T.: "Rheological Studies of Water-Soluble Polymer Solutions with Interacting Solutes," PhD dissertation, Princeton U., Princeton, NJ (1983).
5. Prud'homme, R.K.: "Rheological Characterization of Fracturing Fluids," final reports, PRAC Projects 84–45 and 85–45, API, Dallas (1984–85).
6. Bird, R.B., Armstrong, R.C., and Hassager, D.: *Dynamics of Polymeric Liquids: Vol. 1, Fluid Mechanics*, John Wiley and Sons Inc., New York City (1977) 233–35.
7. Bird, R.B., Stewart, W.E., and Lightfoot, E.N.: *Transport Phenomena*, John Wiley and Sons Inc., New York City (1960) 211–13.
8. Moody, L.F.: *Trans.*, ASME (1944) **66,** 671–84; presented in McCabe, W.L. and Smith, J.C.: *Unit Operations of Chemical Engineering,* McGraw-Hill Book Co. Inc., New York City (1956) 67–69.

SI Metric Conversion Factors

dynes/cm^2	× 1.0*	E−01	=	Pa
°F	(°F−32)/1.8		=	°C
lbf/ft^2	× 4.788 026	E−02	=	kPa
lbm/gal	× 1.198 264	E+02	=	kg/m^3

*Conversion factor is exact.

Appendix D
Turbulent Behavior of Solutions

J.R. Cameron, SPE, Amoco Production Co.
R.K. Prud'homme, Princeton U.

Turbulent Behavior of Newtonian and Inelastic Non-Newtonian Solutions

Dimensional analysis shows that turbulent flow of Newtonian fluids in pipes can be described using the Fanning friction factor, f, correlated with the Reynolds number, N_{Re}, and a parameter descriptive of the wall roughness, e.g., the relative roughness, ϵ/d, where ϵ is the effective height of the protuberances and d is the pipe ID.

For smooth pipes, a commonly used correlation for turbulent flow is the Blasius relation:

$$f \equiv \frac{2\tau_w g_c}{\rho \bar{v}^2} = \frac{d\Delta p_f g_c}{2L\rho\bar{v}^2} = (0.079)/(N_{Re})^{0.25}. \quad \text{(D-1)}$$

Another commonly used correlation is

$$f = (0.046)/(N_{Re})^{0.2}, \quad \text{(D-2)}$$

where

N_{Re} = $\rho d\bar{v}/\mu$,
g_c = gravitational dimensional constant, and
$\Delta p_f/L$ = tubular pressure drop (friction pressure gradient).[1]

The effect of pipe roughness should be considered when fracturing solutions are pumped through a tubing string that is subject to corrosion, scale formation, and particulate erosion (see Sec. 9.4). This is particularly important at higher Reynolds numbers, where roughness becomes more important.

For Newtonian fluids, Moody[2] constructed a friction-factor chart with relative roughness as a parameter (refer to Fig. C-5). In turbulent flow, the wall roughness has no effect on the flow if the surface protuberances do not extend through the laminar sublayer. In this case, flow is hydraulically smooth. As surface protuberances extend through the laminar sublayer into the turbulent region, they increase turbulence. Beyond a certain value of relative roughness, the inertial forces resulting from flow around surface protuberances completely dominate viscous forces, and flow is hydraulically rough with a friction factor independent of N_{Re}. Because the thickness of the laminar sublayer is reduced with increasing N_{Re}, flow may change from hydraulically smooth to hydraulically rough as pump rate increases.[1] For Newtonian fluids in hydraulically smooth turbulent flow, the friction pressure will vary with approximately the 1.8 power of the flow rate; in hydraulically rough flow, it will vary with the square of the flow rate. Because in laminar flow the friction pressure varies with the flow rate to the first power or less (shear-thinning fluids), the transition between various flow regimes is usually quite evident. Fig. D-1 illustrates these trends for a hypothetical Newtonian fluid.

Fracturing solutions, however, are typically non-Newtonian and very often viscoelastic; Newtonian turbulence correlations, such as the Blasius relation, do not apply in these cases. For fluid rheology more complex than Newtonian, it is evident from dimensional analysis that additional dimensionless groups containing the pertinent rheological parameters are required to correlate the friction factor.

For example, the power-law model has two rheological parameter constants: the consistency index, K', and the power-law index, n'. Resulting correlations for inelastic power-law fluids typically involve two dimensionless parameters, n' and the Metzner-Reed generalized Reynolds number, N'_{Re},

$$f = f(N'_{Re},\ n') \quad \text{(D-3)}$$

with

$$N'_{Re} = \frac{\rho d^{n'} \bar{v}^{(2-n')}}{g_c K_p 8^{(n'-1)}},$$

where

$$K_p = K'\left(\frac{6n'+2}{8n'}\right)^{n'}$$

is the pipe-flow-consistency index. (See Eq. 9.3 for N'_{Re} in oilfield units.)

One such correlation is the Dodge-Metzner[3] correlation:

$$\frac{1}{\sqrt{f}} = \frac{4.0}{(n')^{0.75}} \log_{10}[N'_{Re} f^{(1-n'/2)}] - \frac{0.40}{n'^{1.2}}. \quad \text{(D-4)}$$

For the case of inelastic non-power-law behavior (varying n' and K_p), the Dodge-Metzner correlation can still be used, but the n' and K' have to be evaluated at a laminar wall shear stress corresponding to that for turbulent flow. This requires a trial-and-error solution for N'_{Re} because the wall shear stress must be evaluated from an unknown f.

For inelastic pseudoplastic fluids, Hanks and Ricks[4] developed a procedure for generating friction-factor curves as functions of N'_{Re} and n' using a mixing-length model. This supposedly is an improvement over the Dodge-Metzner correlation in that it predicts friction factors for both transitional and turbulent flow conditions and permits the computation of velocity distributions.

In general, these correlations predict a downward displacement of the turbulent portion of the Newtonian-friction-factor curve as n' decreases from 1.0. Thus, purely viscous power-law fluids would exhibit friction pressures varying with flow rate to some power less than 1.8, depending on the value of n'.

Supplementary to Chap. 9.

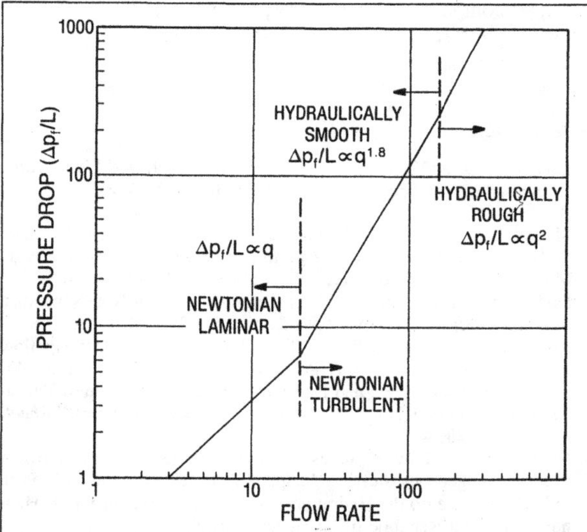

Fig. D-1—Qualitative turbulent flow behavior for a hypothetical Newtonian fluid.

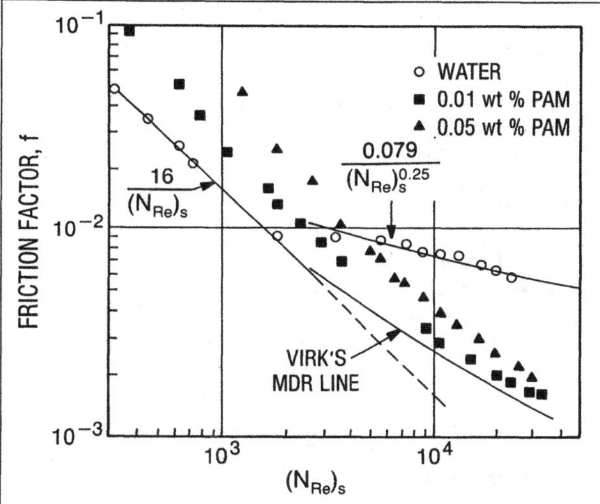

Fig. D-2—Friction factor vs. solvent Reynolds number for 0.01 and 0.05 wt% partially hydrolyzed polyacrylamide solutions. (Data adapted from Ref. 7.)

For inelastic power-law fluids in rough pipes, Szilas *et al.*[5] developed a correlation in terms of the relative roughness and N'_{Re} applicable to the transition region from smooth to totally rough. They found their correlation to be superior to those correlations that did not account for wall roughness when applied to the flow of a pseudoplastic crude oil in a 12-in. [30.5-cm] -diameter oil pipeline.

For inelastic power-law fluids with yield stress, Hanks[6] developed a procedure for calculating the friction factor using the Herschel-Bulkley model parameters n'', K'', and τ'_0 (yield stress), along with d and ρ. This yields a curve of f as a function of N'_{Re}, n'', and the Hedstrom number, defined as

$$N_{He}=\frac{d^2\rho}{\tau'_0}\left(\frac{\tau'_0}{K''}\right)^{2/n''} \quad \text{(D-5)}$$

Hanks' correlation was originally designed for low-Reynolds-number turbulent flow of pseudohomogeneous slurries and may have application to proppant slurries of fracturing solutions if they are inelastic and have a yield stress.

Turbulent Behavior of Viscoelastic Solutions

The majority of fracturing solutions are thought to be drag-reducing and non-Newtonian shear-thinning. Drag reduction occurs when small concentrations of polymer—e.g., <100 ppm by weight—reduce friction pressure at a given turbulent flow rate below that of the solvent alone. Drag reduction can be expressed as a ratio of the polymer-solution friction factor to that of the solvent (f/f_s) at the same solvent Reynolds number.[7] Drag reduction has been observed for the commonly used fracturing solutions made from guar, hydroxypropyl guar (HPG), carboxymethyl HPG, hydroxyethylcellulose (HEC), carboxymethyl HEC, and carboxymethyl cellulose, and for some oil-based solutions. Special turbulent design correlations must be developed for drag-reducing fluids.

The mechanism for drag reduction has not been clearly established, although several ideas have been discussed.[8] Most probably, drag reduction is somehow related to the viscoelasticity imparted to the fluid by addition of the drag-reducing polymer. A descriptive parameter for viscoelasticity is the relaxation time, t_p, which is an indicator of the time required for a polymer molecule to regain its rest conformation after being deformed. Dimensional analysis for the turbulent condition shows that consideration of the fluid relaxation time introduces a new dimensionless group sometimes referred to as the Deborah number:

$$N_{De}=\nu_t t_p, \quad \text{(D-6)}$$

where ν_t is the turbulent-eddy dissipative frequency. Thus, a correlation for a viscoelastic power-law fluid could be written as a function of N_{De}, N'_{Re}, and n'.

To be useful, expressions for N_{De} must be developed. One such relation proposed by Seyer and Metzner[9] that equated stretch rate to ν_t is

$$N_{De}=\frac{\bar{v}}{d}(N'_{Re})^{3/4}t_p. \quad \text{(D-7)}$$

One possible relation for t_p that comes from a simple viscoelastic fluid model (the Maxwell model) involving the measured shear stress, shear rate, and primary normal stress difference is

$$t_p=\frac{\tau_{xx}-\tau_{yy}}{2\tau_w\dot{\gamma}}=\frac{N_1}{2\tau_w\dot{\gamma}}. \quad \text{(D-8)}$$

Expressions for N_{De} show that at constant N_{Re} and fluid velocity, increasing diameter reduces N_{De}.[7,9] Because drag reduction monotonically decreases (i.e., f/f_s increases) with decreasing N_{De}, this is consistent with the experimentally observed decrease in drag reduction as pipe diameter increases.[8-10]

Drag reduction increases as concentration of polymer increases from zero until a maximum is attained at some rather dilute optimum concentration,[9] e.g., 100 ppm by weight for polyacrylamide solutions.[7,11] The existence of a maximum in drag reduction (MDR) has been postulated by Virk *et al.*[12] and is represented by an asymptote on the curve of friction factor vs. solution Reynolds number:

$$f_V=0.42(N_{Re})_s^{-0.55}, \quad \text{(D-9)}$$

where $(N_{Re})_s$ is the Reynolds number based on the solvent properties. At concentrations higher than the optimum, additional polymer will increase the viscosity but may cause little additional increase in the relaxation time. The net effect is to decrease drag reduction. These trends are illustrated in Fig. D-2 for partially hydrolyzed polyacrylamide solutions.[7]

Drag reduction is greatest for long-chain flexible molecules of high molecular weight and relatively short side chains. Molecular conformation changes resulting from changes in chemical environment (e.g., ionic strength for charged macromolecules) can also affect drag reduction.[11]

Several correlations for viscoelastic fluids that incorporate relaxation time have been developed.[7,10,13] These correlations require rheology data, friction pressure data, or both to include the viscoelastic and shear-thinning effects properly. The generality of these correlations is limited by model assumptions. At the higher polymer concentrations used in fracturing solutions, the technique of Darby and Chang[7] may have potential.

Nomenclature

d = pipe ID, in. [cm]
f = friction factor
f_s = solvent friction factor
f_V = Virk's friction factor for maximum drag reduction
g_c = gravitational dimensional constant
K' = power-law index, lbf-sec$^{n'}$/ft^2 [Pa·s$^{n'}$]
K'' = Herschel-Bulkley consistency index
K_p = power-law consistency index for pipe flow, lbf-sec$^{n'}$/ft^2 [Pa·s$^{n'}$]
L = length, ft [m]
n' = power-law index
n'' = Herschel-Bulkley shear-rate exponent
N_{De} = Deborah number
N_{He} = Hedstrom number
N_{Re} = Reynolds number
N'_{Re} = generalized Reynolds number
$(N_{Re})_s$ = Reynolds number based on solvent properties
N_1 = first normal stress difference
Δp_f = friction pressure drop, psi [kPa]
q = flow rate, bbl/min [m^3/min]
t_p = polymer relaxation time
$\bar{v}$ = average fluid velocity
$\dot{\gamma}$ = shear rate
ϵ = roughness
μ = viscosity, cp [mPa·s]
ν_t = turbulent eddy dissipation frequency
ρ = density, lbm/gal [kg/m^3]
τ_{ij} = component of extra stress tensor, $i, j=x, y,$ or z
τ'_0 = yield stress, Herschel-Bulkley model
τ_w = wall shear stress

References

1. Bennett, C.O. and Meyers, J.E.: *Momentum, Heat, and Mass Transfer*, second edition, McGraw-Hill Book Co. Inc., New York City (1974) 163–64.
2. Moody, L.F.: *Trans.*, ASME (1944) **66**, 671–84; presented in McCabe, W.L. and Smith, J.C.: *Unit Operations of Chemical Engineering*, McGraw-Hill Book Co. Inc., New York City (1956) 67–69.
3. Dodge, D.W. and Metzner, A.B.: "Turbulent Flow of Non-Newtonian Systems," *AIChE J.* (June 1959) **2**, No. 5, 189–204.
4. Hanks, R.W. and Ricks, B.L.: "Transitional and Turbulent Pipeflow of Pseudoplastic Fluids," *J. Hydronautics* (Jan. 1975) **9**, No. 1, 39–43.
5. Szilas, A.P., Bobok, E., and Navratil, L.: "Determination of Turbulent Pressure Loss of Non-Newtonian Oil Flow in Rough Pipes," *Rheol. Acta* (1981) **20**, 487–96.
6. Hanks, R.W.: "Low Reynolds Number Turbulent Pipeline Flow of Pseudohomogeneous Slurries," paper C2 presented at the 1978 Fifth Intl. Conference on the Hydraulic Transport of Solids in Pipes, Hannover, Federal Republic of Germany, May 8–11.
7. Darby, R. and Chang, H.D.: "Generalized Correlation for Friction Loss in Drag Reducing Polymer Solutions," *AIChE J.* (March 1984) **30**, No. 2, 274–80.
8. Savins, J.G.: "Drag Reduction Characteristics of Solutions of Macromolecules in Turbulent Pipe Flow," *SPEJ* (Sept. 1964) 203–14; *Trans.*, AIME, **231**.
9. Seyer, F.A. and Metzner, A.B.: "Turbulent Flow Properties of Viscoelastic Fluids," *Cdn. J. Chem. Eng.* (June 1967) **45**, 121–26.
10. Astarita, G., Greco, G. Jr., and Nicodemo, L.: "A Phenomenological Interpretation and Correlation of Drag Reduction," *AIChE J.* (July 1969) **15**, No. 4, 564–67.
11. Sandjani, N.C., Sangster, J.M., and Schreiber, H.P.: "Dependence of Drag Reduction on Polymer Size Distribution and Configuration," *Rheology, Vol. 2: Fluids*, G. Astarita, G. Marrucci, and L. Nicolais (eds.), Plenum Press, New York City (1980) 359–64.
12. Virk, P.S. *et al.*: "The Toms Phenomenon: Turbulent Pipe Flow of Dilute Polymer Solutions," *J. Fluid Mech.* (1967) **30**, 305–28.
13. Seyer, F.A. and Metzner, A.B.: "Turbulence Phenomena in Drag Reducing Systems," *AIChE J.* (May 1969) **15**, No. 3, 426–34.

Appendix E
Numerical Computation of Sand Settlement

Ali Daneshy, SPE, Halliburton Services

Consider a hydraulic fracture with length L, maximum width w_{max}, and height h. Suppose the fracture is divided into small elements, with x denoting the distance between each nodal point and the wellbore and w denoting the fracture width at x (Fig. E-1).

As the fracture propagates away from the wellbore, it reaches x at time t:

$$x_i = A_1 t_i^m \quad \text{(E-1)}$$

or $t_i = (x_i/A_1)^{1/m}$.

The treatment time can be divided into small increments, Δt_n, so that at any time $t=t_n$,

$$t_n = \Sigma \Delta t_n.$$

At each time t_n, the volume of each increment of fracture length, V_{fni}, is computed from

$$V_{fni} = \tfrac{1}{2}(w_{i+1}+w_i)\Big|_{t=t_n}(x_{i+1}-x_i)h. \quad \text{(E-2)}$$

At time $t=t_n$, the point x_i has been exposed to fluid for t_{eni}, where

$$t_{eni} = t_n - t_i.$$

The leakoff volume, V_{Lni}, at $t=t_n$ in Element i is given by

$$V_{Lni} = (x_{i+1}-x_i)\{f(t_{eni})+f[t_{en(i+1)}]\}h. \quad \text{(E-3)}$$

The leakoff function, $f(t)$, is usually determined experimentally and has a shape similar to the curve in Fig. E-2. The mathematical form of $f(t)$ is

$$f(t)=\begin{cases} V_{sp}\sqrt{t/t_{sp}}\,, & t<t_{sp} \\ V_{sp}+2C_w\sqrt{t-t_{sp}}\,, & t\geq t_{sp} \end{cases}.$$

As the fluid leaks into the formation, it travels away from the borehole wall. The distance, Y_{ni}, that the leakoff fluid travels in the formation at $t=t_n$ and $x=x_i$ is

$$Y_{ni} = f(t_{eni})\phi.$$

During each Δt_n, the volume of fluid, V_n, injected into the fracture is

$$V_n = i\Delta t_n,$$

where i is the injection rate. The volume is divided into two components: one occupies the fracture and the other is lost as leakoff. Fig. E-3 shows that the volume lost from V_n as leakoff between t_{m-1} and t_m is given by

$$V_{Lmn} = \tfrac{1}{4}[(x_p-x_r)+(x_i-x_o)]\{f[t_{e(m+1)r}]-f(t_{emp}) + f[t_{e(m+1)p}]-f(t_{emi})\}h, \quad \text{(E-4)}$$

where V_{Lmn} denotes the volume lost from V_n during Δt_m. From mass balance,

$$V_n = \sum_n^m V_{Lmn} + \sum_0^i V_{fni}. \quad \text{(E-5)}$$

Eq. E-5 allows several sets of computations.

1. The location of each fluid increment in the fracture can be calculated by successively increasing i from an initial value of zero until Eq. E-5 is satisfied. Starting with the last increment of injected fluid, one tries to determine the location of the leaking front of each fluid increment. Successive use of Eq. E-5 for preceding fluid increments allows the determination of each fluid increment's location in the fracture.

2. The percent of each fluid increment left in the fracture can be found. If V_n consists of both solid and liquid components (slurry), then Eq. E-5 allows calculation of the increase in the solid concentration as a result of liquid leakoff. Consider an element of fracture at time t containing a slurry and a sand bed (Fig. E-4). At $t+\Delta t$, the top of the slurry column moves down, while the sand bed rises because of particle settling. The height Δh is found from

$$\Delta h = V_t \Delta t. \quad \text{(E-6)}$$

The fluid volume that has lost its sand is $(w_{i+1}+w_i)(x_{i+1}-x_i)\Delta h/2$, which is deposited above the bed during Δt. One can compute the increase in bed height, as well as the location of the slurry in the fracture during the treatment. Although the fluid does not remain stationary in the fracture during Δt, the movement basically replenishes the element.

As long as Δt is selected so that during this period the element contains the same kind of slurry, the fluid inside can be assumed stationary for bed-height computations. The position of sand in suspension in the fracture is determined best by considering the volume element, V_n. If Δt_n is selected so that V_n contains the same concentration and type of sand, then Eq. E-5 yields the location of V_n in the fracture. Eq. E-6 is then used to compute how much sand settles at various x_i. The bed-height/total-deposition profile is found by settling all the sand in suspension above the existing bed at the end of the treatment.

Supplementary to Chap. 10.

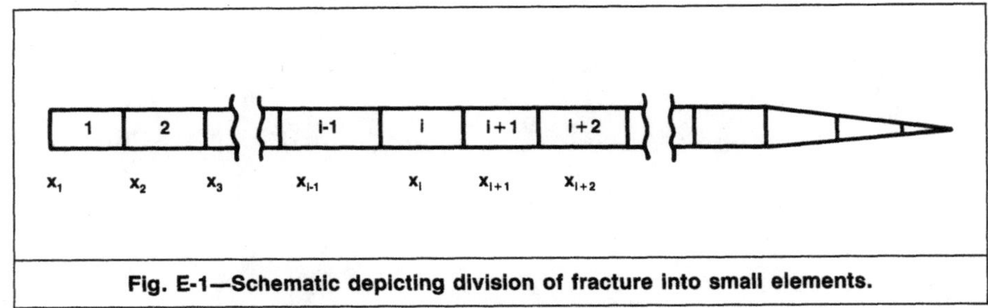

Fig. E-1—Schematic depicting division of fracture into small elements.

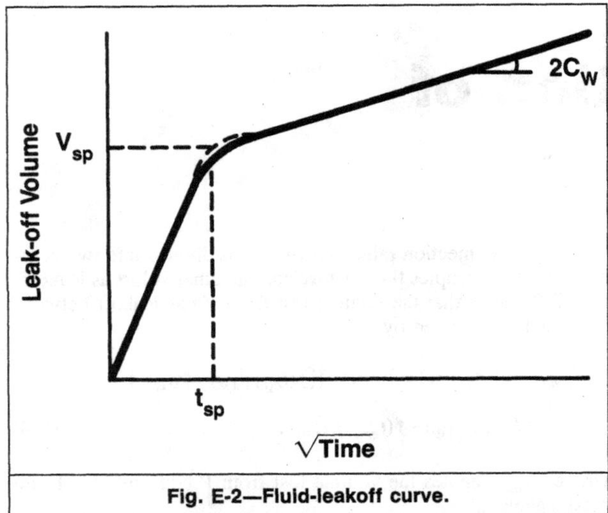

Fig. E-2—Fluid-leakoff curve.

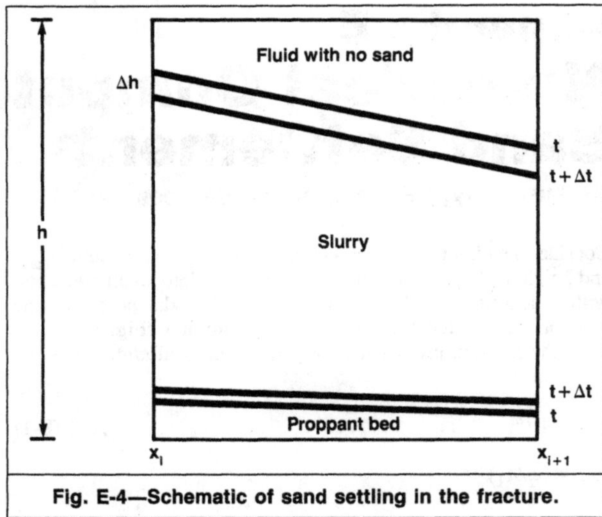

Fig. E-4—Schematic of sand settling in the fracture.

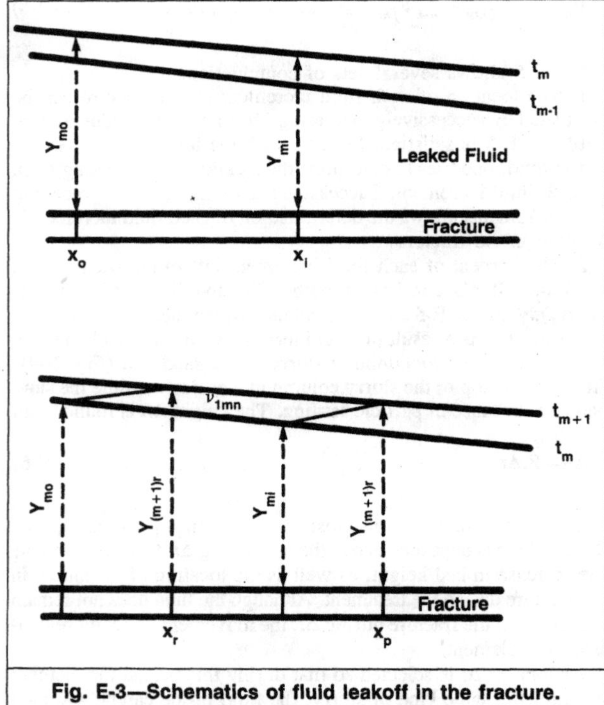

Fig. E-3—Schematics of fluid leakoff in the fracture.

The numerical solution is performed by dividing the fracture into small elements. The treatment time also is divided into small increments so that each increment contains the same slurry.

At a given time $t=t_n$, the location and sand concentration of each of the previous fluid increments in the fracture are calculated with the equations presented in this appendix. Repeating these computations for successive time increments allows the study of sand placement with time.

Nomenclature

A_i = constants
C_w = fluid-loss coefficient
$f(t)$ = leakoff function
h = fracture height, ft [m]
i = injection rate, bbl/min [m^3/min]
L = fracture length, ft [m]
m = constant
t = time
t_n = treatment time
t_{sp} = spurt time, minutes
V_{fni} = volume of each increment of fracture length
V_{Lni} = leakoff volume
V_{sp} = spurt volume, ft/$\sqrt{\text{min}}$ [m/$\sqrt{\text{min}}$]
w = fracture width, ft [m]
w_i = fracture width at Point x_i, ft [m]
w_{max} = maximum fracture width, ft [m]
x = distance between nodal point and wellbore, ft [m]
ϕ = porosity

Appendix F
Equilibrium Bed Height

Ali Daneshy, SPE, Halliburton Services

Equilibrium bed height, h_{eq}, is the maximum height a sand bed can have in a given hydraulic fracture. Babcock *et al.*[1] indicated that

$$v_{eq}=\frac{1.12i}{wh_{eq}}. \quad \text{(F-1)}$$

To calculate h_{eq} from Eq. F-1, v_{eq} must be known. The value is found by the following method.[1]

Let $(v_w)_{eq}$ denote the friction velocity of the particle at equilibrium bed height. This is related to other sand-transport parameters by

$$(v_w)_{eq}=\frac{3.46v_{eq}\left(\dfrac{\rho_f}{\rho_{sp}}\right)^{1/2}}{\left(\dfrac{4v_{eq}r_H\rho_f}{\mu}\right)^{1/2}} \quad \text{(F-2)}$$

for laminar flow,

$$(v_w)_{eq}=\frac{0.2v_{eq}\left(\dfrac{\rho_f}{\rho_{sp}}\right)^{1/2}}{\left(\dfrac{4v_{eq}r_H\rho_f}{\mu}\right)^{1/8}} \quad \text{(F-3)}$$

for turbulent flow, and

$$v/(v_w)_{eq}=0.54\left(\frac{\text{vd}\rho_f}{\mu}\sqrt{\frac{4r_H}{\text{d}}}\right)^{1/2}. \quad \text{(F-4)}$$

Supplementary to Chap. 10.

All parameters in Eq. F-4 are known except $(v_w)_{eq}$ and r_H. The value of r_H is given by

$$4r_H=4\frac{wh}{2(w+h)}\approx 2w. \quad \text{(F-5)}$$

Thus, $(v_w)_{eq}$ can be derived from Eq. F-4. Eq. F-4 is basically for Newtonian fluids; for use with non-Newtonian fluids, μ should be replaced by an apparent viscosity based on the shear stress around the particle. With $(v_w)_{eq}$ known, v_{eq} can be calculated from Eq. F-2 or F-3, depending on whether the fluid is in laminar or turbulent flow. With v_{eq} known, h_{eq} can be computed from Eq. F-1.

Nomenclature

h = height, ft [m]
h_{eq} = equilibrium bed height, ft [m]
i = injection rate, bbl/min [m^3/min]
r_H = hydraulic radius, ft [m]
v = velocity
$(v_w)_{eq}$ = friction velocity of particle at equilibrium bed height
w = fracture width, ft [m]
μ = viscosity, cp [mPa·s]
ρ_f = density of fluid
ρ_{sp} = density of proppant

Reference

1. Babcock, R.E., Prokop, C.L., and Kehle, R.O.: "Distribution of Propping Agents in Vertical Fractures," *Prod. Monthly* (Nov. 1967) 11–18.

Appendix G
Simple Calculation of Fracture Dimensions

D.E. Nierode, SPE, Exxon Production Research Co.

Approximate calculations can be done on a hand calculator or simple computer to arrive at fracture dimensions during pumping. The following equations should be programmed to obtain approximate fracture dimensions for a Perkins-Kern-Nordgren (PKN) or a Geertsma-de Klerk (GdK) fracture. Additional simple calculations are presented for the estimation of final propped fracture dimensions. All expressions for fracture length and volume refer to that quantity for one wing of the total fracture; i.e., fracture length is the distance from the wellbore to one of the fracture tips. Further, all equations contain dimensionalizing constants so that if the dependent variables are put in with the units given in the Nomenclature, the independent variables will also have the units indicated in the Nomenclature. A listing of a FORTRAN 77 program that performs these calculations is contained in the last section of this appendix.

Model Selection

The decision to use the PKN or the GdK model to estimate fracture dimensions can be based on either rock-mechanics or pragmatic considerations. If the decision is based strictly on rock-mechanics considerations, then the GdK model should be used only for shallow wells where fracture height is greater than fracture length; the PKN model should be used in all other situations.[1] The support for this statement involves when it is likely or unlikely that bounding layers adjacent to the fracture will allow slippage. By this line of reasoning, only the PKN model would apply in deep wells because slippage would not be likely in the presence of large overburden forces.

From a pragmatic viewpoint, either model can be applied to any fracturing situation with the realization that only the gross fracture shape is being estimated. When an actual fracture is several hundred feet in height, it usually encounters varying rock properties and stresses. In this situation, the fracture shape is quite irregular in width profile, with largest widths in the lowest-stress, softest areas and smallest widths in the highest-stress, hardest zones. This irregular width profile can be estimated (on the average) by either model as long as an appropriate fluid-loss coefficient is selected. To do this means to neglect the rock-mechanics reasons for selecting one over the other. To neglect the rock-mechanics reasons, one rationalizes that because the details of fracture fluid-loss behavior in an actual treatment are not well known, it makes little sense to select a fracture-dimension model with fluid loss still able to affect predicted dimensions greatly. Instead, one can select a model and then calibrate it to the field with appropriate fluid-loss coefficients. With this line of reasoning, the simplest model to select is the GdK model because subsequent proppant-transport modeling does not introduce additional approximations or difficulties. If the PKN model is selected instead, proppant-transport modeling introduces an approximation where the elliptical cross section is modeled by equivalent-width parallel plates. Regardless of how the model is selected, the following sections present the specific details on implementing both models to do calculations.

Supplementary to Chap. 11.

PKN Dynamic Fracture Dimensions

The following series of equations needs to be solved iteratively on a hand calculator or simple computer. The solution is an approximation to that presented[2,3] so that the effects of non-Newtonian fluids and net sand less than fracture height can be included in the calculations. If calculations are desired for a Newtonian fluid, the calculation is not iterative and the solution defaults to that presented by the authors.[2,3]

Eqs. G-1 through G-10 are solved iteratively with an initial guess for the wellbore maximum fracture width. Calculated values of maximum fracture width are used in subsequent iterations until the calculations converge.

$$L=aL_D, \quad \text{(G-1)}$$

$$w_{wb}=ew_D, \quad \text{(G-2)}$$

$$\mu_e=47{,}880K\left(\frac{80.842q_i}{h_g\bar{w}^2}\right)^{n-1}, \quad \text{(G-3)}$$

$$t_D=\frac{t}{B}, \quad \text{(G-4)}$$

$$L_D=0.5809t_D^{0.6295}, \quad \text{(G-5)}$$

and

$$w_D=0.78t_D^{0.1645}, \quad \text{(G-6)}$$

where

$$\bar{w}=\left(\frac{\pi}{4}\right)^2 w_{wb}, \quad \text{(G-7)}$$

$$B=1.7737\times10^{-4}\left[\frac{(1-\nu)\mu_e q_i^2}{32C^5h_gG}\left(\frac{h_g}{h_n}\right)^5\right]^{2/3}, \quad \text{(G-8)}$$

$$e=5.0872\times10^{-2}\left[\frac{16(1-\nu)\mu_e q_i^2}{C^2h_gG}\left(\frac{h_g}{h_n}\right)^2\right]^{1/3}, \quad \text{(G-9)}$$

and

$$a=7.4768\times10^{-2}\left[\frac{(1-\nu)\mu_e q_i^5}{256C^8h_g^4G}\left(\frac{h_g}{h_n}\right)^8\right]^{1/3}. \quad \text{(G-10)}$$

The algorithm for using Eqs. G-1 through G-10 at a particular pumping time consists of the following steps.

1. Guess an initial maximum wellbore width. Usually 0.10 in. [0.25 cm] will suffice.
2. Calculate overall average fracture width from Eq. G-7.
3. Calculate the effective viscosity from Eq. G-3.
4. Calculate B from Eq. G-8 and t_D from Eq. G-4.
5. Use the t_D value to calculate w_D from Eq. G-6.
6. Use Eq. G-9 to calculate e and Eq. G-2 to calculate maximum wellbore width, and compare with initial guess. Repeat Steps 1 through 6 as necessary until the calculated width value does not significantly change. Use an average of the n and $n-1$ values of maximum wellbore width for the $n+1$ guess.
7. Once the wellbore width value is determined, use Eq. G-10 to calculate a and Eq. G-1 to calculate fracture length.

Eq. G-3 evaluates the viscosity of a non-Newtonian fluid at the wall shear rate determined by the average fracture width with the full, single-fracture-wing injection rate. It is at best an estimate of the effective viscosity value. For very high fluid-loss rates, a lesser value of flow rate would probably be more representative.

Eqs. G-5 and G-6 are log-log linear approximations to the numerical curves presented in Figs. 2 and 3 of Nordgren's[3] paper. These two equations could be improved by slightly more accurate expressions if a more exact curve fit were done, or they could be replaced by the asymptotic expressions given in Ref. 3 for zero fluid loss and very high fluid loss.

Note also that Eqs. G-8 through G-10 were modified from the original presentation[3] to account for a net formation thickness less than the gross fracture height. The power terms in the h_g/h_n grouping were obtained by returning to the original equations and inserting the gross fracture height in fracture-dimension terms and the net thickness in terms that involved fluid loss.

Fracture volume is approximately given by Eq. G-11, which is based on the assumption that the fracture is strictly elliptical in both the vertical and horizontal planes.

$$V=\bar{w}h_gL. \qquad \text{(G-11)}$$

GdK Dynamic Fracture Dimensions

The following series of equations needs to be solved iteratively to arrive at the fracture dimensions according to the solution presented by Geertsma and de Klerk[4] for a vertical fracture with a rectangular cross section in the vertical plane. Fracture shape in the horizontal plane is elliptical. The equations were modified to include the effect of net sand thickness less than gross fracture height.

An initial guess for the final wellbore width at the end of the job is iterated upon until the equations converge, and the lengths and width during the treatment are then calculated from that value.

Eq. G-12 is an explicit expression for fracture length that depends on the final wellbore width at the end of the job, w_{we}:

$$L=\frac{9.308\times10^{-3}}{h_gC^2}q_i\left(\frac{h_g}{h_n}\right)^2\left(\frac{\pi w_{we}}{12}+8V_{sp}\frac{h_n}{h_g}\right)\times\left[\frac{2\alpha_L}{\sqrt{\pi}}-1+e^{\alpha_L^2}\operatorname{erfc}(\alpha_L)\right], \qquad \text{(G-12)}$$

where

$$\alpha_L=\frac{8C\sqrt{\pi t}\left(\dfrac{h_n}{h_g}\right)}{\dfrac{\pi w_{we}}{12}+8V_{sp}\left(\dfrac{h_n}{h_g}\right)}.$$

Eq. G-13 is a simplified expression for the width of the fracture:

$$w_{wb}=0.1295\sqrt{\frac{84(1-\nu)\mu q_iL^2}{\pi Gh_g}}. \qquad \text{(G-13)}$$

Eq. G-14 expresses the volume of one wing of the fracture at an instant for use in the approximate proppant placement calculations that follow.

$$V=\frac{\pi hLw_{wb}}{48}. \qquad \text{(G-14)}$$

To take non-Newtonian fluid behavior into account, Eq. G-3 can be used to calculate a fluid-viscosity value, where average width is given by

$$\bar{w}=\frac{\pi}{4}w_{wb}. \qquad \text{(G-15)}$$

Eqs. G-12 through G-15 are used with the following algorithm.

1. For the final pumping-job-time value, guess an initial value of maximum wellbore width and calculate the length from Eq. G-12.
2. Calculate effective viscosity from Eq. G-3 using Eq. G-15 for average fracture width.
3. Calculate the maximum wellbore width from Eq. G-13 and compare it to the initial value guessed. Iterate until the values converge with the average of the n and $n-1$ values used for the $n+1$ guessed value.
4. After the wellbore width at the end of the job is known, use that value in Eq. G-12 to calculate fracture lengths at lesser values of time. Similarly, use Eq. G-13 for the corresponding values of wellbore width during pumping.

Approximate Propped Fracture Length. The limiting cases for propped fracture length can be approximated from the fracture volumes calculated in the previous section of this appendix. The limits, total suspension and equilibrium bank, were presented in Chap. 10. For total suspension, the propped length is approximately given by

$$L_P=\frac{12(V_{EOJ}-V_{pad})}{h_g\bar{w}}. \qquad \text{(G-16)}$$

Eq. G-16 assumes that no additional fluid loss occurs from the fluid in the pad volume while the proppant-laden fluid is pumped. The propped width from a totally suspending fluid results from a simple volume balance on the amount of proppant pumped:

$$w_P=\frac{12m_P}{\rho_Ph_gL_P}. \qquad \text{(G-17)}$$

Similar expressions for an equilibrium bank condition are given by Eqs. G-18 and G-19:

$$w_P=\left(\frac{\pi}{4}\right)^2\bar{w}_{EOJ} \qquad \text{(G-18)}$$

and

$$L_P=\frac{12m_P}{\rho_Ph_gw_P}. \qquad \text{(G-19)}$$

Approximate Proppant-Carry Distance. The approximate proppant-carry distance in a non-Newtonian fluid is given by Eq. G-21, which is based on Stokes Eq. G-20 for simple settling in a non-Newtonian fluid:

$$v_S=\left[\frac{0.0241\,\bar{d}_P(\gamma_P-\gamma_{fl})}{K}\right]^{1/n}\bar{d}_P \qquad \text{(G-20)}$$

and

$$L_{sc}=\frac{1.123q_i}{\bar{w}V_S}. \qquad \text{(G-21)}$$

```
      IMPLICIT REAL*8(A-H,O-Z)
      DIMENSION WIDTH(11),XLENG(11),VOLUM(11),TIME(11),XLENP(11)
      DIMENSION WIDTE(11),CMNT(20),EFF(11)

C READ IN DATA SETS IN FORTRAN F FORMAT SPECIFYING DECIMAL POINTS
C READ IN COMMENT CARD
1000  READ(5,800,END=1001) (CMNT(I),I=1,20)
C  READ IN ONE WING RATE(BPM), H GROSS(FT), H NET(FT),
C         FLUID LOSS C(FT/ROOT MIN), POISSON'S RATIO(DIMENSIONLESS),
C         POWER LAW n (DIMENSIONLESS), POWER LAW k(LBFSECN/FT2)
      READ(5,801) QI,HG,HN,C,XNU,XN,XK
C READ IN PAD PUMP TIME(MIN), TOTAL JOB TIME(MIN), LBS OF PROPPANT FOR
C         ONE WING(LBS), GRAVITY OF PROPPANT(REL. WATER), GRAVITY OF
C         FRAC FLUID(REL. WATER), AVERAGE DIAMETER OF PROPPANT(INCHES),
C         AND RESERVOIR PERMEABILITY(MD)
      READ(5,801) TPAD,TF,XLBP,SGPR,SGFL,DIAP,XKRES
C READ IN SPURT VOLUME(FT), YOUNG'S MODULUS(PSI), AND/OR SHEAR MODULUS(PSI)
      READ(5,801) VSPT,EMOD,G

C PRINT DATA INPUT STREAM
      WRITE(6,802) (CMNT(I),I=1,20)
      WRITE(6,803) QI,HG,HN,C,VSPT,XNU,XN,XK
      WRITE(6,804) TPAD,TF,XLBP,SGPR,SGFL,DIAP,XKRES,EMOD,G
      IF(EMOD.GT.0.) G=EMOD/(2.*(1.+XNU))

C BEGIN PKN CALCULATIONS FOR 10 TIME STEPS
      W0=.1
      PI=3.14159
      TME=0.
      DTIME=TF/10.
      WRITE(6,805)
      WRITE(6,806)
      DO120K=1,11
      TME=TME+DTIME
      IF(K.EQ.11) TME=TPAD

C BEGIN WIDTH ITERATION LOOP INTERNAL TO TIME LOOP
      DO100I=1,100
      W1=(PI/4.)**2*W0
      IF(XN.EQ.1.D0) GO TO 105
      TERM1=(80.842*QI/HG/W1/W1)**(XN-1.)
      XMUEFF=47880.*XK*TERM1
 105  IF(XN.EQ.1.D0) XMUEFF=47880.*XK
      TERM2=(1.-XNU)*XMUEFF*QI*QI/32./C**5/HG/G*(HG/HN)**5
      B=1.7737E-04*TERM2**(2./3.)
      TD=TME/B
      WD=.78*TD**.1645
      TERM3=16.*(1.-XNU)*XMUEFF*QI*QI/C**2/HG/G*(HG/HN)**2
      E=5.0872E-02*TERM3**(1./3.)
      W2=E*WD
      TERM4=DABS((W2-W0)/W0)
      IF(TERM4.LE.1.E-04) GO TO 200
      W0=(W0+W2)/2.
  100 CONTINUE

C SUCCESSFULLY EXIT WIDTH ITERATION LOOP
  200 TERM5=(1.-XNU)*XMUEFF*QI**5/256/C**8/HG**4/G*(HG/HN)**8
      A=7.4768E-02*TERM5**(1./3.)
      XLD=.4981*TD**.5961
      XL=XLD*A
      TIME(K)=TME
      WIDTH(K)=W2
      XLENG(K)=XL
      VOLUM(K)=W2*(PI/4.)**2*HG*XL/12.
      EFF(K)=VOLUM(K)/(TME*QI)/5.614*100.
  120 CONTINUE

C WRITE FRACTURE GEOMETRY RESULTS IN TABLE
      WRITE(6,807) (TIME(K),WIDTH(K),XLENG(K),VOLUM(K),EFF(K),K=1,10)
```

Fig. G-1—FORTRAN 77 program for performing calculations.

```
C CALCULATE  AND PRINT TOTAL SUSPENSION PROPPED FRACTURE LENGTH
      XLSP=(VOLUM(10)-VOLUM(11))/HG/WIDTH(10)*12.*(4./PI)**2
      XLENP(1)=XLSP
      WRITE(6,808) XLSP

C CALCULATE  AND PRINT TOTAL SUSPENSION PROPPED FRACTURE WIDTH
      WIDTS=XLBP/100.*12./XLSP/HG
      WIDTE(1)=WIDTS
      WRITE(6,809) WIDTS

C CALCULATE  AND PRINT EQUILIBRIUM BANK PROPPED FRACTURE LENGTH
      XLEQ=XLBP/100.*(4./PI)**2/HG/WIDTH(10)*12.
      WRITE(6,810) XLEQ
      XLENP(2)=XLEQ

C CALCULATE  AND PRINT EQUILIBRIUM BANK PROPPED FRACTURE WIDTH
      WIDEQ=WIDTH(10)*(PI/4.)**2
      WIDTE(2)=WIDEQ
      WRITE(6,811) WIDEQ

C CALCULATE AND PRINT STOKES SETTLING VELOCITY AND SAND CARRY DISTANCE
      VSTOK=DIAP*(.0241*DIAP*(SGPR-SGFL)/XK)**(1./XN)
      WRITE(6,812) VSTOK
      SCD=1.123*QI/VSTOK/WIDTH(10)*(4./PI)**2
      WRITE(6,813) SCD

C  START GDK CALCULATIONS
      WRITE(6,814)

C   CALCULATE ITERATIVELY, WELLBORE WIDTH AT FINAL PUMPING TIME
      W0=.1
      W1=W0
      DO250I=1,100
      ALPHL=8.*C*(HN/HG)*(PI*TF)**.5/(PI*W1/12.+8.*VSPT*(HN/HG))
      IF(ALPHL.GT.5.) GO TO 9912
      TERM10=2.*ALPHL/PI**.5-1.+DEXP(ALPHL*ALPHL)*ERFC(ALPHL)
9912  IF(ALPHL.GT.5.) TERM10=2.*ALPHL/PI**.5-1.+1./(ALPHL*PI**.5)
      TERM11=PI*W1/12.+8.*VSPT*(HN/HG)
      XL=2.*QI*TERM11*TERM10/(HG*C*C*HN*HN/HG/HG)*5.584E-02
      WAVG=(PI/4.)*W1
      IF(XN.EQ.1.D0) GO TO 205
      TERM1=(80.842*QI/HG/WAVG/WAVG)**(XN-1.)
      XMUEFF=47880.*XK*TERM1
 205  IF(XN.EQ.1.D0) XMUEFF=47880.*XK
      W2=0.1295*(84.*(1.-XNU)*XMUEFF*QI*XL*XL/G/HG/PI)**.25
      TERM12=DABS((W2-W1)/W2)
      IF(TERM12.LE.1.E-04) GO TO 210
      W1=(W1+W2)/2.
 250  CONTINUE
 210  CONTINUE
      WRITE(6,815) W1
      WRITE(6,816) XL

C CALCULATE FRACTURE DIMENSIONS AT 10 JOB TIMES
      DTIME=TF/10.
      TME=0.
      WWE=W2
      DO300K=1,11
      TME=TME+DTIME
      IF(K.EQ.11) TME=TPAD
      W1=WWE
```

Fig. G-1 continued—FORTRAN 77 program for performing calculations.

If L_{sc} is much larger than the fracture length, then proppant suspension is nearly total. If L_{sc} is much less than fracture length, then an equilibrium bank will likely form.

FORTRAN Program Listing

The program listing in Fig. G-1 performs the calculations in the preceding parts of this appendix. It needs to be compiled with a FORTRAN 77 compiler.

Nomenclature

a = Nordgren length constant, ft [m]
B = Nordgren time constant, minutes
C = fluid-loss coefficient, ft/$\sqrt{\text{min}}$ [m/$\sqrt{\text{min}}$]
$\bar{d}_P$ = average proppant diameter, in. [cm]
e = Nordgren width constant, in. [cm]
G = shear modulus of elasticity, psi [kPa]
h_g = gross fracture height, ft [m]
h_n = net permeable sand thickness, ft [m]
K = power-law constant, (lbf-sec^n)/ft^2 [Pa·s^n]
L = fracture length, ft [m]
L_D = dimensionless fracture length
L_P = propped fracture length, ft [m]
L_{sc} = sand-carry distance, ft [m]
m_P = amount of proppant per wing of fracture, lbm [kg]
q_i = flow rate into one wing of a vertical fracture, bbl/min [m^3/min]
t = job pumping time, minutes

```
C START INTERNAL WIDTH ITERATION LOOP
      DO400I=1,100
      ALPHL=8.*C*(PI*TME)**.5*HN/HG/(PI*WWE/12.+8.*VSPT*HN/HG)
      IF(ALPHL.GT.5.) GO TO 305
      TERM10=2.*ALPHL/PI**.5-1.+DEXP(ALPHL*ALPHL)*ERFC(ALPHL)
 305  IF(ALPHL.GT.5.) TERM10=2.*ALPHL/PI**.5-1.+1./(ALPHL*PI**.5)
      TERM11=PI*WWE/12.+8.*VSPT*HN/HG
      XL=2.*QI*TERM11*TERM10/(HG*C*C*HN*HN/HG/HG)*5.584E-02
      WAVG=(PI/4.)*W1
      IF(XN.EQ.1.D0) GO TO 306
      TERM1=(80.842*QI/HG/WAVG/WAVG)**(XN-1.)
      XMUEFF=47880.*XK*TERM1
 306  IF(XN.EQ.1.D0) XMUEFF=47880.*XK
      W2=0.1295*(84.*(1.-XNU)*XMUEFF*QI*XL*XL/G/HG/PI)**.25
      TERM12=DABS((W2-W1)/W2)
      IF(TERM12.LE.1.E-04) GO TO 410
      W1=(W1+W2)/2.
 400  CONTINUE

C SUCCESSFULLY EXIT ITERATION LOOP
 410  TIME(K)=TME
      WIDTH(K)=W2
      XLENG(K)=XL
      VOLUM(K)=PI*HG*XL*W2/4./12.
      EFF(K)=VOLUM(K)/(TME*QI)/5.614*100.
 300  CONTINUE

C WRITE FRACTURE GEOMETRY RESULTS IN TABLE
      WRITE(6,806)
      WRITE(6,807) (TIME(K),WIDTH(K),XLENG(K),VOLUM(K),EFF(K),K=1,10)

C CALCULATE  AND PRINT TOTAL SUSPENSION PROPPED FRACTURE LENGTH
      XLSP=(VOLUM(10)-VOLUM(11))/HG/WIDTH(10)*12.*(4./PI)
      XLENP(1)=XLSP
      WRITE(6,808) XLSP

C CALCULATE  AND PRINT TOTAL SUSPENSION PROPPED FRACTURE WIDTH
      WIDTS=XLBP/100.*12./XLSP/HG
      WRITE(6,809) WIDTS
      WIDTE(1)=WIDTS

C CALCULATE  AND PRINT EQUILIBRIUM BANK PROPPED FRACTURE LENGTH
      XLEQ=XLBP/100.*(4./PI)/HG/WIDTH(10)*12.
      WRITE(6,810) XLEQ
      XLENP(2)=XLEQ

C CALCULATE  AND PRINT EQUILIBRIUM BANK PROPPED FRACTURE WIDTH
      WIDEQ=WIDTH(10)*(PI/4.)
      WRITE(6,811) WIDEQ
      WIDTE(2)=WIDEQ

C CALCULATE AND PRINT STOKES VELOCITY AND SAND CARRY DISTANCE
      VSTOK=DIAP*(.0241*DIAP*(SGPR-SGFL)/XK)**(1./XN)
      WRITE(6,812) VSTOK
      SCD=1.123*QI/VSTOK/WIDTH(10)*(4./PI)
      WRITE(6,813) SCD
      GO TO 1000
 1001 CONTINUE
  800 FORMAT(20A4)
  801 FORMAT(7F10.5)
  802 FORMAT(1H1,4X,20A4//)
  803 FORMAT(1X,'USER SPECIFIED INPUT'/
     15X,'ONE WING INJECTION RATE, BPM = ',F10.2/
     25X,'GROSS FRACTURE HEIGHT   ,  FT = ',F10.1/
     35X,'NET FRACTURE HEIGHT     ,  FT = ',F10.1/
     45X,'FLUID LOSS COEFFICIENT  ,FT/RM= ',F10.4/
     45X,'FLUID LOSS SPURT VOLUME,    FT= ',F10.4/
     55X,'POISSONS RATIO,DIMENSIONLESS = ',F10.2/
     65X,'POWER LAW EXPONENT, N        = ',F10.3/
     75X,'POWER LAW COEFFICIENT, K     = ',F10.5)
```

Fig. G-1 continued—FORTRAN 77 program for performing calculations.

t_D = dimensionless job time

V = volume of one wing of fracture, ft^3 [m^3]

V_{EOJ} = volume of one wing of fracture at end of pumping, ft^3 [m^3]

V_{pad} = volume of one wing of fracture at end of pad volume, ft^3 [m^3]

v_S = Stokes settling velocity, ft/sec [m/s]

$\bar{w}$ = volumetric average fracture width, in. [cm]

w_D = dimensionless fracture width

$\bar{w}_{EOJ}$ = average propped fracture width at end of pumping, in. [cm]

w_P = propped fracture width, in. [cm]

w_w = fracture width as function of length, in. [cm]

w_{wb} = fracture width at wellbore, in. [cm]

w_{we} = fracture width at wellbore at end of pumping, in. [cm]

```
804 FORMAT(
   15X,'TIME TO INJECT PAD      , MIN = ',F10.1/
   25X,'TOTAL JOB TIME          , MIN = ',F10.1/
   35X,'ONE WING AMT. OF PROPPANT,LBS= ',F10.0/
   45X,'PROPPANT SPECIFIC GRAVITY    = ',F10.2/
   55X,'FLUID SPECIFIC GRAVITY       = ',F10.2/
   65X,'AVERAGE PROPPANT DIAMETER,IN = ',F10.4/
   75X,'RESERVOIR PERMEABILITY , MD  = ',F10.5/
   85X,'YOUNGS MODULUS          , PSI = ',F10.0/
   85X,'SHEAR MODULUS           , PSI = ',F10.0/)
805 FORMAT(/1X,'PKN DATA ANALYSIS:'/)
806 FORMAT(15X,'MAX.',1X,'WELLBORE',1X,'ONE WING',2X,'ONE WING',6X,
   1'FLUID'/8X,'TIME',6X,'WIDTH',7X,'LENGTH',4X,'VOLUME',5X,
   1'EFFICIENCY'/9X,'MIN',7X,'IN',11X,'FT',7X,'FT3',10X,'%'/)
807 FORMAT(5X,F7.1,5X,F5.3,5X,F8.1,3X,F8.1,3X,F8.2)
808 FORMAT(/1X,'PROPPED LENGTH FOR TOTAL SUSPENSION IS ',F8.1,' FT')
809 FORMAT(/1X,'PROPPED WIDTH FOR TOTAL SUSPENSION IS ',F8.3,' IN')
810 FORMAT(/1X,'PROPPED LENGTH FOR EQUILIBRIUM BANK IS ',F8.1,' FT')
811 FORMAT(/1X,'PROPPED WIDTH FOR EQUILIBRIUM BANK IS ',F8.3,' IN')
812 FORMAT(/1X,'STOKES SETTLING VELOCITY IS ',F8.3,' FT/SEC')
813 FORMAT(/1X,'PROPPANT CARRY DISTANCE IS ',F10.0,' FT')
814 FORMAT(/1H1,'GDK DATA ANALYSIS:'/)
815 FORMAT(/1X,'GDK WELLBORE WIDTH AT EOJ IS ',F10.3,' IN')
816 FORMAT(/1X,'GDK LENGTH AT EOJ IS ',F10.3,' FT'/)
    END

C FUNCTION THAT REPRODUCES COMPLIMENTARY ERROR FUNCTION FROM 0 TO 5
      DOUBLE PRECISION FUNCTION ERFC(XVLUE)
      IMPLICIT REAL*8(A-H,O-Z)
      DIMENSION ERFCT(22),ERFXV(20),ZZ4(20)
      DATA ERFCT/0.,.3983175,-.03174029,-.1095944,.1384539,
     1-.04423302,-.06162986,.1909586,-.6024937,2.519318,-6.074964,
     27.480577,-3.886774,.6386388,-.1592837,.03668499,-.0115424,
     3-.01312481,-.005891394,-.002967419,-.0002572478,-.0001350366/
      DATA ERFXV/0.,.1,.2,.3,.4,.5,.6,.7,.8,.9,1.,1.25,1.5,1.75,2.,
     12.5,3.,3.5,4.,5./
      ERFC=SPLN2(ERFXV,ZZ4,ERFCT,XVLUE,20,2)
      RETURN
      END

C CURVE FIT FUNCTION USING SPLINE EQUATIONS- FINDS COEFFS.
      DOUBLE PRECISION FUNCTION SPLN2(X,Y,Z1,X1,N)
      IMPLICIT REAL*8(A-H,O-Z)
      DIMENSION X(50),Y(50),Z1(50),Z2(50)
      DIMENSION A(50,51)
      IF(X1-X(1))200,55,55
   55 DO60I=1,N
      IF(X1-X(I))65,65,60
   60 CONTINUE
   65 I7=I-1
      IF(X1.GT.X(N)) I7=N
      RAB=Z1(1)+Z1(2)*(X1-X(1))
      DO70I=1,I7
      RAB=RAB+Z1(I+2)*((X1-X(I))**3)
   70 CONTINUE
      SPLN2=RAB
      GO TO 100
  200 IF(II.EQ.3) GO TO 300
      SPLN2=Z1(1)+Z1(2)*(X1-X(1))
      GO TO 100
  300 SPLN2=1.+.1*X1
      IF(X1.LE.0.) SPLN2=.40
  100 CONTINUE
      RETURN
      END
```

Fig. G-1 continued—FORTRAN 77 program for performing calculations.

α_L = dimensionless fluid-loss parameter including spurt loss
γ_{fl} = fracture-fluid specific gravity, dimensionless
γ_P = specific gravity of proppant (relative water)
μ = gas viscosity, cp [mPa·s]
μ_e = effective non-Newtonian fracture-fluid viscosity, cp [mPa·s]
ν = Poisson's ratio, dimensionless
ρ_P = proppant bulk density, lbm/ft^3 [kg/m^3]

References

1. Perkins, T.K.: "Discussion of On the Design of Vertical Hydraulic Fractures," *JPT* (Jan. 1973) 93–95.
2. Perkins, T.K. and Kern, L.R.: "Widths of Hydraulic Fractures," *JPT* (Sept. 1961) 937–49; *Trans.*, AIME, **222**.
3. Nordgren, R.P.: "Propagation of a Vertical Hydraulic Fracture," *SPEJ* (Aug. 1972) 306–14; *Trans.*, AIME, **253**.
4. Geertsma, J. and de Klerk, F.: "A Rapid Method of Predicting Width and Extent of Hydraulically Induced Fractures," *JPT* (Dec. 1969) 1571–81; *Trans.*, AIME, **246**.

Appendix H
Treatment Cost and Payout Calculation

D.E. Nierode, SPE, Exxon Production Research Co.

Treatment costs for the purpose of gross treatment sizing are restricted to four major components: fluid cost, proppant cost, pumping charges, and gross equipment charges. For simplicity, detailed costs of tank storage, equipment rental, and other extra charges are lumped into a gross equipment cost.

Specific costs include the following.

Fluids.

Crosslinked, gelled water, 40 lbm/1,000 gal [4.8 kg/m^3]	\$0.50/gal [\$132/m^3].
Gelled oil	\$1.00/gal [\$264/m^3].

Proppants.

Sand	\$0.075/lbm [\$0.165/kg].
Bauxite	\$0.50/lbm [\$1.10/kg].

Pumping Charges.

0 to 5,000 psi [0 to 34.5 MPa]	\$3.75/hhp [\$5.03/kW]
5,000 to 6,000 psi [34.5 to 41.4 MPa]	\$4.05/hhp [\$5.43/kW]
6,000 to 7,000 psi [41.4 to 48.3 MPa]	\$4.65/hhp [\$6.23/kW]
7,000 to 8,000 psi [48.3 to 55.2 MPa]	\$5.25/hhp [\$7.04/kW]
8,000 to 9,000 psi[55.2 to 62.1 MPa]	\$6.15/hhp [\$8.24/kW]
9,000 to 10,000 psi [62.1 to 70 MPa]	\$7.95/hhp [\$10.66/kW]
10,000 to 12,000 psi [70 to 82.7 MPa]	\$9.90/hhp [\$13.27/kW]
12,000 to 15,000 psi [82.7 to 103.4 MPa]	\$13.50/hhp [\$18.10/kW]

Therefore,

$$\text{job cost} = \text{fluid cost} + \text{proppant cost} + \text{pumping charge} + \text{equipment charge}.$$

Payout is calculated from job cost and incremental recovery:

$$\text{payout} = \text{job cost}/(\text{incremental revenue/Mcf} \times \text{incremental Mcf/D}).$$

Supplementary to Chap. 11.

Appendix I
Field Measurement of Fracturing Pressures

K.G. Nolte, SPE, Dowell Schlumberger Inc.

The successful analysis of fracturing pressures requires the collection of quality data, preferably at minimum cost and alteration to general fracturing procedures. These measurements should be continuous recordings of the actual, or accurately inferred, values of bottomhole pressure (BHP) for the pretreatment determination of closure pressure, pressure during the treatment, and the pressure decline after shut-in. For a quantitative analysis of the data, a sensitivity level on the order of 5 psi [35 kPa] and an accuracy level on the order of 25 psi [175 kPa] are generally sufficient. It is desirable to have the injection or flowback rate simultaneously recorded and multiplexed with the pressure, particularly for closure-pressure determinations using the flowback technique discussed later. Also during the treatment, the proppant concentration should be recorded with the pressure and rate. Surface recordings of the injection pressure during a treatment generally introduce large uncertainties into the fracturing BHP because the fluid friction variations down the string and head changes resulting from variations in proppant concentration are greater than the required sensitivity for an accurate analysis of the data. Future advances in predicting these variations, however, could change the benefit of surface injection pressures. The variation in fluid friction is caused by variations in fluid quality and proppant concentrations. An exception to the general applicability of surface injection pressures is the analysis of screenouts (i.e., Sec. 14.5) for which the rate of pressure increase can be larger than the variations in the injection pressure. For perforated completions, the assumption must be made that the friction drop through the perforations either is negligible or can be quantified.

The most direct way to measure BHP is to place the sensor downhole, as illustrated on the left side of Fig. I-1. The disadvantages of this configuration are that a real-time display of the data at the surface (and hence simultaneous recording of other parameters) currently is not possible; the potential loss of the instrument; and wireline work. This configuration, with a wireline sensor through the tubing, is commonly used for calibration treatments (no proppant). An instrument can also be placed downhole in a tubing (without a packer) with the fluid injected down the annulus, as illustrated in the center of Fig. I-1. In general, this configuration is not used because there are no direct advantages over surface recordings through a static fluid column in either the tubing or annulus for applications when the density of the column remains essentially constant. An annulus/tubing arrangement without a packer can be used with the fluid injected down either the annulus or the tubing and the static (backside) pressure on the other string monitored at the surface. Because there is no flow in the static string, BHP is the sum of the surface pressure and the hydrostatic head of the fluid in the static string, p_h (right side of Fig. I-1). However, surface recordings are applicable only if the fracturing pressures are greater than the hydrostatic head of the fluid column. This will generally be the case except for nearly depleted reservoirs. Another limitation of this open-annulus arrangement is no packer capability to protect the casing from fracturing pressures. Placing weighted fluids in the annulus (inject down tubing) can provide some protection. The configuration also creates higher horsepower losses compared with pumping down casing without tubing. Because pressure-decline data are recorded after injection stops, surface measurements can be used for any well configuration in communication with the fracture if the density of the communicating fluid column remains essentially constant. A field-pressure chart can be initiated at shut-in and left to record until flowback. These data can be used to pick the time of closure on the proppant and to provide an inference of fluid efficiency (Appendix J, Eqs. J-16 and J-22). Surface recordings without an open-ended annulus can also provide data during treatments by periodic shut-ins. The instantaneous shut-in pressure (ISIP) provides an estimate of the fluid BHP after a head correction is made. Because of the high rate of pressure decrease at shut-in, however, ISIP's can provide only a lower-bound estimate on the pressure during injection.

The condition of essentially constant density for a static fluid column was cited several times in the prior discussion. This required condition for an accurate determination of BHP from surface pressure can be violated for formations of significant depth and temperature, particularly for use of the tubing as the static column after injection down the tubing with fluid volumes that are not significantly greater than the combined tubular and annular volumes. For these conditions, the much larger annular volume (relative to the tubular volume) will not be significantly cooled. Consequently, the temperature of the fluid in the tubing can quickly recover to near the static temperature gradient and cause a significant density and head change, Δp_h, for the tubing column. Fig. I-2 provides the upper range (i.e., complete recovery to static temperature gradient) for water with various formation depths and temperatures. A significant head change can result in erroneous data for a pressure-decline analysis, as illustrated by the insert in Fig. I-2. This figure can be used as a guide to assess when head changes may be significant and surface pressures may be inappropriate for inference of BHP.

The closure pressure is essential for an analysis of the fracturing pressure because it is the datum for determining the net pressure. The term "closure pressure," p_c, is defined as the fluid pressure at which the fracture closes—i.e., zero width (see Fig. 14.1). This pressure is equal to, and counteracts, the minimum principal stress in the rock that is perpendicular to the fracture plane. The closure pressure reflects a global average of the minimum stress, which is a local quantity and generally is not constant over the zone of interest. The closure pressure generally is less than the breakdown pressure required to initiate a fracture and always less than the pressure required to extend an existing fracture—i.e., fracture-extension pressure. The extension pressure is equivalent to the pressure generally called the "parting pressure" inferred from the break point of a step-rate test (Fig. I-3), which is commonly performed for reservoir flooding purposes. Although parting pressure has a connotation similar to the definition of closure pressure, the parting pressure—or, more precisely, the extension pressure—is

Supplementary to Chap. 14.

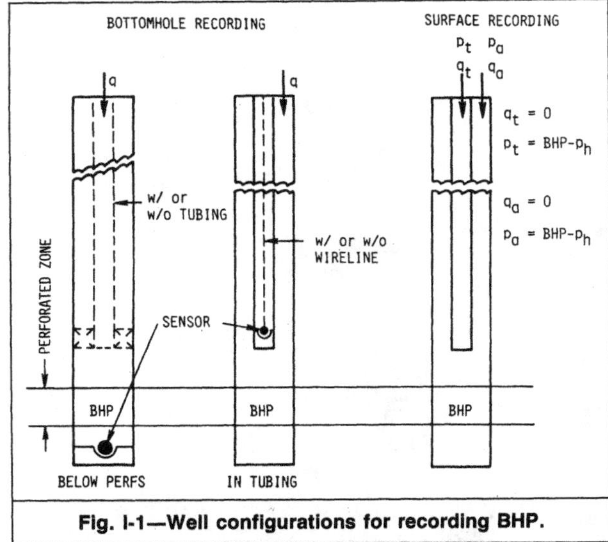

Fig. I-1—Well configurations for recording BHP.

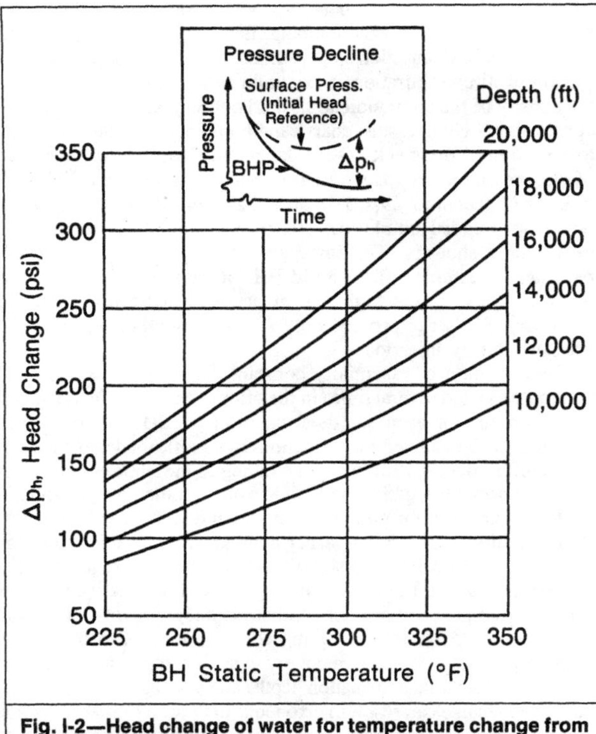

Fig. I-2—Head change of water for temperature change from 75°F to static gradient.

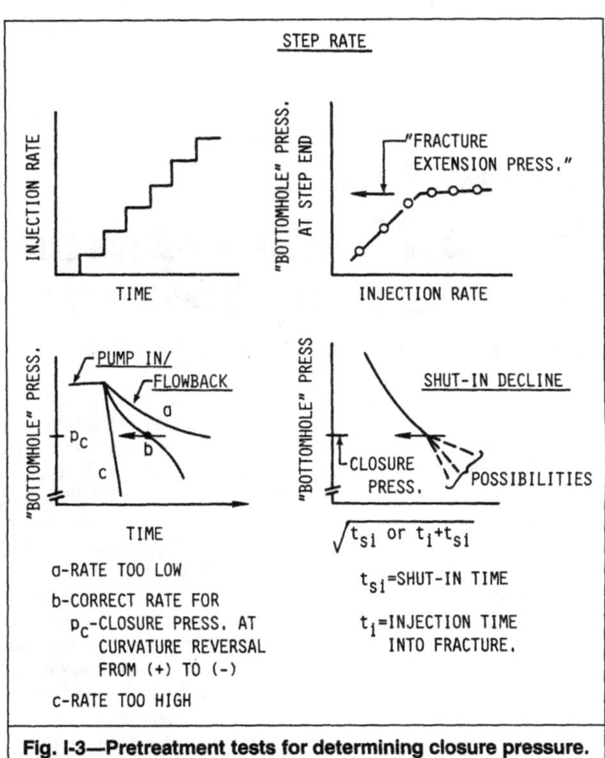

Fig. I-3—Pretreatment tests for determining closure pressure.

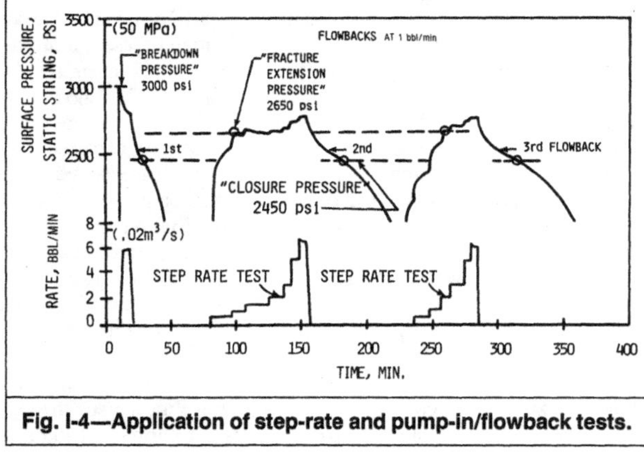

Fig. I-4—Application of step-rate and pump-in/flowback tests.

greater than the closure pressure because of fluid friction in the fracture and a finite resistance to extension. The step-rate test is very useful for determining an initial estimate that is an upper bound for the closure pressure and to ensure that the formation has been fractured. Each injection rate should be maintained for a fixed period of time (e.g., 5 minutes).

An upper bound for the closure pressure can also be approximated by the initial shut-in pressure after an acid breakdown or immediately after the breakdown of a fracture treatment (upper bound because width ≥ 0). This interpretation is similar to the method[1] commonly used for determining the local minimum stress. This method uses a small amount of injected fluid to ensure a minimum width at shut-in for a good approximation of the stress by the ISIP. For permeable sections, however, it is also necessary to ensure that a fracture was created by the small amount of fluid to obtain a representative value of stress. The performance of a valid step-rate test (Fig. I-3) provides confidence that a fracture was created.

A reliable method of determining the closure pressure is the pump-in/flowback procedure[2] performed before the fracture treatment with the interpretation shown in Fig. I-3. For this procedure, fluid of sufficient quantity and rate is injected to create a fracture—e.g., a step-rate test—followed by an immediate flowback at a *constant* rate controlled by an adjustable valve, choke, or preferably a rate controller, and recorded by an accurate low-rate flowmeter. If the flowback rate is within the correct range, the resulting pressure decline will show a characteristic reversal of curvature that must be from positive to negative curvature at the closure pressure. The accelerated pressure decline at the curvature reversal is caused by the flow restriction introduced when the fracture effectively closes. The correct range of flowback rates must be determined by trial and error for any specific field; however, the range is on the order of one-quarter of the prior injection rate to extend a fracture. The effect of flow rates outside the correct range is shown in Fig. I-3. Ungelled water can be used as the fluid for low-permeability formations, while higher-permeability formations may require fracturing fluids[3] to reduce the rate of fluid loss and closing during flowback. When applied in the field, the procedure should be repeated to ensure a consistent indication of closure pressure. For higher-permeability liquid reservoirs, the closure pressure can increase with the additional injection of fluids[3,4]; however, this generally will not occur for lower-permeability gas reservoirs (Fig. I-4).

Fig. I-4 shows a field example of the step-rate test to determine extension pressure, the flowback test to determine closure pressure, and the breakdown pressure. These surface data, requiring a correction for head, were collected by use of water in a tight-gas sand with the well configuration shown on the right of Fig. I-1. Even though only a small amount of the injected fluid was backflowed, a change in extension or closure pressure is not indicated over a period of 5 hours.

Another procedure for determining closure pressure is the analysis of the shut-in pressure decline after a fracture is created that is relatively large compared with the size after breakdown. This procedure is not as reliable as the pump-in/flowback method. Hence, decline data should be used only as a confirmation of, and not as a substitute for, the pump-in/flowback procedure because of the critical role of the closure pressure in the analysis of fracturing pressures. During the decline, the very visible change in rate of decline when the fracture closes on proppant, as in Fig. 14.10, is generally not seen for an unpropped fracture. An unpropped fracture is a necessary condition for determining the formation's closure pressure. For permeable sections, however, the pressure decline approaches a linear relation with the square root of time since shut-in began (Fig. 14.26). When the fracture closes, a distinct change in slope may be seen on this type of plot, as illustrated in Fig. I-3. This figure shows that the break can be either way and depends on the relative relationship of the fracture's variables and those of the reservoir; consequently, it can also show no break. The square-root behavior after closure results if the reservoir's pressure transient is pseudolinear flow induced by the fluid loss during injection and fracture closure. The square-root analysis is useful after a calibration treatment or an injection after a sequence of pump-in/flowback tests as a confirmation of the inferred closure pressure. An interpretation method analogous to the square-root plot for determining closure pressure[5] is a linear plot of pressure vs. $G(t_D)$. This plot (see Fig. 14.28) follows from Eq. 14.64b with $t_D^*=0$. Also, this equation indicates that the slope of the plot is equal to the match pressure, p^*. This method is subject to the same limitations as the square-root analysis for indicating a definitive closure pressure.

The collection and analysis of quality fracturing pressures has been greatly facilitated by the large number of monitoring vans provided by the service companies. These vans are generally equipped with the required instrumentation and computers to record and to analyze the relevant data.

References

1. Haimson, B.C.: "Crustal Stress in the Continental United States as Derived from Hydrofracturing Tests," *The Earth's Crust,* American Geophysical Union Monograph Series, **20**.
2. Nolte, K.G.: "Determination of Fracture Parameters from Fracturing Pressure Decline," paper SPE 8341 presented at the 1979 SPE Annual Technical Conference and Exhibition, Las Vegas, Sept. 23-26.
3. Smith, M.B.: "Stimulation Design for Short, Precise Hydraulic Fractures," *SPEJ* (June 1985) 371-79.
4. Cleary, M.P. *et al.*: "Surface Integral Schemes for Fluid Flow and Induced Stresses Around Fractures in Underground Reservoirs," paper SPE 11632 presented at the 1983 SPE/DOE Low Permeability Gas Reservoirs Symposium, Denver, March 14-16.
5. Castillo, J.L.: "Modified Fracture Pressure Decline Analysis Including Pressure-Dependent Leakoff," paper SPE 16417 presented at the 1987 SPE/DOE Low-Permeability Reservoirs Symposium, Denver, May 18-19.

Appendix J
Derivations and Considerations for Pressure-Decline Analysis

K.G. Nolte, SPE, Dowell Schlumberger

Derivation of Basic Relations

The derivations in this Appendix were originally presented in Ref. 1 unless otherwise noted.

Area Fluid-Loss Relation. The fracture's rate of areal growth can be bounded by either the assumption of negligible fluid loss to the formation (i.e., upper bound) or predominant fluid loss (i.e., lower bound), as given in Sec. 14.3. The upper and lower bounds for the area are

$$f_A = a/A = \tau/t \quad \text{(J-1a)}$$

and

$$f_A = a/A = (\tau/t)^{1/2}, \quad \text{(J-1b)}$$

respectively, or

$$\tau/t = (a/A)^e = f_A^e, \quad \text{(J-2)}$$

where

e = 1, upper bound,
e = 2, lower bound,
A = current area,
t = current time,
a = previous area, and
τ = previous time.

The upper bound used in the following assumes not only negligible fluid loss but also constant fracture width. Because the width generally increases and stores a portion of the injected fluid, the upper-bound exponent in Eq. J-1a is actually less than unity—i.e., $1/e<1$. From Eq. 14.41, noting that area is proportional to either $h_f L$ or r_f^2, these values for each model are

$$1/e < (2n+2)/(2n+3), \quad \text{(J-3a)}$$

$$1/e < (n+1)/(n+2), \quad \text{(J-3b)}$$

and

$$1/e < (4n+4)/(3n+6), \quad \text{(J-3c)}$$

for the PK[2] (Perkins-Kern), KG (Khristianovich-Zheltov[3] and Geertsma-de Klerk[4]), and radial[2,4] models, respectively.

The rate of fluid loss through an incremental area, da, at time t is expressed by the Carter[5] relation as

$$q_\ell(t,\mathrm{d}a) = \frac{2C\mathrm{d}a}{\sqrt{t-\tau(a)}}, \quad \text{(J-4)}$$

Supplementary to Chap. 14.

where C is the fluid-loss coefficient. Integrating Eq. J-4 after substituting Eq. J-2 gives

$$q_\ell(t,a) = \frac{2CA}{\sqrt{t}} \int_0^{f_A} \frac{\mathrm{d}f_A}{\sqrt{1-f_A^e}}$$

$$= \frac{2CA}{\sqrt{t}} \begin{Bmatrix} 2\left(1-\sqrt{1-f_A}\right) \\ \sin^{-1} f_A \end{Bmatrix} \quad \text{(J-5)}$$

$$= \frac{2CA_p}{\sqrt{t_i}} \begin{Bmatrix} 2\sqrt{t/t_i}\left(1-\sqrt{1-f_A}\right) \\ \sin^{-1} f_A \end{Bmatrix}, \quad \text{(J-6)}$$

where the upper expression in the braces is for the upper bound (e=1) and the lower expression is for the lower bound (e=2). Eq. J-6 follows from Eqs. J-5 and J-1 by substituting $a=A_p$, the final permeable area, at $\tau=t_i$ at the end of pumping—i.e., the beginning of shut-in.

If it is assumed that the total area for one face of the fracture, A_f, and A_p retain a constant ratio during fracture propagation, then the following expression can be substituted into the above.

$$A_p = f_p A_f$$

and

$$f_p = \frac{h_p}{h_f} \le 1, \quad \text{(J-7)}$$

where h_f is the total vertical fracture height and h_p is the loss section height for the PK and KG models. Generally, f_p would be unity for the radial or penny model.

Rate and Volume of Fluid Loss. The fluid-loss rate after shut-in ($t \ge t_i$, $a=A_p$; assume $A=A_p$=constant) is found by substituting Eq. J-1 for f_A with $\tau=t_i$ into Eq. J-6 and using Eq. J-7:

$$q_\ell(t,A_p) = \frac{2Cf_pA_f}{\sqrt{t_i}} \begin{Bmatrix} 2\sqrt{t/t_i}\left(1-\sqrt{1-t_i/t}\right) \\ \sin^{-1}\sqrt{t_i/t} \end{Bmatrix}$$

$$= \frac{2Cf_pA_f}{\sqrt{t_i}} \begin{Bmatrix} 2[(1+t_D)^{1/2}-t_D^{1/2}] \\ \sin^{-1}(1+t_D)^{-1/2} \end{Bmatrix}$$

$$= \frac{2Cf_pA_f}{\sqrt{t_i}} f(t_D), \quad \text{(J-8)}$$

where

$$t_D=(t-t_i)/t_i \quad \text{(J-9)}$$

$$=t/t_i-1.$$

The volume of fluid lost during pumping is found by integrating Eq. J-6 over the time of pumping with $f_A=1$ and substituting Eq. J-7:

$$V_\ell(t_i,A_p)=\int_0^{t_i} q_\ell(t,f_A=1)\mathrm{d}t$$

$$=\frac{2\kappa C f_p A_f}{\sqrt{t_i}}\int_0^{t_i}\left\{\begin{matrix}2\sqrt{t/t_i}\\ \pi/2\end{matrix}\right\}\mathrm{d}t=2\kappa C f_p A_f\sqrt{t_i}\left\{\begin{matrix}4/3\\ \pi/2\end{matrix}\right\}, \quad \text{(J-10)}$$

where κ ($\kappa \geq 1$) is a multiple to C that accounts for additional fluid loss *only* during pumping (e.g., spurt loss or opening of natural fractures). These effects, along with the definition of κ, are discussed in more detail later. Eq. J-10 can be expressed as

$$\pi > \frac{V_\ell}{\kappa C f_p A_f \sqrt{t_i}} > \frac{8}{3} \quad \text{(J-11)}$$

for the bounds on the fluid volume lost during pumping.

The volume of fluid lost after shut-in is found by integrating Eq. J-8 from $t_D=0$ (i.e., shut-in), using the definition

$$g(t_D)=\int f(t_D)\mathrm{d}t_D=\left\{\begin{matrix}4/3[(1+t_D)^{3/2}-t_D{}^{3/2}]\\ (1+t_D)\sin^{-1}(1+t_D)^{-1/2}+t_D{}^{1/2}\end{matrix}\right\}$$

as

$$V_{\ell s}=\frac{2Cf_pA_f}{\sqrt{t_i}}\left[t_i\int_0^{t_D} f(t_D)\mathrm{d}t_D\right]$$

$$=2Cf_pA_f\sqrt{t_i}\,[g(t_D)-g(0)], \quad \text{(J-12)}$$

where the upper and lower bounds for $g(t_D)$ are given in the braces of the first expression. Inspection of Eq. J-10 shows that the values in braces are equal to $g(t_D=0)$ in Eq. J-12, or rewriting Eq. J-10 for the volume lost during pumping:

$$V_\ell=2\kappa C f_p A_f\sqrt{t_i}\,g_0, \quad \text{(J-13)}$$

where

$$g_0=g(0)=\left\{\begin{matrix}4/3\\ \pi/2\end{matrix}\right\}.$$

Relations Based on Loss Coefficient and Time

The volume of fluid lost between shut-in and fracture closure is found from Eq. J-12 evaluated at the time for fracture to close after shut-in, t_c, or dimensionless closing time, t_c/t_i. This volume is also equal to the fluid in the fracture at shut-in, V_f—i.e., fracture volume, assuming that no proppant was injected as is the case for a calibration treatment:

$$V_{\ell s}(t_D=t_c/t_i)=V_f$$

and

$$2Cf_pA_f\sqrt{t_i}\,(g_c-g_0)=A_f\bar{w}, \quad \text{(J-14a)}$$

or

$$\bar{w}=2Cf_p\sqrt{t_i}\,(g_c-g_0) \quad \text{(J-14b)}$$

with

$$g_c=g(t_c/t_i);\ g_0=g(0),$$

where $\bar{w}$ is the average fracture width at shut-in. Eq. J-14 implies that $\bar{w}$ can be determined if the loss coefficient, C, can be estimated and the time of fracture closure determined from data after the treatment.

Fluid Efficiency. The introduction of the parameter ρ will facilitate subsequent derivations:

$$\rho \equiv V_f/V_\ell \quad \text{(J-15)}$$

and

$$\eta \equiv V_f/V_i = \rho/(1+\rho), \quad \text{(J-16)}$$

where η is the fluid efficiency.

From the relations from Eqs. J-15, J-14, and J-13—i.e.,

$$\rho=V_f/V_\ell,$$

$$V_f=2Cf_pA_f\sqrt{t_i}\,(g_c-g_0),$$

and

$$V_\ell=2\kappa C f_p A_f\sqrt{t_i}\,g_0, \quad \text{(J-17)}$$

and Eq. J-16, it follows that

$$\rho=(g_c-g_0)/(\kappa g_0)=(g_c/g_0-1)/\kappa,$$

$$1+\rho=1/(1-\eta)=(\kappa-1+g_c/g_0)/\kappa,$$

$$=g_c/g_0, \text{ if } \kappa=1. \quad \text{(J-18a)}$$

Also, substituting Eq. J-18a into Eq. J-16 yields

$$\eta=1-\kappa/(\kappa-1+g_c/g_0)$$

$$=1-g_0/g_c, \text{ if } \kappa=1. \quad \text{(J-18b)}$$

Thus, ρ and η can be determined directly from fracturing data by closure time through the function $g(t_D)$ evaluated at the closure time if $\kappa=1$. Conversely, for the case without proppant, the predicted time for closure can be found from the efficiency—Eqs. J-18—by the inverse of the function g, (g^{-1}), as

$$t_c/t_i=g^{-1}\{g_0[1-\kappa+\kappa/(1-\eta)]\}$$

$$=g^{-1}[g_0(\kappa\rho+1)],$$

or for $\kappa=1$,

$$t_c/t_i=g^{-1}[g_0/(1-\eta)]$$

$$=g^{-1}[g_0(1+\rho)]. \quad \text{(J-19)}$$

The above expressions for time to close assume that the fracture closes completely—i.e., no proppant. If proppant is considered, the effective fracture volume that will close is V_f-V_{pr}, where V_{pr} is the volume of proppant including the porosity of the proppant pack. The value of $g(t_D)$ for closing on the proppant will be denoted as g_c'. Then the equivalent of the expression in Eq. J-17 for the volume of fracture that closes is

$$V_f-V_{pr}=2Cf_pA_f\sqrt{t_i}\,(g_c'-g_0), \quad \text{(J-20)}$$

and from Eqs. J-13 and J-15,

$$V_f = V_\ell \rho = 2\kappa C f_p A_f \sqrt{t_i}\, g_0 \rho. \quad \text{(J-21)}$$

Dividing Eq. J-20 by J-21 yields

$$1 - \frac{V_{pr}}{V_f} = \frac{(g_c' - g_0)}{\kappa g_0 \rho}.$$

Substituting $V_f = \rho V_i/(1+\rho)$ found from Eq. J-16, and solving for ρ when proppant is present, yields

$$\rho = \frac{\kappa f_{Vp} + (g_c' - g_0)/g_0}{\kappa(1 - f_{Vp})},$$

or for $\kappa = 1$,

$$\rho = \frac{g_c'/g_0}{1 - f_{Vp}} - 1 \quad \text{(J-22a)}$$

and

$$f_{Vp} = V_{pr}/V_i = W/[\omega_{pr} V_i (1-\phi)],$$

where

f_{Vp} = volume fraction of proppant pumped (including proppant porosity, ϕ) relative to total slurry volume pumped,
W = proppant weight, and
ω_{pr} = specific weight of proppant material—e.g., 165 lbm/ft^3 [2643 kg/m^3]; 2.65 g/cm^3 for sand.

The efficiency, η, for the case with proppant can be determined from g_c' as

$$\eta = f_{Vp}(1 - \eta') + \eta', \quad \text{(J-22b)}$$

where η' is the apparent efficiency (because of closing on proppant) determined by g_c' with Eq. J-18b. The closure time for a treatment with proppant can be predicted by use of Eq. J-19 with Eq. J-22a based on the values of ρ and f_{Vp} predicted for the treatment.

Fracture Width. Substituting $\kappa \rho g_0 = g_c - g_0$ from Eqs. J-18 into Eq. J-14b yields

$$\bar{w} = 2\kappa C f_p \sqrt{t_i}\, g_0 \rho \quad \text{(J-23)}$$

for the average width at shut-in with ρ from Eq. J-18a for no proppant or from Eq. J-22a for a case with proppant.

Fracture Area and Penetration. Combining the continuity, or material-balance, equation, $V_i = V_\ell + V_f$, and Eqs. J-13 and J-15 gives the following for the fracture area:

$$V_i = V_\ell + V_f = V_\ell(1+\rho)$$

$$= 2\kappa C f_p A_f \sqrt{t_i}\, g_0 (1+\rho)$$

or

$$A_f = \frac{V_i}{2\kappa C f_p g_0 \sqrt{t_i}\,(1+\rho)} = \begin{pmatrix} h_f L \\ h_f L \\ \pi r_f^2 \end{pmatrix} \begin{matrix} \text{PK} \\ \text{KG} \\ \text{radial} \end{matrix}, \quad \text{(J-24)}$$

where the expressions in parentheses are the areas for the three indicated fracture models: h_f and L are fracture height and tip-to-tip length, and r_f is the fracture radius for the radial model. Eq. J-24 can be expressed in terms of time to close by substituting g_c from Eq. J-18a or J-22a for ρ.

Relations Based on Pressure

The average fracture width for each of the standard fracture models can be related to the net fluid pressure (see Eqs. 14.23 and 14.27) by

$$\bar{w} = c_f p_n = c_f(p_w - p_c), \quad \text{(J-25a)}$$

where

$$c_f = \frac{\pi\beta}{2E'} \begin{pmatrix} h_f \\ L \\ \dfrac{32}{3\pi^2} r_f \end{pmatrix} \begin{matrix} \text{PK} \\ \text{KG} \\ \text{radial} \end{matrix}, \quad \text{(J-25b)}$$

and

$\bar{w}$ = average width,
c_f = fracture compliance for width,
E' = plane-strain elastic modulus of formation, and
β = ratio of "average" to wellbore net pressures.

Fluid-Loss Coefficient. The continuity (or material-balance) equation implies that, after shut-in and before the fracture closure, the rate of fluid-loss volume is equal to the rate of change in fracture volume; i.e.,

$$-\frac{dV_f}{dt} = -A_f \frac{d\bar{w}}{dt} = q_\ell(t, A_p).$$

Substituting Eq. J-8, which does not contain κ because pumping has stopped, and Eq. J-25 gives

$$-A_f c_f \frac{dp_n}{dt} = \frac{2 C f_p A_f}{\sqrt{t_i}} f(t_D)$$

or

$$-\frac{dp_n}{dt} = \frac{2 C f_p}{c_f \sqrt{t_i}} f(t_D), \quad \text{(J-26)}$$

where c_f in Eq. J-25b was assumed to be constant, which is consistent with the prior assumption that A_f is constant during closing.

Eq. J-26 implies that the fluid-loss coefficient is proportional to the rate of net pressure decline divided by the dimensionless loss function $f(t_D)$ and that this coefficient can be inferred from the decline data if c_f in Eq. J-25b can also be inferred. The practical use of rate-of-change data is sometimes difficult because of data scatter, while differences over longer lengths of time average out the scatter. Integrating Eq. J-26 from some reference time, t^*, to a later time, t, and using the first expression in Eq. J-12 gives, in terms of dimensionless time, the pressure difference:

$$\Delta p_n(t_D^*, t_D) = \frac{2 C f_p \sqrt{t_i}}{c_f} [g(t_D) - g(t_D^*)]. \quad \text{(J-27a)}$$

Defining

$$G(t_D, t_D^*) = \frac{4}{\pi}[g(t_D) - g(t_D^*)] \quad \text{(J-27b)}$$

and

$$p^* = \Delta p_n(t_D^*, t_D) \text{ when } G(t_D, t_D^*) = 1 \quad \text{(J-27c)}$$

gives

$$p^* = \frac{\pi}{2} \frac{C f_p \sqrt{t_i}}{c_f}. \quad \text{(J-27d)}$$

Substituting these definitions into Eq. J-27a yields

$$\Delta p_n(t_D^*,t_D)=p^*G(t_D,t_D^*). \qquad \text{(J-28a)}$$

Rearranging Eq. J-27d gives

$$C=\frac{2p^*}{\pi f_p\sqrt{t_i}}c_f=\frac{p^*}{f_p\sqrt{t_i}\,E'}\beta\begin{pmatrix} h_f \\ L \\ \dfrac{32}{3\pi^2}r_f \end{pmatrix}\begin{matrix}\text{PK}\\ \text{KG}\\ \text{radial}\end{matrix}. \qquad \text{(J-28b)}$$

Pressure-decline data, after shut-in, permit p^* and C to be determined from Eqs. J-28. The two methods used are based on Eq. J-28a with examples given in Chap. 14. One method is based on a log-log plot-matching technique using pressure differences[6]; the other is based on a linear plot[7] of pressure vs. $G(t_D)$, i.e., G with $t_D^*=0$. Both methods can directly use wellbore pressure instead of net pressure if closure pressure remains essentially constant (discussed later in the Application section). Also, the derivation of Eq. J-28a assumed that the compliance c_f was constant during closure, which is generally approximated for a treatment without proppant—a calibration treatment. When the fracture begins to close on proppant, the compliance begins to decrease and hence invalidates the assumptions for Eq. J-28a and the pressure-decline analysis based on c_f. For applications with proppant, the closure-time analysis for efficiency can provide some information from Eqs. J-22. These expressions can be coupled with the analysis on the basis of the linear plot for Eq. J-28a by using the value of $G(t_D)$ evaluated at the closure time. This value is denoted as G_c' when closure is with proppant and simply G_c without proppant. Using the lower bound (and hence an approximation) for g_0 from Eq. J-13 and the definition of G from Eq. J-27b, we can approximate Eq. J-22a as

$$\rho\approx\frac{\kappa f_{Vp}+(G_c'/2)}{\kappa(1-f_{Vp})} \qquad \text{(J-29a)}$$

in general, or, for $\kappa=1$ (only fluid-loss contribution from C, discussed in the Application section) and $f_{Vp}=0$ (no proppant),

$$\rho\approx G_c/2. \qquad \text{(J-29b)}$$

For the latter case and from Eq. J-16,

$$\eta\approx\frac{G_c}{2+G_c}. \qquad \text{(J-29c)}$$

Fluid Efficiency. Using Eq. J-25a with the value of p_n directly after shut-in denoted as p_s, and after substituting C from Eq. J-28b into Eq. J-13, Eq. J-15 yields

$$\rho=\frac{A_f\bar{w}}{V_\ell}=\frac{\pi p_s}{4\kappa g_0 p^*}=\frac{p_s}{\kappa p^*}\begin{Bmatrix}3\pi/16\\ 1/2\end{Bmatrix} \qquad \text{(J-30)}$$
$$=\eta/(1-\eta),$$

where Eq. J-16 was used and the expressions in braces result for the upper and lower bounds for g_0, respectively, for fracture growth.

Eq. J-30 defines the basic parameter, ρ, and efficiency, η, in terms of shut-in pressure and the decline-match pressure, while Eq. J-18 defines these parameters in terms of dimensionless closure time. Both expressions contain the possible correction for C during pumping, κ. It is important to note that neither of these expressions depends on an assumed fracture model.

Fracture Penetration and Width. Substituting Eq. J-28b for C into Eq. J-24 and multiplying by $2\beta c_f E'/\pi$ yields

$$\frac{V_iE'}{2\kappa g_0 p^*(1+\rho)}=\frac{2\beta c_f E'}{\pi}A_f=\beta\begin{pmatrix} h_f^2L \\ h_fL^2 \\ \dfrac{32}{3\pi}r_f^3 \end{pmatrix}\begin{matrix}\text{PK}\\ \text{KG}\\ \text{radial}\end{matrix}, \qquad \text{(J-31a)}$$

and substituting Eqs. J-16 and J-30 yields

$$\frac{2\eta V_iE'}{\pi p_s}=\beta\begin{pmatrix} h_f^2L \\ h_fL^2 \\ \dfrac{32}{3\pi}r_f^3 \end{pmatrix}. \qquad \text{(J-31b)}$$

These equations permit the fracture penetration—L (assuming h_f can be estimated) or r_f—to be found from the pressure-decline analysis in terms of p^* and ρ, or in terms of efficiency, η, and the net pressure immediately after shut-in, p_s. These equations, unlike those in the prior section, do not depend on the fluid-loss coefficient. If closure pressure can be assumed constant, Eqs. J-25a and J-28a imply that p^* can be found from the bottomhole or surface wellbore pressure because p^* is in terms of differences and a constant closure pressure, p_c; i.e., hydrostatic head is eliminated. These same parameters can be used to define the average fracture width at shut-in by substituting C of Eq. J-28b into Eq. J-23:

$$\bar{w}=\frac{2g_0\kappa\rho p^*}{E'}\beta\begin{pmatrix} h_f \\ L \\ \dfrac{32}{3\pi^2}r_f \end{pmatrix}\begin{matrix}\text{PK}\\ \text{KG}\\ \text{radial}\end{matrix}, \qquad \text{(J-32a)}$$

or directly from Eq. J-25,

$$\bar{w}=\frac{\pi p_s}{2E'}\beta\begin{pmatrix} h_f \\ L \\ \dfrac{32}{3\pi^2}r_f \end{pmatrix}. \qquad \text{(J-32b)}$$

Approximate Proppant Schedule

Approximations for the proppant addition schedule and duration of temperature exposure for fluid scheduling can be obtained[8] from the fluid efficiency, η. Fig. J-1 illustrates an ideal proppant addition schedule in terms of the dimensionless volume injected, V_{iD}, and the dimensionless concentration, c_{prD}, based on the in-situ fracture concentration at shut-in, c_{prf}. Also illustrated are the proppant-free pad, denoted by f_{pad} for the pad fraction, and the definition of the dimensionless variable, ξ, for the time during which proppant slurry is pumped. Assuming a constant injection rate and minimal dispersion of proppant in the fracture, the approximate pad fraction is

$$f_{pad}\approx(1-\eta)^2\approx\frac{1-\eta}{1+\eta}=\frac{1}{1+2\rho}, \qquad \text{(J-33a)}$$

where the first approximation requires a slight correction,[8] the second approximation[9] tends to include the correction, and the final equality is found from Eq. J-16. This expression assumes that $\kappa=1$. When $\kappa>1$ because of a spurt loss, denoted as κ_S, Eq. J-33a must be modified to separate the fluid-loss contribution owing to spurt, which is lost only at the fracture tip and during the pad, from the total fluid loss during pumping. The modified expression is

$$f_{pad}\approx\frac{1}{1+2\kappa_S\rho}+\frac{\kappa_S-1}{\kappa_S(1+\rho)}, \qquad \text{(J-33b)}$$

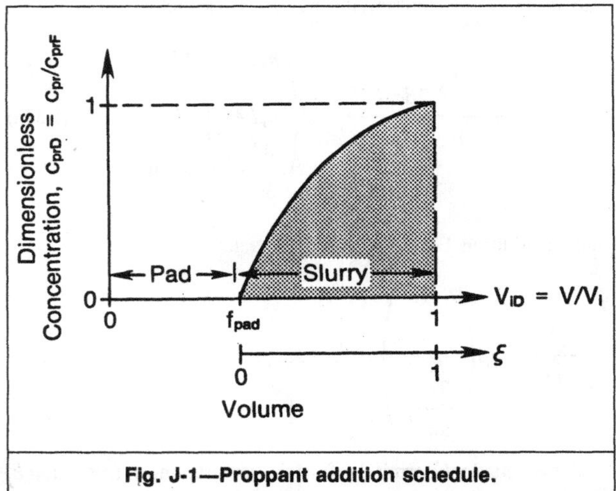

Fig. J-1—Proppant addition schedule.

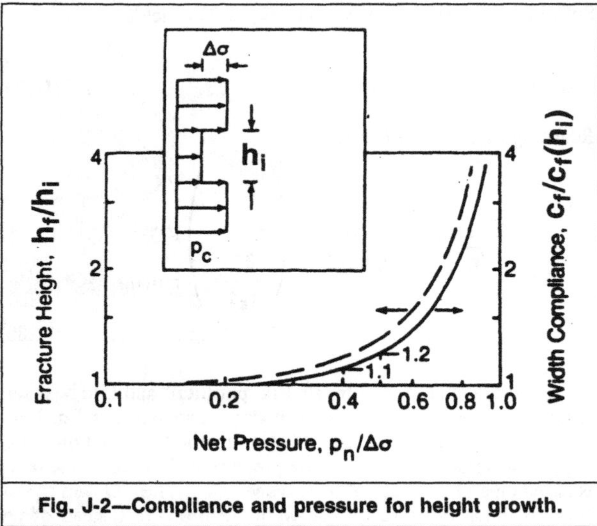

Fig. J-2—Compliance and pressure for height growth.

which follows from the last expression for Eq. J-33a and the definition of κ_S, given in the Application section.

For the assumption that the rate of proppant addition during the slurry portion can be approximated by the power-law relation

$$c_{prD}=c_{pr}/c_{prf}=\xi^{\epsilon}, \quad \text{(J-34a)}$$

it follows that the exponent must equal[8]

$$\epsilon=\frac{1-f_{\text{pad}}-\eta}{\eta}=\frac{1-(1+\rho)f_{\text{pad}}}{\rho} \quad \text{(J-34b)}$$

to satisfy the conditions

$\bar{c}_{pr}=W/V_i$,

$c_{prf}=W/V_f$,

and

$$\bar{c}_{prD}=\eta, \quad \text{(J-34c)}$$

where W and V_i are the total proppant weight and injected volume. The last expression indicates that the shaded area of Fig. J-1 is equal to the fluid efficiency, η.

The expressions in Eqs. J-33b and J-34b permit the proppant schedule to be approximated on the basis of η or ρ, determined from a calibration treatment (i.e., Eqs. J-18a and J-18b or Eq. J-30), if these variables are appropriately altered[8] for the difference in volumes between the stimulation and calibration treatment.

Application

In the following, the applications of the derived expressions are discussed in terms of potential deviations from the assumptions.

Closure Pressure. Two considerations relate to closure pressure: the definition of ρ in terms of the time for the fracture to reach closure pressure—i.e., time to close in Eqs. J-18a and J-22a—and the definition of the terms related to net pressure, p^* and $p_s=p_w-p_c$. If the closure pressure, p_c, is essentially constant, the wellbore pressure, p_w (directly measurable), can be used directly to construct the interpretive plots for p^* from Eq. J-28a. This follows because p^* is defined as differences in net pressure and a constant value of p_c cancels out for the differences. For liquid reservoirs, closure pressure can change significantly[10] during and after a treatment or for subsequent injections. Closure pressure changes generally are insignificant for gas reservoirs, however, because of their relatively high compressibility that retards the reservoir pressure transient from significantly penetrating the reservoir. For this case, the constant value of closure pressure can be determined before the fracture treatment (see Appendix I) and used for subsequent analysis of closure time and net shut-in pressure, p_s.

When closure pressure is expected or found to change significantly, the equations that contain pressure (p^* or p_s, Eqs. J-26 through J-32) cannot be used unless the change in closure pressure with time can be predicted[10,11] and appropriate corrections made.

Pressure-Dependent Loss Coefficient. The above derivations assumed that the fluid-loss coefficient was constant and independent of any pressure variations, whereas it depends on the difference between the fracturing fluid pressure and the reservoir fluid pressure, p_r.[5] The relationship[5] is the square root of the difference for C_{I}, fracturing-fluid viscosity governing; linear dependence on the difference for C_{II}, reservoir fluid governing; and no assumed dependence for C_{III}, fracturing-fluid wall-cake governing. A dimensional analysis[1] indicated that the dependence for C_{III} is the one-sixth power of the pressure difference for the compressible cake resulting from a crosslinked polymer fluid. For an incompressible cake (e.g., hard particulates), the cake permeability would not depend on pressure and the relationship is the square root of the pressure difference. Because the maximum pressure variation during closure is p_s, the maximum error is proportional to $p_s/(p_c-p_r)$. Then a conservative correction—i.e., to obtain largest possible value—for the corrected coefficient, C_c, in terms of the inferred value from the decline analysis is

$$C_c<C\left(1+\frac{p_s}{p_c-p_r}\right)^e, \quad \text{(J-35)}$$

for C_{I}, $e=½$; C_{II}, $e=1$; and C_{III}, $e=½$ (incompressible cake) and ⅙ (compressible cake).

Height Growth. Height growth is generally governed by higher stresses in the boundary formations. For this case, the width is significantly narrower in the boundary formations that are penetrated by height growth (see Fig. 14.7). From the discussion in Sec. 14.3 and Eq. 14.28, the appropriate fracture height, h_f, to use in the prior pressure and dimension relationships is the height of the lower-stress formation, generally the gross reservoir section, without including any additional height growth. Fig. J-2 shows the change in the assumed constant compliance, c_f, of Eqs. J-25 for the idealized case in Fig. 14.7. Fig. J-2 shows that c_f changes by less than 10% for $p_n<0.4\Delta\sigma$ ($\Delta\sigma$=stress difference in formations) or equivalently for height growth of less than 20% of the initial height. The compliance and height are normalized in Fig. J-2 by their initial values in terms of the initial height. Fig. J-2 also shows that the net pressure is limited to about $0.9\Delta\sigma$—i.e., maximum p_n or $p_s=0.9\Delta\sigma$. For cases in which height growth occurs, if the pressure-decline analysis is performed after the net pressure has declined to about one-half of p_s, c_f will remain essentially constant until closure and the analysis will be valid in terms of the initial height. Height growth would appear to violate the assumption of constant $f_p = h_p/h_f$, Eq. J-7. This is not the case if the bounding beds are impermeable (generally shales with no fluid loss) and have higher stress (generally shales). For this case, both the volume of fluid lost and that stored in the narrow width of the boundary formations would be relatively insignificant, and the effective fracture height would not include the height in these formations.

Fluid Loss From Spurt and Opening Natural Fractures. In the prior sections, a distinction was made for fluid-loss coefficient between conditions during pumping (κC) and after pumping stops (C). This distinction was made by multiplying the loss coefficient by the variable κ for expressions based on fluid loss during pumping. Two primary effects that can cause increased loss during pumping are spurt loss and the opening of natural fissures at some value of fluid pressure (see Sec. 14.5). If the decline analysis for p^* is performed below this opening/closing value of pressure, the inferred conditions for fluid loss would not reflect the increase in fluid-loss area, A_{of}, when the fissures were open. The actual loss for this case should be based on the area $A_p + A_{of}$ (A_{of} for only one side of the fracture face).

Fluid loss to spurt occurs when a portion of the formation is first fractured and is generally assumed essentially to occur instantaneously. Therefore, spurt would occur only during pumping and not after shut-in when the fracture penetration is assumed to stop. The spurt loss is not included in the time-dependent loss defined by the loss coefficient. Spurt is defined as volume lost per unit area and has the dimension of width. The amount of volume lost to spurt during a treatment is $2S_\ell A_p$, and $A_p = f_p A_f$. From inspection of Eq. J-13 for the loss volume resulting from time-dependent fluid loss, with the effect of natural fractures opening during pumping included, it follows that

$$\kappa = \left[1 + S_\ell / \left(C\sqrt{t_i}\, g_0\right)\right](A_p + A_{of})/A_p, \quad \text{(J-36)}$$

where C is the fluid-loss coefficient without open fissures or spurt, i.e., as generally determined in the laboratory or as inferred by Eq. J-28b from decline data.

For applications when spurt is expected to be significant (generally a wall-cake-building fluid) the spurt component of κ, κ_S, can be estimated from laboratory data for S_ℓ and C for the appropriate fluid and formation. From Eq. J-36,

$$\kappa_S = \left[1 + (S_\ell / C)/\left(g_0\sqrt{t_i}\right)\right], \quad \text{(J-37)}$$

where g_0 is approximately 1.5 from Eq. J-13 and κ_S depends on the pump time, t_i. This application for Eq. J-37 assumes that the laboratory and in-situ values of the ratio S_ℓ/C are approximately equal.

Nomenclature

a, A = general variables for area
A_f = fracture area
A_{of} = open fissure area
A_p = permeable, or loss, area
c_f = compliance for $\bar{w}$
c_{pr} = proppant concentration, weight/slurry vol
c_{prD} = c_{pr}/c_{prf}
c_{prf} = concentration in fracture at shut-in
C = fluid-loss coefficient
C_c = corrected value of C
e = exponent
E' = plane-strain elastic modulus
f_A = fraction of previous to current areas
f_p = ratio of permeable area to fracture area
f_{pad} = pad fraction
$f(t_D)$ = dimensionless loss-rate function, Eq. J-8
f_{Vp} = volume fraction of proppant
g_c = $g(t_D)$ at closure without proppant
g_c' = g_c with proppant
$g(t_D)$ = dimensionless loss-volume function, Eq. J-12
g_0 = $g(t_D = 0)$
G_c = $G(t_D)$ at closure
$G(t_D)$ = $G(t_D, t_D^*)$ for $t_D^* = 0$
$G(t_D, t_D^*)$ = dimensionless difference function, Eq. J-27b
h_f = vertical fracture height
h_p = permeable section height
L = fracture length, tip to tip
p^* = match decline pressure
p_c = closure pressure
p_n = net pressure, $p_w - p_c$
p_r = reservoir pressure
p_s = net pressure at shut-in
p_w = pressure at wellbore
q_ℓ = fluid-loss rate
r_f = fracture radius
S_ℓ = spurt-loss coefficient
t = time since start of pumping
t_c = closure time after shut-in
t_D = dimensionless shut-in time, $t/t_i - 1$
t_D^* = reference value of t_D
t_i = injection (pumping) time
V_f = volume stored in fracture, end of pumping
V_i = volume injected
V_{iD} = current to final injected volume
V_ℓ = loss volume during pumping
$V_{\ell s}$ = loss volume after shut-in
V_{pr} = proppant volume with voids
$\bar{w}$ = average fracture width, end of pumping
W = proppant weight
β = ratio of average to wellbore net pressures
Δ = difference
ϵ = exponent for rate of proppant addition
η = fluid efficiency, V_f/V_i
κ = correction for C while pumping
κ_S = κ for spurt loss
ξ = variable for slurry volume
ρ = $V_f/V_\ell = \eta/(1-\eta)$
σ = stress
τ = time at which fracture area was created
ϕ = porosity of proppant pack
ω_{pr} = specific weight of proppant

References

1. Nolte, K.G.: "A General Analysis of Fracturing Pressure Decline With Application to Three Models," *SPEFE* (Dec. 1986) 571–83.
2. Perkins, T.K. and Kern, L.R.: "Widths of Hydraulic Fractures," *JPT* (Sept. 1961) 937–49; *Trans.*, AIME, **222**.
3. Khristianovich, S.A. and Zheltov, Y.P.: "Formation of Vertical Fractures by Means of Highly Viscous Liquid," *Proc.*, Fourth World Pet. Cong., Rome (1955).
4. Geertsma, J. and de Klerk, F.: "A Rapid Method of Predicting Width and Extension of Hydraulically Induced Fractures," *JPT* (Dec. 1969) 1571–81; *Trans.*, AIME, **246**.
5. Howard, G.C. and Fast, C.R.: "Optimum Fluid Characteristics for Fracture Extension" (Appendix by R.D. Carter) presented at Mid-Continent District Spring Meeting, Tulsa (1957); also *Hydraulic Fracturing*, Monograph Series, SPE, Richardson, TX (1970) **2**.
6. Nolte, K.G.: "Determination of Fracture Parameters from Fracturing Pressure Decline," paper SPE 8341 presented at the 1979 SPE Annual Technical Conference and Exhibition, Las Vegas, Sept. 23–26.
7. Castillo, J.L.: "Modified Fracture Pressure Decline Analysis Including Pressure-Dependent Leakoff," paper SPE 16417 presented at the 1987 SPE/DOE Low-Permeability Reservoirs Symposium, Denver, May 18–19.
8. Nolte, K.G.: "Determination of Proppant and Fluid Schedules From Fracturing-Pressure Decline," *SPEPE* (July 1986) 255–65; *Trans.*, AIME, **283**.
9. Meng, H.-Z.: "Design of Propped Fracture Treatments," *Reservoir Stimulation*, second edition, M.J. Economides and K.G. Nolte (eds.), Prentice-Hall, New York City (1989) Chap. 8.
10. Smith, M.B.: "Stimulation Design for Short, Precise Hydraulic Fractures," *SPEJ* (June 1985) 371–79.
11. Cleary, M.P. *et. al.*: "Surface Integral Schemes for Fluid Flow and Induced Stresses Around Fractures in Underground Reservoirs," paper SPE 11632 presented at the 1983 SPE/DOE Low Permeability Gas Reservoirs Symposium, Denver, March 14–16.

Appendix K
Examples of Postfracture Formation Evaluation

W. John Lee, SPE, Texas A&M U.

Example 15.1—Productivity-Index-Ratio Increase

A gas well was fractured and then produced at constant bottomhole pressure (BHP) for almost 3 years. Fracture and formation properties include the following:

L_f = 500 ft,
wk_f = 2,200 md-ft,
k = 0.1 md,
A = 320 acres (square),
L_e = 1,867 ft (distance to side of square); r_e, radius of circle with same area=2,106 ft,
r_w = 0.354 ft,
h = 50 ft (net pay *and* propped fracture height),
ϕ = 0.08,
μ = 0.0278 cp, and
c_T = 9.71×10^{-5} psi^{-1}.

From measured rates, pressure drawdown, and fluid properties, the group

$$\frac{q_g\bar{\mu}\bar{z}}{(\bar{p}^2-p_{wf}^2)}$$

was calculated as a function of time. Before fracturing, the value of

$$\frac{q_g\bar{\mu}\bar{z}}{(\bar{p}^2-p_{wf}^2)}$$

was determined to be 5.7×10^{-10} Mscf-cp/D-psia^2 [3.4×10^{-10} std m$^3\cdot$mPa$\cdot$s/d$\cdot$kPa^2] (stabilized after about 266 days). From these data and the tabulation below, do the following.

1. Determine the productivity-index (PI) ratio at stabilization from observed well performance, and compare this result with predictions made with the Prats[1] method, the Tinsley *et al.*[2] chart, and the Holditch[3] modification of the McGuire and Sikora[4] chart.

2. Plot the ratio of postfracture transient PI (Table K-1), J, to stabilized prefracture PI, J_o, vs. time (as in the Morse-Von Gonten[5] method). Compare the shape of the curve with similar curves in the Morse-Von Gonten paper.

Solution—PI-Ratio Increase. The well stabilized at about 490 days; the stabilized PI ratio is about 5.3.

1. A. From Prats' method, we predict

$$J/J_o=\frac{\ln(r_e/r_w)}{\ln(r_e/0.5L_f)}=\frac{\ln(2{,}106/0.354)}{\ln[2{,}106/(0.5)(500)]}=4.08.$$

B. From Tinsley *et al.*'s chart, with $L_f/L_e=500/2{,}106=0.237$,

$$\text{relative capacity}=(0.00159)\frac{wk_f}{k}\frac{h_f}{h}\ln\left(\frac{r_e}{r_w}\right)\sqrt{\frac{10}{A}}$$

$$=(0.00159)\left(\frac{2{,}200}{0.1}\right)\left(\frac{50}{50}\right)\ln\left(\frac{2{,}106}{0.354}\right)\sqrt{\frac{10}{320}}=53.742.$$

From Tinsley's figure, for $h_f/h=1.0$, $J/J_o=2.9$.

C. Using the McGuire-Sikora method yields

$$L_f/L_e=500/1{,}867=0.268.$$

$$\frac{12wk_f}{k}\sqrt{\frac{40}{A}}=\frac{(12)(2{,}200)}{(0.1)}\sqrt{\frac{40}{320}}=9.33\times10^4.$$

From the Holditch modification of the McGuire-Sikora chart,

$$\frac{J}{J_o}\left[\frac{7.13}{\ln(0.472L_e/r_w)}\right]\cong4.8;$$

$$\frac{J}{J_o}=\frac{4.8\ln[(0.472)(1{,}867)/(0.354)]}{7.13}=5.3.$$

The estimate from the modified McGuire-Sikora chart agrees closely with the result from the simulator; Prats' method gives a result in moderate agreement, and the result from Tinsley *et al.*'s chart is significantly different.

2. A plot of transient, postfracture J/J_o is shown in Fig. K-1. The shape of the curve is similar to those presented by Morse and Von Gonten. As a check, note that for t=0.121 days,

$$t_{eD}=\frac{0.0002637kt}{\phi\mu c_T L_e^2}$$

$$=\frac{(0.0002637)(0.1)(0.121)(24)}{(0.08)(0.0278)(9.7\times10^{-5})(1{,}867)^2}=0.00010.$$

From Morse and Von Gonten's Fig. 4, for $L_f/L_e=0.268$, $J/J_o\cong80$; this is quite close to the observed J/J_o of 87.7 at 0.121 days.

Note also that at $t_{eD}\cong0.25$, J/J_o has reached a stabilized value of about 5.3 in Morse and Von Gonten's Fig. 4. At stabilization,

Supplementary to Chap. 15.

TABLE K-1—POSTFRACTURE PRODUCTIVITY INDEXES

t (days)	$\frac{q_g \bar{\mu} \bar{z}}{\bar{p}^2 - p_{wf}^2}$ (Mcf-cp/D-psia²)	J/J_o
0.001	2.97×10^{-7}	521
0.004	1.50×10^{-7}	263
0.013	1.11×10^{-7}	195
0.040	7.59×10^{-8}	133
0.121	5.00×10^{-8}	87.7
0.364	3.18×10^{-8}	55.7
1.093	1.97×10^{-8}	34.6
3.280	1.22×10^{-8}	21.4
9.841	7.82×10^{-9}	13.7
29.52	5.27×10^{-9}	9.25
88.57	3.88×10^{-9}	6.81
195.1	3.30×10^{-9}	5.79
332.1	3.08×10^{-9}	5.40
494.4	3.01×10^{-9}	5.28
580.0	3.00×10^{-9}	5.26
889.8	3.01×10^{-9}	5.28

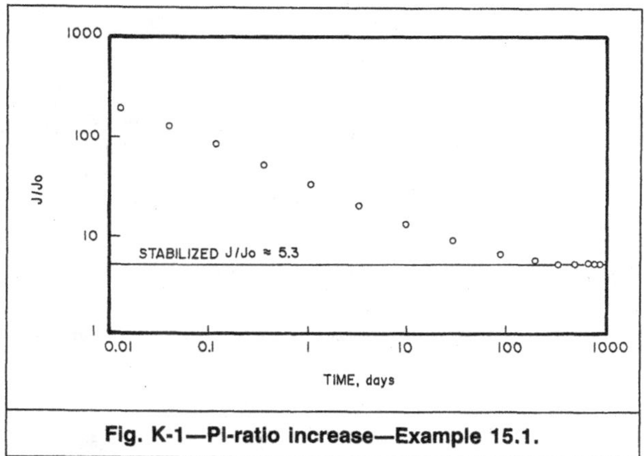

Fig. K-1—PI-ratio increase—Example 15.1.

the time in days is

$$t=\frac{\phi\mu c_T L_e^2}{0.0002637k}\frac{t_{eD}}{(24)}$$

$$=\frac{(0.08)(0.0278)(9.7\times10^{-5})(1{,}867)^2}{(0.0002637)(0.1)}\frac{0.25}{24}=297 \text{ days.}$$

This is roughly comparable to the observed stabilization time of 490 days; in fact, the well had essentially stabilized at about 300 days.

Example 15.2—Predicting Performance With a Constant-Pressure Type Curve

For a gas well centered in a square reservoir (see Table K-2),

k = 0.1 md,
h = 50 ft,
ϕ = 0.1,
L_f = 500 ft,
wk_f = 2,200 md-ft,
A = 320 acres, and
p_i = 6,000 psia.

Assuming that the well produces at a constant p_{wf}=1,000 psia, estimate the cumulative gas produced at times of 1, 2, 3, 4, 6, 8, 10, 12, 14, 16, 18, and 20 years.

Solution—Predicting Performance With a Constant-Pressure Type Curve. In this case,

$$C_r=wk_f/\pi kL_f=\frac{2{,}200}{\pi(0.1)(500)}=14.$$

TABLE K-2—GAS PROPERTIES

p (psia)	B_g (RB/Mscf)	μ_g (cp)	c_T (psia)
6,000	0.666	0.0278	1.04×10^{-4}
5,000	0.746	0.0251	1.13×10^{-4}
4,000	0.876	0.0221	1.62×10^{-4}
3,000	1.116	0.0192	2.48×10^{-4}
2,000	1.656	0.0165	4.16×10^{-4}
1,000	3.428	0.0146	8.55×10^{-4}

We will use the type curve for C_r=10, which is the best approximation available to a curve for C_r=14. This curve is given as Fig. 15.23.

$$L_e=\left[\frac{(320)(43{,}560)}{4}\right]^{1/2}=1{,}867 \text{ ft.}$$

$$L_D=1{,}867/500=3.73.$$

$$p_{av}=(p_i+p_{wf})/2=\frac{6{,}000+1{,}000}{2}=3{,}500 \text{ psia.}$$

Interpolating from the gas properties table, for p_{av}=3,500 psia,

$\bar{\mu}$ = 0.0207 cp,
$\bar{c}_T$ = 2.05×10^{-4} psia^{-1}, and
$\bar{B}_g$ = 0.996 RB/Mscf.

The following relationships can be computed with the definitions of Q_{L_fD} and t_{L_fD}:

$$Q_{L_fD}=\frac{0.8936Q(0.996)}{(0.1)(2.05\times10^{-4})(50)(500)^2(6{,}000-1{,}000)}$$

$$=6.9465\times10^{-7}Q.$$

$$t_{L_fD}=\frac{(0.0002637)(24)t}{(0.1)(0.0207)(2.05\times10^{-4})(500)^2}, \ t \text{ in days}$$

$$=5.966\times10^{-2}t.$$

With these relationships, the constant-pressure type curve for C_r=10 and interpolation between values of L_D, the results given in Table K-3 were calculated.

TABLE K-3—CUMULATIVE PRODUCTION FROM TYPE CURVE

t (years)	t (days)	t_{L_fD}	Q_{L_fD}	Q (Bscf)
1	365	2.18	1.57	2.26
2	730	4.36	2.56	3.69
3	1,095	6.54	3.41	4.91
4	1,460	8.72	4.10	5.90
6	2,190	13.1	5.21	7.50
8	2,920	17.4	5.95	8.57
10	3,650	21.8	6.59	9.49
12	4,380	26.2	7.02	10.1
14	5,110	30.5	7.41	10.7
16	5,840	34.9	7.65	11.0
18	6,570	39.2	7.84	11.3
20	7,300	43.6	8.02	11.6

TABLE K-4—TIMES FOR FLOW REGIMES				
Case	k (md)	L_f (ft)	Time to End of Linear Flow (hours)	Time to Start of Pseudoradial Flow (hours)
1	1	100	0.12	22.8
2	0.01	100	12	2,280 (95 days)
3	0.01	1,000	1,200 (50 days)	228,000 (26 years)

Example 15.3—Predicting Performance With Radial-Flow Type Curve

Consider the gas well described in Example 15.2, with formation properties summarized below and in Table K-2:

k = 0.1 md,
h = 50 ft,
ϕ = 0.1,
A = 320 acres (square, well centered),
p_i = 6,000 psia,
p_{wf} = 1,000 psia (constant), and
r_w = 0.527 ft.

For the cases below, estimate cumulative gas production after 5 and 10 years.

1. Undamaged well, s=0.
2. Damaged well, s=1.834.
3. Stimulated well, s=−3.

Solution—Predicting Performance With Radial-Flow Type Curve. ***Case 1: Undamaged Well.***

$$r_{ea}=\sqrt{A/\pi}=\sqrt{(320)(43{,}560)/(\pi)}=2{,}106 \text{ ft}.$$

$$r_{wa}=r_we^{-s}=r_w=0.527 \text{ ft}.$$

$$r_{eaD}=r_{ea}/r_{wa}=2{,}106/0.527\cong 4{,}000.$$

$$p_{av}=(p_i+p_{wf})/2=(6{,}000+1{,}000)/2=3{,}500 \text{ psia}.$$

Interpolating from Table K-2, for p_{av}=3,500 psia, $\bar{\mu}$=0.0207 cp, $\bar{c}_T$ =2.05×10^{-4} psi^{-1}, and $\bar{B}_g$=0.996 RB/ Mscf.

The relationships between (Q,Q_D) and (t,t_D) are then

$$Q_D=\frac{0.8936QB_g}{\phi\bar{c}_Thr_{wa}^2(p_i-p_{wf})}$$

$$=\frac{(0.8936)Q(0.996)}{(0.1)(2.05\times10^{-4})(50)(0.527)^2(6{,}000-1{,}000)}$$

$$=0.625Q.$$

$$t_D=\frac{0.0002637kt}{\phi\bar{\mu}\bar{c}_Tr_{wa}^2}=\frac{(0.0002637)(0.1)(24)(t,\text{ days})}{(0.1)(0.0207)(2.05\times10^{-4})(0.527)^2}$$

$$=5.370\times10^3t,\ t \text{ in days}.$$

For t=5 years=1,825 days, t_D=9.80×10^6 and Q_D=1.22×10^6 from the type curve. For t=10 years, t_D=1.96×10^7 and Q_D= 2.23×10^6. Thus, for t=5 years,

$$Q=\frac{Q_D}{0.625}=\frac{1.22\times10^6}{0.625}=1.95\times10^6 \text{ Mscf}=1.95 \text{ Bcf},$$

and for t=10 years,

$$Q=\frac{2.23\times10^6}{0.625}=3.57\times10^6 \text{ Mscf}=3.57 \text{ Bcf}.$$

TABLE K-5—PRESSURE/TIME DATA FOR DRAWDOWN TEST	
Δt (hours)	p_{wf} (psia)
0.0	2,684
0.1	2,599
0.2	2,564
0.4	2,514
0.6	2,484
0.8	2,454
1.0	2,434
2.0	2,349
4.0	2,244
6.0	2,164
8.0	2,104
9.0	2,074
10.0	2,049
15.0	1,946
20.0	1,874
25.0	1,820
30.0	1,753
35.0	1,699
40.0	1,650
45.0	1,608
50.0	1,570

Case 2: Damaged Well (s=1.834).

$$r_{wa}=r_we^{-s}=0.527e^{-1.834}=0.0842.$$

$$r_{eaD}=r_{ea}/r_{wa}=2{,}106/0.0842\cong 25{,}000.$$

The relationships between (Q,Q_D) and (t,t_D) are

$$Q_D=\frac{0.8936Q(0.996)}{(0.1)(2.05\times10^{-4})(50)(0.0842)^2(6{,}000-1{,}000)}=24.5Q$$

and

$$t_D=\frac{(0.0002637)(0.1)(24)(t,\text{ days})}{(0.1)(0.0207)(2.05\times10^{-4})(0.0842)^2}$$

$$=2.10\times10^5t,\ t \text{ in days}.$$

For t=5 years=1,825 days, t_D=3.83×10^8 and Q_D=3.87×10^7. For t=10 years, t_D=7.67×10^8 and Q_D=7.23×10^7. Thus, for t=5 years,

$$Q=\frac{Q_D}{24.5}=\frac{3.87\times10^7}{24.5}=1.58 \text{ Bcf},$$

and for t=10 years,

$$Q=\frac{7.23\times10^7}{24.5}=2.95 \text{ Bcf}.$$

Case 3: Stimulated Well (s=−3).

$$r_{wa}=r_we^{-s}=0.527e^{-(-3)}=10.59.$$

$$r_{eaD}=r_{ea}/r_{wa}=2{,}106/10.59\cong 200.$$

The relationships between (Q,Q_D) and (t,t_D) are

$$Q_D=\frac{0.8936Q(0.996)}{(0.1)(2.05\times10^{-4})(50)(10.59)^2(6{,}000-1{,}000)}$$

$$=0.001549Q$$

TABLE K-6—PRESSURE-DRAWDOWN-TEST SUMMARY

Well: linear and pseudoradial analysis
Location: gas-well drawdown test (postfracture)

Net pay, ft	67.000
Wellbore radius, ft	0.36000
Initial reservoir pressure, psia	2,684.0
Initial adjusted pressure, psia²/cp	1,743.7
Bottomhole temperature, °F	190.00
Dry-gas flow rate, Mcf/D	2,955.0
Wet-gas flow rate, Mcf/D	2,955.0
Total porosity, fraction	0.84000×10^{-1}
Water saturation, fraction	0.35000
Water compressibility, psia⁻¹	0.35000×10^{-5}
Formation compressibility, psia⁻¹	0.40000×10^{-5}
Separator-gas specific gravity	0.62000
Gas/condensate ratio, scf/STB	0.00000
Condensate gravity, °API	0.00000
Gas equivalent of condensate, scf/STB	0.00000
Wet-gas specific gravity	0.62000
Number of pressure/time data points	21
Data point for start of middle-time region	17
Data point for end of middle-time region	21
Fluid Properties at Average Pressure	
Average pressure, psia	2,127.00000
Gas viscosity, cp	0.01599
Total compressibility, psia⁻¹	0.0003159
z factor	0.89497
Gas FVF, RB/Mcf	1.37695

TABLE K-7—SEMILOG AND TYPE-CURVE PLOTTING FUNCTIONS

	t (hours)	p (psia)	Δp (psia)	Δt_a (hours)	p_a (psia)	Δp_a (psia)
1	0.0000	2,684.0	0.0000	0.0000	1,744.0	0.0000
2	0.1000	2,599.0	85.00	0.1212	1,646.0	97.42
3	0.2000	2,564.0	120.0	0.2399	1,607.0	136.7
4	0.4000	2,514.0	170.0	0.4736	1,551.0	192.7
5	0.6000	2,484.0	200.0	0.7040	1,518.0	226.0
6	0.8000	2,454.0	230.0	0.9319	1,485.0	258.8
7	1.000	2,434.0	250.0	1.158	1,463.0	280.7
8	2.000	2,349.0	335.0	2.265	1,371.0	372.3
9	4.000	2,244.0	440.0	4.402	1,261.0	482.8
10	6.000	2,164.0	520.0	6.464	1,179.0	564.7
11	8.000	2,104.0	580.0	8.470	1,119.0	625.1
12	9.000	2,074.0	610.0	9.456	1,089.0	654.5
13	10.00	2,049.0	635.0	10.43	1,065.0	678.8
14	15.00	1,946.0	738.0	15.17	966.4	777.3
15	20.00	1,874.0	810.0	19.73	899.8	843.9
16	25.00	1,820.0	864.0	24.16	851.2	892.5
17	30.00	1,753.0	931.0	28.46	792.7	951.0
18	35.00	1,699.0	985.0	32.63	746.2	997.5
19	40.00	1,650.0	1,034.0	36.70	706.1	1,038.0
20	45.00	1,608.0	1,076.0	40.68	671.6	1,072.0
21	50.00	1,570.0	1,114.0	44.57	641.7	1,102.0

and

$$t_D=\frac{(0.0002637)(0.1)(24)(t,\text{ days})}{(0.1)(0.0207)(2.05\times10^{-4})(10.59)^2}=13.30t,\ t\text{ in days.}$$

For t=5 years=1,825 days, t_D=24,300 and $Q_D=4.77\times10^3$. For t=10 years, t_D=48,500 and $Q_D=8.30\times10^3$. Thus, for t=5 years,

$$Q=\frac{Q_D}{0.001549}=\frac{4.77\times10^3}{0.001549}=3.08\text{ Bcf,}$$

and for t=10 years,

$$Q=\frac{8.30\times10^3}{0.001549}=5.36\text{ Bcf.}$$

Example 15.4—Duration of Flow Regimes

For each of the cases below, estimate the end of the linear flow regime and the time at which pseudoradial flow begins. In all cases,

$$\phi = 0.1,$$
$$\bar{\mu} = 0.025\text{ cp,}$$
$$\bar{c}_T = 8\times10^{-5}\text{ psia}^{-1}\text{, and}$$
$$C_r = 100.$$

Note that

$$t_{L_fD}=\frac{0.0002637kt}{\phi\bar{\mu}\bar{c}_T L_f^2}.$$

Assume that pseudoradial flow begins when $t_{L_fD}\cong3$.

1. L_f=100 ft; k=1 md.
2. L_f=100 ft; k=0.01 md.
3. L_f=1,000 ft; k=0.01 md.

Solution—Duration of Flow Regimes. In all cases, the end of the linear flow regime is given by $t_{L_fD}\cong0.016$ or

$$t\cong60\phi\bar{\mu}\bar{c}_T L_f^2/k=(60)(0.1)(0.025)(8\times10^{-5})L_f^2/k$$

$$\cong1.2\times10^{-5}L_f^2/k.$$

Also, in all cases, time to reach pseudoradial flow is $t_{L_fD}\cong3$ or

$$t\cong11{,}400\phi\bar{\mu}\bar{c}_T L_f^2/k$$

$$=(11{,}400)(0.1)(0.025)(8\times10^{-5})L_f^2/k\cong2.28\times10^{-3}L_f^2/k.$$

Table K-4 gives further data.

Example 15.5—Linear and Pseudoradial Analysis of Gas-Well Drawdown Test (Postfracture)

A constant-rate drawdown test was run in a gas well after a small fracture treatment. Known reservoir and fluid properties are summarized below:

p_i = 2,684 psia,
p_{sc} = 15.025 psia,
T = 190°F,
T_{sc} = 60°F,
casing size = 4½ in.,
tubing size = 2⅜ in.,
r_w = 0.36 ft,
ϕ = 0.084,
S_w = 0.35,
h = 67 ft,
γ_g = 0.62,
total depth = 7,520 ft, and
perforated interval = 7,303 to 7,490 ft.

During the test, the well produced 2,955 Mcf/D dry gas. Pressure/time data are summarized in Table K-5.

1. From the appropriate type curve, determine whether linear or pseudoradial flow regimes are present. Analyze those regimes present with the appropriate special graph.
2. Confirm the analysis with type-curve analysis.
3. State your best determinations of k, s, L_f, fracture conductivity, and start and finish of flow regimes, such as linear or pseudoradial.

TABLE K-8—LINEAR-FLOW PLOTTING FUNCTIONS

	$t^{1/2}$ (hours$^{1/2}$)	p (psia)	$t_a^{1/2}$ (hours$^{1/2}$)	p_a (psia)
1	0.0000	2,684.0	0.0000	1,744.0
2	0.3162	2,599.0	0.3482	1,646.0
3	0.4472	2,564.0	0.4898	1,607.0
4	0.6325	2,514.0	0.6882	1,551.0
5	0.7746	2,484.0	0.8391	1,518.0
6	0.8944	2,454.0	0.9653	1,485.0
7	1.000	2,434.0	1.076	1,463.0
8	1.414	2,349.0	1.505	1,371.0
9	2.000	2,244.0	2.098	1,261.0
10	2.449	2,164.0	2.542	1,179.0
11	2.828	2,104.0	2.910	1,119.0
12	3.000	2,074.0	3.075	1,089.0
13	3.162	2,049.0	3.230	1,065.0
14	3.873	1,946.0	3.895	966.4
15	4.472	1,874.0	4.442	899.8
16	5.000	1,820.0	4.915	851.2
17	5.477	1,753.0	5.335	792.7
18	5.916	1,699.0	5.713	746.2
19	6.325	1,650.0	6.058	706.1
20	6.708	1,608.0	6.378	671.6
21	7.071	1,570.0	6.676	641.7

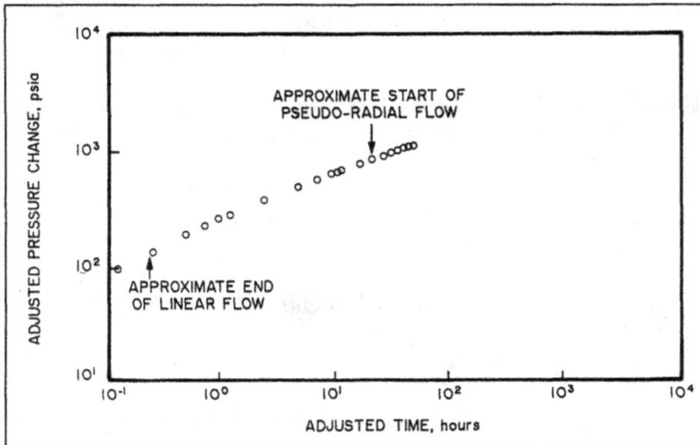

Fig. K-2—Type-curve plot of linear and pseudoradial analysis, postfracture gas-well drawdown—Example 15.5.

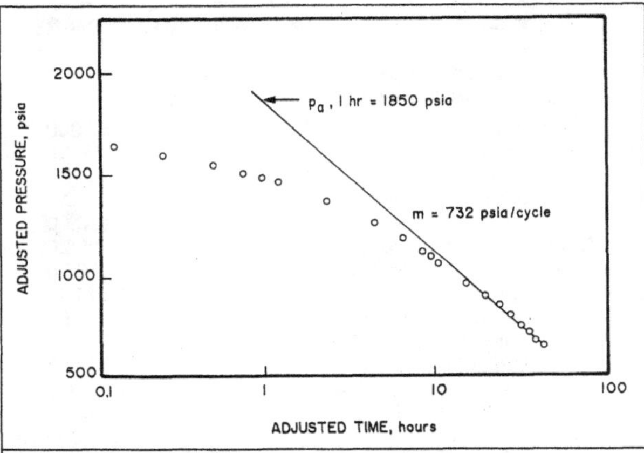

Fig. K-3—Semilog graph of linear and pseudoradial analysis, postfracture gas-well drawdown—Example 15.5.

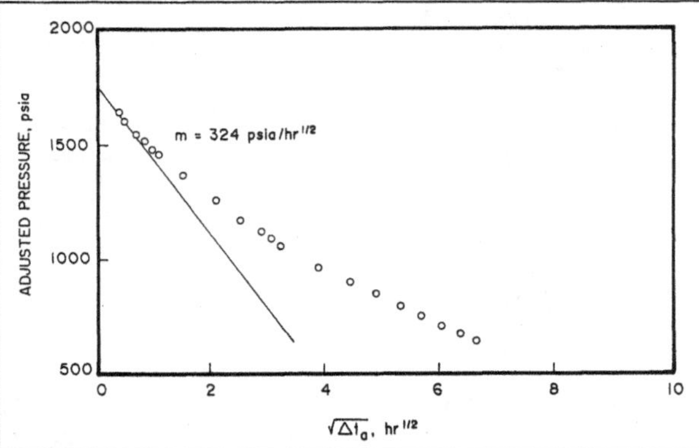

Fig. K-4—Linear-flow plot of linear and pseudoradial analysis, postfracture gas-well drawdown—Example 15.5.

Tables K-6 through K-8 summarize the pressure-drawdown test data.

Solution—Linear and Pseudoradial Analysis of Gas-Well Drawdown Test (Postfracture). ***Preliminary Type-Curve Analysis.*** With a Gringarten *et al.*[6] type curve (Fig. K-2) for infinite-conductivity vertical fractures, a preliminary fit indicates that both linear and pseudoradial flow regimes were dominant at some time during the drawdown test.

Semilog Analysis. Even though a graph of adjusted pressure vs. real (rather than adjusted) time may lead to more accurate analysis of test data than a graph of adjusted time, we will use adjusted time in all analyses presented here (Fig. K-3) to avoid unnecessary confusion. In this case, the difference in results is negligible.

Assuming that pseudoradial flow begins at about $\Delta t_a = 20$ hours ($t_{L_fD} \cong 1.5$),

$m = 732$ psia/cycle (see plot),

$$k = 162.6\frac{q_g \bar{B}_g \bar{\mu}}{mh} = 162.6\frac{(2{,}955)(1.38)(0.016)}{(732)(67)}$$

$=0.216$ md, and

$$s' = 1.151\left[\left(\frac{p_{a,i} - p_{a,1\text{hr}}}{m}\right) - \log\left(\frac{k}{\phi\bar{\mu}\bar{c}_T r_w^2}\right) + 3.23\right]$$

$$= 1.151\left\{\left(\frac{1{,}744 - 1{,}850}{732}\right) - \log\left[\frac{0.216}{(0.084)(0.016)(3.16\times10^{-4})(0.36)^2}\right] + 3.23\right\}$$

$= -4.04.$

Assuming that $wk_f \cong \infty$,

$$L_f = 2r_{wa} = 2r_w e^{-s} = (2)(0.36)e^{-(-4.04)}$$

$= 40.8$ ft.

Linear-Flow Analysis (Fig. K-4). Assuming that linear flow influences the first few data points ($\Delta t_a \leq 0.24$ hours or $t_{L_fD} \leq 0.016$),

$m_L = 287$ psi/hr (see plot),

$$\sqrt{k}L_f = \frac{4.064 q \bar{B}_g}{m_L h}\left(\frac{\bar{\mu}}{\phi \bar{c}_T}\right)^{1/2}$$

$$= \frac{(4.064)(2{,}955)(1.38)}{(287)(67)}\left[\frac{0.016}{(0.084)(3.16\times10^{-4})}\right]^{1/2}$$

$= 21.2.$

For $k = 0.216$ md (semilog analysis), $L_f = 45.5$ ft.

Operator Example 15.5 Test Date Drawdown Test Analyst ____

Well Linear/Pseudo-Radial Analysis Stimulation ____ Date ____

Location ____ Formation ____

Perforations ____ Depth of Bomb ____ ft Pressure Gradient in Wellbore ____ psi/ft

Pressure Datum ____ Pressure Gradient in Reservoir ____ psi/ft

Correction to Pressure Gauge Reading ____ (psi/ft) Wellbore Gradient x ____ (ft) Distance Mid-Perfs to Bomb + ____ (psi/ft) Reservoir Gradient x ____ (ft) Mid-Perfs to Datum + ____ (psia) Atmospheric Pressure = ____ psia

BASIC PROPERTIES

q_g(MCF/D)(wet) 2,955 r_w(ft) 0.36 $\bar{\mu}$ (cp) 0.016

T (°R) 650 S_w(fraction) 0.35 $\bar{z}$ 0.895

h (ft) 67 t_p(hr) - $\bar{c}_T$(psia^{-1}) 3.16 x 10^{-4}

ϕ (fraction) 0.084 t_{pa}(hr) - $\bar{B}_g$(RB/Mcf) = 5.04 $\frac{T\bar{z}}{p_{av}}$ 1.38

γ_g (air = 1.0) 0.62 p_{av} (psia) = $\frac{1}{2}(p_i + p_{wf})$ 2127

PRELIMINARY TYPE CURVE ANALYSIS — START SEMILOG STRAIGHT LINE

Curve Used: Gringarten-Ramey-Raghavan[17]

Matching Parameter: C_De^{2s} = ____ or ____ = ____

Start Semilog Straight Line: Δt_{ae} = ____ hr; Δt = 20.30 hr; $(t_{pa}+\Delta t_a)/\Delta t_a$ = ____

Δt_a = 20.0 hr

End Linear Flow: Δt_{ae} = ____ hr; Δt = 0.20 hr; $(t_{pa}+\Delta t_a)/\Delta t_a$ = ____

Δt_a = 0.24 hr

If $\Delta t_{max} < \Delta t$ at start of semilog straight line, then semilog analysis is not possible.

SEMILOG ANALYSIS

Average drainage area pressure at start of test, p_i = 2,684 psia; $p_{a,i}$ = 1,744 psia at ____ ft depth

$(t_{pa}+\Delta t_a)/\Delta t_a$ at start of MTR = ____ (Buildup) t at start of MTR = 20.30 (Drawdown)

$(t_{pa}+\Delta t_a)/\Delta t_a$ at start of LTR = ____ (Buildup) t at start of LTR = not present (Drawdown)

$m = p_{a2} - p_{a1}$ = ____ - ____ = ____ psia

$$k = \frac{162.6\, q_g \bar{B}_g \bar{\mu}}{mh} = \frac{162.6(2{,}955)(1.38)(0.016)}{(732)(67)} = 0.216 \text{ md}$$

SKIN FACTOR CALCULATION

BUILDUP TEST:

Horner Time Ratio at Δt_a = 1 hr = $\frac{t_{pa}+1}{1}$ = ____

$p_{a,1hr}$ = ____ psia

$p_{a,wf}(\Delta t_a = 0)$ = ____ psia

$$s' = 1.151\left[\left(\frac{p_{a,1hr}-p_{a,wf}}{m}\right) - \log\left(\frac{k}{\phi\bar{\mu}\bar{c}_T r_w^2}\right) + 3.23\right]$$

$$s' = 1.151\left[\left(\frac{(\quad)-(\quad)}{(\quad)}\right) - \log\left(\frac{(\quad)}{(\quad)(\quad)(\quad)(\quad)^2}\right) + 3.23\right] = \_\_\_\_$$

DRAWDOWN TEST:

$p_{a,i}$ = 1,744 psia

Δt_a = 0 hr

$p_{a,1hr}$ at Δt_a = 1 hr = 1,850 psia

$$s' = 1.151\left[\left(\frac{p_{a,i}-p_{a,wf}}{m}\right) - \log\left(\frac{k}{\phi\bar{\mu}\bar{c}_T r_w^2}\right) + 3.23\right]$$

$$s' = 1.151\left[\left(\frac{(1{,}744)-(1{,}850)}{(732)}\right) - \log\left(\frac{(0.216)}{(0.084)(0.016)(3.16\times10^{-4})(0.36)^2}\right) + 3.23\right] = -4.04$$

QUANTITATIVE TYPE CURVE ANALYSIS

Curve Used = Gringarten-Ramey-Raghavan

PRESSURE MATCH POINT

CASE 1: Semilog Analysis Possible
Pre-Calculated Pressure Match Point, Using Known Permeability

$$\Delta p_a = \frac{141.2\, q_g \bar{B}_g \bar{\mu} p_D}{kh} = \frac{141.2(2{,}955)(1.38)(0.016)(1)}{(0.216)(67)} = 637 \text{ psia}$$

CASE 2: Semilog Analysis Not Possible
Permeability From Pressure Match Point

Δp_a(psia) = ____ p_D = ____

$$k = \frac{141.2\, q_g \bar{B}_g \bar{\mu} p_D}{h \Delta p_a} = \frac{141.2(\quad)(\quad)(\quad)(\quad)}{(\quad)(\quad)} = \_\_\_\_ \text{ md}$$

TIME MATCH POINT

Δt_a (hr) = 0.19 (t_D/C_D) = ____ (GRINGARTEN) or t_D = 0.014 (CINCO OR OTHERS)

CASE 1: Gringarten Type Curve
Matching Parameter: C_De^{2s} = ____

$$C_D = \frac{0.0002637\, k}{\phi r_w^2 \bar{c}_T \bar{\mu}}\frac{\Delta t_{ae}}{t_D/C_D} = \frac{0.0002637(\quad)}{(\quad)(\quad)(\quad)(\quad)}\frac{(\quad)}{(\quad)} = \_\_\_\_$$

$s' = 0.5 \ln(C_De^{2s}/C_D) = 0.5 \ln($ ____ $)$ = ____

CASE 2: Cinco or Other Type Curve
Matching Parameter: C_f = ____ or ____ = ____

$$L_f = \left[\frac{0.0002637\, k}{\phi c_{t,av}\mu_{av}}\frac{\Delta t_a}{t_D}\right]^{1/2} = \left[\frac{(0.0002637)(0.216)}{(0.084)(3.16\times10^{-4})(0.016)}\frac{(0.19)}{(0.014)}\right]^{1/2} = 42.7 \text{ ft}$$

$wk_f = \pi C_f L_f k = \pi(100)(42.7)(0.216)$ = 2898 md-ft

PRODUCTION TIME DURATION CHECK

To assure that flow period exceeded duration of wellbore storage during production, calculate $t_{pa,min}$ from

$$t_{pa,min} = \frac{\phi\bar{\mu}\bar{c}_T r_w^2}{0.0002637\, k}(60 + 3.5\, s)\, C_D$$

$$t_{pa,min} = \frac{(\quad)(\quad)(\quad)(\quad)^2}{(0.0002637)(\quad)}[60 + 3.5(\quad)](\quad) = \_\_\_\_ \text{ hr}$$

$t_{pa,actual}$ = ____ hr

If $t_{pa,actual} < t_{pa,min}$, test analysis is invalid. Test must be rerun with longer production period prior to shut-in.

RADIUS OF INVESTIGATION

RADIUS OF INVESTIGATION ACHIEVED IN PRODUCTION PERIOD PRECEDING SHUT-IN PERIOD

$$r_{i,prod} = \left(\frac{k t_{pa}}{948\,\phi\bar{\mu}\bar{c}_T}\right)^{1/2} = \left(\frac{(\quad)(\quad)}{948(\quad)(\quad)(\quad)}\right)^{1/2} = \_\_\_\_ \text{ ft}$$

RADIUS OF INVESTIGATION DURING TEST

Start of Semilog Straight Line

$$r_{i,start} = \left(\frac{k\Delta t_a}{948\,\phi\bar{\mu}\bar{c}_T}\right)^{1/2} = \left(\frac{(0.216)(20.0)}{948(0.084)(0.016)(3.16\times10^{-4})}\right)^{1/2} = 104 \text{ ft}$$

End of Linear flow

$$r_{i,end} = \left(\frac{k\Delta t_a}{948\,\phi\bar{\mu}\bar{c}_T}\right)^{1/2} = \left(\frac{(0.216)(0.24)}{948(0.084)(0.016)(3.16\times10^{-4})}\right)^{1/2} = 11.4 \text{ ft}$$

End of Test

$$r_{i,final} = \left(\frac{k\Delta t_a}{948\,\phi\bar{\mu}\bar{c}_T}\right)^{1/2} = \left(\frac{(0.216)(44.57)}{948(0.084)(0.016)(3.16\times10^{-4})}\right)^{1/2} = 155 \text{ ft}$$

NOTE: Reservoir heterogeneities and boundaries determined from data during shut-in test at $r_i > r_{i,prod}$ are highly questionable. If the drawdown did not reach beyond a certain distance in the reservoir, the buildup will not sample reservoir properties beyond that point.

REMARKS

Fig. K-5—Gas-well test analysis—Example 15.5.

Type-Curve Analysis. Using k=0.216 md, calculate pressure match point.

$$\Delta p_a = \frac{141.2 q \bar{B}_g \bar{\mu} p_D}{kh} = \frac{(141.2)(2{,}955)(1.38)(0.016)(1)}{(0.216)(67)}$$

$$= 637 \text{ psia.}$$

Gringarten *et al.* type-curve matching parameters are Δt_a = 0.19 hour, t_{LfD}=0.014, $L_e/L_f = \infty$,

$$L_f = \left[\frac{0.0002637k}{\phi\bar{c}_T\bar{\mu}}\left(\frac{\Delta t_a}{t_{LfD}}\right)\right]^{1/2}$$

$$= \left[\frac{0.0002637(0.216)(0.19)}{(0.084)(3.16\times10^{-4})(0.016)(0.014)}\right]^{1/2} = 42.7 \text{ ft.}$$

TABLE K-9—BUILDUP TEST DATA

Δt (hours)	p_{ws} (psia)
0.0	2,618
0.033	2,626
0.1167	2,637
0.25	2,657
0.333	2,671
0.4167	2,690
0.5	2,725
0.5833	2,771
0.6667	2,818
0.75	2,889
0.8333	2,942
0.9167	2,985
1.0	3,026
1.25	3,108
1.50	3,166
1.75	3,212
2.0	3,240
2.5	3,284
3.0	3,321
3.5	3,352
4.0	3,377
6.0	3,455
7.0	3,492
8.0	3,528
9.0	3,554
10.0	3,579
12.0	3,625
14.0	3,666
17.0	3,721
20.0	3,776
25.0	3,848
30.0	3,911
35.0	3,970
40.0	4,016
50.0	4,095
60.0	4,166
70.0	4,228
80.0	4,284
100.0	4,376
140.0	4,525
160.0	4,587
180.0	4,633
200.0	4,681
220.0	4,722
240.0	4,756
260.0	4,790
290.0	4,842
320.0	4,892
350.0	4,935
380.0	4,973
410.0	5,010
440.0	5,045
470.0	5,080
504.0	5,109

Summary of results:

k = 0.216 md,
s' = −4.04,
L_f = 45 ft,
$wk_f \geq$ 3,050 md-ft [for $C_r \geq 100$, $wk_f \geq 100(\pi k L_f)$,
$wk_f \geq 100(\pi)(0.216)(45) = 3{,}050$ md-ft],

start of linear flow, $\Delta t_a \leq 0.1$ hour,
end of linear flow, $\Delta t_a = 0.24$ hour ($t_{L_fD} = 0.016$),
start of pseudoradial flow, $\Delta t_a = 20$ hours ($t_{L_fD} \cong 1.5$),
and end of pseudoradial flow, not reached by end of test.
Fig. K-5 summarizes gas-well test analysis.

Example 15.6—Fractured Well With Finite Fracture Conductivity

A pressure-buildup test was run on a fractured well. Well and reservoir properties are summarized below:

ϕ = 0.10,
h = 89 ft,

TABLE K-10—PRESSURE-BUILDUP-TEST SUMMARY

Well: finite-conductivity fracture
Location: postfracture pressure-buildup test

Net pay, ft	89.000
Wellbore radius, ft	0.33000
Production time, hours	4,674.0
Adjusted production time, hours	7,144.7
Pressure at shut-in, psia	2,618.0
Initial reservoir pressure, psia	7,365.0
Initial adjusted pressure, psia	5,919.1
Bottomhole temperature, °F	250.00
Dry-gas flow rate, Mcf/D	940.00
Wet-gas flow rate, Mcf/D	969.98
Total porosity, fraction	0.10000
Water saturation, fraction	0.35000
Water compressibility, psia^{-1}	0.35000×10^{-5}
Formation compressibility, psia^{-1}	0.40000×10^{-5}
Separator-gas specific gravity	0.68200
Gas-condensate ratio, scf/STB	24,544.00000
Condensate gravity, °API	52.000
Gas equivalent of condensate, scf/STB	782.89
Wet-gas specific gravity	0.80052
Number of pressure/time data points	57
Data point for start of middle-time region	50
Data point for end of middle-time region	57
Fluid Properties at Average Pressure	
Average pressure, psia	4,991.50000
Gas viscosity, cp	0.02668
Total compressibility, psia^{-1}	0.0000854
z factor	0.99730
Gas FVF, RB/Mcf	0.71420

T = 250°F,
r_w = 0.33 ft,
γ_g = 0.682 (0.75% CO_2, 0.22% N_2),
p_i = 7,365 psia, and
S_w = 0.35.

Before shut-in for the buildup test, the well produced 940 Mcf/D dry gas for an equivalent time of 4,672 hours. The producing gas-condensate ratio was 24,544 scf/STB, and the condensate gravity was 52°API.

From the buildup test data in Tables K-9 and K-10, estimate the formation permeability and fracture properties. (As explained in the solution, you will encounter a uniqueness problem in the type-curve match of the test data. You may therefore *assume* a permeability of 0.0178 md; its source is explained in the solution.)

Solution—Fractured Well With Finite Fracture Conductivity. ***Horner Analysis (Table K-11).*** The true semilog straight line never becomes apparent, but the slope of a straight line between the last data point and $p_{a,i}$ provides an upper bound on permeability (see Fig. K-6).

$m = 1{,}900$ psia/cycle (from last data; see plot).

$$k_{max} = 162.6 \frac{q_g \bar{B}_g \bar{\mu}}{mh} = \frac{162.6(970)(0.714)(0.0267)}{(1{,}900)(89)}$$

$= 0.0178$ md.

Type-Curve Analysis (Table K-12). A fit is possible for $C_r = 0.2$ on the Cinco-L. *et al.*[7] type curve (Fig. K-7). (The fit is not unique, and curves for other low values of C_r also fit the data. Therefore, we will *force* one fit using $k = 0.0178$ md as a permeability estimate.)

$$\Delta p_a = 141.2 \frac{q_g \bar{B}_g \bar{\mu} p_D}{kh} = 141.2 \frac{(970)(0.714)(0.0267)(1.0)}{(0.0178)(89)}$$

$= 1{,}648$ psia.

TABLE K-11—HORNER PLOTTING FUNCTIONS

	t (hours)	Horner Time	p (psia)	t_a (hours)	Adjusted Horner Time	p_a (psia)
1	0.0000	0.0000	2,618.0	0.0000	0.0000	1,284.0
2	0.3300×10^{-1}	0.1416×10^{6}	2,626.0	0.1671×10^{-1}	0.4275×10^{6}	1,291.0
3	0.1167	0.4005×10^{5}	2,637.0	0.5925×10^{-1}	0.1206×10^{6}	1,301.0
4	0.1667	0.2804×10^{5}	2,646.0	0.8475×10^{-1}	0.8430×10^{5}	1,309.0
5	0.2500	0.1870×10^{5}	2,657.0	0.1274	0.5609×10^{5}	1,318.0
6	0.3330	0.1404×10^{5}	2,671.0	0.1701	0.4201×10^{5}	1,330.0
7	0.4167	0.1122×10^{5}	2,690.0	0.2134	0.3349×10^{5}	1,346.0
8	0.5000	9,349.0	2,725.0	0.2568	0.2782×10^{5}	1,376.0
9	0.5833	8,014.0	2,771.0	0.3010	0.2374×10^{5}	1,417.0
10	0.6667	7,012.0	2,818.0	0.3458	0.2066×10^{5}	1,459.0
11	0.7500	6,233.0	2,889.0	0.3916	0.1825×10^{5}	1,522.0
12	0.8333	5,610.0	2,942.0	0.4383	0.1630×10^{5}	1,570.0
13	0.9167	5,100.0	2,985.0	0.4858	0.1471×10^{5}	1,608.0
14	1.000	4,675.0	3,026.0	0.5340	0.1338×10^{5}	1,646.0
15	1.250	3,740.0	3,108.0	0.6815	0.1049×10^{5}	1,722.0
16	1.500	3,117.0	3,166.0	0.8324	8,585.0	1,775.0
17	1.750	2,672.0	3,212.0	0.9858	7,248.0	1,818.0
18	2.000	2,338.0	3,240.0	1.141	6,262.0	1,843.0
19	2.500	1,871.0	3,284.0	1.455	4,910.0	1,885.0
20	3.000	1,559.0	3,321.0	1.774	4,029.0	1,920.0
21	3.500	1,336.0	3,352.0	2.095	3,411.0	1,949.0
22	4.000	1,170.0	3,377.0	2.420	2,954.0	1,973.0
23	5.000	935.8	3,415.0	3.075	2,324.0	2,009.0
24	6.000	780.0	3,455.0	3.738	1,912.0	2,047.0
25	7.000	668.7	3,492.0	4.410	1,621.0	2,082.0
26	8.000	585.3	3,528.0	5.088	1,405.0	2,116.0
27	9.000	520.3	3,554.0	5.773	1,239.0	2,141.0
28	10.00	468.4	3,579.0	6.463	1,106.0	2,165.0
29	12.00	390.5	3,625.0	7.859	910.1	2,209.0
30	14.00	334.9	3,666.0	9.272	771.6	2,249.0
31	17.00	275.9	3,721.0	11.42	626.5	2,302.0
32	20.00	234.7	3,776.0	13.61	526.1	2,355.0
33	25.00	188.0	3,848.0	17.31	413.7	2,425.0
34	30.00	156.8	3,911.0	21.09	339.7	2,487.0
35	35.00	134.5	3,970.0	24.93	287.5	2,544.0
36	40.00	117.8	4,016.0	28.83	248.8	2,589.0
37	50.00	94.48	4,095.0	36.77	195.3	2,667.0
38	60.00	78.90	4,166.0	44.87	160.2	2,737.0
39	70.00	67.77	4,228.0	53.11	135.5	2,798.0
40	80.00	59.42	4,284.0	61.49	117.2	2,854.0
41	100.0	47.74	4,376.0	78.57	91.94	2,945.0
42	120.0	39.95	4,456.0	96.03	75.40	3,024.0
43	140.0	34.39	4,525.0	113.8	63.78	3,093.0
44	160.0	30.21	4,587.0	131.9	55.17	3,155.0
45	180.0	26.97	4,633.0	150.2	48.56	3,201.0
46	200.0	24.37	4,681.0	168.7	43.34	3,249.0
47	220.0	22.25	4,722.0	187.5	39.11	3,289.0
48	240.0	20.48	4,756.0	206.4	35.62	3,323.0
49	260.0	18.98	4,790.0	225.4	32.70	3,357.0
50	290.0	17.12	4,842.0	254.2	29.10	3,409.0
51	320.0	15.61	4,892.0	283.4	26.21	3,459.0
52	350.0	14.35	4,935.0	312.9	23.83	3,502.0
53	380.0	13.30	4,973.0	342.7	21.85	3,540.0
54	410.0	12.40	5,010.0	372.7	20.17	3,577.0
55	440.0	11.62	5,045.0	402.9	18.73	3,612.0
56	470.0	10.94	5,080.0	433.4	17.49	3,647.0
57	504.0	10.27	5,109.0	468.2	16.26	3,676.0

With this pressure match, a time match point is ($t_{L_fD}=1\times10^{-5}$, $\Delta t_{ae}=0.072$ hours).

$$L_f=\left[\frac{0.0002637k}{\phi\bar{\mu}\bar{c}_T}\left(\frac{\Delta t_{ae}}{t_D}\right)\right]^{1/2}$$

$$=\left[\frac{(0.0002637)(0.0178)(0.072)}{(0.10)(0.0267)(8.54\times10^{-5})(1\times10^{-5})}\right]^{1/2}=385\text{ ft.}$$

$$wk_f=C_r\pi L_f k=(0.2)\pi(385)(0.0178)$$

$$=4.3\text{ md-ft.}$$

This test actually exhibits approximately two cycles of bilinear flow [quarter slope on log-log graph and straight line on $p_{a,ws}$ vs. $(\Delta t_{ae})^{1/4}$; see Fig. K-8]. This bilinear flow continues to $\Delta t_{ae}\cong70$ hours. In bilinear flow, a fixed relationship exists between C_r and k, but because neither quantity is known, we cannot use the value of one quantity to find the other in this specific case. Again, we note that these results merely *bound* the correct results; more precise analysis requires an estimate of formation permeability.

Fig. K-9 summarizes this well analysis.

Example 15.7—Fractured Well, Constant-BHP Production

A prefracture buildup test indicated a formation permeability of 0.0081 md for a well. After a massive hydraulic fracture treatment,

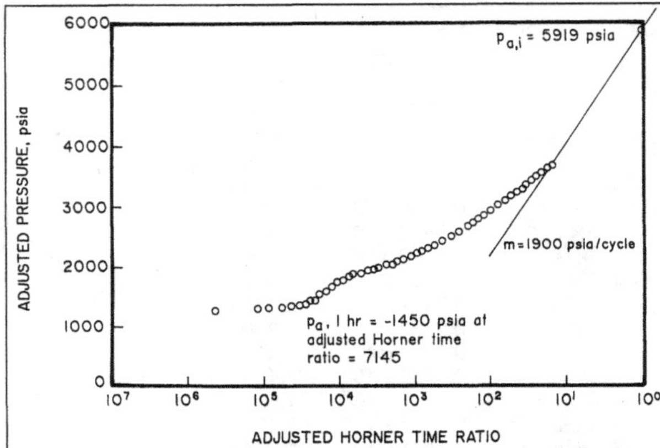

Fig. K-6—Horner graph of fractured well with finite fracture conductivity—Example 15.6.

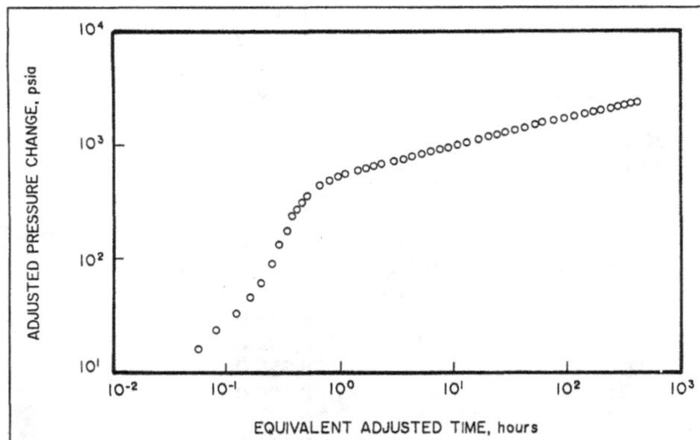

Fig. K-7—Type-curve graph, postfracture test, finite-conductivity fracture—Example 15.6.

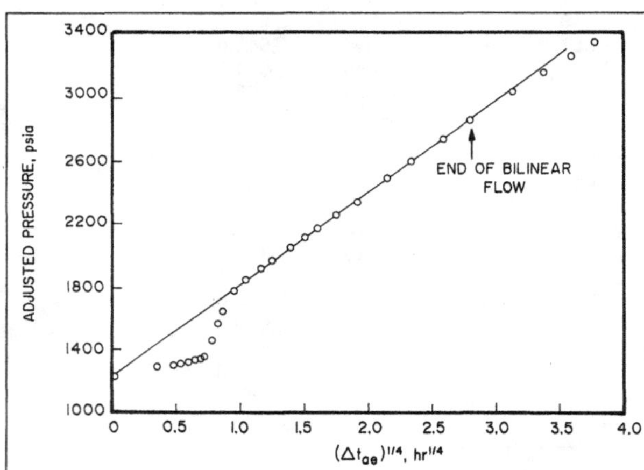

Fig. K-8—Bilinear-flow analysis graph, postfracture pressure-buildup test, finite-conductivity fracture—Example 15.6.

the well produced at constant flowing BHP with the following rate history.

t (days)	q (Mcf/D)
20	625
35	476
50	408
100	308
150	250
250	208
300	192

TABLE K-12—TYPE-CURVE PLOTTING FUNCTIONS

	t_p (hours)	Δp (psia)	t_{ae} (hours)	Δp_a (psia)
1	0.0000	0.0000	0.0000	0.0000
2	0.3300×10^{-1}	8.000	0.1671×10^{-1}	6.880
3	0.1167	19.00	0.5925×10^{-1}	16.34
4	0.1667	28.00	0.8475×10^{-1}	24.08
5	0.2500	39.00	0.1274	33.54
6	0.3330	53.00	0.1701	45.58
7	0.4167	72.00	0.2134	61.92
8	0.4999	107.0	0.2568	92.02
9	0.5832	153.0	0.3009	132.3
10	0.6666	200.0	0.3458	174.4
11	0.7499	271.0	0.3916	238.0
12	0.8332	324.0	0.4383	285.4
13	0.9165	367.0	0.4858	323.9
14	0.9998	408.0	0.5339	361.4
15	1.250	490.0	0.6814	437.1
16	1.500	548.0	0.8323	490.7
17	1.749	594.0	0.9857	533.1
18	1.999	622.0	1.141	559.0
19	2.499	666.0	1.455	600.4
20	2.998	703.0	1.773	635.4
21	3.497	734.0	2.095	664.7
22	3.997	759.0	2.419	688.3
23	4.995	797.0	3.074	724.2
24	5.992	837.0	3.736	762.1
25	6.990	874.0	4.407	797.0
26	7.986	910.0	5.085	831.6
27	8.983	936.0	5.769	856.6
28	9.979	961.0	6.458	880.7
29	11.97	1,007.0	7.850	924.9
30	13.96	1,048.0	9.260	964.4
31	16.94	1,103.0	11.40	1,017.0
32	19.91	1,158.0	13.58	1,071.0
33	24.87	1,230.0	17.27	1,141.0
34	29.81	1,293.0	21.03	1,202.0
35	34.74	1,352.0	24.85	1,260.0
36	39.66	1,398.0	28.72	1,305.0
37	49.47	1,477.0	36.58	1,383.0
38	59.24	1,548.0	44.59	1,453.0
39	68.97	1,610.0	52.72	1,514.0
40	78.65	1,666.0	60.96	1,569.0
41	97.91	1,758.0	77.71	1,660.0
42	117.0	1,838.0	94.75	1,740.0
43	135.9	1,907.0	112.0	1,808.0
44	154.7	1,969.0	129.5	1,870.0
45	173.3	2,015.0	147.1	1,916.0
46	191.8	2,063.0	164.8	1,964.0
47	210.1	2,104.0	182.7	2,005.0
48	228.3	2,138.0	200.6	2,039.0
49	246.3	2,172.0	218.5	2,073.0
50	273.1	2,224.0	245.5	2,125.0
51	299.5	2,274.0	272.6	2,175.0
52	325.6	2,317.0	299.8	2,218.0
53	351.4	2,355.0	327.0	2,256.0
54	376.9	2,392.0	354.2	2,293.0
55	402.1	2,427.0	381.4	2,328.0
56	427.1	2,462.0	408.6	2,363.0
57	454.9	2,491.0	439.4	2,392.0

Well and reservoir data include the following:

p_i = 2,394 psia,
p_{pi} = 4.050×10^8 psia^2/cp,
p_{wf} = 341 psia,
p_{pwf} = 9.210×10^6 psia^2/cp,
γ_g = 0.765,
T = 720°R (260°F),
h = 32 ft,
$\bar{B}_g$ = 2.404 RB/Mscf,
$\bar{\mu}$ = 0.0151 cp,
ϕ = 0.107,
S_w = 0.446,
r_w = 0.365 ft,
c_w = 3.5×10^{-6} psia^{-1},
c_f = 4×10^{-6} psia^{-1}, and
$\bar{c}_T$ = 4.376×10^{-4} psia^{-1}.

Operator Example 15.6 Test Date Post-Fracture Test Analyst ______

Well Finite-Conductivity Fracture Stimulation ______ Date ______

Location ______ Formation ______

Perforations ______ Depth of Bomb ______ ft Pressure Gradient in Wellbore ______ psi/ft

Pressure Datum ______ Pressure Gradient in Reservoir ______ psi/ft

Correction to Pressure Gauge Reading ______ (psi/ft) × ______ (ft) + ______ (psi/ft) × ______ (ft) + ______ (psia) = ______ psia
(Wellbore Gradient; Distance Mid-Perfs to Bomb; Reservoir Gradient; Mid-Perfs to Datum; Atmospheric Pressure)

BASIC PROPERTIES

q_g(MCF/D)(wet) 970	r_w(ft) 0.33	$\bar{\mu}$ (cp) 0.0267
T (°R) 710	S_w(fraction) 0.35	$\bar{z}$ 0.997
h (ft) 89	t_p(hr) 4674	$\bar{c}_T$(psia^{-1}) 8.54 x 10^{-5}
ϕ (fraction) 0.10	t_{pa}(hr) 7145	$\bar{B}_g$(RB/Mcf) = $5.04 \frac{T\bar{z}}{\bar{p}_{av}}$ 0.714
γ_g (air = 1.0) 0.682	p_{av} (psia) = $\frac{1}{2}(p_i + p_{wf})$ 4991.5	

PRELIMINARY TYPE CURVE ANALYSIS – START SEMILOG STRAIGHT LINE

Curve Used: Cinco-Ley[18]

Matching Parameter: C_De^{2s} = ______ or C_r = 0.2

Start Semilog Straight Line: Δt_{ae} = Not present hr; Δt = ______ hr; $(t_{pa}+\Delta t_a)/\Delta t_a$ = ______

Δt_a = ______ hr

Start Boundary Effects: Δt_{ae} = Not present hr; Δt = ______ hr; $(t_{pa}+\Delta t_a)/\Delta t_a$ = ______

Δt_a = ______ hr

If $\Delta t_{max} < \Delta t$ at start of semilog straight line, then semilog analysis is not possible.

SEMILOG ANALYSIS

Average drainage area pressure at start of test, $\bar{p}_i$ = 7,365 psia; $p_{a,i}$ = 5919.1 psia at ______ ft depth

$(t_{pa}+\Delta t_a)/\Delta t_a$ at start of MTR = ______ (Buildup) t at start of MTR = ______ (Drawdown)

$(t_{pa}+\Delta t_a)/\Delta t_a$ at start of LTR = ______ (Buildup) t at start of LTR = ______ (Drawdown)

$m = p_{a2} - p_{a1}$ = ______ - ______ = 1,900 psia (last data)

$$k_{max} = \frac{162.6\, q_g \bar{B}_g \bar{\mu}}{mh} = \frac{162.6(970)(0.714)(0.0267)}{(1,900)(89)} = 0.0178 \text{ md}$$

SKIN FACTOR CALCULATION

BUILDUP TEST:

Horner Time Ratio at Δt_a = 1 hr = $\frac{t_{pa}+1}{1}$ = 7,146

$p_{a,1hr}$ = 1,450 psia

$p_{a,wf}(\Delta t_a = 0)$ = 1,284 psia

$$s' = 1.151\left[\left(\frac{p_{a,1hr}-p_{a,wf}}{m}\right) - \log\left(\frac{k}{\phi\bar{\mu}\bar{c}_T r_w^2}\right) + 3.23\right]$$

$$s' = 1.151\left[\left(\frac{(1,450)-(1,284)}{1,900}\right) - \log\left(\frac{(0.0178)}{(0.10)(0.0267)(8.54\times10^{-5})(0.33)^2}\right) + 3.23\right] = -2.92$$

DRAWDOWN TEST:

$p_{a,i}$ = ______ psia

Δt_a = ______ hr

$p_{a,1hr}$ at Δt_a = 1 hr = ______ psia

$$s' = 1.151\left[\left(\frac{p_{a,i}-p_{a,wf}}{m}\right) - \log\left(\frac{k}{\phi\bar{\mu}\bar{c}_T r_w^2}\right) + 3.23\right]$$

$$s' = 1.151\left[\left(\frac{(\quad)-(\quad)}{(\quad)}\right) - \log\left(\frac{(\quad)}{(\quad)(\quad)(\quad)(\quad)^2}\right) + 3.23\right] = \_\_\_\_$$

QUANTITATIVE TYPE CURVE ANALYSIS

Curve Used = Cinco-Ley

PRESSURE MATCH POINT

CASE 1: Semilog Analysis Possible
Pre-Calculated Pressure Match Point, Using Known Permeability

$$\Delta p_a = \frac{141.2\, q_g \bar{B}_g \bar{\mu} p_D}{kh} = \frac{141.2(970)(0.714)(0.0267)(1.0)}{(0.0178)(89)} = 1,648 \text{ psia}$$

CASE 2: Semilog Analysis Not Possible
Permeability From Pressure Match Point

Δp_a (psia) = ______ p_D = ______

$$k = \frac{141.2\, q_g \bar{B}_g \bar{\mu} p_D}{h\Delta p_a} = \frac{141.2(\quad)(\quad)(\quad)(\quad)}{(\quad)(\quad)} = \_\_\_\_ \text{ md}$$

TIME MATCH POINT

Δt_{ae} (hr) = ______ (t_D/C_D) = ______ (GRINGARTEN) or t_D = ______ (CINCO OR OTHERS)

CASE 1: Gringarten Type Curve
Matching Parameter: C_De^{2s} = ______

$$C_D = \frac{0.0002637\, k\, \Delta t_{ae}}{\phi r_w^2 \bar{c}_T \bar{\mu}\; t_D/C_D} = \frac{0.0002637(\quad)(\quad)}{(\quad)(\quad)(\quad)(\quad)(\quad)} = \_\_\_\_$$

$s' = 0.5 \ln(C_De^{2s}/C_D) = 0.5 \ln(\quad / \quad) = \_\_\_\_$

CASE 2: Cinco or Other Type Curve
Matching Parameter: C_r = 0.2 or ______ = ______

$$L_f = \left[\frac{0.0002637\, k\, \Delta t_{ae}}{\phi c_{t,av}\mu_{av}\; t_D}\right]^{1/2} = \left[\frac{(0.0002637)(0.0178)(0.072)}{(0.10)(8.54\times10^{-5})(0.0267)(1\times10^{-5})}\right]^{1/2} = 385 \text{ ft}$$

$wk_f = \pi C_r L_f k = \pi(0.2)(385)(0.0178) = 4.3$ md-ft

PRODUCTION TIME DURATION CHECK

To assure that flow period exceeded duration of wellbore storage during production, calculate $t_{pa,min}$ from

$$t_{pa,min} = \frac{\phi\bar{\mu}\bar{c}_T r_w^2}{0.0002637\, k}(60 + 3.5\, s)C_D$$

$$t_{pa,min} = \frac{(\quad)(\quad)(\quad)(\quad)^2}{(0.0002637)(\quad)}[60 + 3.5(\quad)](\quad) = \_\_\_\_ \text{ hr}$$

$t_{pa,actual}$ = ______ hr

If $t_{pa,actual} < t_{pa,min}$, test analysis is invalid. Test must be rerun with longer production period prior to shut-in.

RADIUS OF INVESTIGATION

RADIUS OF INVESTIGATION ACHIEVED IN PRODUCTION PERIOD PRECEDING SHUT-IN PERIOD

$$r_{i,prod} = \left(\frac{k t_{pa}}{948\, \phi\bar{\mu}\bar{c}_T}\right)^{1/2} = \left(\frac{(0.0178)(7145)}{948(0.10)(0.0267)(8.54\times10^{-5})}\right)^{1/2} = 768 \text{ ft}$$

RADIUS OF INVESTIGATION DURING TEST

Start of Semilog Straight Line

$$r_{i,start} = \left(\frac{k\Delta t_a}{948\, \phi\bar{\mu}\bar{c}_T}\right)^{1/2} = \left(\frac{(\quad)(\quad)}{948(\quad)(\quad)(\quad)}\right)^{1/2} = \_\_\_\_ \text{ ft}$$

End of Semilog Straight Line

$$r_{i,end} = \left(\frac{k\Delta t_a}{948\, \phi\bar{\mu}\bar{c}_T}\right)^{1/2} = \left(\frac{(\quad)(\quad)}{948(\quad)(\quad)(\quad)}\right)^{1/2} = \_\_\_\_ \text{ ft}$$

End of Test

$$r_{i,final} = \left(\frac{k\Delta t_a}{948\, \phi\bar{\mu}\bar{c}_T}\right)^{1/2} = \left(\frac{(0.0178)(468.2)}{948(0.10)(0.0267)(8.54\times10^{-5})}\right)^{1/2} = 196 \text{ ft}$$

NOTE: Reservoir heterogeneities and boundaries determined from data during shut-in test at $r_i > r_{i,prod}$ are highly questionable. If the drawdown did not reach beyond a certain distance in the reservoir, the buildup will not sample reservoir properties beyond that point.

REMARKS

Fig. K-9—Gas-well test analysis—Example 15.6.

1. Estimate fracture length and fracture conductivity. First, use only pressure; then repeat the analysis using pseudopressure.
2. Comment on the analysis that would have been possible without the prefracture test data.

Solution—Fractured Well, Constant-BHP Production. ***Type-Curve Graph.*** Plot $1/q$ vs. Δt on a type-curve graph and use Agarwal *et al.*'s[8] type curve (see Fig. K-10). Using k=0.0081 md and real pressures, precalculate the rate match point:

$$1/q = \frac{141.2\bar{\mu}\bar{B}_g}{kh(p_i - p_{wf})}\left(\frac{1}{q_D}\right)$$

$$= \frac{141.2(0.0151)(2.404)}{(0.0081)(32)(2,394-341)} \times \left(\frac{1}{1}\right)$$

$$= 9.63\times10^{-3}.$$

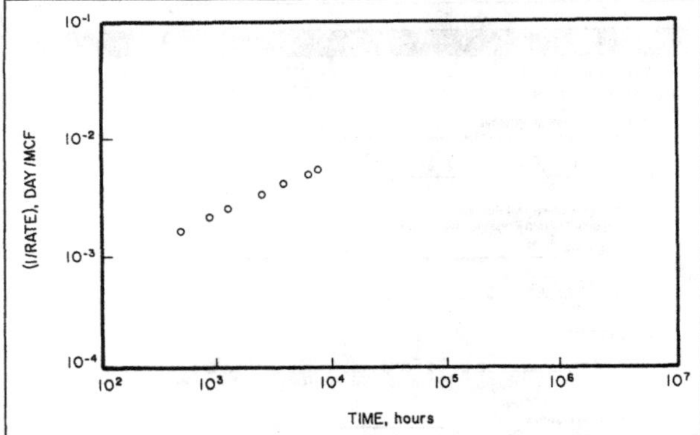

Fig. K-10—Constant-BHP production analysis—Example 15.7.

TABLE K-13—BUILDUP TEST DATA

Δt (hours)	p_{ws} (psia)
0.0	408
0.33	468
0.50	508
0.667	653
1.0	788
1.167	829
1.333	859
1.5	908
2.0	1,034
2.5	1,139
3.0	1,229
3.5	1,315
4.0	1,382
4.5	1,443
5.0	1,492
5.5	1,537
6.5	1,622
7.5	1,689
8.5	1,749
10.5	1,849
11.5	1,888
15.5	2,023
19.5	2,126
23.5	2,199
27.5	2,264
31.5	2,318
35.5	2,365
39.5	2,404
43.5	2,438
47.5	2,470
51.5	2,500
55.5	2,525
63.5	2,570
71.5	2,611
87.5	2,682
95.5	2,704
103.5	2,725
111.5	2,747
119.5	2,765
127.5	2,782
135.5	2,798
143.5	2,811
151.5	2,824
159.5	2,833
164.8	2,837

The match-point parameters are Δt=16,000 hours, t_D=0.13, and C_r=100. Thus,

$$L_f=\left[\frac{0.0002637k}{\phi\bar{\mu}\bar{c}_T}\left(\frac{\Delta t}{t_D}\right)\right]^{1/2}$$

TABLE K-14—PRESSURE-BUILDUP-TEST SUMMARY

Well: fractured well with wellbore storage distortion
Location: postfracture pressure-buildup test

Net pay, ft	65.000
Wellbore radius, ft	0.20380
Production time, hours	272.00
Adjusted production time, hours	464.32
Pressure at shut-in, psia	408.00
Initial reservoir pressure, psia	3,200.0
Initial adjusted pressure, psia	2,656.3
Bottomhole temperature, °F	158.00
Dry-gas flow rate, Mcf/D	1,420.0
Wet-gas flow rate, Mcf/D	1,420.0
Total porosity, fraction	0.11800
Water saturation, fraction	0.25000
Water compressibility, psia^{-1}	0.35000×10^{-5}
Formation compressibility, psia^{-1}	0.40000×10^{-5}
Separator-gas specific gravity	0.65000
Gas-condensate ratio, scf/STB	0.00000
Condensate gravity, °API	0.00000
Gas equivalent of condensate, scf/STB	0.00000
Wet-gas specific gravity	0.65000
Number of pressure/time data points	48
Data point for start of middle-time region	40
Data point for end of middle-time region	48
Fluid Properties at Average Pressure	
Average pressure, psia	1,804.00000
Gas viscosity, cp	0.01524
Total compressibility, psia^{-1}	0.0004498
z factor	0.86072
Gas FVF, RB/Mcf	1.48449

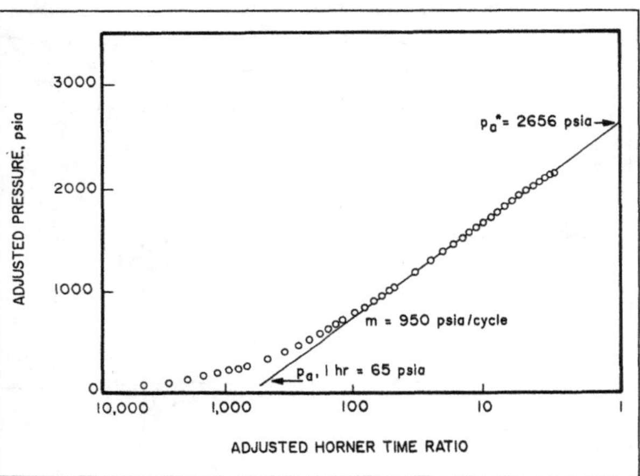

Fig. K-11—Horner graph of postfracture buildup test with wellbore-storage distortion—Example 15.8.

$$=\left[\frac{(0.0002637)(0.0081)(16{,}000)}{(0.107)(0.0151)(4.376\times10^{-4})(0.13)}\right]^{1/2}$$

=610 ft, and

$$wk_f=C_r\pi kL_f=100(\pi)(0.0081)(610)$$
$$=1{,}552 \text{ md-ft.}$$

Using k=0.0081 md and pseudopressures, calculate the rate match point:

$$1/q=\frac{1{,}422T}{kh(p_{pi}-p_{pwf})}\left(\frac{1}{q_D}\right)$$

$$=\frac{1{,}422(720)}{(0.0081)(32)(4.050\times10^8-9.210\times10^6)}\times\left(\frac{1}{1}\right)$$

$$=9.98\times10^{-3}.$$

TABLE K-15—HORNER PLOTTING FUNCTIONS

	t (hours)	Horner Time	p (psia)	t_a (hours)	Adjusted Horner Time	p_a (psia)
1	0.0000	0.0000	408.0	0.0000	0.0000	52.99
2	0.3300	825.2	468.0	0.1027	4,524.0	68.58
3	0.5000	545.0	508.0	0.1610	2,884.0	81.54
4	0.6670	408.8	653.0	0.2278	2,039.0	133.3
5	0.8330	327.5	729.0	0.3050	1,523.0	164.8
6	1.000	273.0	788.0	0.3889	1,195.0	192.6
7	1.167	234.1	829.0	0.4772	973.9	213.2
8	1.333	205.1	859.0	0.5682	818.1	228.3
9	1.500	182.3	908.0	0.6632	701.1	253.6
10	2.000	137.0	1,034.0	0.9699	479.7	327.7
11	2.500	109.8	1,139.0	1.305	356.7	396.8
12	3.000	91.67	1,229.0	1.664	280.0	459.4
13	3.500	78.71	1,315.0	2.043	228.3	523.9
14	4.000	69.00	1,382.0	2.439	191.3	576.5
15	4.500	61.44	1,443.0	2.850	163.9	626.8
16	5.000	55.40	1,492.0	3.273	142.9	667.2
17	5.500	50.45	1,537.0	3.707	126.3	706.9
18	6.500	42.85	1,622.0	4.604	101.9	783.1
19	7.500	37.27	1,689.0	5.535	84.89	845.8
20	8.500	33.00	1,749.0	6.496	72.48	903.7
21	9.500	29.63	1,804.0	7.483	63.05	957.1
22	10.50	26.90	1,849.0	8.494	55.67	1,003.0
23	11.50	24.65	1,888.0	9.524	49.75	1,044.0
24	15.50	18.55	2,023.0	13.82	34.61	1,187.0
25	19.50	14.95	2,126.0	18.33	26.33	1,300.0
26	23.50	12.57	2,199.0	23.00	21.18	1,384.0
27	27.50	10.89	2,264.0	27.81	17.70	1,458.0
28	31.50	9.635	2,318.0	32.73	15.19	1,522.0
29	35.50	8.662	2,365.0	37.74	13.30	1,578.0
30	39.50	7.886	2,404.0	42.84	11.84	1,625.0
31	43.50	7.253	2,438.0	48.01	10.67	1,666.0
32	47.50	6.726	2,470.0	53.25	9.720	1,706.0
33	51.50	6.282	2,500.0	58.55	8.931	1,742.0
34	55.50	5.901	2,525.0	63.90	8.266	1,773.0
35	63.50	5.283	2,570.0	74.76	7.211	1,829.0
36	71.50	4.804	2,611.0	85.79	6.412	1,881.0
37	79.50	4.421	2,648.0	96.98	5.788	1,928.0
38	87.50	4.109	2,682.0	108.3	5.286	1,970.0
39	95.50	3.848	2,704.0	119.8	4.876	1,998.0
40	103.5	3.628	2,725.0	131.3	4.535	2,026.0
41	111.5	3.439	2,747.0	143.0	4.247	2,054.0
42	119.5	3.276	2,765.0	154.7	4.001	2,077.0
43	127.5	3.133	2,782.0	166.5	3.789	2,099.0
44	135.5	3.007	2,798.0	178.4	3.603	2,120.0
45	143.5	2.895	2,811.0	190.3	3.440	2,137.0
46	151.5	2.795	2,824.0	202.3	3.295	2,154.0
47	159.5	2.705	2,833.0	214.3	3.166	2,166.0
48	164.8	2.650	2,837.0	222.3	3.089	2,171.0

The match-point parameters are Δt=15,000 hours, t_D=0.10, and C_r=100. Thus,

$$L_f=\left[\frac{0.0002637k}{\phi\bar{\mu}\bar{c}_T}\left(\frac{\Delta t}{t_D}\right)\right]^{1/2}$$

$$=\left[\frac{(0.0002637)(0.0081)(14{,}500)}{(0.107)(0.0151)(4.376\times10^{-4})(0.10)}\right]^{1/2}$$

$$=662 \text{ ft, and}$$

$$wk_f=C_r\pi kL_f=100(\pi)(0.0081)(662)$$

$$=1{,}685 \text{ md-ft.}$$

Without a prefracture estimate of k, a unique match of these data would be impossible.

Example 15.8—Fractured Well With Wellbore-Storage Distortion

A fractured well had the following well and reservoir properties:

ϕ = 0.118,
h = 65 ft,
r_w = 0.2038 ft,
γ_g = 0.650,
p_i = 3,200 psi, and
S_w = 0.25.

Before shut-in for a buildup test, the well produced at a stabilized rate of 1,420 Mcf/D for an effective time of 272 hours. The data in Tables K-13 and K-14 were recorded in the buildup test. Estimate formation permeability, fracture length, and fracture conductivity for this well.

Solution—Fractured Well With Wellbore-Storage Distortion. ***Horner Graph (Fig. K-11).*** A preliminary match using the Barker and Ramey[9] type curve indicates the approximate start of the semilog straight line at Δt_{ae}=21.92 hours, Δt=23.5 hours, and $(t_{pa}+\Delta t_a)/\Delta t_a$=21.2 (see Table K-15).

m=950 psia/cycle.

$$k=\frac{162.6q_g\bar{B}_g\bar{\mu}}{mh}=\frac{162.6(1{,}420)(1.484)(0.0152)}{(950)(65)}$$

$$=0.084 \text{ md.}$$

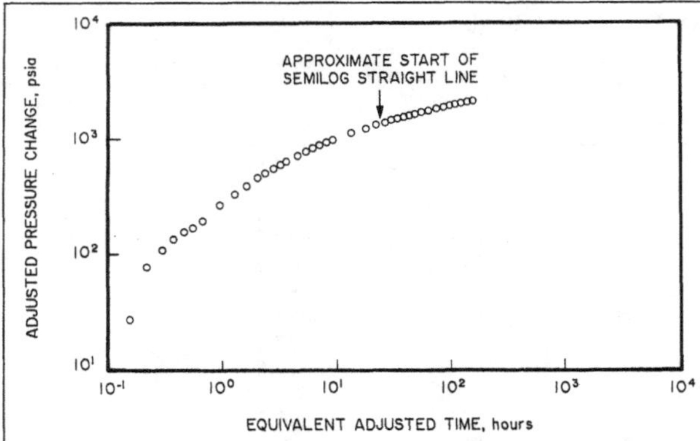

Fig. K-12—Log-log graph of postfracture buildup test with wellbore-storage distortion.

$$s'=1.151\left[\left(\frac{p_{a,1\mathrm{hr}}-p_{a,wf}}{m}\right)-\log\left(\frac{k}{\phi\bar{\mu}\bar{c}_T r_w^2}\right)+3.23\right.$$

$$\left.+\log\left(\frac{t_{pa}+1}{t_{pa}}\right)\right]$$

$$=1.151\left\{\left(\frac{65-52.99}{950}\right)\right.$$

$$-\log\left[\frac{0.084}{(0.118)(0.0152)(4.498\times10^{-4})(0.2038)^2}\right]+3.23$$

$$\left.+\log\left(\frac{464.32+1}{464.32}\right)\right\}=-3.63.$$

Assuming infinite fracture conductivity, the effective wellbore radius should be $L_f/2$.

$$r_{wa}=r_we^{-s}=(0.2038)e^{-(-3.63)}=7.69 \text{ ft}.$$

$$L_f\cong 2r_{wa}=15.4 \text{ ft}.$$

Type-Curve Plot (Fig. K-12). Precalculate a pressure match point using k=0.084 md. For p_D=10,

$$\Delta p_a=\frac{141.2q_g\bar{B}_g\bar{\mu}p_D}{kh}=\frac{(141.2)(1{,}420)(1.484)(0.0152)(10)}{(0.084)(65)}$$

$$=8{,}283 \text{ psia}.$$

The match-point values from the Barker and Ramey type curve ($wk_f=\infty$) are (see Table K-16) Δt_{ae}=1.1 hours, t_D=0.17, C_{fD}= 0.25, and

$$L_f=\left[\frac{0.0002637k}{\phi\bar{c}_T\bar{\mu}}\left(\frac{\Delta t_{ae}}{t_D}\right)\right]^{1/2}$$

$$=\left[\frac{(0.0002637)(0.084)}{(0.118)(4.498\times10^{-4})(0.0152)}\frac{(1.1)}{(0.17)}\right]^{1/2}$$

$$=13.3 \text{ ft}.$$

Summary:
k = 0.084 md,
L_f = 13.3 ft,
s' = −3.63,
C_r ≥ 100, and

TABLE K-16—TYPE-CURVE PLOTTING FUNCTIONS

	t_e (hours)	Δp (psia)	t_{ae} (hours)	Δp_a (psia)
1	0.0000	0.0000	0.0000	0.0000
2	0.3296	60.00	0.1026	15.59
3	0.4991	100.0	0.1610	28.54
4	0.6654	245.0	0.2277	80.29
5	0.8305	321.0	0.3048	111.8
6	0.9963	380.0	0.3886	139.6
7	1.162	421.0	0.4768	160.2
8	1.326	451.0	0.5675	175.3
9	1.492	500.0	0.6623	200.6
10	1.985	626.0	0.9679	274.7
11	2.477	731.0	1.302	343.8
12	2.967	821.0	1.658	406.4
13	3.456	907.0	2.034	470.9
14	3.942	974.0	2.427	523.5
15	4.427	1,035.0	2.833	573.8
16	4.910	1,084.0	3.250	614.2
17	5.391	1,129.0	3.678	653.9
18	6.348	1,214.0	4.558	730.1
19	7.299	1,281.0	5.470	792.8
20	8.242	1,341.0	6.406	850.7
21	9.179	1,396.0	7.364	904.1
22	10.11	1,441.0	8.341	950.4
23	11.03	1,480.0	9.333	990.6
24	14.66	1,615.0	13.42	1,134.0
25	18.20	1,718.0	17.63	1,247.0
26	21.63	1,791.0	21.92	1,331.0
27	24.97	1,856.0	26.24	1,405.0
28	28.23	1,910.0	30.57	1,469.0
29	31.40	1,957.0	34.90	1,525.0
30	34.49	1,996.0	39.22	1,572.0
31	37.50	2,030.0	43.51	1,613.0
32	40.44	2,062.0	47.77	1,653.0
33	43.30	2,092.0	51.99	1,689.0
34	46.09	2,117.0	56.17	1,720.0
35	51.48	2,162.0	64.39	1,776.0
36	56.62	2,203.0	72.41	1,828.0
37	61.52	2,240.0	80.23	1,875.0
38	66.20	2,274.0	87.84	1,917.0
39	70.68	2,296.0	95.22	1,945.0
40	74.97	2,317.0	102.4	1,973.0
41	79.08	2,339.0	109.3	2,001.0
42	83.02	2,357.0	116.0	2,024.0
43	86.81	2,374.0	122.6	2,047.0
44	90.44	2,390.0	128.9	2,067.0
45	93.94	2,403.0	135.0	2,084.0
46	97.30	2,416.0	140.9	2,101.0
47	100.5	2,425.0	146.6	2,113.0
48	102.6	2,429.0	150.3	2,118.0

$$wk_f \geq (100)(\pi)(k)(L_f)=(100)(\pi)(0.084)(13.3)$$
$$= 351 \text{ md-ft}.$$

Fig. K-13 summarizes the gas-well test analysis.

Nomenclature

A = drainage area, ft² [m²]
B = FVF, reservoir vol/surface vol
B_g = 5.04 Tz/p, gas FVF, RB/Mscf [res m³/std m³]
$\bar{B}_g$ = gas FVF evaluated at p_{av}, RB/Mscf [res m³/std m³]
c_f = pore-space compressibility, psi⁻¹ [kPa⁻¹]
c_T = total compressibility, psi⁻¹ [kPa⁻¹]
$\bar{c}_T$ = total compressibility evaluated at p_{av}, psi⁻¹ [kPa⁻¹]
c_w = water compressibility, psi⁻¹ [kPa⁻¹]
C = wellbore storage coefficient, bbl/psi [m³/kPa]
C_D = $0.8936C/\phi c_T h r_w^2$, dimensionless wellbore storage coefficient
C_{fD} = $0.8936C/\phi c_T h L_f^2$, dimensionless wellbore-storage coefficient for fractured wells (slightly compressible liquid)

Operator Example 15.8 Test Date Buildup Test Analyst ______

Well Fractured Well with Storage Stimulation ______ Date ______

Location ______ Formation ______

Perforations ______ Depth of Bomb ______ ft Pressure Gradient in Wellbore ______ psi/ft

Pressure Datum ______ Pressure Gradient in Reservoir ______ psi/ft

Correction to Pressure Gauge Reading ______ (psi/ft) Wellbore Gradient x ______ (ft) Distance Mid-Perfs to Bomb + ______ (psi/ft) Reservoir Gradient x ______ (ft) Mid-Perfs to Datum + ______ (psia) Atmospheric Pressure = ______ psia

BASIC PROPERTIES

q_g(MCF/D)(wet) 1,420 r_w(ft) 0.2038 $\bar{\mu}$ (cp) 0.0152

T(°R) 618 S_w(fraction) 0.25 $\bar{z}$ 0.8607

h (ft) 65 t_p(hr) 272 $\bar{c}_T$(psia^{-1}) 4.498×10^{-4}

ϕ (fraction) 0.118 t_{pa}(hr) 464.32 $\bar{B}_g$(RB/Mcf) $= 5.04 \frac{\bar{T}\bar{z}}{\bar{p}_{av}}$ 1.484

γ_g (air = 1.0) 0.650 p_{av} (psia) $= \frac{1}{2}(p_i + p_{wf})$ 1804

PRELIMINARY TYPE CURVE ANALYSIS - START SEMILOG STRAIGHT LINE

Curve Used: Barker-Ramey[16]

Matching Parameter: C_De^{2s} = ______ or C_{fD} = 0.25

Start Semilog Straight Line: Δt_{ae} = 21.92 hr; Δt = 23.5 hr; $(t_{pa}+\Delta t_a)/\Delta t_a$ = 21.18

Δt_a = 23.0 hr

Start Boundary Effects: Δt_{ae} = not present hr; Δt = ______ hr; $(t_{pa}+\Delta t_a)/\Delta t_a$ = ______

Δt_a = ______ hr

If $\Delta t_{max} < \Delta t$ at start of semilog straight line, then semilog analysis is not possible.

SEMILOG ANALYSIS

Average drainage area pressure at start of test, p_i = 3,200 psia; $p_{a,i}$ = 2,656.3 psia at ______ ft depth

$(t_{pa}+\Delta t_a)/\Delta t_a$ at start of MTR = 21.2 (Buildup) t at start of MTR = ______ (Drawdown)

$(t_{pa}+\Delta t_a)/\Delta t_a$ at start of LTR = not present (Buildup) t at start of LTR = ______ (Drawdown)

$m = p_{a2} - p_{a1}$ = 2,450 - 1,500 = 950 psia

$$k = \frac{162.6\, q_g \bar{B}_g \bar{\mu}}{mh} = \frac{162.6(1,420)(1.484)(0.0152)}{(950)(65)} = 0.084 \text{ md}$$

SKIN FACTOR CALCULATION

BUILDUP TEST:

Horner Time Ratio at $\Delta t_a = 1$ hr $= \frac{t_{pa}+1}{1}$ = 465.32

$p_{a,1hr}$ = 65 psia

$p_{a,wf}(\Delta t_a = 0)$ = 52.99 psia

$$s' = 1.151\left[\left(\frac{p_{a,1hr}-p_{a,wf}}{m}\right) - \log\left(\frac{k}{\phi\mu c_T r_w^2}\right) + 3.23\right]$$

$$s' = 1.151\left[\left(\frac{(65)-(52.99)}{(950)}\right) - \log\left(\frac{(0.084)}{(0.118)(0.0152)(4.498\times10^{-4})(0.2038)^2}\right) + 3.23\right] = -3.63$$

DRAWDOWN TEST:

$p_{a,i}$ = ______ psia

Δt_a = ______ hr

$p_{a,1hr}$ at $\Delta t_a = 1$ hr = ______ psia

$$s' = 1.151\left[\left(\frac{p_{a,i}-p_{a,wf}}{m}\right) - \log\left(\frac{k}{\phi\mu c_T r_w^2}\right) + 3.23\right]$$

$$s' = 1.151\left[\left(\frac{(\ )-(\ )}{(\ )}\right) - \log\left(\frac{(\ )}{(\ )(\ )(\ )(\ )^2}\right) + 3.23\right] = \_\_\_$$

QUANTITATIVE TYPE CURVE ANALYSIS

Curve Used = Barker-Ramey

PRESSURE MATCH POINT

CASE 1: Semilog Analysis Possible
Pre-Calculated Pressure Match Point, Using Known Permeability

$$\Delta p_a = \frac{141.2\, q_g \bar{B}_g \bar{\mu} p_D}{kh} = \frac{141.2(1,420)(1.484)(0.0152)(10)}{(0.084)(65)} = 8283 \text{ psia}$$

CASE 2: Semilog Analysis Not Possible
Permeability From Pressure Match Point

Δp_a(psia) = ______ p_D = ______

$$k = \frac{141.2\, q_g \bar{B}_g \bar{\mu} p_D}{h\Delta p_a} = \frac{141.2(\ )(\ )(\ )(\ )}{(\ )(\ )} = \_\_\_ \text{ md}$$

TIME MATCH POINT

Δt_{ae}(hr) = 1.1 (t_D/C_D) = ______ (GRINGARTEN) or t_D = 0.17 (CINCO OR OTHERS)

CASE 1: Gringarten Type Curve
Matching Parameter: C_De^{2s} = ______

$$C_D = \frac{0.0002637\, k}{\phi r_w^2 c_T \mu}\frac{\Delta t_{ae}}{t_D/C_D} = \frac{0.0002637(\ )(\ )}{(\ )(\ )(\ )(\ )(\ )} = \_\_\_$$

$$s' = 0.5\ln(C_De^{2s}/C_D) = 0.5\ln(\ \ /\ \ ) = \_\_\_$$

CASE 2: Cinco or Other Type Curve
Matching Parameter: C_r = ______ or C_{fD} = 0.25

$$L_f = \left[\frac{0.0002637\, k\, \Delta t_{ae}}{\phi c_{t,av}\mu_{av}\, t_D}\right]^{1/2} = \left[\frac{(0.0002637)(0.084)(1.1)}{(0.118)(4.498\times10^{-4})(0.0152)(0.17)}\right]^{1/2} = 13.3 \text{ ft}$$

$wk_f = \pi C_r L_f k = \pi(100)(13.3)(0.084)$ = 351 md-ft

PRODUCTION TIME DURATION CHECK

To assure that flow period exceeded duration of wellbore storage during production, calculate $t_{pa,min}$ from

$$t_{pa,min} = \frac{\phi\mu c_T r_w^2}{0.0002637\, k}(60 + 3.5\, s)\, C_D$$

$$t_{pa,min} = \frac{(\ )(\ )(\ )(\ )^2}{(0.0002637)(\ )}[60 + 3.5(\ )](\ ) = \_\_\_ \text{ hr}$$

$t_{pa,actual}$ = ______ hr

If $t_{pa,actual} < t_{pa,min}$, test analysis is invalid. Test must be rerun with longer production period prior to shut-in.

RADIUS OF INVESTIGATION CHECK

RADIUS OF INVESTIGATION ACHIEVED IN PRODUCTION PERIOD PRECEDING SHUT-IN PERIOD

$$r_{i,prod} = \left(\frac{kt_{pa}}{948\,\phi\mu c_T}\right)^{1/2} = \left(\frac{(0.084)(464.32)}{948(0.118)(0.0152)(4.498\times10^{-4})}\right)^{1/2} = 226 \text{ ft}$$

RADIUS OF INVESTIGATION DURING TEST

Start of Semilog Straight Line

$$r_{i,start} = \left(\frac{k\Delta t_a}{948\,\phi\mu c_T}\right)^{1/2} = \left(\frac{(0.084)(23.0)}{948(0.118)(0.0152)(4.498\times10^{-4})}\right)^{1/2} = 50.3 \text{ ft}$$

End of Semilog Straight Line Not present

$$r_{i,end} = \left(\frac{k\Delta t_a}{948\,\phi\mu c_T}\right)^{1/2} = \left(\frac{(\ )(\ )}{948(\ )(\ )(\ )}\right)^{1/2} = \_\_\_ \text{ ft}$$

End of Test

$$r_{i,final} = \left(\frac{k\Delta t_a}{948\,\phi\mu c_T}\right)^{1/2} = \left(\frac{(0.084)(222.3)}{948(0.118)(0.0152)(4.498\times10^{-4})}\right)^{1/2} = 156 \text{ ft}$$

NOTE: Reservoir heterogeneities and boundaries determined from data during shut-in test at $r_i > r_{i,prod}$ are highly questionable. If the drawdown did not reach beyond a certain distance in the reservoir, the buildup will not sample reservoir properties beyond that point.

REMARKS

Fig. K-13—Gas-well test analysis—Example 15.8.

C_r = dimensionless fracture conductivity
h = net pay thickness, ft [m]
$J = q/(\bar{p}-p_{wf})$, productivity index, STB/(D-psi) [stock-tank m^3/(d·kPa)]
J_o = productivity index before stimulation, STB/(D-psi) [stock-tank m^3/(d·kPa)]
k = effective formation permeability, md
k_f = propped fracture permeability, md
k_s = permeability of damaged fracture face, md
$L_D = L_e/L_f$, dimensionless distance
L_e = square drainage-area half-width, ft [m]
L_f = fracture half-length, ft [m]
m = slope of semilog straight line, psi/log$_{10}$ cycle [kPa/log$_{10}$ cycle]
m_L = slope of square-root-of-time-plot straight line, psi/hr$^{1/2}$ [kPa/h$^{1/2}$]
p = pressure, psi [kPa]
p^* = pressure obtained when linear portion of Horner graph is extrapolated to Horner time ratio of 1, psi [kPa]

$\bar{p}$ = average static drainage-area pressure, psi [kPa]
p_a = $(\bar{\mu}\bar{c}_T/2p_{av})p_p$, adjusted pressure, psi [kPa]
$\bar{p}_a$ = adjusted pressure evaluated at $\bar{p}$, psia [kPa]
$p_{a,i}$ = adjusted pressure evaluated at p_i, psia [kPa]
p_{av} = $½(\bar{p}+p_{wf})$, arithmetic average pressure, psi [kPa]
$p_{a,wf}$ = adjusted pressure evaluated at p_{wf}, psia [kPa]
$p_{a,ws}$ = adjusted pressure evaluated at p_{ws}, psia [kPa]
$p_{a,1hr}$ = adjusted pressure at 1-hour test time and on semilog straight line, psia [kPa]
p_D = $kh\Delta p/141.2qB\mu$, dimensionless pressure (slightly compressible liquid)
p_i = original reservoir pressure, psi [kPa]
p_p = $2\int_0^p \rho dp/\mu z$, pseudopressure, psia²/cp [kPa²/Pa·s]
p_{pi} = pseudopressure evaluated at p_i, psia²/cp [kPa²/Pa·s]
p_{pwf} = pseudopressure evaluated at p_{wf}, psia²/cp [kPa²/Pa·s]
p_{sc} = standard-condition pressure, psia [kPa]
p_{wf} = flowing BHP at shut-in in buildup test; also flowing BHP in drawdown test, psi [kPa]
p_{ws} = shut-in BHP in buildup test, psi [kPa]
Δp = pressure change from start of transient test, psi [kPa]
Δp_a = change in adjusted pressure from start of transient test, psi [kPa]
q_g = gas flow rate at surface, Mscf/D [std m³/d]
Q = cumulative production, STB [stock-tank m³]
Q_D = $0.8936QB/\phi c_T h r_{wa}^2(p_i - p_{wf})$, dimensionless cumulative production (slightly compressible liquid)
Q_{L_fD} = $0.8936QB/\phi c_T h L_f^2(p_i - p_{wf})$, dimensionless cumulative production referred to fracture half-length (slightly compressible liquid)
r_e = drainage radius, ft [m]
r_{ea} = $\sqrt{A/\pi}$, drainage radius of circle with same area as reservoir, ft [m]
r_i = $(kt/948\phi\bar{\mu}\bar{c}_T)^{½}$, radius of investigation, ft [m]
$r_{i,end}$ = radius of investigation at end of semilog straight line, ft [m]
$r_{i,final}$ = radius of investigation at end of test, ft [m]
$r_{i,prod}$ = radius of investigation in production period preceding shut-in test, ft [m]
$r_{i,start}$ = radius of investigation at start of semilog straight line, ft [m]
r_w = wellbore radius, ft [m]
r_{wa} = $r_w e^{-s}$, apparent wellbore radius, ft [m]
s = skin factor, dimensionless
s' = effective skin factor for gas well, dimensionless
S_w = water saturation, fraction
t = time, hours
t_a = $\bar{\mu}\bar{c}_T t_{ap}$, adjusted time, hours
t_{ae} = equivalent adjusted time, hours
t_{ap} = $\int_0^t dt/\mu c_T$, pseudotime, hr-psia/cp [h·kPa/Pa·s]
t_D = $0.0002637kt/\phi\mu c_T r_w^2$, dimensionless time (slightly compressible liquid)
t_e = equivalent time, hours
t_{eD} = $0.0002637kt/\phi\mu c_T L_e^2$, dimensionless time referred to square drainage-area half-width (slightly compressible liquid)
t_{L_fD} = $0.0002637kt/\phi\mu c_T L_f^2$, dimensionless time referred to fracture half-length (slightly compressible liquid)
t_p = effective producing time, hours
t_{pa} = producing time expressed as adjusted time, hours
$t_{pa,actual}$ = actual producing time before shut-in, hours
$t_{pa,min}$ = minimum producing time required to obtain data undistorted by wellbore storage, hours
Δt = time elapsed from start of transient test, hours
Δt_a = adjusted time elapsed from start of transient test, hours
Δt_{ae} = adjusted equivalent time, hours
Δt_{max} = maximum time elapsed since start of transient test, hours
T = reservoir temperature, °R [K]
T_{sc} = standard-condition temperature, °R [K]
w = fracture width, ft [m]
z = gas-law deviation factor, dimensionless
$\bar{z}$ = gas-law deviation factor evaluated at p_{av}, dimensionless
γ_g = gas specific gravity (air=1.0)
μ = viscosity, cp [Pa·s]
$\bar{\mu}$ = viscosity evaluated at p_{av}, cp [Pa·s]
ϕ = total porosity, fraction

References

1. Prats, M.: "Effect of Vertical Fractures on Reservoir Behavior—Incompressible Fluid Case," *SPEJ* (June 1961) 105–18; *Trans.*, AIME, **222**.
2. Tinsley, J.M. *et al.*: "Vertical Fracture Height—Its Effect on Steady-State Production Increase," *JPT* (May 1969) 633–38; *Trans.*, AIME, **246**.
3. Holditch, S.A.: *Quarterly Low-Permeability Gas Well Research Report for Fall, 1975*, quarterly report, Petroleum Engineering Dept., Texas A&M U., College Station.
4. McGuire, W.J. and Sikora, V.J.: "The Effect of Vertical Fractures on Well Productivity," *Trans.*, AIME (1960) **219**, 401–03.
5. Morse, R.A. and Von Gonten, W.D.: "Productivity of Vertically Fractured Wells Prior to Stabilized Flow," *JPT* (July 1972) 807–11.
6. Gringarten, A.C., Ramey, H.J. Jr., and Raghavan, R.: "Unsteady-State Pressure Distributions Created by a Well with a Single Infinite-Conductivity Vertical Fracture," *SPEJ* (Aug. 1974) 347–60; *Trans.*, AIME, **257**.
7. Cinco-L., H., Samaniego-V., F., and Dominguez, N.: "Transient Pressure Behavior for a Well With a Finite-Conductivity Vertical Fracture," *SPEJ* (Aug. 1978) 253–64.
8. Agarwal, R.G., Carter, R.D., and Pollock, C.B.: "Evaluation and Performance Prediction of Low-Permeability Gas Wells Stimulated by Massive Hydraulic Fracturing," *JPT* (March 1979) 362–72; *Trans.*, AIME, **267**.
9. Barker, B.J. and Ramey, H.J. Jr.: "Transient Flow to Finite Conductivity Vertical Fractures," paper SPE 7489 presented at the 1978 SPE Annual Technical Conference and Exhibition, Houston, Oct. 1–3.

SI Metric Conversion Factors

acres	× 4.046 873	E−01	= ha
°API	141/(131.5+°API)		= g/cm³
bbl	× 1.589 873	E−01	= m³
cp	× 1.0*	E−03	= Pa·s
ft	× 3.048*	E−01	= m
ft³	× 2.831 685	E−02	= m³
°F	(°F−32)/1.8		= °C
in.	× 2.54*	E+00	= cm
psi	× 6.894 757	E+00	= kPa
psi⁻¹	× 1.450 377	E−01	= kPa⁻¹
°R	× 5/9		= K
scf/bbl	× 1.801 175	E−01	= std m³/m³

*Conversion factor is exact.

Author Index

Subject Index*

*Boldface indicates key discussion or plots.

D

E

F

N

O

Q

R

T

U

V

W

X

Y

Z

www.ingramcontent.com/pod-product-compliance
Lightning Source LLC
Chambersburg PA
CBHW080132220326
41598CB00032B/5033

* 9 7 8 1 5 5 5 6 3 0 2 0 1 *